P9-BIM-927

# Foundations of
# ANIMAL DEVELOPMENT

# Foundations of ANIMAL DEVELOPMENT

SECOND EDITION

ARTHUR F. HOPPER
*Professor Emeritus of Zoology*
*Rutgers University*

NATHAN H. HART
*Professor of Zoology*
*Rutgers University*

New York Oxford
OXFORD UNIVERSITY PRESS
1985

Oxford University Press

Oxford London New York Toronto
Delhi Bombay Calcutta Madras Karachi
Kuala Lumpur Singapore Hong Kong Tokyo
Nairobi Dar es Salaam Cape Town
Melbourne Auckland

and associated companies in
Beirut Berlin Ibadan Mexico City Nicosia



Published by Oxford University Press, Inc.,
200 Madison Avenue, New York, New-York 10016



**Library of Congress Cataloging in Publication Data**
Hopper, Arthur F., 1917–
Foundations of animal development.
Bibliography: p.
Includes index.
1. Developmental biology.
2. Embryology.
I. Hart, Nathan H., 1936– II. Title.
QL971.H79 1984 591.3 84-5759
ISBN 0-19-503476-7

Printing (last digit): 9 8 7 6 5 4 3

Printed in the United States of America

# PREFACE
# to the second edition

Our major aim in the second edition of *Foundations of Animal Development,* as in the first edition, is to provide the student with a readable account of the development of animals, integrating descriptive, experimental, and biochemical approaches. Developmental phenomena are examined at various levels of organizational complexity, in a variety of invertebrate and vertebrate organisms. An appreciation of contemporary research in developmental biology requires a firm background in both facts and principles: therefore, much of the content of the first edition has been retained, but revised and updated in light of the current literature.

The text is divided into two parts. The first, Chapters 1 to 13, provides a background in the onset of development (gametogenesis, fertilization) and the subsequent events that lead to the construction of organ primordia and the early morphogenesis of the multicellular organism (cleavage, gastrulation, neurulation, and embryonic axis formation). Substantial treatment is given to the basis of cell differentiation and controls of gene expression and to the various mechanisms that underlie the generation of organ and tissue form. Coverage of the role of the Sertoli cells during sperm maturation has been expanded (Chapter 3). Chapter 4 introduces the concept of cell-to-cell communication and discusses the possible role of gap junctions during oocyte growth. We have expanded sections dealing with gene activity during oogenesis and the mechanisms mediating oocyte maturation. In Chapter 5, on the physiology of reproduction, information has been added on the neuroendocrine control of the estrus and menstrual cycles. The hormonal and nonhormonal factors affecting male fertility and the significance of diethylstilbesterol (DES) in uterine cancer are discussed. We have expanded coverage of the blocks to polyspermy and the roles of calcium and cytoplasmic pH during activation in Chapter 6 (fertilization). Chapter 7 (cleavage and blastulation) now considers electrotonic coupling and its significance in the coordination of cell division, and includes new information on the process of blastulation. In Chapter 8 (gastrulation), the section on gastrulation in fishes has been enlarged, and a discussion of the roles of bottle cells and deep cells during involution in amphibians is presented. In Chapter 9, on the organization and activities of the early embryo, we have expanded the coverage of gene activity and protein synthesis to include the synthesis of histones in sea urchins and amphibians, as well as new information on the development of ascidians and their contribution to our understanding of the significance of cytoplasmic localization in cell lineage specification. In Chapter 10 (neurulation and embryonic induction), the student is introduced to computer simulation as a technique currently being used to study the relationships between cell shape changes and organ formation (i.e., neural tube formation). We have expanded our discussion of attempts to explain the spatial determination and regional segregation of the central nervous system, and we mention experiments designed to determine how the mesodermalizing and neuralizing factors interact with the target ectoderm. Chapter 11 (cell differentiation) has been thoroughly revised to provide a more comprehensive picture of gene expression and cell specialization, including levels at which the synthesis of specific proteins may be controlled within a cell. The techniques of two-dimensional gel electrophoresis and DNA cloning and their applications to studies of gene expression are discussed, particularly the expression of the 5S ribosomal gene in amphibians. The effects of maternal stress and alcohol consumption on fetal development are considered in Chapter 12 on mammalian development. Recent experiments on the differentiation of inner cell mass and trophoblast cells are also presented, as well as information on *in vitro* fertilization and embryo transfer in humans. In Chapter 13, we discuss more thoroughly the structure and organization of the extracellular matrix and their roles in the migration of the neural crest cells. The role of positional information in the development of the limb is treated in detail.

The second part of this book, Chapters 14 to 23, is a detailed analysis of the development of organ systems. Chapter 14 includes new information on cleft palate and epithelio-mesenchymal interactions during tooth development. In Chapter 16, on the respiratory system, the role of collagen in the branching morphogenesis of the lung is reexamined in light of experiments that employ drugs known to interfere with collagen metabolism. We have expanded our coverage of hemoglobin formation and the controls of gene expression during hemoglobin synthesis, the factors regulating cardiac looping (particularly structural and compositional changes in the cardiac jelly), and development of the heart conducting system in Chapter 18. In Chapter 19 (urogenital system), considerable attention is given to the role of the sex chromosomes in the differentiation of the gonad. Material has been added in Chapter 20 (nervous system) on the establishment of pattern in the early differentiation of the nervous system, as well as on the differentiation and derivatives of the neural crest. In the analysis of sensory organ development (Chapter 21), inductive interactions during the development of the nose, inner ear, and eye are considered in more detail. The role of the extracellular matrix in mediating morphogenetic change between lens and optic cup is discussed. Finally, we have added information on the development of the axial skeleton (Chapter 22) and the differentiation of the epidermal appendages of the skin (Chapter 23).

The authors are grateful to the many people who have assisted in the preparation of this second edition. We express gratitude to those who took the time to share their comments and ideas about the first edition with us. We are indebted to friends and colleagues who again generously provided photographs and illustrations, and to publishers who have permitted us to reproduce previously published illustrations.

Finally, we express our thanks and appreciation to the staff at Oxford University Press. We thank Susan Meigs, who coordinated much of the manuscript preparation, and William Curtis, whose prodding and encouragement did much to make this second edition possible.

NATHAN H. HART
*March, 1985* ARTHUR F. HOPPER

# PREFACE to the first edition

The teacher of an introductory course in animal embryology or animal development is faced with the difficult task of determining the content of his course and the methodology of its presentation. By its very nature, the subject matter is difficult to define and delineate. The essence of development is change over time and this is reflected at all levels of structural organization from the molecular to the organismal. In selecting material for such a course, one is also influenced by the students' varied interests and backgrounds. Some students may be terminating their biological training at the level of the college or university, while others may intend to pursue advanced professional training in the disciplines of biology, medicine, dentistry, and so on. Such considerations have influenced our approach to and selection of material for the writing of this textbook. From the very beginning, we realized that it would be impossible to design a format and provide coverage which would satisfy the needs of all teachers of this subject area.

We have endeavored to set forth in direct and readable form an account of the ontogenetic development of organisms through an integration of descriptive, experimental, and biochemical approaches. The focus is the whole embryo and the sequence of events by which the form and structure of its body and constituent organ systems emerge from the superficially simple-appearing egg. Concurrently, the book attempts to set the subject of development in a modern context by recognizing and formulating major questions associated with these events and by analyzing, where appropriate, the underlying mechanisms. There is an emphasis on the basic, morphological aspects of development. Central to our thinking is the belief that insights into the complexities of the developmental process are diminished unless the student is thoroughly familiar with this type of information. Students working with this book will be prepared to study in more detail a specialized area of development which might be assigned by the teacher or to meet the challenge afforded by an advanced course in embryology or developmental biology.

The book may be divided into two parts. The first part gives consideration to the onset of development (gametogenesis, fertilization, and cleavage), the rearrangements of embryonic cells and the organization of the early embryo, tissue interactions and the basis of cellular differentiation, and principles of morphogenesis. Concepts and principles are introduced which provide a background against which the second part of the book, the development of various organ systems, can be examined in detail. Our comprehensive survey of the development of the organ systems includes an analysis of the mechanisms responsible for their form and structure. Emphasis is placed on the development of the mammal, a treatment which reflects an increasing interest in this group by investigators as a source for the study of developmental processes.

The authors are grateful to the many people who have assisted in the preparation of this book. We express particular thanks to friends and colleagues who generously provided photographs and illustrations, and to the publishers who have permitted reproduction of previously published illustrations. The names of these colleagues appear in the legends of the textbook figures. A special word of thanks to Ms. Diane Abeloff, our illustrator, who patiently listened to us and then carefully designed and drew what was requested. We are indebted to Mrs. Elaine Derry who typed the book manuscript and in the process made some helpful suggestions.

Finally, we express our gratitude to the staff at Oxford University Press, especially to Mr. Robert Tilley, whose encouragement did much to bring this book to completion.

*New Brunswick, N.J.* NATHAN H. HART
*June 1979* ARTHUR F. HOPPER

# CONTENTS

## 14
## DEVELOPMENT OF THE FACE, PALATE, ORAL CAVITY, AND PHARYNX 338



## 15
## THE DIGESTIVE TUBE AND ASSOCIATED GLANDS 356



## 16
## THE RESPIRATORY SYSTEM 366



## 17
## THE COELOM AND MESENTERIES 380



## 18
## THE CARDIOVASCULAR SYSTEM 392



## 19
## THE UROGENITAL SYSTEM 444



## 20
## THE NERVOUS SYSTEM 479



## 21
## THE SENSE ORGANS 523



## 22
## DEVELOPMENT OF THE SOMITES 546

Foundations of
ANIMAL DEVELOPMENT

# 1

# INTRODUCTION

The existing individuals of any species have a finite lifespan. Thus, if a species is to persist, each generation of individuals must possess some mechanism whereby it can produce the next generation. The process whereby individuals of a species are perpetuated is called reproduction. In vertebrates the process is concerned with the production of specialized reproductive (germ) cells in the male and the female whose union results in the formation of a fertilized egg *(zygote)*, which will develop into a new individual.

Embryology is the field of biology that concerns the description and experimental analysis of the development of the individual; literally translated it means "the study of embryos." Historically, the discipline was initially an anatomical science with studies primarily recording the changes in the shape and form of animals during their embryonic stages. Today, however, the emphasis is on the processes and causal mechanisms underlying the development of an organism. We now know that the factors underlying and controlling the various changes that occur as the single-celled fertilized egg becomes transformed into a multicellular adult are complex and their analysis requires the application of knowledge and techniques from a wide variety of disciplines, including cytology, physiology, biochemistry, radiation biology, molecular biology, and genetics. Hence, developmental biology (or modern embryology) is a multidisciplinary science concerned with the totality of the processes and mechanisms by which an organism acquires its specialized structural and functional features. Increasingly, we are finding that understanding the basic processes of development demands a sound knowledge of the nature of the information that is stored in the nucleus and cytoplasm of the fertilized egg, and how it is utilized during the course of development.

## Embryology, Development, and Ontogeny

Generally speaking, an embryo is regarded as something in a developmental or rudimentary condition. In biology, we can assign a more specific definition to the embryonic state: It is one that encompasses the development of the individual up to the time of hatching or, in mammalian species, up to the time of birth. In mammalian embryology, the prenatal period may be divided into two stages: (1) the period of the embryo, covering development up until the individual attains the form—but, of course, not the size—of the adult, and (2) the period of the fetus, covering the remainder of the prenatal period until birth.

Dramatic as the events of birth and hatching are, we must not consider them as representing the termination of the development of an individual. To be sure, the change from an aquatic to a gaseous environment does indeed represent, in many species, a dramatic change to which an individual must adjust. However, since the systems that support an individual in its aquatic environment are exactly those that continue, with only minor adjustments, to support it in its gaseous environment, birth and hatching do not represent sharp changes in the development of an organism. They represent only a single point in a process of continuing, progressive change.

The term *development*, then, is not synonymous with embryology but includes the entire lifespan of the individual from fertilization to death. Development is simply defined as progressive change. It includes both prenatal and postnatal stages leading to structural and functional maturity. In turn, these stages are followed, after a certain length of time, by the changes associated with senescence, which lead ultimately to death. Development proceeds in an orderly sequence, and each change leaves the organism in a state different from its previous one and, generally, unable to return to it. Unfortunately, the term development often carries with it the connotation that the progressive change is for the better or toward a higher plane. If we include this meaning in the term, there is then a point in a life history at which development stops. Or rather, there are many points, since the time at which each organ system reaches its maximum development varies considerably. Thus, we should consider that development encompasses the entire lifespan of the individual from fertilization to death, and development is then synonymous with *ontogeny*.

## Ontogeny and Phylogeny

*Ontogeny*, the development of the individual, is only one very small phase of a much larger developmental sequence, *phylogeny*—the historical development of the species, that is, the evolution of the species. Early in the 19th century a German scientist, Johann F. Meckel, proposed that there was a relationship between the embryos of higher forms and the adults of lower forms, the former progressing through stages in which they bore a marked resemblance to the latter. After Charles Darwin's theory on evolution appeared in the mid 1800s, Meckel's views were reconsidered and rephrased in evolutionary terms by another German scientist, Ernest H. Haeckel, who expressed in the succinct phrase "ontogeny recapitulates phylogeny" what has come to be known as the *recapitulation theory*. This implied that the embryos of each species progress through stages in which they resemble the structure of adults belonging to their evolutionary line. Although this theory is a fascinating one—one that a superficial consideration of the embryonic development of the higher animal forms appears to support—it goes beyond the facts. It is not correct to consider that the human embryo is at one stage of its development comparable to a fish because at that time it has developed the fishlike characteristics of five branchial (gill) arches, an undivided heart, and a head kidney.

It is closer to fact to note that certain organs in the embryos of higher forms may in some respects resemble the same organs in the embryos of lower forms, as was indeed proposed by the well-known German embryologist, Karl E. von Baer, in his 1839 monograph on embryology. However, it is still not correct to consider that any embryonic stage of a higher species is comparable to any embryonic stage of a lower one. The embryos of all species are specific unto themselves, and a human embryo is always a human embryo. Even though it at one time develops branchial arches, a human embryo is never the counterpart of a fish, or even a fish embryo.

The development of a number of organs illustrates another relationship between ontogeny and phylogeny. Structures that appear in the embryo as phylogenetic holdovers often function in the embryo before they are replaced by evolutionally more recent acquisitions. Some structures that function in the embryo and are then superseded by later developing structures may degenerate and disappear, or they may be retained as nonfunctional vestigial structures. However, many structures, after giving up their original function, may be retained and diverted to an entirely different function. This is illustrated by the conversion of the excretory (mesonephric) duct of the embryonic kidney of the mammal to a major part of the genital duct (vas deferens) of the adult male. Another example is the conversion of the branchial arches and pouches into a variety of structures in the adult mammal, none of which have anything to do with the respiratory function of these structures in the fish.

Although many evolutionally older parts are lost along the way, it appears that the higher forms are basically conservative and reluctant to abandon completely their phylogenetic inheritance, for—whenever possible—they convert older structures into functional parts of a newer system.

## Preformation and Epigenesis

Interest in the development of animal organisms goes back several thousand years. The earliest studies of embryology, which date to Aristotle's observations on sharks and chicks, were generally concerned with descriptions of the development of different species. Before the advent of the microscope in the 18th century and the concept of the cell theory in the 19th, any real knowledge of the principles of embryology, especially of the earliest stages, was limited. William Harvey, who is most well known for his writings on the circulation of the blood, probably made a more original contribution to science by stating in 1651 that all animals develop from eggs in a manner in which complexity of form gradually appears where uniformity previously existed. Over 100 years later, Harvey's observations were further enlarged on by Kaspar F. Wolff, who described the embryo as first consisting of a formless array of "globules" (cells) that gradually are arranged into organ rudiments and then into organs, more and more complex structures appearing in a stepwise fashion as development proceeds. This concept of development is called *epigenesis.*

In opposition to the epigenetic theory, there arose, also in the middle of the 18th century, the *performation* theory of embryonic development. This theory stated that development consists merely of a growth or unfolding of structures that already exist preformed, but in a miniature state, in the germ cells. The theory emerged with the development of the microscope and its use in the examination of eggs and sperm. Investigators with somewhat fertile imaginations claimed to see fully formed tiny adults in eggs and sperm. Some of the leading microscopists of the time became involved in a controversy over whether the preformed adult resided in the egg (they were called ovists) or in the sperm (spermists). Further analysis and more careful observation saw the end of the preformation theory (at least in the form in which it was proposed in the 18th century), and the gradual acceptance of the concept of epigenesis by most biologists. The fertilized egg is a simple "undifferentiated" cell from which, through an orderly series of developmental transformations, are produced differentiated structures that make up the organs and organ systems of the adult. The bold statement that epigenesis means the development of the complex from the simple may present the erroneous impression that the egg is a simple, uncomplicated cell. As we shall see, it is in reality a highly complex cell—made so by the tremendous synthetic activity that takes place during oogenesis. It contains a multitude of complex proteins and other substances that, in addition to controlling the normal metabolic functions of the cell itself, must also supply the programmed information necessary for the early development of the embryo as well as the building blocks and energy sources used in this process.

## Stages of Development

Development is an ongoing process marked by progressive and continual change in the structural and functional properties of the cells and tissues of an organism. To investigate and examine these changes, it is often convenient to refer to the *developmental stages* of an embryo. A developmental stage represents a particular period of development that is characterized by a unique and particular pattern of events and processes. The processes that underlie these events occur at a given time and take place for a given duration. At any given stage, patterns of change are essential to the expression of the next stage, and so forth. The reader should keep in mind, therefore, that developmental stages are convenient but useful representations of embryos at particular times. We will now describe briefly the major stages of animal development. Additional details on the patterns of change in developing embryos and the mechanisms underlying these changes will follow in later chapters.

### Fertilization

Embryonic development starts with the union of the haploid egg and sperm, and this process of *fertilization* produces the diploid *zygote.* This union not only restores the diploid number characteristic of the species but also provides the stimulus for the beginning of development. Male and female gametes arise from specialized cells of the reproductive organs known as *primordial germ cells*. These cells, which form the basis of continuity between generations, appear to arise from a specialized portion of the egg cytoplasm known as the *germ plasm*. Prior to the fusion of the gametes, the cells of the germ line undergo a maturation process, *gametogenesis*, which is marked by a type of division called *meiosis* in which the diploid number is halved. Both *spermatogenesis*, the development of the sperm, and *oogenesis*, the development of the egg, involve complex structural and functional specializations of cell types. In the case of the egg, there is the storage of nutrient and informational materials, and the formation of accessory envelopes. The condensation of nuclear material, the elimination of excessive cytoplasm, and the formation of locomotor structures mark the development of sperm.

## Cleavage

Fertilization is followed by a period of rapid cell multiplication, a period known as *cleavage*, in which the single-celled zygote is changed into a multicellular structure. The cleavage cells are known as *blastomeres*. During the cleavage stage, cell division is so rapid that the cells do not increase in size between divisions, and thus the embryo remains about the same size and the individual blastomeres become progressively smaller. Cell division, of course, does not stop at the end of the cleavage period but continues throughout development. However, after the cleavage stage, cell division no longer appears as the dominant feature of development. Cleavage patterns vary in different species, a large part of the variation being dependent on the amount and distribution of the inert yolk that may be present in the ovum.

## Blastulation

Cleavage leads to the formation of the *blastula*, which usually consists of a group of cells surrounding a cavity, the *blastocoele*. In the mammal, the embryo at this time is known as the *blastocyst*. The cells of the blastula may be arranged either in a single layer or in a number of layers.

## Gastrulation

*Gastrulation* is a period of cellular movements in which the cells of the blastula are rearranged, some moving into the inside of the embryo and some remaining on the outside. During gastrulation the blastocoele is obliterated and a new cavity, the *gastrocoele*, develops. In most species the gastrocoele may also be called the *archenteron*, marking it as the forerunner of the gut tube. The gastrula is a layered structure, not in the sense that there is more than a single tier of cells but in the sense that the movements of gastrulation give rise to sheets of cells that are known as the *primary germ layers*. These consist of an outer, a middle, and an inner layer, respectively, *ectoderm*, *mesoderm*, and *endoderm*.

## Neurulation and the Establishment of the Organ Rudiments

Through gastrulation, there is little evidence in the embryo of a basic body plan. The basic body plan is fashioned immediately after gastrulation and is constructed along the axis of symmetry. In vertebrate embryos, this is manifest along the craniocaudal axis with the formation of the *neural plate* and *neural tube*, and by the elaboration of other axiate structures such as the somites and the alimentary tract. The paired somites develop lateral to the neural tube from the most medial region of the mesoderm. They will form parts of the skeletal, muscular, and integumentary systems. Lateral to the somites, the mesoderm splits into two layers, one of which becomes associated with the overlying ectoderm and the other with the underlying endoderm. The cavity formed between these two layers of mesoderm is the *coelom*. Thus, immediately after gastrulation we begin to see morphological changes in which the primary germ layers of the gastrula develop into structures that represent the earliest stages in the formation of specific organs. These structures are then called *primary organ rudiments*. The primary organ rudiments usually consist of populations of cells derived from more than one germ layer.

## Organogenesis

The development of the primary organ rudiments is the first stage in *organogenesis* (organ formation).

The formation of the neural tube and the alimentary tract occurs during and shortly after gastrulation. Other organ rudiments develop later. Each rudiment consists of a group of cells with special properties. They are segregated from other cells and groups of cells in the embryo and are destined to develop into specific organs. An organ rudiment is often called an *anlage*. Organogenesis is an extensive period in development during which a number of rather complicated mechanisms interact to form the organ rudiments and in turn guide their differentiation into the adult state, where each organ then becomes capable of performing its particular physiological function. The two major processes associated with organogenesis are *morphogenesis* and *differentiation*. Morphogenesis refers to the various processes by which the form of the embryo and its parts are acquired; these processes include *morphogenetic cellular movements*, particularly of sheets of cells, *growth*, quantitative and qualitative alterations in the *extracellular matrix*, and *cell death*. Differentiation refers to the gradual changes in cells during development and their eventual integration into a structural and functional whole entity (i.e., organ).We will discuss the general principles of organogenesis more fully in Chapter 13.

## Descriptive and Experimental Embryology

As mentioned previously, the earliest studies of embryology were primarily descriptive. It was not until the late 1800s and early 1900s that experimental manipulation of the embryo was introduced as a method of analysis of the underlying factors controlling the events of development. Early studies in Germany by Wilhelm Roux and Hans Driesch on the separation and subsequent development of the early blastomeres of frog and sea urchin embryos formed the basic framework of modern experimental embryology. Although their experiments resulted in two opposing views by which to explain development, both investigators left as a legacy the experimental approach through the manipulation of embryos as a way of revealing cause-and-effect relationships underlying morphogenetic change. Subsequently, in the United States in the 1900s, E. B. Wilson and E. G. Conklin carefully traced the precise contributions of the early blastomeres of invertebrate embryos to the future structure of the animal. The famous German embryologist, Hans Spemann, and his colleagues perfected and applied techniques by which groups of cells could be grafted within and between amphibian embryos; by so doing, they pointed out the importance of cellular rearrangements and interactions between groups of cells during the course of development. Although advances in genetics, biochemistry, microscopy, and molecular biology have skyrocketed our knowledge of the basic mechanisms of development, many of the questions raised in the studies of these eminent experimental embryologists remain as problems still under active investigation today.

### Experimental Analysis of Early Development

The union of the egg and the sperm may seem on the surface a rather simple, although an obviously important, event. However, the process of fertilization presents a number of interesting problems that include, in addition to the mechanism of the development of the gametes, the role of chemical substances in egg–sperm interaction, the mechanism of sperm penetration, the prevention of polyspermy, and the immediate reaction of the egg to sperm penetration. Also, contemporary students of developmental biology have an interest in the role of the egg cell in controlling early events of morphogenesis. There is evidence, for example, that the cytoplasm of some eggs sequesters specific programs of development. As the blastomeres divide and receive the egg cytoplasm, they appear to inherit "instructions" for very specific types of cellular differentiations. The nature of these "instructions," the level in the cell at which they operate, and the mechanisms by which they are distributed to given regions of the embryo are currently challenging problems.

There has been concern with understanding the mechanism of cell division during cleavage and the control of its pattern and rate in different organisms. Since cell division involves both *mitosis* (nuclear division) and *cytokinesis* (cytoplasmic division), the interaction between nucleus and cytoplasm during this period of development has been of particular interest. The unique configuration of the blastula has offered opportunities to study the behavior of individual as well as sheets of cells.

The rearrangements of cells at gastrulation have received considerable attention because they are the most dramatic of cell movements during the course of embryogenesis. The movements may involve sheets of cells, as in the case of invagination and epiboly, and individual cells that detach themselves from epithelia. Questions posed by these complex rearrangements include: what initiates cell movement, how do cells move, and what is the factor(s) that causes cells to stop moving? A consequence of gastrulation is the formation of the primary germ layers, and their formation introduces the problem of how cells become segregated into homogenous tissue fabrics. In addition, germ layer formation is one of the first indications of differentiation in the embryo. The three germ layers are clearly distinguished from one another not only by their respective locations, but also by the specific tissues and organs each layer forms.

*The Germ Layer Concept*

The germ layer concept states that each of the three germ layers is marked to develop into its own specific tissues and organs. Epidermis and neural structures will be formed from ectoderm, the alimentary tract and its accessory organs from endoderm, and muscle, connective tissue, and the urogenital system from mesoderm. Because of the simplicity of this concept and because of its inclusion in all textbooks of embryology, and its reinforcement in the laboratory by the use of colored pencils to denote the germ layer origin of the tissues and the organs, the histological performance of the germ layers is one of the facts longest remembered by most embryology students. However, this emphasis may serve to give a false idea of realism to the germ layers and to imply a specificity not supported by the facts.

The germ layer concept played an important role in comparative embryology when it was shown that the germ layers were similar throughout the entire animal kingdom—although only ectoderm and endoderm appear in some invertebrates—and that their formation and fates were essentially the same in all animal species. During the debate on the theory of evolution in the 1800s, this similarity of origin and fate was used as one of the strongest confirmations of the principles of evolution. However, these earlier concepts assigned to the germ layers an absolute specificity and a reality that does not exist as well as a controlling force that is not warranted. E. B. Wilson noted that the typical relationships between the germ layers and the structures that are derived from them can be experimentally changed, and he stated that if this is the case, these relationships cannot be causally connected and the germ layers themselves do not have any intrinsic morphological value. Experiments conducted to change the fate of the germ layers will be described in a later chapter, where it will be seen that the cells of any germ layer may be induced to overstep the classical boundaries of their own particular layer and to exhibit a variety of potentialities for the formation of tissues and organs of other germ layers. Thus, the germ layers are not regions of fixed specificities; nor are they causally related to the development of particular tissues or organs.

The value of the germ layer concept lies in the fact that germ layer formation is the first overt indication of the beginnings of differentiation. The establishment of the germ layers may be looked on as the embryo's method of sorting out and locating its constituent parts in such a manner that, under normal circumstances, cells in a particular layer will form particular structures. Each germ layer is a region to which the building blocks for later organ formation are consigned. As labels and markers to indicate future differentiation, they aid in our understanding of the events of development. However, the germ layer concept tells us nothing about the underlying factors responsible for these differentiations.

*Induction and Tissue Interactions*

The early development of the vertebrate nervous system has been the subject of investigation for many years. Through extirpation and grafting techniques, it has been shown that the differentiation of the central nervous system is dependent upon the presence of the underlying mesoderm. The influence of one part of the embryo on the fate of another resulting in the differentiation of specific cell types is known as *induction*. The inductive influence of the mesoderm on the overlying ectoderm was considered such an important factor in the development of the axiate pattern of the embryo that Spemann termed the tissue responsible for induction, the *organizer*. The nature of the organizing substance(s) and the mechanism of neural induction are still the subject of considerable investigation, as will be discussed in Chapter 10.

The morphogenesis and differentiation of many

vertebrate organs are dependent upon an inductive interaction between tissues typically of diverse origins (so-called *epithelio-mesenchymal interactions*). These inductions require that cells exchange information by some type of *cell communication*, either by physical contact or by the transmission of diffusable substances.

## Differentiation and Determination

### Differentiation

Development has been defined as progressive change taking place over the lifetime of the individual. Nowhere in the lifespan is progressive change more evident than in the embryonic stages of development, and it is therefore understandable why embryologists are concerned with the causal processes underlying these changes. The progressive changes in the embryo by which its cells acquire their distinctive structures and functions is called *differentiation*. In differentiation, the cells become phenotypically different from other cells and from their precursors.

Although the formation of the germ layers is the first easily recognizable evidence of differentiation, some differentiation does occur before this time. The pregastrular cellular specialization associated with the acquisition of locomotor capability represents an early differentiation of the blastula cells of the teleost. An ultrastructural change that results in the appearance of intercellular junctions between trophoblast (membrane) cells of the mammalian blastocyst is another example of early differentiation.

Cells of multicellular organisms are organized as tissues; tissues then become organized as organs, and organs as organ systems. The term differentiation, therefore, is frequently applied to these different levels of organizational complexity. One can, for example, speak of the course of differentiation of the heart or eye *(organ differentiation)*. Cells that acquire the ability to perform a specific function are known as tissues; their differentiation is often referred to as *histodifferentiation*. For example, the cells lying along the middorsal axis of the amphibian gastrula differentiate into neural tissue whose function is to transmit nerve impulses. Histogenesis of the somites gives rise to connective tissue (dermis and bone) and to muscular tissue. Finally, the morphological and functional properties associated with the differentiated state of a cell are acquired by the process of *cytodifferentiation*. Morphological changes are easily recognizable steps in differentiation. However, recent advances in the analysis of cytodifferentiation indicate that less easily detectable biochemical changes often precede morphological changes.

As we have previously stated, the assembling of cells and embryonic tissues into complex, highly structured and fully functional organs occurs during organogenesis. Morphogenesis and differentiation are integral parts of organ formation. Growth, a component of morphogenesis, is the increase in size of an organism or its parts as the result of the synthesis of protoplasm or substances such as bone, cartilage matrix, or connective tissue fibers that form a part of the extracellular matrix of tissues or organs.

#### *Gene Control of Differentiation*

The structural and functional characteristics of a cell depend on the nature of its proteins. The nature of its proteins in turn depends on the expression of the genetic information contained in the cell's genome. Differentiation, then, is the function of gene activity, the sequence of each protein reflecting the linear sequence of the nucleotide-containing coding units of a gene. This, of course, immediately raises the problem of how cells, each one of which is known to have exactly the same genome, can synthesize different proteins in different tissues of the body. The only logical answer is that the entire genome is not active continually in every cell and only a small part of the DNA is transcribing in any given cell, the activity differing in different cells. In other words, differentiation is the expression of variable gene activity.

This brings us only a step closer to the solution of the mechanism of differentiation, since we are still left with the question why different parts of the genome are turned on or off in different cells. The answer to the question most probably lies in regional differences in the environment, both extracellular and intracellular. It is usually proposed that a cyclic interaction between the environment and the genome occurs, so that the environment specifies which part of the genome will function and the genome in turn modifies the environment. Nucleo-cytoplasmic interaction will be the subject of Chapter 11. Basically, the interplay between nucleus and cytoplasm is considered to be the force that moves the cell along a specific pathway of differentiation and is thus responsible for development.

### Determination

Another concept that often appears in embryology is that of *determination.* At one time in its development every egg is probably totipotent, and indeed in many eggs individual blastomeres of the two-, four-, or even eight-cell stage are capable of forming an entire—although smaller—individual. As development proceeds, the potency of the cells becomes progressively restricted and the lines of development open to them become fewer. That is, they gradually become "determined" to develop in a certain direction. Determination thus involves a series of events that results in a change from an indefinite to a definite condition. An area of the embryo first becomes determined toward some particular formation, and then within this area continuing determination defines the fates of smaller groups of cells until eventually each cell becomes determined to differentiate in a single specific direction.

We should note that determination and differentiation are different processes, and the former precedes the latter. The time period between determination and differentiation varies in different species and organs. In many invertebrates the cells may be fully determined long before any evidence of differentiation can be observed. However, in the early mammalian embryo the interval between the two is short. Differentiation is characterized by some cellular change—morphological, physiological, or biochemical—which is observable to an investigator at some point in development. However, the time at which a cell becomes determined to produce a specific differentiation product usually cannot be measured precisely. Perhaps with more technological advances the act of determination may become itself recognizable, and the determination and differentiation might be considered a single phenomenon of embryonic change.

In many instances determination and differentiation may be changed or reversed. Thus, cells that are positioned in the outer germ layer of the gastrula are determined, under normal conditions, to form ectodermal structures only. However, experimental manipulation of the embryo may allow these cells to form structures characteristic of either of the other two germ layers. That is, determination may not be irrevocable. In addition, even differentiated cells may be made to dedifferentiate and redifferentiate into cells of entirely different characteristics, and thus differentiation may also be reversible.

Although determination can be simply visualized as progressive restriction of potencies, and although in this sense determination is a single underlying principle in the development of every organ and its parts, the causative agents behind the events of determination are multiple. That is, the mechanism of the determination of a region of the frog egg to become a limb bud is entirely different from that which determines what part of this limb bud will become muscle, which is in turn certainly different from the determination that a muscle cell will synthesize actin and myosin.

## Topographical Terminology

In anatomy it is important that we use well-defined terms by which we may relate parts of the body to other parts. Such terms as "under," "over," "up," "down," and so on vary according to the position of the body and are generally not used in descriptive anatomy.

The terms *ventral, dorsal, cranial,* and *caudal* are in common usage and should require no definition. However, there are some terms used in human anatomy that may be confusing. For the purpose of description in human anatomy, the body is assumed to be in what is known as the anatomical position: erect, with the hands at the sides and the palms directed forward. In this position, the normal direction in which a person would move is called *anterior,* the opposite direction being *posterior.* Thus, anterior is synonymous with ventral and posterior with dorsal. Two other terms are also used: *superior* or toward the head, synonymous with cranial, and *inferior* or toward the feet, synonymous with caudal. *Rostral* is a term that means toward the cranial end of the nervous system.

# 2

# THE REPRODUCTIVE SYSTEM AND THE GERM CELLS

"Ex ova omnia" is a statement made by the physiologist William Harvey in the 17th century. The *ovum* from which everything arises is the mature female *germ cell*. It usually develops into a new individual only after uniting with a male germ cell, the *spermatozoon*. Germ cells are highly specialized cells that are generally considered to be formed from stem cells called *primordial germ cells*, set aside from the body (somatic) cells early in development. There is convincing evidence in a number of species that the primordial germ cells are unique in their possession of a specific cytoplasmic component known as the *germ (germinal) plasm*. The germ plasm undergoes a precocious segregation into only a few cells of the early embryo, which then become the primordial germ cells, the stem cells for the germ line in each new individual.

The germ plasm presents a number of scientific problems such as its chemical nature, the method of its segregation into the primordial germ cells, and the mechanism of its action. These will be discussed in the chapter on the development of the reproductive system. In addition, it presents a philosophical problem. If we consider that the germ plasm is the factor that determines the germ line in each new individual and then becomes a part of every germ cell formed by the individual, we are faced with the question of its immortality. Since each individual, in turn, is the product of germ cells from his previous generation, this specialized material, the germ plasm, is the substance that links each individual to both his previous and his future generations. The germ plasm is thus self-perpetuating and lives on indefinitely and in each generation is retained in a temporary body before it passed on to the next generation. H. Bergson (1944) aptly expressed this concept:

*Life is like a current passing from germ to germ through the medium of a developed organism. It is as if the organism itself were only an excrescence, a bud caused to sprout by the former germ endeavoring to continue itself in a new germ. The essential thing is the continuous progress indefinitely pursued, an invisible progress, on which each visible organism rides during the short interval of time given it to live.*

The formation of mature male and female gametes from their respective stem cells is known as gametogenesis. Spermatogenesis refers to the production of spermatozoa, and oogenesis refers to the production of ova. Although gametogenesis is not, strictly speaking, a part of the study of embryology, a knowledge of its processes is fundamental. This,

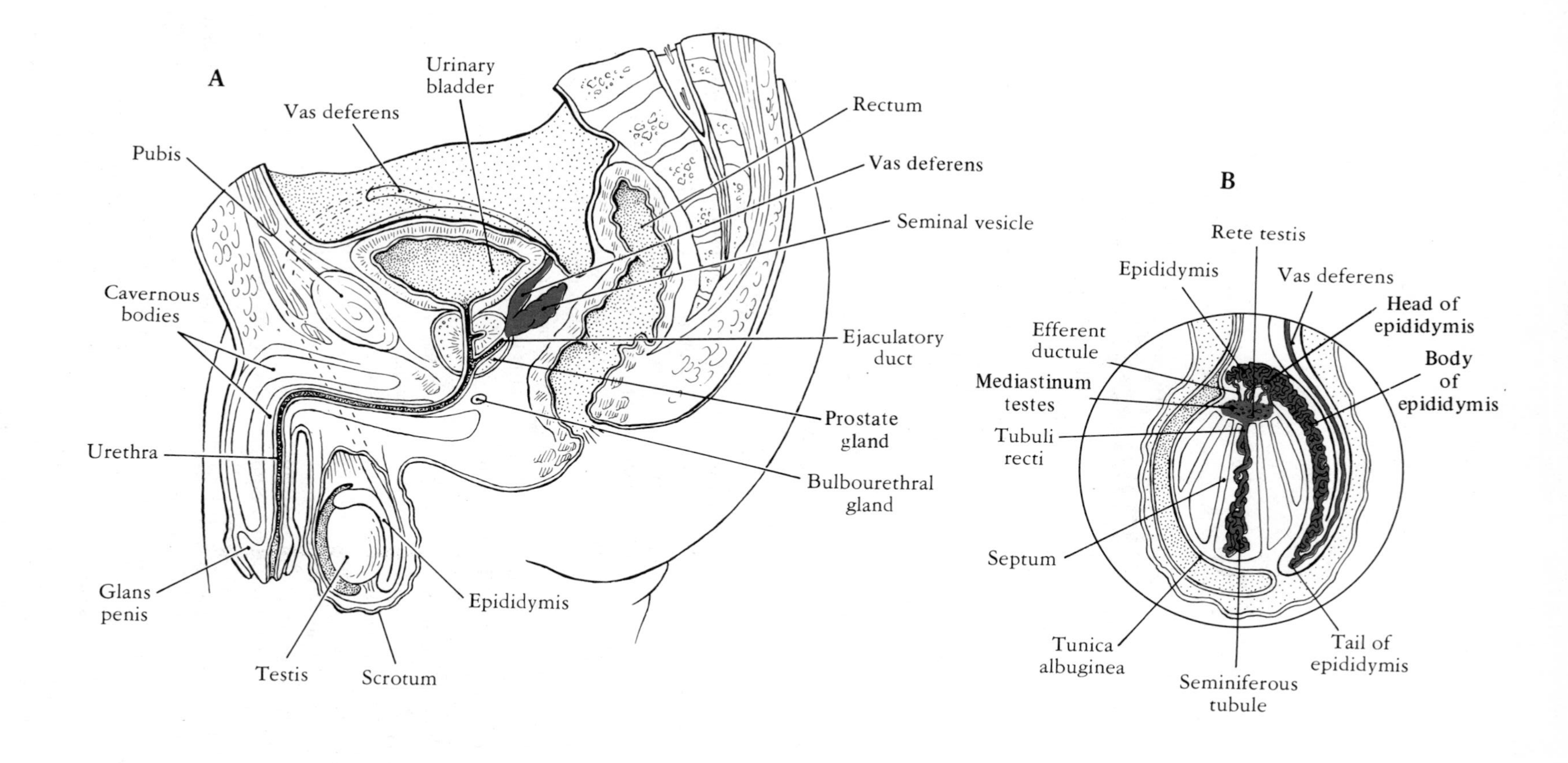


**FIG. 2–1.** A, sagittal section of the pelvic region of the male; B, sagittal section of the testis and scrotal sac.

in turn, leads us to a brief consideration of the anatomy of the male and female reproductive systems where these processes take place.

## The Male Reproductive System

The male reproductive system (Fig. 2–1 A,B) consists of the *primary sex organs* (gonads), the *testes*, and a set of *accessory sex organs* that includes the tubes and tubules through which the sperm pass to the outside, various glands that empty their secretions into these passageways, and a copulatory organ, the *penis*, by means of which sperm are deposited in the female reproductive tract. Not part of the reproductive system, but certainly associated with reproduction, are the body modifications characteristic of the different sexes, the *secondary sexual characteristics*. These include in the male such features as the higher coloration in fishes and birds, the cock's comb in the chicken, the thumb pad in the frog, and the enlarged larynx, the beard, and the pattern of pubic and cranial hair in man.

### The Testes

Although the testes develop in the abdominal cavity, they are, in the adults of most mammals, suspended in a sac, the *scrotum*, outside of the abdominal cavity. They perform a dual function: (1) the production of sperm and (2) the synthesis and secretion of male sex hormones, called *androgens*. *Testosterone* is the principal testicular hormone. Spermatogenesis takes place within a set of tortuous tubes, the *seminiferous tubules*, which are lined with *germinal epithelium*. Located between the seminiferous tubules are cells known as *interstitial* or *Leydig cells* that produce the androgens. They are responsible for the maintenance of the accessory sexual structures and the continued expression of the secondary sexual characteristics.

### The Reproductive Tract

The path by which the sperm reach the outside may be traced by reference to Figure 2–1 A,B. Ducts, nerves, and blood vessels enter and leave the testis at its posterocephalic margin, a region known as the *mediastinum*. Radiating out from the mediastinum are connective tissue septa that divide the testis into over 200 lobules each containing a number of highly convoluted seminiferous tubules. As the seminiferous tubules approach the mediastinum, those in each lobule unite to form straight tubules, *tubuli recti*. The tubuli recti enter the mediastinum and form a branched network, the *rete testis*. From the rete, a number of ducts, the *vasa efferentia*, carry the sperm into a single common duct, the *epididymis*. The convoluted epididymis, over 15 feet long in man, lies along the posterior aspect of the testis. It may be divided into head (caput), a body (corpus), and a tail (cauda). The epididymis, in turn, connects with the *vas deferens* which passes craniad out of the scrotal sac through the inguinal canal into the abdominal cavity. At the base of the prostate gland the vas deferens is joined by the duct of the *seminal vesicle*. The common duct, now called the *ejaculatory duct*, runs between the lobes of the prostate and opens into the *prostatic urethra*—the portion of the urethra surrounded by the prostate gland. The ejaculatory ducts from either side converge as they run through the prostate gland and open into the urethra in close proximity to each other. Secretions from the seminal vesicles and the prostate form a large part of the ejaculate. From the point of entrance of the vas deferens into the urethra, the urethra in the male is a common pathway for both the urinary and the reproductive systems.

## The Female Reproductive System

The female reproductive system consists of the gonads, the *ovaries*, and the accessory sex organs, the *uterine tubes*, *uterus*, and *vagina*. The external genitalia consist of the *labia minora*, the *labia majora*, and the *clitoris* (Fig. 2–2).

### The Ovaries

The ovaries are paired organs lying within the pelvic cavity on either side of the uterus, attached to the posterior surface of the *broad ligaments* by a mesentery known as the *mesovarium*. The broad ligaments are folds of peritoneum that extend from the sides of the uterus to the sides of the pelvic walls. Each ovary is also attached to the uterus by an *ovarian ligament*.

Beneath the outer peritoneal covering of the ovary, the serosa, is a single layer of cuboidal or low columnar germinal epithelium. Below the germinal epithelium, subdivided by a network of connective tissue, are nests of undeveloped *primary fol-*

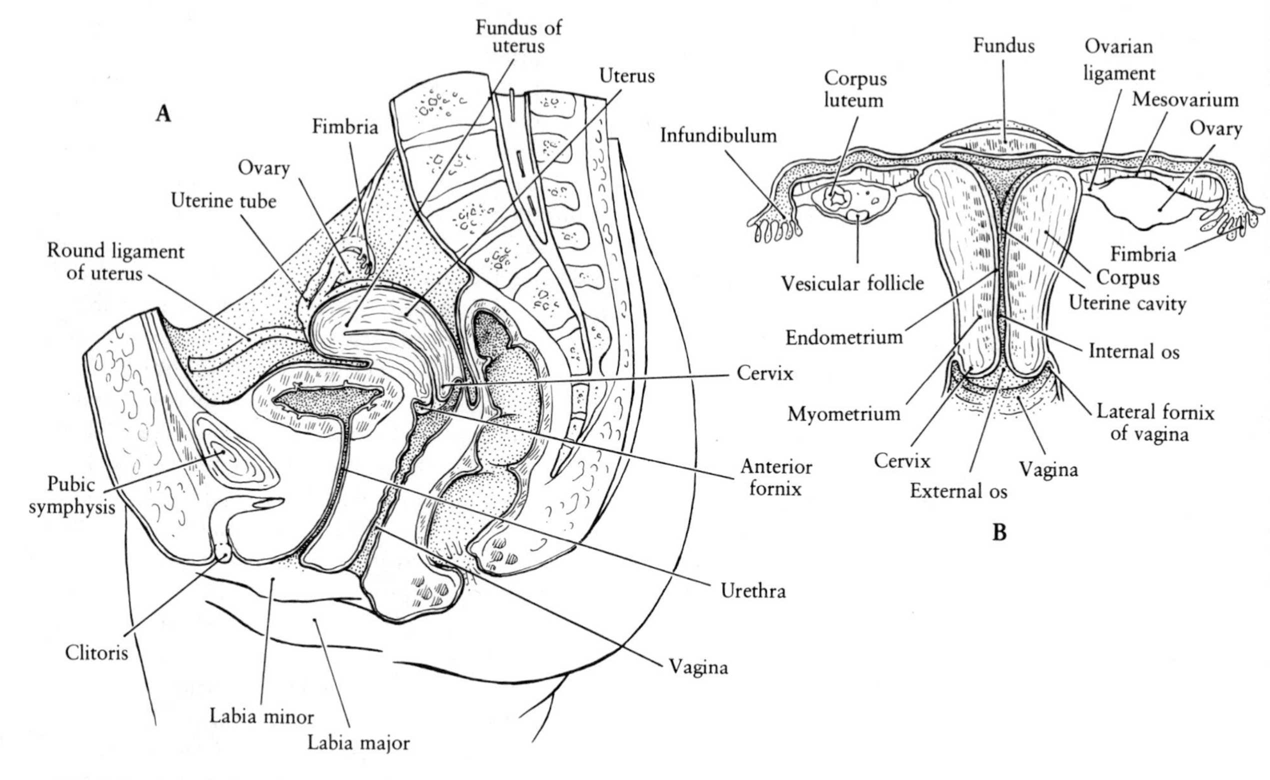


**FIG. 2–2.** A, sagittal section through the pelvic region of the female, B, posterior view of the ovaries, uterine tubes, and uterus.

*licles, developing follicles,* and *corpora lutea.* The sequence of development of the ovarian follicles and the formation of the corpora lutea after ovulation will be described in a later chapter.

During the embryonic development of the ovary, cells from the germinal epithelium move into the ovarian cortex to form the primary follicles. It has been estimated that there may be as many as 400,000 primary follicles in the human ovary at the time of birth. While there is a possibility that some germ cells may be formed from the germinal epithelium postnatally, it is more probable that all of the follicles that develop to maturity after puberty are those that are already present in the ovaries at the time of birth. Since in humans only a single ovum is normally released from the ovaries each month during the reproductive lifetime of the individual, it is apparent that a tremendous number of ova never mature but instead undergo *atresia.*

## The Reproductive Tract

The open, fimbriated ends of the uterine tubes *(oviducts)* encircle one pole of each ovary. Each uterine tube follows a C-shaped course back toward its opening into the uterus (Fig. 2–2 B). The uterus is a thick-walled muscular organ situated between the bladder anteriorly and the rectum posteriorly. The muscular layer *(myometrium)* makes up the bulk of the uterine wall. The thinner inner part of the wall *(endometrium)* is covered by a simple columnar epithelium and contains numerous tubular uterine glands. At *menstruation,* portions of the endometrium are sloughed off. This pear-shaped organ can be divided into, in order from its cranial to its caudal end, the *fundus,* the *corpus,* the *isthmus,* and the *cervix.* The rather small uterine cavity opens through the cervix into the vagina. The cervix of the uterus projects downward into the vagina so that a recess, the *fornix,* is formed between the two. Since the cranial end of the uterus is tipped anteriorly, the posterior fornix is deeper and the posterior wall of the vagina is longer than the anterior. The slit-shaped cavity of the vagina opens into the *vestibule* between the labia minora. The urethra opens into the vestibule anterior to the opening of the vagina.

Sperm are deposited in the vagina and pass

through the reproductive tract of the female to the upper part of the uterine tubes where, if an ovum is present, fertilization takes place. The fertilized ovum then passes back down through the uterine tube and into the uterus where it implants and develops.

## Cell Division in Somatic Cells and Germ Cells

### Mitosis

All of the body or somatic cells are normally diploid. That is, they have two full sets of homologous chromosomes, having received one set from each parent. Each cell contains the same number of chromosomes, a number characteristic of the species (46 in humans). Cell reproduction is accomplished by mitotic divisions in which each daughter cell receives a chromosome complement that is exactly the same as that of the parent cell.

Cell division has been studied in innumerable plant and animal cells and is known to consist of a fairly complicated process divided into different phases progressing from *prophase* through *metaphase* and *anaphase* to *telophase* (Fig. 2–3). In prophase, (Fig. 2–3 A,B) the chromosomes become increasingly coiled and condensed, and at the end of prophase, individual chromosomes may be seen with the aid of the light microscope. Each chromosome is seen to be double, made up of two *chromatids*. During prophase, the nuclear membrane begins to break down and the nucleolus disappears. In metaphase (Fig. 2–3 C), a *spindle* develops whose fibers radiate from *centrioles* located at opposite poles of the cell. The chromosomes become aligned in a plane passing through the middle of the cell, the metaphase plate. Spindle fibers attach to the *centromere (kinetochore)* of each chromosome. At anaphase (Fig. 2–3 D), the centromeres divide and the sister chromatids move toward opposite poles with the centromeres leading the way. Each of the chromatids may now be called a daughter chromosome. Movement to the opposite poles is completed at telophase (Fig. 2–3 E), when the nuclear membrane develops, the nucleolus reappears, the chromosomes uncoil, and the spindle disappears. Cell division, *cytokinesis*, divides the telophase cell into two daughter cells, each with identical genetic material (Fig. 2–3 F).

The actual process of mitosis usually occupies only a small part of what is known as the *cell cycle*. The major part of the cell cycle is spent in what has been termed the *interphase*, that is, the period between mitotic divisions, a period when the individual chromosomes are not visible. However, interphase is not a period of inactivity—as the term resting phase, which was once its synonym, implied—but one of considerable metabolic and synthetic activity.

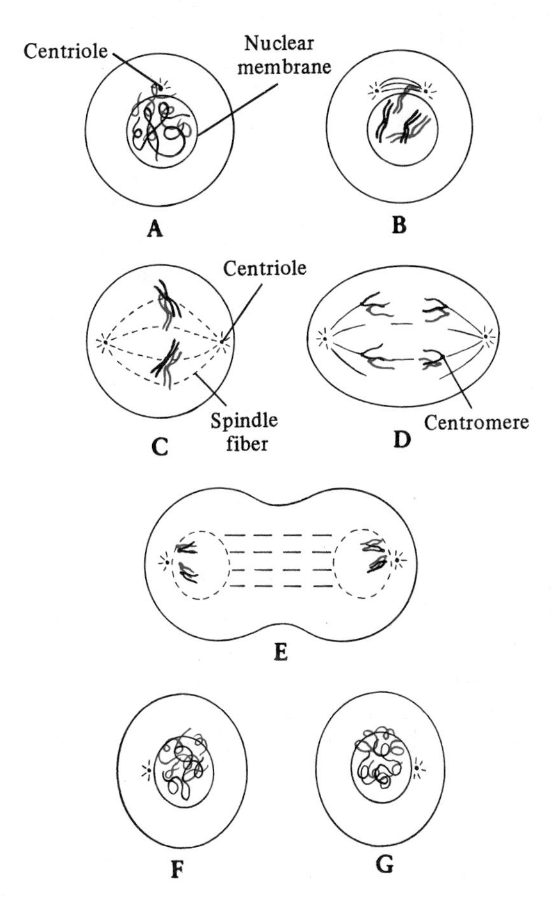


**FIG. 2–3.** Mitotic division in a cell with two pairs of chromosomes. For description see text.

The technique of autoradiography has been an extremely useful tool in the analysis of the cell cycle (Fig. 2–4) and in particular in demonstrating that one phase of the cycle shortly before mitosis is the time when the cell synthesizes DNA. This phase is the S phase, and during this period—and this period only—is a radioactive precursor of DNA, tritiated thymidine, taken up by the cell. A short postsynthetic $G_2$ phase occurs between the S phase and mitosis (M phase). A $G_1$ phase follows cell division. In most cells this is the longest phase of the cycle even in such rapidly dividing cells as those of the

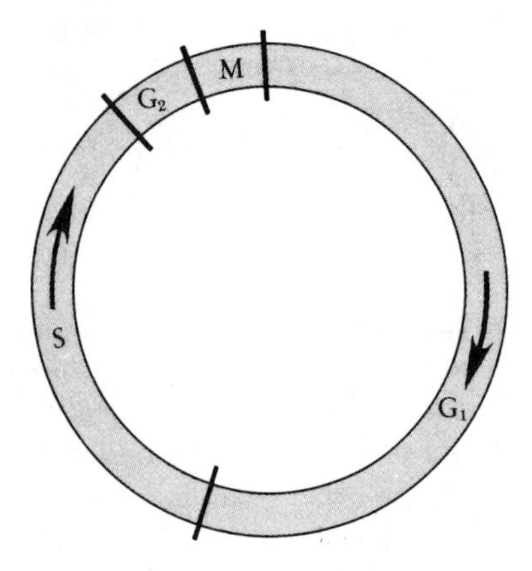


FIG. 2–4. Diagram of the cell cycle. S, synthetic phase; M, mitosis; $G_1$ and $G_2$, presynthetic and postsynthetic phases, respectively.

lining of the intestine. Synthesis of DNA indicates that chromosome replication is occurring; thus, shortly before mitotic division—during the $G_2$ phase—each nucleus contains an amount of DNA that is double that of the presynthetic cell. This material is then halved at mitosis and the diploid value restored in each daughter cell (Fig. 2–5 A).

## Meiosis

Although the stem cells of the germ line, *spermatogonia* and *oogonia*, are diploid cells and divide mitotically to reproduce themselves, the mature germ cells that are derived from them contain only half the number of chromosomes characteristic of the somatic and stem cells of the species. They are haploid. At fertilization, the union of the 1N or haploid cells restores the diploid number. If the germ cells were not haploid, the total number of chromosomes would, of course, double in each generation.

During the maturation of the germ cells, the stem cells, *gonia*, undergo a series of events that involves two divisions of the nucleus, which results in the halving of the chromosome number, since during these two nuclear divisions the chromosomes duplicate themselves only once. The divisions are called *maturation* or *meiotic* divisions, although it is only in the prophase of the first division that the chromosomes behave differently from the way they do in ordinary mitosis.

Prior to any maturation division the gonia go through a period of growth that results in a considerable increase in size for most oogonia but only a small increase for spermatogonia. At the end of this growth period, the germ cells are *primary oocytes* and *primary spermatocytes*. These cells now undergo their first meiotic divisions.

FIG. 2–5. Diagram illustrating the amount of DNA present in the cell nucleus during the different stages of the cell cycle in mitosis (A) and meiosis (B).

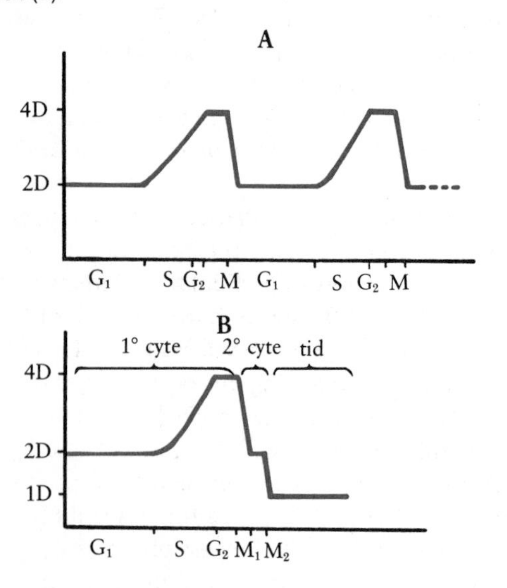


### *Meiosis I*

Meiosis I is characterized by a long and complicated prophase that is divided into a number of stages (Fig. 2–6). The first stage is *leptotene*, when the chromosomes condense into long, threadlike, loosely coiled structures. Although the leptotene chromosomes are not visibly double, DNA synthesis studies show that each chromosome has already duplicated itself in the preceding S phase

The next stage is *zygotene*, when homologous segments of maternal and paternal chromosomes—and thus the chromosomes themselves—pair. Pairing of homologous chromosomes is known as *synapsis*, and it results in a nucleus that appears to have only the haploid number of chromosomes although neither the number of chromosomes nor the amount of DNA has changed.

In the next stage, *pachytene*, the shortening and thickening of the chromosomes that was begun in zygotene continues. Each synaptic pair is called a *bivalent*. It is now possible to determine that each bivalent consists of four chromatids, forming what is known as a *tetrad*.

In the next stage, *diplotene*, the individual chromatids are more apparent. In synapsis, in leptotene, homologous chromosomes pair over their entire

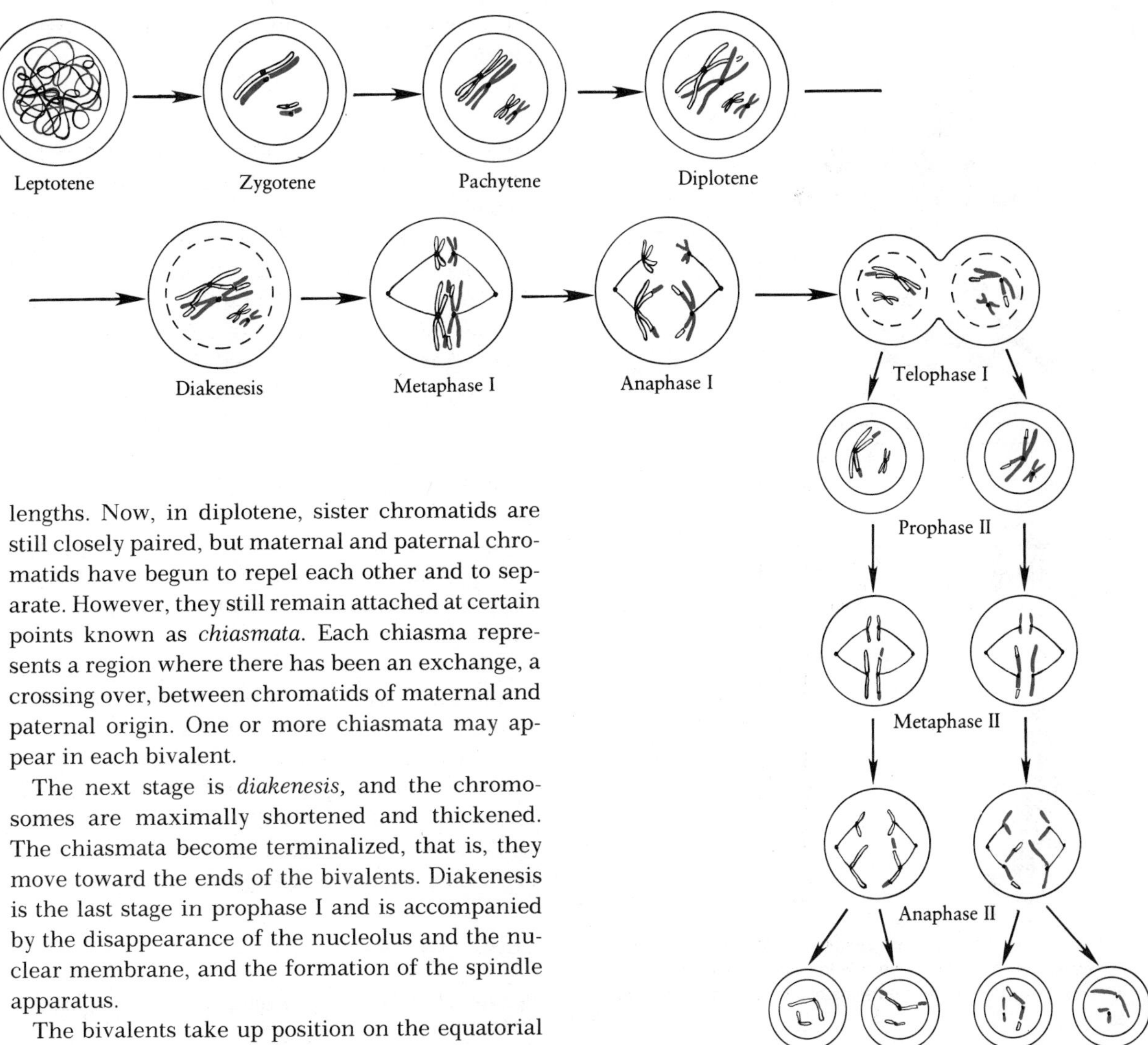


FIG. 2-6. Meiotic divisions in a germ cell with two pairs of chromosomes. (For description see text.)

lengths. Now, in diplotene, sister chromatids are still closely paired, but maternal and paternal chromatids have begun to repel each other and to separate. However, they still remain attached at certain points known as *chiasmata*. Each chiasma represents a region where there has been an exchange, a crossing over, between chromatids of maternal and paternal origin. One or more chiasmata may appear in each bivalent.

The next stage is *diakenesis,* and the chromosomes are maximally shortened and thickened. The chiasmata become terminalized, that is, they move toward the ends of the bivalents. Diakenesis is the last stage in prophase I and is accompanied by the disappearance of the nucleolus and the nuclear membrane, and the formation of the spindle apparatus.

The bivalents take up position on the equatorial plate of the spindle during metaphase, and spindle fibers from opposite poles attach to each homologous chromosome at its kinetochore. During anaphase, homologous chromosomes move toward the opposite poles. Anaphase I ends with a haploid set of chromosomes at each pole. However, each is a double structure consisting of two chromatids, known as a *dyad*.

Telophase I is short and usually abortive, and the cell moves rapidly into the next division. When a short interphase does occur it is only a transition stage, and no DNA synthesis or chromosome duplication takes place. Thus, following the first maturation division, each daughter cell contains a haploid number of chromosomes. However, each chromosome is double—as a result of DNA synthesis in the previous S phase—and the amount of DNA present is characteristic of a diploid cell (Fig. 2–5 B). Each duplicated chromosome is of either maternal or paternal origin, except where crossing over has produced interchanges.

At the end of the first meiotic division, each primary spermatocyte has produced two functional daughter cells, *secondary spermatocytes*. However, each primary oocyte produces only a single *second-*

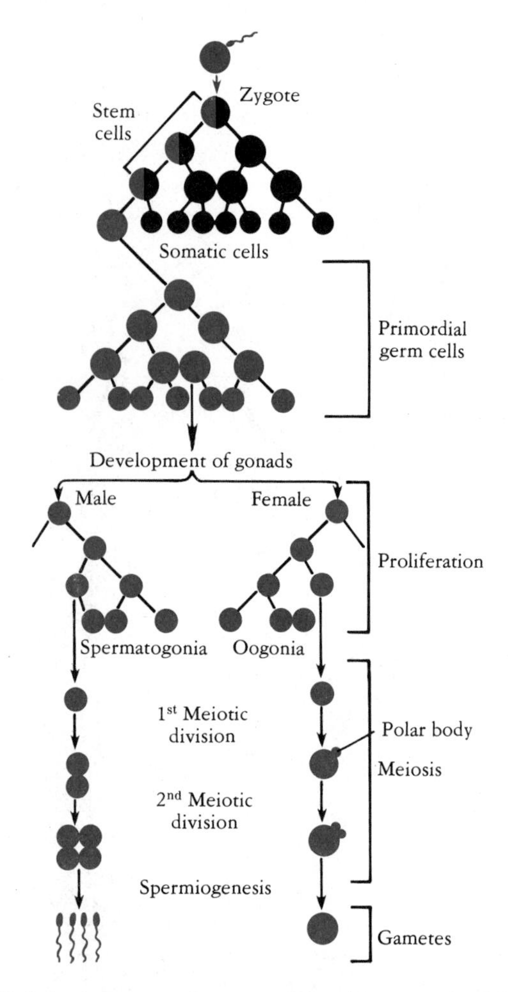


**FIG. 2–7.** A diagrammatic representation of the separation, following fertilization, of the germ line cells from the somatic line cells and the proliferation and meiotic divisions of the male and female germ cells.

*ary oocyte* and a small nonfunctional *polar body* (Fig. 2–7). The secondary spermatocytes and oocytes now continue with the second maturation division.

*Meiosis II*

Although the second maturation division is preceded by an abortive interphase in which there is no S phase, the nuclear division is exactly the same as that which occurs during normal mitosis. If an interphase stage is lacking, no prophase occurs; and the dyads align along the equatorial plate of the spindle and separate into two units, each of which is consigned to a daughter cell. Each daughter cell now contains the haploid number of chromosomes and also an amount of DNA characteristic of a haploid nucleus (Fig. 2–5 B).

Each secondary spermatocyte produces two *spermatids*, which will develop into two spermatozoa in a process called *spermatid maturation* or *spermiogenesis*. Although each secondary oocyte also produces two daughter cells, again only a single one is functional, the other forming another small polar body.

A diagrammatic representation of the separation of the germ line cells from the somatic line cells and the proliferation and meiotic divisions of the male and female germ cells is shown in Figure 2–7. Each spermatogonium forms four mature sperm while, as will be described in the next chapter, each oogonium forms only a single mature ovum.

## References

Bergson, H. 1944. Creative Evolution. New York: Random House.

Smith, L. D. 1975. Germinal plasm and primordial germ cells. 33rd Symposium, The Society for Developmental Biology. Eds., C. L. Markert and J. Papaconstantinou. New York: Academic Press.

# 3

# SPERMATOGENESIS

In the previous chapter we noted that the seminiferous tubules of the testes were lined with germinal epithelium. The cells of the germinal epithelium are made up of male germ cells in all stages of transition from spermatogonia to mature sperm. A cross section of a testis will show as its main feature seminiferous tubules cut in many different planes. For a study of spermatogenesis, tubules cut in cross section should be selected. Three such tubules from a human testis are represented in Figure 3–1.

## Spermatid Development

The most immature cells, the spermatogonia, are found forming a few layers closest to the basement membrane of the tubule. Two types of spermatogonia are found in the testis: type A and type B. Type A spermatogonia may be further subdivided in the human testis into those with pale and those with dark nuclei, but essentially type A spermatogonia are a population of cells that represent the stem cells of the germ line and divide mitotically. However, the pale type A spermatogonia produce two daughter type B spermatogonia that may be recognized histologically by their larger, spherical, lightly staining nuclei containing flakes and granules of chromophilic chromatin. These histological changes then define spermatogonia that have taken the beginning step in the irreversible process of differentiation that will end up in the formation of mature spermatozoa. This process is divided into stages by giving names to the different cell types that can be recognized in the seminiferous epithelium, but we should realize that the differentiation process is in reality a continuous modification involving a diversity of morphological and functional changes taking place within the germ line lasting over a period of months.

Successive steps in the formation of spermatozoa will be seen progressing from the periphery of the tubule to its lumen. The first step, the formation of *primary spermatocytes*, does not involve cell divisions but results from the growth of type B spermatogonia to about double the size of the original cells. Each primary spermatocyte now goes through the two maturation or meiotic divisions, first forming two *secondary spermatocytes* and from these four *spermatids*, each with the haploid number of chromosomes. Since the secondary spermatocytes divide rapidly without going through an interphase, they are often difficult to locate in sections of the seminiferous tubules.

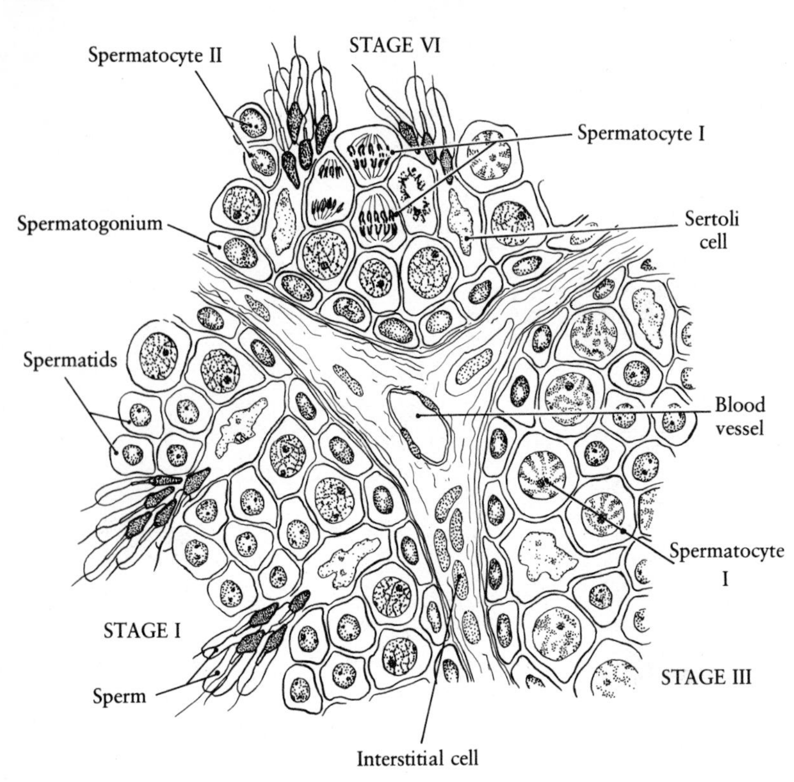


**FIG. 3–1.** Diagram of the seminiferous tubules in man.

## Spermatid Maturation

After the formation of the four spermatids from each of the original spermatogonia, spermatogenesis is still not complete. The task of the mature spermatozoon is to carry to and introduce into the egg the genetic material of the male parent. Spermatids are not adapted to carry out this function and must undergo extensive morphological and biochemical modification before they are capable of doing so. This process is called spermatid maturation or *spermiogenesis*. It does not involve any cell division but is concerned with getting rid of those structures not necessary for the task at hand and modifying those structures that are retained to form a highly specialized cell geared for motility and penetration. Spermatozoa are spermatids "stripped for action." The remodeling of the spermatid will be followed by considering in turn the parts of the mature sperm (Fig. 3–2) and the processes by which they are formed. A diagram of a mature sperm, which includes the cellular components of the spermatid that contribute to its formation, is shown in Figure 3–3.

### The Head

The nucleus of the spermatid becomes the major part of the *head* of the sperm, forming a small compact mass of chromatin in a shape characteristic of the species. The human sperm is oval when viewed on its flat surface and lanceolate when viewed on its edge. The amount of nuclear material appears to be smaller in the sperm than in the spermatid, but this is due to a condensation of the chromatin rather than any reduction in amount. In sections prepared for electron microscopy, the condensed chromatin in the mammalian sperm appears homogeneous or coarsely granular, although freeze-etching studies reveal a structural order in the form of lamellae parallel to the flattened surfaces.

### The Acrosome

The *acrosome* appears as a membrane-bounded vesicle covering the anterior end of the head. It is of almost universal occurrence in all sperm so far examined with the exception of some fishes and some insects. It shows a wide variety of shapes and

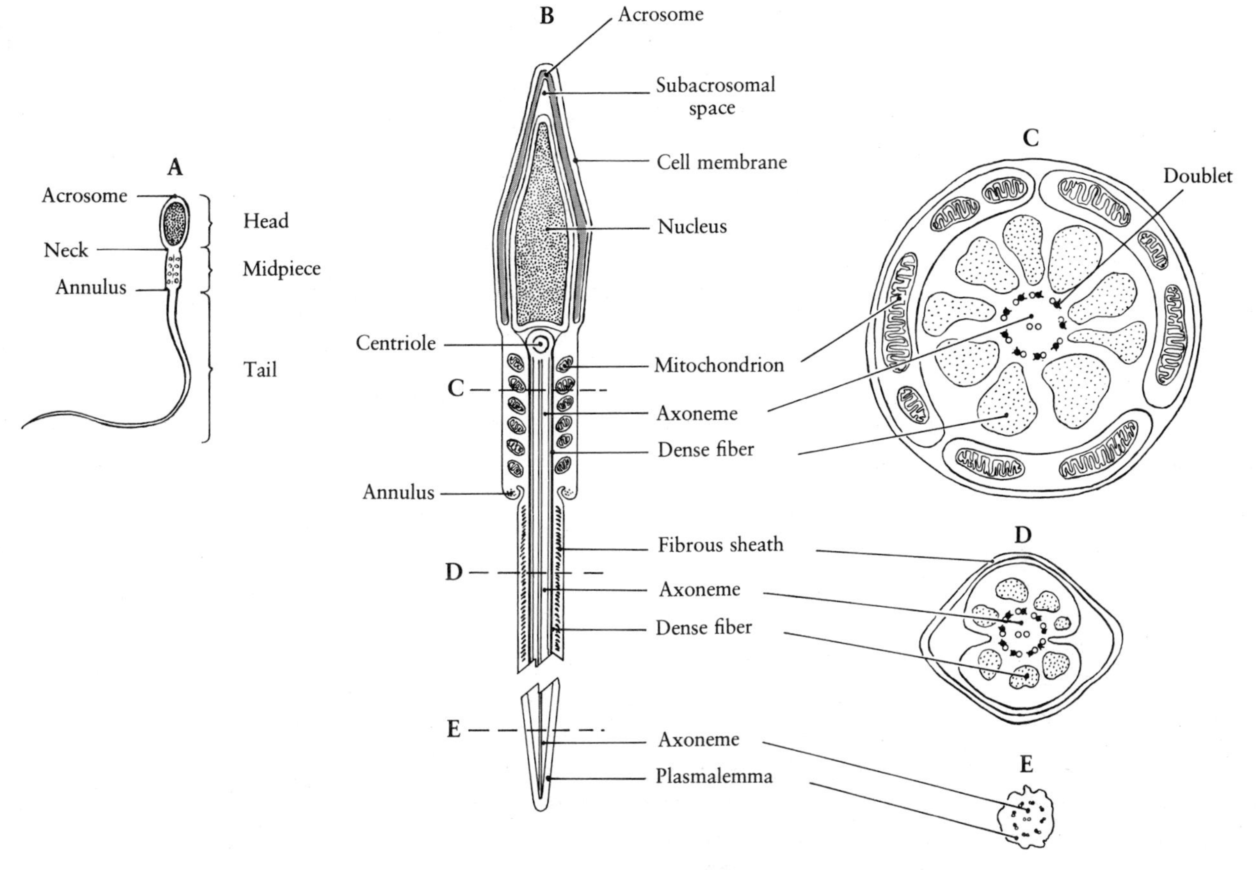


**FIG. 3–2.** A, diagram of a typical mammalian sperm; B, structural features, C, D, and E, cross sections through the middle piece, principal piece, and end piece at the levels indicated. (After J. Lash and J. R. Whittaker, 1974. Concepts of Development. 1st ed. Sinauer Associates, Sunderland, Mass.)

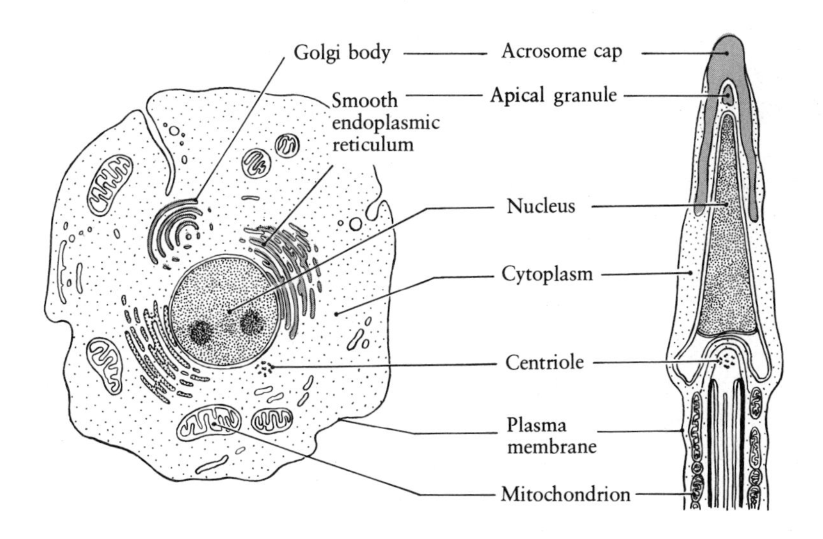


**FIG. 3–3.** A generalized diagram of the role of the cellular elements of the spermatid in the formation of the structures of the mature sperm. (After R. Hadek, 1969. Mammalian Fertilization. Academic Press, New York.)

sizes in different species. The acrosome develops in the Golgi complex, which becomes oriented close to the anterior end of the head early in spermatid maturation. Frequently the acrosome first appears as a proacrosomal vesicle between the Golgi complex and the nucleus (Fig. 3–4). It is not surprising that the acrosome is associated with the Golgi complex, since the acrosome is a specialized secretion granule containing lytic enzymes and the Golgi apparatus is known to be involved in the formation of secretion granules. The action of these acrosomal lytic enzymes will be described in the section on fertilization. As the original acrosome vesicle enlarges, it spreads over the anterior surface of the nucleus (Fig. 3–5) and eventually covers somewhat more than the anterior half. An acrosomal granule forms a small distinct element in the human sperm, and the acrosomal vesicle forms what is often referred to as the *anterior head cap*, although the term *acrosomal cap* better reflects its origin.

## Neck and Flagellum

The *neck* is the small region just posterior to the head and connects it to the *middle piece*. The important developmental processes occurring in the neck region are concerned with the centrioles and the formation of the *flagellum*. The spermatid has two centrioles located just below the plasma membrane, a *proximal centriole* closest to the nucleus, and a *distal centriole*. The flagellum develops in association with the distal centriole. The centrioles move from their original peripheral position toward the nucleus, and the proximal centriole assumes a position indenting the posterior pole of the nucleus (Fig. 3–6). Early in spermiogenesis the

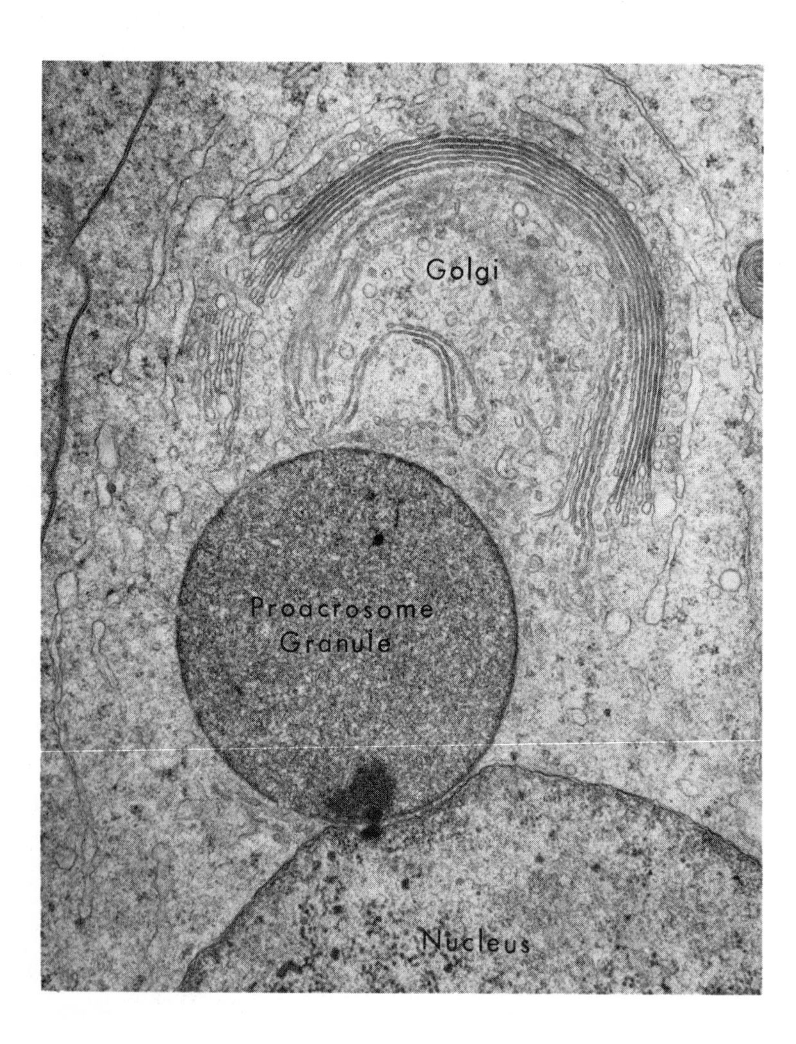


**FIG. 3–4.** Formation of the acrosome in an insect, showing the relation to the Golgi apparatus. (From D. M. Phillips, 1974. Spermiogenesis. Academic Press, New York.)

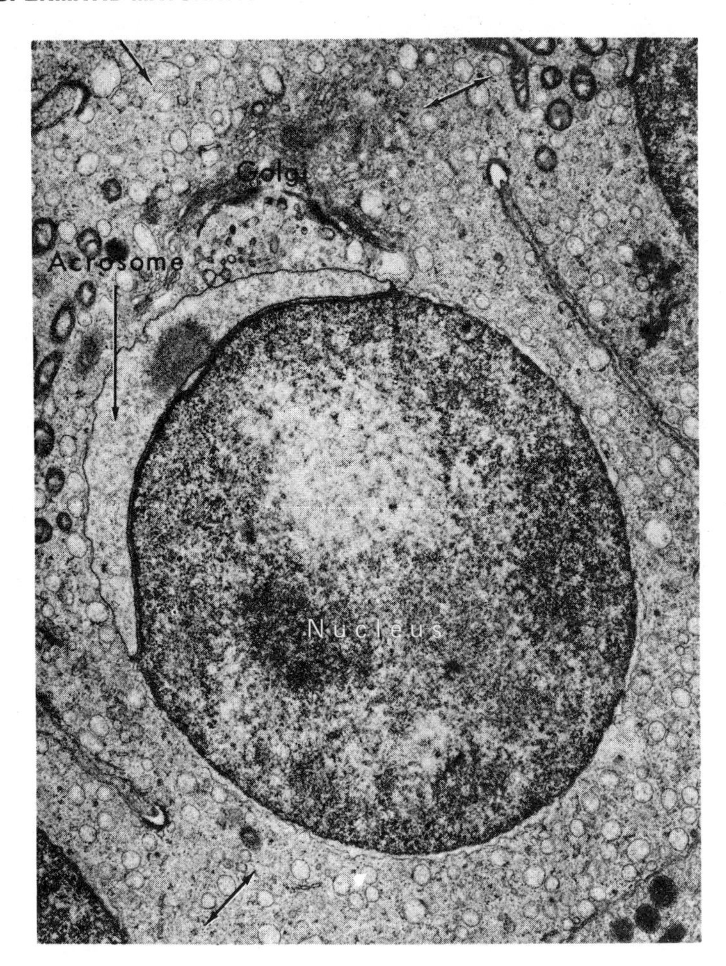


**FIG. 3–5.** An early stage in the formation of the acrosome in the rabbit. (From D. M. Phillips, 1974. Spermiogenesis. Academic Press, New York.)

plasma membrane becomes associated with an electron-dense material lying just distad of the centriole pair. It is called the *chromatid body.* This structure, of as yet unknown origin, will form the *annulus.* As the centrioles and the central body migrate toward the nucleus, the plasma membrane retains its connection with the developing annulus and forms an involution around the flagellum (Fig. 3–7). The annulus later moves distally and marks the end of the middle piece. It is sometimes called the *ring centriole.* This is a rather unfortunate term, since the annulus, although closely associated with the centrioles, does not develop from them; and electron microscopy studies show that it does not have the structure characteristic of a centriole.

The flagellum develops a central axial structure, the *axoneme,* which consists of two central single tubules surrounded by nine pairs of tubules or doublets. This structure is characteristic of the axial filaments of flagellae and cilia developed as locomotor organs in a variety of animals. Surrounding the axoneme in the mammalian sperm are nine outer *dense fibers,* each associated with one of the nine doublets. A fibrous sheath forms the outer covering (Fig. 3–2).

## The Middle Piece

The *middle piece* consists of the flagellum surrounded by a mitochondrial element. The mitochondrial sheath is in the form of a helical structure located outside of the dense fibers. The middle piece extends from the neck to the annulus (Fig. 3–2).

## The Principal Piece and the End Piece

The *principal piece* and the *end piece* run from the annulus to the tip of the flagellum. The axoneme is present as the core in both of these regions. In the

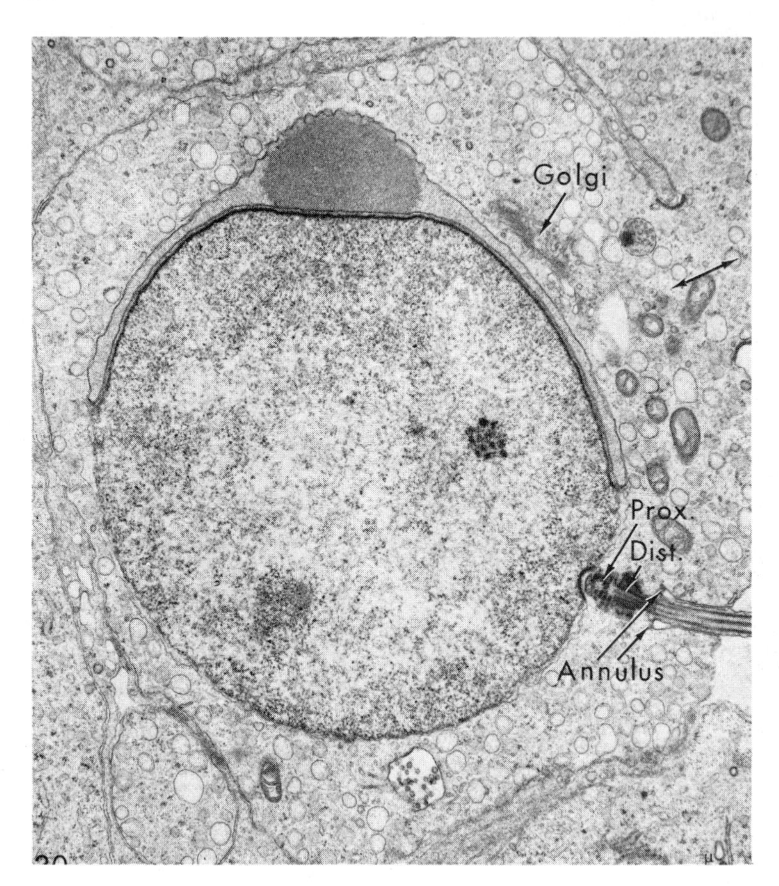


**FIG. 3–6.** Centrioles and annulus of the maturing rabbit spermatid at a stage later than that shown in Figure 3–5. (From D. M. Phillips, 1974. Spermiogenesis. Academic Press, New York.)

principal piece the nine-plus-two tubular structure is surrounded by seven dense fibers instead of nine as in the middle piece. The fibrous sheath that surrounds the dense fibers shows two longitudinally running columns on opposite sides of the flagellum, each making contact with a doublet (Fig. 3–2). The dense fibers of these doublets are the two that are missing from the original nine found in the middle piece. Progressing toward its tip, the flagellum gradually tapers as the dense fibers and the fibrous sheath become thinner and eventually disappear. Approximately the last 5 microns of the tail, the end piece, consist of the axoneme covered by only a thin layer of cytoplasm and its plasmalemma (Fig. 3–2 E).

### The Cytoplasm

The surface of the spermatozoon is covered by only a very thin layer of cytoplasm lying under the plasma membrane, and thus a large volume of the cytoplasm of the spermatid is eliminated during spermiogenesis. This residual cytoplasm containing the leftover inclusions and organelles not used in the formation of the constituent parts of the spermatozoon is eliminated simply by being relegated to a cytoplasmic baglike projection, the *residual body* (Fig. 3–7 E), which is eventually sloughed off from the spermatid. A second method of cytoplasmic elimination takes place through the formation of a special type of cell complex between the spermatids and the surrounding supporting cells that will be described in the section on the Sertoli cell later in this chapter.

## The Mature Sperm

The mature sperm is thus a specialized cell that is efficiently packaged to carry the paternal DNA condensed in its head to the vicinity of the ovum. As a locomotor organ, it uses the flagellum, with special contractile elements, the dense fibers, to which energy is supplied by the mitochondria of the middle piece. Once contact with the ovum is made, the acrosome functions to penetrate the egg membranes, and the male nucleus moves into the cytoplasm of

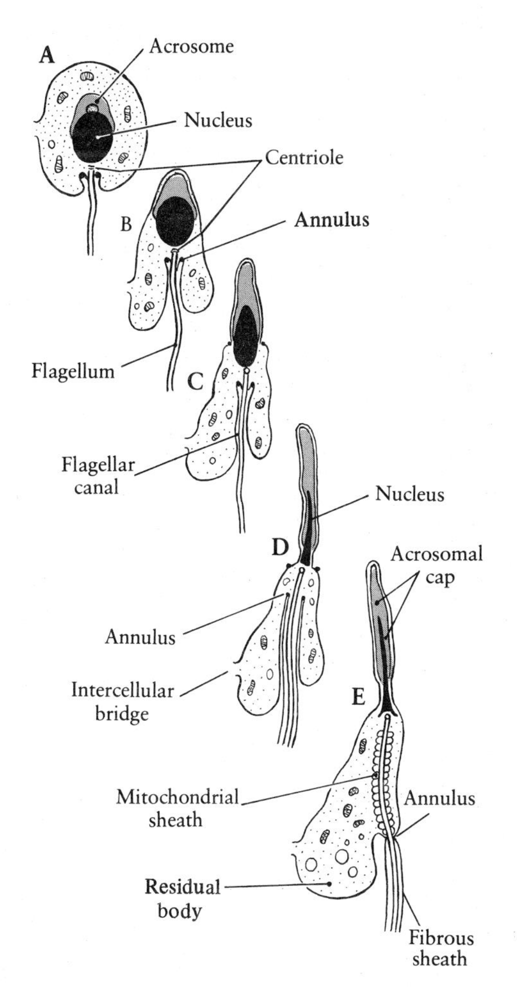


**FIG. 3-7** Diagram of spermatid maturation in the guinea pig. (After D. W. Fawcett et al., 1971. Dev. Biol. 26: 220.)

the egg. As part of the package, the sperm centriole moves into the cytoplasm, and there it functions in forming the achromatic spindle apparatus of the first division of the zygote nucleus.

## Sertoli Cells

In addition to the germinal elements, the seminiferous tubules also contain a population of nongerminal cells that have been variously called nurse cells, sustentacular cells, and supporting cells, but more commonly *Sertoli cells*. The close association of the Sertoli cells with the germinal cells suggests that they must be in some way implicated in the process of the maturation of the germ cells. In fact they play a more important role than originally suspected, not only in relation to germ cell maturation but also as a part of the endocrine system, and recent studies on the morphology and function of the Sertoli cells represent one of the major advances in recent years in the field of reproductive physiology.

### Sertoli–Germ Cell Interrelations

Spermatogonia are located around the basal portion of the seminiferous tubule, and as they mature successive stages are found progressively closer to the center of the tubule until mature sperm are liberated into the lumen. Ultrastructural studies depicting the morphology of the Sertoli cells and their relationships to the differentiating germ cells have shown that the Sertoli cells are responsible for the directed movement of the differentiating germ cells toward the lumen of the seminiferous tubule as well as the eventual release of the mature spermatozoa into the lumen.

Morphologically the Sertoli cell has been likened to a tree with its trunk resting on the basal lamina of the seminiferous tubule, extending from there into the lumen of the tubule with branches (actually sheets of tissue) that pass laterally between the nearby germ cells (Figs. 3–8, 3–9). In the basal region of the tubules these branches may make contact with the branches of adjacent Sertoli cells. These contacts have been described as forming desmosome-like junctions, occluding junctional complexes, that prevent substances that may pass freely through the basal lamina of the seminiferous tubule and between the germ cells from moving any further toward the lumen than the level at which the complexes are formed, in the region just above the spermatogonia and the preleptotene primary spermatocytes. A barrier, the blood–testis barrier, is thus formed relegating these types of germ cells to a basal compartment and all other types of germ cells to the intermediate and adluminal compartments (Fig. 3–9). The blood–testis barrier may be likened to the blood–brain barrier but is morphologically different in that the blood–brain barrier is formed by specializations of the capillary walls rather than specialized occluding junctions of cells outside of the vascular system.

Desmosome-like junctions are also formed between Sertoli cells and germ cells. They are poorly developed between Sertoli cells and type A spermatogonia, are highly developed and conspicuous between Sertoli cells and type B spermatogonia and preleptotene spermatocytes, but gradually become

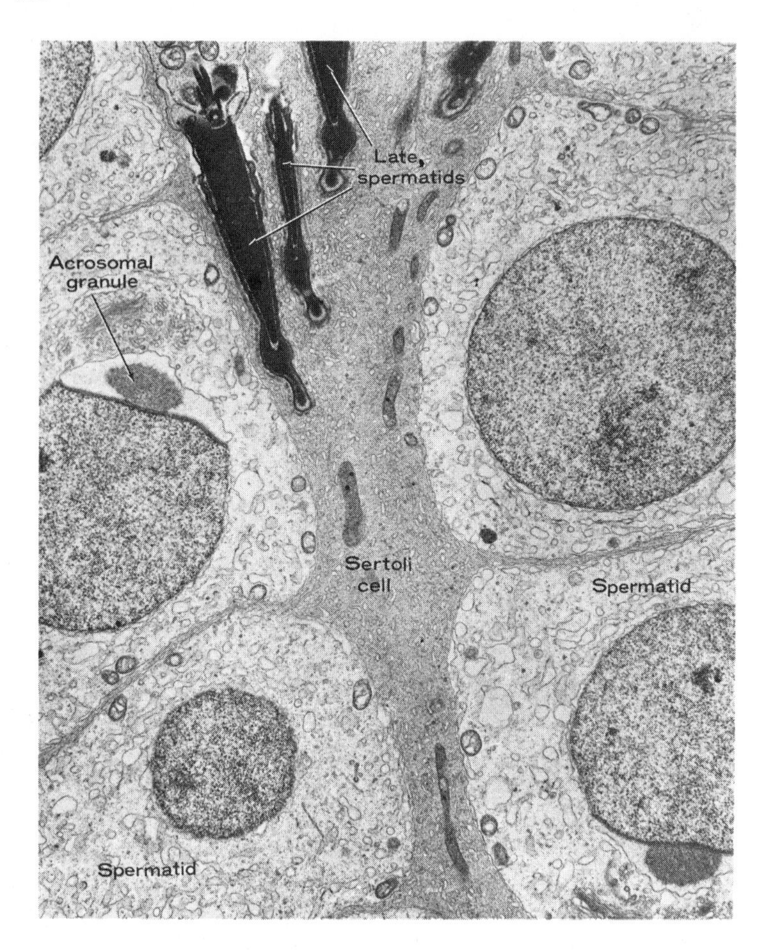


**FIG. 3-8.** Relationship of a Sertoli cell to maturing spermatids. Sertoli-cell processes passing between younger spermatids. Advanced spermatids found in recesses at the luminal end of a Sertoli cell. (Courtesy of D. W. Fawcett.)

fewer and less conspicuous between dividing meiotic cells and are rarely seen in stages later than the young spermatids. It is proposed that the junctional complexes would serve to move the germ cells toward the lumen of the seminiferous tubule as the Sertoli cells undergo configurational changes of their cytoplasm that move the complexes toward the luminal region of the cell carrying the attached germ cells with them.

As germ cell maturation progresses to the stage of spermiogenesis, the elongating spermatids become positioned in deep recesses of the main stem of the Sertoli cell, their heads oriented toward the basal lamina of the seminiferous tubule and their tails toward the lumen (Fig. 3–9). A different type of specialization is found in the surface cytoplasm of the Sertoli cells that line these recesses. Known as *ectoplasmic specializations,* they are characterized by actin or actinlike fibers attached to the plasma membrane and to underlying saccules of the endoplasmic reticulum (Fig. 3–9 C). It is generally concluded that these Sertoli cell specializations play a major role during the release of the spermatozoon, although the exact nature of the process is not clear. There is a gradual loss of ectoplasmic specializations as the recesses in which the spermatids are lodged become shallower to a point where only the tip of the spermatid head is surrounded by Sertoli cell cytoplasm. The breakdown of the specializations could be responsible for the filling up of the recesses and the subsequent movement of the spermatids toward the lumen or could result in the gradual loss of some kind of holding connection between the Sertoli cell cytoplasm and the spermatid head.

Finally, an interesting type of interrelationship between Sertoli cells and spermatids appears during the last few days before sperm release. Structures called *tubulobulbar complexes* develop at this time. They first appear when only the heads of the spermatids are associated with the Sertoli cells, when the deep recesses and the ectoplasmic spe-

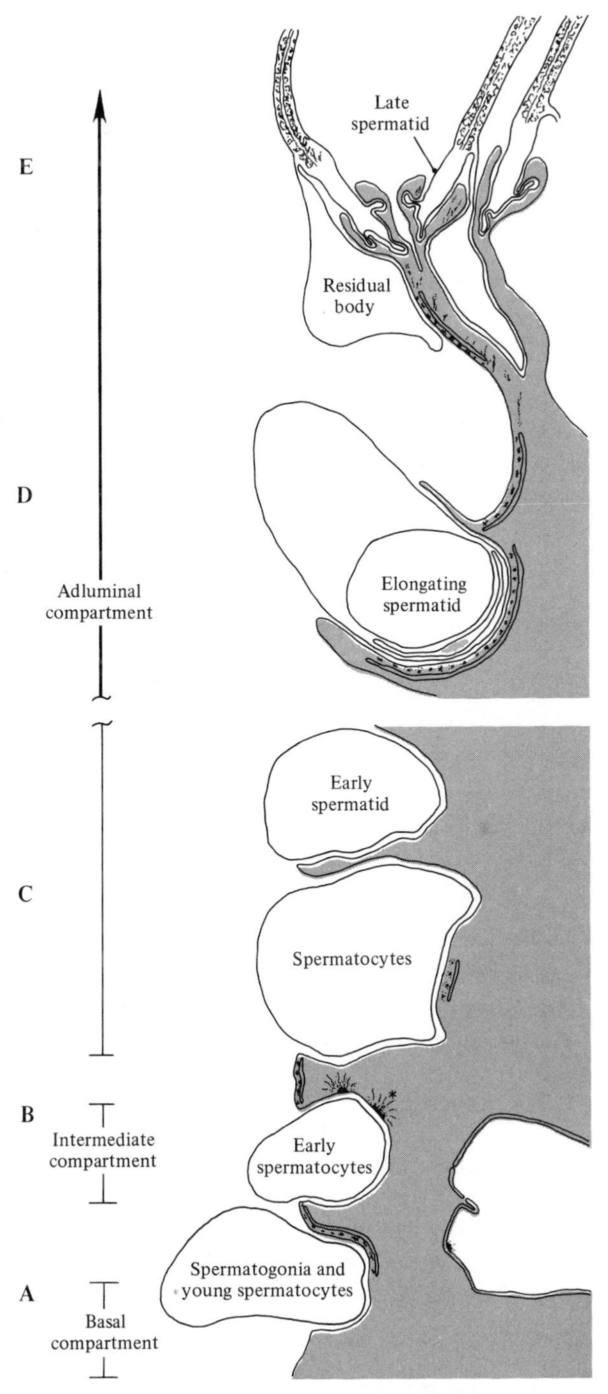


**FIG. 3–9.** Diagrammatic representation of the association between a Sertoli cell (colored) and the developing germ cells. The Sertoli cell extends from the basal lamina to the lumen of the seminiferous tubule. Sheetlike branches extend laterally from the "trunk" of the treelike cell partially enclosing the maturing germ cells. A, basal compartment containing spermatogonia and young spermatocytes; B, intermediate compartment with young spermatocytes; C, D, E, adlumenal compartment with recesses in which the maturing germ cells are lodged. The recesses become shallower toward the apex of the cell. Asterisks—desmosome-like junctional complexes between adjacent Sertoli cells and between the Sertoli cell and the germ cells; arrowheads—ectodermal specializations; arrow–tubulobulbar complexes. (After L. D. Russel, 1980. Gamete Res. 3:179.)

cializations are disappearing. They consist of extensions of the cytoplasm of the spermatid in the region of the acrosome into apposing pits in the Sertoli cell. Initially the pits are shallow, but they later deepen, and the spermatid extensions in turn lengthen (Fig. 3–9 D,E). A bulbar region is usually found somewhere along the tubule and hence the name tubulobulbar complex. They vary in number, because the Sertoli cell continually degrades them by phagocytizing the older complexes at the same time new ones are being formed. Thus, one function of the tubulobulbar complex is the elimination of excess spermatid cytoplasm. This cytoplasm, unlike that eliminated with the residual body, contains no inclusions or organelles.

As the spermatid is gradually released, less and less of its head is associated with the Sertoli cell, but the tubulobulbar complexes still persist when only the tip of the head lies in the Sertoli cytoplasm. This then suggests that they may act as mooring devices, and only upon their complete disappearance does the spermatid finally lose all association with the Sertoli cell and move into the lumen of the seminiferous tubule (Fig. 3–10). The spermatid flagellum, immobile at this time, plays no role in the release process.

## Other Sertoli Cell Functions

In addition to their interrelationships with the germ cells in which they form the cytoskeleton of the seminiferous tubule, establish a blood–testis barrier, provide a means of translocation of the germ cells from the basal region of the tubule toward the lumen, and form an anchoring mechanism before the ultimate release of the spermatid, the Sertoli cells have other equally important functions. One is the secretion of a fluid that transports the sperm from the seminiferous tubule to the epididymis. This fluid contains substances important to sperm survival, including a peptide that inhibits the premature breakdown of the acrosome and a variety of steroid hormones. Sertoli cells also secrete *inhibin*, responsible for the inhibition of the secretion of pituitary gonadotrophins, and a hormone of the fetal testis, *anti-Müllerian hormone*,

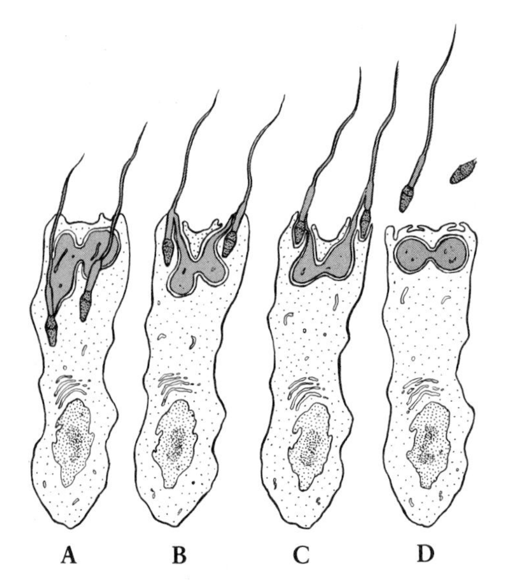


**FIG. 3-10.** Stages in the release of sperm by a Sertoli cell. A, spermatid deep within the cytoplasm of a Sertoli cell, spermatid cytoplasm between the spermatid head and the lumen of the seminiferous tubule; B, beginning of sperm extrusion, with cytoplasmic remnant retained in a Sertoli cell; C, extrusion continued, with sperm connected to the cytoplasmic remnant by a narrow neck; D, sperm free, with cytoplasmic remnant remaining in a Sertoli cell. (From D. W. Fawcett, 1975. Handbook of Physiology, Section 7, Endocrinology Vol. 5. D. W. Hamilton and R. O. Greep, eds. The American Physiological Society, Bethesda, Md.)

which is of importance in the development of the male reproductive ducts. These functions will be considered in later chapters.

## Cycle of the Seminiferous Tubule

Examination of cross sections of the rodent testis early showed that associations of cells with fixed compositions occurred regularly. That is, the cell types (stages in spermatogenesis and spermatid maturation) did not occur in a haphazard fashion, but various stages of spermatid maturation were always associated with certain types of spermatogonia and spermatocytes. These consistent cell associations seen at any point in time then make up specific stages in a progressing wave of maturation passing down the seminiferous tubule, one following the other in an orderly fashion, making up what is called the cycle of the seminiferous tubule. Although the histological appearance of the human seminiferous tubule was at first thought not to fit into this pattern and to show a haphazard arrangement of germinal elements, it was subsequently shown that there are, in fact, six types of cell associations in man and thus six stages in the cycle of the seminiferous tubule. These stages continually succeed each other in an orderly sequence. The fact that in man these stages may sometimes show irregularities made the original definition of the human cycle difficult.

## Male Infertility

In about 25 percent of the couples who are unable to conceive, infertility is attributable to defective production of spermatozoa. Thus, it is important to establish criteria for the evaluation of the ejaculate. Such criteria usually consider semen volume as well as sperm density, motility, and morphology. While no rigid line of demarcation between fertile and infertile males may be drawn at any specific level for any of the above criteria, a number of guidelines are considered indicative in semen analysis. Although 3 milliliters is given as the mean semen volume, there does not seem to be too much relation between volume and ease of conception; consequently, semen volume is probably the least reliable criterion of fertility.

The sperm count, expressed as either or both the number of sperm per ejaculate or the number of sperm per milliliter (sperm density) has also been considered as a measure of potential fertility. The American Fertility Society in 1971 proposed 40 million sperm/ml and 125 million sperm/ejaculate as normal minimum numbers. However, males with less than 20 million sperm/ml have been known to be fertile, and in one study on 2000 men with counts between 20 and 40 million sperm/ml, 47 percent proved to be fertile. Another study by Zukerman et al. (1977) found significant differences in fertility only when sperm counts dropped below 10 million/ml and 25 million/ejaculate, and they suggested that only when counts were below this range should they be considered a major factor in infertility.

Sperm motility is important, although muscular contractions and beating of the cilia of the female reproductive passages are responsible for most of the movement of the sperm from the vagina to the uterine tubes. Under their own power sperm move much too slowly (about 10–60 $\mu$m/second) to traverse this distance in the allotted time. However, the sperm's own motility is important in passing through the cervical mucus and the narrow utero-

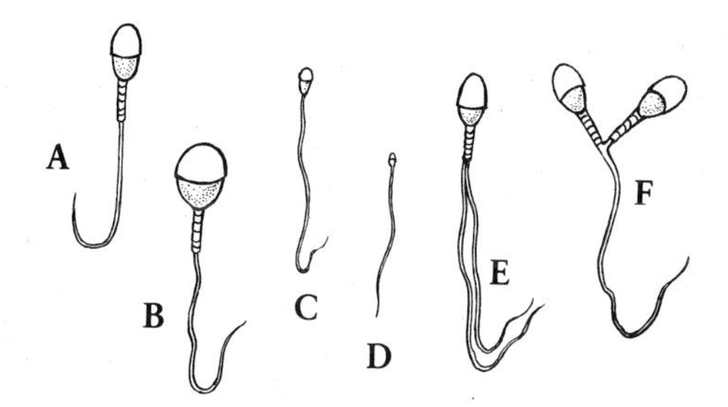


FIG. 3-11. Some types of abnormal sperm.

tubal junction, and, once in the vicinity of the ovum, vigorous movement is necessary to penetrate the membranes surrounding the egg. Active forward progression rather than pure mobility is what counts. Sperm acquire their mobility as they pass through the epididymis. In the head of the epididymis, the type of movement may vary from none at all to mild, or wild unprogressive thrashing. Normal forward movement appears in the middle of the body of the epididymis, when both the flexibility of the flagellum and the arc of its movement are reduced.

Not all sperm are normal morphologically, and a variety of abnormal sperm are seen on examination of any semen sample (Fig. 3–11). If the percentage of abnormal forms exceeds 20 percent, impaired fertility may be expected.

In evaluating male fertility it is thus not possible to set absolute standards. The only absolute standard for sterility, since it takes but a single sperm to fertilize an egg, would be a semen sample with no motile sperm. However, certain minimum standards in semen volume, sperm concentration, mobility, and morphology may be considered in classifying individuals as potentially fertile or subfertile.

## References

Clermont, Y. 1963. The cycle of the seminiferous epithelium in man. Am. J. Anat. 112:35–52.

Fawcett, D. W. 1972. Observations on cell differentiation and organelle continuity in spermatogenesis. In: International Symposium on the Genetics of the Spermatozoan. Eds., R. A. Beatty and S. Gluecksonn-Waelsh. Edinburgh, New York: Department of Genetics, University of Edinburgh.

Fawcett, D. W. 1975. Ultrastructure and function of the Sertoli cell. In: Handbook of Physiology-Endocrinology, V, Section 7, pp. 21–25. Eds., D. W. Hamilton and R. O. Greep. Baltimore: Williams and Wilkins.

Koehler, J. K. 1970. A freeze-etching study of rabbit spermatozoa with particular reference to head structures. J. Ultrastruct. Res. 33:598–610.

Zukerman, Z., L. J. Rodriguez-Regau, K. D. Smith, and E. Steinberger. 1977. Frequency distribution of sperm counts in fertile and infertile males. Fertil. Steril. 28:1310–1313.

# 4

# OOGENESIS

A single cell, the oocyte, represents the connecting link between the ongoing generation and the next generation. As such, it is a highly specialized cell that is capable of expressing and maintaining the characteristics of the species. Oogenesis constitutes a continuum of events and processes involved in the origin, growth, and differentiation of the oocyte or egg cell from stem cells in the ovary. The transformations from stem cell to mature ovum are complex, embracing cellular, molecular, and physiological phenomena.

The egg cell in all animals is nonmotile and quite large by comparison to the somatic cells of the body. The importance of a careful examination of the growth and differentiation of the oocyte was clearly articulated by E. B. Wilson, an eminent cytologist, in 1896 when he stated that embryogenesis begins in oogenesis. Ample evidence currently available suggests that the development of the early embryo is in great measure controlled by the egg cell. In other words, the egg contains virtually all of the information required to direct the morphological and molecular aspects of development. How and in what form this developmental information is synthesized and stored in the egg cell is an area of active investigation.

In addition to storing developmental information, the ovum also carries the nutrients (i.e., yolk) to provide building materials and energy for the support of the early embryo. The synthesis and packaging of these materials is a major process in oogenesis. The amount and distribution of yolk vary in eggs of different animal species. The yolk of the hen's egg, for example, represents most of the cell and weighs about 55 grams. Large quantities of yolk in the egg influence events during and following fertilization, including patterns of cleavage and gastrulation.

An equally important aspect of oogenesis is the extent to which the programming of developmental information and the packaging of yolk materials are regulated by factors outside of the oocyte. We know that the growth and differentiation of the oocyte are under endocrine control (Chapter 5). However, the nature of the mechanisms by which hormones act upon the oocyte remains to be fully determined.

In the pages that follow, we describe pertinent aspects of egg formation. We will see that oogenesis yields a highly differentiated cell specialized to undergo meiosis, to participate in fusion with the male gamete, and to store information and materials that are essential to the events of early embryonic development.

## Organization of the Developing Egg

Descriptions of oogenesis often seem confusing because of the variety of terms used to refer to the "egg cell." During its growth and differentiation within the ovary, the egg cell is an *oocyte*. In order to emphasize homological relationships with the male sex cell, it is more appropriately termed the *primary oocyte*.

Following the growth and differentiation phase, the primary oocyte enters a stationary phase until the resumption of meiosis (maturation). The *secondary oocyte* is a product of the first meiotic division. The oocyte at the completion of maturation is technically the *ovum* or *mature egg cell*.

As in the development of the male gamete, oocytes arise from stem cells in the ovary called *oogonia*. Oogonia are often difficult to identify in ordinary histological sections of the ovary because of their size and proximity to the germinal epithelium of this organ. With the exception of fishes and amphibians, oogonia of most vertebrates have a limited capacity for self-renewal. Indeed, the oogonia stop dividing and enter meiotic prophase before the onset of oocyte growth and differentiation. In the human female, for example, the ovary at birth has few if any oogonial cells; most are converted into a finite population of primary oocytes with no further ability to proliferate. Oogonia are typically small cells (10–20 μm in diameter) but have the same basic structure and complement of organelles as any eukaryotic cell. The transformation of the stem cell into the primary oocyte is accompanied by an enlargement of the nucleus and a rearrangement of the chromatin material.

Before a primary oocyte begins its growth phase, its chromatin material condenses and proceeds to the terminal stage of the first meiotic prophase (diplotene stage). The period of growth and differentiation of the primary oocyte of most organisms is manifest in an enormous increase in size, a change that may take place over a time interval of several days or several years. During this time the oocyte becomes surrounded by accessory cells termed *follicle cells*. Follicle cells are specialized somatic cells derived from the germinal epithelium of the ovary. As pointed out below, they play an important role in the normal metabolic activity of the oocyte and in the regulation of meiosis. Initially, only a single layer of follicle cells invests the primary oocyte. These *unilaminar follicles* (or primordial follicles) characterize the human female ovary during the postembryonic period of sexual immaturity (Fig. 4–1). Many of these follicles may remain in this state for several decades, or until recruited for growth.

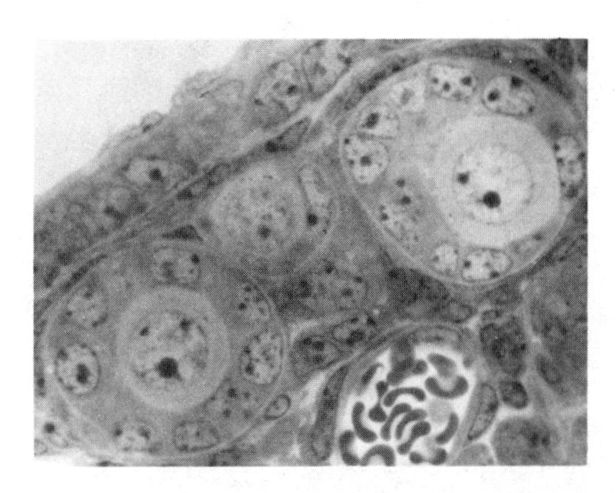

**FIG. 4–1.** Quiescent oocyte in a primary follicle in the cortex of a mature mouse ovary. (From L. Zamboni, 1970. Biol. Reprod. Suppl. 2:44.)

With the onset of sexual maturity in the adult mammal, there is a rapid growth of the oocyte and proliferation of the follicle cells (Fig. 4–2). The simple column-shaped cells of the primary follicles multiply, first forming a double *(secondary follicle)* and finally a stratified cuboidal epithelium around the oocyte.

The rapid proliferation of the follicle cells is under the control of gonadotropins secreted by the pituitary gland. At about the time that the growth of the primary oocyte ceases (140 μm in diameter in the human female), small, fluid-filled irregular spaces appear between the follicle cells. These gradually coalesce to form a single, large cavity, the *follicular antrum*, filled with a fluid known as the *liquor folliculi* (Fig. 4–3). Autoradiographic studies have shown that the elaboration of the liquor folliculi, rich in polysaccharides, is due to the secretory activity of the follicle cells. The continued increase in the volume of the liquor folliculi separates the primary oocyte, ensheathed by several layers of follicle cells, from the remaining follicle cells.

The primary oocyte with its halo of follicle cells is termed the *cumulus oophorus;* it is suspended in the follicular fluid by a stalk of follicle cells and occupies an eccentric position in the antral cavity (Fig. 4–3). The large (20–25 mm in diameter) and vesicular follicle is known as the *Graafian follicle*. During the major growth phase of the mouse oocyte, there is a dramatic increase in size from about

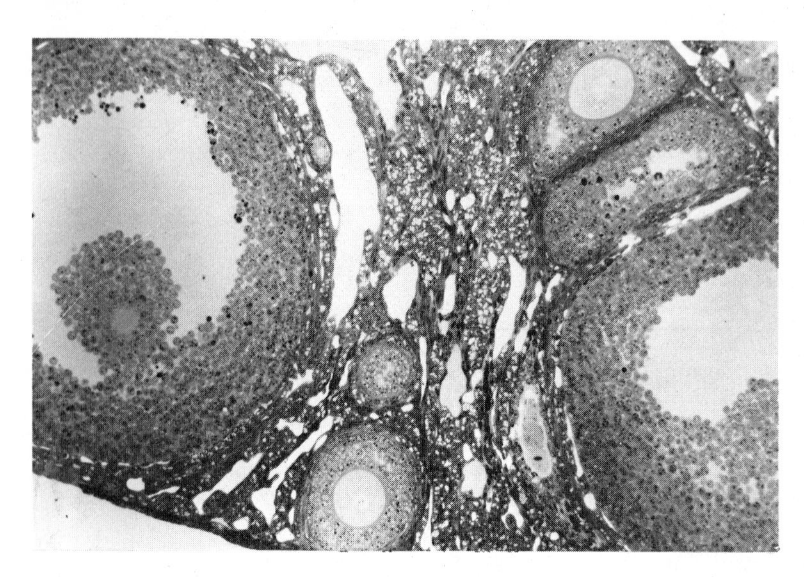

**FIG. 4–2.** A section through the mouse ovary to show follicles in various stages of maturation. A young Graafian follicle with the forming cumulus oophorus can be seen to the left. (From L. Zamboni, 1972. Oogenesis. J. Biggers and A. Schuetz, eds. University Park Press, Baltimore.)

20 μm to 70 μm in a period of about two weeks. The oocyte remains at full size for the next five days while the whole follicle enlarges until ovulation or atresia.

Figure 4–4 summarizes in diagrammatic form the various changes in the primary oocyte and its associated follicle cells that lead to the formation of the mammalian Graafian follicle. Note that the antral cavity of the Graafian follicle is invested by follicle cells arranged in a stratified cuboidal epithelium. These constitute the structural layer known as the *stratum granulosum.* Beyond the stratum granulosum, the connective tissue of the ovary typically condenses around the growing follicle as the *theca folliculi.* The theca capsule will differentiate into inner and outer layers.

**FIG. 4–3.** Formation of the antral cavity in a mouse follicle. (From L. Zamboni, 1970. Biol. Reprod. Suppl. 2:44.)

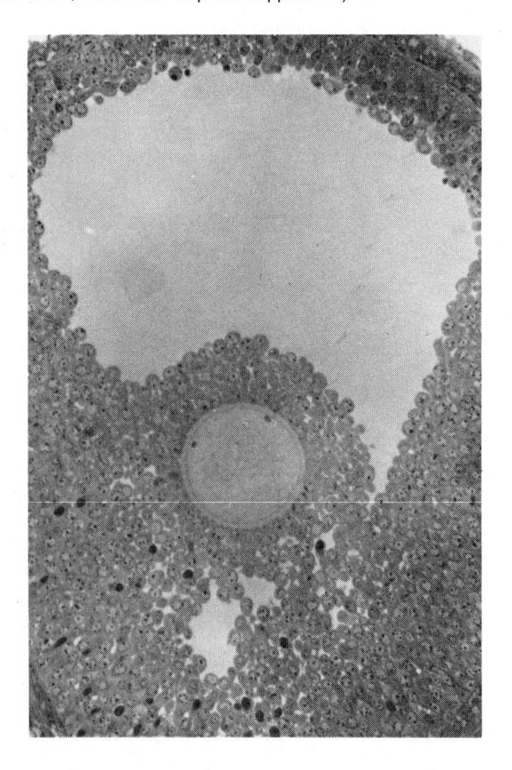

As the antral cavity enlarges with fluid accumulation, the Graafian follicle pushes closer to the surface of the ovary. Rupture of the wall of the Graafian follicle results in the release of the oocyte with its surrounding 12 to 15 layers of follicle cells. The stratum granulosum and the theca folliculi then rearrange themselves to form the *corpus luteum.* Those Graafian follicles that are not ovulated will degenerate, a process known as *atresia.*

Close examination of the primary oocyte in the Graafian follicle shows that, in addition to follicle cells, it is surrounded by a transparent, noncellular envelope, the *zona pellucida* (Fig. 4–5). The zona pellucida is approximately 15 microns in thickness and constructed of glycoproteins. Recent studies by Bleil and Wassarman indicate that the glycoprotein matrix is synthesized by the oocyte during its growth phase. Both scanning and transmission electron microscopy show the zona pellucida to be porous. A narrow fluid-filled space, the *perivitelline space,* can be identified between the zona pellucida and the plasmalemma of the oocyte.

Our understanding of the structural changes in

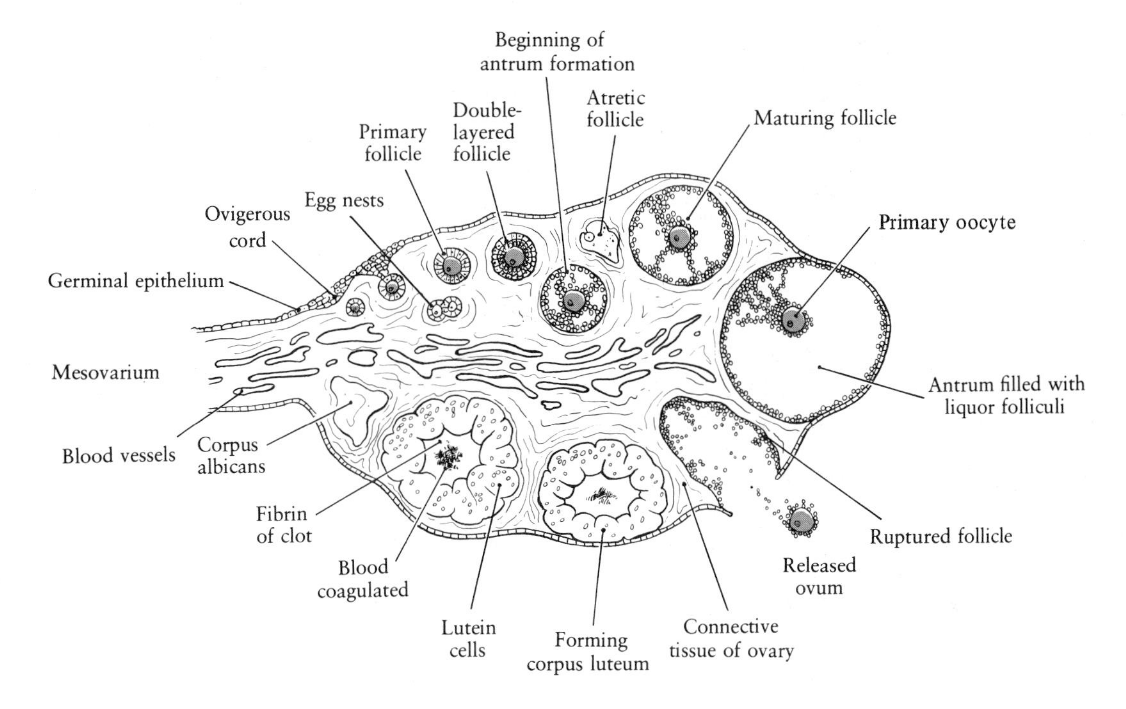


**FIG. 4–4.** Schematic diagram of the mammalian ovary showing the sequence of events in the origin, growth, and eventual rupture of the Graafian follicle. Start at the mesovarium and follow in a clockwise direction. Note also the formation and regression of the corpus luteum.

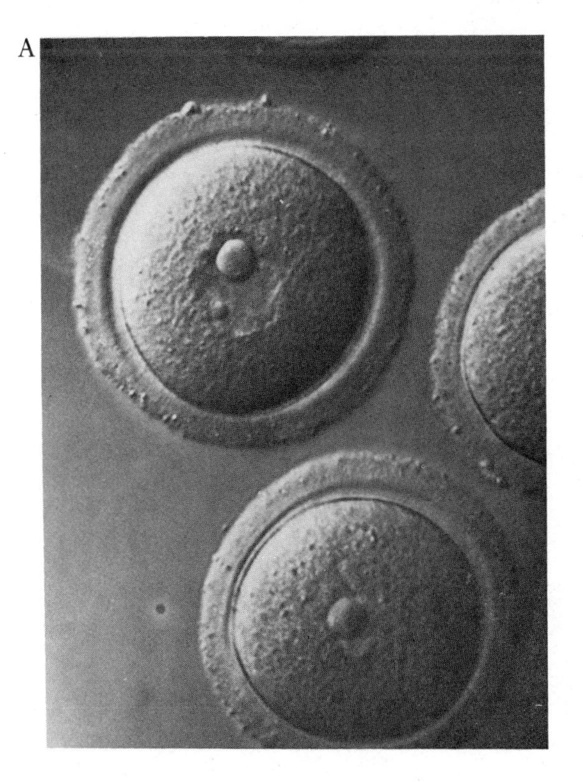


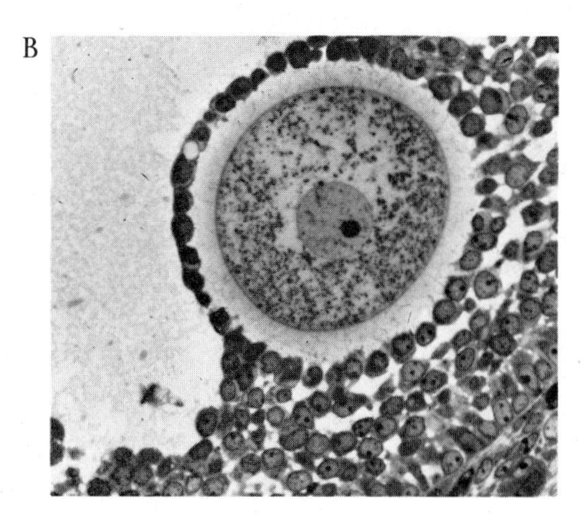


**FIG. 4–5.** A, intact oocyte under Nomarski differential-interference microscopy showing the zona pellucida (From J. Bleil and P. Wassarman, 1980. Dev. Biol. 76:185); B, human oocyte in antral follicle. Note the zona pellucida and the surrounding cumulus cells. Organelles are distributed throughout the oocyte cytoplasm. (From L. Zamboni, 1972. Oogenesis. J. Biggers and A. Schuetz, eds. University Park Press, Baltimore.)

the oocyte, as well as the interrelationships between the oocyte and the surrounding environment of follicle cells, during oogenesis has been greatly enhanced by the use of the transmission electron microscope and various labeling techniques. Typically, the young oocyte of the primary follicle is rather simple in appearance and structural organization (Fig. 4–6). The cytoplasm tends to be rather transparent and granular. The granules are particles of ribonucleoprotein. In the mammalian oocyte, the nucleus is large and spheroidal in shape. Most of the organelles of the cell are clustered in a limited region around the nucleus. Electron-microscopic studies have shown that the perinuclear organelles consist mostly of *endoplasmic reticulum,* closely packed *mitochondria, lysosomes,* and a prominent *Golgi complex.*

The mitochondria are frequently associated with a granular, electron-dense material. These "mitochondrial clouds" or Balbiani bodies were once believed to be involved with the formation of yolk, but current evidence does not support this view. The granular material around the mitochondria is very similar to *nuage material,* which occurs in the pores of the nuclear membrane. A particularly interesting organelle, the *annulate lamella,* has been observed in the young human oocyte (Fig. 4–7). Annulate lamellae are stacks of parallel, paired membranes interrupted at regularly spaced intervals by pores or annuli. Similar structures have been found in the sea urchin oocyte in association with ribosome-like particles. Presumably, these annulate lamellae arise by blebbing activity of the nuclear membrane. The functional significance of these structures during oogenesis is not known.

The primary oocyte undergoes marked and progressive biochemical and morphological differentiations during its growth. The nucleus becomes substantially enlarged; it is in this state commonly referred to as the *germinal vesicle.* The mitochon-

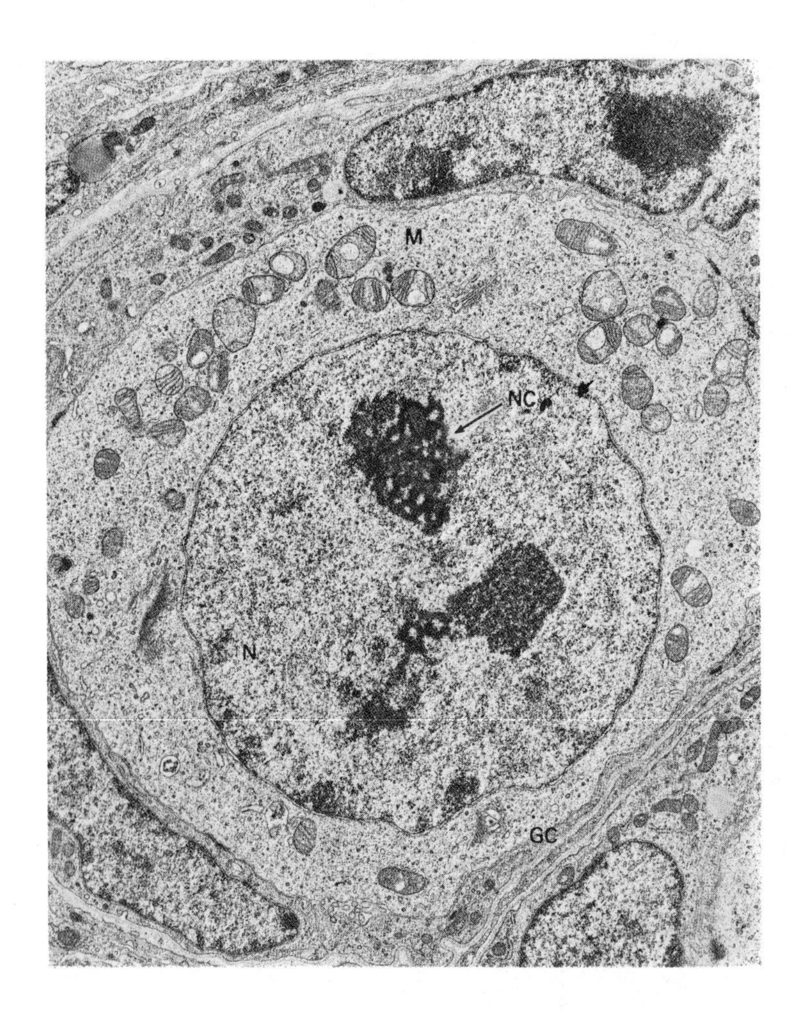


**FIG. 4–6.** Electron micrograph of a young oocyte of the mouse ovary showing the distribution of Golgi material (GC), mitochondria (M), and endoplasmic reticulum. N, nucleus; NC, nucleolus. (Courtesy of E. Anderson.)

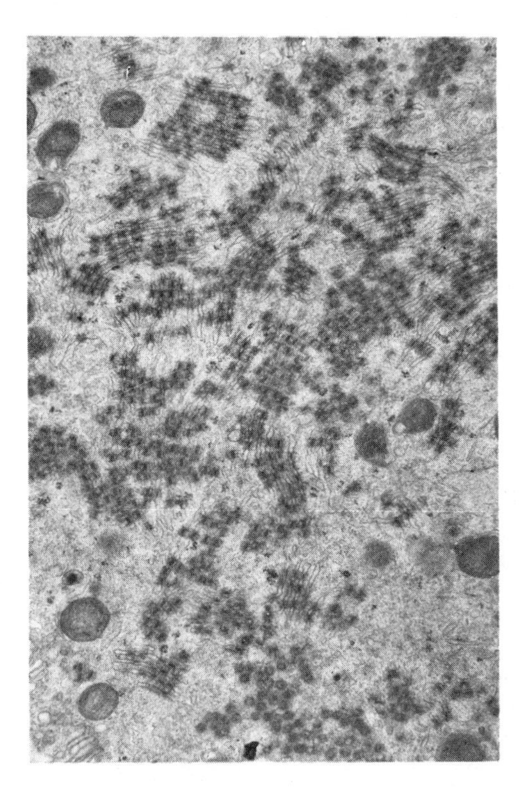

**FIG. 4–7.** Annulate lamellae in the cytoplasm of the human oocyte. (From L. Zamboni, 1972. Oogenesis. J. Biggers and A. Schuetz, eds. University Park Press, Baltimore.)

dria increase in number and become more uniformly distributed throughout the cytoplasm. The elements of the endoplasmic reticulum become abundant and pronounced. Their association with numerous ribosomes is functionally associated with the dramatic increase in protein synthesis, which is known to take place during the growth stage of the primary oocyte.

One of the most dynamic transformations in the cytoplasm of the maturing oocyte involves the Golgi complex. The Golgi complex has been implicated in a variety of animals in the formation of a population of vesicles known as the *cortical granules.* These spheroid-shaped bodies have been identified in the peripheral cytoplasm of the mature egg of frogs, echinoderms, teleost fishes, bivalve molluscs, and several mammalian species (including the rabbit, hamster, and human [Fig. 4–8]). They are apparently absent in the ova of urodeles, insects, birds, and several mammals (e.g., the rat and the guinea pig).

Cortical granules are membrane-bound organelles; they range in size from about 0.5 $\mu$m (sea urchin), to 2 $\mu$m (frogs), to 20 $\mu$m (some fishes) (Fig. 4–8). Several investigators have recently attempted to isolate and characterize the contents of the cortical granules in sea urchin eggs. Results suggest that the cortical granules are rich in soluble, high-molecular-weight, PAS (periodic acid Schiff)-positive glycoproteins and several enzymes, including $\beta$1,3-glucanohydrolase, protease, and peroxidase. With regard to the glycoproteins, the proteinaceous component of the granule is probably synthesized in the endoplasmic reticulum of the oocyte and the carbohydrate moiety in the Golgi complex.

The modification of the Golgi complex into the cortical granules appears to occur throughout the cytoplasm of the oocyte. In forms like the rabbit and human, the Golgi complex is observed to divide into aggregates of vesicular and tubular elements (Fig. 4–9 A). These become filled with an electron-dense material. The mature cortical granule, approximately 300 to 500 millimicrons in diameter, is produced through the coalescence of these vesicles with their contents. These then migrate to the extreme periphery of the oocyte where they are organized into several layers beneath the oolemma (Fig. 4–9 B). The fusion of vesicles pinched off from the Golgi complex also appears to be the method of cortical granule formation in other invertebrate and vertebrate organisms.

Ultrathin sections of sea urchin oocytes reveal that the organization of the cortical granule is very complex. The mature cortical granule is limited by

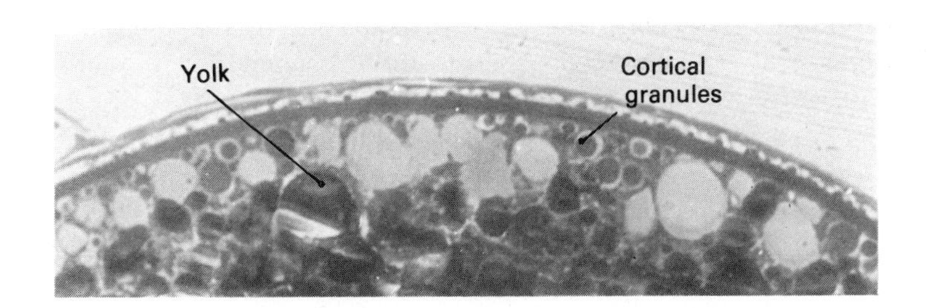


**FIG. 4–8.** A section through the mature egg of the zebra fish (teleost) showing the cortical granules.

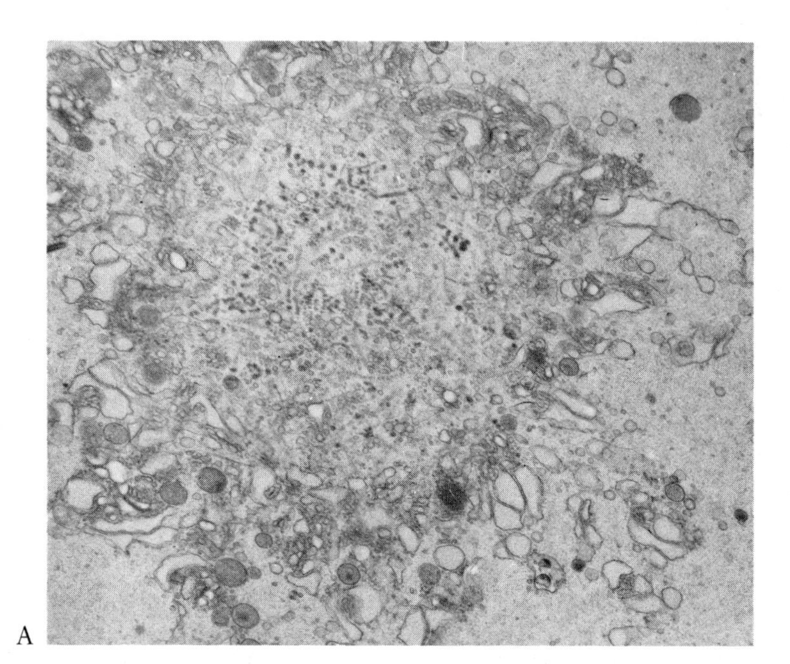

A

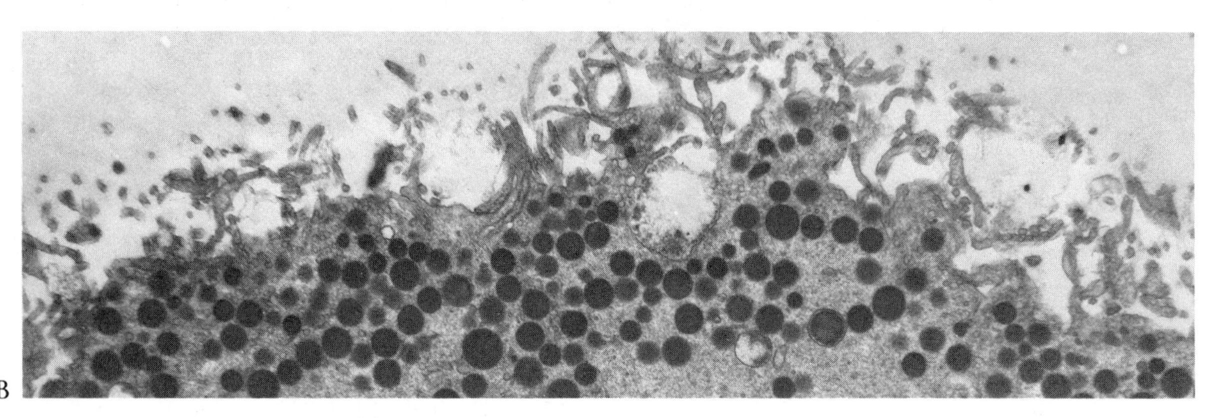

B

**FIG. 4–9.** A, cortical granule formation in the Golgi complex of a human oocyte; B, cortical granules located in the cortex of the human oocyte. (From L. Zamboni, 1972. Oogenesis. J. Biggers and A. Schuetz, eds. University Park Press, Baltimore.)

a unit membrane approximately 50 Å in thickness. Some studies suggest that the cortical granule membranes are bonded to the inner surface of the egg cell membrane. The contents of the cortical granules of echinoderm eggs show considerable variation in appearance. Frequently, the contents are arranged as concentric layers of electron-dense material alternating with layers of electron-light material (Fig. 4-10 A). In other eggs, the granule contents consist of a large, centrally located electron-dense mass (250–500 Å) and several lighter hemisphere-shaped globules (200–300 mμ in diameter) (Fig. 4–10 B). By contrast, the cortical granules in fishes and amphibians are homogeneous and finely granular in appearance (Fig. 4–11).

The cortical granules of the fully grown oocyte are associated with and part of a peripheral layer of cytoplasm known as the *cortical cytoplasm* or *cortex*. The cortex of the animal egg is a layer of cytoplasm of variable thickness, typically less than five microns, lying directly beneath the plasmalemma. Its properties appear to be quite different from the cytoplasm found throughout the rest of the cell. If sea urchin ova, for example, are subjected to moderate centrifugation, the inclusions of the interior of these cells, such as the mitochondria, ribosomes, and yolk particles, are movable and easily displaced; that is, the organelles appear to be suspended in a liquidlike cytoplasm. By contrast, the cortical granules and the cytoplasm in which they are embedded remain intact and undisturbed. It is now known, after study by a variety of techniques,

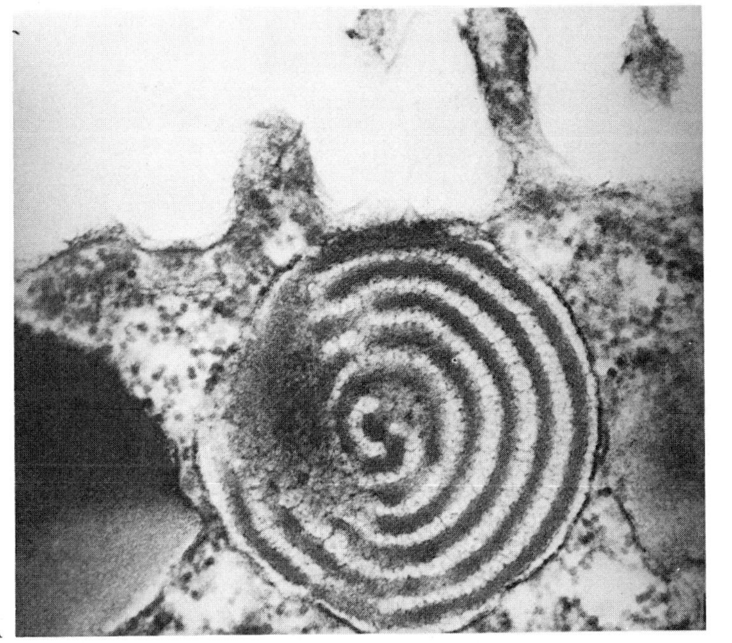


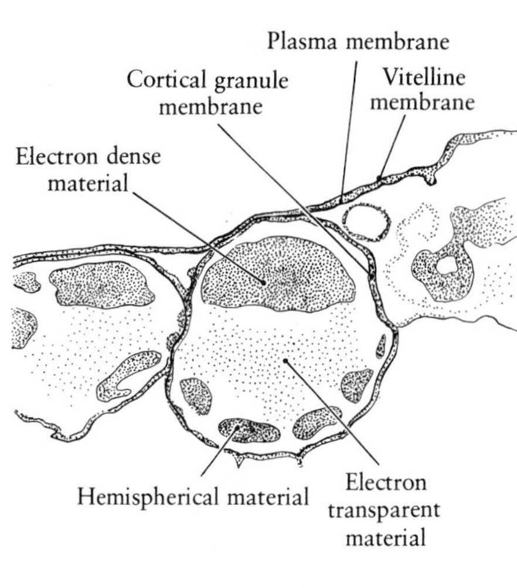


**FIG. 4-10.** A, a micrograph of a section through a cortical granule of *Strongylocentrotus purpuratus.* (Courtesy of V. Vacquier.) B, reconstruction from electron micrographs of a section through the cortical granule of the egg of the sea urchin, *Clypeaster.* (After Y. Endo, 1961. Exp. Cell Res. 25: 383.)

that the cortex exists in a gel-like state and exhibits viscoelastic mechanical properties. Indeed, the rigidity of this cortical layer appears to maintain the spherical shape of the egg. Contractile proteins presumably mediate the state of gelation of the cortex. We shall have more to say about the role of the cortex in morphogenesis in later chapters. It is a highly dynamic complex capable of rapid change in response to both internal and external stimulation.

**FIG. 4-11.** A cortical granule (CG) in the mature oocyte of the frog. Note also the distribution of pigment granules (PG) and yolk platelets (YP). (From N. Kemp and N. Istock, 1967. J. Cell Biol. 34:111.)

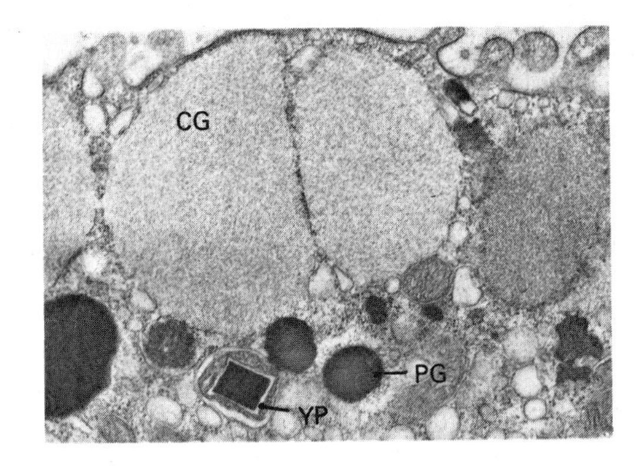


The cortex of the egg cell appears to be an important site for the localization and storage of information that is critical to the development of the future embryo. Needless to say, the identification of the nature of this cortical information and determination of its influence on the processes of embryogenesis have been major challenges to the developmental biologist.

The advanced primary oocyte emerges as a highly differentiated cell with a variety of organelles and inclusions, such as mitochondria, ribosomes, membrane complexes (endoplasmic reticulum, the Golgi complex), pigment granules, and yolk particles (see section that follows on vitellogenesis), embedded in a liquidlike cytoplasm. Many of these cellular constituents become localized in specific areas of the oocyte during its growth and maturation, giving a distinct *polarity* to the cell. For example, there is in most animals an enormous increase in the number of ribosomes during the growth phase of the oocyte. However, instead of being arranged at random in the ooplasm, they are distributed along a gradient in decreasing numbers from one pole of the cell to the opposite pole. The distribution of mitochondria follows a similar pattern. Also, in the oocytes of many vertebrates, particularly those with moderate to large amounts of yolk (frogs, fishes, birds), the yolk inclusions or particles are smaller and more loosely packed at one end of the cell; they progressively become larger and more tightly packed toward the opposite end of the cell. This stratification permits us to refer to the region of the oocyte containing the nucleus and few yolk inclusions—but rich with ribosomes and mitochondria—as the *animal pole*. The opposite or yolk-filled region of the oocyte is termed the *vegetal pole*. The imaginary line passing from the animal pole to the vegetal pole is the *animal-vegetal axis*.

Hence the position of the nucleus and the distribution of other organelles define a distinct polarity to the egg. Although polarity of the egg has been known for some time, factors regulating its expression are poorly understood. The factors do, however, appear to be intrinsic to the egg itself. When *Xenopus* oocytes are stripped of their follicular envelopes and grown in culture, pigment granules are initially distributed throughout the entire cell; they subsequently become localized in the animal hemisphere.

The architectural organization of the primary oocyte in many animal species is frequently visible in regional differences in the appearance of the ooplasm. In the ova of many amphibians, dark brown or black pigment granules are located in the cortical cytoplasm of the animal hemisphere (Figs. 4–11, 4–12 A). This is in sharp contrast to the vegetal hemisphere, which appears colorless or white because of the densely packed mass of yolk and few pigment granules. Although the transition from dark to light is macroscopically distinct, there is an intermediate or *marginal zone* where the pigment granules are intermediate in intensity. The mature egg of the tunicate or ascidian, *Styela*, possesses yellow pigment granules uniformly distributed throughout the cortical cytoplasm (Fig. 4–12 B). In several species of mollusc *(Dentalium, Ilyanassa)*, the fully grown oocyte shows three distinct cytoplasmic areas (Fig. 4–12 C): a yolk-free, clear, unpigmented zone of transparent cytoplasm at one pole of the cell *(vegetal polar plasm)*, a narrow, pigment-free zone of cytoplasm at the other pole of the cell *(animal polar plasm)*, and an intervening zone of cytoplasm filled with yolk particles and pigment granules. Similar animal and vegetal polar plasms are observed in the ovum of the snail *(Limnaea)*. These cytoplasmic regions apparently differentiate just as the ovum leaves the ovary. Additionally, six

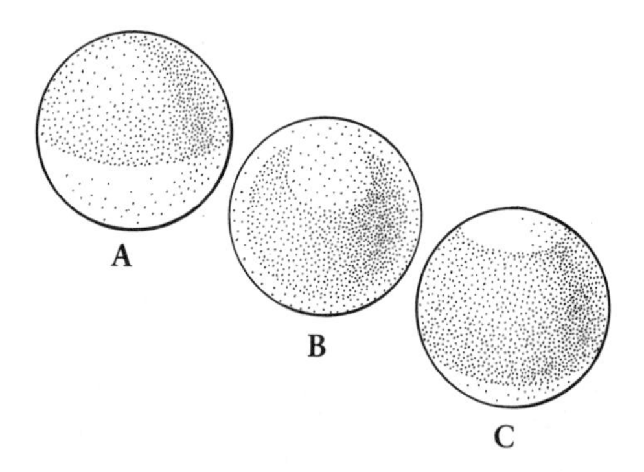


**FIG. 4–12.** Superficial views of the mature oocyte of the frog (A), tunicate (B), and mollusc (C) to show regional differences in the cytoplasm.

lenticular-shaped patches of cytoplasm, known as *subcortical accumulations*, have been described by Raven as occurring at the equator of the snail oocyte. Their significance is very much in doubt. However, the positions of the subcortical accumulations directly reflect the locations of six follicle cells that surround the oocyte during its development, thus suggesting that the organization of the ooplasm is, in part, brought about or dictated by cells accessory to the primary oocyte.

Since pigment granules are perhaps the most visible expression of oocyte differentiation, it is logical to ask if they play a key role in the future development of the embryo. It is unlikely that pigment granules are very important to the growth and differentiation of the embryo. However, we can consider that the uneven distribution of pigment granules may very well reflect a more subtle organization of the ooplasm into qualitatively different regions that subsequently are vital to the specific differentiations of the embryo.

In most animal species the oocyte develops in close association with a population of accessory or follicle cells. These accessory cells have been shown to be essential to the complete structural and physiological maturation of the oocyte. They participate in such activities of the ovary as transportation of nutritive materials into the oocyte, hormonal stimulation of the maturation of the oocyte, and formation of noncellular envelopes that surround and invest the oocyte. The importance of the follicular association is indicated by the fact that only those oocytes having successfully completed meiosis are capable of being fertilized and giving rise to normal development.

Throughout its growth phase, the oocyte is invested by one or more layers of follicle cells. The close functional relationship between the primary oocyte and the surrounding follicle cells is suggested by several structural specializations. One of the earliest structural changes to occur in the growing oocyte of many invertebrates and vertebrates is the formation of numerous fingerlike projections from the surface of the oolemma. These *microvilli,* which increase substantially the surface area of the oocyte, develop rapidly and uniformly over the whole surface of the oocyte. They interdigitate with similar cytoplasmic extensions of the investing follicle cells (Figs. 4–13, 4–14). With the light microscope, this zone of projecting and overlapping microvilli gives a distinct striated appearance to the region of the primary follicle beyond the oolemma. This region of the primary follicle is termed the *zona radiata* in the mammals.

Electron-microscopic studies provide strong evidence that the microvilli function in the exchange and transport of substances between the follicle cells and the primary oocyte. Oocytes actively engaged in the formation of proteins and yolk often show small inpocketings at the bases of these microvilli. Subsequently, after sequestering fluids, these pockets become pinched off and form membrane-bound vesicles within the ooplasm. This process of "cell drinking" is referred to as *micropinocytosis.* By the time the oocyte has reached its full size, micropinocytosis is greatly diminished or no longer visible. The microvilli of the oocyte and the extensions of the follicle cells then retract and are withdrawn.

In the chick, the follicle cells are drawn out into club-shaped projections known as *lining bodies* (Fig. 4–15). These push deep into the cytoplasm of the oocyte but never apparently penetrate its oolemma.

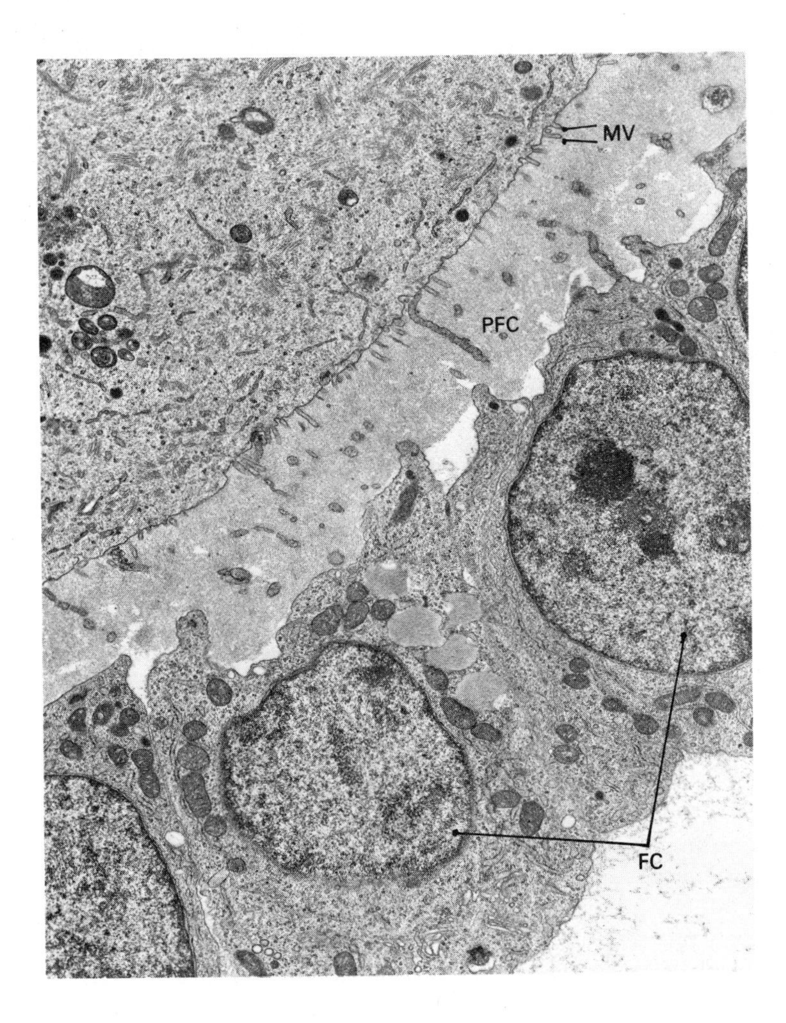


**FIG. 4–13.** An electron micrograph showing microvilli (MV) extending from the surface of the mouse oocyte and interdigitating with projections (PFC) from the investing follicle cells (FC). (Courtesy of E. Anderson.)

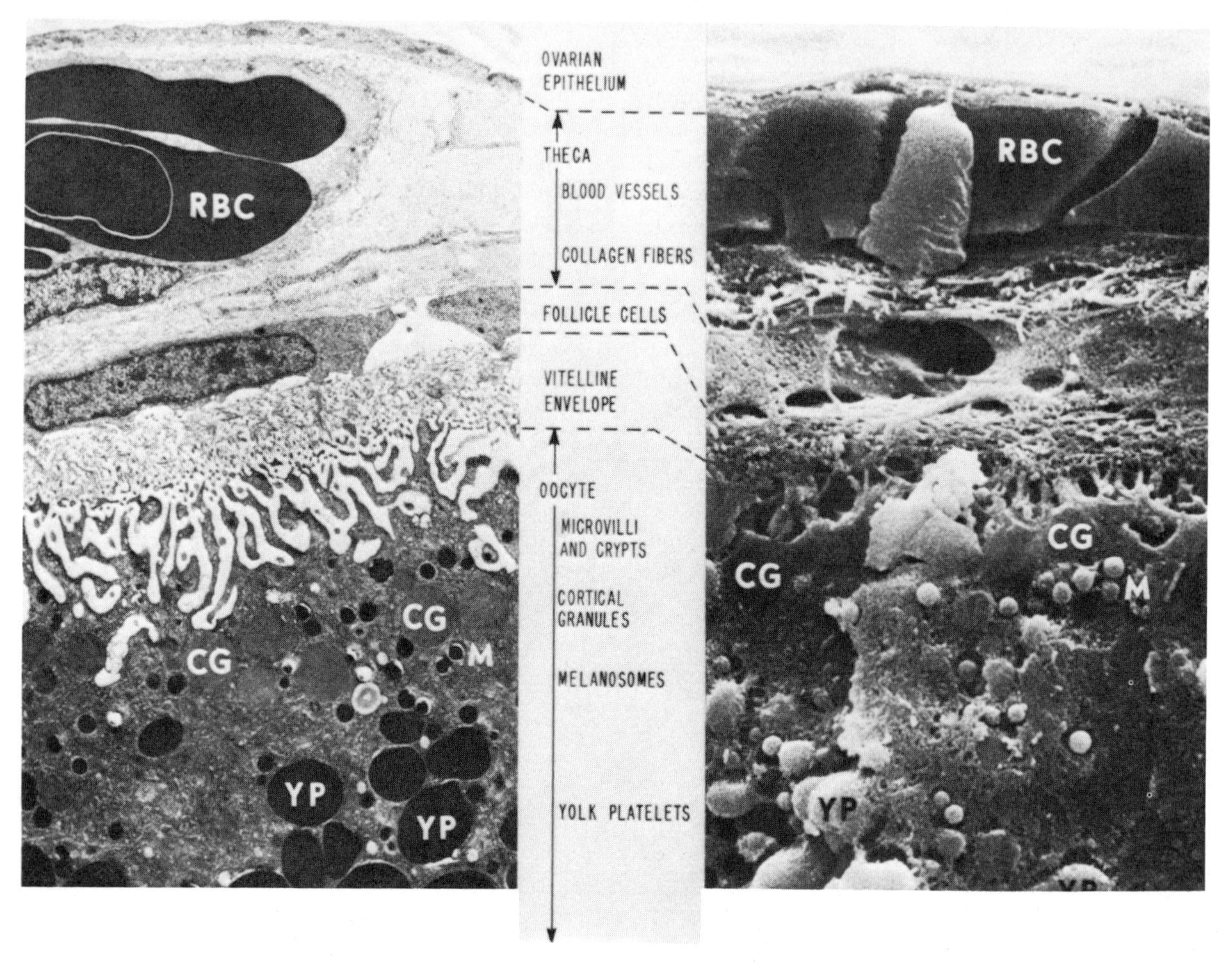


**FIG. 4–14.** TEM micrograph of the architectural relationships between the developing oocyte of *Xenopus* and the adjacent follicular investments. (From J. Dumont and A. Brummett, 1978. J. Morphol. 155:73.)

There is some evidence to suggest that these lining bodies may be nipped off and engulfed by the oocyte. Their fate within the oocyte remains a mystery. Similar specializations of follicle cells have been observed in the primary oocytes of some turtles.

It has long been proposed that follicle cells regulate growth of the mammalian oocyte by providing nutrients to it. Recent studies using electron microscopy and freeze–fracture techniques show the presence of specialized junctional complexes, known as *gap junctions,* between follicle cells and an oocyte during the growth phase. These gap junctions, maintained throughout the period of zona pellucida formation and proliferation of the follicle cells, are small and numerous, and occur where the processes of the follicle cells contact the oolemma (Fig. 4–16). These regions of membrane specialization are thought to contain channels that allow the functional transfer of small molecules between the two cell types. For example, fluorescein dye iontophoretically injected into an oocyte spreads to the surrounding follicle cells. The movement of metabolites from one cell directly into another cell is termed *metabolic cooperativity* or *metabolic coupling.* A variety of experiments have been performed that clearly suggest that metabolic cooperativity exists between follicle cells and the oocyte that they enclose. Heller and his colleagues have demonstrated in a series of radiolabeling experiments that greater than 85 percent of the metabolites present in follicle-enclosed oocytes were originally taken up by the follicle cells and transferred to the oocyte by gap junctions. Conversely, less label

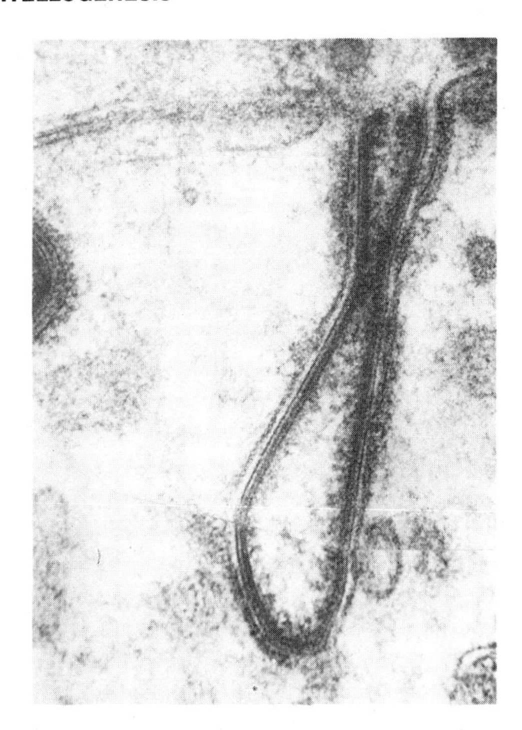

FIG. 4–15. Electron micrograph of a longitudinal section through a projection of an ovarian follicle cell (hen) termed the lining body. The lining body indents but does not penetrate the cell membrane of the oocyte. (From R. Bellairs, 1971. Developmental Processes of Higher Vertebrates. University of Miami Press, Coral Gables, Fla.)

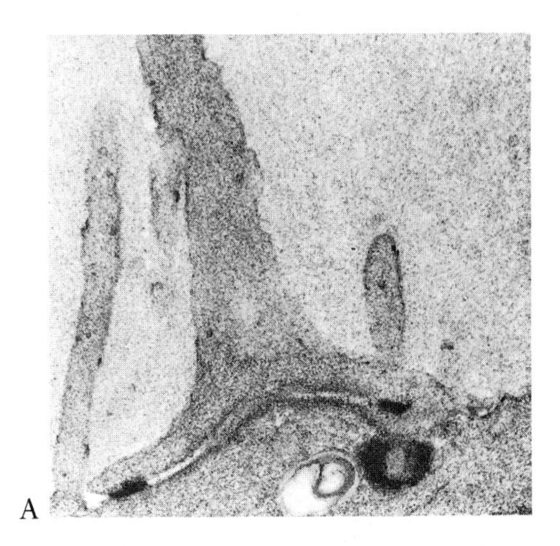


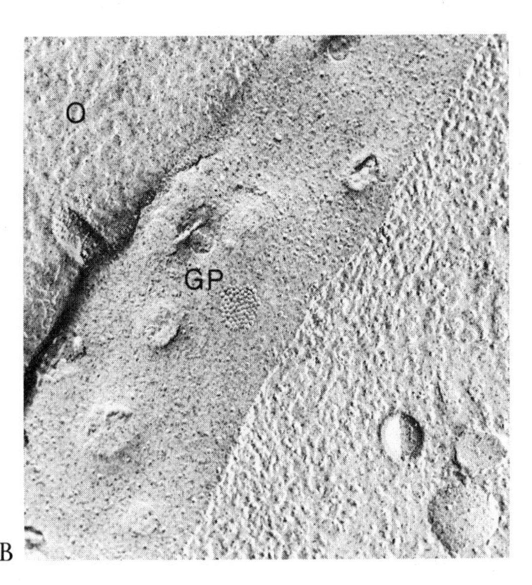


FIG. 4–16. A, an electron micrograph of a follicle cell process establishing gap junctional contact with the oocyte. B, freeze–fracture replica of mouse oocyte depicting a gap junction between the oocyte (O) and a granulosa cell process (GP). (From E. Anderson and D. Albertini, 1976. J. Cell Biol. 71:680.)

is detected in oocytes denuded of their follicle cells. As ovulation approaches, there is a decrease in the number of gap junctions and the extent of metabolic coupling. The naturally occurring union between the follicle cells and the oocyte provides an ideal system for the study of the role of intercellular communication between two different cell types.

In addition to follicle cells, some insects, molluscs, and annelids rely on a special system of accessory cells, known as *nurse cells,* for the growth and differentiation of the oocyte. Unlike follicle cells, nurse cells are derived from the same oogonium that gives rise to the primary oocyte; they are connected to the oocyte by direct intercellular cytoplasmic bridges because of incomplete cytokinesis. In *Drosophila,* for example, a stem cell undergoes four mitotic divisions. Of the 16 cells, one will become the oocyte and the remaining 15 will surround it as the nurse cells. Macromolecules, yolk material, and organelles such as ribosomes traverse the intercellular bridges into the unfertilized egg. The extent of this transfer can be dramatic. It has been estimated that in some insects as many as $2 \times 10^{10}$ ribosomes are transferred from the nurse cells into the oocyte during oogenesis.

## Vitellogenesis

Growth and increase in size of the primary oocyte are conspicuous features of oogenesis. Although an increase in volume of cytoplasm contributes to the growth of the oocyte, much of the change is due to the deposition and stockpiling of foodstuffs. These nutrients are stored in the egg to sustain development until the newly formed individual can care for itself. The growth phase of the primary oocyte may span a considerable period of time. In some frogs, for example, the development of the mature ovum occurs over an interval of approximately three years, beginning shortly after the metamorphosis of the larva into a young froglet. During this time the oocyte increases from a cell size of about 50 microns to one of about 1500 microns in diameter. This corresponds to an increased volume of approximately 27,000-fold. However, most of the change in size occurs during the third year when deposition of yolk within the cell is particularly intense.

By contrast, the growth of other oocytes may be much more rapid and the whole process completed over a shorter time interval. The size of the hen's egg changes very rapidly during the 14-day period immediately preceding ovulation. The volume of the oocyte increases some 200-fold, with most of the yolk being laid down within the six-day period before the egg leaves the ovary (Fig. 4–17).

The eggs of mammals are considerably smaller than those of either amphibians and birds. The mouse egg, for example, increases from about 20 microns to 70 microns during its growth phase.

The phase of oogenesis during which nutritive material or *yolk* is deposited and accumulated within the primary oocyte is termed *vitellogenesis*. For all animals studied, this vital step or process begins after the oocyte enters the first prophase stage of nuclear maturation or meiosis.

The term yolk or *deutoplasm* as used in the literature appears to have several meanings. Yolk is often used to refer to the reserve foodstuffs in mature ova, including fat droplets and glycogen granules. In the strict sense of its definition, however, yolk refers to reserve materials present in special cytoplasmic inclusions laid down during oogenesis and present in the developing embryo. The chief constituents of these yolk inclusions or bodies are phospholipids, proteins, and carbohydrates. Conjugated lipoproteins typically predominate in the yolk, and these may make up more than 80 percent of the egg's dry weight. Depending on the animal, the structural organization and composition of yolk will vary. In forms like cephalochordates and echinoderms, most of the yolk is proteinaceous (*proteoid yolk)* and distributed throughout the ovum as fine *granules*. By contrast, less than 50 percent of the dry weight of the mature egg cell is yolk protein in the amphibian. Lipids constitute about 25 percent of the dry weight of the fully grown amphibian egg and are distributed in the cytoplasm in the form of inclusions known as *lipochondria*. Most of the protein yolk in the amphibian oocyte is found in the form of large, flattened crystalline bodies termed *yolk platelets*. Similar inclusions with a crys-

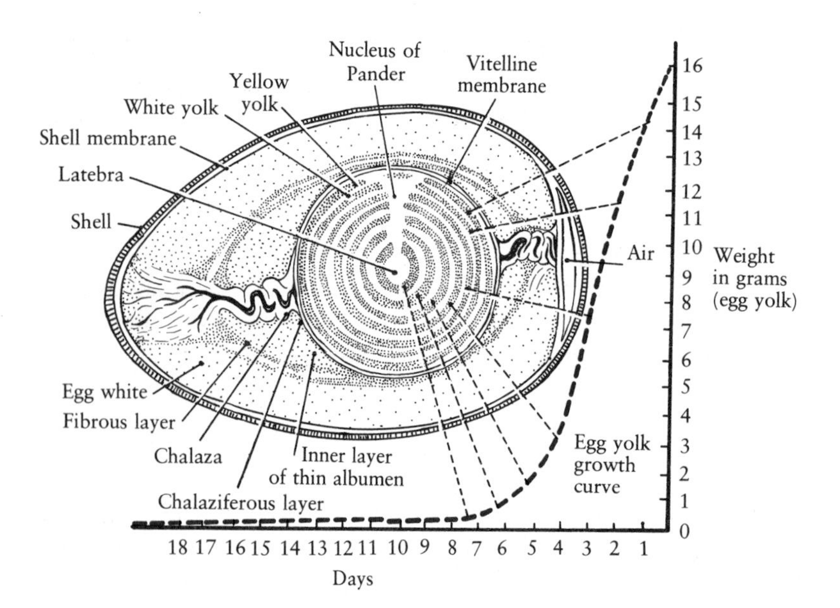


**FIG. 4–17.** Diagram showing the structural organization of the hen's egg at the time of laying. The graph indicates the rate of growth of the egg, measured by weight of yolk in grams, during the 18 days preceding oviposition (laying). (After E. Witschi, 1956. Development of Vertebrates. W. B. Saunders Company, Philadelphia.)

talline structure are identifiable in the oocytes of cyclostomes, elasmobranchs, and bony fishes. Avian yolk is a combination of water (48.7%), proteins (16.6%), phospholipids (32.6%), and carbohydrates (1.0%). Vitamins $A_1$, $B_1$, $B_2$, and D are also present. Most of the yolk is in liquid form. Approximately 25 percent of the yolk is organized as *yolk globules* or *yolk spheres*.

It is not surprising that differences in the amount and distribution of yolk within the mature ovum exist among members of the animal kingdom. The egg of the bird has a large and generous supply of yolk, sufficient to provide for rapid and complete development. The chick is, in fact, a small adult upon emergence from the shell and fully capable of taking care of its own needs. Ova with large amounts of yolk are termed *polylecithal* or *megalecithal* and can be found in teleost fishes, elasmobranch fishes, reptiles, and birds. Generally, the active cytoplasm in a polylecithal egg is restricted to a thin layer on the surface of the animal pole; this will thicken into a disc of cytoplasm *(blastodisc)* containing the maternal nucleus. By contrast, the yolk in the ovum of a frog or salamander is moderate in amount *(mesolecithal),* sufficient to carry the developing embryo only to the larval or tadpole stage. The tadpole is then able to secure enough food for its growth and development into a small frog. The yolk in polylecithal and mesolecithal ova tends to be localized at the vegetal pole, a condition referred to as being *telolecithal*. The amount of the yolk in the ova of echinoderms, lower chordates (cephalochordates, urochordates), and mammals is small (*oligolecithal* or *microlecithal*) and rather evenly distributed throughout the ooplasm (*isolecithal* or *homolecithal*). Arthropods, especially insects, have an unusual distribution of yolk in their ova. The yolk lies in the interior of the cell, surrounding a mass of cytoplasm containing the nucleus. A thin layer of cytoplasm also lies just beneath the surface of the cell. Such an egg is termed *centrolecithal*.

Within recent years, studies by a number of investigators have greatly increased what we know concerning the sources of the constituents of the yolk inclusions and the mechanisms by which yolk is packaged within the primary oocyte. We now know more, for example, about the sites of synthesis for the precursors of protein and carbohydrate yolk as well as the organelles charged with fashioning the yolk into a cytoplasmic inclusion. In general, yolk production results from precursors synthesized either within *(autosynthetic)* or outside of *(heterosynthetic)* the primary oocyte. A variety of organelles including the endoplasmic reticulum, the Golgi complex, and the mitochondria have been identified as possible sites for the assembly of yolk into an inclusion within the oocyte.

Techniques employing the electron microscope and radioactively labeled amino acids have provided clear evidence that the precursors of protein yolk in both invertebrates and vertebrates are commonly manufactured outside of the primary oocyte and subsequently sequestered within it. If the ovaries of insects are exposed to a tritiated amino acid, such as [$^3$H]leucine, radioactivity is initially detected in the accessory cells surrounding the oocyte. Shortly thereafter, the radioactive label is present throughout the cytoplasm of the oocyte. Similar labeling methods have been used to trace the source of yolk in amphibians and birds. Here, the yolk protein precursor, known as *vitellogenin,* is synthesized and secreted by the liver under the influence of estrogen. Vitellogenin, characterized as a large lipoglycophosphoprotein, is transported in the plasma of the bloodstream to the follicular epithelium around the oocyte. It passes through the follicle cells and is incorporated into the oocyte by micropinocytosis or endocytosis (Fig. 4–18). In the case of *Xenopus,* the endocytosed vesicles fuse with each other to form the primordial yolk platelets. The contents of these bodies will become crystalline to form the mature yolk platelets. The vitellogenin is cleaved proteolytically into two major yolk proteins, *lipovitellin* and *phosvitin.* The major steps in vitellogenesis appear to be quite similar in all vertebrates (Fig. 4-19). Anderson has observed a very similar process of uptake of yolk protein in eggs of the American cockroach (Fig. 4-20). Exogenous proteins and polysaccharides appear to be initially adsorbed onto specialized areas of the egg plasma membrane containing a sticky, matted substance or *bristle coat.* These areas then become invaginated and pinched off to form *coated vesicles.* The coated vesicles are precursors of the yolk bodies.

Observations on the eggs of other animals during oogenesis clearly point to the machinery of the oocyte as the site of yolk synthesis. The yolk bodies in the ova of the crayfish and lobster are formed in the cisternae of the smooth endoplasmic reticulum. Dense granules considered to be precursors of the yolk protein are initially visible in the cisternae of the rough endoplasmic reticulum near the oocyte

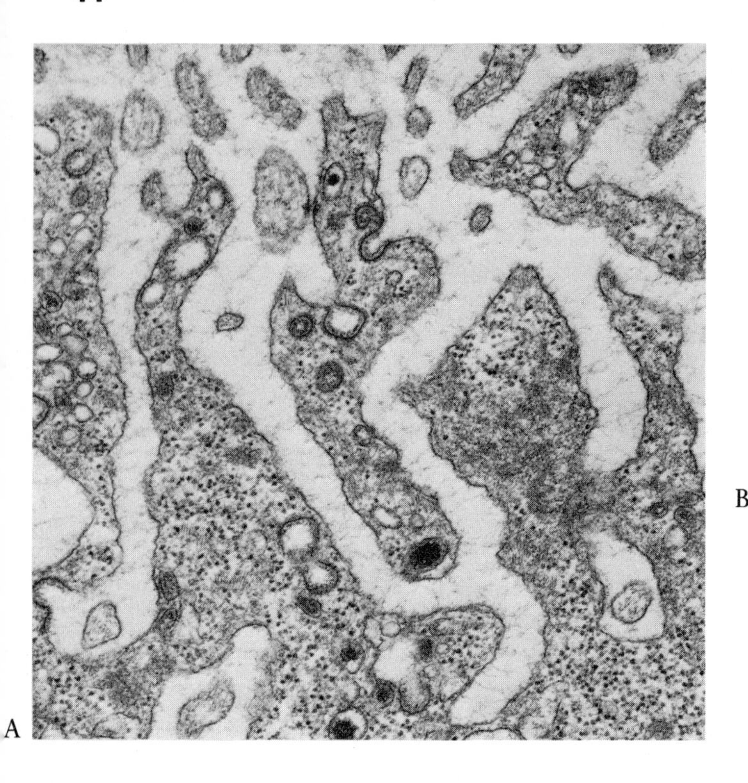


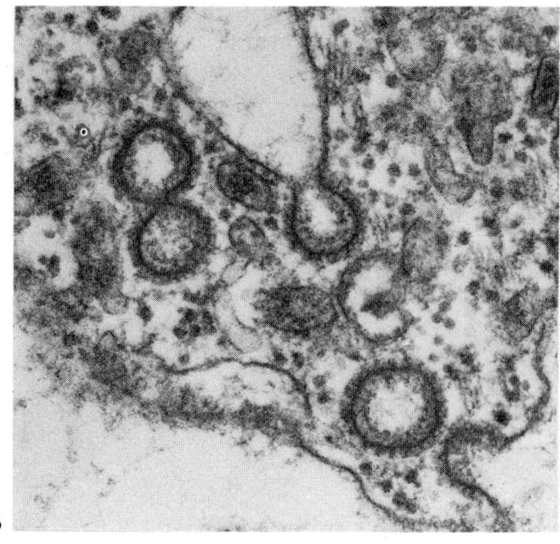


**FIG. 4–18.** A, precursors of yolk are taken up into coated vesicles by the process of endocytosis in eggs of *Xenopus* (Courtesy of K. Selman.) B, higher magnification of the coated vesicles at the egg surface during vitellogenesis. (From R. Wallace et al., 1983. Molecular Biology of Egg Maturation. Ciba Foundation Symposium 98, p. 228.)

**FIG. 4–19.** A diagrammatic scheme showing the basic steps in vitellogenesis as it typically occurs in growing vertebrate oocytes.

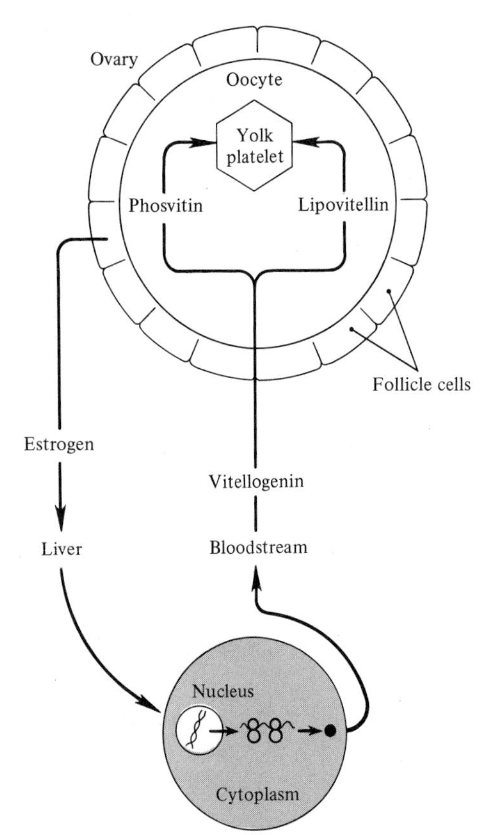


**FIG. 4–20.** A diagrammatic section through the oolemma of the cockroach oocyte to show the formation of yolk bodies. The surface of the oolemma is thrown into a number of microvilli (MV). Areas between the microvilli show a sticky, matted substance termed bristle coat (BC). Micropinocytic or coated vesicles form, some of which are small ($CV_1$) and some which are large ($CV_2$). The large vesicles fuse to form dense-cored yolk bodies (YB). The small coated vesicles fuse to form tubular structures; some of these fuse with larger coated vesicles to form yolk bodies (YB). Exogenous proteins and polysaccharides are probably absorbed onto the bristle coat and then taken into the ooplasm by micropinocytosis. (From E. Anderson, 1969. J. Microsc. 8:721.)

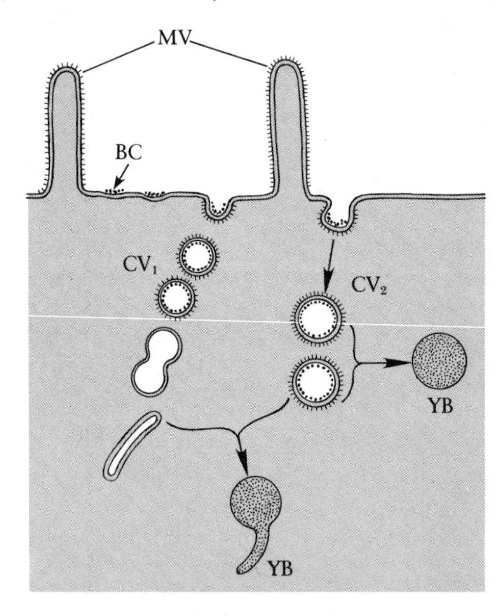

nucleus. After being transported peripherally through a series of interconnecting tubes to the smooth endoplasmic reticulum, the granules aggregate and are then pinched off into the ooplasm as yolk bodies (Fig. 4–21). Hence, the endoplasmic reticulum functions as the site of both the synthesis of yolk precursors and their assembly into an inclusion.

The Golgi complex of the oocyte appears to be a center of yolk formation in such diverse forms as the horseshoe crab, the hydrozoan jellyfish, and the killifish (a teleost). Whether the cisternae of the Golgi complex synthesize as well as assemble the constituents of the yolk bodies is still unclear. The yolk bodies are initially visible as a collection of small vesicles adjacent to the Golgi apparatus. There is a close association between the Golgi bodies and the rough endoplasmic reticulum in the eggs of these animals during vitellogenesis. It is likely that the proteins of the yolk bodies are synthesized on the ribosomes of the endoplasmic reticulum and then are transported to the saccules of the Golgi. Here the carbohydrate moiety is added. Electron micrographs of amphineuran (Mollusca) oocytes actively engaged in vitellogenesis show that vesicles arise by evagination from the endoplasmic reticulum and then appear to be transported to the Golgi complex.

**FIG. 4–21.** Differentiation and growth of the crayfish oocyte showing the role of the endoplasmic reticulum in the formation of proteinaceous yolk. The structures shown are nucleus (N), nuclear pores (NP), Golgi complex (GC), agranular endoplasmic reticulum (AER), differentiated stacks of rough-surfaced endoplasmic reticulum (ERS), intercommunicating smooth-surfaced cisternae (ICC), intracisternal granules, (ICG), aggregates of intracisternal granules (ACG), immature yolk bodies (IYB), and mature yolk bodies (YB). (From H. Beams and R. Kessel, 1963. J. Cell Biol. 18:621.)

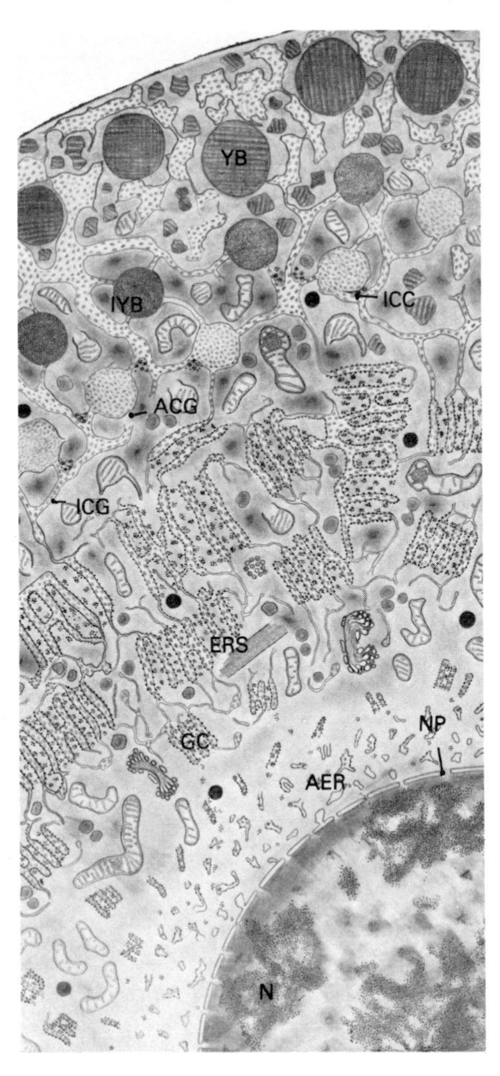


The formation of yolk bodies is complex, and there is evidence that many animal species may synthesize yolk precursors both inside and outside of the oocyte. The formation of protein yolk in a dual fashion is amply illustrated in both amphibians and fishes. In addition to the fusion of micropinocytotic vesicles, the yolk bodies are formed as large crystalline inclusions inside of modified mitochondria (Fig. 4–22). Because the oocyte mitochondria in *Rana* appear to contain all the components required for protein synthesis, it is probable that the yolk proteins in this frog are synthesized as well as packaged in the mitochondria. By contrast, although yolk platelets are known to be associated with mitochondria in *Xenopus*, the yolk proteins appear to be synthesized solely in the liver. Following release from the mitochondria, the yolk platelets of the frog egg are large, membrane-bound, ovoid crystalline structures that are flattened in one plane. Each platelet consists of a proteinaceous core surrounded by a superficial granular coat of polysaccharide. Analysis of the platelet shows the presence of two prominent yolk proteins, phosvitin (molecular weight of 35,000) and lipovitellin (molecular weight of 40,000). Two molecules of phosvitin are joined to each molecule of lipovitellin to form the organizational unit of the yolk crystal.

The mature egg of the mammal is often described as having no yolk. Within the framework of our loose definition of this term, yolk is definitely present in mammalian eggs in the form of lipid droplets and glycogen granules. Whether distinct inclusions of a protein-lipid nature are part of the

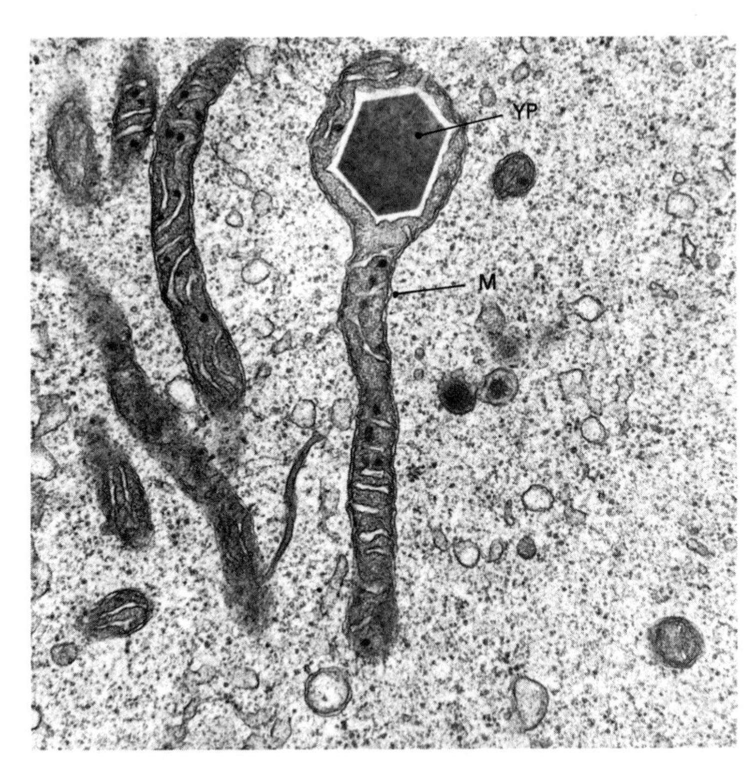


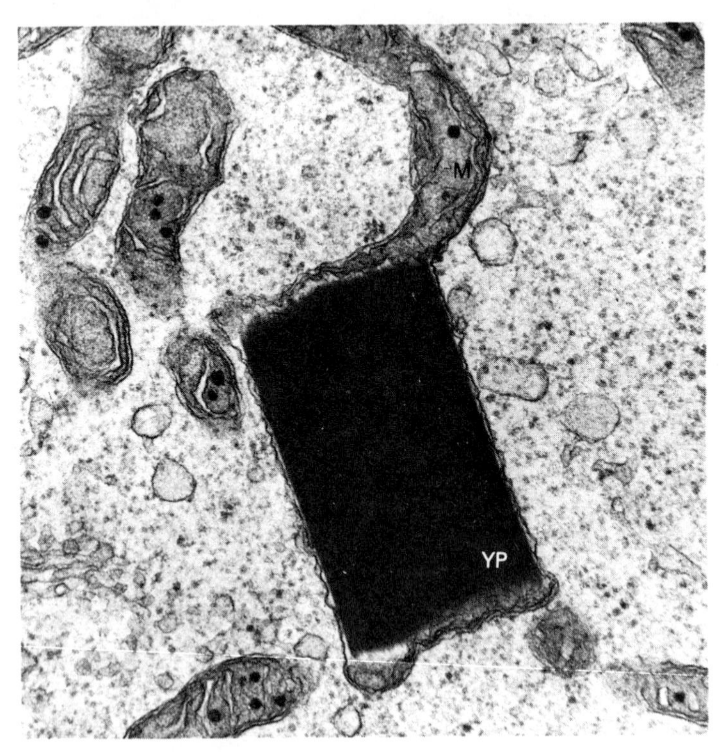


**FIG. 4–22.** Two views of the yolk platelet (YP) in varying degrees of growth inside the mitochondrion (M) of a frog. (From R. Kessel, 1971. Z. Zellforsch. Mikrosk. Anat. 112:313.)

organization of the egg is still an open question. In the mouse, rat, and hamster, multiple stacks of fibrous material appear in the cytoplasm of young oocytes. These continue to accumulate within the ooplasm until the oocyte is mature. Interestingly, these fibrous stacks disappear during the preimplantation period of the embryo. Perhaps this fibrous material is being used as an energy source by the embryo during the interval between fertilization and implantation. Large yolklike vesicles filled with flocculent material have also been identified in the ova of rabbit, ferret, and sheep.

## Gene Activity During Oogenesis

Most of our discussion on oogenesis to this point has been directed at structural changes occurring at the surface and within the cytoplasm of the oocyte. Also, we have seen that the oocyte is a depository for products synthesized in the liver (or fat body of insects); these hepatic proteins serve the function of nutrition or provide growth factors and building blocks for the early developing embryo. Additionally, however, many products are synthesized within the developing oocyte. Some of these are important to preparing the machinery of the egg for DNA, RNA, and protein synthesis. The growing oocyte also regulates the synthesis and storage of messenger RNA that is required immediately after fertilization.

The nucleus of the primary oocyte is a major site of biosynthetic activity. As already pointed out, the nucleus becomes enormously large and swollen during vitellogenesis to form the germinal vesicle. Examination of the germinal vesicle shows its major components to be the *nuclear envelope*, the *nuclear sap* or *nucleoplasm*, the *chromosomes*, and the *nucleoli*.

The large size of the nucleus of the growing oocyte has been of decided advantage in studies designed to probe the structure and function of this organelle during oogenesis. With the amphibian oocyte, it is a rather simple procedure to isolate the nucleus manually using a pair of fine forceps. Ultrathin sections of the isolated nucleus show the nuclear envelope to consist of two concentric membranes separated by a space of 100 Å to 300 Å. The envelope is interrupted periodically by *nuclear pores* or *annuli*. Close analysis indicates that these pores are not simple openings in the nuclear envelope. Each pore is complex and appears to be occupied by a granular-type material. Commonly, aggregates of dense fibrous material can be seen projecting through the pores, suggesting that these annuli act to control the passage of material from nucleus to cytoplasm. Experiments in which ions and low-molecular-weight proteins have been injected into amphibian oocytes give clear evidence that the nuclear envelope is very selective in its permeability. Hence the nuclear envelope would appear to function as a semipermeable membrane and thereby regulate the free diffusion of ions and macromolecules into the cytoplasm.

The nucleoplasm is fluid and acts to suspend the chromosomes and the nucleoli. Both of these structures actively engage in the synthesis of ribonucleic acid (RNA) during the growth and differentiation of the oocyte. As in the case of the primary spermatocyte, distinct and characteristic changes in the configuration of the chromosomes become visible in the primary oocyte. In extremely young oocytes, for example in the ovaries of newborn mice or tiny *Xenopus* toads prior to metamorphosis, the chromosomes are observed to be in leptotene or early zygotene state of meiosis. By four days of age in mice, the chromosomes are in the diplotene stage of meiosis. They will remain in such a configuration until the time of ovulation.

Shortly after the onset of vitellogenesis, the arrested chromosomes attain a high degree of extension and assume what is commonly known as the lampbrush configuration (Fig. 4–23). Thin threads or loops characteristically branch out at right angles to the long axis of each chromosome, thus suggesting the appearance of lampbrushes used for cleaning petroleum lamps before the invention of the electric light. Such lampbrush chromosomes have been observed in the nuclei of oocytes in a wide variety of both vertebrates and invertebrates, including man (Table 4–1). Although widespread among animal oocytes, not all lampbrush chromosomes are of the same size and therefore equally amenable to experimental analysis. Lampbrush chromosomes are rather small in *Xenopus;* they are much larger in the newt *(Triturus).*

The structure of the typical lampbrush chromosome as proposed by Gall and Callan is shown in Figure 4–23. The main axis of each chromosome consists of two homologous chromatids, which are closely paired. The chromatid is considered to be a single, long fiber of double-stranded deoxyribonucleic acid (DNA) and tightly packed protein. The

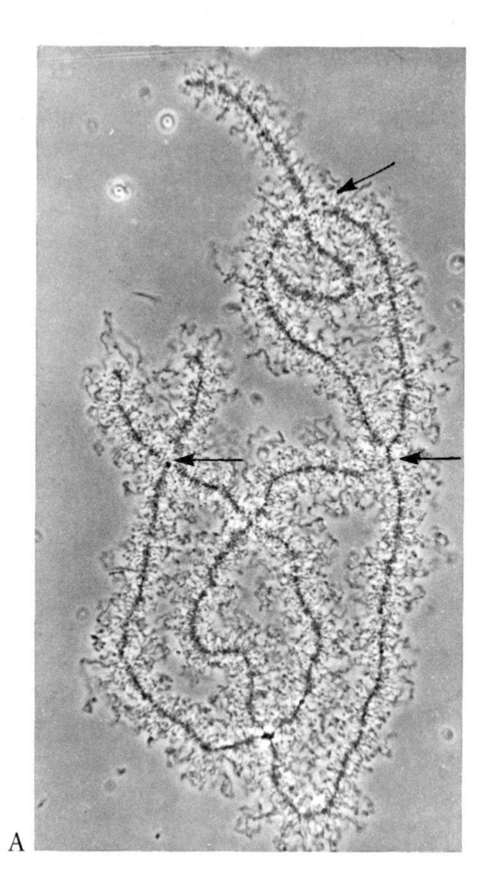


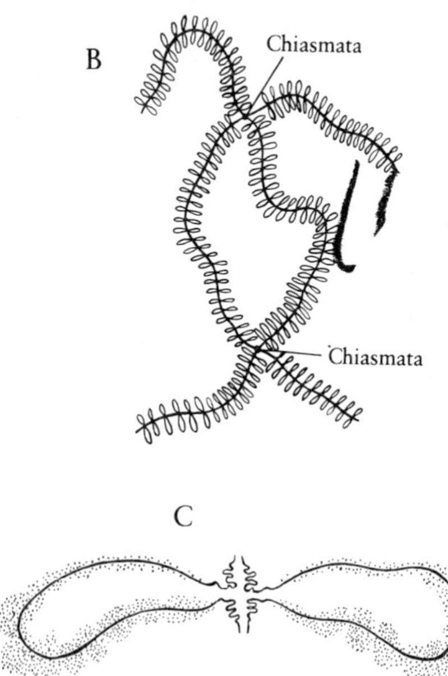


TABLE 4–1 Occurrence of Lampbrush Chromosomes in Animal Oocytes and the Duration of the Lampbrush Stage

| *Species and Affiliation of Animals in Which Lampbrush Chromosomes Have Been Reported* | *Estimated Duration of Lampbrush Stage (Where Available)* |
|---|---|
| Deuterostome | |
| Chaetognath | |
| Arrow worm | |
| Echinoderm | |
| Sea urchin | |
| Chordate | |
| Cyclostome | Several months in lamprey |
| Shark | |
| Teleost | |
| Amphibian | |
| Urodele | About seven months in *Triturus* |
| Anuran | Four to eight months in *Xenopus*, 30–40 days in *Engystomops* |
| Reptile | Some months in lizards |
| Bird | Three weeks in chick |
| Mammal | Perhaps years in man |
| Protostome | |
| Mollusk | |
| Gastropod | |
| Cephalopod | |
| Insect | |
| Orthopteran | Three months in cricket |

From E. Davidson and B. Hough, 1972. In: Oogenesis. Eds. J. Biggers and A. Schuetz. Baltimore, University Park Press.

paired loops represent uncoiled segments of the DNA fiber on adjacent regions of the sister chromatids. The base of each loop is thickened into a swelling known as a *chromomere;* it is believed that the chromomere serves as the site from which the DNA of the loop originates. In the salamander *(Triturus)*, there are some 20,000 loops in the whole set of chromosomes with the average loop being about 50 microns in length. Thus, there are about 5000 loops per haploid set of chromosomes. It has been estimated that *Plethodon*, also a salamander, may have as many as 10,000 loops per chromosomal set.

**FIG. 4–23.** A, a phase-contrast photomicrograph of an isolated lampbrush chromosome from the oocyte nucleus of the salamander, *Triturus viridescens*—note the multiple, paired loops extending from the axes of the homologous chromatids, which are held in a tetrad configuration by chiasmata (arrows); B, a diagrammatic sketch of a pair of lampbrush chromosomes joined by two chiasmata; C, an interpretation of chromosome structure as two continuous chromatids. (Courtesy of J. Gall.)

The configuration of the lampbrush chromosome is suggestive of intensive and widespread gene activity in the nucleus. When lampbrush chromosomes are manually isolated and exposed in a culture medium to a radioactive precursor of RNA ([$^3$H]uridine), the results of autoradiography indicate that most of the label is detected along the length of the lateral loops. Newly synthesized proteins can also be found over the chromosomal loops. Hence the lateral loops of lampbrush chromosomes are active sites of protein synthesis, and each loop contains a matrix of ribonucleoprotein. As shown in Figure 4–23 C, however, this matrix is not uniformly distributed along a loop, but rather increases in thickness from one end of the loop to the other end. Miller and his co-workers have used the electron microscope to construct the relationship between RNA synthesis and loop morphology. They found that the matrix consists of innumerable fibrils arranged in increasing length around the loop. The fibrils represent nascent or growing RNA transcripts being synthesized continuously along the loop of DNA. The entire loop is considered to be a *transcriptional unit*. Recognizing that the length of the DNA of a single loop in *Triturus* is about 50 microns, it is estimated that the RNA transcripts total in length some $5 \times 10^4$ to $10 \times 10^4$ nucleotides. The lampbrush chromosomes are very active during vitellogenesis in producing *nuclear* or *heterogeneous RNA*. Presumably, these large transcripts, DNA-like in base composition and highly complex in their base sequences, give rise to the mRNA molecules of the egg's cytoplasm. How this processing occurs is still unclear.

There is extensive gene activity and accumulation of RNA in the egg during its growth phase. In *Xenopus*, for example, the growth phase of the egg lasts about 6 months; during this time the primary oocyte increases to a diameter of 1.5 mm and accumulates about 4 micrograms of ribosomal RNA (rRNA) and 40 to 80 nanograms (ng) of *3′-polyadenylic (Poly A) RNA*, most of which appears to be mRNA. Mature eggs of *Strongylocentrotus* contain a total of 3 nanograms of RNA with approximately 50 to 100 picograms per egg being RNA that probably serves as maternal messenger RNA. Maternal messenger RNA is egg RNA that the embryo utilizes to synthesize proteins on polyribosomes after fertilization, or unfertilized egg RNA that supports protein synthesis in an in vitro translation system. Most of the translatable mRNA in the unfertilized egg contains 3′-polyadenylic acid (or "tails" that are about 50–120 nucleotides long). Little of this mRNA is associated with the polyribosomes of the mature oocyte. It is primarily stored in the form of *ribonucleoprotein particles* and unavailable for translation. This type of RNA is often referred to as *masked messenger RNA*.

Because the patterns of biosynthesis in the fertilized egg remain essentially those determined by maternal mRNA, there has been considerable interest in the extent to which the egg genome is transcribed during oogenesis and how much information is actually stored in the egg. The majority of the mRNA molecules appear to be synthesized from *nonrepeating* or so-called *single-copy* sequences of DNA. By measuring the extent of the DNA to which these RNA molecules are complementary (by hybridization techniques), one can estimate the diversity of the population of the RNAs. It has been estimated that the RNA of the fully grown oocyte of *Xenopus* is complementary to about 1.2 percent of the nonrepeated portion of the genome. This represents a complexity or total sequence length of unfertilized egg RNA of about $27 \times 10^6$ nucleotides. Similar complexity values for *Strongylocentrotus* and *Arbacia* fall in the range of $30 \times 10^6$ to $37 \times 10^6$ nucleotides. From this data, it can be further calculated that this population of RNA molecules could code for some 25,000 to 40,000 different polypeptides. This represents the amount of information that is programmed into the egg to sustain the early development of the embryo. There are several specific proteins for which maternal mRNAs are known to be stored. These proteins include *tubulins, actin,* and *histones*.

Most of the RNA stored in the egg is not informational. It is a mix of low-molecular-weight transfer RNAs (4S RNA), 5S RNA, and ribosomal RNA. Ribosomal RNA makes up more than 95 percent of these RNAs. It is synthesized in the nucleoli of the primary oocyte nucleus. Nucleoli are extrachromosomal bodies of rRNA genes set aside within the nucleus for the specific production of rRNA. The ribosomal genes are referred to as *rDNA*. The oocytes of different animal species show variable numbers of nucleoli. In most invertebrates, the nucleolus is a large, single, spheroid-shaped organelle. By contrast, the oocytes of vertebrates commonly have hundreds of nucleoli of various sizes distributed just inside the nuclear envelope. There are about 600 nucleoli in the germinal vesicle of *Triturus* and about twice this number in *Xenopus*.

Ultrathin sections of a typical nucleolus show a bipartite structure with a granular cortex surrounding a central fibrillar core (Fig. 4–24 A). In *Triturus*, the core has been isolated from the granular cortex, dispersed, and examined under the electron microscope. The core consists of thin circular axial fibers along which, at regularly spaced intervals, are groups of 80 to 100 short-to-long fibrils (Fig. 4–24 B). If treated with selected enzymes known to digest DNA, RNA, and protein, the results show that each core fiber is composed of DNA, the fibrils are RNA, and both nucleic acids are coated with protein. Elegant studies by Gall, MacGregor, and Miller permit the view that the fibrils are growing chains of rRNA being transcribed from segments of DNA. Indeed, each segment of DNA is a single gene (i.e., transcriptional unit) coding for rRNA. Hence, as many as 80 to 100 precursor rRNA molecules are being synthesized simultaneously on each rDNA gene.

Each rRNA molecule sediments at 40S. This type of RNA will later be processed into smaller (18S) and larger (28S) subunits that go into the assembly of a ribosome. The synthesis of 28S and 18S RNA takes place during vitellogenesis.

Since vertebrate oocyte nuclei contain many nucleoli when compared to somatic cell nuclei, there must be many copies of genes coding for nucleolar rRNA. Indeed, the considerable accumulation of rRNA in the eggs of such organisms as *Triturus* and *Xenopus* is due to the *selective amplification* of rDNA genes. Amplification is generally thought to occur during the pachytene stage of meiosis I and refers to the repetitive copying of rDNA sequences; the copies organize to form the nonchromosomal nucleoli. It has been demonstrated in amphibian oocytes that the extra genes originate from specific, condensed regions (caps or nucleolar-organizing regions) of the chromosomes.

The significance of gene amplification is clearly demonstrated in the amphibian egg. At the end of the growth period, the large cytoplasmic mass is estimated to be populated with about $1.1 \times 10^{12}$ ribosomes. In the absence of amplification, it would take the rDNA genes, working at maximal transcriptional rate, some 500 years to produce as much ribosomal material as is synthesized during the normal growth period. Ribosomal gene amplification apparently does not occur during oogenesis in sea urchins. The high rate of synthesis suffices to produce in a few months of oogenesis the population of ribosomes (about $4 \times 10^{18}$) required by the egg. It is interesting to note that 5S RNA, which later functions in conjunction with the ribosomes and is therefore required in amounts similar to 18S and 28S RNA, is not synthesized on amplified RNA genes. The synthesis of 5S RNA begins earlier than either 18S or 28S RNA, and therefore this type of RNA has a longer time during which to accumulate in the egg.

## Oocyte Maturation

When the primary oocyte has reached the end of its growth phase, it enters a stationary phase that persists until ovulation. Oocytes by this time have lost the ability to divide and proliferate. At about the time of ovulation, the fully grown egg is ready to undergo maturation, a process that makes the cell fertilizable. Although maturation and ovulation occur almost simultaneously in a number of animal species, these two processes do not appear to be causally linked. In mammals, for example, fully grown oocytes can be induced in vitro by gonadotropins to complete maturation within their follicles and spontaneously develop into oocytes.

The term *oocyte maturation* is used in various ways by different investigators. We will use oocyte maturation to refer to the progression of events from the breakdown of the germinal vesicle (i.e., resumption of meiosis) through the completion of meiosis. It includes activities within the nucleus as well as in the cytoplasm. The reinitiation of meiosis is marked by the breakdown of the germinal vesicle (GVBD) and the condensation of the chromatin material. Reorganization of the chromatin material involves the regression of the loops of the lampbrush chromosomes. In forms such as *Chaetopterus*, amphibians, and mammals (rat), there is a fascinating undulation of the nuclear envelope just prior to its disintegration. With continued maturation, the condensed, synaptic chromosomal pairs migrate to a position beneath the egg plasma membrane. Here, an achromatic figure is formed with the spindle fibers oriented perpendicularly to the surface of the oocyte. The primary oocyte is then divided unequally into a large *secondary oocyte* and a small *polar body (polocyte)*. The second reduction proceeds in a similar fashion, resulting in the production of a mature egg cell (ovum) with a haploid set

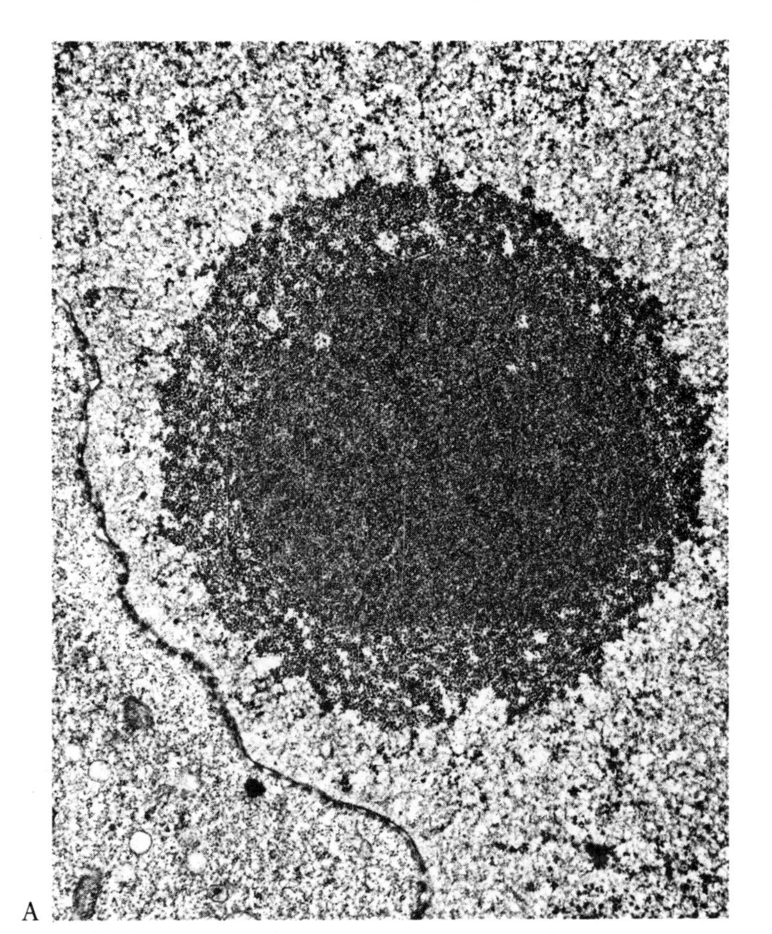



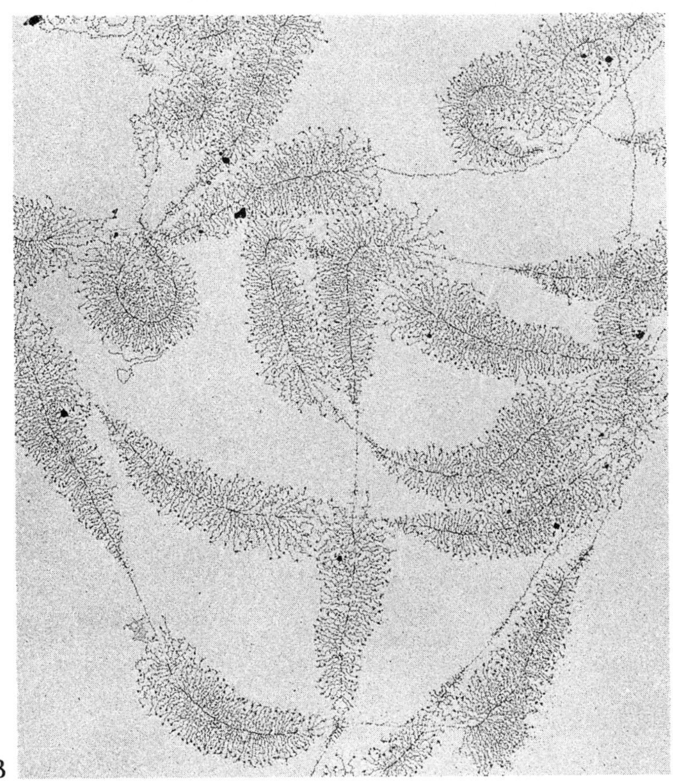



**FIG. 4–24.** A, an ultrathin section of an extrachromosomal nucleolus from the oocyte of a salamander to show a compact fibrous core surrounded by a granular cortex (From O. L. Miller and B. Beatty, 1969. J. Cell Physiol. 74 Suppl. 1:225); B, a portion of the dispersed core of the extrachromosomal nucleolus showing the active (RNA-producing fibrils) and non-active segments of the DNA. (From O. L. Miller, B. Beatty, and B. Hamkalo. 1972. Oogenesis. J. Biggers and A. Schuetz, eds. University Park Press, Baltimore.)

of chromosomes. The haploid nucleus of the egg is called the *female pronucleus*. The first polar body may divide to form two polocytes. Theoretically, at least, each primary oocyte produces a mature egg cell and three abortive polar bodies during meiosis (Fig. 4–25).

Oocytes of different species are fertilizable at different stages of maturation. With the exception of sea urchins and coelenterates, completion of the reduction divisions is dependent on the entrance of the sperm into the egg. Eggs of sea urchins are mature (i.e., meiosis is completed) before ovulation and fertilization. In *Ascaris* (nematode) and *Nereis* (marine annelid), both reduction divisions occur after the sperm cell has been incorporated into the egg cell. By contrast, the spindle apparatus for the first reduction division in *Styela* is formed at the time that the egg is discharged into the seawater; the meiotic process is then arrested in metaphase I and not continued until penetration of the egg by the male gamete. In *Amphioxus* and most vertebrates, the meiotic process is initiated in the ovary but then arrested in metaphase of the second reduction division. It is unclear why different animal species show this variability in the timing of sperm entry into the egg relative to the progression of meiosis.

**FIG. 4–25.** Reduction divisions in an oocyte. A, an oocyte before the onset of the meiotic divisions showing two tetrads in the germinal vesicle (nucleus) and the achromatic figure; B, C, D, first meiotic divisions; E, the formation of first polar body with achromatic figure in preparation for second meiotic division; F, G, second meiotic division including division of the first polar body; H, meiosis completed.

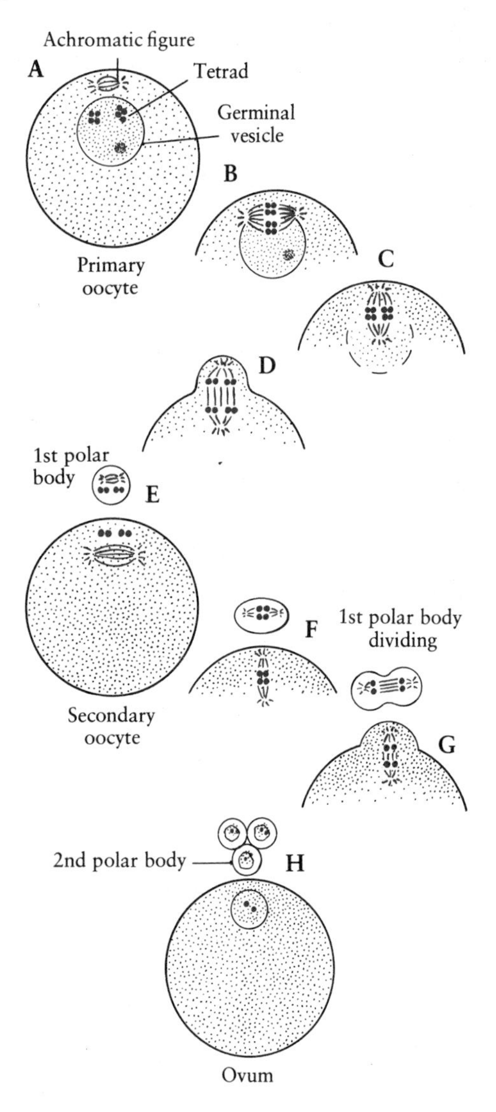


The importance of germinal vesicle breakdown for continued development of the egg cell was pointed out at the beginning of this century in classical experiments by Delage on oocytes of *Asterias* (starfish). By shaking ripe ova in a vial, Delage discovered that each cell could be fragmented into two spherical halves. If fragmentation occurred when the germinal vesicle was still intact, it was determined that the nucleated half (i.e., with the germinal vesicle) would cleave after the addition of sperm. The nonnucleated (i.e., without the germinal vesicle) half failed to cleave following exposure to a sperm suspension. However, if fragmentation of mature ova occurred just after dissolution of the germinal vesicle membrane, both resulting halves cleaved following the addition of sperm. These studies provided the first experimental evidence that the ability of the egg cytoplasm to cleave was dependent on the nucleoplasm of the germinal vesicle. Similar results have been obtained using the eggs of other invertebrates, such as *Nereis* and *Arbacia*.

The mechanism(s) underlying oocyte maturation are currently the subject of considerable investigation, particularly on eggs of amphibians, starfish, and mammals (mouse, rat). The initiation of maturation in vertebrates is dependent on stimulation by hormones produced by the pituitary gland (*gonadotropins*) and the follicle cells of the ovary. A gonadotropin-like polypeptide hormone (termed *radial nerve factor* or *gamete-shedding substance*) is released by the radial nerves and stimulates maturation in the eggs of the starfish. The hormones produced by the nervous system act directly on the ovary.

The sequence of events during maturation and the signals underlying the regulation of meiosis have been extensively studied in amphibians, particularly in eggs of *Rana* and *Xenopus*. Experiments with fragments of the ovary containing immature oocytes established the role of the pituitary gland in triggering maturation. If pieces of ovary tissue are cultured with fragments or extracts of the pituitary gland, the eggs undergo GVBD, and the chromosomes condense and proceed to the second meiotic metaphase. Although initially it was thought that the pituitary gonadotropin acted directly on the oocyte itself, it is now known that the targets of the hormone are the follicle cells. If fully grown oocytes are denuded of their follicle cells and then exposed to pituitary extracts, the germinal vesicle remains intact and the eggs show none of the other manifestations of maturation. However, when the same follicle-free oocytes are treated with selected concentrations of *progesterone*, there is full resumption of meiosis. Hence the follicle cells mediate the action of the pituitary hormone by releasing a steroid hormone, presumably progesterone, which then stimulates the maturation of the oocyte. The changes that occur in the follicle cells in response to the action of the pituitary hormone are partially known. If ovarian fragments containing immature oocytes are treated with either actinomycin D (inhibitor of transcription) or puromycin (inhibitor of translation) and then exposed to the gonadotropin, there is no evidence of maturation. These results are interpreted to mean that messenger RNA and protein appear to be synthesized by follicle cells in response to pituitary hormone stimulation. These activities are probably related to the conversion of *pregnenolone* to progesterone in oocytes of *Xenopus*. By contrast, if fully grown, immature oocytes devoid of follicle cells are exposed to actinomycin D and then treated with progesterone, maturation is not inhibited; the process, however, is blocked when actinomycin D is replaced with puromycin. Progesterone, therefore, appears to function as a natural *maturation* (meiosis)-*inducing factor* in amphibians. Also, oocyte maturation does not require the synthesis of new mRNA, but it does require the synthesis of protein. The protein appears to be translated from some preexisting mRNA. In *Rana pipiens*, this synthesis is initiated within 5 to 6 hours after gonadotropin stimulation of follicle-enclosed oocytes. By 18 to 24 hours after treatment, the level of protein synthesis has increased by a factor of 10. The increase in the synthesis of proteins during maturation is a general phenomenon among oocytes of animal species. Certain proteins synthesized at specific stages are essential to the progression of the events of maturation.

The nature and general action of mediators involved in the maturation of starfish and mammalian oocytes have also been studied in some detail. Extensive experiments by a number of investigators, particularly by Kanatani and his colleagues, have provided considerable insight into the control of echinoderm maturation and ovulation. When aqueous extracts of the radial nerves are injected into starfish gonads with gravid ovaries, eggs are immediately shed and meiosis reinitiated. The factor that induces gamete shedding is a simple peptide with a low molecular weight (about 2100) and has been shown in numerous experiments to act on the follicle cells surrounding the egg. It functions by stimulating the follicle cells to release a *meiosis-inducing substance* (MIS). The MIS has been purified from oocyte-free ovarian tissue and identified as *1-methyladenine*(1-MA). If 1-MA is added to naked starfish oocytes, this simple purine base induces dissolution of the germinal vesicle envelope within 30 minutes. The process of meiotic maturation then proceeds directly to the completion of the formation of the second polar body.

As in lower vertebrates and echinoderms, resumption of meiosis and ovulation in mammalian eggs are dependent on prior events in the follicle cells. Oocytes enclosed by follicle cells can be stimulated to maturation by *luteinizing hormone* both in vivo and in vitro. When oocytes are freed of their follicular envelopes, many tend to mature spontaneously. However, these oocytes lack complete developmental potentiality, clearly indicating that maturation is highly dependent on the follicle cells. The pituitary gonadotropin (i.e., luteinizing hormone) appears to interact with some intrafollicular factor that stabilizes the oocyte in the first prophase of meiosis. Interestingly, naked oocytes can be progressively blocked from spontaneous maturation by the graded addition of follicle cells from the cumulus oophorus. Results of other experiments support the view that follicular steroid hormones, such as progesterone and *17β-estradiol*, are also involved in inducing mammalian oocyte maturation. However, the mechanism of action of the steroid hormones has yet to be clearly elucidated.

How do meiosis-inducing substances act at the level of the oocyte? Do they act directly on the germinal vesicle? Meiosis-inducing substances, such as progesterone and 1-MA, are ineffective if microinjected into a fully grown egg, suggesting that they act on or near the surface of the oocyte (i.e., on the plasma membrane or the subjacent cortical cytoplasm). In the case of 1-MA, the receptor appears to be present on the surface of the oocyte; it is heat stable and insensitive to proteolytic enzymes such as trypsin. The binding of the MIS to its receptor appears to promote the production or activation of a cytoplasmic factor that directly induces GVBD and the subsequent processes of meiotic maturation. This factor, termed the *maturation-promoting factor* (MPF), has been identified in MIS-stimulated eggs of *Xenopus, Rana,* sturgeon, and starfish. MPF appears in the cytoplasm of amphibian eggs shortly before GVBD and then increases in activity. In starfish oocytes, MPF can be detected within 13 minutes of stimulation by 1-MA. If cytoplasm from MIS-stimulated eggs is injected into immature oocytes, meiosis is completed normally. The chemical nature of MPF is not well known. In both amphibians and starfish, the MPF appears to be a phosphoprotein; it is heat labile, sensitive to proteases, and can be extracted from oocyte homogenates using a medium containing inhibitors of phosphatases. With respect to the specificity of MPF, amphibian MPF extracted from progesterone-treated oocytes of *Bufo* or *Xenopus* can induce starfish oocyte maturation. Conversely, starfish MPF brings about maturation of *Xenopus* eggs.

The steps leading to the production of MPF are still imperfectly understood. However, Figure 4–26, based primarily on studies with amphibian oocytes, summarizes the critical steps believed to occur that result in MPF activity. The production of functional MPF appears to require the release of free calcium ions from bound form in the cortex, and the synthesis of protein. The importance of the calcium ion has been documented using a number of techniques. When calcium ions are iontophoretically (i.e., by microelectrodes) introduced into the cortical cytoplasm of amphibian oocytes, maturation is induced without exposure to progesterone. If oocytes are first injected with a calcium-specific chelating agent, such as ethylene glycol bis($\beta$-aminoethylether)-$N,N'$-tetraacetic acid (known as EGTA), and then bathed in progesterone, there is no evidence of maturation. Oocytes exposed to a Ringer's solution containing high concentrations of calcium or to the ionophore A23187, an antibiotic that promotes the movement of divalent cations across biological membranes, resume meiosis in the absence of progesterone. It has been proposed that the ionic calcium binds immediately to a surface protein (the *calcium-dependent regulatory protein*), which, following a conformational change, activates an enzyme known as *phosphodiesterase*. Phosphodiesterase catalyzes the breakdown within a cell of cyclic adenosine monophosphate (cAMP) to 5′-cAMP.

Since MIS-induced maturation of amphibian oocytes is blocked by inhibitors of protein synthesis, the process leading to the initial increase in MPF activity requires protein synthesis. An *initiator protein* is thought to be continuously synthesized and degraded in the egg before the onset of maturation. This phosphoprotein is produced as a result of the activity of an enzyme known as *cAMP-dependent protein kinase.* As long as the level of this protein kinase activity remains high, the block to maturation is maintained. The key, therefore, to triggering the maturational response appears to be the availability of cAMP. The concentration of cAMP in the egg is regulated by the activity of two enzymes: phosphodiesterase (which lowers cAMP) and *adenyl cyclase* (which promotes the formation of cAMP from ATP). When the cAMP level is low or depleted, the initiator protein becomes active, functions as a *cAMP-independent protein kinase,* and catalyzes the phosphorylation of inactive MPF precursors stored in the oocyte. Once the MPF becomes active, it acts to break down the germinal vesicle.

As previously pointed out, oocytes following GVBD typically become arrested in either metaphase I or metaphase II of meiosis until fertilization. There is good evidence that during oocyte maturation a self-inhibiting factor blocking further reduction division is set up in the cytoplasm. If cytoplasm from mature, unfertilized eggs of *Rana* is injected into one of the blastomeres of a two-celled embryo, the cycle of the injected blastomere will become arrested in metaphase. This factor inhibiting further meiosis after GVBD is termed the *cytostatic factor*. In addition to amphibians, cytostatic factor has also been identified in mammalian eggs. The cytostatic factor appears in the cytoplasm in-

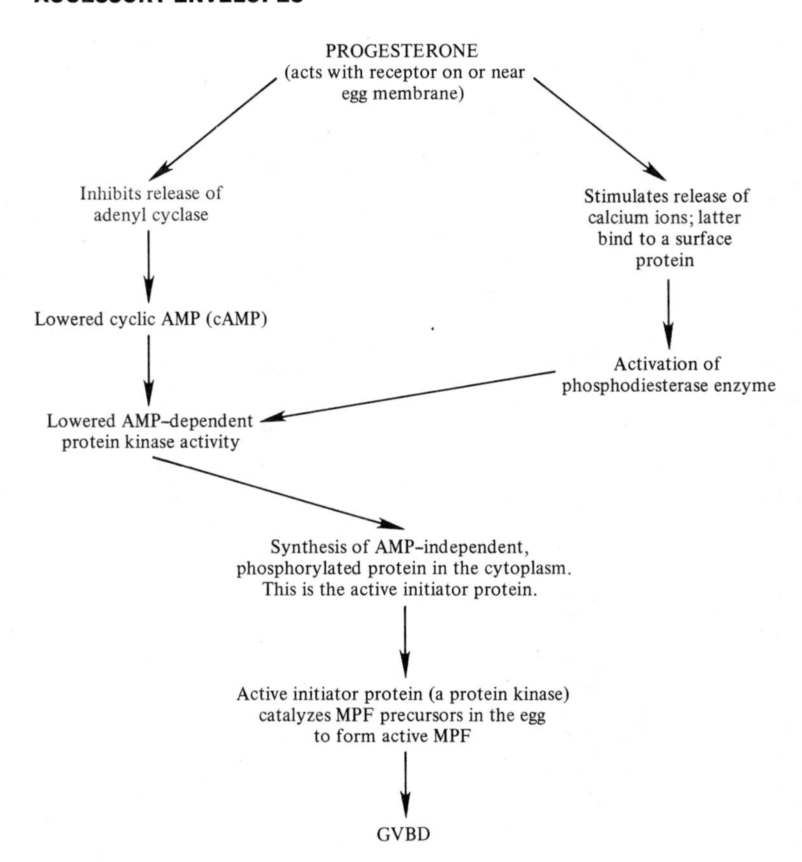


**FIG. 4–26.** Possible steps leading to the production of active MPF in progesterone-stimulated oocytes of *Xenopus*.

dependently of GVBD. If fully grown oocytes are enucleated and then treated with progesterone, cytostatic factor activity is detected.

## Ovulation

The release or discharge of the mature egg through the surface of the ovary is the event known as ovulation. In the mammal, as the Graafian follicle swells, it pushes against the surface of the ovary and causes the latter to bulge locally (Fig. 4–27). At this site the wall of the ovary gradually becomes thinner, attenuated, and avascular. Rupture of the ovarian surface and the wall of the Graafian follicle releases the ovum, surrounded by its investment of follicle cells, with the antral fluid into the peritoneal cavity. Although the egg cell is technically liberated into the peritoneal cavity, the favorable positioning in the mammal of the infundibulum of the oviduct virtually assures its reception of the ovulated egg.

The mechanism that stimulates the rupture of the ovarian follicle is still imperfectly understood. Although a sudden elevation in the pressure of the antral fluid was an early hypothesis, it is now not deemed critical to the ovulatory step. There is some evidence to suggest that lytic enzymes, produced under the influence of pituitary hormones, may act to weaken the follicle wall. More recently, studies with the electron microscope have shown myofilaments, similar to those in striated muscle, to be present in cells of the connective tissue of the ovary and in the cells of the theca of several mammalian species. These contractile cells may play a functional role in ovulation, but just how and to what extent they are necessary is not clear.

## Accessory Envelopes

As in the case of the mammalian egg cell, ovulated eggs of most animal species are surrounded by one or more acellular envelopes or membranes. *Pri-*

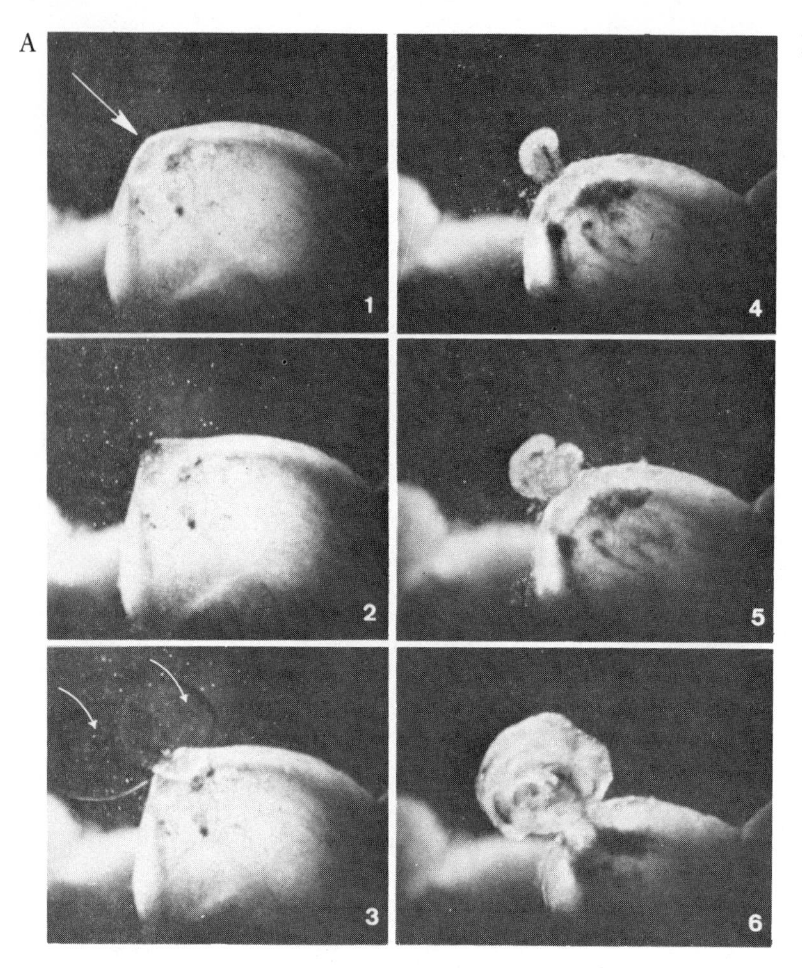


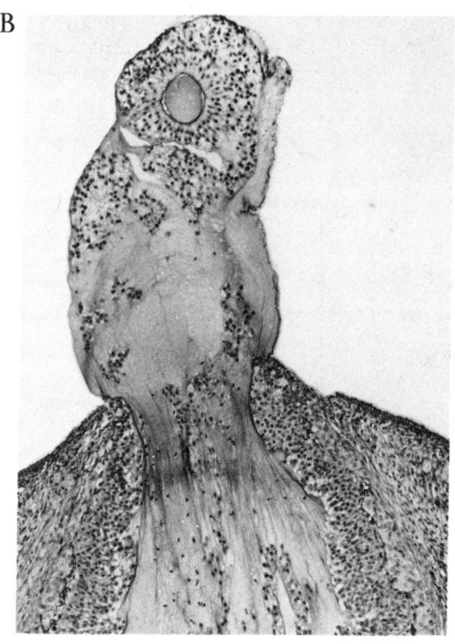


**FIG. 4–27.** A, photographs showing the sequence of ovulation in the rabbit. The arrow in (1) points to the stigma or site where follicle will rupture along the wall of the ovary. The stigma represents an attenuation of the stratum granulosum and the theca. The arrows in (3) show the expulsion of thin follicular fluid. The egg is finally shed with a surrounding halo of follicle cells (6); B, a micrograph of the cumulus oophorus with follicular fluid being discharged from the ovary in the rabbit. (Courtesy of R. Blandau.)

*mary egg envelopes* are those produced by the oocyte itself or in cooperation with the neighboring follicle cells during oogenesis. *Secondary egg envelopes* are those added to the egg as they pass through the oviduct.

Examples of primary egg envelopes include the *vitelline envelope* (membrane) enclosing the egg of sea urchins, cephalochordates, amphibians, and birds. In sea urchins, the vitelline membrane is about 30 Å in thickness, bound tightly to the oolemma, and stains positively for acid mucopolysaccharides. It is surrounded by a *jelly coat*. The comparable envelope in mammals is the zona pellucida; as stated previously, it has recently been shown to be synthesized by the oocyte during its growth phase. The *chorion* of the fish egg is a tough, relatively thick, fibrous membrane. In many species, the chorion shows a single, funnel-shaped opening, the *micropyle*, through which spermatozoa gain access to the surface of the egg. Eggs of insects are surrounded by a vitelline membrane and a second, thicker envelope (called the chorion) that appears to be secreted by the follicle cells.

Secondary envelopes are investments that are secreted by the oviducts after the egg is ovulated. These include the jelly coats surrounding the eggs of amphibians, the leatherlike capsule of the shark egg, and the hard calcareous shells of the eggs of reptiles and birds. Without question the most complicated membranes are found in bird eggs. The innermost membrane is the vitelline membrane. Although initially granular in appearance, a complex system of fibers is deposited in this envelope just before ovulation. Once ovulation has taken place, other membranes are added by the secretory activity of cells of the genital tract (Fig. 4–28). Following the addition to the vitelline membrane of a complex fibrous layer of material, the egg white or albumen is added. Further down the oviduct, two fibrous shell membranes are added to surround the

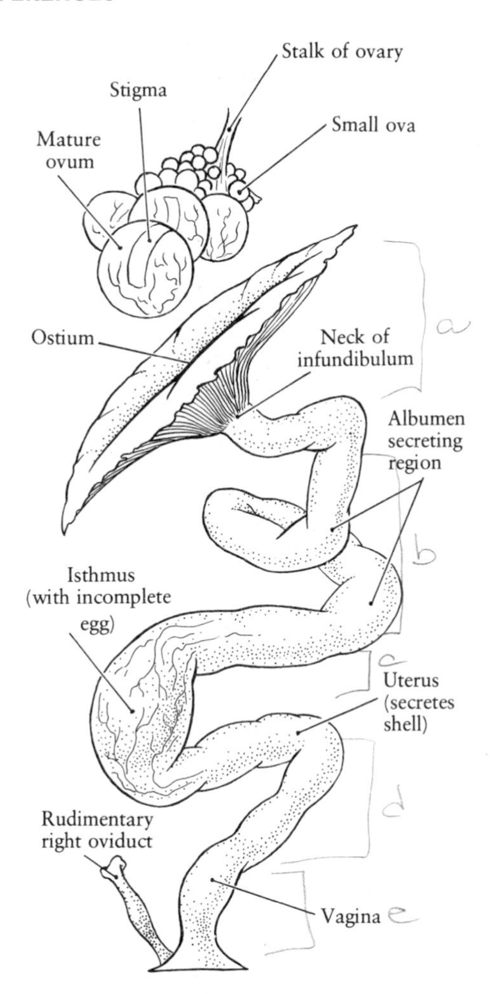


**FIG. 4–28.** Ovary and left oviduct of the domestic fowl.

albumen. A calcareous shell is laid down directly onto the outer shell membrane following entrance of the ovum into the uterus. The calcareous membrane is composed chiefly of calcium carbonate, the source of which appears to be the long bones of the laying hen. The hard shell is pierced by fine pores and is therefore permeable to respiratory gases, water vapor, and the like. The relationships between these five accessory membranes are summarized in Figure 4–17.

Egg envelopes serve a variety of functions, including protection of the egg against mechanical injury and as an additional source of nourishment. As will be seen in the chapter on fertilization, the primary egg envelopes are a physical barrier to the entry of sperm into the egg. The zona pellucida of mammals may carry the receptors for sperm and provide a substratum to direct the extension of follicle cell processes to the oocyte surface.

## References


Anderson, D. and L. Smith. 1978. Patterns of synthesis and accumulation of heterogeneous RNA in lampbrush stage oocytes of *Xenopus laevis* (Daudin). Dev. Biol. 67:274–285.

Anderson, E. and H. Beams. 1960. Cytological observations on the fine structure of the guinea pig ovary with special reference to the oogonium, primary oocyte, and associated follicle cells. J. Ultrastruct. Res. 3:432–446.

Anderson, E. 1969. Oogenesis in the cockroach, *Periplaneta americana,* with special reference to the specialization of the oolemma and the fate of coated vesicles. J. Microsc. 8:721–738.

Anderson, E. and D. Albertini. 1976. Gap junctions between the oocyte and companion follicle cells in the mammalian ovary. J. Cell Biol. 71:680–686.

Bachvarova, R., M. Baran, and A. Tejblum. 1980. Development of naked growing mouse oocytes *in vitro.* J. Exp. Zool. 211:159–169.

Beams, H. and R. Kessel. 1963. Electron microscope studies on developing crayfish oocytes with special reference to the origin of yolk. J. Cell Biol. 18:621–649.

Bellairs, R. 1971. Developmental Processes of Higher Vertebrates. Coral Gables, Fla: University of Miami Press.

Bellé, R., J. Boyer, and R. Ozon. 1982. Carbon dioxide reversibly inhibits meiosis of *Xenopus laevis* oocyte and the appearance of the maturation promoting factor. Dev. Biol. 90:315–319.

Biggers, J. D. and A. W. Schuetz, eds. 1972. Oogenesis. Baltimore: University Park Press.

Bleil, J. and P. Wassarman. 1980. Structure and function of the zona pellucida: identification and characterization of the proteins of the mouse oocyte's zona pellucida. Dev. Biol. 76:185–202.

Brower, R. and R. Schultz. 1982. Intercellular communication between granulosa cells and mouse oocytes: existence and possible nutritional role during oocyte growth. Dev. Biol. 90:144–153.

Capco, D. and W. Jeffery. 1982. Transient localization of messenger RNA in *Xenopus laevis* oocytes. Dev. Biol. 89:1–12.

Cloud, J. and A. Schuetz. 1979. 1-Methyladenine induction of oocyte (starfish) maturation: inhibition by procaine and its pH dependency. J. Exp. Zool. 210:11–16.

Dekel, N., T. Lawrence, N. Gilula, and W. Beers. 1981. Modulation of cell-to-cell communication in the cumulus–oocyte complex and the regulation of oocyte maturation by LH. Dev. Biol. 86:356–362.

Delage, Y. 1901. Etudes experimentales chez les Echinodermes. Arch. Zool. Exp. Gen Ser 9:285–326.

Dolecki, G. and L. Smith. 1979. Poly $(A)^+$ RNA metabolism during oogenesis in *Xenopus laevis.* Dev. Biol. 69:217–236.

Doree, M. 1981. 1-Methyladenine induced stimulation of protein phosphorylation and $Na^+$ pump does not require the presence of the nucleus. J. Exp. Zool. 217:147–150.

Doree, M., K. Sano, and H. Kanatani. 1982. Ammonia and other weak bases applied at any time of the hormone-dependent period inhibit 1-methyladenine-induced meiosis reinitiation of starfish oocytes. Dev. Biol. 90:13–17.

Dumont, J. 1978. Oogenesis in *Xenopus laevis* (Daudin). VI. The route of injected tracer transport in the follicle and developing oocyte. J. Exp. Zool. 204:193–218.

Dumont, J. and A. Brummett. 1978. Oogenesis in *Xenopus laevis* (Daudin). V. Relationships between developing oocytes and their investing follicular tissues. J. Morphol. 155:73–98.

Endo, Y. 1961. Changes in the cortical layer of sea urchin eggs at fertilization as studied with the electron microscope. I. *Clypeaster japonicus*. Exp. Cell Res. 25:383–397.

Gall, J. G. and H. G. Callan. 1962. $^3$H-uridine incorporation in lampbrush chromosomes. Proc. Natl. Acad. Sci. U.S.A. 48:562–570.

Heller, D., D. Cahill, and R. Schultz. 1981. Biochemical studies of mammalian oogenesis: metabolic cooperativity between granulosa cells and growing mouse oocytes. Dev. Biol. 84:455–464.

Kanatani, H. 1983. Nature and action of the mediators inducing maturation of the starfish oocyte. In: Molecular Biology of Egg Maturation. Ciba Foundation Symposium 98, pp. 159–170.

Kanatani, H., H. Shirai, K. Nahanishi, and T. Kurokawa. 1969. Isolation and identification of meiosis inducing substance in starfish, *Asterias amurensis*. Nature (Lond.) 221:273–274.

Kemp, N. and N. Istock. 1967. Cortical changes in growing oocytes and in fertilized or pricked eggs of *Rana pipiens*. J. Cell Biol. 34:111–121.

Kessel, R. G. 1971. Cytodifferentiation in the *Rana pipiens* oocyte. II. Intramitochondrial yolk. Z. Zellforsch. Mikrosk. Anat. 112:313–331.

Kishimoto, T., S. Hirai, and H. Kanatani. 1981. Role of germinal vesicle material in producing maturation-promoting factor in starfish oocyte. Dev. Biol. 81:177–181.

Kraumeyer, J., N. Jenkins, and R. Raff. 1978. Messenger ribonucleoprotein particles in unfertilized sea urchin eggs. Dev. Biol. 63:265–278.

Longo, F. J. and E. Anderson. 1974. Gametogenesis. In: Concepts of Development. J. Lash and J. R. Whittaker, eds. Stamford, Conn.: Sinauer Associates.

Massover, W. 1971. Intramitochondrial yolk crystals of frog oocytes. J. Cell Biol. 48:266–279.

Masui, Y. and H. Clarke. 1979. Oocyte maturation. Int. Rev. Cytol. 57:186–282.

Masui, Y., P. Meyerhof, M. Miller, and W. Wasserman. 1977. Roles of divalent cations in maturation and activation of vertebrate oocytes. Differentiation 9:49–57.

Meijer, L. and P. Guerrier. 1981. Calmodulin in starfish oocytes. I. Calmodulin antagonists inhibit meiosis reinitiation. Dev. Biol. 88:318–324.

Meyerhof, P. and Y. Masui. 1979. Properties of a cytostatic factor from *Xenopus laevis* eggs. Dev. Biol. 72:182–187.

Miake-Lye, R., J. Newport, and M. Kirschner. 1983. Maturation-promoting factor induces nuclear envelope breakdown in cycloheximide-arrested embryos of *Xenopus laevis*. J. Cell Biol. 97:81–91.

Miller, O. L., B. Beatty, and B. Hamkalo. 1972. Nuclear structure and function during amphibian oogenesis. In: Oogenesis. J. Biggers and A. Schuetz, eds. Baltimore: University Park Press.

Moor, R., C. Polge, and S. Willadsen. 1980. Effect of follicular steroids on maturation and fertilization of mammalian oocyte. J. Embryol. Exp. Morphol. 56:319–335.

O'Connor, C. and L. Smith. 1979. *Xenopus* oocyte cAMP-dependent protein kinases before and during progesterone-induced maturation. J. Exp. Zool. 207:367–374.

Opresko, L., H. Wiley, and R. Wallace. 1979. The origin of yolk DNA in *Xenopus laevis*. J. Exp. Zool. 209:367–376.

Raven, C. P. 1970. The cortical and subcortex of cytoplasm of the *Lymnaea* egg. Int. Rev. Cytol. 28:1–44.

Richter, J. and R. McGaughey. 1979. Specificity of inhibition of steroids of porcine oocyte maturation *in vitro*. J. Exp. Zool. 209:81–90.

Ruderman, J. and M. Schmid. 1981. RNA transcription and translation in sea urchin oocytes and egg. Dev. Biol. 81:220–228.

Samson, D. and A. Schuetz. 1979. Progesterone induction of oocyte maturation in *Rana pipiens:* reversibility of cycloheximide inhibition. J. Exp. Zool. 208:213–220.

Schatz, F. and D. Ziegler. 1979. The role of follicle cells in *Rana pipiens* oocyte maturation induced by $\Delta^5$-pregnenolone. Dev. Biol. 73:59–67.

Schuetz, A. and D. Samson. 1979. Protein synthesis requirement for maturation promoting factor (MPF) initiation of meiotic maturation in *Rana* oocytes. Dev. Biol. 68:636–642.

Smith, L. D. and R. E. Ecker. 1969. Role of the oocyte nucleus in physiological maturation in *Rana pipiens*. Dev. Biol. 19:281–309.

Wallace, R., L. Opresko, H. Wiley, and K. Selman. 1983. The oocyte as an endocytic cell. In: Molecular Biology of Egg Maturation. Ciba Foundation Symposium 98, pp. 228–249.

Wasserman, W. 1982. The role of the maturation-promoting factor in controlling protein synthesis in *Xenopus* oocytes. Dev. Biol. 90:445–447.

Wilson, E. B. 1896. On cleavage and mosaic-work. Wilhelm Roux' Arch. Entwicklungsmech. Org. 3:19–26.

Wu, M. and J. Gerhart. 1980. Partial purification and characterization of the maturation-promoting factor from egg of *Xenopus laevis*. Dev. Biol. 79:465–477.

# 5

# THE PHYSIOLOGY OF REPRODUCTION

The primary function of the gonads is the production of eggs and sperm. However, both the ovary and the testis also synthesize and secrete hormones whose function is to maintain the reproductive system in a state best suited to promote the development, delivery, and union of the germ cells that the gonads produce, and, in the mammal, if pregnancy occurs, to provide a suitable environment for the conceptus.

The sex hormones are given the general names of *androgens* (male producing) and *estrogens* (female producing). Regardless of the sex, both types of hormones are produced in all individuals, although the male produces a preponderance of androgens and the female a preponderance of estrogens.

## Female Reproductive Activity

Reproductive activity in the female shows a cyclic pattern of rather complex, interrelated behavior of the ovary, the pituitary gland, the hypothalamus, and the reproductive tract. The ovary and the reproductive tract show periods of activity and inactivity that are marked by and under the control of the cyclic activity of the hypothalamus and the pituitary.

### The Estrous Cycle

In nonprimate mammals, the hormonal interplay results in the female periodically reaching a state in which she is receptive to the male. When in this condition, she is said to be in heat or in *estrus*, and the cycles are thus termed *estrous cycles*. Estrous cycles in different species vary in length, but essentially they all show the following phases:

1. *Diestrus*—a period of quiescence, which in some species may be prolonged into an extended seasonal period of sexual inactivity termed *anestrus*. During the diestrus, the ovarian follicles are small, the reproductive tract is shrunken and anemic, and the glands of the uterine lining, the endometrium, are collapsed.
2. *Proestrus*—a period of reawakening of reproductive activity just prior to estrus. The ovarian follicles are maturing, the uterine lining is growing rapidly, and its vascularity is increasing.
3. *Estrus*—the period of receptivity and the time when ovulation occurs. Ovulation is usually spontaneous but in a few species occurs only when induced by copulation. This stage is the continuation and culmination of proestrus; all the manifestations of proestrus reach their peaks in estrus.

4. *Metestrus*—the period following estrus and ovulation when the follicles develop into corpora lutea. If fertilization does not take place, the corpora lutea regress and the reproductive tract returns to its state of quiescence, losing its vascularity and motility.

Thus, the estrous cycle pivots about a periodic attainment of sexual receptivity at which time ovulation takes place. Ovulation is the result of the maturation of the Graafian follicle, and the events leading up to ovulation are concerned with follicular growth. This period (proestrus and the beginning of estrus up to the time of ovulation) may be called the *follicular phase* of the cycle. Following ovulation, the corpus luteum becomes the dominant feature of the ovary and, until it regresses, marks the *luteal* phase of the cycle.

In the rat and the mouse, two favorite laboratory animals, estrous cycles are continuous throughout the year. Each cycle lasts four to six days, the estrous phase lasting about 12 hours at which time ovulation occurs spontaneously. These animals are thus continuous breeders. This is the exception to what we see in the natural environment where reproductive activity is generally restricted to a particular season. Seasonal breeders generally reflect a response to the environment in that sexual activity takes place at a time when environmental conditions are most likely to be kindest to the pregnant mother and the newborn offspring. Most mammals in temperate climates bear their young in the spring or summer when conditions for the survival of the young are optimal. Seasonal breeders may be *monestrous* (one cycle per year) or *polyestrous* (a number of consecutive cycles during the breeding season or more than one breeding season per year).

*Hormonal Control of the Estrous Cycle*

Cyclic activity in the female is controlled by trophic hormones *(gonadotrophins)* secreted by the anterior lobe of the pituitary gland. Following hypophysectomy, all cyclic activity stops but may be restored by injections of the appropriate pituitary extracts. The pituitary gonadotrophins influencing ovarian activity are *follicle-stimulating hormone* (FSH), *luteinizing hormone* (LH), and *luteotrophic hormone* (LTH). The ovary itself secretes *estrogens* and *progesterone* in response to the stimuli of the trophic hormones of the pituitary. These ovarian hormones control the cyclic changes in the reproductive tract that periodically prepare it for the anticipated reception of the fertilized egg.

The blood levels of the ovarian hormones in turn influence the release of the pituitary gonadotrophins. The feedback from the ovary to the pituitary may be positive (resulting in an increase in the release of the trophic hormone) or negative (resulting in a decrease). Although a small part of this ovarian feedback may be ascribed to a direct effect on the pituitary gland itself, the major pathway is through the brain, specifically the *hypothalamus*, a subdivision of the diencephalon. The concept of neural control over gonadotrophin release dates back to the 1940s when it was proposed that specific substances secreted by neurons in the hypothalamus were carried to the pituitary gland by way of a *hypothalamic-hypophyseal portal system.*

The hypothalamus is the region of the diencephalon lying closest to the pituitary gland. In fact, a part of the pituitary gland—its posterior lobe—is formed by a ventral evagination of the hypothalamus, to which it remains attached by a stalk. However, it is the anterior lobe of the pituitary, formed by an evagination of the roof of the oral cavity, which secretes the gonadotrophins. The hypothalamic control over gonadotrophin output is mediated through what are termed neural releasing factors. A specific *LH-releasing factor* (LH-RF) was reported in 1960, and shortly thereafter an FSH-releasing factor (FSH-RF) was demonstrated. More recently it has been established that there is a single compound, a decapeptide, that stimulates the pituitary to release both FSH and LH. This is called LH-releasing hormone (LHRH) or LH-FSH releasing hormone (LH-FSHRH), or more simply gonadotrophin-releasing hormone (GnRH).

There are two types of hypothalamic mechanisms controlling the cyclic release of the pituitary gonadotrophins. The first, a tonic control mechanism, stimulates a continuous release of gonadotrophins in quantities sufficient to maintain follicular growth and estrogen secretion. Gonadotrophin-releasing hormone needed for the tonic control is produced in an area called the *hypophysiotrophic area* (HTA) of the hypothalamus. The HTA is in the median basal region of the hypothalamus (MBH) and includes the *arcuate* and *ventromedial nuclei* (Fig. 5–1). This region contains neurons that secrete GnRH. The releasing factors are transported along the axons and released at the axon terminals, from where they pass in the portal circulation to

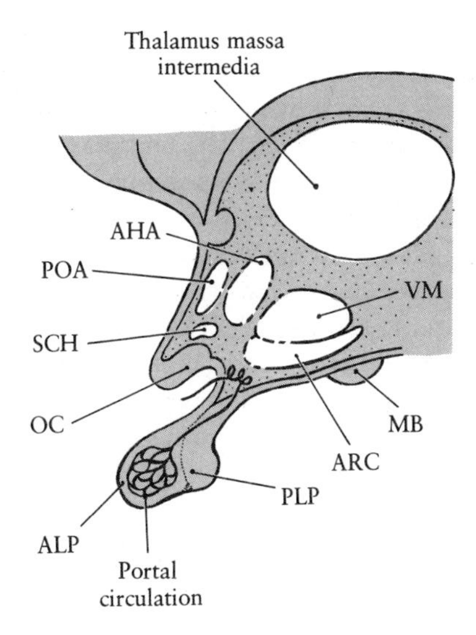


**FIG. 5–1.** Diagram of hypothalamic areas secreting tonic and cyclic-releasing factors controlling pituitary gonadotrophins. AHA, anterior hypothalamic area; ALP, anterior lobe of pituitary; ARC, arcuate nucleus; MB, mammillary body; OC, optic chiasma; PLP, posterior lobe of pituitary; POA, preoptic area; SCH, superchiasmatic nucleus; VM, ventromedian nucleus.

the anterior lobe of the pituitary gland. In addition to the tonic mechanism, a cyclic mechanism also exists. It functions periodically to secrete sufficient additional GnRH to induce surges of LH and FSH secretion by the pituitary. These surges occur during estrus, at the time when the follicle is mature, and are the stimulus for ovulation. A different, more rostral region, including the *preoptic* and the *anterior hypothalamic areas* (Fig. 5–1) contains these secretory neurons. Their axons impinge on the neurons of the HTA. Surgical interruption of neural connection between the MBH and the more rostral regions of the hypothalamus results in the disappearance of the preovulatory surge in LH and a consequent failure of ovulation.

The interplay between the pituitary, the ovary, and the hypothalamus is still not completely understood; in addition, there are, of course, species differences. However, a generalized pattern, although one certainly not applicable to all species, may be described. The early development of the ovarian follicle proceeds without any trophic influence from the pituitary. However, development beyond the stage at which the ovum is surrounded by four layers of granulosa cells is dependent on gonadotrophins. In the absence of a negative estrogen feedback to the hypothalamus at the time when the follicle is first developing, increasing amounts of FSH are synthesized and released by the pituitary, stimulating the further development of the follicle. When the theca interna forms, its cells synthesize and release estrogens. Follicle-stimulating hormone alone is not capable of stimulating estrogen secretion. Luteinizing hormone is also necessary. The preovulatory follicle also secretes small amounts of progesterone (Fig. 5–2). The dependence of follicular growth on FSH and of estrogen production on the added stimulus of LH has led to the concept that FSH is primarily a morphogenic hormone and LH is primarily a steroidogenic hormone. The secretion of increasing amounts of estrogen by the maturing follicle exerts a positive feedback effect on the rostral regions of the hypothalamus whose excitatory signals then result in an increased release of GnRH and in turn a surge of gonadotrophin release—both FSH and LH. The LH surge is higher than the FSH, and the two may not be completely coordinated (Fig. 5–2). The LH surge, in spontaneous ovulators, induces ovulation. In the rat, the surge occurs in the afternoon of proestrus and is followed by ovulation about 10 hours later. Luteinizing hormone is secreted in amounts considerably higher than needed, as ovulation can be induced by LH levels much lower than those that normally occur. If LH is blocked, FSH alone can induce ovulation, although under these conditions, the corpus luteum does not develop. Following their preovulatory peaks, both FSH and LH fall to levels comparable to those present before the surge.

During the follicular phase of the cycle, estrogens stimulate uterine growth, while during the luteal phase, progesterone stimulates the estrogen-primed uterus to differentiate into an organ optimally prepared to receive and nourish the fertilized

**FIG. 5–2.** Schematic diagram of serum concentrations of LH, FSH, estradiol, and progesterone during the estrous cycle of the rat.

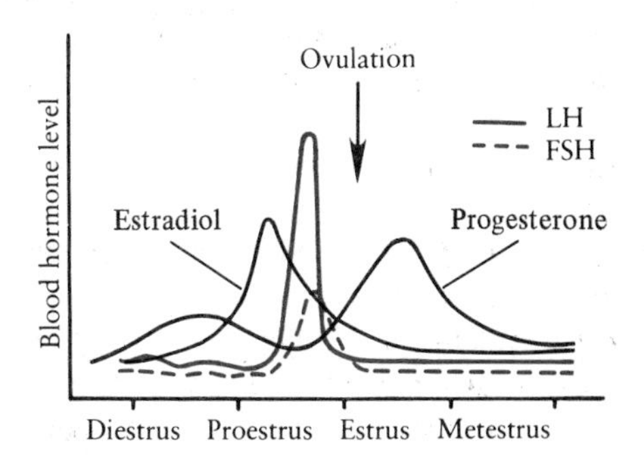

egg. Estrogens are primarily growth hormones, and progesterone is primarily a differentiation hormone.

The control of the development of the corpus luteum and the secretion of progesterone presents an excellent example of the great diversity in the mechanisms controlling the reproductive cycles in different species. The corpus luteum of the pig is completely autonomous and functions without any trophic stimulus. The guinea pig needs hypophyseal support for only the first three or four days and after this is autonomous. The sheep and the human female require the continual trophic stimulus of LH, while rats and mice require not only LH but also the LTH, *prolactin*.

The length of the luteal phase also shows considerable species variation. In the rat, mouse, and hamster, this phase of the cycle is short, and the corpus luteum regresses after only a brief and insignificant existence. On the other hand, in most domestic animals, such as the sheep, the cow, and the horse, the luteal phase is the longest part of the cycle.

The mechanisms controlling the regression of the corpus luteum are not completely known but may involve a decline in LTH, the appearance of a luteolytic factor, or a combination of both. Considerable species difference again exists, as has been demonstrated by numerous experiments on the effects of hypophysectomy and/or hormone administration on the maintenance of the corpus luteum. An important part of the evidence dates back to Leo Loeb who, in 1923, showed that hysterectomy prolonged the lifespan of the corpus luteum in the guinea pig. Studies from other laboratories have supported Loeb's observations and shown that in a number of species, including cows, sheep, and pigs, the uterus exerts a lytic effect on the corpus luteum. Hysterectomy removes the lytic principle and thereby prolongs the life of the corpus luteum. If only one horn of the uterus is removed, the corpus luteum on the operated side will persist while that on the unoperated side will regress at the normal time. The action is thus a local one. There is some indication that the lytic effect is produced only after the uterus has been exposed to progesterone for a certain length of time. If sheep are given continuous injections of progesterone before they are induced to ovulate, the corpus luteum regresses almost as soon as it is formed. Thus, there appears to be in some species a local utero-ovarian interaction whereby progesterone secreted by the mature corpus luteum induces the synthesis of a lytic agent in the uterine endometrium that, in turn, then "murders" the corpus luteum. The uterus is then implicated as playing an important role in controlling the periodicity of the estrous cycle. However, the evidence is still circumstantial, and it has not been demonstrated unequivocally by extraction, purification, and injection that a specific luteolytic substance exists. Perhaps the best demonstration is seen in the fact that uterine flushings from sows in days 14 to 18 of a 20-day estrous cycle will destroy luteal cells grown in vitro, while flushing at other stages of the cycle will not.

After it was reported that prostaglandin $F_{2\alpha}$ ($PGF_{2\alpha}$) caused luteolysis in the rat, investigation of its role in the regression of the corpus luteum received a big impetus. The following were determined: (1) it is present in the uterine tissue; (2) when its synthesis is blocked, luteal activity is prolonged; and (3) its direct administration causes luteolysis in a number of species. The site of action of $PGF_{2\alpha}$ is not known. It has been proposed that it might inhibit the uptake of LH by acting directly on the luteal cells to impair the activity of the LH receptor complex. Inhibition of LH stimulation would result in a functional luteolysis (i.e., the loss of the ability to secrete progesterone). Morphological degeneration of the corpus luteum follows later, comprising a complex series of events such as leukocytic infiltration, cellular degeneration, and eventual resorption.

## The Menstrual Cycle

Cyclic phenomena associated with the reproductive activity of the female also occur in primates. The cycles are called *menstrual* cycles. Their overt manifestation is a periodic *menstruation,* every three to five weeks, associated with a physiological breakdown of the uterine lining (the endometrium) resulting in a blood-stained vaginal discharge, the *menses*.

Menstrual cycles are quite comparable to estrous cycles and are controlled by the same interplay of ovarian, pituitary, and hypothalamic hormones as are the estrous cycles. The menstrual cycle has a preovulatory follicular phase that ends in spontaneous ovulation and is followed by a postovulatory luteal phase, as does the estrous cycle. The major differences between the two are menstruation and the absence of any periodicity of sexual activity in the primates. In the nonprimate mammals, a period of sexual receptivity at which time ovulation occurs, provides a mechanism by which mating takes place only at the time of the cycle most likely to result in fertilization. A considerably higher percentage of fertile mating is then expected, much higher than in the primates, whose time of ovulation bears no relation to mating.

The menstrual cycle is considered to begin on the first day of menstruation. This may be, however, somewhat of an unfortunate choice because, as we shall see, menstruation actually represents the terminal stage of the cycle. The follicular stage of the cycle starts immediately after menstruation. During this stage, the follicles mature under the influence of FSH and small amounts of LH (Fig. 5–3). The follicle secretes large amounts of estrogens and little, if any, progesterone (Fig. 5–4). The estrogens stimulate the rapid growth of the uterine endometrium. New uterine glands are formed that, however, during the follicular phase, do not branch and contain only small amounts of glycogen. Ovulation occurs spontaneously approximately in the middle of the cycle. About 24 hours before ovulation, LH titers start to increase and reach a sharp preovulatory peak in midcycle apparently due to the influence of increasing amounts of estrogen secreted by the vesicular follicle. The surge in LH secretion lasts for one to three days and is associated with a corresponding increase in FSH secretion. The FSH surge, however, is neither so pronounced nor consistent as the LH surge. Following their midcycle peaks, the levels of both LH and FSH fall to levels lower than those seen during the preovulatory stage.

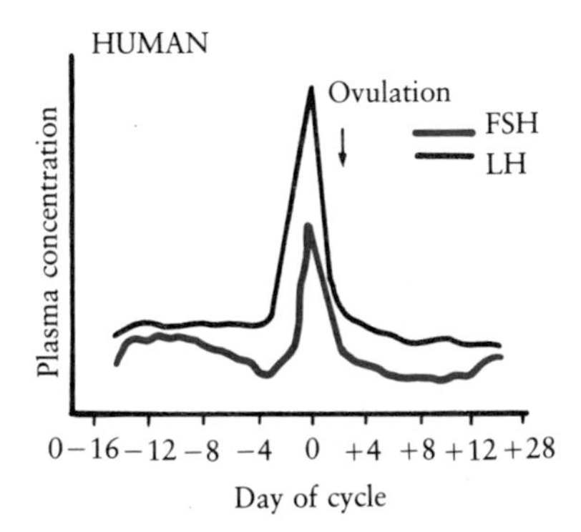


FIG. 5–3. FSH and LH levels during the menstrual cycle determined by radioimmunoassay of plasma and urinary extracts. (From V. C. Stephans, 1969. J. Clin. Endocrinol. 29:904.)

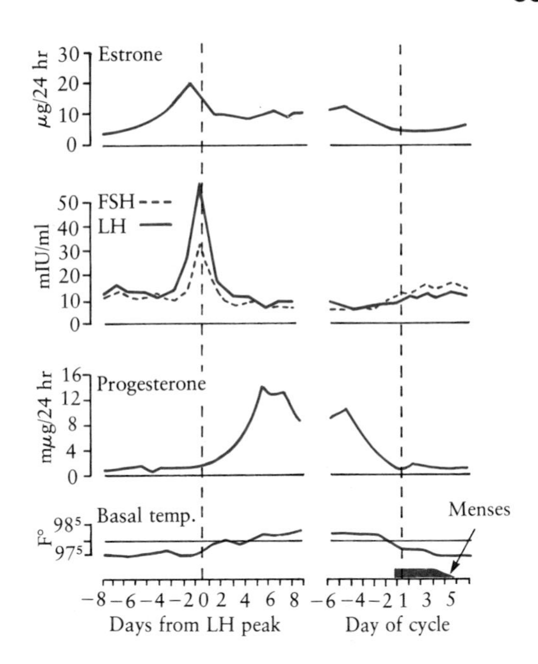


FIG. 5–4. Basal temperature, serum FSH and LH, and urinary estrone and progesterone concentrations related to LH midcycle peak and days from menstruation. (After U. Goebelsmann et al., 1969, J. Clin. Endocrinol. 29:1222.)

Following ovulation, the corpus luteum develops and, under the stimulation of LH, secretes increasing amounts of progesterone (Fig. 5–4). In the luteal phase, the reproductive tract, already primed by estrogen, differentiates further under the additional stimulation of progesterone. Development at this time involves cellular changes associated with secretion of the uterine glands rather than with growth. The glands become serrated and branched, the nuclei of their cells assume a basal position characteristic of secretory epithelia, and glycogen is secreted into the lumen of the glands. Secretory activity reaches its maximum about the middle of the luteal phase, at about the time implantation will take place if fertilization occurs during that cycle. If fertilization does not occur, functional degeneration of the corpus luteum begins about 8 to 10 days after ovulation. In the human, hysterectomy has no influence on the lifespan of the corpus luteum, and evidence is against a uterine luteolytic agent causing regression of the corpus luteum. One may then propose the concept of an inherent lifespan of the corpus luteum (certainly an easy way out) or search for other regulatory mechanisms. One suggestion is that regression may be the result of an intraovarian action of steroid hormones, possibly estrogens.

Following the regression of the corpus luteum, the uterine endometrium reacts to the subsequent low levels of progesterone and estrogen by undergoing degenerative changes. The uterine glands involute. Constriction of the muscular walls of certain arteries produces local areas of restricted blood supply *(ischemia)*, which leads to necrosis. This breakdown of the endometrial lining results in an extravasation of blood and cellular debris into the uterine lumen and the appearance of a vaginal discharge marking the end of the old cycle—or the beginning of the new. Since the cyclic buildup of the uterus is solely in preparation for the reception of the fertilized egg, menstruation has been aptly characterized as the reaction of a disappointed uterus.

*Hormonal Control of the Menstrual Cycle*

The hypothalamus and its GnRH are just as important in maintaining gonadotrophin secretion by the pituitary gland in the control of the menstrual cycle in the primate as they are in the control of the estrous cycle in the nonprimate mammal. However, a series of experiments on the rhesus monkey (a representative higher primate) has demonstrated that there are fundamental differences in the control of these two types of cycles (Knobil, 1980). As in the rat, the pattern of the tonic basal control of gonadotrophin release is directed by the MBH. However, in definite contrast to the rat, surgical interruption of the nervous connection between the MBH and the more rostral regions of the hypothalamus result neither in the loss of the preovulatory surge of LH and FSH nor in the failure of ovulation. The neuroendocrine mechanisms that govern the tonic secretion as well as the cyclic surge of gonadotrophins are dependent only on the MBH, mainly the arcuate nucleus. The arcuate nucleus releases a pulse of GnRH approximately once every hour, stimulating the pituitary in turn to secrete FSH and LH. The immature follicles respond to these gonadotrophins by increasing in size and secreting increasing amounts of estrogen (estradiol). When the level of estradiol, at midcycle, exceeds a threshold of 150 pg/ml for at least 36 hours, it now acts directly on the pituitary causing the preovulatory surge of FSH and LH necessary for subsequent ovulation. That the estrogen acts directly on the pituitary and not by way of the MBH may be demonstrated by maintaining monkeys with radiofrequency-induced lesions of the arcuate nucleus with hourly infusions of synthetic GnRH, simulating the normal pattern of GnRH release. Since, in these experiments, a careful control of the time of administration and the amount of GnRH is necessary to maintain ovarian function, it is evident that there is an absolute requirement for GnRH. With the correct pattern of GnRH established, the follicle develops normally and secretes increasing amounts of estrogen, and ovulation then occurs normally. Since the arcuate nucleus has been eliminated, the estrogen must act directly on the pituitary to provide the LH surge necessary for ovulation. The menstrual cycle is thus pelvic (ovary) controlled, whereas the estrous cycle is brain (hypothalamus) controlled. The menstrual cycle depends on and is timed by the ebb and flow of the ovarian hormones acting on the pituitary gland. The action of GnRH, necessary for proper ovarian function, although it is obligatory, is only permissive. The MBH is essentially unregulated and is not subject to cyclic impulses from more rostral hypothalamic centers during normal menstrual cycling.

*Ovulation*

There is no external manifestation of ovulation in humans, and it is therefore difficult to pinpoint the time of its occurrence. This is unfortunate, because in any attempt to increase (or decrease) the chance of pregnancy, it is important to be able to predict the time of ovulation with some degree of accuracy. The fertilizable life of the human ovum is no more—and probably less than—48 hours, and that of the sperm is only slightly longer.

One of the earliest methods of attempting to determine the time of ovulation was by daily measurement of early-morning body temperature. Temperature changes are, of course, subject to wide variation depending on a number of circumstances, but, in general, it has been shown—following the examination of many temperature graphs—that the temperature is lower during the preovulatory period than following ovualtion (Fig. 5–4). In some cases there may even be a small drop in temperature just preceding ovulation. The entire range of the temperature change is not more than 1 to 1.5 degrees, and the efficiency of predicting the time of ovulation by this method is not without its pitfalls.

The cyclic secretion of estrogens and progesterone results in characteristic changes of the reproductive tract. In the nonprimate mammal, examination of smears obtained by inserting a cotton swab into the vagina (the *vaginal smear technique*) allows these changes to be used very easily to determine what stage of the estrous cycle the animal is in. However, vaginal cytology in humans has been of little predictive value in determining the time of ovulation, although changes have been shown to occur. Considerable training and experience may enable an examiner to distinguish in general between the estrogenic and progestational phases of the cycle, but this method is of limited value in predicting the time of ovulation.

Another method used to determine changes in hormone levels indirectly has been by noting their effects on the vascularity of the rat ovary. Some investigators have claimed considerable success with this method.

Recently, advances have been made in the methods available for the direct determination of hormone levels in the blood, such as radioimmunoassay. The use of these methods has given a clear-cut picture of changes in hormone levels during the menstrual cycle. For example, the time of the preovulatory surge in LH presents an excellent indication of the midcycle occurrence of ovulation.

Pregnancy is an obvious positive evidence of ovulation and in the case of isolated coitus or artificial insemination may be used to indicate the approximate time of ovulation. Most studies report that insemination results in pregnancy when it takes place around midcycle. Direct observations of the ovary through an endoscope introduced into the pelvic cavity through the posterior wall of the vagina *(culdoscopy)* also have confirmed the fact that ovulation occurs near midcycle. It is probable that in the large majority of cases ovulation normally takes place within a rather short period covering from three days before to two days after midcycle. However, deviation from this mean is often considerable, and indeed ovulation may occur on any day of the cycle. In addition, the length of the menstrual cycles in many women is not regular—particularly in younger women and those approaching menopause—and thus the midcycle time may vary from one cycle to the next. Of course, the precise midcycle time for any given cycle cannot be determined until after the cycle is completed.

Once it was determined that the menstrual cycle and ovulation depended on an interplay between ovarian hormones and pituitary gonadotrophins, it became evident that it might be possible to control ovulation by the administration of exogenous hormones. Hence the appearance in the mid 1950s of the "pill," a hormonal contraceptive. Essentially, the rationale for the mechanism of action of the pill was the negative feedback of the ovarian hormones on the hypothalamic–pituitary system that would block the release of FSH and LH. Blockage of FSH secretion by estrogen inhibits follicular growth and in turn ovulation, while inhibition of LH secretion by progestins directly blocks ovulation by preventing the necessary preovulatory LH surge.

When first introduced, the pill contained large doses of estrogen and progestin, a shotgun approach. Since then, many refinements have been made. A major advance was lowering the amount of or completely eliminating estrogen. The large doses of estrogen first employed were found to result in a number of undesired side effects. The major one was the formation of leg vein clots with the consequent risk of embolisms. Other side effects were a small increase in blood pressure and (rarely) the production of gallstones. The likelihood of the occurrence of adverse effects varies with a number of conditions, becoming highest in women over 35 who smoke. Although it was at one time suspect, the pill probably does not increase the risk of cancer and may, in fact, lower it.

When both an estrogen and a progestin are given in combination, they prevent ovulation, but in addition, the progestin has other contraceptive effects. This led to the introduction of the "minipill," a pill without estrogen and containing only small amounts of a progestin. The minipill generally does not prevent ovulation but suppresses normal uterine endometrial development, preventing implantation. It also adversely affects cervical viscous fluidity, inhibiting sperm transport.

The isolation and the determination of the structural formula of GnRH opened up a new approach to contraception: the use of of GnRH antagonists or analogs. The analogs compete with GnRH for pituitary binding sites and in turn lead to lowered blood levels of FSH and LH. In the rat and hamster they suppress spontaneous ovulation, and in the rabbit they inhibit the ovulatory reflex resulting from cervical stimulation. GnRH analogs completely suppress the postcastration rise in gonadotrophins in the rhesus monkey. In humans, GnRH antagonists injected intramuscularly have been shown to inhibit the preovulatory surge in FSH and LH, preventing ovulation. Also, when injected for the first three days of menstruation, they result in a decreased concentration of FSH and defective follicular growth, followed by a shortened luteal phase and a consequent failure of the uterine endometrium to develop into a structure adequate to support implantation. However, the clinical use of these compounds has been hindered by the necessity of resorting to parenteral administration. Recently, an orally active analog has been reported to be effective in preventing ovulation in the rat (Nekola et al., 1982).

*The Diethylstilbestrol (DES) Story*

The hyperplastic effect of estrogens on the uterine endometrium made them possible suspects in the induction of uterine cancer following the use of any estrogen-containing pill. It is now concluded that the risk, if any, is extremely small. However, estrogen replacement therapy in postmenopausal women (female forever) does present a much greater risk, although many factors including the dose and the length of treatment as well as individual differences, must be taken into account. An unexpected chapter in the story of reproductive tract cancer occurred in the 1970s when cases of adenocarcinoma of the vagina appeared in a number of young women. These cases were unique because of the nature of the cancer—which in the vagina is classically squamous cell cancer—and the age of the patients (15–22), as opposed to the usual over-50 ages. The cause of the vaginal adenocarcinomas in the young women was for some time a mystery, until it was discovered that all of their mothers had been given DES during their pregnancies. The mothers had histories of bleeding and pregnancy loss, and they were given DES to prevent prematurity and toxicity of pregnancy. Although the link between vaginal cancer of a daughter and a mother taking DES while pregnant is unquestionable, it is difficult to understand how this synthetic, nonsteroidal estrogenic compound produces such a specific effect. It has been reported that female mice exposed transplacentally to DES on days 9 to 16 of gestation show reduced fertility as adults (McLachlan et al., 1982). They noted that DES stimulated both Müllerian and Wolffian duct derivatives in the fetus. In the female, the Wolffian duct derivatives become cystic and hyperplastic, and may then cause morphological abnormalities of the reproductive tract.

Fortunately, only a small number of vaginal cancers have appeared in the daughters of an estimated 2 million women who took DES while pregnant, and the risk for exposed female fetuses is about 1 in 1000. However, a larger number (about 30%) show clinically detectable changes in the vagina of both an anatomical and a histological nature. Even in the presence of vaginal and cervical DES-induced abnormalities, however, these women are still able to have normal pregnancies and deliveries.

## Hormones of Pregnancy

### Human Chorionic Gonadotrophin (HCG)

In the normal menstrual cycle in the absence of pregnancy, the corpus luteum functions in steroidogenesis for only a limited period of less than 10 days before it undergoes regression. However, if fertilization occurs, the corpus luteum does not regress but continues to function throughout pregnancy. The factor responsible for the maintenance of the corpus luteum is introduced by the developing embryo, which begins to implant in the uterine wall about one week after fertilization—about in the middle of the luteal phase of the cycle. Implantation takes place when the embryo is in the blastocyst stage. It consists of a thin-walled vesicle (the trophoblast) enclosing a fluid-filled cavity (the blastocyst cavity) and a small mass of cells at one pole, which will ultimately form the embryo proper. The trophoblast proliferates rapidly after implantation and actively invades the maternal uterine tissues, forming the fetal part of the placenta. The trophoblast, which forms the embryonic membrane called the chorion, secretes a hormone that in the human is called *human chorionic gonadotrophin* (HCG). It is the luteotrophic action of this hormone that maintains the corpus luteum. Human chorionic gonadotrophin is detectable in the maternal blood during the second week of pregnancy, reaches a peak during the 7th to 12th weeks, and then drops sharply to levels about one-fifth to one-tenth of peak values until term (Fig. 5–5).

Human chorionic gonadotrophin resembles LH in many of its actions but also has some of the properties of LTH, one of which is to stimulate the continued secretion of progesterone by the corpus luteum of pregnancy. The precise physiological role of HCG has not yet been determined, although a

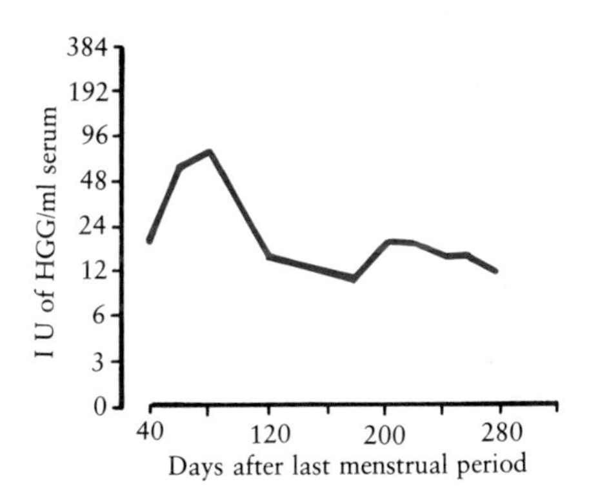


**FIG. 5-5.** Ranges in values of urinary excretion of HCG during pregnancy. (From S. Brody and G. Carlstrom, 1965. J. Clin. Endocrinol. 25:792.)

number of possible functions have been suggested. These include the stimulation of placental steroid synthesis and the stimulation of the growth of the fetal adrenal gland. In addition, HCG may play a role in sex differentiation and in altering the immunological reactivity of the maternal tissues by local immunosuppressive action on the maternal leukocytes in the region of the invading trophoblast.

*Pregnancy Tests*

Whatever the physiological function of HCG may be, there is no doubt of its value to the obstetrician who uses its presence as a basis for pregnancy tests. It might well be stated that HCG is an invention of the obstetrician in order to have available a highly reliable method for the early determination of the fact of pregnancy. Pregnancy tests use the presence of HCG in the urine of the pregnant female to produce specific effects on the reproductive systems of a number of different species. With our characteristic penchant for getting things done in a hurry, each new pregnancy test that is developed gives us the answer in a shorter period of time. The *Ascheim-Zondek* tests use immature female mice that respond to twice daily subcutaneous injections of pregnancy urine for three days by showing hemorrhagic follicles or corpora lutea when examined on day 5. The endpoint of the rabbit *(Friedman)* test is ovulation one to two days after a single intravenous injection of pregnancy urine. The female South African clawed toad responds by ovulating in 12 hours, while the male frog will show sperm in the cloaca in 6 hours if the test is positive.

The more recently developed serological tests are by far the most rapid; the results are available in 10 to 15 minutes. Serological tests involve the capacity of the HCG in pregnancy urine, when mixed with a solution containing HCG antibody, to block this antibody. When the solution containing the blocked antibody is mixed with another solution containing HCG antigen, no precipitate forms. In view of the fact that the pregnant female can now determine in a quarter of an hour that she has eight and a half months before the final outcome, we have here an excellent example of the old saying "hurry up and wait." While noting that each new pregnancy test tends to give the answer in a shorter time, we should also point out the much more important aspect that the newest tests are easier to make, more foolproof (the rabbit is an extremely excitable animal that may ovulate for reasons best known to itself), and definitely more reliable.

## Human Chorionic Somatomammotrophin (HCS)

This hormone, detected in the human placenta in the early 1960s, has been shown to have both a somatotrophic and lactogenic effect. It has been variously called human placental lactogen (HPL), chorionic growth hormone-prolactin (CGP), and human chorionic somatomammotrophin (HCS). It is detectable at the sixth week of pregnancy and, unlike HCG, does not peak and then fall, but increases progressively to reach a maximum at about 35 weeks. The increase in HCS parallels the increase in placental mass, and the ratio of HSC to placental weight remains constant throughout pregnancy. Its physiological role is not known. It has been suggested that it might function in diverting glucose from maternal to fetal tissues, where the energy requirements are met almost exclusively by the metabolism of glucose.

## Human Chorionic Thyrotrophin (HCT)

Extracts of the placenta yield a thyrotrophic hormone that has immunological properties relating it to the thyroid-stimulating hormone (TSH) of the adult pituitary. It reaches its highest level of concentration during the first two months and then declines progressively. Its role is also in doubt, although it may be responsible, in part, for the early development of the fetal thyroid gland, whose follicles develop and become functional two or three weeks before the fetal pituitary begins to secrete TSH.

### Luteinizing Hormone–Releasing Factor (pLRF)

The human placenta contains fairly large amounts of a decapeptide that is biologically, biochemically, and immunologically similar to the luteinizing hormone–releasing factor of the hypothalamus. The amount of pLRF increases from 12 to 16 weeks and plateaus from then until parturition. Its role has not been established. Synthetic LRF stimulates the release of HCG from the placenta in vitro, and it has been suggested that pLRF may act in the same manner in vivo. The time of its appearance in the placenta, however, does not coincide with the peak time of the secretion of HCG.

### Steroid Hormones

Although the trophic hormones discussed above are synthesized directly by the placenta, many of the steroid hormones synthesized during pregnancy require interplay between fetal and placental tissues. Progesterone is one exception to this and like the placental trophic hormones may be synthesized independently of the fetal tissues other than those contributing to the placenta. The synthesis of other steroid hormones requires the functioning of what has been called the *fetoplacental unit* or *complex*. Many steroid hormones are formed during pregnancy, including estrogens, progestogens, androgens, and corticoids. We will consider only estrogens and progesterone.

*Progesterone*
Steroid synthesis in the placenta and the fetoplacental unit involves the same pathways as in the adult, starting with the 2-carbon *acetate* molecule to synthesize the 27-carbon *cholesterol* with its steroid configuration. Although cholesterol is found in the placenta, it has been shown that the placenta cannot synthesize this compound itself and must get it from the maternal circulation or from the fetus. The fetal liver rapidly converts acetate to cholesterol. In the placenta, cholesterol is converted into progesterone.

*Estrogens*
In the synthesis of the three main estrogens—*estrone, estradiol*, and *estriol*—some of the steps occur in the fetal liver and adrenal, and some in the placenta. There is a back-and-forth transfer of compounds from fetus to placenta. In each location, a step or steps takes place until finally, through the mutual effort, the definitive compound is formed. Neither the fetus nor the placenta alone is capable of synthesizing the product.

There is a progressive rise in the urinary excretion of estrogens throughout pregnancy. Levels are low during the first trimester, increase after the 10th to 12th week, and show a sharp rise during the last 3 to 4 weeks. Some 20 different estrogens have been isolated from late-pregnancy urine. Of these, estriol is the most abundant.

## Parturition

After the fetus has developed to term, there remains still one final process, birth or *parturition*. Parturition has been extensively investigated, and although a number of processes have been suggested as important precursors of this event, parturition has not been shown to be the result of any single causative agent or process. Rather, it is more probable that many processes may be involved. They lead by a series of interrelated events to the final common pathway, which is the contraction of the smooth muscle of the uterine wall resulting in the expulsion of the fetus.

Throughout gestation the mass of the fetus increases, and the purely mechanical stimulus of stretching the uterine walls that results may be a factor. However, more emphasis on biological factors, particularly hormonal levels, has been stressed. This is not to say that hormonal events are the only ones involved, although it is probable that they are the most important—and certainly the most interesting.

Analysis of changes in estrogen and progesterone levels late in gestation and the effects of these hormones on the response of the uterine muscle to the posterior pituitary hormone, *oxytocin*, have suggested a possible mechanism for the induction of labor. Throughout most of gestation, the uterus accommodates to the expanding fetus and remains relatively quiescent. During this time, the uterus is dominated by progesterone, and substances such as oxytocin that normally stimulate uterine muscular contraction are prevented from doing so. At term, progesterone levels fall and estrogen levels peak, and now the uterus becomes estrogen dominated and sensitive to muscle stimulants. Analysis of hormone levels in a number of different species, particularly the sheep and goat, shows that serum progesterone levels fall and estrogen levels rise just prior to the onset of labor, thus supporting this proposal. However, in the human, analysis of hormone levels has yielded controversial results, and evidence is not so satisfactory nor so conclusive as it is in other species.

An additional facet was introduced when it was suggested, mainly from work on the sheep, that the signal for the initiation of labor came from the fetus. The signal is a rise in *corticosteroids* secreted by the fetal adrenal. Experimental evidence shows that fetal hypophysectomy or adrenalectomy prolong pregnancy, while maternal hypophysectomy or adrenalectomy have no effect on the onset of labor. Fetal adrenocorticotrophic levels rise during the last half of pregnancy and the fetal adrenal cortex doubles in size during the last 10 days of pregnancy. Not only is labor preceded by a rise in fetal corticosteroids, but also injection of corticosteroids will induce labor. Liggins (1972) proposed that one function of the fetal corticosteroids in the initiation of labor could be to inhibit progesterone either by a direct inhibition of the progesterone effect on uterine muscle or by an inhibition of progesterone synthesis or metabolism. Increases in fetal corticosteroids would also increase the synthesis of estrogens. Again, the evidence on fetal hormone levels in humans, which is necessarily scant, is not conclusive and neither supports nor denies this hypothesis.

Prostaglandins present another link in the chain. In the sheep, the onset of labor is associated with an increase in $PGF_{2\alpha}$ in the maternal placenta, uterine myometrium, and uterine venous blood. The fetal adrenal may also be a factor in this event. Secretions from the fetal adrenal increase the rate of synthesis of prostaglandins in maternal tissues. Also, infusion of the pregnant ewe with prostaglandins will induce labor. The site of formation of the prostaglandins that may play a role in labor is not known, but the uterus and the maternal placenta are suggested as possibilities. Prostaglandins could function by increasing uterine muscle contractability by lowering the threshold of response to oxytocin. In line with the estrogen and progesterone changes at parturition, estrogen is known to stimulate prostaglandin release while progesterone has an inhibitory effect.

We can consider the initiation of labor as the result of a series of events leading to a final process, contraction of the uterine musculature. These events are not necessarily the same in all species. In any one species, a certain event or process may be more important than in another. It is possible to list events that have been described as occurring in some species, but not in all, in a sequence to suggest a possible chain of events leading up to parturition.

1. Throughout most of pregnancy, the uterus is under the influence of progesterone secreted by the placenta. Progesterone decreases uterine sensitivity to muscle stimulators and maintains the uterus in a quiescent state.
2. As pregnancy progresses, more and more ACTH is secreted by the fetal pituitary. The fetal adrenal cortex responds to this by showing an increase in mass and finally, near term, an increase in the output of corticosteroids.
3. Fetal corticosteroids decrease the synthesis, inhibit the action, or change the metabolism of progesterone, and increase the synthesis of estrogens and prostaglandins.
4. Fall in progesterone activity and rise in estrogens result in the release of the uterine muscle from progesterone inhibition. This allows the muscle to respond to muscle stimulants such as oxytocin.
5. Prostaglandins, also under progesterone and estrogen control, are released, mainly from the maternal placenta, in increasing amounts; and they also act to lower the threshold of response to agents stimulating muscle contraction. The uterine muscle now begins the contractions characteristic of labor.

All of these events have been described in one species or another as occurring during or prior to parturition. They appear to be arranged in a logical time sequence, but whether or not they all occur in humans is not known. In this concept, the dramatic onset of labor, which we associate with the beginning of uterine contractions, is considered as only the final step of a gradual process taking place over the final weeks of pregnancy. It is not necessarily associated with sharp fluctuations in hormonal levels. It is characterized by gradual progressive change in the levels of a number of different hormones of the fetus and the placenta, occurring as a chain of interrelated events climaxing in the onset of uterine contractions.

## Male Reproductive Activity

In nature, the basic pattern of reproductive activity in the male is also cyclic. Mature sperm are produced only during the breeding season. In domestic animals and in laboratory animals, spermatogenesis continues throughout the year, and mature sperm are always present in the testes and the epididymis.

It was first demonstrated in 1927 by P. E. Smith, and has been repeatedly confirmed since, that both the spermatogenic and secretory functions of the testes are under the control of pituitary gonadotrophins. The same pituitary trophic hormones that are known to be active in the female are also found to be present in the male, with FSH functioning in the support of spermatogenic activity and LH, sometimes referred to as *interstitial cell-stimulating hormone* (ICSH), functioning in the support of secretory activity (Fig. 5–6).

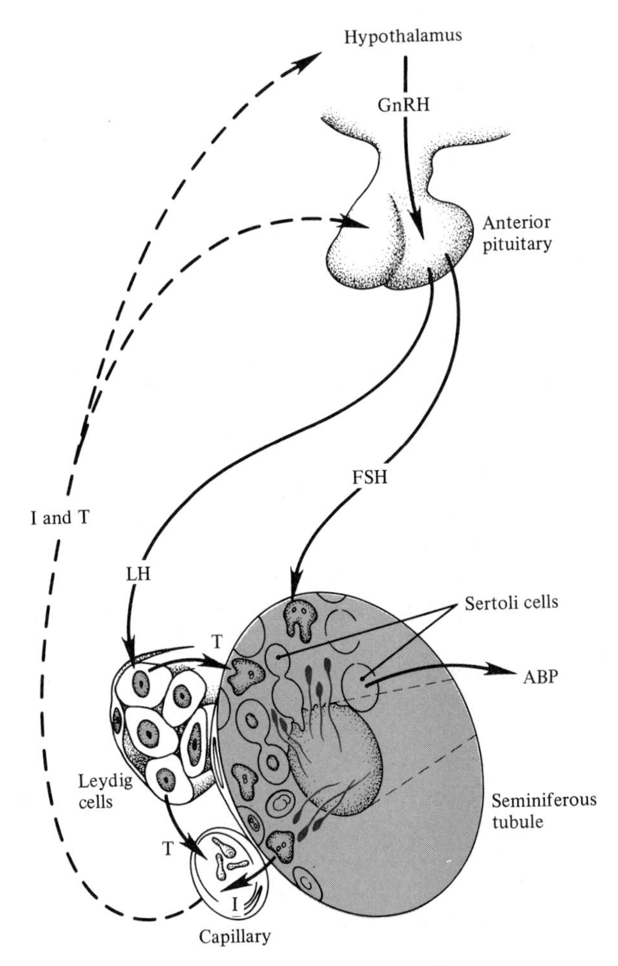


**FIG. 5–6.** Diagram of the hormonal control of spermatogenesis and testosterone secretion. LH stimulates the Leydig cells to secrete testosterone (T), which (1) stimulates spermatogenesis and (2) passes into the bloodstream and maintains the sex glands and ducts. Testosterone reaching the brain exerts a negative-feedback effect on gonadotrophin production. FSH stimulates the Sertoli cells to secrete an androgen-binding protein (ABP). Inhibin (I), secreted by the Sertoli cells, exerts a negative-feedback effect on the hypothalamus and the pituitary.

### Control of Spermatogenesis

*Hormonal Factors*

Although spermatogonia are present in the testes of the newborn and undergo some mitotic activity in prepuberal males, the formation of mature sperm does not take place until the age of puberty—12 to 15 years in man. FSH acting on the seminiferous tubules is necessary for the initiation of spermatogenesis in puberal males. However, its target cell and exact function are not known. In the adult, FSH acts on the Sertoli cells to stimulate secretion of an androgen-binding protein (ABP) that may have an important function in maintaining a high intratubular concentration of testosterone. In the mature testes, spermatogonial divisions and maturation to the prophase of the primary spermatocyte are independent of any hormonal control. Development from this stage through both meiotic divisions and the early and middle stages of spermatid maturation are dependent on adequate levels of pituitary and testicular hormones. Spermatogenesis takes place in an androgen-rich environment maintained by the reciprocal interaction among the hypothalamus, the pituitary, and the testis. Hypophysectomy results in an immediate cessation of sperm production, and, conversely, gonadotrophins will stimulate spermatogenesis in prepuberal males. Both LH and FSH are important, but in the adult LH appears to play the major role. It acts on the Leydig cells to stimulate the secretion of androgens, which then provide the necessary androgen-rich environment for the germ cells. FSH, in contrast, acts directly on the seminiferous epithelium and is apparently most important during the later stages of spermatogenesis. Whether it has a direct effect on the germ cells themselves in addition to its stimulation of the secretion of ABP by the Sertoli cells is not known.

A negative-feedback effect of testosterone on the hypothalamus regulates the secretion of LH. In humans, the blood levels of LH do not fluctuate greatly. A negative-feedback control over FSH secretion also exists. It is due to an as yet unidentified compound produced in the seminiferous tubules (and also in the ovaries) and named *inhibin*. Determining which cells are responsible for the secretion of inhibin and what its exact mode of action is are difficult because inhibin has not yet been defined chemically, nor is there any assay for it. Inhibin was at first thought to be formed by the germ cells—particularly the spermatids—but it is now believed to be a product of the Sertoli cells. Its major action is to provide a negative feedback at the level of the hypothalamus and the pituitary, lowering the secretion of FSH (Fig. 5–6). It has been proposed that inhibin also acts at the level of the gonad, significantly decreasing the uptake of FSH by the immature testis.

*Nonhormonal Factors*

The testes are normally found in the scrotal sacs outside of the body cavity. In some seasonal breeders the testes move out of the scrotal sacs and into the abdominal cavity during the nonbreeding season. When in this location, spermatogenic activity ceases. Positioning of the testes within the abdominal cavity *(cryptorchidism)* may be experimentally produced in laboratory animals. This operation results in a cessation of spermatogenic activity, and if the condition is prolonged, all of the epithelial cells of the seminiferous tubules, with the exception of the Sertoli cells, degenerate. The reason for this can be found in the rather small temperature difference (2.2°C) between the abdominal cavity and the scrotum. Testicular transplants to locations within the abdominal cavity result in a loss of spermatogenic activity, but transplants to the anterior chamber of the eye or to the scrotum function normally. Heat applied directly to the testes is also effective in inhibiting sperm development.

The seminiferous epithelium is a tissue in which the cells divide continually. It is thus subject to the injurious effects of many agents that adversely affect rapidly dividing cell populations. Interest in the effect of toxic substances on spermatogenesis increased in the late 1970s, when it was shown that the pesticide dibromochloropropane (DBCP) produced infertility in exposed workers, particularly in those exposed to the higher indoor levels associated with the manufacturing process. Many other substances are known to be testicular toxins causing lowered sperm counts, morphologically abnormal sperm, and reduced sperm motility. These include lead, cadmium, chlorinated flame retardants, and a number of other pesticides in addition to DBCP.

Radiation is particularly injurious to rapidly dividing tissues and thus to the seminiferous epithelium, but the doses one would normally receive are well below those that affect fertility. The annual dose to the gonads from natural radiation is only about 0.1 rads and from medical diagnostic procedures is also about 0.1 rads. The spermatogonia are the most sensitive, and the farther alone the line the germ cells progress from spermatogonia to mature sperm, the more they become resistant to radiation. Thus, moderate doses in the 50-rads to 100-rads range affect mainly the spermatogonia and result in a reduction in sperm count starting at the time the germ cells that were spermatogonia at the time of irradiation would have matured into active sperm—about one to two months in most mammals. If the dose is high enough to destroy large numbers of spermatogonia, the individual then goes through three successive periods: (1) a postirradiation fertile period when the unaffected cells past the gonial stage continue to mature and function, (2) a postirradiation sterile period at the time when the destroyed spermatogonia should have matured, and finally (3) a postirradiation fertile period resulting from the multiplication and maturation of those irradiated spermatogonia that were not killed by the irradiation.

While considering the effects of various chemical and physical agents on fertility, we should also indicate the important and serious consequences that may result from the mutagenic activity of these agents. The amount of exposure necessary to produce mutations in the germ line is a great deal less than that necessary to cause cell death or morphological abnormalities in the exposed cells. However, assessing the risk of genetic damage as the result of exposure to any mutagenic agent and in turn its effect on future generations is an extremely difficult, if not impossible task, particularly when dealing with low levels of exposure.

Dietary factors are also important. Inanition or protein depletion depress spermatogenic activity. This effect, however, is not direct but is mediated through a depression in gonadotrophin secretion. Vitamin A and E deficiencies also depress spermatogenesis, the depression in the case of vitamin A being reversible and that of vitamin E irreversible.

Other pituitary-mediated depressors of spermatogenesis are confinement and estrogen, the latter through its negative-feedback effect. A number of nonsteroidal compounds are also known to inhibit gonadotrophin secretion.

Some disease organisms also have an adverse effect on the testes. One in particular is the causative agent of infectious mononucleosis, a disease quite common in college students. This disease has been known to result in a 10-fold reduction in the number of sperm per ejaculate as well as a marked increase in the percentage of abnormal sperm.

## Control of Fertility in the Male

Concern about population growth has led to a search for a means of regulating fertility in the male as well as in the female. Although there has been a great amount of effort directed at the possibility of the control of fertility in the female, a relatively small amount has been directed at the control of male fertility. This should not be laid at the door of male chauvinism but considered in its proper biological perspective. In the female, we are faced with the control of a single event in a cyclic process: either the inhibition of the monthly release of a single oocyte or the production of an endometrial environment unsuitable for implantation. In the male, however, we are faced with the problem of interfering with the continual process of the development and maturation of a seminiferous epithelium that produces millions of sperm every day of its functional life.

### *Hormonal Control*

Controlling spermatogenesis through the use of hormones has been an ongoing endeavor but has not yet yielded a satisfactory solution. One might expect that the female pill would be equally effective in the male, lowering LH and FSH production and consequently depriving the seminiferous epithelium of the high levels of testosterone necessary for spermatogenesis. It is. Estrogens and/or progestins will inhibit sperm production, but these hormones in turn "feminize" the male and may induce gynecomastia, a mild but uncomfortable enlargement of the breasts. More important, lowered testosterone levels result in a loss of libido—which is in itself an effective but certainly undesirable contraceptive method.

### *Nonhormonal Control*

Many nonhormonal compounds have been found that are effective in inhibiting spermatogenic activity in experimental animals, and many of them have been tested in men. Compounds acting on almost every step in spermatogenesis have been tested. Alkylating agents, such as nitrogen mustard and esters of sulfonic acid, are radiomimetic drugs that inhibit mainly the division of the spermatogonia. Some heterocyclic compounds (nitrofurans, dinitropyrroles) have been shown to inhibit meiotic divisions, while others (diamines) exert their influence at the level of spermatid maturation. Prerequisite for the use of any antifertility agent in men is that its action is reversible, causing no permanent damage, and that it does not have any undesirable or toxic side effects. Unfortunately, most of the tested compounds are active at levels close to toxic, and many of them produce undesired side effects, among them antibuse action (intolerance to even low levels of alcohol).

The most recent entry into the field comes from China and is called gossypol. It is a phenolic compound extracted from the seeds, stems, and roots of the cotton plant, where it functions to protect the plant against insect pests. It is highly effective in men in inhibiting spermatogenesis and reducing the sperm count. It also increases the percentage of abnormal sperm and markedly decreases sperm motility (from 83% to 4%). Gossypol's effect on sperm motility may be mediated through a decrease in ATPase activity. Electron-micrographic studies of sperm from gossypol-treated monkeys show a breakdown of the A tubules of the peripheral doublets and a disintegration of the spokelike links between the central doublet and the B tubules of the peripheral doublets (Shandilya et al., 1982). The Chinese researchers report that gossypol produces oligospermia by acting directly on the germ cells. Other investigators report that it acts to block the response of the Leydig cells to LH, inhibiting testosterone synthesis at a number of steps along the synthetic pathway. The levels of testosterone are reduced below those necessary to maintain spermatogenesis but not far enough to impair the action of testosterone on the libido. Gossypol's effects on the testis are reversible, and it has no effect on body, seminial vesicle, or prostate weights.

## Control of Secretory Activity

It has been known since "days of old," from castration experiments on men and animals, that the testis is responsible for the maintenance of the male reproductive tract and the male sexual characteristics. However, knowledge of the mechanisms of this control, the hormones involved, the cells responsible for the synthesis of these hormones, and the control of this synthesis has been only recently acquired.

The Leydig cells (interstitial cells) of the testis are the cells responsible for the secretion of the male hormone. These cells appear in clusters between the seminiferous tubules (Fig. 3–1). In the late 1920s it was established that the androgenic hormone secreted by the Leydig cells was testosterone and that these cells were under the trophic influence of the anterior pituitary gonadotrophin, LH. In hypophysectomized animals, LH will maintain the synthesis and secretion of testosterone by the Leydig cells. Testosterone acts locally on the seminiferous tubules and also passes into the bloodstream, acting to maintain the sex glands and ducts, and the secondary sexual characteristics.

### Differentiation of the Hypothalamic Neuroendocrine System

We will discuss, in the chapter on the development of the reproductive system, the importance of the gonadal hormones of the fetus in the differentiation of the sex ducts and the external genitalia. Gonadal hormones secreted during the perinatal period are also important in the differentiation of the hypothalamic centers that function in the secretion of the gonadotrophin-releasing hormone, which in turn regulates the release of the pituitary gonadotrophins that control the cyclic behavior of the female reproductive system and maintain the spermatogenic and secretory activity in the male.

The neuroendocrine activity of the male and female brains differs functionally. We have seen that during the estrous cycle, a basic tonic release of GnRH controlled by the neurons of the MBH is periodically subjected to a cyclic impulse from more rostral hypothalamic centers, resulting in an LH and FSH surge and ovulation. An early advance in the understanding of the mechanism that sets up this female type of brain function was the finding that female rats injected with testosterone shortly after birth, when they reach puberty never run estrous cycles and are permanently sterile. Later experiments established other pertinent facts. The brain of females castrated at birth still functions in a female pattern. But the brain of a castrated male also functions in a female pattern. Also, normal adult males can never be induced to show an LH surge no matter what exogenous hormones are administered. We may then draw the following conclusions: (1) The female-like cyclic activity of the hypothalamus is the basic pattern, and the male-like activity is superimposed on an inherently female-like brain, resulting in the overriding of the female-like cyclic activity; in addition, (2) the pattern imprinted on the perinatal brain is permanent. The facts appear to point clearly to the androgens of the testis of the newborn as the controlling agents. However, an apparent contradiction arose when it was found that estradiol was more potent than testosterone in masculinizing the female brain and also that antiestrogen compounds inhibited masculinization. This contradiction was resolved when it was considered that testosterone is metabolized in the brain and might be aromatized to estradiol there, as it is in other tissues. It could then exert a masculinizing effect. Support for the conclusion that estradiol formed from testosterone in the brain is the responsible compound comes from the fact that in castrated males the nonaromatizable androgenic compound dihydrotestosterone fails to masculinize the brain. Why, then, does the newborn female not masculinize her own hypothalamus? Is it because the ovary of the newborn female is synthetically inactive? No. Plasma levels of estradiol are very high in the neonatal female. This dilemma was resolved when it was discovered that the blood of the newborn rat contains an estrogen-binding protein. In the female, the estrogen secreted into the bloodstream is bound by this protein and prevented from reaching the brain. In the male, the testosterone passes freely to the brain and into the hypothalamic neurons, where it is converted to estradiol and then exerts its masculinizing effect.

Thus, secretion of testicular hormones during a temporally restricted critical perinatal period organizes a permanent functional response on a preoptic–hypophysiotrophic system that is sexually indifferent before the time of the exposure. This stimulus abolishes the capacity of the male hypothalamus ever to impose a cyclic pattern of LH and FSH release on the pituitary, as is characteristic of the female hypothalamus. The hypothalamus, at the same time, is programmed to control other postpuberal sexual responses concerned with characteristic male and female sexual behavior and aggression.

## References

Assheton, R. 1898. The segmentation of the ovum of the sheep with observation of the hypothesis of hypoplastic origin for the trophoblast. Quart. J. Microsc. Science 41:205–261.

Diczfalusy, E. 1969. Steroid metabolism in the foetoplacental unit. Excerpta Med. Int. Congr. Ser. 183:65–109.

Gallagher, T. F. and F. C. Koch. 1929. The testicular hormone. J. Biol. Chem. 84:495–500.

Igarashi, M. and S. M. McConn. 1964. A hypothalamic follicle stimulating hormone-releasing factor. Endocrinology 74:446–456.

Ingram, D. L. 1953. The effect of hypophysectomy on the number of oocytes in the adult albino rat. J. Endocrinol. 9:307–311.

Klopper, A. 1974. The hormones of the placenta and their role in the onset of labour. MTP International Review of Science. Reproductive Physiology Series One, 8th ed. R. O. Greep, ed.

Knobil, E. 1980. The neuroendocrine control of the menstrual cycle. Recent Prog. Horm. Res. 36:53–87.

Liggins, G. C. 1972. Endocrinology of the foeto-maternal unit. In: Human Reproductive Physiology, pp. 138–197. R. P. Shearman, London: Blackwell Scientific Publications.

Liggins, G. C. 1973. Hormonal Interactions in the Mechanism of Parturition. In: Endocrine Factors in Labour, pp. 119–139. A. Klopper and J. Gardner, eds. Cambridge: Cambridge University Press.

Loeb, L. 1923. The effect of extirpation of the uterus on the life and function of the corpus luteum of the guinea pig. Proc. Soc. Exp. Biol. Med. 20:441–443.

McConn, S. M., S. Taleisnik, and H. M. Friedman. 1960. LH-releasing activity in hypothalamic extracts. Proc. Soc. Exp. Biol. Med. 104:432–434.

McLachlan, J. A., R. R. Newbold, H. C. Shah, M. D. Hogan, and R. L. Dixon. 1982. Reduced fertility in female mice exposed transplacentally to diethylstilbestrol (DES). Fertil. Steril. 38:364–371.

Nekola, M. V., A. Horvath, L. J. Ge, D. H. Cory, and A. V. Schally. 1982. Suppression of ovulation in the rat by an orally active antagonist of luteinizing hormone–releasing hormone. Science 218:160–161.

Schally, A. V., A. Arimura, A. J. Kastin, H. Matsuo, Y. Bara, T. W. Redding, R. M. G. Nair, L. Debelyuk, and W. F. White. 1971. Gonadotropin-releasing hormone: one polypeptide regulates secretion of lutenizing and follicle-stimulating hormones. Science 173:1087–1094.

Shandilya, L., T. B. Clarkson, M. R. Adams, and J. C. Lewis. 1982. Effects of gossypol on reproductive and endocrine functions of male cynomolgus monkeys (*Macaca fascicularis*). Biol. Reprod. 27:241–252.

Smith, P. E. 1927. The disabilities caused by hypophysectomy and their repair. J.A.M.A. 88:158–161.

Steinberger, E. and A. Steinberger. 1974. Hormonal control of testicular function in mammals. In: Handbook of Physiology-Endocrinology, IV, Part 2, pp. 325–345. R. O. Greep and E. B. Astwood, eds. Baltimore: Williams and Wilkins.

# 6

# FERTILIZATION

Fertilization is the process by which the sperm initiates and participates in the development of the egg. Penetration of the egg by the sperm results in the initiation of a series of reactions that indicate the beginning of the development of the zygote. This phase of fertilization is known as *activation.* Following penetration and activation of the egg, the male and female pronuclei approach each other and fuse to form the zygote nucleus. Fusion of the nuclei is termed *amphimixus* or *syngamy.* Although we may then ask whether we should consider the act of fertilization as occurring at the time the sperm penetrates the egg or at the time of amphimixus, it is best to consider all of the events starting with the approach of the sperm to the egg and ending with the fusion of the pronuclei under the heading of the fertilization process. The entire process takes only a few hours or less in most invertebrates and lower vertebrates (fish, amphibians), but in mammals the time required is usually 12 hours or longer.

Fertilization has three distinct results. It activates the egg. It restores the diploid number and provides the male side of the genetic material to the system. In many species, it supplies the central body necessary for cell division, although there is some question whether or not this occurs in the mammal.

Fertilization may take place either inside or outside the body of the maternal parent. Interestingly, in those forms in which fertilization takes place outside of the body, we may correlate the number of eggs shed with the mating behavior of the parents. Where there is little close contact between the male and the female and the sperm are liberated merely in the general proximity of the ova, the female often releases many thousands of eggs. Such is the case in many invertebrates. Even a vertebrate such as the codfish sheds over a million eggs at a single spawning. In other forms, the eggs may be deposited in a specific location to await fertilization. Here the number of eggs may be reduced to 1000 or 2000. In species characterized by a more intimate relationship between the male and the female, where the male clasps the female forcing the eggs out and fertilizing them as they are shed, the number of eggs released may be a hundred or less. This type of behavior is seen in the Siamese fighting fish and in some amphibia.

For those forms that do not shed their eggs into an aquatic environment, internal fertilization is necessary. Following fertilization, the young may be carried within the mother and born at an advanced stage of development, a process known as *viviparity.* Alternatively, the fertilized egg may be enclosed in protective envelopes and laid and allowed to develop outside of the mother, a process known as *oviparity*. Internal fertilization requries the development of a copulatory apparatus by means of which the sperm are deposited within the genital tract of the female. This process is termed *semination* and should not be confused with fertilization. Internal fertilization is characteristic of all forms above the amphibians, although it is not restricted to these species—as anyone who has raised tropical fish well knows.

## Egg–Sperm Interacting Substances

Since, by definition, fertilization begins with the approach of the sperm to the egg, we must first consider what is occurring at this time. This involves a study of the chemical substances that are carried or secreted by the eggs and the sperm and that play various roles in the fertilization process. The first comprehensive theory of fertilization involving the interplay of egg and sperm substances was introduced by F. R. Lillie (1919). He proposed a quite ingenious scheme with a major role assigned to an egg substance that he called *fertilizin*—thus labeling his theory as the *fertilizin theory*. The theory was based on his observations of the effect of "egg water"—obtained by allowing sea urchin eggs to stand in seawater—on sperm introduced into the water after the eggs were removed.,Lillie proposed that egg water had three effects on sperm: attraction, activation, and agglutination. He postulated that fertilizin was a diffusible component of the egg surface and that fertilization was a species-specific reaction between fertilizin in the egg and a complementary factor, *antifertilizin*, present in the sperm head.

### Egg Substances—Attraction

Although attractive substances are often of importance in plants, particularly in ferns and mosses—where one such substance has been given the appropriate name of sirenin—there is considerable evidence that the eggs of most animals do not exert any attractive influence on the sperm. Anyone seeing the large numbers of sperm clustered around the egg has no difficulty understanding how someone could propose that eggs attract sperm. However, experimental results that have been advanced in support of attraction have proved to be inconclusive insofar as the eggs of most animals are concerned. In a number of species of *Hydrozoa*, however, the gonangia (the egg-containing organs) are known to secrete a chemical substance that both activates and attracts the sperm. Sperm in the vicinity of the gonangia—or in the vicinity of extracts of the gonangia—exhibit directional turning toward the attractive substance, which is a low-molecular-weight, heat-stable basic peptide. Chemotaxis in these hydroids was at first thought to be species specific and therefore of possible importance in discouraging cross-fertilization, but was later shown not to be so. Sperm chemotaxis toward eggs has also been described in the tunicates (Chordate).

### Egg Substances—Activation

Sperm are normally active cells. Thus, an increase in activity as the result of exposure to egg substances is somewhat difficult to evaluate. However, there is little doubt that sperm activity is increased by egg secretions in many species. This becomes especially evident if sperm are allowed to age after their release before exposure to egg substances. Aged sperm are almost immotile, but when exposed to egg water, they show marked and easily noticeable increases in movement. Oxygen consumption may be used as an indication of increased activity and a fourfold increase in oxygen uptake has been reported in sea urchin sperm exposed to egg water. Again, increases in oxygen consumption are more apparent when aged sperm are used.

Attempts have been made to determine the chemical composition of the egg substance that increases sperm activity. The activating substance and the agglutinating substance are normally bound together but may be split by a number of physical and chemical methods. Good evidence suggests that these two substances are chemically different. Chemical analysis of fertilizins in general indicate that they are glycoproteins of high molecular weight that yield upon hydrolysis a number of amino acids, one or a few monosaccharides, and considerable sulfate. Since the amino acids and the monosaccharides vary from species to species, we properly speak of fertilizins rather than a single compound, fertilizin.

### Egg Substances—Agglutination

Although not of universal occurrence, sperm-agglutinating substances have been found in the eggs of a variety of species including echinoderms, molluscs, annelids, tunicates, and many vertebrates.

The reaction is generally species specific. When cross-agglutination does occur, it is usually between closely related species where some degree of cross-fertilization takes place naturally. When sperm are exposed to homologous egg water, they immediately lose their motility and clump together in groups, a reaction for which the agglutinating factor is responsible. In the sea urchin, agglutination reverses itself spontaneously. This has been explained on the basis of a mutual multivalence theory using concepts employed in antigen–antibody reactions. The reacting molecules are pictured as being multivalent with respect to their combining groups (Fig. 6–1). Large groups of sperm are thus built up as the fertilizin molecules combine with the sperm and bind them together (Fig. 6–1 A). Reversal occurs as the result of the splitting of the multivalent agglutinin into univalent molecules attached to individual sperm (Fig. 6–1 B). After this spontaneous reversal, the sperm have their reacting groups occupied and as a result of this, although they are still motile, they have lost their fertilizing power and also cannot be agglutinated again by fresh egg water. In favor of this explanation is the fact that agglutinating egg water may be made nonagglutinating by treatment with heat, ultraviolet or x rays, or proteolytic enzymes. Sperm exposed to egg water so treated, although they are not agglutinated and retain their motility, lose their fertilizing ability. The bonds in the fertilizin molecule broken by these treatments, which thus converts the compound into a univalent configuration, are assumed to be the same as those broken by the sperm, either mechanically or enzymatically, in spontaneous reversal.

**FIG. 6–1.** Reversible agglutination. A, multivalent fertilizin molecules agglutinate sperm; B, motile sperm with receptors covered by univalent agglutinin.

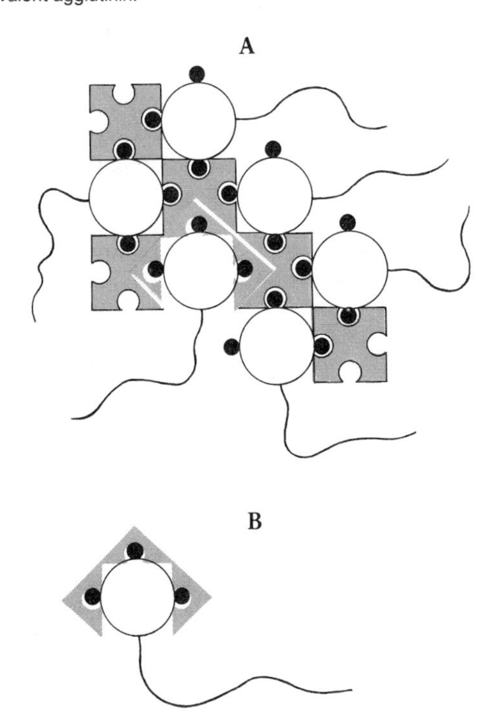


The agglutinin is a component of the egg jelly, although dejellied demembranated sea urchin eggs retain a cytoplasmic agglutinin, *cytofertilizin,* with properties identical to the jelly component. The agglutinins are acid polysaccharide–amino acid complexes with a high sulfate content.

Recently, it has been proposed that in the sea urchin, *Strongylo-centrotus purpuratus,* agglutination is actually the swarming of freely moving sperm toward a common focus. The clusters containing motile sperm always remain spherical in shape even as more sperm join up, implying that the sperm are continually changing their positions. Sperm motility is necessary for this reaction and, if inhibitors of sperm motility are added, agglutination does not occur. However, this proposal is not supported by investigation in the starfish, *Asterias amurensis,* where antimotility agents have no effect on agglutination. Before the classical fertilizin–antifertilizin isoagglutination scheme—which is supported by a wealth of experimentation—can be questioned seriously, much more evidence must be introduced.

## Sperm Substances—Antifertilizin

Antifertilizin, the sperm substance with which fertilizin reacts, can be assayed by its ability to neutralize the agglutinating action of fertilizin. Antifertilizin also has the capacity to agglutinate suspensions of eggs, and in this process a precipitation membrane is formed on the jelly coat. Chemical analysis of antifertilizin extracted from sea urchin sperm show it to be an acidic protein of low molecular weight containing about 16 percent nitrogen.

## Sperm Substances—Membrane-Penetrating Substances

Sperm have been shown to carry a number of substances other than antifertilizin. Many eggs are surrounded by tough membranes that, although they serve to protect the egg, also act as barriers to the penetration of the sperm. In order to reach the inside of the egg, sperm must have the means of overtaking these barriers. This is accomplished by a number of different kinds of penetrating agents carried by the sperm head.

*Lytic Substances in Some Invertebrates*

In general, the membranes of invertebrate eggs are not so complicated as those of the mammals, and the sperm do not need to secrete lytic substances in order to penetrate through them to the surface of the egg. However, this is not true of all species. The keyhole limplet, *Megathura crenulata*, has a tough membrane that is resistant to strong acids but is broken down in seconds by lytic agents of the sperm. Sperm lysins have also been demonstrated in a number of other species including the mussel (*Mytilus*), the abalone (*Haliotis*), and the polychaete worm (*Hydroides*). In some species, the sperm leave distinct holes in the membranes after they have passed through. However, in other species—those of the echinoderms for example—the results are not so clear-cut, and there is evidence both for and against the presence of lytic enzymes, which aid in sperm penetration.

*Sperm Penetration in Mammals*

As has been described in Chapter 4, the mammalian egg, after ovulation, is enclosed in its plasma membrane outside of which are found the noncellular zona pellucida and a number of layers of follicle cells forming the corona radiata and the cumulus oophorus. A sperm must pass through all three of these membranes, and it does so by virtue of three different enzymes that it carries. The outer follicle cells are interspersed in a ground substance that is a mucopolysaccharide containing considerable hyaluronic acid. Hyaluronic acid is a component of the ground substance of all connective tissue. It can be broken down by *hyaluronidase,* an enzyme similar to one found in snake venom. When carbon particles and hyaluronidase are injected into an animal subcutaneously, the carbon particles penetrate for considerably greater distances than when they are injected alone. For this reason, the term "spreading factor" has been applied to hyaluronidase. Hyaluronidase has been extracted from testicular tissue and has been localized on the head of the sperm, specifically in the acrosome. Release of hyaluronidase from the acrosome breaks down the intercellular connections of the cumulus cells and allows the sperm to penetrate between the cells to the corona layer.

Sperm, in turn, penetrate the corona radiata by releasing a second enzyme, called *corona-penetrating enzyme* (CPE). The mechanism of penetration also involves an attack on the ground substance. Although CPE has not been chemically identified, there is no question that it is different from hyaluronidase.

The release of hyaluronidase and CPE occurs by means of a process called membrane vesiculation. It is accomplished by multiple fusions of two membranes situated in close apposition to each other. The process is diagrammed in Figure 6–2. Individual breakthroughs and fusions result in a fenestrated double-walled membrane and, as the process continues, the result is the development of individ-

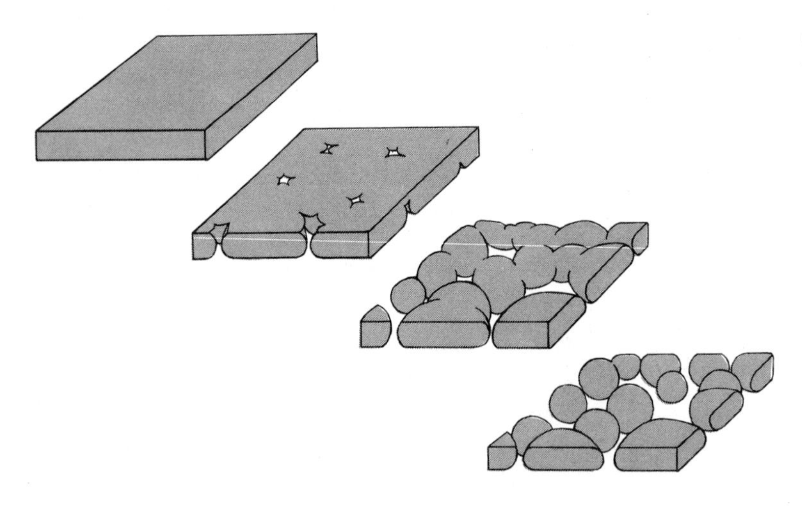

**FIG. 6–2.** Process of vesiculation between two apposed membranes.

ual separated double-walled vesicles. The membranes involved in the case of the sperm are the plasma membrane on the outside and the outer acrosomal membrane on the inside. Electron microscope studies have shown that sperm outside of the cumulus oophorus show no morphological difference from epididymal sperm. However, EM photographs of sperm penetrating the corona radiata and of sperm up against the zona pellucida show small individual vesicles over the anterior half of the sperm previously enclosed in the plasma and outer acrosomal membranes (Figs. 6–3, 6–4). At the equator of the sperm the outer acrosomal membrane becomes continuous with the plasma membrane; thus, the sperm is still enclosed in a single continuous membrane. Over the posterior half of the head, posterior to the location of the broken-down acrosome, this membrane is the plasma membrane, while over the anterior half it is the inner acrosomal membrane. Vesiculation and the formation of the composite sperm covering is illustrated diagrammatically in Figure 6–5. Vesicula-

**FIG. 6–3.** Vesiculation of the acrosomal membrane and the sperm plasma membrane over the anterior portion of the mouse sperm. Over the posterior region the membranes are still intact. (From J. M. Bedford, 1968. Am. J. Anat. 123:329.)

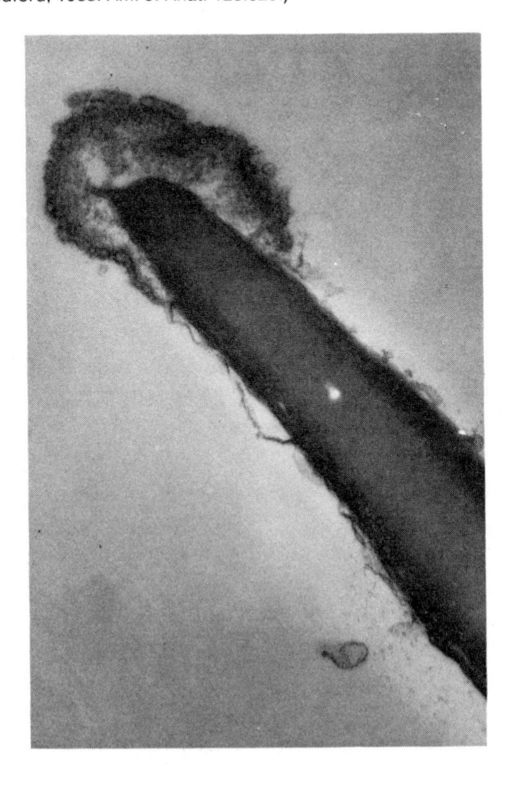

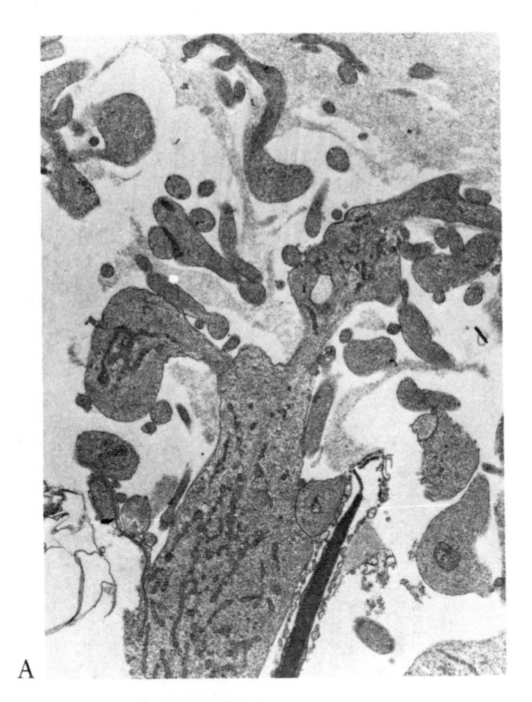


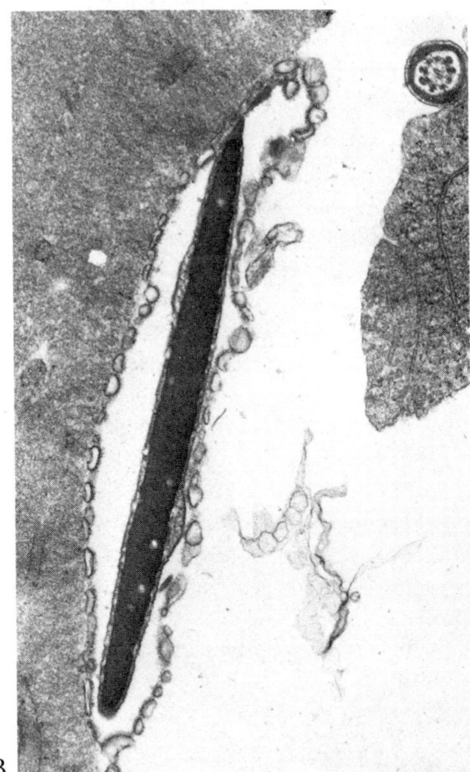


**FIG. 6–4.** A, vesiculation of the anterior acrosomal cap of the rabbit sperm as it passes through the corona cells toward the zona pellucida; B, vesiculation of the acrosome of the rabbit sperm lying against the zona pellucida. (From J. M. Bedford, 1968 Am. J. Anat. 123:329.)

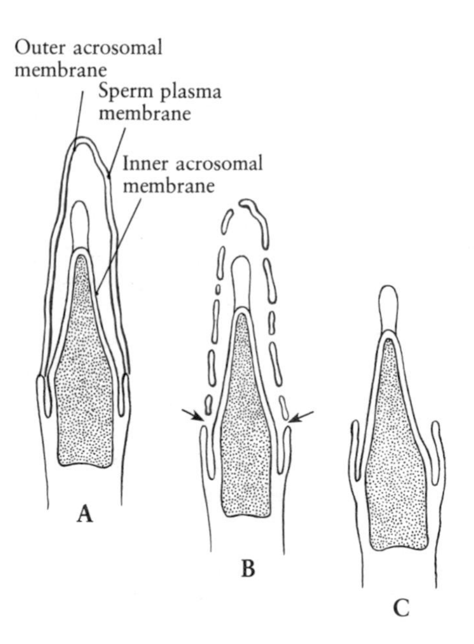


FIG. 6-5. Diagrammatic representation of vesiculation resulting in release of the acrosomal contents and exposure of the inner acrosomal membrane. At the arrow, fusion of outer acrosomal and sperm plasma membranes.

tion is considered to be a process that results in a slow progressive release of the acrosomal enzymes.

The final barrier is the zona pellucida. This calls into action another enzyme, which is presumed to be located on the inner arosomal membrane. Slits are seen in the zona pellucida after sperm penetration and are indicative of an enzymatic breakdown of the zona substance. This enzyme shows a close similarity to pancreatic trypsin as determined by inhibitor studies, amino acid analyses, and immunodiffusion tests, a quite remarkable finding in view of the different origins and functions of these enzymes. The enzyme has been called a number of different names including trypsinlike enzyme (TLE) and acrosin. The International Committee on Biochemical Nomenclature has proposed the use of the name *acrosomal proteinase,* which is indicative of both its origin and its proteolytic effect on the zona pellucida.

## Capacitation

Penetration of the egg membranes thus occurs by means of the action of various enzymes released as the acrosome undergoes vesiculation. However, before the acrosomal reaction can occur, sperm must undergo a process known as *capacitation.* In most nonprimate mammals, mating is timed so that it takes place before ovulation and sperm are present in the uterine tubes for some time before the arrival of the ovum. This turns out to be a necessary condition of the fertilization process because, if animals are mated at the time of ovulation or if freshly ejaculated sperm are placed in the uterine tubes, fertilization does not occur. Capacitation represents a change in the sperm that makes them capable of fertilzing the ovum. Capacitation also affects the metabolic activity of the sperm and results in sperm "activation," one aspect of which is increased flagellar movement. The length of time necessary for capacitation varies in different species, being one hour in the mouse, six hours in the rabbit, and seven hours in man.

The question then arises, "Why do sperm have to be capacitated?" The answer is related to the all-important acrosome reaction. Sperm that have not been capacitated will not undergo an acrosome reaction and are incapable of penetrating the egg membranes. In sperm that have not undergone capacitation, the breakdown of the acrosome does not occur because, as sperm pass through the male reproductive tract, they are exposed to what has been termed a *decapacitating factor* (DF). Decapacitating factor is found throughout the male reproductive tract and, in fact, capacitated sperm may become decapacitated when reexposed to DF in epididymal fluid or seminal plasm. Capacitation thus invovles the removal or inactivation of a DF imposed on the sperm in the male reproductive tract.

Some physiological significance of decapacitation is seen in that it (1) prolongs the life of the sperm, (2) prevents sperm from penetrating the lining cells of the male and female reproductive tracts through which they pass, (3) prevents sperm agglutination, and (4) prevents phagocytosis of the sperm in the female reproductive tract.

It has also been shown that sperm must reside within the female reproductive tract in order to bring about the conversion of the enzymatically inactive proacrosin present in the ejaculated sperm to acrosin. Wineck et al. (1979) have demonstrated that the uterine factor which promotes this conversion is a glycosaminoglycan.

## Fusion of the Gametes

After the breakdown of the acrosome and the release of the enzymes contained therein have resulted in the sperm penetrating through the outer membranes of the egg down to the egg plasma membrane, the next step is the passage of the sperm through the plasma membrane into the egg cytoplasm. However, this does not involve a simple penetration of the sperm through the plasma membrane but rather a fusion of the sperm and the egg plasma membranes to form a common membrane. This is followed by the movement of the sperm nucleus, neck, and midpiece into the egg cytoplasm. Electron microscope studies of fertilization show that in the mammal a side-to-side contact is made (Fig. 6–6). The region that establishes and maintains contact is the postacrosomal part of the sperm. This is also the area where initial fusion of the sperm and the egg plasma membranes occurs, eventually enclosing the egg and the sperm in a single common mosaic membrane (Fig. 6–7). The process of egg–sperm membrane fusion in the mammal differs markedly from that in many invertebrates, as will be described in the following section.

The microvilli of the egg plasma membrane with which the sperm membrane fuses are important to the egg–sperm fusion process, since fusion normally occurs only at those regions of the egg where microvilli are present. In many mammals microvilli are lacking over a fairly extensive area of the egg surface around the region where the second polar body is extruded, and fusion rarely occurs in this region.

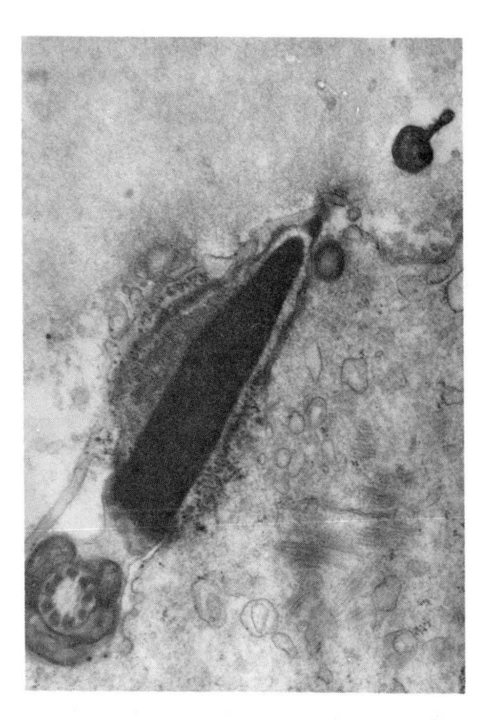

**FIG. 6–7.** Fusion of the sperm and egg plasma membranes establishing a continuity between the gametes. (From M. Stefanini et al., 1969. J. Submicrosc. Cytol. 1:1.)

**FIG. 6–6.** Head of mouse sperm lying parallel to the surface of the egg. The rows of small vesicles are the remnants of the vesiculated sperm plasma membrane and the outer acrosomal membrane. (From M. Stefanini et al., 1969. J. Submicrosc. Cytol. 1:1.)

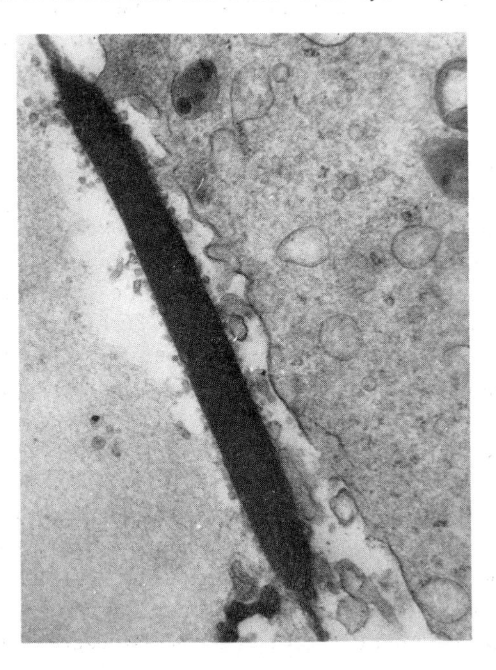

Although the acrosomal membrane is not involved in egg–sperm fusion, the acrosomal process is a necessary part of the fusion process. Unreacted sperm (sperm in which the acrosomal reaction has not taken place) cannot penetrate the egg membranes and thus cannot fertilize the egg. However, even if unreacted sperm are brought into contact with the egg plasma membrane of corona radiata- and zona pellucida-free eggs, no membrane fusion and hence no fertilization takes place. Thus, the acrosomal reaction, in addition to making available the substances necessary for egg membrane penetration, also brings about changes in the sperm plasma membrane making it capable of binding with sperm receptors on the egg surface. The nature of these changes is unknown although the reason a lowered pH (below 6.1) prevents fertilization is considered to be its effect in blocking the sperm

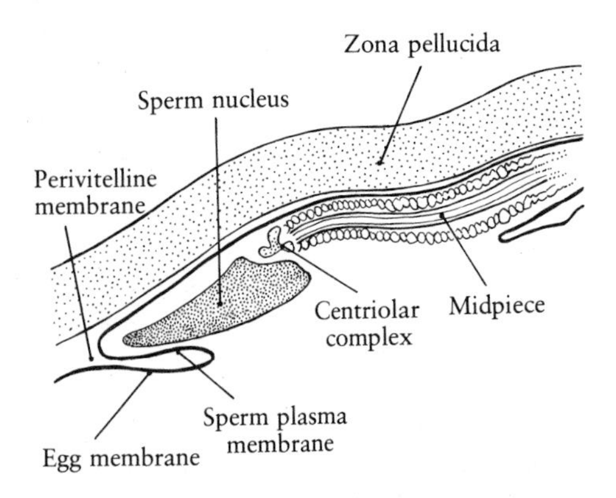


**FIG. 6–8.** Diagram of a rabbit sperm nucleus and midpiece with mitochondrial spiral inside of the zona and egg vitelline membrane. (After D. Szöllösi and H. Ris, 1961. J. Biophys. Biochem. Cytol. 10:275.)

membrane-bound molecules that participate in egg–sperm fusion.

Very soon after the sperm enters the egg cytoplasm its nuclear membrane disappears and its previously dense chromatin material begins to become loose in texture and filamentous in form. The entire sperm is incorporated into the egg cytoplasm (Fig. 6–8), but only the head forms the male pronucleus. The midpiece and the entire tail (with the possible exception of the centriole) degenerate.

## Morphological Aspects of Fertilization in Some Invertebrates

Electron microscope studies of fertilization in many invertebrate species have given a somewhat different picture from that described for the mammal. The first accurate analysis of the structure of the acrosome and the morphology and significance of the acrosomal reaction were provided by the Colwins in the early 1960s in studies on two species of invertebrates, *Hydroides* (a marine annelid) and *Saccoglossus* (a hemichordate). Since then, numerous studies have shown that there is a similarity of acrosomal structure and function in a variety of invertebrates.

### Fertilization in the Sand Dollar (Echinarachnius Parma)

In general, fertilization in the echinoderms involves at least three events. The first is the reaction between the egg jelly coat and the sperm plasma membrane resulting in the acrosome reaction. The second is the adhesion of an acrosomal granule-coated process to the egg vitelline membrane, and the third is the fusion of the acrosomal membrane and the egg plasma membrane.

#### *Acrosomal Structure*

The acrosomal region (Figs. 6–9 A; 6–10 A) consists of a membrane-bound acrosomal vesicle and periacrosomal material. The acrosomal vesicle, the most anterior part of the sperm, lies just beneath the sperm plasma membrane. It is almost spherical but somewhat flattened posteriorly. Within the vesicle is a granule separated from the anterior half of the vesicle membrane by an EM-lucid area but continuous with the basal part of the vesicle membrane through a more EM-dense region. Periacrosomal material surrounds the vesicle, but the bulk is contained in a depression on the anterior surface of the nucleus.

#### *The Acrosomal Reaction*

The acrosomal reaction, induced by extracellular egg substances, takes place as the sperm approaches the egg. It may easily be induced by egg water. The reaction consists of a breakdown of both the sperm plasma membrane, and the acrosomal vesicle membrane and their fusion to each other in a circular area below the middle of the acrosomal vesicle (Figs. 6–9 B,C; 6–10). The region of this fusion, which the Colwins have termed the "rim of dehiscence," is evident in the unreacted sperm. In the process of this fusion, the anterior halves of both the sperm plasma membrane and the acrosomal vesicle membrane are lost, and the contents of the acrosomal vesicle are exposed. The basal part of the acrosomal vesicle membrane everts to begin the formation of an acrosomal tubule. This tubule elongates rapidly carrying with it the adherent acrosomal vesicle material (Figs. 6–9 C,D; 6–10 D,E).

Microfilaments at the base of the acrosomal tubule are apparent early in its formation. They have also been described in other invertebrates and found to contain the contractile protein, actin, as a principal component.

The initiation of the acrosome reaction by egg water and the fact that this reaction is species specific recalls the process of sperm agglutination and the fertilizin–antifertilizin system, and suggests a possible relation between the sperm-agglutinating

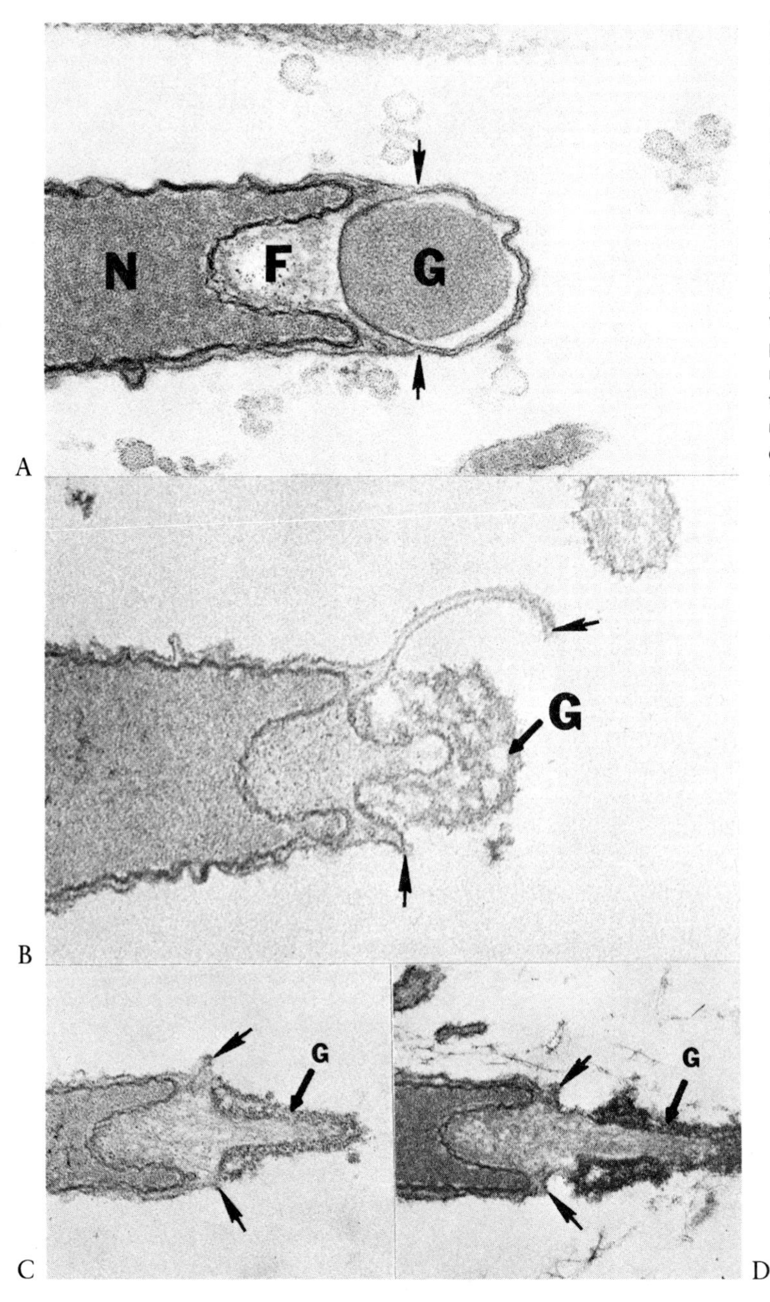


**FIG. 6–9.** Electron micrographs of the unreacted acrosomal region and the acrosomal reaction of the sperm of *Echinarachnius parma*. A, unreacted acrosomal region. The acrosomal vesicle contains an acrosomal granule (G) enclosed in a complete acrosomal membrane. Periacrosomal material surrounds the vesicle with the bulk contained in a fossa (F) at the anterior end of the nucleus (N). Arrows show the region of future dehiscence and fusion of the sperm plasma membrane and the acrosomal vesicle membrane; B, fusion of the sperm plasma membrane and the acrosomal vesicle membrane; C, D, elongation of the acrosomal tubule with the former contents of the acrosomal vesicle adhering to it. (From R. G. Summers et al., 1975. J. Biophys. Biochem. Cytol. 10:275.)

and the acrosome reaction processes. Both activities are retained after dialysis or alcohol precipitation, are destroyed by heating, radiation, and proteolytic digestion, and are proportional to fucose and sialic acid content. Comparisons support the view that egg fertilizin initiates both sperm agglutination and the acrosome reaction. However, the presence of calcium is required for the acrosome reaction in all species of sea urchins (except *Clypeaster japonica*) but is generally not required for agglutination. It has been proposed (Metz, 1978) that the fertilizin–antifertilizin reaction that results in sperm agglutination—a process that does not require the presence of calcium—then, in turn, produces changes in sperm membrane permeability allowing the influx of the calcium necessary for the fusion of the sperm plasma and acrosome membranes.

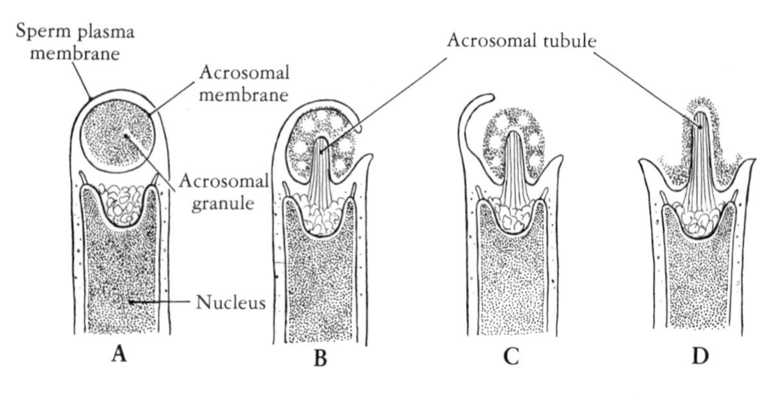


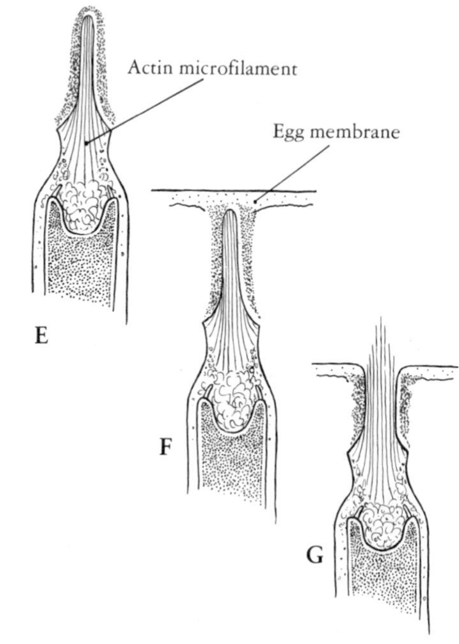


**FIG. 6–10.** Schematic representation of the unreacted acrosomal region in *Echinarachnius parma,* A, and the acrosomal reaction, B–G. B, C, the breakdown and fusion of the sperm plasma membrane and the acrosomal membrane and the beginning of the formation of the acrosomal tubules; D, E, the elongation of the acrosomal tubule with the adherent granule material; F, G, fusion of the egg and sperm membranes. (After R. G. Summers et al., 1975. J. Biophys. Biochem. Cytol. 10:275.)

*Attachment and Penetration*

The acrosomal tubule penetrates through the egg jelly to reach the vitelline membrane. The acrosomal vesicle material adherent to the tubule is the first sperm product to reach the vitelline membrane, and it forms a morphological complex with this membrane, a process termed *primary gamete binding.* Because the species-specific reaction between eggs and sperm is a function of some component of the vitelline membrane and the surface of the sperm, the sperm substance responsible for this reaction should be a part of the acrosomal vesicle material adherent to the acrosomal tubule. A protein called *bindin* has been isolated from the acrosomal vesicle of the sea urchin and has also been localized by its immunological reaction with rabbit anti-bindin on the surface of the acrosomal process (Vacquier, 1979). The bindin receptor on the egg microvilli is a glycoprotein.

*Membrane Fusion*

Following primary gamete binding, a second and final step in the process of the movement of the sperm into the egg occurs. This is the fusion of the egg and sperm membranes (Fig. 6–10 F,G). This fusion takes place between the egg plasma membrane and the acrosomal tubule. The acrosomal tubule consists partly of the acrosomal vesicle membrane and the sperm plasma membrane. The end result is the formation of a common egg–sperm plasma membrane, a system in which the egg and sperm nuclei now lie within the same covering. Following membrane fusion, egg cytoplasm may flow up around the sperm to form the classical fertilization cone.

The movement of the sperm nucleus into the main mass of the egg cytoplasm is not completely understood. It may be merely a passive event as when two droplets of water coalesce, or it may be due to contraction of actin filaments, or it may be due to movements of the fluid plasma membrane toward the point of egg–sperm membrane fusion.

Markedly similar series of events have been described in other species (Figs. 6–11, 6–12). Sperm and egg plasma membrane fusion, rather than penetration of the sperm through the egg plasma membrane, would appear to represent a widespread and fundamental pattern of fertilization.

## Responses of the Egg to Fertilization

From the discussion given in the preceding paragraphs, it is evident that successful penetration by

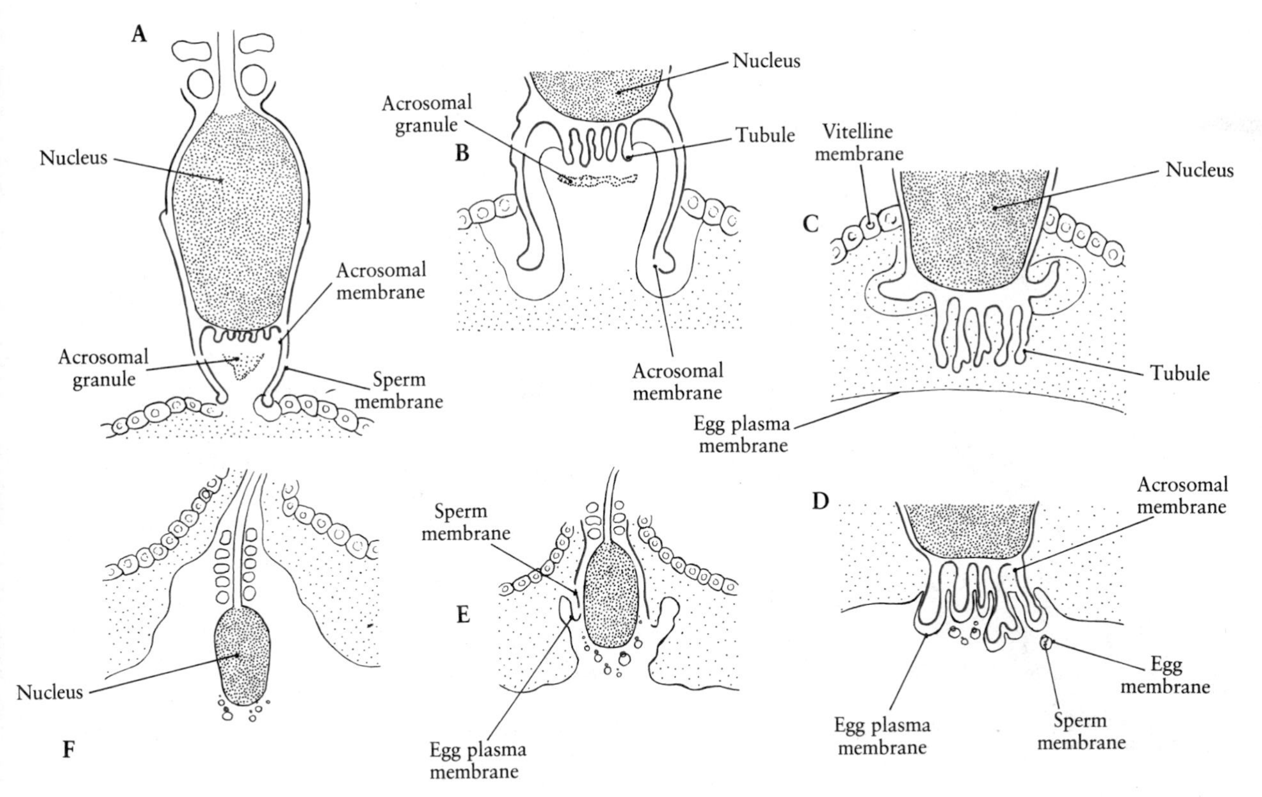


**FIG. 6–11.** Diagrams of the egg–sperm interaction in *Hydroides*. A, the dehiscence and fusion of the acrosomal membrane and the sperm plasma membrane as the sperm contacts the outer egg membrane. The acrosomal granule lies within the EM-lucid cavity of the acrosomal vesicle; B, C, the eversion of the acrosomal membrane to form acrosomal tubules accompanied by the breakdown of the acrosomal granule; D, E, the acrosomal tubules pass through the vitelline membrane and interdigitate with the egg plasma membrane. The two membranes fuse and egg cytoplasm rises around the sperm nucleus to form the fertilization cone. Persisting acrosomal remnant marks the point of the sperm plasma membrane; F, the sperm moves into the egg cytoplasm. (From A. L. Colwin and L. H. Colwin, 1961. J. Biophys. Biochem. Cytol. 10:211.)

the spermatozoon into the egg's cytoplasm requires a very specific interaction between the plasma membranes of the male and female gametes. The fusion of the plasmalemmas of the egg cell and the spermatozoon is a critical step in the fertilization process. First, it appears to be a prerequisite for the eventual incorporation of a functional sperm nucleus into the egg cytoplasm. Second, it typically is followed by the activation of the egg cell, thereby triggering the development of the embryo. *Egg activation* refers to the cascade of morphological, physiological, molecular, and metabolic changes that take place in the egg cell in response to contact or fusion with the spermatozoon. Some of these changes occur at the surface of the egg cell and result in rearrangements of the organization of the cortical granules and cortical cytoplasm. Other changes occur in the interior of the egg cell, including the fusion of male and female pronuclei to form the zygote nucleus. The absence of a successful interaction between egg and sperm results in death, since the unfertilized egg will soon degenerate.

Over the past several years, the results of intensive investigations, principally from the laboratories of Epel, Steinhardt, and Vacquier using sea urchins and sand dollars, have begun to shed light on the extent of the morphological, molecular, and biochemical changes in the egg evoked by its fusion with the sperm cell. Epel (1977), for example, indicates that the activation process in the West Coast sea urchin *(Strongylocentrotus purpuratus)* can be divided into a group of changes occurring within the first 60 seconds after sperm–egg contact and a group of changes beginning at about 5 minutes after activation (Fig. 6–13). The early responses of

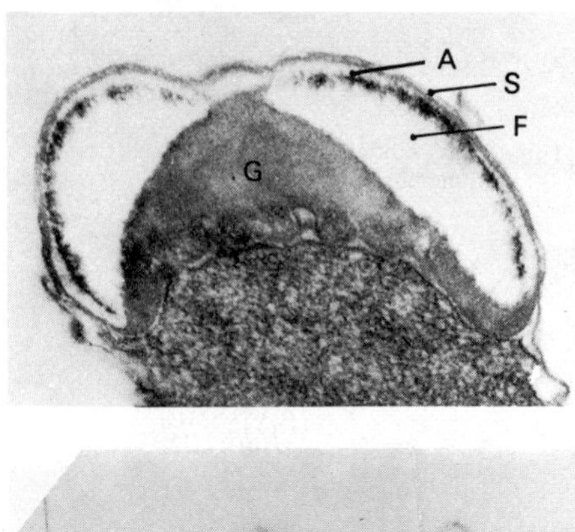


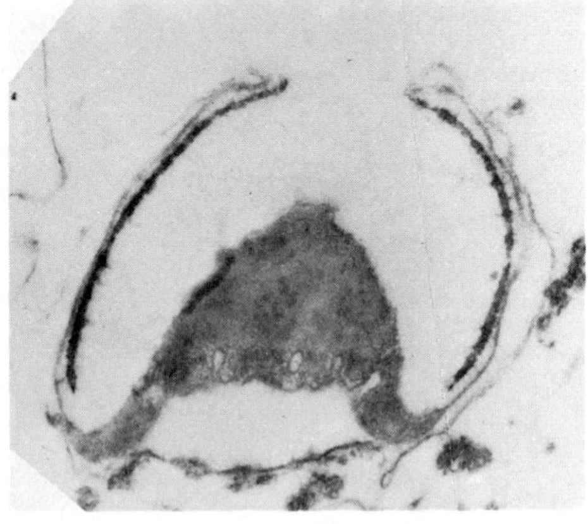

**FIG. 6–12.** Electron micrograph of sections through a sperm of *Hydroides*. A, the unreacted sperm. The acrosomal membrane (A) and the sperm plasma membrane (S) are continuous around the rim of the opening. Fine granular material (F) lines the acrosomal membrane. An acrosomal granule (G) lies within the cavity of the acrosomal vesicle; B, stage comparable to Figure 6–11A. (From A. L. Colwin and L. H. Colwin, 1961. J. Biophys. Biochem. Cytol. 10:231.)

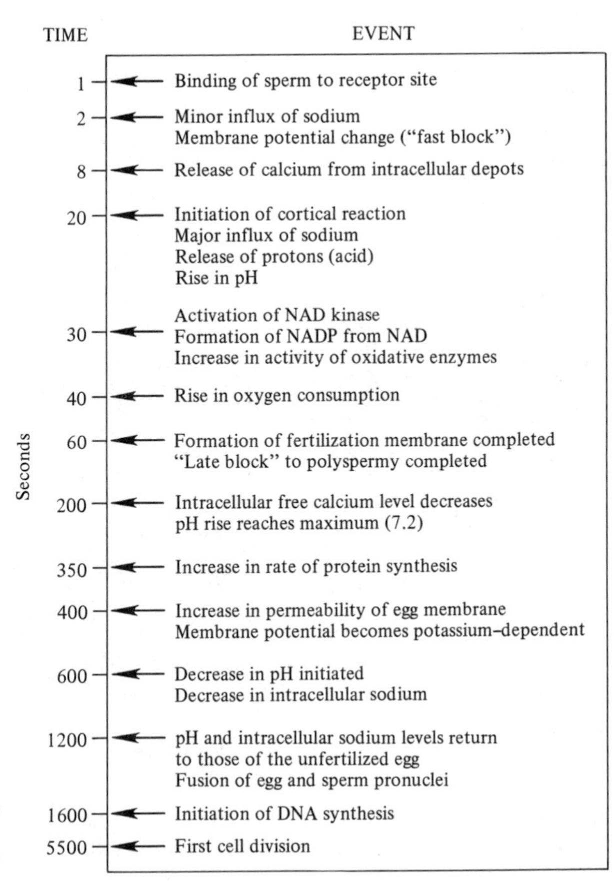


**FIG. 6–13.** The temporal sequence of morphological and biochemical events following fertilization of the sea urchin egg.

the egg include alterations in the intracellular concentrations of certain ions, the conversion of metabolically important coenzymes, and the release of stored enzymes into the cytoplasm. Late responses include, the synthesis of proteins and DNA, both events being critical to embryonic development. Additionally, studies have been and continue to be focused on the mechanism(s) underlying the activation process and the causal relationships between the early and late responses of the fertilized egg. We will first examine some of the major responses of the fertilized egg.

It should be pointed out that understanding the role of the sperm cell in the activation process is complicated by the observation that egg cells can be stimulated experimentally to start development in the absence of the male gamete. *Experimental* or *artificial activation* of unfertilized eggs can be induced by electric currents, acids, bases, and pricking with a needle. The development of artificially activated eggs is usually quite limited. There are also cases, among both invertebrate and vertebrate animals, when the mature egg cell is spontaneously activated and develops into the adult organism in the absence of the sperm cell. This condition, referred to as *natural parthenogenesis* or *virginal reproduction,* is part of the life cycle of several free-living and parasitic organisms, including species of rotifers, crustacea, annelids, insects, and lizards. Domestically, there is a strain of turkey whose eggs regularly develop without a spermatozoon. However, the eggs of birds do not typically develop by parthenogenesis. In several species of rotifers, arthropods, and lizards, the parthenogenetic populations consist of only females. The egg from which a naturally parthenogenetic individual develops is haploid. Although adult organisms of some species produced by parthenogenesis have the haploid number of chromosomes, there are mechanisms in a number of other species by which the diploid number of chromosomes is restored prior to adulthood. In some lizards, for example, this is accomplished by doubling the number of chromosomes just before meiosis. The diploid state may also be attained after the embryo starts to develop by the fusion of pairs of nuclei of cleaving cells.

## The Cortical Reaction

A characteristic response of most eggs to penetration by a spermatozoon (or other activating stimulus) is a dramatic rearrangement of the cortical granules. The cortical reaction is the first visible, morphological sign of egg activation and refers to the breakdown (i.e., exocytosis) of cortical granules at the surface of the egg. Although it has been extensively studied by both light and electron microscopy, the complete role of the cortical reaction in the development of the fertilized egg is still to be determined.

When eggs of the sea urchin are inseminated and viewed under the phase-contrast microscope, one of the first visible changes is the rupture of the cortical granules and the discharge of their contents at the egg surface (Fig. 6–14). A consequence of the released cortical granule material is the elevation of a membrane and the formation of the fluid-filled *perivitelline space* (Fig. 6–14 C,D). The process of

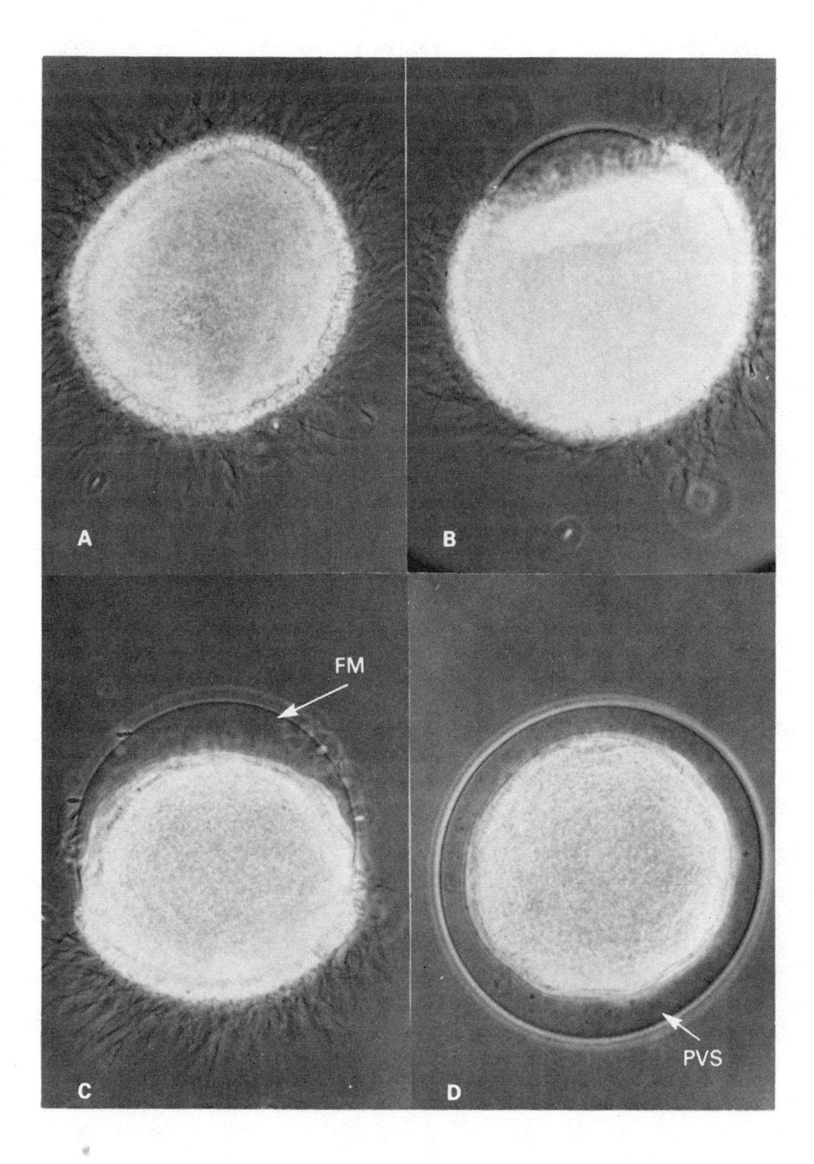


**FIG. 6–14.** Eruption of cortical granules and elevation of the fertilization membrane (FM) as seen by light microscopy in eggs of *Lytechinus pictus* (echinoderm). A, 10 seconds after insemination the sperm are bound over the entire egg surface; B, 25 seconds after insemination; C, 35 seconds after insemination the cortical reaction begins to propagate over the egg; D, at 50 seconds the vitelline layer is completely elevated as the fertilization membrane. Note the cortical granule material in the perivitelline space (PVS). (From V. Vacquier and J. Payne, 1975. Exp. Cell Res. 82:227.)

granule dissolution begins at the site of successful sperm attachment to the egg cell and propagates itself in wavelike fashion around the egg during the next 20 to 30 seconds. The reaction of the cortical granules is completed within a minute or so of activation. The structural details of the cortical granule reaction as revealed by electron microscopy are shown in Figures 6–15 and 6–16. Following activation, there is an approximation and fusion of the cortical granule membranes with the plasma membrane of the egg. A rupture or perforation, formed perhaps because of the digestive action of an enzyme within the cortical granule, subsequently develops at the site of fusion between these two membranes. The contents of the cortical granules swell and are then discharged at the egg cell surface. The contents of the cortical granules include enzymes, structural proteins, and mucopolysaccharides. One enzyme, *vitelline delaminase,* dissolves away the protein that binds the vitelline membrane to the unfertilized egg plasma membrane. As the mucopolysaccharides of the perivitelline space swell, a fluid pressure is generated that causes the vitelline membrane to lift away from the egg surface. Structural proteins derived from the contents of the cortical granules harden the vitelline membrane and transform it into the *fertilization membrane* (envelope) or *activation calyx. Peroxidase* and *glucanase* enzymes, also of cortical granule origin, may function in the hardening process by cross-linking the

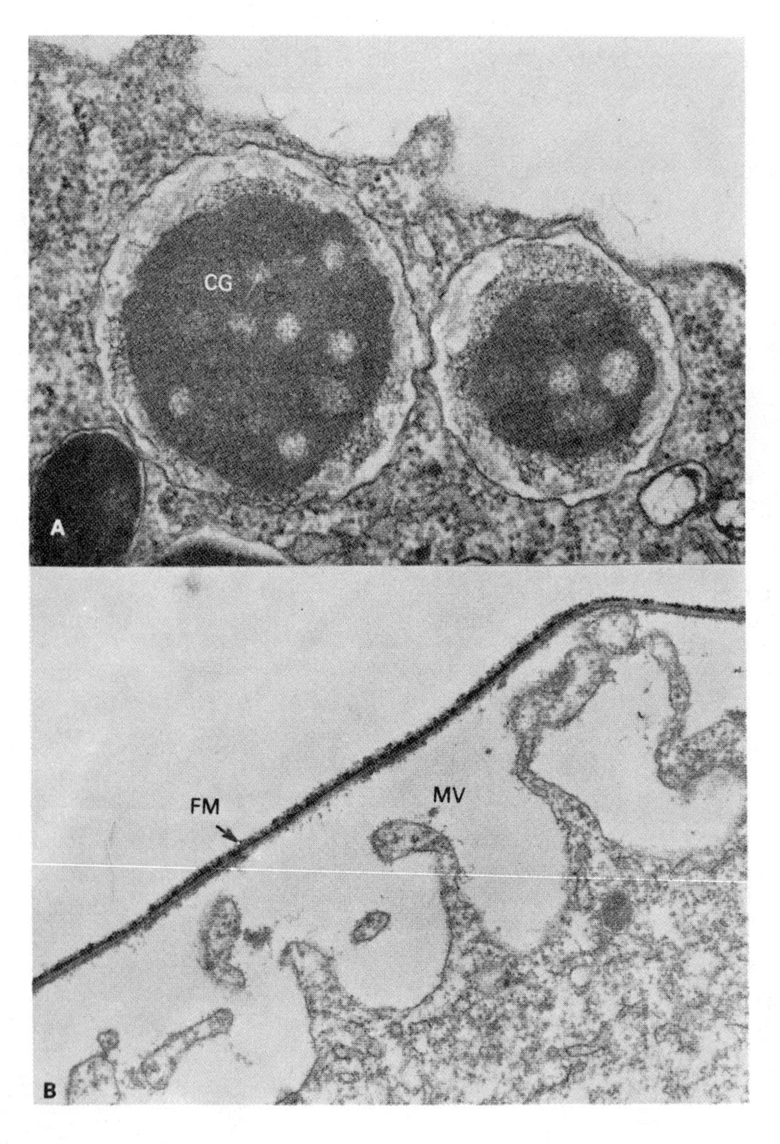


**FIG. 6–15.** The echinoderm egg during the cortical reaction as seen in section. A, electron micrograph of the cortical granules (CG) of *Dendraster excentricus.* Note the dense material surrounded by less dense fibrous material in each granule; B, electron micrograph of the egg surface of *Dendraster* after breakdown of the cortical granules. The cortical granule contents have united with the vitelline layer of the egg to form the fertilization membrane (FM). Note that the surface of the egg is extended into microvilli (MV). (From V. Vacquier, 1975. Exp. Cell Res. 90:465.)

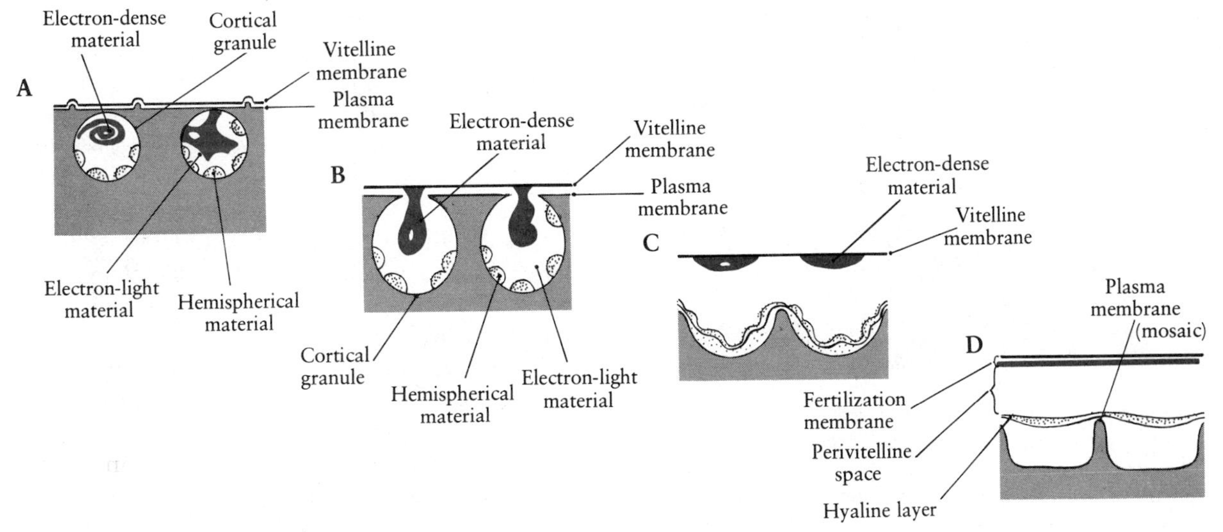


**FIG. 6–16.** Diagrammatic summary of the formation of the fertilization membrane, the hyaline layer, and the plasma membrane of the fertilized egg in the sea urchin. (From Y. Endo, 1961. Exp. Cell Res. 25:518.)

molecules of the structural proteins. Another enzyme has been identified in cortical granule exudate, and it appears to play an important role in the block to polyspermy (see below).

During the exocytosis of the cortical granules, a viscous and transparent material is released, remains closely associated with the egg plasma membrane, and spreads and flows to form the *hyaline layer* (Fig. 6–16 C,D). The chief constituent of this layer is a major protein termed *hyalin*. Hylander and Summers (1982) have shown by immunocytochemical techniques that hyalin is stored in the homogeneous component of the cortical granules of *Strongylocentrotus*. Recent studies by Hall and Vacquier (1982) suggest that hyalin may not be the only constituent of the hyaline layer. They identify several glycoproteins, distinct from hyalin, which persist as a fibrous layer (called the *apical lamina*) surrounding the sea urchin embryo after removal of the hyalin. The intact hyalin layer remains as a matrix around the embryo through late larval life. It serves as a rigid enclosure during early development to keep dividing blastomeres together. At the late blastula stage, the embryonic cells are seen to be tightly attached to the hyaline layer by microvillar processes. The tight cellular adhesion to the hyaline layer may also be important in directing later morphogenetic events of the developing embryo.

As a consequence of cortical granule breakdown, the limiting membranes of the cortical granules fuse with the plasma membrane of the egg. Hence, the plasmalemma of the egg at fertilization develops a mosaic composition whose components are derived from the original plasmalemma of the ovum and the membranes circumscribing the cortical granules. It has been a widely held view that the cortical granule membranes become permanently integrated into the mosaic surface of the fertilized egg (Fig. 6–16). However, recent studies with sea urchin and fish eggs suggest that surface membrane is continuously being retrieved in the form of *coated vesicles* back into the cytoplasm of the recently fertilized egg. The fate of this internalized membrane is not known.

The expulsion of cortical granule contents at the egg surface upon fertilization is found in many groups of animals, including frogs, teleost fishes, and some mammals. In the frog, extrusion of the cortical granules is initiated several minutes after contact with the sperm cell (Fig. 6–17). The breakdown of the cortical granules proceeds as a wave from the site of sperm entry and takes one to three minutes to traverse the egg. As in the sea urchin, the cortical reaction is accompanied by the elevation of the vitelline membrane. Cortical granule material associates with the vitelline membrane and converts it into a fertilization membrane. In teleost fishes, the spermatozoon reaches the egg surface after passage through a preformed opening, the micropyle, in the chorion at the animal pole. The cortical reaction begins at the micropyle and

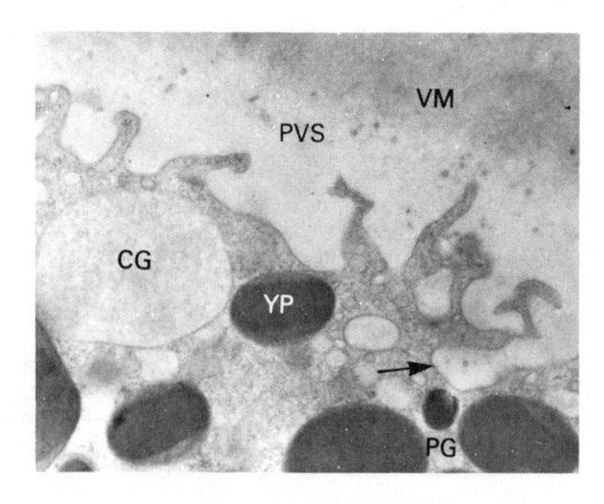


**FIG. 6–17.** The extrusion of the contents of cortical granules in the frog's egg. The contents of some of the cortical granules have been liberated into the perivitelline space (PVS). Note the erupted cortical granule at the tip of the arrow. A large granule (CG) just before extrusion of its contents is seen to the left. The underside of the vitelline membrane (VM) appears to have a layer of diffuse material whose density is very much like that of the cortical granules. PG, pigment granule; YP, yolk platelet. (From N. Kemp and N. Istock, 1967. J. Cell Biol. 34, 111.)

quickly passes around the cell. As in sea urchins and frogs, the chorion lifts off the egg surface in the wake of cortical granule breakdown. Surface views of the egg during the cortical reaction show that the central contents of the cortical granules are released intact into the perivitelline space (Fig. 6–18). They subsequently break down, and their contents become part of the perivitelline fluid.

A cortical reaction is also observed in the fertilized eggs of many mammals, including rat, hamster, and human. The contents of the cortical granules are liberated into a preformed perivitelline space between the zona pellucida and the egg cell. No new membrane is elevated after fertilization as in sea urchins and frogs. The perivitelline space does gradually increase in volume, but this is probably due to the shrinkage of the egg itself following fertilization.

Although the release of cortical granule contents onto the egg surface during the cortical reaction might be considered typical in the animal kingdom, there are several animal species whose cortical granules show an unusual response with fertilization. For example, in the surf clam *(Spisula)* and the mussel *(Mytilus)*, the cortical granules remain completely intact and undisturbed following fertilization. The cortical granules in the barnacle *(Barnea)* egg spontaneously break down, independently of fertilization, and their contents are liberated into the interior of the cell.

The variability in the response of the cortical granules to sperm penetration (or experimental activation) of the egg cell has made it difficult to determine the significance of the cortical reaction in the developmental process. For those organisms in which the cortical granules release their contents at the egg surface upon activation, the cortical reaction has been implicated in prevention of multiple sperm entry into the egg cell at fertilization. It has already been pointed out that the cortical granules contain structural proteins that harden the fertilization membrane of echinoderm eggs; a similar change is probably effected in the chorion of fish eggs following fertilization. This envelope then

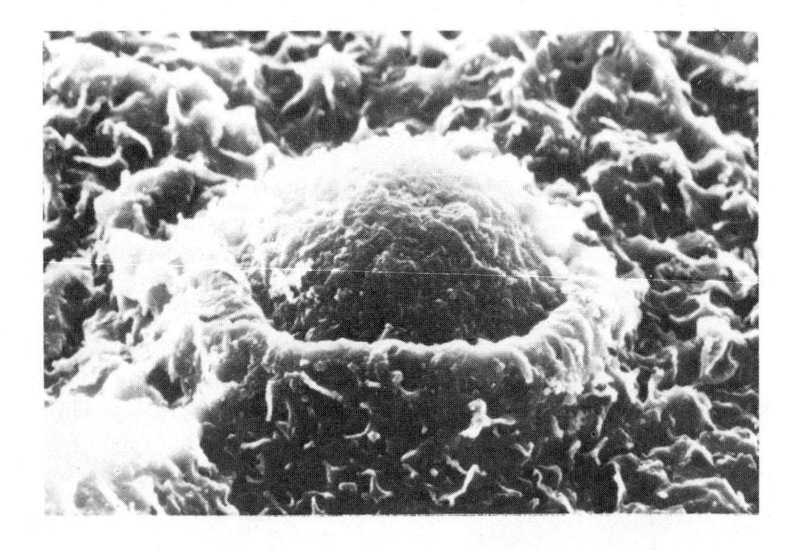

**FIG. 6–18.** Scanning electron micrograph showing the discharge of a cortical granule from the egg surface in the zebra fish.

functions to form a protective environment for the developing embryo until hatching.

Finally, it must be remembered that there are major groups of animals (insects, salamanders, birds) whose eggs lack cortical granules. Hence there is no cortical reaction in response to fertilization.

## Blocks to Polyspermy

Only one sperm cell normally enters the egg *(monospermy)* and fuses with the female pronucleus in most animal organisms. One of the most important physiological responses of the monospermic egg at fertilization, therefore, is the reaction that prevents penetration into the egg by additional sperm. The regulation of the entry of the sperm into the egg is a critical process in the initiation of embryonic development. The entrance of more than one sperm into the egg *(polyspermy)* disturbs development and typically leads to the eventual death of the embryo. Initially, it was proposed that the elevation of the fertilization envelope, dependent on the exocytosis of the cortical granules, acted as a physical barrier to the entrance of many sperms into the egg. However, the cortical reaction and the elevation of the fertilization membrane may take one to several minutes. Hence it is now known that the time course of these two events is too slow to account for complete protection against multiple sperm entry. Studies by Rothschild and Schwann in the 1950s suggested that the block to polyspermy was a diphasic (i.e., a two-step) process. The first step was postulated to occur very rapidly or within the first several seconds of activation. Kinetic analysis of fertilization rates indicated that this step reduced the receptivity of the egg to a second sperm by about 80 percent. The second step was attributable to the completion of the cortical reaction with the subsequent elevation of the fertilization membrane. The experimental basis for this was the observation that inhibition of the cortical reaction typically resulted in the production of polyspermic eggs. Results of a number of morphological and electrophysiological studies support the view that several processes contribute to the prevention of polyspermy in animal eggs. These are often grouped as *early* (rapid) and *late* (slow) *blocks to polyspermy.*

The correlation between the exocytosis of cortical granules and the late block to polyspermy has been demonstrated morphologically in elegant studies by Tegner and Epel (1973). Since the interaction of egg and sperm is a surface phenomenon, a "sperm's-eye" view of the sequence of events at fertilization can be obtained using the scanning electron microscope. Hundreds of spermatozoa attach to the outer surface of the vitelline membrane within the first 25 seconds following insemination of *Strongylocentrotus* eggs (Fig. 6–19 A, B). Indeed the sperm appear to bind progressively to the vitelline surface during this time. They bind to an array of regularly spaced projections (receptors) on the vitelline membrane (Fig. 6–19 D). The cortical reaction is initiated 25 seconds after insemination. During the cortical reaction, there is a progressive detachment of the supernumerary spermatozoa from the vitelline membrane coincident with cortical granule breakdown (Fig. 6–19 C). The detachment phase is mediated by the release from the cortical granules of a protease that enzymatically digests the sperm–egg binding sites (except for the fertilizing sperm). The lift-off of the fertilization membrane also effectively removes the receptors for sperm binding away from the egg surface. Artificial agents that induce polyspermy in the sea urchin, such as *nicotine* and *soybean trypsin inhibitor,* do so by interfering with the cortical reaction, specifically by inhibiting proteolytic activity and the wave of sperm detachment.

Results of studies with other animal eggs suggest that the contents of the cortical granules play a similar role in effecting a slow block to polyspermy. In the hamster, for example, observations on artificially inseminated eggs indicate that many spermatozoa bind to the zona pellucida. However, following the discharge of the cortical granules, the zona pellucida is conspicuously devoid of spermatozoa. The number of spermatozoa attaching to the zona pellucida of the mouse is significantly reduced by 5 minutes postfertilization. The alteration of the zona pellucida to prevent penetration and further attachment by spermatozoa in mammals is known as the *zona reaction.* This reaction is mediated by a component of the cortical granules (a protease) whose properties are similar to those of the proteolytic enzyme bringing about sperm detachment in the sea urchin.

In organisms studied to date, the time required to initiate the cortical reaction is too long to guarantee a monospermic condition. The cortical reaction in the sea urchin egg, for example, can only account

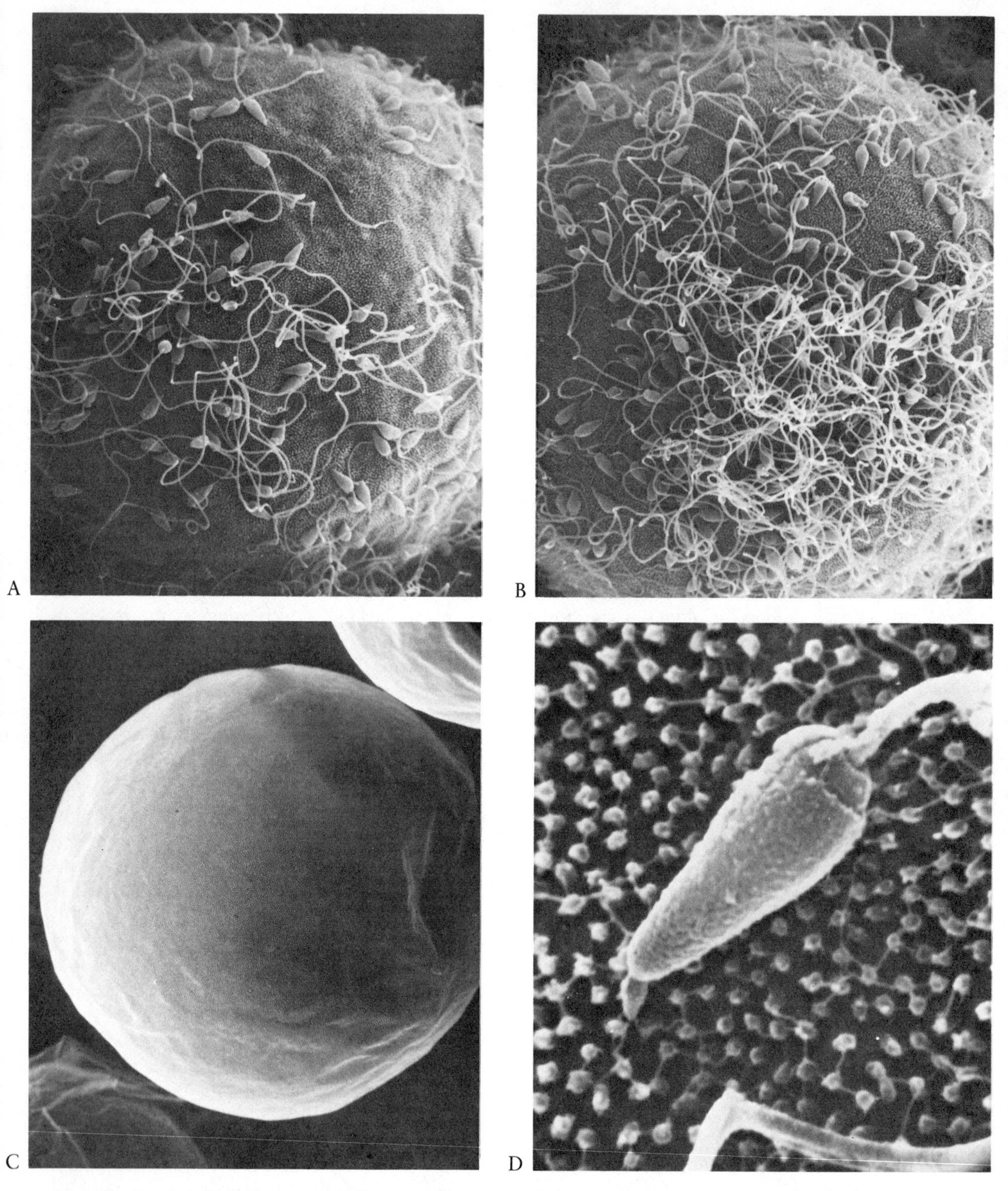


**FIG. 6-19.** Scanning electron micrographs by Tegner and Epel of the sperm attachment-detachment sequence on the surface of *Strongylocentrotus purpuratus* eggs. The number of sperm bound to the egg surface increases until 25 seconds after insemination. The cortical reaction is then initiated and the sperms begin to detach. A, approximately 5 to 10 seconds after insemination; B, approximately 15 seconds after insemination; C, 3 minutes after insemination showing the hardened fertilization membrane; D, a spermatozoon having undergone the acrosome reaction and attached to the vitelline surface by an acrosomal process. Note the projections of the surface of the fertilization membrane. (From M. Tegner and D. Epel, 1973. Science 179:685. Copyright 1973 by the American Association for the Advancement of Science.)

for protection against polyspermy after 25 seconds following fertilization. Hence other mechanisms must be operative to block multiple sperm entry prior to the onset of cortical granule breakdown. There is now substantial evidence that an early electrical block to polyspermy exists during the interval between the first successful sperm–egg interaction and the completion of the cortical reaction. Jaffe (1976) placed microelectrodes across the plasma membrane of dejellied eggs of *Strongylocentrotus* and recorded a negative potential of −70 millivolts (mV) (i.e., interior of egg is negative). During and after the addition of a sperm suspension, the egg membrane depolarizes to a plateau at −30 to + 20 mV (Fig. 6–20 A). This fast depolarization is referred to as the *activation* or *fertilization potential.* This elevated potential lasts for about one minute; the egg membrane then returns to a negative reading. Eggs with activation potentials of greater than 0 mV were never observed to be polyspermic; those that reached a plateau less positive than −10 mV were sometimes polyspermic. These results suggest that the entry of more than one sperm is prevented by the shift of the membrane potential in the positive direction. Indeed, when Jaffe artificially raised the unfertilized egg membrane to a potential more positive than +5 mV sperm did not fertilize eggs (Fig. 6–20 B). Yet many sperm were seen to bind to the egg surface, but no fertilization membrane formed and no action potential was ever recorded. The rapid electrical depolarization of the sea urchin egg plasma membrane is mediated by the influx of sodium ions. Eggs inseminated in low-sodium (choline substituted) seawater are consistently polyspermic.

**FIG. 6–20.** Membrane potentials and fertilization in eggs of the sea urchin *Strongylocentrotus.* The top trace is voltage against time; the bottom line is current against time. The dashed line indicates 0 millivolts. A, the resting potential across the unfertilized egg membrane measures −70 millivolts. When sperm are added, the egg membrane depolarizes to a plateau ranging from −30 to +20 mV. B, when current is applied to hold the unfertilized egg membrane at a potential more positive than +5 millivolts, sperm do not fertilize the cell. No electric event resembling an activation potential develops. As soon as the current is turned off, fertilization occurs. (After L. Jaffe, 1976. Nature 261:68.)

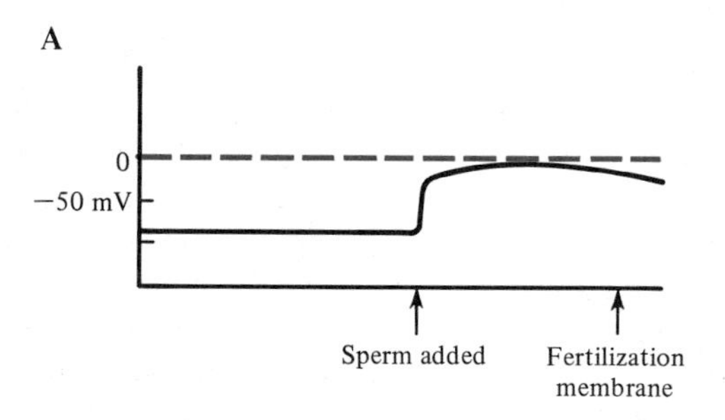


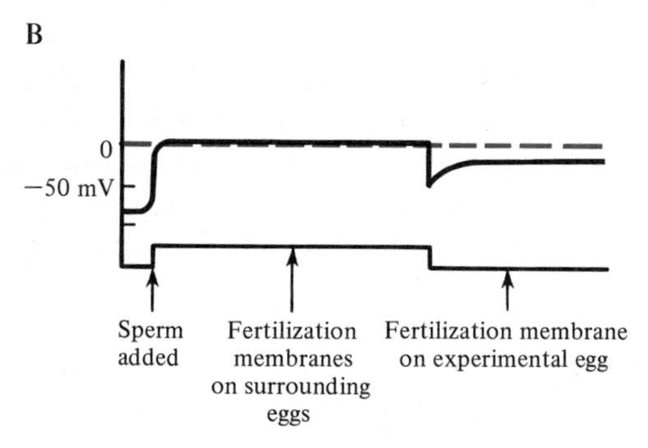


A fast electrical block to polyspermy at the level of the plasma membrane appears to operate in animals other than sea urchins, including the marine worm, *Urechis,* the starfish, *Asterina,* and the anurans, *Rana pipiens* and *Xenopus laevis.* At fertilization in *Rana,* the membrane potential quickly shifts from about −30 mV to approximately +5 mV; it remains positive for some 20 minutes and then returns to its prefertilization level. When the membrane potential of unfertilized eggs is held at or above +1 mV, there is no fertilization upon the addition of sperm (Cross and Elinson, 1980). Grey et al. (1982) report that eggs of *Xenopus* become polyspermic when fertilized by natural mating under conditions that inhibit the formation of the fertilization potential.

An electrical block at the plasma membrane, however, is not common to all animal species. Voltage clamp experiments have shown that fertilization can occur over a wide range of membrane potential levels in the medaka *(Oryzias)* egg. Monospermy in this fish is probably ensured by a mechanical block, as sperm have been observed to approach the plasma membrane in single file through the micropyle. Changes in the electrical properties of the plasma membrane upon fertilization are also not apparent in eggs of the hamster, mouse, rabbit, and *Ciona* (ascidian). Recent experiments with mouse eggs show that the membrane potential remains constant at −41 mV, except for a small oscillation at 5 minutes postinsemination, during a 60-minute interval after fertilization (Fig. 6–21). There is a slow depolarization of the plasma membrane of the rabbit egg following insemination, but its small amplitude suggests that this re-

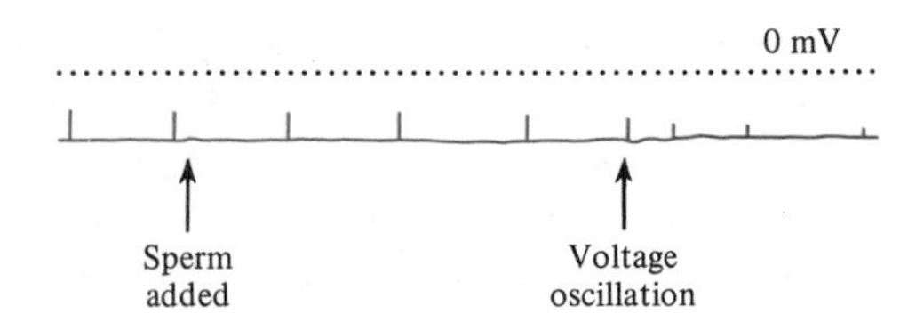


**FIG. 6–21.** Trace showing little change in membrane potential (voltage) following fertilization of the mouse egg. (From L. Jaffe et al., 1983. Dev. Biol. 96:317.)

sponse is insufficient to form a block to polyspermy. It appears that eggs of humans and the hamster rely primarily on the zona reaction to prevent multiple sperm penetration. The mouse represents a group of mammals in which, in addition to the zona block, there is a functionally important block at the level of the plasma membrane. The nature of this fast block is still unclear.

There are undoubtedly other processes that contribute to the prevention of polyspermy. For example, fertilization of the sea urchin egg causes the production and release of *hydrogen peroxide.* The possibility has been raised that egg-derived hydrogen peroxide might act to prevent polyspermy by inactivating or destroying other sperm surrounding the egg. Enzymes that break down hydrogen peroxide, such as *catalase,* cause sea urchin eggs to become polyspermic. Also, catalase induces polyspermy when it is present prior to the complete elevation of the fertilization membrane. It is presently believed that the egg responds to the fertilizing sperm by releasing hydrogen peroxide, which then reacts with a putative peroxidase associated with supernumerary sperm to inactivate them. Polyspermy prevention in the mouse, in addition to the two blocks at the zona pellucida and the plasma membrane, involves restriction of the number of sperm reaching the site of fertilization and the elimination of supernumerary sperm from the egg cytoplasm.

In contrast to the examples cited above, the eggs of many insects, amphibians (salamanders), reptiles, and birds are normally entered by several spermatozoa at fertilization. Although these animals have avoided the difficulties of excluding extra sperm from penetrating the egg, mechanisms have been developed to prevent more than one sperm nucleus from joining with the female nucleus. A particularly interesting example is offered by the egg of the salamander. In the salamander *(Triturus),* as many as nine spermatozoa enter the egg cytoplasm at fertilization. For the first 2 hours and 20 minutes after insemination, the centriole of each sperm head enlarges into a conspicuous aster (Fig. 6–22 A). As soon as one sperm nucleus fuses with the nucleus of the egg, the remaining sperm nuclei begin to degenerate. Elegant experiments by Fankhauser and his colleagues with the egg of *Triturus* suggest that the newly formed zygote produces a substance that spreads throughout the cytoplasm and specifically acts to induce degeneration of the supernumerary nuclei (Fig. 6–22 B).

## Reorganization of Constituents of the Egg Cell

With penetration of the spermatozoon into the cytoplasm of the egg cell, there are in many animal species dramatic shifts and noticeable changes in the organization of the cytoplasmic constituents. In some cases, it is evident that these changes have a profound influence on the organization and future development of the embryo. The rearrangement of ooplasmic constituents at fertilization is best illustrated in those animal egg cells whose cytoplasm bears distinctive features or regional specializations (i.e., pigment granules). The amphibian egg cell, you will recall, is typically covered in its upper two thirds by a heavily pigmented layer; the lower third of the egg cell is light in color. At fertilization in *Rana,* the spermatozoon enters the egg at a point 20 to 30 degrees from the midregion of the animal pole. The site of sperm entry on the egg surface after 20 minutes is indicated by a circular area free of microvilli (Fig. 6–23). Within 5 to 10 minutes, the pigmented cortical cytoplasm opposite the site of sperm entrance moves upward toward the animal pole (Fig. 6–24 A). As a consequence of this cytoplasmic movement, a distinct zone of grayish cytoplasm is revealed. This area is known as the *grey crescent* (Fig. 6–24 B). A sudden increase in the rigidity of the cytoplasm surrounding the sperm aster may be important in bringing about the shift of the cytoplasm and the formation of the grey crescent.

The importance of this dramatic displacement and rearrangement of the cortical cytoplasm, which results in the expression of a structure not previously present (i.e., grey cresent), is severalfold. First, the position of the grey crescent marks the future dorsal side of the embryo and the site of the dorsal lip of the blastopore at gastrulation. The

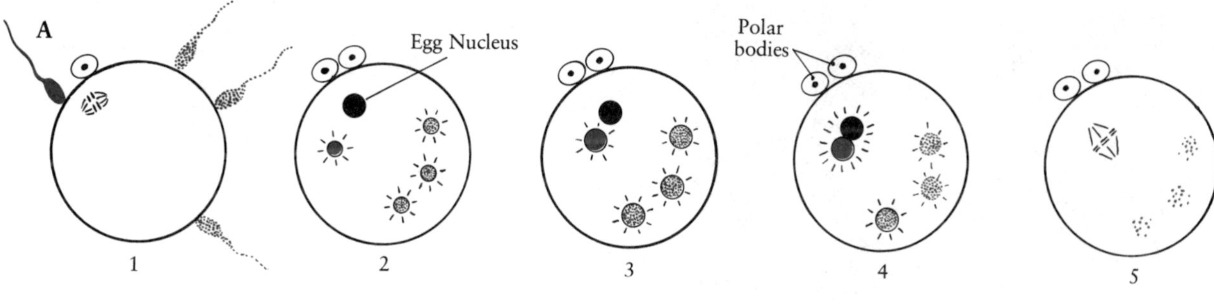


region opposite the grey crescent marks the future ventral side of the embryo. Second, the grey crescent establishes the axis of bilateral symmetry of the future embryo. Thus, the cleavage plane that bisects the grey crescent divides the embryo into right and left halves. In the majority of cases, this will be the first cleavage plane. It is interesting to note that the grey crescent also forms in the frog egg following artificial activation by pricking with a needle, but its expression bears no apparent relationship to the site of activation. Yet, studies of Roux in 1885 clearly showed that when the point of entry of sperm was controlled at the surface, the grey crescent consistently developed opposite the point of entry into the egg. He determined this by placing drops of sperm onto fine silk threads that were attached to selected sites in the animal hemisphere. We can conclude, therefore, that the action of the spermatozoon in part is to strengthen and accentuate one of serveral meridians of bilateral symmetry built into the unfertilized egg. In short, the main directional axes of the future amphibian embryo (dorsoventral; left-right), probably fixed in the cortical cytoplasm, are organized and determined by the site of sperm penetration. Similar observations on the relationship between fertilization and symmetrization have been made on the eggs of many molluscs and annelids. Third, the grey crescent appears to initiate the gastrulative process and to play a fundamental role in the determination of embryonic structures (Chapter 10).

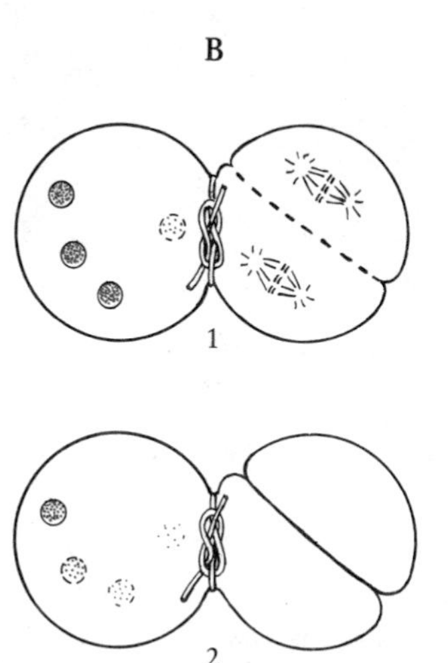


**FIG. 6–22.** Polyspermy in the salamander egg. A, the spermatozoa enter the egg cell in the metaphase of the second meiotic division (1). The nuclei of all spermatozoa enlarge during the next 2 hours and 20 minutes (2,3). After fusion of the male nucleus with the female nucleus, the supernumerary sperm begin to regress (4). As the first cleavage division occurs, suppression of most supernumerary male nuclei is evident (5); B, the experiment of Fankhauser suggesting that the fused male and female nuclei release an inhibitory factor causing the destruction of accessory spermatozoa. The egg of *Triturus* was constricted before the beginning of cleavage (1). In the half on the right, the male and female nuclei have fused and cleavage initiated. Subsequently, in the left half of the egg, the nuclei of the sperm furthest from the constriction are still large; those nearest to the right half of the egg are degenerating. (After G. Fankhauser, 1948. Ann. N.Y. Acad. Sci. 49:684.)

The development of the urochordate *(Styela)* egg is another good example of fundamental change in cytoplasmic organization that becomes visible following fertilization. As seen from the surface, the unfertilized egg cell of *Styela* shows no evidence of bilateral organization. Its surface is covered by a layer of cortical cytoplasm containing yellow pigment granules. Following the entrance of the spermatozoon, there is a rapid movement of the yellowish cytoplasm toward the vegetal pole. Shortly, the yellowish cytoplasm reorganizes as a crescent-shaped area on the side of sperm entrance and just below the equator of the egg cell (Fig. 6–25). Subsequently, a light-grey, crescent-shaped zone of cytoplasm develops opposite this yellow crescent. These cytoplasmic displacements give a distinct bilateral organization to the egg cell. Below the yellow and light-grey crescents, the vegetal pole remains as the *vegetal cytoplasm* (slate grey in color). The

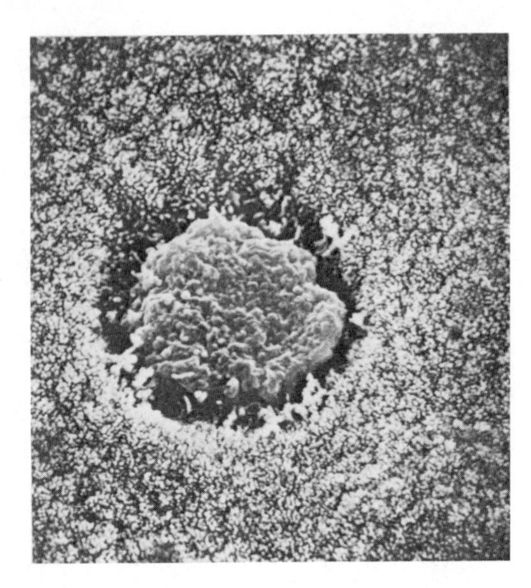

FIG. 6–23. A scanning electron micrograph showing a small circular structure on the egg surface of the *Rana* egg 20 minutes after insemination. It has been identified as the site of sperm entry into the egg. (From R. Elinson and M. Manes, 1978. Dev. Biol. 63:67.)

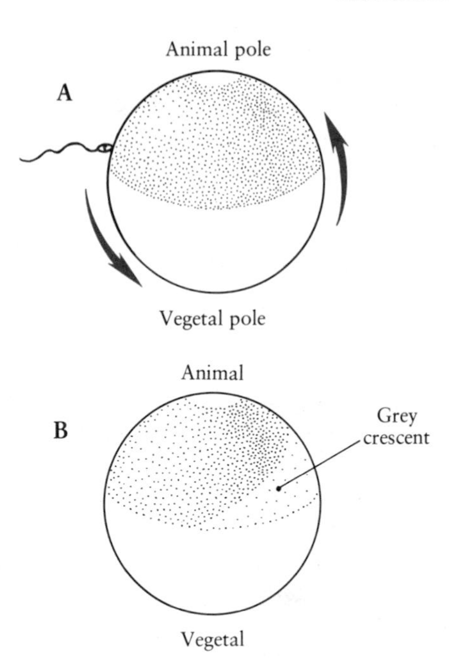


FIG. 6–24. Formation of the grey crescent in the egg of the amphibian. A, the direction of cortical cytoplasmic movements (arrows); B, location of the grey crescent opposite to the site of sperm entry.

transparent-appearing cytoplasm at the animal pole is known as the *animal plasm*. As we will discuss in Chapter 9, these visible plasms contain *morphogenetic determinants* that will become segregated into particular cell lines during cleavage and eventually influence their developmental pathways. Similar to the case of the grey crescent of amphibians, the dramatic *ooplasmic segregation* in the ascidian egg appears to be mediated by contraction of the cortex. Jeffrey and Meier (1983) have recently proposed that the motive force underlying these cytoplasmic movements is provided by a network of actin filaments associated with the plasma membrane. The spatial reorganization of the ascidian egg may also involve components of the cell surface. *Lectins* are proteins, typically derived from plants, that interact with specific carbohydrate groups on cell surfaces. An antibody prepared against a given lectin can then be conjugated with a visible staining marker, such as fluorescein iso-

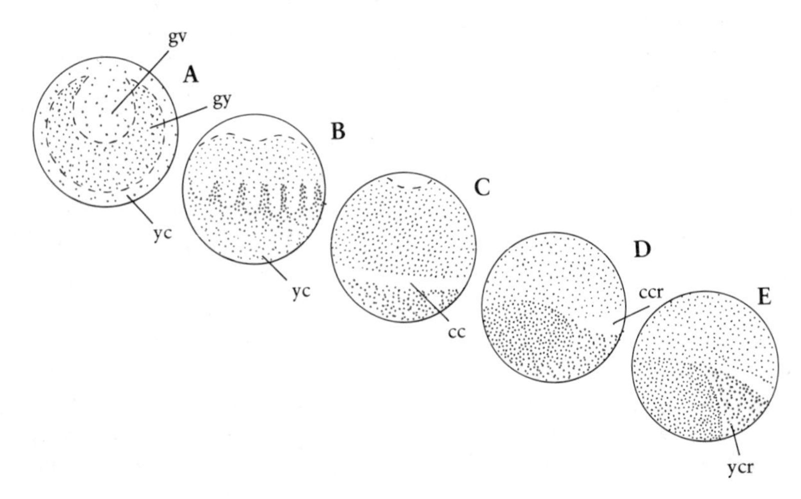


FIG. 6–25. Reorganization of the cytoplasm in the egg of the tunicate, *Styela partita,* following fertilization. A, ripe, unfertilized egg showing germinal vesicle (gv), central mass of gray yolk (gy), and peripheral layer of yellow cytoplasm (yc); B, five minutes after fertilization showing streaming of yellow cytoplasm toward lower pole where sperm enters cell; C, consolidation of yellow cytoplasm at vegetal pole. Note the clear cytoplasm (cc) beneath yellow cap; D, E, later stages in the formation of the yellow crescent (ycr) and the clear cytoplasmic crescent (ccr). The yellow crescent marks the future posterior end of the embryo. (From E. G. Conklin, 1905. J. Acad. Nat. Sci. Philadelphia 13, 1.)

thiocyanate (FITC). When exposed to a cell, the antilectin–FITC conjugate will bind to that chemical group(s) on the surface against which the antibody has been made. Unfertilized eggs of *Ascidia* when stained with fluorescein-conjugated *Dolichos* anti-A lectin show a patchy fluorescence. However, following the formation of the second polar body, the fluorescence becomes localized in a single patch over the vegetal pole. It appears to accompany the cytoplasmic movements of the yellow plasm.

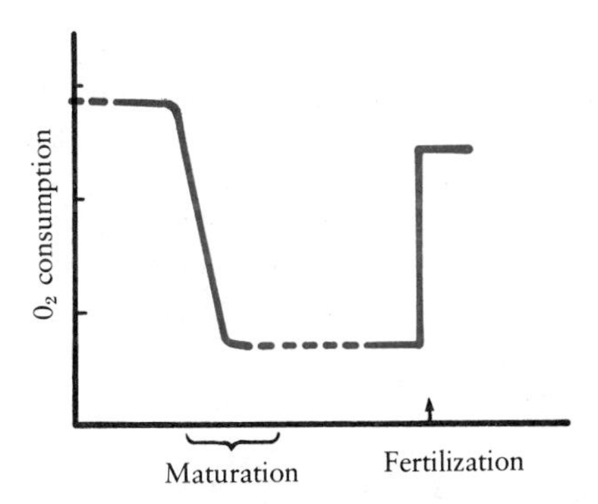


**FIG. 6–26.** Changes in the rate of oxygen consumption during maturation and at the time of fertilization in the egg of the echinoderm *(Paracentrotus).* (From P. Lindahl and H. Holter, 1941. C. R. Tra. Lab. Carlsberg, Ser. Chim. 24:49.)

## Metabolic and Synthetic Activities

The act of fertilization (or activation) transforms a physiologically and metabolically quiescent cell into one showing a series of complicated and rapid changes. The fertilized egg cell shows striking alterations in the permeability of its plasmalemma, in the viscosity of its cytoplasm, utilization of oxygen, activities of proteolytic and oxidative enzymes, and in the synthesis of proteins. The essence of fertilization lies in understanding why the unfertilized egg becomes metabolically "shut down" during late oogenesis and what mechanisms are invoked that suddenly "derepress" this inhibitory state. Many of the biochemical and physiological changes that constitute the program of fertilization in the sea urchin egg are shown in Figures 6–13 and 6–41. The data in these figures are based on several papers by David Epel (1977, 1978, 1979).

An initial effort at understanding the program of egg activation came with the discovery by Otto Warburg some 60 years ago that the rate of oxygen consumption in the sea urchin egg increased markedly within minutes of fertilization. Later studies by Lindahl and Holter with *Paracentrotus* (sea urchin) eggs confirmed this observation (Fig. 6–26). Shortly after the breakdown of the germinal vesicle, the rate of oxygen utilization sharply decreases and remains at a low level until fertilization. At this time it shows a precipitous rise. Similar changes in respiratory metabolism have been observed in teleost oocytes (Fig. 6–27). It would appear that an increase in oxygen consumption is a key reaction in the fertilization process. However, elevation in the rate of oxygen consumption upon fertilization is by no means a universal occurrence among animal species. Little change in the level of oxygen activity can be detected between unfertilized and fertilized eggs in the frog. In some marine annelids there actually appears to be a decrease in the rate of respiraton in the fertilized egg. It is interesting to note that the direction and magnitude of respiratory change at the time of fertilization may be dependent on the stage of nuclear maturation. Generally, only those eggs with reduction divisions completed will show an increase in respiration at the time of fertilization. Eggs of other species, penetrated by spermatozoa before completion of nuclear maturation, may not show any change in respiration until the process of maturation is terminated. It appears that the fertilized egg can regulate the rate of oxygen utilization, although the direction and magnitude of change vary with different animal species.

**FIG. 6–27.** Rate of oxygen consumption in the course of the maturation and fertilization of teleost eggs *(Oryzias).* (From A. Monroy, 1965. Chemistry and Physiology of Fertilization. Holt, Rinehart and Winston, New York.)

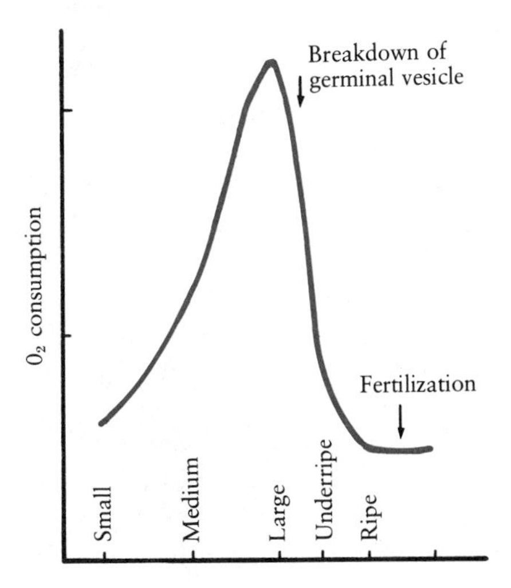

The intriguing question remains as to why the mature, unfertilized egg in so many animal species shows a low rate of respiration. Since oxygen is typically tied to the degradation of carbohydrates, it is possible that the physiological state of the unfertilized egg reflects a deficiency and/or inhibition of essential oxidative enzymes. Also, perhaps there are insufficient levels of oxidizable substrates in the cytoplasm. In both frogs and sea urchins, it has been established that carbohydrate metabolism in the egg consists of the Embden-Meyerhoff glycolytic pathway, the pentose phosphate shunt, the Krebs tricarboxylic acid (TCA) cycle, and a cytochrome system. The unfertilized egg apparently contains all of the necessary enzymes to catalyze the step-by-step reactions during carbohydrate metabolism. Yet, in the echinoderm egg, oxidative enzymes associated with the pentose phosphate cycle, such as glucose 6-phosphate dehydrogenase and 6-phosphogluconate, rise sharply in activity following fertilization. There is also a rise in NAD kinase, an enzyme that is responsible for the phosphorylation of the coenzyme NAD (nicotinamide adenine dinucleotide) to NADP (nicotinamide adenine dinucleotide phosphate), and glycogen phosphorylase. The increase in glucose 6-phosphate dehydrogenase plus elevated levels of NADP and glucose 6-phosphate could account for the increased respiration of eggs. It is probable that during oogenesis many of these enzymes are prevented from interacting with their substrates because they become bound to either structural proteins or some cellular component of the egg cytoplasm. Several investigators have described the presence of proteolytic enzymes (proteases) in sea urchin eggs and observed marked increases in their activities following insemination. Such proteases might act in the presence of calcium to release these oxidative enzymes from their inactive states.

Among the metabolic changes that take place upon activation of the repressed, unfertilized egg is a rise in the synthesis of proteins. This has been demonstrated by comparing the relative rates of the uptake and incorporation of radioactively labeled amino acids into protein between unfertilized and fertilized eggs (Fig. 6–28). Although the egg cell does become more permeable to protein precursors upon fertilization, these differences in amino acid incorporation between unfertilized and fertilized eggs are not due to differences in plasma membrane permeability. For example, the same differences in the rates of protein synthesis can be shown using homogenates of eggs rather than whole cells. The extent of the shift in protein synthesis upon fertilization is variable. In sea urchins, the rise in protein synthesis is dramatic and rapid; a 5- to 30-fold increase in the rate of synthesis occurs within the first 5 to 10 minutes after insemination. This response in echinoderms has been generally considered to be a late event in the program of fertilization (Fig. 6–13). However, recent analyses using a quantitative computer model simulating the rise in protein synthesis of the echinoderm egg suggest that several of the processes influencing this activity are in fact initiated within zero to two minutes of fertilization (Raff et al., 1981). Protein synthesis, therefore, may be an earlier response of the fertilized egg than previously suspected. In the surf clam *(Spisula)*, the change in the rate of protein synthesis is considerably less, being on the order of a twofold to fourfold increase within 10 minutes of fertilization. By contrast, most studies with mouse eggs show little significant increase in overall protein synthesis upon insemination. That the newly synthesized proteins are vital to the early development of the embryo can be demonstrated by the use of inhibitors known to block the formation of proteins *(translational inhibitors)*. The incubation of eggs in such translational inhibitors as puromycin and cycloheximide shortly after fertilization quickly arrests further development.

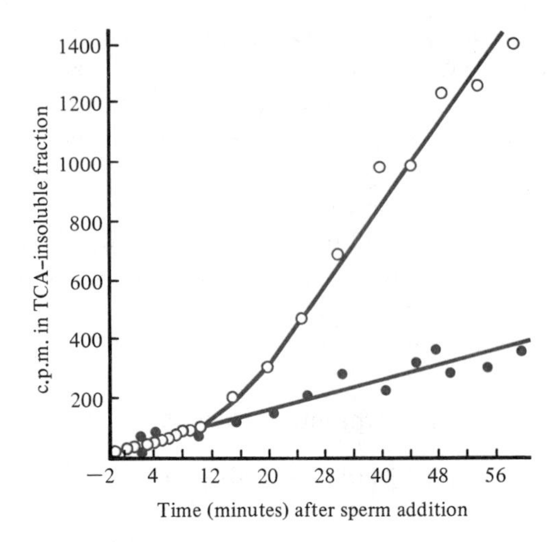


**FIG. 6–28.** Protein synthesis in unfertilized (dots) and fertilized (open circles) eggs of sea urchins as measured by the incorporation of [$^{14}$C]leucine. (From D. Epel, 1967. Proc. Natl. Acad. Sci. U.S.A. 57:899.)

What is the nature of the mechanism(s) that suddenly leads to the increase in protein synthesis at fertilization? Before examining in greater detail the question of protein synthesis at activation, it will be useful to review the basic steps in the production of a protein (Figs. 6–29, 11–1). Proteins are combinations of amino acids whose organizational arrangement is determined by the linear sequence of nucleotide bases of chromosomal DNA. The initial step, therefore, in protein synthesis is the transcription of DNA to form RNA. A special group of three enzymes, termed *DNA-dependent RNA polymerases*, mediate this process. RNA polymerase II is required for the transcription of structural genes, while RNA polymerase III functions in the production of transfer RNA and 5S RNA. The RNA polymerase molecule attaches to the portion of the DNA to be copied; as it moves along the DNA, the latter becomes uncoiled. In the presence of the RNA polymerase, the base sequences of DNA are copied as a linear sequence of ribonucleotides complementary to one strand of DNA. Hence the result of the synthesis of the RNA molecule is a chain in the form of a continuous sequence of triplets of nucleotide bases *(codons)*. As the RNA is transcribed, it becomes associated with protein to form a *ribonucleoprotein complex*, termed the *primary transcript*. The RNA chain has a distinct polarity that is important for the orientation of the molecule during translation. One end of the chain terminates by a 5′-carbon atom and is associated with a phosphate group (called the 5′ end), while the other end contains a 3′ -carbon and is associated with a hydroxyl group (called the 3′ end).

The primary transcript (*nuclear* or *heterogeneous RNA*) produced by transcription may undergo substantial modification before becoming a functional messenger RNA molecule. One change typically includes the addition of several adenylate residues to the 3′ hydroxyl end of the RNA, a process referred to as *polyadenylation*. Further details of posttranscriptional processing of the primary transcript are considered in Chapter 11 (see also Figure 11–1). The functional messenger RNA is then used in the

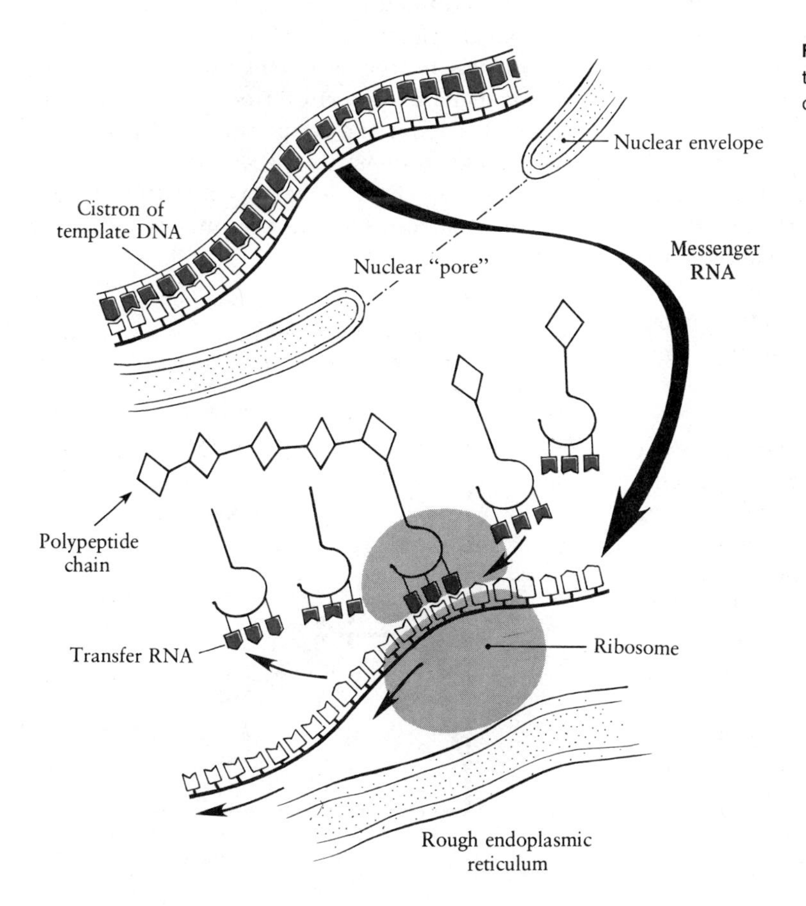


**FIG. 6–29.** Diagrammatic drawing showing the key steps in protein synthesis. (See text for details.)

cytoplasm to construct a protein during *translation*. Translation is generally recognized to consist of three steps: *initiation, elongation,* and *termination*. All polypeptides are synthesized beginning at the 5′ end and progressing to the 3′ end of the messenger molecule. A specific type of tRNA, termed methionyl $tRNA_1^{Met}$, initiates polypeptide synthesis. After combining with the small ribosomal subunit, this $tRNA_1^{Met}$ molecule recognizes an initiating codon (AUG) for methionine at the 5′ end of the mRNA. The large ribosomal subunit is then added to complete the ribosome and the *initiation complex*. The initiation complex begins translation by specifying the placement of the amino acid methionine at the beginning of the polypeptide chain. Amino acids are progressively added (elongation) to the chain in an order specified by the linear sequence of mRNA codons. The amino acids are coupled by peptide bonds. Each amino acid is transported to the ribosome by a specific tRNA molecule. One end of the tRNA molecule is occupied by a triplet sequence *(anticodon)* complementary to a specific codon in the mRNA molecule, while the other end bears the specific amino acid for which the mRNA triplet must code. The aminoacyl–tRNA complex is formed by a reaction involving ATP and activating enzymes known as *aminoacyl synthetases*. After an amino acid has been added to the growing polypeptide chain, its tRNA is released from the ribosome. Elongation of the polypeptide will continue until the ribosome encounters a *termination codon* at the 3′ end of the mRNA. The protein chain and the ribosome are then released into the cytoplasm. The process of polypeptide formation is usually carried out by numerous ribosomes attached to the same mRNA strand. The complex of mRNA and the ribosomes is termed a *polyribosome* or *polysome*.

It is apparent that there are several levels at which control over the sudden increase in protein synthesis upon fertilization might be exerted. For example, the increase could be due to an accelerated rate of transcription leading to the production of mRNA, or to the sudden availability of mRNA previously stored in the egg. There is now a large body of evidence from both classical embryology and more recently from biochemical studies to show that the primary mechanism triggering the rise in protein synthesis is the great increase in the availability of mRNA molecules for translation. The modulation of protein synthesis at the translational level has been experimentally shown for several species, including the sea urchin, surf clam, and mouse. It would be useful, perhaps, to mention several key experiments demonstrating that the elevated protein synthesis is not due to the production of mRNA made available through transcription. When Gross and his colleagues in the early 1960s preincubated *Arbacia* eggs in actinomycin D (a transcriptional inhibitor) and then added a sperm suspension, they observed that these same eggs exposed to a radioactive amino acid label rapidly synthesized protein for about three hours (Fig. 6–30). Similar *chemical enucleation* experiments with the same inhibitor have been carried out with amphibian and fish eggs; results clearly show that early protein synthesis is unaffected by transcriptional inhibitors. Also, when the unfertilized egg of the sea urchin is centrifuged under given experimental conditions, it can be divided into anucleate (without nucleus) and nucleate (with nucleus) halves. Neither half incorporates a radioactive amino acid into protein. If artificially activated, however, both egg halves, including the one without a nucleus, rapidly form protein. Puromycin effectively terminates all polypeptide formation in activated anucleate and nucleate egg halves.

Templates for the early embryonic proteins are mRNAs that are transcribed from the maternal nucleus and stored in the cytoplasm during oogenesis. These same mRNAs are either not translated or translated very inefficiently by unfertilized eggs. The maternal mRNAs are presumably coated or

**FIG. 6–30.** Protein synthesis (measured by the uptake of [$^{14}$C]valine) in fertilized eggs of *Arbacia* without (control) and with (experimental) actinomycin D. Note that actinomycin D does not inhibit protein synthesis immediately after fertilization. (From P. Gross et al., 1964. Proc. Natl. Acad. Sci. U.S.A. 51:407.)

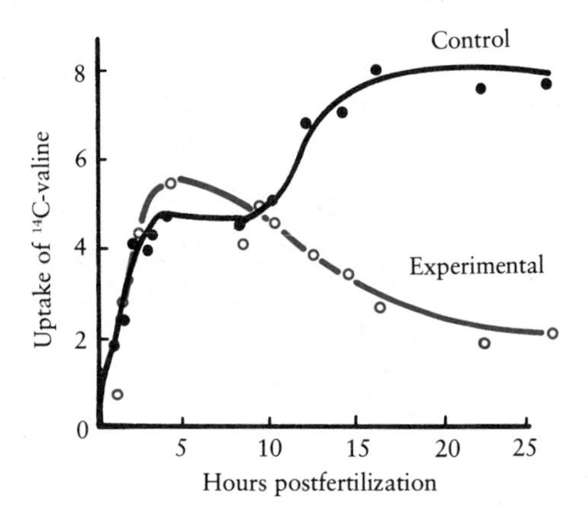

masked by phenol-soluble macromolecules and therefore exist as messenger ribonucleoprotein complexes (mRNPs). The "mask" is proposed to function as an inhibitor of translation. The mRNAs are then recruited into polysomes upon fertilization, and protein synthesis is increased. The maternal mRNAs become increasingly available for translation, apparently as a result of the "unmasking" of the mRNP complexes in the cytoplasm. The results of many studies with sea urchin eggs have supported the so-called *masked-messenger hypothesis*. The major data in support of this hypothesis come from studies showing that the in vitro translational activity of crude mRNA preparations is less than that of the RNA purified from the mRNPs themselves. However, these same studies have shown that protein synthesis in the sea urchin is very complex, and whether masking is the only mechanism of translational control in eggs remains to be determined. For example, Raff and coinvestigators have compared the rates of ribosome transit during protein synthesis in both fertilized and unfertilized eggs. They observe that the *translational efficiency* for protein synthesis in unfertilized eggs is 0.23 polypeptides completed per mRNA strand per minute, while in the zygote it increases to 0.58 molecules completed per mRNA strand per minute. Hence, the rise in protein synthesis could, at least in part, be explained by the twofold to threefold increase in translational efficiency of messages that are already active in the mature oocyte.

## Formation and Fusion of Pronuclei

The important structures of the spermatozoon that typically enter the egg cytoplasm are the nucleus, with its condensed chromatin material, and the centrosome (proximal and distal centrioles). In some animals, particularly the mammals, other constituents of the midpiece and/or the tail piece of the spermatozoon may also come to lie within the egg's cytoplasm. It is not known what role(s), if any, that the latter components of the spermatozoon play in the subsequent development of the egg cell. In mammals, electron microscope studies have shown that the mitochondria of the midpiece and the axial filaments of the tail piece undergo degeneration shortly after being incorporated into the egg's cytoplasm.

The nuclei of both male and female gametes undergo marked structural transformations after the incorporation of the spermatozoon into the egg's cytoplasm. These changes lead to the formation of the *male* and *female pronuclei*. The formation of the male pronucleus from the male nucleus begins with the breakdown of the sperm nuclear envelope and entails the reorganization of the sperm chromatin and the establishment of a pronuclear envelope. In the sea urchin, the spermatozoon, shortly after incorporation into the egg, rotates approximately 180 degrees and comes to lie lateral to its point of entrance. Concomitantly, the sperm nuclear envelope becomes vesicular and the chromatin material begins to disperse or decondense (Fig. 6–31 A). Subsequently, vesicles aggregate along the periphery of the dispersing chromatin and form the pronuclear envelope (Fig. 6–31 B). These structural changes occur very rapidly (within the first six minutes of insemination), leaving the male pronucleus considerably smaller than its counterpart, the female pronucleus (Fig. 6–32). Since meiosis is complete at the time of fertilization in echinoderms, the transformation of the egg nucleus into the female pronucleus occurs well in advance of the development of the male pronucleus. In organisms not having completed meiosis by the time of fertilization, development of the male pronucleus may or may not occur simultaneously with elaboration of the female pronucleus. For example, in the mussel, the dispersal of chromatin and the formation of the pronuclear envelope of the male pronucleus are completed within 18 minutes of insemination. This is well in advance of the completion of meiosis. By contrast, the fusion of vesicles along the margin of the dispersed chromatin material in the surf clam occurs at the completion of meiosis and at the onset of the development of the female pronucleus (approximately 40 minutes after insemination). Interestingly, both the female and male pronuclei are about of equal size. It has been suggested that the egg's cytoplasm contains some substance(s) that may regulate the synchronous development of male and female pronuclei in the surf clam.

Formation of the sperm *aster* accompanies the development of the male pronucleus. Generally, both the proximal and distal centrioles are incorporated into the egg. They tend to remain oriented at right angles to each other. The regions around each centriole gradually become packed with short, radiating microtubular fibers and endoplasmic vesicles of the smooth variety. The microtubules form

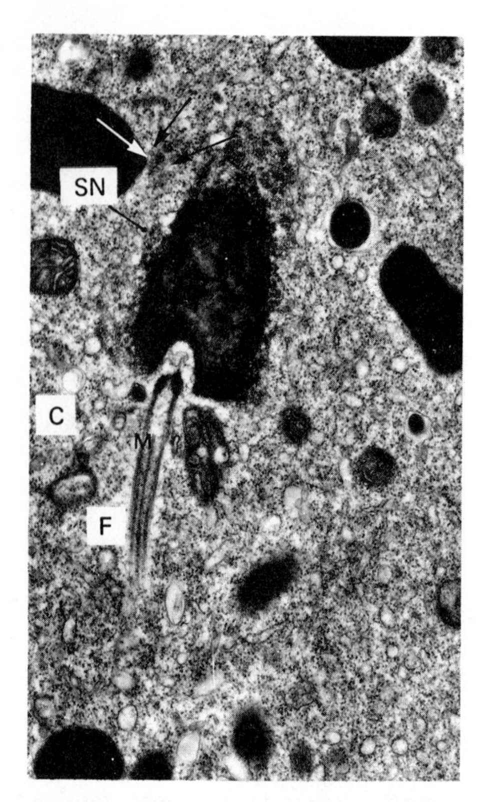


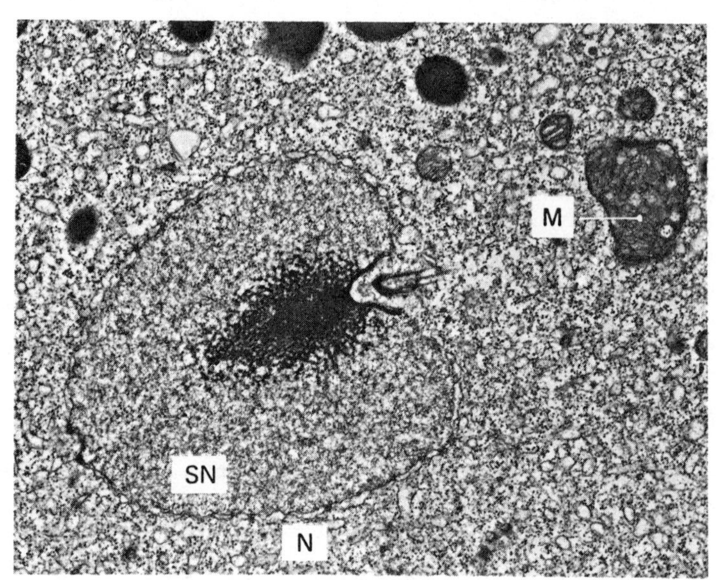


**FIG. 6–31.** A, the beginning stage of the dispersal of sperm chromatin (arrows) in the sea urchin; B, the heart-shaped male pronucleus delimited by its own pronuclear envelope in the sea urchin. C, centriole; F, remains of spermatozoon flagellum; M, remains of mitochondrial body of male gamete; N, nuclear membrane; SN, nucleus of spermatozoon. (From F. Longo and E. Anderson, 1968. J. Cell Biol. 39:339.)

the *astral rays* yielding the sunburst arrangement characteristic of the aster. The sperm aster develops from the radiating array of microtubules of the proximal centriole. As the sperm aster continues to enlarge, the distal centriole becomes dissociated from the male pronucleus. The precise role of the sperm aster in egg development is still largely an unresolved problem. In forms such as the sea urchin and the frog, the sperm aster eventually divides to form the poles of the first *cleavage amphiaster*. It is assumed that the egg centrosome loses its capacity for division during the process of nuclear maturation. The ability, then, of the fertilized egg to undergo division is dependent on the activities of the sperm aster. However, the fact that cleavage can follow parthenogenetic activation of eggs in several animal species supports the contention that elaboration of the first cleavage amphiaster is not always dependent on a sperm centriole. Zamboni maintains that in the mammal, no centrioles are present at either pole of the first cleavage spindle. Hence, the sperm centriole either degenerates or simply does not particpate in the first cleavage division.

As the male and female pronuclei undergo structural and physiological changes, they move toward each other in the cytoplasm. The mechanisms underlying the movement of the sperm and egg pronuclei through the cytoplasm are largely unknown. Recently, however, Schatten and his co-workers have used time-lapse video microscopy to trace the pronuclear movements in the sea urchin. This technique has shown that movements of the pronuclei can be grouped into three phases. Beginning at about 5 minutes after sperm–egg fusion, the sperm aster starts to move into the cytoplasm at a rather slow rate (about 2.5 $\mu$m/minute$^{-1}$) At approximately 8 minutes after fusion, the female pronucleus moves rapidly toward the sperm aster at a speed of about 15 $\mu$m per minute$^{-1}$; hence, the two pronuclei appear to approximate each other within an average time of about 70 seconds. Finally, the paired pronuclei migrate centripetally at a slow

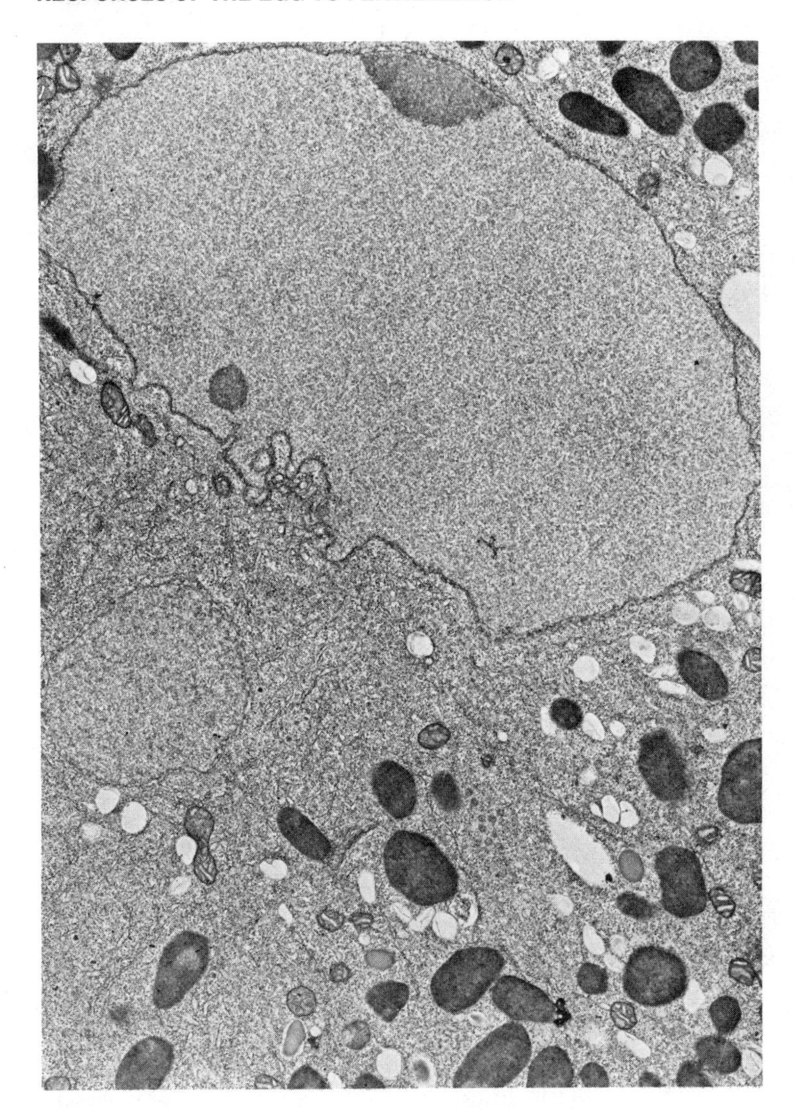

**FIG. 6–32.** The approach of the male and female pronuclei in the fertilized egg of *Arbacia*. Note the cytoplasmic projections of the female pronucleus. (From F. Longo and E. Anderson, 1968. J. Cell Biol. 39:339.)

speed (about 2 $\mu$m/minute$^{-1}$) and become located at the center of the cell. Microtubules are known to be involved in these movements. Inhibitors of microtubule assembly, such as *colcemid*, prevent fusion of the pronuclei. Using an antitubulin immunofluorescence probe, Schatten has carefully followed the pronuclear movements and determined that the microtubules of the sperm aster are capable of transmitting both pushing and pulling forces (Fig. 6–33).

The behavior of the pronuclei following their association in the center of the cell varies among animal species. In some species, particularly those in which meiosis has been completed prior to fertilization, the pronuclei actually fuse to form a diploid zygote nucleus. The pronuclei in other species only approximate each other; their pronuclear envelopes dissolve and the chromosomes immediately become aligned on a common chromosome plate. An example of the latter condition is found in mammals. The two pronuclei of the rabbit egg become tightly apposed to each other in the center of the cell. As they approach each other, numerous projections are elaborated on their proximal surfaces (Fig. 6–34). Some internuclear communication and exchange of material may result from contacts observed between adjacent projections of both pronuclei. Following the interdigitation of the two pronuclei, the chromatin material begins to condense. There follows breakdown and vesiculation of the pronuclear envelopes. The chromosomes of the two pronuclei then move together and assume

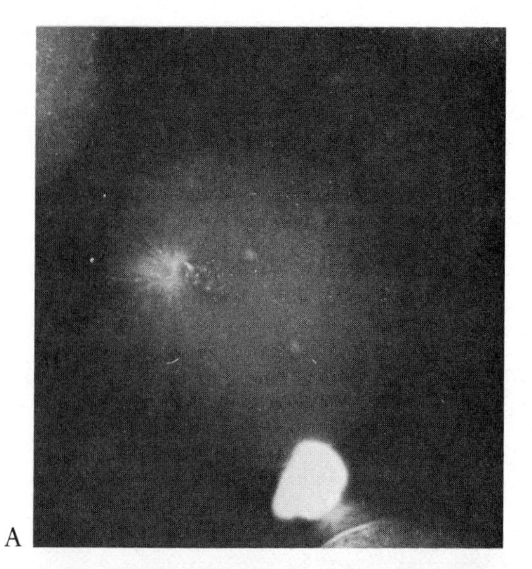


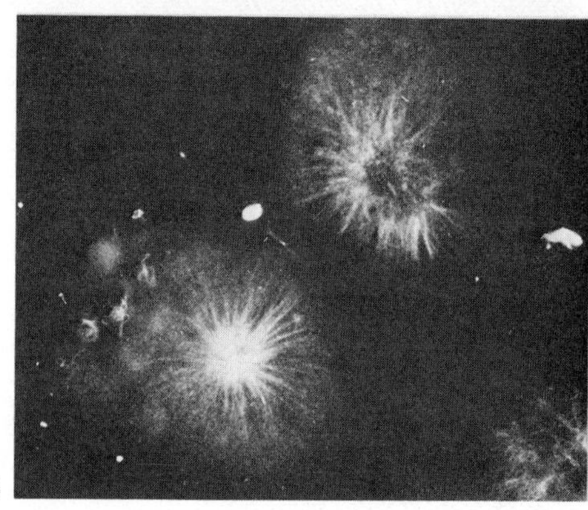


**FIG. 6–33.** Movements of the pronuclei following fertilization of the sea urchin egg have been analyzed using an antitubulin immunofluorescence technique. A, the micrograph shows the expansion of the sperm aster during the centripetal movement of both male and female pronuclei after they have been brought into apposition (about 20 minutes postinsemination); B, the sperm aster expands and confines the zygote nucleus in the center of the cell. (From T. Bestor and G. Schatten, 1981. Dev. Biol. 88:80.)

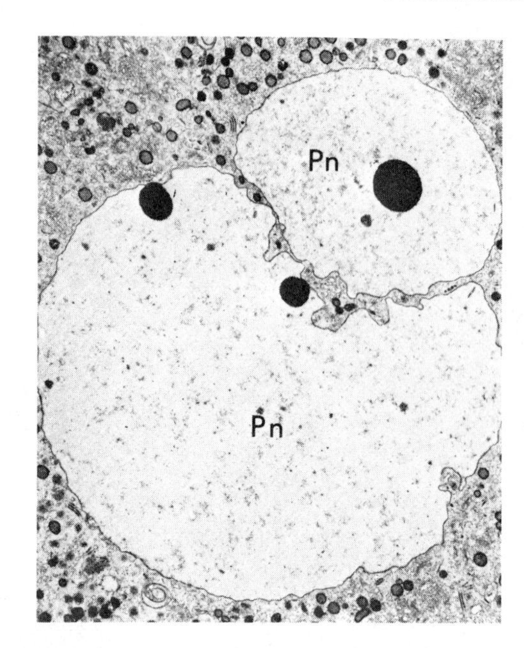


**FIG. 6–34.** The approach of the male and female pronuclei in the rabbit. Note the irregularities in the adjacent pronuclear membranes. Pn, pronucleus. (From B. Gondos et al., 1972. J. Cell Sci. 10:61.)

positions on the first cleavage spindle. No true zygote nucleus is produced. The stages of development of the fertilized rabbit egg between the approximation of pronuclei and the completion of the first cleavage division are shown in Figure 6–35. A similar pattern of pronuclear fusion is present in the rat, mouse, hamster, mussel, and surf clam.

In contrast to mammals, the envelopes of the pronuclei actually fuse in eggs of echinoderms (Fig. 6–36). The pronuclei contact each other at one site, fuse, and then form a narrow bridge of cytoplasm to unite the two bodies. Subsequently there is a distinct mixing of nucleoplasmic contents. Following formation of this zygote nucleus, the male chromatin diffuses throughout the nucleoplasm and becomes indistinguishable from that of the female.

There is also variability among animal organisms as to the time when DNA synthesis is initiated for the first cleavage division. DNA synthesis in some sea urchins, such as *Arbacia,* does not begin until after pronuclear fusion. However, DNA synthesis begins in sand dollars, amphibians, and mammals at the pronuclear stage of development.

## The Mechanism(s) of Egg Activation

The responses of the egg to fertilization presumably result from the fusion (or binding) of the plasma membranes of the sperm and the egg. How is this contact or fusion transduced to activate the development of the egg? Are the early and late phases of the program of fertilization initiated by the same mechanism or process? We will address these questions using data primarily gathered from studies with echinoderm eggs.

As already pointed out, the earliest detectable response to the fertilizing sperm is a change in plasma membrane potential, a shift important in

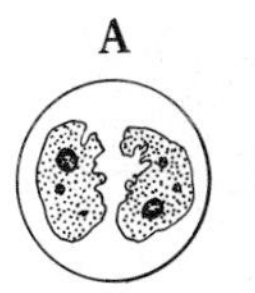

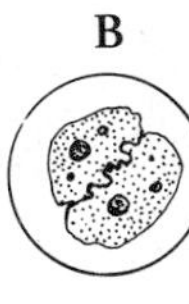

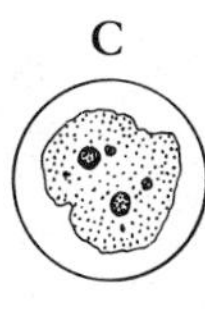

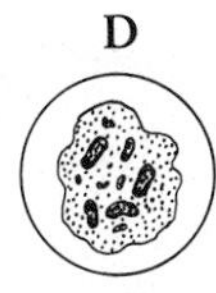

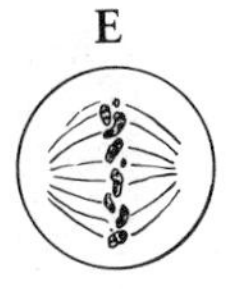

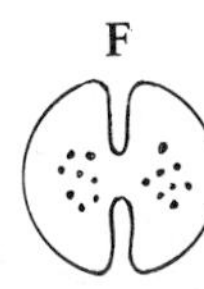


**FIG. 6–35.** Stages in the development of the fertilized rabbit egg between 18 and 22 hours after mating. A, approach of male and female pronuclei; B, close association of pronuclei with flattening of opposing surfaces; C, communication between nuclei; D, breakdown of pronuclear membranes; E, metaphase spindle of first cleavage; F, first cleavage division. (After B. Gondos et al., 1972. J. Cell Sci. 10:61.)

the fast block to polyspermy. The ionic basis of the generation of this membrane potential appears to be a small influx of sodium from the seawater. However, the change in membrane potential per se does not appear to trigger development. If the membrane potential of an egg is experimentally elevated, one cannot elicit early responses such as the cortical reaction. Also, the exocytosis of cortical granules can be induced in eggs bathed in sodium-free seawater.

An important clue to the understanding of the mechanism of egg activation was the observation by Mazia (1937) that the concentration of free calcium increased after fertilization. Since this observation, free cytosolic calcium in fertilized eggs has been directly measured and visualized with special image-intensifying techniques. The concept that free calcium might be related to egg activation therefore has had a long history. An important question, however, is whether the free calcium in the cytoplasm is a primary cause of or a result of the activation process. An approach that allows one to distinguish between these two alternatives has been to elevate the intracellular level of free calcium in eggs artifically using *ionophores*. Ionophores are antibiotic drugs that make biological membranes selectively permeable to certain ions. The *ionophore A23187* is specific for the transport of divalent ions such as $Ca^{++}$ and $Mg^{++}$, but particularly for the former ion. Unfertilized sea urchin eggs exposed to micromolar amounts of A23187 undergo cortical granule exocytosis and elevation of the fertilization membrane. The A23187 compound will also activate eggs bathed in a calcium-free medium, clearly suggesting that the primary source of the calcium is intracellular. The nature of the storage sites is not known, but the endoplasmic reticulum may sequester calcium. The activation of eggs using the A23187 ionophore has now been demonstrated for a wide variety of organisms, including starfish, teleost fishes, amphibians, and mammals. Calcium buffers injected into sea urchin eggs are also capable of triggering activation.

A direct and elegant demonstration of the increase in cytosolic calcium upon fertilization comes from experiments by Gilkey and co-workers using the large, transparent egg of the ricefish *(Oryzias)*. They injected *aequorin*, a protein isolated from jellyfish, into the eggs of *Oryzias*. Aequorin is a calcium-dependent luminescent protein that emits light in the presence of free $Ca^{++}$. When such eggs are fertilized, there is a 10,000-fold increase in luminescence when compared to unfertilized eggs (Fig. 6–37). The luminescence is primarily visible during the interval of the cortical reaction, beginning at the animal pole and sweeping around the egg as a band of light. The pulse of light is transient. Gilkey believes that the large pulse of free calcium

**FIG. 6–36.** Fusion of the male and female pronuclei in the echinoderm egg. (From F. Longo and E. Anderson, 1968. J. Cell Biol. 39:339.)

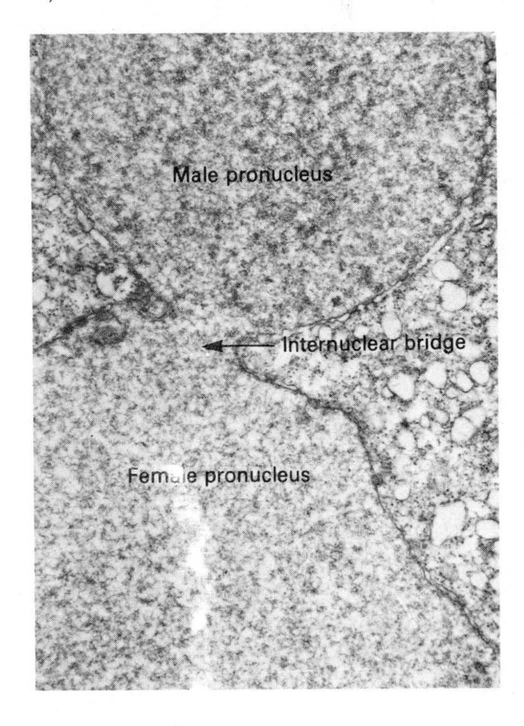

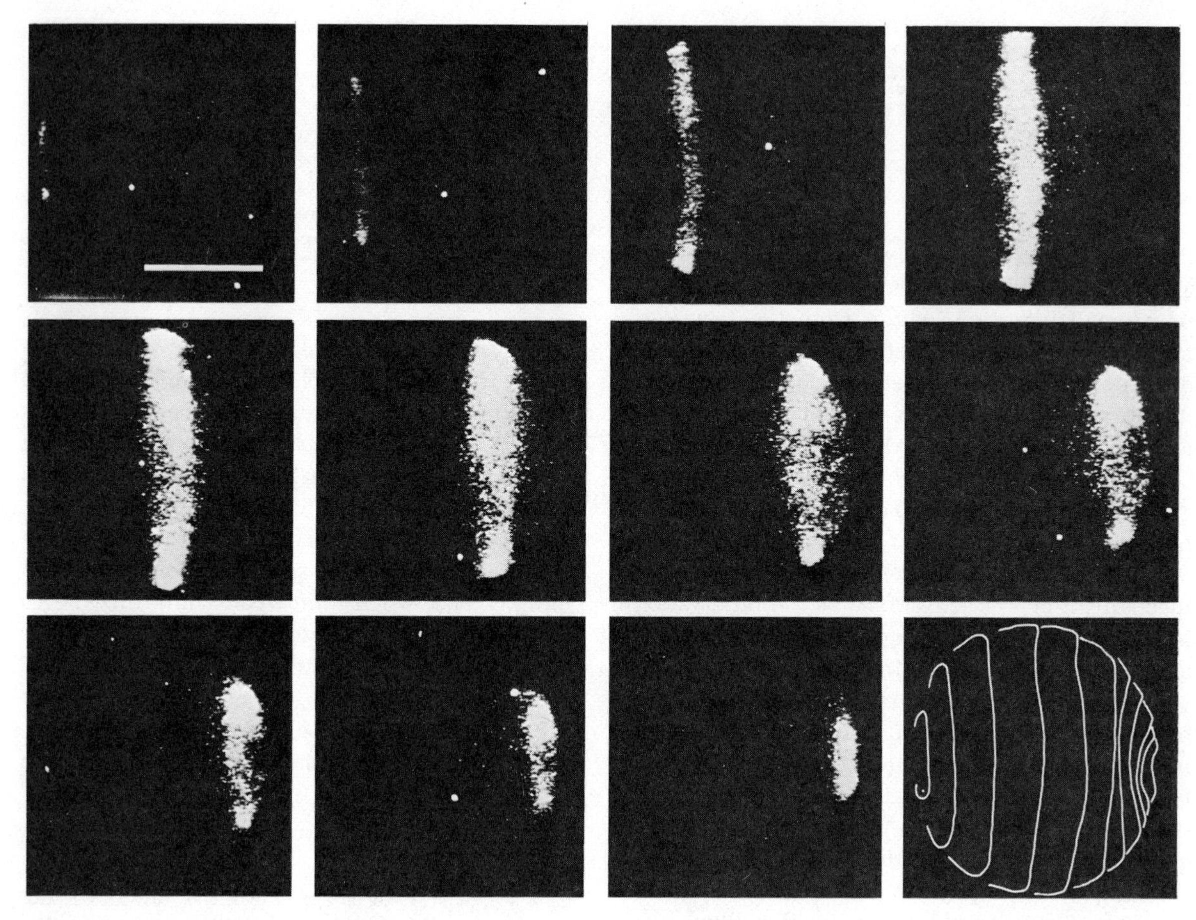

**FIG. 6-37.** A free calcium wave propagating across the cytoplasm of a sperm-activated medaka *(Oryzias)* egg. Successive photographs are 10 seconds apart. The last panel is a tracing of the leading edges of the 11 illustrated wavefronts. The micropyle of the fish egg is to the left. (From J. Gilkey et al., 1978. J. Cell Biol. 76:448.)

is propagated across the cytoplasm by a chain reaction involving a calcium-stimulated calcium release mechanism. The level triggering the calcium explosion has been calculated to be about 3 $\mu M$ $Ca^{++}$; the propagation rate of the calcium wave is about 10 $\mu$m/second. Similar experiments have been conducted with aequorin-loaded eggs of the sea urchin, *Lytechinus*. Direct calibration of the light emitted by the aequorin indicated a peak calcium level of about 5 $\mu M$.

Hence, a transient increase in intracellular free calcium triggers the development of the echinoderm egg. Consequences of this dramatic rise in calcium are cortical granule exocytosis and the elevation of the fertilization membrane. How the sperm–egg fusion is transduced into the wave of calcium release is still undetermined.

A key toward complete understanding of egg activation came with the demonstration that the late metabolic and biosynthetic responses of the egg took place quite independently of the earlier events. When eggs of the sea urchin are bathed in dilute solutions of either ammonium hydroxide or ammonium chloride (pH 9), late events such as potassium conductance and DNA synthesis are initiated, but these same eggs show neither cortical granule exocytosis nor fertilization membrane elevation. For the late responses to occur, there appears to be an obligatory sodium requirement during the first 5 to 10 minutes after fertilization. Metabolic activation of eggs is inhibited when eggs are exposed to sodium-free water between 1 and 5 minutes. When sodium is restored to the bathing medium, development of the eggs immediately resumes.

Between one and four minutes after fertilization,

sea urchin eggs release acid into the seawater. This *fertilization acid* is produced as a consequence of the efflux of $H^+$ (protons) from the egg cells, a step that is dependent on the presence of extracellular sodium. On the basis of their experiments, Johnson et al. (1976) indicate that the hydrogen flux is mediated through a sodium–proton exchange transport system across the plasma membrane; that is, sodium ions move into the cell and protons move out of the cell. That a sodium–$H^+$ exchange is involved in metabolic activation has been shown by fertilizing eggs in normal seawater and within the first minute transferring them to sodium-free seawater. Such eggs undergo cortical granule exocytosis, induced by the calcium pulse, but they do not release fertilization acid. If sodium is now added to the external medium, fertilization acid is discharged from the eggs; the rate of release of the acid depends on the concentration of the sodium (Fig. 6–38). Additionally, when fertilized eggs are treated with *amiloride,* a drug that inhibits the passive transport of sodium, acid release is markedly reduced. The efflux of $H^+$ raises the intracellular pH of egg homogenates by 0.3 pH units (Fig. 6–39).

A crucial question is now raised: Are the late metabolic responses triggered by the loss of sodium ions or by the rise in intracellular pH? Several experiments point to the importance of the increase in intracellular pH as the activating stimulus. First, activation of the biosynthetic activities can be prevented by reducing the extracellular pH, a manipulation that impedes acid efflux. Second, ammonia will initiate $H^+$ efflux in the absence of external sodium. The capacity of ammonia to substitute for sodium is shown in the data of Figure 6–40. When eggs are fertilized and transferred at one minute to sodium-free seawater containing either 50 m*M* sodium or 5 m*M* ammonia ($NH_4Cl$), they show cell division, although at a somewhat slower rate than control eggs bathed in regular seawater. Fertilized eggs transferred to sodium-free seawater do not cleave. These data are also interpreted to mean that the metabolic activities of the unfertilized egg are

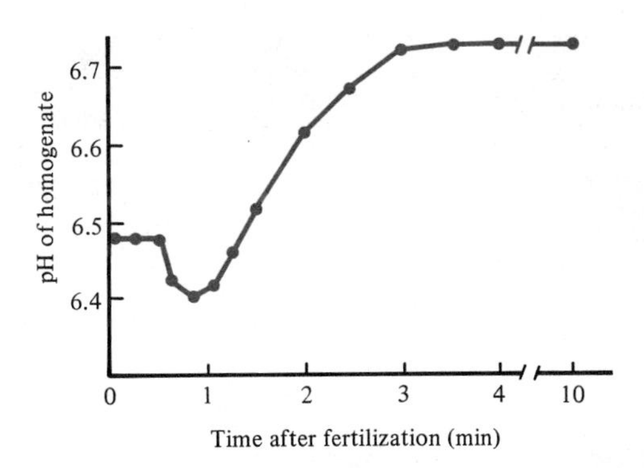


**FIG. 6–39.** The pH of egg homogenates of *Strongylocentrotus* at various times after fertilization. (From J. Johnson et al., 1976. Nature 262:661.)

**FIG. 6–38.** Acid efflux in eggs of sea urchins as a function of sodium concentration. The acid release was measured by the drop in pH of the seawater surrounding the eggs. (From J. Johnson et al., 1976. Nature 262:661.)

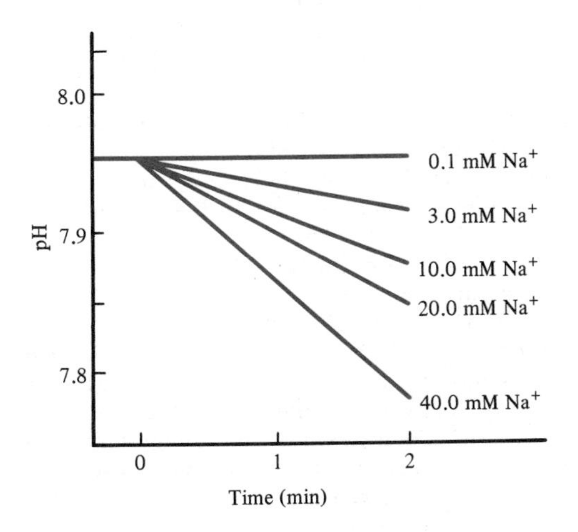


**FIG. 6–40.** Capacity of ammonia to substitute for sodium in activating the late metabolic responses of the echinoderm egg. The metabolic response being evaluated is cell division. Eggs were fertilized in seawater and, at 1 minute after fertilization, briefly washed in sodium-free seawater. They were then transferred to: seawater (A), seawater containing either 50 mM sodium (B) or 5 mM $NH_4Cl$ (C), and sodium-free seawater (D). Note that eggs under conditions A, B, and C undergo cleavage, while those under condition D do not. (From J. Johnson et al., 1976. Nature 262:661.)

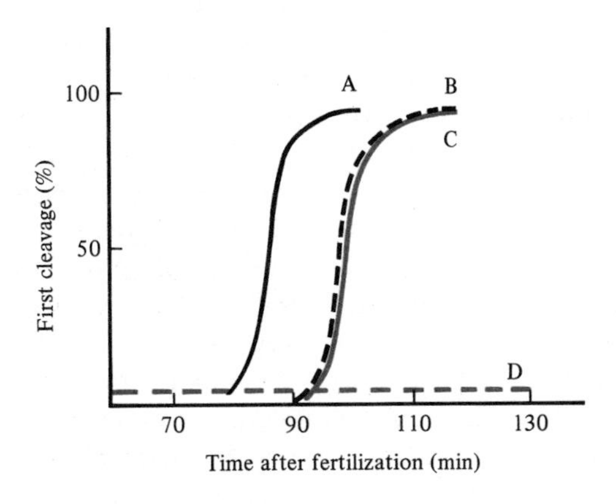

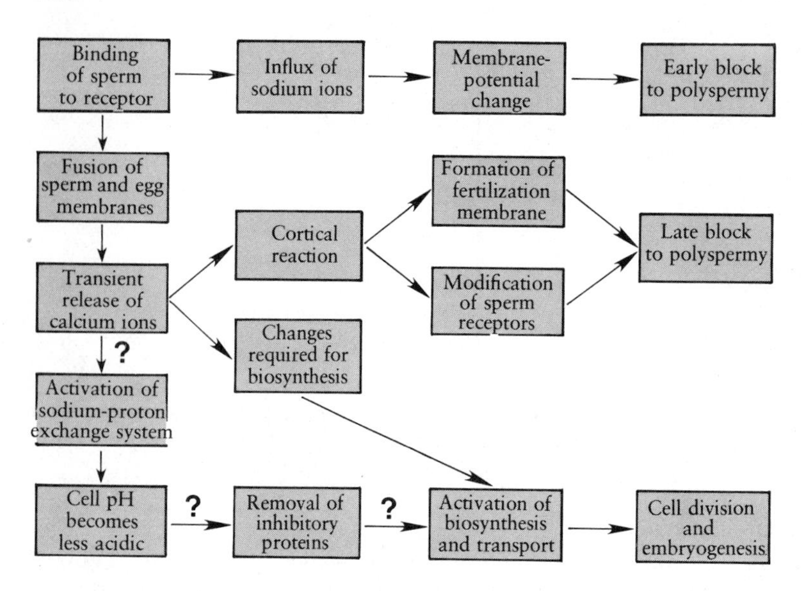


**FIG. 6–41.** A flow diagram summarizing the suggested relationships among the early calcium-dependent changes and the late pH-dependent changes in the fertilized egg of *Strongylocentrotus*. A question mark indicates that the relationship between the activities is not well established. (From D. Epel, 1977. Sci. Am. 237, 129.)

suppressed because the cell maintains a low pH. The mechanism that initiates the sodium–$H^+$ exchange transport system is not known. A glycoprotein, released from the egg surface at the time when the ion exchange begins, is thought to function as an inhibitor of the transport system.

In summary, the activation process in the echinoderm egg appears to involve two distinct steps, both of which are required to transform the unfertilized egg rapidly into a metabolically active cell. The first phase is calcium dependent; the second phase requires an increase in cytoplasmic pH (Fig. 6–41). Although the events of the program of fertilization have been extensively studied in the sea urchin, many of the causal relationships within as well as between these two phases of activation are still not clear. For example, does the early pulse of free cytosolic calcium play any role in triggering the sodium–proton exchange transport system? Does the release of bound calcium function in the activation of enzymes important in metabolism and biosynthesis of the egg? An equally important question is whether the mechanisms proposed for the activation of the sea urchin egg apply to eggs of other invertebrates and vertebrates. Calcium is an important regulator of a number of cellular processes, and it is not surprising that this ion has been shown to be critical to the activation of a variety of egg types. Jaffe (1983) has recently reviewed much of the literature on calcium and egg activation in a wide spectrum of organisms. He suggests that all deuterostome eggs utilize intracellular calcium as a source for activation, while protostome *(Urechis, Nereis)* eggs are activated by calcium entering the cytosol from the external medium, presumably in response to egg membrane depolarization. As for the late events, proton efflux has been reported to occur upon fertilization of the eggs of *Urechis*. Further research should provide additional insight into the extent to which the regulatory mechanisms proposed for the echinoderm egg are found in other organisms.

## References

Anderson, E. 1968. Oocyte differentiation in sea urchin, *Arbacia punctulata,* with particular reference to the origin of cortical granules and their participation in the cortical reaction. J. Cell Biol. 37:514–539.

Austin, C. R. 1960. Capacitation and the release of hyaluronidase from spermatozoa. J. Reprod. Fertil. 1:310–311.

Austin, C. R. 1968. Ultrastructure of Fertilization. New York: Holt, Rinehart and Winston.

Barros, C. and C. R. Austin. 1967. *In vitro* fertilization and the sperm acrosome reaction in the hamster. J. Exp. Zool. 11:317–323.

Barros, C. and R. Yanagimachi. 1971. Induction of zona reaction in golden hamster eggs by cortical granule material. Nature (Lond.) 233:268–269.

Bedford, J. M. 1968. Morphological aspects of sperm capacitation in mammals. In: Advances in Bioscience, 4. G. Raspe, ed., New York: Pergamon Press.

Bestor, T. and G. Schatten, 1981. Anti-tubulin immunofluorescence microscopy of microtubules present during the pronuclear movements of sea urchin fertilization. Dev. Biol 88:80–91.

Boldt, J., H. Schuel, R. Schuel, P. Dandekar, and W. Troll. 1981. Reaction of sperm with egg-derived hydrogen peroxide helps prevent polyspermy during fertilization in the sea urchin. Gamete Res. 4:365–377.

Colwin, A. L. and L. H. Colwin. 1964. Role of gamete membranes in fertilization. In: Cellular Membranes in Development. 22nd Symposium of the Society for the Study of Development and Growth. M. Locke, ed., New York: Academic Press.

Cross, N. 1981. Initiation of the activation potential by an increase in intracellular calcium in eggs of the frog, *Rana pipiens*. Dev. Biol. 85:380–384.

Cross, N. and R. Elinson, 1980. A fast block to polyspermy in frogs mediated by changes in the membrane potential. Dev. Biol. 75:187–198.

Dale, B., A. de Santis, and G. Ortolani. 1983. Electrical responses to fertilization in ascidian oocytes. Dev. Biol. 99:188–193.

Decker, S. and W. Kinsey. 1983. Characterization of cortical secretory vesicles from sea urchin eggs. Dev. Biol. 96:37–45.

Denny, P. and A. Tyler. 1964. Activation of protein biosynthesis in nonnucleate fragments of sea urchin eggs. Biochem. Biophys. Res. Commun. 14:245–259.

Elinson, R. and M. Manes, 1978. Morphology of the site of sperm entry in the frog egg. Dev. Biol. 63:67–75.

Endo, Y. 1961. Changes in the cortical layer of sea urchin eggs at fertilization as studied by the electron microscopy. I. *Clypeaster japonicus*. Exp. Cell Res. 25:383–397.

Epel, D. 1967. Protein synthesis in sea urchin eggs. A "late" response to fertilization. Proc. Natl. Acad. Sci. U.S.A. 57:899–906.

Epel, D. 1977. The program of fertilization. Sci. Am. 237:129–138.

Epel, D. 1978. Mechanisms of activation of sperm and egg during fertilization of sea urchin gametes. Curr. Top. Dev. Biol. 12:186–246.

Epel, D. 1979. Experimental analysis of the role of intracellular calcium in the activation of the sea urchin egg at fertilization. In: The Cell Surface: Mediator of Developmental Processes, pp. 169–185. S. Subtelny and N. Wessells, eds. New York: Academic Press.

Fankhauser, G. 1948. The organization of the amphibian egg during fertilization and cleavage. Ann. N.Y. Acad. Sci. 49:684–708.

Gilkey, J., L. Jaffe, E. Ridgway, and G. Reynolds. 1978. A free calcium wave traverses the activating egg of the medaka, *Oryzias latipes*. J. Cell Biol. 76:448–466.

Goldenberg, M. and R. Elinson. 1980. Animal/vegetal differences in cortical granule exocytosis during activation of the frog egg. Dev. Growth Differentiation 22:345–356.

Gondos, B., P. Bhiraleus, and L. Conner. 1972. Pronuclear membrane alterations during approximation of pronuclei and initiation of cleavage in the rabbit. J. Cell Sci. 10:61–78.

Grainger, J., M. Winkler, S. Shen, and R. Steinhardt. 1979. Intracellular pH controls protein synthesis rate in the sea urchin egg and embryo. Dev. Biol. 68:396–406.

Grey, R., M. Bastiani, D. Webb, and A. Schertel, 1982. An electrical block is required to prevent polyspermy in eggs fertilized by natural mating of *Xenopus laevis*. Dev. Biol. 89:475–484.

Gross, P., L. Malkin, and W. Moyer, 1964. Template for the first proteins of embryonic development. Proc. Natl. Acad. Sci. U.S.A. 51:407–414.

Hall, H. and V. Vacquier, 1982. The apical lamina of the sea urchin embryo: major glycoproteins associated with the hyaline layer. Dev. Biol. 89:168–178.

Hylander, B. and R. Summers, 1982. An ultrastructural immunocytochemical localization of hyalin in the sea urchin egg. Dev. Biol. 93:368–380.

Jaffe, L. 1976. Fast block to polyspermy in sea urchin eggs is electrically mediated. Nature (Lond.) 261:68–71.

Jaffe, L. 1983. Sources of calcium in egg activation: a review and hypothesis. Dev. Biol. 99:265–276.

Jaffe, L., A. Sharp, and D. Wolf. 1983. Absence of an electrical polyspermy block in the mouse. Dev. Biol. 96:317–323.

Jeffrey, W. and S. Meier. 1983. A yellow crescent cytoskeletal domain in ascidian eggs and its role in early development. Dev. Biol. 96:125–143.

Jenkins, N., J. Kaumeyer, E. Young, and R. Raff. 1978. A test for masked message: the template activity of messenger ribonucleoprotein particles isolated from sea urchin eggs. Dev. Biol. 63:279–298.

Johnson, J., D. Epel, and M. Paul. 1976. Intracellular pH and activation of sea urchin eggs after fertilization. Nature (Lond.) 262:661–664.

Kemp, N. and N. Istock. 1967. Cortical changes in growing oocytes and in fertilized or pricked eggs of *Rana pipiens*. J. Cell Biol. 34:111–121.

Lillie, F. R. 1919. Problems of Fertilization. Chicago: University of Chicago Press.

Longo, F. 1981. Morphological features of the surface of the sea urchin *(Arbacia punctulata)* egg: oolemma–cortical granule association. Dev. Biol. 84:173–182.

Longo, F. and E. Anderson. 1968. The fine structure of pronuclear development and fusion in the sea urchin, *Arbacia punctulata*. J. Cell. Biol. 39:339–368.

Longo, F. and E. Anderson. 1969. Cytological aspects of fertilization in the lamellibranch, *Mytilus edulis*. II. Development of the male pronucleus and the association of the maternally and paternally derived chromosomes. J. Exp. Zool. 172:97–120.

Longo, F. and E. Anderson. 1970. An ultrastructural analysis of fertilization in the surf clam, *Spisula solidissima*. II. Development of the male pronucleus and the association of maternally and paternally derived chromosomes. J. Ultrastruct. Res. 33:515–527.

McClay, D. and R. Finke, 1982. Sea urchin hyalin: appearance and function in development. Dev. Biol. 92:285–293.

Metafora, S., L. Fellicetti, and R. Gambino. 1971. The mechanism of protein synthesis activation after fertilization of sea urchin eggs. Proc. Natl. Acad. Sci. U.S.A. 68:600–604.

Metz, C. M. 1978. Sperm and egg receptors involved in fertilization. Curr. Top. Dev. Biol. 19:107–147.

Monroy, A. 1965. Chemistry and Physiology of Fertilization. New York: Holt, Rinehart and Winston.

Monroy, A., R. Maggio, and A. Rinaldi. 1965. Experimentally induced activation of the ribosomes of the unfertilized sea urchin egg. Proc. Natl. Acad. Sci. U.S.A. 54:107–111.

Moon, R., M. Danilchik, and M. Hille. 1982. An assessment of the masked message hypothesis: sea urchin egg messenger ribonucleoprotein complexes are efficient templates for *in vitro* protein synthesis. Dev. Biol. 93:389–403.

Moy, G., G. Kopf, C. Gache, and V. Vacquier. 1983. Calcium-mediated release of glucanase activity from cortical granules of sea urchin eggs. Dev. Biol. 100:267–274.

Nishioka, D. and N. McGwin. 1980. Relationships between the release of acid, the cortical reaction, and the increase of protein synthesis in sea urchin eggs. J. Exp. Zool. 212:215–223.

Ortolani, G., D. O'Dell, and A. Monroy. 1977. Localized binding of *Dolichos* lectin to the early ascidian embryo. Exp. Cell Res. 106:402–404.

Raff, R., J. Brandis, C. Huffman, A. Koch, and D. Leister, 1981. Protein synthesis as an early response to fertilization of the sea urchin egg: a model. Dev. Biol. 86:265–271.

Schuel, H. 1978. Secretory functions of egg cortical granules in fertilization and development: a critical review. Gamete Res. 1:299–382.

Schuel, H. and R. Schuel. 1981. A rapid sodium-dependent block to polyspermy in sea urchin eggs. Dev. Biol. 87:249–258.

Stambaugh, R. and M. Smith. 1974. Amino acid content of rabbit acrosomes; proteinase and its similarity to human trypsin. Science 186:745–746.

Tegner, M. and D. Epel. 1973. Sea urchin sperm–egg interactions studied with scanning electron microscopy. Science 179:685–688.

Vacquier, V. D. 1979. The interaction of sea urchin gametes during fertilization. Am. Zool. 19:839–849.

Vacquier, V. 1981. Dynamic changes in the egg cortex. Dev. Biol. 84:1–25.

Vacquier, V. and J. Payne. 1975. Methods for quantitating sea urchin sperm-egg binding. Exp. Cell Res. 82:227–235.

Whitaker, M. and R. Steinhardt. 1983. Evidence in support of the hypothesis of an electrically mediated fast block to polyspermy in sea urchin eggs. Dev. Biol. 95:244–248.

Wineck, T. J., R. F. Parrish, and K. F. Polakoski. 1979. Fertilization: a uterine glycosaminoglycan stimulates the conversion of sperm proacrosin to acrosin. Science 203:553–554.

Wolf, D. 1978. The block to polyspermy in zona-free mouse eggs. Dev. Biol. 64: 1–10.

Yu, S. and D. Wolf. 1981. Polyspermic eggs can dispose of supernumerary sperm. Dev. Biol. 87:203–210.

Zanewald, L. J. and W. L. Williams. 1970. A sperm enzyme that disperses the corona radiata and its inhibition by DF. Biol. Reprod. 2:363–368.

# 7

# CLEAVAGE AND BLASTULATION

Soon after fertilization the zygote becomes rapidly converted into a population of cells. The series of cell divisions by which this transformation is achieved is known as *cleavage* or *segmentation*. The cells formed by this process are termed *cleavage cells* or *blastomeres*. The period of cleavage is considered to extend from fertilization to the time of the formation of a multicellular embryo known as the *blastula*. Generally, there is little evidence of growth during this time, and the shape of the embryo does not visibly change.

Cleavage is a process that is easily observed in a variety of animal egg cells with the assistance of a microscope. The cleavages of the fertilized egg cell are typical animal cell divisions in that mitosis or *nuclear division* is immediately followed by *cytokinesis* or *cytoplasmic division*. However, there are several characteristic differences between the cell divisions of cleavage and those found in later stages of embryonic development and in dividing tissues of the adult animal. The consecutive divisions of cleaving cells are not separated by intervening periods of growth. Hence, the total cytoplasmic volume of the embryo during this period of development remains approximately constant. As a consequence, the blastomeres of the embryo become smaller and smaller until each reaches a cell size characteristic of the species. The absence of a distinct growth phase between successive cleavages also gradually changes the ratio of nuclear to cytoplasmic volume. Segmentation begins with a cell that has a low nucleus to cytoplasm volume ratio when compared to the cells of the adult organism. In the mature sea urchin egg, the volume ratio of nucleus to cytoplasm is 1:500. By the 64-celled stage, this same ratio is 1:12. At the end of cleavage, the proportion is approximately 1:6.

The rate or rhythm of cleavage is measured by the time interval between two consecutive divisions *(generation time.)* Eggs cleave at rates characteristic of the species, but this rate may be greatly influenced by temperature. Generation times at a given temperature are easily determined during early cleavage, because the first several divisions of the egg tend to occur simultaneously in all blastomeres. Zebra fish eggs cleave about every 15 minutes at 26°C. The early cleavages of the killifish *(Fundulus)* are separated by an interval of about 45 to 60 minutes. By contrast, there may be 10 to 12 hours at 37°C between cleavages in young mouse embryos. It is interesting to note that forms like fishes and frogs reach the blastula stage at a lower temperature more rapidly than typical mammals at a higher temperature. Several studies, particularly those involving hybridization experiments between echinoderm species with different generation times at cleavage, have indicated that the cytoplasm of the fertilized egg cell carries a factor that is important in determining the rate at which cleavages occur.

With the exception of mammalian embryos, the successive mitotic division of blastomeres is synchronous during the early stages of cleavage. Two cells produce four cells, four cells produce eight cells, and so forth. Coordination of this doubling of cells at each mitotic division is imperfectly understood, but recent studies suggest that the dividing cells may be responding at precisely the same time to some common signal mediated by intercellular pathways. Bennett and his colleagues have shown that mechanically dissociated blastomeres of *Fundulus* and *Ambystoma* (the Mexican axolotl) embryos will reassociate and become electrotonically coupled (i.e., current measurements by intracellular electrodes show that two cells behave as if connected by low resistance). The different structures that might mediate *electrotonic coupling* between the blastomeres include direct cytoplasmic continuity by bridges and gap junctions. As pointed out in Chapter 4, freeze–fracture techniques show a gap junction to be composed of a cluster of small channels on a membrane that allow the passage of ions and small molecules between cells (Fig. 7–1 A). Under the transmission electron microscope, this junction type exhibits a characteristic septilaminar organization (Fig. 7–1 B). Coupling at the very earliest stage of multicellular development may arise from cytoplasmic continuity resulting from

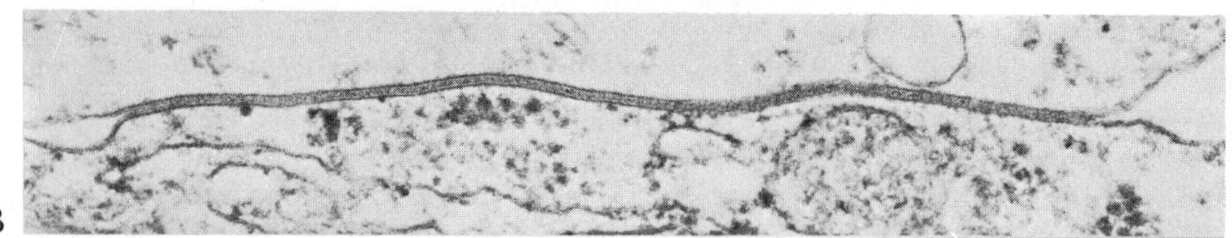


**FIG. 7–1.** Electron micrographs of a gap junction between axolotl blastomeres as seen in a freeze–fracture replica (A) and in an ultrathin section (B). The diameter of the junction is about 1.3 microns. Note the septilaminar appearance of the gap junction in B. (From R. Hanna et al., 1980. Am. J. Anat. 158:111.)

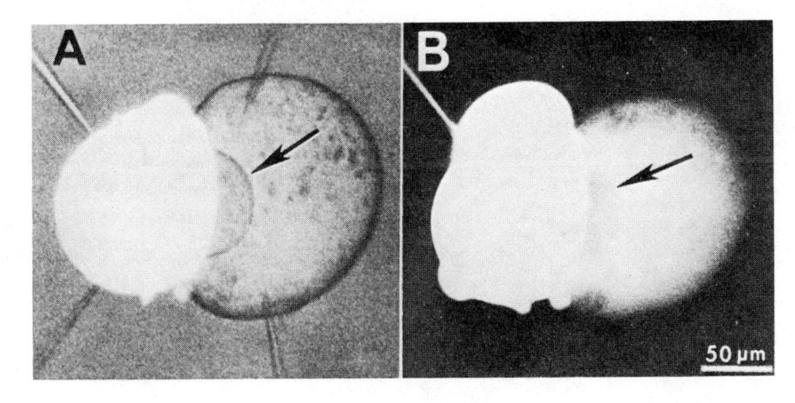


FIG. 7–2. Intercellular movement of dye and intercellular junctions in *Fundulus* blastomeres. A, mixed transmitted light and fluorescence micrograph of two coupled cells penetrated by four electrodes, one containing the fluorescent dye Lucifer yellow; B, fluorescence micrograph showing spread of Lucifer dye into second blastomere. (From M. Bennett et al., 1981. Am. Zool. 21:413.)

mitoses with incomplete cleavage. Four-cell mouse embryos appear to be coupled in this fashion. More typically, however, cells of cleaving embryos appear to be coupled by gap junctions. Do these junctions allow direct cell-to-cell transmission of small molecules? The answer to this question is probably positive. When the fluorescent tracer Lucifer yellow is microinjected into one of two coupled blastomeres of *Fundulus*, there is a gradual spread of the dye into the second blastomere (Fig. 7–2). Although the idea that gap junctions transmit some qualitative or quantitative message that triggers cell division at a given time is attractive, direct evidence for this function during cleavage remains to be forthcoming.

## Patterns of Cleavage

As pointed out in the chapter on oogenesis, animal egg cells not only show a range in size variation, but distinct differences in the amount and distribution of yolk or yolky cytoplasm. Indeed, from species to species the differences in the relative amounts of yolk and active cytoplasm influence the pattern by which the egg cell is progressively subdivided into daughter cells. The correlation between cleavage pattern and yolk was recognized by G. W. Balfour (1894), who observed that the rate at which the cleavage furrow passed through the egg was inversely proportional to the amount of yolk within the cell. Both mitosis and cytokinesis are progressively impeded with increasing amounts of yolky cytoplasm.

The egg cells of many annelids, molluscs, nematodes, echinoderms, cephalochordates, amphibians, and mammals contain small to moderate amounts of yolk. The volume of such an egg cell is always divided into complete daughter cells, a pattern of cleavage that is termed *holoblastic*. In reptiles, birds, and elasmobranch and teleost fishes, the egg cell possesses a larger amount of yolk heavily concentrated at one pole. Only the active cytoplasm with the nucleus, never the yolk, undergoes cleavage. Such a pattern of cleavage is termed *meroblastic* or *discoidal*.

It is advisable at this time, as a basis for the understanding of later events of development, to examine in detail selected animals that illustrate these cleavage patterns.

### Cephalochordate—Amphioxus

The egg cell of *Amphioxus* at the time of fertilization is approximately 120 microns in diameter. The first cleavage appears as a furrow that completely encircles the cell some 60 to 90 minutes after insemination. The furrow subsequently divides the egg into two equal blastomeres along a meridional plane that extends from animal to vegetal pole (Fig. 7–3 A,B). This plane along which the egg is cleaved represents the axis of bilateral symmetry of the adult organism. The second cleavage occurs 45 minutes later and is also meridional, but at right angles to the first furrow (Fig. 7–3 C). The two blastomeres are divided into four blastomeres. The third cleavage is horizontal, at right angles to the first two cleavage furrows, and slightly above the equators of the four blastomeres. Upon completion of the furrowing process, eight blastomeres are produced. Although the upper quartet of blastomeres is only slightly smaller than the lower quartet, the former are often called *micromeres* and the latter *macromeres* (Fig. 7–3 D). The fourth cleavage is meridional (Fig. 7–3 E). The fifth cleavage is horizon-

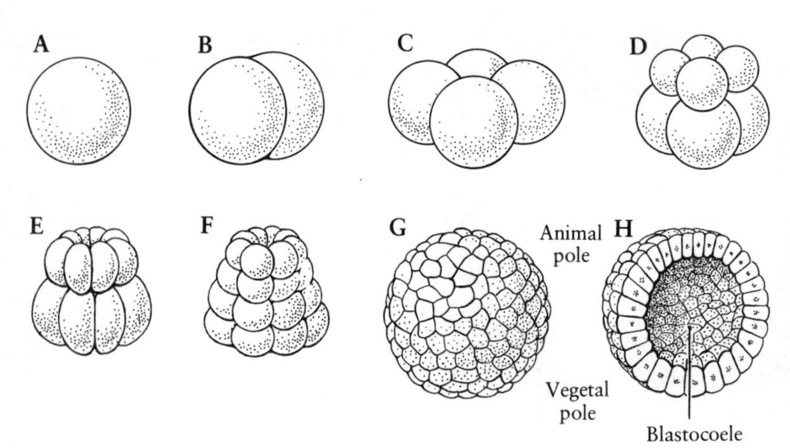


FIG. 7–3. Cleavage and blastulation in the egg of *Amphioxus.* A, fertilized egg; B, 2-celled stage; C, 4-celled stage; D, 8-celled stage; E, 16-celled stage; F, 32-celled stage; G, blastula, surface view; H, blastula, hemisection. (From A. F. Huettner, 1972. Fundamentals of Comparative Embryology of the Vertebrates. Copyright 1949 by Macmillan Publishing Co., Inc., renewed by M. R. Huettner, R. A. Huettner, and R. J. Huettner.)

tal, dividing the eight micromeres and eight macromeres simultaneously into a total of 32 blastomeres (Fig. 7–3 F). During succeeding cleavages, the larger macromeres in the vegetal hemisphere tend to divide more slowly than the smaller micromeres in the animal hemisphere. By the end of the period of cleavage, the cephalochordate embryo is organized as a hollow sphere or blastula whose cells enclose a fluid-filled cavity termed the *blastocoele* (Fig. 7–3 H). The 200 or so blastomeres are structured as a simple epithelium in which the larger cells are located in the vegetal hemisphere and the smaller cells in the animal hemispehre (Fig. 7–3 H). This pattern of cleavage is often referred to as *equal holoblastic*.

## Echinoderm—Sea Urchin

The fertilized egg of the sea urchin is approximately 90 to 100 microns in diameter and roughly sperhical. The first cleavage generally occurs within 60 to 90 minutes of fertilization (Fig. 7–4 A,B). The furrow is meridional and cuts the egg cell into two equal-sized blastomeres. The second cleavage is meridional, but at right angles to the first. The third pair of cleavages is visible about three hours after fertilization. Each is horizontal and results in a cluster of four animal (upper) and four vegetal (lower) blatomeres (Fig. 7–4 D). The cleavages of the fourth division are quite unusual. The four blastomeres of the animal hemisphere divide meridionally, but the four blastomeres of the vegetal hemisphere divide horizontally. The result is a 16-celled embryo with the following layers: an upper zone of eight medium-sized blastomeres *(mesomeres)*, a middle zone of four very large blastomeres (macromeres), and a bottom zone of four very small blastomeres (micromeres) (Fig. 7–4 E). This set of cleavages provides the basis for our understanding that different regions of the echinoderm egg have unique properties that influence the early expression of cell types (Chapter 9). In two subsequent cleavages, first the mesomeres and then the macromeres divide equatorially to form an embryo of 64 cells. A cavity, termed the blastocoele, appears very early in cleavage between the blasto-

FIG. 7–4. Cleavage and blastulation in the egg of the sea urchin, *Paracentrotus.* A, fertilized egg; B, 2-celled stage; C, 4-celled stage; D, 8-celled stage; E, 16-celled stage; F, 32-celled stage; G, blastula, surface view; H, blastula, hemisection. Mes = mesomere; Mac = macromere; Mic = micromere.

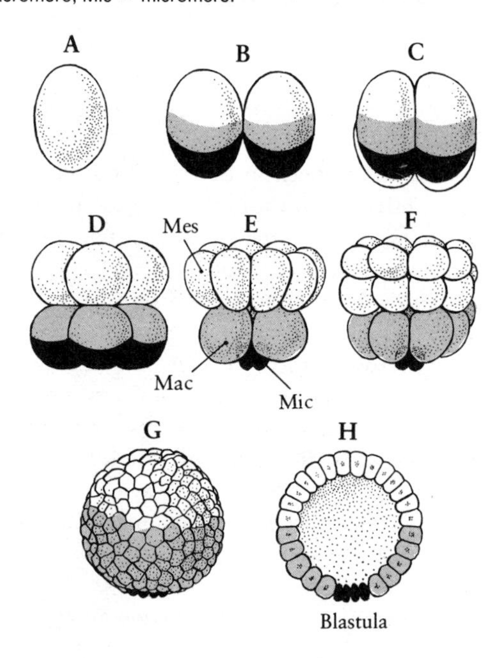

meres. It continues to enlarge as the blastomeres increase in number and decrease in size.

## Amphibia—Frog or Salamander

A good example showing the effect of the amount and distribution of yolk on the pattern of segmentation is seen in the amphibian egg. Here the yolk is distributed along the animal–vegetal axis with the greatest concentration at the vegetal pole. Cleavage is initiated at the animal pole some two to three hours (temperature and species dependent) after fertilization. The first two furrows are meridional and at right angles to each other, producing four cells of approximately equal size (Fig. 7–5 A–C). These furrows do not appear simultaneously around the circumference of the egg, as in echinoderms; each starts at the animal pole and gradually extends toward the vegetal pole, cutting deeper into the egg as it does so until two cells are formed. The first furrow generally passes through the grey crescent so that each of the resulting blastomeres contains a portion of this cytoplasm. Because of the unequal distribution of the yolk within the egg, the third pair of cleavages is displaced markedly toward the animal pole (Fig. 7–5 D). The larger blastomeres or macromeres produced by this division are confined to the vegetal hemisphere and the smaller blastomeres or micromeres to the animal hemisphere of the embryo. After about the fourth or fifth cleavages, the regularity of the segmentation process is lost and blastomeres tend to divide at different rates. Blastomeres containing yolk-laden platelets tend to divide more slowly than those with less yolk. For a short period of time, the blastomeres are packed together and the embryo takes on the appearance of a cluster of mulberries. Such a configuration is referred to as a *morula stage*. Narrow spaces appear between the blastomeres. These gradually coalesce into a single cavity or blastocoele. As with *Amphioxus* and the sea urchin, there emerges with continued cleavages a hollow, spherically shaped ball of cells, the blastula (Fig. 7–5 G). A hemisection through the late blastula shows the blastocoele eccentrically displaced toward the animal pole (Fig. 7–5 H). The cells of the animal hemisphere are small and organized into two or more layers; they form the roof of the blastocoele. The cells of the vegetal hemisphere are large, yolk laden, and form the floor of the blastocoele. The pattern of cleavage in amphibians is often termed *unequal holoblastic*, a phrase that recognizes the no-

**FIG. 7–5.** Cleavage and blastulation in the egg of a typical amphibian. A, fertilized egg; B, beginning of 2-celled stage; C, 4-celled stage; D, 8-celled stage; E, 16-celled stage; F, 32-celled stage, G, blastula, surface view; H, blastula, hemisection. (From A. F. Huettner, 1972. Fundamentals of Comparative Embryology of the Vertebrates. Copyright 1949 by Macmillan Publishing Co., Inc., renewed by M. R. Huettner, R. A. Huettner, and R. J. Huettner.)

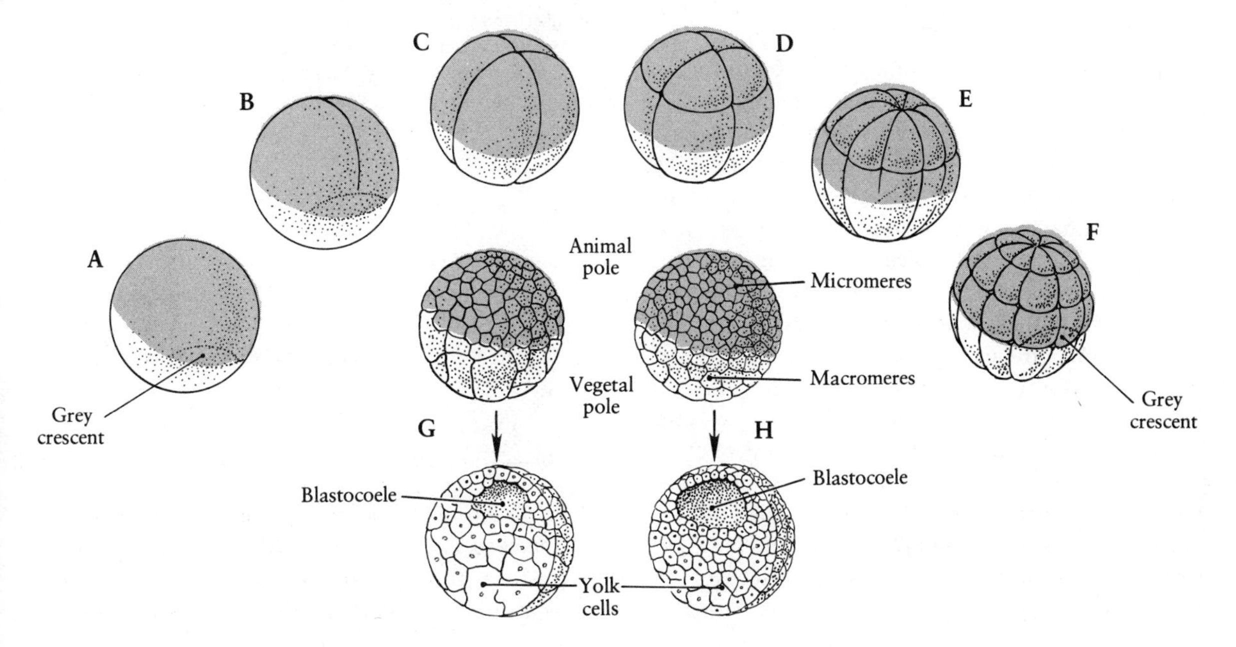

ticeable contrast in the size of the blastomeres of the blastula.

## Osteichthyes—Bony Fishes

The egg cells of bony fishes have more yolk than those of the amphibians, and consequently the active cytoplasm along with the nucleus tends to be located at the animal pole. Representative species within this group of vertebrates show a range in cleavage pattern from incomplete holoblastic to meroblastic. In a primitive bony fish such as *Amia* (the bowfin), meridional cleavages start at the animal pole, but they are greatly retarded in their efforts to reach the vegetal pole because of the large yolk mass (Fig. 7–6 A). Indeed, subsequent divisions are initiated in the animal pole well before preceding furrows cut through the yolk toward the vegetal pole. As in the frog and the salamander, however, the whole egg is divided into daughter blastomeres.

More typical in higher bony fishes (as well as in elasmobranch fishes) is the cleavage pattern illustrated in Figure 7–6 B for the zebrafish *(Brachydanio)*. Within 25 minutes of fertilization, the cytoplasm of the egg cell is entirely segregated from the yolk and appears as a distinct cap or *blastodisc* at the animal pole (Figs. 7–6 B,1; 7–7 A). Only the blastodisc becomes cellularized during cleavage. Initially, the cleavages are meridional or vertical, and all of the resultant blastomeres lie in the same plane with their lower surfaces formed by the yolk below (Figs. 7–6 B,2–4; 7–7 B,C). The sixth cleavage is horizontal and yields an upper set of blastomeres completely separated from neighboring blastomeres, while the lower set of blastomeres retains a connection with the yolk mass. Continued divisions in the horizontal plane add more cells with completed boundaries. Within four hours of fertilization in the zebrafish, the blastula stage is reached. The compact mass of small blastomeres elevated above the yolk mass is termed the *blastoderm* (Fig. 7–6 B,G). A hemisection through the blastoderm shows the presence of a blastocoele (Fig. 7–6 B, 7). Although the blastocoele is largely lined by distinct blastomeres, its floor is formed by a syncytial cytoplasm in which numerous nuclei can be detected. This layer, known as the *yolk syncytial layer* or *periblast*, is formed as follows. At the end of cleavage, the lowermost blastomeres, produced as a result of horizontal divisions, are still continuous with the

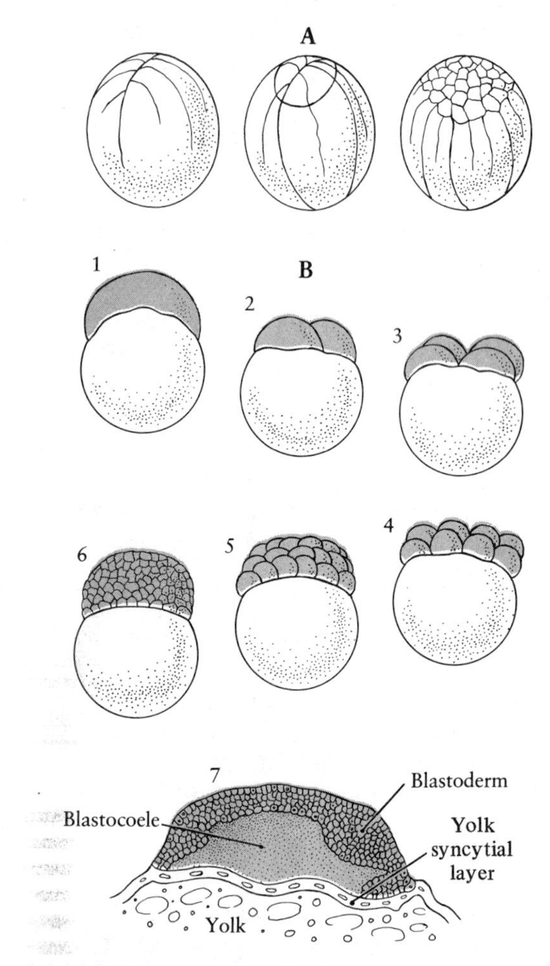


**FIG. 7–6.** Patterns of cleavage in bony fishes. A, holoblastic cleavage in a primitive bony fish such as *Amia*. (From E. Korschelt, 1936. Vergleichende Entwicklungsgeschichte der Tiere. G. Fisher, Jena); B, Meroblastic cleavage in advanced bony fish such as *Brachydanio*. (1) egg 25 minutes after fertilization; (2) two-celled stage; (3) four-celled stage; (4) eight-celled stage; (5) late cleavage; (6) blastula, lateral view; (7) blastula, hemisection. (After K. Hisaoka and H. Battle, 1958. J. Morphol. 102:311.)

yolk mass (i.e., they lack lower boundaries). The lateral boundaries between these blastomeres disappear and their cytoplasms and nuclei join together to form a distinct layer on top of the yolk mass.

## Aves—The Bird

Segregation of the active cytoplasm from the yolk in the bird is completed by the end of oogenesis. Therefore, the whitish blastodisc, approximately three millimeters in diameter, is already organized

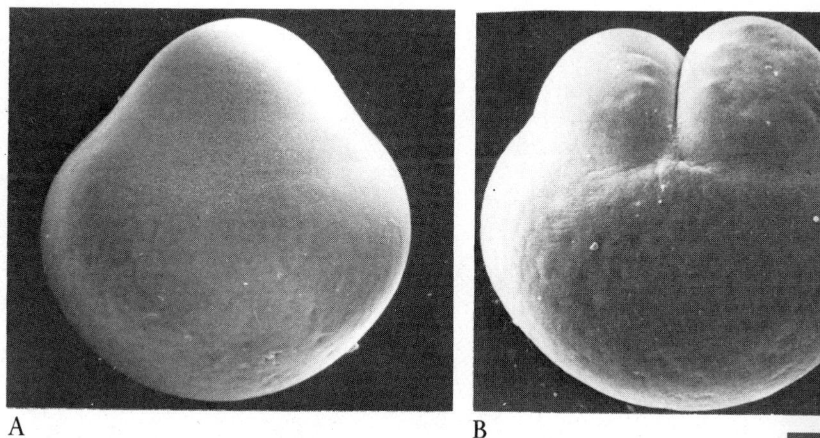


at the time of fertilization and clearly marks the animal pole of the cell. With successful fertilization of the ovum in the oviduct, the first cleavage furrow appears some three to five hours after ovulation. The first cleavage furrow is considered to be meridional; it bisects the blastodisc into two partially separated blastomeres (Fig. 7–8 A,G). The second cleavage consists of two meridional furrows, each one of which is at a right angle to the first cleavage furrow (Fig. 7–8 B). The third set of cleavages are also vertical and tend to be variable in position, but typically are parallel to the plane of the first cleavage furrow. The fourth set of cleavages are considered to be vertical and occur in such a fashion that eight *central cells* are separated from eight *marginal cells* (Fig. 7–8 D). The centrally located cells have upper and lateral surfaces, but, lacking a lower surface, are continuous with the yolk below (Fig. 7–8 H). The marginal cells are incomplete peripherally since the furrows do not extend to the margins of the blastodisc; they lack a lower surface and therefore possess only a pair of lateral surfaces (Fig. 7–8 D,H). From this point onward, the succession of cleavages becomes asynchronous and irregular. Three types of furrows are evident: (1) vertical furrows that extend peripherad toward the margin of the blastodisc; (2) vertical furrows that cut across the inner ends of radiating furrows, thereby producing peripheral boundaries to marginal cells; these blastomeres then become part of the centrally located cells; (3) horizontal furrows that occur below the surface (and parallel to it) and establish lower boundaries to the cells. Horizontal cleavages begin sometime after the 32-celled stage.

By the time the embryo or blastoderm consists of 60 to 100 cells (Fig. 7–8 I), it is organized as a mass

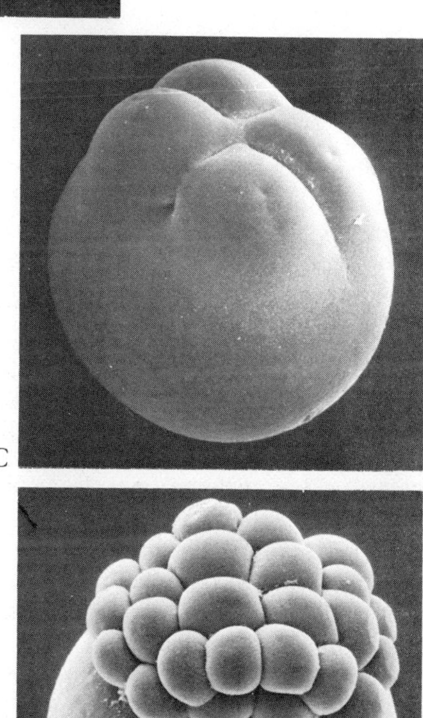


**FIG. 7–7.** Scanning electron micrographs of the blastodisc stage (A), 2-celled stage (B), 4-celled stage (C), and 32-celled stage (D) in *Brachydanio.* (From H. Beams and R. Kessel, 1976. Am. Sci. 64, 279. Reprinted by permission of American Scientist, Journal of Sigma Xi, The Scientific Research Society.)

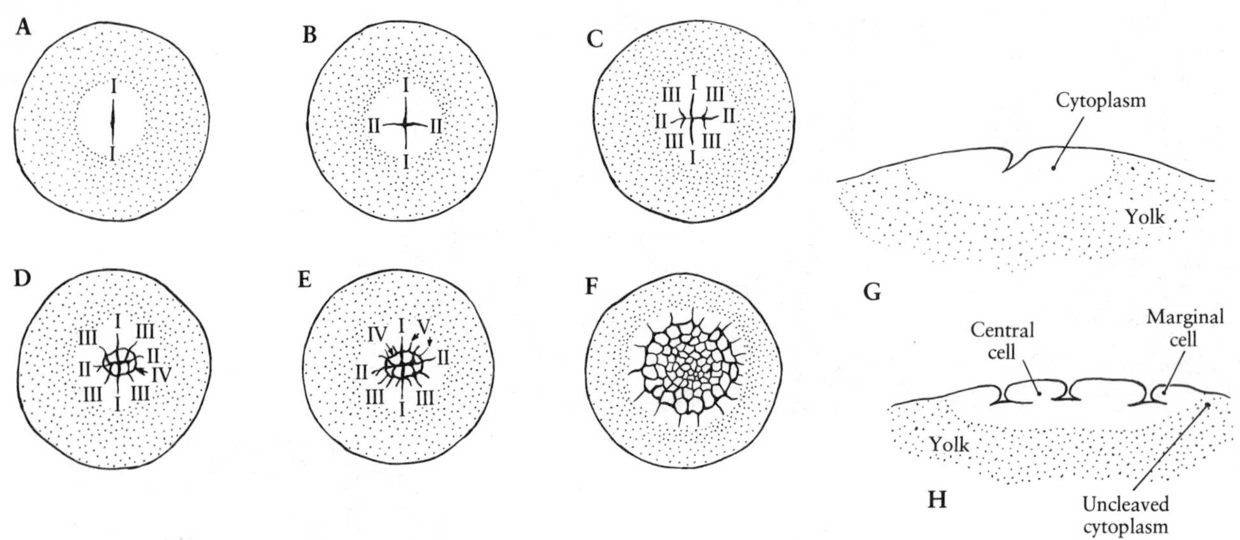


of centrally located cells, actively dividing, that lie over a fluid-filled cavity. Beyond the centrally located cells, vertical cleavages continue to add more central cells, thereby advancing peripherally the segmentation of the blastodisc. Horizontal cleavages progress centrifugally and establish lower boundaries for the more peripherally located cells. Some of the nuclei formed by these horizontal divisions have been observed to wander yolkward and peripherally into uncleaved portions of the cytoplasm. This nucleated cytoplasm becomes evident below the blastoderm. It forms the periblast. Eventually, the periblast beneath the margins of the blastoderm becomes cellularized and further adds to the expanding population of blastomeres with completed boundaries.

At the end of cleavage, the chick embryo appears as a discoidal cap of cells atop the yolk (Fig. 7–8 J). This blastoderm is five to six cells thick in the center but only one to two cells deep at the periphery. With transmitted light under the microscope, the blastoderm can be divided into two recognizable regions. There is a central region, or *area pellucida;* it appears translucent because its blastomeres are separated from the yolk by a fluid-filled cavity. By contrast, the peripheral part of the blastoderm lies against the yolk, thus rendering it more opaque (*area opaca*). Only the cells of the area pellucida will contribute to the construction of the embryo.

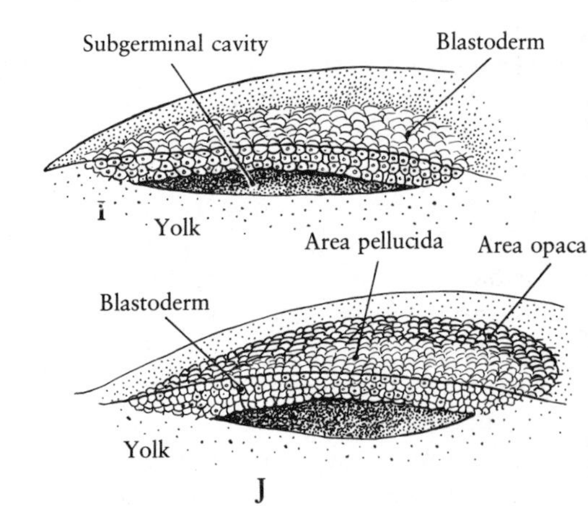


**FIG. 7–8.** Cleavage and blastulation in the egg of the bird. A, 2-celled stage, polar view; B, 4-celled stage, polar view; C, 8-celled stage, polar view; D, 16-celled stage, polar view; E, 32-celled stage, polar view; F, early morula, polar view; G, 2-celled stage, hemisection; H, 8-celled stage, hemisection; I, early blastula, hemisection; J, advanced blastula, hemisection. I–V, temporal sequence of cleavage furrows. (A–F, from Foundations of Embryology by B. Patten and B. Carlson. Copyright © 1974 by McGraw-Hill Inc. Used with permission of McGraw-Hill Book Company; H–J, after A. F. Heuttner, 1972. Fundamentals of Comparative Embryology of the Vertebrates. Copyright 1949 by Macmillan Publishing Co., Inc., renewed by M. R. Huettner, R. A. Huettner, and R. J. Huettner.)

## Mammals—The Pig and Monkey

With the exception of the eggs of primitive mammals (i.e., duckbill platypus and spiny anteater), which have a large amount of yolk and divide in meroblastic fashion, the ova of marsupial and placental mammals are exceedingly small. Their size reflects the scanty amount of yolk that is manufactured and stored in the cell during oogenesis. The ovum is only about 70 microns in diameter in the mouse and approximately 120 microns in the dog.

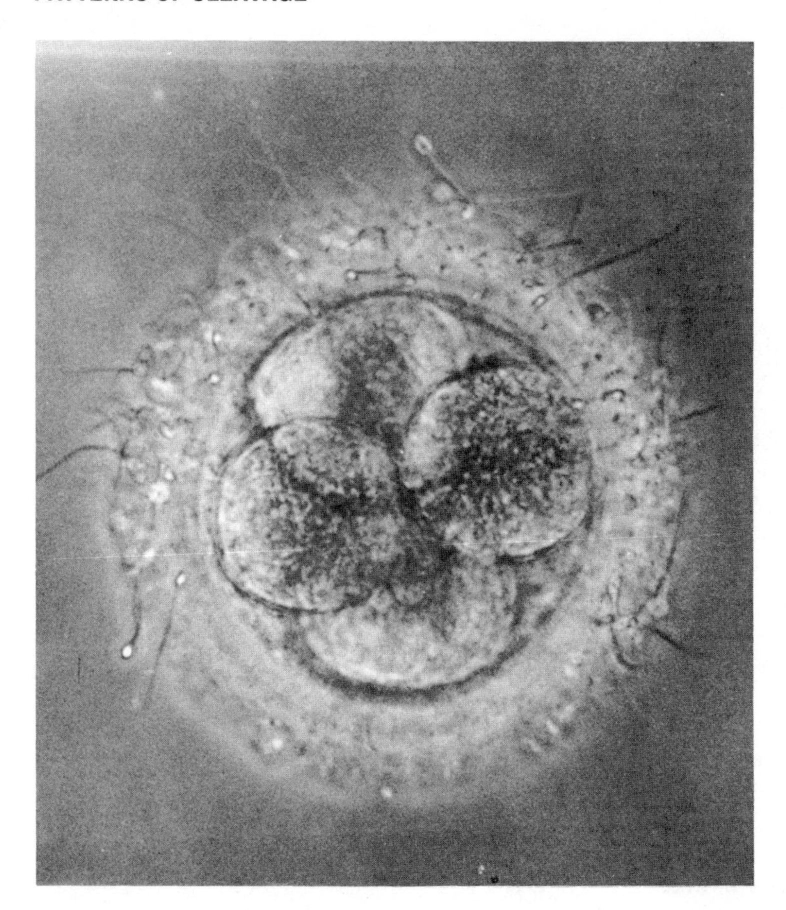

**FIG. 7–9.** Human egg at the four-celled stage. Note the spermatozoa in the zona pellucida. (From R. Edwards and R. Fowler, 1970. Sci. Am. 223:44.)

As one might expect, the segmentation of the fertilized egg is holoblastic, and all blastomeres are more or less of equal size (Fig. 7–9).

Cleavage of the mammalian egg is initiated in the upper end of the oviduct some 24 to 25 hours after insemination. In the mouse, rabbit, and monkey, the first cleavage furrow appears to be meridional, extending along an imaginary axis from animal to vegetal pole. By contrast to the oligolecithal eggs of other invertebrate and chordate organisms, the mammalian egg lacks synchronization of mitoses of blastomeres from the two-celled stage onward. Consequently, it is not unusual to find three-celled, five-celled, and seven-celled stages in the mammalian embryo (Figs. 7–10, 7–11). Examine carefully the sequence of cleavage and blastulation in the egg

**FIG. 7–10.** Early cleavage of the egg of the monkey *Macacus rhesus.* A, two-celled stage; B, three-celled stage; C, four-celled stage; D, five-celled stage; E, six-celled stage; F, seven-celled stage. pb1, pb2, polar bodies 1 and 2; zp, zona pellucida. (After W. Lewis and C. Hartmann, 1933. Carnegie Contributions to Embryology 24:189.)

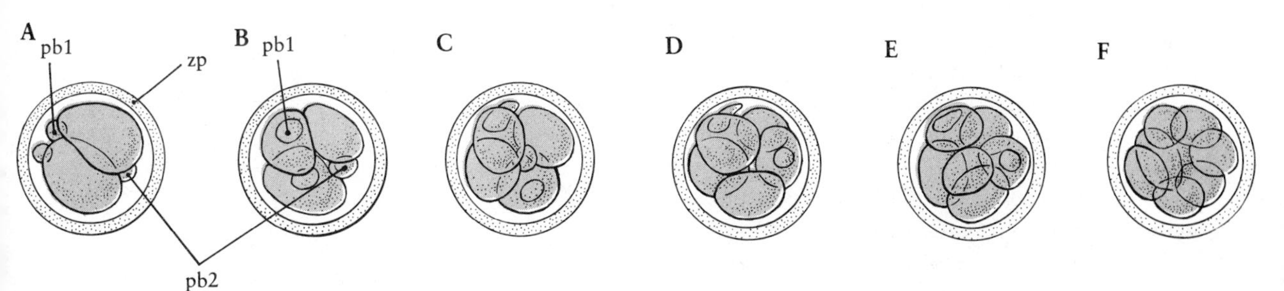

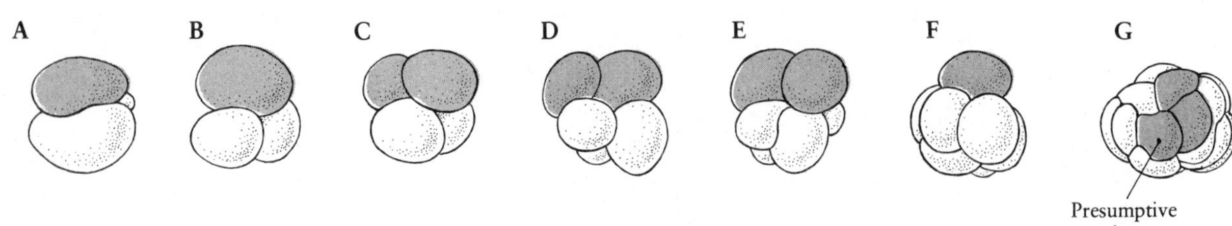


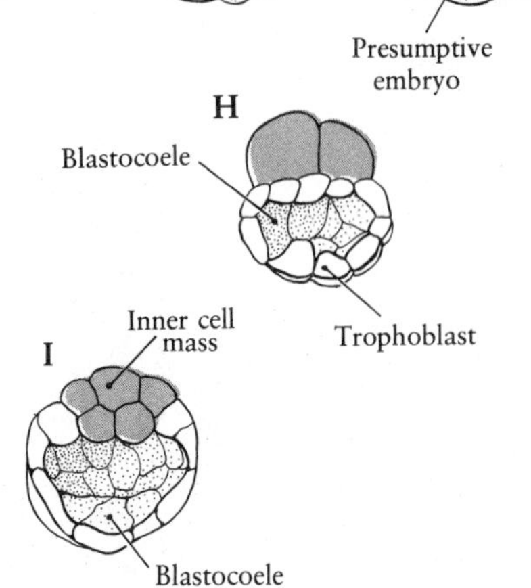


**FIG. 7–11.** Cleavage and blastulation of the egg of the pig based on models reconstructed in wax. A, two-celled stage; B, three-celled stage; C, four-celled stage; D, five-celled stage; E, six-celled stage; F, eight-celled stage, G, morula stage; H, formation of the blastocoele; I, early blastocyst. (After C. Heuser and G. Streeter, 1929. Carnegie Contributions to Embryology 20:1.)

of the pig (Fig. 7–11). Note that one of the first two blastomeres is more precocious than the other in the onset of the second cleavage, thus giving rise to a three-celled stage. From this stage onward, one can divide the embryo into a part that divides more rapidly and a part that divides more slowly (Fig. 7–11 C–I). If the fates of the cells comprising these two areas are traced into the blastula stage, it can be shown that the more rapidly dividing cells will form the *trophoblast* and the more slowly dividing cells the *formative* or *inner mass cells.*

The result of the segmentation of the fertilized egg is the production of a solid sphere of cells or morula. In such forms as the bat, the superficial cells are organized at this stage into a loose epithelial layer (Fig. 7–12 A). This outer layer is the trophoblast and will contribute to the formation of the extraembryonic membranes of the embryo, make contact with the uterine wall during implantation, and mediate the supply of nourishment to the embryo from the maternal body by way of the placenta. The cells lying in the interior constitute the inner cell mass. The inner cell mass contributes directly to the construction of the various parts of the embryo proper.

Before entrance into the uterine portion of the oviduct, the morula gradually becomes transformed into the *blastocyst* (Fig. 7–12 B). Crevices increasingly appear between the inner cell mass and most of the cells of the overlying trophoblast. As the fluid-filled spaces progressively coalesce, a large cavity or blastocoele is formed with displacement of the inner cell mass toward the future dorsal side of the embryo. At this site, the inner cell mass remains connected to the inner surface of the trophoblast. Continued uptake of water and of fluids from the oviduct into the blastocoele noticeably increases the volume of the blastocyst. Under these conditions, the zona pellucida is greatly stretched. Eventually, this membrane ruptures and the blastocyst is set free in preparation for implantation into the wall of the uterus.

## Insects

Segmentation of the fertilized egg of insects occurs in a most unusual fashion in that mitotic division of the nucleus is not followed by division of the cytoplasm. The nucleus of the recently fertilized egg lies in the center of the cell surrounded by a small volume of nonyolky cytoplasm. Mitosis of the nucleus produces a number of nuclei, each of which

**FIG. 7–12.** The inner cell mass and trophoblast at the morula (A) and early blastocyst (B) stages of the bat. (From J. Brachet, 1935. Traité d'Embryologie des Vertebres, 2nd ed. Revised by A. Dalcq and P. Gerard. Masson, Paris.)

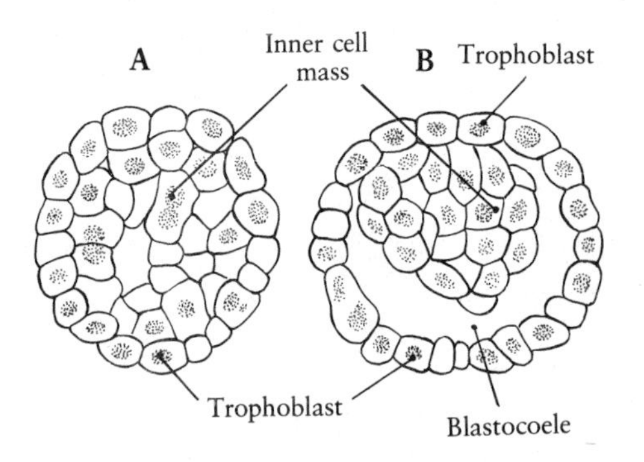

becomes embedded in an undivided mass of cytoplasm. Surrounded by a halo of cytoplasm, each nucleus then migrates outward toward the surface of the embryo. The nuclei then spread out in a common undivided layer of cytoplasm just beneath the surface. Subsequently, each nucleus becomes separated from its neighbor by the inward growth of membranes from the surface of the embryo. These cells will gradually become completely separated from the yolk. It has been estimated that in *Drosophila* some 3500 nuclei become compartmentalized into cells by this method. This type of cleavage is termed *superficial cleavage*, a phrase recognizing that only the surface layer of the egg becomes divided into blastomeres.

## Blastulation

Irrespective of the pattern by which it occurs, cleavage converts the single-celled, fertilized egg into a multicellular embryo. During the early stages of cleavage in oligolecithal and mesolecithal eggs, the blastomeres often tend to cluster together and assume the configuration of the morula. Subsequently, the rate of cell division slows down and the blastomeres then become displaced into a tightly organized peripheral layer(s) surrounded by a fluid-filled cavity. The process by which the blastomeres become arranged into a peripheral epithelium enclosing the blastocoele is known as blastulation. Although imperfectly understood, the process of blastulation has received some attention in embryos of sea urchin and starfish. The formation of the blastocoele and its enlargement in sea urchins are probably related to an osmotic pressure exerted by the blastocoelic fluid. The uptake of water from the environment during and following the cleavage stage forces the blastomeres outward where they attach to the hyaline layer. Specialized junctional complexes *(septate desmosomes)* appear to supplement the hyaline layer in holding cells together in the blastula. An increasing adhesiveness and surface tension between dividing blastomeres also appear to be important factors in maintaining the peripheral position of the cells of the blastula.

Recently, Dan-Sohkawa and Fujisawa (1980) have examined the mechanism of blastulation using denuded eggs (i.e., without fertilization membranes) of the starfish, *Asterina pectinifera*. Such eggs provide an experimental system in which the behavior of cells undergoing blastulation can be studied without restrictions that might be imposed by the fertilization membrane. One must keep in mind, however, that caution must be exercised in the application of such results to the normal developing system. The normal blastulation process for starfish embryos is show in Figure 7–13. The first few divisions of the denuded egg produce blastomeres that are loosely arranged and virtually unconnected. After about the eighth cleavage, however, the blastomeres start to pack together and organize themselves into a tightly packed sheet of cells. At the $2^9$ or $2^{10}$-celled stages, the free edge of the cell sheet turns up and forms itself into an irregularly shaped, hollow blastula. These "closing movements" of blastulating cells are correlated with the appearance of specialized junctional complexes (septate desmosomes) between blastomeres. A basement membrane is then deposited around the blastula. The behavior shown by the blastomeres of the starfish embryo is similar to that of an isolated epithelial sheet. The blastomeres gradually acquire the ability to recognize each other as being of the same type, initiate the development of structures that will hold them together (desmosomes), and then form themselves into a closed, hollow sphere.

## The Cleaving Cell

The cleavage of a blastomere involves a mitosis or nuclear division followed immediately by cytokinesis or physical separation into two daughter cells. In holoblastic organisms, each blastomere tends to round up just prior to division (Fig. 7–14). By the late anaphase stage of mitosis, the blastomere elongates in a plane parallel to the mitotic spindle apparatus. Consequently, the surface at the equator of the dividing blastomere is noticeably flattened. This condition is exaggerated in telophase when the diameter of the blastomere actively decreases with the formation of the cleavage furrow.

When one spherical blastomere divides into two daughter cells with a similar shape, the combined diameters and surface areas of the two descendant cells exceed that of the parent cell. The formation of the two blastomeres results in a substantial increase in surface area. The cleavage process, therefore, produces a large increase in the total cell surface area of the embryo, requiring the formation of additional surface membrane. Rather than being manufactured de novo, this membrane is more likely mobilized from preexisting sources within

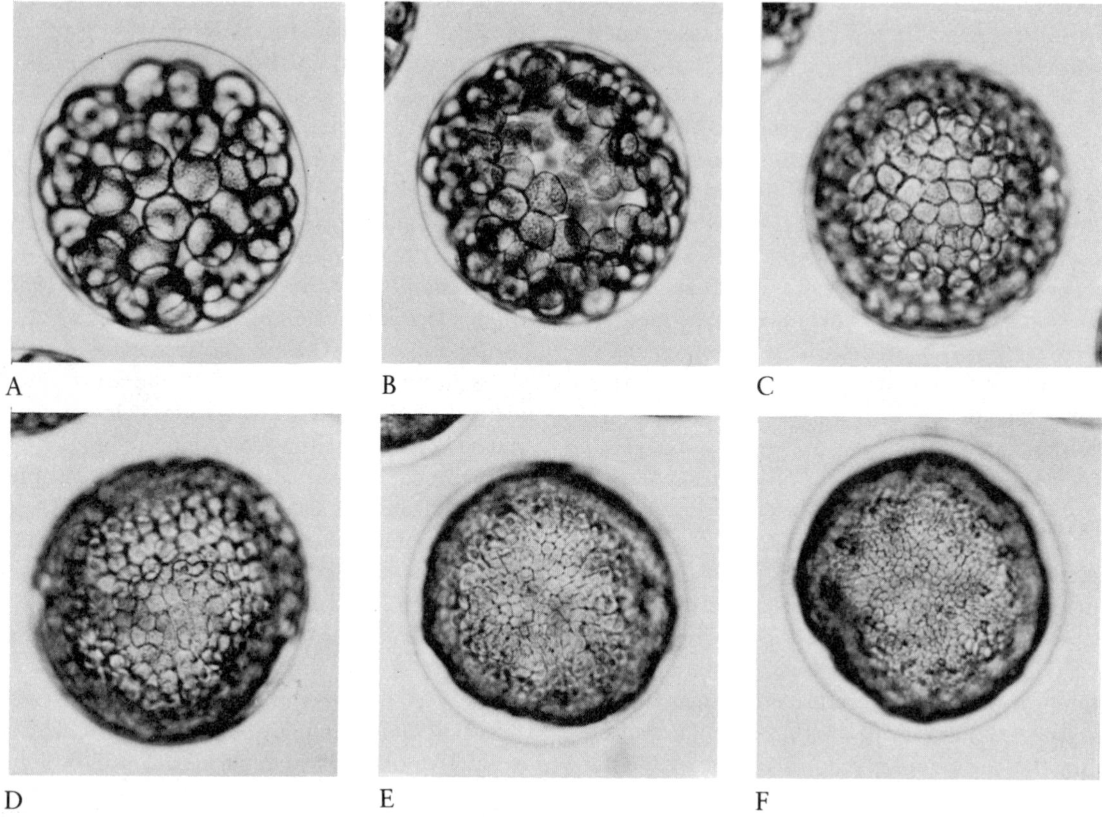


**FIG. 7–13.** Normal blastulation in embryos of the starfish, *Asterina pectinifera.* A, $2^6$-celled stage; B, $2^7$-celled stage; C, $2^8$-celled stage; D, $2^9$-celled stage; E, $2^{10}$-celled stage; F, $2^{11}$-celled stage. Time from A to F is about four hours. (From M. Dan-Sohkawa and H. Fujisawa, 1980. Dev. Bio. 77:328.)

the dividing blastomeres. In some amphibians, for example, it appears that membrane-bound cytoplasmic vesicles fuse with the leading edge of the cleavage furrow to form the required new membrane surface. Whether the microvilli or folds that typically cover the undivided egg surface play any role in providing new membrane is unclear.

Because of their size, shape, and transparency (particulary invertebrate eggs), cleaving eggs have been viewed as ideal subjects for the study of the fundamental processes of cell divison. Investigators have been particualrly interested in the role(s) of the visible parts of the blastomeres in the cleavage process and in the physicochemical basis for the separation of one cell into two cells.

The most visible structural component of the cleaving cell is the mitotic apparatus (Fig. 7–15). This complex includes the chromosomes, spindles, asters, and centrioles. Under the light microscope, each centriole appears as a rod-shaped particle embedded in a clear area of cytoplasm, the *centrosome.* With the electron microscope, each centriole is, in fact, a pair of hollow cylinders with the long axis of one being at right angles to the long axis of the other. A typical cylinder consists of nine peripherally arranged fibrils, each of which is a triplet of microtubules. Although the precise mechanism remains unclear, centrioles are believed to self-duplicate from precursor molecules in the cytoplasm just prior to the onset of a cleavage. Each centriole serves as a site for the organization of protein subunits into a linear array of microtubules and vesicular elements that compose the spindle and astral fibers. The centriole plus the astral fibers form the "sunburstlike" structure known as the *aster*. Chromosomes are engaged by the spindle fibers running between the asters and are drawn toward the centrioles during the mitotic phase of the cell division cycle. The size of the mitotic apparatus varies among different animal species. In general, the size of the mitotic apparatus is proportional to the volume of cytoplasm that it occupies. There is abun-

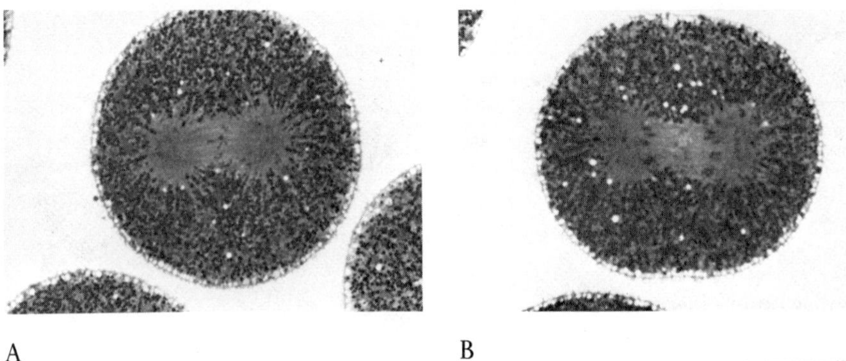


dant experimental evidence to show that the mitotic apparatus is essential for normal cytokinesis.

The location of the mitotic apparatus within the cytoplasm of a dividing blastomere appears to determine the position of the cleavage furrow as well as the time of its appearance. When the mitotic apparatus is in the center of a spherical blastomere, the furrow tends to appear simultaneously throughout the circumference of the cell. However, if the mitotic apparatus is naturally or experimentally displaced to one side of the blastomere, the furrow appears first in the nearest surface and then in the more distant surface. The types of cleavage found among invertebrate and vertebrate organisms clearly demonstrate that the furrowing of the cytoplasm in dividing blastomeres bears a structural relationship to the orientation of the mitotic apparatus. In forms like echinoderms, amphibians, and urochordates, the cleavage spindles of the mitotic apparatus tend to be oriented with their long axes perpendicular or parallel to the axes of cell polarity. The plane of furrowing is always perpendicular to the axis of the mitotic spindle. Such a division is said to be *meridional* or *radial*, because the resultant blastomeres are organized in a radially symmetrical pattern around a polar axis. An example of radial cleavage is shown in Figure 7–16 A. Note that in the first two cleavages the long axis is perpendicular to the axis of polarity (i.e., the animal–vegetal axis); the furrows are vertical and parallel to it. Radial cleavage produces tiers of cells arranged directly one above the other. A pecularity of the radial type of cleavage is seen in some invertebrate and chordate embryos. Here, for example, two of the first four blastomeres may be larger than the other two, thereby establishing a distinct axis of bilateral symmetry in the embryo. This arrangement of the blastomeres may be further evident with subsequent divisions. Cleavage of this type is referred to as being *bilaterally symmetrical.*

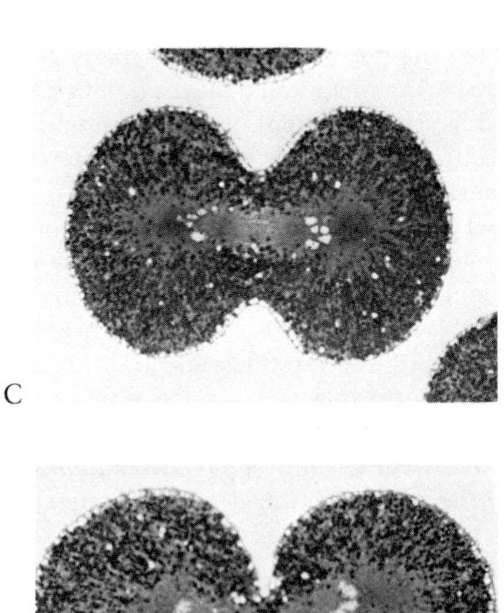


**FIG. 7–14.** Cleavage in sea urchin eggs. Midlongitudinal sections (1 μm) of a cleaving *Strongylocentrotus purpuratus* egg with fertilization membranes and hyaline layers removed. The dark cytoplasmic granules are yolk granules. The yolk-free zone contains the mitotic apparatus. A, metaphase: chromosomes are condensed (dark) and line up at the equator; B, early telophase: chromosomes have separated and furrowing has commenced; C, midtelophase; D, late telophase: cleavage is nearly complete. Cleavage is complete within 15 minutes. (Courtesy of C. Asnes and T. Schroeder.)

An entirely different relationship between the orientation of the mitotic spindle and the axis of cell polarity is visible in the dividing eggs of many molluscs, annelids, and nemertean worms (Fig. 7–16 B). After three divisions, it is noted that the upper quartet of blastomeres is shifted in the same

A

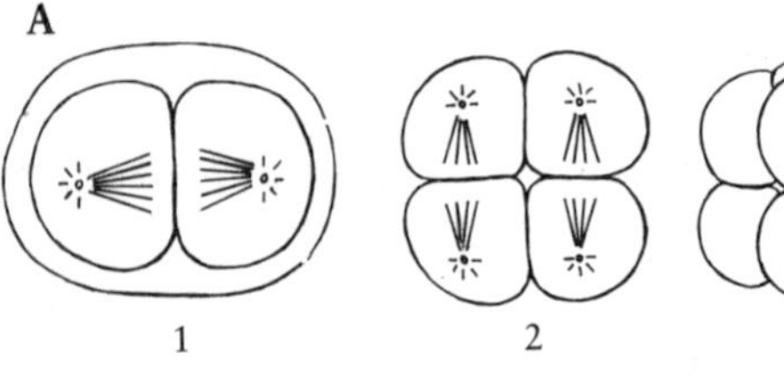
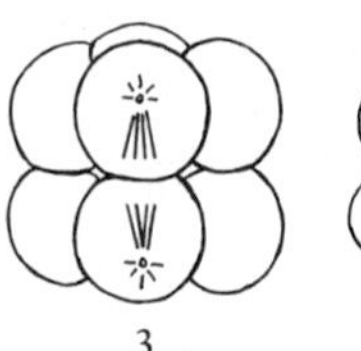
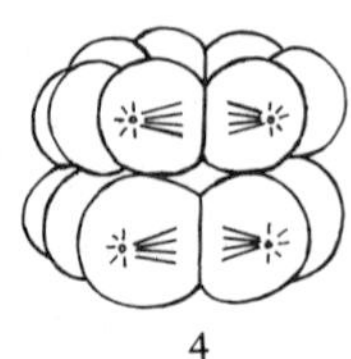

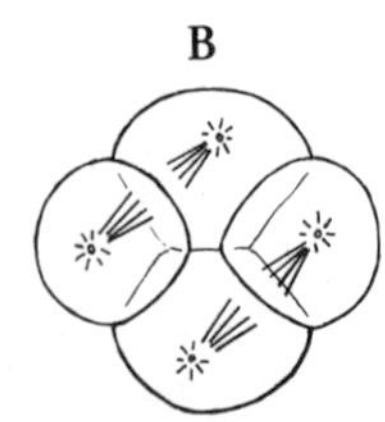
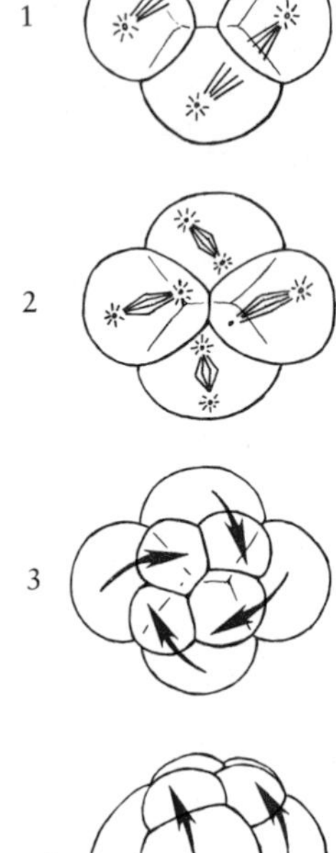

direction such that each upper blastomere lies over the junction of two adjacent blastomeres of the lower quartet of cells. Note that this is in sharp contrast to the condition in the sea urchin where, at the same stage, the upper tier of blastomeres lies directly over the lower tier of blastomeres. This arrangement in molluscs and annelids is due to the fact that the four cleavage spindles of the third cleavage division are tipped at an oblique angle (something other than 90° or 180°) to the axes of the blastomeres and coordinately oriented in the pattern of a spiral. Consequently, although the division furrows are at right angles to the axes of the mitotic apparatus, the resulting upper tier of blastomeres comes off in spiral fashion. This *spiral type* of cleavage is continued with subsequent divisions. If the blastomeres appear to rotate in a clockwise direction, the spiral is *dextral;* if the blastomeres appear to rotate in a counterclockwise direction, the spiral is *sinistral.* The coiling, either dextral or sinistral, of the shells of snails reflects the spiral pattern of cleavage. Results of genetic crosses between populations of snails showing different directions of shell coiling have demonstrated that the orientation of cleavage planes is programmed into the egg cytoplasm under the influence of maternal genes.

**FIG. 7–16.** A, an example of radial cleavage as seen in an echinoderm. Note the orientation of the cleavage furrow to the mitotic apparatus. (1) 2- celled stage; (2) 4-celled stage (viewed from animal pole); (3) 8-celled stage (lateral view); (4) 16-celled stage (lateral view; (5) 32-celled stage (lateral view). B, an example of spiral cleavage as seen in a mollusc. (1) 4-celled stage with spindles of second division still visible; (2) 4-celled stage with metaphase spindles in place for tyhe third division; (3) 8-celled stage (animal pole view); (4) 8-celled stage (lateral view). Arrows indicate direction of spiral. (From E. Korschelt, 1936. Vergleichende Entwicklungsgeschichte der Tiere. G. Fischer, Jena.)

**FIG. 7–15.** The most visible structural component of the cleaving cell is the mitotic apparatus. The mitotic apparatus includes the chromosomes, spindles, asters, and centrioles.

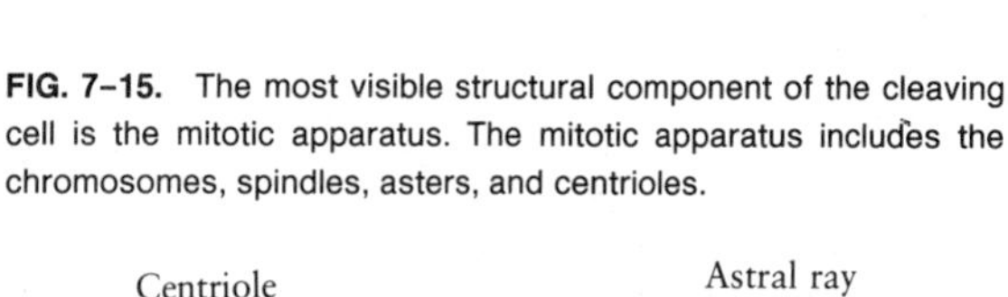
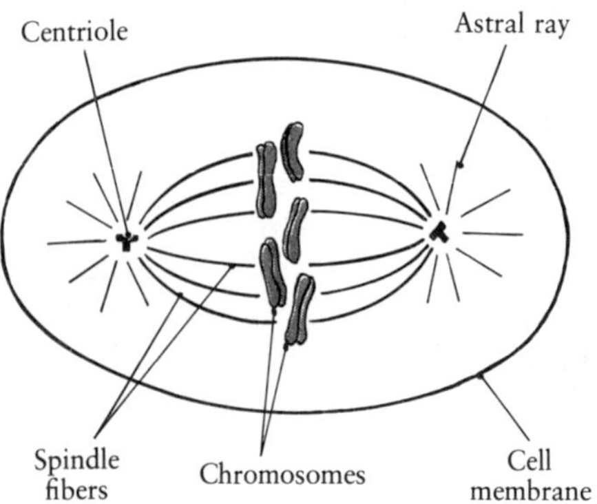


Since the events of mitosis and cytokinesis overlap in the cell cycle, it has been suspected that these two processes are causally connected. There has always been, for example, intense interest in the possibility that the form and arrangement of the mitotic spindle with its asters might function in the physical separation of a blastomere into two daughter cells. There is now substantial experimental evidence to indicate that the mitotic apparatus and other division-related organelles do not play a role

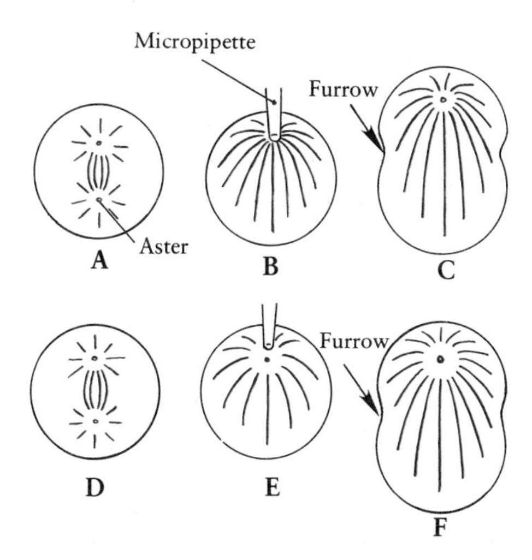


**FIG. 7-17.** A–C, Successive stages of cleavage when the mitotic apparatus is removed by aspiration. A, before the spindle is removed; B, spindle removed; C, furrow still appears in predetermined position; D–F, successive stages of cleavage when the mitotic spindle is displaced by removing part of the egg protoplasm. D, before displacement; E, after displacement; F, furrow still appears at predetermined position (From Y. Hiramoto, 1956. Exp. Cell Res. 11:630.)

in the physical establishment and functioning of the cleavage furrow. Y. Hiramoto removed the entire mitotic apparatus by aspiration with a micropipette shortly after metaphase in an echinoderm cleaving egg (Fig. 7–17 A–C). The division of the experimental cell was unaffected by this manipulation. Even if the mitotic apparatus was displaced by removal of part of the egg cytoplasm, the cleavage furrow still appeared in the normal position (Fig. 7–17 D–F). He further demonstrated, by injecting an oil droplet into the sea urchin egg at anaphase to displace the mitotic apparatus, that furrow formation was normal (Fig. 7–18). Additional studies by Rappaport on echinoderm eggs support the view that the formation and functioning of the division mechanism is independent of mitosis. For example, the division of the egg continues following the initial appearance of a cleavage furrow even if the cytoplasm between the mitotic apparatus and cell surface is stirred vigorously or if the cell contents are displaced by alternate compression at the subpolar regions of the cell. Repeated skewering of the mitotic apparatus fails to prevent cleavage. Two important conclusions can be drawn from these experiments. First, the position where the cleavage furrow will appear at the surface is fixed by anaphase of the cell cycle. Second, the furrowing process itself is independent of the mitotic apparatus.

However, other experiments clearly indicate that the mitotic apparatus is instrumental in setting up the conditions that result in the development of the furrowing mechanism. The mitotic apparatus, for example, cannot be removed from a cleaving blastomere prior to anaphase without affecting subsequent division. Questions on the nature of this interaction and how long the mitotic apparatus must interact with the surface in order to induce a permanent, functional furrow have only rudimentary answers, despite more than a century of speculation and investigation. It has always been assumed

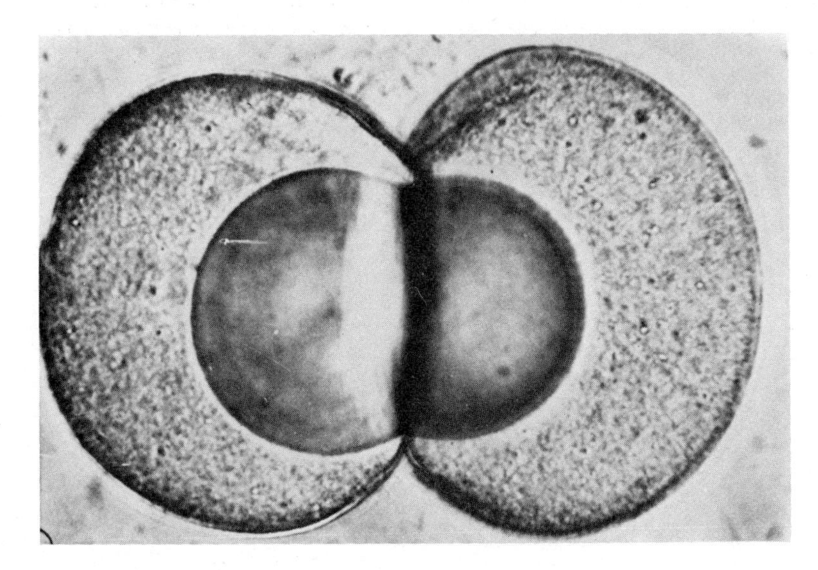

**FIG. 7-18.** Injection of a large oil droplet into a dividing sea urchin egg at anaphase disrupts the mitotic apparatus but does not block the development of the cleavage furrow. (From Y. Hiramoto, 1965. *J. Cell Biol.* 25:161.)

that the interaction between the cell surface and the mitotic apparatus is completed immediately before the latter becomes indispensable. Rappaport, however, has shown that there is a distinct time interval between the establishment of the cleavage furrow and the beginning of furrowing activity. This was done by aspirating the mitotic apparatus from one pair of cells at the second cleavage, and timing the interval between this manipulation and the appearance of the furrow in the control cell. When the mitotic apparatus is removed four minutes or less before control furrowing, the experimental cell still cleaves. However, if the operation is carried out five minutes or more before control division, cleavage does not take place in the experimental cell. Hence, the mitotic apparatus must interact with the cell surface up until a given time before the furrowing process; only after this time does the mitotic appartus become dispensable. Also, the development of a successful furrow appears to require an interaction between the mitotic apparatus and the subsurface in some areas but not in other areas of the blastomere. Elegant experiments have recently been conducted by Rappaport and Rappaport, in which blocks were placed between different regions of the mitotic apparatus and the surface in flattened sea urchin and sand dollar eggs. When oil droplets, slit punctures, or needles were positioned between the mitotic apparatus and the equatorial plane, the furrowing process was affected (Fig. 7–19). The magnitude of the effect was related to the size of the block (i.e., larger or smaller oil droplet). When the same blocks were placed outside the equatorial plane, furrowing was normal. These results are consistent with other experiments indicating that furrowing is dependent on the mitotic apparatus only at the equatorial surface of the cell.

## The Mechanism of Cleavage

The capacity of a cleavage furrow to continue its activities despite isolation from and disruption of subsurface structures clearly suggests that the site of the division mechanism resides in the surface of the blastomere. Within recent years, several theories have been proposed to account for the ability of a blastomere to divide itself into two halves at a highly predictable time and in a precisely predictable pattern. These theories include: the *polar expansion theory*, the *contractile ring theory*, and the *polar relaxation theory*. The polar expansion concept explains segmentation as due to the active

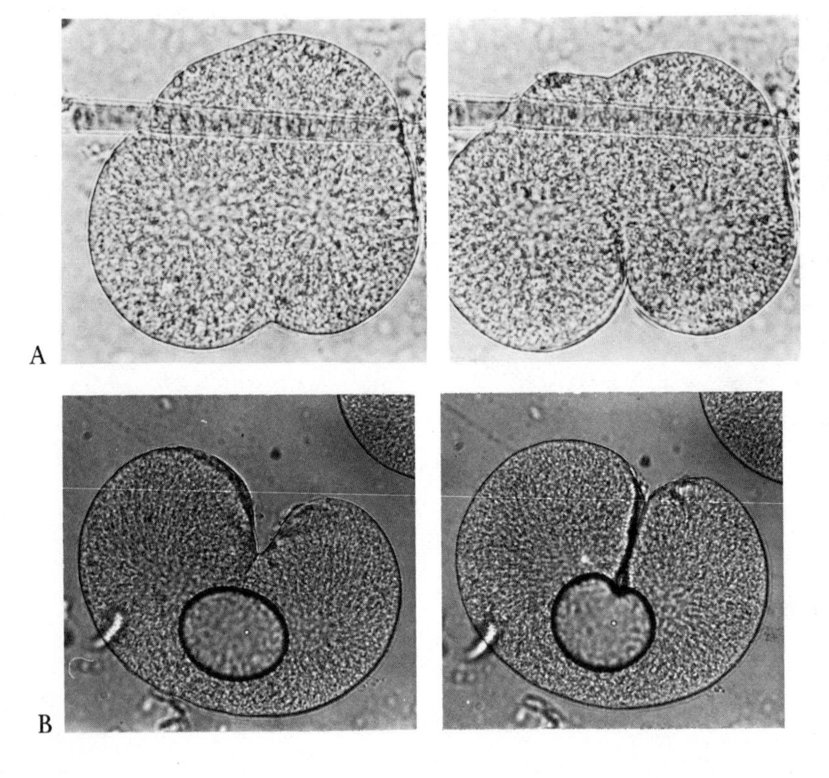


**FIG. 7–19.** Effects of various blocks positioned between the mitotic apparatus and the surfaces in flattened eggs of a sea urchin and a sand dollar. A, a 7.5-micron diameter needle inserted perpendicular to the equatorial plane of a dividing cell can mechanically halt an active furrow in a sea urchin; B, a large oil droplet (60 microns in diameter) placed in the equatorial plane will interfere with further progress of the furrow (sand dollar). (From R. Rappaport and B. Rappaport, 1983. J. Exp. Zool. 227:213.)

growth and expansion of the polar regions of the blastomere, resulting in the passive inward dipping (i.e., furrowing) of the equatorial surface. The polar relaxation theory proposes that segmentation occurs because the polar surfaces of the blastomere relax, allowing the equatorial surface of the cell to contract actively. The contractile ring (or *equatorial constriction*) concept states that a blastomere divides because of the presence of a band of contractile material at the equator of the cell.

All three theories recognize that cleavage is a rather brief event with furrowing itself being preceded by dramatic changes in the physical and/or chemical properties of the cell surface. In the polar relaxation and equatorial constriction theories, it is additionally recognized that the entire cell surface or cortical cytoplasm becomes contractile just before the onset of furrowing. Eventually, the contractility is heightened at the equator of the cell, either through an increase at the equator of the cell (contractile ring) or a decrease at the poles of the cell (polar relaxation).

Most current data support the hypothesis that the cleaving cell divides because of the assembly of a specialized equatorial ring of contractile material located in the cortical cytoplasm. Evidence for the presence of contractile material comes from several sources. With the light microscope, the progress of the cleavage furrow is inhibited when two microneedles are inserted into the path of a dividing echinoderm egg (Fig. 7–20). Presumably the microneedles act as a barrier and block the constricting action of the contractile ring. Observations with the electron microscope reveal the presence of an electron-dense band of microfilaments at the base of cleavage furrows in a variety of animals, including squids, salamanders, jellyfish, echinoderms, and mammals (Fig. 7–21). In echinoderms, this band (about 0.1 μm thick and 10 μm wide) consists of numerous microfilaments, each of which is about 50 Å to 70 Å in diameter. These cytoplasmic filaments have been shown by various labeling techniques to contain the contractile protein actin. Within the furrow, the microfilaments form linear arrays oriented parallel to the equatorial surface and parallel to the plane of division. These microfilaments are presumed to be the visible expression of the contractile ring that cuts the blastomere into two daughter cells. Support for this interpretation comes from the correlation between the temporal appearance of the furrow and the morphological

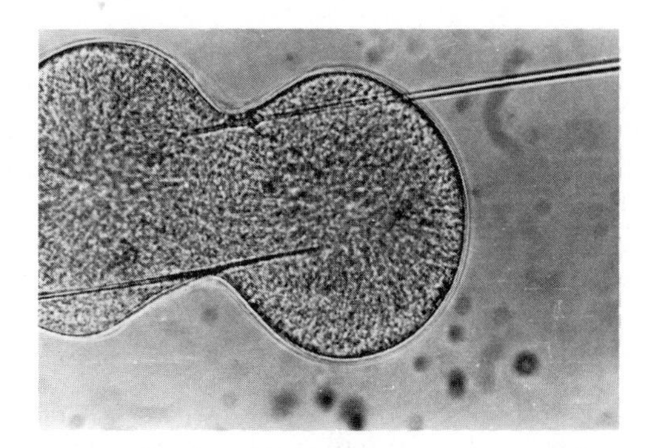

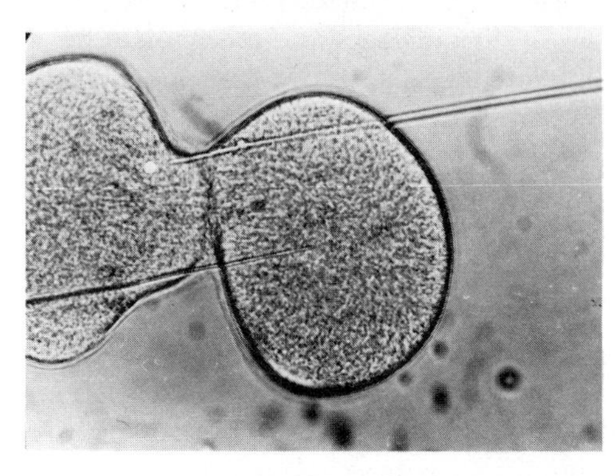

**FIG. 7–20.** Cleavage of the sand dollar *(Echinarachnius)* egg with two microneedles placed through the plane of division. A, the appearance of the dividing egg shortly after positioning of the needles; B, as the furrow deepens, the needles are pulled closer together. The furrow typically fails to divide the egg into two daughter cells if the needles are left in place. (From R. Rappaport, 1966. J. Exp. Zool. 161:1.)

**FIG. 7–21.** Microfilaments form a band in the furrow plane between two forming blastomeres in a rat egg. (After D. Szöllösi, 1970. J. Cell Biol. 44:192.)

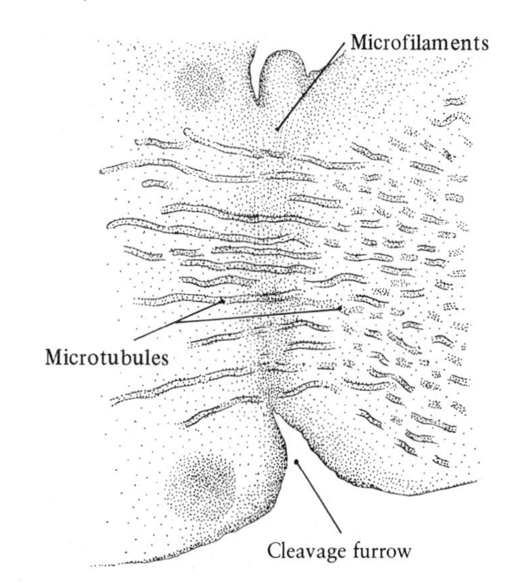

presence of the specialized microfilamentous band. Also, cytochalasin B, the antibiotic from molds that is known to dissolve microfilaments into their subunits, severely disturbs cytokinesis. Eggs treated with cytochalasin B before cleaveage will show nuclear division but no physical separation into daughter cells. Eggs treated with this compound after physical division is initiated show either arrest or reversal of the furrowing process.

Although a variety of studies have demonstrated the existence of contractile proteins in the cleavage furrow of dividing eggs, the mechanism by which contraction of the microfilamentous band is achieved and translated into a furrow is not known. It has been suggested that contraction may occur by the actin filaments sliding toward each other, much as in the actomyosin filament model proposed to account for the shortening of vertebrate skeletal muscle. Presumably, actin filaments of the cleavage furrow, attached to the overlying plasmalemma, slide past each other as a result of their association with molecules of myosin. The cell progressively is separated into two parts as its circumference along the contractile ring diminishes. The process has been likened to the closure of the opening into a purse by "drawing the strings." An actomyosin basis for the cleavage contractile mechanism is supported by several lines of evidence. Actin or actinlike proteins have been isolated from the cortical cytoplasm of sea urchin eggs. When combined with rabbit muscle myosin, the complexed protein shows many of the properties of ordinary actomyosin extracted from voluntary muscle tissue. Calcium, a critical ion in skeletal muscle contraction, appears to be essential to the furrowing process. Agents that chelate or complex calcium are know to inhibit the capacity of the egg to form furrows. Also, the diameter of the furrow microfilaments is very similar to that recorded for actin threads. Rappaport has determined that the microfilamentous band in echinoderm cleaving eggs has a capacity for isometric contraction of approximately 1.25 to $2.4 \times 10^5$ dynes/cm$^2$. Actomyosin threads during isometric contraction exert a tension of about $2.45 \times 10^5$ dynes/cm$^2$. The tension developed during cytokinesis is clearly visible in the appearance of the cleavage furrow (Fig. 7–22). Other indirect evidence that the cleavage process depends on an actomyosin contraction includes the observation by Mabuchi and Okuno that antibodies prepared against myosin and injected into starfish eggs inhibit cleavage. Although myosin has been

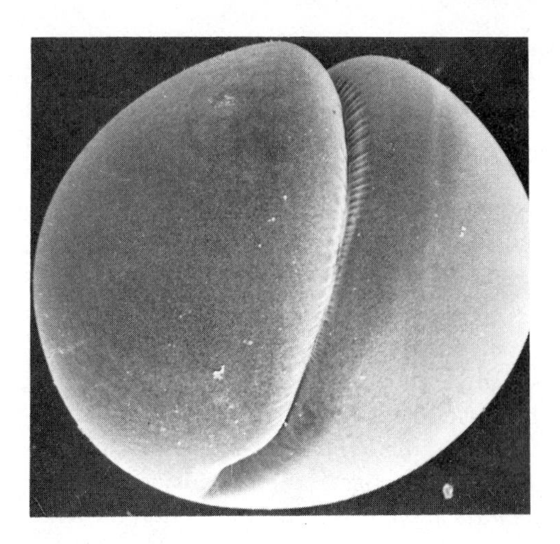

**FIG. 7–22.** A scanning electron micrograph of the first cleavage in the egg of a frog to show the tension lines in the furrow. (From H. Beams and R. Kessel, 1976. Am. Sci. 64, 279. Reprinted by permission of American Scientist, Journal of Sigma Xi, The Scientific Research Society.)

isolated from the cortices of cleaving eggs, there is little evidence that it is organized into filaments.

An alternative and older hypothesis to explain the chemical basis for the cleavage contractile mechanism has been offered by Sakai. He has isolated a KCl-soluble protein from a water-insoluble residue of homogenized sea urchin eggs. This protein can be made to form threads that contract and relax in the presence of certain metal ions, such as calcium, and oxidants of -SH (sulfhydryl) groups. The ability of these proteinaceous threads to shorten varies with the cleavage cycle, increasing as metaphase approaches and decreasing just prior to the onset of furrow formation. Interestingly, the -SH levels in the cortical cytoplasm show a similar fluctuation during the division cycle. Since the total -SH content of the sea urchin protein remains constant over the division cycle, Sakai proposes the existence of another -SH-containing protein whose level fluctuates in a fashion opposite to that of the cortical protein. This protein, precipitated from the water-soluble fraction of homogenized eggs in the presence of calcium, is postulated to be associated with the mitotic apparatus. Contraction of the cortex in the furrow is presumed to be the result of the in vivo oxidation of -SH groups and a complicated exchange reaction between the protein of the mitotic apparatus and the contractile protein of the cortical cytoplasm.

When does the egg surface or cortex develop the capacity to furrow? How are the changes initiated

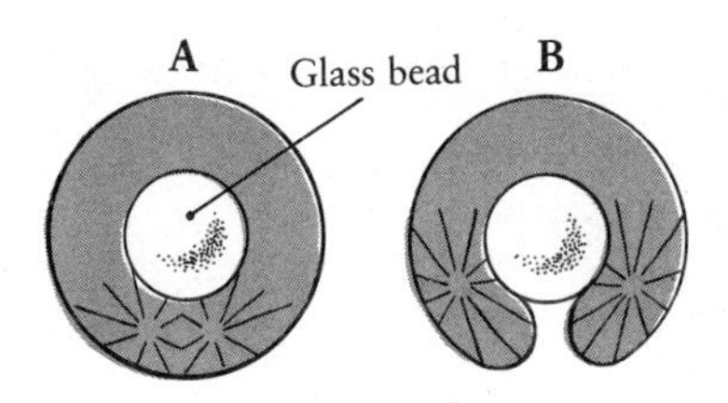

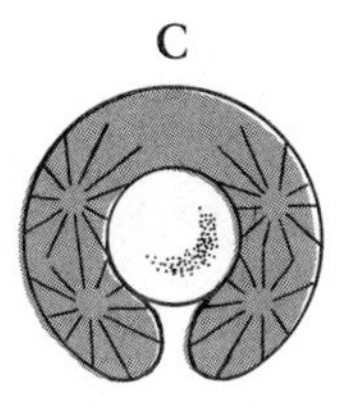

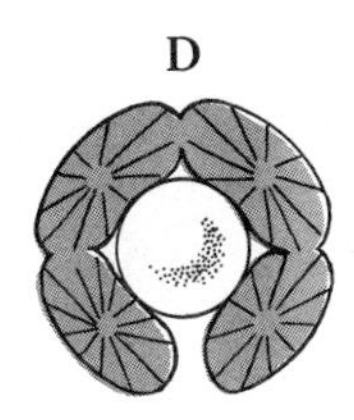


**FIG. 7–23.** Cleavage of a torus-shaped echinoderm egg. A, before the first cleavage; B, after the first cleavage; C, before the second cleavage to show a mitotic apparatus in each arm of the binucleate cell; D, beginning of third cleavage to show that a furrow has formed between two asters not joined by a spindle, resulting in uninucleate cells. (From R. Rappaport, 1971. Int. Rev. Cytol. 31:169.)

that lead to the required modifications in the properties of the equatorial surface of the cleaving cell? What is the nature of the stimulus and the stimulation process? Answers to these questions are still very much incomplete. *Xenopus* oocytes become capable of furrowing, and possibly of normal cleavage, sometime after the breakdown of the germinal vesicle. Mixing of nuclear and cytoplasmic plasms during egg maturation appears to initiate a change in the egg cortex that gives the egg an ability to furrow. However, we have previously seen that there is also required for furrow formation an alteration in the physical properties and behavior of the surface by the mitotic apparatus. It now appears that changes in the cortex of dividing echinoderm eggs are induced as a consequence of stimulation by the paired asters of the mitotic apparatus. Assessment of the role of the asters and other components of the mitotic apparatus has been accomplished by the geometrical alteration of the blastomere before metaphase (Fig. 7–23). When an echinoderm egg is converted into a torus-shaped structure by a glass bead, the first cleavage produces a horseshoe-shaped binucleate cell (Fig. 7–23 A,B). Two mitotic apparatuses form in the arms of the horseshoe-shaped cell in anticipation of the second cleavage division (Fig. 7–23 C). Only the astral rays are present in the bend of the torus-shaped cell. Successive cleavages appear not only at right angles to the mitotic apparatuses but also across the asters at the horseshoe bend to produce uninucleate cells (Fig. 7–23 C,D). Hence, furrow formation does not require the presence of either the mitotic spindle or chromosomes in the plane of division. Asters alone would appear to bring about the surface differentiation necessary for the development of a furrow in echinoderm eggs. Studies by Heidemann and Kirschner (1981) have shown that asters stimulate furrow formation in frog eggs. However, since some furrowing could also be induced in unfertilized eggs by an activation stimulus, there is the suggestion in amphibians that astral structures may not be strictly required for the formation of the cleavage furrow.

When asters of a cleaving cell form, their rays penetrate to all regions of the cell. At the equatorial surface, the rays of one aster overlap with the rays of the other aster (Fig. 7–24). It has been postulated that the astral rays probably subject the equatorial surface to greater influence or stimulatory activity because of their overlapping configuration. It would be convenient to propose that the astral fibers add or remove some critical substance that results in the differentiation of surface changes leading to the formation of the microfilamentous band. Since the 1930s, a number of investigators have identified centrifugable, division-related substances in a variety of cell types. However, these cleavage-initiating substances appear to play more of a role in nuclear division than cytoplasmic division. Unfortunately, ultrastructural studies have thus far failed to provide decisive information on the concept of crossed astral rays.

In summary, the factors implicated in controlling the furrowing process in cleaving animal eggs include the mitotic apparatus, subcortical components, and a band of cortical microfilaments. How

**FIG. 7–24.** The equatorial stimulation pattern of cleavage as proposed by Rappaport. The equatorial surface of a cleaving cell is subjected to greater stimulatory activity because of the overlapping of the astral rays at this site. (From R. Rappaport, 1971. Int. Rev. Cytol. 31:169.)

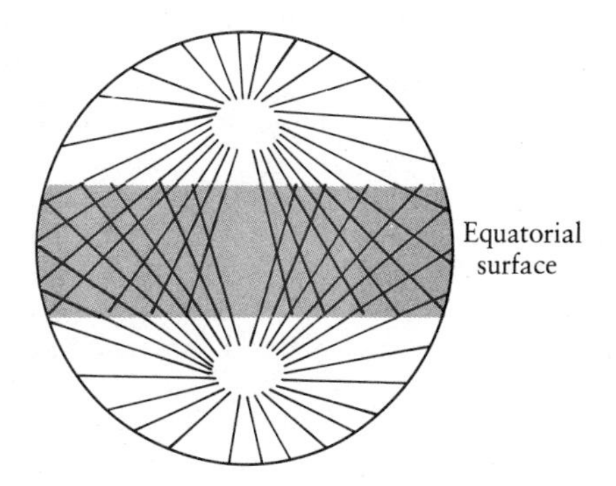

these various parts of the cell are integrated to induce furrowing requires their dissociation and experimental manipulation. As technical difficulties in manipulating the components and events of the dividing cell are overcome, a clearer picture of cell cleavage and its relationship to astral growth and differentiation of the egg surface will emerge.

## The Significance of Cleavage

The period of cleavage is a time during which the fertilized egg becomes transformed into a multicellular embryo. During this time the nucleus to cytoplasm ratio volume is adjusted to that characteristic of the adult somatic cell. Since DNA synthesis and cell division are prominent features of cleavage, this developmental period is particularly susceptible to exogenous agents that damage chromosomes and nucleic acids, such as x irradiation and ultraviolet radiation.

Additionally, cleavage provides the embryo with sufficient cell numbers to permit systematic movement and rearrangements of cells during the next major phase of development (gastrulation) in anticipation of the complex multilayered structure of the adult organism. Indeed, the blastula (or blastoderm) can be visualized as consisting of populations of cells of presumptive organ-forming areas that are topographically organized into distinct zones. These presumptive organ-forming areas will be considered in greater detail in subsequent chapters.

## References

Arnold, J. 1969. Cleavage furrow formation in a telolecithal egg *(Loligo pealii)*. 1. Filaments in early furrow formation. J. Cell Biol. 41:893–904.

Asnes, C. and T. Schroeder. 1979. Cell cleavage: ultrastructural evidence against equatorial stimulation by aster microtubules. Exp. Cell Res. 122:327–338.

Beams, H. and R. Kessel. 1976. Cytokinesis: a comparative study of cytoplasmic division in animal cells. Am. Sci. 64:279–290.

Bennett, M., and D. Goodenough. 1978. Gap junctions, electrotonic coupling and intercellular communication. Neurosci. Res. Program Bull. 16:373–486.

Bennett, M., D. Spray, and A. Harris. 1981. Electrical coupling in development. Am. Zool. 21:413–427.

Conrad, W. and R. Rappaport. 1981. Mechanisms of cytokinesis in animal cells. In: Mitosis/Cytokinesis, pp. 365–396. A. Zimmerman and A. Forer, eds. New York: Academic Press.

Dan-Sohkawa, M. and H. Fujisawa. 1980. Cell dynamics of the blastulation process in the starfish, *Asterina pectinifera*. Dev. Biol. 77:328–339.

Edwards, R. and R. Fowler. 1970. Human embryos in the laboratory. Sci. Am. 223:44–54.

Hanna, B., P. Model, D. Spray, M. Bennett, and A. Harris. 1980. Gap junctions in early amphibian embryos. Am. J. Anat. 158:111–114.

Heideman, S. and M. Kirschner. 1981. Induced formation of asters and cleavage furrows in oocytes of *Xenopus laevis* during *in vitro* maturation. J. Exp. Zool. 204:431–444.

Hiramoto, Y. 1956. Cell division without mitotic apparatus in sea urchin eggs. Exp. Cell Res. 11:630–636.

Hiramoto, Y. 1965. Further studies on cell division without mitotic apparatus in sea urchin egg. J. Cell Biol. 25:161–167.

Hiramoto, Y. 1971. Analysis of cleavage stimulus by means of micromanipulation of sea urchin eggs. Exp. Cell Res. 68:291–298.

Inoué, S. 1981. Cell division and the mitotic spindle. J. Cell Biol. 91:131s–147s.

Kobayakawa, Y. and H. Kubota. 1981. Temporal patterns of cleavage and the onset of gastrulation in amphibian embryos from eggs with reduced cytoplasm. J. Embryol. Exp. Morphol. 62:83–94.

Mabuchi, I. and I. Okuno. 1977. The effect of myosin antibody on the division of starfish blastomeres. J. Cell Biol. 74:251–263.

Ohta, T. and T. Iwamatsu. 1980. Initiation of cleavage in *Oryzias latipes* eggs injected with centrioles from sea urchin spermatozoa. J. Exp. Zool. 214:93–99.

Perry, M., H. John, and N. Thomas. 1971. Actin-like filaments in the cleavage furrow of the newt egg. Exp. Cell Res. 65:249–253.

Rappaport, R. 1961. Experiments concerning the cleavage stimulus in sand dollar eggs. J. Exp. Zool. 148:81–89.

Rappaport, R. 1966. Experiments concerning the cleavage furrow in invertebrate eggs. J. Exp. Zool. 161:1–8.

Rappaport, R. 1967. Cell division: direct measurement of maximum tension exerted by furrow of echinoderm eggs. Science 156:1241–1243.

Rappaport, R, 1979. Cytokinesis in animal cells. Int. Rev. Cytol. 31:169–213.

Rappaport, R. 1981. Cytokinesis: cleavage furrow establishment in cylindrical sand dollar eggs. J. Exp. Zool. 217:365–375.

Rappaport, R. 1983. Cytokinesis: furrowing activity in nucleated endoplasmic fragments of fertilized sand dollar eggs. J. Exp. Zool. 227:247–253.

Rappaport, R. and B. Rappaport. 1983. Cytokinesis: effects of blocks between the mitotic apparatus and the surface on furrow establishment in flattened echinoderm eggs. J. Exp. Zool. 227:213–227.

Sakai, H. 1968. Contractile properties of protein threads from sea urchin eggs in relation to cell division. Int. Rev. Cytol. 23:89–112.

Schroeder, T. 1968. Cytokinesis: filaments in the cleavage furrow. Exp. Cell Res. 53:272–276.

Szöllösi, D. 1970. Cortical cytoplasmic filaments of cleaving eggs: a structural element corresponding to the contractile ring. J. Cell Biol. 44:192–209.

# 8

# Gastrulation

We have described cleavage and blastulation as processes that offer little in the way of morphological differentiation but serve to provide the building blocks for future tissue and organ construction. The expression of organs and perfection of tissues await further development. The blastula is a mass of cells showing little or no axiate pattern, and the arrangement of the cells bears no apparent relationship to the body plan of the future organism. Gastrulation is the period of development concerned with the sorting out and movement of the building blocks to the various regions of the embryo where they will be used. It involves a dynamic series of orderly morphogenetic movements that rearrange and reorganize large groups of cells into patterns in which they can take part in the formation of the various tissues and organs of the body. Following gastrulation the axiate pattern of the organism is easily recognizable.

Gastrulation also results in the formation of a multilayered embryo. In some species, such as *Amphioxus*, gastrulation results in the formation of a two-layered embryo. In others, such as the frog, a three-layered condition results. However, in those forms in which a two-layered condition is the initial consequence of gastrulation, reorganization of the inner layer soon takes place and results in the formation of a three-layered embryo. These three layers are known as the primary germ layers: ectoderm, endoderm, and mesoderm. The presumptive endoderm and mesoderm, located on the surface of the egg in the blastula, are moved to the inside of the gastrula. The result is essentially an outer tube of ectoderm surrounding an inner tube of endoderm with a tube of mesoderm placed between them, a structure characteristic of the general vertebrate body plan. Each of these layers will develop into specific parts of the organism. The ectoderm differentiates into epidermis and neural tissue, the endoderm into the lining of the gastrointestinal tract and the respiratory system, and the mesoderm into urogenital structures, circulatory system, connective tissue, and muscle.

A word of caution is in order on the use of the words gastrula and gastrulation. As can be concluded from the paragraph above, it is not possible to give a definition of gastrula in the structural sense that will apply to all chordates. Some have an archenteron at the end of the gastrulation, while others do not. Gastrula is also difficult to define in the chronological sense. Even if we consider that the embryo in its period of active morphogenetic movements is a gastrula, these cellular displacements are not initiated or necessarily completed at the same time during development. For these reasons, Ballard recommends that the term gastrulation be phased out as far as vertebrates are concerned. However, the term will be retained here for purposes of convenience. We will use gastrulation to refer to "that ensemble of processes during post-cleavage early development which when complete has brought the cells which will form the various organs to the places where these organs are to form." (Trinkaus, 1976). The "ensemble of processes" are morphogenetic movements.

## The Cell Movements of Gastrulation

Certain general types of cell movements occur during gastrulation. To a large extent the amount of yolk influences the type and amount of movement, and not all types are seen in all eggs. *Invagination* is the major movement in most eggs with small amounts of yolk. Invagination is the inpushing or inpocketing of an unbroken sheet of cells at one region of the blastula. It has been compared to placing the thumbs on the surface of a hollow rubber ball and pushing in with sufficient force to obliterate the original cavity (the blastocoele), establishing a new cavity (the gastrocoele). This movement is best exemplified in the sea urchin and in *Amphioxus*. The term invagination has also been applied to the movement of the cells through the primitive streak of the chick embryo. *Involution* is the rolling in of cells over a rim. The cells that involute are replaced by cells on the surface, which move toward the point of involution so that a continual stream of cells passes over the rim into the interior. This is the type of movement characteristic of eggs with moderate amounts of yolk such as those of the amphibians. However, if we consider the rubber ball analogy, it is apparent that involution must necessarily also accompany invagination. A third major movement is that of the cells on the surface of the blastula. This is called *epiboly*. Cells move over the surface toward the region of invagination or involution. At the completion of gastrulation, the epibolic movements have resulted in the spread of the presumptive ectoderm over the entire surface of the embryo. In most embryos the region at which the cells move into the inside is limited to a small part of the embryo. Thus, as cells from outlying areas approach this region they must show a considerable amount of *convergence*. Conversely, once inside, the cells move away from the point of entry, a movement of *divergence*. The region where cells move to the inside is known as the *blastopore* in *Amphioxus* and the amphibians. In avian species it is known as the *primitive streak*.

## Gastrulation in *Amphioxus*

Gastrulation at its simplest occurs in those species in which cellular movements are not restricted by the presence of large amounts of inert yolk. Such is the case of *Amphioxus*. The blastula of *Amphioxus* has been described in Chapter 7 as a single layer of columnar cells surrounding a large cavity, the blastocoele (Fig. 8–1 A.) An animal and a vegetal pole may be recognized on the basis of the larger cells at the vegetal pole. The first indication of gastrulation is a flattening of the blastula at the vegetal pole (Fig. 8–1 B). This flattened plate of cells, presumptive endoderm, then gradually folds inward, invaginating into the blastocoele converting the spherical blastula into a cup-shaped gastrula (Fig. 8–1 C,F). The invaginated cells move toward the surface ectodermal cells at the animal pole gradually obliterating the blastocoele and forming a new cavity, the *gastrocoele* or *archenteron*. The gastrocoele opens to the outside by way of the *blastopore*, an opening that becomes progressively smaller as gastrulation and neurulation continue (Fig. 8–1,2). The gastrula has a double-layered wall, the two layers being continuous with each other at the lips of the blastopore. The inner layer will form endoderm, mesoderm, and notochord, while the outer layer will form epidermis and neural tissue. During the formation of the gastrula, the embryo rotates through an arc of about 120 degrees.

The cup-shaped gastrula now undergoes elongation in the craniocaudal axis. One region of the elongated embryo develops a flattened surface marking the dorsal side (Fig. 8–2). A cross section through the gastrula at this time (Fig. 8–3 A) shows an outer layer of cells flattened at one side surrounding an inner layer of larger cells arranged around a cavity, the gastrocoele. The flattened dorsal region of the outer layer marks the site of the formation of the nervous system. The inner layer is made mostly of presumptive endoderm but contains a middorsal strip of presumptive notochord bounded on either side by strips of presumptive mesoderm (Fig. 8–3 A,B).

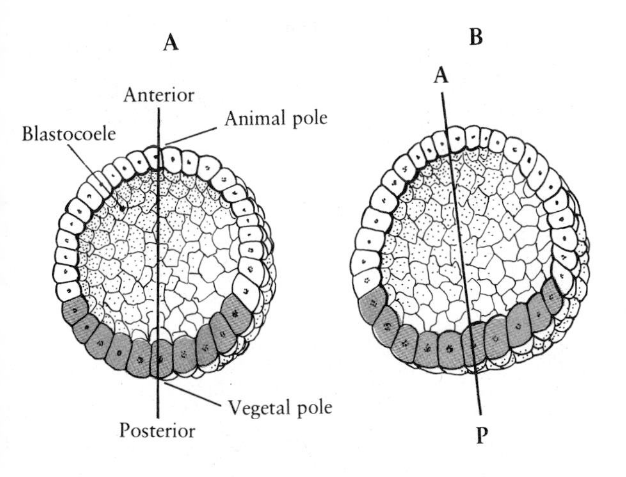


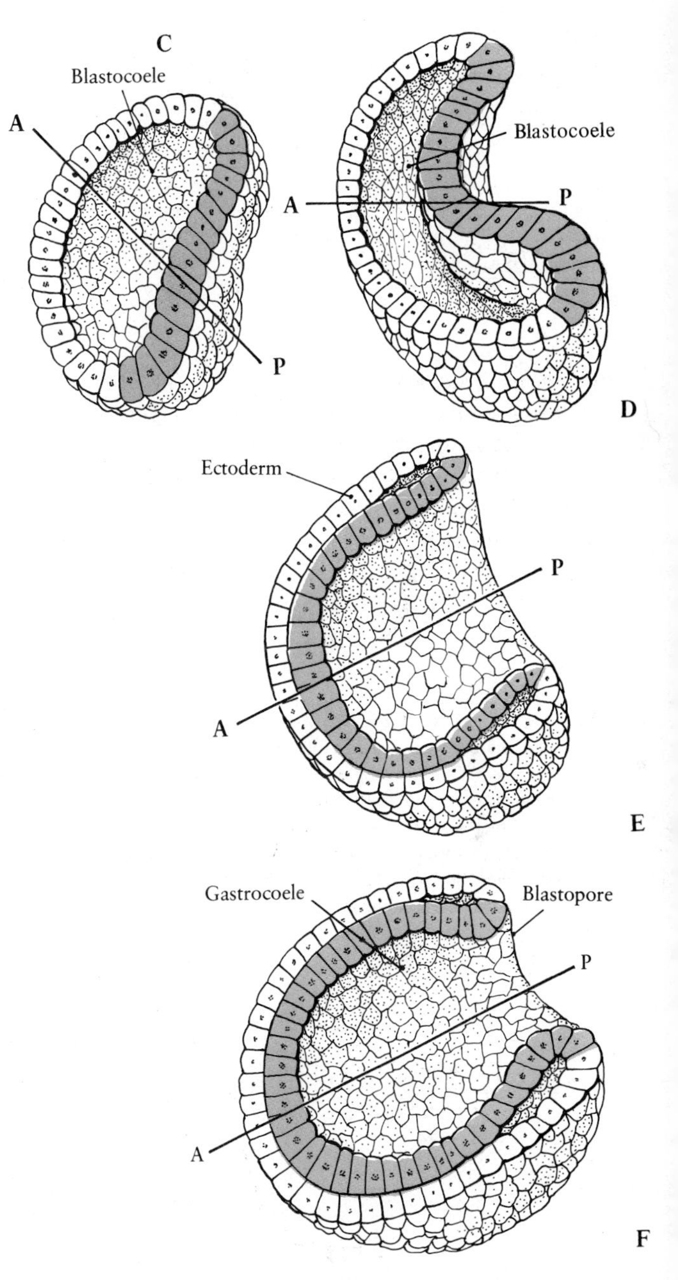


FIG. 8–1. Gastrulation in *Amphioxus* showing change in polarity. Cells at the posterior pole flatten and invaginate toward the animal pole, eventually obliterating the blastocoele and replacing it with a new cavity, the gastrocoele. (From A. F. Huettner, 1972. Comparative Embryology of the Vertebrates. Copyright 1949 by Macmillan Publishing Co., Inc. Renewed 1977 by M. R. Huettner, R. A. Huettner and R. J. Huettner.)

## Development of the Organ Rudiments

Further development involves the differentiation of the primary organ rudiments by local proliferation and folding of particular regions of the inner and outer layer of cells.

*The Nervous System*

The presumptive nervous system is in the form of a flattened longitudinal plate of cells, the *neural plate*, on the dorsal surface of the embryo (Fig. 8–3 A). This longitudinal plate separates from the surrounding cells and begins to sink below the surface, and at the same time the cells, that border the neural plate begin to grow dorsally over it as two folds of epidermis (Fig. 8–3 B). The edges of these two folds approach each other and then fuse, completely covering the neural plate with a sheet of epidermis (Figs. 8–3 C,D; 8–4). This process commences in the region of the blastopore and progresses cranially. As the neural plate sinks below the surface, its lateral edges begin to fold upward and join dorsally, converting the neural plate into the neural tube (Fig. 8–3 E,F). The folding does not immediately close off the neural tube either cranially or caudally but leaves an anterior opening to the ouside, the *anterior neuropore*, and a posterior opening into the gastrocoele, the *neurenteric* canal (Fig. 8–4). This canal arises because the epidermis that covers the posterior part of the neural tube is derived from tissue ventral to the blastopore, which thus not only covers the neural tube but also the blastopore. The epidermis shuts off the opening of the gastrocoele to the outside but leaves it in communication with the cavity of the neural tube. The neurenteric canal persists for only a short time.

The gastrocoele reestablishes an opening to the outside posteriorly and also forms one anteriorly—the anal and oral openings—but only at a much later stage of development.

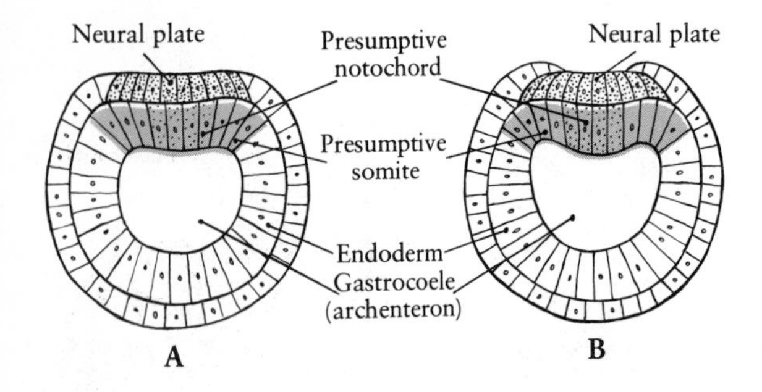


*Notochord, Mesoderm, and Intestine*

These are all formed from the inner layer of cells surrounding the gastrocoele (Fig. 8–3). The gastrocoele developes three longitudinally running outpocketings, one along the midline and the other two dorsolaterally on either side of the first. When these are completely pinched off, the result is the formation of four tubelike structures (Fig. 8–3 E,F). The middorsal tube will form the notochord, the two dorsolateral tubes will form the mesoderm, and the tube remaining after the others have pinched off will become the alimentary tract. The mesodermal bands soon show transverse divisions forming segmentally arranged somites. From the somites, mesoderm extends laterally and ventrally between the gut and the overlying ectoderm forming the lateral mesoderm. The lateral mesoderm splits into two sheets surrounding a cavity, the *coelom*. The inner of these sheets becomes associated with the gut and the outer with the ectoderm. They are named the *splanchnic* and the *somatic* mesoderm. The combined endoderm and mesoderm is called

**FIG. 8–2.** Gastrulae of *Amphioxus.* A, elongation in a craniocaudal direction, the blastopore representing the future caudal end of the embryo; B, ectoderm growing over the blastopore as the neural tube develops. Cavity of the neural tube is connected to the gastocoele by way of the neurenteric canal. (From A. F. Huettner, 1972. Comparative Embryology of the Vertebrates. Copyright 1949 by Macmillan Publishing Co., Inc. Renewed 1977 by M. R. Huettner, R. A. Huettner, and R. J. Huettner.)

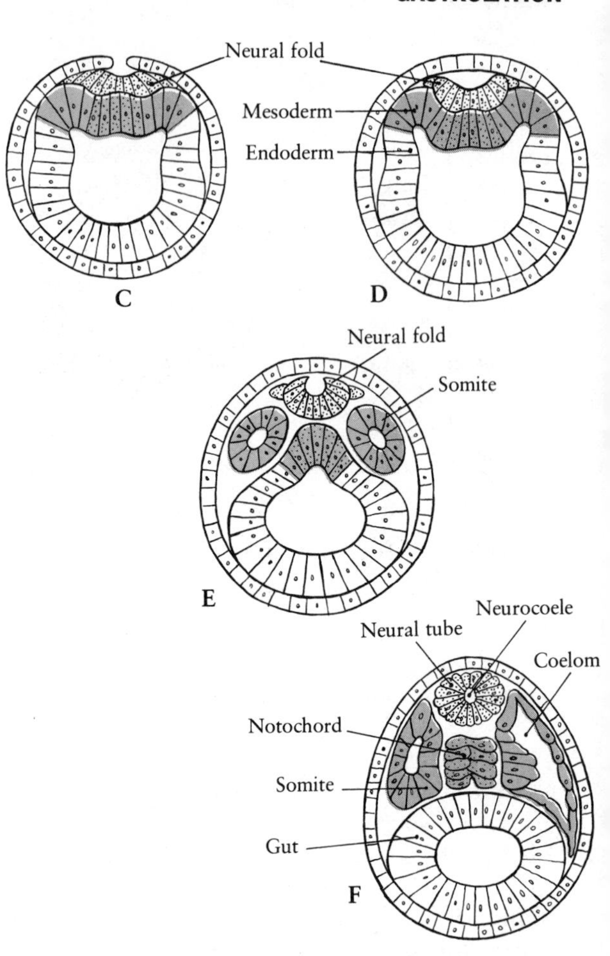


**FIG. 8–3.** Transverse sections through the *Amphioxus* embryo during the early differentiation of the embryonic axis showing the formation of the neural tube from the outer layer of cells and the endoderm, notochord, and mesoderm from the inner layer of cells. (From A. F. Huettner, 1972. Comparative Embryology of the Vertebrates. Copyright 1949 by Macmillan Publishing Co., Inc. Renewed 1977 by M. R. Huettner, R. A. Huettner, and R. J. Huettner.)

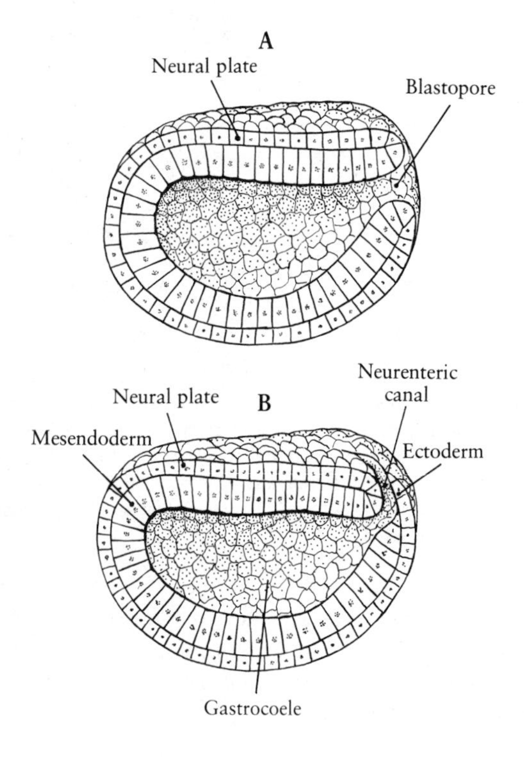

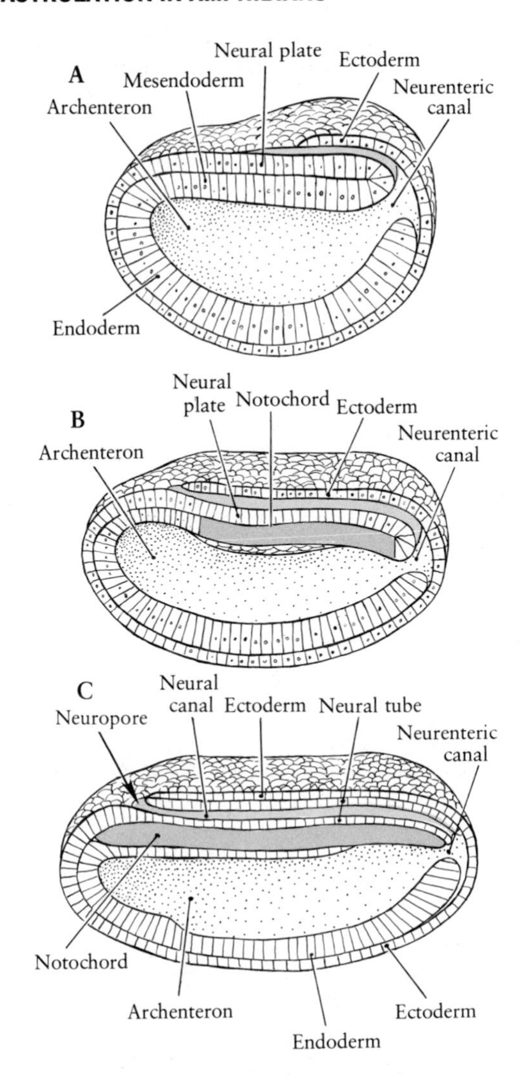


**FIG. 8–4.** Midsagittal sections of *Amphioxus* during the formation of the neural tube. These diagrams should be used in conjunction with the transverse sections in Figure 8–3. (From A. F. Huettner, 1972. Comparative Embryology of the Vertebrates. Copyright 1949 by Macmillan Publishing Co., Inc. Renewed 1977 by M. R. Huettner, R. A. Huettner, and R. J. Huettner.)

the *splanchnopleure,* and the combined ectoderm and mesoderm, is called the *somatopleure.*

## Gastrulation in Amphibians

The end result of gastrulation in the amphibians is the same as it is in *Amphioxus*—the conversion of an undifferentiated blastula into a trilaminar gastrula in which the inner sheets of cells are destined to form endodermal and mesodermal structures, and the outer layer of cells is destined to form ectodermal derivatives. However, because of the large amount of yolk in the amphibian egg, the movements by which these events are brought about differ from the relatively simple type of gastrulation in *Amphioxus.*

The first indication of gastrulation is the appearance of a small groove on the surface of the blastula just ventral to the position of the grey crescent (Fig. 8–5 A). This slit is the beginning of the blastopore and represents the area where the surface cells are beginning to move into the inside. The slit-shaped blastopore increases in length and becomes sickle-shaped, then semicircular, then horseshoe-shaped, and finally the ends of the horseshoe met to form a circular blastopore surrounding a plug of endodermal material, the *yolk plug* (Fig. 8–5 B–F). The region where the blastopore first forms is called the *dorsal lip* of the blastopore, and the region of the completion of the circular blastopore is the *ventral lip* of the blastopore. Between them, around the periphery, are the *lateral lips.*

An extensive series of vital stain studies by W. Vogt (1929) presented a comprehensive and coherent study of the complicated cellular movements that occur during amphibian gastrulation. Reference to presumptive fate maps and to the figures representing the results of vital staining experiments is essential to the understanding of gastrulation in the amphibians. However, the results of the vital staining experiments should not be misinterpreted. Because particular areas on the surface of the blastula may be shown to undergo specific movements that result in their becoming a part of certain tissues or organs does not mean that the blastula is a mosaic of discrete regions differing from one another morphologically. These experiments merely reveal that under normal conditions of undisturbed development certain regions of the blastula will develop in accordance with their position within the whole. These regions have a certain fate. However, the blastula is not a mosaic of predetermined organs. Vital stains reveal nothing of the intrinsic properties or potencies of the regions they mark, and the vast majority of the regions of the blastula have the potency of differentiate into a much greater variety of structures than the particular fate indicated by the vital stain. These potencies, not revealed by the technique of vital staining, may be analyzed by transplantation and isolation experiments as will be described in a later chapter.

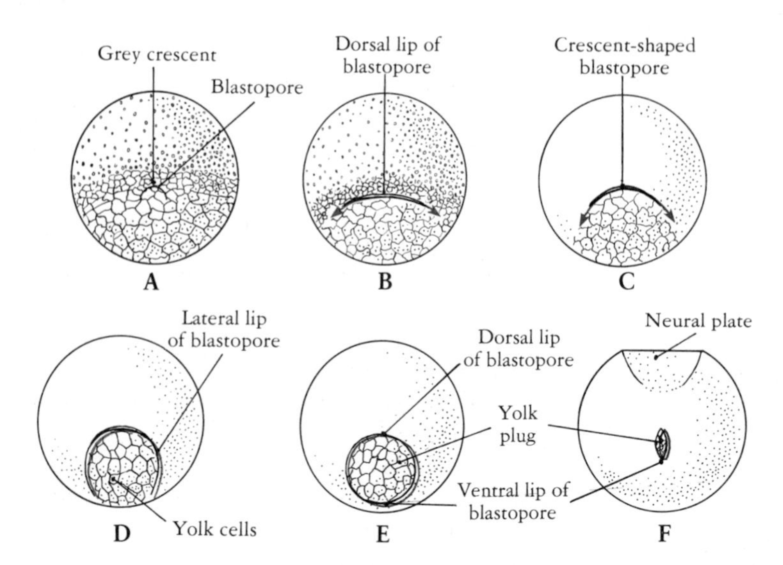


**FIG. 8–5.** Diagram of the development of the blastopore in the frog seen from the caudal aspect.

Examination of an amphibian presumptive fate map (Fig. 8–6) gives an excellent indication of the general type of movement that takes place during gastrulation. One important line on the map is that which separates the invaginating material, prospective endoderm and mesoderm, from the noninvaginating material, prospective ectoderm. The prospective endoderm is seen as a disk-shaped region surrounding the vegetal pole. Between it and the ectoderm is a girdle of mesoderm, often called the *marginal zone*. The mesoderm of the grey crescent region is mostly presumptive notochord. Lateral to the presumptive notochord in the marginal zone are the presumptive somite, lateral, and tail mesoderm, respectively. It is quite obvious from the lcoation of the blastopore that the presumptive areas marked out on the surface of the blastula must undergo considerable movement during gastrulation, both on the surface and on the inside. The general pattern of convergence of the marginal zone over the surface of the blastula toward the blastopore and its elongation underneath the surface is illustrated in Figure 8–7. Of particular interest are those structures that, at the end of gastrulation, are oriented along the craniocaudal axis, particularly the notochord and the nervous system. In the early blastula, they have a mediolateral orientation, perpendicular to the future craniocaudal axis (Fig. 8–8).

**FIG. 8–6.** Generalized presumptive fate map of the amphibian showing the position of the presumptive areas on the surface of the early gastrula at the time of the first appearance of the blastopore. A, caudal view; B, lateral view.

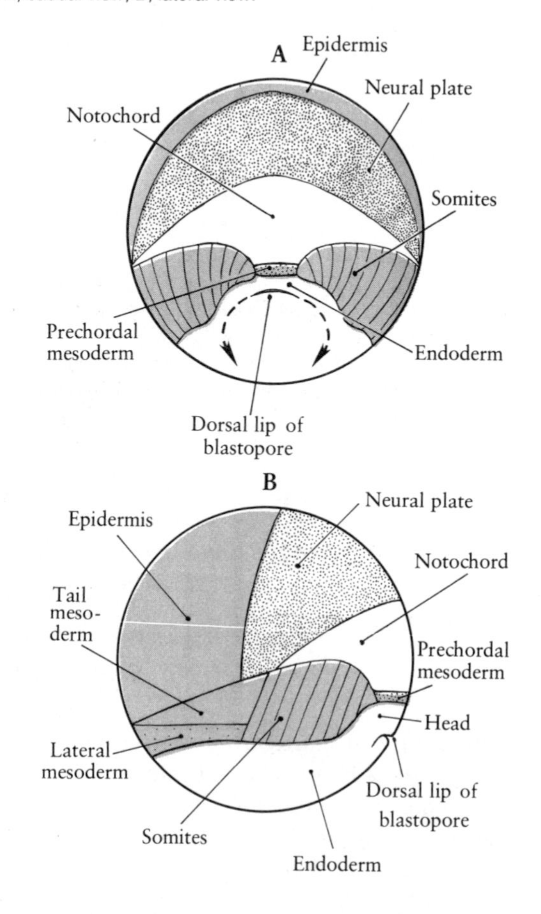


## Movements of the Presumptive Mesoderm

Separation of the mesoderm from the endoderm follows somewhat different patterns in the anurans

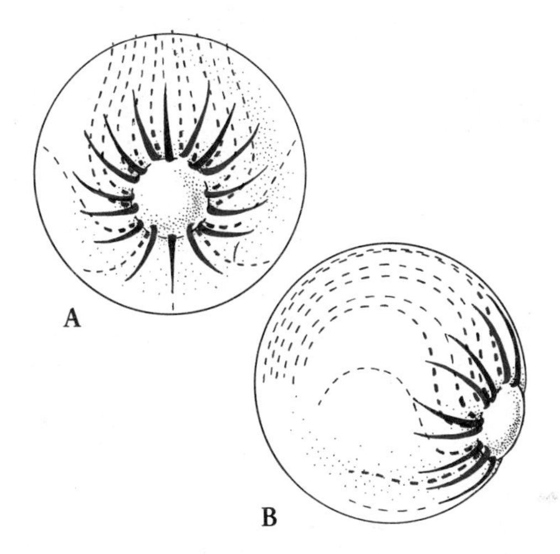


**FIG. 8-7.** Scheme of convergence of the mesoderm toward the blastopore (heavy lines) and its elongation in a craniocaudal direction (dotted lines) once it has involuted. (After W. Vogt, 1929.)

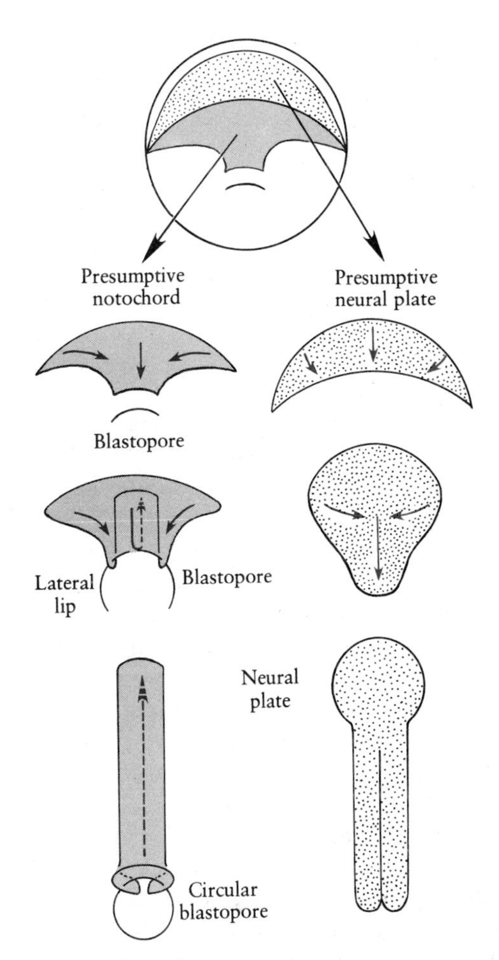


**FIG. 8-8.** Diagram of the changes in shape of the presumptive chordamesoderm and neural plate regions during gastrulation. Solid lines represent surface movements, and dotted lines represent movements after involution. For both regions there is a change from a side-to-side orientation before gastrulation to a craniocaudal orientation after gastrulation. (From Embryology, revised and enlarged edition by Lester George Barth. Copyright 1949 and 1953 by Holt, Rinehart and Winston, Inc. Reprinted by permission of Holt, Rinehart and Winston.)

and the urodeles. In the anurans, the mesoderm does not split off from the endoderm until late in gastrulation, after invagination, whereas, in the urodeles this separation occurs during invagination and the two germ layers move into the interior of the gastrula as separate units. The gastrular movements associated with the formualtion of the mesoderm and the endoderm described in this section refer primarily to those occurring in the urodeles as originally presented in Vogt's comprehensive vital staining studies.

Two important facts in regard to the formation of the blastopore have already been presented. First, the blastopore develops in the region of the presumptive endoderm; and second, it shows a definite sequential growth pattern progressing from an original slit-shaped structure through stages to a circular structure.

The first statement indicates that the presumptive endoderm must be the first material to involute over the dorsal lip of the slit-shaped blastopore (Fig. 8–6). It is followed by the cells lying just below the presumptive notochord, which next involute over the dorsal lip and move cranially to form the *prechordal plate.* This is the first mesoderm to involute and it is, in turn, followed by the presumptive notochord.

The second statement about the development of the blastopore indicates that there is a time sequence in the involution of the presumptive areas corresponding to the stepwise formation of the blastopore. That is, the first mesoderm to move into the interior is that which involutes over the dorsal lip at the beginning of the formation of the blastopore to form prechordal plate and notochord. As the lateral lips develop, somite material moves over the surface and involutes over the lateral lips and is followed by lateral mesoderm. Finally, tail mesoderm involutes over the ventral lip at late stages in gastrulation. The smaller amount of material involuting over the lateral, and particularly the ventral, lips is indicated by the narrowness of the marginal zone in the region of these presumptive areas (Fig. 8–6).

It may help to visualize the cell movements dur-

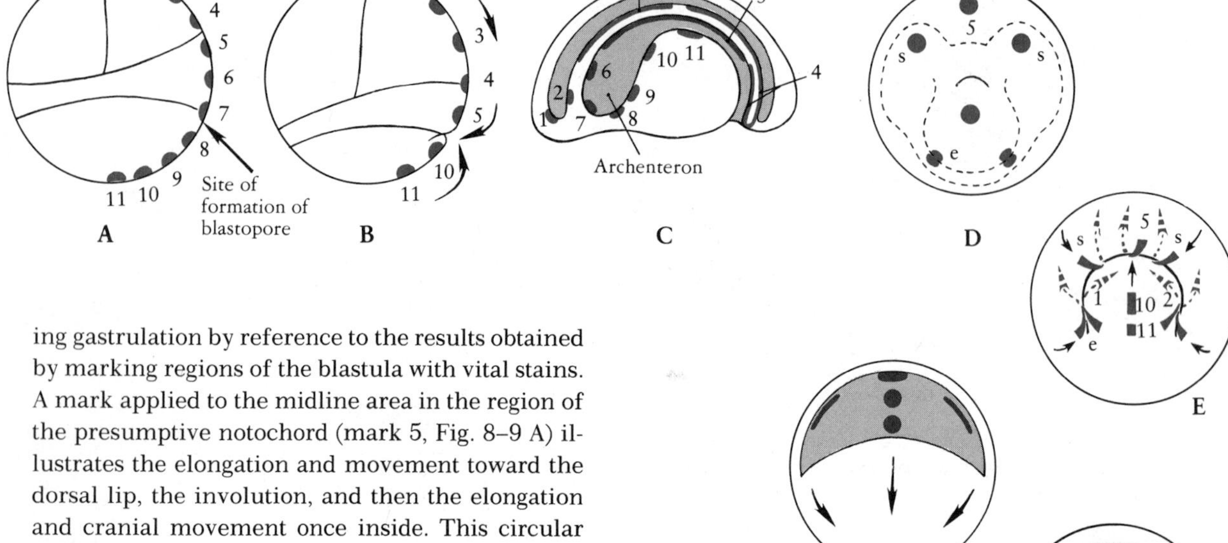


FIG. 8–9. Vital staining experiments on anuran gastrulae. A–C, movements of marks placed along the middorsal region of the embryo running from the animal to the vegetal pole and passing through the middle of the presumptive chordamesoderm (refer to presumptive fate map, Fig. 8–6). A, before gastrulation; B, midgastrula—marks 6–9 have passed over the rim of the blastopore; C, completion of gastrulation. D, E, movement of marks placed on presumptive mesoderm and presumptive endodermal material seen from the caudal aspect; F, G, surface movements of marks placed on presumptive neural plate tissue. (From V. Hamburger, 1960. A Manual of Experimental Embryology. The University of Chicago Press.)

ing gastrulation by reference to the results obtained by marking regions of the blastula with vital stains. A mark applied to the midline area in the region of the presumptive notochord (mark 5, Fig. 8–9 A) illustrates the elongation and movement toward the dorsal lip, the involution, and then the elongation and cranial movement once inside. This circular mark ends up on an elongated group of cells forming the roof of the archenteron (Fig. 8–9 C). Marks placed on the presumptive somite material (mark s, Fig. 8–9 D) also converge toward the dorsal lip, involute, and continue to converge slightly as they move cranially. A single small circular mark will stain adjoining regions of a number of somites, indicating again the elongation that takes place in this material, particularly after it involutes.

Although we consider notochord, somite, lateral, and tail mesoderm as separate presumptive regions, they are, of course, not physically separated from each other during gastrular movements. The entire mass of mesoderm involutes as a continuous sheet of cells, which may be called the mesodermal mantle (Fig. 8–10). The mesodermal mantle shows its greatest development and undergoes its largest movements in the presumptive notochord and somite regions.

As gastrulation continues, the gastrocoele replaces the blastocoele. This cavity is formed directly ventral to the earliest involuting mesodermal material, and thus the chordamesoderm during gastrulation forms the first (although only temporary) roof of the archenteron before it is replaced by the dorsal growth of the endoderm (Figs. 8–11, 8–12).

## Movements of the Presumptive Endoderm

As previously mentioned, the blastopore first appears in the region of the presumptive endoderm, and presumptive endoderm is the first material to involute over the dorsal lip (Fig. 8–6). Inside of the gastrula this material moves cranially and forms the floor and walls of the archenteron and the roof of the most cranial end of the archeteron. However, by far the largest amount of endoderm enters the interior of the gastrula in another manner. A mark placed in the middle of the presumptive endoderm illustrates this type of movement (mark 10, Fig. 8–9). The mark approaches the area of the dorsal lip of the blastopore and sinks into the interior below the dorsal lip. The movement of the presumptive endoderm, which is made up of large cells contain-

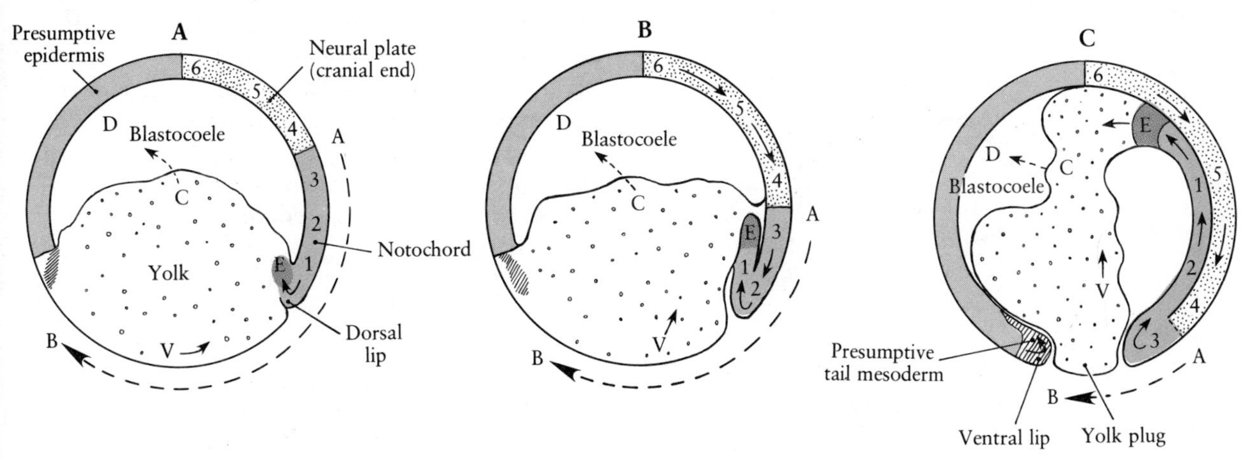


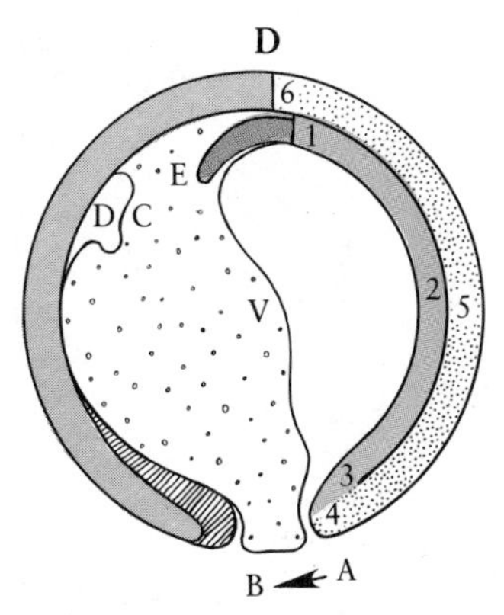


**FIG. 8–11.** Diagram of midsagittal sections of anuran embryos from the time of the first appearance of the blastopore to the yolk-plug stage. Groups of cells are differentiated and marked so that their movements may be followed. (From Embryology, revised and enlarged edition by Lester George Barth. Copyright 1949 and 1953 by Holt, Rinehart and Winston, Inc. Reprinted by permission of Holt, Rinehart and Winston.)

ing considerable amounts of inert yolk, is much less extensive and less active than that of the mesoderm. Marks 9 to 11 (Fig. 8–9), which sink in and become a part of the floor of the archenteron, show considerably less stretching than their counterparts (marks 4 and 5), which form the temporary roof of the archenteron.

The marks placed along a semicircular line through the middle of the gray crescent in Figure 8–9 A stain a continuous line of tissue on the surface of the blastula. When the blastopore forms between marks 7 and 8, these areas reach the interior through different routes: mark 7 involuting over the dorsal lip, mark 8 sinking in below the dorsal lip. Nevertheless, these marks still stain a continous line of tissue on the interior of the gastrula. The prechordal plate and the notochord also involute over the dorsal lip, trailing behind mark 7, while marks 9 to 11 follow behind mark 8. Inside of the gastrula they still form a continuous sheet of tissue, a part of which forms the roof of the archenteron—the presumptive chordamesoderm—and a part of which forms the floor of the archenteron—the presumptive endoderm. Figure 8–11 also illustrates the movements of the presumptive chordamesoderm and endoderm, as seen in midsagittal section, and diagrams the replacement of the blastocoele by the gastrocoele. A careful study of Figure 8–11 should give a clear indication of the cellular move-

**FIG. 8.–10.** Movement of the mesodermal mantle after involution. Arrows represent direction of movement and dotted lines represent the location of the leading edge of the mantle at four successive stages. (From V. Hamburger, 1960. A Manual of Experimental Embryology. The University of Chicago Press.)

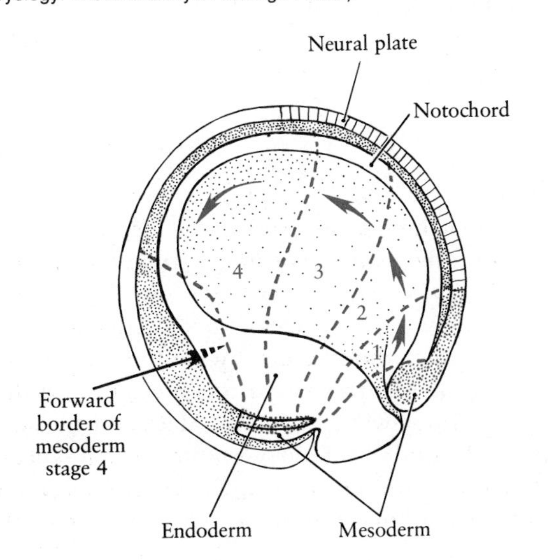

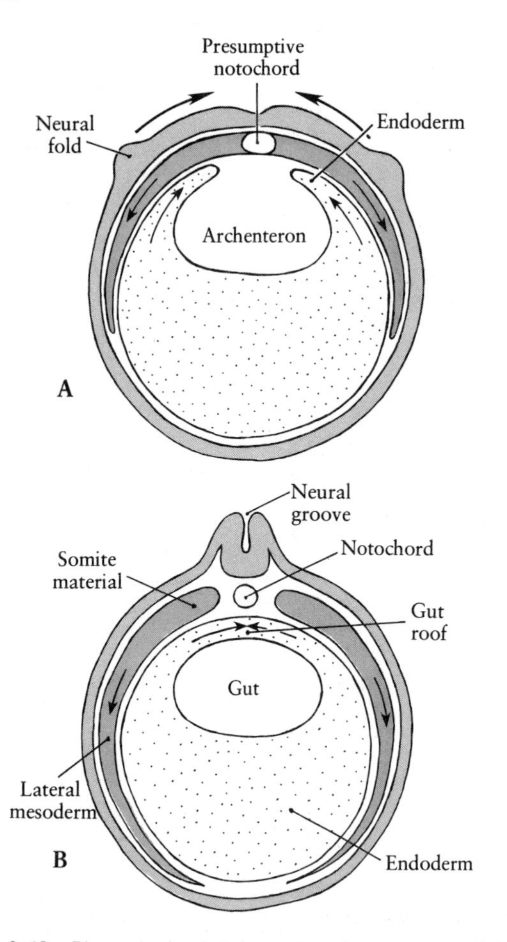


**FIG. 8–12.** Diagrams showing the progressive movement of the mesoderm and endoderm inside the amphibian gastrula. A, the chrodamesoderm forms the temporary roof of the archenteron. The ventral movement of the mesoderm and the dorsal movement of the endoderm are indicated by the dotted arrows; B, the dorsal movement of the endoderm has resulted in the formation of a gut cavity completely surrounded by endoderm. (From Embryology, revised and enlarged edition by Lester George Barth. Copyright 1949 and 1953 by Holt, Rinehart and Winston, Inc. Reprinted by permission of Holt, Rinehart and Winston.)

ments that are taking place in the regions of the dorsal and ventral lips of the blastopore.

A mark placed exactly where the lateral lip of the blastopore will develop (mark e; Fig. 8–9 D) illustrates the manner in which the endoderm and the mesoderm are separated at the time they pass into the interior of the gastrula. Half of the mark will involute over the lateral lip of the blastopore as a part of the mesoderm mantle to form the lateral mesoderm, but the other half will invaginate with the endodermal mass to form a part of the wall of the archenteron. Regions that are in contact with each other on the surface of the blastula may thus have very different presumptive fates.

Once inside the gastrula, the movements of the presumptive endoderm and mesoderm are quite different. Although both tissues elongate as they move cranially, the mesoderm also shows a ventral migration (Fig. 8–12). The endoderm, in contrast, moves dorsally as well as cranially (Fig. 8-12). Originally, the endoderm forms only the floor and the walls of most of the archenteron, but its dorsal movement up the sides of this cavity is continued to a point where the two lateral sheets of endoderm meet dorsally, and thus the archenteron becomes completely lined with endoderm (Fig. 8–12 B). The ventral movement of the mesoderm carries it between the ectoderm and the endoderm, where it also finally forms a complete layer.

Recently, Løvtrup (1975) reviewed Vogt's vital staining experiments and presented a different interpretation of their significance. He proposed that the material on the surface of the blastula below the limit of invagination consists only of the notochord and endoderm, and these should be the only features shown on fate maps referring to the superficial layer. He concluded that the presumptive mesoderm was located in the same area pictured by Vogt but was represented by a ring of small spherical cells beneath the surface. His contention was that when presumptive mesoderm cells were stained, the stain had actually marked these cells below the surface layer. Keller (1976) has reported vital staining experiments of *Xenopus*. He also concludes that only presumptive endoderm is located on the surface of the blastula below the limit of invagination and that all of the presumptive mesoderm and also the notochord are located in a deeper layer below the surface.

## Movements of the Prospective Ectoderm

Although all of the ectoderm always remains on the surface of the gastrula, this tissue also undergoes extensive movement during gastrulation. As the mesodermal mantle passes to the inside, the ectoderm remaining in contact with it also moved over the surface of the gastrula toward the blastopore. And also, as the mesoderm converges toward the midline, so does the ectoderm. Marks on the surface in the midline of the presumptive neural plate move toward the dorsal lip and undergo elongation as they do so (marks m,n, Fig. 8–9 F,G). They re-

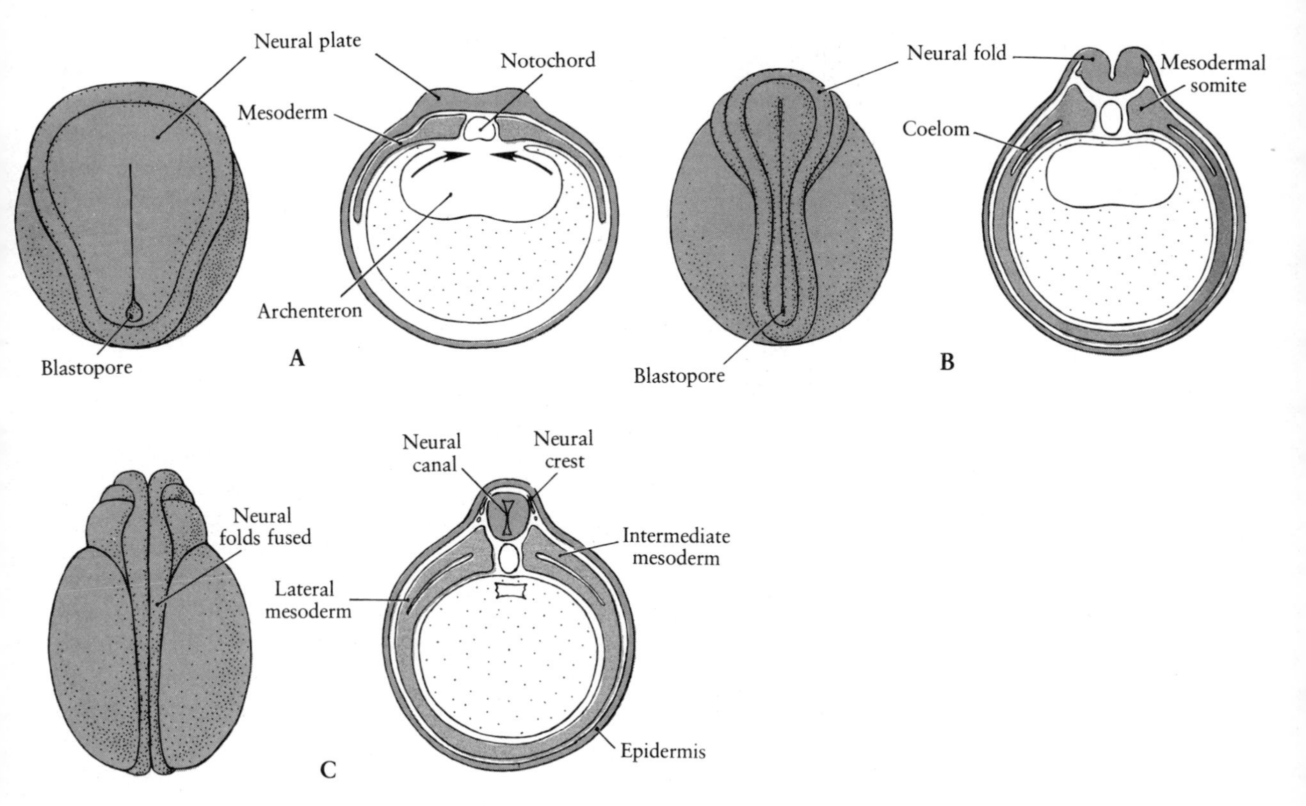


**FIG. 8–13.** Three stages in neurulation in the amphibian embryo. The drawings on the left in A, B, and C represent dorsal views of whole embryos at successively later stages. The drawings on the right represent cross sections of the embryos on the left.

main in the midline. Marks of the lateral wings of the presumptive neural plate show movements of both elongation and convergence (Fig. 8–9 F,G).

The remainder of the ectoderm, the presumptive epidermis, also undergoes extensive movement, retaining its contacts with the presumptive neural plate and with the presumptive mesoderm and moving in conformity with these contacts. Its type of movement could be compared with the opening of a fan.

## Formation of Organ Rudiments

The chordamesoderm develops into notochord and mesoderm. A middorsal rod of cells separates from the remainder of the mesoderm to form the primordium of the notochord (Figs. 8–12, 8–13). The portion of the mesoderm adjacent to the notochord develops a series of transverse fissures, which forms it into longitudinally oriented bands of somites, one band on either side of the notochord (Figs. 8–12, 8–13). The transverse separations occur only in the most dorsal part of the mesoderm, the more lateral and ventral regions remaining unsegmented. Adjoining the somites, between them and the lateral mesoderm, the intermediate mesoderm, which is presumptive nephrogenous tissue, develops. Lateral to the intermediate mesoderm, the mesoderm separates into somatic and splanchnic layers surrounding a cavity, the embryonic coelom (Fig. 8–13 B).

The presumptive neural plate at the end of gastrulation is an elongated oval area overlying the cranial end of the archenteron, the prechordal plate, the notochord, and the somites. Soon the edges of the plate thicken and are raised above the surface of the embryo to form the neural folds (Fig. 8–13 B). As the neural folds continue to elevate they meet in the middorsal line to form the neural tube (Fig. 8–13 C), which will form the brain anteriorly and the spinal cord posteriorly. The cavity of the neural tube is the primordium of the ventricular system of the brain and the central canal of the spinal cord.

Fusion of the neural folds occurs first in the region of the future hindbrain and progresses cranially and caudally. The neural plate is much broader in the head region; and at the time of the formation of the neural folds, the cranial end can be recognized by its greater size and earlier differentiation (Fig. 8–13).

As the neural folds elevate, they carry with them the adjacent epidermal cells; and when the neural folds fuse, the epidermis grows over the neural tube to form its epidermal covering. As this occurs, the longitudinal band of cells that originally formed the outer limit of the neural plate is cut off and appears on either side in the pocket between the neural folds and the epidermis (Fig. 8–13). This is the *neural crest.*

## Gastrulation and the Formation of the Organ Rudiments in Fishes

The process of gastrulation in fishes is not as well understood as it is in such vertebrates as the amphibians and the birds. The construction of fate maps of late-blastula embryos and the examination of morphogenetic movements underlying the formation of germ layers is still far from complete. Many of the studies on gastrulation in fishes date back to the 1930s and 1940s, when techniques for tracing the movements of cells were limited to the use of vital dyes. More recent studies by Betchau and Trinkaus (1978) on *Fundulus* and by Ballard (1973 a,b; Ballard and Ginsburg, 1980) on *Salmo* and *Acipenser* (the freshwater sturgeon) illustrate the complexity of gastrulation and the remarkable contrast between the morphogenetic movements and fate maps among the bony fishes.

In primitive bony fishes (*Polypterus* and *Acipenser*) and in the lungfishes, the fertilized egg cleaves in holoblastic fashion and produces a blastula whose organization is very similar to that of the frog. The late blastula of *Acipenser* consists of a small, irregular segmentation cavity displaced toward the animal pole; several hundred micromeres, organized into several layers, lie above the cavity, while about 50 macromeres constitute the whole vegetal hemisphere. Between these two regions is the marginal zone; here the cells are about one-half to one-fifth the diameter of the macromeres (Fig. 8–14 A). Ballard and Ginsburg (1980), by placing a large series of small spots of vital Nile blue sulfate dye on late-blastula embryos, have constructed a fate map of the sturgeon (Fig. 8-15). The marginal zone cells produce principally the endoderm of the roof of the archenteron and the macromeres the endoderm of the floor of the archenteron. Dye solution markings confirm that the entire lining of the gut is derived from the transfer of cells from the surface layer of the blastula to the interior. Above the zone of prospective endoderm is a complete band of cells that will form notochord and somite mesoderm; most of the somite mesoderm, however, lies beneath the surface layer of the blastula. Head mesoderm and intermediate and lateral plate mesoderm are located completely inside the surface of the blastula. This condition is similar to that in the late blastula of *Xenopus,* where only prospective ectoderm and endoderm cells are in its surface layer. The neural plate in *Acipenser* is represented by cells both at and immediately inside the surface of the blastula.

What happens in *Acipenser* when the morphogenetic movements of gastrulation are initiated? The movements of cells of the embryonic surface can be charted, as in frogs and salamanders, by recording the successive positions and shapes of particular dye spots at regular intervals of time. The first indication of gastrulation is the appearance of a shallow furrow in the transition area between the marginal zone and the large macromeres of the vegetal hemisphere (Fig. 8-14 B). Quickly, a dorsal lip

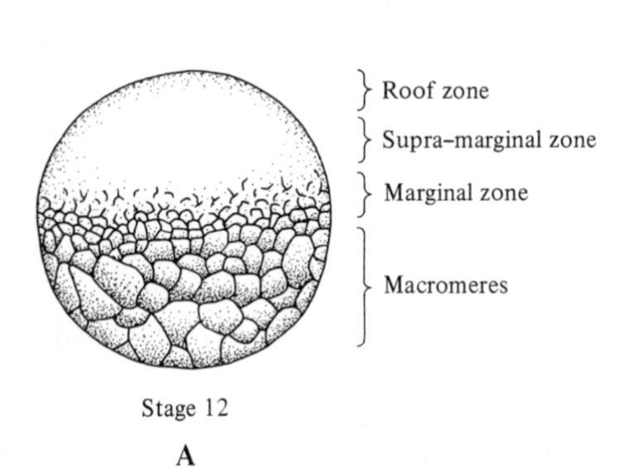


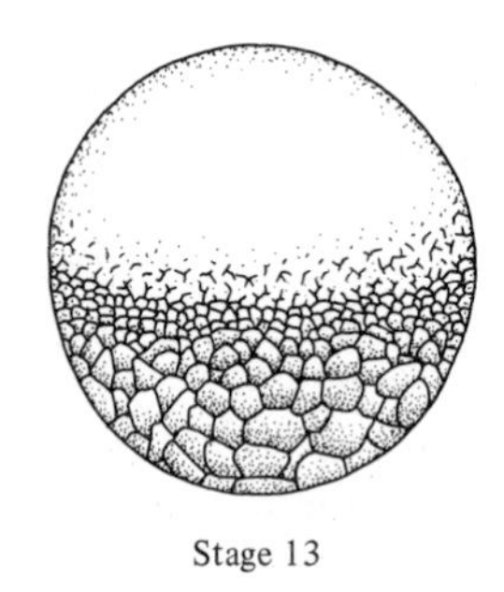


**FIG. 8–14.** Lateral views of *Acipenser* embryos at blastula (A) and gastrula (B) stages. (From W. Ballard and A. Ginsburg, 1980. J. Exp. Zool. 216:69.)

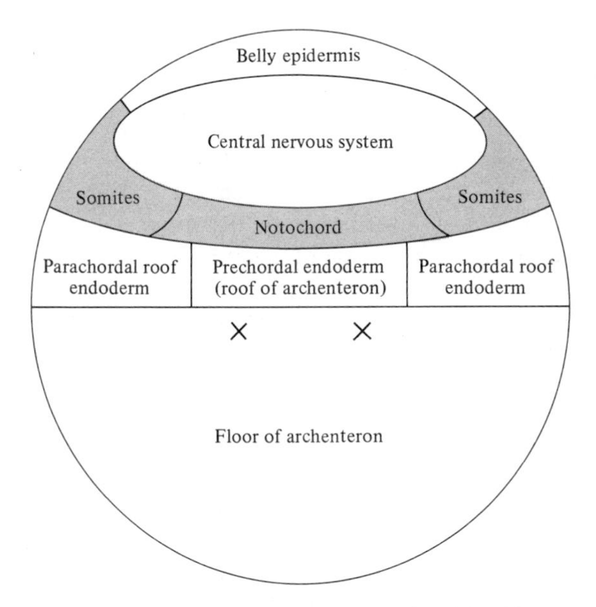


**FIG. 8–15.** Fate map of *Acipenser* as seen in dorsal view. The area between the *x*'s marks the site of the initiation of gastrulation. Most of the prospective somite mesoderm and prospective cells of the nervous system lie beneath the surface. (After W. Ballard and A. Ginsburg, 1980. J. Exp. Zool. 213:69.)

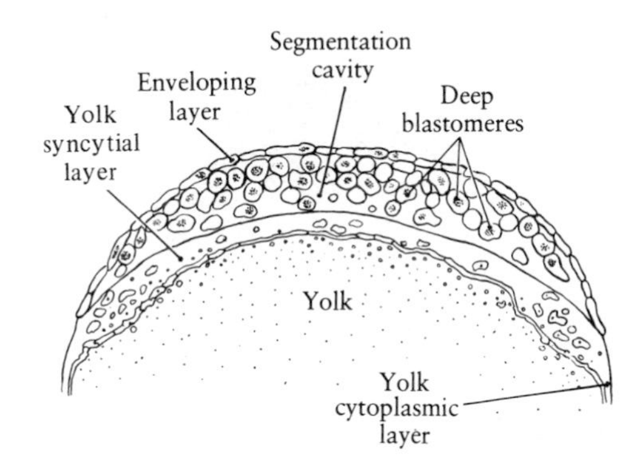


**FIG. 8–16.** Diagram of the blastoderm of a typical bony fish *(Fundulus).* (After T. Lentz and J. P. Trinkaus, 1967. J. Cell Biol. 32:121.)

develops and the presumptive endoderm of the roof of the archenteron rolls over it by involution to become internalized. As the invaginating groove spreads laterally to form the circular blastopore, more of the external marginal cells disappear from the surface. Concurrently, the dorsal lip rolls downward to enclose passively the macromeres of the vegetal hemisphere, a consequence of the epibolic stretching of epidermal and neural cells on the surface.

As previously described, the late-stage blastoderm of teleost fishes consists of a cellularized mass lying upon a multinucleated *yolk syncytial layer* (periblast) that is continuous peripherally with the yolk cytoplasmic layer (Fig. 8–16). The yolk syncytial layer is closely associated with the uncleaved yolk. Analysis of the blastoderm shows that it also consists of a superficial layer of enveloping blastomeres, a tightly coherent sheet one cell in thickness, and deeper-lying, rounded blastomeres (Fig. 8–16). Only the blastoderm is responsible for the formation of the various parts of the embryo. The yolk syncytial layer, the yolk mass, and the cytoplasmic layer enclosing the yolk are extraembryonic, since they make little if any contribution to the formation of the embryo.

Our knowledge of the potencies or fate of the cells of the late-stage blastoderm of fishes is scarce and, in the specific case of teleosts, a matter of considerable controversy. For example, until recently only two fate maps had been published for all teleost fishes. The early studies of Oppenheimer on *Fundulus* and Pasteels on *Salmo* (trout), in which spots of vital dye placed upon the surface of the pregastrula embryo were traced during gastrulation, suggested that both prospective endoderm and mesoderm were located on the surface of the blastula. They hypothesized that these two populations of cells were wheeled into place in the interior by a movement of invagination along the *germ ring,* the latter being a temporary thickening of the entire outer edge of the blastoderm. The germ ring forms as the time of gastrulation approaches. In other words, the edge of the blastoderm was postulated to function as a blastopore through which the cells of endoderm and chordamesoderm were invaginated, much as in the frog and salamander.

More recent investigations on *Salmo,* particularly those by Ballard (1973 a,b), have questioned the accuracy of teleost fate maps and challenged the traditional view that the germ layers are formed during gastrulation as a result of invagination. Using the technique of implanting chalk particles at selected sites on the surface layer and in the deeper portions of the blastoderm, Ballard has rigorously charted the movements of cells during gastrulation. A fate map based on his observations is shown in Figure 8–17. In contrast to the case of *Acipenser,* the fate map of *Salmo* is very complex and three-dimensional, with zones of cells overlapping and at different depths in the blastoderm. Surprisingly, analysis of the fate map shows that the cellular en-

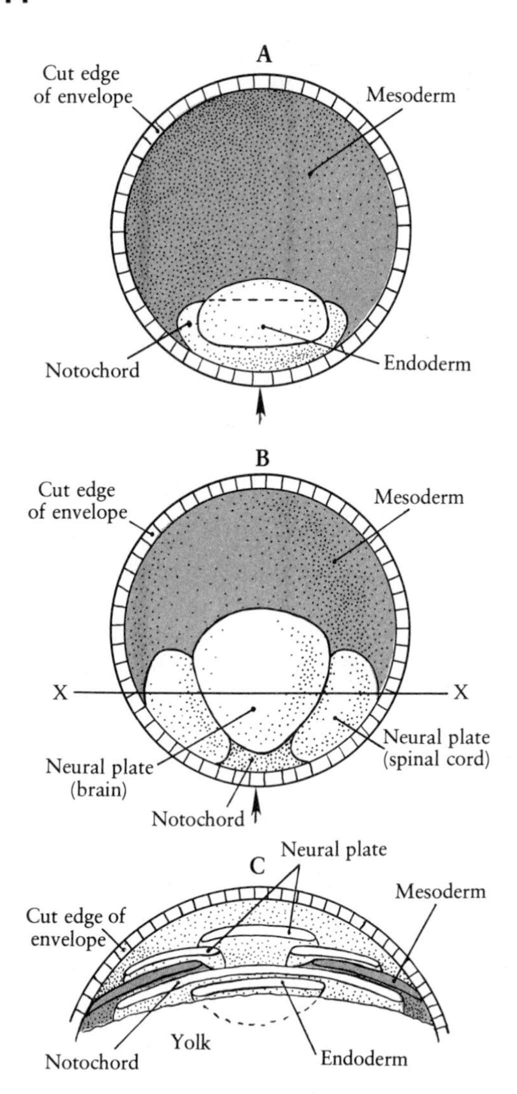


**FIG. 8–17.** Three-dimensional fate map of the trout blastoderm as shown (A) from below and (B) from above. The arrow marks the axis of symmetry. A, the area of prospective notochord underlies the general mesodermal sheet, and is itself underlaid by the area of prospective endoderm; B, the neural areas overlie the general mesodermal sheet. The cellular envelope lies superficial to all other areas but is only shown at the margin of the blastoderm; C, a typical transverse section through the posterior end of the blastoderm at level of X in (B). (After W. Ballard, 1973b. J. Exp. Zool. 184:49.)

velope that covers the entire surface of the blastoderm will only give rise to the ectoderm. A broad sheet of cells forms the prospective mesoderm and will contribute to the somites of the trunk and tail, to the lateral plates, and the heart. It is continuous throughout the "middle portion" of the blastoderm, tending to be the only cell type in the nonaxial portion of the blastoderm. Along the axial portion of the blastoderm, at different levels, are cells contributing to the notochord, the nervous system, and the endoderm. All of these prospective cell groups are confined to the "posterior" half of the blastoderm.

Hence, the studies of Ballard lead to the conclusion that the prospective endoderm and chordamesoderm are already present in the interior of the blastoderm at the time of gastrulation. The endodermal and mesodermal layers of the embryonic trunk (i.e., germ layer formation) do not arise by invagination or ingression, but by the rearrangement and convergence of blastomeres beneath the enveloping layer.

At the onset of gastrulation, the high mound of blastoderm cells flattens slightly. From the floor of the blastoderm, cells begin to disengage and drift centrifugally along the yolk syncytium. These cells accumulate as a diffuse *hypoblast* at the perimeter of the blastoderm to form the germ ring, and within the posterior quadrant of the blastoderm, which will become the *embryonic shield.* The cells of the hypoblast will form the endoderm and are initially organized as a sheet beneath the shield. Other disengaged internal cells *(nubbin cells)* aggregate on the rim of this quadrant to form a terminal node, the first external indication of the plane of bilateral symmetry of the embryo. These movements leave the central portion of the blastoderm to consist of the enveloping layer and a thin zone of consolidated deep blastomeres; cells of these two regions constitute the *epiblast.* The node cells, probabaly representing prechordal plate mesoderm, appear to form a firm association with the yolk syncytical layer; this anchorage may be a significant factor in coordinating the convergent movements of the deep blastomeres that result in the assembling of the embryonic axis and the definition of the primary germ layers.

Over the next several days, the blastoderm will spread by epiboly of the germ ring over the yolk mass and differentiate the embryo. These movements of cells have been studied by implanting chalk particles within the germ ring (Figs. 8–18, 8–19). The epiblast spreads over the yolk sphere by epiboly, leaving in its wake the ectodermal and mesodermal layers of the yolk sac. The hypoblast cells show two component movements. They not only spread epibolically over the yolk sphere, but they also converge from both sides toward the future axis of symmetry in which the nubbin lies. The result of convergence movements of the hy-

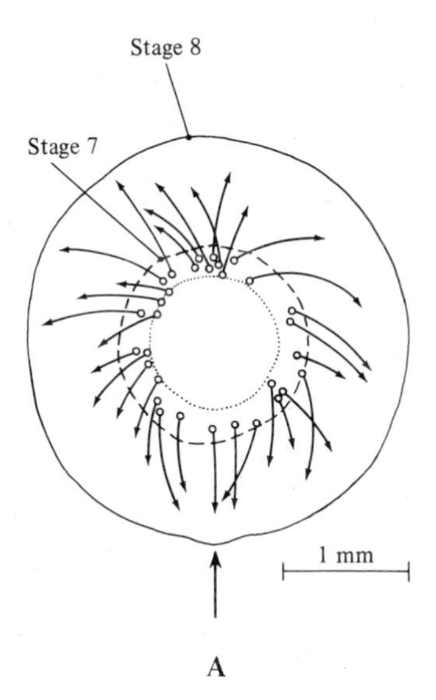


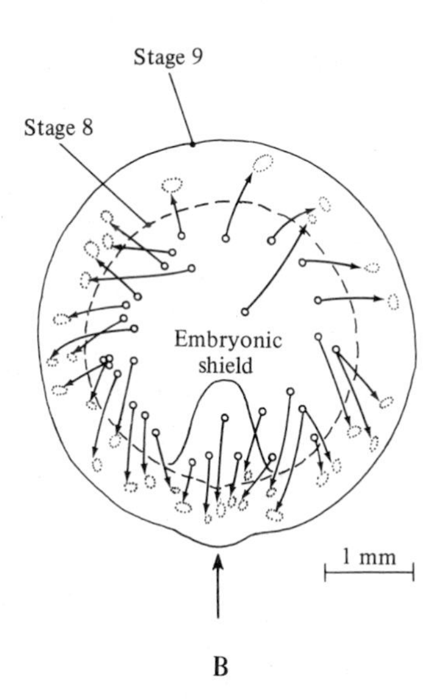


**FIG. 8–18.** Movements of the deep cells of the germ ring between stages 7 and 8 (A) and stages 8 and 9 (B) in the *Salmo* (trout) embryo. In both A and B, the margin of the blastoderm of the younger stage is shown within the contour of the blastoderm margin of the older stage. Chalk particles (indicated by small circles) were placed on the surfaces of deep cells of the germ ring at stages 7 (A) and 8 (B); the location of these particles was then examined at stages 8 (A) and 9 (B). The arrows indicate the direction and approximate extent of the movement of the chalk particles. (From W. Ballard, 1973a. J. Exp. Zool. 184:27).

poblast is the production of a broad, crescent-shaped thickening termed the embryonic shield (Fig. 8–20). It is within the embryonic shield that the primary organs of the embryo are laid down in anteroposterior progression, including the neural tube, the notochord, and the somites (Fig. 8–20). The germ ring does not appear to contribute any cells to either the notochord or the central nervous system. Hypoblast cells are the chief source of somite and lateral plate mesoderm.

The first structure to become visible in the teleost embryo is the *axial strand*, so named because it contains cells giving rise to two axial organs, the notochord and the neural tube (Fig. 8–20). The notochord will separate from the axial strand and adjacent mesodermal cells, a process completed by the time that the germ ring is approximately at the equator of the yolk sphere. The mesodermal cells converge on either side of the notochordal tissue and segment to form somites (Fig. 8–20). They form in the trout at the rate of about one pair per hour at 10°C. The more lateral wings of the mesoderm will be drawn toward the midline from the margins of the blastoderm as epiboly continues; these will contribute to the kidneys and lateral plates.

The process of neurulation in bony fishes differs in many details from the neural fold methods of other vertebrate organisms. In the trout, for example, the neural cells, following their separation from the axial strand, thicken above the notochord to form an elongated *neural keel*. Anteriorly, the presumptive brain portion of the keel presses down against the yolk syncytium. The ventricles of the brain and the central canal of the spinal cord are subsequently formed by the separation of cells within the neural keel.

Gastrulation and the formation of primary organ rudiments in the elasmobranch fishes are often described as following the same pattern as that observed in bony fishes. However, the recent investigations of Ballard showing extraordinary diversity among bony fishes in cleavage and patterns of morphogenetic cells movements raise serious questions regarding similarities in the process of gastrulation between the two major groups of modern fishes. Using the shark *(Scyllium)* as an example, it is generally presumed that the presumptive organ-forming areas are laid out on the surface of the blastoderm (Fig. 8–21 A). The notochord, mesoderm, and much of the endoderm appear to involute over the posterior margin of the blastoderm during gastrulation. In essence, the posterior edge of the blastoderm acts as a dorsal lip area. The result of gastrulation in the shark is the production of an embryo with germ layers arranged as shown in Figure 8–21

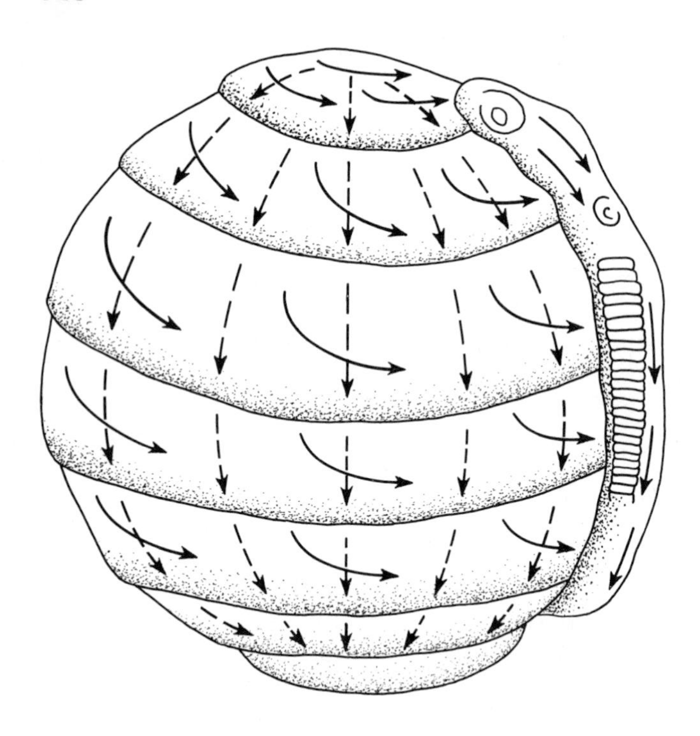

**FIG. 8–19.** Diagrammatic illustration of the movements of the epiblast (deeper cells of germ ring) and hypoblast cells during yolk sac and embryo formation in *Salmo.* The epiblast cells (dashed lines) move by epiboly around the yolk mass while attached to the yolk syncytial layer. The hypoblast cells (solid arrows) descend with the germ ring but also continually converge toward the body axis to form somites and lateral plates. (From W. Ballard, 1973a. J. Exp. Zool. 184:27.)

B. Similar to the amphibian and amniote embryo, the neural plate in the shark is rolled into the neural tube.

## Gastrulation and the Formation of the Organ Rudiments in Birds

Gastrulation in birds involves the same processes that take place during gastrulation in *Amphioxus* and the amphibians, and produces the same end results—a multilayered embryo with well-defined organ primordia. However, in the highly telolecithal avian egg, in which the active cytoplasm is limited to the blastoderm, the yolk plays no part in gastrulation—except to complicate the process—and all of the events of gastrulation take place only in the blastoderm. Only in later stages does the growth of the extraembryonic membranes encompass the yolk and make it available for the nutrition of the developing embryo.

Integrated cellular movements of the blastoderm cells are the essential elements of gastrulation. These movements, as previously described, involve the progression of cells over the surface of the blastula toward a region where they will be able to find their way to the inside and establish a three-layered embryo in which the organ primordia form and begin to differentiate. In the chick, the region where the cells move to the inside is the primitive streak; avian gastrulation involves essentially the formation and the regression of the primitive streak and the progressive movements of cells associated with these processes. It is now widely accepted that all of the embryonic endoderm and mesoderm is of gastrular origin (i.e., they pass through the primitive streak).

### Formation of the Endoderm

A major and presistent controversy in chick embryology involves the formation of the endoderm. Various factors have contributed to this controversy, not the least of which is the fact that at the time of laying the egg is already a two-layered system with an outer layer of cells, termed the *epiblast*, underlain, at least in its future posterior region, by an inner layer of cells termed the *hypoblast*. It has been proposed that the epiblast is presumptive ectoderm and mesoderm and that the hypoblast is presumptive endoderm—thus denying the involvement of the primitive streak in the formation of the endoderm. The controversy concerns two points; (1) the origin and movement of the inner layer of cells, and (2) the prospective fate of the inner layer

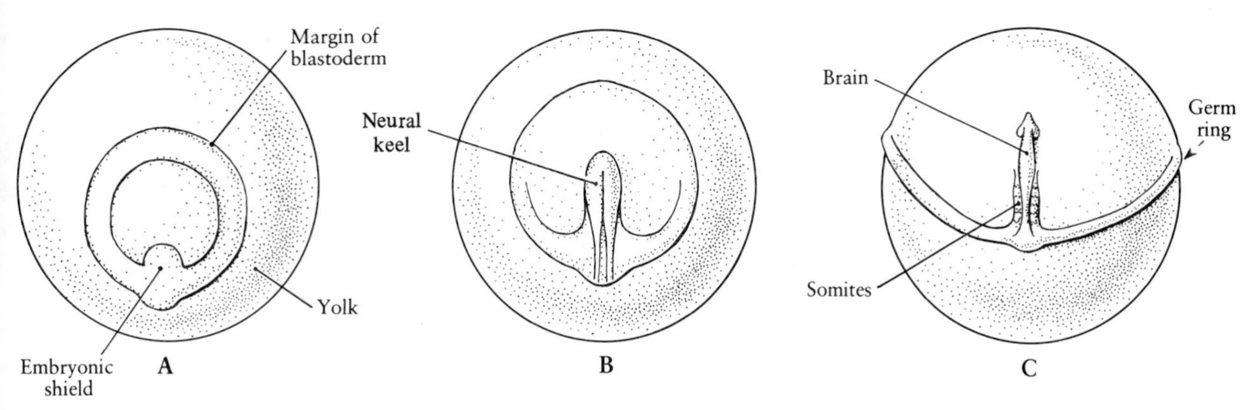


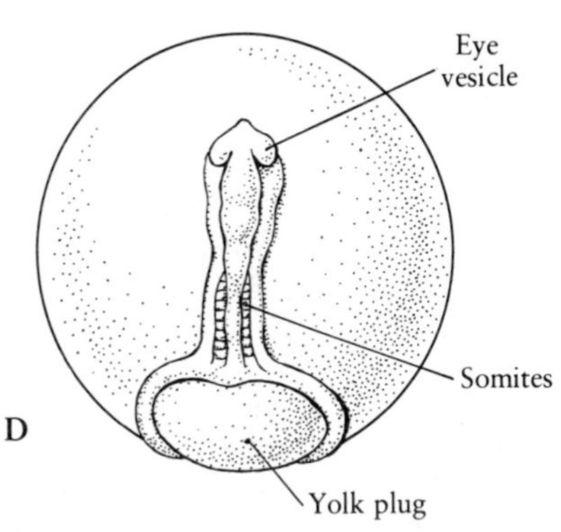


**FIG. 8–20.** Stages in the development of the trout *(Salmo).* A, beginning of gastrulation and formation of the embryonic shield; B, formation of the neural plate; C, formation of neural tube and somites; D, overgrowth of the yolk by germ ring nearly completed: (From B. I. Balinsky, 1975. An Introduction to Embryology, 4th edition. W. B. Saunders Company, Philadelphia.)

of cells, mainly its role in the formation of the embryonic endoderm.

*Origin and Movement of the Inner Layer of Cells*

Over a number of years, many investigators have proposed and supported the formation of the hypoblast simply by a separation of the deeper yolk laden cells from the superficial cells of the blastoderm, the separation beginning mainly at the future posterior end of the embryo and moving anteriorly. Whether this separation is by a cutting off or delaminating of a more or less continuous sheet of cells (Fig. 8–22 C) or by a separating of individual cells and their subsequent joining to form a continuous hypoblast layer (Fig. 8–22 B)—a process sometimes termed *polyinvagination*—the result is essentially the same: the formation of a two-layered embryo (8–22 D). Before the formation of the hypoblast, a cavity develops below the blastoderm, a cavity best referred to as the *subgerminal cavity* (Fig. 8–22 A). After the appearance of the hypoblast, the cavity between the epiblast and the hypoblast is usually termed the blastocoele, although whether it is homologous to the amphibian blastocoele is a matter of question.

In regard to the question about the origin of the endoderm, it is appropriate to consider the techniques that have been and are being used to solve this problem. Methods of investigation are varied, and technical difficulties are always present. Early investigators attempted to predict directions of cell migration from histological sections by using cell position and shape as clues, an obviously difficult piece of detective work. Later, vital staining and the application of carbon or other particles were used to mark cells and follow morphogenetic movements. Marking procedures may be done in ova or in vitro, using a number of different methods of explantation and orientation of the blastoderm. Difficulties arise because vital stains often tend to diffuse; particles may not remain associated with the same group of cells to which they are applied; larger particles may behave differently than smaller; morphogenetic movements may be restricted depending on the method of transplant, the medium, and the layer in contact with the substrate; and development itself may be inhibited in transplanted blastoderms. Cell labeling with radioisotopes and subsequent transplantation of a labeled graft to an unlabeled host (Fig. 8–23) was introduced in the early 1950s and has proved to be a highly valuable technique.

On the basis of a number of studies it is now generally concluded that the presumptive endoderm does not come entirely from the separation of a hypoblast from an epiblast layer but is built up from

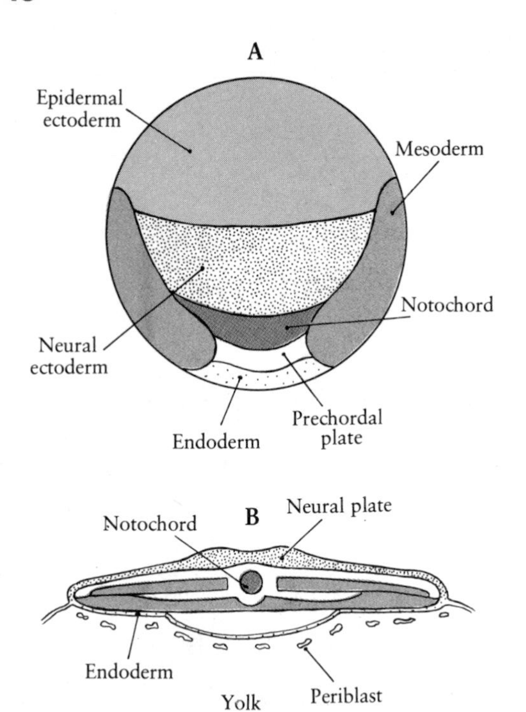


**FIG. 8–21.** A, presumptive organ-forming areas in the blastoderm of the shark *(Scyllium)* embryo, top view; B, section through the shark embryo at the end of gastrulation. (From Comparative Embryology of the Vertebrates by O. E. Nelson. Copyright © 1953 The Blakiston Co., Inc. Used with permission of McGraw-Hill Book Company.)

cells from two different sources. The first is from the posterior germ wall in the area opaca in the region of the future posterior end of the embryo. Cells spread out from this point of origin in a fan-shaped pattern (Fig. 8–24). These cells are destined to form only the extraembryonic endoderm. Some investigators call this layer the *endophyll*. The endophyll (or hypoblast) should be distinguished from the second source of endoderm cells, which are of gastrular origin and will form the embryonic endoderm. At the time of the formation of the endophyll, these presumptive, embryonic endoderm cells are still a part of the surface layer of cells, the epiblast. In conjunction with the term endophyll, the surface layer may be called *ectophyll*. The formation of the embryonic endoderm will be described in relation to the formation and function of the primitive streak.

*Development of the Primitive Streak*

Morphogenetic movements that are concerned with the formation of the primitive streak appear within a few hours after incubation and probably begin as soon as the temperature of the blastoderm cells reaches the normal level for growth (38.5°C). After three to four hours of incubation, a thickening on one quadrant of the area pellucida, the future pos-

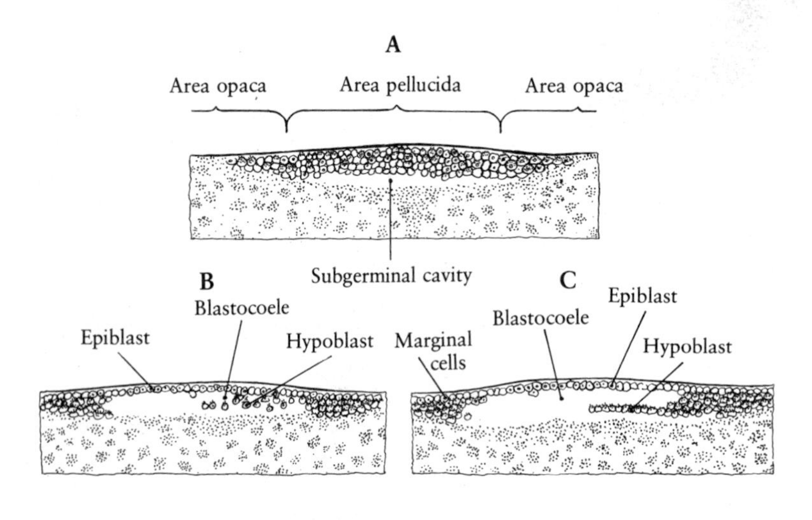


**FIG. 8–22.** Two proposed methods of formation of the hypoblast. A, cross section through the blastoderm before the separation of the hypoblast (shaded) cells from the epiblast cells by B, polyinvagination or C, delamination. Both methods result in the formation of a two-layered embryo, D.

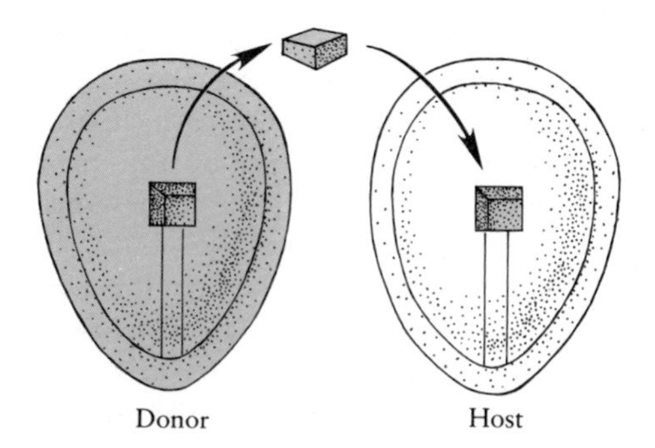


**FIG. 8–23.** Diagram illustrating the method of tracing cell movements by transplanting grafts from embryos labeled with tritiated thymidine (stippled) into an unlabeled host.

**FIG. 8–24.** Diagram to show the fanlike cell movements of the hypoblast away from its point of origin. The solid arrows represent endoderm cell movements, and the dotted arrows show the expansion of the area pellucida and the area opaca. (After N. T. Spratt, Jr. and H. Haas, 1960. J. Exp. Zool. 144:139.)

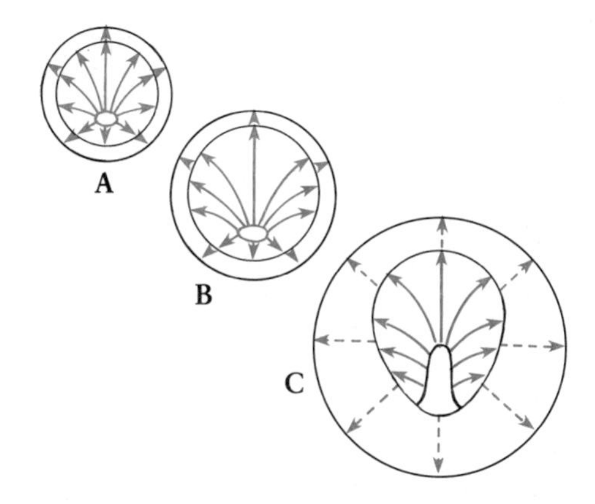


terior quadrant, represents the beginning of the formation of the primitive streak (Fig. 8–25 A). Within a few hours after its first indication, the thickening becomes more pronounced and begins to show an elongation in the future craniocaudal axis (Fig. 8–25 B). By the end of the first half day of incubation, the rather indefinite caudal thickening has developed into a well-defined fingerlike process that extends about halfway across the area pellucida. This may be called the intermediate streak stage (Fig. 8–25 C). During the time the primitive streak is developing, the area pellucida is increasing in size as the blastoderm cells begin to spread over the yolk. At the intermediate streak stage, the enlarging area pellucida is no longer circular but shows a posterior projection. By 18 to 19 hours of incubation, the streak has reached its maximum length, extending about three quarters of the way across the area pellucida. It ends anteriorly in a depression, the *primitive pit*, which is surrounded by an elevated area, *Hensen's node* (Fig. 8–25 D).

The increase in thickness of the caudal end of the blastoderm, which is the first indication of the formation of the primitive streak, is the result of an active migration of cells toward this region. Marking studies have not produced uniform results, but it appears that the part of the blastoderm that contributes to the primitive streak is rather small, probably not more than the posterior quarter of the area pellucida. A presumptive fate map of the prestreak blastoderm indicates that all of the material to be invaginated through the primitive streak occupies a relatively small area in the posterior region of the blastoderm (Fig. 8–26). The movement of the cells of the posterior half of the blastoderm as determined by marking experiments is illustrated in Figure 8–27. Here it is seen that a mark placed approximately in the middle of the blastoderm at six hours of incubation moves away from the developing streak, is never incorporated into the streak, and never invaginates (Fig. 8–27 A). A transverse mark across the blastoderm, just posterior to the midline, illustrates that the surface cells in the center of the blastoderm, just posterior to the

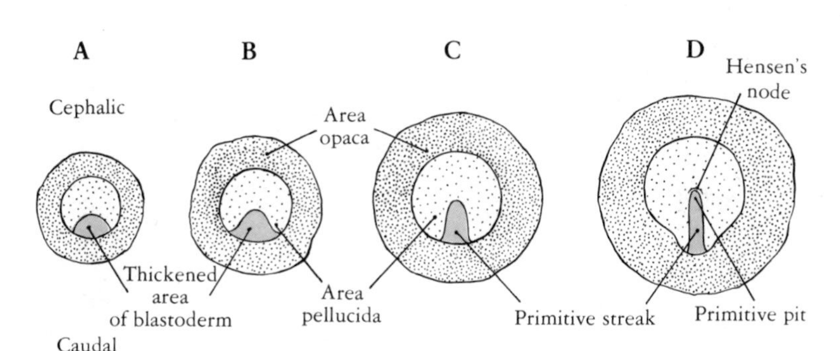


**FIG. 8–25.** Chick embryos showing the early development of the primitive streak. A, 3 to 4 hours of incubation; B, 7 to 8 hours of incubation (early streak); C, 10 to 12 hours of incubation (mid-streak); D, 18 to 19 hours of incubation (definitive streak). These diagrams cover Hamburger and Harrison stages one to four. (After V. Hamburger and H. L. Hamilton, 1951. J. Morphol. 88:49.)

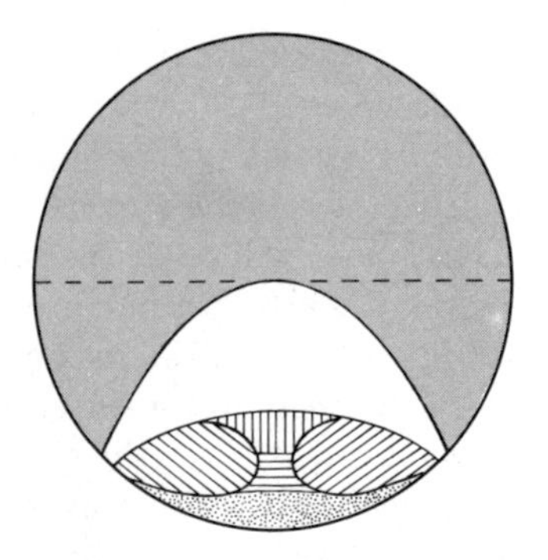

**FIG. 8-26.** Presumptive fate map of the two- to three-hour incubation stage. White crescent, neural material; vertical lines, chorda; horizontal lines, head mesoderm; inclined lines, somite mesoderm; dots, lateral mesoderm. (From M. E. Malan, 1953. Arch. Biol. 64:149.)

midline, move away from the developing streak (Fig. 8–27 B). Marks slightly posterior to the midline but lateral to the developing streak also move away from the streak. The more lateral one goes, the less the movement; the most lateral regions show little, if any, displacement. Only when a transverse mark is placed at some distance posterior to the midline do we find that these cells move toward and become incorporated into the primitive streak (Fig. 8–27 C) before they invaginate to form mesoderm. The marks still on the surface at 17 hours represent presumptive mesoderm moving toward the primitive streak prior to invaginating.

As the primitive streak develops, the area pellucida continues to increase in size by overgrowing the yolk in all directions. However, it grows more rapidly in a posterior direction and soon assumes a pear-shaped configuration (Fig. 8–28). During this time the primitive streak continues to elongate until it reaches its maximum length. It has been suggested that the elongation of the primitive streak is due to the incorporation of new material at both its anterior and posterior ends. Recent marking experiments do not confirm this hypothesis. Not only do marks placed on the blastoderm anterior to the developing streak never become incorporated into the streak, but marks placed on either the posterior or anterior ends of the streak remain in these areas as the streak elongates. These facts point to an elongation of the primitive streak as a stretching of the streak at about its midpiont, at an area where it first appears between the area pellucida and the area opaca. This concept is illustrated in Figure 8–28, indicating that the point of origin of the primitive streak ends up in the middle of the early streak and remains in this site. This concept very nicely gives an understanding of how the primitive streak elongates and how the area pellucida gradually changes from a circular to a pear-shaped structure.

*Function of the Primitive Streak*

The primitive streak is an invagination area (Fig. 8–29). Although it has been suggested that it acts as a blastema (a group of rapidly dividing undifferentiated cells that will develop into differentiated structures), there is convincing evidence to the contrary: (1) The mitotic index is no higher in the primitive streak than in other areas of the blastema; (2) staining and marking experiments show that cells move into the primitive streak, move to

**FIG. 8-27.** Making experiments showing the movement of the cells of the epiblast during the early development of the primitive streak. A, a mark in the middle of the blastodisc at the beginning of primitive streak formation does not become incorporated into the definitive primitive streak; B, a transverse mark posterior to the midline shows the incorporation of only a small amount of material into Hensen's node; C, a transverse mark near the posterior end of the area pellucida at three hours colors almost all of the primitive streak at 17 hours. (From M. E. Milan, 1953. Arch. Biol. 64:149.)

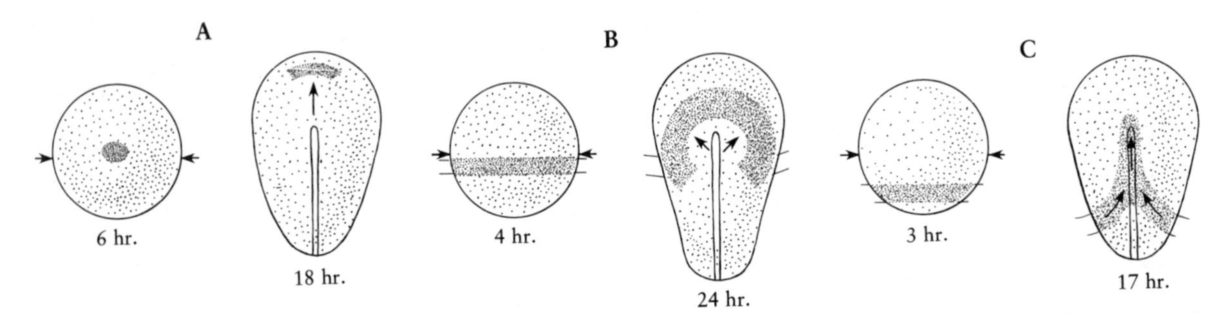

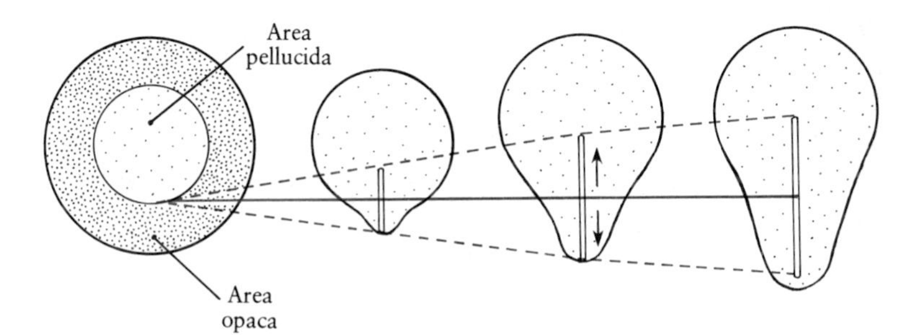


**FIG. 8–28.** Diagram of the formation and elongation of the primitive streak. The primtive streak is considered to originate at the posterior border of the area pellucida and the area opaca. this region then becomes the middle of the primitive streak in later stages as elongation takes place in both directions from the midpoint. (From L. Vakaet, 1962. J. Embryol. exp. Morphol. 10:38.)

the inside, and then move away from the streak; and (3) cells implanted into the primitive streak move quickly out of the streak and then are replaced. Marking experiments indicate that invagination takes place mainly at the sides of the streak, with very little occurring at either extremity. Invagination at the two ends occurs most actively during the formation of the streak, ceases at the node region at the definitive streak stage (18–19 hours of incubation), and ceases in the posterior part of the streak at the headfold stage (23–25 hours of incubation).

## Cellular Movements During Gastrulation

*Lower Layer*

The lowest layer is derived from two sources. One of these has already been described as being laid down before the formation of the primitive streak, originating from the posterior germ wall and forming the extraembryonic endoderm.

The process of invagination through the primitive streak involves a coordinated movement of surface cells toward the primitive streak and a downward (inward) movement of these cells at the primitive streak, which is followed by lateral and anterior movement away from the streak underneath the surface cells (Fig. 8–29). The lips of the primitive streak, the same as those of the blastopore of the amphibian egg, are occupied by a continually changing population of cells that remain in the primitive streak only temporarily as they move into the inside and then move away from the streak region. The movement of the surface cells toward the streak and their movement once inside are diagrammed in Figure 8–29 B.

Both marking and transplant experiments have given evidence that presumptive embryonic endoderm cells have a gastrular origin and are invaginated through the anterior part of the early streak. The cells that are to form the embryonic endoderm penetrate into the lower layer of cells, the hypoblast (endophyll), and push these cells laterally and anteriorly, replacing them. Thus, the deep layer of cells in the region of the anterior part of the streak is now made up of cells of gastrular origin. Figure 8–30 illustrates the position of the presumptive embryonic endoderm around the anterior end of the primitive streak during the early stages of streak formation and its disappearance from the surface layer by the time the streak reaches its maximum length. The embryonic endoderm and the endo-

**FIG. 8–29.** The function of the primitive streak. A, cells reach the primitive streak and move inward, leaving the surface layer. Some of these cells replace cells of the already present hypoblast (endophyll), and some migrate away from the streak area betweeen the epiblast and the hypoblast to form the mesoderm; B, movements of the surface cells on the left side (solid lines) and movements of the inner cells invaginated over the primitive streak (dotted lines) on the right side.

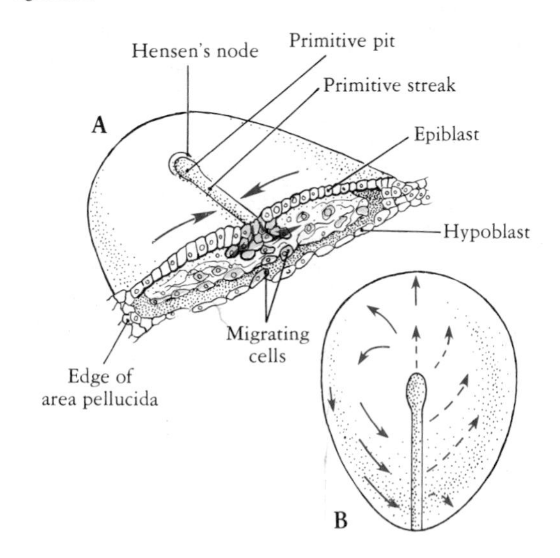

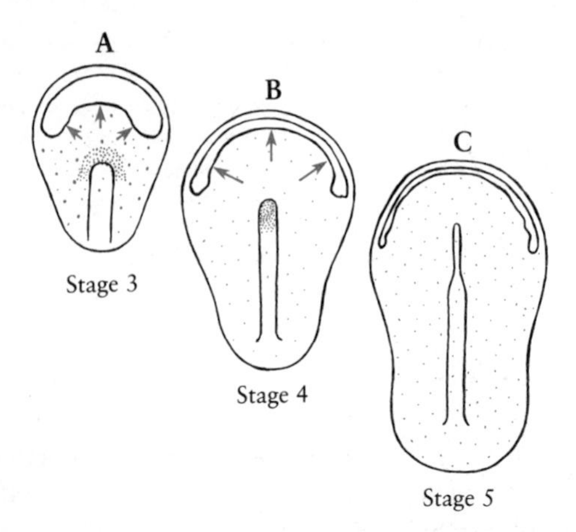


**FIG. 8–30.** Gastrular origin of the embryonic endoderm. A, at stage 3, some of the endoderm has invaginated and moved laterally and anteriorly (fine dots), some is in the process of invaginating (closely spaced fine dots), and some is still on the surface as part of the epiblast (heavy dots); B, at stage 4, the endoderm has spread under the surface in all directions, and a small amount is still in the process of invagination. At this stage there is no longer any endoderm on the surface. C, at stage 5. (From G. Nicolet, 1971. Adv. Morphog. 9:234.)

derm of germ wall origin form a continuous sheet of cells.

*The Mesoderm Layer*

The entire middle layer, which includes the mesoderm and the notochord, is invaginated through the primitive streak. The process begins as the streak first forms and continues during its regression. At the time of the maximum development of the primitive streak, most of the extraembryonic mesoderm, the prechordal plate, the head mesoderm, and the presumptive cardiac vesicles have moved to the inside. The notochordal material is condensed in the node, and the somite and lateral plate mesoderm, and some of the extraembryonic mesoderm, are still on the surface of the blastoderm (Fig. 8–31). During the formation of the head process and the regression of the primitive streak, the notochordal material condensed in the node is laid down and the somite, lateral plate, and extraembryonic mesoderm invaginate in succession. Com-

**FIG. 8–31.** A, fate map at the primitive streak stage. Surface material is shown on the left as solid lines and invaginated material on the right as dotted lines. The order in which the mesoderm invaginates is seen to be prechordal plate, notochord, head mesoderm (HM), somite mesoderm, lateral plate mesoderm (LM), and extraembryonic mesoderm (EM). Forebrain (FB), midbrain (MB), hindbrain (HB), and spinal cord (SC) are formed from the ectoderm around the anterior end of the primitive streak. B, at the head process stage. (From G. Nicolet, 1970. J. Embryol. exp. Morphol. 23:79.)

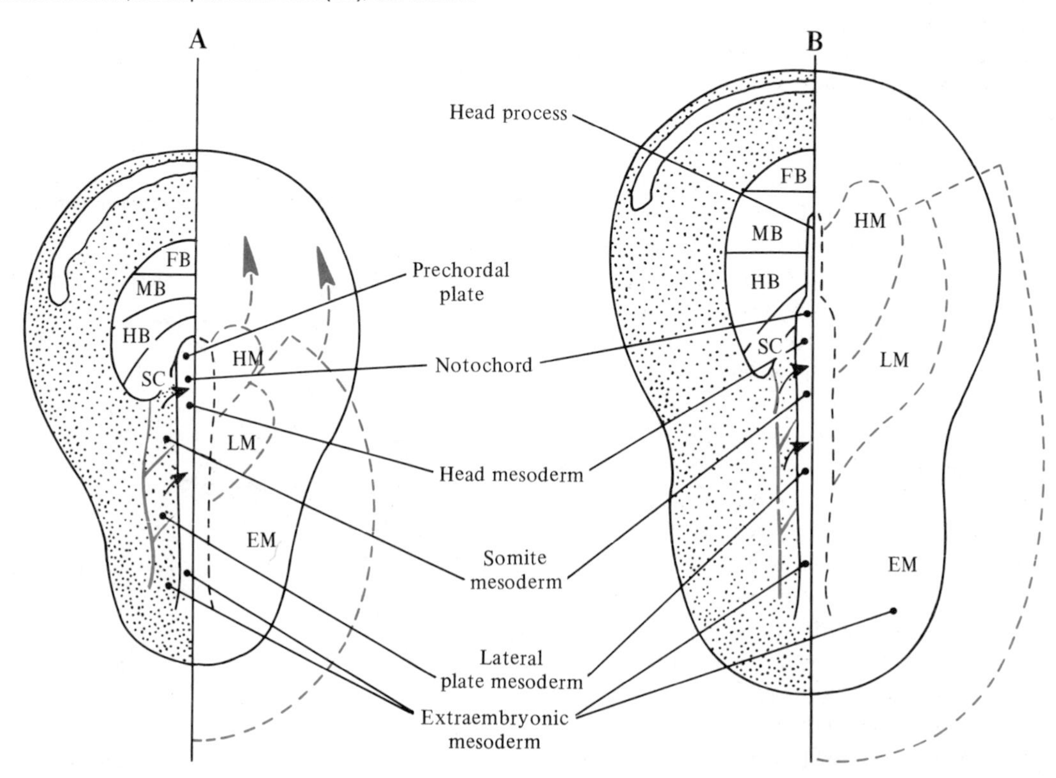

paring this fate map (Fig. 8–31) with that of the amphibian (Fig. 8–6) shows that the presumptive areas are similarly arranged, although there is a wide lateral dispersion of the areas in the amphibian and a marked restriction of the areas close to the primitive streak in the chick. In the chick egg a large part of the mesoderm is concerned with the formation of the extraembryonic membranes. The chronology of invagination is similar in the two species: (1) The invagination of the foregut and prechordal plate precedes that of the notochord; (2) the notochord forms a large part of the node, as it does of the dorsal lip of the blastopore, which is considered to be the homolog of the node; and (3) the anterior somite, posterior somite, and lateral plate mesoderm invaginate in succession. The cells, which then remain on the surface, will form neural and epidermal tissues.

## Regression of the Primitive Streak and the Formation of the Organ Rudiments

The laying down of the organs of the embyronic axis takes place during the regression of the prim-

FIG. 8–32. Graphic representation of the regression of the primitive streak and the increase in the length of the notochord. (From N. T. Spratt, Jr., 1947. J. Exp. Zool. 104:69.)

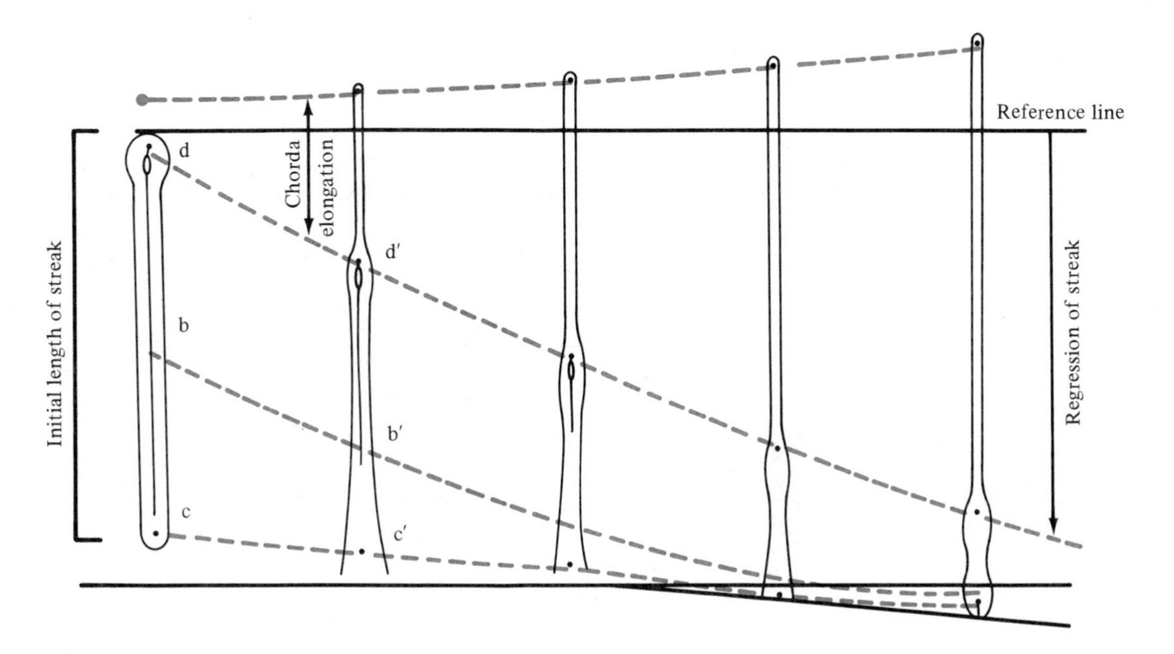

itive streak. Streak regression takes place shortly after it reaches its maximum development. A graphic summary of streak regression is shown in Figure 8–32. Transverse marks placed across the area pellucida during streak regression indicate that the entire area pellucida caudal to Hensen's node undergoes regression, although it is much more rapid in the area of the primitive streak (Fig. 8–33). The most posterior end of the streak does not contribute anything to embryonic structures.

The anterior end of the streak consists of presumptive prechordal plate and notochord, which are laid down as the streak regresses (Fig. 8–34). During this regression, cells leave Hensen's node and the primitive streak immediately posterior to it and condense to form the tissue of the notochord. The presumptive neural material, located in front of and lateral to the node, elongates and stretches in a craniocaudal direction, and the lateral parts coverge toward the midline. Neurulation and organ formation take place in a craniocaudal sequence. As the node retreats, the neural plate, underlain by the notochord, develops in front of it (Fig. 8–34 C).

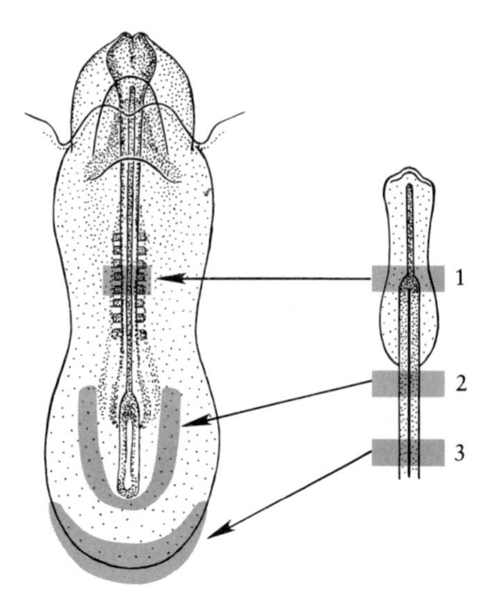


**FIG. 8–33.** Marking experiments on the regression of the primitive streak. The greatest regression is seen in the midline, but there is a regression of lateral areas also. The posterior end of the streak (mark 3) does not form any part of the embryo. (From G. Nicolet, 1971. J. Embryol. exp. Morphol. 23:79.)

**FIG. 8–34.** Formation of the embryonic axis anterior to the regressing primitive streak. A, definitive streak, 18–19 hours, stage 4; B, head process, 19–22 hours, stage 5; C, beginning of somite formation, 23–24 hours, stage 7 +; D, four-somite stage.

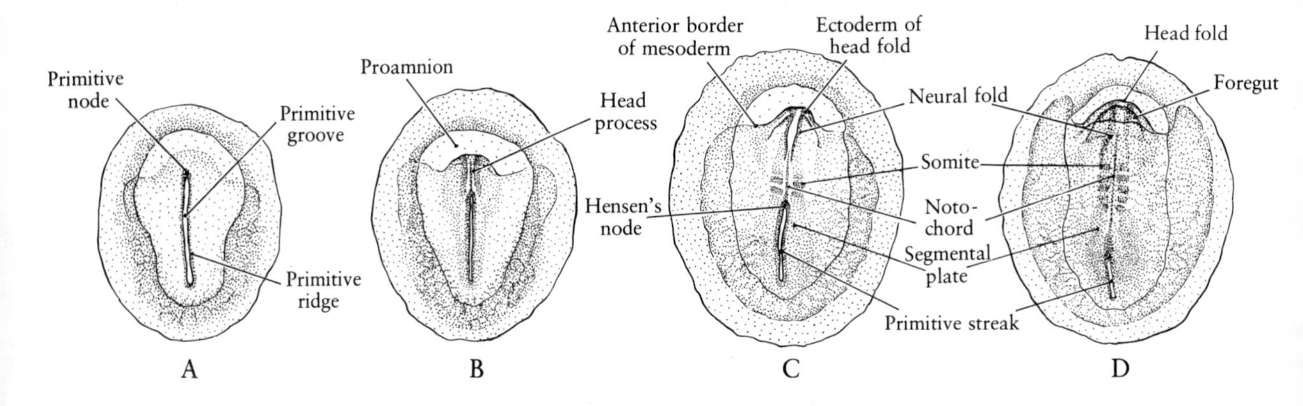

*Formation of the Somites*
Laterad of the developing notochord, the mesoderm can be divided into two populations of cells. One, the most medial, represents presumptive somite material. The cells form a loosely arranged columnar epithelium attached to the basal lamina of the overlying epiblast in a region that is coextensive with the developing neural plate. The cells are joined to each other at their basal ends. A more lateral, loosely arranged group of cells presents the flattened appearance of typical mesenchyme, which does not form cellular junctions either to the overlying presumptive epidermis or between themselves. They will form the lateral mesoderm.

As the neural plate condenses toward the midline, so do the presumptive somite cells below it. As they do so, they lose their loose arrangement and become tightly opposed, virtually eliminating the intercellular spaces. New intercellular junctions are formed at the apical ends of the cells. At intervals along this group of cells, the *segmental plate*, gaps appear between the plate and the overlying neural plate. These gaps are the beginning of the formation of the intersomitic furrows and occur at regularly spaced intervals running in a craniocaudal direction where the somite cells break their connections with and are "released" from the neural plate cells (Fig. 8–34 C,D).

As the neural plate folds up to form the neural tube, all connection between the somites and the neural tube is lost. Later, the somites become enclosed in a loose network of fibers which also establishes a fibrous connection to the notochord.

*The Mechanism of Somite Formation.* The mechanisms involved in the early morphogenesis of the chick somites have been the subject of numerous investigations. In the 1950s, Spratt presented a widely accepted explanation that proposed that two "somite-forming centers" develop in a region just posterior and lateral to Hensen's node (Fig. 8–35 A). Although they could not be distinguished morphologically, on the basis of extirpation and transplantation experiments they were pictured as two small areas of the blastoderm that, as they regressed with the primitive streak, were responsible for the induction of the somites. Spratt reported that transections of the blastoderm, either anterior or lateral to the centers, had no effect on somite formation, whereas if the transections were posterior or medial to the centers, no somites were formed. Other

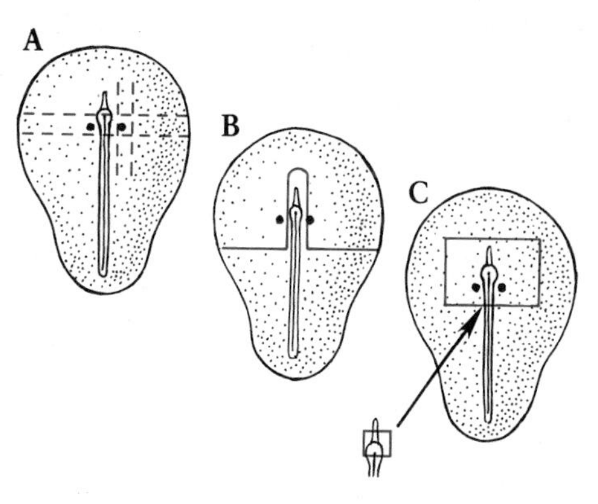


**FIG. 8–35.** Diagrams of three experiments on the role of the somite centers. A (Spratt, 1954), if the blastoderm is cut either anterior or lateral to the somite centers (asterisks), somite development is normal. If the cut is posterior or medial to the centers, no somites develop. B (Bellairs, 1963), the somite centers are removed but the entire primitive streak and Hensen's node are retained. Although lacking somite centers, somite development is normal. C (Nicolet, 1970), the entire anterior end of the primitive streak is removed. No somites will develop. However, if a portion of Hensen's node is implanted at the anterior end of the cut streak, somites will develop. (From B. H. Lipton and A. G. Jacobson, 1974. Dev. Biol. 38:91.)

investigators also reported that if the somite centers were removed by excising the entire anterior end of the primitive streak, then the somites did not develop. However, it was also shown that if the somite centers and not the primitive streak were removed (Fig. 8–35 B), or if the anterior part of the streak and the centers were removed and a Hensen's node implanted at the anterior end of the cut streak (Fig. 8–35 C), then the somites developed normally. The existence of somite-forming centers is thus questionable, and other mechanisms for somite formation must be sought out. A number have been proposed, including induction by the overlying neural plate or by the notochord or some influence brought about by the regression of the primitive streak.

A series of experiments by Lipton and Jacobson (1974) has presented a comprehensive view of somite formation that, in fact, implicates the neural plate, the regression of the primitive streak, and the notochord as all playing a role in the process. In one experiment, the blastoderm was cut longitudinally so that one part contained all of the primitive streak as well as Spratt's proposed somite-forming centers (Fig. 8–36 A). In a second, they removed a wedge-

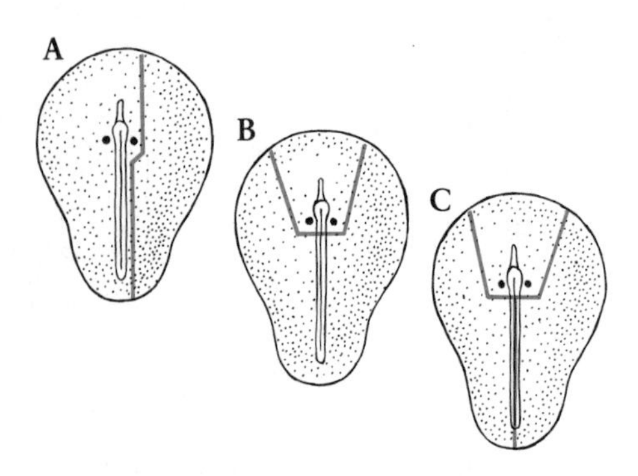


**FIG. 8-36.** Diagrams of three studies of Lipton and Jacobson on the mechanism of somite formation. For explanation, see text. (From B. H. Lipton and A. G. Jacobson 1974. Dev. Biol. 38:91.)

shaped section of the blastoderm containing the anterior end of the primitive streak and the adjacent somite centers (Fig. 8–36 B); and in a third, they removed the same wedge-shaped piece as in the second and then extended a cut posteriorly through the middle of the primitive streak (Fig. 8–36 C). In a high percentage of cases in all three experiments, the pieces lacking the anterior end of the primitive streak and the somite-forming centers developed somites. In examining the blastoderms cultured after the above-mentioned operations were performed, four important observations were made. First, there was no notochordal tissue in any of the cultures, since Hensen's node was completely removed in all of the experiments. Second, the fragments that formed somites always contained some neural plate material. Third, only those embryos in the second experiment that ended up showing a U-shaped configuration as the result of the formation of a longitudinal split down the middle, after the wedge was removed, formed somites. Fourth, although somites developed after 10 hours of incubation, over the next 14 to 20 hours most of them broke down, dispersing laterally. Based on these four observations, we may reach four conclusions. From the first, neither Hensen's node, the notochord, nor the somite-forming centers are necessary for somite formation. From the second, neural plate material is necessary. From the third, a longtiudinal splitting of the embryo is important. From the fourth, the presence of the notochord may be necessary for the stabilization of the somites once they are formed.

Thus, we are presented with the following view of somite formation. As the presumptive somite cells migrate laterally away from the primitive streak, they become closely associated with that region of the epiblast that will form the neural plate and, through this association, are imprinted with a prepattern of segmentation. This segmentaiton, however, is only realized when the primitive streak regresses and splits the mesoderm into right and left halves. This shearing action of the regressing primitive streak can be simulated by mechanically cutting the mesoderm longitudinally. Interestingly, in normal somite formation, as the streak regresses, the somites appear at some distance in front of Hansen's node, at the level where the notochord is developing, and they are formed progressively in a craniocaudal sequence. Yet, when the blastoderm is split mechanically, the entire length of the mesoderm is divided into right and left halves simultaneously and, following this, all of the somites along the length of the split form at the same time. Finally, recalling that the somites develop fibrous connections to the notochord, we may postulate that, although the notochord has no function in somite formation, it may be necessary for the stabilization of the somites once they are formed. Lacking these connections the somites tend to disperse laterally.

*Establishment of the Foregut, Midgut, and Hindgut*

While the neural folds are developing, an elevation of the entire blastoderm at the cranial end of the embryo appears. This forms a process that extends forward over the underlying ectoderm. This is the *head fold* (Fig. 8–34 C,D; 8–37). The pocketlike recess between the head fold and the ectoderm is the *subcephalic pocket*. Craniad of the head fold, the blastoderm contains no mesodermal layer in a region known as the *proamnion*. Since the entire thickness of the blastoderm is included in the head fold, the endoderm is elevated also and is pushed into the elevation as a shallow pocket beneath the neural plate. This finger-shaped pocket is the foregut, and the opening leading into it is the *anterior intestinal portal* (Fig. 8–37 C). A hindgut and a *posterior intestinal portal* develop at a later time (60–70 hours) in conjunction with the formation of the tail fold. Between the foregut and the hindgut, the median sheet of endoderm, lying underneath the notochord and the somites and on the top of the yolk, is the roof of the open midgut. As development pro-

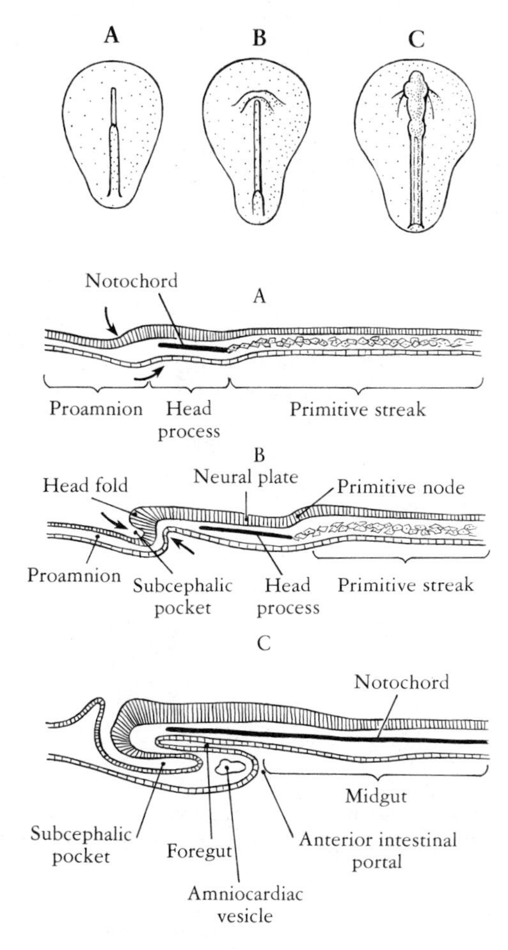


**FIG. 8–37.** Diagrams of midsagittal sections of the chick embryo during the formation of the embryonic axis. Arrows indicate foldings that lead to the formation of the head fold and the foregut. A, head process, 20–22 hours; B, head fold, 23–24 hours; C, 28–29 hours.

ceeds, continued downward and inward folding of the endoderm of the midgut area, accompanied by cranial and caudal growth of the head fold and the tail fold, increases the size of the foregut and hindgut regions and progressively decreases the size of the open midgut, which retains its connection to the yolk sac by a narrow stalk.

Figure 8–38 shows the appearance of an embryo during early somite formation. The cross sections illustrate the connection between the closed foregut and the open midgut by way of the anterior intestinal portal. The mesoderm is differentiating into somite and lateral mesoderm between which the intermediate mesoderm is seen.

# The Mechanisms of Gastrulation

## Morphogenetic Cell Movements

Gastrulation is a developmental event that involves the massive translocation of embryonic cells after the blastular stage is completed. The future primordia of mesodermal and endodermal structures, such as muscle and gut, are removed from the surface to new positions within the embryo. Their place at the surface of the embryo is taken by an actively spreading population of cells (ectodermal) that will contribute to the skin, nervous system, and sense organs. Hence, this crucial phase of development is characterized by movements of cells that involve the whole embryo. As a consequence, the external shape of the embryo begins to change and groups of cells are topographically placed in locations that anticipate the primitive body plan of the organism. The embryo acquires distinct anteroposteriority and bilateral symmetry. Since the movements of these cells assist in the creation of new form and shape, they have been termed *morphogenetic movements*.

Using a variety of techniques, including carbon particle and tritiated thymidine labeling, vital dye marking, and microsurgical extirpation, we know that cells move during gastrulation, and we know where they go. An appreciation for the systematic and highly coordinated movement of embryonic cells at this time during development is now being obtained using time-lapse cinemicrography of sea urchin and amphibian gastrulas as well as chick embryos at the primitive streak stage. This technique is providing insight into the roles that changes in cell shape and cell motility play in the gastrulative process.

The displacement and rearrangement of cells by massive movements appear to be accomplished by the coordinate, motile interaction of collections of individual cells. To understand better the cellular mechanisms underlying gastrulation, it is necessary to answer several questions: What are the paths of movement of the different populations of cells? What are the activities of the cells that bring about these movements? How do changes in cellular behavior generate the mechanical forces required to move cells? How is the behavior of these cells controlled? Two major approaches are being used to establish the mechanisms by which cells

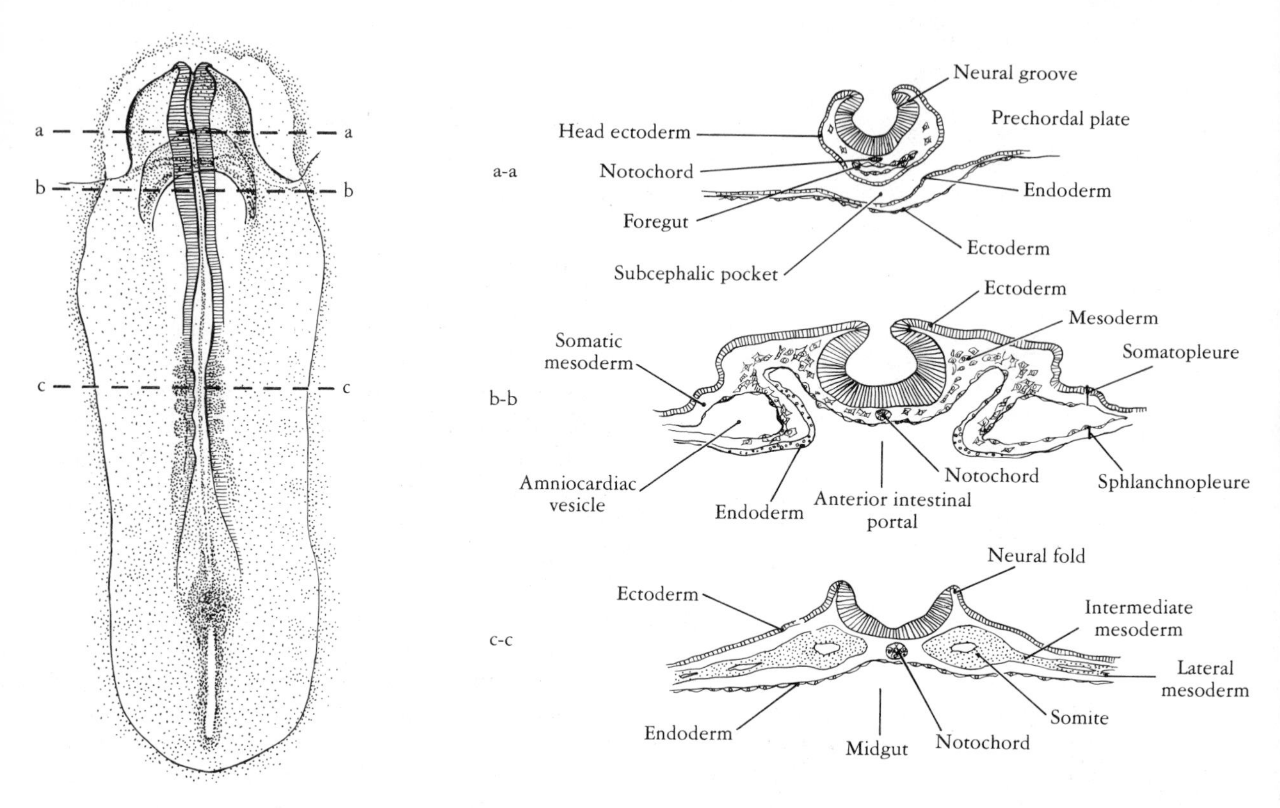


**FIG. 8-38.** Whole mount of a five-somite embryo with transverse sections pictured at the levels indicated: a–a, through the head fold showing the closed foregut and subcephalic pocket; b–b, through the anterior intestinal portal; c–c, through a pair of somites.

move during gastrulation and to determine how the activities of individual cells are coordinated to give mass cell movement. These include the examination of patterns of cell behavior in different regions of living embryos and the behavior of dissociated cells in culture. The former technique is preferable in principle, but generally it is impractical, since embryos are often opaque and difficult to observe. Much of what is currently known regarding the mechanisms of cell locomotion comes from the tissue culture of cells. However, when cells of gastrulating embryos are disrupted and subsequently cultured for study, it must be remembered that modes of behavior may be generated in vitro that have no consequence for normal gastrulation.

The small, transparent holoblastic eggs of echinoderms have been particularly useful in studying the mechanics of the cellular activities responsible for the rearrangement of cells and the changes in form during gastrulation. At about 12 hours after the start of development in the sea urchin, some 40 cells (micromeres) migrate from the vegetal pole into the hollow, fluid-filled blastocoele (Fig. 8–39 A,B). Movement is accomplished by means of numerous long *pseudopodia,* some up to 30 microns in length, being thrown out from the cell surfaces into the interior of the blastula. When these pseudopodia contact and adhere to the inner wall (ectoderm) of the blastula, shortening or contraction of these thin processes pulls the cell body toward the point of attachment (Fig. 8–39 C). Electron micrographs through these migratory cells, destined to become the primary mesenchyme cells, show large numbers of microtubules distributed parallel to their long axes, thus indicating that microtubules are probably responsible for the development and activity of the pseudopodia. The primary mesenchyme cells will later lay down the skeletal system of the embryo.

Gastrulation begins by a small flattening of the columnar-shaped cells of the vegetal plate, which gradually extends into the blastocoele as the archenteron (Fig. 8–39 D). Just prior to this infolding, the cells of the vegetal plate lose contact with their neighbors, round up at their inner ends, and show strong pulsatory activity. This pulsatory activity ap-

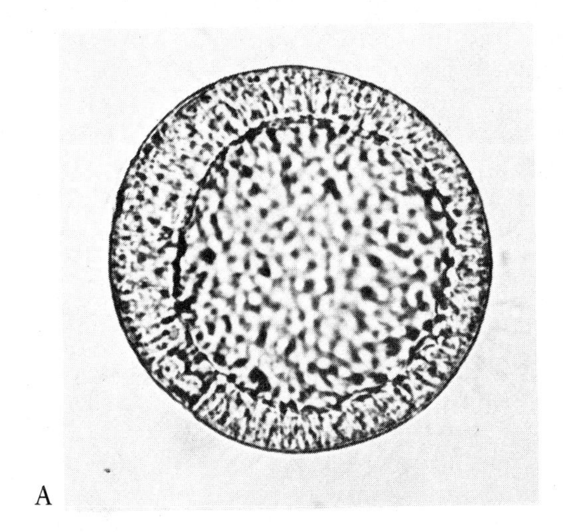


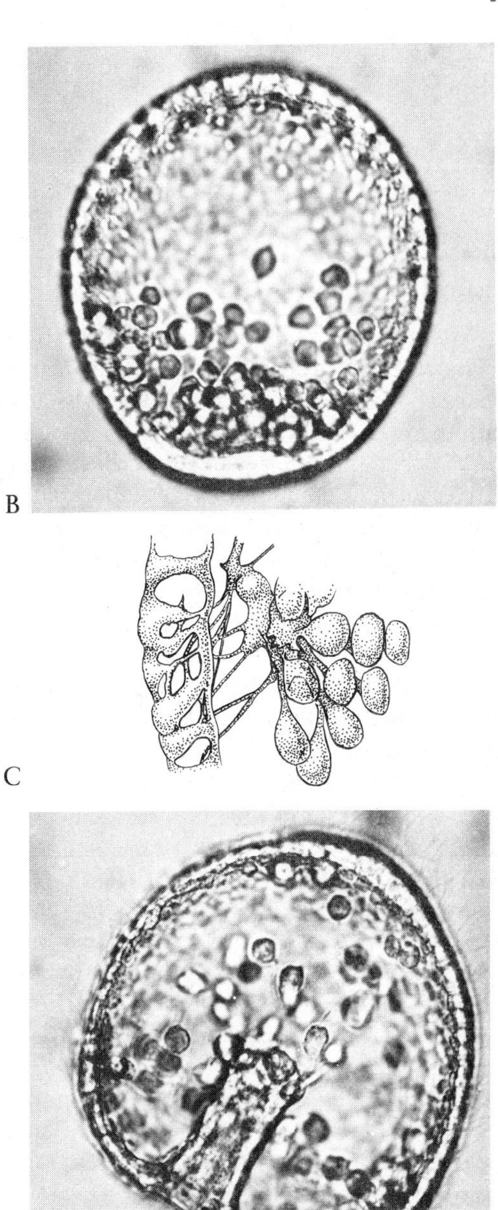


FIG. 8–39. A late blastula of the sea urchin (A) is shown just before the migration of primary mesenchyme cells into the blastocoele (B). A sketch of the attachment of the pseudopods of the primary mesenchyme to the thickened ectoderm is shown in C. Gastrulation involves the flattening and invagination of cells of the vegetal plate (D). Note the secondary mesenchyme cells at the tip of the gut. (A, B, and D from G. Karp and M. Solursh. 1974. Dev. Biol. 41:110; C, after T. Gustafson and L. Wolpert, 1967. Biol. Rev. 42:442.)

pears to cause the initial invagination that brings the tip of the infolded cellular layer about one-third of the way across the blastocoele. The second phase of gastrulation begins, after a pause, with the appearance of intense pseudopodial activity of the cells at the tip of the archenteron. Thin, long pseudopodia, shot out from these cells, form stable contacts with the inner surface of the gastrular wall. Subsequent contraction of the pseudopodia pulls the archenteron further into the interior of the embryo. The attachment of the pseudopodia to the inner wall also appears to be important in maintaining the structural integrity of the archenteron during gastrulation. If the pseudopodia are detached experimentally, the archenteron may evert to the outside so that the embryo becomes an exogastrula.

Amphibian embryos have also been particularly popular material for a variety of investigations on the gastrulative process. Their cells are large and the movements of cells are more clearly visible than in most other vertebrate embryos. Gastrulation involves a complex of morphogenetic movements, including the spreading of the superficial and deep regions of the marginal zone vegetally and their coordinate involution to form the roof of the archenteron and the chordamesodermal cell mantle. The onset of gastrulation in frogs or salamanders is marked by an infolding of the presumptive endodermal cells on the borderline of the vegetal region just below the center of the grey crescent area. The upper rim of the groove represents the dorsal lip of the developing blastopore.

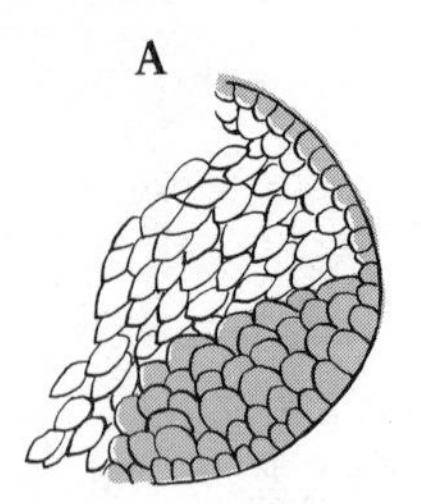

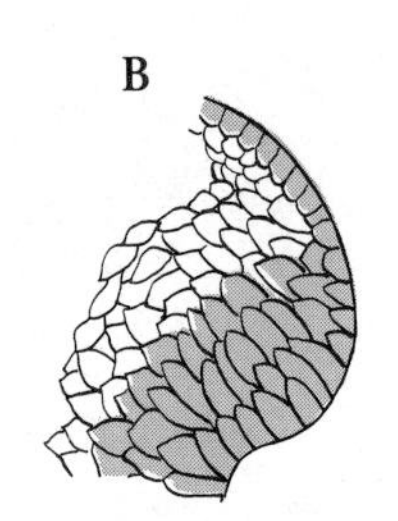

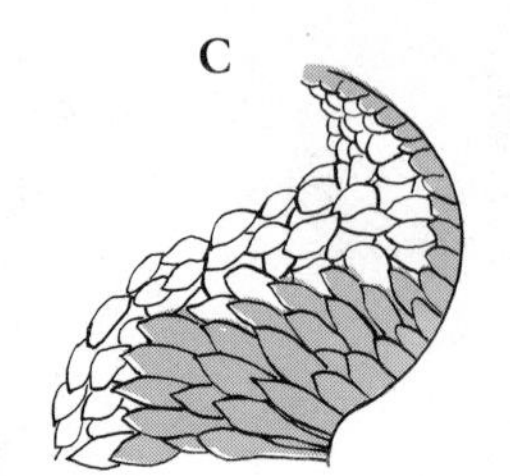

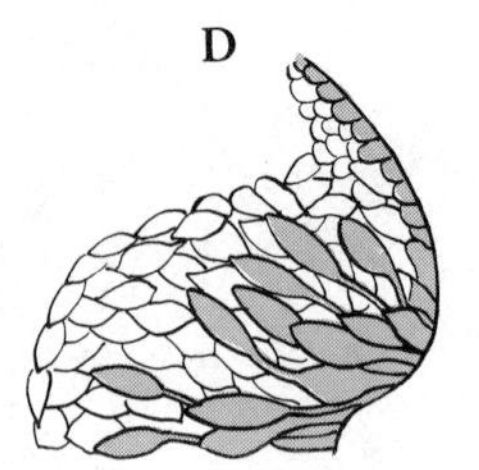

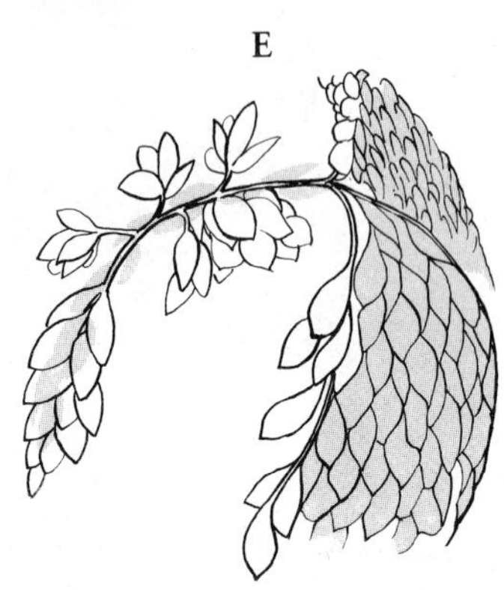


**FIG. 8–40.** A–D, successive changes in the shape of blastoporal cells during gastrulation in the amphibian as seen with the light microscope (After J. Holtfreter, 1943b. J. Exp. Zool. 94:261); E, a part of the dorsal lip of the blastopore showing cells held together by the surface coat as proposed by Holtfreter. (After J. Holtfreter, 1943a. J. Exp. Zool. 93:251.)

The cells that initiate the formation of the blastopore are observed to change dramatically in shape.

Analysis of the shape and pattern of cells at and within the blastoporal groove has provided a basis for understanding the mechanics of gastrulation in both frogs and salamanders. As a cell approaches and becomes part of the blastopore, it becomes transformed from a cuboid-shaped cell to a flask-or bottle-shaped cell. Bottle-shaped cells were initially described in elegant studies using the light microscope by Rhumbler (1902) and Holtfreter (1944) and more recently by Baker (1965), Nakatsuji (1975), and Keller (1981) using the electron microscope. During flask cell formation, the outer or apical end of the cell becomes long and attentuated while the inner or basal end becomes rotund (Figs. 8–40, 8–41). These shape changes are postulated to generate the stress in the superficial sheet of endodermal cells, resulting in a bending or invagination that leads to the formation of the blastoporal groove. During the invagination process, the outer surfaces of these migratory cells always maintain firm connections with neighboring cells on the surface of the embryo despite being greatly stretched. Although the process has been studied by a number of investigators, it is not known what stimulates bottle cell formation and how the activities of these cells are coordinated to form the complete circular blastopore.

The idea that changes in the shapes of cells at the blastopore play an essential part in the folding inward and subsequent movement of sheets of endoderm and mesoderm grew out of light microscope studies, particularly those of Holtfreter in the 1940s. Since Holtfreter made significant contributions to our understanding of the structure and migration of presumptive mesodermal and endodermal cells during amphibian gastrulation, the salient features of his hypothesis regarding this subject should be briefly mentioned. He isolated cells of the blastopore and observed that, when cultured on glass, each at its basal end attached, spread, and then led other cells in movement. When explanted onto a fragment of endodermal cells, blastoporal cells moved into the interior of the mass and simulated the formation of a blastoporal groove (Fig. 8–53). Additionally, he postulated that the driving force initiating the formation of the blastopore was located within the presumptive bottle cells of the endoderm. Presumably, the inner ends of these cells migrate inward in response to the alkaline pH of the blastocoelic fluid, thus creating the blastoporal invagination. The distal ends of these cells remain firmly attached to a specialized layer of material on the surface of the gastrula, termed the *surface coat* (Fig. 8–40 E). As each blastoporal cell moves inward, it is drawn out into the shape of a flask. The presumptive notochordal and mesodermal cells are drawn over the lips of the blastopore in sheetlike fashion, because the flask cells exert a pulling force on these cells by way of the surface coat. Hence, the tighty joined endodermal, notochordal, and mesodermal cells at the surface function to integrate into a coordinated system the pulling activites of the independently migrating bottle cells. The presence of surface coat is essen-

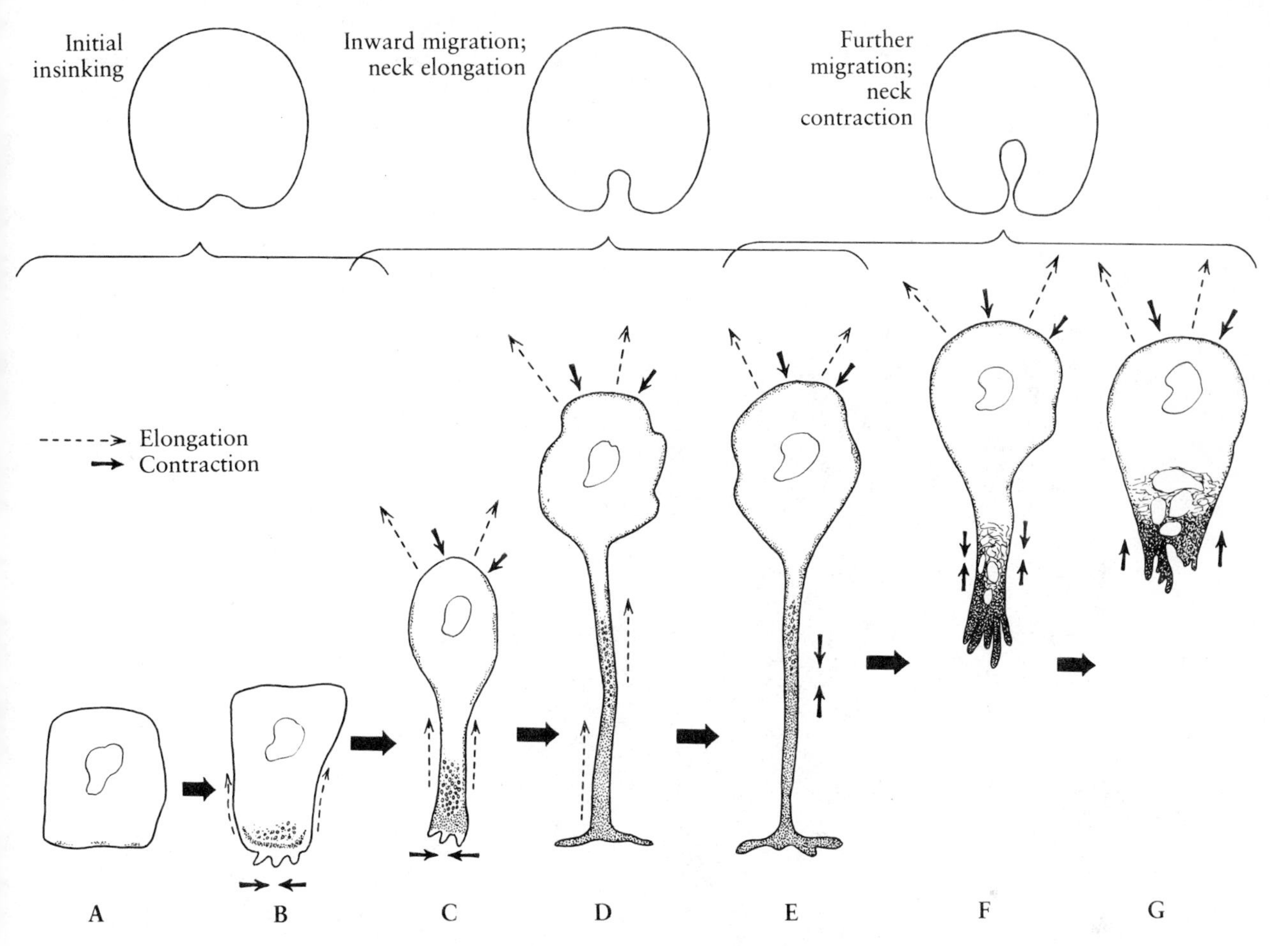


**FIG. 8–41.** Theoretical scheme for the transformation of a blastoporal cell during gastrulation as proposed by Baker. A, the cell as it appears prior to gastrulation; B, the distal end of the cell contracts causing an initial insinking and formation of the blastoporal groove. Note the shape (wedge) of the cell. C–D, the proximal surface of the cell migrates inward, deepening the blastoporal groove. The distal end of the cell continues to contract and the neck of the cell elongates. E–G, the neck of the cell contracts after maximum elongation of the cell is reached. Note that this shortening furthers invagination and pulls adjacent cells into the groove. (From P. Baker, 1965. J. Cell Biol. 24:95.)

tial to Holtfreter's theory of gastrulation. Yet positive and incontrovertible evidence for the existence of such a coat or covering of elastized material remains to be forthcoming. Electron-microscopic studies have not generally supported the notion that there is a continuous extracellular covering for the young amphibian embryo. However, the connections between adjacent blastoporal cells by way of microvilli would effectively serve the same purpose as the proposed surface coat.

Results of electron-microscopic studies, particularly those of Baker and Nakatsuji, have allowed the construction of a plausible model to explain the general mechanism of bottle cell formation and cell movement through the blastopore (Fig. 8–41). According to this model, each cell moving through the blastopore passes through a sequence of shapes that resemble a cube, a wedge, and a flask (Figs. 8–42, 8–43). The distal ends of cells invaginating over the blastopore are observed to contain a layer of electron-opaque material that has a filamentous appearance, suggestive of a contractile function (Fig. 8–43). It is proposed that alternate shortening (contraction) and elongating (expansion) of this dense layer in the distal ends of these cells accounts for their transformation into flask-shaped cells and their consequent movement through the blastopore (Fig. 8–41). Some contractile activities in the neck regions of flask cells may also be partially responsible for cell movement. As invaginating cells con-

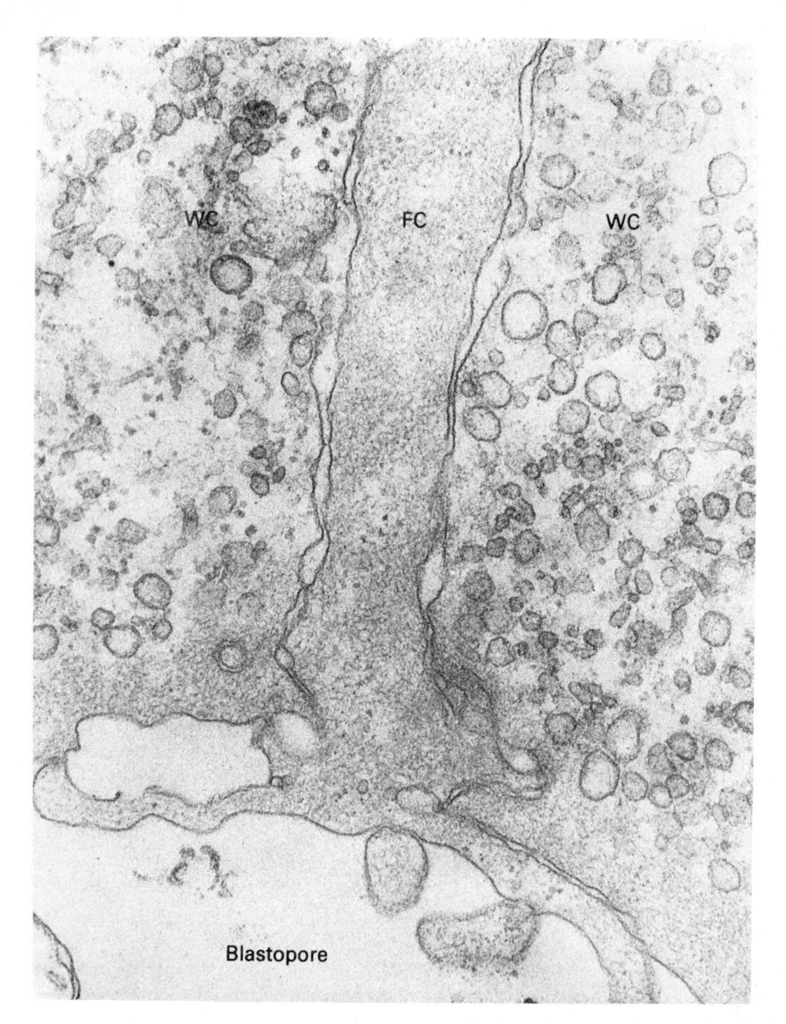


**FIG. 8-42.** The distal end of a flask cell (FC) of an amphibian gastrula, showing adjacent wedge cells (WC) with areas of dense material. (From P. Baker, 1965. J. Cell Biol. 24, 95.)

tract distally and migrate inward, neighboring cells are drawn in behind. Temporary bonds or junctions between adjacent cells of the blastopore appear to coordinate the movements of invagination. The cells, then, appear to move as a supercellular unit into the interior in great measure because of the pulling activities of the bottle cells. Other factors may also be involved in the orderly, integrated involution of the chordamesodermal cells. Among these are the inherent tendency for chordamesodermal cells to stretch in an anteroposterior direction. Following movement through the blastopore, the endodermal cells of the roof of the archenteron and the mesodermal cells in many amphibian embryos make contact with the inner surface of the blastocoelic wall (i.e., overlying presumptive ectoderm). Where contact is made with the blastocoelic wall, the cells are observed to form pseudopodia. Continued forward movement of both endoderm and the mesodermal mantle appears to be by an active migration of individual cells over the inner surface of the wall of the blastocoele.

Do bottle cell formation and active cell migration account for the locomotive forces underlying the morphogenetic movements of endoderm and mesoderm in all amphibians? Recent studies by Keller indicate that bottle cells in *Xenopus* embryos do not appear to play the role often assigned to them. If bottle cells are completely extirpated in the early gastrulating embryo, there is visible truncation of the archenteron, but there are no abnormalities of involution of the mesodermal cells. It would appear, therefore, that the bottle cells forming in the endoderm are not necessary for either the extension or convergence of mesodermal cells toward the blastopore or their involution. Even if the superficial cells destined to form the first bottle cells are removed before any shape changes, the meso-

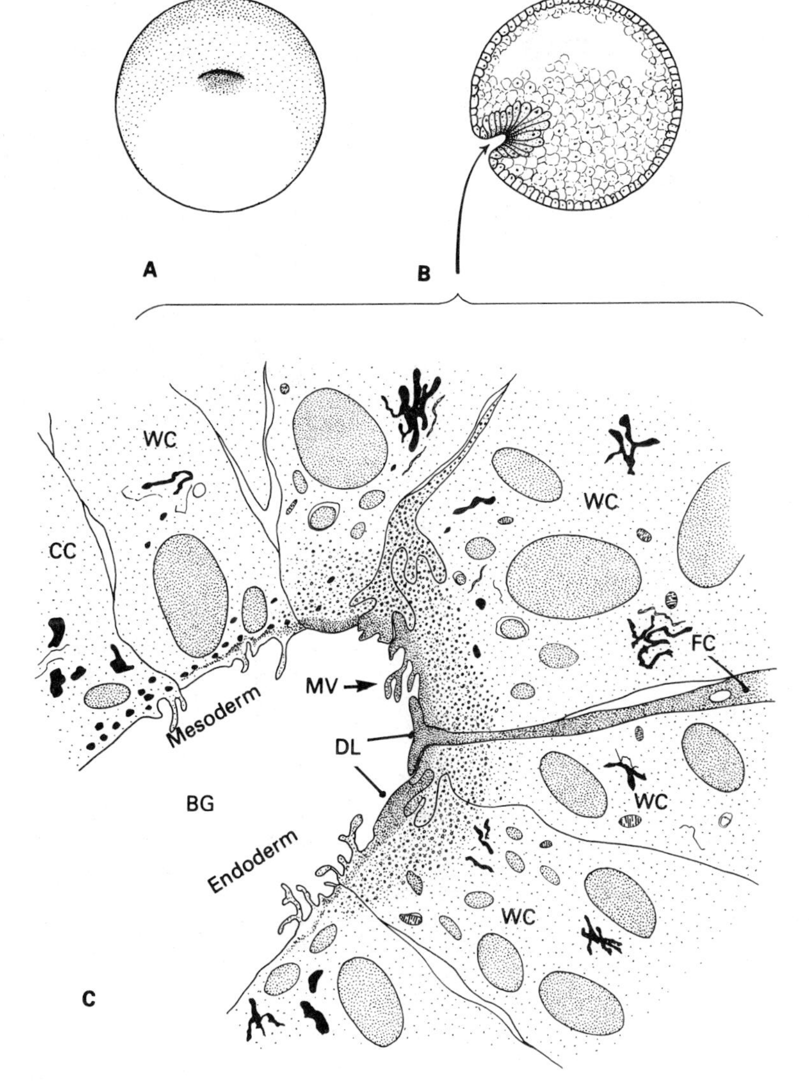


**FIG. 8–43.** A, early gastrula of the frog with blastoporal lip just beginning to form; B, a sagittal section through the blastoporal groove showing wedge and flask cells; C, a reconstruction from electron micrographs of a parasagittal section through the blastoporal groove at a stage represented by B. Note the flask (FC), wedge (WC), and cuboidal (CC) cells that line the groove. The distal surfaces of these invaginating cells contain a dense layer (DL). BG, blastoporal groove; MV, microvilli. (From P. Baker, 1965. J. Cell Biol. 24, 95.)

dermal cells seem to move in normal fashion. Failure of bottle cell removal to disrupt significantly the gastrulative process suggested that the behavior of cells in the deep region of the marginal zone was critical to involution. To test this possibility, Keller (1981) grafted to the dorsal lip superficial and deep cells from various regions of the gastrula and then assessed their ability to undergo involution (Fig. 8–44). A patch of superficial and deep cells does not show normal extension, and all involution is blocked in the region of the graft. If only the superficial layer of the animal region is grafted to the marginal zone, involution of both deep and superficial cells occurs, including the graft. When the deep region of the blastopore is replaced with a patch of cells from the deep region of the animal pole, involution is arrested when the grafted patch reaches the lip of the blastopore. Hence, cells of the deep region appear to have unique properties that are essential for involution. In contrast, the cells of the superficial layer, including the bottle cells, do not have special properties and would appear to be passively moved into the interior to form the lining of the archenteron. The deep cells form the mesodermal cell stream and use the inner surface of the overlying marginal zone as a substratum upon which to migrate. The major features of the model proposed by Keller to explain gastrulation in *Xenopus* are summarized in Figure 8–45.

Because their eggs are often transparent, teleost

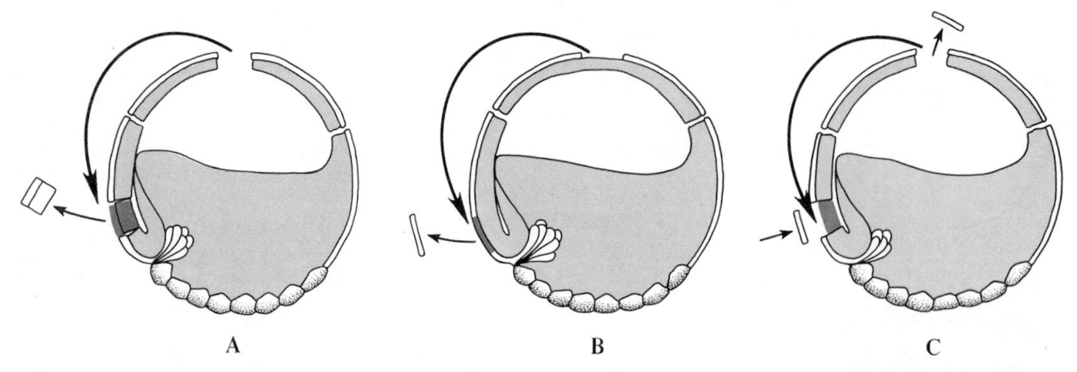


FIG. 8–44. To test the competence of superficial and deep cells from various regions of the gastrula *(Xenopus)* to undergo involution, cells of the superficial and deep layers of the animal region (A), the superficial layer alone (B), or the deep region alone (C) were extirpated and grafted to the dorsal marginal zone. Only cells of the deep region appear to have properties that are essential for involution. (From R. Keller, 1981. J. Exp. Zool. 216:81.)

fishes have been used in analyzing morphogenetic movements at gastrulation. Following cleavage, the blastoderm flattens and, because of an intrinsic capacity to spread, moves by epiboly over the yolk sphere toward the vegetal pole. Three separate structures appear to participate in the spreading movements: the enveloping layer of the blastoderm or the cohesive epithelium that forms the outer surface of the blastoderm, the deep cells of the blastoderm, and the yolk syncytial layer which sits atop the fluid yolk (Fig. 8–16). With the onset of gastrulation in *Fundulus*, the yolk syncytial layer spreads

FIG. 8–45. A model of the mechanism proposed to underlie gastrulation in some amphibians *(Xenopus).* During early gastrulation (A), the superficial cells flatten, spread, and divide. Deep cells of the marginal zone, held together by interdigitations (small arrows), move toward the vegetal pole (large arrows). Upon reaching the blastopore, the deep cells undergo a change in shape and behavior leading to the involution. The involuted cells form the mesoderm (M) and migrate toward the animal pole using the inner surface of the overlying marginal zone as a substratum. In the second half of gastrulation (B), the deep cells flatten and spread as the marginal cells (large arrows) continue to extend toward the vegetal pole. The mesodermal cells continue to migrate, and the overlying superficial layer of cells (including the bottle cells) is pulled toward the animal pole to form the roof of the archenteron. The bottle cells, which formed a thick compact mass in the early gastrula (A), contribute to much of the periphery of the archenteron by the end of gastrulation (C). (From R. Keller, 1981. J. Exp. Zool. 216:81.)

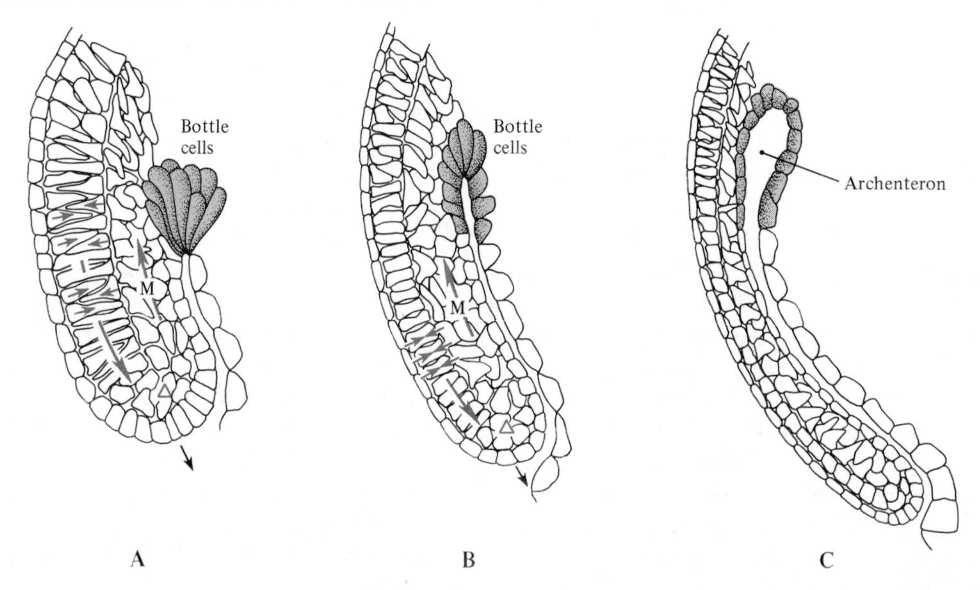

over the yolk in advance of the epibolic movements of the cellular blastoderm. Subsequently the blastoderm moves over the underlying yolk syncytial layer until the entire egg is enclosed. Several types of experiments, including the complete removal of the enveloping layer and deep cells, point to the conclusions that the blastoderm acquires this intrinsic capacity for epiboly during the late-blastula stage and that the blastoderm requires for its spreading movements a natural substratum, the yolk syncytical layer. What is the nature of the forces responsible for the spectacular movements of the blastoderm? At one time it was believed that the blastoderm was pushed over the yolk as a consequence of the division of its cells. However, young fish embryos treated with colchicine, a reagent known to block cellular mitoses, still show expansion of the blastoderm over the yolk mass.

Transmission and scanning electron microscopy of *Fundulus* embryos by Trinkaus and his co-workers have provided considerable insight into the mechanism of teleost epiboly. When the marginal cells of the expanding enveloping layer are carefully examined, they are observed to be attached to the yolk syncytial layer. If this marginal attachment is severed, the blastoderm retracts immediately; this suggests that the blastoderm is under considerable tension and the epibolic movements depend on a firm attachment between marginal cells and the yolk syncytial layer. A major question in the analysis of epiboly is whether the enveloping layer spreads actively over an expanding syncytial substratum or moves passively because of its adhesion to the independently moving yolk syncytial layer. In contrast to movements of fibroblast cells over a flat substratum in culture, where contacts need to break and form anew constantly, the edges of the marginal cells of the enveloping layer during spreading are firmly connected to the syncytial layer by special junctional complexes, consisting of a mixture of *tight* and *close junctions*. Since these marginal contacts are very stable, the epiboly of the enveloping layer would appear to be passive and in response to a pull exerted by an independently moving yolk syncytial layer. What keeps the enveloping layer intact during the stretching process? The electron microscope shows that the cells of the enveloping layer are bound together by tight junctions in which the outer leaflets of opposing plasmalemmas are fused by primitive desmosomes. Cell contact is also seen in the interdigitation of fingerlike cytoplasmic projections of adjacent enveloping layer cells (Fig. 8–46).

If spreading of the enveloping layer is not due to the locomotor activities of its marginal cells, but rather to its firm adhesion to the yolk syncytial layer, then the key to understanding epiboly resides in the properties of the syncytial layer. Prior to the onset of epiboly, the external yolk syncytial layer is about 400 microns wide and smooth in contour. When epiboly begins, the yolk syncytial layer progressively narrows and in the region of the margin of the enveloping layer is thrown into numerous folds. An interesting observation is that convolution of the surface of the external yolky layer is accompanied by a thickening of the cortical microfilamentous network that fills the cytoplasm of the folds. The possibility is raised, therefore, that a contractile force resides in the cortex of the external yolk syncytial layer; it acts to throw the surface of this layer into folds, to narrow it, and to stretch the blastoderm toward the vegetal pole by exerting tension on the attached margin of the enveloping layer. Further support for this viewpoint comes from the observation that folding and narrowing of the syncytial layer take place even if the blastoderm is detached. In contrast to the cells of the enveloping layer, which spread as an epithelial sheet, the deep blastomeres of the blastoderm appear to translocate by an entirely different mechanism. Prior to the onset of gastrulation, both time-lapse cinemicrographic and ultrastructure studies show that the deep blastomeres change shape constantly, although no actual movement is in evidence. Each cell forms protruding, rounded, and transparent bulges known as *lobopodia* (Fig. 8–47). With the onset of epiboly, however, the lobopodia begin to form adhesive contacts with other deep cells as well as with cells of the overlying enveloping layer. As this occurs, each lobopodium becomes stretched into a thin cytoplasmic projection or *filopodium*. A firm contact with another cell apparently provides traction, for, when the filopodium shortens, the cell is pulled in the direction of the adhesion. Hence, the deep blastomeres appear to move as individual units by crawling over each other. Although their movements are initially random, they subsequently migrate in unidirectional fashion toward the margins of the blastoderm to participate in the formation of the germ ring. Eventually they converge dorsally and posteriorly to form the embryonic shield. It should be pointed out at this time that the deep

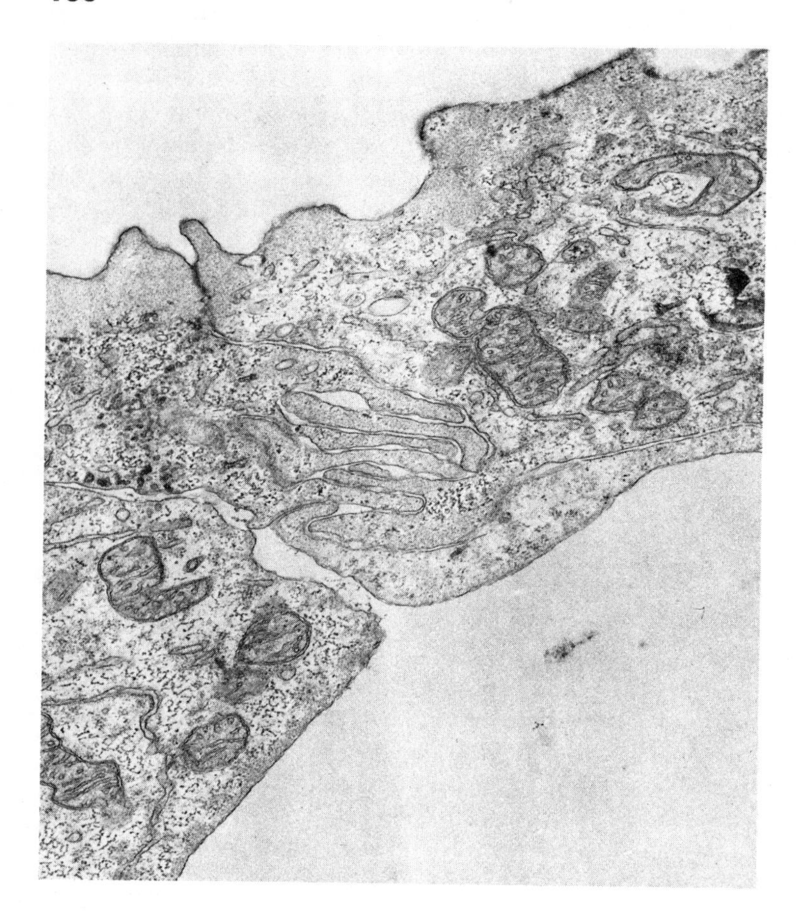

**FIG. 8–46.** Enveloping layer cells of *Fundulus* showing several cytoplasmic projections interdigitating with one another. (From J. Trinkaus and T. Lentz, 1967. J. Cell Biol. 32:139.)

**FIG. 8–47.** A deep cell of the early *Fundulus* gastrula showing contact with another blastomere by its lobopodium (LP). (From J. Trinkaus and T. Lentz, 1967. J. Cell Biol. 32:139.)

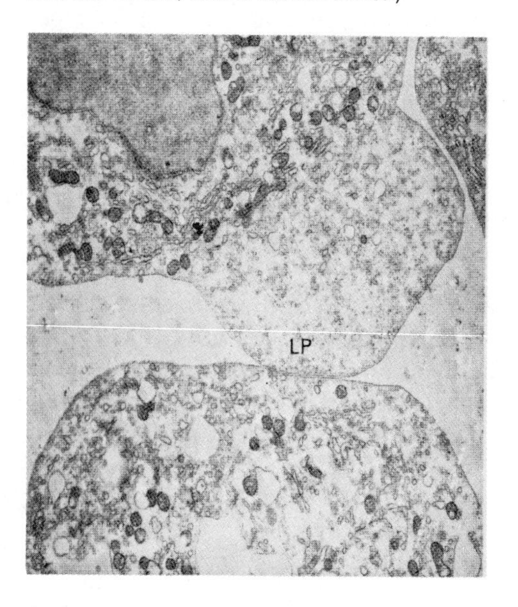


cells, which give rise to the tissues of the teleost embryo, do not appear to play an important role in epiboly.

The key to our complete understanding of the mechanisms underlying the cellular movements of the enveloping layer in teleost fishes is knowing more about the activities of the yolk syncytial and the yolk cytoplasmic layers during epiboly. These layers of the gastrulating embryo are confluent and serve as a specific substratum for the spreading blastoderm. The yolk syncytial layer expands in area and diminishes in thickness, but how this is accomplished remains elusive. Apparently, the cytoplasm of the yolk syncytial layer flows around the yolk, and as it does so the cytoplasm of the yolk cytoplasmic layer joins and mixes with it. Thick microfilaments (10 to 12 nm) have been identified under the electron microscope in the yolk syncytial layer beneath the marginal contact with cells of the enveloping layer; they run parallel to the contact and circumferentially relative to the whole egg. Interestingly, their marginal location and circumferential arrangement coincides with the peripheral constriction of the egg that takes place with spreading of the blastoderm. These filaments might be the motive force in epiboly and bring about the vegetal movement of the yolk syncytial layer with its attached blastoderm.

Studies of the mechanisms of morphogenetic movements during gastrulation and early morphogenesis of the chick have been directed in great measure at the spreading of the margin of the blastoderm, the immigration or invagination of the chordamesodermal and endodermal cells through the primitive streak, and the expansion of the hypoblast. The epibolic movements involved in spreading of the blastoderm over the yolk occur quite independently of those leading to the formation of the germ layers.

The area opaca of the chick blastoderm, which consists of extraembryonic ectoderm, mesoderm, and endoderm, spreads as an epithelial sheet beneath the vitelline membrane to encompass the yolk during the first several days following incubation. It is well known that the chick blastoderm fails to spread normally if removed from the yolk and cultured on an agar or plasma clot. New (1959), however, demonstrated that spreading of the area opaca is perfectly normal when the blastoderm is cultured with its own vitelline membrane (Fig. 8–48). Other experiments by New have given clear evidence that the cells of the margin of the area opaca (the so-called *margin of overgrowth*) are attached to the inner surface of the vitelline membrane and use it as a substratum upon which to move. Although the area opaca in vitro will adhere to the outer surface of the vitelline membrane, expansion is only possible upon the inner surface of the vitelline membrane, suggesting that the latter is uniquely constructed to promote movement. Textural and/or the chemical composition of the vitelline membrane may be important to blastoderm expansion. When a chick blastoderm is cultured on the vitelline membrane from a turkey, it expands at a rate comparable to that in the normal system. Expansion does not occur if duck vitelline membrane is used. Duck and turkey vitelline membranes are morphologically similar, but chick and duck membranes appear to be chemically different. As in the case of teleost epiboly, it was once thought that centrally located sites of intense cellular proliferation were the motive force underlying the epibolic spreading of the chick blastoderm. In essence, movement was proposed to be effected by a "push from behind." Although there is cell division within the blastoderm during epiboly, principally near the margins of the blastoderm, there is little experimental evidence to support such a proposal. The area opaca will continue to spread upon the vitelline membrane when blastoderms are cultured in the presence of known mitotic inhibitors, such as colchicine.

**FIG. 8–48.** When the chick blastoderm is explanted normally onto its vitelline membrane (A), its adhesive margin (AM) attaches to the inner surface of the vitelline membrane (B). Normal expansion of the blastoderm then follows (C). If the blastoderm is inverted and placed on the vitelline membrane (D), the adhesive surface of the blastoderm curls under to bring it into contact with the vitelline membrane (E). Expansion in this case leads to the formation of the hollow vesicle (F). BE, blastoderm edge; LA, liquid albumin; VM, vitelline membrane. (After D. A. T. New, 1959. J. Embryol. Exp. Morphol. 7:149.)

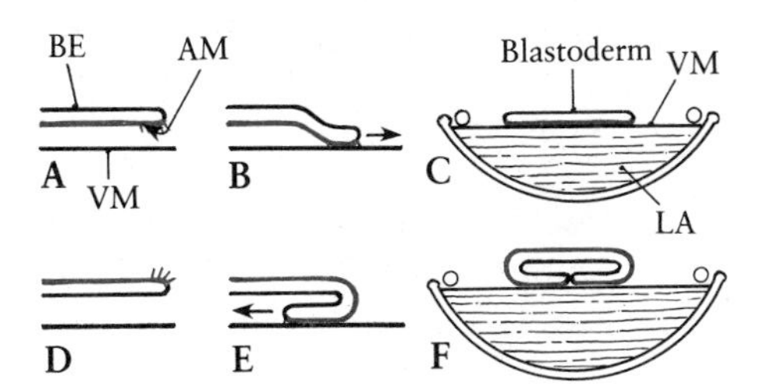


An alternative hypothesis, namely that the marginal cells of the area opaca act by pulling the rest of the blastoderm over the yolk, is supported by evidence from studies with the electron microscope. In contrast to the other cells of the chick blastoderm, the cells at the margin of the blastoderm are flattened and attached to the overlying vitelline membrane by protrusions and electron-dense thickenings known as plaques. The distalmost cells of the marginal zone are ectodermal, and they are observed to form long, thin processes or filopodia during spreading activity (Fig. 8–49). These processes act as locomotor organs and upon contraction effectively pull the marginal cells with them. Hence, the marginal cells appear to advance as an epithelial sheet with the nonmarginal cells being pulled along in rather passive fashion. The pull created by the marginal cells puts the rest of the cells of the blastoderm under tension during movement. As with the enveloping layer of cells in the teleost blastoderm, specialized junctional complexes between the nonmarginal cells keep these cells together as a tight cohesive sheet. This prevents their being pulled apart under the tension of epiboly.

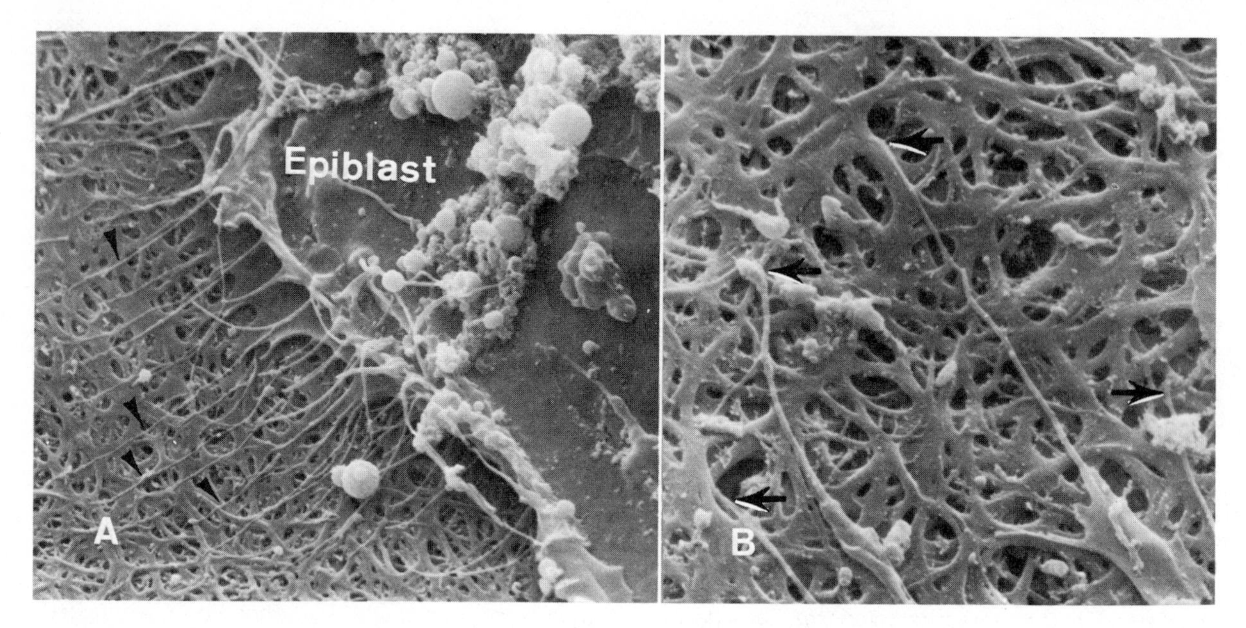


FIG. 8–49. A, during the overgrowth of the yolk by the blastoderm in the chick embryo, the margins of the epiblast cells show numerous filipodia extending across to the vitelline membrane; B, a higher magnification of several filipodia in contact with the vitelline membrane. (From E. Chernoff and J. Overton, 1977. Dev. Biol. 57:33.)

As noted previously, the embryonic body of the chick starts to take shape with the formation of the primitive streak and the segregation of the germ layers. The morphogenetic movements of cells leading to the formation of the primitive streak are initiated shortly after the incubation temperature reaches 38.5°C. The directions in which cells of the epiblast of the bilaminar blastoderm migrate during the formation of the primitive streak are rather well know, chiefly through the carbon and carmine particle studies of Spratt and Haas. Although the movements of these particiles have given a rather accurate picture of the movements of epiblast cells toward the primitive streak, the mechanism by which cells of the epiblast converge and form the streak is far from clear. Many ingenious hypotheses have been put forth to account for the manner in which the primitive streak develops. The presence of numerous points of contact (i.e., tight junctions) between surfaces of adjacent cells suggests that endoderm and chordamesoderm migrate as an epithelial sheet. There is additional evidence that the hypoblast plays an important role in directing the movements of cells of the epiblast and in causing their deformation to form the primitive streak. If the hypoblast from a blastoderm at the early primitive streak stage is removed and rotated 90 degrees to its original axial position, the primitive streak will continue to form, but it bends in the direction of the original anterior end of the endoderm.

The primitive streak is an invagination center. It is not surprising, therefore, to find that movement of epiblast cells through the streak appears to be very similar to that described for the movement of chordamesodermal and endodermal cells through the blastopore of the gastrulating amphibian embryo. As cells approach the streak, many of them elongate and become irregular in shape; they then become progressively elongated and flask shaped. Both gap and tight junctions are particularly evident between cells in the region close to the upper surface of the streak. Cytoplasmic microtubules and bundles of microfilaments (4–6 nm in diameter) have been identified in cells of the primitive streak. However, the role that either of these cytoskeletal components might play in providing the motive forces for changing cell shape resulting in the translocation of cells through the streak is not clear.

The movement of mesodermal cells from the base of the primitive streak outward beneath the epiblast is a steady and continuous process (Fig. 8–

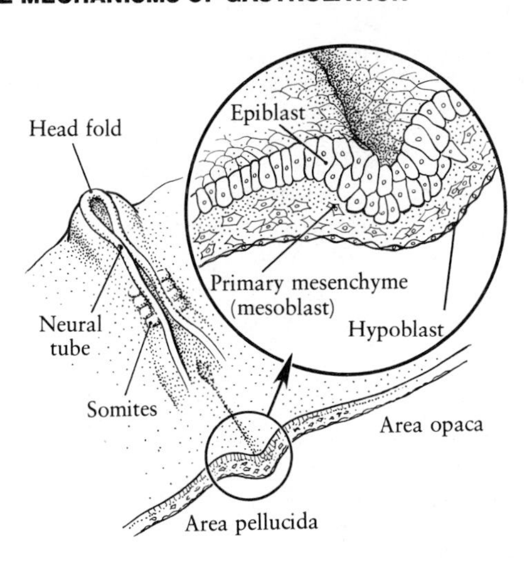


**FIG. 8–50.** Diagram of the gastrulating chick embryo at the four-somite stage. The embryo consists of head fold, open neural groove and neural folds, somites, and primitive streak. A transverse section through the primitive streak shows the cells of the area pellucida and the area opaca. The inset shows the cells of the epiblast and the laterally migrating mesoderm. The hypoblast is a thin epithelium below the mesoderm. (From E. Hay, 1968. Epithelial-Mesenchymal Interactions. R. Fleishmajer and R. E. Billingham, eds. Williams and Wilkins, Baltimore. Copyright © 1968, Williams and Wilkins Company, Baltimore.)

50). Each mesodermal cell is highly polarized during migration, with the basal end of the cell (containing the nucleus) leading off and the apical end (containing the Golgi complex) trailing behind. The advancing edge of the mesoblast cells elaborates numerous filopodia; each filopodium contains filaments about 50 Å in diameter and extends into and adheres to the adjacent epithelial surfaces of the overlying epiblast and the underlying hypoblast. Also, broader cytoplasmic projections (ruffled membranes?) mark the leading edges of thse mesodermal cells. The filamentous appearance of the filopodia suggests that these organs may be contractile and hence effect movement by pulling. Earlier studies indicated that mesodermal cells migrate as individual units away from the primitive streak. However, electron-microscopic studies now clearly show that the mesodermal cells during migration are connected to each other as well as to cells of the epiblast and hypoblast by a variety of junctional complexes, leaving the impression that movement away from the primitive streak is in the form of a loosely arranged sheet of interconnected cells (Fig. 8–51).

The hypoblast of the late-stage chick blastoderm is presumptive to the extraembryonic endoderm of the yolk sac. Its cells are loosely packed and show signs of movement well in advance of the formation of the primitive streak. Cells of the hypoblast move forward and radially, apparently using the undersurface of the overlying epiblast as a specific substratum. What causes this migration and whether it is active or passive is still very much a source of controversy. The proposal that hypoblast cells are pushed passively by the action of a center of rapid cell division in the posterior part of the blastoderm has not been adequately tested. It is equally possible that movements of hypoblast cells are initiated with changes in such cell parameters as motility and adhesiveness.

In all gastrulating embryos, cells move with precision and order from one location to another. Although the mechanisms of most morphogenetic movements accomplishing this task are still obscure, it is clear from the previous examples that such properties of cells as surface adhesiveness, the ability to change shape (deformation), and the ability to form organs of locomotion (i.e., ruffled membranes and filopodia) are important to the event of translocation. It is extremely important for the reader to appreciate that movements, and, indeed, the larger picture of gastrulation, involve the interplay of cell motility, cell shape, and cell contact behavior. For example we have seen several cases in which a cell must adhere to a substratum before it can move from one place to another. For the cell to leave one location, it must be free of contacts with adjacent cells (i.e., lose surface adhesiveness). The formation of filopodia in cells that move as individual units, or the formation of ruffled membranes at the margins of cell masses moving as epithelial sheets, necessitates contact with a substratum and consequent changes in cell shape.

Many fundamental questions can still be raised about morphogenetic movements at gastrulation. Why do cells move in their particular directions? Do cells of the gastrula have a capacity to read and respond to some chemical and/or molecular "road map" to guide them? What signals initiate and terminate such changes in cell properties as adhesiveness and motility? Recent advances in cell membrane structure and in the surface chemistry of adult cells may point the way to answering these questions on gastrulating embryos.

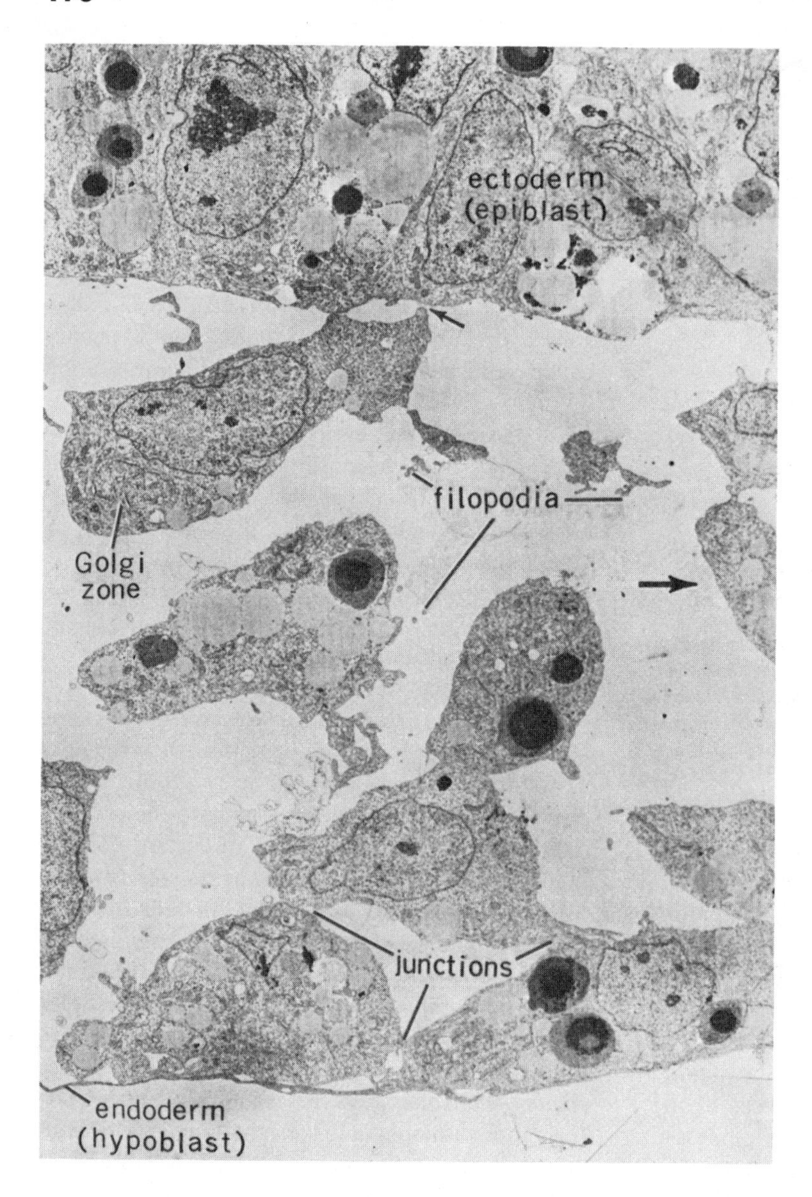


**FIG. 8–51.** Each mesodermal cell in the chick embryo, following gastrulation through the primitive streak, moves with "basal" end forward and "apical" end trailing behind. Filipodia form along the leading edge of the cell. Note the contacts of the mesodermal cells with adjacent cells of the ectoderm and endoderm. The arrow indicates the direction of mesodermal cell movement. (From R. Trelstad et al., 1967. Dev. Biol. 16:78.)

## Cellular Rearrangement and Cellular Segregation

The transformation of the fertilized egg into a multicellular system in which groups of cells are shifted into arrangements that foreshadow the tissue patterns of the adult organism occurs principally between the blastula and neurula stages of development. Irrespective of whether the postgastrula embryo is round (frog) or flat (chick), the morphogenetic movements of gastrulation segregate sheets of cells known as the primary germ layers. The germ layers are always arranged such that the mesoderm occupies a position between the outer ectoderm and the inner endoderm, suggesting that cells associate spatially in a highly selective manner.

Much of the history of embryology has been concerned with the nature of the mechanisms controlling the rearrangement and segregation of groups of cells that result from morphogenetic movements not only during gastrulation but also during the period of organ formation. For example, why do the mesodermal cells, which form a loosely packed association, stay together after gastrulation as a homogeneous layer insinuated between ectoderm and endoderm? Are the germ layers of the postgastrular embryo organized the way they are because of special properties of their constituent cells, or are they passively channeled into their topographical positions because of some overriding influence of the whole embryo? The rearrangement of cells and the sorting out or segregation of cells into specific tissues or organs are inextricably related to changes in cell shape, cell contact behavior, and cell motility.

Unfortunately, our current knowledge of the roles of cell shapes, cell contact behavior, and cell motility in the organization of cells of the embryonic germ layers as well as in the segregation of organ tissues is based largely on observations using artificial or in vitro techniques. Fragments of embryos or isolated organ primordia, or reaggregated suspensions of isolated embryonic cells, are allowed first to interact in the culture medium, and the cellular patterns of reassociation are then observed. A routine procedure is illustrated for amphibian cells in Figure 8–52. A fragment from the embryo is excised with glass needles and placed into a potassium hydroxide solution (pH 9.6–9.8). At this pH, the tissue fragment disaggregates or dissociates into individual single cells. Treatment of similar fragments with trypsin or solutions containing EDTA (ethylenediaminetetraacetic acid, a chelating agent) without calcium and magnesium ions will also bring about *disaggregation*. Cells prepared in this way from a variety of embryonic tissues (i.e., gastrular ectoderm, endoderm, mesoderm; neural plate) are then brought together in selected combinations in a culture medium at a pH of 8.0. The cells are then observed to reaggregate.

Early studies by Holtfreter pioneered and elucidated much of the basis of current views on the rearrangement and self-isolation of embryonic cells. Experiments by Holtfreter using fragments of embryos, individual cells in culture, and reaggregated suspensions of embryonic cells led him to believe that different kinds of cells have different inherent associative properties or *affinities*. For example, when he isolated a fragment of presumptive endoderm from a frog embryo and placed it in culture, the fragment initially rounded up into a compact aggregate and then spread on the bottom of the culture dish. When an isolated dorsal lip of the blastopore was then added to the endodermal preparation, the mesodermal cells not only remained segregated but formed a distinctive invagination into the endoderm (Fig. 8–53). On the basis of an extensive series of experiments utilizing various combinations of embryo fragments and disaggregated cells, it gradually became apparent that tissues and groups of cells recognized and adhered to only like tissues and cells. Holtfreter introduced the concept of *selective cellular affinities* to explain these patterns of behavior between different kinds of cells, although the mechanism underlying the behavior was unknown to him.

When ectodermal and endodermal cells of the gastrula are dissociated and combined in vitro, there is initially random movement and indiscriminate union between diverse cell types. Gradually, however, there is sorting out and self-isolation in the reaggregate, leading to the formation of homogeneous layers of superficial ectoderm and internal endoderm (Fig. 8–54 A). Although both germ layers are capable of forming an epithelium, a permanent association between the two layers is lacking. A mosaic of cells is seen during the first 10 hours in an aggregate formed by the recombination of ectoderm and mesoderm. The mesodermal cells then slip to the interior of the aggregate to become surrounded by a thick layer of ectoderm (Fig. 8–54 B). The inner mesodermal tissue may then differen-

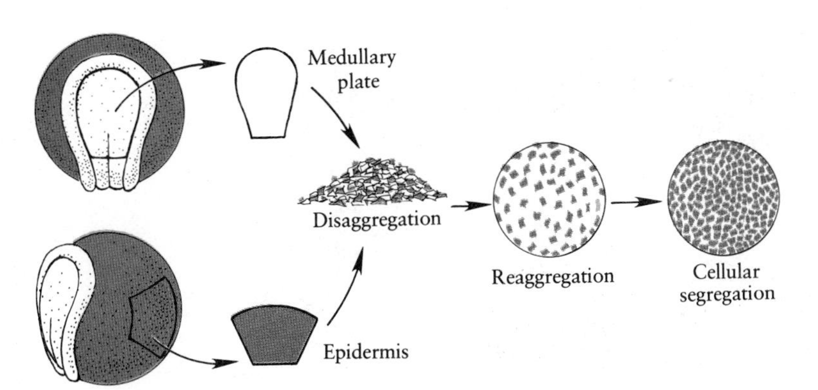


**FIG. 8–52.** A diagrammatic summary of the technique for the disaggregation of embryonic tissues (medullary plate and epidermis) and their reaggregation for studies designed to examine cellular segregation. (From P. Townes and J. Holtfreter, 1955. J. Exp. Zool. 128:53.)

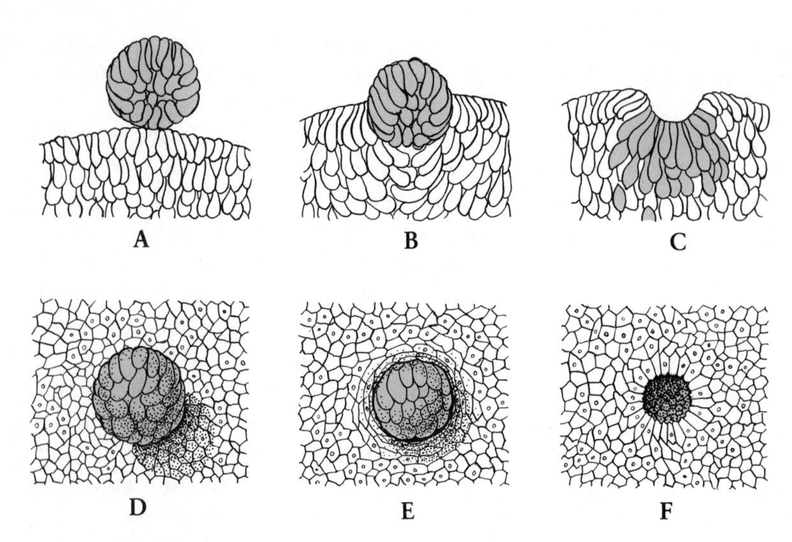


FIG. 8–53. A piece of the dorsal lip of the blastopore will sink into a piece of endoderm, forming a distinct invagination. A, B, and C are cross sections; D, E, and F are surface views of A, B, and C. (From J. Holtfreter, 1944. J. Exp. Zool. 95:171.)

tiate into blood vessels, mesenchyme, and so on while the outer ectodermal tissue becomes a thin epidermis. When endoderm and mesoderm are combined as single cells, the reaggregate shows that the endoderm forms an external covering to the entire preparation (Fig. 8–54 C). This pattern of cell segregation, contrary to the normal topographical arrangements of these two layers, resembles that of a total exogastrula. If all three germ layers are dissociated into single cells and then mixed, an aggre-

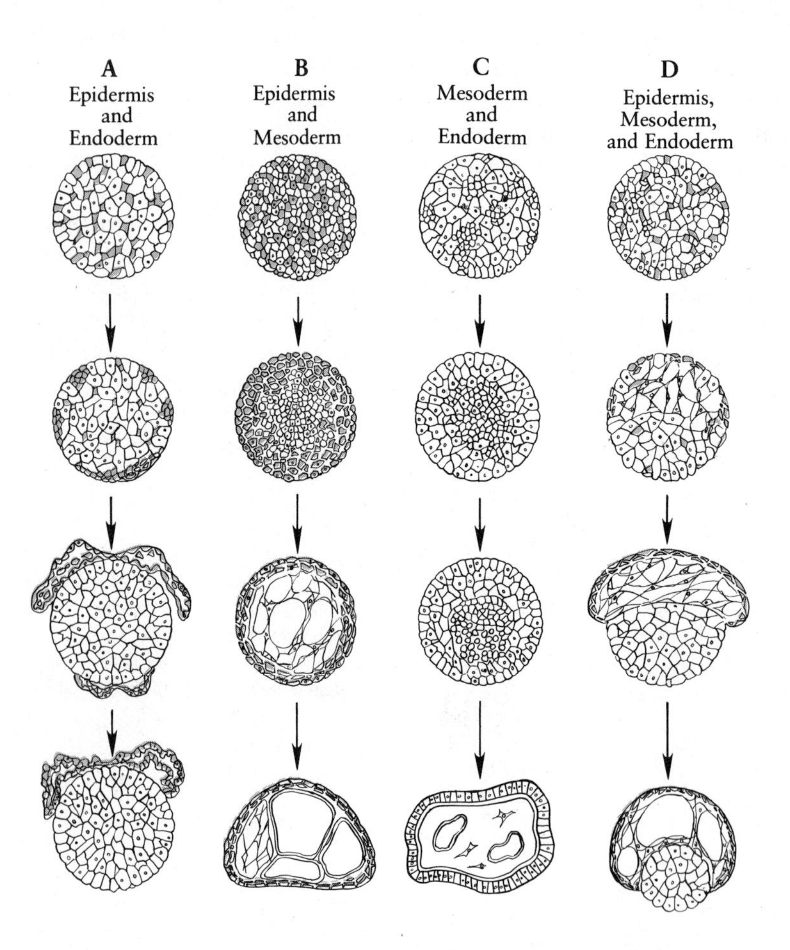


FIG. 8–54. Rearrangement and segregation of disaggregated and reaggregated embryonic cells (amphibian). A, combined epidermal and endodermal cells; B, combined epidermal and mesodermal cells; C, combined mesodermal and endodermal cells; D, combined epidermal, mesodermal, and endodermal cells. (From P. Townes and J. Holtfreter, 1955. J. Exp. Zool. 128:53.)

gate forms in which the mesoderm holds together the ectoderm at the surface and the endoderm at the interior (Fig. 8–54 D). As with other cell combinations, the rearrangement of germ cells in this aggregate involves the segregation and association of like cells and their placement into topographical relationships similar to those produced by the normal movements of ectoderm, mesoderm, and endoderm at gastrulation. The sorting-out process requires cell motility, such as the inward movements of the mesoderm and the outward spreading of the ectoderm, and the adhesion of like cell types to yield histologically specific tissue layers. This pattern of cell behavior is also observed in mixed aggregates of embryonic cells taken from other species as well as in combinations of dissociated cells or tissue fragments prepared from organ primordia. If two or more organs, such as chick heart and liver, are dissociated and the cell suspensions mixed, movements of cells are followed by segregation and self-assembly in which a central mass of heart cells becomes surrounded by a layer of liver cells (Chapter 13).

Although the experimental investigations of Holtfreter did not reveal the basis for cellular rearrangement and sorting out, he clearly demonstrated that the preferential association of like cells or groups of cells is important to germ layer formation, and to tissue construction and organization. Since his observations, efforts have been directed at understanding the forces that underlie differences in affinities among cells and in their abilities to sort out into highly predictable associations. We will consider briefly several hypotheses that attempt to account for the adhesion of like cells into a cohesive unit and the positioning of these units in patterns peculiar to each combination.

*Chemotaxis* has often been invoked to account for cell positioning and cell self-assembly. Might not cells segregate themselves in response to differential concentrations of metabolites, migrating along a gradient either toward or away from the highest point of concentration? For example, certain cells in an aggregate might be attracted toward the periphery in response to a higher concentration of oxygen. The evidence to support this hypothesis is weak and unconvincing, primarily because the behavior of cells in an aggregate does not generally conform to that dictated by the principle of chemotaxis.

Curtis (1961) postulates that the sorting out of vertebrate cells is primarily due to differences in the time *(timing hypothesis)* at which diverse cell types become adhesive and hence can be joined together. He assumes that the cell surface becomes modified by the process of dissociation. Each cell loses its adhesiveness and becomes migratory. Diverse cell types within an aggregate gradually reacquire their adhesive qualities, but at different times. The combined effects of random movements, rising adhesiveness, and reduced shear (i.e., a force produced by cells moving over each other, which tends to promote motility; its effects are least at the surface of an aggregate) trap cells at the surface of an aggregate. In this manner, all cells of one type become trapped at the surface with other less adhesive cells being herded toward the center of the aggregate. Some evidence is availabe to support the timing hypothesis. Experiments have been conducted in which the position of the germ layers in an aggregate can be altered by varying the times at which they are dissociated before being recombined in culture. A strong argument against the hypothesis is that it requires an artifactual effect (i.e., loss of adhesiveness owing to dissociation) to explain the normal topographical relationships of cells in a mixed aggregate.

Spreading and sorting out of trypsin-dissociated embryonic tissues in mixed aggregates have been extensively studied in the laboratories of Steinberg and Moscona. Steinberg and colleagues observe that there appears to be a hierarchy of preference for the external or internal position of like cells in mixed aggregates of cells. The hierarchy predicts topographical relationships between cell types and follows a transitive rule: If cell type A surrounds cell type B, and cell type B surrounds cell type C, then cell type A will surround cell type C. Random motility and quantitative differences in general adhesiveness between cell types form the basis of the *differential adhesiveness hypothesis;* they are proposed to play central roles in the self-isolation and organization of cells into functional units.

Steinberg believes that an aggregate may be treated as if it were a multiple-phase system of immiscible liquids. When a cell touches other cells in culture, it may remain in contact with that cell(s) or move away to make contact with other cells. What determines how the cell responds to its initial contact with another cell? Suppose that an aggregate is composed of two cell types, A and B, and that A cells cohere more strongly than B cells. When cell A adheres to cell A (or cell B), it is said to possess a certain amount of free energy, which is used to make and maintain the contact. For the aggregate to achieve thermodynamic stability or equilibrium, the free energy of A and B cells must be at a minimum. This can only be obtained when there is maximum adhesion of cell surfaces (i.e., when the adhesive contacts on the cell surface are used to a maximum extent). Hence, cells will tend to exchange weaker for stronger bonds of adhesion. If A-B adhesions are intermediate in strength between A-A and B-B adhesions, but weaker than the average of the two, then there will be a continual exchange of A-B adhesions for A-A adhesions until most of the A cells cohere. Since cells tend to maximize their contacts over as much of their surface as possible, the more adhesive A cells will move to the interior of the aggregate, thereby excluding the less adhesive B cells to the surface of the aggregate. Hence, the more cohesive A cells will form the internal component of the aggregrate, surrounded by and embedded in a continuous layer of less cohesive B cells. Since the B cells at the surface of the aggregate do not use their adhesiveness to maximal capacity, the surface area of the aggregate is reduced to a minimum (i.e., a sphere).

Steinberg and his co-workers have developed several lines of experimental evidence to support the hypothesis that the sorting out of cells in embryonic systems, including gastrulation, requires motility and differences in the frequency of adhesive sites on cell surfaces. For example, when cells of embryonic tissues are dissociated and then separately allowed to reaggregate, a rounded aggregate of cells will form. If each aggregate is then subjected to a small centrifugal force and evaluated for its ability to resist changes in shape, flattening is observed in those aggregates whose cells typically segregate to the exterior in mixed cell cultures; resistance to deformation is most noticeable in those aggregates whose cells occupy the center of mixed cell cultures.

The differential adhesion hypothesis provides an explanation for the sorting patterns of cells based on the relative strengths with which those cells cohere with themselves. It does not explain how cells selectively recognize and adhere to each other. Studies from the laboratory of Moscona suggest that the specificity of cell adhesion is probably related to specific *cell ligands* or carbohydrate-containing molecules located on or near the outer surface of cells. Moscona believes that cell recognition and sorting out are due to qualitative differences between embryonic cells in the configuration of their surface molecules. Plasma membrane constituents commonly include monosaccharide sugars linked to proteins (glycoproteins) and lipids (glycolipids). Side chains of sugars of variable lengths are envisioned as projecting from the glycoproteins into the environment of the cell. To test the hypothesis that cells associate by virtue of chemically specific attachment sites, membranes from embryonic tissues have been prepared and efforts made to isolate and characterize glycoproteins that promote cell aggregtion. Moscona and his colleagues have demonstrated remarkable success in this direction. They have isolated, for example, a specific factor from embryonic neural retina that rapidly promotes the aggregation of dissociated cells from the same tissue source. Presumably, the basis for other histospecific cellular associations lies in similar cell-aggregating factors. We will return to the subject of cell recognition and cell sorting when consideration is given to factors underlying the construction of organs (Chapter 13).

How do cell recognition and segregation relate to gastrulation and the arrangement of the three germ layers? Remember that the cells of the ectoderm, mesoderm, and endoderm become spatially segregated in a highly predictable pattern as a consequence of the movements of gastrulation. In analyzing gastrulation in amphibians, Holtfreter demonstrated that the mesoderm becomes surrounded by endoderm, thus indicating the greater cohesiveness of the mesoderm. Also, the endodermal tissue generally takes up an internal position in combination with ectoderm. Thus, there appears to be a hierarchy of cohesiveness between germ layers (from greatest to least): mesoderm, endoderm, and ectoderm. How does one account for the fact that the relationship between the endoderm and mesoderm in reaggregate experiments is exactly the reverse of the relationship of these two germ layers in the intact, gastrulating embryo? In studying the results of recombination experiments, one should remember that the relationship of endoderm and mesoderm occurs with cells obtained from the germ layers *after* the latter have achieved their normal topographical relationships in the intact embryo. Hence, it is likely that there are changes in the cellular adhesiveness of endoderm and mesoderm during gastrulation (i.e., the cellular adhesiveness of mesoderm increases). It can also be concluded that an ectodermal covering is required for the normal shifting of the mesoderm over the endoderm. In the absence of an ectodermal covering, the mesoderm will exhibit its "abnormal" behavior of invaginating into the endodermal mass.

## References

Baker, J. 1965. Fine structure and morphogenetic movements in the gastrula of the tree frog, *Hyla regilla*. J. Cell Biol. 24:95–116.

Ballard, W. 1973a. Morphogenetic movements in *Salmo gairdneri*. J. Exp. Zool. 184:27–48.

Ballard, W. 1973b. A new fate map of *Salmo gairdneri*. J. Exp. Zool. 184:49–75.

Ballard, W. W. 1976. Problems of gastrulation: real and verbal. Biol. Sci. 26:36–39.

Ballard, W. 1981. Morphogenetic movements and fate maps of vertebrates. Am. Zool. 21:391–399.

Ballard, W. and A. Ginsburg. 1980. Morphogenetic movements in acipenserid embryos. J. Exp. Zool. 213:69–116.

Betchau, T. and J. Trinkaus. 1978. Contact relations, surface activity, and cortical microfilaments of marginal cells of the enveloping layer and of the yolk syncytial and yolk cytoplasmic layers of *Fundulus* before and during epiboly. J. Exp. Zool. 206:381–426.

Chernoff, E. and J. Overton. 1977. Scanning electron microscopy of chick epiblast expansion on the vitelline membrane. Dev. Biol. 57:33–46.

Curtis, A. S. G. 1961. Timing mechanisms in the specific adhesions of cells Exp. Cell Res. Suppl. 8:107–122.

Granholm, N. and J. Baker. 1970. Cytoplasmic microtubules and the mechanism of avian gastrulation. Dev. Biol. 23:563–584.

Gustafson, T. and L. Wolpert. 1967. Cellular movement and contact in sea urchin morphogenesis. Biol. Rev. 42:442–498.

Hamburger, V. 1960 A Manual of Experimental Embryology. Chicago: University of Chicago Press, pp. 47–60.

Hamburger, V. and H. L. Hamilton. 1951. A series of normal stages in the development of the chick embryo. J. Morphol. 88:49–92.

Holtfreter, J. 1943a. Properties and functions of the surface coat in amphibian embryos. J. Exp. Zool. 93:251–323.

Holtfreter, J. 1943b. A study of the mechanics of gastrulation. Part I. J. Exp. Zool. 94:261–318.

Holtfreter, J. 1944. A study of the mechanics of gastrulation. Part II. J. Exp. Zool. 95:171–212.

Johnson, K. 1970. The role of changes in cell contact behavior in amphibian gastrulation. J. Exp. Zool. 175:391–428.

Karp, G. and M. Solursh. 1974. Acid mucopolysaccharide metabolism, the cell surface, and primary mesenchyme cell activity in the sea urchin embryo. Dev. Biol. 41:110–123.

Keller, R. E. 1976. Vital dye mapping of the gastrula and neurula of *Xenopus laevis*. II. Prospective areas and morphogenetic movements of the deep layer. Dev. Biol. 51:118–137.

Keller, R. 1980. The cellular basis of epiboly: an SEM study of deep-cell rearrangement during gastrulation in *Xenopus laevis*. J. Embryol. Exp. Morphol. 60:201–234.

Keller, R. 1981. An experimental analysis of the role of bottle cells and the deep marginal zone in gastrulation of *Xenopus laevis*. J. Exp. Zool. 216:81–101.

Lentz, T. and J. Trinkaus. 1967. A fine structural study of cytodifferentiation during cleavage, blastula, and gastrula stages of *Fundulus heteroclitus*. J. Cell Biol. 32:121–138.

Lipton, B. H. and A. G. Jacobson, 1974. Experimental analysis of the mechanisms of somite formation. Dev. Biol. 38:91–103.

Løvtrup, S. 1975. Fate maps and gastrulation in amphibia—a critique of current views. Can. J. Zool. 53:473–479.

Malan, M. E. 1953. The elongation of the primitive streak and the localization of the presumptive chordamesoderm of the early chick blastoderm studied by means of coloured marks with Nile blue sulphate. Arch. Biol. 64:149–182.

Monroy, R., B. Baccetti, and S. Denis-Donin. 1976. Morphological changes of the surface of the egg of *Xenopus laevis* in the course of development. IV. Scanning electron microscopy of gastrulation. Dev. Biol. 49:250–259.

Moscona, A. A. 1968. Cell aggregation: properties of specific cell-ligands and their role in the formation of multicellular systems. Dev. Biol. 18:250–277.

Nakatsuji, N. 1975. Studies on the gastrulation of amphibian embryos: cell movement during gastrulation in *Xenopus laevis* embryos. Wilhelm Roux's Archiv. Dev. Biol. 178:1–14.

New, D. A. T. 1959. Adhesive properties and expansion of the chick blastoderm. J. Embryol. Exp. Morphol. 7:146–164.

Nicolet, G. 1971. Avian gastrulation. Adv. Morphog. 9:231–261.

Pasteels. J. J. 1937. Etudes sur la gastrulation des vertébrates méroblastiques. III. Oiseaux. IV. Conclusions générales. Arch. BIol. 48:381–488.

Peter, K. 1938. Die Entwicklung des Endoderms beim Hühnchen. Z. Mikrosk-Anat. Forsch. 43:362–415.

Roth, S. 1968. Studies on intercellular adhesive selectivity. Dev. Biol. 16:602–613.

Roth, S., E. McGuire, and S. Roseman. 1971. Evidence for cell-surface glycosyltransferases: their potential role in cellular recognition. J. Cell Biol. 51:536–547.

Spratt, N. T., Jr., and H. Haas. 1965. Germ layer formation and the role of the primitive streak in the chick. I. Basic architecture and morphogenetic tissue movements. J. Exp. Zool. 158:8–38.

Steinberg, M. 1964. The problem of adhesive selectivity in cellular interactions. In: Cellular Membranes in Development, pp. 321–366. M. Locke, ed. New York: Academic Press.

Townes, P. and J. Holtfreter. 1955. Directed movements and selective adhesion of embryonic amphibian cells. J. Exp. Zool. 128:53–120.

Trelstad, R., E. Hay, and J. Revel. 1967. Cell contact during early morphogenesis in the chick embryo. Dev. Biol. 16:78–106.

Trinkaus, J. and T. Lentz, 1967. Surface specializations of *Fundulus* cells and their relation to cell movements during gastrulation. J. Cell Biol. 32:139–153.

Trinkaus, J. 1976. On the mechanism of metazoan cell movements. In: *The Cell Surface in Animal Embryogenesis and Development*. Cell Surface Rev. 1:225–329. G. Poste and G. Nicolson, eds. New York: Elsevier/North-Holland Biomedical Press.

Vakaet, L. 1962. Some new data concerning the formation of the definitive endoblast in the chick embryo. J. Embryol. exp. Morphol. 10:38–57.

Vakaet, L. and C. Vanroelen. 1982. Localization of microfilament bundles in the upper layer of the primitive streak stage chick blastoderm. J. Embryol. exp. Morphol. 67:59–70.

Vogt, W. 1929. Gestaltunganalyse am Amphibienkern mit ortlicher Vitalfarbung. Vorwort ueber Wege Zeil. I. Methodik and Wirkungweise der ortlicher Vitalfarbung mit Agar als Farbtrager. Wilhelm Roux' Arch. Entwicklungsmech. Org. 120:385–706.

Yehudit, A. and H. Eyol-Giladi. 1981. Interaction of epiblast and hypoblast in the function of the primitive streak and the embryonic axis in the chick, as revealed by hypoblast rotation experiments. J. Embryol. exp. Morphol. 59:133–144.

# 9

# THE EARLY EMBRYO—ITS ORGANIZATION AND ACTIVITIES

## Preformation and Epigenesis

Each living organism can be viewed as consisting of a time series of systems in which fundamental parts and processes are arranged in an orderly temporospatial pattern. The tissues and organs of these systems in turn have their own characteristic structure and function. If the history of each organism is traced back through time, we know that this structural and functional complexity that organisms demonstrate originates with the fertilized egg, a deceptively simple and apparently unstructured cell. The visible steps by which the fertilized egg becomes transformed into the embryonic and adult stages, and the processes underlying these gradual transformations, constitute a meaning for the term development. Generally, multicellular animals with sexual reproduction show remarkable similarity in the broad steps by which these transformations occur (i.e., fertilization, cleavage, gastrulation, etc.).

Humans in general, and the scientist in particular, have always been interested in the forces or mechanisms underlying the precise, predictable, and ordered expressions of development. Historically, two viewpoints have been invoked to explain the events of development: preformation and epigenisis. Prior to the late 1800s, each theory was based primarily on comparative and descriptive observations of embryos. The theory of preformation denied that development resulted in an increased level of order and postulated instead that the complexity associated with embryogenesis represented the gradual growth and enlargement of preexisting organization. The preformationists of the 17th and 18th centuries held that the complete adult organism was present in each egg (these people were called ovists) or sperm (these people were called animaculists or spermists), but in miniaturized state. Development was simply a matter of an increase in size.

The classical idea of preformation was doomed to a rather short existence, as the ontogenetic process was subject to careful observation. Using the chick for study, Kaspar Friedrich Wolff (1733–1794) concluded that the egg contained no future parts of the embryo. He saw only the relatively simple, unstructured material that we know as protoplasm. Hence, the order and complexity associated with the development of the embryo must emerge gradually from a simple, formless cell by formative and synthetic processes. The process by which the adult properties arise anew during ontogeny was termed epigenesis.

With improvement in optical equipment and perfection of the light microscope in the 19th century, it became obvious that the organism did not exist in miniaturized form in either the egg or the sperm. As originally proposed, therefore, the theory of preformation was abandoned. However, the concept itself did not die. August Weismann (1834–1914) proposed that the various parts and organs of the body of the embryo were represented by linearly arranged *particles* and *determinants* in the nucleus of the fertilized egg. He envisaged development as an expression of these invisible, but preformed elements, with each one determining a specific part of the embryo.

During the last half of the 19th century, it gradually became apparent that simple observations on how an embryo developed were inadequate to allow for a choice between the theories of preformation and epigenesis. Wilhelm Roux (1850–1924) dramatically changed the approach to embryogenesis by shifting the emphasis from an observational or descriptive to an experimental one. He founded the discipline of *analytical embryology* (developmental mechanics). Assuming that the concept of preformation as postulated by August Weismann was correct, he reasoned that destruction of one of the first two blastomeres of an embryo should produce an individual lacking certain parts and organs. Roux performed the simple experiment of destroying with a hot needle one of the first two blastomeres of the egg of the frog *(Rana)* (Fig. 9–1A). The surviving blastomere developed as a half-embryo, thus suggesting that each cell at the two-celled stage possessed half of the parts necessary to form a complete embryo. Although these studies by Roux were subsequently repeated by others, including the famous German embryologist Hans Spemann, and shown to be erroneous, his experimental approach pointed toward a new direction for embryological thought.

Hans Driesch (1867–1941) followed up the work of Roux by studying the development of the early-cleaving blastomeres in echinoderms. By vigorous shaking, he was able to separate completely the first two blastomeres of the *Echinus* egg (Fig. 9–1 B). Each blastomere subsequently formed a ciliated blastula that gastrulated and developed into a pluteus larva somewhat smaller in size than normal. The ability of each blastomere to form a whole embryo led Driesch to view the developing egg as a "harmonious equipotential system"; that is, parts of the egg, represented by the two blastomeres, had an equal ability to reorganize and form a whole embryo. The results by Driesch emphasized the epigenetic aspect of early development and created a new intellectual framework for the study of ontogeny. Driesch failed to understand the basis for epigenesis, and in later life invoked the *principle of vitalism* to explain the totipotency of the mature egg cell.

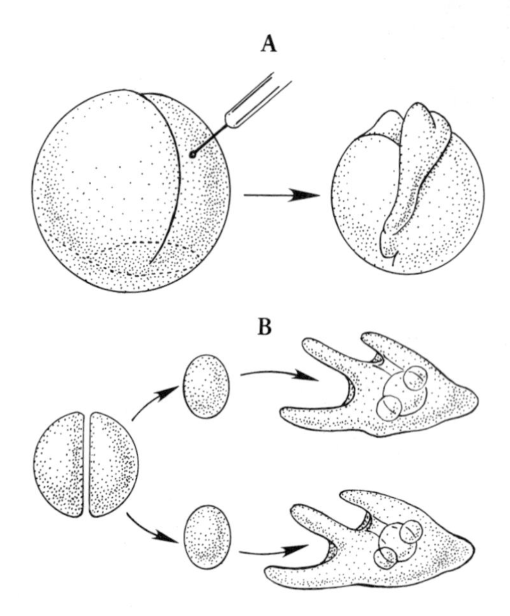


**FIG. 9–1.** Preformation and epigenesis. A, half-embryo produced in experiments by W. Roux following destruction of one blastomere with a hot needle at the two-celled stage in the frog. B, development of two normal-appearing embryos (plutei larvae) in the sea urchin produced by H. Driesch following separation of the first two blastomeres by shaking in seawater.

The mechanisms involved in the progressive increase in complexity of diverse cell types and in the emergence of their proper spatial organization have continued to occupy the attention of investigators since the early studies of Roux and Driesch. Within the past several years, the results of several studies in developmental biology have allowed us to reconcile the concepts of preformation and epigenesis. The gradual transformation of the egg and the emergence of the structured embryo appear to require information that is prelocalized in the egg (preformation) as well as newly formed information arising from interactions between cells (epigenesis). The information for total development of an organism is present in the fertilized egg cell in the form of its genetic material. In a real sense, the nucleotide sequence of the chains of DNA form an invisible but preformed code. Under the direction of this code, constituents of the cytoplasm as well as the cortex of the egg become fashioned into the recognizable parts of the embryo. How these various components of the fertilized egg interact to produce an organism with its various tissue and organ systems will be discussed here and in chapters to follow.

## Gene Activity and Protein Synthesis

Through gastrulation, there is little change in the mass of the embryo and little to indicate the future shape of the embryo. Yet the early embryo is confronted with many tasks that are important to its gradual transformation into a structural and functional whole. These tasks are accomplished in a progressive, predictable, and orderly series of steps.

As described in the chapter on cleavage, there is a rapid increase in cell numbers shortly after fertilization. Every nuclear division requires an increase in DNA and an increase in those proteins associated with the DNA of the chromosomes. Cleaving cells also require structural proteins to form the microtubules of the mitotic apparatus and the contractile proteins involved in cytokinesis.

In all cleaving cells, the synthesis of DNA occurs during the S (synthetic) phase of the cell division cycle. Typically, the frequency of DNA synthesis or the number of times per unit time that the genome is replicated tends to decrease through gastrulation (i.e., the length of the cell cycle increases). The genome of midcleavage frog nuclei is replicated approximately once per hour. By gastrulation, however, the genome is replicated only about once per day. In early cleavage stages of the sea urchin, the number of nuclei doubles every one to two hours and the DNA, amounting to $1.8 \times 10^{-12}$ grams per diploid nucleus, doubles during an S period lasting 10 to 12 minutes at 16°C. This rate of DNA synthesis is about 60 times greater than that in most other eukaryotic cells. The duration of the S phase in the frog is about 15 minutes at midcleavage. This lengthens to approximately six hours during gastrulation. Generally, then, it is at the end of cleavage that mitotic activity of the embryo decreases and permits a typical cell cycle to become detectable. The general pattern of change in the synthesis of DNA during oogenesis and the early development of the frog embryo is summarized in Figure 9–2.

**FIG. 9–2.** Changes in nucleic acid syntheses during oogenesis and early development of the frog embryo. (From J. Gurdon, 1968. Essays Biochem. 4:25.)

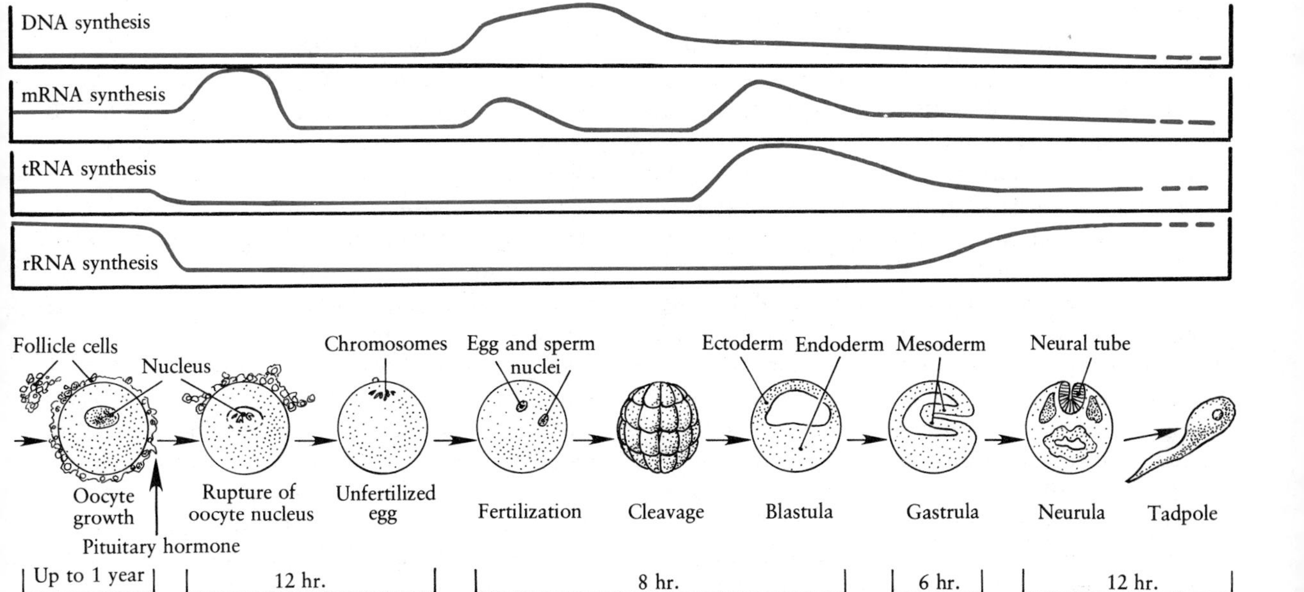

Cleaving eggs appear to be capable of synthesizing DNA at a rapid rate because the precursors for this acid, as well as the essential catalyzing enzymes, are already present in the cytoplasm of the fertilized egg. *DNA polymerase*, the enzyme that promotes the construction of the new polynucleotide chain on the existing DNA template, is present in substantial amounts in the egg's cytoplasm. In the sea urchin, it has been shown that the egg starts with a large amount of DNA polymerase. The localization of the polymerase gradually changes as it becomes progressively concentrated in the nucleus with each successive division cycle. How this polymerase migration occurs remains an unanswered question.

The sources of the precursors for the DNA molecule have been postulated to be severalfold. In the sea urchin, thymidine is probably provided as a result of the transformation and breakdown of ribonucleotides in the cytoplasm of the egg. *Ribonucleotide reductase*, an enzyme that converts ribonucleotides into deoxyribonucleotides, is known to be present in substantial amounts in the cytoplasm of developing sea urchin eggs. There is also evidence in both sea urchins and frogs that DNA can be synthesized from low-molecular-weight precursors. When glycine, an amino acid that can be used for the synthesis of purine groups in the DNA molecule, is labeled ([$^{14}$C-]glycine) and the eggs of sea urchins (or frogs) exposed to it, the radioactive carbon atoms are actively incorporated into DNA in the eggs.

We have previously seen in Chapter 4 that the egg stockpiles a number of products synthesized under the control of its own genome. These include all three major species of RNA (mRNA, tRNA, and rRNA), yolk, mitochondria, and various structural and enzymatic proteins. Upon fertilization there is a dramatic increase in the rate of protein synthesis, and, by cleavage, the embryo begins to synthesize in its nuclei a high-molecular-weight heterogeneous RNA. Some of this RNA is clearly of the messenger type. In efforts to understand the course and control of the early development of the embryo, the task of the developmental biologist has been to articulate the relative importance of maternally stored and embryonically synthesized products.

The synthesis of RNA in the early embryo is generally described as being weak and limited in comparison with later stages of development. The active incorporation of radioactively labeled precursors of RNA, such as [$^{3}$H]-uridine, has been shown to occur in most animals by gastrulation (Fig. 9–2). Most of the RNA synthesized in the embryo before the onset of gastrulation appears to be of the mRNA class. Embryos from different classes of organisms differ somewhat in the stage of development at which the different species of RNA begin to be transcribed. In the sea urchin, activation of the DNA-like mRNA-synthesizing machinery is not apparent until the embryo is at the 2- to 4-celled stage. Between the 8- and 16-celled stages, there is a dramatic increase in embryonic RNA output. It is interesting that the nucleotide sequences of most of the mRNA molecules are identical to the base sequences of maternally synthesized messenger RNA molecules, indicating that the genes being transcribed are the same as those during oogenesis. Some mRNA transcripts have base sequences complementary to DNA segments not previously transcribed, but there are few copies per cell of these new kinds of mRNAs. The onset of mRNA synthesis appears to be slightly later in the amphibian embryo with transcription being detected at the midblastular stage (Fig. 9–2).

Transfer RNA in most embryos is detected after the initiation of mRNA synthesis and shortly before the onset of gastrulation. Ribosomal RNA is not produced in embryos of sea urchins and frogs until gastrulation. The onset of this type of RNA synthesis at this particular time can be correlated with the appearance of visible nucleoli within the nucleoplasm of cells of the gastrula. Some investigators are of the opinion that rRNA synthesis occurs earlier than gastrulation in the sea urchin, but that the rapid accumulation of DNA-like RNA (probably of the mRNA type) prevents rRNA detection. The pronounced delay in rRNA synthesis in both sea urchin and frog is apparently related to the stockpile of ribosomes built up in the cytoplasm during oogenesis. By contrast, rRNA production occurs much earlier (early cleavage stages) in mammalian embryos because few ribosomes are packaged in the cytoplasm of the unovulated egg.

The appearance of different species of RNA at different times during early embryogenesis indicates that the genes controlling their synthesis are being independently activated and regulated. There is some evidence to suggest that the activity of these RNA genes may be regulated by different inhibitors in the egg's cytoplasm. For example, it has been shown that nuclei of the endoderm of *Xenopus* neurulae actively produce rRNA. When such a nucleus is isolated and injected into an activated egg whose own nucleus has been killed, synthesis of rRNA suddenly is arrested. Presumably, a factor in the cytoplasm blocks transcriptive activity of rRNA genes. However, if such an egg is allowed to develop, rRNA synthesis is initiated at the gastrula stage (i.e., at the time that it would appear in a normally fertilized egg).

Although the synthesis of RNA begins shortly after the activation of the egg, transcription of the embryonic genome does not appear to be essential to development prior to the stage of gastrulation. If fertilized eggs are treated with actinomyocin D, an agent known to block transcription, they show all outward signs of being able to develop in normal fashion, including a normal rate of protein synthesis. How the embryonic mRNAs are prevented from being translated is not clear; presumably, they are coated or masked by some substance that prevents their translation until a later time in development. In both frogs and sea urchins, embryos in which transcription has been experimentally blocked with actinomycin D will develop to the blastula stage. At this time further development of the embryos is arrested (i.e., actinomycin D–treated embryos fail to gastrulate). These experiments with early embryos exposed to actinomycin D permit several conclusions. First, the early events of morphogenesis (i. e., cleavage and blastulation) and associated patterns of protein synthesis are not dependent on newly synthesized RNA of the embryonic genome. Second, gastrulation is arrested in the absence of transcription of RNA from the embryonic genome.

It has already been pointed out that there is a marked elevation in the synthesis of proteins following fertilization of sea urchin egg cells (Chapter 6). The rate of polypeptide formation climbs sharply through early cleavage and then levels off. Indeed, within minutes of fertilization, it has been estimated that in the sea urchin the rate of protein synthesis increases by 10- to 20-fold when compared to the unfertilized egg cell. Just prior to gastrulation, there is a second surge in polypeptide synthesis. By the gastrula stage the rate of embryonic protein synthesis is about 100 times that of unfertilized eggs. The change in the rate of protein synthesis elicited by fertilization may not be as dramatic in other animal embryos (such as frogs). However, a burst of intense protein formation during gastrulation appears to be shared by all organisms. Hence, on the basis of the elevated levels of RNA and protein syntheses, gastrulation is a developmental period of large-scale gene activity.

As one might expect, the normal development of the early embryo requires the continual input of newly synthesized proteins. Treatment of embryos after fertilization with the classical inhibitors of protein synthesis, such as puromycin and cycloheximide, immediately interrupts cell division and brings development to a standstill. It is apparent, therefore, that newly appearing proteins are essential to normal embryogenesis. A number of questions can be raised about these early embryonic proteins. What are these proteins, and what is their amino acid composition? Are these proteins fashioned from mRNAs synthesized during oogenesis (maternal RNA transcripts) or from mRNAs synthesized after fertilization (early embryonic RNA transcripts)? What roles do these proteins play during early embryogenesis?

The sea urchin, and more recently the mouse, have been favorite subjects for the analysis of proteins and the dual nature of genetic programming for the early biochemical and morphological events of development. How is it possible to detect differences between proteins synthesized on maternal RNA templates from those produced on embryonic RNA templates? An initial method to distinguish between these two possibilities employed hybrid embryos between two species. The technique requires that developing embryos of the two species show morphological differences, and it assumes that only those RNAs underlying these differences are being produced. Following the penetration of the foreign sperm of one species into the egg of the second species, it is possible to determine when the hybrid embryo becomes paternal-like and therefore when its embryonic genome activated. Such morphological studies with frog and sea urchin hybrids have shown paternal gene expression at gastrulation, suggesting that prior development depends primarily on maternal mRNA. However, results of other studies using more sensitive analytical and molecular techniques now indicate that embryonic transcription clearly begins prior to gastrulation. Some of the newly synthesized transcripts may be translated into proteins with an important role in early development.

The rise in the synthesis of proteins after fertilization is accompanied by a corresponding increase in the number of polyribosomes. Each polyribosome consists of three to seven ribosomes per strand of mRNA. By the blastula stage the number of polyribosomes within the cytoplasm of the constituent blastomeres increases dramatically. Analysis of this population of polyribosomes shows that two classes are present: *heavy polyribosomes,* with an average of 23 ribosomes per mRNA, and *light polyribosomes,* with an average of 9 ribosomes per mRNA strand. Both heavy and light polyribosomes are active in the incorporation of labeled amino acids into protein. The light or *s-polyribosomes* appear to be totally responsible for the manufacture of *histones.*

Histones are small, low-molecular-weight proteins that are associated with the DNA of chromatin material. Not unexpectedly, therefore, the rate of histone synthesis increases is parallel with the rate of embryonic DNA synthesis during early embryogenesis. Approximately 50 percent of the proteins formed during cleavage appear to be histones. As the synthesis of DNA gradually declines, approximately 10 hours after fertilization in sea urchins, so also does the rate of histone synthesis. Hence, histone protein synthesis is activated shortly after fertilization. Histones are probably formed on maternal, preformed mRNA templates as well as on newly synthesized embryonic RNA templates. If sea urchin eggs are treated with actinomycin D before fertilization and the polyribosome population analyzed at selected time intervals after insemination, it can be shown that there is a considerable reduction in the number of s-polyribosomes and in the synthesis of histones. Presumably, the reduction in histones indicates that newly synthesized embryonic RNA is required to make these nuclear proteins. Direct evidence of the involvement of the embryonic genome in the production of this protein has come from studies designed to extract and analyze the mRNA of the histone-synthesizing s-polyribosomes of fertilized eggs. However, the fact that histones are produced in actinomycin-treated embryos is strong evidence that preformed mRNA for histones is present in the unfertilized egg and translated after fertilization. It has been estimated that two-thirds of the histone formed during cleavage is translated from material templates. Following their formation, the histones migrate from the cytoplasm of the blastomere into the nucleus.

In contrast to eggs of echinoderms, the oocytes of *Xenopus* have an abundant pool of preformed histones as well as histone mRNAs. Similar to the sea urchin, the embryonic genome of this amphibian transcribes histone RNAs shortly after fertilization. The advantage of a large stockpile of preformed histone proteins becomes evident in *Xenopus* when it is realized that the rate of DNA synthesis is very rapid and actually exceeds the rate of histone synthesis through the blastula stage.

Histones have received more attention than other embryonic proteins because their special properties have made them more amenable to biochemical study. There are other proteins, however, synthesized after fertilization that appear to utilize preformed, maternal RNA templates. These include the *tubulin proteins* that participate in the formation of the microtubules of the mitotic apparatus of the cleaving cell, the enzyme *ribonucleotide reductase,* and the *hatching enzyme.* Hatching enzyme appears in homogenates at the midblastula stage of sea urchin embryos, and is used to digest away the fertilization membrane, thus setting free the swimming larva. There is some evidence to suggest that the microtubular proteins are also present in large amounts in the unfertilized egg, but these are not assembled into macromolecular structures until after fertilization.

The development of embryos utilizing preformed mRNAs synthesized as a result of maternal gene activity is a surprising and remarkable accomplishment. As efforts to identify these proteins derived from transcription during oogenesis continue, questions are raised about the mechanisms regulating the translation of proteins known to be formed on maternal templates. Are all species of mRNA translated upon fertilization, or are only particular mRNAs selectively recruited into polyribosomes? For ribonucleotide reductase, tubulin, and certain classes of histones, there is strong evidence in sea urchins that regulation of translation is qualitative. That is, their preformed mRNA templates are selectively recruited and translated at highly predictable times (stage specific) during early development. Ribonucleotide reductase, for example, can first be detected at one hour after insemination, even in the presence of actinomycin D. Tubulin shows a marked increase in its rate of synthesis just before hatching when compared to its rate during cleavage, suggesting that there is a more efficient qualitative control over the tubulin templates. Whether the increased rate of protein synthesis after fertilization is due entirely or partially to the translation of qualitatively new species of mRNA is far from clear. In studies that have compared overall profiles of the population of proteins in unfertilized and fertilized eggs, the patterns of protein synthesis have been reported to be very similar. Brandhorst and his colleagues, for example, have used two-dimensional polyacrylamide gel electrophoresis (see Chapter 11 for technique) to compare patterns of radioactively labeled proteins synthesized in oocytes, fertilized eggs, and early embryos of sea urchins and molluscs. This method identifies proteins of homogenates on the basis of their isoelectric points and molecular weights. Individual proteins appear as radioactive spots on the gel. By careful analysis of the several hundred protein spots detected at various developmental stages, they concluded that the patterns of protein synthesis appear to be very similar before and after fertilization in both sea urchins and molluscs. If translation was being strictly regulated by a qualitative mechanism, one would expect that the proteins produced in the unfertilized egg would be different from those appearing after fertilization. We are left with the conclusion that many proteins are probably formed exclusively on randomly activated mRNA molecules. The increased rate of protein synthesis would be under quantitative regulation and due to elevated translation of the same species of mRNA that is translated in the unfertilized egg.

An extensive maternal program unfolds after fertilization in sea urchin, amphibian, and mouse embryos. This information, formed as gene products of oogenesis, becomes supplemented with and gradually replaced by gene products synthesized by the developing embryo. How long are the mRNAs of maternal origin utilized during early development? It is generally believed that the mRNAs synthesized in the oocyte begin to break down and become unavailable for translation on polyribosomes sometime after the onset of cleavage. In the mouse, for example, about 40 percent of the bulk maternal RNA is degraded by the 2-celled stage; another 30 percent is lost by the blastocyst stage. Since the mouse embryo synthesizes RNA at a high rate per cell from the 2-celled stage onward, there appears to be a pronounced shift from the use of maternal to embryonic-derived RNA beginning with the embryo of 2 cells. A major change in the qualitative pattern of protein synthesis between the 2-celled and the 8-celled stages of development tends to support this view. Embryonic mRNAs begin to be rapidly produced between the 8-celled and 16-celled stages in the sea urchin. As previously noted, many of the mRNA transcripts produced by the embryo are identical to the mRNAs synthesized by the egg during oogenesis. Indeed, although the gastrula and pluteus stages of the sea urchin appear to have fewer species of mRNA, the nucleotide sequences of these messages are also similar to those of the oocyte.

In addition to maternal mRNA molecules, the fertilized egg starts development with preformed proteins manufactured during oogenesis. The role of these proteins in early development is far from clear, although some have been implicated in determining the fates of certain cells of the embryo. A preformed protein that is critical to gastrulation has been described in the Mexican axolotl, *Ambystoma mexicanum*. In one form of the axolotl, there is a *mutant gene o* that is inherited as a simple recessive and has a maternal effect on offspring. Females homozygous for the *o* mutation yield eggs that, following fertilization by either a wild type or heterozygous sperm for *o*, become arrested in development just prior to gastrulation. This developmental block can be removed by injecting cytoplasm from normal eggs into fertilized eggs of females recessive for the *o* gene. Also, nucleoplasm obtained from the germinal vesicle can restore the maternal deficiency. This suggests that the normal *o* gene produces a substance, presumably a protein, in the germinal vesicle that is essential to gastrulation. Since embryos of *o*/*o* females show little RNA synthesis at the end of cleavage when compared to normal embryos, it is believed that the *o* substance acts as a regulatory factor by triggering transcription and the production of RNAs necessary for gastrulation.

In summary, the embryo prior to gastrulation is primarily concerned with chromosomal replication, cell division, and protein synthesis, utilizing as a primary information source many of the mRNAs manufactured during oogenesis and stored within the egg cell. At given periods of time following fertilization, specific mRNA molecules are selected for translation into proteins. These proteins are vital to the construction of new DNA and cell membrane. Gradually, the stockpile of maternal mRNAs is replaced by an increasing population of messengers transcribed from the embryo's own genome.

## The Equivalence of Nuclei

Development of the embryo begins with the rapid conversion of the single, fertilized egg into a population of cells. Although the cells of the early embryo are initially alike, they gradually diverge phenotypically and give rise to specialized cell types with very specific properties. By the end of gastrulation, the cells are broadly fixed or determined with respect to their fate or specific destiny. Some cells will become ectodermal cells, some endodermal cells, and others mesodermal cells. In this connection, one might propose that the origin of differences between these cells and the specification of embryonic cell fate could be attributed to differences in either their nuclei or the cytoplasm inherited by the embryonic blastomeres during the process of cleavage.

In his germ plasm theory, A. F. L. Weismann postulated that every part of the embryo was represented by a separate determinant or particle located on the chromosomes of the nucleus of the sex cell. These determinants were then distributed to different blastomeres during cleavage, thus accounting for the fate of the various blastomeres. Differentiation of a given blastomere could only take place in accordance with the type of determinants present. Although we now know that Weismann's proposal is untenable from a genetic point of view, it is still important to ask whether the cells of the early embryo, and hence the structures of the embryo fashioned from them, become different because of differences in their nuclei. That is, might not the early specialization of cells involve the selective elimination of genes and/or the differential distribution of genetic material to blastomeres during segmentation of the fertilized egg?

Several lines of evidence suggest that all cells of the embryo possess nuclei containing all of the genetic information necessary for normal cell differentiation and expression. The studies by Driesch on isolated blastomeres of the sea urchin, the experiments by Spemann on delayed nuclear supply in the egg of *Triturus*, and the nuclear transfer studies of Briggs and King, as well as by Gurdon and colleauges, support this concept.

The view by Weismann that blastomere specialization resulted from the selective distribution of determinants during cleavage was initially, as we previously noted, contradicted by the studies of Driesch. Development of complete but slightly smaller than normal embryos from isolated blastomeres of the 2-celled and 4-celled stages of sea urchin and frog eggs indicated that the cells of the very early embryo, at least, have all of the information necessary to support total development.

To investigate more fully the characteristics and potentialities of cleavage nuclei, Spemann (1928) devised the elegant technique of *delayed nuclear implantation* or supply (Fig. 9–3). Using the fertilized egg of *Triturus,* he constricted the fertilized cell into two halves with a specially designed hair loop just prior to cleavage. By constricting the egg still further through tightening of the loop, the zygote nucleus was displaced toward one pole of the egg. A thin cytoplasmic bridge remained as the only connection to the non-nucleated portion of the cell (Fig. 9–3 A). Cleavage occurred but was restricted entirely to the nucleated portion of the egg (Fig. 9–3 A). After approximately the fourth division, the nuclei were quite small in size. One of these nuclei crossed the cytoplamic bridge into the nonsegmented portion of the original egg (Fig. 9–3 B,C). Immediately, this half of the egg began to cleave. When Spemann then tightened the hair loop, the original fertilized egg was separated into two distinct cleaving halves. In most cases, each cleaving half was observed to gastrulate and develop into a normal embryo (Fig. 9–3 D,E).

It is important to point out that both types of embryos initially possessed about one half of the original egg cytoplasm. Yet, the development of one type of embryo was directed by only $\frac{1}{16}$ of the original egg nucleus, while development of the second type of embryo was directed by $\frac{15}{16}$ of the original egg nucleus. Spemann was one of the first investigators to provide experimental support for the concept that the nuclei of the early cleaving embryo are equivalent to each other as well as to the nucleus of the fertilized egg. That is, these nuclei contain all of the nuclear information required for the achievement of normal development.

A more direct method for testing the genetic equivalence of the nuclei of cleaving blastomeres and, indeed, of cells in general has been that of *nuclear transplantation.* The objective of a nuclear transplant is to insert the nucleus of a blastomere or specialized cell type into an unfertilized, but ac-

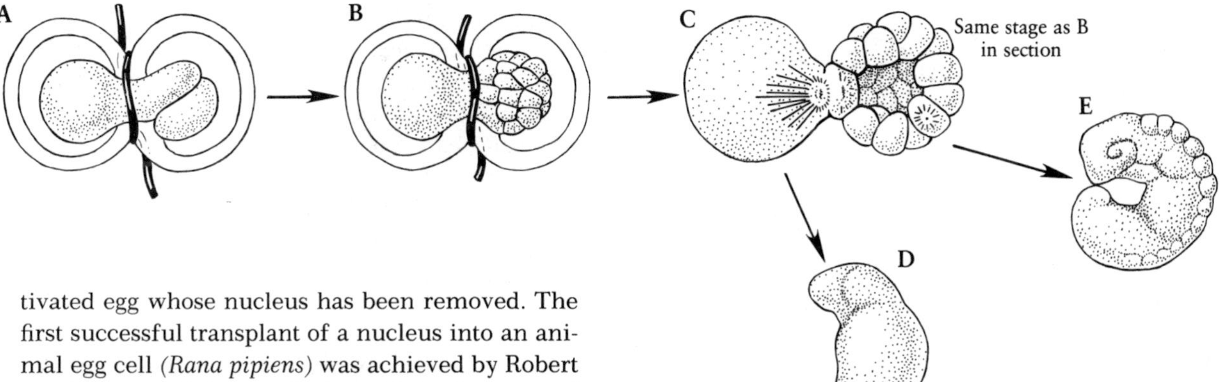


FIG. 9-3. The delayed nuclear implantation experiment of Spemann. A, a hair loop placed around the fertilized egg of the salamander *(Triturus)*—cleavage has begun in the nucleated half of the egg; B, a nucleus has passed across the cytoplasmic bridge into the undivided portion of the egg; C, same as in B, but in section; D–E, normal-appearing embryos develop from the half with the zygote nucleus and the half with the delayed nuclear supply. (From H. Spemann, 1938. Embryonic Development and Induction. Yale University Press, New Haven.)

tivated egg whose nucleus has been removed. The first successful transplant of a nucleus into an animal egg cell *(Rana pipiens)* was achieved by Robert W. Briggs and Thomas J. King of the Institute of Cancer Research in Philadelphia. The technique basically involves three steps, which are summarized in Figure 9–4. The first step involves the removal of the egg nucleus. A mature egg cell is artifically activated by pricking with a drawn glass needle. With the aid of a dissecting microscope, the female nucleus can then be observed as it approaches the cell surface to complete the second maturation division. It is then flipped out of the cell with a second glass needle. The second step requires the preparation of the nucleus to be transplanted *(donor nucleus)*. A cell, a blastomere for example, is dissociated from its neighbors in a medium typically lacking calcium or magnesium ions. The most difficult step is the placement of the donor nucleus into the enucleated egg. This requires a high level of manual skill and endless patience. Briggs and Kings discovered that the nuclear transplantation is best accomplished by sucking a single donor cell into a micropipette whose bore is slightly smaller than the diameter of the cell. As the cell is drawn into the micropipette, the cell membrane ruptures, thus liberating the nucleus. The nucleus with a halo of its own cytoplasm is then injected into the cytoplasm of the enucleated or recipient cell. If the membrane of the donor cell fails to break in the transplant process, the donor nucleus will not respond to the cytoplasm of the egg.

To determine if nuclei remain equivalent to one another during development and retain the same genetic information, Briggs and King used donor nuclei from various stages of development of the leopard frog, *Rana pipiens*. Nuclei transplanted from blastula cells resulted in about 80 percent of the transfer embryos undergoing cleavage, with the majority of these continuing to develop into tadpoles. However, when the donor nuclei were obtained from progressively later stages, the resulting transfers showed increased disturbances in development with fewer transfers reaching the tadpole stage. For example, only about 65 percent of the transfers reach the blastula stage when the donor nucleus (endodermal) is contributed by a late-gastrula embryo. This figure drops to about 33 percent when the donor nucleus (endodermal) is supplied by an embryo in the neurula stage. Additionally, many of these transplant embryos subsequently become arrested in gastrula or prehatching stages. These experiments with *Rana* suggest that nuclei do gradually become modified and apparently restricted in their capacity to promote normal development.

Wishing to test the permanence and stability of the presumed changes in the nucleus, Briggs and King undertook the difficult and tedious task of cloning nuclei *(serial nuclear transplantation)*. In a serial transplant, a donor nucleus is isolated from an abnormal or arrested embryo and injected into an enucleated egg. At the blastula stage, nuclei are again isolated from blastomeres and injected into a new population of enucleated eggs. This procedure can be repeated several times, the result being a group of individuals all having an identical set of genes in their nuclei. Figure 9–5 summarizes in

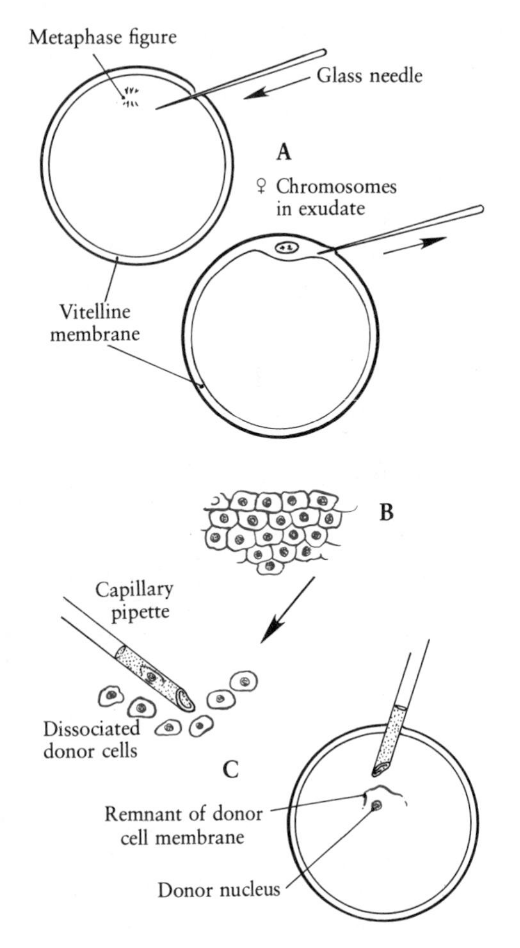


**FIG. 9–4.** A diagrammatic summary of the basic steps involved in nuclear transplantation. A, activation and removal of the nucleus of the ripe egg cell; B, aspiration into a capillary pipette of a dissociated donor cell; C, injection of the donor nucleus and donor cell debris into the enucleated egg.

diagrammatic fashion the basic steps in serial transplantation as carried out in amphibians.

With the leopard frog, it has been shown that the original recipient eggs, injected with nuclei isolated from the endoderm of late-gastrula embryos, developed into a variety of embryo types, ranging from arrested gastrulas to normal embryos (Fig. 9–5). However, when one blastula of the original recipient generation served as a donor for transfers to a new group of enucleated eggs, the new clone of embryos displayed a very uniform type of development. Most of the embryos of this generation gastrulated normally but later showed marked morphological deficiencies, particularly in ectodermal structures. Continued serial transplantation of blastular nuclei from the original recipient generation showed no change in the type of deficiencies produced. The nuclear alterations appeared to be stable and specific. Hence, these nuclear cloning experiments support the hypothesis that some of the nuclei had experienced a stable, reproducible change, expressed in the inability of these nuclei to support normal development of an enucleated egg cell (i. e., ectodermal differentiation). It appears that different patterns of abnormal embryogenesis can be correlated with the origin of the donor nucleus.

Does the observation that *Rana* nuclei appear to become irreversibly altered during the course of development, expressed in a failure to support total development, mean that the nuclei have experienced gene loss or elimination? Not at all. Indeed, the failure of later stage embryonic nuclei to promote and sustain total development is probably related to chromosomal damage precipitated by an incompatibility between the donor nucleus and the host cytoplasm. When nuclei of spermatogonial cells are injected into enucleated, activated eggs of *Rana pipiens*, most of the nuclear transfers never develop to the tadpole stage. Analysis of these arrested embryos shows marked chromosomal abnormalities. It must be remembered that donor nuclei isolated from gastrular and neurular stages of development, as well as from juvenile and adult tissues, are being obtained from cells with significantly reduced rates of mitosis. Sudden transfer of such a nucleus into the cytoplasm of an egg, which is programmed for rapid division, can produce physical damage to chromosomes and subsequent embryonic deficiencies because of its failure to divide rapidly enough.

Although the nuclei from later developmental stages of *Rana pipiens* appear to show severe restrictions after transplantation, experiments carried out by Gurdon and his colleagues using *Xenopus* (African clawed frog) have yielded different interpretations (Fig. 9–6). A particularly useful feature in *Xenopus* is that a mutant strain exists that possesses one nucleolus per nucleus, while the normal, wild-type strain has two nucleoli per nucleus. When the mutant strain is used as a source of donor nuclei, one can always distinguish the division products of the transplanted nucleus.

Gurdon and his colleagues have concentrated on testing the capacity of nuclei from fully differentiated cells to support total development. When nuclei of fully differentiated larval intestinal epithe-

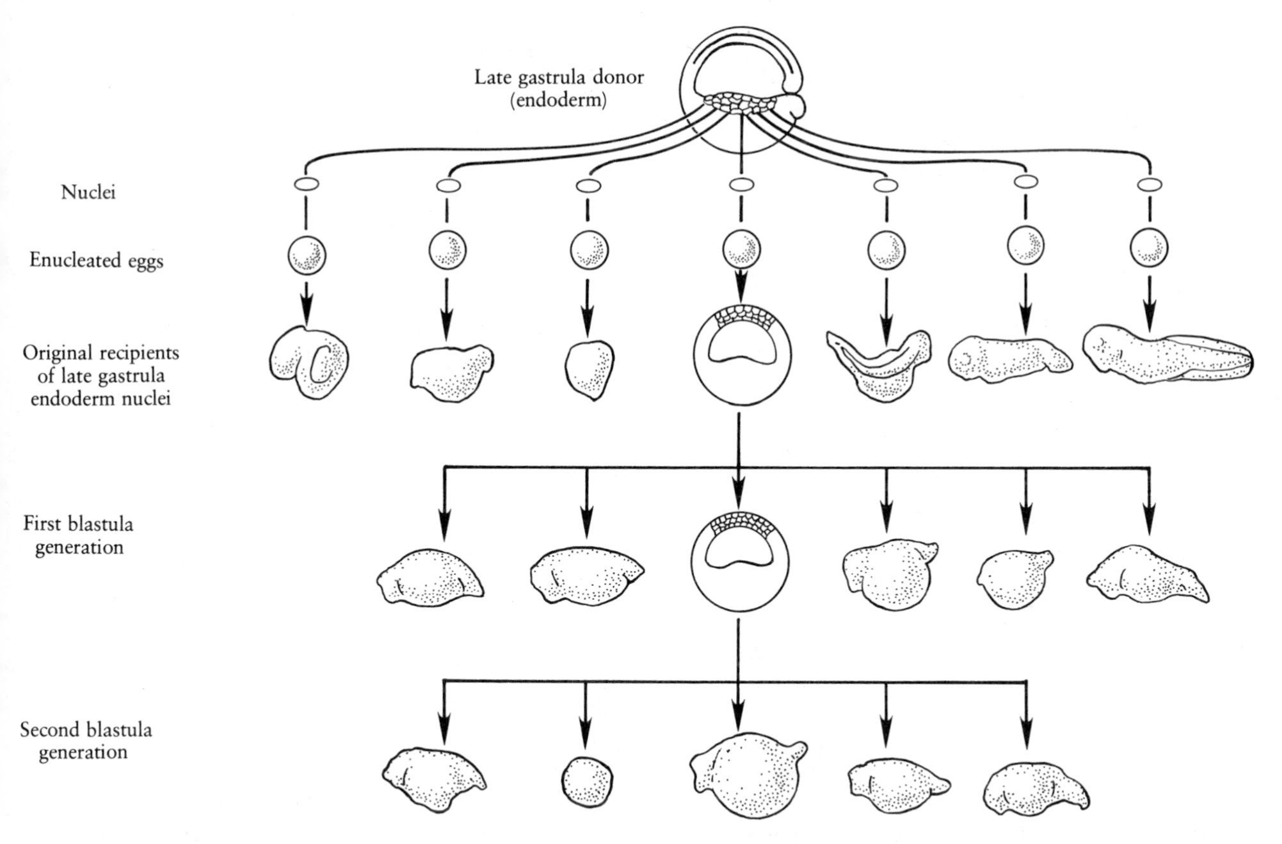


**FIG. 9–5.** Diagram illustrating the basic steps involved in the serial transplantation of endodermal nuclei. Donor nuclei are initially obtained from the presumptive anterior midgut region of the late gastrula. They promote the various types of development shown for the "original recipients" following transplantation into enucleated eggs. One of the original recipients, sacrificed at the blastula stage, provides nuclei for a single clone of individuals (first and second blastula generations). The first and second blastula generations show more uniform types of development. (After T. King and R. Briggs, 1956. Cold Spring Harbor Symp. Quant. Biol. 21:271.)

lial cells from the mutant strain of *Xenopus* are used as donors, approximately 1.5 percent of the total number of transfers develop into normal adult frogs. All cells in these embryos possess a single nucleolus in their nuclei. Indeed, both female and male *Xenopus* frogs, fertile and completely normal, have been raised from eggs into which intestinal nuclei had been placed. It would appear, then, that many of the transplanted nuclei retain all the genetic information required for the development of all cell types. About 20 percent of the transfers reach the stage at which there are distinct neuromuscular responses, demonstrating that intestinal nuclei retain the genes required for the differentiation of several very specific cell types (nerve and muscle). Donor nuclei from cultured adult skin cells also have the capacity to support the differentiation of many cell types. Gurdon and his colleagues argue that no irreversible changes occur in the nucleus during development, because some transfers always show *totipotency*.

The inability of some transplanted nuclei to support normal development does appear to increase as the cells from which they are taken become differentiated. In particular, DiBerardino and her colleagues have concluded that there are severe disturbances in the number and shape of the chromosomes in these nuclear-transplant embryos. In other words, the failure of many nuclear transplants to develop into more complete embryos may be limited by the nuclear-transplant technique itself. The origin of these chromosomal abnormalities in *Xenopus* is attributable to an incompatibility between the very slow rate of division of differentiating cells (serving as donor nuclei) and the rapid rate of division of the egg (recipient cell).

Although there are differences in the results of nuclear transfer experiments with *Rana* and *Xeno-*

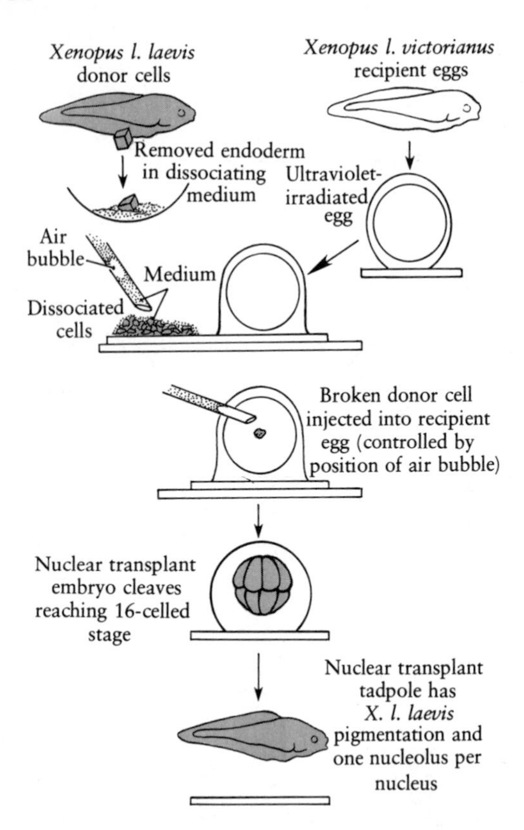


**FIG. 9–6.** The principal stages involved in the transplantation of nuclei in the amphibian, *Xenopus laevis.* Donor nuclei are obtained from a strain of subspecies, *Xenopus laevis laevis,* which has only one nucleolus per nucleus. The young tadpoles of this species have many pigment cells in their bodies. Recipient eggs are from another subspecies, *Xenopus laevis victorianus.* The nuclei of diploid cells in this subspecies have two nucleoli, and the tadpoles have no body pigment. The nuclear-transplant tadpoles have the characteristics of the nuclear, not the cytoplasmic, parent. (From J. Gurdon, 1966. Endeavor 25:95.)

*pus,* it is known that the nuclei of both organisms become smaller, divide more slowly, and may be susceptible to greater damage by injection as development progresses. Whether nuclei do, in fact, experience stable and irreversible changes during embryogenesis remains an open question. However, because nuclei from fully differentiated tissues have been shown in several experiments to sustain normal development, the conclusion appears to be warranted that nuclei are totipotent and contain a full gene complement. Certainly, none of the nuclear-transfer experiments supports the hypothesis that blastomeres and the specialized cell types derived from them become different because of selective gene loss.

## Significance of the Egg Cytoplasm and the Egg Cortex

Studies employing techniques of nuclear transplantation, primarily with amphibian eggs, have given a rather clear indication that the basis for differences in the destinies of cells of the early embryo cannot be ascribed to differences in their nuclei. With few exceptions (see Chapter 11), every nucleus in the multicellular organism contains the same genetic information and is fully equivalent to the nucleus of the fertilized egg. By logic, then, one must examine the possibility that the specification and differentiation of blastomeres and their descendant cells are somehow related to the organization of the cytoplasm of the egg. In other words, the basic features of the embryonic body plan (i.e., cell diversity, spatial organization of cells) are traceable to patterns of cytoplasmic constituents that are present in the egg and early stages of development. Presumably, substances *(morphogens)* localized in the egg cytoplasm, and segregated into appropriate cells of the developing embryo during cleavage, trigger the process of differential gene activity leading to the differences between embryonic cell types. The basis of cell diversity and spatial organization, therefore, is the unequal distribution of regional differences in the egg's cytoplasm during cell division and the influence of these cytoplasmic components on genetically equivalent nuclei. In the section that follows, we will discuss several examples of how the cytoplasmic organization of the egg is related to the spatial pattern of cell differentiation in the developing embryo. As you will see, a wide variety of different embryos has been employed in attempts to understand the relationship between cell specification and the cytoplasmic composition of the egg.

You will recall that shortly after fertilization in the egg of the frog a distinct, crescent-shaped zone of grey-colored cytoplasm appears at the lower margin of the animal hemisphere approximately opposite to the site of sperm penetration (Fig. 9–7). This grey crescent can also be identified in the uncleaved eggs of salamanders. In most cases, the first cleavage furrow coincides with the plane of bilateral symmetry and passes through the center of the grey crescent. Typically, therefore, each of the first two blastomeres receives approximately half of the grey crescent material. Spemann noted that if he placed a hair ligature on the fertilized egg in such

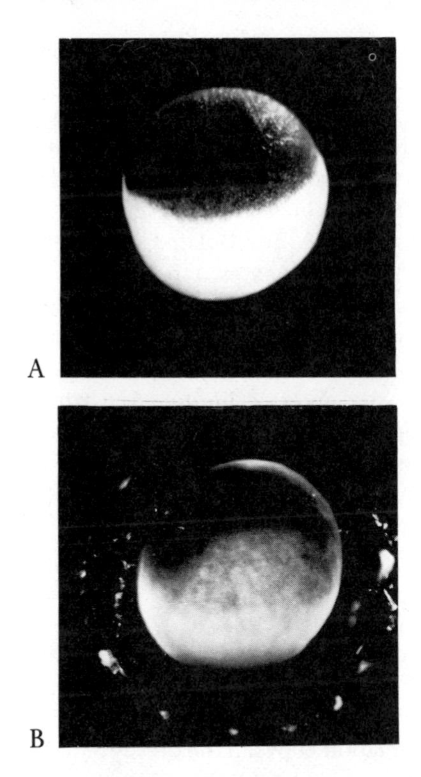


**FIG. 9–7.** Formation of the grey crescent in *Xenopus laevis*. A, mature egg following isolation from the ovary; B, same egg after fertilization showing the grey crescent and the fertilization membrane. (Courtesy of J. Brachet.)

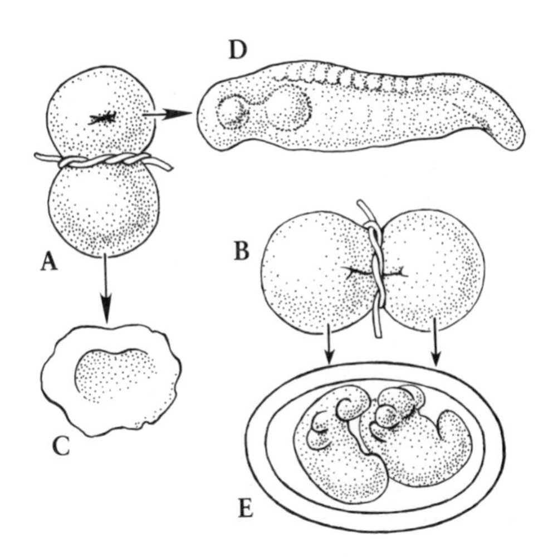


**FIG. 9–8.** Experiments of Spemann involving constriction of the salamander's egg in the frontal (A) and medial (B) planes. After frontal constriction, only the egg half containing the grey crescent cytoplasm developed into a complete embryo (D); the half lacking the grey crescent developed abnormally (C). After medial constriction (B), both halves, each of which contains grey crescent cytoplasm, developed into complete embryos (E). (From H. Spemann, 1938. Embryonic Development and Induction. Yale University Press, New Haven.)

a way that constriction divided the grey crescent in half, the resulting two blastomeres, when isolated and cultured, developed into two normal embryos (Fig. 9–8). However, if the ligature was placed and tightened so that only one blastomere received grey crescent material, the results were quite different. The blastomere with the grey crescent developed into a normal embryo; however, the blastomere lacking grey crescent produced only an epithelial ball of cells with little visible differentiation, particularly of axial structures such as notochord and nervous system. Spemann showed that these differences were not dependent on which blastomere received the original egg nucleus, and thus, whatever controls the future differentiation of the embryo must be associated with some cytoplasmic material contained in the grey crescent of the fertilized egg.

Since the early experiments of Spemann, a number of investigators have attempted to define more precisely the nature of the grey crescent material and its role in amphibian embryogenesis. Curtis (1962), for example, surgically removed the grey crescent from fertilized eggs of *Xenopus* and observed that cleavage was unaffected. However, the embryos failed to gastrulate and showed little differentiation of cell types. More recently, ultraviolet irradiation has been used to probe the properties of the frog's egg and its grey crescent. Irradiation of the vegetal half, but not the animal half, of fertilized but uncleaved eggs typically results in a delay of gastrulation and produces defects in neural morphogenesis. When this part of the egg is exposed to ultraviolet irradiation within 60 minutes after insemination, grey crescent formation is inhibited and disturbances in development similar to those noted above are produced. The grey crescent normally forms between 90 and 120 minutes postinsemination. These studies suggest that an ultraviolet-sensitive, cytoplasmic substance, critical to the development of neural structures, becomes regionally localized at the marginal zone (future dorsal side) of the egg of the frog several hours after fertilization. As one might expect, ultraviolet-irradiated eggs will show a reasonably normal pattern of development if supplied with unaltered grey crescent material. This has been done by injecting marginal zone cytoplasm from nonirradiated eggs into ultraviolet-irradiated eggs.

Studies with ascidian embryos have probably provided the strongest foundation for the notion that there are *morphogenetic substances* or *determinants* localized in the egg cytoplasm and that their segregation during early development relates to different pathways of cell differentiation. A spectacular and classical example of this localization phenomenon is the egg of *Styela*. Recall that before the onset of furrow formation, the fertilized egg displays four areas of recognizable plasms. The animal hemisphere is characterized by clear, transparent cytoplasm while the vegetal hemisphere cytoplasm is slate grey in color and packed with coarse yolk granules. Just below the equator of the cell, there is a light grey, crescent-shaped zone of cytoplasm and a yellow, crescent-shaped zone of cytoplasm (Fig. 9–9). This "yellow plasm" is characterized by an abundance of yellow pigment granules and numerous mitochondria.

The visible distinctions between these plasms allowed E. G. Conklin (and others) to trace and map out the relationships between the various pigmented regions of the egg cytoplasm and the tissues ultimately formed from these areas. A diagram of the cell lineage fate map of *Styela* is reproduced in Figure 9–10. The clear cytoplasm of the animal hemisphere becomes segregated in the ectodermal cells of the embryo and tadpole larva, the slate grey cytoplasm is inherited by the endodermal cells, the light grey cytoplasm becomes localized in cells of the nervous system and the notochord, and the yellow plasm is distributed to cells differentiating into muscle tissue and mesenchyme. As early as the 64-celled stage, the distribution of these different kinds of egg cytoplasm into their respective cell lineages has been completed.

**FIG. 9–9.** Localization of various pigmented plasms in the ascidian *Styela partita* at the 2-celled (A), 8-celled (B), and 64-celled stages (C). cc = clear cytoplasm representing ectodermal cells; dg = dark grey cytoplasm representing endodermal cells; lg = light grey cytoplasm representing neural plate (NP) and notochord (N); yc = yellow cytoplasm representing muscle cells (M) and mesenchymal cells (ME). (After E. G. Conklin, 1905. J. Acad. Nat. Sci. Philadelphia 13:1.)

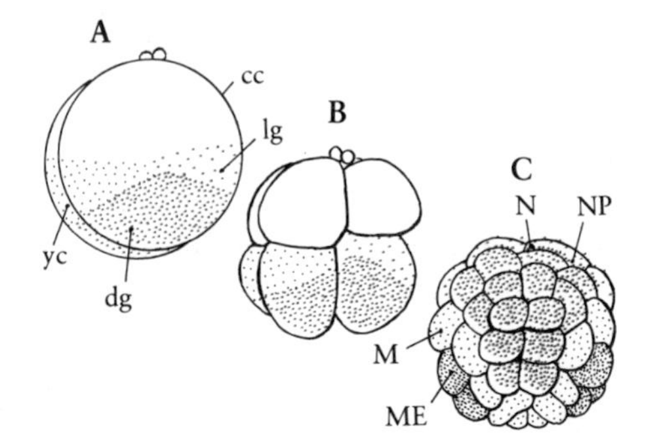


The arrangement of these four cytoplasmic areas defines a distinct bilateral organization for the uncleaved egg. The first cleavage furrow always divides the egg cell along the plane of bilateral symmetry, thus providing portions of the four plasms to the first two blastomeres (Fig. 9–9 A). If the first two blastomeres are isolated and cultured, each blastomere develops as a half-embryo. At the 4-celled stage, the anterior two blastomeres can be separated from the posterior two blastomeres. As one would expect on the basis of the segregated plasms, the anterior blastomeres when cultured form nervous tissue, notochord, ectoderm and endoderm, but not muscle or other mesodermal tissues. By the 8-celled stage, the yellow plasm is specifically confined to the posterior pair of the lower quartet of blastomeres (Fig. 9–9 B). These two blastomeres will give rise to the muscle and mesenchyme cells of the larva.

What is the evidence that these various cytoplasmic plasms underlie the differentiation of the cells in which they become distributed? Certainly, the correlation between the presence of specific visible cytoplasmic materials and particular tissue differentiations is an indirect argument in favor of the existence of morphogenetic determinants. Experimentally, a cause-and-effect interdependence between plasms and structural parts of the embryo is suggested by centrifugation studies. If eggs of *Styela* are subjected to moderate centrifugation after fertilization but before cleavage, there results a redistribution of the visible cytoplasms. Larvae may develop from these centrifuged eggs, but there are atypical distributions of muscle, nerve, and endodermal cells. Interestingly, the muscle cells develop in that position where the yellow plasm comes to lie as a result of the centrifugation.

Recent studies by Whittaker and his colleagues with eggs of *Ciona intestinalis* provide convincing evidence that cytoplasms segregated into specific blastomeres can be correlated with the appearance of tissue-specific proteins. Three enzymes have been identified during the course of the development of *Ciona*. The enzyme *acetylcholinesterase* (AChE) appears only within the tail muscles of larvae, *alkaline phosphatase* only within the endodermal cells of the gut, and *tyrosinase* (dopa oxidase) in two giant black melanocytes of the brain. Refer to the ascidian lineage fate map (Fig. 9–10), and note that the B4.1 cell pair carries the potential to produce larval muscle tissues. When these cells are removed and placed in culture, they will give rise

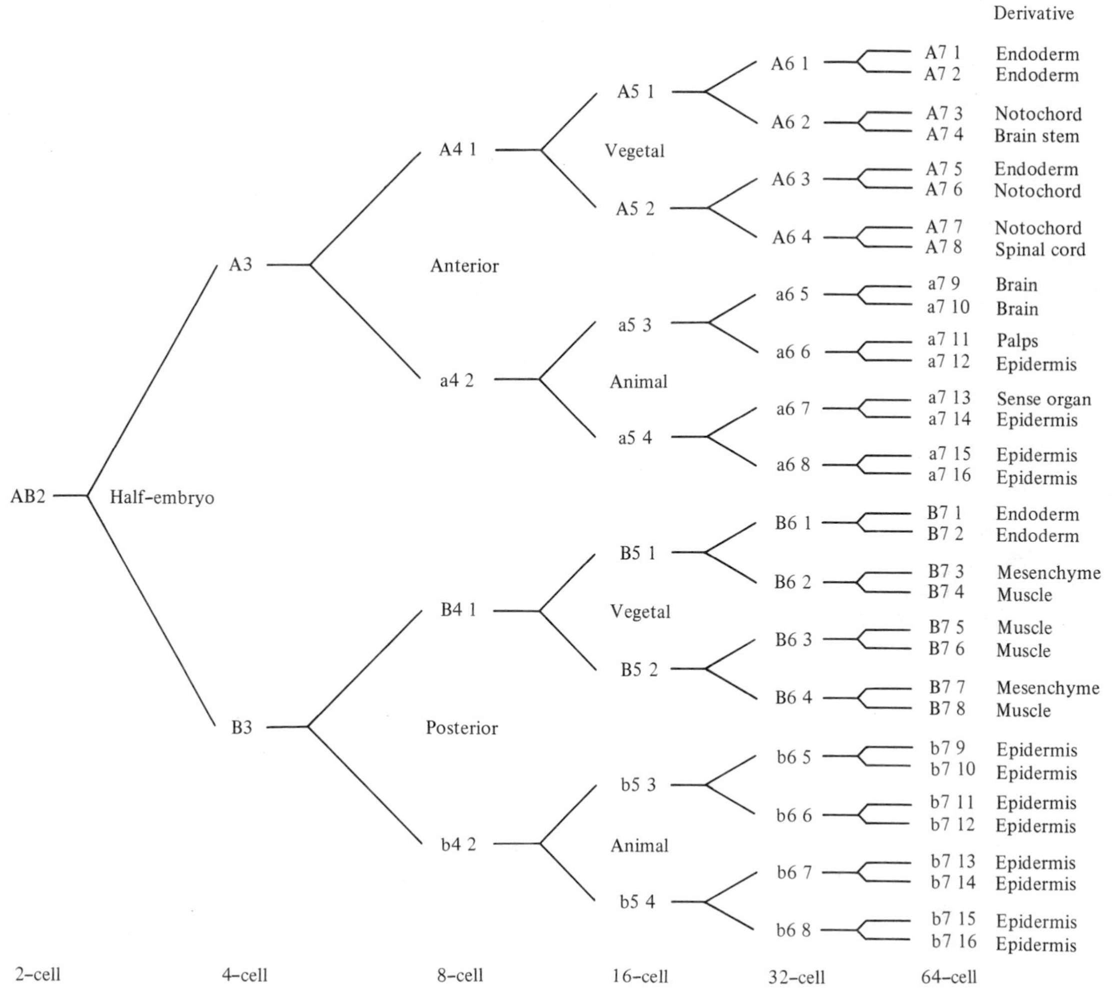


**FIG. 9–10.** Lineage fate map of a bilateral half of an ascidian embryo as reconstructed from data of Conklin (1905) and Ortolani (1955, 1957, 1962). (From J. R. Whittaker, 1979. The 37th Symp. Soc. Dev. Biol., p. 29.)

to cells that stain positively for AChE some 15 hours after fertilization (Fig. 9–11 B, C). If the B4.1 blastomere pair is destroyed in the embryo at the 8-celled stage, the remaining 6 blastomeres fail to produce any tissues containing AChE. Hence, the enzyme develops in the cells according to the cell number and pattern of muscle tissue lineage. The expression of the enzyme occurs independently of cleavage. If *Ciona* embryos are treated with cytochalasin B at different stages, cytokinesis (but not nuclear division) is immediately blocked and further development is arrested. However, AChE can be localized in these embryos at precisely the time that the enzyme appears in untreated control embryos. A similar series of experiments has shown that only the 4 vegetal blastomeres at the 8-celled stage have the potential to give rise to endoderm cells containing alkaline phosphatase. As in the case of AChE, this enzyme can be localized in cytochalasin-arrested embryos. Additional evidence for the existence of morphogenetic determinants has come from experiments in which the fates of ascidian cells have been changed by altering the cytoplasmic relationships. Whittaker, for example, compressed an embryo of *Styela* during the third cleavage and observed that the resulting division furrow was meridional rather than equatorial. As a consequence, yellow crescent material was distributed to 4 instead of 2 blastomeres of the 8-celled stage. If such embryos are released from compression and prevented from undergoing further division by treatment with cytochalasin B, histospecific

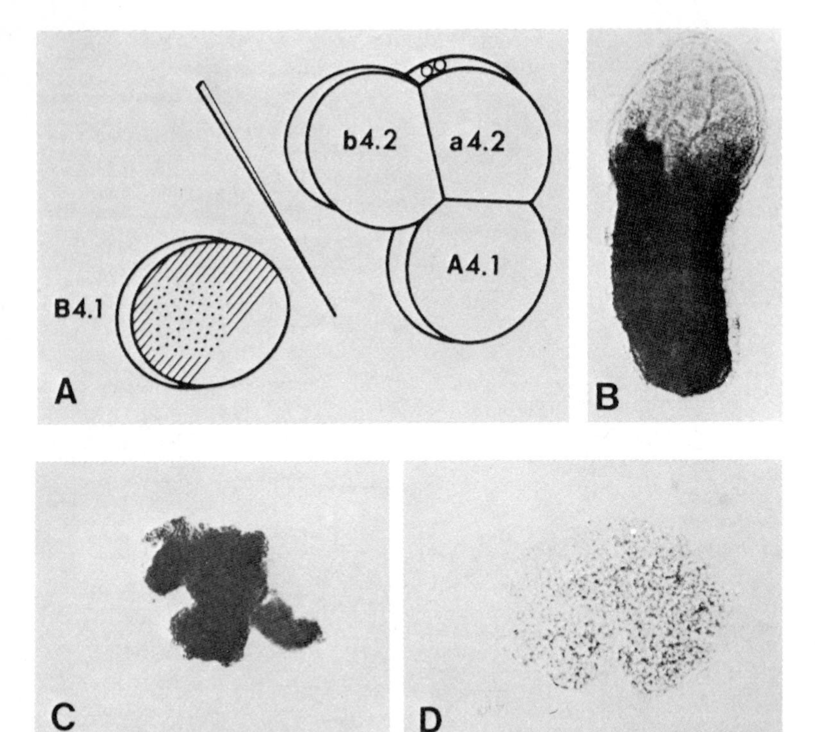


**FIG. 9–11.** Diagram illustrating the lateral orientation of the ascidian *(Ciona)* embryo at the eight-celled stage and the surgical separation of the B4.1 blastomere pair (A). The B4.1 pair is presumptive to muscle (diagonal lines) and mesenchyme (dotted) tissues; B, localization of acetylcholinesterase (AChE) activity in the early tailbud larva; C, localization of AChE activity at the middle tail bud stage in progeny cells of the single B4.1 cell pair; D, AChE activity in the dechorionated control embryo. (From J. Whittaker, 1977. Dev. Biol. 55:196.)

acetylcholinesterase appears in 3 or 4 blastomeres instead of just the 2 muscle lineage cells in control embryos. The yellow crescent, therefore, has a functional autonomy in relation to muscle differentiation that is evident by its displacement to other cellular locations.

Cytoplasmic localization phenomena have been investigated in a number of other invertebrates, including ctenophores *(Mnemiopsis)*, nemertine and annelid wourms *(Cerebratulus, Arenicola)*, and molluscs *(Crepidula, Ilyanassa, Dentalium)*. In *Cerebratulus*, there is a specific determinant for a ciliated-like apical tuft of the *pilidium larva*. This determinant is localized throughout the vegetal region of the vegetal hemisphere of the fertilized egg. Gradually, successive cleavages localize the *tuft determinant* in the animal hemisphere of the embryo. If the process of localization in *Cerebratulus* is compared to either the yellow crescent of ascidians or the grey crescent of amphibians, it appears to be slower and not correlated with any visible cytoplasmic reorganization. Other experiments, particularly with cytochalasin B and ethyl carbamate (which inhibits cleavage by suppressing aster formation), clearly suggest that a normal cell cycle is necessary for localization of the apical tuft determinant.

The spirally cleaving eggs of many molluscs, including those of *Ilyanassa* and *Dentalium*, are distinguished by visible regional differences in their cytoplasm. The ooplasm of *Dentalium*, for example, becomes reorganized and segregated into three distinct zones shortly after fertilization: a clear layer of cytoplasm at the animal pole, a broad granular layer at the equator of the cell, and a transparent layer of cytoplasm at the vegetal pole (Fig. 9–12 1). Just prior to the first cleavage division, the vegetal cytoplasm becomes localized in a large transient protrusion known as the *polar lobe* (Figs. 9–12 2; 9–13). The embryo at this time has the appearance of three cells and is referred to as the *trefoil stage*. During the first division of the egg, the polar lobe remains connected to the egg by a thin bridge of cytoplasm. By the two-celled stage, it is observed that the polar lobe becomes associated with one of the daughter blastomeres, the so-called *CD blastomere* (Figs. 9–12 3; 9–13). A second polar lobe forms in similar fashion from the vegetal region of the CD blastomere just prior to the second cleavage division. At the four-celled stage, the polar lobe plasm

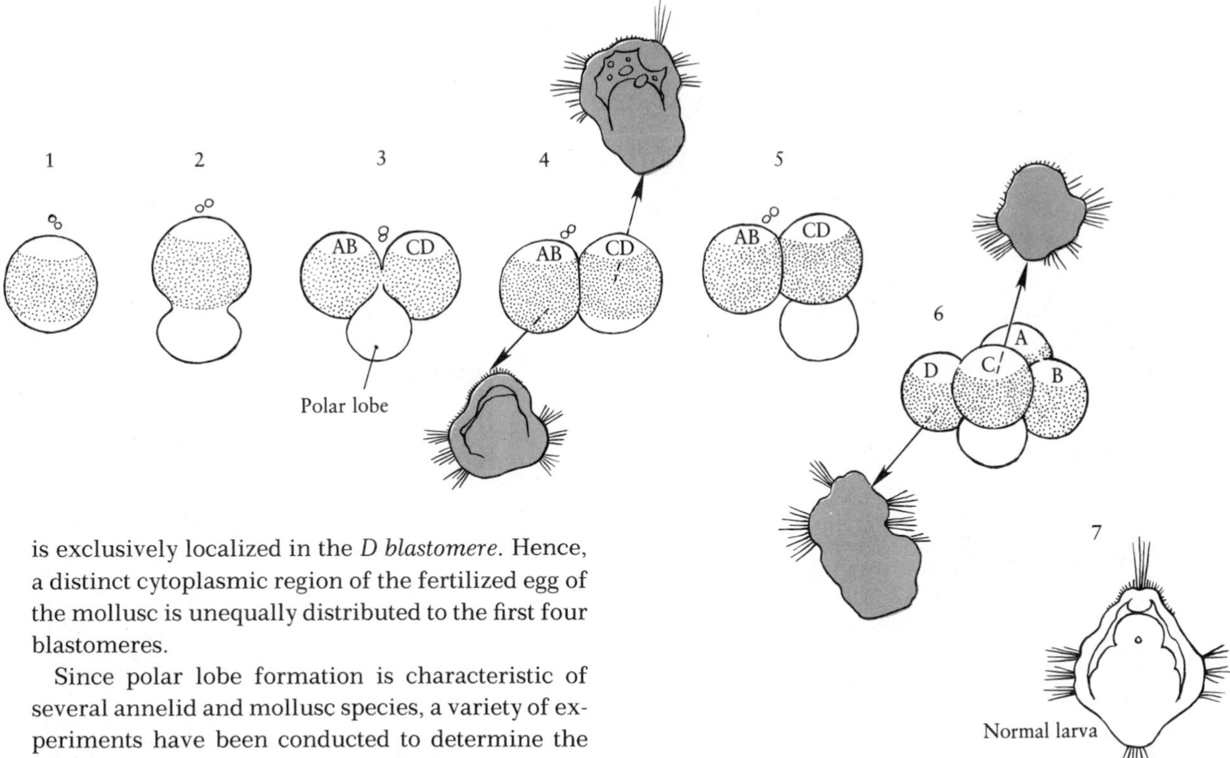


is exclusively localized in the *D blastomere*. Hence, a distinct cytoplasmic region of the fertilized egg of the mollusc is unequally distributed to the first four blastomeres.

Since polar lobe formation is characteristic of several annelid and mollusc species, a variety of experiments have been conducted to determine the role of the polar lobe plasm during embryogenesis. The role of the polar lobe has been evaluated by the culture of isolated blastomeres, extirpation or removal of selected blastomeres, and removal of the polar lobe. If the AB and CD blastomeres of *Dentalium* are separated and cultured, only the CD blastomere containing the polar plasm yields a complete *trochophore larva*. The AB blastomere forms a defective larva that typically lacks structures of mesodermal origin (Fig. 9–12 4). At the four-celled stage, only the D blastomere is able to develop as a complete, though smaller than normal larva (Fig. 9–12 6). Experiments in which the polar lobe is severed from the egg just prior to furrow formation lends further support to the view that the polar plasm is related to the differentiation of several adult structures, particularly those of mesodermal origin. Removal of the polar lobe at the first cleavage in *Dentalium* and *Ilyanassa* produces a larva that lacks an apical tuft of cilia and such structures as a velum, foot, shell, heart, and intestine. If the vegetal 60 percent of the lobe is extirpated, an apical tuft still appears in the larva, suggesting that the cytoplasm required for tuft expression is located in the "animal region" of the polar lobe. Removal of the vegetal one-third of an unfertilized egg yields results similar to those produced by removal of the polar lobe at the first cleavage division. Hence, the polar lobe appears to contain morphogenetic determinants that are set aside as early as the unfertilized egg stage and responsible for the expression of lobe-dependent structures.

FIG. 9–12. Cleavage of the mollusc egg *(Dentalium)* and the larvae that develop from isolated blastomeres. Only those blastomeres receiving polar lobe material develop normally. The letters indicate the blastomere from which each larva develops. (From E. B. Wilson, 1904. J. Exp. Zool. 1:1.)

The polar lobe of annelids and molluscs probably possesses several cytoplasmic determinants that become progressively segregated during cleavage until each reaches its own target embryonic cell. By removing the D blastomere at successively later stages of development, Clement was able to establish precisely the times at which the morphogenetic determinants originally present in the egg are segregated into appropriate descendant cells. Extirpation of either the D blastomere (leaving blastomeres ABC) or the 1D blastomere (leaving blastomeres ABC + 1d) produces embryos similar in appearance to those in which the polar lobe is removed at the trefoil stage. At the 16-celled stage,

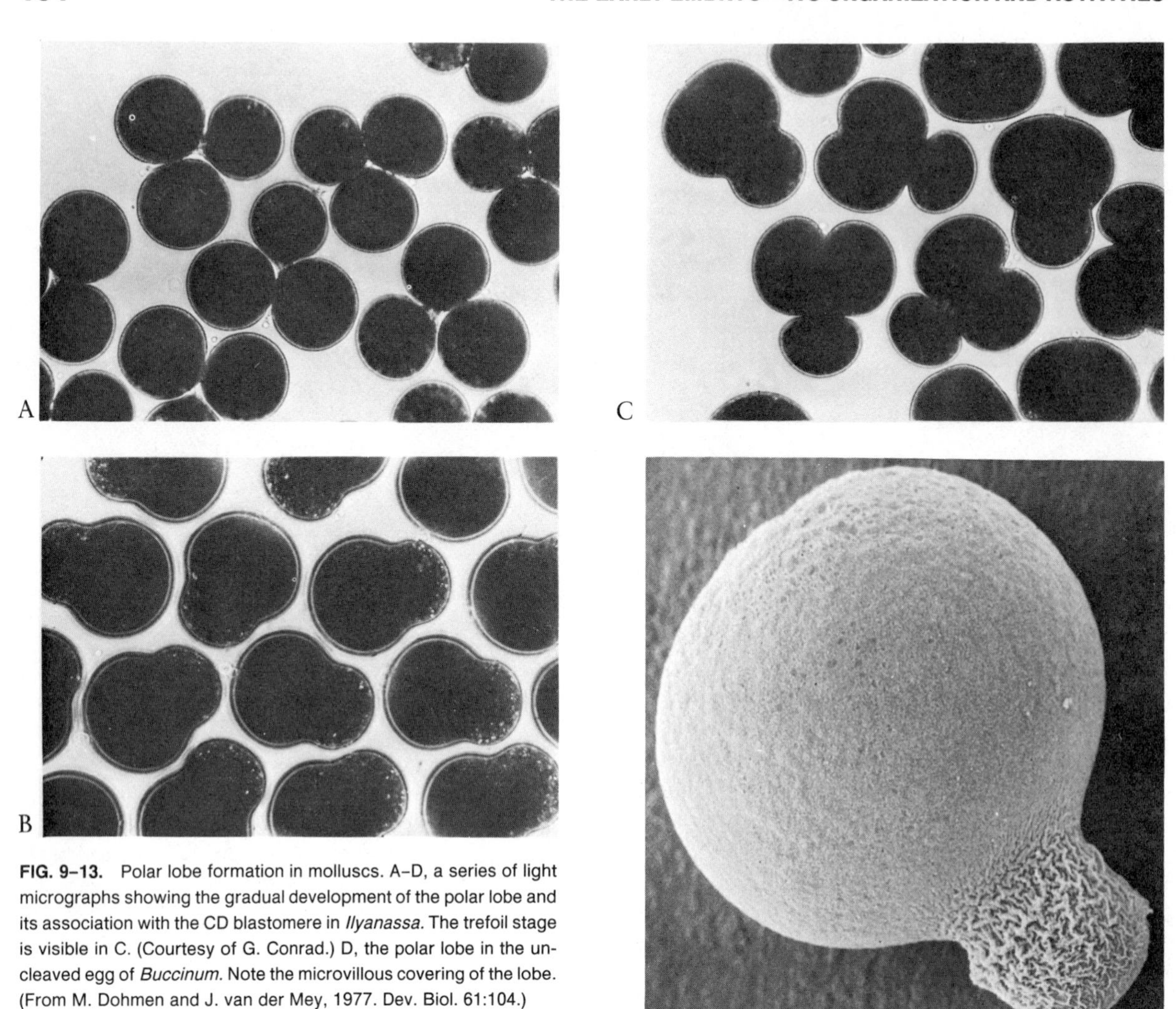


**FIG. 9–13.** Polar lobe formation in molluscs. A–D, a series of light micrographs showing the gradual development of the polar lobe and its association with the CD blastomere in *Ilyanassa.* The trefoil stage is visible in C. (Courtesy of G. Conrad.) D, the polar lobe in the uncleaved egg of *Buccinum.* Note the microvillous covering of the lobe. (From M. Dohmen and J. van der Mey, 1977. Dev. Biol. 61:104.)

removal of the 3D blastomere (leaving blastomeres ABC + 1d + 2d + 3d) yields an embryo with velum, eyes, shell, and foot, but without a heart and intestine. At the 32-celled stage, extirpation of the 4D blastomere has little qualitative effect on tissue differentiation, and the resulting embryo is normal except for its small size. The morphogenetically important polar lobe determinants appear to be shunted into the 3d and 4d blastomeres. The 4d blastomere will give rise to the *mesentoblast cells* or those cells involved in forming organs of mesodermal origin (i. e., heart, intestine).

The ascidian egg and the polar lobe have been used in a number of cytological and biochemical studies designed to yield information on the nature and subcellular localization of morphogenetic determinants. Although morphogenetic determinants appear to be real, we know little about their composition and how they regulate the expression of specific tissues. In *Ciona,* a provocative suggestion is that the determinants for the histospecific proteins that appear during larval development may be maternal messenger RNA (mRNA) molecules. Results of experiments with known inhibitors of RNA synthesis indicate that maternal mRNA for the alkaline phosphatase enzyme is present in the unfertilized egg and segregated into the intestinal endodermal cell lineage of the embryo. However, since there is no direct proof that actinomycin D inhibits RNA synthesis during pregastrular stages, the correlation between alkaline phosphatase synthesis and the presence of preformed maternal mRNA must be considered tentative. An alternative possibility is that the determinant is a "selective translation factor" whose segregation into the endodermal cells regulates the expression of mRNA specifying the alkaline phosphatase enzyme. The nature of the determinants for acetylcholinesterase and tyrosinase remains unknown.

A common technique in efforts to understand the role of cytoplasmic morphogens in embryogenesis has been to displace the constituents of the egg by centrifugation. When a relatively "simple" egg cell, such as that of the sea urchin, is centrifuged at moderate speeds for several minutes, the components of the fluid plasm separate into layers according to their specific gravities, but the gel-like, viscous peripheral cytoplasm or cortical cytoplasm (cortex) is unaffected by the centrifugal field. Generally, the pigment and yolk granules accumulate as distinct layers at the centrifugal pole (i. e., the point farthest from the axis of the centrifuge rotor) of the cell while the lipid droplets gather in a layer at the centripetal pole (Fig. 9–14 A). The transparent cytoplasm and nucleus organize between these other stratifications. It was Morgan who first demonstrated that, independently of this rearrangement of egg cytoplasm and pigment distribution, the centrifuged eggs of the sea urchin would cleave in accord with their original axiate pattern, gastrulate by invagination in the region where the micromeres form (which marks the vegetal pole), and develop into normal larvae (Fig. 9–14 B).

Similarly by centrifugation of uncleaved *Ilyanassa* eggs, the clear plasm of the animal hemisphere can be drawn into the vegetal lobe plasm, or the lobe plasm can be forced to "exchange positions" with the clear plasm (Fig. 9–15). Despite this redistribution, a polar lobe will form in the vegetal hemisphere, even though it contains hyaline cytoplasm derived from the animal pole of the egg cell. Clement (1976) has shown that these eggs continue to develop and form lobe-dependent structures, such as an apical tuft, indicating that determinants responsible for organ expression are not contained in the bulk of the cytoplasm. In *Dentalium*, the embryo lacks an apical tuft if the polar lobe is removed after centrifugation. This is clear evidence that the determinant for apical tuft formation is not displaced by centrifugation. Results of ablation and centrifugation experiments are consistent with the view that lobe determinants are positioned in the cortex. Cytologically, the polar lobe of *Bithynia* (a gastropod), comprising less than 1 percent of the total egg volume, shows a unique structure known

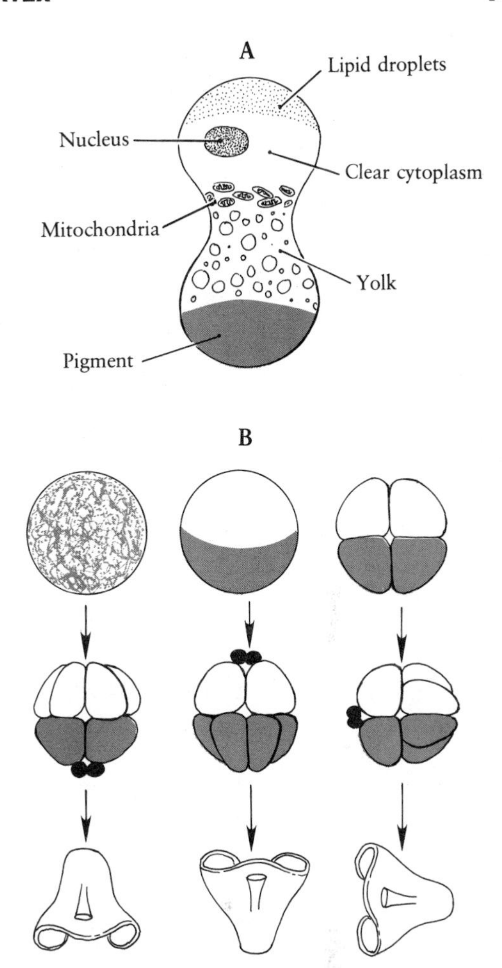


**FIG. 9–14.** A, stratification of the sea urchin *(Arbacia)* whole egg following moderate centrifugation (From E. Harvey, 1936. Biol. Bull. 71:101); B, centrifuged eggs will develop normally despite the dislocation of cytoplasmic constituents. (After T. Morgan, 1927. Experimental Embryology. Columbia University Press, New York.)

as the *vegetal body* (Fig. 9–16). The vegetal body consists of numerous small membrane-bound vesicles, and it stains intensely for RNA. It is segregated in the CD blastomere at first cleavage. The distribution of this organelle is unaffected by centrifugation. The development of *Bithynia* embryos lacking the vegetal body (i. e., lobeless) is similar to that reported for lobeless embryos of *Dentalium* and *Ilyanassa*. The embryos will cleave and gastrulate in the same fashion as lobed embryos. However, after gastrulation, the embryos swell into balloonlike structures with the endodermal tissue lying against the superficial ectodermal cells. Some of the endodermal cells may even project to the exterior or exogastrulate. (Fig. 9–17). However, no

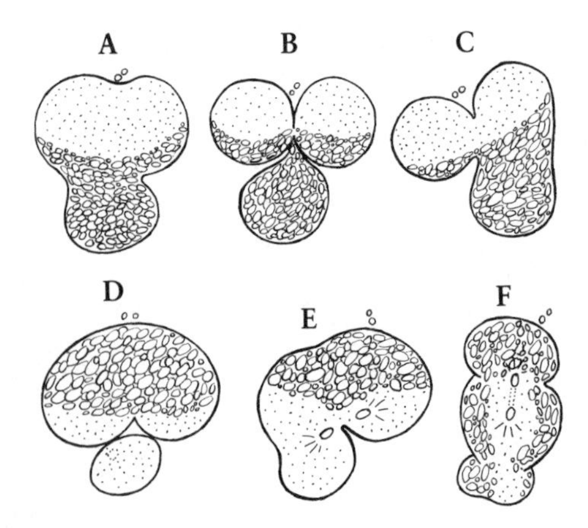


**FIG. 9-15.** A–C, normal cleavage of the egg of the mollusc, Ilyanassa; D–F, cleavage in the centrifuged egg of *Ilyanassa.* Note that centrifugation displaces the yolk-rich vegetal plasm to the animal pole. (From T. Morgan, 1927. Experimental Embryology. Columbia University Press, New York.)

mesodermal tissue is observed in these lobeless gastropod embryos. Unfortunately, there is no direct evidence for a functional relationship between the vegetal body and the known morphogenetic determinants of the polar lobe. At the molecular level, the polar lobe of *Ilyanassa* has been the object of considerable study. There is some evidence that the polar lobe houses a specific set of maternal messenger RNAs or factors that select a specific group of messenger RNAs for translation. The advantage of the polar lobe plasm is that investigators can now examine more carefully how these determinants regulate the expression of the different tissues of the early mollusc embryo.

Other experiments over the past several years with eggs of different animal species point to the cortical cytoplasm as an important sequestering site of developmental determinants. If eggs of forms such as *Lymnaea* (snail), *Styela*, and *Arbacia* are permitted to stand after disruption of their normal organization through moderate centrifugation, the special plasms and inclusions do not remain in their new positions. They reorganize throughout the egg and tend toward a restoration of their original arrangement. Hence, it follows that the fluid cytoplasm with its regional variations is probably organized in relation to differences in the immovable portion of the cell or egg cortex. We have previously mentioned that the yellow crescent of the ascidian embryo appears to be localized in the ve-

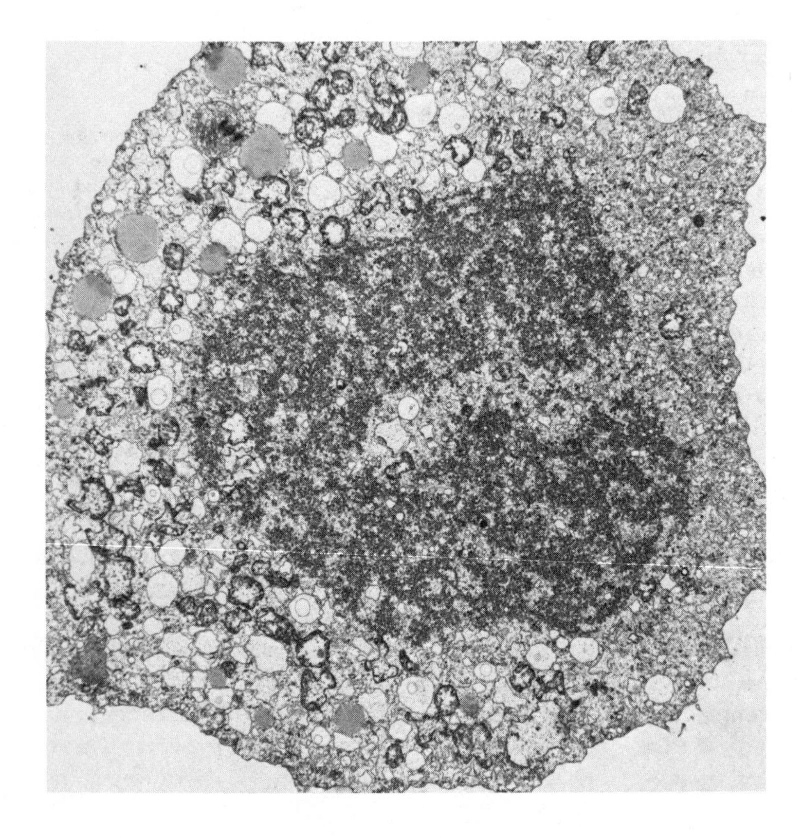

**FIG. 9-16.** Electron micrograph of the first polar lobe with the vegetal body in *Bithynia tentaculata* (gastropod). (From M. Dohmen and N. Verdonk, 1974. J. Embryol. Exp. Morphol. 31:423.)

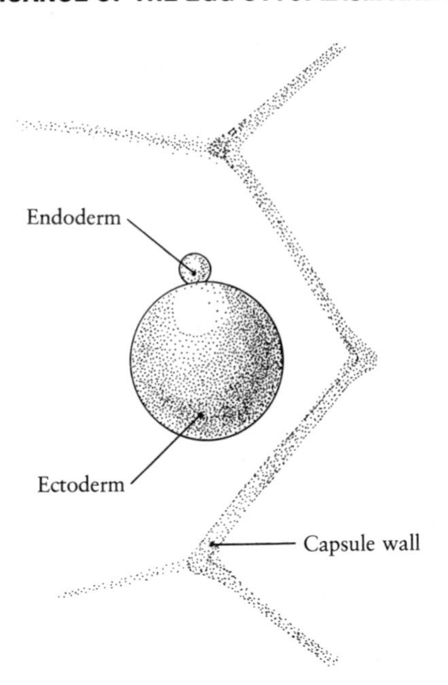


FIG. 9–17. Sketch of an 11-day-old living lobeless embryo showing partial exogastrulation. (After J. Cather and N. Verdonk, 1974. J. Embryol. Exp. Morphol. 31:415.)

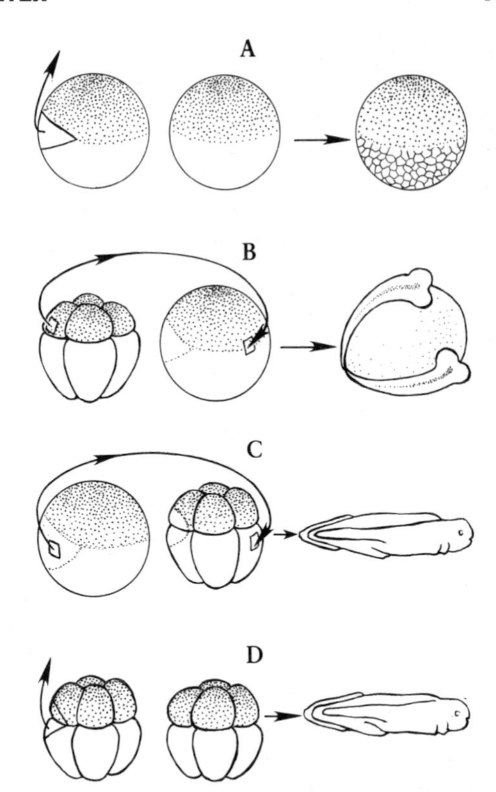


FIG. 9–18. Grey crescent transplantation experiments of Curtis in *Xenopus*. A, uncleaved egg from which cortical grey crescent has been excised fails to gastrulate; B, transplant of cortical grey crescent from eight-celled stage onto ventral side of a second egg produces an embryo with two sets of axial structures; C, transplant of cortical grey crescent from uncleaved egg to eight-celled embryo does not result in the induction of a second embryonic axis; D, extirpation of cortical grey crescent at eight-celled stage produces a normal embryo. (From A. Curtis, 1962. J. Embryol. Exp. Morphol. 10:410.)

getal pole because of the presence of *Dolichos biflorum* lectin-binding sites on the plasma membrane (Chapter 6).

The cortical cytoplasm of the grey crescent of amphibian eggs has been studied rather extensively, but its importance as a site of developmental information is subject to differences of opinion. Classical experiments of Brachet and Pasteels pointed out that localized destructions of the grey crescent of the frog's egg are followed by abnormalities of such axial structures as the neural tube, notochord, and somites. Experiments by Curtis (1962) on *Xenopus* showed that removal of the cortex of the grey crescent (approximately 0.5–3.0 $\mu$m in thickness) did not disturb mitosis or cytokinesis of the fertilized egg; however, the formation of the blastopore and the development of axial structures was inhibited (Fig. 9–18 A). After perfecting a technique for the isolation of cortical cytoplasm material, he transplanted grey crescent cortex to the ventral side of a second or host egg cell. When the grey crescent was obtained from the 8-celled or younger stages, the resulting embryo developed with two sets of axial structures (Fig. 9–18 B). An egg cell receiving a transplant of cytoplasm taken from beneath the grey crescent developed into an embryo with only a single set of axial structures. Interestingly, the 8-celled *Xenopus* embryo no longer responded to a transplant of additional crescent taken from a younger stage (Fig. 9–18 C). Also, extirpation of the grey crescent at this stage no longer interrupted the process of gastrulation (Fig. 9–18 D). These results suggested that a change in cortical organization occurs during the second or third cleavages. Grey crescent cortex implanted by Tompkins and Rodman (1971) into the blastocoele of *Xenopus* embryos provided additional support for the localization of morphogenetic determinants in the cortex. Recently, Malacinski and colleagues carried out experiments similar to those of Curtis using eggs of *Ambystoma* (salamander). They could find no evidence that the grey crescent cortex of the fertilized egg contained morphogenetic information critical to subsequent development. Rather, the blastomeres on the future dorsal side of the cleaving embryo appeared to acquire progressively the ability to initiate gastrulation and influence cell and tissue development. Until further studies are done, the concept that there is localization of information in the amphibian grey crescent must be viewed with reservation. We will have more to say about the role of the grey crescent in the organization of the embryo in the next chapter.

Many questions remain about the cortical cytoplasm as a source of developmental information. How widespread in the animal Kingdom is the cortical control of embryogenesis? What is the nature of specific morphogenetic substances or determinants, and what is their pattern of organization in the egg cortex? When is the pattern of morphogenetic information established in the egg cortex? Attempts to isolate and identify chemically active morphogenetic determinants in the egg cytoplasm have been a difficult and unrewarding task. Horstadius and his colleagues have made efforts in this direction using unfertilized egg homogenates of the sea urchin (see next section). Investigators at the Zoological Laboratory of the University of Utrecht have identified Feulgen-positive granules (probably DNA) in the cortical region where the first polar lobe will form in *Dentalium*. These granules may have a template function, specifying those morphogenetic determinants responsible for the formation of the apical tuft and lobe-dependent mesodermal structures. More recently, Malacinski has isolated an axial-specific determinant from the germinal vesicle of unfertilized eggs of frogs, which he suggests becomes part of the grey crescent following germinal vesicle breakdown (see Chapter 10).

## Cytoplasmic Gradients and Morphogenetic Substances: The Sea Urchin

It would be convenient to view the differentitation of the tissue and organ rudiments of the early embryo as being solely controlled by a system of morphogenetic determinants that exhibit a certain spatial organization and become sequentially segregated into different cell lineages. In such an embryonic system, the developmental potential of each blastomere becomes progressively restricted. Such a cell, if isolated, would never develop qualitatively into tissues other than those it is expected to form. Yet the early experiments of Driesch on isolated sea urchin blastomeres clearly indicate that regional differences in the egg cytoplasm and the egg cortex are only some of the factors required for normal development. Isolated sea urchin blastomeres possess a remarkable ability to organize and form themselves into whole embryos. Hence, each blastomere in isolation gives rise to structures not normally part of its expression in the intact embryo. Such a cell with unlimited potential (i.e., it can form all the constituent cells of the adult organism) is referred to as being *totipotent*. Driesch recognized that embryonic cells could have different properties during development. The expected fate of an embryonic cell under normal conditions of development is referred to as its *prospective significance* (i.e., *prospective fate*). The *prospective potency* of an embryonic cell is its developmental potential at any given stage. The existence of an embryonic system in which a part can produce a whole raises interesting questions about the relationships between the prospective significance and the prospective potency of a cell. Driesch postulated that the expected fate of a cell depends on its position in the whole embryo. The ability of a part to realize the potential of a complete embryo is called *regulation*, and it characterizes the generation of form in a number of animal species. The basis of regulation is still imperfectly understood, but insight into questions raised by this phenomenon has come from analysis of the developing sea urchin embryo.

You will recall from the chapter on cleavage that the fertilized egg of the sea urchin divides in holoblastic fashion (Fig. 9–19). The embryo at the 16-celled stage consists of eight mesomeres in the animal hemisphere and eight blastomeres in the vegetal hemisphere, four of which are very large (macromeres) and four of which are small (micromeres) (Fig. 9–19 D). By the 64-celled stage, the eight mesomeres have proliferated into two tiers of 16 cells each. These are termed the $An_1$ and $An_2$ (animal) layers. Below the animal hemisphere, there are three recognizable layers of cells: $Veg_1$ (8 cells), $Veg_2$ (8 cells), and Mic (micromeres, 16 cells) (Fig. 9–19 F). Continued division of cells yields a ciliated blastula some six hours after fertilization. At approximately 10 hours after fertilization, the blastula is released from the confines of the fertilization membrane, and the embryo becomes free-swimming. Long, stiff cilia make up the apical tuft

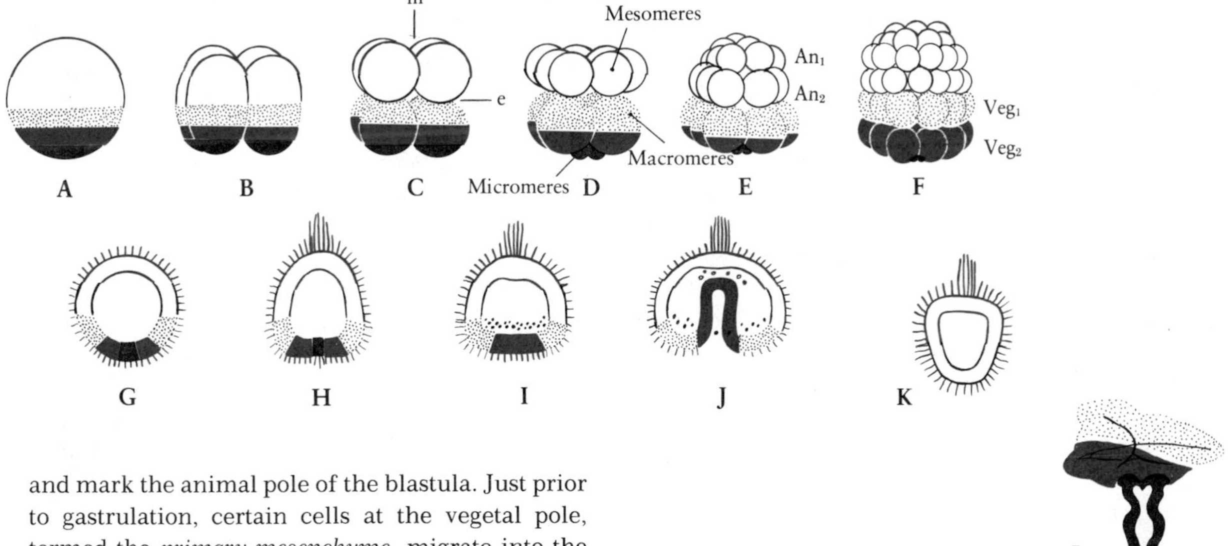


**FIG. 9–19.** Diagram of the normal development of the sea urchin. *Paracentrotus lividus*. A, uncleaved egg; B, 4-celled stage; C, 8-celled stage; D, 16-celled stage, E, 32-celled stage, F, 64-celled stage; G, young blastula; H, later blastula with apical tuft; I, blastula after migration of primary mesenchyme; J, gastrula with archenteron or gut cavity; K, dauerblastula produced by isolation of the four animal cells at the 8-celled stage; L, exogastrula produced by the isolation of the four vegetal cells at the 8-celled stage. $An_1$, animal layer 1; $An_2$, animal layer 2; $Veg_1$, vegetal layer 1; $Veg_2$, vegetal layer 2. (From S. Horstadius, 1939. Biol Rev. 14:132.)

and mark the animal pole of the blastula. Just prior to gastrulation, certain cells at the vegetal pole, termed the *primary mesenchyme*, migrate into the interior of the blastocoele (Fig. 9–19 I). They are descendants of the micromeres and will form the skeletal system. Gastrulation occurs by the invagination of cells largely derived from $Veg_2$. The gastrula subsequently produces *secondary mesenchyme* from the tip of the gut (Fig. 9–19 J). Following the formation of the gut, skeletal system, and the mouth, the free swimming organism is known as the *pluteus larva*.

If the different layers of blastomeres are followed through gastrulation, a specific fate can be assigned to each (Fig. 9–19). The $An_1$, $An_2$, and $Veg_1$ cells form the ectoderm of the larva, contributing also to the apical tuft, mouth, ciliary bands, and anal arms. The $Veg_2$ cells form the endodermal lining of the gastrointestinal tract or *archenteron*. The Mic cells form primary mesenchyme, the major derivative of which is the skeletal system with its calcareous spicules.

Since the first two cleavages in the sea urchin are radial, each resultant blastomere receives all the major regions of the egg cytoplasm (i.e., presumptive ectoderm, endoderm, and mesoderm or mesenchyme) (Fig. 9–19). However, the third set of cleavages is equatorial or horizontal. For the first time during development, therefore, blastomeres exist that clearly reflect differences in the animal and vegetal material received. The upper four cells contain primarily animal material, and the four lower cells contain primarily vegetal material.

The restriction in the totipotency of the blastomeres by the eight-celled stage is shown by examining the development of isolated half-embryos. If the upper quarter of blastomeres at the eight-celled stage is isolated by an equatorial separation (Fig. 9–19 C; along plane e) and subsequently cultured, the result is the production of a ciliated, hollow ball of cells known as a *dauerblastula* (Fig. 9–19 K). Neither an archenteron nor a skeleton is formed in the dauerblastula, and development does not proceed beyond the blastula-like stage. Such an embryo lacking endodermal and mesodermal differentiations is said to be *animalized*. The vegetal four cells yielded by the equatorial separation differentiate as an ovoid-shaped embryo with a disproportionately large archenteron, few spicules, and little ectodermal specialization (i.e., mouth, cilia). Often, the gut cavity is evaginated outward rather than inward, a condition known as *exogastrulation* (Fig. 9–19 L). Such an embryo is said to be *vegetalized*. By contrast, if the eight-celled embryo is separated into two four-celled halves by a meridional divison (Fig. 9–19 C; along plane m), each half develops into a complete, though slightly smaller than normal larva.

These same patterns of expression are also evident in the manipulated, unfertilized egg. The unfertilized egg can be cut equatorially into upper and lower halves with a glass needle and each half can subsequently be fertilized. Each egg fragment cleaves, but the upper fragment develops as a dauerblastula while the lower fragment develops as a vegetalized embryo. In short, the unfertilized egg appears to consist of two halves, one animal and one vegetal, each of which possesses properties that the other lacks.

The interpretation of the early development of the sea urchin egg, as advanced in extensive studies by Horstadius, Needham, and Runnström, is that the direction in which a blastomere will differentiate is under the control of two mutually antagonistic gradients throughout the embryo. Each gradient is viewed as having an organizational center. One gradient exerts a maximum influence at the animal pole; the other gradient exerts a maximum influence at the vegetal pole. Each gradient diminishes in influence from its organizational center. The differentiation of the various parts of the embryo is directly related to an interaction between these animal-vegetal and vegetal-animal gradients. Defects and anomalies in embryonic development can be traced to naturally or artificially induced disturbances in the balance of these gradients.

Support for the paired-gradient concept derives from several types of experiments, including the microsurgical separation and culturing of blastomeres, the production of differential reduction gradients by dyes, the alteration of patterns of carbohydrate and protein metabolism, and the isolation of morphogenetic agents from unfertilized egg cells. Presently, the interrelationships between the proposed cytoplasmic gradients, the metabolism of the egg itself, and the presence of morphogenetic determinants in the cytoplasm are unclear.

Horstadius (1939) found that the sea urchin egg is particularly well suited to examining a variety of developmental problems, including questions relating to the existence of cytoplasmic gradients. By using fine glass needles and very narrow-diameter pipettes in conjunction with a calcium-free surgical medium, Horstadius was able to separate blastomeres at selected stages of development (particularly at the 32- and 64-celled stages). He then combined specific groups of blastomeres and observed their development in a small depression of celluloid. If the double-gradient hypothesis is correct, then differentiation of various combinations of blastomeres should be predictable on the basis of the relative amounts of animal and vegetal components. In other words, if a proper balance of animal and vegetal material is the main requirement for normal development, then it would be possible to bring about this state by using novel combinations of blastomeres.

The influence of $Veg_1$, $Veg_2$, and the micromeres upon the differentiation of isolated animal embryo halves is shown in Figure 9–20. As previously noted, an isolated animal embryo half (i.e., $An_1$ + $An_2$ blastomeres) develops into a hollow, ciliated ball of cells often with an enlarged apical tuft (Fig. 9–20 A). If a layer of $Veg_2$ cells is added to the animal half, the result is a blastula with cilia, an apical tuft, archenteron, and mouth cavity. A perfect pluteus larva is obtained from this combination of blastomeres (Fig. 9–20 C). The $Veg_2$ cells have the ability to check apical tuft size and stimulate differentiation of a ciliated band and mouth cavity. Additionally, the presence of $Veg_2$ cells causes the differentiation of skeletal tissue, the latter being normally produced by the micromeres. An animal half to which has been added a layer of $Veg_1$ cells produces an imperfect pluteus larva (Fig. 9–20 B). However, a normal larva can be obtained if only four micromeres are combined with the animal half, thus showing the greater vegetalizing influence of the micromeres (Fig. 9–20 E). The micromeres provide the required vegetal influence to balance the animal half, thus restoring the balance between animal and vegetal ratios compatible with normal development. Regulation is evident in the sense that structures normally derived from $Veg_1$ and $Veg_2$ blastomeres are formed from animal cells under the vegetalizing influence of micromere material.

Additional evidence that normal sea urchin development requires a balance of animal and vegetal factors or substances comes from experiments in which various numbers of micromeres are cultured with other cellular layers (Fig. 9–21). One or two micromeres added to a layer of $An_1$ cells reduces the animalizing influence, but does not permit a normal pluteus larva to develop. However, the addition of four micromeres to the $An_1$ layer results in a normal pluteus larva. If $An_2$ cells are used, fewer micromeres must be added to achieve nor-

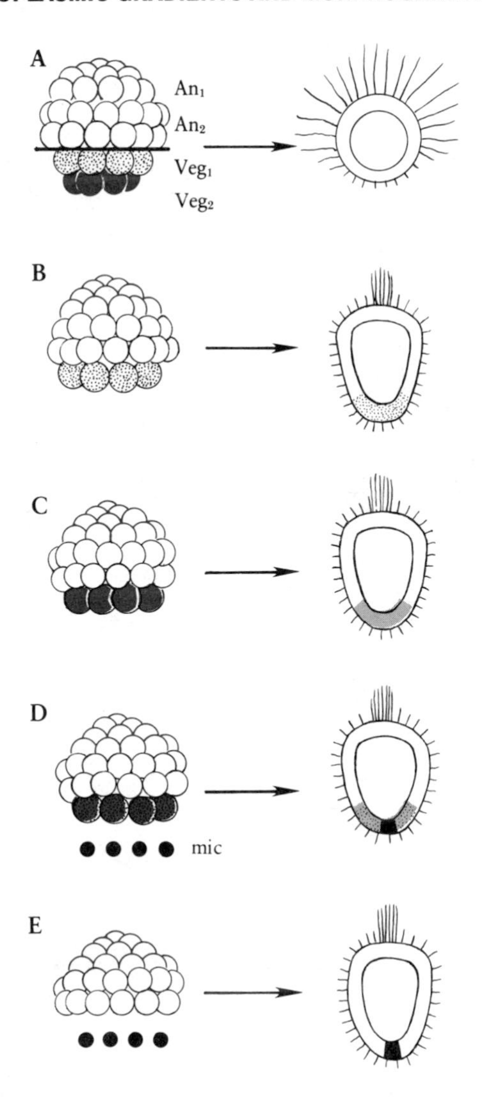


**FIG. 9–20.** Diagram of the influence of $Veg_1$, $Veg_2$, and the micromeres on the differentiation of isolated animal halves in *Paracentrotus*. A, isolated animal half; B, animal half + $Veg_1$; C, animal half + $Veg_2$; D, animal half + $Veg_2$ + micromeres; E, animal half + micromeres. (From S. Horstadius, 1939. Biol. Rev. 14:132.)

mal development. Micromeres added to $Veg_1$ and $Veg_2$ layers, as expected, tend to exaggerate vegetal characteristics.

In experiments first done by Ernest F. G. Herbst dating back to 1893, it has been demonstrated that a wide range of chemical treatments can modify the direction of the development of sea urchin embryos. When lithium chloride is added to the medium surrounding fertilized eggs, the resultant embryos show signs of excessive gut development and exogastrulation. Such agents, which cause an embryo to develop as if it were an isolated vegetal half-embryo, are referred to as *vegetalizing agents*. These agents, then, appear to bias differentiation in the same direction as that observed by combining an isolated half-embryo with micromeres. Other vegetalizing agents include certain amino acids (valine and tyrosine) and metabolic inhibitors such as dinitrophenol and chloramphenicol. On the other hand, trypsin, chymotrysin, and sodium thiocyanate cause fertilized eggs to develop as dauerblastulas, thus acting as *animalizing agents*. In view of the heterogeneity of these animalizing and vegetalizing agents, some investigators have suggested that these substances act by upsetting biochemical equilibriums in the egg.

There is also evidence to indicate that the proposed pair of cytoplasmic gradients may be closely linked to the metabolism of the sea urchin egg. When embryos (blastulas or gastrulas) are placed in a chamber with Janus green dye and sealed from an external source of oxygen, the oxygen of the seawater is gradually exhausted. The Janus green then acts as an acceptor of electrons from oxidative phosphorylation and becomes reduced. As it does so, the dye progressively changes in color from grayish blue to red and then to colorless. The pattern of reduction of the dye in late blastulas begins near the vegetal pole and spreads laterally and toward the equator of the embryos (Fig. 9–22). When the advancing border of the reduced dye reaches the equator, a second reduction front appears at the animal pole and proceeds toward the vegetal hemisphere. When the same experiments are conducted using isolated vegetal halves (Fig. 9–23 B), an early and intense reduction center is noted at the vegetal pole; it then moves toward the animal hemisphere. However, no secondary gradient is visible in the animal end of the embryo. A similar observation has been made on animal halves (i.e., no second reduction front appears after the first one forms) (Fig. 9–23 A). Hence, each gradient appears to have its own reduction potential, suggesting that the gradients are qualitatively different. It is interesting to note that when micromeres are implanted into either an isolated animal half or into the side of a whole embryo, they precipitate their own center of reduction. Such implants also stimulate a second archenteron to develop. These results on reduction gradients strongly suggest the existence of a double-gradient system in the sea urchin embryo.

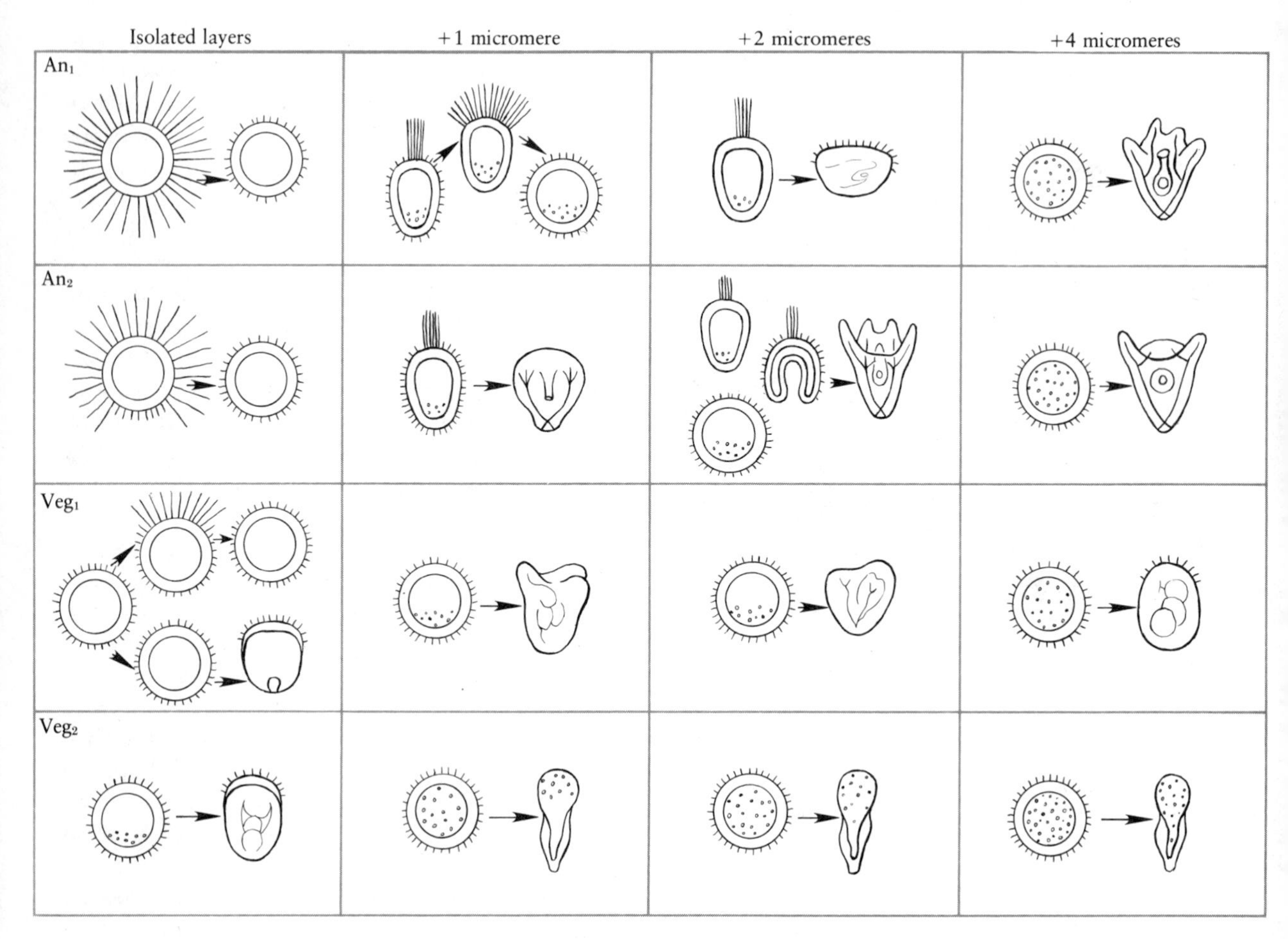


**FIG. 9-21.** Diagram of the development of the layers $An_1$, $An_2$, $Veg_1$, and $Veg_2$ following isolation (left column) and the addition of one, two, and four micromeres. (From S. Horstadius, 1939. Biol. Rev. 14:132.)

**FIG. 9-22.** Change in the color of Janus green showing reduction gradients in a later blastula (A, 1–6) and a gastrula (B, 1–6). Numbers indicate time of observations. Large dots, blue color; small dots, red color; no dots, colorless. (From S. Horstadius, 1952. J. Exp. Zool. 120:421.)

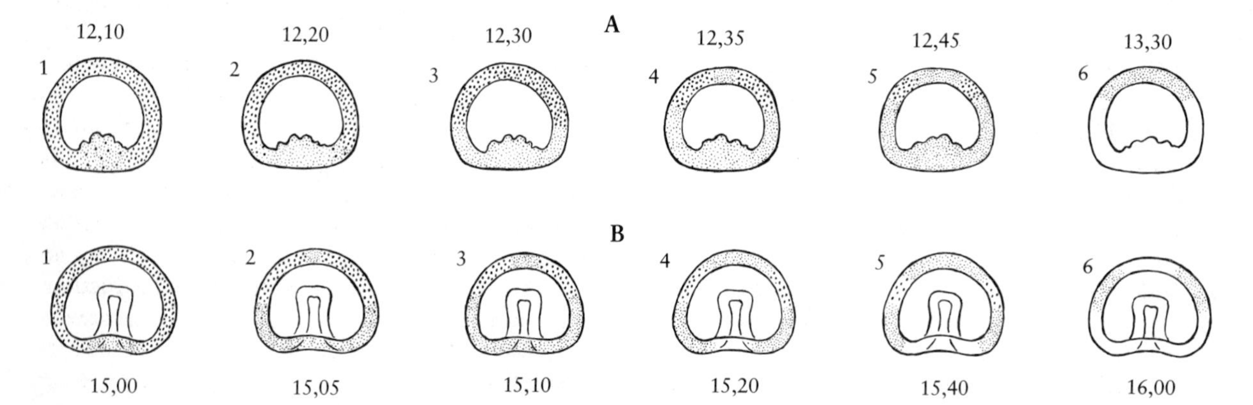

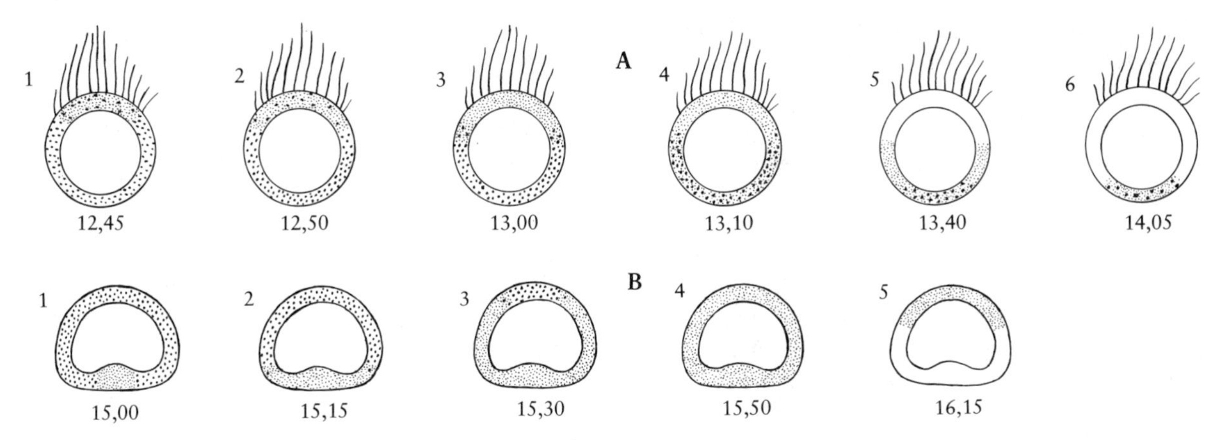


**FIG. 9–23.** Reduction of Janus green in isolated animal (A, 1–6) and vegetal (B, 1–5) halves of the sea urchin embryo. Numbers indicate time of observations. Symbols as in Figure 9–22. (From S. Horstadius, 1952. J. Exp. Zool. 120:421.)

Runnström and others believe that these reduction gradients are, in fact, manifestations of metabolic gradients in the cytoplasm of the egg. Studies designed to measure the incorporation of [$^{14}$C]leucine and [$^{14}$C]valine into protein in isolated animal and vegetal sea urchin halves indicate that most of the protein synthesis occurs in the vegetal hemisphere. By contrast, the bulk of carbohydrate metabolism appears to be conducted in the animal hemisphere. That the animal–vegetal gradients may be systems of metabolic reactions tends to be supported by the actions of vegetalizing and animalizing agents. Lithium chloride, a vegetalizing agent, disturbs the respiration of the embryo by either uncoupling the oxidative phosphorylating reactions or inhibiting the actions of certain glycolytic enzymes. Dinitrophenol appears to act in a similar fashion. Vegetalization, then, is essentially a suppression or disturbance in the oxidative processes of the embryo (i.e., the animalizing influence of the animal hemisphere is affected). Why structures developing at the animal pole are more sensitive to these agents than structures derived from the vegetal pole remains unknown.

Animalizing agents, such as trypsin and polysulfonated organic compounds (Evans blue), appear to interfere with the formation of functional proteins. Although the mechanism by which they act is not understood, it is suggested that acidic groups of some of these compounds immobilize the proteins by joining to functional side groups. The exaggeration of ectodermal structures under the influence of these agents appears to be due to disturbances in the vegetal–animal gradient.

Treatment of sea urchin embryos with various chemical compounds has been helpful in establishing the presence of gradients and probable biochemical basis for their existence. The two principal systems, one centered in the animal hemisphere and the other in the vegetal hemisphere, control regulation and morphogenesis along the animal–vegetal axis. If these two antagonistic systems exist as overlapping gradients of developmental potencies, what is the chemical basis for the determination of animal and vegetal properties? Are there specific substances in the sea urchin egg with animal and vegetal properties that act as morphogenetic determinants? Generally, attempts to demonstrate the presence of specific cytoplasmic components in the early sea urchin embryo have not been successful. Horstadius, Josefsson, and Runnström in a series of papers (1967, 1969) have reported on their efforts to isolate and identify animalizing and vegetalizing substances in unfertilized egg and early-cleavage stages of the sea urchin. By using sophisticated techniques of centrifugation and column chromatography, some fractions were obtained that exhibited strong animalization of whole eggs and vegetal embryo halves. Other fractions vegetalized animal halves but failed to affect whole eggs. Absorption curves of animalizing fractions using ultraviolet light indicated that there may be several animalizing agents. One fraction strongly resembles the amino acid tryptophan, while another fraction resembled a nucleotide structure.

## Regulative and Mosaic Patterns of Development

Our discussion of developing eggs of tunicates, molluscs, sea urchins, and amphibians tends to suggest that there are two broad patterns controlling early embryogenesis among animals. In most molluscs, annelids, and tunicates, the fate of a blastomere is typically fixed or determined by the onset of segmentation. Cleavage parcels out unequally to various blastomeres the morphogenetic substances localized in patchwork fashion in the egg. Each blastomere can only express its potential in terms of the morphogenetic determinants that it has inherited. When isolated, each blastomere shows a striking capacity for autonomous self-differentiation, but its developmental potential is never qualitatively more than that predicted by its prospective fate. If certain blastomeres are deleted from the developing embryo, the embryo remains deficient in those structures normally formed by the extirpated cells. Eggs that reflect this pattern of development are termed *mosaic* or *determinate*.

By contrast to mosaic eggs and embryos, the early blastomeres of echinoderm and vertebrate embryos are labile and totipotent, capable of regulation and developing into normal embryos when experimentally isolated. In the sea urchin, meridional halves of the blastula are capable of regulating themselves into well-proportioned embryos. Each meridional half, therefore, forms a considerably larger repertoire of cell types than would normally be expected. Eggs that reflect this pattern of development are referred to as being *regulative* or *indeterminate*.

This apparent dichotomy led to the classical view that animal eggs can be categorized as either "mosaic" or "regulative" in their developmental character. Typically, cytoplasmic localization was purported to be important in the development of mosaic eggs, but not to regulative eggs. It is now clear that this distinction between mosaic and regulative eggs is inaccurate and by no means absolute. Two examples can be used to illustrate this point. Horstadius has shown that the various layers of the 16-celled stage of *Cerebratulus* (a nemertean worm), following isolation, will differentiate only into what is expected. When blastomeres are dissected from the embryo and recombined, each blastomere cell line develops as it would have in the intact embryo; there is no evidence of interactions between animal and vegetal components as in the sea urchin. In short, the embryo behaves as a perfect mosaic. Yet, if a nucleated fragment of the unfertilized egg is inseminated, it develops into a qualitatively normal embryo. This is indicative of regulative development. Recall that each of the first two blastomeres of the frog's egg can in isolation compensate for the loss of the other blastomere and subsequently form a whole embryo. By contrast, at the four-celled stage, only the two blastomeres containing grey crescent material can yield normal embryos. Since the grey crescent represents an area of localized morphogenetic information responsible for the axial organization of the embryo, the egg can be considered to be mosaic.

Differences between mosaic and regulative patterns of development probably reflect differences in the time at which events involved in the determination of parts of the embryo occur. In some organisms the localization of cytoplasmic determinants that control the direction of later development takes place by fertilization. In other organisms, definitive localization may not occur until after several cleavages. The development of most embryos depends on both autonomous self-differentiation of blastomeres that occurs as a result of localized morphogenetic determinants, and on interactions between neighboring blastomeres. The capacities for self-differentiation predominate in forms such as molluscs and ascidians early in development, and patterns of localization are quite fixed. In forms such as sea urchins, the initial patterns of localization are labile, as Horstadius demonstrated in his artificial blastomere recombination experiments. The regulative abilities of these eggs depend on cell-to-cell interactions. Hence, differences in the significance of blastomere interactions are a measure of the regulative ability of an embryonic system.

## References

Arnold, J. and L. Williams-Arnold. 1974. Cortical-nuclear interactions in cephalopod development: cytochalasin B effects on the informational pattern in the cell surface. J. Embryol. Exp. Morphol. 31:1–25.

Brachet, J. and E. Hubert, 1972. Studies on nucleocytoplasmic interactions during early amphibian development. I. Localized destruction of the egg cortex. J. Embryol. Exp. Morphol. 27:121–145.

Brandhorst, B. 1976. Two-dimensional gel patterns of protein synthesis before and after fertilization of sea urchin eggs. Dev. Biol. 52:310–317.

Brandhorst, B. and K. Newrock. 1981. Post-transcriptional regulation of protein synthesis in *Ilyanassa* embryos and isolated polar lobes. Dev. Biol. 83:250–254.

Briggs, R. and T. J. King. 1952. Transplantation of living nuclei from blastula cells into enucleated frogs' eggs. Proc. Natl. Acad. Sci. U.S.A. 38:455–463.

Cather, J. and N. Verdonk. 1974. The development of *Bithynia tentaculata* (Prosobranchia, Gastropoda) after removal of the polar lobe. J. Embryol. Exp. Morphol. 31:415–422.

Cather, J. and N. Verdonk. 1979. Development of *Dentalium* following removal of D-quadrant blastomeres at successive cleavage stages. Roux's Arch. Dev. Biol. 187:355–366.

Clement, A. C. 1976. Cell determination and organogenesis in molluscan development: a reappraisal based on deletion experiments in *Ilyanassa*. Am. Zool. 16:447–453.

Curtis, A. S. G. 1962. Morphogenetic interactions before gastrulation in the amphibian *Xenopus laevis*—the cortical field. J. Embryol. Exp. Morphol. 10:410–422.

Curtis, A. S. G. 1963. The cell cortex. Endeavor 22:134–137.

DiBerardino, M. A. and N. Hoffner. 1970. Origin of the chromosomal abnormalities in nuclear transplants, a revaluation of nuclear differentiation and nuclear equivalence in amphibians. Dev. Biol. 23:185–209.

Dohmen, M. and N. H. Verdonk. 1974. The structure of a morphogenetic cytoplasm present in the polar lobe of *Bithynia tentaculata* (Gastropoda, Prosobranchia). J. Embryol. Exp. Morphol. 31:423–433.

Driesch, H. 1892. The potency of the first two cleavage cells in echinoderm development. Experimental production of partial and double formations. In: Foundations of Experimental Embryology, pp. 38–50. B. H. Willier and J. M. Oppenheimer, eds. Englewood Cliffs, N.J.: Prentice-Hall.

Dohmen, M. and J. van der Mey. 1977. Local surface differentiations at the vegetal pole of the eggs of *Nassarius reticulatus, Buccinum undatum* and *Crepidula fornicata* (Gastropoda, Prosobranchia). Dev. Biol. 61:104–113.

Geilenkirchen, W., N. H. Verdonk, and L. Timmermans. 1970. Experimental studies on morphogenetic factors localized in the first and second polar lobe of *Dentalium* eggs. J. Embryol. Exp. Morphol. 23:237–243.

Gurdon, J. B. 1968. Transplanted nuclei and cell differentiation. Sci. Am. 219:24–35.

Gurdon, J. 1979. The cytoplasmic control of gene activity. Endeavor 25:95–99.

Horstadius, S. 1939. The mechanics of sea urchin development, studied by operative methods. Biol. Rev. 14:132–179.

Horstadius, S. 1952. Induction and inhibition of reduction gradients by the micromeres in the sea urchin egg. J. Exp. Zool. 120:421–436.

Horstadius, S., L. Josefsson, and J. Runnström. 1967. Morphogenetic agents from unfertilized eggs of the sea urchin, *Paracentrotus lividus*. Dev. Biol. 16:189–202.

Jackle, H. 1980. Possible control of tubulin synthesis during early development of an insect. J. Exp. Zool. 214:219–227.

Jeffrey, W. and D. Capco. 1978. Differential accumulation and localization of maternal poly(A)-containing RNA during early development of the ascidian, *Styela*. Dev. Biol. 67:152–166.

Josefsson, L. and S. Horstadius. 1969. Morphogenetic substances from sea urchin eggs. Isolation of animalizing and vegetalizing substances from unfertilized eggs of *Paracentrotus lividus*. Dev. Biol. 20:481–500.

King, T. and R. Briggs. 1956 Serial transplantation of embryonic nuclei. Cold Spring Harbor Symp. Quant. Biol. 21: 271–290.

Kobel, H., R. Brun, and M. Fischberg. 1973. Nuclear transplantation with melanophores, ciliated epidermal cells, and the established cell-line A-8 in *Xenopus laevis*. J. Embryol. Exp. Morphol. 29:539–547.

Malacinski, G. 1974. Biological properties of a presumptive morphogenetic determinant from the amphibian oocyte germinal vesicle nucleus. Cell Differ. 3:31–44.

Malacinski, G., H. Chung, and M. Asashima. 1980. The association of primary embryonic organizer activity with the future dorsal side of amphibian eggs and early embryos. Dev. Biol. 77:449–462.

Manes, M. and R. Elinson. 1980. Ultraviolet light inhibits grey crescent formation on the frog egg. Roux's Arch. Dev. Biol. 189:73–76.

Petzoldt, U., P. Hoppe, and K. Illmensee. 1980. Protein synthesis in enucleated fertilized and unfertilized mouse eggs. Roux's Arch. Dev. Biol. 189:215–219.

Raff, R. 1982. Maternal mRNA utilization in the control of protein synthesis by sea urchin embryos. Cell Differ. 11:305–307.

Raven, C. P. 1970. The cortical and subcortical cytoplasm of *Lymnaea* egg. Int. Rev. Cytol. 28:1–44.

Rosenthal, E., T. Hunt, and J. Ruderman. 1980. Selective translation of mRNA controls the patterns of protein synthesis during early development of the surf clam, *Spisula solidissima*. Cell 20:487–494.

Roux, W. 1888. Contributions to the developmental mechanics of the embryo. On the artificial production of half-embryos by destruction of one of the first two blastomeres, and the later development (postgeneration) of the missing half of the body. In: Foundations of Experimental Embryology, pp. 2–37. B. H. Willier and J. M. Oppenheimer, eds. Englewood Cliffs, N.J: Prentice-Hall.

Spemann, H. 1938. Embryonic Development and Induction. New Haven: Yale University Press.

Thompkins, R. and W. Rodman. 1971. The cortex of *Xenopus laevis* embryos: regional differences in composition and biological activity. Proc. Natl. Acad. Sci. U.S.A. 68:2921–2923.

Timmermans, L., W. Geilenkirchen, and N. H. Verdonk. 1970. Local accumulation of Feulgen-positive granules in the egg cortex of *Dentalium dentale L*. J. Embryol. Exp. Morphol. 23:245–252.

Verdonk, N. H., W. Geilenkirchen, and L. Timmermans.

1971. The localization of morphogenetic factors in uncleaved eggs of *Dentalium*. J. Embryol. Exp. Morphol. 25:57–63.

Whittaker, J. 1977. Segregation during cleavage of a factor determining endodermal alkaline phosphatase development in ascidian embryos. J. Exp. Zool. 202:139–154.

Whittaker, J. 1979. Cytoplasmic determinants of tissue differentiation in the ascidian egg. In: Determinants of Spatial Organization. The Thirty-Seventh Symposium of the Society for Developmental Biology. S. Subtelny and I. Konigsberg, eds. New York: Academic Press.

Whittaker, J. 1982. Muscle lineage cytoplasm can change the developmental expression in epidermal lineage cells of ascidian embryos. Dev. Biol. 93:463–470.

Whittaker, J., G. Ortolani, and N. Farinella-Ferruzza. 1977. Autonomy of acetylcholinesterase differentiation in muscle lineage cells of ascidian embryos. Dev. Biol. 55:196–200.

Wilson, E. 1904 Experimental studies in germinal localizaton. II. Experiments on the cleavage-mosaic in *Patella* and *Dentalium*. J Exp. Zool. *1:* 197–268.

# 10

# NEURULATION AND PRIMARY EMBRYONIC INDUCTION

Gastrulation is a developmental process that involves the movement of the primordia of presumptive internal structures from the surface to specific positions in the interior of the embryo. These morphogenetic movements and the subsequent events of neurulation are responsible for the first visible signs of a distinct embryonic axis and the organization of the basic body plan of the embryo. The expression of the ectoderm, mesoderm, and endoderm by the end of gastrulation, as well as the subsequent development of the neural tube and neural crest, constitute an early building plan of the embryo that requires dramatic changes in the physical and chemical properties of embryonic cells. The nuclear-transplantation experiments discussed in the previous chapter provide evidence that the same genetic material is present in all embryonic cells throughout the course of development. Hence, the initial changes in embryonic cells that lead to the formation of cell types such as ectoderm, mesoderm, and endoderm can be regarded as due to the differential expression of the same genome.

As pointed out in the previous chapter, results of various microsurgical and tissue culture techniques have shown that the fate of individual embryonic cells is controlled, in large part, by extrinsic factors. These factors include the changing fluid microenvironment of the cells and the intimate associations with adjacent or neighboring populations of cells. The interdependent relationship between adjacent populations of cells, which is the basis for the early differentiation of most vertebrate tissues and organs, is best illustrated in vivo by removing portions of the embryo. When this is done, remaining intact areas of the embryo may fail to differentiate properly or be abnormally expressed. In contrast to earlier developmental stages in which embryonic cells are remarkably adaptable, these same cells show more limited capacities of expression by the end of gastrulation. The production of different cell types is one of the immediate consequences of the ordered, morphogenetic movements of gastrulation.

## Neurulation and the Event of Neural Induction

One of the key interactions between cells of vertebrate gastrulas that has occupied the attention of embryologists for many years, perhaps longer than any other event in development, is that of neural induction. In this process, the chordamesoderm, with passage through the blastopore or primitive streak, acts on the overlying adjacent ectoderm to stimulate the latter to form a thickened plate of cells, the *neural plate*. Failure of the chordamesoderm to migrate properly into a position adjacent to the cells that respond to its influence leads to defects in the central nervous system. The dependence of one group of cells (ectoderm) on another group of cells (chordamesoderm) for the expression of the differentiated state (neural tissue) is an example of an *embryonic induction*. The cell group that is the source of the stimulation of influence is the *inductor* or *inducer*.

Neural induction is an epigenetic process that depends on the capacity of the chordamesoderm to induce the ectoderm coming in contact with it during the process of gastrulation. Since the early days of the classical school of German embryologists (Hans Spemann and his colleagues) and their discovery of the importance of the dorsal lip of the blastopore in the tissue differentiations of the early embryo, the succeeding decades have seen numerous efforts at establishing the basis of neural induction and the subsequent morphogenesis of the central nervous system. Although the event of neural induction has been for many years the most extensively studied of tissue interactions, its mechanism still eludes us. Presently, this area of research appears to be entering a new and exciting phase as contemporary ideas and techniques are being applied to solve this old problem. The event of neural induction may be used here to illustrate some of the basic problems associated with the study of tissue interactions in general. We shall in the following pages review the famous grafting studies of Spemann and give consideration to some of the approaches currently being employed to examine the mechanisms of neural induction. Before doing this, the events of neurulation itself should be examined in greater detail.

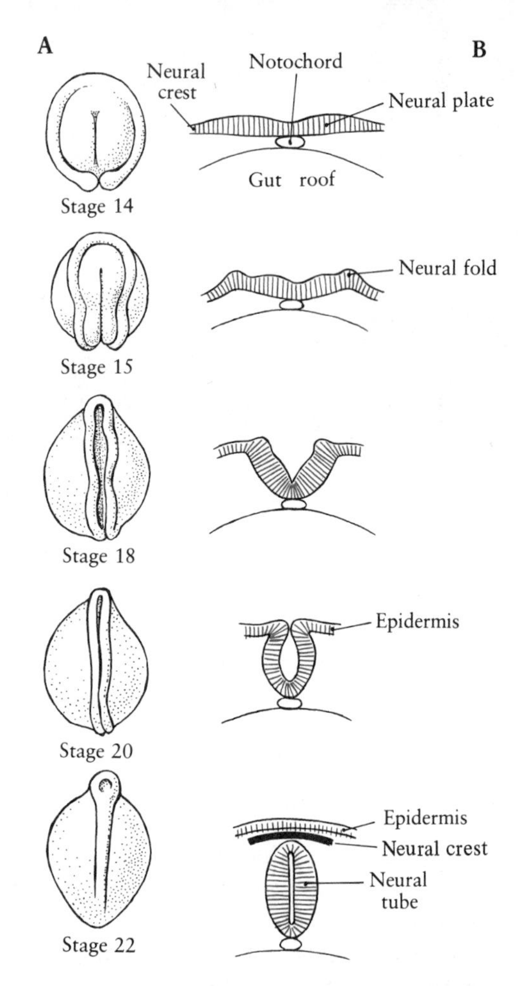


**FIG. 10-1.** A, the sequence of stages in the neurulation of an amphibian such as *Ambystoma;* B, diagrammatic cross sections of stages shown in A. (After B. Kallen, 1965. Organogenesis. R. H. DeHaan and H. Ursprung, eds. Holt, Rinehart and Winston, New York.)

## Neurulation, Microtubules, and Microfilaments

Neurulation is the process whereby the neural or medullary plate becomes transformed into the dorsal, hollow neural tube. During this time the primordia of brain, spinal cord, and neural crest become internalized and completely segregated from the rest of the surface ectoderm. In forms such as the frog and the salamander, the neural tube is completed throughout its length at much the same time (Fig. 10–1). By contrast, neural tube formation in amniote embryos, beginning in advance of the completion of the gastrulaton process, occurs by the sequential, anterior-to-posterior fusion of the neural folds in the wake of primitive streak regression (Fig. 10–2). Despite this visible difference among vertebrate embryos, the mechanics of the transformation of a flattened plate of neural cells into a tube appear to be essentially the same. Changes in the shape and size of neural plate cells apparently play an important role in the morphogenetic movements required in the conversion of plate to tube.

Of all the vertebrates, amphibian embryos have been the most extensively studied with respect to

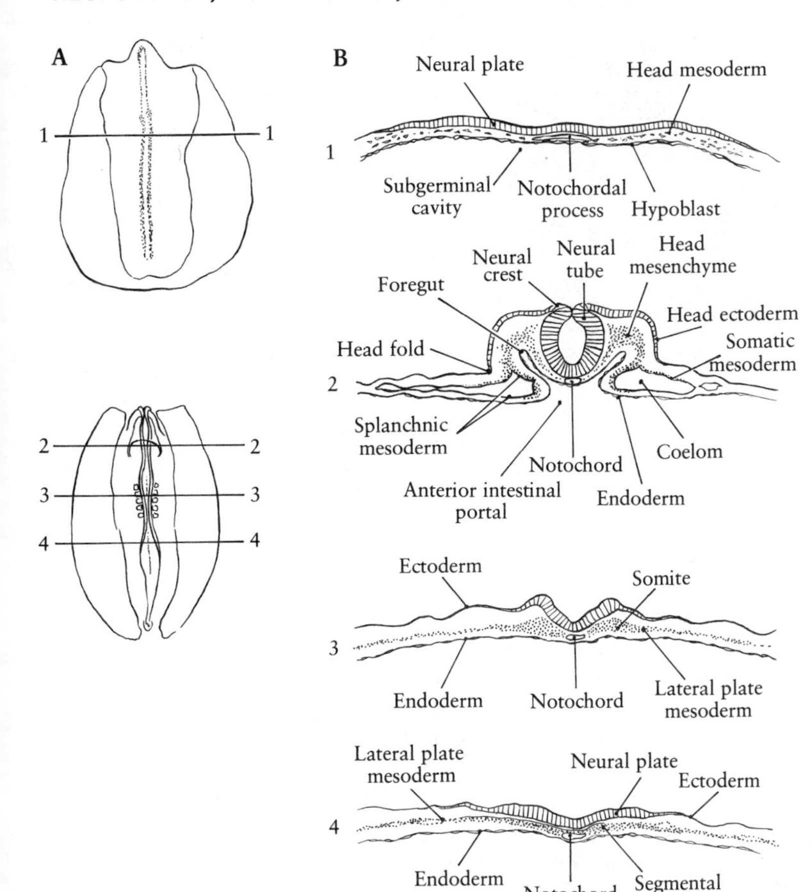


**FIG. 10–2.** A, the sequence of stages in neurulation of the chick embryo; B, diagrammatic cross sections of stages shown in A. (Reprinted with permission of Macmillan Publishing Company, from Atlas of Descriptive Embryology by W. W. Mathews. Copyright © 1972 by Willis W. Mathews.)

neurulation. As much as 50 percent of the outer ectodermal layer contributes to the formation of the neural plate in the amphibian. The overall shape and extent of the territory of the cells forming the plate are presumably related to the effective boundaries of the underlying inducing chordamesoderm. Initially, the neural plate appears in the dorsal hemisphere of the embryo as a flattened, compact zone of cells (Fig. 10–3 A). The plate then assumes the shape of a keyhole, being wider anteriorly where the future brain will develop (Fig. 10–3 B). Transverse sections through the early neural plate show that it consists of either a single layer *(Triturus)* or several layers *(Xenopus)* of cuboid-shaped cells (Fig. 10–3 C). The margins of the neural plate then become thickened and raised into the *neural folds*. The neural folds will then converge in the dorsal midline and fuse to form the *neural tube* (Figs. 10–1; 10–5). At the same time that the neural folds approximate each other, there is a general lengthening of the embryo along the anteroposterior axis.

The morphogenesis of neural plate and neural tube formation has been rather extensively studied experimentally in amphibians, particularly in salamanders, where the entire process concerns the transformation of a simple epithelium. During the formation of the neural plate, the cells of the neural epithelium become elongated but do not appear to change their volume. Consequently, the surface area of the embryo occupied by the neural cells during this shape change decreases and therefore results in the concentration of the neural ectoderm along the dorsal midline. The elongation and stretching of the neural plate cells appear to be intrinsic properties. Holtfreter, for example, has shown that isolated neural plate cells retain their columnar shape and even continue to elongate when cultured. If plate cells are isolated prior to elongation, the explant undergoes shape changes

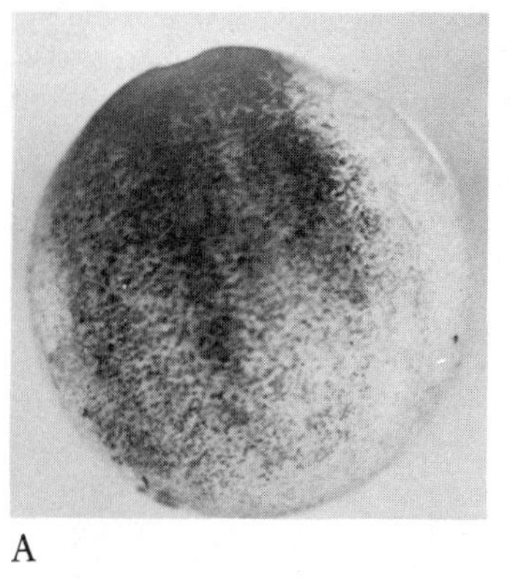


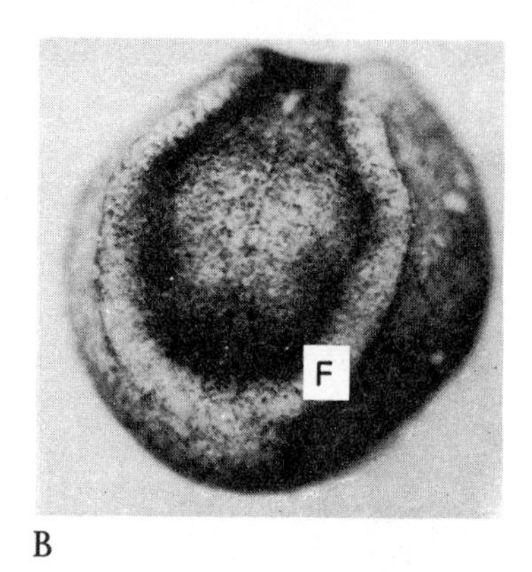


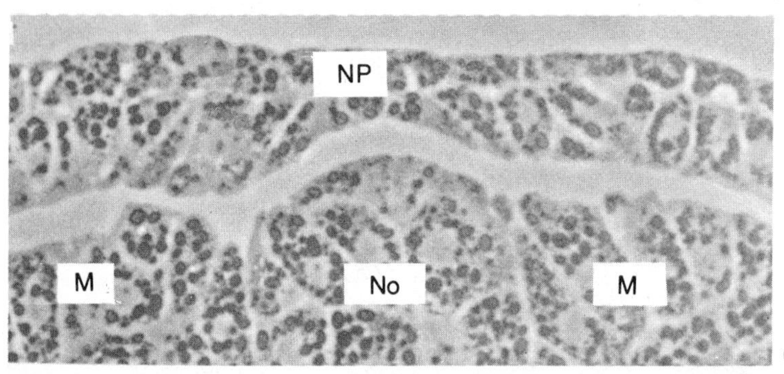


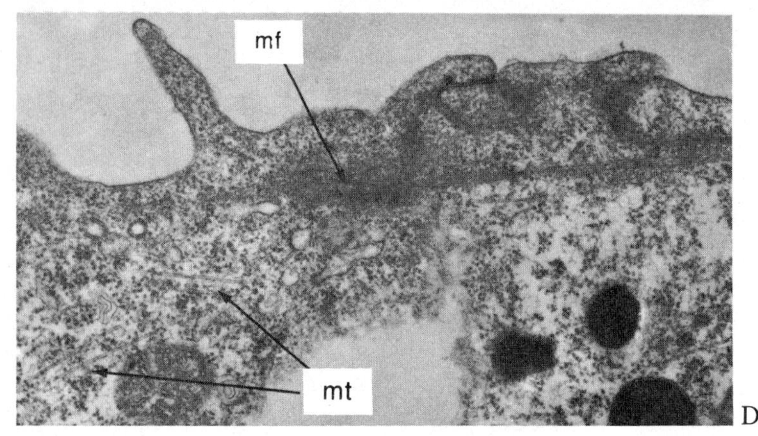


**FIG. 10–3.** A, light micrograph showing the early neural plate of *Ambystoma* (axolotl); B, light micrograph showing the beginning of the formation of the neural folds in *Ambystoma;* C, cross section through the early neural plate of *Xenopus;* D, cross section through the neural plate of *Xenopus* showing subapical microfilaments (mf) and microtubules (mt) in the medially situated cells. F, neural folds; M, muscle; No, notochord; NP, neural plate. (A and B, Courtesy of R. Brun; C and D, from P. Karfunkel, 1971. Dev. Biol. 25:30.)

characteristic of those observed in the intact embryo. In addition to form-building because of forces within the neural plate itself, there is also evidence that neurulation is dependent on movements of the underlying chordamesoderm. The isolated neural plate in culture never develops the keyhole shape unless explanted with notochordal tissue. The neural plate in the intact embryo shows a tight adhesion to the median notochordal strip of mesoderm. Vigorous stretching movements of the notochordal cells along the anteroposterior axis are thought to alter the shape of the overlying plate and produce its characteristic keeled appearance. Whether these same movements play any role in converting the neural plate into a tube remains unclear.

Although the neural plate in *Xenopus* is multilayered, the process of neurulation is very similar to that observed in salamanders. Dramatic changes in the shapes of cells occur at all levels of the plate (Figs. 10–4; 10–6). The medially situated cells of the upper layer become taller and wedge-shaped, their narrowed ends abutting upon the neurocoele. The cells of the lower layer become more columnar

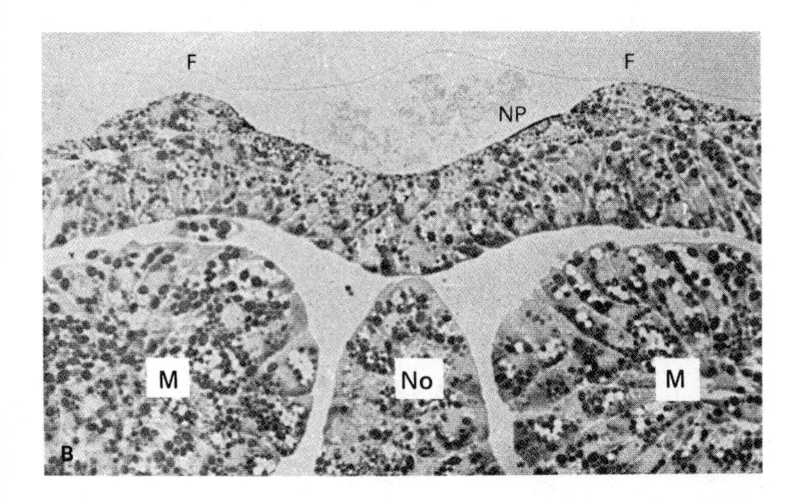


**FIG. 10–4.** Cross section through the early neural plate (NP) stage of *Xenopus*. F, neural fold; M, muscle; No, notochord. (From P. Karfunkel, 1971. Dev. Biol. 25:30.)

in shape with their long axes oriented in a dorsoventral plane. Upon approximation and fusion of the neural folds in the middorsal line (Fig. 10–5), the long axes of the deep cells are always radially positioned with respect to the lumen of the neural tube. The changes in the orientation and shape of the neural cells during neurulation in *Xenopus* are diagrammatically summarized in Figure 10–6.

How does one explain this remarkable and dramatic transformation of a sheet of cells into a tube? Historically, it was initially believed that regional differences in mitotic rate shaped the plate and early tube. The process of neurulation in amphibians appears to be brought about by distinct changes in cell shape. Electron-microscopic observations show that elongation of the neural cells to form the columnar epithelium characteristic of the neural plate is correlated with changes in the microtubular cytoskeleton (Figs. 10–3 D; 10–6). Schroeder, Karfunkel, and others have noted that the cells of the early neural plate contain numerous microtubules, approximately 240 Å in diameter and randomly oriented in the cytoplasm. Microtubules of similar size and orientation are present in cells of the adjacent presumptive epidermis. Gradually, the microtubules become arranged parallel to the axes of elongating cells; in *Xenopus*, this change occurs primarily in the medial superficial cells of the neural plate (Fig. 10–6). Following elongation to form the neural plate, the cells of the neural epithelium become constricted at their apical ends; the cells become increasingly flask-shaped and convert the plate into a curved tube. This change in cell shape is brought about by microfilaments. Microfilaments initially appear in the upper cells of the neural plate *(Xenopus)* as a dense band just below the exposed or free outer surface of each cell (Fig. 10–6). As the neural plate becomes tubelike, the microfilaments are observed as thickened bundles encircling the apex of each cell. Since microfilaments of neighboring cells are joined to the same desmosome, there is an extensive, interconnected microfilamentous apparatus throughout the neural epithelium.

You will recall that microtubules and microfilaments have been previously associated with morphogenetic events involving changes in cell shape and cell movement (i.e., blastopore and primitive streak formation; cytokinesis of cleaving blastomeres). Although both microtubules and microfilaments can be correlated with changes in cell shape during amphibian neurulation, it is still not clear how these cytoskeletal elements generate the tensile and bending forces underlying cell shape changes. It has been proposed that the microfilaments, initially arranged in circumferential bundles parallel to the neural plate cell surfaces, act by virtue of contracting, thereby narrowing the apical ends of the simple neural epithelium in the salamander or the median superficial layer of cells in *Xenopus*. The consequence of this contractile activity is a progressive change in the shape of each cell (i.e., from cuboid, to wedge, to flask), the result of which is an imposition of curvature on the previously flattened neural plate. Since the basal or lower surfaces of the cells are less constricted than the apical surfaces, the effect is a rolling upward and inward of the margins of the neural plate. This rolling of the neural plate into the neural tube has been likened to drawing tight the strings of a purse.

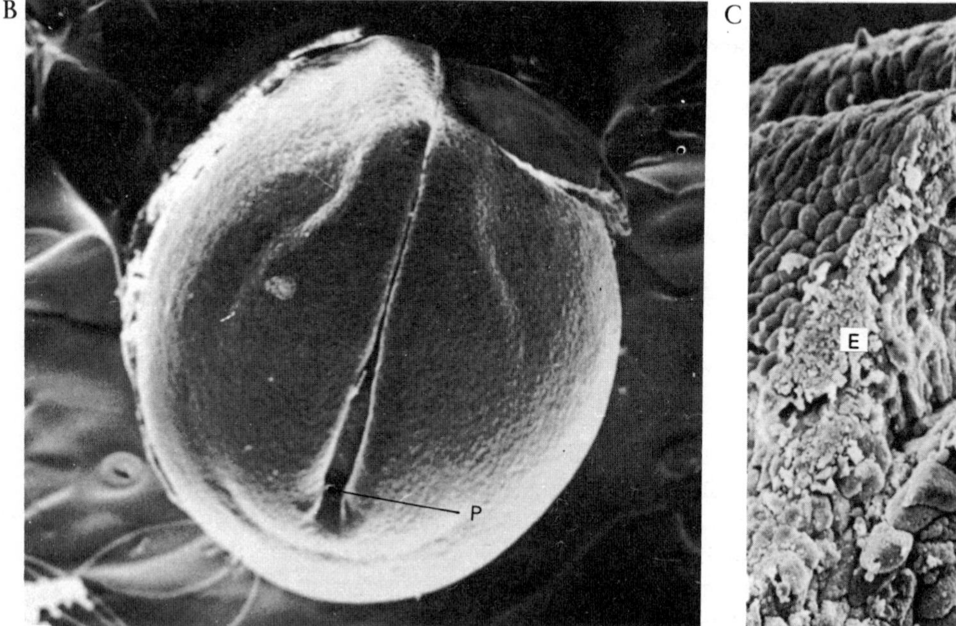


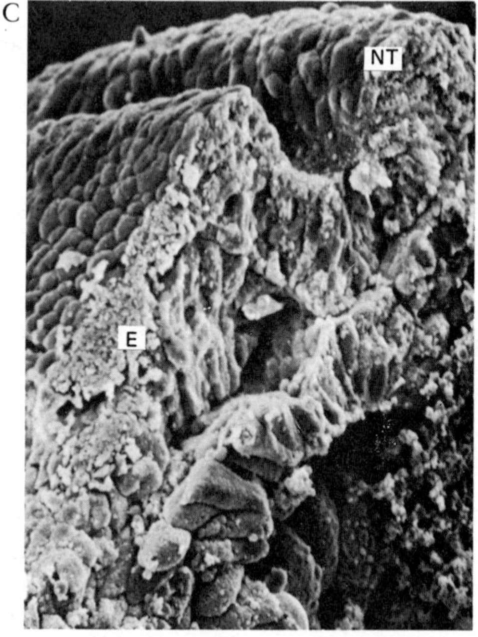


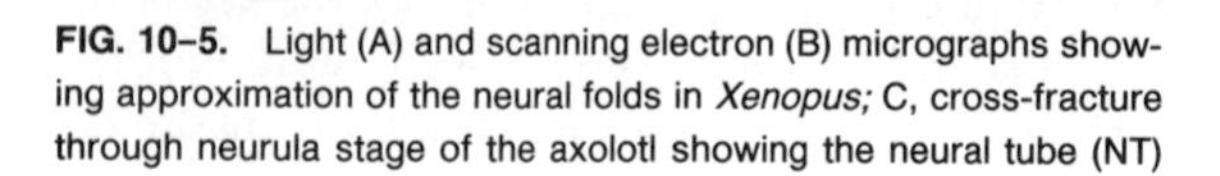

**FIG. 10–5.** Light (A) and scanning electron (B) micrographs showing approximation of the neural folds in *Xenopus;* C, cross-fracture through neurula stage of the axolotl showing the neural tube (NT) being formed. E, epidermis; P, anterior neuropore. (A and C, Courtesy of R. Brun; B, From D. Tarin, 1972. J. Anat. 111:1.)

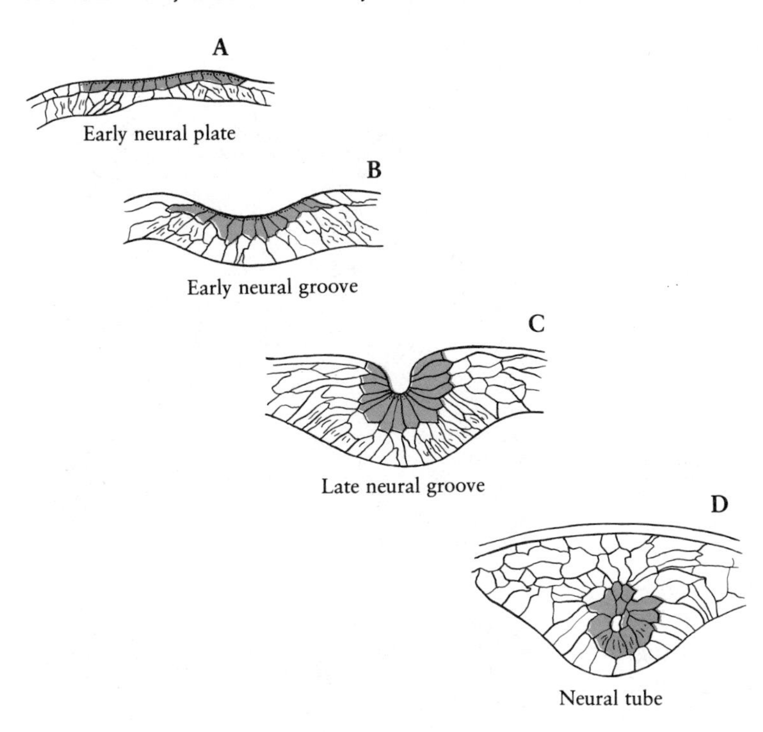


**FIG. 10–6.** Tracings of shapes of neural plate cells during neurulation in *Xenopus*. Note that only the superficial, medial cells form the boundary of the neurocoele. The dots in the cells represent microfilaments. Straight lines inside cells indicate microtubules. Note their orientation during tube formation. (After P. Karfunkel, 1971. Dev. Biol. 25:30.)

The microtubules appear to be responsible for the first step in neurulation or the elongation of ectodermal cells to form the flat neural plate. Although most investigators feel that microtubules play an active role in the elongation of ectodermal cells, there is uncertainty as to how this is accomplished. The directional addition and polymerization of tubulin subunits onto the ends of microtubules could cause elongation of a cell by pushing out on its apical and basal ends. A similar effect could be produced by the sliding of microtubules past one another. Another possibility is that microtubules might bring about elongation by the active transport of cytoplasm toward the basal end of the cell. Once the shape of the cell has been altered, the extension could be stabilized by the addition of new subunits to the microtubules.

That changes in cell shape are critical to the neurulation process has been shown by treating amphibian embryos with various agents or drugs whose actions are known to interfere with the assembly or maintenance of microfilaments and microtubules. For example, if frog embryos are exposed at the early neural plate stage (i.e., prior to cell elongation) to vinblastine, a compound that precipitates microtubular protein and thus destroys the integrity of microtubules, wedge-shaped and flask-shaped cells never appear, and neural folds fail to form. If the cells are treated after elongation, they fail to maintain their shape, thus offering direct experimental proof that microtubules are responsible for the elongation of cells of the neural plate. Vinblastine also disrupts microfilaments. It not only prevents the apical ends of cells of the neural plate from contracting, but it also causes them to lose their constricted state if treatment is initiated after tube formation. The drug cytochalasin B, a fungal metabolite, also interferes with microfilament formation. Treatment of amphibian embryos during neurulation with this drug causes an opening up and subsequent flattening of the neural tube.

A troublesome question in studies of amphibian neurulation is whether the shape changes of the neural plate can autonomously form the neural tube without interactions with adjacent tissues. That is, do parts of the embryo other than the neural plate actively contribute to neurulation? Using a computer model of neurulation, Odell and his colleauges found that a wave of apically constricting neural cells is sufficient to transform the neural plate into a neural tube. However, results of other studies suggest that this may not be the case. Movements of neurulation continue in the embryo even if the neural plate is removed. This phenomenon has usually been interpreted as being due to wound healing and not related to neural tube formation. However, when embryos of *Ambystoma* are exposed to either *colchicine* or *nocodazole* at the neural plate stage, they show formation and apposition of the neural folds. Yet these same drugs on close examinaton eliminate the formation of wedge-shaped cells in the plate. Results of these various studies suggest that the epidermis produces a force by spreading that, in addition to the forces generated by apical contraction, is important to neural fold and tube formation (Fig. 10–7).

With the exception of teleost fishes, the process of the formation of the neural tube is quite the same in all vertebrates. In the chick, the epithelium of the neural plate consists of a single layer of very tall columnar cells. During the transformation of the neural plate into the neural tube (Fig. 10–8), a deep *neural groove* forms between a pair of prominent neural folds. The edges of the neural folds gradually approach one another and fuse. During this event, the cells lying along the neural groove become elongate and wedge-shaped, their apices appearing narrower and more convex. At the ultrastructural level, bundles of microfilaments are very prominent in the subapical cytoplasm of these cells. As in the case of amphibian embryos, bundles of microtubules are oriented parallel to the long axes of cells during neural plate formation. The changes in the shape of cells as the neural tube forms are summarized for the chick in Figure 10–9.

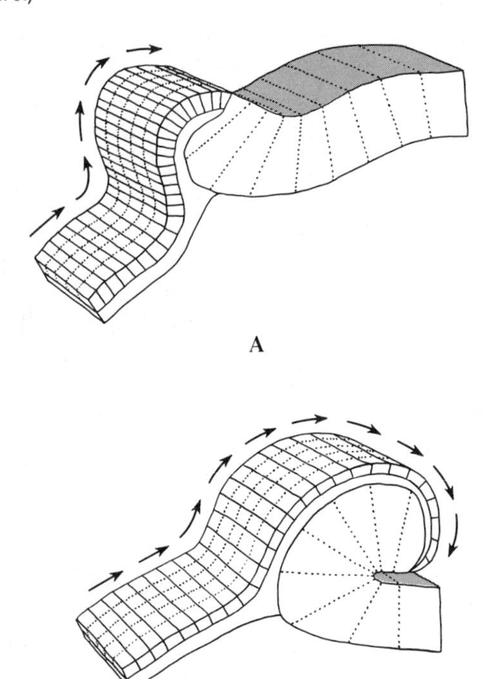


**FIG. 10–7.** Neurulation in *Ambystoma*. A, at the beginning of neural fold formation, the cells of the superficial epidermis are columnar; B, as the neural folds are elevated, the epidermal cells become more flattened. It is sugeested that this cell shape change in the epidermis produces a spreading force that is important in neural tube formation. (From R. Brun and J. Garson, 1983. J. Embryol. Exp. Morphol. 74:275.)

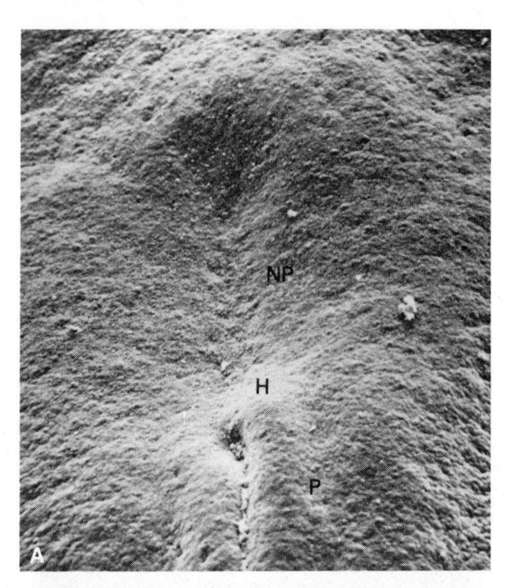


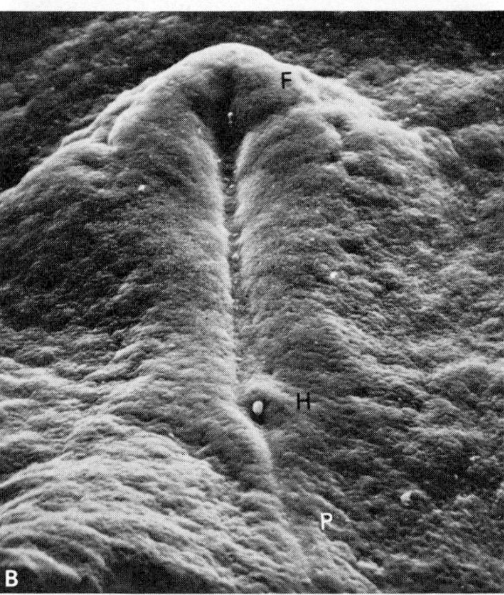


**FIG. 10–8.** A survey picture showing the late neural plate (NP) stage of the chick; B, formation of the neural folds in the chick. F, neural folds; H, Hensen's node; P, lateral folds of primitive streak. (From P. Portch and A. Barson, 1974. J. Anat. 117:341.)

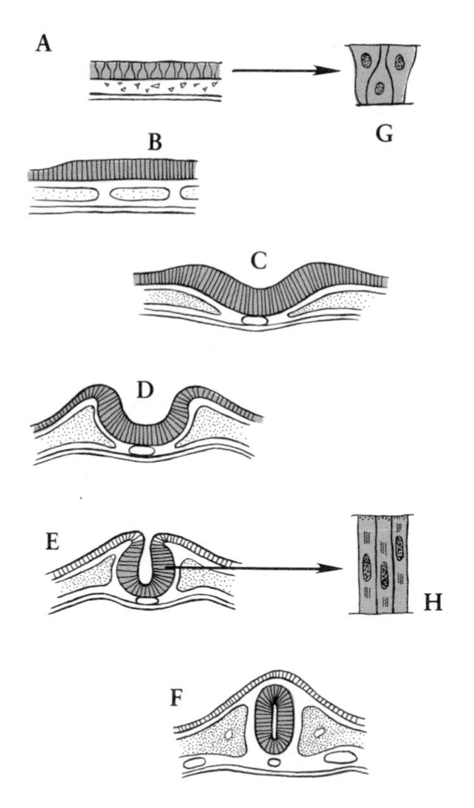


FIG. 10–9. Diagrammatic drawings of the cells of the chick neural plate, neural folds, and neural tube during neurulation. A, the preplate blastoderm; B, the neural plate; C, early neural fold formation; D, the early neural groove; E, the late neural groove; F, the neural tube; G, the appearance of individual cells of the preplate blastoderm; H, the cells of the neural groove are elongate with microtubules arranged parallel to the cells' long axes and microfilaments (dots) at the apices of cells. (After P. Karfunkel, 1972. J. Exp. Zool. 181:289.)

The active role of both microtubules (cell elongation) and microfilaments (cell constriction) during neurulation is supported by experimental manipulations with chick neurulas. When chick embryos at the neural fold stage are cultured with either colchicine or colcemid, the cells of the neural folds round up, and subsequently the neural folds fall back toward the yolk. These results suggest that microtubules are at least essential to the stabilization of the elongated cell shape. Similarly, if cells of the neural plate are treated with cytochalasin B before becoming apically constricted, these same cells fail to contract and the neural folds fail to elevate.

It may be concluded, at least in amphibian and chick embryos, that the curvature of the neural plate and the elevation and approximation of the neural folds during neurulation is a two-step process dependent on the presence of microtubules and microfilaments (Fig. 10–10). First, the ectodermal cells undergo elongation to form a neural plate with an active role played by microtubules. Second, constriction of the neural plate cells at their apical ends is microfilament mediated, resulting in the initial invagination of the neural plate and the subsequent formation of the neural tube. Fusion of the neural folds and their separation from the adjacent cells of the ectoderm produces the neural tube. Whether the correlation between the presence of the cytoskeletal structures and the transformation of the neural epithelial cells into a tube applies to all vertebrate embryos remains to be established.

The induction of the neural plate in the chick embryo appears to involve the de novo formation of microtubules and microfilaments, because neither of these are present prior to neural plate formation. It has not been determined whether the formation of these cytoskeletal structures is due to

FIG. 10–10. Neurulation is a two-step process dependent on the transformation of a plate of cells (A) into elongate neural cells by microtubules (B) and their apical constriction by microfilaments to convert into a tube (C).

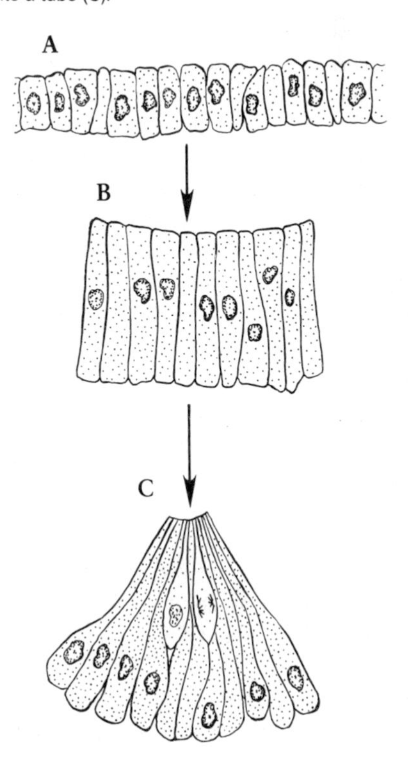

the polymerization of preformed subunits or the synthesis and polymerization of new subunits.

## Determination and Spemann's Primary Organizer

The techniques of transplantation or tissue grafting and culturing of embryos have been successfully employed in examining several of the fundamental questions relating to the significance of cellular rearrangements and the differentiation of tissue and organ primordia in the early embryo. Small pieces of tissue may be surgically removed from preselected sites of embryos at various stages of development and inserted into prepared cuts or wounds on the same or different embryos. If the graft is placed on the same embryo from which it is excised, the transplant is referred to as being *autoplastic*. Grafts can as easily be made between individuals of the same species *(homoplastic transplant)* or between different species of the same genus *(heteroplatic transplant)*. The embryo providing the graft is the *donor;* the embryo receiving the graft is the *host*.

It is important to know in a transplantation experiment which tissues of the embryo are derived from the graft and which tissues are from the host. An induction, for example, can be said to occur under the influence of a graft if host cells participate in the formation of structures not normally formed by them. A distinction is readily made if donor and host cells are sufficiently different from each other. Heteroplastic transplants are often used because differences in cell size, cell-staining properties, or cell inclusions (i.e., pigment granules) commonly exist between donor and host cells, thus facilitating interpretation of transplant experiments.

In the early 1920s Spemann investigated the question of when specific areas of the vertebrate embryo become firmly committed to specific developmental pathways. Using embryos of the salamander *(Triturus taeniatus)*, he placed a small piece of presumptive ectoderm from an early-gastrula stage into the presumptive neural plate area of another embryo of the same stage of development (Fig. 10–11 A). The result was that the graft developed in accordance with the cells around it and gave rise to the neural plate cells of the host (Fig. 10–11 A). The reciprocal transplant, or graft of presumptive

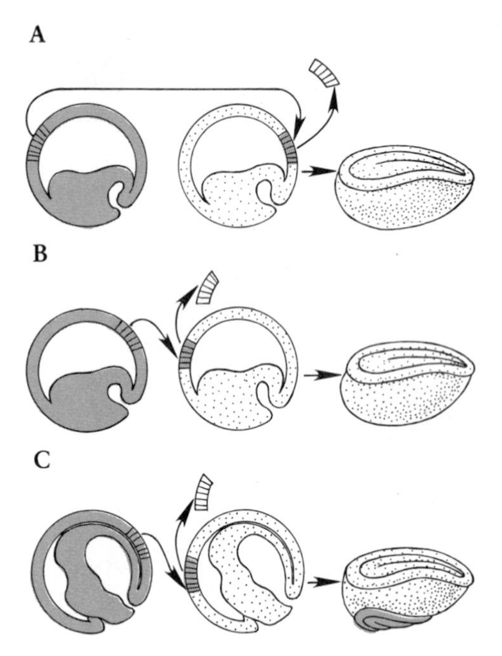


**FIG. 10–11.** A transplant of presumptive ectoderm (epidermis) from early newt gastrula placed into presumptive neural plate of another early newt gastrula will develop in accordance with its new surroundings (i.e., neural plate); B, transplant of presumptive neural plate from early newt gastrula placed into presumptive ectoderm (epidermis) of another early newt gastrula will develop in accordance with its new surroundings (i.e., epidermis); C, the transplant of neural plate from late newt gastrula placed into presumptive ectoderm (epidermis) of early newt gastrula develops in its new surroundings in accordance with its expected fate (i.e., neural plate).

neural plate cells placed into the anteroventral ectodermal (epidermal) area (Fig. 10–11 B), also developed in conformity with the surrounding cells and contributed to the integumentary cells of the host (Fig. 10–11 B). If ectodermal or presumptive neural plate tissue was grafted to the marginal or chordamesodermal area of the early gastrula, Spemann observed that the transplant moved through the blastopore with the morphogenetic movements of the host embryo and differentiated in accordance with its new mesodermal environment, giving rise to such mesodermal derivatives as notochord, somites, and kidney tubules. Heteroplastic transplants between *Triturus cristatus* and *Triturus taeniatus* embryos at the early-gastrula stage have yielded similar results (Fig. 10–12).

From reciprocal transplants between ectoderm and neural plate areas, as well as those involving the grafting of chordamesoderm and endoderm to other embryonic areas, it can be concluded that the

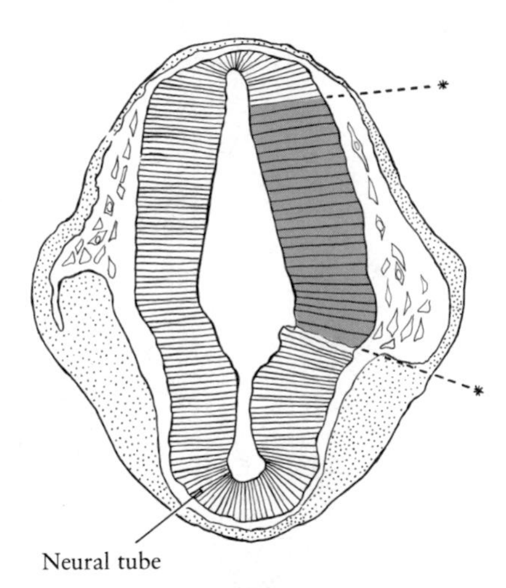


**FIG. 10–12.** A cross section through the anterior end of a *Triturus taeniatus* embryo formed from a graft with presumptive epidermis from *T. cristatus* placed into the neural plate region. The region of the forebrain (shown between the asterisks) of *T. taeniatus* has developed from *T. cristatus* presumptive epidermis. (From H. Spemann, 1938. Embryonic Development and Induction. Yale University Press, New Haven.)

cells of the early gastrula possess the ability to differentiate into a variety of cell types. The presumptive neural cell can develop into a skin or somite cell when placed in the appropriate environment. Presumptive somite cells can differentiate into skin cells. We refer to this lability or capacity to differentiate in several directions of an early embryonic cell as being its *prospective potency*. The normal or expected fate of a cell is its *prospective significance*.

Suppose reciprocal transplants between epidermal ectoderm and presumptive neural plate areas are executed using late-stage gastrulas as tissue donors. The results are very different when compared to those with early-stage gastrulas as tissue donors. In all cases, the grafts from late gastrulas differentiate in accordance with the presumptive areas from which they were excised before transplantation. For example, a transplanted piece of neural plate when placed into a cut in the skin ectoderm of an early gastrula will sink beneath the surface and form recognizable neural structures (Fig. 10–11 C). A graft of epidermal tissue when placed into the presumptive neural area will differentiate as epidermis only. These results serve to establish that the neural plate and the epidermis have lost their abilities to develop into any derivatives other than neural and integumentary structures, respectively.

Hence, by the end of gastrulation, the potencies of presumptive epidermal and neural cells become restricted to their expected fate or prospective significance. This restriction of cell lability or channeling of developmental potential is known as the process of *determination*. We speak of the nervous system or the epidermis as being determined by the end of gastrulation. Similarly, chordamesodermal and endodermal cells also become broadly determined or limited with respect to cell type. The expression of further change within these areas (i. e., the determination of individual structures and parts within the ectoderm, neural plate, chordamesoderm, and endoderm) is dependent on subsequent and successive determinative steps.

The early grafting experiments of Spemann provided strong evidence that determination of the epidermal and neural tissues was related to the rearrangement of cells brought about by the process of gastrulation. Hilde Mangold, a student of Spemann's, carried out a variety of heteroplastic transplant experiments in an effort to define the exact times of determination of the different parts of the gastrula. Her efforts established that the chordamesoderm (mainly notochord and paraxial or somitic mesoderm), which shifts forward beneath the ectoderm with the morphogenetic movements of gastrulation, plays a decisive role in the determination of the epidermal and neural areas. She excised a piece of the dorsal lip of the blastopore from an early gastrula of *Triturus cristatus* and placed it onto the lateral lip of the blastopore of an early gastrula of *Triturus taeniatus* (Fig. 10–13 A). The transplanted tissue was subsequently incorporated into the interior of the host embryo. When the host embryo developed further, there was found a nearly complete second set of organs, giving to the host the appearance of conjoint twins (Fig. 10–13 B, C). Sections through the host showed that the secondary embryo lacked the anterior part of the head. There was, however, a neural tube on either side of which were somites and kidney tubules, A notochord and gut tube were identified beneath the neural tube (Fig. 10–13 D). All of these structures appeared to be arranged into a unified whole. Working from differences in pigmentation of cells between the two species in the heteroplastic transplant (i.e., *T. taeniatus* cells contain dark pigment granules), Mangold determined that most of the neural tube, part of the somites, the kidney tubules, and the inner ear primordia of the secondary embryo were derived from host tissues. The notochord

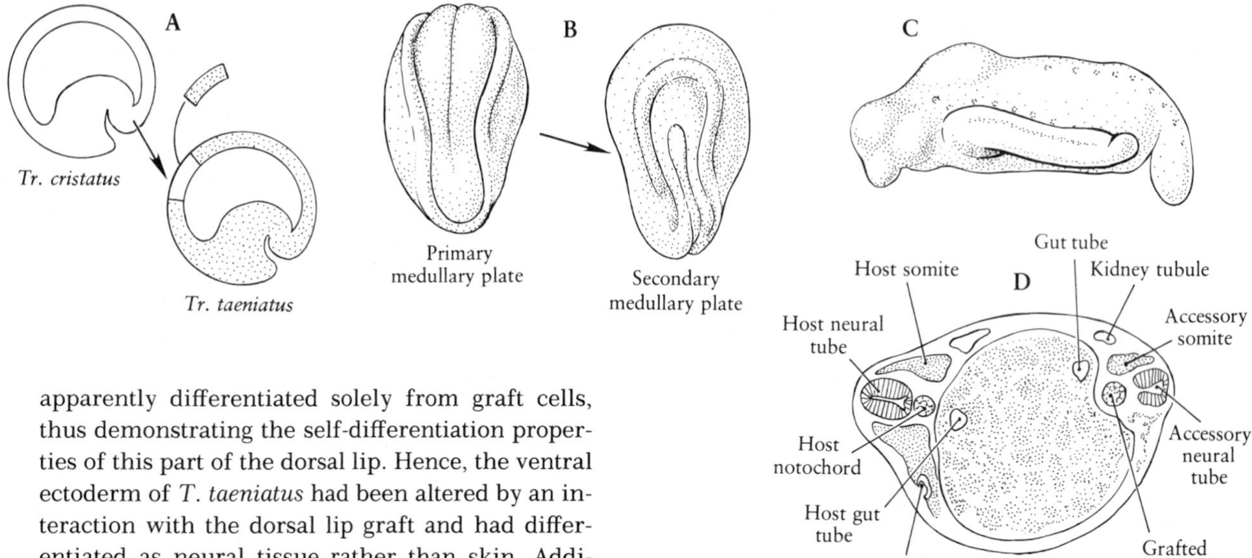


**FIG. 10–13.** A, diagram of Mangold's experiment in which dorsal lip of the early gastrula of *Triturus cristatus* was placed onto the lateral aspect of an early gastrulating embryo of *T. taeniatus;* B, the host embryo is induced by the dorsal lip transplant to form a secondary neural plate; C, subsequently, the host embryo shows a secondary embryo on its belly which includes ear vesicle, neural tube, and somites; D, transverse section through *Triturus* embryo showing structures of host and secondary embryo. (From H. Spemann, 1938. Embryonic Development and Induction. Yale University Press, New Haven.)

apparently differentiated solely from graft cells, thus demonstrating the self-differentiation properties of this part of the dorsal lip. Hence, the ventral ectoderm of *T. taeniatus* had been altered by an interaction with the dorsal lip graft and had differentiated as neural tissue rather than skin. Additional evidence for the role of the dorsal lip in neural differentiation comes from studies in which early amphibian gastrulae have been exposed to hypertonic media. Under these conditions, the mesoderm and endoderm surprisingly become displaced and migrate away from rather than into the interior of the embryo. This unusual morphogenetic process is referred to as *exogastrulation*. The ectoderm in the absence of a close association with the chordamesoderm shows no evidence of neural development.

Because of the ability of the dorsal lip of the blastopore (or the roof of the primitive archenteron) to stimulate from host cells the development of a nearly whole secondary embryo after transplantation, Spemann termed the dorsal lip of the blastopore the *primary organizer* (or simply organizer). Under the influence of the organizer, a secondary embryo, possessing the essential features of axial organization, can be induced to form from tissues normally having distinctly different prospective fates. For example, the transplanted dorsal lip induced presumptive epidermis of the host to form a neural plate. It also influences the mesoderm and endoderm, stimulating the formation of structures from host tissues as a result of inductions. The results of these interspecific transplantation experiments led Spemann to assign two properties to the dorsal lip. First, the dorsal lip can induce neural differentiation in the overlying ectoderm, tissue whose normal prospective fate is that of epidermis. Second, the dorsal lip self-differentiates as the notochord, which serves as a center around which the major embryonic axis with associated structures is organized. The process that leads to the early segregation of the central nervous system from the rest of the ectoderm is often referred to as *primary induction,* a phrase reflecting that this is the first apparent cellular interaction during vertebrate development.

The organizer or primary embryonic inductor appears to be traceable to that region of the fertilized egg previously identified as the grey crescent. Since the early studies of Spemann, the possible causal relationship between the grey crescent cytoplasm and the primary embryonic organizer has been the subject of considerable experimentation. Complete extirpation of the grey crescent at the time of fertilization results in an embryo lacking neural and mesodermal structures. Destruction of the grey crescent by controlled exposure to ultraviolet irradiation produces results similar to those by extirpation. As prevously pointed out, transplantation of the cortex of the grey crescent onto the ventral side of another embryo yields a second embryo with a second set of axial organs (Fig. 9–18). We are able to conclude from these experiments that the phenomenon of primary determination or morphogenetic potential for determining the pattern of axial organization appears to be related to this early cytoplasmic localization of some substance(s) in the grey crescent. Whether it is built into the grey crescent cortex of the zygote, however, is controversial in light of findings by Malacinski and his colleagues. Using organizer implants from various stages (zygote to gastrulation) of *Xenopus* and *Ambystoma* embryos, they concluded that cortical implants of the zygote grey crescent were unable to induce a secondary set of axial structures. Rather, cells on the future dorsal side of the embryo appear to acquire gradually the ability to induce axial morphogenesis. It is clear from differences in the results of various grafting experiments that continued investigation of the association between primary organizer activity and the egg cortex is in order.

In the amphibian blastula and early gastrula, the region that has inducing capabilities appears to coincide spatially with the region of cells that becomes invaginated and internalized during gastrulation. This is a rather large region, and it might be expected that different regions of the organizer induce different structures along the axis of the embryo (i.e., brain and spinal cord). Spemann, in the late 1920s, demonstrated that the inductor influence of the blastoporal lip of an early gastrula differed from that of the same tissue of an advanced gastrula.

Recall that the portion of the organizer that is first to become invaginated at the rim of the dorsal lip of the blastopore will reach furthest forward, coming to lie beneath the anterior end of the presumptive neural plate. The posterior region of the organizer is temporally last to turn in at the dorsal lip of the blastopore. It comes to lie beneath the presumptive spinal cord of the embryo. These temporal relationships between different regions of the organizer or chordamesoderm and the overlying ectoderm are summarized in Figure 10–14. By excising pieces of the dorsal lip at different times during gastrulation and transplanting them into cuts prepared on the surface of early gastrulas, it is possible to determine if regional or topographical differences exist in the organizer with respect to inductive capacities. When the dorsal lip of the early gastrula is transplanted, it acts primarily as a *head inductor*. It induces in the host embryo only head structures such as the formation of brain, eyes, and nose rudiments (Fig. 10–15 A). If the dorsal lip of the late gastrula is used as a donor graft, only trunk

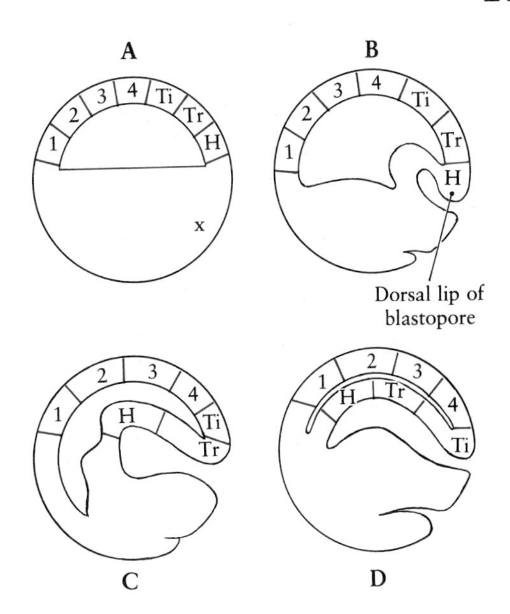


**FIG. 10–14.** Diagram to show in longitudinal section the relative positions of the inductors within the chordamesoderm during different stages of amphibian development. A, late blastula; B, early gastrula; C, midgastrula; D, late gastrula. 1–4, different regions of the ectoderm; H, head inductor; Tr, trunk inductor; Ti, tail inductor; x, future position of the dorsal lip of the blastopore. By transplanting the dorsal lip of the blastopore at different times during gastrulation, the regional specificity of the chordamesoderm can be demonstrated.

**FIG. 10–15.** A, if a transplant is made of dorsal lip at the early-gastrula stage *(Triturus),* the host develops a second head; B, if the transplant of the dorsal lip is taken from a late gastrula, the host embryo develops an accessory tail. (After H. Mangold, 1932. Naturwissenschaften 20:371.)

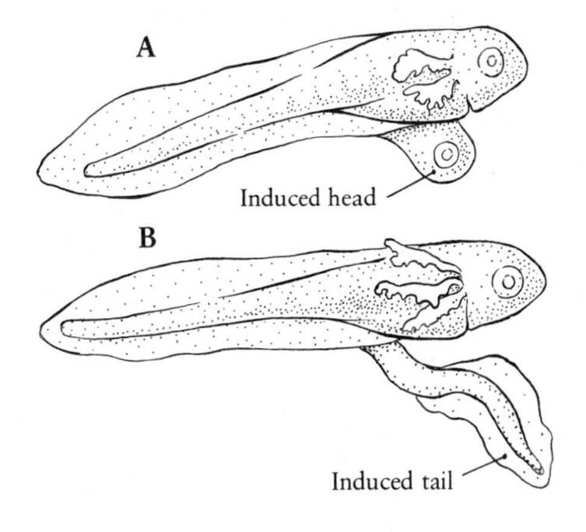

and tail structures are induced (Fig. 10–15 B). The graft acts as a *spinocaudal inductor*. Further experimental analysis of the head inductor indicates that it can be divided into an *archencephalic inductor* (stimulating formation of forebrain, optic cups, and nose primordia) and a *deuterencephalic inductor* (stimulating formation of hindbrain and inner ear vesicles).

The inducing capacity of the blastopore thus changes during gastrulation as the chordamesoderm invaginates and moves to the inside of the embryo. Consequently, the organizer is viewed as having regional specificity; that is, it has regions that induce only certain parts or structures of the embryo. The presence of inductor tissue, however, does not assure that an induction will take place in host tissues. Several studies have demonstrated that the reacting cells, or those cells being induced, must be in a particular physiological state to differentiate under the influence of the inductor. This state of receptiveness to determinative stimuli is referred to as *competence*. The nature of this physiological state is poorly understood. However, competence in any inductive interaction is always required for a morphologically specific response to a particular stimulus. Under the influence of the chordamesoderm, the differentiation of the neural tissues is an expression of the *primary competence* of the ectoderm. The ectoderm gradually loses its capacity to react to the inductive influence of the chordamesoderm between the gastrula and neurula stages of development. This can be demonstrated by transplanting the dorsal lip beneath the flank ectoderm of successively later stages of amphibian embryos. The decrease in neural competence with aging of the ectoderm has also been tested by isolating fragments of gastrula ectoderm, maintaining them for given intervals of time in culture, and then exposing each fragment to a strong inductor. Ectoderm that is not exposed to a chordamesodermal influence, or ectoderm exposed to the inductor at a very late stage of gastrulation, will differentiate as epidermis.

The loss of neural competence by the ectoderm with time does not prevent responsiveness by this germ layer to new and different inductive stimuli. Although the postgastrular ectoderm cannot be stimulated to form neural tissue, it does become competent to respond to other inductors.

The appearance after gastrulation of a new responsiveness to other inductors is known as *secondary competence*. For example, the primary optic vesicle of the forebrain and the hindbrain induce the overlying ectoderm to form the eye lenses and the inner ear vesicles, respectively. Hence, the gradual and progressive elaborations of embryonic structures in all germ layers would appear to require successive states of competence along with a succession of inductors.

The inductive processes observed by Spemann and others in the embryogenesis of the amphibian appear to be essentially the same in other anamniote and amniote embryos. The inducing action of the chordamesoderm has been experimentally tested in several vertebrate groups, including the cyclostomes, bony fishes, birds, and mammals. Waddington and his collaborators have succeeded in inserting grafts of the anterior half of the primitive streak of the chick, which includes Hensen's node, beneath a prepared cut in the ectoderm of the area pellucida (Fig. 10–16). This part of the primitive streak induces a secondary embryonic axis

**FIG. 10–16.** A portion of the anterior half of the primitive streak of the chick when transplanted beneath host ectoderm of a second embryo will induce an accessory head.

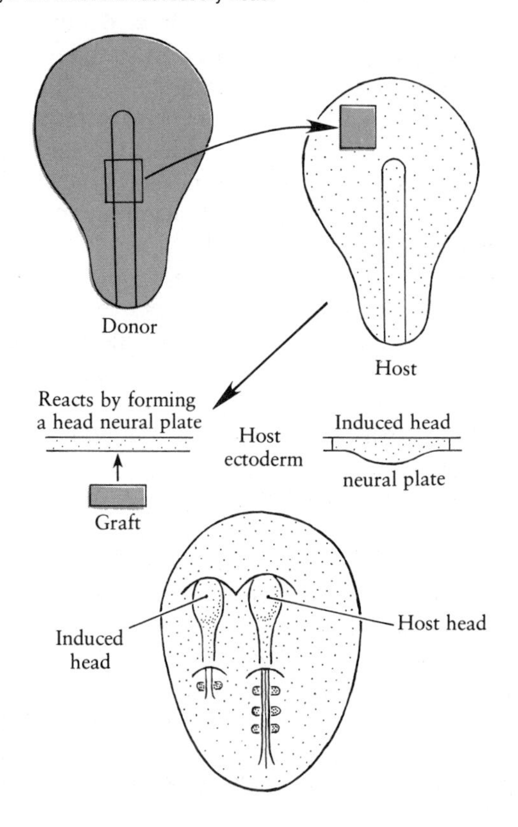

consisting of neural tube, notochord, and somites (Fig. 10–16). The posterior half of the primitive streak lacks the ability to stimulate neural plate formation in host ectoderm, presumably because the prospective fate of the cells migrating through this part of the streak is that of extraembryonic mesoderm. As with the dorsal lip of the amphibian gastrula, the anterior end of the primitive streak shows regional specificity with respect to inductive capacities. This has been demonstrated by removing Hensen's node at progressively later stages of primitive streak regression and transplanting it beneath host ectoderm.

The organizing action of the bird primitive streak does not appear to be species specific. The anterior end of the primitive streak of the duck can be isolated and grafted beneath the ectoderm of a chick embryo at a comparable stage of development (Fig. 10–17). The duck graft induces the host ectoderm to form a secondary neural tube (Fig. 10–17).

Very inadequate information is currently available on the chordamesoderm and its influence on the overlying ectoderm in embryos of reptiles and mammals. Presumably, the ectoderm undergoes a spatial and temporal determination upon stimulation by the chordamesoderm cells, following the latter's displacement through the primitive streak, and becomes specified into different regions of the neural tube. Some work done with rabbit embryos shows that the ectoderm is competent to form neural tissue.

The fact that the primary organizer or chordamesoderm is capable of inducing the overlying ectoderm to form a neural plate explains several additional observations. We now know why there is a correlation between the width of the neural plate and the width of the primitive gut roof in different groups of vertebrates. Also, our understanding of the role of the organizer in primary determination has served to explain the basis for a number of teratological or abnormally developed types known as anterior, posterior, and cross-doubled embryos (*duplicitas anterior*, *duplicitas posterior*, and *duplicitas cruciata*, respectively). If the egg of a salamander is partially constricted with a hair loop along the plane of bilateral symmetry during the period of gastrulation, the invaginating chordamesoderm becomes split into two tongues of tissue upon contact with the strand of hair. Each finger-shaped piece of chordamesoderm then pushes forward on either side of the constriction at the blastopore. The resulting Y-shaped chordamesodermal sheet induces a pair of neural folds anteriorly. The resulting embryo appears to be doubled with two more or less perfectly formed anterior ends joined to a single posterior end. Doubling of both anterior and posterior ends of an embryo can also be obtained by grafting halves of gastrulas together in such a way that their organizers either converge or diverge anteriorly. If the Y-shaped chordamesoderm converges anteriorly, the embryo will be doubled at its posterior end.

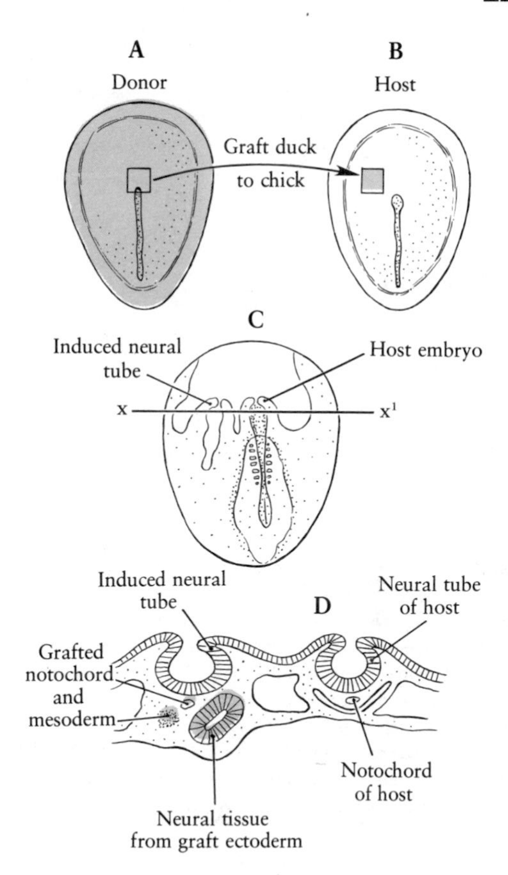


**FIG. 10–17.** Diagrammatic drawing to show the induction of an accessory neural tube as a result of grafting notochordal tissue (Hensen's node) from a duck donor to a chick host. A, duck showing the location of the graft; B, chick host showing the site of the graft transplant; C, chick host showing induced accessory neural tube after 31.5 hours following graft transplant; D, section (transverse) at the level of x–x′ in C to show induced neural tube. (From Foundations of Embryology by B. Patten and B. Carlson. Copyright © 1974 by McGraw-Hill, Inc. Used with permission of McGraw-Hill Book Company.)

The production of crossed-doubled embryos experimentally is particularly interesting. Two gastrula halves, each having a dorsal lip, are brought

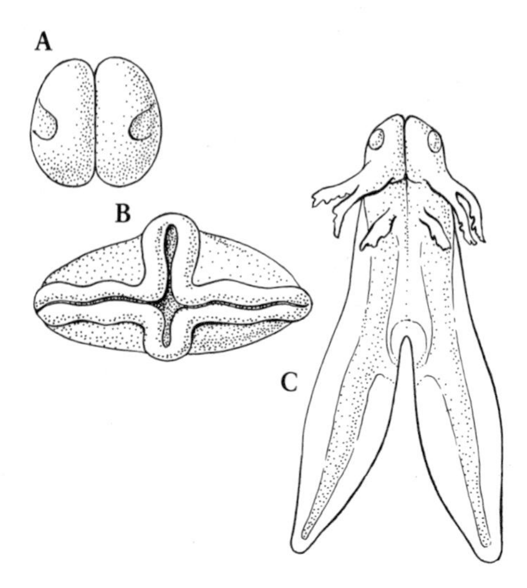


**FIG. 10–18.** A, two dorsal gastrula halves of a salamander grafted together so that the directions of invagination of their blastopores are opposed; B, the cross-shaped appearance of the chordamesoderm following invaginations of the two gastrula halves; C, the resulting embryo showing crossed-doubling or duplicitas cruciata. Each half-gastrula has produced a posterior trunk region with spinal cord, but two heads and brains are formed at right angles to the axis of the trunks, each formed partly from both half-gastrulae. (After J. Huxley and G. DeBeer, 1934. Elements of Experimental Embryology. The Cambridge University Press, New York.)

together with their sites of invagination opposite to each other (Fig. 10–18 A). The chordamesoderm of each gastrula-half moves in through its own blastopore. As the primitive archenteron of one gastrula meets the primitive archenteron of the other gastrula, each diverges to either side. The whole chordamesodermal complex takes on the appearance of a cross (Fig. 10–18 B). The portions of the cross at right angles to the plane of chordamesodermal migration are actually composites of the organizers of both gastrulas. Neural folds arise in the ectoderm over the cross-shaped organizer complex, producing a double embryo whose two heads and brains are at right angles to the axis of the trunks. (Fig. 10–18 C). Each half-gastrula produces a posterior trunk region with spinal cord. However, each head and each brain are partly formed from both gastrulas.

## Induction—A Chemical Process?

The discovery of the organizer-inductor property of the dorsal lip of the amphibian blastopore, and later of the anterior half of the primitive streak, provided a tremendous impetus to the discipline of experimental embryology. Particularly in the 1930s and 1940s, there was a frantic rush to uncover the nature and basis for neural induction. Questions regarding the interaction between cells of the chordamesoderm and cells of the overlying ectoderm were numerous. Was contact between the chordamesoderm and the ectoderm required in order for a successful induction to occur? Was it necessary for the chordamesodermal tissue to be viable and intact? Was the inductor a chemical substance, and, if so, could it be isolated and characterized? Where was the inductor located? What was the minimum period of time required for the inductor to stimulate the ectoderm to form neural tissue? The organizer concept of Spemann seemed to provide a basis for understanding embryonic organization. Although there has been an accumulation of substantial data since the articulation of this concept, the mechanism of primary embryonic induction continues to be elusive.

The whole issue of neural induction was, and to a large extent still is, complicated by several rather surprising discoveries. First, the dorsal lip of the blastopore need not be living and its cells intact in order to induce the formation of a neural plate in the amphibian embryo. One can kill or denature the dorsal lip by heating, freezing, or treatment with alcohol. When a dorsal lip treated in this fashion is subsequently implanted into an early gastrula, neural induction still occurs and the host embryo forms a secondary embryonic axis.

Second, a wide variety of tissues (known as abnormal, foreign, or heterogeneous inductors because they do not actually induce the neural plate in the living embryo), other than the chordamesoderm, are good neural inductors. Indeed, almost any adult tissue, including liver, kidney, gut, and skin, can induce competent ectoderm to form a neural plate and tube under the proper experimental conditions. Even pieces of the neural plate and neural tube are effective inducers. We are thus led to believe that the inductive principle or agent is nonspecific and either widely distributed in tissues or that various substances are effective inducing agents.

Efforts at determining which tissues are good inductors have been facilitated by the perfection of additional grafting techniques. In the original transplantation experiments by Spemann with the primary organizer, the dorsal lip was placed in a wound cut on the surface of the host embryo. Owing to its innate tendency to undergo invagination, the graft slips beneath the surface to become located in a position to influence the ectoderm (Fig. 10–19 A). Unfortunately, adult tissues do not possess the capacity for self-invagination. Mangold, therefore, developed a method whereby the tissue to be tested is placed directly into the blastocoele through a cut made in the roof of a late blastula or early gastrula (Fig. 10–19 B). As gastrulation progresses, the implanted tissue is pressed against the inner surface of the ectoderm by the invaginating chordamesoderm and endoderm. This method makes it possible to test the action of tissues older than the host and to determine the action of killed inductors. Subsequently, Holtfreter developed the *sandwich* or *explantation method* for the study of the inductive activity of various tissues. The inductor to be tested, often in the form of a centrifuged tissue homogenate, is placed between two strips of competent ectoderm excised from the animal hemisphere of a late blastula or early gastrula (Fig. 10–19 C). The free edges of the ectoderm fuse together, thus completely enclosing the inductor material.

Third, inductions can be experimentally produced with tissues from phylogenetically distant species. For example, tissues from *Hydra* and the guinea pig serve as effective inductors of neural plate in the salamander embryo.

**FIG. 10–19.** Three methods of exposing competent gastrula ectoderm to induction. A, transplantation of part of the dorsal lip onto the ventral, marginal surface of a second gastrula where it invaginates and forms a secondary archenteron; B, transplantation of part of the dorsal lip of the blastopore into the blastocoele through a slit in the animal hemisphere; C, inductor, such as centrifuged tissue homogenates, "sandwiched" between two pieces of presumptive ventral ectoderm.

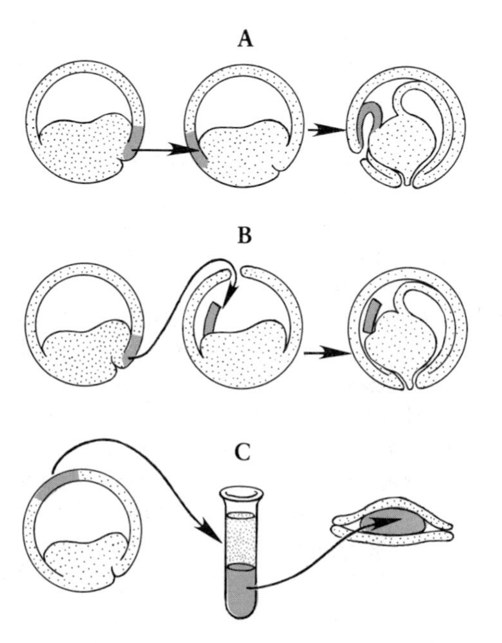


Fourth, strong acids and alkalies, as well as salt and weak organic dyestuff solutions, can induce isolated, competent amphibian ectoderm to form neural tubes. It goes without saying that neural induction in vivo is clearly not initiated by sharp fluctuations in the pH of either the ectoderm or the chordamesoderm. However, since such diverse substances have inductive capacities, with many acting on the reacting tissue by sublethal cytolysis (i.e., they cause cell injury through the destruction of cell membranes), some investigators believe that the key to understanding primary induction lies in the ectoderm. They argue that the process of neural induction involves the unmasking of some preexisting, inactive substance in the ectoderm, which then triggers the specific program of neural differentiation. Induction as an event of *evocation* will be examined in a later section of this chapter.

There are several ways by which the dorsal lip inductor tissue might exert its influence on the responding ectoderm. First, the dorsal lip might provide the ectoderm with a specific type of information that instructs it to differentiate in a specific direction. Such an interaction between two cell populations is said to be of the *instructive type;* it restricts or limits the developmental pathways available to the responding tissue. Second, the dorsal lip might induce differentiation of the ectoderm by providing an essential ingredient that then permits the responding cells to differentiate along an already predetermined pathway. Such an interaction between two cell populations is said to be of the *permissive type*. There are now several lines of evidence to indicate that neural induction is an example of an instructive interaction, produced by substances that are chemical in nature and diffusible. Definition of the properties and characteristics of these chemical substances that function as inductive messengers during primary determination is largely based on *foreign inductors*. The diffusion hypothesis was originally set forth by Mangold. Whether direct cell contact between ectoderm and chordamesoderm layers, as initially proposed by Weiss in 1950, plays any role in primary embryonic induction is still a matter of question (see below).

The *transfilter technique* has been particularly useful in testing whether inductive interactions are dependent on diffusible molecules or cell-to-cell contact. A typical transfilter assembly is shown in

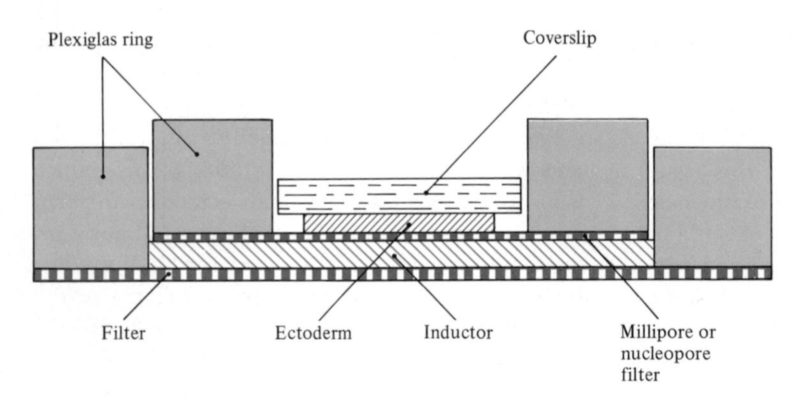


**FIG. 10–20.** Diagram of a typical transfilter assembly. The inductor tissue and competent ectoderm are placed on opposite sides of a Millipore or Nucleopore filter. An additional filter is often placed on the bottom of the culture dish to support the inductor tissue.

Figure 10–20. In this method, inductor and responding tissues are separated by a nitrocellulose filter of a given thickness with pores of a specified diameter. By comparing the patterns of differentiation of the responding tissue in the presence or absence of the inductor tissue, one can evaluate the role of diffusible signals during induction. Saxén (1961) showed in *Triturus* that when competent gastrula ectoderm and dorsal lip were separated by a membrane filter (20 μm thick, 0.8 μm pore size diameter) and then cultured for 24 hours, the ectodermal explants differentiated into definite neural structures. Ectoderm cultured in the absence of inductor tissue failed to become determined and formed typical epidermis. Toivonen (1979) extended these studies on *Triturus* to include Nucleopore filters (0.05 μm pore size diameter) and dialyzing membranes with pores of only 12,000 daltons. After 18–22 hours of exposure to the dorsal lip tissue and 8–10 hours of culture in isolation, a high percentage of the ectodermal explants showed differentiation into neural structures irrespective of the membranes employed. Studies of this type have not been limited to amphibian embryos. Gallera, for example, observed that neural induction occurred in the chick embryo when chordamesodermal inductor and competent ectoderm were separated by a Millipore filter of average pore size and thickness.

Evidence from transfilter experiments suggests that primary induction is the result of diffusion of an inductor substance across the extracellular space between the ectoderm and the chordamesoderm. This explanation of neural induction is not universally shared. With the transfilter technique, it is possible that projections from one or both of the interacting tissues might pass through the pores. An inductive interaction might result from direct contacts between the two tissues. Efforts to resolve this question have included examination of the filters during the culture period with the transmission electron microscope. Many of these studies report few if any cellular processes penetrating into the pores. However, small granules about the size of ribosomes and apparently originating from the chordamesoderm are frequently seen to a depth of 2–4 microns in the filter pores. It is interesting to note that Kelly and Tarin report the presence of granules in the interzone between the chordamesoderm and ectoderm during induction in *Xenopus* (Fig. 10–21). Many of these granules are sensitive to ribonuclease treatment, a property that aligns them to the active component of foreign inductors (see below). At present, it is not known if the RNA-containing granules actually participate in the tissue interaction, whether they are transferred from one tissue to another, and what role the RNA might play in the inducing system.

Whether cell contacts mediated by cell processes can be definitely excluded as being of importance in primary neural induction requires further investigation. It must be remembered that the evidence indicating no requirement for cell-to-cell contact is based primarily on in vitro conditions. The interrelationships between the interacting tissues may be quite different in the intact embryo. Recent transmission electron micrographs of midgastrulae of *Triturus* not only confirm the presence of an interspace between the chordamesoderm and the overlying ectoderm, but also show the space to be traversed by small cell projections originating from both tissues (Grunz and Staubach, 1979). In addition, intimate membrane contacts were observed between ectoderm and chordamesoderm during the period of primary embryonic induction. Also, there is some evidence that the mechanism of

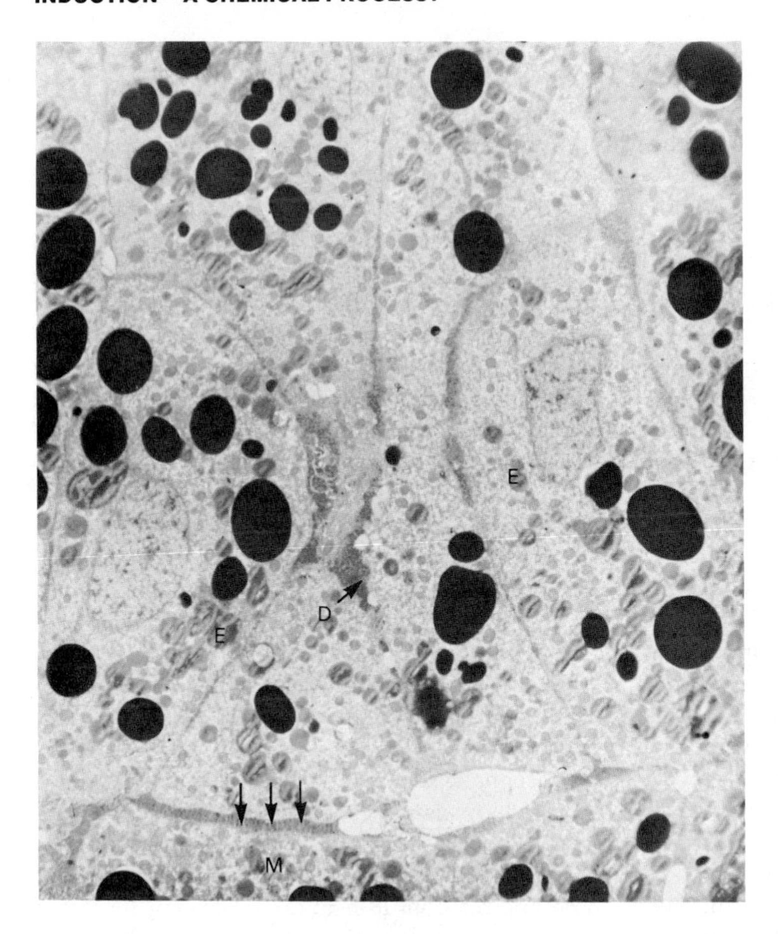


**FIG. 10-21.** The ectoderm and chordamesoderm junction during neural induction in *Xenopus*. Aggregates of dark granules (D) are seen between the ectoderm (E) and chordamesoderm (M) layers. The arrows indicate the position of the chordamesodermal boundary. (From D. Tarin, 1972. J. Anat. 111:1.)

transfer of inducing information may not be the same during the whole period of the development of the central nervous system. Toivonen (1979), for example, reports that the general neuralization of the ectoderm can occur without cell contact between the interacting tissues, whereas the inductive action causing the appearance of the different regions of the central nervous system from the preneuralized ectoderm is mediated by cell-to-cell contact.

Direct experimental evidence that the neural inductor is probably a diffusible molecule released from the chordamesoderm originated with classical studies by Niu and Twitty in 1953. They removed the dorsal lip of the blastopore from a salamander embryo and placed the tissue fragment in a small quantity of saline on the surface of a coverslip. The coverslip was then inverted over the hollow of a depression slide and sealed to prevent dessication. After 7 to 10 days, a fragment of competent ectoderm from the gastrula stage was added to the culture medium, now "conditioned" by the presence of dorsal lip tissue. Following a period of 24 hours, nerve and pigment cell types were detected in the medium (Fig. 10–22). If fragments of competent ectoderm were cultured in drops withdrawn from the original culture and containing no

**FIG. 10–22.** Chromatophores formed from gastrula ectoderm of *Triturus* after introduction into a 10-day-old culture conditioned by tissue presumptive to mesodermal structures of the tail and posterior part of the trunk. (After M. Niu and V. Twitty, 1953. Proc. Natl. Acad. Sci. U.S.A. 39:985.)

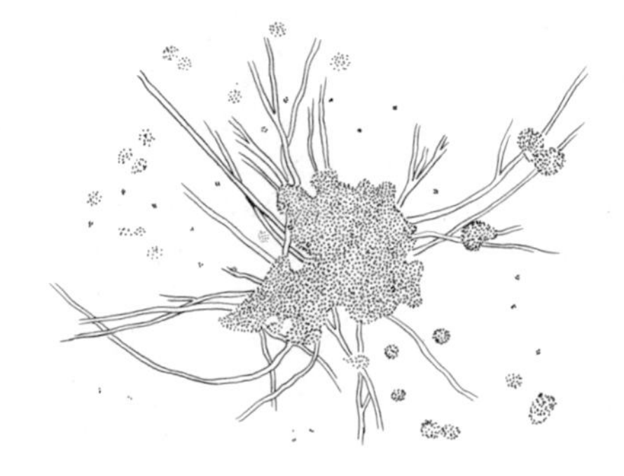

dorsal lip cells, nerve and pigment cells still differentiated from the ectodermal tissue. Competent ectoderm placed in Holtfreter's solution lacking dorsal lip tissue remained undifferentiated. If the chordamesoderm remained in the culture medium for 12 to 15 days, competent ectoderm incubated in drops of this conditioned medium differentiated into pigment, nerve, and precursor muscle cells. Niu and Twitty concluded that the chordamesoderm releases different substances into the medium. One appears early and induces neural tissue, while the other appears later and induces mesodermal structures. Niu later found the conditioned medium to be rich in RNA and strongly suggested that this nucleic acid was the natural inducer.

During the last two decades, a number of studies have been conducted in an effort to analyze carefully the nature of inducing substances in gastrulating embryos and to determine if such substances actually pass from the chordamesoderm to the overlying ectoderm. Unfortunately, it has not been possible to obtain in sufficient quantities extracts of dorsal lip or primitive streak that will allow for detailed chemical analysis. However, a variety of adult tissues are known to display strong inductive capacities, imitating the action of the natural primary organizer. Extracts and pellets of these tissues have been made and the inducing power of the preparations subjected to various testing procedures. With few exceptions, many adult tissues following heat or alcohol treatment are able to induce rather specific differentiations in uncommitted ectoderm. Yamada and his colleagues in Japan use alcohol-treated guinea pig liver to induce primary archencephalic structures, such as brain vesicles and nose primordia. Guinea pig kidney following alcohol treatment is primarily a spinocaudal inductor; it stimulates the expression of spinal cord, notochord and somites. Alcohol-treated guinea pig bone marrow induces almost exclusively the mesodermal structures found in the trunk of the embryo, these being notochord, somites, and kidney tubules.

Extracts of embryos have been used in similar kinds of studies. Tiedemann and associates have been successful in isolating two extracts from 9- to 12-day chick embryos, one of which induces uncommitted gastrula ectoderm of *Triturus* to form head structures *(neuralizing factor)*. The other extract is called the *vegetalizing factor* because it induces endodermal and mesodermal differentiations in gastrula ectoderm, tissues normally derived from the vegetal part of the embryo. Certain foreign inductors, therefore, demonstrate not only an ability to induce a general differentiation in competent ectoderm (i.e., neural tissue, mesodermal tissue), but also a morphological response with pattern and regional specialization (i.e., forebrain, spinal cord, and somites). Presumably these inductor tissues or fractions prepared from them release chemical substances that bias the developmental pathway of the ectoderm. But, how do the inducing substances act to restrict spatially determination and regional segregation of the ectoderm?

The concept that neural induction may be due to the interaction of two chemically distinct inductive agents emerged from studies in which the inducing properties of adult tissues were experimentally altered. For example, guinea pig liver tissue continues to be an effective inductor of archencephalic structures even if heated to 70°C to 90°C. However, the spinocaudal inducing activity of guinea pig kidney is markedly reduced following similar heat treatment; yet, in some cases, the tissue induced archencephalic structures. This suggests that guinea pig kidney tissue contains two factors or substances, one sensitive to heat (and inducing such spinocaudal structures as notochord and somites) and one resistant to heat (and inducing forebrain structures). Other experiments employing these foreign (or heterotypic) inductors permit one to conclude that there are basically two responses during the primary induction process: *neuralization* and *mesodermalization*. Neuralization ultimately leads to the formation of cranial neural structures (forebrain, optic vesicle), whereas mesodermalization culminates in the expression of purely mesodermal structures (notochord, somites). Toivonen and Saxén proposed several years ago that two substances, extracted from a variety of tissues, were responsible for neuralization and mesodermalization: a *neuralizing agent* (thermostable and soluble in organic solvents such as petroleum ether) and a *mesodermalizing agent* (highly thermolabile and insoluble in organic solvents). The expression of midcranial and postcranial structures, such as hindbrain and spinal cord, resulted from the combined action of the primary neuralizing and mesodermalizing agents.

Support for the "combined effect" hypothesis was obtained in elegant experiments by Toivonen and Saxén using pellets prepared from guinea pig

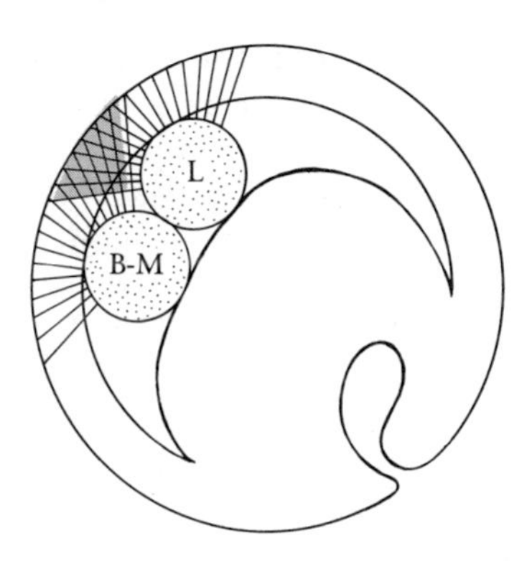


**FIG. 10–23.** Simultaneous implantation of a neuralizing agent (L, liver) and a mesodermalizing agent (B–M, bone marrow) into the blastocoele of a salamander gastrula. Deuterencephalic and spinocaudal structures are induced where the effects of the two agents overlap. (From L. Saxén and S. Toivonen, 1962. Primary Embryonic Induction. Prentice-Hall, Englewood Cliffs, N.J.)

tissues. Two pellets, one prepared from guinea pig liver (neuralizing action) and the other from guinea pig bone marrow (mesodermalizing action), were implanted simultaneously into the blastocoele of a young salamander gastrula (Fig. 10–23). The liver pellet alone induces only archencephalic structures. The bone marrow pellet alone induces only mesodermal structures. Together, however, the two pellets induce deuterencephalic and spinocaudal structures (hindbrain, inner ear vesicles, spinal cord) in the ectoderm, presumably because of the interaction of the diffusible, active components being released by the pellets. Conversely, a deuterencephalic inductor can be chromatographically separated into two components, one of which can only induce neural structures and the other mesodermal structures.

Are the proposed neuralizing and mesodermalizing agents two distinctly different substances or merely two different states of the same substance? Partial answers to these questions have been obtained from centrifuged fractions of heterotypic tissues tested on isolated, competent ectoderm (Fig. 10–24). Generally, both ribonucleoprotein (PNP—pentose nucleoprotein) and protein fractions of these tissues have proved to be effective inductors. The active component of tissues inducing archencephalic and deuterencephalic structures appears to be a ribonucleoprotein (Fig. 10–24). Purification of the PNP by ultracentrifugation does not disturb its inducing capacity. Importantly, treatment of PNP fractions with ribonuclease does not reduce the original inducing capability of the preparation, thus suggesting that the nucleic acid moiety of the active agent is not the essential component of the inductor. However, archencephalic and deuterencephalic structures fail to develop if PNP fractions are treated with proteolytic enzymes such as pepsin. Indeed, there appears to be an enhancement of the differentiation of nonaxial structures in the presence of the treated fractions. The vegetalizing factor of Tiedemann has recently been purified by the techniques of isoelectric focusing and sodium dodecyl sulfate (SDS) acrylamide electrophoresis. The active component has a molecular weight of 30,000 and is inactivated by heat and proteolytic enzymes.

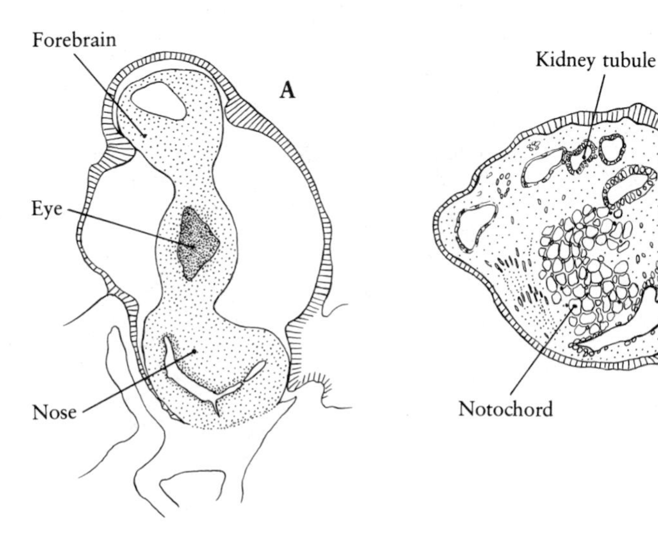


**FIG. 10–24.** A, nose, forebrain, and eye-type structures with pigment (archencephalic induction) induced by liver PNP in isolated *Triturus* ectoderm; B, notochord, somites, and kidney tubules are induced in isolated ectoderm by the non-nucleoprotein fraction of bone marrow. (After T. Yamada, 1958. Experientia 14:81.)

In short, the inductors of such axial structures as neural tube, notochord, and somites are proteins that may or may not be coupled with RNA. The mesodermalizing agent appears to be a protein with a low molecular weight that can be eluted from a chromatographic column at a pH of 7.0. The neuralizing agent appears to be ribonucleoprotein that can be eluted from a chromatographic column at a pH of 5.6. The presence of inducing activity in tissue ribonucleoprotein following ribonuclease treatment suggests that the nucleic acid component is not informationally essential in the inductive process. Recent studies by Niu conflict with this conclusion. He claims that RNA extracted from calf testis (germ-cell RNA), when added to postnodal pieces of the definitive primitive streak of the chick, will induce a range of embryonic axial differentiations.

Assuming that proteins of the neuralizing and mesodermalizing (vegetalizing) agents are responsible for neural induction in the intact gastrulating embryo, where are these factors located, and how are they organized to determine the specific direction of differentiation taken by the ectoderm? The model of induction put forth by Toivonen and Saxén proposes that these substances are located in the chordamesodermal mantle or roof of the archenteron (Fig. 10–25). The regional differentiations along the axis of the embryo appear to be controlled by a balance between the neuralizing substance and the mesodermalizing substance, each of which is distributed in the form of a gradient. The dual-gradient concept emerged from studies in which neural and predominantly mesodermal inductor cells were mixed semiquantitatively in various ratios and the cellular combinations then analyzed with respect to the types of neural and spinocaudal structures induced (Fig. 10–26). Forebrain was expressed when the combination consisted of only neural inductor cells. Progressive caudalization of neural structures was associated with a relative increase in the amount of the mesodermalizing inductor. Hence, the neuralizing agent is organized equally along the anteroposterior axis of the chordamesoderm with its highest concentration in the dorsal midline. It declines in concentration toward the lateral and ventral parts of the embryo. The mesodermalizing agent is absent from the extreme anterior region of the embryo. It begins in the presumptive hindbrain region of the embryo and progressively increases in concentration toward the tail.

The anterior end of the chordamesodermal layer (i.e., presumptive prechordal plate) produces only neuralizing agent and induces the ectoderm to form archencephalic structures. Slightly further in the posterior direction, the ectoderm is exposed to neuralizing agent and a small amount of mesodermalizing substance. This combination induces hindbrain and inner ear vesicles. As the relative amount of mesodermalizing to neuralizing sub-

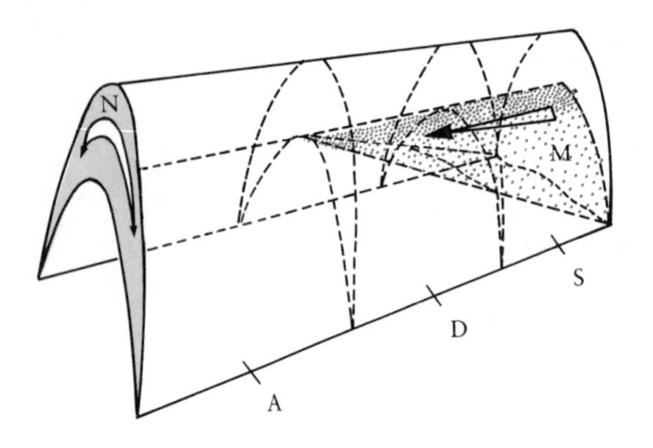


**FIG. 10–25.** The two-gradient model of Saxén and Toivonen. The chordamesoderm contains both a neuralizing agent (N) and a mesodermalizing agent (M). The neuralizing agent is distributed in gradient fashion from dorsal to ventral. The mesodermalizing agent increases in concentration from the level of the future hindbrain to the tail of the embryo. A, D, S, levels of archencephalic, deuterencephalic, and spinocaudal inductions, respectively. (From L. Saxén and S. Toivonen, 1962. Primary Embryonic Induction, Prentice-Hall, Englewood Cliffs, N.J.)

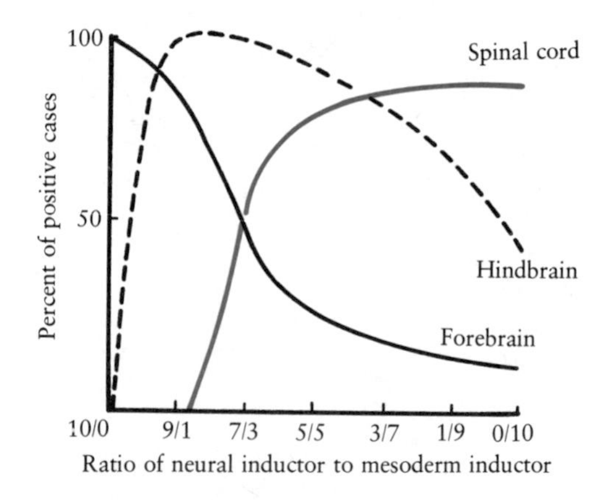


**FIG. 10–26.** Percentage of neural structures belonging to the central nervous system induced in experiments in which two types of foreign inductors were mixed in different ratios. (From L. Saxén and S. Toivonen, 1961. J. Embryol. Exp. Morphol. 9:514.)

stance increases in the anterior-to-posterior direction, trunk and tail structures are induced. The gradient of the neuralizing substance may also be responsible for the dorsal-to-ventral differentiations of such ectodermal derivatives as neural crest and sensory placodes (lens and nose primordia). Where the gradient of the neuralizing agent progressively diminishes, the ectoderm becomes epidermis. In the chordamesodermal mantle, the gradient of the mesodermalizing agent induces the differentiation of somites and lateral plates.

Nieuwkoop (1973) has put forth a slightly different hypothesis to explain the events of neural induction. He proposes that the ectoderm is initially determined to be a forebrain differentiation as a result of the activation of neural principles within its cells. A nonspecific agent from the tip of the chordamesoderm induces this alteration in the overlying ectoderm as it slides forward during gastrulation. Subsequently, the activated ectoderm becomes transformed under the influence of a second agent released from the chordamesoderm. The differentiation of a given region of ectoderm is the result of a balance between the initial archencephalic tendency and the length of time that this layer has been exposed to the transforming agent. Since the posterior ectoderm is first to be influenced by the transforming agent, it is exposed for a longer period of time than the anterior ectoderm and therefore differentiates as trunk and spinal cord. The *activating* and *transforming agents* of Nieuwkoop are probably equivalent to the neuralizing and mesodermalizing agents, respectively, of Toivonen and Saxén. Both induction models are quite similar if one assumes that neuralizing and mesodermalizing agents are released at different rates or at different times from the chordamesodermal mantle.

That regionalization of the central nervous system is complex and cannot be due simply to a single inductive event has been experimentally shown by Saxén and his colleagues using the techniques of disaggregation and reaggregation (Fig. 10–27). Competent ectodermal cells from an amphibian gastrula were isolated and exposed to either guinea pig liver (culture A) or guinea pig bone marrow (culture B). Approximately 24 hours later, the inductors were removed from each culture, and the ectodermal cells disaggregated into single cell populations. When the suspensions were allowed to reaggregate, the cells in culture A formed forebrain and optic cups while those in culture B formed spinal cord, muscle blocks, kidney tubules, and notochord. However, if the two types of induced suspensions were mixed and cultured as a combined aggregate, hindbrain and inner ear vesicles were readily formed. It can be concluded that these structures were determined after the initial period of induction and before the regionalization of the central nervous system was stabilized. If presumptive forebrain cells from young amphibian neurulas are cultured in mixed aggregates of axial mesodermal cells, they can be "transformed" into hindbrain and spinal cord structures. Interestingly, a gradual increase in the ratio of axial mesodermal to forebrain cells results in the gradual caudalization of the central nervous system derivatives (Fig. 10–28). Results of these types of experiments are offered as additional evidence for the two-gradient hypothesis.

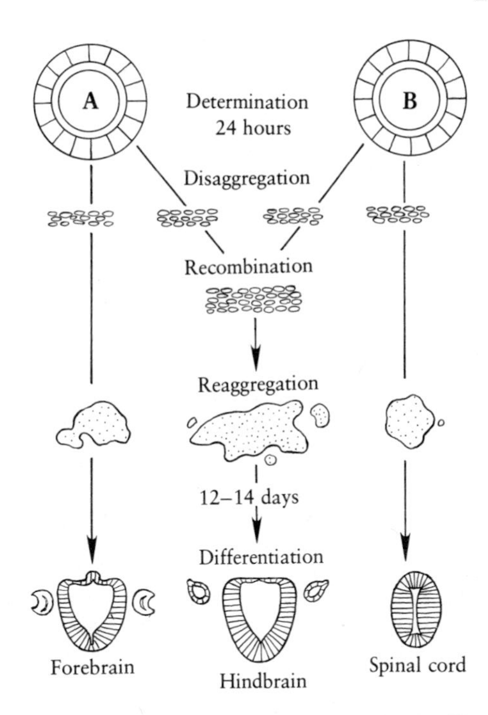


**FIG. 10–27.** Experimental procedure demonstrating the differentiation of the central nervous system in cultures of competent ectoderm exposed to guinea pig liver (A) and bone marrow (B) utilizing the techniques of disaggregation and reaggregation. (From L. Saxén, S. Toivonen, and T. Vainio, 1964. J. Embryol. Exp. Morphol. 12:333.)

These experiments strongly indicate that primary embryonic induction is a multistep process. The ectoderm of the gastrula is multipotent and developmentally flexible, capable of being converted into neural, mesodermal, and endodermal

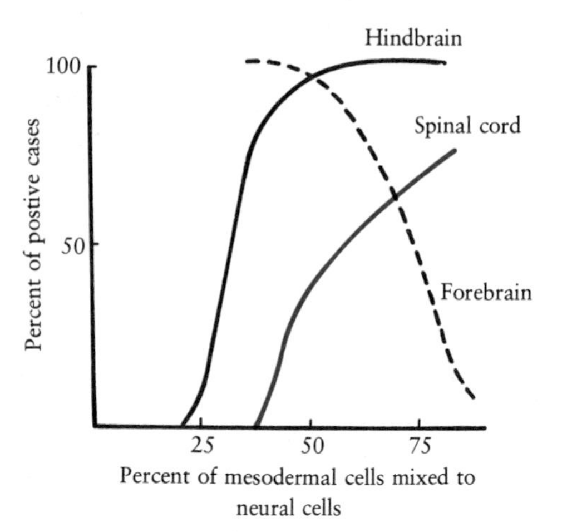


**FIG. 10–28.** Percentage of neural derivatives (forebrain, hindbrain, and spinal cord) formed when cells of the axial mesoderm are mixed with neural cells of the forebrain region (neurulae). (From S. Toivonen and L. Saxén, 1968. Science 159:539. Copyright 1968 by the American Association for the Advancement of Science.)

derivatives. During a short initial phase, the ectoderm of the embryo under the influence of the dorsal lip or its equivalent is probably determined in either a neural or mesodermal direction. Without further influence, cells determined in the neural direction will become forebrain structures. However, continued mesodermal influence alters their futher morphogenetic pattern into caudal-neural structures. This secondary interaction, which affects the specificity of morphogenetic expression, operates in a quantitative fashion and acts to stabilize the initial determination.

It must be remembered that much of what we know concerning primary embryonic induction is based on studies using foreign or heterotypic tissue inductors. There remains, of course, the obvious question whether active components extracted from heterotypic inductors are the same as those functioning during normal embryonic induction. Some progress toward the goal of isolating, identifying, and characterizing natural inductors is currently being made. From crude extracts prepared from different regions of *Xenopus* gastrulas, Deuchar has determined that those from the dorsal lip give the highest percentages of neural induction. Others have succeeded in separating whole extracts of *Xenopus* gastrulas into protein, RNA, RNA-protein, and DNA-protein fractions. When tested with competent ectoderm, certain protein and high-molecular-weight RNA-protein fractions induced hindbrain, spinal cord, and mesodermal structures.

Several additional studies have established a relationship between the inducing capacity of the dorsal lip and ultraviolet (UV)-sensitive cytoplasmic components that are present in the amphibian egg prior to cleavage division. Malacinski (1974) has shown that crude extracts of the germinal vesicle of *Rana,* following injection into the blascocoele of a normal embryo, produce an embryo with an enlarged neural plate (Fig. 10–29). The substance, termed the *axial structure determinant* (ASD), is characterized as being thermolabile, insensitive to ribonuclease, but sensitive to trypsin. When the vegetal hemisphere of frog eggs shortly after fertilization is subjected to UV irradiation, the embryo's capacity for normal neural tube and axis formation is destroyed. The UV-sensitive target appears to be restricted to the grey cresent (dorsal) region of the egg. Yet, these lesions in UV-treated frog embryos can be corrected by the microinjection of crude extracts of germinal vesicle plasm. Chung and Malacinski (1975) experimentally have shown that localized UV damage to the egg subsequently

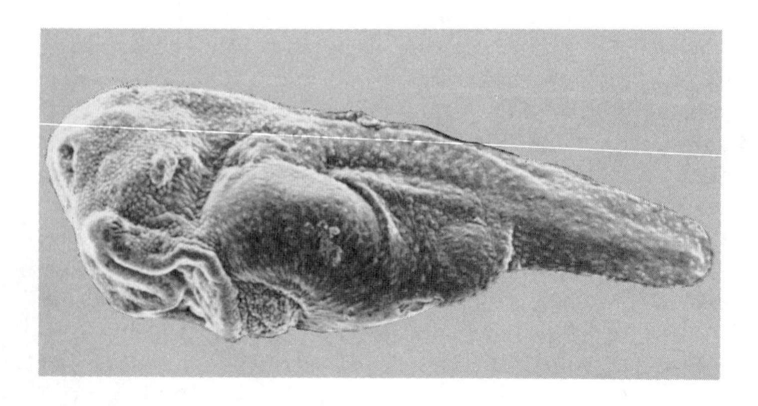

**FIG. 10–29.** Microinjection of germinal vesicle plasm into the blastula of a frog embryo results in the development of an enlarged neural plate and a macrocephalic embryo. (From G. Malacinski, 1974. Cell Differ. 3:31.)

affects the inducing capacity of the dorsal lip during gastrulation. The design of their experiments is shown in Figure 10–30. Approximately 75 to 90 minutes after fertilization in *Rana*, eggs are UV-irradiated at the vegetal hemisphere. At the early gastrula stage, their dorsal lips can be grafted to the surface of normal embryos (Fig. 10–30 B). The recipient embryos are then examined at the muscular response stage for evidence of secondary neural induction. Dorsal lips of irradiated gastrulae when transplanted to normal host embryos fail to induce a secondary set of axial structures. Virtually all of the UV-irradiated embryos that receive a dorsal lip from a UV-irradiated donor embryo as a replacement show defects in neural development. Grafting of a lip from a UV-irradiated embryo onto a normal embryo results in a high percentage of the recipients developing abnormal neural morphology. However, when UV-irradiated embryos receive a dorsal lip from a normal embryo, approximately half of the UV-irradiated recipients develop normally, thereby suggesting that there is correction of the neural lesion. The importance of the dorsal lip is clearly indicated by experiments in which ventral marginal zone (i.e., opposite to the dorsal lip) tissue is exchanged between irradiated and nonirradiated embryos (Fig. 10–30 C). We can conclude from these experiments that UV irradiation of the vegetal hemisphere of the precleaved amphibian egg destroys a component of the cytoplasm that is critical to neural induction by the dorsal lip. Attempts are being made to understand better what cellular processes are affected by irradiation to result in the disturbances of neural development. Malacinski and colleagues suggest that ultraviolet irradiation has at least two effects that would account for reduced neural development. It appears not only to inhibit the cell movements of gastrulation, but also to reduce the inducing capacity of the dorsal lip.

## Transmission and Mode of Action of the Inductor

Primary embryonic induction is an epigenetic process that depends on the chordamesodermal cells of the archenteron roof stimulating the neuralization of the overlying ectoderm. There are two schools of thought, although not necessarily with an equal number of adherents, regarding the loca-

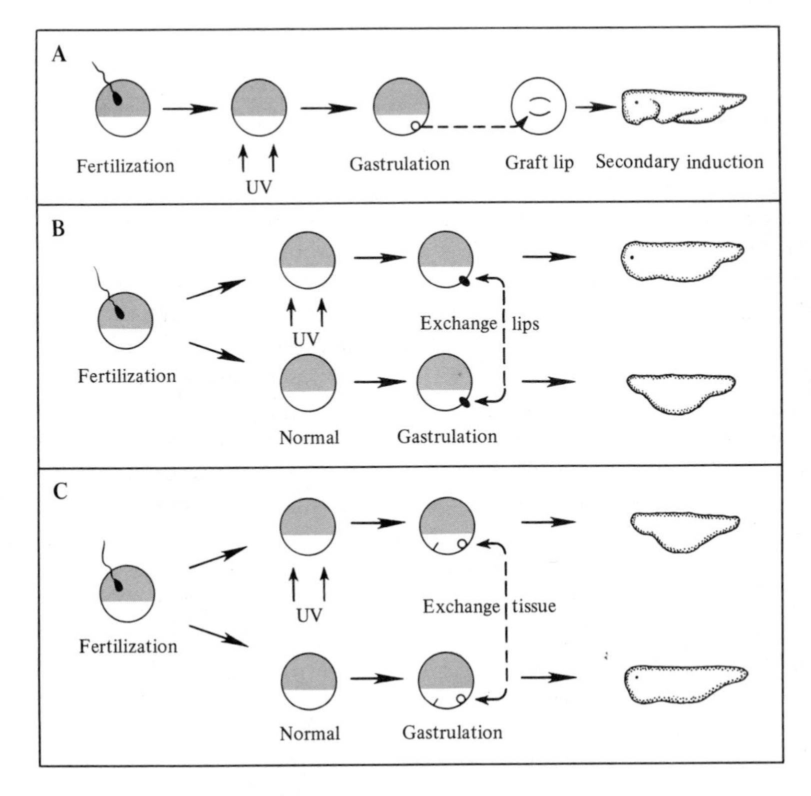


**FIG. 10–30.** Experiments demonstrating the effect of UV irradiation on the inducing capacity of the dorsal lip. A, at 75 to 90 minutes after fertilization, eggs are irradiated with UV light at the vegetal hemisphere. At the early-gastrula stage, their dorsal lips are transplanted to the ventral surface of normal recipient gastrulae. The extent of secondary induction is subsequently scored. B, assay of the ability of normal (non-UV-irradiated) dorsal lip to promote neural induction in irradiated embryos. Dorsal lips were exchanged between irradiated and normal embryos; note that UV-irradiated dorsal lips do not induce neural structures. C, determination of the effects of UV irradiation of the vegetal hemisphere on the ability of the ventral marginal zone to affect subsequent neural induction. (From H. Chung and G. Malacinski, 1975. Proc. Natl. Acad. Sci. U.S.A. 72:1235.)

tion of the substance(s) responsible for neural induction within the embryo. Since a wide spectrum of agents and treated tissues can induce competent ectoderm by sublethal cytolysis to form neural structures, there are those, particularly Waddington and his colleagues, who are inclined to believe that the inductor is resident within the cytoplasm of the reacting cells. According to them, induction involves the unmasking, probably by some nonspecific signal *(indirect evocator)*, of the inductor molecule *(direct evocator)* in the ectoderm. During normal development, activation of the direct evocator as a result of its interaction with the underlying chordamesoderm triggers the transformation of the ectodermal cells into neural cells. This view suggests that the ectoderm is already determined to some degree. Additional support for this position on neural induction comes from the phenomenon of *autoneuralization*. Competent ectoderm in salt solutions will form a neural tube in the absence of any inducer substance (i.e., self-neuralization). This suggests that all factors for neuralization are present in the responding ectoderm and simply await activation.

Chiefly from studies in vitro, in which the induction process has been analyzed using selected filters interposed between interacting tissues, most investigators are of the opinion that primary embryonic induction is a case of the transmission of specific inductive stimuli across the ectodermal–chordamesodermal junction. Unfortunately, the characterization of the inductive substances has been limited, as previously pointed out, and concrete evidence for the passage of substances from the chordamesodermal mantle to the overlying ectoderm in the intact embryo is disappointing and still largely lacking. Experiments have been conducted in which competent ectoderm has been cultured with chordamesodermal tissue tagged with radioactive amino acids or fluorescently labeled proteins. Because label is detected in the ectoderm following induction, the results from such experiments are suggestive of a signal transfer between inducer tissue and the reacting ectoderm. However, there is still some question as to whether the labeled compounds were essential to the induction process. Also, it must be pointed out that induction does not always occur with the transfer of these labeled compounds. Recently, attempts have been made to isolate material from the extracellular space between the neural plate and the chordamesoderm of early neurulae of *Triturus*. A phenol-extracted protein from this extracellular material showed neural-inducing activity.

Assuming that diffusible signals are involved in neural induction, how do they act on the target ectoderm, and what are the responses of the ectoderm? At the cell level, two primary sites to be considered are the plasma membrane and the intracellular compartment. Although investigators have only recently begun to examine the role of the ectodermal membranes, there are already data to suggest that these may be important for the specificity of neuralization. Tiedemann and his colleagues (1978) have designed several experiments to determine whether inducing factors simply interact with the outer plasma membranes of ectodermal cells or must be incorporated into the cells. To prevent the incorporation of either the neuralizing factor or the vegetalizing (mesodermalizing) factor into competent ectoderm, they covalently bonded the extracts of these factors to either cellulose or Sepharose particles. Pellets of factor–matrix particles were then implanted into blastocoeles of young *Triturus* gastrulas. When the vegetalizing factor was bound to the matrix, the inducing ability of the fraction was markedly reduced. However, the inducing ability of the neuralizing fraction under similar conditions remained unchanged. Although there are concerns about the technical design and interpretation of these experiments (for example, might not the vegetalizing factor be inhibited by the binding procedure itself? Does adsorption of a factor to the matrix limit its availability to the target cell?), the possibility is raised that the mechanism of action of these two factors may be quite different. The data suggest that the vegetalizing factor must be incorporated into the target ectodermal cells, whereas the neuralizing factor exerts its influence on the outer plasma membranes.

Other studies in various preliminary stages suggest that the neuralizing inductive signal requires for its transmission distinct membrane receptors and/or a normal molecular conformation of the target cell surface membrane. Exciting and potentially useful probes in analysis of the possible role of the ectodermal cell surface in normal embryonic induction are lectins. Lectins, as previously pointed out, are plant proteins that bind selectively to specific carbohydrate residues of the cell surface. They can be visualized under the light microscope by fluorescence and under the transmission electron microscope by staining through coupling with ferritin or horseradish peroxidase. Ectodermal cells from young amphibian gastrulas possess receptors for several lectins, including soybean lectin (SBA), garden pea lectin (PSA), and concanavalin A (Con A), indicating that the surface carbohydrates contain chiefly α-D-mannose and α-D-galactose. Whether the neuralizing factor interacts with one or any of these surface molecules binding the lectins to elicit neural differentiation is unclear. The experimental approach to this question has been to treat competent isolated ectoderm with selected concentrations of lectin and then assay for induction. SBA and PSA in *Pleurodeles* (amphibian) appear to have no inducing effect. Indeed, if SBA- or PSA-treated ectoderm is then exposed to natural inducer tissue, there is little evidence of induction either. This finding suggests that these lectins probably disturb the structural integrity of the cell surface and thereby inhibit the passage of the inductive factor. By contrast, there is evidence that Con A receptors may be important to the specificity of the neural inductive process. Con A appears to have a strong neural-inducing effect on the presumptive ectoderm of newt gastrulas. Autoradiography of [$^3$H]-Con A explants shows that lectin distribution is restricted to the inner surface of the ectoderm. Subsequently, some of the tritiated label can be identified in the cytoplasm of the ectodermal cells, suggesting that the incorporation of Con A may be a requirement for neural induction. However, this contradicts the observation that Con A has a strong inductive effect even if bound to Sepharose particles. Further studies are necessary to characterize more fully the molecules on the target cell surface. It appears likely, however, that certain sugar-containing complexes along the inner surface of the ectoderm above the roof of the archenteron are involved in the initiation of the neural-inducing mechanism.

What are the responses of the ectoderm to the neural signal? Some insight into this question has been obtained by determining if synthetic processes are stimulated into action during gastrulation and by establishing if these activities are essential to neural differentiation. We now know on the basis of a variety of studies that the genes of the embryo, after being largely inactive through the period of cleaveage and blastulation, begin to express themselves during gastrulation; they subsequently assume greater control over the processes directing the development of the embryo. There is increased nuclear activity at gastrulation as shown in the appearance of new proteins, enzymes, and nucleic acids. Much of this evidence has been obtained from experiments employing hybrids formed between two species. As one might expect during a period of major gene activity, there is positive evidence that some of the messenger RNAs synthesized during gastrulation are different from those present in the cytoplasm of the egg or the cells of the blastula. Structural genes previously inactive would appear to become active at this particular time of development. The observation that sequences of DNA are being transcribed for the first time at gastrulation was determined using the elegant technique of *DNA-RNA hybridization* (Fig. 10–31). Likewise, supportive components in the chain of events leading to the formation of proteins, such

**FIG. 10–31.** The purpose of DNA–RNA hybridization is to permit the investigator to determine whether mRNA synthesized at one stage of development (i.e., gastrulation) is the same as or different from mRNA formed at another stage of development (i.e., cleavage). Advantage is taken of the known binding specificity between whole sequences of bases of DNA and complementary bases of RNA. The principal steps in the technique are shown here with two transcriptive sites. A, double-stranded DNA is separated by heating into two single DNA strands; B, mRNA from the cleavage stage is added to one of the single DNA strands and hybridizes with the complementary section of the DNA; C, mRNA from the gastrula stage is then added, and only those mRNAs that are different from the cleavage stage will attach to the DNA. The proportion of mRNA molecules that fail to hybridize gives an indication of the similarity in the mRNAs between the two stages.

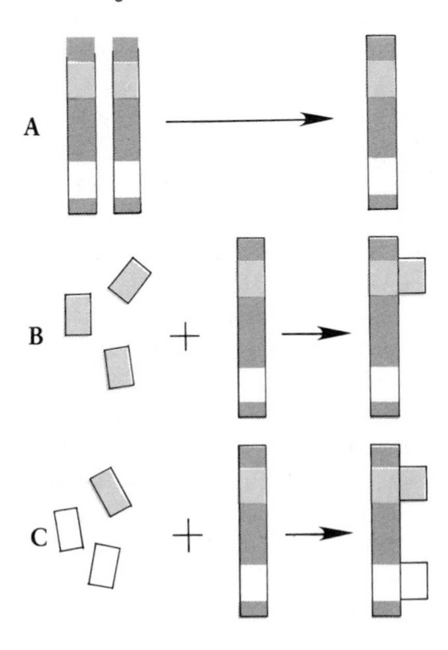

as tRNA and rRNA, are known to be transcribed from nuclear genes at this time.

Since the activation of genes during gastrulation and the resulting synthesis of proteins are manifestations of the developing embryo, it is important to ask whether these activities are associated with the ectoderm and critical to the event of neural induction. A number of biochemical changes, apparently essential to neural differentiation, are detectable in the neural ectoderm immediately following induction. In *Triturus*, for example, there is a substantial increase in the incorporation of amino acids into proteins in isolated gastrula ectoderm induced by the vegetalizing factor. Also, the neural plate of *Triturus* contains several antigens not present in the ordinary eipdermis. Using immunoelectrophoretic methods, Stanisstreet and Deuchar (1972) have shown that one of the earliest features of neural differentiation in *Xenopus* in response to induction by dorsal gastrula mesoderm is an increase in the concentration of three antigenic components when compared to ventral or noninduced ectoderm. Indeed, this particular biochemical change in the neural ectoderm is detected prior to any visible sign of the neural plate.

Neural induction, therefore, like any other mechanism that controls differentiative processes in embryonic cells, appears to act by initiating the synthesis of new proteins in the target cells. Is the presence of new protein in the neural ectoderm evidence of gene activity? Are these proteins essential for successful induction to occur? Partial answers to these questions are available based on experiments in which inhibitors have been used to block key steps in the synthetic activities of the neural ectoderm.

If the dorsal lips of young *Triturus* gastrulas are cultured in either actinomycin D (transcriptional inhibitor) or puromycin (translational inhibitor) and then the treated tissue confronted with competent ectoderm, induction still occurs and the ectoderm responds by forming neural structures. This suggests that neither the transcription of mRNA nor the synthesis of proteins that may take place in the chordamesoderm at the time of gastrulation is essential for induction to occur. However, if the dorasl lip is explanted with the adjacent ectoderm and then treated with sufficient concentrations of actinomycin D to block synthesis of mRNA, the ectoderm will not differentiate into neural plate. Hence, for the reacting cells to move in the direction of neuralization, the formation of mRNA by transcription of nuclear DNA is required. Without the active participation of the nuclei of the target cells, neural differentiation will not take place. The action of the neural inductor is not simply to transform the ectoderm into neural tissue, but rather to set into motion a series of nuclear-dependent events that will lead to neural differentiation.

The mechanism by which the proposed diffusible, chemical substances act on the overlying ectoderm to stimulate a specific program of nuclear activity is still a subject of considerable speculation. The primary inductors resemble hormones in that they act as intercellular messengers. Accordingly, it has been postulated that the neuralizing factor might act on the membranes of target cells to stimulate release of a secondary messenger molecule (possibly 3′,5′-monophosphate or cAMP); the mesodermalizing factor, entering the target cell by simple diffusion, might complex with an intracellular cytoplasmic-specific binding protein or receptor. Presumably, the secondary messenger molecule or the inductor–receptor complex passes into the nucleus where it selectively binds to chromatin material, thereby derepressing specific genes for transcription of mRNA. Unfortunately, there have been few studies to test the validity of these hypotheses. The cAMP secondary messenger hypothesis has been tested by Wahn and his associates (1975). They excised presumptive epidermis from the dorsal lip stage of several amphibian species and cultured the tissue fragments with various nucleotide compounds, including 3′,5′-monophosphate (dibutyryl cAMP). Although their results varied somewhat from one species of amphibian to another, in all cases several cell types characteristic of neural differentiation were recognizable in those cultures treated with dibutyryl cAMP. Their experiments do not prove per se that cAMP is the normal agent mediating the effects of the primary inductors. Further work is required to determine if, in fact, there is a rise in cAMP activity in presumptive neural plate cells during primary induction. Experiments by other investigators have failed to confirm the inducing ability of cyclic nucleotides.

L. G. Barth and L. T. Barth, long interested in the problems of embryonic induction, have offered a hypothesis that attempts to associate local physiological factors, such as changes in ion concentrations, with consequent modification of gene activity leading to differentiation. They argue strongly that the actual process of induction is initiated by an alteration in cell membrane properties of the target ectoderm which, in turn, results in the release and redistribution of cations within the cells. Their hypothesis also holds that there are no net changes in total ion concentration, but only a change in the ratio of bound to free ions in the cells of the early gastrula. Their observations are based on an extensive series of experiments in which it could be demonstrated that induction of nerve and pigment cells in the presumptive epidermis of the frog gastrula was dependent on the concentration of the sodium ion in the culture medium. Nerve and pigment cells are induced when the culture medium contains a sodium concentration of 88 m*M;* no inductions of these cell types occur at a lower concentration (such as 44 m*M*). Barth and Barth interpreted their data to mean that normal embryonic induction requires an internal supply of ions that are released from bound storage sites during late gastrulation. Whether free cations may regulate gene expression in the target cells of ectoderm by freeing segments of DNA from their histones is still very unclear. Sodium might compete with neighboring groups on DNA during induction, thereby resulting in a series of anabolic processes leading to the differentiation of neural tissue. Since the regulation of many metabolic processes and ionic concentrations is cAMP dependent, it is conceivable that the adenosine 3′,5′-monophosphate may play a role in the modification of the ratio of bound to free cations during induction.

Most of the investigators studying neural induction consider only the initial reactions of the target cells. Yet there is some evidence that communication between the target ectodermal cells may be an important aspect of the inductive process. Deuchar (1970) has shown that a minimum number of cells is necessary for neural differentiation in *Xenopus*. The success of neuralization appears to depend in this frog on an adequate number of ectodermal cells. No neural differentiation was visible when less than 10 ectodermal cells from the neural plate region of late gastrulas were cultured with mesodermal cells. However, groups of fewer than 10 cells, when first cultured with dorsal lip and then seeded into a larger ectodermal mass, were able to stimulate the latter to form neural structures. The quality of the neural differentiation, therefore, possibly depends on some second-stage interaction involving the transfer of molecules between the ectodermal cells.

Future investigators will undoubtedly concentrate on the molecular alterations and interactions that take place at the surface as well as inside the target ectoderm during primary embryonic induction. Solving the "old problem" of induction depends on knowing the complete chain of events from synthesis and transmission of inducing signals to initiation of the determinative and differentiative steps of neuralization.

## References

Barbieri, F., S. Sanchez, and E. Del Pino. 1980. Changes in lectin-mediated agglutinability during primary embryonic induction in the amphibian embryo. J. Embryol. Exp. Morphol. 57:95–106.

Barth, L. G. and L. T. Barth. 1969. The sodium dependence of embryonic induction. Dev. Biol. 20:236–262.

Brun, R. and J. Garson. 1983. Neurulation in the Mexican salamander *(Ambystoma mexicanum):* a drug study and cell shape analysis of the epidermis and the neural plate. J. Embryol. exp. Morphol. 74:275–295.

Burnside, B. 1973. Microtubules and microfilaments in amphibian neurulation. Am. Zool. 13:989–1006.

Chung, H. and G. Malacinski. 1975. Repair of ultraviolet irradiation damage to a cytoplasmic component required for neural induction in the amphibian egg. Proc. Natl. Acad. Sci. U.S.A. 72:1235–1239.

Deuchar, E. 1970. Neural induction and differentiation with minimal numbers of cells. Dev. Biol. 22:185–199.

Duprat, A., L. Gualandris, and P. Rouge. 1982. Neural induction and the structure of the target cell surface. J. Embryol. Exp. Morphol. 70:171–187.

Gallera, J., G. Nicolet, and M. Baumann. 1968. Induction neurale chez le oiseaux à travers un filtre millipore: etude au microsope optique et électronique. J. Embryol. Exp. Morphol. 9:439–450.

Grunz, H. and J. Staubach. 1979. Cell contacts between chorda-mesoderm and the overlying neuroectoderm (presumptive central nervous system) during the period of primary embryonic induction in amphibians. Differentiation. 14:59–65.

Karfunkel, P. 1971. The role of microtubules and microfilaments in neurulation in *Xenopus laevis*. Dev. Biol. 25:30–56.

Karfunkel, P. 1972. The activity of microtubules and microfilaments in neurulation in the chick. J. Exp. Zool. 181:289–302.

Kelly, R. O. 1969. A electron microscopic study of chordamesoderm–neuroectoderm association in gastrulae of a toad, *Xenopus*. J. Exp. Zool. 172:153–179.

Malacinski, G. 1974. Biological properties of a presumptive morphogenetic determinant from the amphibian oocyte germinal vesicle nucleus. Cell Differ. 3:31–44.

Malacinski, G., A. Brothers, and H. Chung. 1977. Destruction of components of neural induction system of amphibian egg with ultraviolet irradiation. Dev. Biol. 56:24–39.

Malacinski, G., H. Chung, and M. Asashima. 1980. The association of primary embryonic organizer activity with the future dorsal side of amphibian eggs and early embryos. Dev. Biol. 77:449–462.

Manorama, J., J. Janacek, J. Born, P. Hope, H. Tiedemann, and H. Tiedemann. 1983. Neural induction in amphibians. Transmission of a neuralizing factor. Roux's Arch. Dev. Biol. 192:45–47.

Nieuwkoop, P. D. 1973. The "organization center" of the amphibian embryo: its origin, spatial organization, and morphogenetic action. Adv. Morphog. 10:1–39.

Niu, M. and V. Twitty. 1953. The differentiation of gastrula ectoderm in medium conditioned by axial mesoderm. Proc. Natl. Acad. Sci. U.S.A. 39:985–989.

Odell, G., M. Oster, P. Alberch, and B. Burnside. 1981. The mechanical basis of morphogenesis. I. Epithelial folding and invagination. Dev. Biol. 85:446–462.

Portch, P. and A. Barson. 1974. Scanning electron microscopy of neurulation in the chick. J. Anat. 117:341–350.

Saxén, L. 1980. Neural induction: past, present, and future. Curr. Top. Dev. Biol. 12:409–418.

Saxén, L. and S. Toivenen. 1961. The two-gradient hypothesis in primary induction. The combined effect of two types of inductors mixed in different ratios. J. Embrol. exp. Morphol. 9:514–533.

Saxén, L., S. Toivonen, and T. Vainio. 1964. Initial stimulus and subsequent interactions in embryonic induction. J. Embryol. Exp. Morphol. 12:333–338.

Saxén, L. and S. Toivonen. 1962. Primary Embryonic Induction. Englewood Cliffs, N.J.: Prentice-Hall.

Schroeder, T. E. 1973. Cell constriction: contractile role of microfilaments in division and development. Am. Zool. 13:949–960.

Spemann, H. 1938. Embryonic Development and Induction. New Haven, Conn.: Yale University Press.

Spemann, H. and H. Mangold. 1924. Über Induktion von Embryonalanlagen durch Implantation artfremder organisatoren. Wilhelm Roux' Arch. Entwicklungsmech. Org. 100:599–638.

Stanisstreet, M. and E. Deuchar. 1972. Appearance of antigenic material in gastrula ectoderm after neural induction. Cell Differ. 1:15–18.

Takata, K. K. Yamazaki, and R. Ozawa. 1981. Use of lectins as probes analyzing embryonic induction. Roux's Arch. Dev. Biol. 190:92–96.

Tarin, D. 1971. Scanning electron microscopical studies of the embryonic surface during gastrulation and neurulation in *Xenopus laevis*. J. Anat. 109:535–547.

Tarin, D. 1972. Ultrastructural features of neural induction in *Xenopus laevis*, J. Anat. 111:1–28.

Tiedemann, H. and J. Born. 1978. Biological activity of vegetalizing and neuralizing inducing factors after binding to BAC-cellulose and CNBr-Sepharose. Roux's Arch. Dev. Biol. 184:285–299.

Toivonen, S. 1979. Transmission problem in primary induction. Differentiation 15:177–181.

Toivonen, S. and L. Saxén. 1968. Morphogenetic interaction of presumptive neural and mesoderm cells mixed in different ratios. Science 159:539–540.

Waddington, C. H. 1966. Principles of Development and Differentiation. New York: Macmillan.

Waddington, C. H. and G. A. Schmidt. 1933. Induction by heteroplastic grafts of the primitive streak of birds. Wilhelm Roux' Arch. Entwicklungsmech. Org. 128:522–563.

Wahn, H., L. Lightbody, T. Tchen, and J. Taylor. 1975. Induction of neural differentiation in cultures of amphibian undetermined presumptive epidermis by cyclic AMP derivatives. Science 188:366–368.

Yamada, T. 1958. Induction of specific differentiation by samples of proteins and nucleoproteins in the isolated ectoderm of *Triturus* gastrulae. Experientia 14:81–87.

# 11

# CELL DIFFERENTIATION

The earliest stages in the life of the embryo are visible in the multiplication of cells and the organization of these cells into a multicellular blastula. These morphogenetic events are accompanied by mitotic spindle assembly, the formation of cell membranes, and the synthesis of nucleic acids and various proteins, including actin, histones, and tubulin. As observations from several animal groups have shown, the morphological and biochemical activities of the early embryo up to gastrulation proceed in the absence of information generated by the embryonic genomes. At about the time of gastrulation, the embryo enters an extended period of time during which cell lineages are clearly established and a vast array of new, specialized cell types appear that will become ordered into tissues and organs. Hence, cells descendant from a single, complex cell, the fertilized egg, gradually come to differ from one another and form tissues and organs performing specialized functions.

The development of specialized cell types or tissues is a central theme of embryogenesis. The process by which cells acquire their morphological and biochemical properties to perform special functions in the whole organism is known as cell differentiation. The differentiated cell, then, is a unique functional phenotype; it typically synthesizes a distinctive protein or set of proteins (e.g., hemoglobin in the red blood cell; actin and myosin in the muscle cell) and may possess a characteristic assemblage of cytoplasmic organelles (e.g., microtubules of the flagellated spermatozoon). The fully differentiated cell is one that typically has lost its ability to proliferate and is very stable; that is, the changes leading to the *differentiated state* tend to be permanent and irreversible. However, there are conditions under which differentiated cells may show minor variations in the appearance of their phenotype; these fluctuations in phenotype are commonly referred to as *modulations*. The transition between the undifferentiated cell (or one without immediate specialized structure and function) and the differentiated cell does not necessarily mean a transformation from a simple to a more complex state. Red blood cells and cells of the adult lens, for example, lack nuclei and cytoplasmic organelles such as mitochondria. They are less complex than most embryonic cells with respect to these structural properties.

Since the discovery of the role of DNA in determining the synthesis of RNA and protein, it is now widely accepted that the structural and functional parameters of a differentiated cell are the result of gene activity. The fundamental significance of gene action in development and differentiation is not disputed. It follows, then, that cell types could arise during development either as a consequence of the presence of different genes or by the controlled expression of genes that are always present whether or not they are expressed. Evidence from various types of experiments now lead us to believe that only a small fraction of the DNA is transcribed in a differentiated cell. Most of the genetic information is regarded as being repressed or inhibited. An interesting problem in differentiation, therefore, is how large segments of the genome become silenced as a cell passes from the undifferentiated to the differentiated state. We will review the experimental basis for the concept that cell specialization involves the expression of selected portions of the genome in different cell types.

Without question, one of the major problems faced by the developmental biologist is the regulation of gene expression during differentiation or how different portions of the genome are utilized to produce diverse cell types. If, for example, liver-specific proteins are encoded by a set of genes that must be transcribed in the appropriate cell type, then how is this set of genes activated? What mechanisms determine that the liver-specific genes are to be transcribed into mRNA? Are other genes transcribed as well that result in RNA sequences that do not appear appropriate to the specific differentiated function of the liver cell? How does this set of cell-specific genes become activated during the course of development? Most studies on gene regulation during development have been limited to certain terminally differentiated tissues or cell types, particularly those with easily identifiable biochemical markers. However, differentiation is an ongoing process in the embryo, and cells pass through stages when they have the potential for several pathways of development. As differentiation proceeds, the potential for alternative pathways becomes progressively restricted and is eventually lost. It is unlikely, therefore, that the genes involved in the final stages of cellular differentiation are the same as those associated with the initial steps in the specification and determination of cell lines. Although the cytoplasm undoubtedly plays an important role in controlling gene activity during the early stages of development, little is known of the molecular and genetic basis of differentiation at these times. Cloning techniques are now being employed to examine the regulation of genes at the very early stages of development, or long before there is terminal differentiation of tissues. Sargent and Dawid (1983) have identified a special class of polyadenylated RNAs that is differentially expressed in gastrula embryos of *Xenopus laevis*. These RNAs do not generally occur in unfertilized eggs and blastulae, and may encode proteins that are critical to the process of gastrulation.

The appearance of a specific protein(s) within a cell implies that its corresponding structural gene has been activated and transcribed. Particularly at the terminal stages of cellular differentiation, regulation of gene activity would appear to be primarily at the level of transcription. Recall, however, that there are a series of complex steps intervening between the initial step in information processing (i.e., transcription) and the appearance of a functional protein (Fig. 11–1). At any of these steps, controls could be exerted to regulate the expression of genes. For example, there is now considerable evidence for a number of proteins that they are encoded by cytoplasmic mRNAs smaller than the primary transcripts (i.e., nuclear or heterogeneous RNA). How is the cytoplasmic mRNA fashioned from a larger precursor molecule of RNA? Does a cell translate all mRNAs in its cytoplasm, or are there mechanisms that control access to the ribosome, thereby biasing the spectrum of proteins synthesized?

Although answers remain incomplete to questions regarding how changes in the pattern of expression of nuclear genes are regulated, advances in the technology of handling genes will permit experiments with embryonic and adult tissues whose results should enhance understanding of how cell lines become progressively restricted during development and how diverse cell types are produced. This chapter is devoted to a general discussion of cell differentiation and the mechanisms controlling gene expression during the development of a cell's phenotype. Aspects of cell differentiation will also be found in the chapters dealing with individual organ systems.

## Constancy of the Cell Genome

As previously pointed out (Chapter 9), results from transplantation studies, particularly with amphibian embryos, indicate that nuclei of various cell types in most animal organisms appear to contain the same genetic information. The formation of

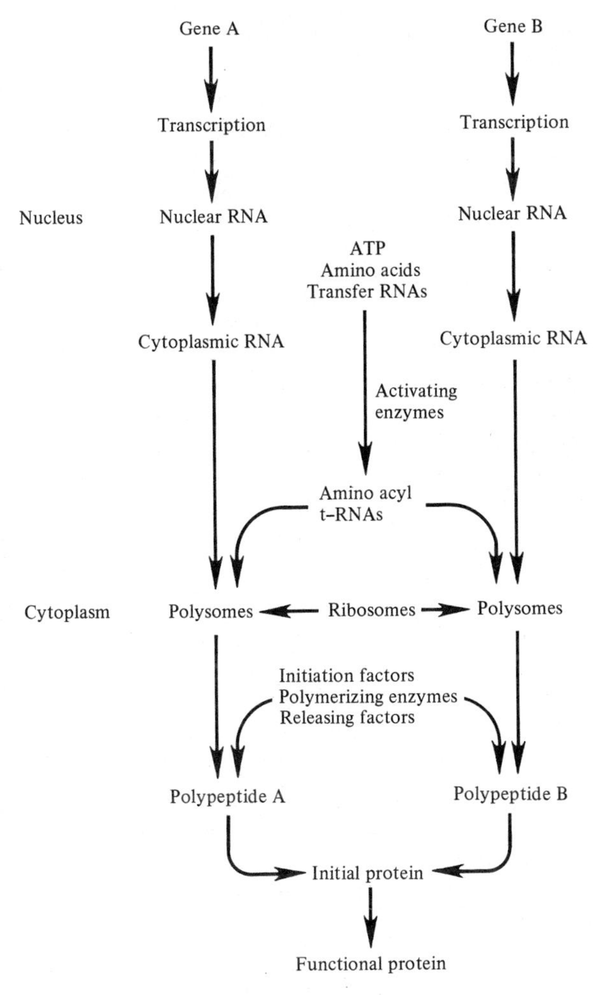


**FIG. 11–1.** A summary of the flow of genetic information during the synthesis of a typical protein involving two different polypeptides. Control of the expression of one or both genes could be exerted at any step.

specialized cell types, therefore, must depend on mechanisms controlling which genes are to be active or inactive within the nucleus. The nucleus is obviously important in differentiation, because it is the location of the genetic information that specifies the instructions for the synthesis of tissue-specific proteins. To this day, nuclear transplantation is the most stringent, functional test of the genetic potentialities of living nuclei; the manipulation of embryonic and adult nuclei has allowed critical examination of the concept of nuclear equivalence.

One of the earliest and most elegant demonstrations of the role of the nucleus in controlling cell type expression comes from the grafting and nuclear-transplantation experiments conducted by Hämmerling using *Acetabularia. Acetabularia* is a marine, unicellular green alga that is regionalized into an apical cap (up to 1 cm in diameter), a stem or stalk (4–6 cm in length), and a caudal portion or rhizoid (Fig. 11–2). The rhizoid anchors the cell to the substratum and houses the nucleus. Various species of *Acetabularia* show variation in the size and shape of the cap. In *A. crenulata,* the cap is lobate in shape with many fingerlike projections, while the cap of *A. mediterranea* is smooth and circular (Fig. 11-2 A).

The large size of this single cell permits portions of the cytoplasm to be manipulated and the nucleus transplanted or grafted to another cell of the same or a different species. Such experimental manipulations allow an evaluation of the nucleus and cytoplasm in regulating the morphogenesis of this cell. For example, if the cap is surgically removed from the rest of the cell, a new cap is quickly regenerated (Fig. 11–2 B). This can be successively repeated, a new cap being formed as long as there remains a nucleated stem. However, a capless stem severed from the rhizoid (i.e., an enucleated stem) will regenerate only once (Fig. 11–2 B). Additionally, if an enucleated stem is cut into three pieces, each fragment shows a different regenerative capacity. These observations indicate that cap regeneration is under nuclear control and requires substances passing from the nucleus into the cytoplasm. Presumably, these substances are organized in an apical–basal gradient in the cytoplasm with the highest concentration near the cap.

Hämmerling evaluated the role of the nucleus in determining the specificity of the cell's characteristics by conducting grafting experiments between two species of *Acetabularia* with known differences in cap morphology. When the rhizoid with its nucleus of *A. mediterranea* is grafted to the stalk of *A. crenulata* from which the cap has been removed, the first cap to regenerate is an intergrade having features characteristic of the caps of both species (Fig. 11–2 C). If the cap of the interspecific hybrid is now severed, the regenerated cap resembles that of the parent that supplied the nucleated fragment (Fig. 11–2 C). Since the initial cap of an interspecific hybrid is intermediate in form, it is presumed that the cytoplasm contains cap-determining information stored from both species. The absence of cap features typical of one of the species after the initial regeneration indicates that cap-determining

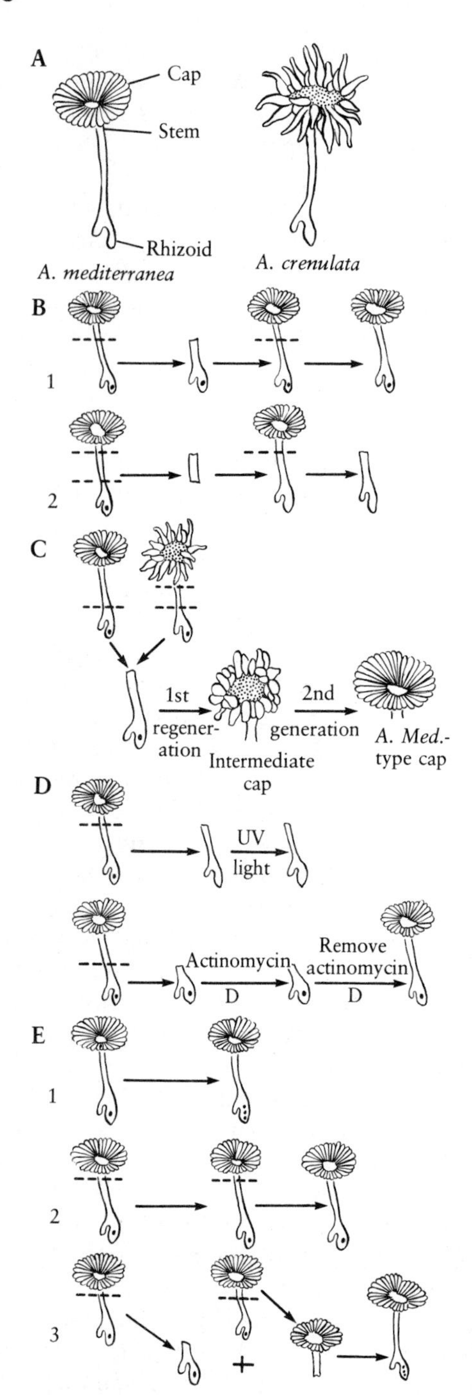


substances of the graft species are exhausted after one generation.

Information flowing from the nucleus appears to be responsible for cap formation. The morphogenetic substances produced by the nucleus and released into the cytoplasm are probably in the form of stable RNA molecules. RNA synthesis has been shown to be essential in cap regeneration. Exposure of nucleated stalks to UV irradiation or treatment with transcriptional inhibitors such as actinomycin D prevent the formation of a cap (Fig. 11–2 D). The same nucleated rhizoid washed free of actinomycin D shows a resumption of RNA synthesis and the regeneration of a cap. Presumably, stable informational mRNAs produced by the nucleus are released into the stem where they are stored. Upon activation, these informational molecules are translated into the proteins required for cap morphogenesis. Although these molecules are supplied by transcription of nuclear DNA, their expression is apparently regulated by control mechanisms in the cytoplasm.

Various processes in *Acetabularia* also demonstrate that activities within the cytoplasm strongly influence nuclear activity. Normally, the primary nucleus in this organism divides successively to produce numerous small secondary nuclei when its cap reaches a maximum size. These secondary nuclei are carried by protoplasmic streaming into the cap, where each forms a cyst. After a period of encystment, each secondary nucleus matures into a single gamete. If the cap of *Acetabularia* is excised before reaching its normal size, the divisions of the primary nucleus do not occur (Fig. 11–2 E). Division of this organelle to produce daughter nuclei will be delayed until a cap size of maximum dimensions has been produced. Conversely, division of the primary nucleus can be induced prematurely by grafting an enucleated fragment having a mature cap onto a young nucleated rhizoid without a cap (Fig. 11–2 E).

**FIG. 11–2.** Nucleocytoplasmic interactions in the unicellular alga, *Acetabularia*. A, the structural organization of the cell in two species of *Acetabularia*. Note the differences in the shape of the cap between *A. mediterranea* and *A. crenulata*. B, a nucleated fragment from which the cap has been removed will repeatedly regenerate a new cap (1). A stem from which the cap has been removed will regenerate a new cap for only one generation (2). C, a uninucleate graft formed by an *A. mediterranea* rhizoid and a capless *A. crenulata* stem will regenerate a cap intermediate in characteristics between the two species. The cap of the second generation will have the characteristics of the species donating the nucleus. D, cap regeneration is sensitive to UV irradiation and actinomycin D; E, the nucleus divides when the cap reaches maximum size (1). If the cap is removed before reaching the size, the nucleus never divides (2). Grafting of a mature cap to a young rhizoid prematurely induces division of the nucleus (3).

The development and perfection of the technique of transplanting nuclei from differentiated cells into the cytoplasm of uncleaved eggs have provided a firm answer to the question whether the many specialized cells of the adult organism contain the entire genome (i.e., a complete set of genes). You will recall that the typical nuclear transplantation technique involves the microinjection of a previously isolated nucleus from some selected tissue into an activated, enucleated frog's egg. Amphibian eggs are convenient host cells because of their large size and ability to remain viable after manipulation. When nuclei from the skin cells of the web of the adult frog's foot, for example, are isolated and microinjected into eggs, many of the nuclear transfers develop to the heartbeat and muscular response stages of the tadpole. Hence, somatic cell nuclei of specialized cell types possess genes necessary for the differentiation of a number of cell types, such as lens, muscle, and nerve. Experiments of this type support the view that the nucleus contains genes directing the differentiation of many cell types, but they also demonstrate that permanent genetic changes do not necessarily accompany the process of cell differentiation; in other words, unexpressed genes in certain cell types can be activated under appropriate conditions. Other evidence offered to support the concept of nuclear equivalence includes the identification of gene products in cells where they are not normally detected. For example, crystallins are the unique proteins of the adult lens fiber cells. Such proteins, however, have been confirmed to be present in other tissues, such as the iris when experimentally induced to form lens fiber cells. The conclusion from these various experiments is that the nuclei of fully differentiated cells can promote the differentiation of a diverse number of cell types. Differences between cell types must therefore depend on differences in the preferential selection of which genes are to be activated and which genes are to be inhibited. Hence, the process of differentiation is not typically accomplished by the loss of genes from some cells and the retention of other genes in other cells.

There are of course some exceptions to the concept that cells differentiate on the basis of the differential regulation of the same full set of genes that was present in the fertilized egg. There are examples during the development of certain insects and nematodes in which gene loss appears to be a prerequisite to somatic cell differentiation. Before the turn of the century, Boveri observed in *Ascaris* that differences in the origin of the somatic and germ cell lines could be traced back to the two-celled stage. As shown in Figure 11–3, the first cleavage division in *Ascaris* is equatorial, although each blastomere has a nucleus similar to that of the zygote. With the second division, however, a large portion of the chromosomal material of the animal hemisphere blastomere is pinched off into the cytoplasm where it degenerates. The two daughter cells (as well as the descendant cells) formed as a result of *chromosome diminution* will contribute the somatic cells of the embryo and the adult organism. The blastomere in the vegetal pole divides normally. However, with the following division of these two blastomeres, one undergoes diminution to yield one somatic cell and one germ cell. A similar process occurs with the fourth division. At the

**FIG. 11–3.** Chromosomal diminution and the determination of germ cells in *Ascaris,* A, the 2-celled stage with germlike stem cell (GC) and a presumptive somatic cell (PS); B, an advanced 2-celled stage in which chromatin is eliminated at the equator of the presumptive somatic cell; C, the 4-celled stage with eliminated chromatin on surface of two somatic cells; D, third cleavage division with second chromosome diminution in the precursor somatic cell; E, the 12-celled stage with third chromatin diminution; F, the 32-celled stage with the fourth chromatin diminution in progress. Note the single primordial germ cell. (From T. Boveri, 1925. The Cell in Development and Heredity by E. B. Wilson. The Macmillan Co., New York.)

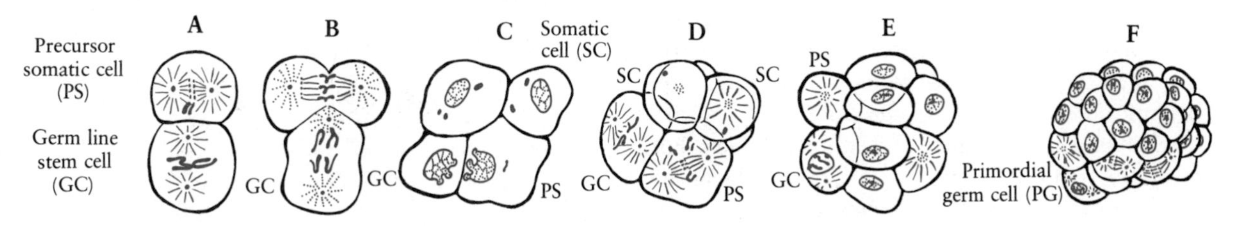

16-celled stage, therefore, only two blastomeres have the full genome. One of the blastomeres will retain its full genomic complement; it is the primordial germ cell and will, beginning with the 32-celled stage, give rise by mitosis to all the germ cells of the adult.

Also, as pointed out in Chapter 4, there are examples of differentiated cells in which the number of genes in the genome is higher than that typically expected. Variation in the number of copies of each gene through amplification is evident in those genes coding for ribosmal RNA in eggs of amphibians and in certain insect follicle cells whose genes specify the proteins of the egg envelope (i.e., *Drosophila*).

## Cytoplasmic Control of Nuclear Activity

In addition to demonstrating that differentiated cells have the same full set of genes that is present in the fertilized egg, nuclear transplantation studies have been employed to show that major changes in gene expression can be induced experimentally, particularly by manipulation of the cell's cytoplasm or by the microinjection of a nucleus of a given cell type into the cytoplasm of a second cell type performing a function quite distinct from the donor cell. The genome of a cell appears to respond to and be controlled by substances or determinants in the cytoplasm. Ultimately, therefore, the differentiation of cell types is a selective response by a complete set of genes to certain *cytoplasmic* or *gene-controlling factors*.

You will recall from Chapter 9 that the eggs of many invertebrate and vertebrate animal species have determinants localized in the cytoplasm. The segregation of these cytoplasmic factors or determinants can be correlated subsequently with the appearance of particular types of cellular differentiation in the embryo. A number of observations lead to the conclusion that nuclear activity under the influence of these cytoplasmic localizations is necessary for the differentiation of early cell types. First, alterations by either extirpation or centrifugation in the spatial arrangement of various cytoplasmic constituents of eggs produce embryos with predictable morphological and cellular defects. For example, removal of the anuclear polar lobe at the first cleavage division in *Ilyanassa* produces embryos lacking mesodermal cells and other lobe-dependent structures. Extirpation of the grey crescent material of the fertilized egg blocks the appearance of neural and notochordal cells in frog embryos. Removal of the yellow cytoplasm of the egg of *Styela* disturbs mesodermal differentiation. Second, suppression of specific activities of the nucleus by treatment of embryos with transcriptional inhibitors, such as actinomycin D, at specific times during development inhibits the subsequent expression of given cell types.

The germinal plasm of amphibian eggs and the pole plasm of insect eggs are perhaps the best examples of morphogenetic determinants influencing nuclear activity. In both groups of organisms, a very special type of cytoplasm containing visible *polar granules* is located near the vegetal pole of the unfertilized egg. This cytoplasm becomes segregated during cleavage into a few vegetal cells, which eventually migrate to the gonad and as primordial germ cells give rise directly to the gametes. The correlation between this specialized part of the cytoplasm and the determination of a given cell type has been demonstrated by manipulation of the germinal plasm either by centrifugation or by the introduction of accessory nuclei into the egg. If, for example, the fertilized egg of *Ascaris* is centrifuged just before cleavage so that each of the first two blastomeres receives germinal plasm, then after five divisions there will be 30 somatic and 2 germ line stem cells (compare Figs. 11–3 and 11–4). The implication from such an experiment is that the germinal plasm influences the resident nucleus of a blastomere to divide unequally. Also, if nuclei not normally found in the polar region of the egg are

**FIG. 11–4.** Redistribution of the germ cell determinants by centrifugation of the *Ascaris* egg before cleavage typically produces two primordial germ cells (PG). Note chromatin diminution in the somatic cells (SC). A and B are two examples of such centrifuged eggs. (From M. Hogue, 1910. Wilhelm Roux' Arch. Entwicklungsmech. Org. 29:109.)

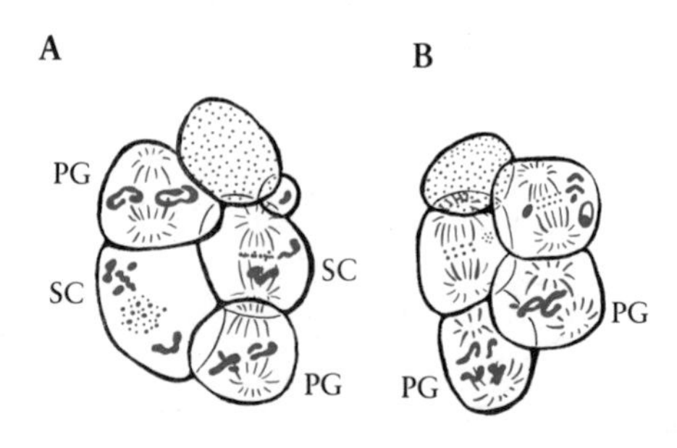

forced into this RNA-rich area, they become part of a population of blastomeres protected from chromosome diminution.

A very distinctive germinal plasm is found at one pole of the egg of *Drosophila* and other dipteran insects (Fig. 11–5). The egg of insects is typically oblong in shape and following fertilization undergoes an incomplete form of cleavage. Recall that the zygote nucleus is initially located in the center of the cell. Here it starts to divide, but without a corresponding division of the cytoplasm (Fig. 11–5 B,C). After several such nuclear divisions, the individual nuclei migrate toward the periphery of the cell. The nuclei that arise at one pole of the cell enter a thin disc of cytoplasm with germinal determinants (Fig. 11–5 C,D). Only those nuclei that come under the influence of the germinal plasm will differentiate as germ line stem cells. It appears that the polar granules of the germinal plasm are storage sites of mRNA that will be used in the synthesis of proteins determining the fates of these cells. If the germinal plasm is eliminated or destroyed by either heat or UV irradiation, the eggs will develop normally, but the resulting adults will lack gametes and be sterile. The best direct evidence that the pole plasm brings about the differentiation of germ cells comes from experiments conducted by Illmensee and Mahowald (1974) on *Drosophila*. The injection of polar plasm into UV-irradiated eggs results in embryos that will produce normal individuals (i.e., adults will be fertile). Also, they have demonstrated that anterior cells of the embryo, which normally differentiate into ectodermal structures, can be converted into germ stem cells by injecting polar plasm into the anterior region of a second (mutant) host embryo.

FIG. 11–5. Determination of the germ cells in a typical beetle. A, recently fertilized egg; B, cleavage showing division of nuclei without corresponding division of cytoplasm; C, nuclei arriving at the periphery initiate blastoderm formation; D, nuclei that invade germ cell plasm become blastomeres differentiating as primordial germ cells. (After R. W. Hegner, 1911. Biol. Bull. 20:237.)

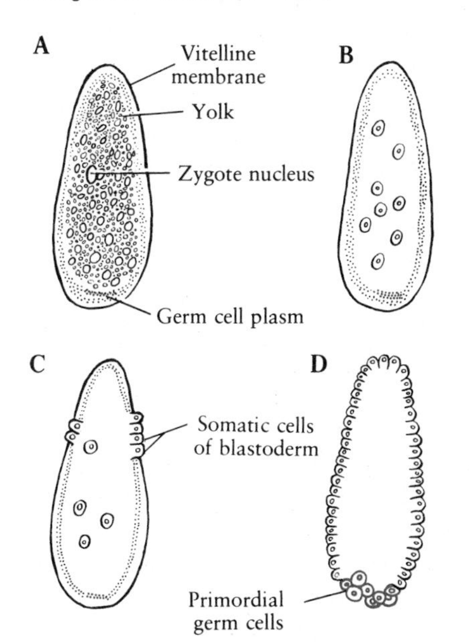


A powerful approach to the study of the role of the cytoplasm in the regulation of nuclear (gene) activity is that of nuclear transplantation. Introduction of the nucleus of a cell type performing a given function into the cytoplasm of another cell type with its own unique function and whose nucleus has been removed permits an analysis of the effect of the foreign cytoplasmic environment on the activities of the donor nucleus. For example, the chief activity of the nucleus following fertilization of the egg is intense DNA synthesis. When an adult frog nucleus from a brain cell, which actively synthesizes RNA (mainly rRNA) but not DNA, is injected into an artifically activated, enucleated frog's egg, the synthesis of RNA is observed to terminate and the nucleus then engages in its own synthesis of DNA (Fig. 11–6). The change in the activity of the nucleus takes place very rapidly. The synthesis of RNA is not detected until the transfer embryo reaches a stage when this nucleic acid normally appears. If a nucleus from the same brain tissue is injected into a ripe ovarian oocyte, whose primary activity is RNA but not DNA production, the nucleus responds by synthesizing only RNA (Fig. 11–6). Also, nuclei of mature red blood cells, which synthesize neither DNA nor RNA, commence to produce both DNA and RNA following their transfer into the cytoplasm of a rapidly dividing cell population.

Other experiments, particularly from the laboratory of Gurdon and his co-workers, demonstrate that the cytoplasm of the egg appears to contain gene-regulating factors that selectively control the activity of different classes of genes within the nucleus (Table 11–1). As pointed out previously, there is a changing but predictable temporal pattern of RNA synthesis during development. Typically, the predominant class of RNA synthesized through midcleavage is nuclear. Transfer (4S) RNA is detected by late cleavage and ribosomal RNA by the

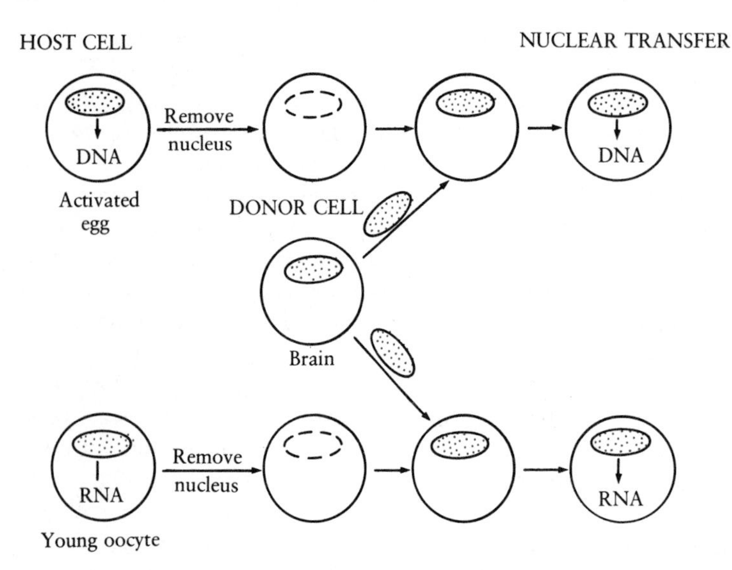


**FIG. 11–6.** Diagrammatic illustration of studies by Gurdon and others to show that a frog brain nucleus following transplantation into an enucleated, activated egg (which promotes DNA synthesis) or a young oocyte (which promotes RNA synthesis) produces a gene product dictated by the synthetic activity of the host cell.

end of gastrulation. When the nucleus of a frog neurula (which synthesizes mainly rRNA) is transplanted into an enucleated frog's egg, the nucleus behaves as if it is reprogrammed to function like the fertilized egg nucleus. RNA synthesis ceases immediately and cannot be detected until mid-cleavage, at which time the transfer embryos display the nuclear, transfer, and ribosomal pattern of RNA synthesis observed in control embryos. Hence, the cytoplasm must contain a factor(s) that temporally inhibits or shuts down the activity of genes specifying one product, in this case rRNA. Also, the cytoplasmic gene-regulating factors must be heterogeneous, because the different genes responsible for the synthesis of different RNAs are selectively activated. The gene-regulating substances are probably specific for each animal species. For example, the transplantation of a neurula nucleus of *Xenopus* into an enucleated egg of the frog *Discoglossus* results in the expression of only some of the classes of RNA found in normal *Xenopus* eggs. The cytoplasm of the eggs of *Discoglossus* apparently lacks certain cytoplasmic factors required to activate the RNA genes in the donor nucleus.

TABLE 11–1 Summary of nuclear transfer experiments in *Xenopus laevis* demonstrating an influence of living cytoplasm on nuclear activity[a]

| | *Synthetic activity of embryos*[b] | | | |
|---|---|---|---|---|
| | *DNA* | *nRNA* | *tRNA* | *rRNA* |
| Neurula cell (donor nuclei) | − | ++ | ++ | ++ |
| Nuclear-transplant embryos | | | | |
| Uncleaved egg (1 hour after transfer) | ++ | −− | −− | −− |
| Midblastula (7 hours after transfer) | + | ++ | −− | −− |
| Late blastula (9 hours after transfer) | + | ++ | ++ | −− |
| Neurula | − | ++ | ++ | ++ |

[a]Single neurula nuclei were transplanted to enucleated eggs, which were labeled with [$^3$H]uridine for one to two hours at various stages during their subsequent development. For details of experiments, see Gurdon and Woodland (1969).

[b]Symbols: −−, no detectable synthesis; −, ≈ 10 percent of nuclei active; +, ≈ 50 percent of nuclei active; ++, rapid synthesis in nearly all nuclei.

From J. B. Gurdon. 1969. In: Communication in Development. A. Lang, ed. 28th Symposium of the Society of Developmental Biology. New York: Academic Press.

The results of nuclear transplantation experiments described above clearly show that egg cytoplasm can alter gene expression. But, does a cell, such as an egg, contain cytoplasmic substances that can control independently the activity of individual genes within the nucleus? To be able to detect changes in the activity of individual genes using the nuclear transplantation method, one must be able to identify the products of individual genes and determine whether the gene activity is due to host or donor components. In a typical experiment designed to determine if the cytoplasm can regulate the activity of individual genes, nuclei are isolated from a selected tissue type and injected into oocytes. The oocytes are then incubated in a medium containing a radioactive amino acid. At the end of the incubation period, the cells are frozen. The labeled proteins are the markers of gene activity in the transfer embryo. These are extracted from the eggs and analyzed by the very sensitive method of two-dimensional gel electrophoresis. Following the separation of proteins on the basis of electrical charge and molecular weight, their location on the gel is revealed either by staining with a protein stain or by special film fluorographic techniques that detect areas of radioactivity. The small protein molecules are visualized as spots. To guarantee that there will be enough protein for analysis and to eliminate any possible confusion with the background expression of the genes of the oocyte itself, as many as 200 nuclei are injected into an oocyte. The technique is very useful when one selects nuclei from tissues whose electrophoretic pattern of proteins is distinguishable from that of the occyte. As one might expect, when the proteins of any two different tissues are compared, some of the identifiable proteins are common to both tissue types (these are the so-called *housekeeping proteins*), while others are unique (these are the so-called *luxury proteins*) to one or both tissue types (Fig. 11–7). The luxury proteins are the expression of individual genes and are of particular interest to the investigator.

Several cell types have been used to demonstrate the shutdown of individual genes in nuclear transfer eggs. Human HeLa cells in culture synthesize some 20 different major proteins. These proteins are easily distinguishable on gels from those synthesized by the oocyte of *Xenopus*. If isolated HeLa nuclei are injected into an oocyte of *Xenopus*, there is no visible expression of HeLa genes. Between three and five days after injection, 3 of the 20 HeLa proteins are synthesized in amounts sufficient to be detected. No further HeLa proteins are expressed in the transfer embryos. It appears, therefore, that 17 recognizable HeLa genes are either turned off or function at such a low level that their products cannot be detected. The expression of 3 proteins, however, clearly suggests that the oocyte cytoplasm can selectively activate a group of HeLa genes.

Unfortunately, experiments such as these with HeLa cells do not provide unequivocal evidence for the reprogramming of nuclear genes. Failure of most of the HeLa genes to synthesize products in the oocyte cytoplasm could be an artifact of the nucleocytoplasmic combination. A better test of the concept that the cytoplasm can selectively reprogram nuclear activity would be a demonstration of the switching on of genes previously inactive in the nucleus of a somatic cell following transfer to oocyte cytoplasm. *Xenopus* oocytes synthesize about 15 major proteins not detectable in a culture line of *Xenopus* kidney cells. One would predict that the genes specifying the oocyte proteins might be activated upon transfer of the kidney nuclei into the oocytes. This experiment has been carried out, but the extensive accumulation of stored mRNA produced by the oocyte genes prevents determination of whether any of the oocyte-specific proteins originate from the transplanted kidney nuclei. However, the oocyte-specific proteins of eggs of *Pleurodeles waltlii* (a salamander) have been shown by electrophoresis to be clearly distinguishable from those of *Xenopus*. Taking advantage of this observation, De Robertis and Gurdon (1977) injected kidney nuclei of *Xenopus* into oocytes of *Pleurodeles* to determine if the unexpressed *Xenopus* oocyte genes in the former could be activated (Fig. 11–8). Within three to seven days following injection, three new proteins are detected in a two-dimensional gel that have size and charge characteristics of proteins normally produced in *Xenopus* oocytes. In other words, inactive oocyte-specific genes of *Xenopus* kidney cells become active upon transfer into the cytoplasm of the salamander egg. Also, eight proteins that are normally synthesized by the cultured kidney cells, but that are typically absent in oocytes of *Xenopus*, could not be resolved in the salamander oocytes. This observation confirms results obtained from HeLa transplantation experiments showing that genes can be turned off in oocyte cytoplasm.

Transplantation studies of this type in which normally inactive genes can be selectively activated support the view that expression of individual protein-coding genes of the nucleus is controlled by cytoplasmic regulatory molecules. In all nuclear transfer experiments, the pattern of gene expression typically reflects the pattern of gene activity characteristic of the host cell. Similar results have been obtained from *somatic cell hybridization* experiments. The technique of somatic cell hybridization involves the fusion of two cells, thereby permitting the confrontation of two genomes in different functional states. The properties of the resulting hybrid cells allow conclusions about the commitment and the capacity of a genome to carry out a given pattern of gene expression. Cell hybridization is usually carried out with cells of permanent lines often derived from tumors. Many of

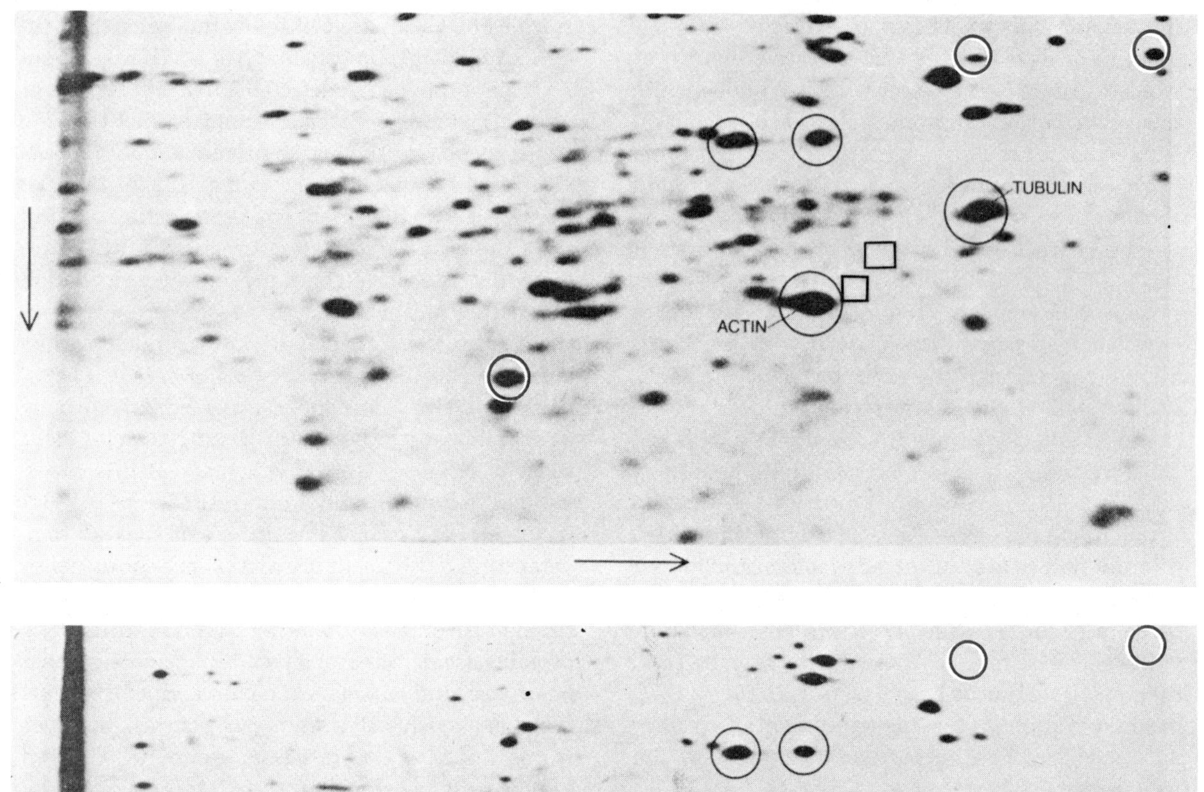


**FIG. 11–7.** Two-dimensional electrophoresis is a sensitive technique for the separation of proteins. The proteins are extracted from selected tissues, such as oocytes (A) and cultured kidney cells (B) of *Xenopus*, and then separated on the basis of electric charge (horizontal arrows) and then according to their molecular weight (vertical arrows). "Housekeeping" proteins (black circles) are synthesized by both cell types. "Luxury" proteins are synthesized specific to each cell type. Oocyte-specific proteins are made by the oocytes but not by the kidney cells (colored circles) while kidney-specific proteins are synthesized by kidney cells but not by oocytes (squares). (From E. De Robertis and J. Gurdon, 1979. Sci. Am. 241:74.)

these cell lines are characterized by the expression of functions specific to the tissue of origin. For example, fused cells formed from the union of mouse leukemic lymphoblast cells and hyperdiploid rat hepatoma cells produce both rat and mouse albumin. Normally, this protein is synthesized only by the rat hepatoma cells. The cytoplasm of this cell type apparently produces a diffusible substance(s) that activates the previously unexpressed mouse albumin gene to synthesize a liver protein. The activation of "silent" genes to express liver-specific proteins has also been observed in crosses of hep-

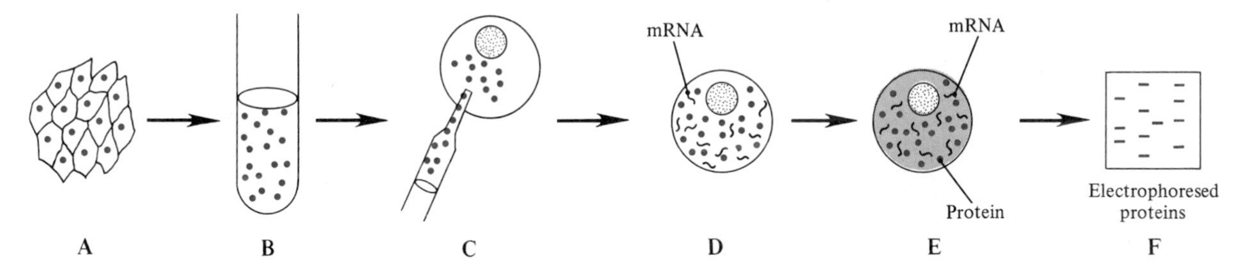


**FIG. 11–8.** Diagrammatic summary of an experiment performed by De Robertis and Gurdon (1977) showing that the cytoplasm of the salamander oocyte can reprogram the genes of kidney cells to form oocyte-specific proteins. Kidney tissue cells are cultured (A) and their nuclei subsequently isolated (B). These kidney cell nuclei (about 200) are injected into a salamander oocyte (C). Messenger RNA is transcribed for several days (D). The addition of a radioactive amino acid is followed by protein synthesis for about six hours (E). When the proteins are extracted and electrophoresed, genes characteristic of the frog oocyte are expressed (F). (After E. De Robertis and J. Gurdon, 1977. Proc. Natl. Acad. Sci. U.S.A. 74:2470.)

atoma cells with fibroblasts, lymphocytes, and melanoma cells. As one might guess, however, the loss of differentiated functions is routinely observed in hybrids between expressing and nonexpressing cells of the same ploidy. However, functions previously lost can be reexpressed in somatic hybrids, suggesting that the disappearance of function is probably due to a temporary block in the final expression of differentiation and not to an alteration in the whole genome. The various observations made on interactions in hybrid cells clearly suggest that there are two distinct levels of regulation in mammalian somatic cells. One level is concerned with genomic commitment (i.e., determination) and the other level with the expression of tissue-specific proteins.

Recent advances in techniques for the isolation and handling of genetic material now make possible promising approaches to identify the molecules that regulate gene activity. Given that the cytoplasm of eggs and oocytes must contain substances that regulate gene expression, then attempts could be made to identify these cytoplasmic components by reisolating injected nuclei after the nuclei have changed their biosynthetic activities. An approach toward this end has been to inject into the nucleus of an oocyte DNA or a large number of copies of a single gene along with an RNA precursor, guanosine triphosphate, labeled with radioactive phosphorus. The rationale behind injecting DNA or genes into frog oocytes is straightforward. If the oocyte contains regulatory molecules that are gene specific and bind to injected DNA, then these molecules could possibly be identified by reextracting the purified genes. A critical test of the value of such an approach is a demonstration that genetic material injected into oocytes can be transcribed accurately. De Robertis and Gurdon (1977) used purified DNA of simian virus 40 (SV 40) injected into *Xenopus laevis* oocytes for such an analysis. They showed that the viral DNA transcribes continuously over a period of several days and produces a quantity of SV 40-specific RNA that is roughly proportional to the amount of injected DNA. The viral RNA is then translated into viral proteins, thus clearly suggesting that at least some of the RNA transcribed from the viral DNA is correctly processed, transported into the oocyte cytoplasm, and translated by the protein-synthesizing machinery of the oocyte. The expression of the 5S ribosomal RNA gene following injection into *Xenopus* oocytes is further evidence that factors in the oocyte appear to recognize specific sites on the DNA, which then initiate a specific program of transcription. The haploid *Xenopus laevis* genome contains approximately 20,000 copies of oocyte 5S rRNA genes. The DNA coding for the 5S ribosomal RNA can be extracted from red blood cells of *Xenopus borealis*, purified, and then microinjected into oocytes of *Xenopus laevis*. When this is done, substantial amounts of RNA are synthesized having biochemical characteristics of the 5S rRNA of *X. borealis*. The selectivity of the expression of the DNA is indicated by the observation that only RNA polymerase type III promotes 5S rRNA synthesis. The amount of DNA specifying the 5S rRNA can be increased by *recombinant DNA techniques* (i.e., gene cloning). The key steps in the cloning of a gene are outlined in Figure 11–9. The gene to be isolated is selectively excised from the chromosome with spe-

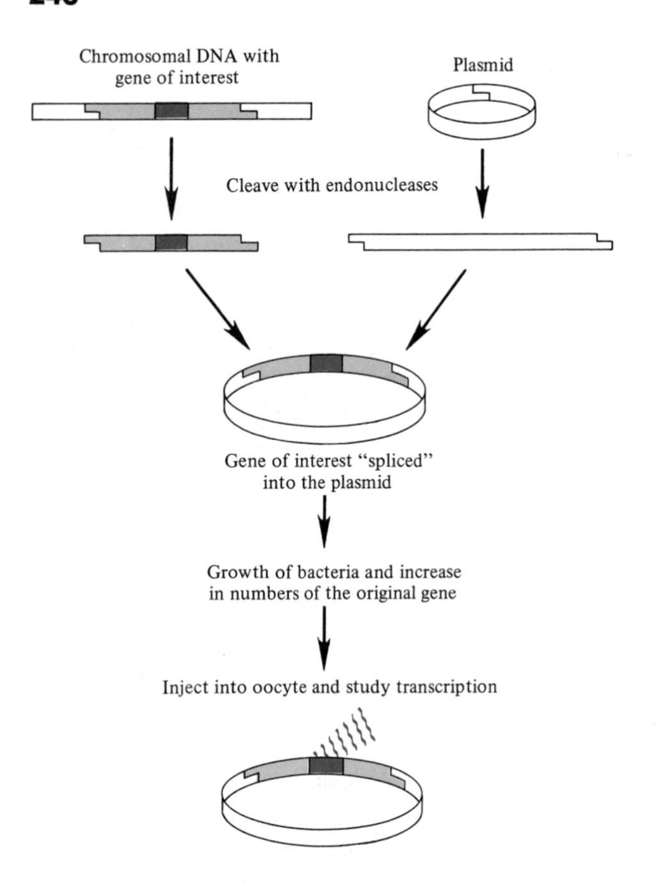


**FIG. 11–9.** The microinjection of cloned genes into oocytes is a technique being used to analyze the regulation of the expression of DNA. Chromosomal DNA carrying a gene of interest, such as the 5S RNA gene, is isolated with restriction enzymes (endonucleases). This DNA segment is then inserted into a circular molecule of bacterial DNA (known as a plasmid) from which a similar segment of DNA has been removed with the same endonucleases. The "sticky ends" of the chromosomal piece allow it to be religated or "spliced" into the plasmid. The gene multiplies with the growth of the bacteria. Genes cloned in this way can then be microinjected into frog oocytes and the subsequent transcription process studied.

cific *restrictive enzymes* known as *endonucleases*. These enzymes act at very specific sites along the length of the DNA. A consequence of this enzymatic action is that the ends of the chromosomal DNA segment containing the gene(s) of interest are very sticky. This DNA segment is then inserted into a circular molecule of bacterial DNA (known as a *plasmid*) from which a similar-length segment of DNA has been removed with the same endonucleases. The recombinant DNA is then allowed to replicate many times as the bacteria multiply. The DNA is then collected and microinjected into an oocyte. For the 5S rRNA cloned gene, it has been shown by Bogenhagen et al. (1982) that a region associated with the gene is necessary and sufficient to direct an accurate initiation of transcription. The region apparently controls transcription through its interaction with a 40,000-dalton protein that acts as a *positive* (activation) *transcription factor*. RNA polymerase III and at least two other components are also necessary for transcription of the 5S rRNA cloned gene.

The techniques associated with gene cloning offer great promise in future studies of gene regulation. They should, when coupled with the electron microscope, help identify regions of the DNA that are important in gene regulation. Since DNA injected into an oocyte nucleus is known to become associated with endogenous nucleoproteins to form stable chromatin, this DNA can be reextracted and examined for RNA and/or proteins that show specific binding to regions of the DNA. Also, since it is now possible to assemble chromatin in vitro, one might use labeled proteins to assemble chromatin and inject such complexes into oocytes. The use of an in vitro chromatin assembly with the native in vivo transcriptional machinery should provide additional insight into the problem of gene regulation.

## Cytological Evidence for Differential Gene Activity

Nuclear grafting and transplantation experiments lead to the conclusion that the structural and functional differences arising between cells during the course of development reflect differences in the utilization of genetic information. Evidence that variable gene activity can indeed be correlated with cell type comes from a variety of studies, but particularly from those in which alterations of the physiological activities of cells commence with visible changes in the gross organization of chromosomes. An outstanding example relating change in functional activity with differential gene expression is the *giant* or *polytene chromosomes* found in certain tissues of larvae and adult insects, including *Drosophila* and *Chironomus* (Fig. 11–10). Since the late 1800s, it has been known that these giant chromosomes are found in a number of larval tissues, including salivary glands, Malpighian tubules, seminal vesicles, and the gut. The large size of each of these chromosomes reflects certain peculiarities in the growth of the larval tissues, namely that the tissues grow by hypertrophy rather than by ordinary cell division. The homologous chromosomes remain permanently paired during interphase of cell division. Repeated DNA replication yields numerous chromatids that remain attached to each other. Consequently, a mature polytene chromosome may contain some 1000 strands of DNA and be some 10,000 times the "diameter" of a typical interphase chromosome. As can be seen from Figure 11–10, each polytene chromosome has a striking striated appearance, consisting at irregular intervals of alternating regions of darker and lighter bands. When an individual chromatid is examined under the light microscope, it shows periodically along its length regions that are highly coiled and condensed. Since all of the chromatids remain perfectly aligned, all of the coiled areas *(chromomeres)* are in register and thereby produce a band across the chromosome. The banding pattern is so characteristic for every tissue in which polytene chromosomes have been described that each band is given a specific designation.

The bands of polytene chromosomes have been the focus of considerable genetic and cytological research because of their identification with specific gene loci and cellular functions. Under normal conditions, the level of polyteny is sufficient only in salivary glands of larvae to permit detailed study of the chromosome organization and behavior. In various regions of a polytene chromosome, bands frequently show an enlarged or swollen appearance, producing what is termed a *chromosomal puff* (Fig. 11–10 B). A puff represents an uncoiled band and therefore appears lighter than adjacent regions of the chromosome. Very large, swollen puffs are termed *Balbiani rings*. Since the early studies of Beerman in the 1950s, there have been experimental data to support the hypothesis that puffs and Balbiani rings are sites of intense transcriptional activity and gene expression. When giant chromosomes are isolated and incubated in radioactive precur-

**FIG. 11–10.** A, reconstructions of the puffing sequence in the proximal arm of chromosome III in *Drosophila*. The numbers refer to specific bands of this polytene chromosome (After M. Ashburner, 1967, Chromosoma 21, 398); B, a photograph of chromosome III in *Chironomus pallidivitatus* showing a large or giant puff (Balbiani ring). (From U. Grossback, 1973. Cold Spring Harbor Symp. Quant. Biol. 38, 619.)

A

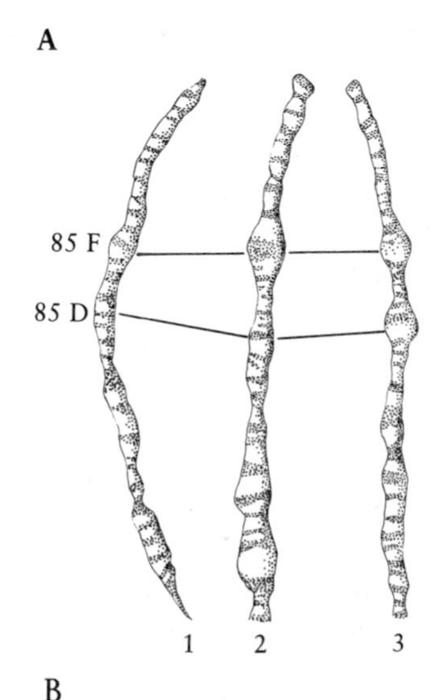


B

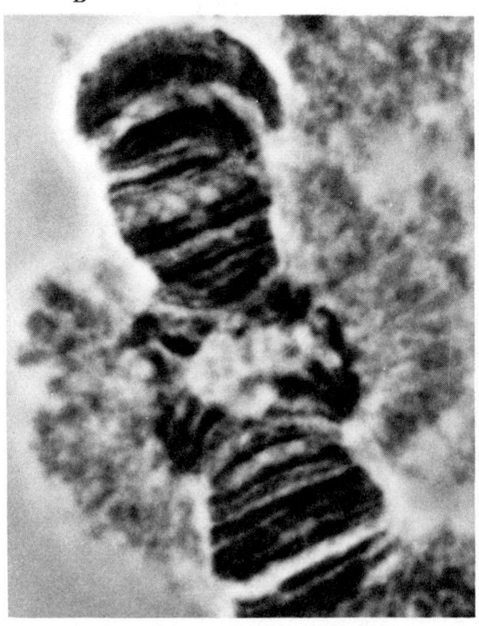

sors of RNA and protein, there is rapid uptake and incorporation of label into the puffs. In the case of RNA, this incorporation can be inhibited with either actinomycin D or *α-amanitin*. Regression of a puff is accompanied by a corresponding decrease in the uptake of labeled uridine or amino acid. Hence, chromosomal puffs are believed to be sites of forming ribonucleoprotein. The transcription process appears to be facilitated by uncoiling of segments of DNA, perhaps making the transcriptive sites more accessible to the action of RNA polymerase.

Patterns of puff formation and regression are particularly evident during insect metamorphosis, when there are dramatic changes in the structural and physiological activities of cells. Studies of the puffing patterns of polytene chromosomes, particularly in *Drosophila* and *Chironomus* (silkworm), indicate that gene expression is regulated in both time and space. *Hemimetabolous insects,* such as cockroaches and locusts, pass through several larval stages known as *molts* before they assume the adult form. In *holometabolous insects,* such as beetles, moths, and flies, the larva is transformed into a *pupa* before reaching adulthood. The major molts from larval to pupal stages and from pupal to adult stages are *metamorphoses.* The stages between larval molts are *instars* (i.e., intermolt periods). It is during the final instar (i.e., just prior to pupation) that there are striking and remarkable changes in the puffing pattern of the salivary giant chromosomes. Ashburner, Berendes, and Cherbas have studied in considerable detail the pattern of puffing, and hence the pattern of gene activity, both in vitro and in vivo during this critical phase of development. In the fruit fly there are about 10 prominent puffs that can be identified in salivary gland chromosomes prior to the molt forming the pupa. As formation of the pupa begins, there is a rapid regression of the "intermolt puffs" and the appearance of a small number of "early puffs" at new locations along the chromosome. The "early puffs" reach their maximal size within three to four hours, regress, and then are replaced by a very large number (about 100) of "late puffs." These regress at particular times during the next 10 to 20 hours. It has been estimated that some 200 puffs are expressed during the last instar (third) and pupal stages of *Drosophila* development.

The function of the giant salivary gland cells is the synthesis of a proteinaceous secretion *(glue polypeptides),* which is stored in membrane-bound vesicles for most of the last instar stage. This secretion is discharged by exocytosis into the lumen of the salivary gland some four to eight hours before the onset of metamorphosis. The salivary gland secretion of *Chironomous pallidivittatus* has drawn the attention of several investigators, because its synthesis is dependent on the regulation of gene activity in two spatially separate populations of cells (Fig. 11–11). Most of the gland cells produce a clear serouslike secretion. The polytene chromosomes of these cells show three unique Balbiani rings (BR1, BR2, BR3), suggesting that their expression presumably reflects the activation of genes directing the synthesis of the clear secretion proteins. BR1 and BR2 produce what is termed *giant mRNA;* the translation units are on the order of $10^6$ daltons. However, the four gland cells in the vicinity of the secretory duct produce a granular secretion whose composition is the same as the clear secretion except for the presence of an additional protein. These cells show Balbiani rings BR1, BR2, and BR3 as well as an additional puff. The fourth puff in these cells appears, therefore, to have the genetic information necessary for the synthesis of the protein that makes the secretion granular.

The function of the intense transcriptive activity that occurs before and during metamorphosis is far from clear. The so-called glue polypeptides manufactured by the salivary glands may play a role in attaching the case in which the pupa develops to an appropriate substratum. However, there are many

**FIG. 11–11.** The salivary gland of *Chironomus pallidivittatus* shows a group of four cells near the secretory duct. These cells produce the protein that makes the gland secretion granular. (From W. Beerman, 1961. Chromosoma 12:1.)

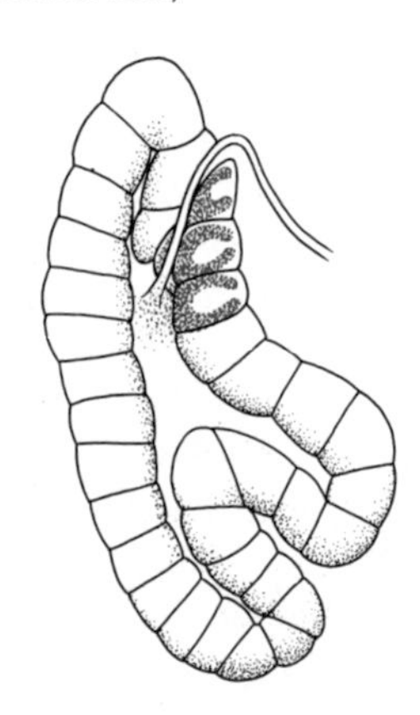

more puffs than identifiable salivary gland specific proteins. Some investigators have suggested that the increased transcription could reflect the production of enzymes involved in either the eventual lysis of cells that accompany metamorphosis or the storage of hemolymph proteins.

Molts and metamorphic events in insects are controlled by neuroendocrine hormones produced by several major organs, including the brain (Fig. 11–12). A hormone synthesized by the neurosecretory cells of the brain is released upon stimulation into the hemolymph or tissue fluid. With high titers of this hormone, the *prothoracic glands* (or ecdysial glands) of the thorax respond by releasing a steroid hormone called *ecdysone* (or *20-hydroxyecdysone*). Ecdysone is the insect molting hormone. The brain hormone also causes the *corpora allata* to release another hormone called *juvenile hormone;* it is the larval growth hormone.

Ecdysone and juvenile hormone interact in a complex way to determine the specific patterns of metamorphic change (Fig. 11–12). Ecdysone stimulates the epidermal cells of the integument to lay down a new cuticle and to release several hydrolytic enzymes that destroy the old cuticle. If the titer of the juvenile hormone is high in the tissue fluids, the cuticle laid down under the influence of ecdysone will be larval in character. When the concentration of juvenile hormone is low, as is the case during the last larval instar stage of holometabolous insects, the epidermis in the presence of ecdysone is triggered to form a pupal cuticle instead of a larval cuticle. Hence, ecdysone and variable titers of juvenile hormone act in concert to control the temporal and phenotypic expression of the epidermis. A variety of experiments provide support for this view. For example, a strip of pupal epidermis following transplantation into a young larva will secrete a larval cuticle when the larva molts. If the last instar stage is treated with juvenile hormone, the larva will undergo another molt, but produce a larval cuticle. We will have more to say below about the role of ecdysone in the regulation of gene transcription.

**FIG. 11–12.** The sequence of metamorphic events in a typical holometabolous insect. Ecdysone and variable titers of juvenile hormone act in concert to control the temporal and phenotypic expression of the epidermis (i.e., type of cuticle). P(L), larval proteins; P(P), pupa proteins; P(A), adult proteins. (From G. Tombes, 1970. An Introduction to Invertebrate Endocrinology. Academic Press, New York.)

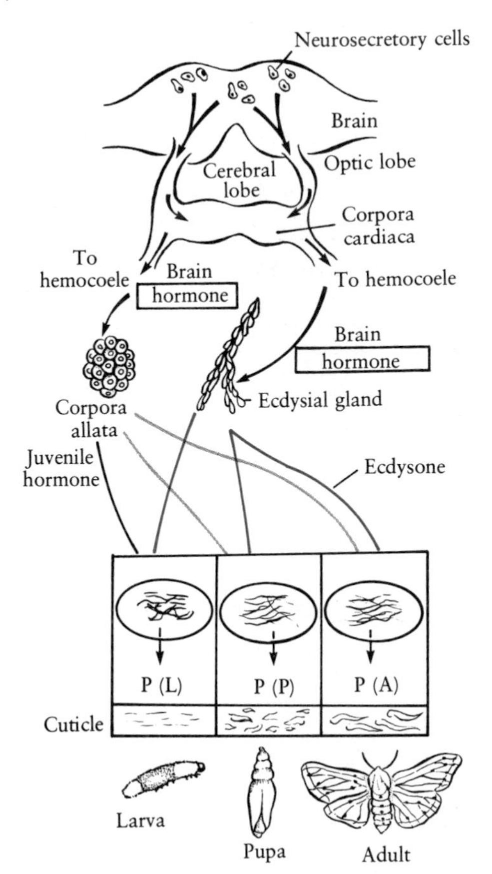


## Mechanisms of Control of Gene Expression

Both nuclear transplantation and somatic cell fusion studies show that the expression of nuclear genes is influenced by positive signals from the cytoplasm. The nature of these cytoplasmic components remains to be determined. However, the functional properties of a cell are altered by other factors that may in turn act to regulate gene activity and expression during development. These factors include signals produced by contact or diffusion between cells (cell-to-cell interactions), ions, cyclic nucleotides, and hormones. We have already referred in other chapters to the possible nature and function of inducer molecules in the differentiation of certain embryonic systems, such as the nervous system. Needless to say, difficulties encountered in isolating and identifying such molecules have precluded their effective use in analyses of cellular differentiation. Hormones, however, have been rather extensively used to examine the regulation of gene activity, particularly at the level of transcription, during the process of differentiation. There are several reasons for their widespread use. First, hormones are known to initiate or modulate gene activity in target cells at some distance from their site of production. They have dramatic but measurable effects on the state of differentiation of target cells. *Erythropoietin,* for example, is a hormone whose titer in the circulating blood is controlled by the level of oxygen tension. It stimulates the formation of erythrocytes and the synthesis of the characteristic red blood cell protein (globin) in the fetal and adult mammal (Chapter 18). In reptiles and birds, the liver is modified under elevated levels or prolonged exposure to estrogen into an organ that produces the precursors to the bulk of the egg-specific yolk proteins. Second, hormones can be administered with relative ease either to the intact animal or to organ cells isolated and cultured in vitro.

## Examples of Transcriptional Control

Direct hormonal control of transcription has been demonstrated in both invertebrate and vertebrate organ systems. The changing pattern of puffs in polytene chromosomes of certain insect tissues appears to be a promising model for analysis of regulation at the molecular level of transcription. A number of experimental manipulations point to the importance of ecdysone (or ecdysteroid hormones related to 20-hydroxyecdysone) in the control of puffing activity. When the salivary glands of mature *Drosophila* larvae are isolated and cultured, the addition of ecydsone provokes a discharge of the secretory product into the gland lumen. At the same time, the hormonal stimulation leads to a sequence of changes in chromosomal puffs identical to that occurring in the organism during normal metamorphosis (Fig. 11–13). Clever injected ecdysone into the last larval stage of *Chironomus tetans* and observed an immediate, premature puff within 15 to 30 minutes on the first chromosome at a site designated as I-18C (Fig. 11–14). This was closely followed by the appearance of a second puff on chromosome IV (Fig. 11–14; site IV-2B). A second group of puffs is expressed between 6 and 48 hours after injection of the hormone (Fig. 11–14). One of these puffs (8A on chromosome I) appears and regresses within this interval of time. Ecdysone-stim-

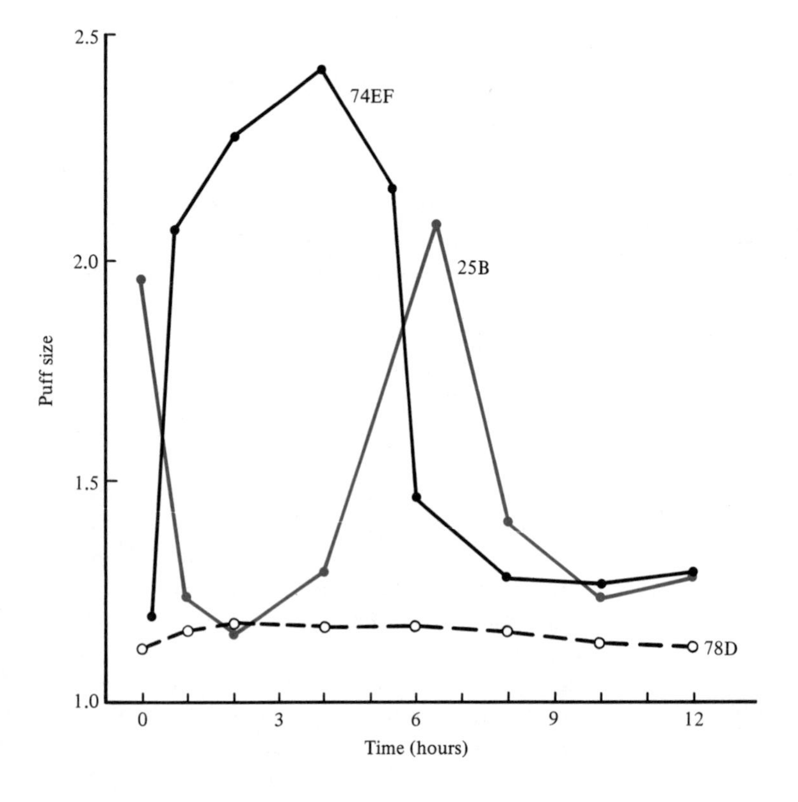


**FIG. 11–13.** Cultured salivary glands of *Drosophila* when exposed to 20-hydroxyecdysone pass through a cycle of puffing chracteristic of the larval and pupal stages of development. Puff responses can be divided into those that regress (25B), those that are activated (74EF), and those that are activated only after longer periods (78D). Puff size is the ratio of puff diameter to the diameter of an adjacent constant region. (After data from M. Ashburner, 1972. Chromosoma 38:255.)

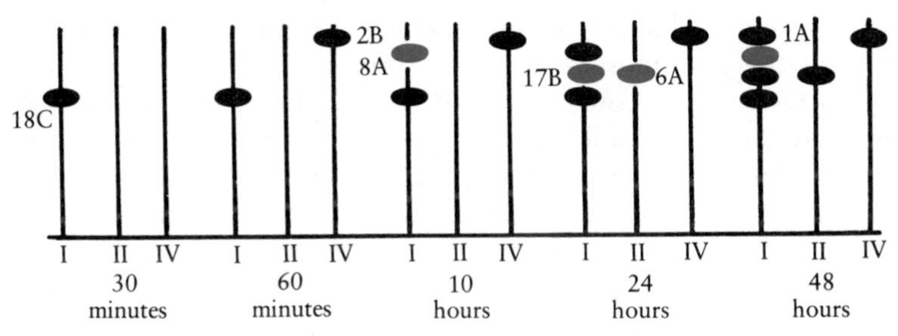


**FIG. 11–14.** Injection of ecdysone into the larvae of *Chironomus* produces a puffing sequence in the chromosomes (I, II, IV) of the salivary glands. (See next for further details.) (After U. Clever, 1964. Science 146:794. Copyright 1964 by the American Association for the Advancement of Science.)

ulated puffs at sites I-18C and IV-2B can be blocked with actinomycin D. Larvae injected with puromycin or cycloheximide followed by treatment with ecdysone show normal puff formation on the first and fourth chromosomes, but later puffs fail to appear. These observations suggest that the transcriptive activity of early puffs is essential for the sequential activation of later puffs. Presumably, proteins synthesized as a result of RNA synthesis at sites I-18C and IV-2B act as initiators of subsequent patterns of gene activity. On the basis of evidence from these and other studies, ecdysone appears to act in the following fashion at the level of the nucleus. It initially induces the transcription of a small group of primary gene loci leading to the formation of a set of "early" puffs. Concurrently, the action of the hormone results in an inhibition of a secondary set of primary loci and regression of "intermolt" puffs. Stimulation of the transcription of "late" puffs and inhibition of the transcription at "early" puff sites appear to depend on gene products of the "early" puffs.

A similar pattern of gene activity is observed in the epidermis of the tobacco hornworm *Manduca* during metamorphosis. During the larval stages, the epidermis produces larval proteins and the blue pigment protein, *insecticyanin*. Upon metamorphosis into the pupa, ecdysone acts in the absence of juvenile hormone to cause the epidermis to lose its ability to make larval products and become committed to the differentiation of a pupal cuticle. Evidence for the inhibition of larval gene activity is the loss of translatable larval-specific mRNAs for insecticyanin and other larval cuticular proteins. The loss of these mRNAs does not occur simultaneously, suggesting that the larval-specific genes may not be coordinately regulated. The activation of most genes producing pupal cuticular mRNAs and proteins requires a subsequent exposure to more ecdysteroid hormone.

The appearance and regression of puffs in polytenic tissues of insect larvae at specific stages of development are strong visual evidence that ecdysone can regulate genes at the level of transcription. The basis of structural and functional changes in tissues with nonpolytenic chromosomes is presumably similar, with uncoiled sequences of DNA serving as templates for the synthesis of mRNA. How does the ecdysone act within a cell to induce puff formation? What is the molecular basis for the selective response of different puff sites to ecdysone? The challenge of working out the precise pathways within the cell by which ecdysone alters nuclear activity is a difficult one, and answers to these questions are yet to be determined. There is some evidence to suggest that ecdysone, similar to vertebrate steroid hormones (see below), complexes with an intracellular receptor before becoming concentrated at distinct chromosomal sites. Thus, one site of ecdysone hormone action appears to be the chromosome itself. However, the same hormone may act at other sites, such as the cell membrane. Dibutyryl adenosine monophosphate (cAMP) has also been shown to induce or repress puff formation in larvae of *Drosophila* at different stages of development, thus suggesting that the gene expression induced in these tissues under the influence of ecdysone may in fact be regulated by the cAMP acting as a *secondary messenger*.

Hormones are signals that are also known to activate a program of differentiation in target cells of certain vertebrate organs, such as the liver, mammary gland, and oviduct. The expression of genes in these target cells responding to circulating steroid hormones appears to be controlled directly at the level of transcription.

Several hormones act in concert to bring about the functional differentiation of the alveolar epithelial cells of the mammary gland. The epithelial cells synthesize and secret several *milk proteins*, the principle ones being *caesin* and the enzyme $\alpha$-lactalbumin. Normally, the alveolar sacs do not arise in the intact animal until the second half of pregnancy. At this time, there is a rapid proliferation and differentiation of the alveolar epithelium from a primordial milk duct system in response to the presence of three hormones, *insulin, hydrocortisone,* and *prolactin*. When mammary gland tissue from young mice is cultured with these three hormones, alveolar cells will differentiate, showing dramatic increases in caesin-specific mRNA and secretory proteins (Fig. 11–15). By withholding one or several of these hormones from the organ culture, the precise role each hormone plays in the differentiation of the alveolar epithelium can be assessed (Fig. 11–15 A,B). Insulin appears to stimulate extensive DNA synthesis and the proliferation of mammary gland epithelial cells, but it has little effect on the ability of the cells to synthesize and secrete the milk proteins. However, unless it is present and stimulates cell division, further functional differentiation under the influences of hydrocortisone and prolactin will not take place. The role of hydrocortisone has yet to be fully established, but it must be present at some time when insulin acts on the alveolar epithelium to enable prolactin to elicit synthetic activity and the production of the milk proteins. Only prolactin is capable of triggering the synthesis and secretion of functional milk proteins associated with the terminally differentiated cell type. By using inhibitors of DNA synthesis, such as *cytosine arabinoside,* Smith and Vonderhaar have shown that blocked explants of mammary glands of mature virgin mice fail to produce either caesin or active $\alpha$-lactalbumin. Interestingly, electron-microscopic examination of these cells indicates that they are cytologically fully able to synthesize proteins. These observations support the view that DNA synthesis is an indispensable requirement for virgin mouse epithelial cell differentiation in vitro. By contrast, actively proliferating mammary gland cells from pregnant mice or from postlactational nonpregmant mice can be stimulated by prolactin to synthesize milk proteins in vitro in the absence of DNA synthesis. In the mammary gland system, therefore, the synthesis of DNA is a required step prior to the full terminal differentiation of the alveolar epithelium. The significance of this nucleic acid requirement to the differentiation of the terminal cell type is not clear. One possibility is that

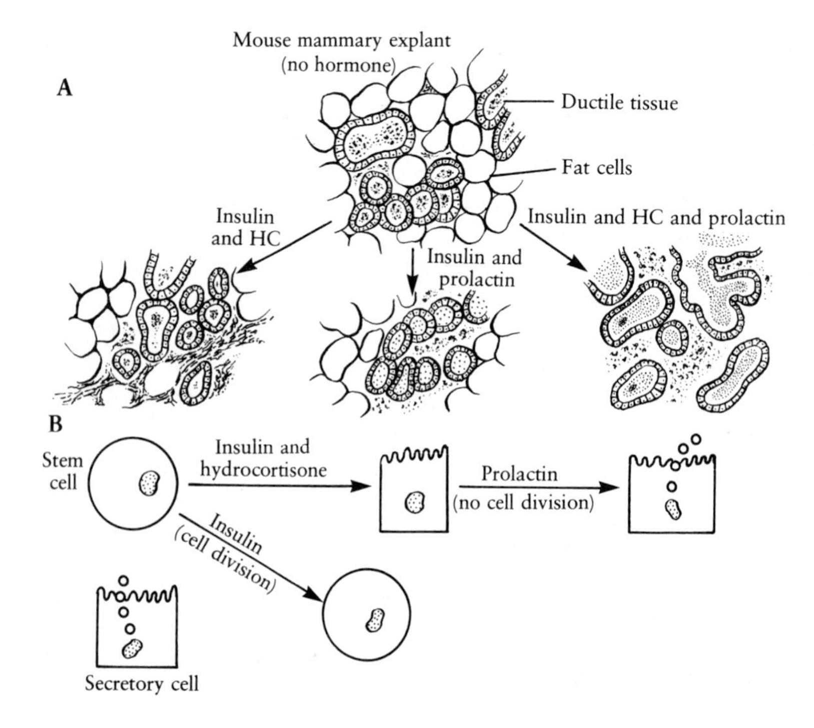


**FIG. 11–15.** The effect of hormones on the differentiation of alveoli in the mammary gland of the mouse. A, appearance of sections of mammary gland explant after exposure to selected combinations of hormones; B, scheme to show the interactions of hormones in the differentiation of gland tissue. Only the presence of all three hormones will lead to a functional secretory epithelium. HC, hydrocortisone. (From R. Turkington, 1968. In: Current Topics in Developmental Biology, Vol. 3. A. Moscona and A. Monroy, eds. Academic Press, New York.)

the mandatory DNA synthesis reflects an amplification of caesin genes that is required to support the high rate of RNA and protein syntheses characteristic of this active tissue.

The role of steroid hormones in controlling gene expression has also been rather extensively studied in the chick oviduct. *Estrogen* stimulates chick oviduct (magnum portion) cells to proliferate and synthesize the so-called egg white proteins. The major egg white proteins, *ovalbumin, conalbumin, ovomucoid,* and *lysozyme,* are formed in all cells of tubular glands in the oviductal epithelium. The effects of selected sex steroids on gland cell differentiation and egg white protein synthesis are commonly analyzed using recently hatched chicks. Here, the epithelium of the oviduct is a uniform, simple layer and shows little evidence of specialized cell types. When estrogen is administered daily by injection to newly hatched (4–5 days) chicks for a period of 10 days, the tubular glands show morphological differentiation as early as 96 hours in the epithelial tissue. By 7 to 8 days, ovalbumin and lysozyme can first be detected in the cells of the tubular glands. After 9 days of continuous treatment with estrogen, another cell type, the *goblet cell,* differentiates in the luminal epithelium. A single injection of *progesterone* into the estrogenized chick then stimulates the goblet cells to synthesize the well-known protein called *avidin.* At least two other steroids, glucocorticoids and androgens, are also known to induce the synthesis of egg white proteins.

Another approach to studying the effects of steroid hormones on the synthesis of egg white proteins has been to interrupt hormone treatment for a given period of time. For example, a chick is typically given a daily dose of estrogen for 10 days *(primary hormone stimulation).* If the treatment is then stopped for 10 days or so, the production of egg white proteins rapidly declines. Reinjection of estrogen *(secondary hormone stimulation)* reinitiates the synthesis of ovalbumin, but now within three to four hours of hormone administration. There are several advantages to using the "hormone-withdrawn" chick. The oviduct of the secondarily stimulated chick is much larger than the oviduct of the chick receiving primary hormone stimulation; this makes the events of ovalbumin synthesis more accessible for analysis. Also, if progesterone is injected into the chick after termination of estrogen treatment, ovalbumin synthesis is reinitiated but much more rapidly than that induced by estrogen alone. Progesterone does not normally elicit egg white protein production on its own.

The increase in the synthesis of ovalbumin and conalbumin under primary and secondary stimulation with estrogen is the direct result of the accumulation of protein-specific messenger RNA molecules (Fig. 11–16). If RNA is extracted from oviduct tissue stimulated by estrogen and then added to a cell-free protein-synthesizing system, the presence of ovalbumin can subsequently be detected. No ovalbumin is found if the extracted RNA is obtained from non-hormone-treated chick oviduct tissue. In addition to estrogen, the expression of the ovalbumin gene appears to require a *somatomedin-like peptide hormone.* Insulin is an equally active substitute for this peptide hormone (Fig. 11–17). Interestingly, the accumulation of conalbumin mRNA is not enhanced by the peptide hormone. Hence, although ovalbumin and conalbumin genes are both transcribed in the same cell in response to the same steroid hormones, the expression of ovalbumin genes requires an additional regulatory signal. The mRNA specifying pure ovalbumin has now been isolated and used in vitro with *reverse transcriptase* to construct strands of *complementary DNA (cDNA);* the DNA strands are essentially genes

**FIG. 11–16.** Comparison of the rate of ovalbumin synthesis with the amount of ovalbumin mRNA per cell after secondary stimulation with either estrogen or progesterone. (From G. McKnight et al., 1975. J. Biol. Chem. 250:8105.)

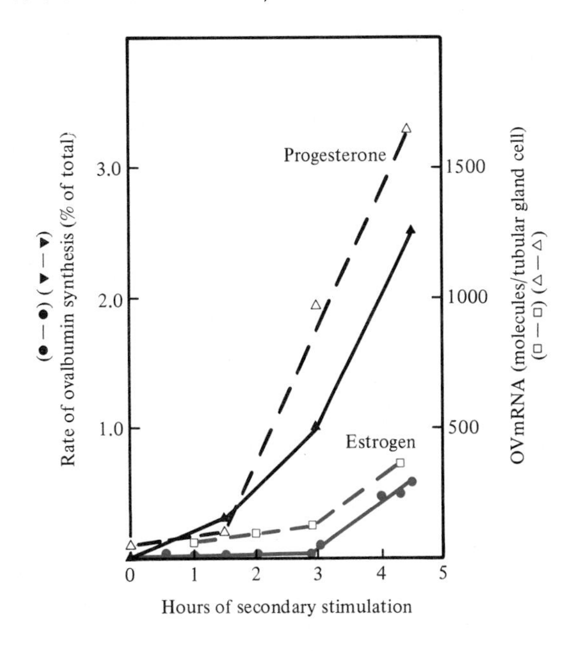

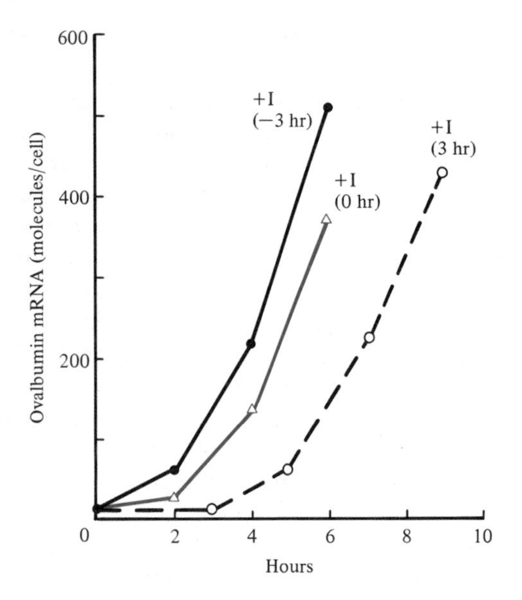


**FIG. 11–17.** The effect of insulin on the accumulation of ovalbumin mRNA. Withdrawn oviduct was induced by the addition of estrogen and insulin following a preincubation period in insulin (●), estrogen (○), or with no hormone (△). (From M. Evans et al., 1981. Cell 25:187.)

coding for ovalbumin. With ovalbumin cDNA as a probe, it has been possible with hybridization analysis to determine how much ovalbumin-specific mRNA is made during hormonal stimulation. It has been estimated that injection of estrogen results in the accumulation of about 50,000 molecules of ovalbumin mRNA per gland cell after 18 days. This number drops to 60 molecules of ovalbumin-specific mRNA per cell if estrogen treatment is discontinued.

The close temporal correlation between the accumulation of mRNA molecules and the detection of ovalbumin proteins strongly suggests that ovalbumin genes are being regulated at the level of transcription. What is the mechanism by which hormones interact with cells of target organs to alter gene activity? Does a hormone, for example, interact directly with chromatin material? The interpretation from most studies that have examined these questions is that target cells respond to hormones by at least two different mechanisms. For steroid hormones such as estrogen, progesterone, hydrocortisone, and androgen, each hormone is postulated to diffuse into the cell from the blood circulation and complex with a specific receptor protein in the cytoplasm to form a *hormone–receptor complex* (Fig. 11–18 A). Nontarget cells do not bind the hormone. Hence, the early development of target organs must involve the differentiation of specific hormone receptor proteins. Presumably, different steroid hormones acting on the same target cells complex with different cytosolic receptor proteins. After undergoing some unspecified modification, the hormone–receptor complex is translocated through the nuclear membrane into the nucleus where it binds to an acceptor site on the chromatin. How the activated hormone–receptor complex increases the transcription rate of specific genes is not known, but it appears to bind near the gene to be activated. Attempts have recently been made to isolate and purify steroid receptors from the chick oviduct. O'Malley and co-workers describe a protein with a high affinity for *[$^3$H]progestin;* it is believed to consist of two dissimilar subunits. Movement of the hormone–receptor complex from cytoplasm to nucleus has been demonstrated by adding labeled hormone, such as [$^3$H]progestin, to the cytosol fraction of a target organ, such as chick oviduct. When the hormone–receptor complex is incubated with nuclei isolated

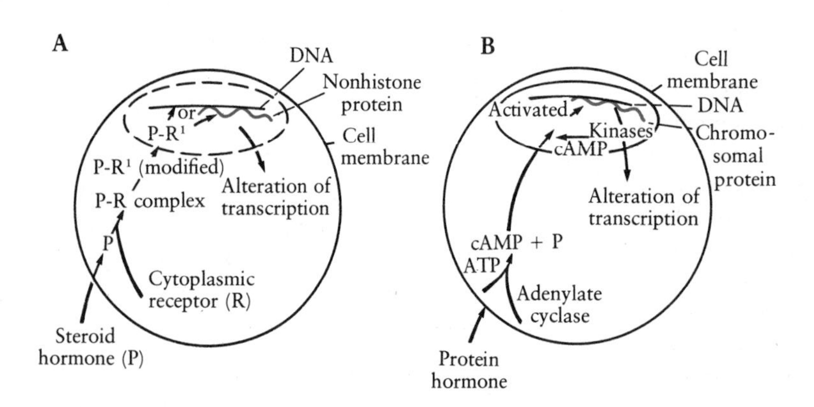


**FIG. 11–18.** Diagrams showing proposed pathways by which steroid (A) and protein (B) hormones act on target cells. (A, from G. Stein et al., 1974. Science 183:817. Copyright 1974 by the American Association for the Advancement of Science.)

from the tubular glands of the oviduct, an uptake of the label by the nuclei is consistently observed. Little uptake of hormone is noted if nuclei are incubated in the hormone alone. Also, nontarget cell nuclei will not concentrate hormone–receptor complexes.

Protein hormones effect their actions on target cells by a totally different mechanism (Fig. 11–18 B). Molecules of the hormone complex with special receptor sites on the cell surface, an event that stimulates the release of membrane-bound adenylate cyclase into the cytoplasm. Adenylate cyclase than catalyzes the conversion of ATP into cAMP (3′,5′-adenosine monophosphate) and pyrophosphate. The cyclic nucleotide is proposed to activate certain protein enzymes *(kinases)* within the nucleus, which subsequently mediate the phosphorylation of chromosomal proteins. The uptake of phosphate by chromatin proteins is an early event that is often observed after hormonal stimulation of target cells and preceding extensive gene activation. Prolactin appears to act on the alveolar epithelial cells of the mammary gland in this fashion. Indeed, the function of hydrocortisone in the mammary gland system described above may well be to stimulate the synthesis of prolactin receptor sites on surface membranes of newly formed gland cells. Insulin appears to act with a membrane-bound receptor to generate an intracellular signal that interacts specifically with the ovalbumin gene. Since the peptide hormone requirement can be obviated by altering intracellular levels of cAMP, there is a possible role of a cAMP-dependent protein kinase in the activation of the ovalbumin gene.

## Chromosomal Proteins and Gene Transcription

Alterations in gene expression are correlated with the uptake by the nucleus of cytoplasmic proteins, cyclic nucleotides, ions, and hormone–receptor complexes. These apparently bind to chromatin material within the nucleus, suggesting that each functions to influence the pattern of transcription. Chromatin, however, is a complex of DNA, large amounts of chromosomal protein, and small amounts of RNA. Since the luxury proteins characterizing different cell types are presumably specified by different regions of the genome, specific regulatory molecules must be present at the level of the chromatin to determine which regions of the genome are to be either expressed or silenced. Unfortunately, the mechanism(s) by which genes are selected and regulated in cells of higher organisms is still largely unknown. Considerably more information is available on gene control systems in viruses and bacteria. Control of gene expression through regulation of DNA transcription was first shown at the molecular level for the lactose operon in *Escherichia coli.* In *E. coli,* a specific repressor has been isolated and shown to regulate transcription by binding to a specific site on the bacterial DNA. Small molecules such as lactose, in turn, control the binding affinity of the repressor by inducing alterations in its conformational organization. The strategy in studying eukaryotic gene regulation has been to follow the successful approaches used in examining gene regulation in bacteria.

For several years now, there has been considerable interest in the role of nuclear chromosomal proteins in selective gene transcription. Historically, the proposed regulatory proteins have been divided into two classes: the histones and the nonhistones or nuclear acidic proteins. Histones comprise about 70 percent of the nuclear proteins of adult nuclei and are best defined as proteins with a positive charge and rich in such basic amino acids as arginine and lysine. Initially, because of their known preference to bind tenaciously to DNA and demonstrated function to inhibit the ability of DNA to serve as a template for RNA synthesis, it was thought that histones were specific repressors of gene transcription. An attractive hypothesis was that histones might function in the differentiation of cell types during embryogenesis by silencing selected DNA sequences. Although new histone proteins appear during certain critical stages of sea urchin and amphibian development, there are really very few developmental changes that can be correlated with variations in the appearance of basic proteins. Also, since histones exhibit remarkable uniformity among tissues and cells (crude histone preparations can only be fractionated into five major classes), they apparently lack sufficient specificity to recognize and influence the activity of specific genes. The absence of sufficient heterogeneity among histones suggests that these proteins probably act as regulatory molecules of gene transcription in some nonspecific way.

As early as the 1940s, Stedman and Stedman proposed that the nonhistone chromosomal proteins were potential regulators of specific gene activity.

Acid nuclear proteins constitute approximately 30 percent of the total nuclear proteins, are frequently rich in aspartic and glutamic acids, and, like histones, are synthesized in the cytoplasm of the cell. In contrast to the histones, the nonhistone proteins display tremendous structural and functional differences in different cell types. They are proportionally higher in transcriptively active regions of the genome; by contrast, inactive portions of the chromatin lack nonhistone proteins. The most direct evidence that nonhistone proteins play an important role in tissue-specific transcription comes from *chromatin reconstitution experiments*. The technique permits an evaluation of the activity of the DNA of a given cell type, measured by the synthesis of a tissue-specific RNA, in the presence of either isolated histones or nonhistones from another cell type (Fig. 11–19). In famous experiments by Gilmour and Paul at the Institute for Cancer Research in Glasgow, rabbit thymus and rabbit bone marrow cells were dissociated into DNA, histone, and nonhistone fractions (Fig. 11–19). When all of these components of the thymus were reconstituted, the RNA that formed behaved just like normally synthesized RNA. When DNA and histones from thymus and bone marrow were pooled together and the chromatin fully reconstituted with the addition of thymus nonhistone proteins, the RNA synthesized had all of the characteristics of thymus RNA. This occurred even though DNA and histones were available from the bone marrow. When the same types of DNA and histones were recombined with nonhistones from bone marrow, the RNA synthesized from the combined chromatin behaved like bone marrow RNA. Similar techniques have been applied to other cell systems in efforts to determine whether nonhistone chromosomal proteins play a role in transcriptional specificity. Isolated chromatin from fetal mouse liver produces mRNA specifying the globins of hemoglobin; brain chromatin does not. However, if brain DNA is reconstituted in the presence of liver nonhistone chromosomal protein, globin mRNA can be detected.

Further support for the view that nonhistone proteins function as acceptor sites for initiators of changes in gene activity, such as hormone–receptor complexes, comes from reconstitution studies using chick oviduct chromatin. When nonhistone chromosomal protein is extracted from the chromatin of estrogen-stimulated oviduct and recombined with DNA and histones obtained from nonstimulated oviduct, the reconstituted chromatin shows ovalbumin mRNA synthesis equivalent to that of normally stimulated oviduct chromatin. Use of radioactive [$^3$H]progestin shows quite clearly that

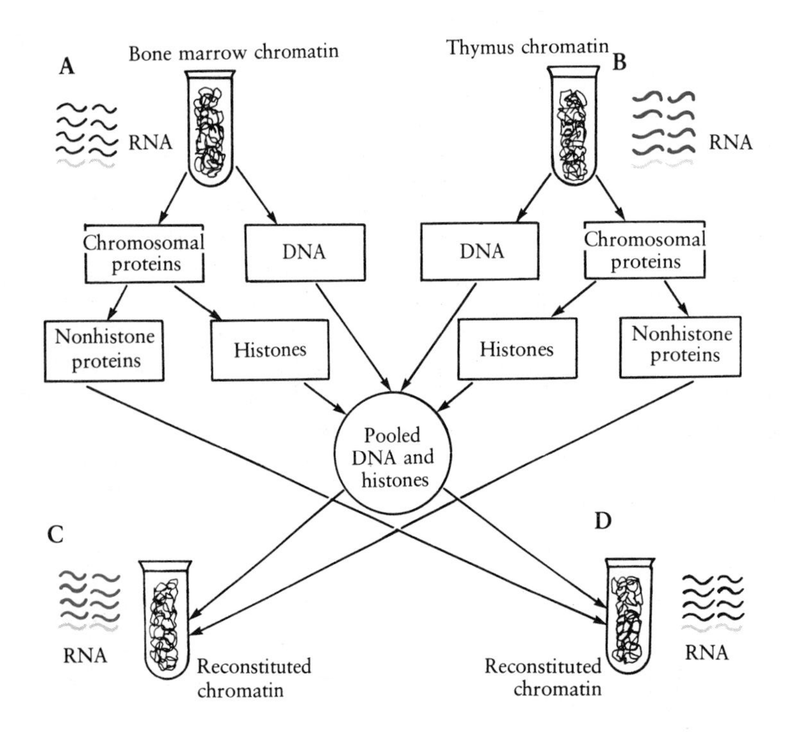


**FIG. 11–19.** The technique of chromatin reconstitution permits an analysis of the role of nonhistone proteins in tissue-specific transcription. This figure illustrates experiments conducted by R. Gilmour and J. Paul using rabbit thymus and rabbit bone marrow chromatin. Chromatin from bone marrow and thymus can be dissociated into DNA, histones, and nonhistones (A, B). When DNA and histones from both tissues are pooled and combined with nonhistone proteins from thymus, the reconstituted chromatin synthesizes RNA characteristic of thymus (C). When DNA and histones from both tissues are pooled and combined with nonhistone proteins from bone marrow, the reconstituted chromatin synthesizes RNA, which behaves like bone marrow RNA (D). (From G. Stein et al., 1974. Science 183:817. Copyright 1974 by the American Association for the Advancement of Science.)

the binding of the hormone–receptor complex follows the nonhistone proteins in reconstitution experiments.

Nonhistone chromosomal proteins appear to play an essential role in regulating the transcription of individual genes during the differentiation of cell types, but how is the control exercised? Although several models have been proposed to account for the molecular mechanism by which this regulation is effected, the histone displacement model set forth by Stein and his colleagues is consistent with a number of observations regarding the behavior of both histone and nonhistone proteins. Histones bind vigorously to DNA, and apparently they function to silence large blocks of genes. Therefore, synthesis of specific proteins in response to initiators of differentiation would require the removal of histones from selected DNA sites.

A striking property of the nonhistone proteins is their extensive phosphorylation or capacity to incorporate phosphate rapidly. The uptake of radioactive phosphate is particularly evident during the synthetic or S phase of the cell cycle, a time when genes also appear to be more actively transcribed. It is proposed that specific nonhistone proteins complex with specific sites of DNA repressed by histone protein (Fig. 11–20 A). When the nonhistone protein becomes phosphorylated, its negatively charged phosphate groups are repelled by the negatively charged DNA (Fig. 11–20 B,C). Consequently, there follows displacement of the nonhistone–histone protein complex from the DNA, allowing a specific region to be transcribed into RNA in the presence of RNA polymerase (Fig. 11–20 D,E). As to the precise control over the utilization of specific regions of the genome during differentiation, there is still uncertainty as to the relationship between displacement of the nonhistone–histone complex and the activation of selected structural genes. Britten and Davidson suggest that repetitive nucleotide sequences in the genome constitute *integrating regulatory transcriptional units.* Each unit is presumably responsive to changes in the behavior of the nonhistone–histone complex. The transcripts of the integrating regulatory transcriptional units are proposed to be cell specific.

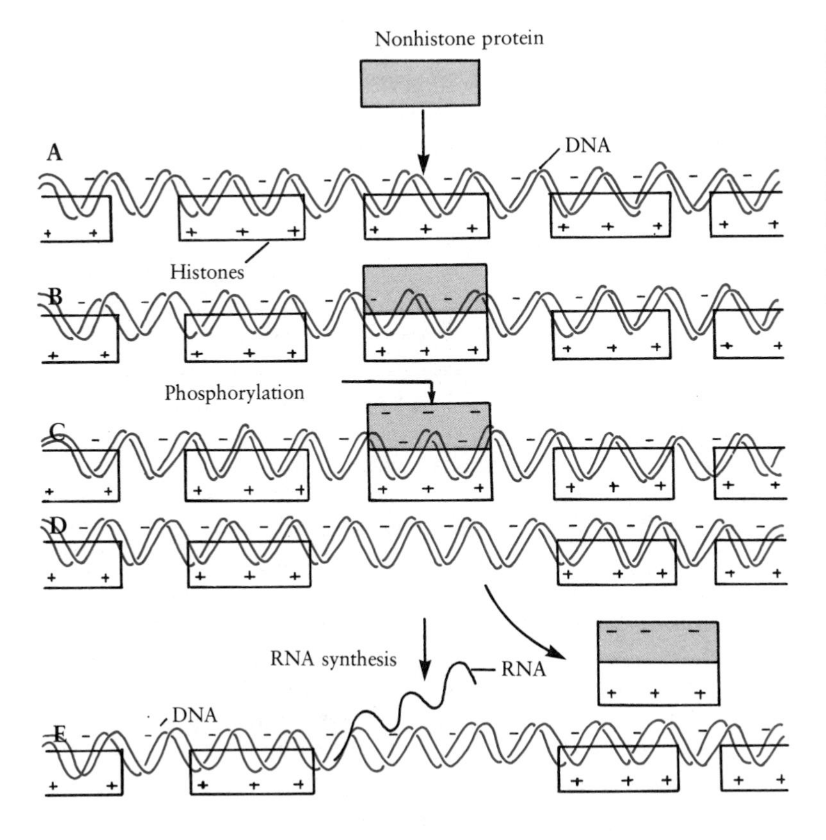


FIG. 11–20. A model showing regulation of gene activity by nonhistone protein. A, a nonhistone protein recognizes a specific site on DNA that is repressed by histone protein; B, the nonhistone protein binds to the site; C, the nonhistone protein undergoes phosphorylation and becomes negative; D, the nonhistone–histone complex is displaced from the negatively charged DNA, leaving a segment of the DNA available for the transcription of RNA (E). (From G. Stein et al., 1975. Sci. Am. 232:47.)

They function as regulatory molecules by triggering the expression of adjacent structural genes containing complementary nucleotide sequences.

## Examples of Posttranscriptional Controls

Control at the initial step in the synthesis of tissue-specific proteins (i.e., at the level of RNA synthesis) could result in the production of different kinds of messenger RNA molecules and thereby of different proteins. Certainly, experimental evidence from dipteran polytene chromosomes and the tissue specificity of RNA transcribed from isolated chromatin indicate that the programming of which genes are to be "on" or "off" in a cell operates at the level of the gene. However, the extent to which gene action is solely controlled at this source by the direct limitation of the transcription of genes into RNA has yet to be fully determined for the eukaryotic genome. The notion that genes for specialized proteins are transcribed only in their specialized cells clearly suggests that transcription is tightly regulated. However, there are cases in which RNA sequences appear to be synthesized that do not seem appropriate to the specific differentiated function of a cell. For example, studies using cDNA probes specific for globin RNA indicate the presence of globin in nonerythroid tissues. Observations of this nature raise the possibility that all genes may be transcribed in all cell types, and that the differentiated state of cells is primarily the result of selective controls operating after transcription. It is also a possibility, therefore, that all regulation of gene expression occurs by the interaction of controlling elements with the initial gene transcript. At the beginning of this chapter, it was pointed out that there are a number of steps between the transcription of a gene and the appearance of a functional protein in the cytoplasm. These steps may be within the nucleus, affecting the primary transcript immediately after transcription or during its transport into the cytoplasm, and/or within the cytoplasm at the level of the ribosome during translation. At any of these points in the machinery of protein synthesis, posttranscriptional regulatory factors could prevent, delay, or affect the rate of messenger RNA translation, and thereby control cell phenotype.

The immediate product of transcription is probably not the functional message that is translated into the functional protein. There is substantial evidence to indicate that the primary gene transcript is a large molecule that becomes altered into a smaller message molecule. The modification of the primary gene transcript into a functional message is known as *processing*. Processing may include methylation (i.e., the addition of methyl groups to the ribose moieties of nucleotide bases) and the addition, deletion, or alteration of nucleotide sequences. One of the clearest examples of a posttranscriptional control at the level of processing is the formation of 28S and 18S ribosomal RNA molecules. In *Xenopus*, as we have already noted in the discussion of ribosomal gene amplification during oogenesis, the 28S and 18S components of RNA are synthesized simultaneously and arise from a complex precursor molecule that sediments at 40S. The processing of the nucleolar 40S molecule appears to take place in the nucleus. It involves methylation and the enzymatic excision by *nucleases* of excess nucleotides, leaving the 28S and 18S components. The 28S and 18S molecules then combine with proteins and 5S RNA to form the ribosomes. In the case of 5S RNA, however, the initial transcript is the final RNA product.

Another type of RNA that undergoes extensive processing is transfer or 4S RNA. The processing of the transcript coding for the transfer RNA specific for the amino acid tyrosine in bakers' yeast (*Saccharomyces cerevisiae*) has been examined in some detail by De Robertis and Olson following cloning of the gene in plasmids. The gene is of interest because it contains a 14-nucleotide segment inserted into the nucleotide sequence coding for the transfer RNA molecule (Fig. 11–21). In addition to the nucleotide sequence adjacent to the coding region, there is a leader sequence of additional nucleotides preceding the tRNA segment (i.e., at the 5′ end of the transcript). Processing involves removal of the leader sequence and addition of the nucleotide sequence -C-C-A at the 3′ end of the transcript. Several methylated nucleotides are added. The intervening sequence is then excised and the two cut ends spliced or ligated to yield the functional tRNA molecule. The processing of tRNA appears to take place in both the nucleus and the cytoplasm.

Cytoplasmic mRNA capable of associating with ribosomes and acting as a template for protein synthesis appears typically to arise from the processing of larger molecular weight nuclear RNA precursor transcripts. Such precursor transcripts have been identified for the cytoplasmic mRNAs specifying mouse $\beta$ globin, sea urchin histone, and chicken ovalbumin. As with ribosomal transcripts, the pro-

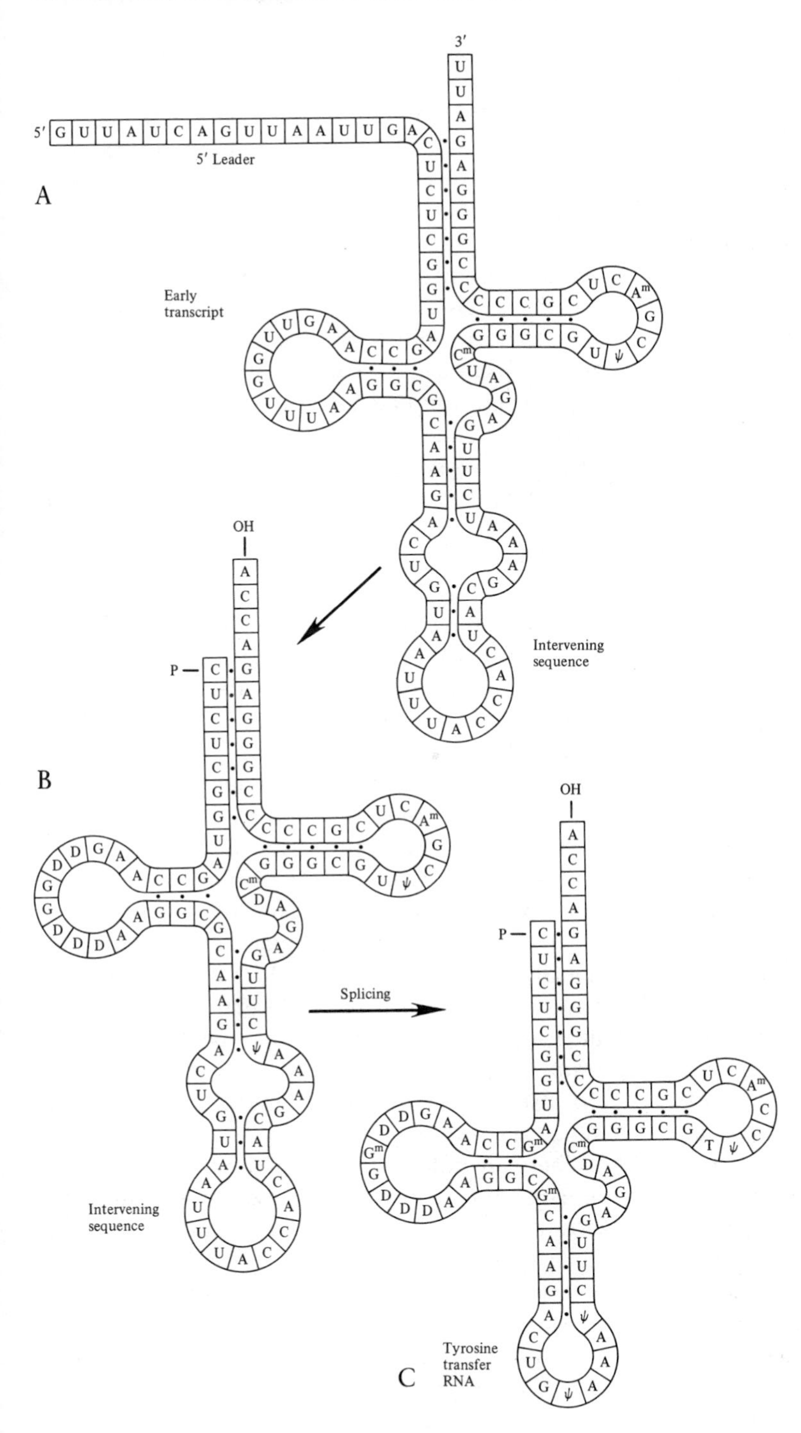


**FIG. 11–21.** Diagrammatic summary of the processing of yeast tyrosine transfer RNA. A, the primary transcript shows a leader sequence at the 5′ end and a 2-nucleotide trailing sequence at the 3′ end of the molecule. There is also a 14-nucleotide intervening sequence. B, processing removes the leader sequence and modifies the 3′ end, adding the nucleotides CCA. The tyrosine tRNA is finally formed when the intervening sequence is excised and the cut ends are spliced together (C). Note that the primary transcript has three unusual nucleotides: a pseudouridylic acid (ψ), methylated adenine ($A^m$) and methylated cytosine ($C^m$). Another unusual nucleotide that appears is dihydrouridylic acid (D). (From E. DeRobertis and J. Gurdon, 1979. Sci. Am. 241:74.)

cessing of these precursor mRNA transcripts may involve the addition of nucleotides, the removal of unwanted sequences, or the modification of existing nucleotides. Because processing reactions are imperfectly understood, all possible points of post-transcriptional control are not known. Typically, however, processing involves both ends of the nuclear RNA transcript. Most mRNAs are modified at the 5′ end of the molecule by the addition of a methyl group(s) to produce the so-called *5′ cap*. This cap may function to protect the structural integrity of the 5′ end of the mRNA molecule and initiate the process of translation itself. At the opposite or 3′ end of the transcript, many mRNAs add a string of 150–300 adenylate residues *(polyadenylation)*. These so-called Poly(A) tracts are not transcribed, and their significance in gene expression is very unclear, particularly in light of the observation that some mRNAs, such as those coding for histones in mammals and sea urchins, lack Poly(A) but are quite functional. Investigators have suggested that Poly(A) tracts may assist in the transport of the mRNA from nucleus to cytoplasm, or somehow increase the efficiency of producing protein on the polyribosome. The nuclear transcript that is "capped" and polyadenylated may require further processing before becoming functional. For example, the nuclear ovalbumin primary transcript is coded by some 6000 base pairs of DNA; however, the cytoplasmic mRNA specifying ovalbumin is only about one-third as long. The conalbumin primary gene transcript contains about 10,000 base pairs, but the cytoplasmic mRNA is less than one-fourth as long. Processing of these molecules, therefore, must involve excision and splicing of nucleotide sequences.

Another posttranscriptional level of possible control of differential gene expression is that of translation. The existence of "translational controls" implies that the rate at which proteins are synthesized from mRNA is different for two kinds of messages within the same cell, for the same message in different cells, or for the same message at different stages in the development of a cell. Each of these conditions would result in differences in the types of proteins synthesized by the cells. There are a number of systems of differentiating cells in which translational control of gene expression has been reported or suspected, observations based partially on inhibitor studies in which estimates of the times of transcription and translation are made relative to the time of appearance of the functional protein. For example, the period of time during which actinomycin D suppresses gene expression is taken as the interval of transcription, while the period of sensitivity to puromycin or cycloheximide is taken as the period of time of message translation. Several to many hours may intervene between the periods of transcription and translation. Recall from Chapters 4 and 6 that the unfertilized animal egg synthesizes and stores untranslated cytoplasmic mRNA in masked form. The mRNA molecules are then selectively utilized after fertilization of the egg. This delay in the recruitment and utilization of mRNA by the early embryo is one of the best examples of gene regulation at the level of translation.

In addition to delay, another factor affecting protein synthesis at the level of translation is the stability of cell-specific mRNA molecules. The stability of mRNA is measured in terms of its half-life, or the time required for 50 percent of the mRNA population existing at any given time to become degraded. In general, cell-specific mRNA molecules of eukaryotic, fully differentiated cells are quite stable. For example, the half-life of ovalbumin mRNA of estrogen-stimulated oviduct cells is about 4 to 5 days. When the estrogen treatment is withdrawn, the half-life changes and the decay of ovalbumin mRNA is very rapid. The increased longevity of cell-specific mRNA molecules makes each template available for a number of rounds of protein synthesis, thus biasing the pattern of proteins produced and the direction of the cell type. Each ovalbumin mRNA may be translated approximately 50,000 times during its lifespan. Hence, differences in the rate of decay between different mRNAs would clearly affect the spectrum of proteins synthesized by the cell. With time and continued differentiation, there is an accumulation of the more stable mRNA molecules. This is precisely what happens during the differentiation of the adult erythrocyte (Chapter 18). The mRNA specifying globin increases and accumulates as the mRNA of nonglobin protein declines. The basis of the differences in stability between different messages is still unclear. However, capping may be important in conferring stability on a mRNA transcript. If this is the case, then regulation of capping may indirectly affect gene expression in a cell.

There are other examples of differentiating cells in which the expression of messengers at different rates appears to be regulated by components of the translational machinery itself. In addition to tRNA and the subunits of the ribosome, successful protein synthesis requires factors that mediate the initiation of translation and govern the rate at which amino acids are assembled in linear fashion into protein. Proteins known as *initiation factors* begin a complex series of steps leading to the formation of an *initiation complex* that permits the translocation of the message along the ribosome. Differences in the nucleotide sequences ahead of the coding region of a polypeptide or the presence of message-specific initiation factors, therefore, may affect the efficiency or rate at which mRNA binds to the ribosome and hence the speed of translation. An example frequently cited to illustrate this type of regulation is the synthesis of the $\alpha$- and $\beta$-globin chains contained in hemoglobin of the mammalial reticulocyte (precursor cell to the erythrocyte). The $\beta$-globin chain has been determined to be synthesized on larger polysomes (four to five ribosomes) than the $\alpha$-globin chain, which is made on polysomes with about three to four ribosomes. Since the $\alpha$ and $\beta$ polypeptides are produced in approximately the same amount in the reticulocyte, the smaller number of nascent $\alpha$ chains must be compensated for by either a more rapid movement of ribosomes along the $\alpha$ mRNA or by more $\alpha$ messengers in the cytoplasm, each molecule of which associates less frequently with an attached ribosome. Experimental evidence from studies using inhibitors of protein synthesis suggests that the latter explanation is probably correct. Normal reticulocytes appear to contain more $\alpha$-globin mRNA than $\beta$-globin mRNA but the $\alpha$-globin mRNA appears to initiate protein synthesis only 60 percent as frequently as the $\beta$-globin mRNA. The result is that the ribosomes associate more rapidly with the $\beta$-mRNA transcripts than with the $\alpha$-mRNA transcripts.

It would be convenient to postulate that the differences between the gene transcripts for $\alpha$ and $\beta$ globin in initiating translation reflect differences in message-specific initiating factors. No such message-specific factors have yet been identified for these two transcripts. However, there is some evidence that message-specific initiating factors do exist in other cell types; these may function to mediate rates of translation. For example, if myosin mRNA is isolated from chick and added to the translational machinery in a cell-free system from chick erythroblast, there is little translation and myosin cannot be detected in the medium. When initiation factors from muscle cell ribosomes are added to this cell-free translation system, translation of myosin mRNA is enhanced. Equivalent factors prepared from the ribosomes of erythroblast cells have little effect on myosin mRNA translation. It would appear, therefore, that at least some differentiated cells have specialized initiation factors that recognize cell-specific mRNA and somehow make these same messages more available for translation. Conversely, these same cells must contain nonspecific initiation factors that enable them to translate messages shared in common with other cell types (i.e., for housekeeping proteins). In short, message-specific initiation factors would have the effect of promoting the translation of messages characteristic of the specialized cell type.

When a message has been translated, the newly synthesized molecule released from the polysome may not necessarily be functional. The control of gene expression may continue beyond the translational step. Posttranslational modification has been examined for a number of proteins, including the $\alpha$-crystallin (lens), the egg white proteins, collagen, light- and heavy-chain immunoglobulin, and various pancreatic enzymes and hormones. Typical modifications include the elimination of amino acids from one end of the polypeptide chain, and/or the alteration of individual amino acids by covalent binding to various residues through methylation, phosphorylation, or hydroxylation. In Chapter 4 it was noted that the precursor to the amphibian yolk proteins, lipovitellin and phosvitin, is a high-molecular-weight complex of protein, lipid, and carbohydrate. It is produced by the liver and secreted into the circulating blood. The vitellogenin in the ovary is processed into three polypeptide chains. Two of these are lipid rich and join to form lipovitellin. The other polypeptide chain becomes phosphorylated and forms phosvitin.

## The Stability of the Differentiated Cell

Once an animal cell has become terminally differentiated, it appears to be a stable entity and generally incapable of being transformed into other cell types. A nerve cell, for example, under normal conditions is not able to become a muscle or epithelial cell. This suggests that the specific controls of gene expression within the cell are somehow stabilized during the differentiation process. Maintenance of the differentiated state of a cell requires that selected genes remain stably active while other genes are permanently repressed for long periods of time. This aspect of cell differentiation has received little attention. However, some insight into the question of how a cell maintains its differentiated state comes from studies with the 5S RNA gene of *Xenopus*. When cloned 5S RNA genes are added to a nuclear extract of oocytes, they assemble into stable, active *transcriptional complexes*. Each complex is formed by the binding of 5S RNA to the 5S-specific transcriptional factor and to other unidentified factors. After RNA polymerase III recognizes the complex, repeated transcription of the gene may occur for several hours. It is probable that all components of the transcriptional complex remain associated with the template following termination and release of the primary transcript. The 5S-specific transcriptional factor is an essential component of the complex for the 5S RNA gene. If histones are added to 5S DNA genes in the absence of the 5S-specific factor, the 5S DNA forms a stable but transcriptionally inactive complex. It is believed that such stable, inactive complexes normally block access of the positive transcriptional factor to 5S DNA. Hence, transcriptionally active complexes would be maintained by strong protein–protein and protein–DNA interactions. The model proposed for the transcriptional control of the 5S RNA gene may help in understanding the basis of the inheritance of the same set of active and inactive genes in a cell type following its embryonic determination.

Although the concept of the stability of the differentiated state is widely accepted, there are reports in the literature suggesting that a terminally differentiated cell can lose its structural and functional properties, become committed to a new developmental pathway, and synthesize an array of macromolecules characteristic of a totally new cell type. We have pointed out previously that nuclei can be isolated from adult cells, microinjected into enucleated eggs, and become reprogrammed to support the differentiation of a variety of cells, including muscle and nerve. However, the process of *metaplasia* or *transdifferentiation* involves the complete transformation of one fully differentiated cell into another differentiated cell. Metaplasia, therefore, would appear to require a loss of those controls over developmental determination as well as those regulating the acquisition of the structural and functional properties of the differentiated state. Examples of true metaplasis are to be distinguished from cases in which cells may lose their phenotypic but not genotypic expression of the differentiated state. The phenotypic properties (i.e., biochemical markers of the differentiated state) of several cell types are known to disappear if such cells are dissociated from tissues and grown in culture with media of selected composition. The cells appear to become dedifferentiated and undifferentiated. For example, when pigment cells are cultured under certain conditions, the melanosomes with their pigment disappear from the cytoplasm. However, it can be shown by various techniques that these cells never lose the capability of synthesizing melanin, suggesting that the absence of pigmentation may be due to a factor that blocks some step in the cascade of controls. These fluctuations in the phenotype of the differentiated cell are often referred to as modulations of the differentiated state.

One of the frequently cited examples of metaplasia is the regeneration of the lens in the adult salamander. If the lens of the urodele is removed, the dorsal iris, normally a derivative of the neural ectoderm, thickens, and subsequently (by day 9 or 10 postoperation) its fully differentiated melanocytes begin to lose their melanin and pigmented appearance. Shortly thereafter, these same cells show DNA and RNA synthesis, the formation of lens fiber cells, and finally the accumulation of lens-specific proteins (crystallins) in the cytoplasm. Crystallins are not normally formed by cells of the dorsal iris. Since no stem cells have been identified in the iris epithelial cells that might be precursors of the fiber cells, the adult iris cells apparently undergo dedifferentiation and then become redirected to a new differentiating pathway specifying the expression of lens protein. Interestingly, when dorsal iris cells are isolated and cultured, they never show the capacity to differentiate lens material. However, iris epithelial cells grafted into the vicinity of the retina of a lensectomized host will regenerate into a lens. The nature of the stimulus provided by the neural retina in this lens differentiation is not known. Lens-specific crystallins also accumulate in long-term cultures of chick embryonic neural and pigmented retinas. Embryonic cells of these tissue layers are apparently able to rechannel their differentiation to form lens fiber cells with their characteristic proteins. Cells isolated and cultured from 3½-day chick retinas will form lenslike structures containing a high concentration of crystallin mRNA. The level of this messenger declines in parallel with a decline in the lens-forming potential of retinal cells. These data are interpreted to mean that the mechanism underlying transdifferentiation involves alterations in the state of determination of cultured neural retina cells.

## Cell Division and Differentiation

During the development of most tissues, there is typically an initial period in which cells proliferate and increase in number. This is followed by a period during which the same cells then undergo change and terminal differentiation. The acquisition of the differentiated state is commonly accompanied by a loss of proliferative activity. The possible role of cell division in the subsequent process of cellular differentiation has been the subject of considerable controversy for many years. Holtzer and co-workers have promulgated the concept that there is a distinct causal relationship between cell proliferation and cell differentiation. They propose that cells must pass through a *critical mitosis* or cell cycle (quantal) before they are competent to acquire and perform their specific differentiative function. During the critical cell division, the genome of the cell is postulated to become reprogrammed.

The critical mitosis theory has been invoked to explain the differentiation of the cell types of a number of organ systems, including the precursors of muscle cells (see Chapter 22), cartilage cells, red blood cells, and mammary gland epithelial cells. DNA synthesis appears to be an absolute requirement for the final differentiation of these cells, but other events that precede or accompany the proposed critical cell division remain very unclear. Smith and Vonderhaar (1981) have recently examined the relationship between DNA synthesis, cell proliferation, and functional differentiation in explant cultures of mouse mammary gland. They found that over half of the epithelial cells in cultured gland explants, upon stimulation by insulin, hydrocortisone, and prolactin, incorporated [$^3$H]thymidine into their nuclei during a 72-hour incubation period. However, when explants were incubated with the same hormones and various concentrations of colchicine, only 5 to 6 percent of the epithelial cells were found to enter mitosis. A large number of the epithelial cells were found to be positive for caesin. The conclusion from these observations is that a critical cell division is not required for the acquisition of the functional differentiation of virgin mouse mammary gland. As previously mentioned, however, the synthesis of DNA is mandatory for the production of milk proteins by the epithelial cells.

## References


Ashburner, M. 1972. Patterns of puffing activity in the salivary gland chromosomes of *Drosophila*. VI. Induction by ecdysone in salivary glands of *D. melanogaster* cultured *in vitro*. Chromosoma 38:255–281.

Beerman, W. 1961. Ein Balbiani-Ring als Locus einer Speicheldrüsen-mutation. Chromosoma 12:1–25.

Bogenhagen, D., M. Wormington, and D. Brown. 1982. Stable transcription complexes of *Xenopus* 5S RNA genes: a means to maintain the differentiated state. Cell 28:413–421.

Britten, R. and E. Davidson. 1969. Gene regulation for higher cells: a theory. Science 165:349–357.

Brown, D. and J. Gurdon. 1978. Cloned single repeating units of 5S DNA direct accurate transcription of 5S RNA when injected into *Xenopus* oocytes. Proc. Natl. Acad. Sci. U.S.A. 75:2849–2853.

Cherbas, L. and P. Cherbas. 1981. The effects of ecdysteroid hormones in *Drosophila melanogaster* cell lines. Adv. Cell Culture 1:91–124.

Clayton, R., I. Thomson, and D. de Pomerai. 1979. Relationship between crystallin mRNA expression in retina cells and their capacity to redifferentiate into lens cells. Nature 282:628–629.

Clever, U. 1964. Actinomycin and puromycin: effects on sequential gene activation by ecdysone. Science 146:794–795.

Clever, U. 1966. Gene activity patterns and cellular differentiation. Am. Zool. 6:33–41.

Davidson, E. and R. Britten. 1979. Regulation of gene expression: possible roles of repetitive sequences. Science 204:1052–1059.

de Pomerai, D. and M. Gali. 1982. A switch for transdifferentiation in culture: effects of glucose on cell determination in chick embryo neuroretinal cultures. Dev. Biol. 93:531–538.

De Robertis, E. 1983. Nucleocytoplasmic segregation of proteins and RNAs. Cell 32:1021–1025.

De Robertis, E. and J. Gurdon. 1977. Gene activation in somatic nuclei after injection into amphibian oocytes. Proc. Natl. Acad. Sci. U.S.A. 74:2470–2474.

De Robertis, E., G. Partington, R. Longthorne, and J. Gurdon. 1977. Somatic nuclei in amphibian oocytes: evidence for selective gene expression. J. Embryol. Exp. Morphol. 40:199–214.

De Robertis, E. and J. Gurdon. 1979. Gene transplantation and the analysis of development. Sci. Am. 241:74–82.

Edström, J., H. Sierakowska, and K. Burvall. 1982. Dependence of Balbiani Ring induction in *Chironomus* salivary glands on inorganic phosphate. Dev. Biol. 91:131–137.

Evans, M., L. Hager, and G. McKnight. 1981. A somatomedin-like peptide hormone is required during the estrogen-mediated induction of ovalbumin gene transcription. Cell 25:187–193.

Gronemeyer, H. and O. Pongo. 1980. Localization of ecdysterone on polytene chromosomes of *Drosophila melanogaster*. Proc. Natl. Acad. Sci. U.S.A. 77:2108–2112.

Grossbach, U. 1973. Chromosome puffs and gene expression in polytene cells. Cold Spring Harbor Symp. Quant. Biol. 38:619–627.

Gurdon, J. 1968. Changes in somatic cell nuclei inserted into growing and maturing amphibian oocytes. J. Embryol. Exp. Morphol. 20:401–414.

Gurdon, J. 1976. Egg cytoplasm and gene control in development. Proc. R. Soc. Lond. (Biol.) 198:211–247.

Gurdon, J. and H. Woodland. 1968. The cytoplasmic control of nuclear activity in animal development. Biol. Rev. 43:233–267.

Gurdon, J. and H. Woodland. 1969. The influence of the cytoplasm on the nucleus during cell differentiation, with special reference to RNA synthesis during amphibian cleavage. Proc. R. Soc. Lond. (Biol.) 173:99–111.

Hager, L., G. McKnight, and R. Palmiter. 1980. Glucocorticoid induction of egg white mRNAs in chick oviduct. J. Biol. Chem. 255:7796–7800.

Hämmerling, J. 1963. Nucleocytoplasmic interactions in *Acetabularia* and other cells. Annu. Rev. Plant Physiol. 14:65–92.

Hegner, R. 1911. Experiments with chrysomelid beetles. III. The effects of killing parts of the eggs of *Lepitinotersa decemlineata*. Biol. Bull. 20:237–251.

Holden, J. and M. Ashburner. 1978. Patterns of puffing activity in the salivary gland chromosomes of *Drosophila*. IX. The salivary and prothoracic gland chromosomes of a dominant temperature-sensitive lethal of *D. melanogaster*. Chromosoma 68:205–227.

Holtzer, H. 1976. Lineages, quantal cell cycles, and generation of cell diversity. Quant. Rev. Biophys. 8:523–559.

Illmensee, K. and A. Mahowald. 1974. Transplantation of posterior pole plasm in *Drosophila*. Induction of germ cells at the anterior pole of the egg. Proc. Natl. Acad. Sci. U.S.A. 71:1016–1020.

McKnight, G., P. Pennequin, and R. Schimke. 1975. Induction of ovalbumin mRNA sequences by estrogen and progesterone in chick oviduct as measured by hybridization to complementary DNA. J. Biol. Chem. 250:8105–8109.

McMahon, D. 1974. Chemical messengers in development: a hypothesis. Science 185:1012–1021.

Mertz, J. and J. Gurdon. 1977. Purified DNAs are transcribed after microinjection into *Xenopus* oocytes. Proc. Natl. Acad. Sci. U.S.A. 74:1502–1506.

Moen, R. and R. Palmiter. 1980. Changes in hormone responsiveness of chick oviduct during primary stimulation with estrogen. Dev. Biol. 78:450–463.

O'Malley, B. and A. Means. 1976. The mechanism of steroid hormone regulation of transcription of specific eukaryotic genes. Prog. Nucleic Acid Res. Mol. Biol. 19:403–419.

Ono, T. and M. Getz. 1980. Levels of ovalbumin messenger RNA sequences in non-oviduct tissues of the chicken. Dev. Biol. 75:481–484.

Renkawitz, R., H. Beug, T. Graf. P. Matthias, M. Grez, and G. Schutz. 1978. Expression of a chicken lysozyme recombinant gene is regulated by progesterone and dexamethasone after microinjection into oviduct cells. Cell 31:167–176.

Riddiford, L. 1982. Changes in translatable mRNAs during the larval–pupal transformation of the epidermis of the tobacco hornworm. Dev. Biol. 92:330–342.

Sargent, T. and I. Dawid. 1983. Differential gene expression in the gastrula of *Xenopus laevis*. Science 222:135–139.

Schwartz, R., C. Chang, W. Schrader, and B. O'Malley. 1977. Effect of progesterone receptors in transcription. Ann. N.Y. Acad. Sci. 286:147–160.

Shigeru, S. and D. Brown. 1982. Contact points between a positive transcription factor and the *Xenopus* 5S RNA gene. Cell 31:395–405.

Smith, G. and B. Vonderhaar. 1981. Functional differentiation in mouse mammary gland epithelium is attained through DNA synthesis, inconsequent of mitosis. Dev. Biol. 88:167–179.

Stein, G., J. Stein, and L. Kleinsmith. 1975. Chromosomal proteins and gene regulation. Sci. Am. 232:47–57.

Stein, G., J. Stein, L. Kleinsmith, W. Park, R. Jansing, and J. Thomson. 1976. Nonhistone chromosomal proteins and histone gene transcription. Prog. Nucleic Acid Res. Mol. Biol. 19:421–445.

Subtelny, S. 1974. Nucleocytoplasmic interactions in development of amphibian hybrids. Int. Rev. Cytol. 39:35–88.

Trendelenburg, M. and J. Gurdon. 1978. Transcription of cloned *Xenopus* genes visualized after injection into oocyte nuclei. Nature 276:292–294.

Tsai, S., S. Harris, M. Tsai, and B. O'Malley. 1976. Effects of estrogen on gene expression in chick oviduct. J. Biol. Chem. 251:4713–4721.

Turkington, R. 1968. Hormone-dependent differentiation of mammary gland *in vitro*. In: Current Topics in Development Biology, Vol. 3, pp. 199–218. A. A. Moscona and A. Monroy, eds. New York: Academic Press.

Vonderhaar, B. and G. Smith. 1982. Dissociation of cytological and functional differentiation in virgin mouse mammary gland during inhibition of DNA synthesis. J. Cell Sci. 53:97–114.

# 12

# EARLY HUMAN DEVELOPMENT, EMBRYONIC MEMBRANES, AND PLACENTATION

## Early Human Development

Because of the obvious difficulties in obtaining material, the study of the early development of the human presents many gaps in the series of embryos available for observation. Nevertheless, by comparison of the available human material with other mammalian, particularly primate, species, the early development of the human ovum can be seen to differ little from that in other primate species. The timing, although necessarily based on assumed times of fertilization, in comparison with other species and studies of in vitro development, is probably quite accurate within the limits of biological variability.

### Development through the First Week: Cleavage and Blastocyst Formation

Four examples of in vivo cleaving human ova have been described—2, 12, 58, and 107 cells. The two-cell embryo (Fig. 12–1), removed from the uterine tube, shows two blastomers of about equal size surrounded by an incomplete zona pellucida inside of which a polar body is visible. The embryo's diameter is 179 $\mu$m. On the basis of the time of coitus and the estimated time of ovulation (based on uterine histology), this two-cell embryo is considered to be about 36 to 40 hours postfertilization.

A 12-cell stage with a diameter of 173 $\mu$m was also recovered from the uterine tube, but unfortunately this specimen was lost during its examination and no photographs are available. At this stage, the blastomers differed in size. There was one large central blastomere with a diameter approximately double that of the average of the other 11 blastomers. Presumably the large blastomere was embryonic in potential—destined to form embryonic structures—and the smaller ones were *trophoblastic* cells—destined to form extraembryonic membranes. This embryo was estimated to be about 72 hours in age.

The 58-cell embryo (Fig. 12–2 A), still surrounded by its zona pellucida, shows a continuing divergence in cell specialization with 5 large *formative* (embryonic) cells and 53 small trophoblastic cells. The formative cells are more central in position and represent the beginning of the development of the *inner cell mass* (ICM) or *germ disc* which will locate at one pole of the embryo, the embryonal pole. Coalescing intercellular spaces represent the beginning of the formation of the segmentation cavity or the blastocoele. The age of this

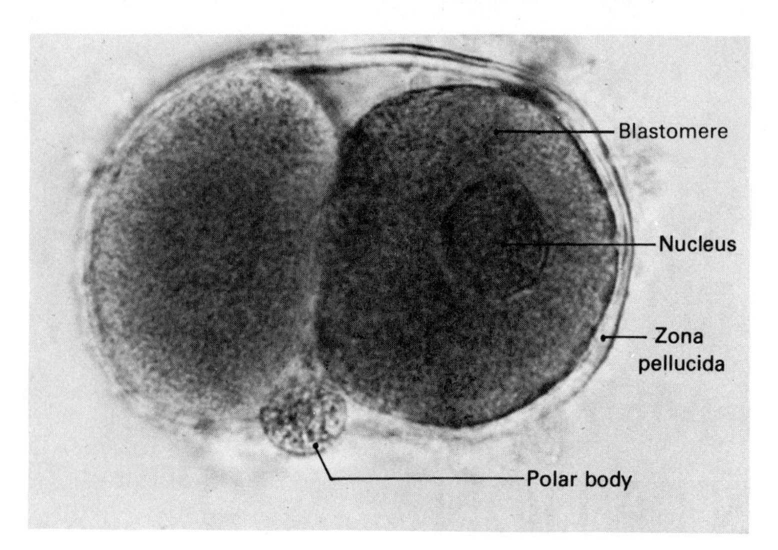


**FIG. 12–1.** Two-celled human embryo, about 36–40 hours postfertilization. (Courtesy of the Carnegie Institution of Washington, Department of Embryology, Davis Division.)

embryo, which was found free in the uterine cavity, was approximately 96 hours.

The 107-cell embryo (Fig. 12–2 B) consists of 8 formative cells and 99 trophoblastic cells. The larger more vacuolated formative cells make up an excentrically located inner cell mass at the embryonal pole of what may now be termed the *blastocyst*. The trophoblast cells are flattened and elongated and form a complete covering for the blastocyst. The zona pellucida has disappeared. This embryo, also found free in the uterine cavity, was estimated to be four and one half days old.

## Implantation

Implantation is the process by which the developing mammalian egg penetrates the outer layers of the uterine wall and, by means of its expanding trophoblast cells, begins to establish a relationship with the maternal tissues that will result in the formation of the placenta. Development of the placenta is a necessary and crucial step, since it is this organ—partly of maternal and partly of fetal origin—that must function in providing the nutritive, excretory, and respiratory needs of the developing fetus until parturition. The intimacy of contact between fetal and maternal tissues varies in different species and will be considered in a separate section on comparative placentation. Implantation and placentation are the functions of extraembryonic trophoblast cells, not a part of the development of the embryo proper, which is formed from the inner cell mass. However, it is an integral part of development and does concern the growth of tissue derived from the fertilized egg; thus, we will consider the early stages of implantation in conjunction with the progressive stages of embryonic development rather than describing them separately.

The youngest human embryo in the process of implantation (Fig. 12–3) is judged to be seven to seven and one half days of age. The embryonic pole of the blastocyst has eroded the uterine epithelium and is starting to penetrate into the endometrium. The blastocyst, in this section, has collapsed so that the abembryonal trophoblast lies up against the under surface of the germ disc partially obscuring the blastocoele—probably not a normal occurrence. The trophoblast wall varies from a thin mesothelial or membranous structure at the abembryonal region to a thick proliferating mass of cells at the embryonal pole. The trophoblast in the embryonal region, which is actively invading the maternal tissues, can be separated into an inner cellular layer that lines the segmentation cavity, the *cytotrophoblast*, and an outer syncytial mass, the *syncytiotrophoblast*. The embryonic disc at this time shows the beginning of the development of a bilaminar form. It consists of a layer of columnar cells, the ectoderm (or epiblast, since the cells in this layer will also contribute to the formation of the mesoderm), beneath which is a layer of irregularly arranged polyhedral cells, the endoderm, adjoining the blastocoele. Clefts may be seen between the epiblast and the overlying trophoblast. These represent the beginning of the formation of the *amnionic cavity*, which will result from the coalescence of these clefts. The precocious formation of the *amnion* by cavition or delamination should be

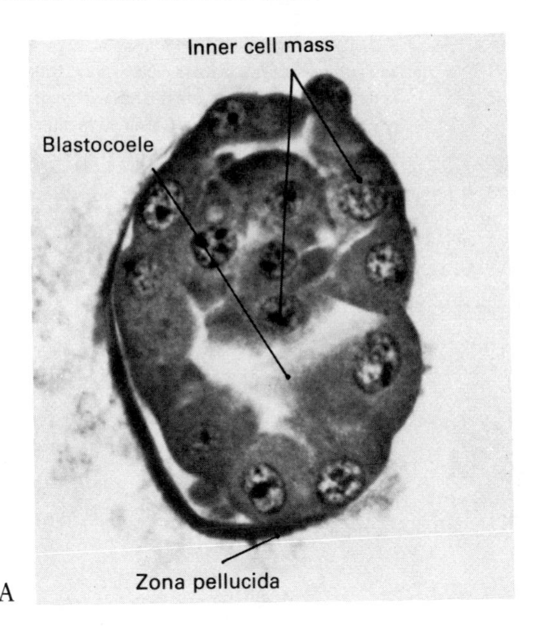


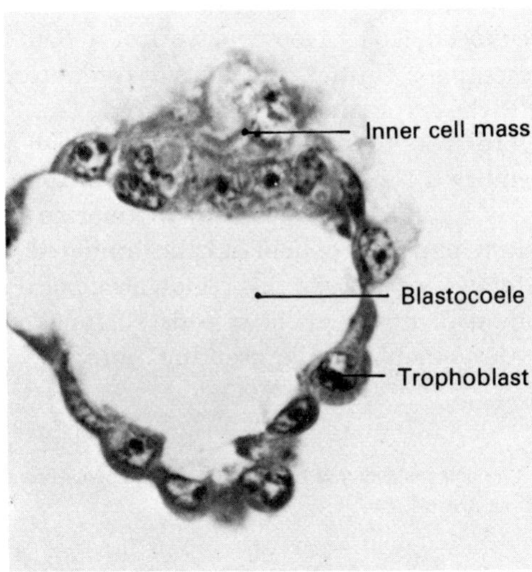


**FIG. 12–2.** A, 58-cell human embryo, about 96 hours. Formation of the inner cell mass and blastocoele; B, 107-celled human embryo, about four and a half days. Unilaminar blastocyst with inner cell mass at the embryonal pole. (Courtesy of the Carnegie Institution of Washington, Department of Embryology, Davis Division.)

compared with its formation by the growth and folding of sheets of cells as described for the chick and the pig in a later section of this chapter.

## Development through the Second Week

The 11-day embryo (Fig. 12–4) is almost completely embedded within the uterine tissues and the uterine epithelium has closed in over it. The continuation of the invasive process is seen in the marked expansion of the trophoblast cells. The cytotrophoblast cells still appear as a single layer around the blastocoele and are easily distinguished from the rapidly expanding syncytiotrophoblast cells. The increase in the number of nuclei in the latter layer is not due to division of the syncytiotrophoblast nuclei themselves, but to cell divisions taking place in the cytotrophoblast. Tritiated thymidine is taken up only by the cytotrophoblast cells. Over a period of time after a pulse label, however, radioactivity may be detected in the syncytiotrophoblast, establishing the pathway of differentiation of the syncytiotrophoblast from the cytotrophoblast. Spaces or lacunae have developed as the result of the coalescence of cavities appearing in the syncytiotrophoblast, and this stage of trophoblast development is termed the *lacunar* or *previllous* stage. The uterine endometrium in contact with the fetal syncytiotrophoblast is undergoing a decidual reaction forming the *decidua* with its characteristic larger pale cells containing glycogen and lipid droplets in the cytoplasm. The term decidua is indicative of the fact that this part of the uterine lining will be cast off at parturition as the so-called afterbirth. Glandular enlargement and vascular congestion are apparent in the uterine tissue. The embryo proper has also undergone significant change. A continued differentiation of epiblast and endoderm is apparent. The amniotic cavity has developed further, and cells apparently differentiated from the overlying cytotrophoblast have formed its roof, giving rise to a complete membrane, the amnion. A thin layer of cells attached to the embryonic endoderm forms a second cavity within the blastocoele, the *primary yolk sac* cavity. This layer of cells, variously termed *Heuser's membrane, exocoelomic membrane,* or *primary yolk-sac membrane,* is considered to be formed from extraembryonic mesothelial cells of the cytotrophoblast, since no embryonic mesoderm has as yet developed. The blastocoele may now be called the *exocoelomic cavity,* and the blastocyst is now considered to be in its bilaminar stage.

At 12 days (Fig. 12–5) the trophoblast has expanded still further, and the syncytiotrophoblast shows an intercommunicating system of lacunar spaces. Maternal blood and engulfed endometrial cells in the lacunar spaces present evidence of the active erosion of uterine tissue by the fetal trophoblast. The cytotrophoblast shows areas where it has started to extend into the syncytiotrophoblast, the beginning of the formation of the *chorionic villi.*

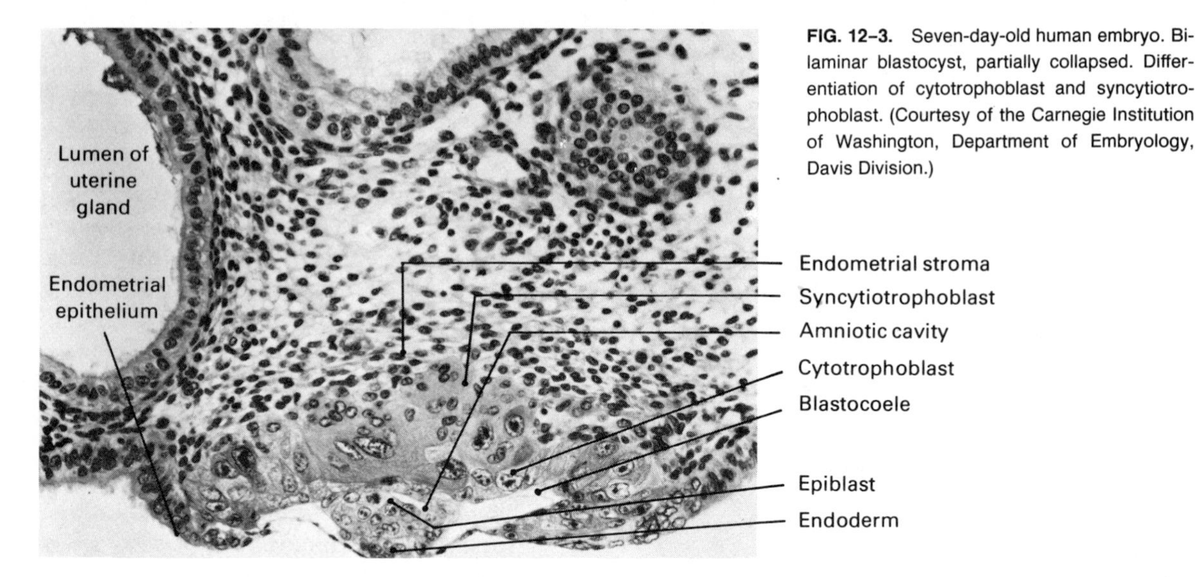


**FIG. 12–3.** Seven-day-old human embryo. Bilaminar blastocyst, partially collapsed. Differentiation of cytotrophoblast and syncytiotrophoblast. (Courtesy of the Carnegie Institution of Washington, Department of Embryology, Davis Division.)

The embryo is still in a bilaminar stage. Some cells from the borders of the endoderm layer have begun to migrate away from the embryonic disc into the inner surface of the yolk sac. These extraembryonic endoderm cells are the third layer of cells surrounding the primary yolk sac and thus represent the beignning of a trilaminar blastocyst. We thus have a bilaminar embryo but an early trilaminar blastocyst. A space has appeared between the primary yolk sac and the cytotrophoblast, and this space is occupied by a loose network of stellate cells representing a further contribution of the trophoblast to the extraembryonic mesoderm.

Figure 12–6 A represents a section through a 13-day embryo. The abembryonal region of the blastocyst is completely enclosed by the uterine endometrium, and the trophoblast is beginning to show proliferative changes in this region also. Local proliferation of cytotrophoblast cells surrounded by syncytiotrophoblast, representing *primary stem*

**FIG. 12–4.** Eleven-day-old human embryo. Bilaminar blastocyst. Lacunar spaces in rapidly expanding syncytiotrophoblast. (Courtesy of the Carnegie Institution of Washington, Department of Embryology, Davis Division.)

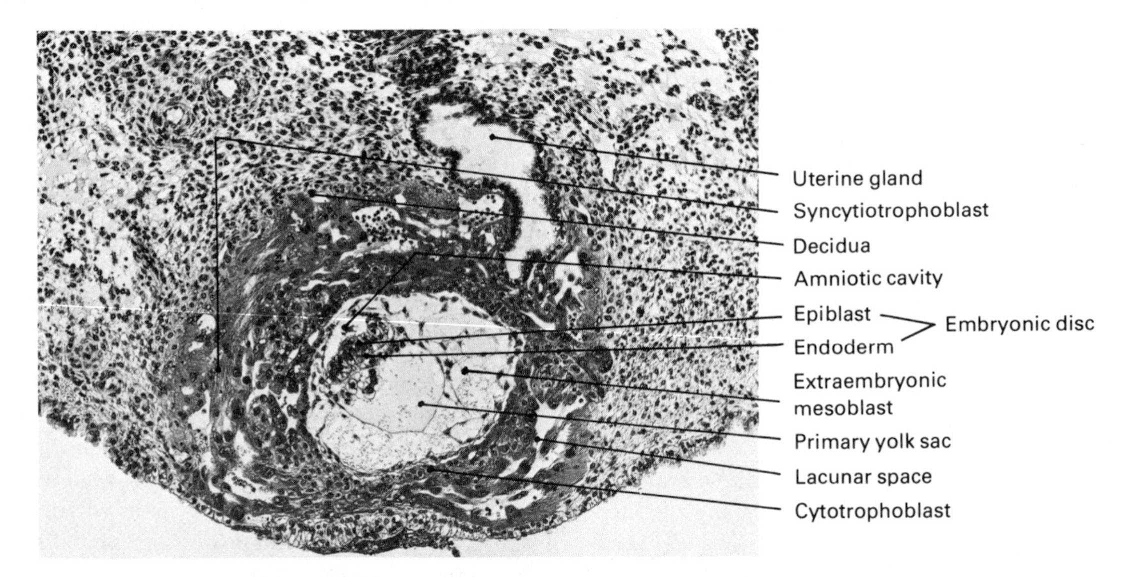

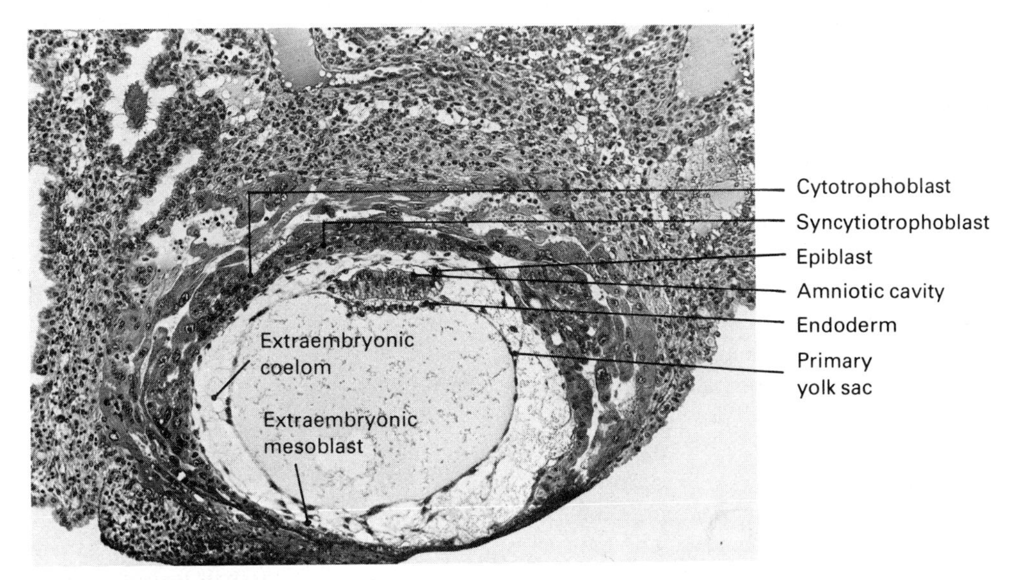


**FIG. 12–5.** Twelve-day-old human embryo. Late bilaminar blastocyst. Intercommunicating spaces in the syncytiotrophoblast. (Courtesy of the Carnegie Institution of Washington, Department of Embryology, Davis Division.)

**FIG. 12–6.** A, drawing of a 13-day-old human embryo. Trilaminar blastocyst. Amnion, chorion, and yolk sac developed. Body stalk present; B, cross section through the middle of the embryonic disc; C, section through the caudal end of the embryonic disc. Embryonic mesoderm differentiating from the early primitive streak.

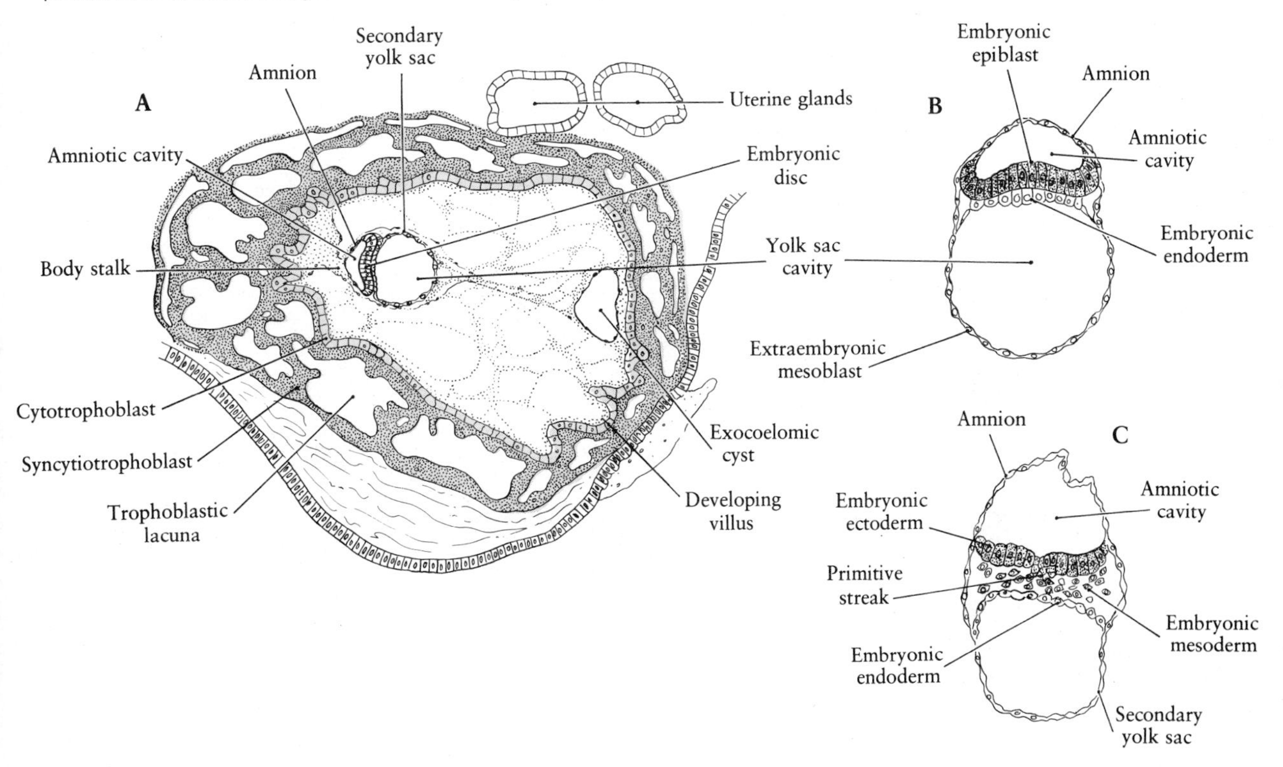

*villi*, are more apparent than in the previously described embryo. The continued proliferation of cells from the endodermal layer of the embryonic disc has now completely lined a cavity much smaller than the primary yolk-sac cavity. These endodermal cells are covered by mesoblast, and together they make up the *secondary* or *definitive yolk sac*. Apparently, large portions of the primary yolk sac are pinched off in the process of forming the smaller definitive yolk sac. Exocoelomic cysts found in what is now the extraembryonic coelom are considered to be these pinched-off remnants of the primary yolk sac. The extraembryonic coelom thus splits the extraembryonic mesoderm into two layers. One, the outer layer of the yolk sac and the other, the inner layer of the cytotrophoblast—which may now be called the chorion. The mesoderm also forms the outer layer of the amnion.

The embryo is still in the form of a flat bilaminar disc. At the caudal end of this disc, a band of mesoderm stretches across the extraembryonic coelom connecting the embryo and the amnion to the cytotrophoblast. This is the *body stalk*. At the cranial end of the embryonic disc an area of columnar endoderm cells forms the prechordal plate, the future site of the mouth. A cross section through the middle of the embryonic disc (Fig. 12–6 B) shows the typical pseudostratified columnar ectoderm, the loosely connected cuboidal endoderm, and the amnionic and yolk-sac cavities. A cross section near the caudal end of the embryo (Fig. 12–6 C) shows the beginning of the formation of a third layer, the embryonic mesoderm, between the ectoderm and the endoderm. This is occurring in the region of the forming primitive streak.

Figure 12–7 is a diagrammatic review of the development of the human embryo during the first two weeks, showing the stages described, their estimated age, and their position within the female reproductive tract.

## Embryonic Membranes

Fishes and amphibians, collectively called *anamniotes*, deposit large numbers of eggs that develop rapidly to an independent free-swimming stage in an aquatic environment. Because of the rapid development and the aquatic environment, these eggs do not require complex auxillary membranes to assist them in respiration, excretion, and nutrition during their embryonic stages. However, the eggs of the anamniotes may form protective coverings such as the chorion of the fishes and the jelly membranes of the amphibians. In addition, fishes also develop a membrane, which is termed a yolk sac. It overgrows and surrounds the yolk and, when vas-

**FIG. 12–7.** Composite of the first two weeks of human development showing the age of the embryo and its general location within the female reproductive tract. (After R. F. Gasser, 1975. Atlas of Human Embryos. Harper and Row.)

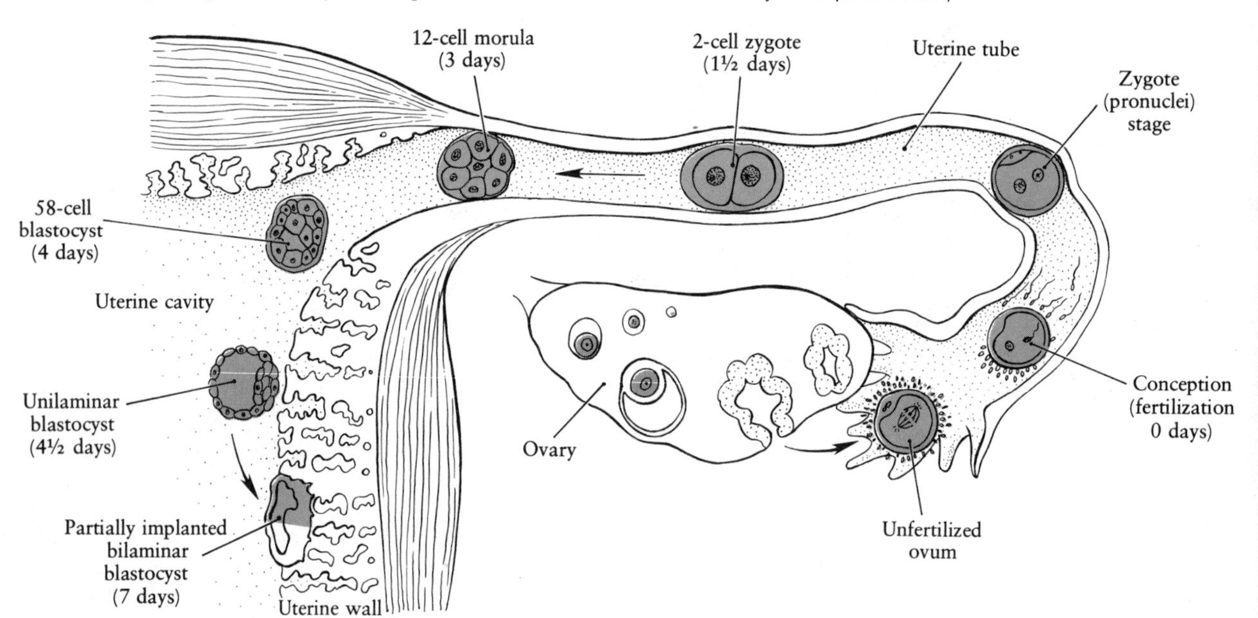

cularized, makes the yolk available to the embryo. The yolk sac is formed by the growth of the edges of the blastoderm (as described in Chapter 8), which consists of ectoderm, mesoderm, and periblast. It does not have any endoderm and therefore is not homologous to the yolk sac of higher forms, although it does serve the same function.

In reptiles and birds that lay their eggs on land, development takes place within a system in which the embryo is largely closed off from the outside environment by the eggshell. This closed *(cleidoic)* system necessitates the development of structures whose sole function is to protect the embryo and provide for its nutritive, excretory, and respiratory needs. These structures are called extraembryonic membranes. They function only during development; they do not form any part of the embryo itself. Four membranes are developed in reptiles and birds, the *serosa (chorion), amnion, yolk sac,* and *allantois.* The first two develop from somatopleure; the last two, from splanchnopleure.

In the mammal, development takes place in the uterus but the same embryonic membranes are formed, although their function may be different. The placenta, which is partly of embryonic and partly of maternal origin, is the major extraembryonic membrane in mammals.

## The Yolk Sac

In the chick, the splanchnopleure and somatopleure extend out in all directions over the yolk beyond the region in which the body of the embryo is forming. Where these layers lie beyond the limits of the embryo proper, they are termed extraembryonic. At first, there is no definite boundary between the extraembryonic and the embryonic splanchnopleure and somatopleure, which form continuous sheets. However, as the embryo takes shape, a series of folds appear all around it and grow downward to undercut it and separate it from the underlying yolk. These folds, which then define the limits of the embryonic body, are known as the *body folds.* The first to appear is the *head fold* (Fig. 12–8). During the second day of incubation it becomes continuous with the *lateral body folds* devel-

**FIG. 12–8.** Schematic diagrams of sagittal sections of the chick showing the formation of the head and tail folds and the extraembryonic membranes. A, 24 hours; B, 72 hours; C, 5 days; D, 9 days. (From B. Patten and B. M. Carlson, 1974. Foundations of Embryology. McGraw-Hill Book Co., New York.)

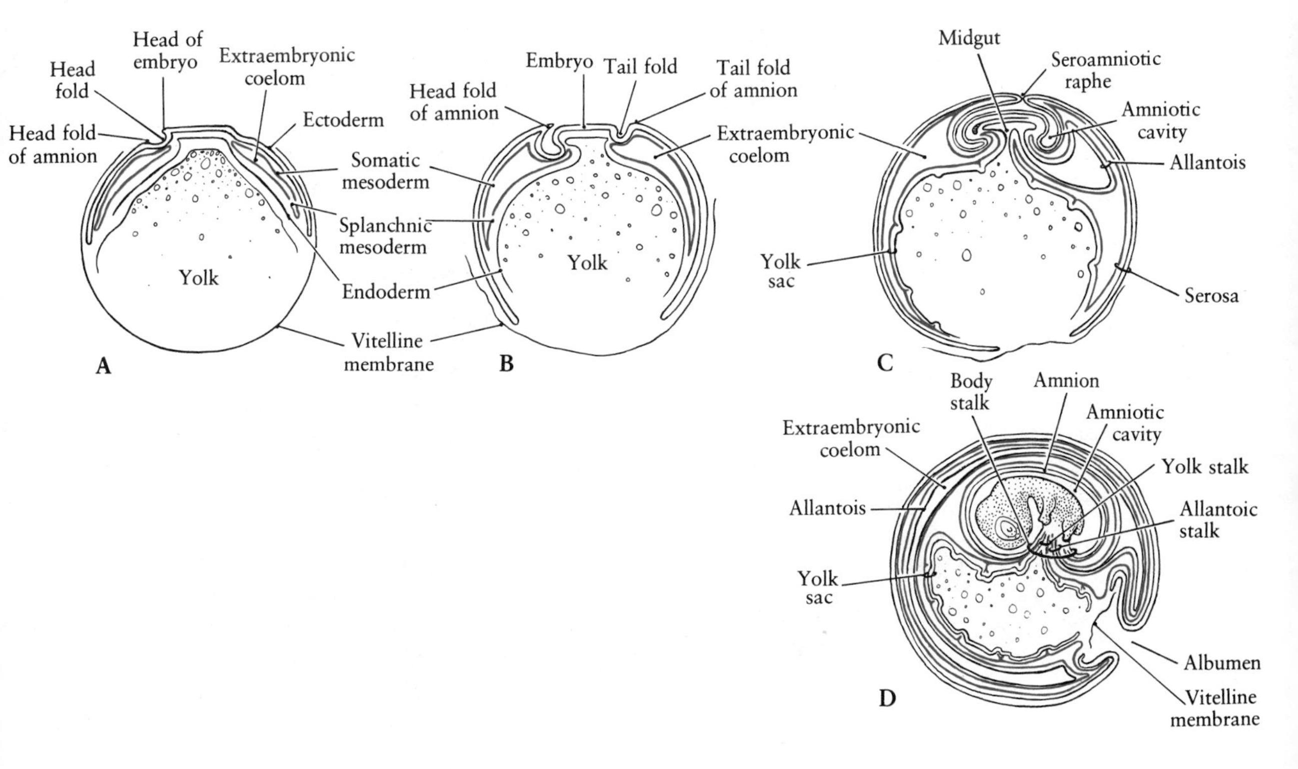

oping along both sides of the embryo (Fig. 12–9). During the third day a *tail fold* begins to undercut the caudal end of the embryo (Fig. 12–8 B). The net result is a progressive separation of the embryo from the yolk, establishing the shape of the body and indicating the boundary between the embryonic and extraembryonic structures.

When the head and tail folds have undercut the embryo, the embryonic gut may be divided into three regions: a closed foregut in the head fold, a closed hindgut in the tail fold, and an open midgut between them, whose floor is the yolk. The extraembryonic splanchnopleure of the yolk sac is continuous with the splanchnopleure that forms the lining of the gut of the embryo. The extraembryonic splanchnopleure gradually grows over the surface of the yolk in all directions and encloses it in a covering membrane, the yolk sac. As the body folds continue to undercut the embryo, the foregut and hindgut become progressively longer, and the midgut becomes progressively shorter. The connection between the splanchnopleure of the yolk sac and the midgut thus becomes restricted to a narrow band of tissue termed the *yolk stalk* (Fig. 12–8 D). The mesoderm of the yolk sac forms a highly vascular network that establishes connection with the circulatory system of the embryo by way of the vitelline blood vessels. By 40 hours of incubation, blood is circulating through the vascular network of the yolk sac. Vascularized projections of the yolk sac penetrate into the yolk and aid in its breakdown and absorption.

**FIG. 12–9.** Cross sections through the early chick embryo showing the relationship of the body folds to the seroamniotic folds during the formation of the serosa and the amnion.

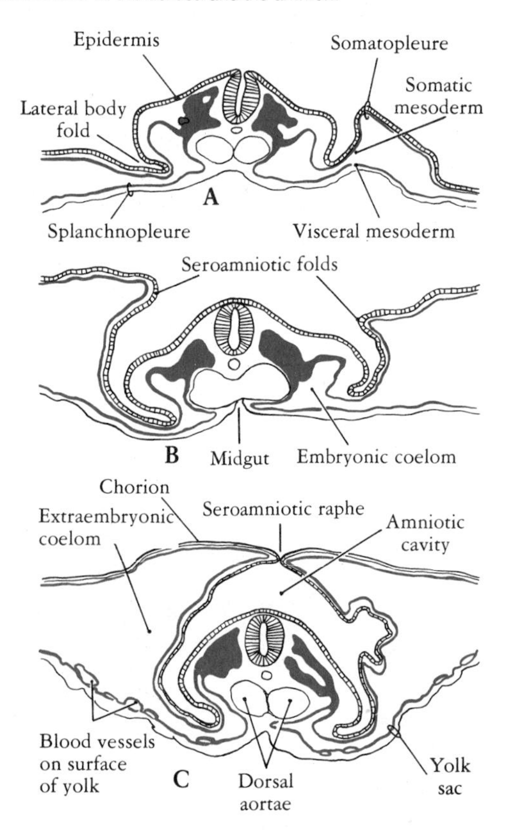


## The Amnion and Serosa

The amnion and serosa arise together as the *seroamniotic folds*, upwardly directed folds of the somatopleure just outside of the body folds (Figs. 12–8 and 12–9). The head, tail, and lateral seroamniotic folds meet and fuse dorsally at the *seroamniotic raphe* and completely enclose the embryo. When the somatopleure fuses, the embryo is now covered by two double-layered membranes. The inner membrane, lined with ectoderm and covered with somatic mesoderm, is the amnion, and its cavity is the amniotic cavity. The outer membrane, lined with somatic mesoderm and covered with ectoderm, is the serosa and its cavity is the extraembryonic coelom, which is continuous with the embryonic coelom. Following its formation, the amniotic cavity becomes filled with fluid and the embryo is immersed in its own aquarium. This acts as a protective device, preventing dessication, distributing shocks equally over the entire surface of the embryo, and forming an isolated chamber where growth and positional changes may occur without restriction.

## The Allantois

The last of the four embryonic membranes to develop appears late in the third day of incubation. The allantois differs from other membranes in that it develops as an extension of a part of the embryo itself, specifically as an evagination of the splanchnopleure of the hindgut (Figs. 12–8 and 12–10). During the fourth day, this evagination is carried out into the extrembryonic coelom between the amnion and the serosa, where it continues to grow until it entirely fills up this space (Fig. 12–8 C,D). The allantoic sac remains attached to the hindgut by the allantoic stalk, which is traversed by the al-

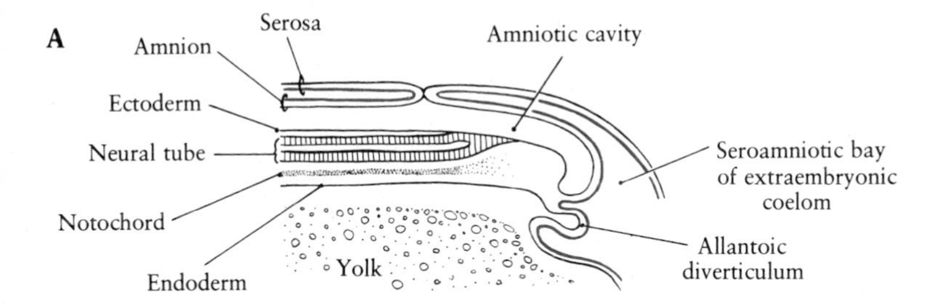


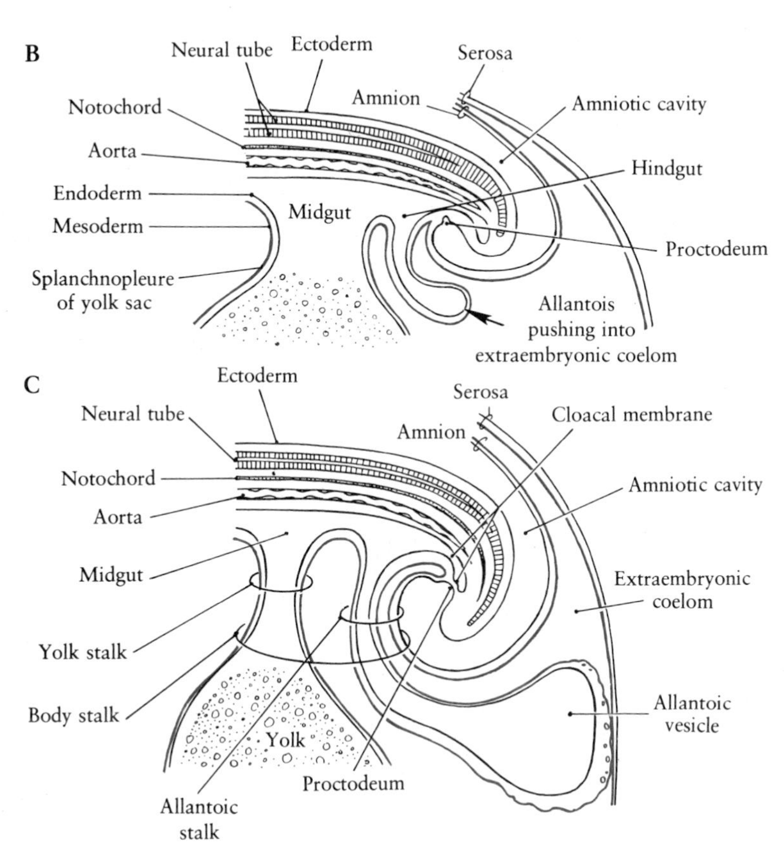


**FIG. 12-10.** Diagrams of sagittal sections through the caudal end of the chick embryo showing the formation of the allantois and its relationship to the extraembryonic coelom and the serosa. A, 72 hours; B, 3½ days; C, 4½ days. (From B. Patten and B. M. Carlson, 1974. Foundations of Embryology. McGraw-Hill Book Company.)

lantoic blood vessels going to and from the embryo. The yolk stalk and the allantoic stalk lie alongside of each other in the body stalk.

The splanchnic mesoderm of the allantoic diverticulum becomes closely associated with the somatic mesoderm of the serosa. When the former becomes vascularized, a rich vascular network is formed just beneath the entire surface of the egg shell. This network provides the mechanism for the exchange of respiratory gases between the outside air and the fetal red cells. The combined serosa and allantois thus function in this respect as the embryonic lung. The similarity of this condition in the chick to the chorioallantoic placenta of the mammal will be apparent when we consider the latter.

In addition to its respiratory responsibility, the allantois also has an excretory function, serving as a large reservoir for the storage of wastes. During the early stages of development, urea is the main breakdown product of protein metabolism in the chick embryo. This compound, however is soluble in water, so that large amounts of water would be necessary to keep the urea below toxic levels if it were to continue to be formed throughout development. The breakdown product of protein during the later stages of development, however, is uric acid. This compound is relatively insoluble in water, so that large quantities can be stored as crystalline material in the allantoic sac without ill effects to the embryo.

## Placentation

In describing the first two weeks of development we have seen how the blastocyst loses its zona pellucida, attaches to the uterine wall, and then rapidly becomes completely embedded within the uterine endometrium. The early period of implantation is characterized by the rapid proliferation of the trophoblast cells, particularly the syncytial layer, which actively erode the uterine endometrium and its glands and blood vessels to establish lacunar spaces lined with syncytiotrophoblast and filled with maternal blood. Blood moves into the lacunar spaces from the open ruptured ends of the branches of the uterine spiral arteries.

The normal implantation site is in the posterior wall of the uterus. Implantation in other areas of the uterus may lead to abnormal development, particularly if it occurs near the cervix. When implantation occurs outside of the uterus, this is known as an *ectopic pregnancy*. Less than 2 percent of all pregnancies are ectopic, the large percentage of these occurring in the uterine tube. Tubal pregnancy usually results in rupture of the tube and severe hemorrhaging during the second month of pregnancy. Less commonly, implantation may occur in the ovary or in the mesenteries of the abdominal cavity. Rarely does any extrauterine pregnancy come to full term.

Early in the third week of development the cytotrophoblast penetrates into the bases of the syncytiotrophoblast columns lining the lacunae and forms primary villi (Fig. 12–11 A). The entire trophoblast is proliferating rapidly to keep pace with the growth of the blastocyst. However, mitotic activity in the trophoblast still remains restricted to the cytotrophoblast, which supplies the nuclei for the expansion of the syncytiotrophoblast as well as for its own proliferation into the syncytiotrophoblast columns. The cytotrophoblast cells in the developing villi are called *Langhans cells*.

Shortly after the formation of the primary villi, mesoderm penetrates them and converts them into *secondary villi* (Fig. 12–11 B). The mesoderm is probably of extraembryonic origin derived from the trophoblast. By the end of the third week, capillaries have developed in situ in the mesodermal cores of the villi, which are now termed *tertiary villi* (Fig. 12–11 B). During this period, the embryonic circulatory system is becoming established; early in the fourth week communication between the embryonic umbilical arteries and veins and the vessels of the chorion, by way of channels in the body stalk, provides a pathway of circulation of blood between the fetus and the placenta. The vessels that are found in the body stalk develop from the allan-

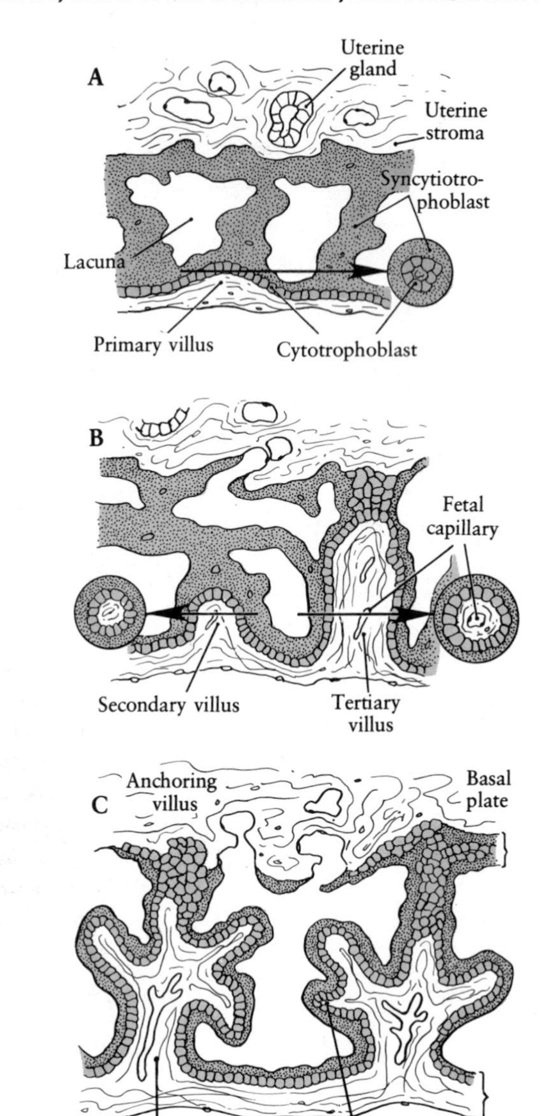


**FIG. 12–11.** Development of chorionic villi. A, primary villus; B, secondary villus with mesodermal core and tertiary villus with fetal capillary; C, anchoring villi connecting chorionic and basal plates of the placenta.

toic mesoderm (Fig. 12–12). Although the allantois itself is rudimentary, its splanchnic mesodermal component forms these important vessels. Because both the chorion and the allantois contribute to the formation of the placenta, it is called a *chorioallantoic* placenta. When the heart begins to beat during the fifth week, the placenta now functions to supply the embryo's needs, which have been taken care of prior to this time by diffusion only.

A further development of the placenta is marked by the proliferation of the cytotrophoblast at the tips of the villi into the overlying syncytiotrophoblast until it reaches the maternal endometrium (Fig. 12–11 C). Cytotrophoblast cells from adjacent villi make interconnections and form a thin cytotrophoblast shell, which marks the line of contact between the fetal and the maternal parts in the placenta. This development starts at the embryonal pole but soon occurs over the entire circumference of the placenta. Cytotrophoblast and syncytiotrophoblast in contact with the maternal decidua form the *basal plate* of the placenta. The fetal area from which the villi develop is termed the *chorionic plate*. Between the basal plate and the chorionic plate, the coalesced lacunae form the *intervillous space*. This space is still lined with syncytiotrophoblast. The villi crossing the intervillous space and attaching to the basal plate are called *anchoring villi* (Fig. 12–11 C).

As the placenta continues to develop, the villi branch extensively, and each main stem forms

**FIG. 12–12.** Embryo and its membranes at about the end of the first month. The allantois is rudimentary but the allantoic mesoderm has formed the body stalk blood vessels.

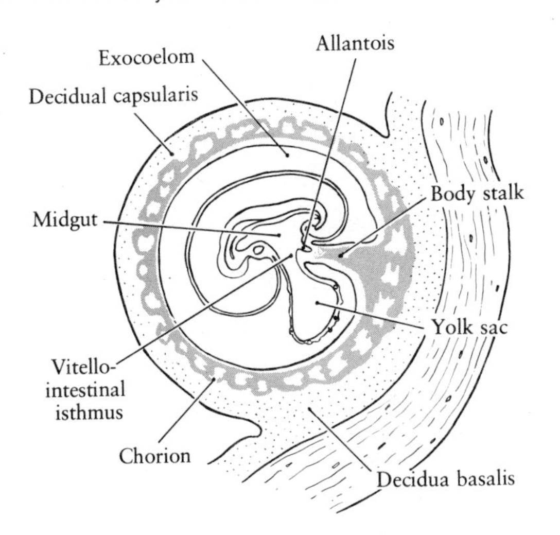


branches that end freely in the intervillous space as well as branches that acquire secondary connections to the basal plate and thus increase the number of the anchoring villi. Each main-stem villus thus resembles the root system of a bush. There are estimated to be from 60 to 200 stem villi in the mature placenta, and their branches form a dense network of villi filling in the intervillous space. A new technique for obtaining fetal cells for prenatal analysis by removing a small sample of the chorionic villi is now being tested. A thin catheter is inserted into the pregnant woman's uterus and a small plug of tissue from the end of one or more villi is removed by suction. This technique, known as *chorionic villus biopsy*, has two advantages over *amniocentesis* (removal of a small amount of amniotic fluid which includes a small number of fetal cells). Firstly, it can be performed as early as the middle of the second month of pregnancy, whereas amniocentesis cannot be performed until the sixteenth week. In addition, the tissue removed is made up of rapidly dividing fetal cells that can be analyzed immediately for chromosomal and biochemical defects. By contrast, the fetal cells in the sample of amniotic fluid are so few in number that they must be grown in tissue culture for two weeks before there are enough of them to analyze.

The amount of tissue separating the fetal capillaries from the intervillous space—mesenchyme, cytotrophoblast, and syncytiotrophoblast in that order—becomes progressively reduced throughout gestation, accompanied by a great increase in the number and a considerable reduction in the cross-sectional diameter of the individual villi. The capillaries increase in size and move from a central to a peripheral position directly under the trophoblast. After the first trimester the cytotrophoblast, which at first forms a complete layer in each villus, partially disappears, while the syncytiotrophoblast thins out considerably. Although at term the capillaries may often appear to bulge into the intervillous space covered only by an attenuated layer of synctiotrophoblast, the Langhans cells never completely disappear from the villi, despite the fact that they are difficult to find in histological sections.

We may list a number of changes that occur during the development of the placenta that apparently increase the efficiency of placental transport: (1) the ratio of the surface to volume of the villi increases, (2) the thickness of the syncytiotrophoblast decreases and the cytotrophoblast becomes discontinuous, (3) the amount of villous connective tissue decreases relative to the amount of trophoblast, and (4) the number of villous capillaries increase in number and move closer to the surface of the villious. These changes increase the absorptive surfaçe and reduce the amount of tissue between the maternal and the fetal systems and thus should increase the efficiency of transport, especially for those substances that pass through the placenta by simple diffusion.

During their early development, the villi cover the entire circumference of the chorion (Fig. 12–12). During the second month, however, they become restricted to a disk-shaped region centered about the embryonal pole. This area is termed the *chorion frondosum* (leafy chorion). The remainder of the villi atrophy and form the *chorion laeve* (smooth chorion) (Fig. 12–13).

At the margin of the disc-shaped chorion frondosum, the chorionic plate and the basal plate fuse, and this part of the placenta is made up of several layers of cytotrophoblast cells enclosing occasional degenerating villi. The chorion laeve is lined internally with avascular chorionic mesoderm associated with the extraembryonic coelom. By the end of the third month the expansion of the amnion obliterates the extraembryonic coelom and brings together the avascular mesoderm of the amnion with that of the chorion. What is termed an *amniochorion* is thus formed (Fig. 12–13 B).

Changes in the fetal chorion are also associated with changes in the surrounding maternal tissues. On the basis of its position in relation to the embryo, the maternal decidua is divided into three areas: a *decidua basalis* over the embryonal pole and the chorion frondosum, a *decidua capsularis* over the abembryonal pole between the blastocyst and its chorion laeve and the uterine cavity, and a *decidua parietalis,* all of the remainder of the uterine decidual tissue (Fig. 12–13). Before the differentiation of the chorion frondosum and the chorion laeve, the decidua basalis and the decidua capsularis are similar in structure. As the functional villi become restricted to the embryonal region of the chorion and the embryo and its membranes increase in size, the decidua capsularis becomes stretched and begins to degenerate. Continued expansion brings the decidua capsularis into contact

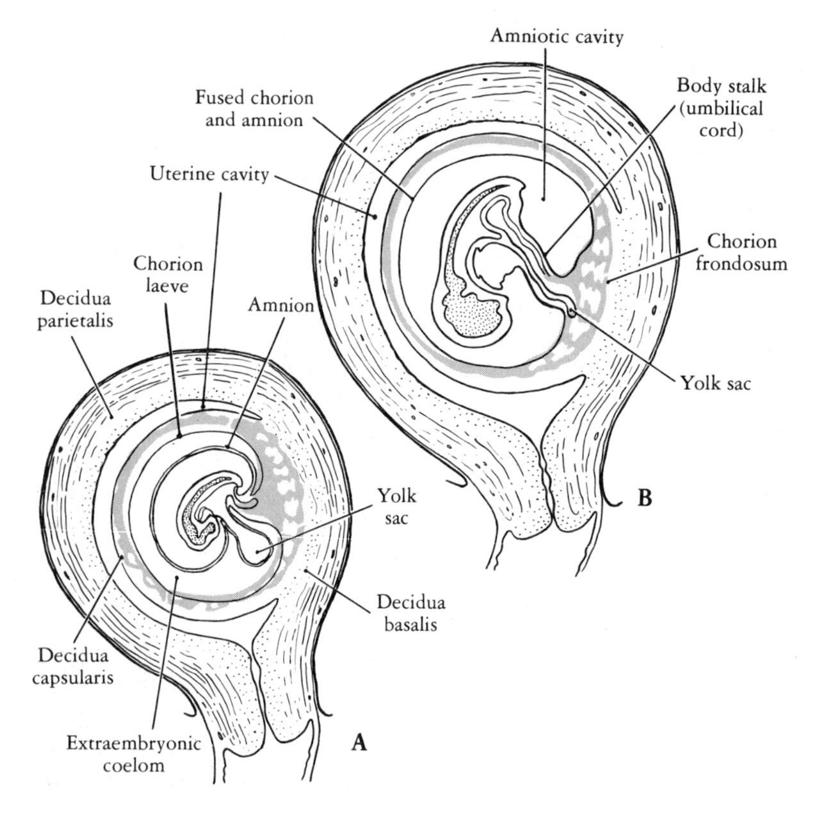


**FIG. 12–13.** A, the uterus, the fetus, and its membranes at about the middle of the second month. Note remnant of the yolk sac; B, about the middle of the third month.

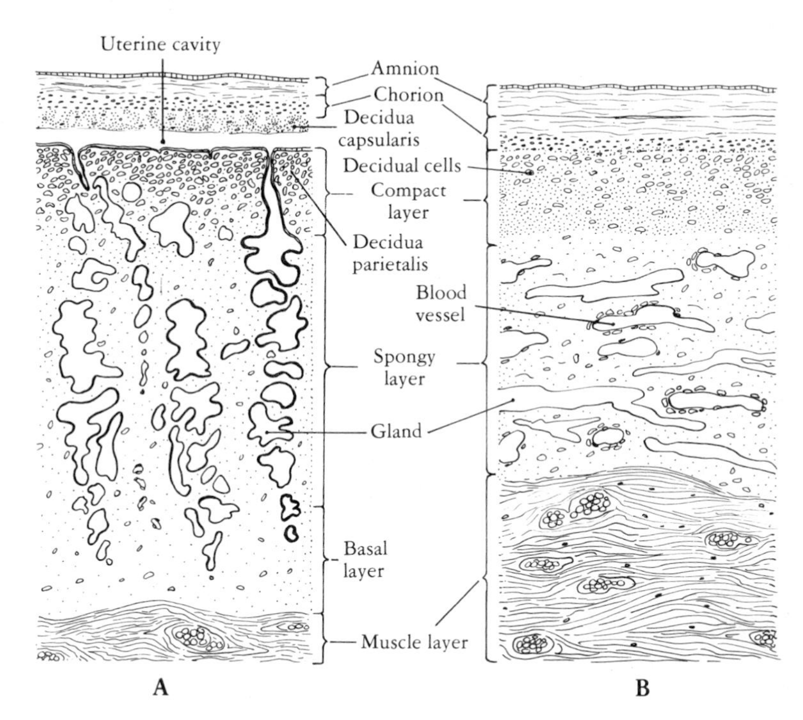


FIG. 12–14. Relation of the decidua capsularis to the decidua parietalis. A, at three months, before fusion; B, at six months, after fusion and obliteration of the uterine cavity. (From L. B. Arey, 1974. Developmental Anatomy. W. B. Saunders Company, Philadelphia.)

with the decidua parietalis of the opposite wall of the uterus. By the end of the first trimester, the uterine cavity has been obliterated, the epithelial layers of the opposing decidua capsularis and decidua parietalis have fused and degenerated, and the remaining endometriums of these two parts of the uterus are no longer distinguishable (Fig. 12–14).

## Placental Circulation

We have described the intervillous space as the region of the disc-shaped placenta between the chorionic and basal plates occupied by the syncytium-covered chorionic villi bathed by maternal blood. In fact, this space is partially subdivided into compartments formed by the growth of a number of septa, the *decidual septa*, projecting from the basal plate into the intervillous space but not reaching the chorionic plate. These septa with a core of maternal tissue are covered by the syncytiotrophoblast. If the placenta is viewed from the maternal side, some 15 to 20 slightly bulging domes of *cotyledons* mark this partial compartmentalization of the intervillous space. Each compartment or cotyledon contains its own stem and anchoring villi and villous network. Even though there is continuous communication between compartments in the region of the chronic plate, the circulation of the maternal blood is concerned mainly with what takes place in each of the cotyledenary areas.

### *Maternal Circulation*

Maternal blood reaches the placenta through the coiled or spiral arteries of the uterine endometrium. From the second month on, these arteries open directly into the intervillous space often at the apices of projections. In the intervillous space, the maternal blood is of course outside of the maternal capillaries and in contact with the syncytiotrophoblast. Within this space there are no anatomical channels to conduct the blood around the villi and back to the maternal veins. The blood entering the intervillous space from the open ends of the spiral arteries is propelled by the maternal blood pressure toward the chorionic plate, since the maternal pressure is much higher than that in the intervillous space. In this space, the blood runs into the irregular projections of the septa and the network of villi that slow it down, reduce its pressure, and spread it throughout the intervillous space. The blood is pushed along by the inflow of additional blood and eventually moves out of the space at the basal plate into the open ends of the endometrial veins. The amount of blood flowing through the placenta increases about 10-fold during the last two trimesters of pregnancy owing to an increase in the diameter of the spiral arteries and dilation of their openings into the intervillous space.

*Fetal Circulation*

Fetal blood reaches the chorionic plate by way of the umbilical arteries passing from the fetus through the body stalk and from here into the capillaries of the villi. Its return to the fetus is through the umbilical veins of the body stalk. It was at one time proposed that the fetal blood within the villous capillaries flowed parallel but in the opposite direction from the maternal blood. Although this relationship of countercurrent flow would provide the greatest potential for exchange, the arrangement of the fetal villi precludes the existence of such a system in the human. Fetal blood flows from artery to vein through the capillaries of a single villus. Single villi are tiny compared to the intervillous space and are oriented in all varieties of directions. Maternal blood in its movement through the intervillous space passes many villi in which the capillary blood flow may be oriented differently in each villus. They type of arrangement of blood flow has been termed *multivillous flow* (Fig. 12–15).

## Comparative Placentation

*Membranes Involved*

As noted previously, the human placenta in which the allantois, specifically the allantoic mesoderm, contributes to the vascularization of the chorion is an example of a chorioallantoic placenta. This type of placenta predominates in all mammals above the marsupials. The most primitive mammals, the monotremes, which include the fascinating Australian duck-billed platypus, are egg layers and of course do not develop a placenta. Most marsupials evidence a specialized type of prenatal development in that their young are born at very immature stages after going through a short gestation period. Because during development the yolk sac is the first embryonic membrane to form, in these species with such short gestation periods it is the only membrane that has time to develop. As it appears, it assumes a relationship with the chorion to form what is called a *yolk-sac placenta,* or what may be better termed a *choriovitelline placenta.* In most other mammals—man is an exception—a portion of the yolk sac fuses with the chorion, and its splanchnic mesoderm vascularizes a temporary choriovitelline placenta. However, this disappears when the exocoelom extends into the area and imposes itself between the yolk sac and the chorion. In some species (e.g., rabbits, mice, bats) a part of the choriovitelline placenta may persist and supplement the chorioallantioc placenta.

FIG. 12–15. Schematic diagram of the relation of fetal (F) to maternal (M) blood flow in the human placenta.

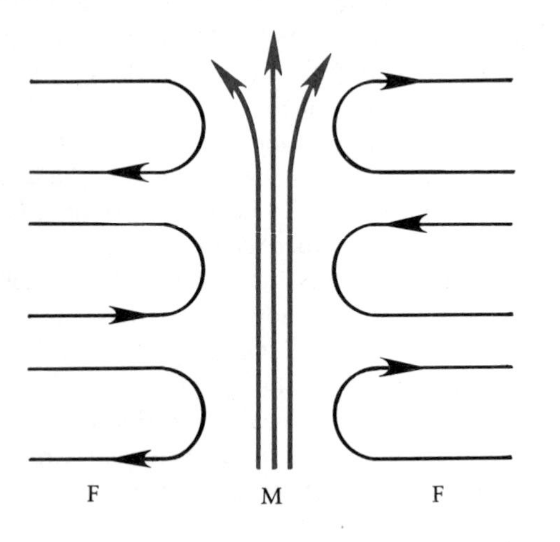


*Placental Shape*

The region of the chorion that becomes vascularized and persists in the mature placenta shows considerable variation and provides a method of classification based on the distribution pattern of the functional villi. Thus, man and most other primates with a cup-shaped region of functional villi have what is called a *discoid placenta.* Bats and rodents also have discoid placentas. *Diffuse placentas,* in which the villi cover the entire surface of the chorion, are found in the lemur, the sow, and the horse (Fig. 12–16 A). In most artiodactyls (e.g., cattle, sheep, deer) villi appear as prominent clusters of cotyledons covering the entire surface of the placenta but separated from each other by areas of smooth chorion forming a *cotyledenary placenta* (Fig. 12–16 B). A *zonary placenta,* in which the villi form a girdle around the middle of the chorion, is characteristic of the carnivores (Fig. 12–16 C).

*Fine Structure of the Placenta*

This basis of classification, with five types, is based essentially on enumerating the number of layers separating the maternal and the fetal blood. The names reflect first the maternal and then the fetal components of the placenta (Fig. 12–17).

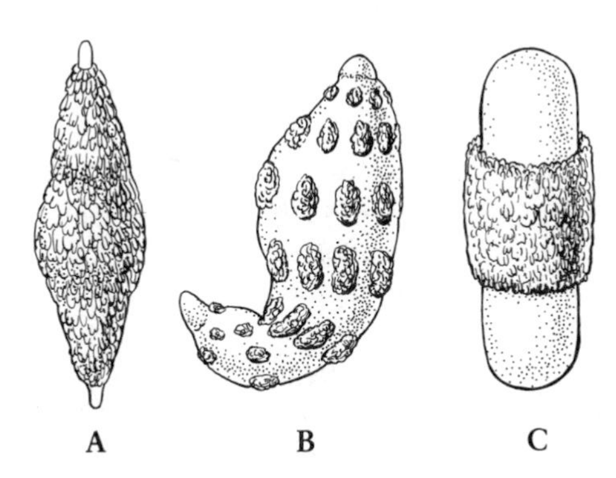


**FIG. 12-16.** Types of placentas based on the gross distribution of villi. A, diffuse (sow); B, cotyledonary (cow); C, zonary (dog).

1. In the *epithelialchorial* type, found in the sow and the horse, the fetal villi of the chorioallantoic placenta rest against the intact uterine epithelium. Extensions of the villi may fit into depressions in the epithelium of the uterus.

2. Cattle and most ruminants have a *syndesmochorial*-type placenta, where uterine erosion occurs but is kept to a minimum.The chorionic villi penetrate through the uterine epithelium into the connective tissue.

3. A further step in the elimination of the amount of tissue between fetus and mother is seen in the *endothelialchorial* placenta, where the invading trophoblast lies up against the endothelium of the maternal capillaries. Carnivores have this type of placenta.

4. In man and most primates, we have seen that the endothelium of the maternal vessels is broken down and the fetal villi are found in the intervillous space bathed in maternal blood released from the open ends of the maternal vessels. The chorionic villi still retain all of their tissue—trophoblast plus connective tissue and capillary endothelium—and hence this type of placenta is termed *hemochorial.* There are two types of hemochorial placentas. One is that found in man and most higher primates with a single communicating intervillous space containing the villous network—a *villous* type. In the second type, *labyrinthine,* the trophoblastic columns are fused and the maternal blood flows in delimited channels. This type of characteristic of some insectivores, rodents, and bats.

5. The closest approach to an intermingling of fetal and maternal blood is found in the *hemoendothelial* type of placenta. Here, in some areas of the mature placentas of some rodents, the trophoblast and the connective tissue overlying the fetal capillaries disappear and only the fetal capillary endothelium separates the two systems.

While the above five classifications appear rather straightforward and apparently present an easily understood picture of progressive thinning of the layers between maternal and fetal blood systems, a number of reservations must be mentioned.

1. The placenta changes with age. As we have seen from the description of the development of the human placenta, the amount of tissue between the fetal and maternal systems diminishes progressively.

2. The placenta is not uniform throughout, and considerable variation in the kind and amount of

**FIG. 12-17.** Schematic drawing of placental classification on the basis of the tissue between the maternal and fetal blood. A, epithelialchorial; B, syndesmochorial; C, endothelialchorial; D, hemochorial; E, hemoendothelial.

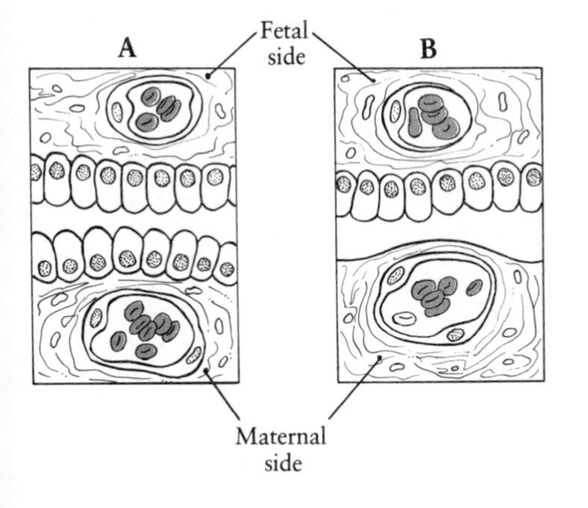


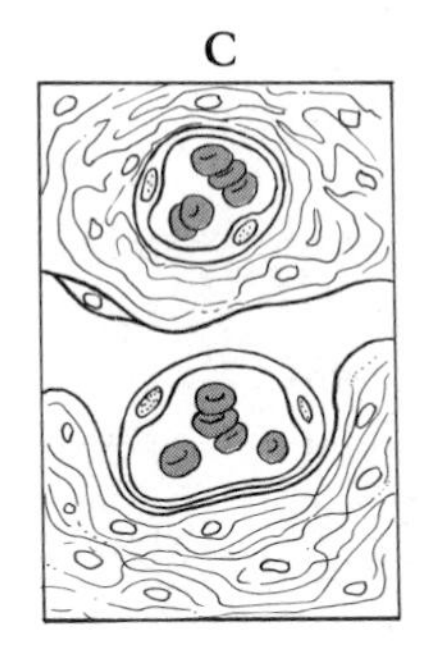


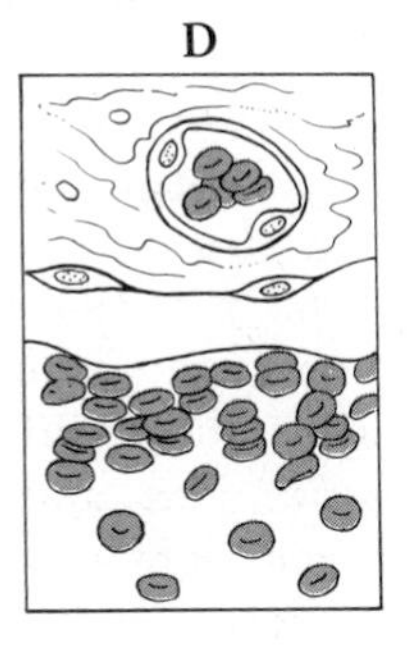


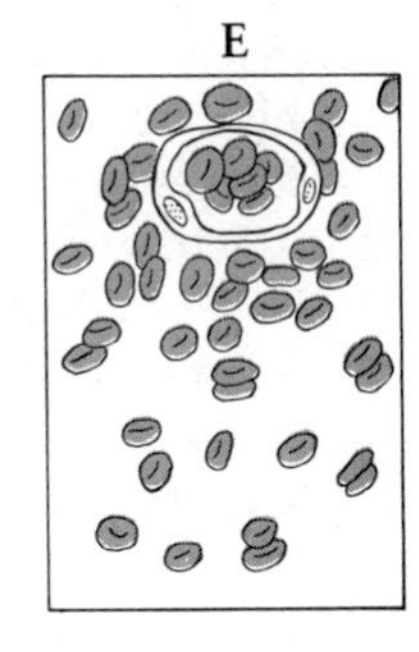

intervening tissue exists in different regions in any single placenta.

3. The classification places emphasis on the placental barrier and in so doing may suggest that the less the barrier the greater the efficiency of placental transport. While this may be true for some compounds that pass through by simple diffusion, it neglects the physiological function of the placenta during transport.

In addition, more recent electron microscope studies reveal that the classificaiton may not be quite so clear-cut or accurate as heretofore suggested. The syndesmochorial placenta of the sheep and the goat is more like an epithelialchorial type, since remnants of the epithelium persist in some parts of the placenta. The term *vasochorial* has been suggested for some of the endothelialchorial placentas, since some connective tissue is probably necessary to support the maternal capillaries. This connective tissue may be fetal trophoblast connective tissue rather than maternal. In the endothelialchorial placenta of the cat and dog, however, maternal connective tissue persists, and these placentas should then be classified as syndesmochorial. Again, the hemoendothelial placenta is really hemochorial, since at least one layer of trophoblast persists throughout gestation.

## Placental Transfer

Exchange between fetus and mother through the placenta is of course essential for normal intrauterine development In this capacity the all-importnat placenta serves as a substitute for the fetal lung, kidney, and gastrointestinal tract. We have already seen that it also serves as an endocrine organ.

We often speak of the "placental barrier." This places an emphasis on the placenta as an organ that restricts exchange between fetus and mother. While the placenta does act as a barrier in many instances, this terminology fails to focus attention on the real function of the placenta, which is to provide a mechanism for fetal–maternal exchange.

Substances pass through the placental membrane by four mechanisms: simple diffusion, facilitated diffusion, active transport, and pinocytosis. These mechanisms are not unique to the placenta but are mechanisms that occur and have been studied in other epithelial membranes such as those of the intestine and the kidney.

*Simple Diffusion*

Diffusion—movement of molecules from a region of higher to a region of lower concentration—is responsible for the placental transfer of many important compounds including the respiratory gases, oxygen and carbon dioxide.

In the transfer of respiratory gases, the placenta approaches the efficiency of the lung. Fetal erythrocytes have a high hemoglobin content of a type known as fetal hemoglobin that has a higher affinity for oxygen than adult hemoglobin. At any given local oxygen tension, fetal blood will contain more oxygen than maternal. In addition, the passage of metabolites from fetus to mother lowers the pH of the maternal blood and facilitates the transfer of oxygen from mother to fetus. Generally, the amount of oxygen reaching the fetus is dependent largely on the rate of blood flow, and fetal hypoxia is usually the result of factors that diminish either the fetal or the maternal blood flow. Although the thickness of the interposed tissues plays a role in transfer by diffusion, other factors are also important. These include molecular size, concentration gradient, rate of uterine and fetal blood flow, and the amount of the substance metabolized by the placenta during transfer.

*Facilitated Diffusion*

Some substances cross the placenta in response to concentration gradients but do so at a rate more rapid than that which would result from the laws of simple diffusion. Glucose of maternal origin, the primary source of energy for fetal metabolism, is one such substance. In facilitated diffusion, compounds combine with a carrier molecule in a process that requires no endogenous energy source supply from the placenta.

The protein carrier combines with the glucose molecule, forming a lipid-soluble complex that crosses the placental membrane with the concentration gradient faster than that which the substrate alone would accomplish in what may be called a "rapid downhill transfer" system. Fructose, which has the same molecular weight as glucose, crosses the placenta much more slowly, since it does not enjoy the services of a carrier.

*Active Transport*

This type of transfer requires the presence in the placenta of an active energy-requiring transport system by means of which a compound may be transported across the membrane against a gradient. In terms of a carrier system, this means an energy-consuming chemical alteration of the carrier substance in order to effect a transport in an "uphill" direction. Some amino acids and divalent ions are transported by this method.

*Pinocytosis*

This involves the surrounding and engulfing of the compound to be transported by the plasma membrane of the cell and its transfer intact by what appears as a vacuole in the cytoplasm. Large molecules such as $\gamma$-globulins and lipoproteins may use this system.

*Transfer of Some Specific Types of Compounds*

*Electrolytes:* Electrolytes cross the placenta in significant quantities. Most univalent ions, such as $Na^+$, $K^+$, and $Cl^-$, appear to cross by simple diffusion. On the other hand, the divalent ions, such as $Fe^{++}$, $Ca^{++}$, $P^{++}$, and $Zn^{++}$, are present in higher concentrations in the fetal than in the maternal system, a distribution reflecting an active-transfer system.

*Proteins:* The fetus synthesizes most of its proteins from amino acids transferred across the placenta. There is a net transfer from mother to fetus and a higher concentration of amino acids in the fetal than in the maternal blood, suggesting an active-transfer system. Additional evidence supporting transplacental movement by means of an active transport system is: (1) the placenta shows a stereospecificity (the L isomers usually being transported more rapidly than the D isomers); (2) the different amino acids compete for transfer; and (3) the rate of transfer diminishes in the presence of energy-uncoupling inhibitors. Nitrogenous end products of amino acid and protein metabolisms, such as urea and creatinine, however, occur in equal concentrations in fetal and maternal blood, indicating their transfer by simple diffusion.

Many high-molecular-weight polypeptides cross the placenta slowly or not at all.This is true of the pancreatic hormone, insulin, and the pituitary trophic hormones, ACTH and TSH. Maternal serum proteins apparently cross the placenta intact. $\gamma$-Globulins cross in large quantities. However, only small amounts of $\alpha$ and $\beta$ globulins cross. The serum proteins are probably transferred by pinocytosis.

*Carbohydrates:* Pentoses and hexoses cross at the same rate and proportional to the concentration gradient—indicating simple diffusion. High-molecular-weight carbohydrates, such as inulin and some dextrans, do not cross, but those with molecular weights less than about 650 cross readily. Fructose and other sugars with similar molecular weights cross more slowly than glucose.

*Vitamins:* The water-soluble vitamins, riboflavin, ascorbic acid, thiamin, and $B_{12}$, are present in the fetal blood in concentrations higher than in the maternal, while the reverse is true for the fat-soluble vitamins. Because of the lipid nature of membranes, this is unexpected One reason for the higher concentration of water-soluble vitamins in the fetal blood may be their conversion by the fetus to an impermeable product—possibly by binding to a protein.

*Lipids:* Lipid precursors such as acetate and free fatty acids cross readily and participate in lipid synthesis in the fetus. Intact lipids cross slowly, if at all, and are usually broken down in the placenta and resynthesized in the fetus. Cholesterol passes only slowly and phospholipids not at all. The immobility of the latter two lipids is probably due to their metabolism in the placenta, phospholipids being hydrolyzed to phosphate and cholesterol being used in steroid synthesis.

*Drugs:* Birth defects occur with alarming frequency. Although some have been shown to have a genetic basis, the cause for many others remains unknown. The thalidomide incident of the early 1960s focused attention on the fact that drugs taken during pregnancy may have damaging effects on the fetus. The fact that this tranquilizer produced an easily recognizable bizarre effect, *phocomelia,* in almost 100 percent of the cases when taken between days 34 and 45 of pregnancy was striking evidence that drugs may be teratogenic. If thalidomide had produced a more common defect—such as cleft lip—in a smaller percentage of cases, its effect may well have gone unnoticed. We might then consider that teratogenic drugs or chemicals might well be the unknown cause of many birth defects simply because the defects are not specific nor readily directly related to particular drugs. If this is true of a large number of drugs, both prescribed and self-administered, despite our lack of knowledge of specific effects, it leads to the sensible proposal that any drug, unless indicated on serious medical grounds, should be avoided during pregnancy. This injunction can be made even though, for most agents, the defects are not specifically known; they may be varied and inconsistent; they may occur in only rather sensitive individuals in low incidence; and they do not universally follow drug administration. Despite the uncertainty about the effects of drugs, they are taken in large quantities, often by pregnant women. In one recent study in England involving 1369 pregnant women, it was reported that 97 percent were taking some form of a prescribed drug and 65 percent some form of a self-administered drug. These included analgesics antihistamines, vitamins, diuretics, tranquilizers, antibiotics, laxatives, cough medicines, and appetite suppressors.

It is beyond the province of this book to attempt to document the evidence implicating specific compounds in the production of fetal defects. Much of the evidence is as yet inconclusive and controversial. Our knowledge of drug effects on the human fetus is necessarily slim and often circumstantial. Only a few drugs have been linked with any degree of certainty to specific birth defects; nevertheless clinicians and researchers are rightly concerned about the effects of indiscriminate use of drugs during pregnancy.

*Alcohol:* There is one drug, alcohol, whose effects on the fetus have been rather extensively studied with results that merit some consideration. That alcohol has an adverse effect on the fetus has been proposed for some time, and in fact Aristotle wrote that drunken mothers produce "morose and languid" children. At the turn of the century, a British study reported that female alcoholics had a record of stillbirths and infant deaths 56 percent greater than their nonalcoholic relatives.

It has now been well documented that the offspring of chronic alcoholics show a remarkable similarity of facial charactertistics, growth deficiencies, and psychomotor disturbances that has been termed the *fetal alcohol syndrome* (FAS) (Jones and Smith, 1973). The facial features include short eye slits, low nasal bridge, short nose, indistinct philtrum (the ridges running between the nose and the upper lip), small chin, and flat midface. The growth deficiencies are in weight, height, and head circumference. They are of prenatal onset and are not corrected postnatally. The central nervous system defects include retarded mental and motor development, hyperactivity, small brain size, and brain malformations. A few children with FAS may have normal intelligence, but most have significant mental handicaps; the average IQ of such children reported in studies from Germany, France, Sweden, and the United States was 68 (Streissguth et al., 1980). The etiology of the syndrome has been demonstrated in studies on mice, rats, guinea pigs, dogs, and monkeys, where alcohol given to pregnant females produced in their offspring all of the developmental defects characteristic of FAS in humans.

FAS occurs in about 1 of every 750 births and ranks as one of the most common forms of mental retardation with a known etiology. It should be noted, however, that FAS is associated with chronic maternal alcoholism and with heavy drinking during pregnancy, and almost certainly the effect of alcohol on the developing fetus is directly related to the amount consumed. The effects produced range from small birthweight deficits and minor emotional disturbances at one end, to the full-blown FAS at the other. It is not known whether a threshold exists below which no defects are produced, and the "safe level" of alcohol consumption has not been established. Birthweight decrements in humans have been reported at levels corresponding on the average of two drinks per day.

*Disease Organisms:* Fortunately, the placenta presents a true barrier to most disease organisms. Although viruses may cross the placenta, only two viral infectious agents, rubella and cytomegalovirus, have been shown to act on the fetus in the production of congenital defects.

Of these two, rubella has been thoroughly studied since it was first recognized in the early 1940s. At first it was considered effective only during the first trimester, when it was shown to produce cataracts, heart lesions, hearing defects, microcephaly, and mental retardation in order of decreasing frequency. Although the defects are most frequent when infections occur in the first month, the incidence may still reach 6 percent for infections during the third, fourth, and fifth months.

Cytomegaloviruses, so named because of the characteristic enlargement of the cells containing the virus, are also able to cross the placenta. Congenital defects produced include microcephaly, microphthalmia, blindness, hepatomegaly, and sometimes encephalitis.

Other viruses may cross the palcenta, but they do not lead to developmental anomalies. These include the agents of polio, smallpox, vaccinia, and mumps, which may produce chronic infections in the fetus.

A number of nonviral organisms are also known to result in fetal infection. This is evidenced as a pathological expression of the disease in the fetus and the newborn rather than as any developmental defect. One such infection is syphilis, which is incurred by the fetus following transmission from the mother through the placenta. The fetus has little resistance to the spirochetes that live and multiply rapidly and are found in practically all of the fetal tissues in numbers far greater than in the adult. Approximately 25 percent of the infected fetuses die before term, and a high percentage are born prematurely.

*Erythroblastosis fetalis:* Even though the tissues of the mother and the fetus, which are immunologically incompatible, develop side by side without tissue destruction—possibly due to some trophoblastic barrier intervening between the two—there is one situation in which the barrier is shown to be defective. This involves cases in which the blood cell antigens of the fetus and the mother are different. Most studied and publicized is Rh incompatibility, where an Rh-positive fetus is carried by an Rh-negative mother. Rh-negative individuals are double recessives. Thus, the fetus, if both parents are Rh negative, presents no problem. However, the fetus of an Rh-negative mother and an Rh-positive father could be either Rh positive or negative depending on well-known genetic laws and probabilities. If the fetus is Rh positive, and red blood cells get through the placenta into the maternal circulation, they can trigger an immune response from the Rh-negative maternal system, which will produce antibodies against the foreign antigen. These antibodies, $\gamma$-globulins, will pass through the placenta and into the fetal circulation; the resulting antigen–antibody reaction will destroy fetal red cells, lead to anemia and jaundice, and cause liver and spleen enlargement. Because of the extensive red cell destruction, this disease is also called *hemolytic disease of the fetus*. Since there is no connection between maternal and fetal circulatory systems, fetal red cells must reach the maternal circulation through some defect in the placenta. Placental transfer of red cells is now well documented; in a recent study, fetal red cells were found to be present in 71 percent of all pregnant women. However, the numbers present were small and, since the probability of sensitization varies with the amount of fetal blood in the maternal circulation, in most cases the amount of fetal blood that crosses the placental barrier is usually less than that necessary to initiate a primary maternal response. Whereas 15 milliliters of fetal blood result in about a 65 percent chance of antibody production, less than 0.2 milliliters represents only a 3 percent chance. Thus, the probability of sensitization of an Rh-negative mother by an Rh-positive fetus is low during a first pregnancy but is much increased during a second pregnancy if Rh incompatibility is present.

There are factors other than the number of fetal red cells, that affect the maternal antibody response. One is the suppression of the maternal response by the elevated levels of steroids present during pregnancy. Also, if the fetal cells are incompatible with the mother's ABO blood group antigens, their survival time in the maternal system is limited and may not be long enough to initiate an immune response. New techniques for prediction, prevention, diagnosis, and treatment have significantly decreased the serious consequences of Rh incompatibility. The use of $Rh_o(D)$ immunoglobulin as an immunodepressant has proved to be very successful in the prevention of sensitization, and it is administered rountinely where an Rh incompatibility exists. In a standard dose (300 $\mu$g), $Rh_o(D)$ Ig will neutralize 15 milliliters of fetal blood. Even before the introduction of these measures, only 1 in 20 to 26 Rh-incompatible matings resulted in involvement of the fetus.

*Stress:* Since there is never any interconnection of the nervous systems of the mother and the fetus—thus rendering impossible any transmission of nervous impulses from mother to fetus—it is correctly stated that maternal thoughts and impressions can have no effect on any aspect of fetal development, inlcuding that of the nervous system. Thoughts, impressions, and moods—be they good or bad—are not soluble in the maternal blood plasma, nor are there any biological or physical mechanisms by means of which they can be transmitted across the placental membranes. The "old wives' tales" of children born with moles because their mothers dreamed of toads, or with strawberry rashes because their mothers loved strawberries are just that—old wives' tales.

However, we must not fail to take into account the fact that some stressful events to which a pregnant female may be exposed may indeed have effects on the development of the fetus. Animal experiments have demonstrated that subjecting pregnant females to repeated rather severe stress such as overcrowding, conditioned anxiety, or a combination of restraint, bright light, and increased temperature will produce both functional and morphological abnormalities in the offspring. Pregnant rats that had learned to anticipate and fear electric shocks produced offspring that themselves showed symptoms of anxiety and were less able to take care of themselves and more closely bound to their mothers. Overcrowding pregnant females produced offspring that showed less spontaneous activity than normal and that were slower to respond to strange stimuli. Although behavioral defects are more commonly produced, both cleft palate and harelip may occur as the result of prenatal stress. The effects on the offspring must be due to some substances formed in the mother as a result of the stress, which then cross the placenta and exert their effects on the fetus. The logical ones to suspect are the maternally produced stress hormones such as adrenocorticotrophic hormone of the pituitary, and epinephrine and corticosteroids of the adrenal gland.

There is well-documented evidence that postpuberal sexual behavior of both males and females may be adversely modified as the result of the effects of prenatal stress on the neuroanatomical and biochemical organization of the parts of the brain that control these functions. In male rats, the masculinization of the central nervous system components mediating normal male behavior, which is a necessary event for male sexual functioning, is inhibited by prenatal stress. This is apparently due to the absence of the required surge in plasma testosterone on days 18 and 19 of pregnancy (Ward and Weisz, 1980). Female offspring of stressed mothers experience disruptions in estrous cycling and fewer conceptions, more spontaneous abortions, and longer pregnancies than offspring of nonstressed females. These results are attributed to the influence of the stressful event on the exchange of adrenal and gonadal hormones between mother and fetus, and also on the balance of these hormones within the fetus at a critical time of hypothalamic differentiation.

Thus, while in the normal course of events the thoughts, ideas, and impressions of the pregnant female have no influence on the development of the fetus, some severe stressful conditions may be forceful enough to result in the prolonged production of stress hormones in the mother. These hormones then pass freely across the placenta and in the fetus may exert adverse effects, especially if they arrive at certain critical periods of development.

*Passive Immunity*

The adult acquires immunity over a period of time, responding to the challenges of invading organisms by establishing specific defense mechanisms mediated in large part through serum antibodies. The fetus is not only to a large extent effectively—but not completely—isolated from these challenges but also is not capable of responding to them by synthesizing any significant amounts of antibodies. Thus, at birth, the fetus is projected into a potentially hostile environment filled with microbial challengers, at a time when it has not had the opportunity to develop its defense against these challengers. Fortunately, the fetus is provided with a defense mechanism that it in a sense has borrowed from its mother by means of the passage of maternal antibodies across the placental membrane. In this instance, the introduction of maternal antibodies into the fetal circulation is not undesirable, as in the case of Rh antibodies, but is instead of definite advantage to the fetus. Almost any infection that the mother has contracted—measles, mumps, whooping cough, influenza—will cause her to develop antibodies against these disease organisms that , after they cross the placenta, provide the newborn with a built-in defense mechanism. This passive immunity lasts for several months of postnatal life, during which time the infant progressively acquires its own immunity in response to the challenges offered to it.

## Fetal–Maternal Immunological Reactions

Grafts or transplants that are genetically alien to a host are quickly rejected. Yet the fetus is genetically distinct from the mother and, even though the fetal placental unit is in intimate contact with the immunocompetent maternal tissues, it is not rejected. It has been suggested that the uterus is an immunologically privileged site. However, the occurrence of ectopic pregnancies in such places as the uterine tube or the ovary, where the fetus is not rejected either, does not support this. In fact, the decidual reaction of the uterus appears to resist the invasive tendencies of the fetus, and this invasiveness is much more apparent in ectopic sites. A balance must be achieved between the aggressive invasive activities of the fetal trophoblast and the tendencies of the maternal organism to rid itself of a foreign graft. One mechanism that may act to prevent the rejection of the histoincompatible fetus resides at the site of contact between the fetal trophoblast and the maternal decidua. Decidua formation depends on the reaction of a uterus primed with estrogen during the follicular phase of the menstrual cycle and then stimulated by progesterone during the luteal phase. An estrogen-primed uterus alone will not react to an implanting trophoblast to form a decidua, and under these conditions the trophoblast's invasive capabilities are increased, and it may even perforate the uterus.

In fact, it has been suggested that there is actually no significant contact between maternal and fetal tissues. This may be due to the fact that the trophoblast surrounds itself with an extracellular acid mucopolysaccharide deposit of "fibrinoid," which effectively isolates it from the maternal tissues. This pericellular layer is a mucoprotein rich in tryptophan, sialic acid, and hyaluronic acid. The foreign antigens of the fetal tissues are contained by this layer, and, in addition, there is a high negative charge at the surface of the mucoprotein layer that causes an electrostatic repulsion of the maternal lymphocytes.

A number of different hormones, including progesterone, estrogen, HCG, and HCS, have been considered to be immunodepressants, exerting a suppressive effect on the maternal cytotoxic cells. Progesterone is the dominant hormone of the maternal blood serum during pregnancy, and it is present in much higher concentrations (12–50 times) in the placenta. A variety of studies have been reported on the effects of progesterone on the immune system, and a number present evidence for the concept of high local placental concentrations as a key factor in immunodepression at the fetal–maternal interface (Siiteri and Stites, 1982). Suppression of induced granulomas and the prolongation of xenoplastic grafts are found in the immediate vicinity of progesterone implants. Progesterone affects immune cell functions, inhibiting human T-lymphocyte activation by plant lectins and mitogens. In vitro studies with mouse cells have shown that the action of progesterone is to block the recognitive and proliferative phases of the T-lymphocytes. The generation of cytotoxic cells directed at foreign grafts is blocked by adding progesterone at this stage of the immune response; however, if the cells are allowed to reproduce before progesterone is added, no amount of progesterone given at this stage will prevent target destruction.

Estrogens have also been shown to have effects on the immune system. However, the action of estrogens in the fetal–maternal immune response, although the concentration of estrogens is higher in the placenta than in the maternal circulation, is questionable. In vitro studies have demonstrated inhibition of lymphocyte activation only at high nonphysiological doses.

The early appearance of HCG and its subsequent rapid increase in the placenta early in pregnancy have made it a possible candidate as an immunosuppressive substance. Although some early studies supported this assumption, later investigations with purified HGC did not produce the same immunosuppressive effects. Contaminants in the unpurified HGC are implicated, but the nature of the contaminants has not yet been identified.

HCS has also been tested, but it does not have any important immunosuppressive activity. It thus appears that progesterone in concentrations produced locally at the maternal–fetal interface is the key hormone in suppressing a maternal immune response and thus plays an important role in allowing the fetus, nature's allograft, to survive.

## Analysis of Early Mammalian Development

With relatively few exceptions, it was not until the 1960s that experiments on the early stages of the mammalian embryo were reported in the literature. The main reason for this was that early attempts at culturing and manipulating mammalian embryos outside of the maternal environment met with limited success. An additional factor involved the low rate of egg production and the difficulty of obtaining embryos. This left unanswered or speculative the question of regulative versus determinative development as the causal basis of the establishment of developmental patterns in mammalian systems. The mammalian embryo is of particular interest in this respect because, although the egg undergoes holoblastic cleavage, the embryo itself develops from only the small number of blastomeres that give rise to the inner cell mass, while the remainder of the blastomeres form the trophoblast. The development of these two cell types presents an early gross morphological manifestation of differentiation that can be used to study the processes involved in the commitment of cells or groups of cells to divergent lines of differentiation. At what time and under what conditions does a commitment to one of the populations occur? Once the two populations are morphologically recognizable, are they irrevocably determined so that cells of one line can no longer differentiate into tissues of the other? Once the two populations have become determined, are all of their parts determined, or can there still be regulation within a population?

Embryologists in the late 19th century and the first half of the 20th century generally supported the view that development in the mammal was determinative and that the fate of the individual blastomeres was decided in early cleavage. As early as 1875, Van Beneden (cited by Assheton, 1898) contended that in the 2-cell stage of the rabbit one cell was larger with hyaline cytoplasm and was destined to form the epiblast (ICM). Later investigators also reported size differences, which they associated with early determination of prospective ICM and trophoblast, in such a large number of cases that inequality of the blastomeres was considered typical of cleavage in the majority of mammalian embryos. In addition, other investigators published results from histological and histochemical studies that seemed to indicate a progressive segregation of cytoplasmic materials into specific blastomeres conveying polarity to the early cleavage stages and determining the potential fate of the blastomeres at least by the 8-cell stage.

However, beginning in the 1960s the refinement of in vitro culture techinques and the development of better manipulative procedures produced a greater amount of experimentation under controlled conditions and modified and contradicted earlier conclusions. The more recent experiments have contributed the general concepts that the mammalian egg is indeed regulative, that the early blastomeres are totipotent probably to the 8-cell stage, and that the basis of determination of which blastomeres will form the inner cell mass and which will form trophoblast is the position they assume in the morula stage rather than differential distribution of cytoplasmic determinants.

## Development of Individual Blastomeres

One of the obvious tests of the potency of any blastomere is to determine what it is capable of developing into when isolated from the other blastomeres and cultured in vitro. The logical sequence to this test, if the isolated blastomere develops into an apparently normal blastocyst, is to implant it into a foster mother, bring it to term, and even raise it to adulthood. A number of studies have shown that single blastomeres of the 2- or 4-cell mouse or rabbit embryo are capable of regulation in culture to form normal blastocysts composed of inner cell mass and trophoblast. Blastocysts from one-half or one-quarter embryos have been transferred to foster mothers and brought to term, and the offspring were normal and fertile. In this type of experiment, all but one of the blastomeres are destroyed and removed from inside of the zona pellucida. The surviving blastomere is then cultured and transferred to the foster mother still within the zona pellucida, since implantation will not take place in zona-free embryos. Since the surviving blastomere is chosen at random, this experiment still does not prove that each and every one of the eight blastomeres is capable of this performance. However, it has been shown that ¼ (one blastomere from a 4-cell embryo) and ⅛(one blastomere from an 8-cell embryo) blastomeres, when removed from their zonas and cultured in vitro, are all capable of differentiating trophectoderm cells, and many of them are capable of forming apparently normal blastocysts with an ICM covered by trophectoderm. The ¼ blastomeres developed about twice as many blastocysts as the ⅛. It is difficult to follow the development of all of the ¼ and ⅛ blastomeres from a single embryo. Of three experiments in which this was accomplished with ¼ embryos, the best result was the differentiation of three normal blastocysts and one trophoblastic vesicle. The one experiment in which all of the ⅛ blastomeres from an 8-cell embryo were followed produced only a signle normal blastocyst. The smaller number of blastocysts developed from ⅛ blastomeres might suggest that these cells have lost some of their developmental capacities by this stage. However, a greater susceptibility to damage and a greater difficulty in manipulating embryos with larger numbers of blastomeres could also account for the result. In these experiments there was no indication of isolated blastomeres from any one embryo developing complementary structures—as one would expect if segregaton of cytoplasmic factors into individual blastomeres was responsible for differentiation. In fact, it was demonstrated that every cell in the embryo up to the 8-cell stage has the potentiality of developing into a trophoblast. In addition, about one third of the single blastomeres developed from 4-cell embryos and 11 percent of the single blastomeres developed from the 8-cell embryo formed more or less normal blastocysts with an inner cell mass covered by a trophoblast.

Although the crucial experiment of following the development of all of the ⅛ blastomeres of the eight-cell embryo to term is lacking, the conclusion that the fate of blastomeres in the 8-cell mouse embryo is not as yet determined is probably warranted. In the mouse, blastulation begins at approximately the time the fertilized egg has undergone five cleavage divisions, and its onset is not affected by experimentally reducing or increasing the number of blastomeres dividing. This would particularly affect an isolated ⅛ blastomere, which would begin blastulation after only two more divisions, at which time it might then not have a sufficient number of cells to provide the physical environment for normal blastulation. This conclusion is supported by the fact that both ¼ and ⅛ blastomeres will contribute to both the ICM and the trophectoderm when aggregated with enough blastomeres from another embryo to restore the normal cell number at the time of blastulation.

In the rabbit, not only ½ and ¼ blastomeres but also ⅛ blastomeres have been cultured, implanted, and brought to term. The rabbit blastulates about two cell generations later than the mouse, and the greater potentiality of the ⅛ blastomeres in the rabbit may reflect that the increase in the number of cells in the ⅛ blastomeres at the time of blastulation provides a more appropriate cellular environment than that found in the mouse.

Similar results have recently been reported for the sheep (Willadsen, 1981), where ⅛ blastomeres are also capable of developing into viable offspring. Interestingly, one set of monozygotic triplets devel-

**FIG. 12–18.** Monozygotic lambs produced from four pairs of blastomeres of an eight-cell embryo. The dead lamb (lower right) was born alive but was trapped in the amnion. (From S. M. Willadsen. 1981. J. Embryol. exp. Morphol. 65:165–172.)

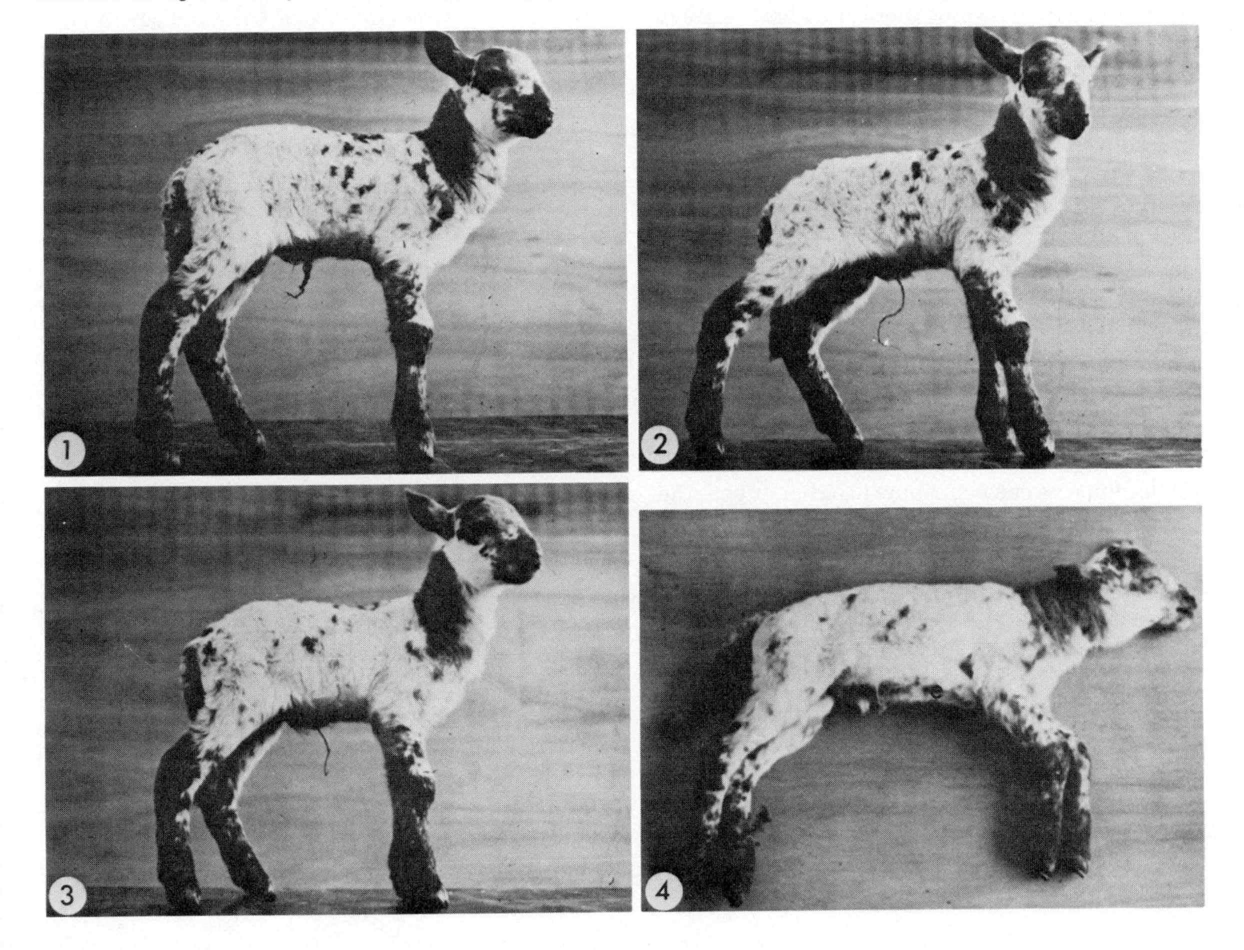

oped from three ¼ blastomeres of a 4-cell embryo, and one set of monozygotic quadruplets developed from four pairs (⅜) of blastomeres from an 8-cell embryo (Fig. 12–18), proving without any doubt that all four of the 4-cell sheep blastomeres are totipotent.

## Disaggregation and the Development of Chimeras

The literature of antiquity presents many examples of legendary chimeras such as the centaur, half man and half horse, and the griffon, a lion's body with the legs and head of an eagle. An actual chimera is a composite organism in which the different cell populations are derived from more than one fertilized egg. Chimeras may result from the combination or transplantation of adult tissues or organs (often called secondary chimeras) or from the combination of cell populations in early stages of development (primary chimeras). Mammalian embryonic chimeras were first studied in the mouse by disaggregating the cells of two different embryos and then pooling them and allowing them to reaggregate. Alternatively, cells may be disaggregated and then injected singly or in groups into the blastocoele of an intact blastocyst. Various markers—antigenic, biochemical, chromosomal, radioactive—may be used to distinguish between the two different sources of cells in assaying the eventual fates of the host and donor cells. Chimeras have been used extensively to test the potential for differentiation of embryonic cells and tissues. Chimeras have been established in which single blastomeres of the 4- and 8-cell mouse embryo are combined with cells of mice of different strains and examined at the blastocyst stage. The results, as we have already noted, are that each of these 4- and 8-cell blastomeres can contribute to either or both the ICM and the trophectoderm, supporting the conclusion that in the early cleavage stages there has been no segregation of morphogenetic factors and that in these early cleavage stages development is still of an epigenetic regulative nature.

Regulative activity may also be demonstrated by disaggregation studies in which zona-free eggs are disaggregated and pooled and then groups of cells from the pooled mass are removed and cultured. Complete disassociation of embryos from the eight-cell to the early blastocyst stage is followed by reaggregation and development to morphologically normal blastocysts in 90 percent of the cases of pooled cells of the same developmental age and 77 percent of the cases of pooled cells from eggs of different developmental ages (4-cell and 16-cell; 4-cell and 32-cell). Pooling of embryos results in a varied mixture of blastomeres, and it is unlikely that blasomeres in the reaggregate take up positions corresponding to any original polarity. Thus, any spatial or gradient-forming relationships between the original blastomeres are completely destroyed. It is apparent that any polarity present in the uncleaved egg and the early embryo is unnecessary for differentiation up to the blastocyst stage.

Regulative development can also be demostrated through the formation of chimeras by fusing the cells from two or more embryos. Again, extensive rearrangment of the blastomeres will result. The development of a high percentage of normal blastocysts (Fig. 12–19) under conditions in which migration and rearrangement or sorting out of fused cells to reestablish a particular pattern or location of specific blastomeres is impossible, also supports the conclusion that cytoplasmic segregation is not responsible for early differentiation. It is not even necessary to fuse embryos of the same age. A good percentage of 8-cell embryos fused with late morulas or early blastocysts can regulate for these chronological differences as well as develop into morphologically normal, but giant, blastocysts (Fig. 12–20).

It is also possible to produce interspecific chimeras. Aggregations of rat and mouse and mouse and vole blastomeres are viable and will implant in foster mothers, but they do not survive to term. A significant advance in the investigation of chimeras took place when the technique of using immunosurgery to lyse the trophectoderm of the blastocyst and free the ICM intact and unharmed was introduced. Using this method, injection chimeras may be formed by injecting the immunosurgically liberated ICM into an intact blastocyst. Injection rat–mouse chimeras have been formed that will implant but, again, only a few survive to term, and those that do are runted. The amount of rat tissue in the chimeras is small, and it appears that there is a selection against rat cells as development progresses. Considerably better results have been reported with injection chimeras of the laboratory mouse *(Mus musculus)* and a wild species of a South Asian mouse *(Mus caroli)* (Rossant and Frels, 1980). The chimeras show a high rate of normal development, and the newborn contained enzymes of both species in all tissues examined.

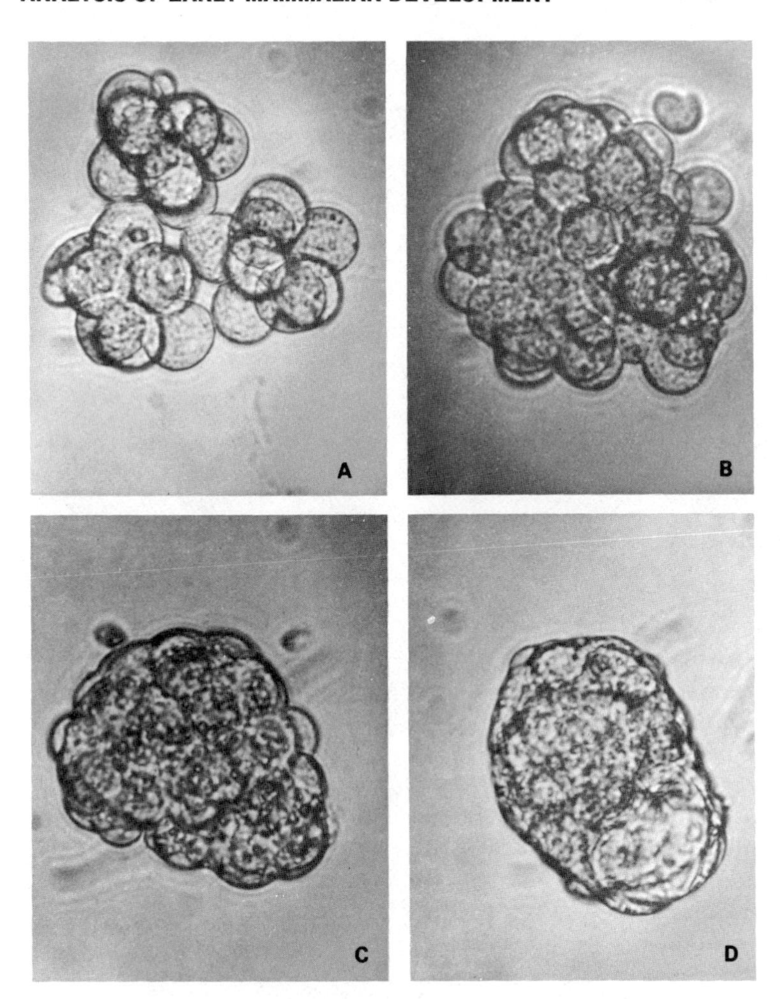


**FIG. 12–19.** In vitro aggregation of three eight-celled mouse embryos. A, at 1 hour after fusion, a few blastomeres from each embryo are in contact; B, at 8 hours, a single compact embryo is beginning to form; C, at 19 hours, a morula three times larger than normal; D, at 25 hours, a small blastocoele has formed. (From M. S. Stern and I. B. Wilson, 1972. J. Embryol. exp. Morphol. 28:247.)

Injection chimeras present the opportunity of implanting a chimera with the trophoblast of one species into the uterus of the other species. *M. caroli* ICMs injected into *M. muculus* blastocysts will develop normally in the *M. musculus* uterus but do not survive to term in the *M. caroli* uterus (Rossant, et al., 1982). The trophoblast cells of the type of the foster mother species present a barrier that protects the foreign ICM cells from destruction by the maternal immune system.

The possible uses of interspecific and intraspecific chimeras in the fields of experimental embryology and developmental genetics are numerous and expanding rapidly.

## Differentiation of the Trophoblast and the Inner Cell Mass

By three and a half days of development, the fully expanded blastocyst is made up of two distinct populations of cells, the inner cell mass and the trophectoderm (trophoblast), committed to different directions of development. On the basis of their experiments on culturing isolated blastomeres from early mouse embryos, Tarkowski and Wroblewska (1967) proposed that the factor that determines which of the two routes of development any blastomeres takes depends on its position in the developing embryo. Those cells that remain on the outside develop into trophectoderm. Those cells that occupy the interior form the innner cell mass by virtue of their isolation from influences of the external environment. The initially totipotent cells of the early cleavage stages, as they develop into the morula, recognize their position as either "outside" or "inside" and develop trophectoderm or inner cell mass accordingly. A variety of experiments have since been reported to support this hypothesis. One method of testing the inside–outside hypothesis is to construct chimeras in which an attempt is

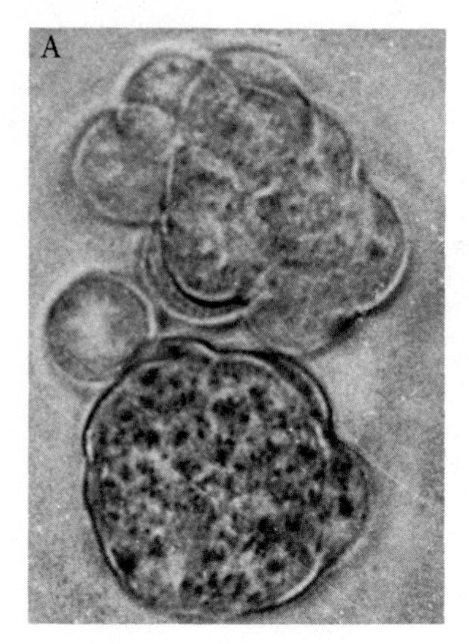


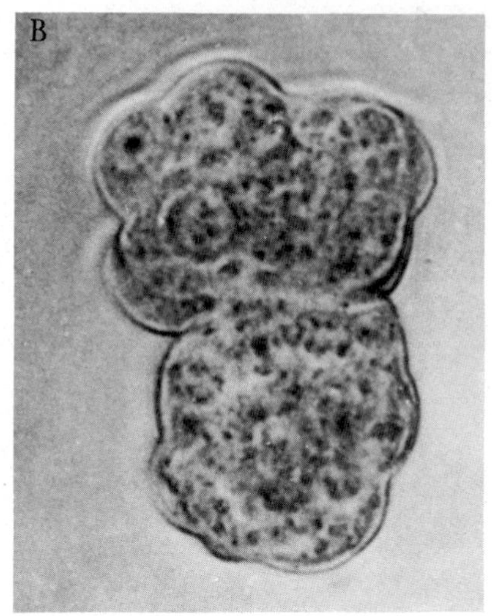


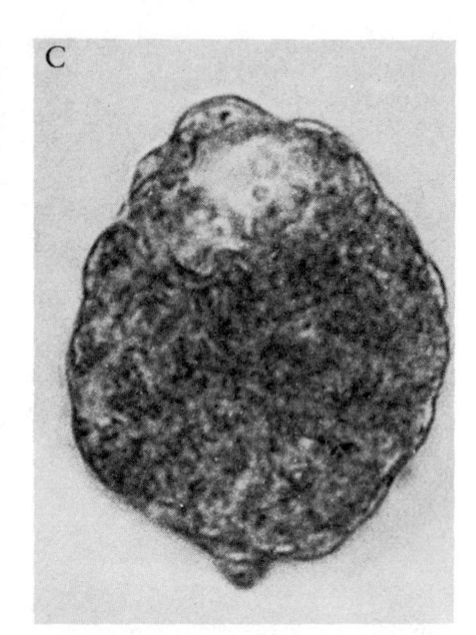


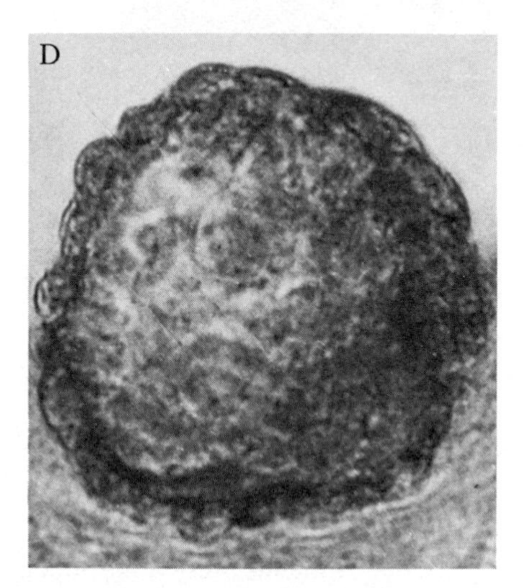


**FIG. 12–20.** Fusion of developing mouse eggs of different ages. A, an eight-celled egg joined with a morula; B, A after 4 hours in culture; C, A after 21 hours in culture. Cavitation is apparently starting in the older partner. D, blastocyst developed from A after 45 hours in culture. (From C. L. Markert and R. M. Peters, 1978. Science 202:56. Copyright 1978 by the American Association for the Advancement of Science.)

made to place individual blastomeres or groups of blastomeres in either an inside or an outside position. In one such experiment (Hillman et al., 1972) tritiated thymidine-labeled donor cells from 4- and 8-cell embryos were aggregated on the outside of cells from unlabeled embryos (Fig. 12–21 A–C). In the resulting blastocysts, over 90 percent of the labeled cells were always found in the trophectoderm. It is difficult, when single labeled blastomeres are reaggregated on the inside of two or four unlabeled blastomeres (Fig. 12–21 D,E), to be sure that they will retain an inside position during further cleavage divisions. When ¼ blastomeres are reaggregated inside of larger numbers of cells (Fig. 12–21 F,G), less and less label is found in the trophectoderm and more and more in the ICM. When an entire 8- to 16-cell labeled embryo is "trapped" inside of 16 unlabeled 8- or 16-cell embryos (Fig. 12–21 H), labeled cells of the blastocyst were found only in the ICM in over 50 percent of the cases, the remainder showing some label in the trophectoderm as well as the ICM.

If the three-and-a-half-day blastocyst consists of two populations of cells that have been affected by their different positions within the embryo to embark on two different lines of development, can we determine when this commitment first takes place and can we find any differences between the cells that might indicate that two different cell populations are differentiating? We have seen that the blastomeres may be considered totipotent up to the 8-cell stage at a time when all of the cells are still exposed to the outside. When the 8-cell stage is first formed, the blastomeres are spherical and their cell outlines are easily seen. Within a few hours a porcess known as *compaction* takes place. Compaction involves substantial changes in the organization of the individual blastomeres themselves and in their relation to each other. The spherical cells become more tightly apposed to each other and flatten out, and tight junctions are formed between adjacent

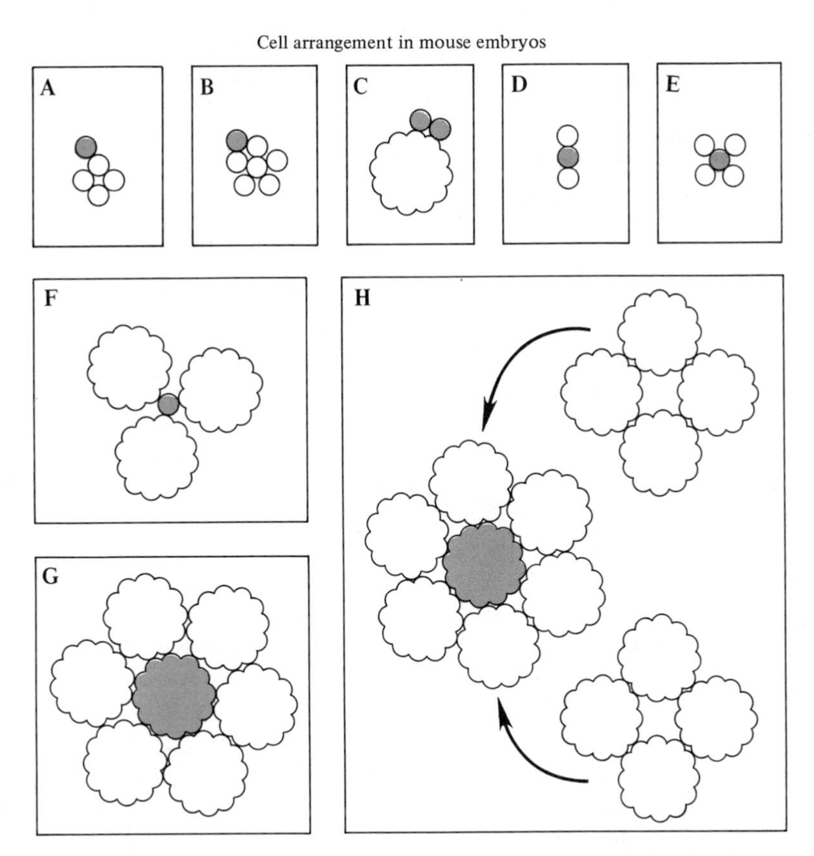


**FIG. 12-21.** Cell aggregations in which labeled (black) blastomeres are arranged in attempts to have them assume either an outside (A–C) or an inside (D–H) position. In H, a labeled 8- to 16-cell embryo is surrounded by six unlabeled 8- to 16-cell embryos, and four more 8- to 16-cell embryos are placed on the top and on the bottom of the aggregation. (From N. Hillman et al., 1972. J. Embryol. exp. Morphol. 28:263–278.)

cells. The outlines of the individual cells become indistinct.

What is of considerable significance is that at this time differences in both surface and internal organization of the individual blastomeres become apparent. These differences cannot be detected in the 2-, 4- or early 8-cell stages but appear at the time compaction occurs. Johnson et al. (1981) proposed that this reorganization, which establishes an axial polarity in the individual blastomeres, represents the first step in the divergent differentiation of inside and outside cells. The cellular polarization is easily seen in the rat using vital stains. The stain appears in each cell as a column running from the nucleus, which has now assumed a more basal position, to the outer periphery of the cell. EM studies reveal that this staining reaction is due to a redistribution of cell organelles, which are now largely restricted to the same area of the cytoplasm between the nucleus and the periphery that stains with the vital dyes. These organelles consist of most of the mitochondria and the agranular endoplasmic reticulum, a small part of the Golgi apparatus, and a heterogeneous assortment of vesicles. Cell surface differences also become apparent when either the whole embryo or the individual disaggregated blastomeres are labeled with a variety of fluorescent ligands. Surfaces that are part of the periphery bind more ligand than internal surfaces. The pattern of ligand binding is reflected in the distribution of the microvilli on the cell surface. Microvilli that are uniformly distributed in the 2-, 4-, and early 8-cell stages become polarized in a pattern that corresponds to that of the ligand binding. In fact, the polarization detected by ligand binding may be the result of the presence of an increased number of ligand-binding sites at the periphery of the cell due to the amount of cellular membrane that the polarized microvilli contribute to this region. When the polarized blastomeres of the 8-cell embryo divide, the 16-cell embryo is made up of two distinct populations of cells, polarized and nonpolarized. SEM and fluorescent-labeling studies show that the polarized cells are on the outside and the nonpolarided cells are generally on the inside, indicating that the polarization established in the 8-cell embryo is continued in later stages and might well be the basis for the establishment of the two different cell populations in the blastocyst. Depending on the plane of cleavage, the two daughter cells (2/16 cells) will be made up of either one polarized and one nonpolarized cell, if the mitotic spindle orients parallel to the axis of polarity, or of two polarized cells, if it does not. The first may be called a differentiating division, the second a generative division. Generally, the first cells to divide are more likely to produce one polarized and one nonpolarized daughter cell. At the 16-cell stage only a few cells are found completely inside the embryo. Inside and outside cells from the 16-cell stage on show different degrees of adhesiveness, the more adhesive cells being on the inside (Kimber et al., 1982). The adhesiveness of the inner cells allows them to move over each other in the formation of the ICM, but there is probably no movement of cells in either direction between the inside and outside populations.

At the three-and-a-half day fully expanded blastocyst stage, the two distinct populations of cells, trophectoderm and ICM, are fully committed to divergent pathways of development. Trophectoderm cells may be dissected free and cultured or transplanted, and under no condition will they form anything but trophectoderm. Conversely, isolated inner cell masses never develop any trophectoderm. The trophectoderm of a blastocyst may be lysed away, leaving only inner cell mass tissue, by reacting the blastocyst with antispecific antisera. When this inner cell mass free of trophectoderm is cultured in vitro, it forms only a compact cell mass with no trophectoderm evident.

Although the three-and-a-half-day blastocyst is committed to two divergent pathways of development, we still have to find out at what point in development this commitment occurs. Whether or not the polarization of the 1/8 blastomeres and the formation of inside and outside cells during subsequent divisions indicates an irreversible determination of these cells may be studied by separating inside from outside cells at various stages from the morula to the early blastocyst and testing their developmental potential. We have already seen that at three and a half days the inner cell mass cannot form any trophectoderm. However, at the morula stage, inner cells differentiate into miniature blastocysts, which show typical trophectoderm as well as inner cell mass. In fact, inner cell masses isolated from cavitating blastocysts only some three or four hours away from the three-and-a-half-day fully expanded blastocysts stage can also form normal blastocysts with typical trophectoderm. Thus, the commitment to trophectoderm or inner cell mass occurs between three and three and a half days, just at the time the trophectoderm is developing, and both cell lines are apparently determined at the same time.

Although the inner cells in the morula and developing blastocyst are not as yet determined, it has been shown that they have already begun to differentiate in a direction that distinguishes them from the outer cells. The inner cells have a distinct pattern of alkaline phosphatase activity and a higher labeled thymidine uptake as early as the morula stage. In the early blastocyst they show evidence of fluid accumulation and will not phagocytose melanin granules—both properties that distingush them from the outer potential trophectoderm cells. Of the five groups of tissue-specific proteins that are present in the embryonic cells of the mature blastocyst, one is found in the 16-cell stage and two more in the morula, beginning to accumulate fluid. This stepwise divergent differentiation finally reaches a point when the inner cell mass is determined to follow a path of development in which its cells are destined to form only embryonic and some extraembryonic structures and in which the potential to form trophectoderm is lost. The two cell lines are distinct.

## Differentiation within the Trophoblast

By day 4½ the trophoblast shows regional differences. The mural cells—those cells not lying above the inner cell mass—stop dividing but continue to duplicate their DNA, forming large cells, primary giant cells, with many times the 2N amount of DNA. The polar cells continue to divide. They form the ectoplacental cone above the inner cell mass. This is the region of the rodent trophoblast that will form the embryonic part of the placenta. As the ectoplacental cone increases in thickness, secondary giant cells appear in the region furthest away from the underlying inner cell mass. The regional differentiation of the trophoblast has been ascribed to inductive influences from the inner cell mass. Trophoblast cells lose the capacity to divide and become giant cells unless they are exposed to influences from the inner cell mass. Secondary giant cells are formed in the ectoplacental cone by progeny of the dividing polar cells, which are pushed to peripheral regions where they are no longer influenced by the inner cell mass. The nature of the inductive influence is not known.

The region between the ectoplacental cone and the embryonic ectoderm of the ICM forms the extraembryonic ectoderm of the chorion. The differentiation of giant cells, the cells of the ectoplacental cone, and the cells of the extraembryonic ectoderm represent pathways of divergent differentiation. However, the cells of the ectoplacental cone and the extraembryonic ectoderm still retain the potential for giant cell formation at least up to eight and a half days of development. Giant cells, in contrast, cannot revert to dividing diploid cells and cannot contribute to either of the other two cell types.

## Regulation within the Inner Cell Mass

We may now return to the inner cell mass cells. Their fate is to form the egg cylinder that will form the embryo proper, the amnion, the yolk sac, and part of the chorioallantoic placenta. However, regulation within these prospective fates is still possible. A number of experiments demonstrate that the cells of the inner cell mass are not committed to specific developmental pathways but that a high potentiality for regulation within the system is still present.

1. When part of the inner cell mass is removed, a normal offspring may result.
2. Whole blastocysts, when cut in half, will develop into fetuses that are morphologically normal.
3. An inner cell mass from a three-and-a-half-day blastocyst may be transplanted to a host three-and-a-half-day blastocyst, where it will aggregate with the host inner cell mass and produce a normal chimeric offspring.
4. When a single, marked inner cell mass cell is injected into a host blastocyst, the progeny of this cell are widely distributed when the resulting 11½- to 17½-day embryos are examined. Donor cells have been found in the embryo proper and in its membranes and in derivatives of all three germ layers.

## Further Development of the Inner Cell Mass

As the inner cell mass develops from three and a half to four and a half days, a monolayer of cells appears on its blastocoelic surface, which may be distinguished from the other cells of the inner cell mass on the basis of their morphology, the formation of tight junctions, and the presence of cytoplasm rich in rough endoplasmic reticulum. This layer is the *primitive endoderm,* which will contribute to the formation of the yolk sac and the placenta. The remaining inner cell mass, the *primitive ectoderm,* will form all of the fetal germ layer material.

The prospective fates of the primitive endoderm and primitive ectoderm are different, but are their potencies also restricted? They are. Culturing primitive endoderm yields only parietal and visceral endoderm of the extraembryonic membranes, and indicates that it is at four and a half days committed to only a single developmental direction. Conversely, the primitive ectoderm cells will form only embryonic ectoderm, endoderm, and mesoderm, and cannot form primitive endoderm. However, within the framework of these potentialities, the primitive ectoderm is not as yet determined to develop in any specific direction. Up to the primitive streak stage, the ectoderm can still form both endoderm and mesoderm. At the time the embryonic endoderm develops—in the primitive streak stage—the ectoderm then loses its capability to form more endoderm. Later, when the mesoderm develops, it also loses its capability to form more mesoderm. There is a progressive restriction of developmental potencies.

Apparently, at the time certain cells in the inner cell mass become committed to some specific line of development, the remaining cells lose the capability of forming any more cells of this line. When the trophectoderm cells appear, the inner cell mass loses its potential to form more trophectoderm; and when the primitive endoderm differentiates, the inner cell mass then loses its potential to form more primitive endoderm; and so on down the line. Each step means the formation of a determined cell line and the restriction of potency in the as yet undetermined cells to form any more cells of the determined line. A scheme of development of the mouse embryo is diagrammed in Figure 12–22, showing in both the inner cell mass and the trophectoderm the same type of progressive restriciton of developmental potencies.

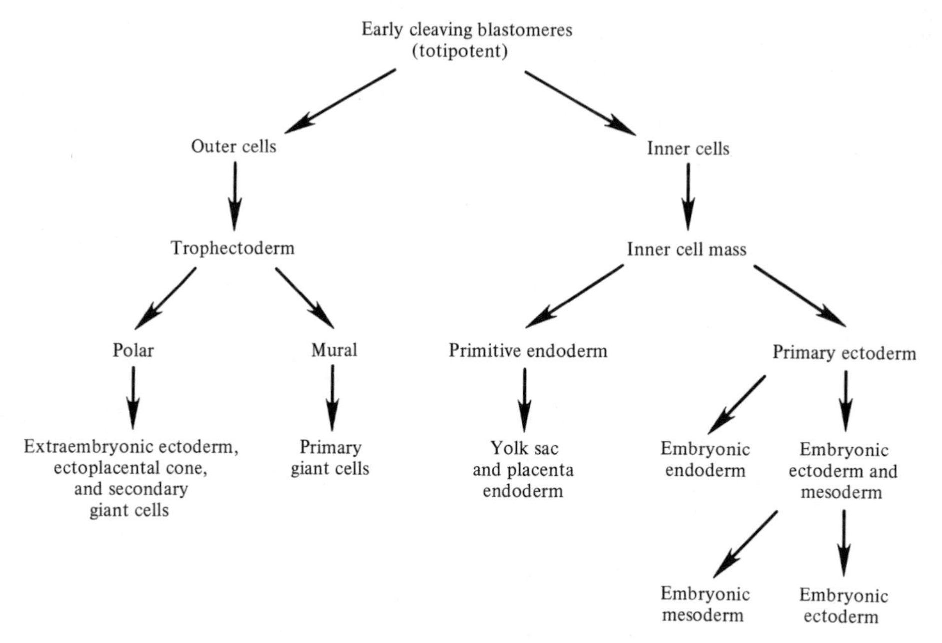


**FIG. 12–22.** Schematic diagram of the progressive steps in the development of the mouse embryo. (After R. L Gardner, 1975. In: The Developmental Biology of Reproduction, C. L. Markert and J. Papaconstantinan, eds. Academic Press, New York.)

It is questionable if all the conclusions drawn from the analysis of differentiation in one mammal can be applied to all of the others. However, there appears to be no question that the mammalian egg may be considered to be a regulative system in which neither polarity nor the differential distribution of cytoplasmic factors are important for future development. The inner–outer determination of inner cell mass and trophoblast differentiation and the regulative properties of the inner cell mass cells appear to be general mammalian characteristics. However, the inductive influence of the inner cell mass on the differentiation of the polar trophoblast has not been completely substantiated in all the species studied.

## In Vitro Fertilization, Culture, and Transfer of the Human Embryo

In vitro fertilization and transfer of the conceptus into the uterus of a foster mother have become increasingly common experimental procedures in mammals over the past two decades, and some of the results from the use of these procedures have been described in the previous sections. In the human, however, owing in large part to the scarcity of available material, in vitro fertilization (IVF) and embryo culture experiments have been relatively few in number. In addition, the technical difficulties and the ethical, moral, and legal considerations of embryo transfer (ET) were such that it was not until 1978 that a report of the first human oocyte recovered from the ovary, fertilized in vitro, cultured to the 8-cell stage, implanted in the uterus, and delivered by cesarean section was published (Steptoe and Edwards, 1978), when the birth of Louise Joy Brown took place in London in July of that year. A report of a second successful birth was published in 1981 by a group of Austrialian gynecologists (Lopata et al., 1981).

In vitro fertilization and culture of human oocytes was first reported in the early 1940s (Rock and Menkin, 1944). One hundred thirty-five of over 800 human follicular oocytes isolated from surgical material were observed after exposure to sperm. Of these, one 2-celled and one 3-celled embryo developed after 22 to 27 hours of culture. Sections of the 2-celled embryo appeared quite normal when compared to the 2-celled embryo recovered from the uterus (Fig. 12–1). Since then, there have been a number of reports on in vitro fertilization and culture of human oocytes, and with the continued refinement in the technqiues of obtaining the oocytes and the improvement of the culture media, the percentage of successful experiments has increased markedly. The original culture media for mammalian eggs were simple clots formed by mixing chick embryo extract and blood plasma. Since that time considerable effort has gone into the modification of tissue culture media. Pyruvate and lactate are used as an energy source, and bicarbonate is used for buffering. The addition of crystallized bovine serum as a protein source significantly improves the fertilization rate, although its exact function is not known. Control of oxygen tension and pH is critical to successful fertilization.

Although in normal fertilization millions of sperm are deposited in the vagina, the actual number that reach the vicinity of the oocyte in the uterine tube is considerably fewer. Attrition takes place mainly at the narrowest parts of the female reproductive tract—the uterotubal junction and the isthmus of the uterine tube. The long passage from the vagina to the upper third of the uterine tube serves as an obstacle course that gives the more active and vigorous sperm an advantage, and most of the inactive and abnormal spern do not complete the journey. In in vitro fertilization, only a portion of the total ejaculate is used to inseminate the oocyte. In this case sperm are eliminated purely by chance, and the selective effect of the passage through the female reproductive tract is eliminated. Some method for the selection of the more active and normal sperm in in vitro fertilization would be advantageous.

For a variety of reasons, successful fertilization does not always follow in vitro insemination. Criteria for successful fertilization are thus important. We might expect that the first indication of successful fertilization would be the cleavage of the egg, and this criterion is used frequently. In fact, cleavage is not a reliable criterion. First, it does not reveal whether or not the sperm nucleus has actually participated in the process, since a sperm may merely stimulate parthenogenetic development of the oocyte. Evidence that is cytological and genetic—if the gametes are genetically dissimilar—is necessary to prove that the male nucleus is indeed performing its normal role. In addition, what may appear to be normal cleavage when viewed uncritically or when seen in photographs in published reports may actually by the fragmentation of an unfertilized egg. Many fragmenting eggs are practically indistinguishable from normal cleaving eggs.

It is interesting in this respect to consider the tale of a "human embryo" whose photograph appeared (1) in a textbook in 1960, described as a morula, (2) upside down in a textbook of embryology in 1971, described as a morula at three and a half days, and (3) in a scientific journal in 1973, described as a living human blastocyst five days after fertilization about to implant—and which, in reality, was nothing but a fragmented unfertilized oocyte.

Trouson et al., in 1982, recorded the timing and the appearance of the normal cleavage stages of the human embryo developing in vitro. Examples of these stages are shown in Figure 12–23. The 1-cell stage, 12 to 31 hours after insemination, has two visible pronuclei, and often two polar bodies may be identified. The 2-cell stage is reached between 27 and 43 hours (mean, 35.6 hours), the 4-cell stage between 36 and 65 hours (mean, 45.7 hours), and the 8-cell stage between 45 and 73 hours (mean, 54.3 hours), and the 16-cell stage between 68 and 85 hours. There is thus considerable overlap between the stages. Those embryos with the shorter time intervals between successive stages appear to be the most viable, and embryos with the longer intervals may have reduced viability and may be more likely to become arrested as development continues.

Some of the problems involved in IVF and ET in humans are those concerned with (1) timing the menstrual cycle so that at the time the oocytes are collected—usually by laparoscopy—they are in the correct stage of development, (2) providing the proper conditions for in vitro fertilization and culture, and (3) perfecting the technique of transplanting the conceptus into the uterine lumen.

In the first two reported cases of live births, the oocytes were removed from the preovulatory follicles of the women running natural menstrual cycles. The preovulatory surge of LH is used as an indicator and the oocyte is removed at a chosen time interval after the LH surge—usually 27 to 28 hours. Ultrasound scanning of the ovaries may be used to determine if the oocyte is still present in the follicle, if it is the proper size, and if it is accessible to operation. In one study (Lopata et al., 1981), various times, from 24 to 30 hours after the first LH surge were chosen for oocyte recovery. At 24 hours 50 percent of the operations were successful; at 26 hours 72 percent, at 28 hours 64 percent, and at 30 hours 60 percent. None of the oocytes recovered at 24 hours began cleavage following insemination. The highest percentage of cleavage (78%) was seen

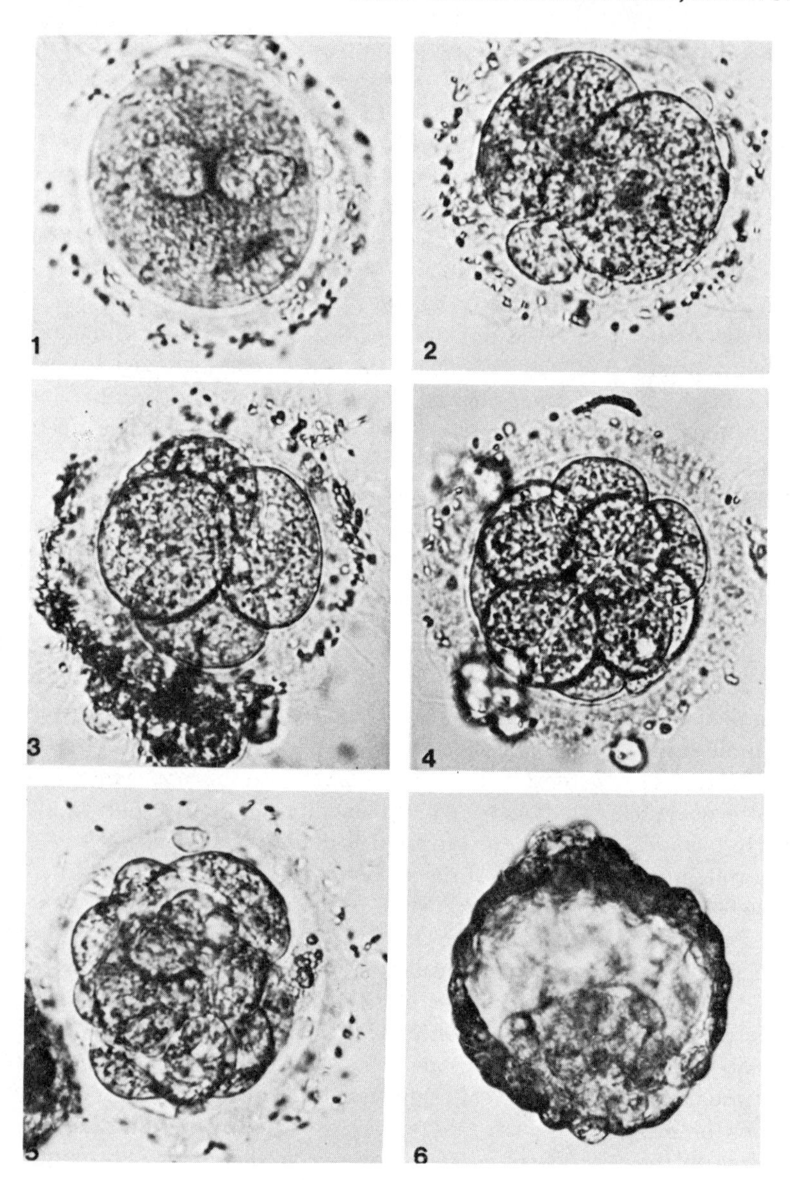

**FIG. 12–23.** Stages in the in vitro development of the human embryo. (1) pronuclear stage; (2) 2-cell embryo; (3) 4-cell embryo; (4) 8-cell embryo; (5) morula stage; (6) blastocyst hatched out of its zona pellucida. (From A. O. Trounson et al., 1982. J. Reprod. Fertil. 64:285.)

in oocytes aspirated at 28 hours, and of the two pregnancies that ensued following 14 embryo transfers, both were from oocytes recovered at 28 hours and transplanted at the 8-cell stage. The percentage of pregnancies was extremely low; two pregnancies and one live birth in a total of 69 patients.

Although both of the first two human births recorded were from oocytes recovered during normal, spontaneous ovulatory cycles, better results have been realized by increasing the yield of preovulatory oocytes by artifically controlling the menstrual cycles. The earliest attempts to increase the yield of oocytes by inducing the development of multiple follicles with a drug (clomiphene citrate) that stimulates follicular development and by controlling the final maturation of the oocyte with human chorionic gonadotrophin were not successful, in that they did not result in the establishment of pregnancies. More recently (Trounson et al., 1981), an increase in the number of oocytes recovered, embryos developed, and pregnancies occurring was recorded in a group of 20 women given clomiphene and HCG compared to a group of 20 women allowed to run natural spontaneous ovulatory cycles. In the controlled group 31 mature oocytes were recovered and 27 embryos developed; in the 16 patients receiving embryo transfers, 3 pregnancies resulted. The uncontrolled group yielded only 13 oocytes and 11 embryos developed; in the 11 patients receiving embryo transfers, no pregnancies resulted. The increase in the number of oocytes recovered using clomiphene to produce multiple follicle development and HCG to determine more accurately the optimum time for oocyte aspiration was significant, but the rate of pregnancies resulting after embryo transfer was low—only 11 percent—indicating either that embryos grown in vitro, despite their apparently normal appearance, have reduced viability, or that the transfer techniques and subsequent treatment need improvement.

Lowered embryonic viability might be due to the fact that oocyte maturation was not completed at the time of fertilization. When the oocytes are recovered, they are at unknown intervals before what would have been the normal time of ovulation and these intervals certainly vary from oocyte to oocyte. There is no indicator available that can be used to measure the state of oocyte maturation, and there is thus no way to ascertain if the oocytes are completely mature at the time of recovery and insemination. This suggested that delaying insemination for a period of time after oocyte recovery might allow time for complete maturation and thereby increase normal fertilization, embryonic development, and successful pregnancy after embryo transfer. In fact, delayed insemination has been shown to have a significant effect in increasing the percentage of normally cleaving embryos, from approximately 30 percent when insemination was not delayed, to 90 percent when insemination was delayed for 5–5½ hours, and to 70 percent after 6–6½ hours delay (Trounson et al., 1983). In addition, the only embryos that survived transfer were those in the groups in which insemination was delayed for from 5 to 6½ hours. There was a significant increase in successful pregnancies when twin embryos were transferred. While the transfer of 26 single embryos resulted in only two pregnancies, the transfer of seven pairs of twin embryos resulted in five pregnancies, a pregnancy rate of 36 percent for the transfer of twin embryos. Two of the transfers of twin embryos resulted in twin fetuses. If all transfer embryos have an equal potenital for devleopment, it would certainly be expected that the transfer of two embryos instead of one would yield a higher rate of pregnancy.

The advantage of transferring more than one conceptus has also been reported in the Eastern Virginia Medical School program (Jones et al., 1982). In this program, stimulated and controlled ovulations were also used to obtain more than one fertilizable oocyte per cycle and to control the timing of follicular growth. In 1981 the pregnancy rate for the transfer of single embryos was 13 percent;

**FIG. 12–24.** Photographs of two human embryos fertilized and cultured in vitro just before they were transferred into the uteri of their donors. A, 4-cell; B, 8-cell. (Courtesy of Howard W. Jones, Jr., M.D., Eastern Virginia Medical School.)

A

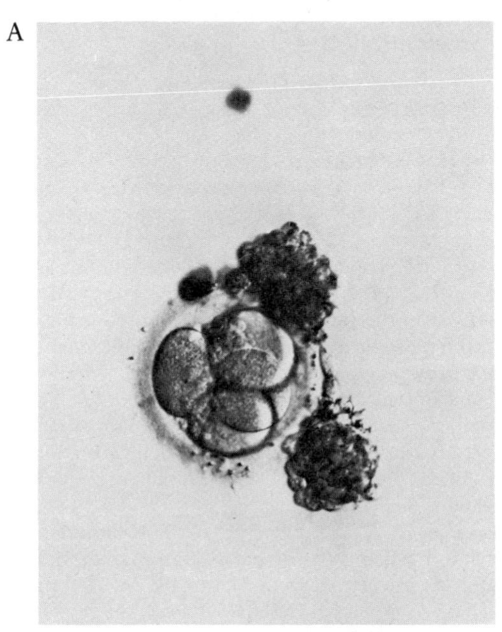

B

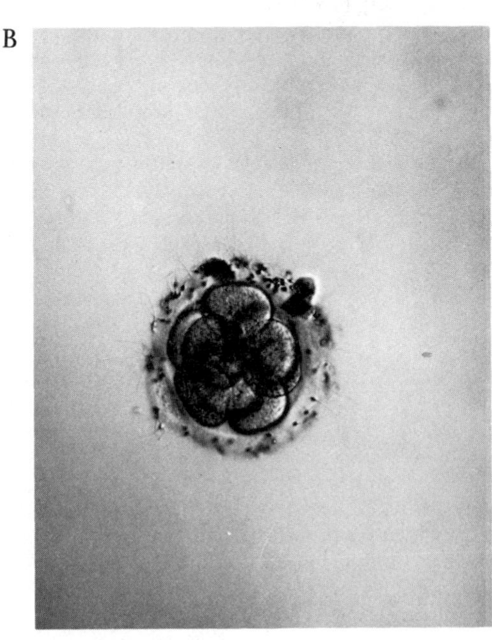

for twin embryos it was 31 percent and for triplet embryos 50 percent. Although the actual numbers are small, they do indicate a definite advantage of multiple-over single-embryo transfers. Two embryos fertilized and cultured in the laboratory of this group are shown in Fig. 12–24. The 4-cell and 8-cell embryos were transferred into the uteri of the donors shortly after these photographs were made.

Although at present the precentage of births occurring after IVF and ET is low, it is increasing every year as more is learned about the processes involved and as the techniques are perfected. IVF and the culture and transfer of human embryos is now an established clinical practice.

## References


Bartels, H., W. Moll, and J. Metcalfe. 1962. Physiology of gas exchange in the human placenta. Am. J. Obstet. Gynecol. 84:1714–1730.

Dalcq, A. M. 1957. Introduction to General Embryology, pp. 103–198. London: Oxford University Press.

Garnder, R. L. 1975. Analysis of determination and differentiation in the early mammalian embryo using intra- and interspecific chimeras. In: The Developmental Biology of Reproduction, pp. 207–236. C. L. Markert and J. Papaconstantinau, eds. New York: Academic Press.

Hillman, N., M. L. Sherman, and C. Graham. 1972. The effect of spatial arrangment on cell determination during mouse development. J. Embryol. exp. Morphol. 28:263–278.

Johnson, M. H., A. H. Handyside, and P. R. Braude. 1977. Control mechanisms in early mammalian development. In: Development in Mammals. M. H. Johnson, ed., New York: North-Holland.

Johnson, M. H., H. P. M. Pratt, and A. H. Handyside. 1981. The generation and recognition of positional information in the preimplantation mouse embryo. In: Cellular and Molecular Aspects of Implantation, pp. 55–74. S. R. Glasser and D. W. Bullock, eds. New York: Plenum Press.

Jones, H. W., and G. S. Jones, M. C. Andrews, A. Acosta, C. Bundren, J. Garcia, B. Snadow, L. Veeck, C. Wilkes, J. Witmyer, J. E. Wortham, and G. Wright. 1982. The program for in vitro fertilization at Norfolk. Fertil. Steril. 38:14–21.

Jones, K. L. and D. W. Smith. 1973. Recognition of the fetal alcohol syndrome in early infancy. Lancet 2:999–1001.

Kimber, S., M. A. H. Surani, and S. C. Barton. 1982. Interactions of blastomeres suggest changes in cell surface adhesiveness during the formation of inner cell mass and trophectoderm in the preimplantation mouse embryo. J. Embryol. exp. Morphol. 70:133–152.

Lopata, A., I. W. H. Johnston, I. J. Hoult, and A. L. Speirs. 1981. In vitro fertilization in the treatment of human infertility. In: Bioregulation of Reproduction, pp. 411–426. G. Jagiella and H. J. Vogel, eds. New York: Academic Press.

Mintz, B. 1964. Synthetic processes and early development in the mammalian egg. J. Exp. Zool. 157:85–100.

Moore, N. W., C. E. Adams, and L. E. A. Rowan. 1968. Developmental potential of single blastomeres of the rabbit egg. J. Reprod. Fertil. 17:527–537.

Nelson, M. M. and J. O. Forar, 1971. Association between drugs administered druing pregnancy and congenital abnormalities of the fetus. Br. Med. J. 1:523–527.

Ramsey, E. M. 1973. Placental vasculature and circulation. In: Handbook of Physiology, Endocrinology II, Part 2. R. O. Greep, ed. Baltimore: Williams and Wilkins.

Rasmussen, D. M. 1968. Syphilis and the fetus. In: Intra-Uterine Development. A. C. Barnes, ed. Philadelphia: Lea and Febiger.

Rock, J. and M. F. Menkin. 1944. In vitro fertilization and cleavage of human ovarian eggs. Science 100:105–106.

Rossant, J. 1977. Cell commitment in early rodent development. In: Development in Mammals, ed. M. H. Johnson, New York: North-Holland.

Rossant, J. and W. T. Frels. 1980. Interspecific chimeras in mammals: successful production of live chimeras between *Mus musculus* and *Mus caroli*. Science 208:419–421.

Rossant, J., V. M. Mauro, and B. A. Croy. 1982. Importance of trophoblast genotype for survival of interspecific murine chimeras. J. Embryol. exp. Morphol. 69:141–145.

Seeds, A. E. 1968. Placental transfer In: Intra-Uterine Development. A. C. Barnes, ed. Philadelphia: Lea and Febiger.

Seidel, F. 1960. Die Entwicklungsfähigkeiten isolierter Furchungszellen aus dem Ei des Kaninchens, *Oryctolagus cuniculus*. Wilhelm Roux' Arch. Entwicklungomech. Org. 152:44–130.

Sever, J. L. 1971. Viral infections and malformations. Fed. Proc., Fed. Am. Soc. Exp. Biol. 30:114–117.

Siiteri, P. C. and D. P. Stites. 1982. Immunologic and endocrine relationships in pregnancy. Biol. Reprod. 26:1–14.

Solter, D. and B. B. Knowles. 1975. Immunosurgery of mouse blastocysts. Proc. Natl. Acad. Sci. U.S.A. 72:5099–5102.

Steptoe, P. C. and R. G. Edwards. 1978. Birth after the reimplantation of a human embryo. Lancet 2:366.

Stern, M. S. 1972. Experimental studies on the organization of the preimplantation mouse embryo. II. Reaggregation of disaggregated embryos. J. Embryol. exp. Morphol. 28:255–261.

Stern, M. S. and I. B. Wilson. 1972. Experimental studies on the organization of the pre-implantation mouse embryo. I. Fusion of asynchronously dividing eggs. J. Embryol. exp. Morphol. 28:247–254.

Streissguth, A. P., S. Landesman-Dwyer, J. C. Martin, and D. W. Smith. 1980. Teratogenic effects of alcohol in humans and laboratory animals. Science 209:328–329.

Tarkowski, A. K. 1959. Experiments on the development of isolated blastomereres of mouse eggs. Nature (Lond.) 184:1286–1287.

Tarkowski, A. K. and J. Wroblewska. 1967. Development of blastomeres of mouse eggs isolated at the 4- and 8-cell stage. J. Embryol. Exp. Morphol. 18;155–180.

Trounson, A. O., J. F. Leeton, C. Wood, J. Webb, and J.

Wood. 1981. Pregnancies in humans by fertilization in vitro and embryo transfer in the controlled ovulatory cycle. Science 212:681–682.

Trounson, A. O., L. R. Mohr, C. Wood, and J. F. Leeton. 1982. Effect of delayed insemination on in vitro fertilization, culture and transfer of human embryos. J. Reprod. Fertil. 64:285–294.

Ward, I. L. and J. Weisz. 1980. Maternal stress alters plasma testosterone in fetal males. Science 207:328–329.

Willadsen, S. M. 1981. The developmental capacity of blastomeres from 4- and 8-cell sheep embryos. J. Embryol. exp. Morphol. 65:165–172.

Wynn, R. M. 1973. Fine structure of the placenta. In: Handbook of Physiology, Endocrinology II, Part 2. R. O. Greep, ed. Baltimore: Williams and Wilkins.

# 13

# PRINCIPLES OF MORPHOGENESIS AND PATTERN FORMATION

By the end of gastrulation, each germ layer, occupying a different topographical position in the embryo, is broadly committed to develop into specific tissue and organ primordia. In both amniote and anamniote embryos, the basic body plan at this time is that of a tube within a tube. The inner, endodermal tube is separated from the outer or ectodermal tube by coelomic tubes of mesoderm. The free surface of each tube is initially organized as a sheet of cells or an *epithelium*. An epithelium is a monolayer of cells whose tight-knit cohesiveness is maintained by specialized intercellular junctions and extracellular materials. An important property of embryonic epithelia is the capacity for their individual cells to change shape. Alterations in the shapes of individual cells are translated and integrated into morphogenetic movements involving many cells. Transformation of epithelia in multicellular animal embryos by modifications in the shape of cells is an important instrument in the establishment of definite form in many organs.

Organogenesis can be considered to occur during that period of time when organs and organ systems become fashioned from germ layer organ rudiments. Most organs develop from rudiments that consist of an epithelial tissue and a mesenchymal tissue. During the ontogeny of an organ, there is the development of three-dimensional structure involving the arrangements of cell populations in a precise and organized manner (i.e., morphogenesis) and the specialization of cell types (i.e., cytodifferentiation). The twin processes of morphogenesis and cytodifferentiation must occur if tissues and organs are to acquire their proper structural and functional features. The importance of these two events is readily apparent in most cases of organ development. For example, it would be futile for the embryo to differentiate cells capable of producing bile (liver) or secretory enzymes (pancreas) without the development of a network of small ductules and ducts (morphogenesis) to carry the products to the small intestine.

The precise relationships between the differentiation of the functional cell types of an organ and the development of that organ's morphology (i.e., shape, form) are still imperfectly understood. Since the synthesis of specific molecules within a differentiating cell is dependent on its gene activity, it is natural to ask whether the genes controlling macromolecular synthesis and the genes controlling morphogenesis are functionally coupled or independently regulated. Primarily on the basis of studies using embryonic pancreas, thyroid gland, and glands of the oviduct, it appears that specific steps in the development of the morphology of an organ can be temporally correlated with the differentiation of its unique cell types. For example, Rutter and his colleagues have examined the relationships between cytodifferentiation and morphogenesis in the development of the mouse pancreas. The dorsal pancreas in the mouse arises as a hollow diverticulum from the endoderm; this epithelial rudiment pushes into the investing mesodermal tissue (Chapter 15). Subsequently, fingerlike groups of cells develop from the epithelial outpocketing that branch to form a labyrinth of ductules and ducts. Special cell clusters termed acini arise at the terminal ends of the ductules. These differentiate as exocrine cells and will produce the digestive enzymes of the pancreatic juice. Using sensitive biochemical assays for the detection of pancreatic proteins, Rutter was able to identify, although at very low levels, the first pancreatic proteins at about the time that the initial form of the gland was taking shape. Increased levels of the enzymes were subsequently detected several days later when ducts, ductules, and acini were clearly formed. Hence, the morphogenetic phase of an organ can be correlated with levels of differentiation of its cell types.

Of the morphogenetic and cytodifferentiative phases that characterize organ development, much of the research to date has been centered on how individual cells acquire their special and unique properties. Increasingly, however, attention is being directed toward undertanding the factors influencing and underlying morphogenesis. The processes involved in this phase of organ development include *growth, cell movements, localized cell death,* and alterations in the *extracellular matrix*. It should be emphasized that these determinants of organ shape may not be completely separate. For example, growth in its broadest sense refers to an increase in biomass resulting from cellular hypertrophy, hyperplasia, and production of extracellular materials. It is rarely uniform, resulting in differental rates of size change. Differential rates of growth can be due to the net additon or subtraction of tissue, including extracellular components, thereby leading to a remodeling of the organ. Similar changes of form may appear as a consequence of stress generated by mechanical forces organized within the extracellular matrix. Differential growth processes appear to dominate over slower morphogenetic changes. In the remainder of this chapter and in later chapters addressing the development of specific organs, we will discuss the various types of morphogenetic processes active in organ formation.

## Morphogenesis of Embryonic Epithelia by Changes in Cell Shape

Most of the internal organs of the vertebrate organism, such as lungs, pancreas, salivary glands, and so on, originate from primordia consisting of two dissimilar cell populations. One population of cells is epithelial in its organization. These cells line tubes, ducts, cavities, and so forth and differentiate into specific functional components of organs. The other population of cells, generally of mesodermal origin, is less regularly or uniformly organized and will contribute to the connective and supportive functions of organs. Generally, a diverse number of organ systems require interactions between the epithelium and the mesenchyme in order for morphogenesis to occur. These tissue interactions are commonly referred to as *epitheliomesenchymal interactions*.

The form and shape associated with many internal organs relate to the localized transformation of the epithelial component of primary organ rudiments. Commonly, the intitial phase in the formation of an organ is expressed as a thickening in an embryonic epithelial layer, as in the case of the neural plate. Within the same embryonic epithelial layer, the primordia of organs may be paired (lens and nasal placodes) or in multiples (hair follicles, mammary glands). Localized thickenings in an epithelium may be a consequence of localized cell division and/or rearrangements of cells due to shape changes. Interestingly, the mechanism underlying the same cell shape change may be quite different. For example, both the neural plate (Chapter 10) and the lens placode (Chapter 21) are constructed of elongated cells containing microtubules oriented parallel to their length. Although the organization of microtubular elements appears to be critical to neural plate formation, it may not be the major factor determining the change in cell shape in the lens placode. An increase in cell volume appears to produce cell elongation in the lens placode.

Thickenings within an epithelial layer may be directed outward or inward. Often, the thickening may secondarily lose its connection with the epithelial layer and remain solid (thyroid gland) or acquire an internal cavity (Fig. 13–1 A). For example,

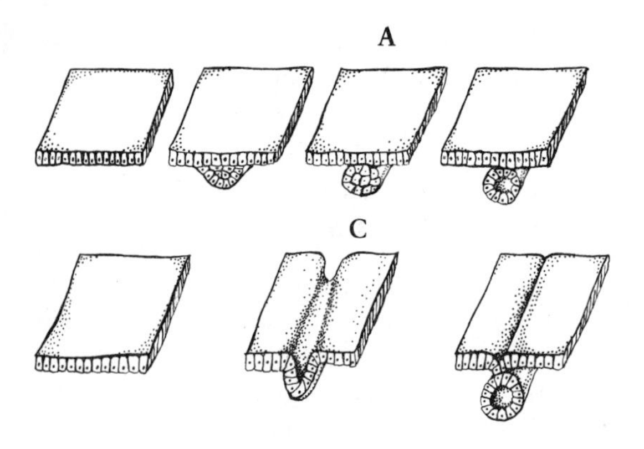


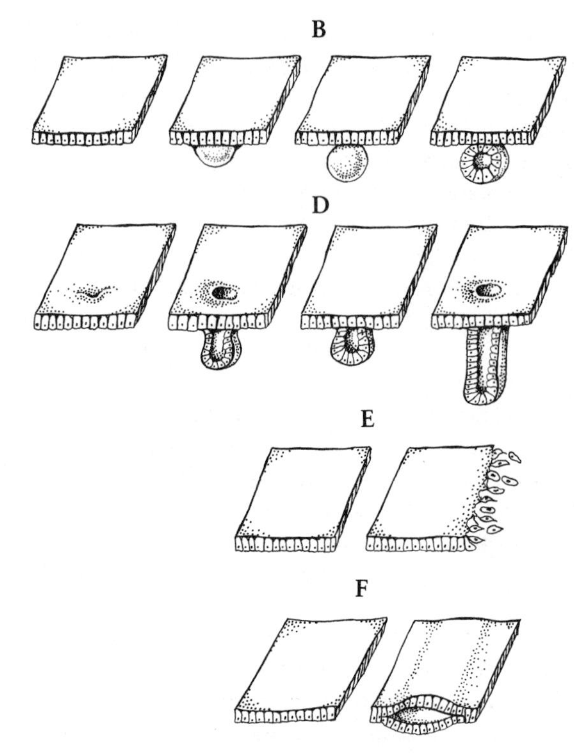


**FIG. 13–1.** A diagram showing various types of transformations in an embryonic epithelium that may occur during the formation of an organ. A, development of a tube from a solid, longitudinal thickening of the epithelium; B, development of a vesicle from a localized thickening of the epithelium; C, a longitudinal groove in an epithelium giving rise to a hollow tube; D, a localized pit may give rise to a vesicle or to a tubule continuous with the epithelial surface; E, disaggregation of an epithelium to produce mesenchyme-like cells; F, splitting of an epithelial layer into two layers (i.e., coelom formation). (From B. I. Balinsky, 1975. An Introduction to Embryology. W. B. Saunders Company, Philadelphia.)

in frogs and bony fishes, the lens arises as a localized, thickened group of ectodermal cells that separates from the adjacent epidermis and slips beneath the surface as a solid cellular mass. Here an internal reorganization of cells leads to the formation of a hollow body or lens vesicle (Fig. 13–1 B).

The folding and migration of an epithelial sheet, either by invagination or by evagination, is perhaps the most common method in the formation of organs. Recall that gastrulation in echinoderms and cephalochordates, as well as the initial indentation of the endoderm in amphibian gastrulas, is essentially the infolding of a simple epithelial layer of cells. The reverse of invagination—evagination—is utilized in constructing various diverticula from the germ layer tubes formed as a consequence of gastrulation. The gut tube, for example, evaginates locally to form the visceral pouches, the thyroid gland, the laryngotracheal groove, and the pancreatic and liver primordia.

Two types of epithelial foldings are evident in organ formation. If the infolding of the epithelial sheet is linear, there results a longitudinal groove or furrow (Fig. 13–1 C). Complete separation from the epithelial layer produces a closed tube. If separation is incomplete, then the tube remains connected to the epithelial layer by an opening (i.e., union of trachea with gastrointestinal tract). The infolding of a portion of the epithelial layer may be quite localized, producing a depression or pocket-like structure (Fig. 13–1 D). Separation of these blisterlike formations yields epithelial-lined, hollow vesicles. The lenses of the eye and the inner ear primordia of amniote embryos are fashioned in this manner.

The folding and movement of epithelial cell sheets produce the variety of geometrical configurations that are associated with the distinctive form of an organ. It has been known for some time that such movements of cell populations are generally accompanied by changes in the shapes of individual cells within the sheet. A major problem in understanding the morphogenetic processes in cell populations is how local cell shape changes are generated and coordinated into the movements of embryonic epithelia during organ construction. Although a number of investigators have attempted to correlate observed cell shape changes with various morphogenetic patterns, the mechanism(s) responsible for generating alterations in cell shape that underlie movement is subject to differences of opinion. A review of the literature suggests that there are at least two basic models to explain morphogenetic patterns. One model (genetic) postulates that the individual cells of a sheet are genetically programmed to carry out a sequence of instructions directing each movement as well as its precise timing. Each cell is envisioned as having an autonomous program of movement and some signal or "clock" for activating the program. The other model (mechanical) proposes that various morphogenetic patterns such as evagination and invagination are regulated solely by mechanical forces generated locally by each cell. The coordination of epithelial cell shape changes is presumed to be accompanied by the propagation of mechanical contraction waves. If, for example, each cell can deform its neighbors sufficiently, then a wave of cell shape change can be propagated across the epithelial layer. Evidence supporting the idea that large-scale epithelial movements are effected solely by mechanical forces comes from current understanding of the organization of the cell's cytoplasm.

One of the problems in studying morphogenesis of embryonic organs is the complexity of shape changes that take place. Generally, two types of cell shape are commonly recognized during movements of epithelia. First, individual cells may elongate, as in the early formation of such organs as the neural plate, lens, thyroid gland, and pancreas. Second, individual cells may narrow or constrict at either their apical or basal ends. Changes in the cytoplasmic organization of epithelial cells correlate with changes in cell shape. Each epithelial cell is equipped with a structural framework or cytoskeleton consisting of microtubules, *actin filaments*, and *intermediate filaments* arranged in a complex, three-dimensional meshwork of various combinations. The fibrous meshwork, organized so as to constitute an "excitable" viscoelastic medium, can sustain various stresses and transduce chemical energy into mechanical energy. Microfilaments are generally held responsible for changing the shape of epithelial cells through their contractile activity. They can be temporally and spatially correlated with the regional constriction of epithelial cells. Experiments with cytochalasin B, an antibiotic that disrupts the structural integrity of microfilaments in some cell types, offer indirect evidence that these organelles are important to generating shape changes associated with constriction. Bundles of microfilaments are considered essential to the formation of a variety of organs, including the neural tube, the lens vesicle, and the pancreatic diverticulum. It might seem rather surprising that nonmuscle embryonic cells have contractile properties. Yet, numerous biochemcial studies have shown both actin and myosin to be present in many nonmuscle cell types. There is compelling evidence that actin, considered to be the major protein of microfilaments, extracted from nonmuscle cell types behaves physiologically and biochemically like actin of skeletal muscle. Although there is the suspicion that motility and deformations of epithelial sheets during organ formation are mediated by musclelike proteins, there needs to be careful assessment of the properties of these "motility proteins" and their exact form of distribution within filaments of nonmuscle cells.

As pointed out in Chapter 10, microtubules are long, cylindrical, unbranched structures that generate changes in cell shape by virtue of their orientation and distribution within epithelial cells. Microtubules are almost always observed to be oriented parallel to the long axis of lengthening cells (i.e., lens placode and neural plate formation). Also, it has been shown that inhibitors of microtubule assembly, such as colchicine and colcemid, frequently disturb cell elongation. Although the organization of microtubular elements during organ construction is temporally and spatially associated with the elongation process, other factors may underlie the capacity of cells to become elongate.

Cell elongation and deformation by constriction are not the only changes involved in the development of organ form. Organ shape can be generated by an increase in size due to cell division, shifts in position of cells relative to their neighbors, and differential increases in surface area by cell spreading. It is often difficult to determine which cellular parameters are important in the generation of epithelial form because several of these processes may occur at the same time. A potentially useful technique in identifying the basis of the key changes in the formaiton of an epithelial organ is *computer simulation*. In computer modeling of organogenesis, changes in the shape of a developing organ, such as cell height and cell width, are measured and a computer program then written that will "simulate" the shape of the organ primordium. The computer simulation permits the identificaiton of local sites of morphogenetic activity and shape changes that might not be readily visible by inspection of the whole primordium. The investigator can then design experiments to test the possibilities of

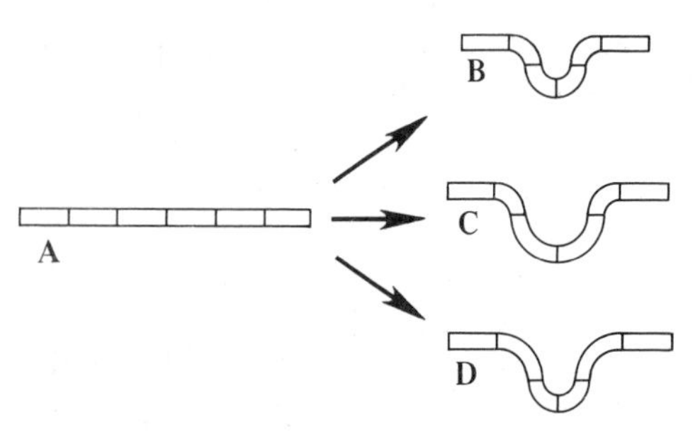


**FIG. 13–2.** Computer representation of three possible ways that a flat epithelial sheet (A) could be shaped into a pitlike primordium (B). Constriction could occur at both apices and bases at sites of bending. If the apical surface does not change (C), the basal portion must increase at the pit and constriction must occur at the basal margins of the primordium. The primordium will be larger than in B. If the basal surface does not change (D), constriction will occur at the apices in the pit and expansion of the apical surface will take place at the margin, leading to a deeper structure than in B. (From S. Hilfer and E. Hilfer, 1983. Dev. Biol. 97:444.)

how shape changes are generated. Figure 13–2 shows a computer representation of three possible ways by which a flat, epithelial sheet could be shaped into a pitlike primordium.

## Individual Cells and Their Movements

Although the spreading, folding, and branching of cohesive sheets of cells are instrumental in rearranging cells and thereby giving shape to an organ, individual cells or clusters of cells are an important source of materials for organ formation. Mesenchymal cells, primordial germ cells, and neural crest cells migrate, often over considerable distances from their point of origin, to accumulate at specific sites in the embryo where they will aggregate to give rise to organ rudiments or disperse and participate in organ construction. Cells may migrate as individual units or as clusters, crawling over a substratum that may be the basal lamina underlying an epithelium or the extracellular matrix between epithelia. Properties of embryonic substrata are important in determining when cells begin to move, where cells move, and when cells stop moving.

Both mesenchyme and neural crest cells are examples of individual cell types produced by the breaking up or disaggregation of epithelial tissues (Figs. 13–1 E; 13–3). Neural crest cells arise dorsally from the epithelial cells at the lateral edges of the neural plate. They detach from the ectodermal epithelium shortly after closure of the neural tube across the midline and then become organized about the neural tube as an axial ribbon of tightly clustered cells. Prior to this segregation, presumptive neural crest cells are morphologically distinct from presumptive ectodermal cells at all stages of neural tube development; they are, however, very similar to cells of the neural epithelium. As the neural crest cells become segregated and discon-

**FIG. 13–3.** A, SEM showing events preceding neural crest migration. During neural fold elevation, the presumptive ectoderm (E) and presumptive neural epithelium become closely apposed. Note that the mesoderm is excluded from the area of apposition. (From K. Tosney, 1982. Dev. Biol. 89:13.) B, diagram of a cross section through an early vertebrate embryo showing the two alternative pathways for the migration of neural crest cells. One (A) dorsolateral stream leads to the formation of epidermal pigment cells. The other (B) stream leads to the formation of posterior root ganglia, autonomic nervous system ganglia, and cells of the medulla of the adrenal gland.

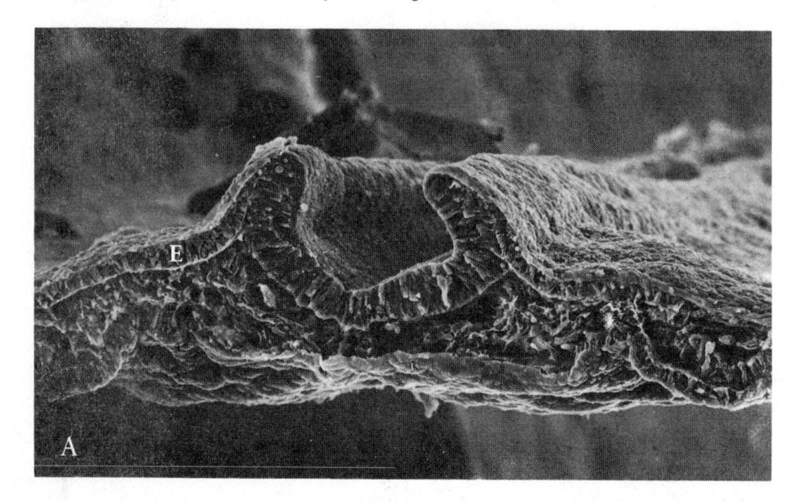


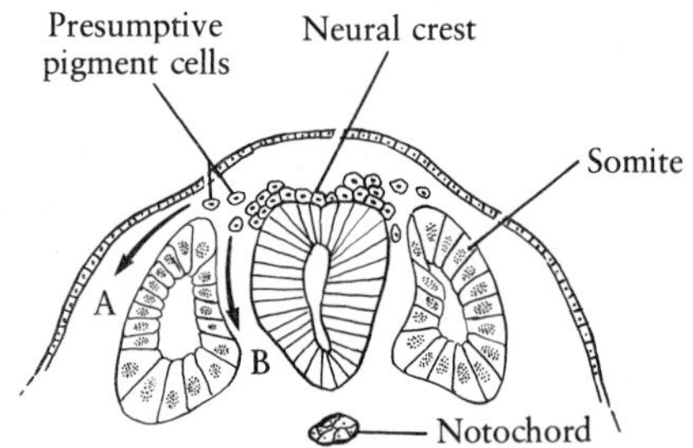

nected from the apical surface of the neural epithelium, they show extension of cell processes along their basal surfaces and disruption of the adjacent basal lamina. Subsequently, the neural crest cells migrate away from the different levels of the axis in two highly specific pathways and become localized at particular sites in the embryo (Fig. 13–3 B). A dorsal stream of cells moves laterally and glides beneath the overlying ectoderm (skin). A second stream of cells migrates laterally and ventrally to a position beneath the neural tube and the somites. A variety of techniques, including vital staining, grafting of cells with known nuclear markers, and transplantation of [$^3$H]thymidine-labeled neural tube fragments, have provided useful information on the migratory pathways and diversity of fate of crest cells. Pigment cells of the dermis, nerve cells of the spinal ganglia, ganglia of the sympathetic and parasympathetic divisions of the nervous system, and the cartilages of the embryonic pharynx are all derivatives of neural crest cell populations.

Since neural crest cells follow different pathways during cellular migration, invade and aggregate in specific regions of the embryo, and differentiate into such varied cell types, they afford a unique opportunity to study mechanisms of cell migration, cell proliferation, cell adhesion, and cell differentiation during embryogenesis. Do individual cells migrate to specific sites because of their intrinsic genetic properties, or do they utilize preformed, directional pathways provided by other tissues or the extracellular material? Why do cells stop at a specific target site in the embryo? Are they directed to a specific target because of the special properties of that site? Several hypotheses have been put forth to account for the remarkable precision by which neural crest cells reach their target sites; these include the ideas that crest cells may preferentially respond to some chemcial substance produced by a target site *(chemotaxis)*, migrate over a gradient of adhesiveness in the substratum *(haptotaxis)*, or be directed by a structurally oriented substratum *(contact guidance)*.

Recent data obtained from several laboratories indicate strongly that crest cells are not predetermined to migrate to specific destinations. Rather, it appears clear that the environment through which crest cells move influences the initial migration, the extent of the migration, and the final cell distribution. Neural crest cells move initially away from the neural tube into a large cell-free space. During this migration, the cells appear to remain as a confluent multicellular layer. Transmission and scanning electron micrographs of amphibian, chick, and mouse embryos show that this space is filled with a complex meshwork of extracellular fibrillar bundles (about 50–100 nm in diameter) and electron-dense granules (Fig. 13–4). Several macromolecular components have been identified in this matrix. These include *glycosaminoglycans, fibronectin,* and *collagen* (see following section on extracellular matrix). In amphibian embryos (axolotl), the fibrils are oriented and aligned in such a way as to suggest that crest cells are guided in their movements away from the neural axis; however, the meshworks of fibrils do not appear to show any prevailing orientation that would guide and direct cell movements in chick embryos. Crest cells in the chick embryo tend to orient toward areas abundant in matrix components. For example, the majority of the processes extending from crest cells adjacent to the mesencephalic portion of the brain adhere to the extracellular materials of the ectodermal undersurface where fibrillar bundles and electron-dense bodies are concentrated.

Neural crest cells have an inherent motility. However, they also require for their migration a cell-free space rich in extracellular materials. The properties of the routes of migration and the roles of the different macromolecular components of the matrix in allowing or guiding crest cell movements have yet to be clearly articulated. There is abundant evidence indicating an important role for the environment in influencing crest migration. When postmigratory, neural crest-derived cells are isolated and reimplanted onto the ventral migratory route of a younger embryo, they actively migrate. However neural crest cells implanted into older embryos do not migrate as extensively as when implanted into the same region of a younger host. Hence, the extracellular environment may change with time and thereby affect the extent of migration. The migratory pathways appear to interact uniquely with neural crest-derived cells, suggesting that the matrix also determines the selection of cell types to migrate along the neural crest pathways. Somite and fibroblast cells, for example, fail to migrate when grafted onto the ventral pathway. If early neural crest cells from the cranial and trunk levels of the neural axis are exchanged, the cells migrate in pathways characteristic of their origin rather than site of implantation.This suggests that populations of neural crest cells from different axial levels of the embryo may have different prop-

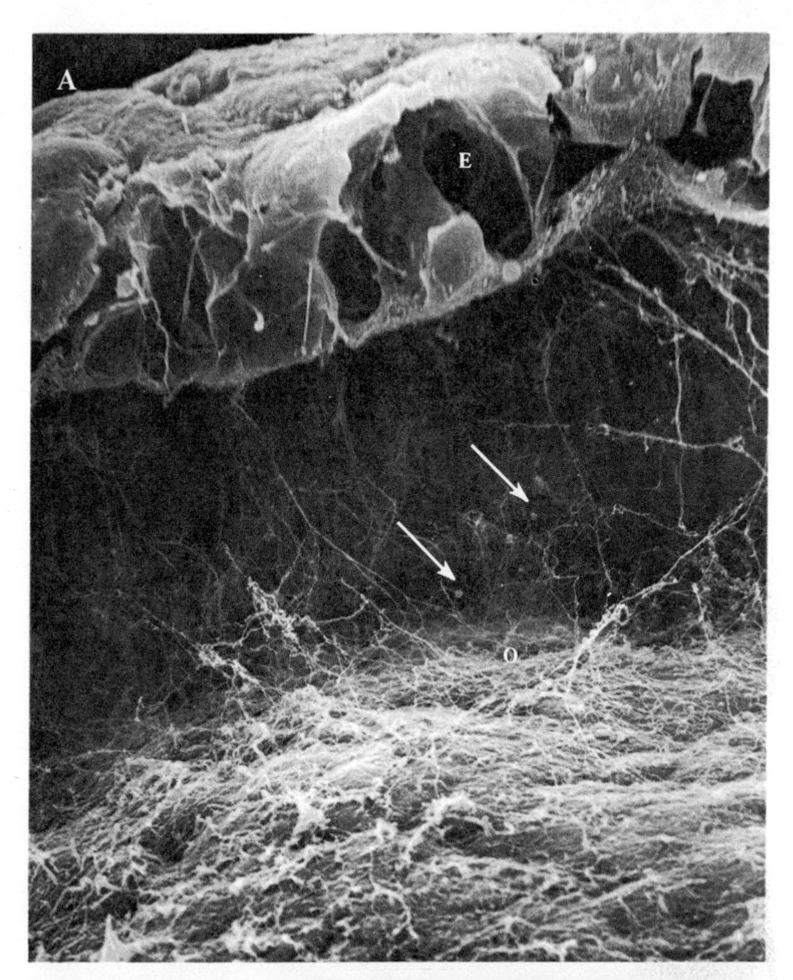


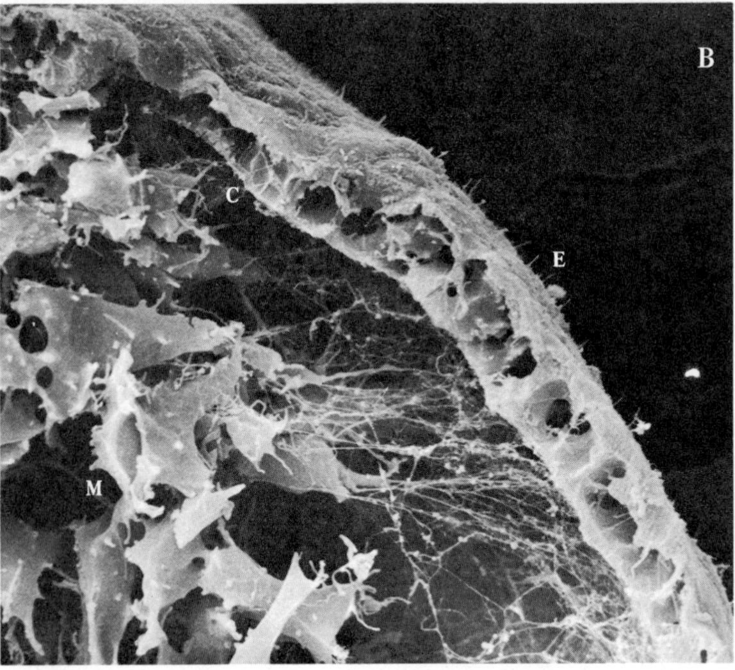


**FIG. 13–4.** Neural crest migration. A, crest cells migrate into a cell-free space filled with fibrils and often studded with interstitial bodies (arrows). At the level of the anterior mesencephalon, these fibrils extend between the ectoderm (E) and the optic epithelium (O). B, SEM showing crest migration at the level of the lateral mesencephalon. Note the paucity of extracellular matrix; crest cells appear to disrupt the extracellular matrix as they migrate into it. As the crest cells migrate, the fibril bundles appear to become oriented toward the direction of crest population movement. C, crest cells; M, mesodermal cells. (From K. Tosney, 1982. Dev. Biol. 89:13.)

erties. Neural crest movements would appear to be the result of complex interactions between the cells and their environment.

Neural crest cells migrate over long distances and remain in well-defined extracellular pathways until they reach their target site. After their arrest, the cells continue to proliferate and thereby induce the aggregation of cells into organ rudiments. Certain components of the extracellular matrix have been implicated to play different roles in the migration and aggregation of crest cells. Hyaluronic acid, for example, is actively synthesized and released by crest cells; it is postulated to produce the cell-free space between the tissues through which the cells must pass. Collagen is also present in the fibrils of the cell-free space. In culture, neural crest cells attach to collagen substrates. The attachment is mediated by the adhesive molecule known as fibronectin. Indirect immunofluorescence studies show that migrating crest cells generally do not contain fibronectin; however, much of the fibrillar network along the crest pathways contains high amounts of fibronectin. When cultured neural crest cells grown on collagen substrates are treated with various concentrations of fibronectin, the molecule appears to stimulate motility of the cells and even serve as a chemoattractant for them. It is hypothesized that fibronectin is synthesized and secreted by tissues surrounding the migratory pathways. If produced in a regional and temporal pattern, conditions (for example, a gradient of adhesiveness) might be set up that could account for the observed directionality of neural crest cell movement. Cells would tend to migrate toward sites where they can adhere more strongly. Conversely, the disappearance of fibronectin from the extracellular environment at the end of the migratory pathway would appear responsible for the aggregation of crest cells at the target site. By removing fibronectin from the extracellular matrix, the environment may alter the predominantly cell-to-substrate adhesion during migration to a cell-to-cell adhesion during aggregation. Fibronectin, therefore, may be an important informational macromolecule that, when complexed with collagen, provides a suitable substratum for migration and adhesion. It should be pointed out that all migrating neural crest cells are not in contact with a collagen substrate. Some are in contract only with other crest cells. Therefore, some crest cells may move passively as a result of *contact inhibition* or mutual repulsion.

Mesenchyme cells are individual, stellate cells that migrate away from mesodermal epithelia and aggregate at specific target sites in the embryo. Here, they differentiate as rudiments of organs (e.g., cartilage and bone cells of the skeletal system) or as connective tissues for the support of the epithelial components of organs. The movements of individual mesenchyme cells have been studied in several systems, particularly at gastrulation in the sea urchin embryo using cinemicrography (Chapter 8). Prior to the onset of gastrulation, the primary mesenchyme cells of the vegetal plate begin locomotor activity by pulsating and forming hemispherical protrusions called blebs. Following their separation from neighboring cells as well as from each other, presumably because of a decrease in cell-to-cell adhesiveness, the mesenchyme cells round up and move into the blastocoele. Each cell sends out long, thin dendritic pseudopodia or filopodia whose tips probe the inner surface of the blastocoele wall. When an adhesive contact with the blastocoele wall is made (i.e., at the junction between ectodermal cells), the filopodium contracts and thus pulls the cell along. The movement of these primary mesenchyme cells along the inner surface of the ectoderm appears to be rather like an inchworm, accomplished by the filopodia alternately making and breaking contacts with the cellular substratum. During gastrulation there develops a voluminous, fibrillar matrix beneath the basal lamina of the ectodermal cells. The matrix appears to be constructed of glycosaminoglycans and perhaps collagen. The extracellular materials may regulate and guide the movements of the mesenchyme cells along the inner blastocoelic surface by controlling both adhesion and cell motility.

Studies on the migratory behavior of mesenchyme cells in the intact vertebrate embryo have been rather limited. One organ that has received considerable attention is the eye of the chick. During the third day of avian embryonic development, the lateral head ectoderm overlying the lens vesicle becomes the *cornea epithelium*. The basal cells of this two-layered epithelium gradually become columnar and rich in secretory organelles. Between four and five days, the cornea epithelium deposits a *primary stroma* abundant in collagen fibrils between itself and the lens vesicle. During this time neural crest cells migrate between the primary stroma and the anterior lens to establish the corneal monolayer endothelium (i.e., blood vessels).

The presumptive endothelial cells, although somewhat flattened, appear to move into the eye by the extension of numerous filopodia. The blood vascular tissue initiates the synthesis of hyaluronate, a high-molecular-weight polymer of a disaccharide carbohydrate, which causes increased hydration and swelling of the stroma. As the stroma swells (beginning at day 7), mesenchyme cells derived from the neural crest and destined to give rise to the corneal fibroblasts invade the primary stroma. These fibroblasts will contribute additional collagen that will characterize the *secondary* or *mature corneal stroma*. During their migration, the spindle-shaped fibroblasts extend numerous branched filopodia. Locomotion appears to be accomplished by the flow of cytoplasm from the cell body into certain filopodia and not into others. There is a striking correlation between hyaluronate synthesis and movement into the stroma by the neural crest cells. Cell migration into the stroma ceases as the enzyme *hyaluronidase* appears in the cornea tissue The stroma then dehydrates under the influence of thyroid hormone and becomes transparent. The cornea provides a useful example of the regulation of individual cell movements by alterations in the composition of the extracellular matrix.

A particularly interesting observation is that these corneal fibroblasts, when isolated and cultured in vitro on a glass substratum, behave like typical fibroblasts. Ordinary fibroblasts are the predominant cell type in connective tissues; they differentiate from mesenchyme cells. When observed in culture under phase microscopy, fibroblasts glide across the glass surface without filopodia formation or internal cytoplasmic flow. During movement, the fibroblast is in contact with the substratum only at its leading edge (i.e., the direction of movement) and its trailing edge. The leading edge of the cell is drawn out into a broad, thin, fanlike membrane known as a *lamellipodium*. Locomotion depends on the advance of the lamellipodium over the substratum. Localized protrusions or *ruffles* along the margin of the lamellipodium undulate and intermittently make and break contacts with the surface over which the cell moves. The cell typically moves in the direction from which a major lamellipodium extends from its surface. As the leading edge of this lamella advances, the cell becomes elongate in that direction because its trailing edge remains adherent to the substratum for a time.

Differences in the migratory behavior of the chick eye fibroblast cells, when observed in situ and in vitro, point out the importance of the substratum and its influence on cell form. When these same cells were cultured on a collagen gel, they assumed a spindle shape, formed filopodia, and behaved the same way as in vivo. This lends support to the suggestion that a collagen gel may very well simulate the embryonic environment in which the cells normally move.

Contact and adhesion with a substratum are important features in the morphogenetic movements of individual cells whether in vitro or in vivo. As in the case of the folding of epithelial sheets, the locomotion of cell types is invariably accompanied by changes in cell form. A variety of special surface modifications appear to serve as locomotor "organs" and effect the movement of a cell from one place to another place. These structures include filopodia, lamellipodia, and blebs or lobopodia (Chapter 8). Since these structures appear to move the cell by pulling it along, one might suspect that the basis of cell form changes is contractile and effected by the same intracellular organelles previously identified in the filopodia of primary mesenchyme cells, in the lobopodia of *Fundulus* deep cells, and in the lamellipodia of fibroblast cells. The availability of antibodies to actin and other major cytoplasmic structural proteins has enabled investigators to use indirect immunofluorescence to follow the distribution of these molecules under conditions of individual cell movement. Using this technique with a variety of fibroblast cells, Lazarides has demonstrated the presence of actin filaments in the membrane ruffles of lamellipodia. The actin filament pattern observed with immunofluorescence corresponds to the microfilament pattern observed with electron microscopy. Also, microtubules have been seen to extend from the main part of the fibroblast into the main body of a lamellipodium, thus indicating that they are probably the structural supporting elements of this type of membrane extension. The conclusion is warranted, therefore, that individual cells move by contractility that is effected through a complex intracellular system of filaments.

## The Extracellular Matrix

Within recent years, there has been an increased awareness of the importance of the extracellular environment or matrix in both morphogenetic and cytodifferentiative phases of organ formation. It has been pointed out previously that organogenesis frequently involves interactions between cells from diverse embryonic territories. Such interactions are the result of spatially and temporally coordinated events involving cell migration, proliferation, adhesion and differentiation, and so on. A variety of substances, including a complex network of macromolecules, water, and ions, is found between the interacting tissues. Collectively, these constitute the *extracellular matrix*. As seen from our previous discussion of neural crest migration and the development of the cornea, these materials of the extracellular matrix appear to exert an influence over such morphogenetic processes as migration and short-range inductive interactions between groups of cells. This medium also acts as a vehicle through which a variety of chemical messengers, such as nucleic acids, proteins, and hormones, pass and function to alter the activities of neighboring cells.

The extracellular matrix fills all the embryonic space except that occupied by cells and lumina. Its components are becoming better understood in relation to cell and tissue interactions because of technological advances in the analysis of proteins and carbohydrates. Typically, the matrix is organized into a distinct basement membrane or basal lamina beneath epithelia and a series of surface-associated fibrous materials. Basal laminae are seen under the transmission electron microscope as sheets of dense and filamentous substance. The chief macromolecular constituents of the extracellular matrix are *collagen, glycosaminoglycans (GAGs),* and *glycoproteins*. Collagen is the most common structural protein found in tissues of higher vertebrates. The collagen molecule, *tropocollagen,* consists of three subunits called $\alpha$ chains. Each $\alpha$ chain has a molecular weight of about 95,000 and contains slightly more than 1000 amino acids. Based on other criteria, five types of $\alpha$ chains have been identified; known combinations of these chains produce four types of collagen (called type I, type II, etc.). Each of the collagen types is fibrous in form; fibers vary in both length and thickness. Glycosaminoglycans, common constituents of the extracellular matrix, are large molecules containing linear polymers of monosaccharides covalently linked to polypeptides (Fig. 13–5 A). The protein is generally considered to be a minor component. Common GAGs include *hyaluronate, chondroitin sulfate, heparan sulfate,* and *keratan sulfate*. With the possible exception of hyaluronate, most glycosaminoglycan chains are attached laterally to a pro-

A

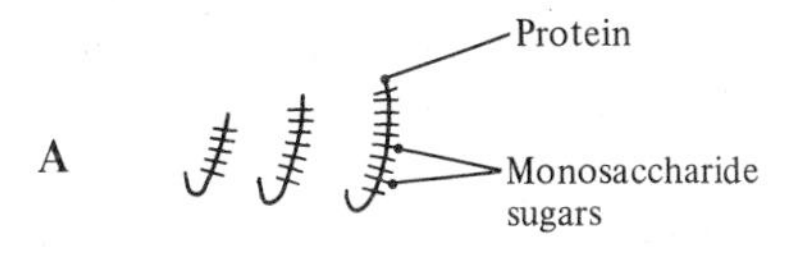


B

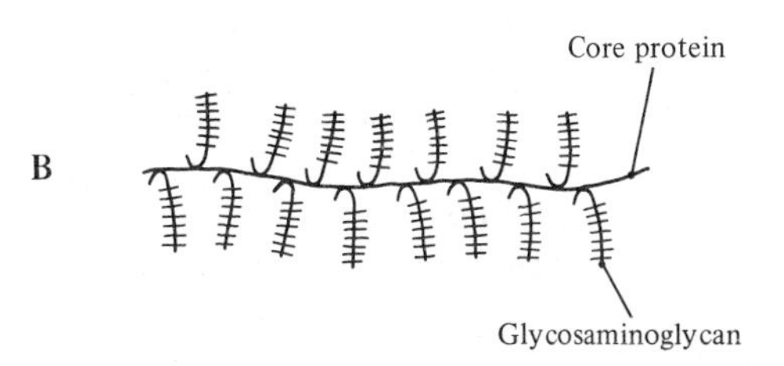


C

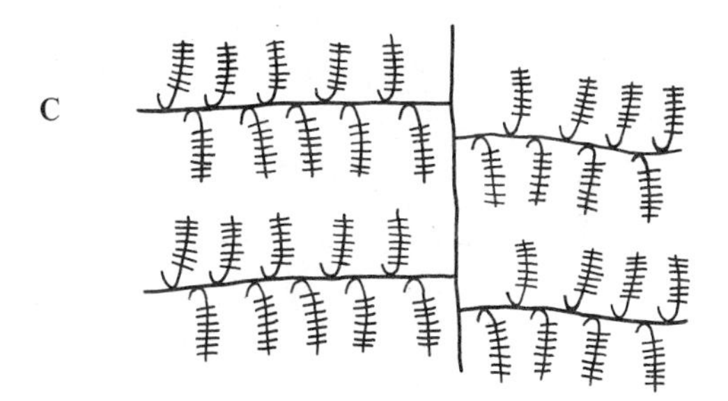

**FIG. 13–5.** Diagrammatic representation of the structure of glycosaminoglycan (A) and proteoglycan (B) molecules. Proteoglycans may form complex aggregates (C).

tein core to form another macromolecule known as a *proteoglycan* (Fig. 13–5 B). Many proteoglycans are capable of forming multimolecular aggregates of very high molecular weight (Fig. 13–5 C). Glycoproteins are molecules containing polysaccharides covalently linked to protein, but, in contrast to GAGs, the sugar component is relatively small. Two predominant glycoproteins have been identified in the extracellular matrix: *laminin* and *fibronectin*. Laminin is an important glycoprotein of the basal lamina. Fibronectin is a molecule with multiple sites for binding to other macromolecules of the extracellular matrix; it is associated with basal laminae and areas beneath epithelia. The possible in vivo functions of glycoproteins are being studied intensively. Fibronectin is of special interest. It appears to mediate and promote the attachment of different cell types, such as neural crest and muscle, to collagen, regulate cell shape, and guide cell migration. Muscle cells in vitro, for example, adhere to, align with, and move along oriented matrices of artificial fibronectin.

The embryonic extracellular matrix is a complex and dynamic entity. Combinations of collagen, proteoglycans, and highly ordered arrays of GAGs produce a diversity of fibrous complexes over and between epithelial surfaces. Matrices appear to exhibit spatial and temporal variation in the dimensions of these meshworks and in the composition of their constituents. Alterations in the distribution and in the concentration levels of one or several components presumably produce profound changes in the matrix and the morphogenetic activities influenced by it. For example, there is a marked increase in the synthesis of GAGs just prior to the emigration of primary mesenchyme cells from the vegetal plate during echinoderm gastrulation. A sudden burst of hyaluronate synthesis precedes neural crest-derived mesenchyme cell movement into the developing cornea. Localized changes in the synthesis of GAGs appear to be an important mechanism in the regulation and facilitation of cell migration. Alternatively, cessation of movement of presumptive fibroblast cells into the cornea is linked to a decline in hyaluronic acid production and an increase in the synthesis of hyaluronidase. Observations such as these are particularly provocative, because they suggest that the migratory activities of embryonic cells, and hence the stability of tissue configurations of organs, are controlled by localized fluctuations in the composition of the extracellular matrix. Information about the extracellular matrix and its role in the development of most organ systems is still very inadequate. As the properties and composition of the matrix of a developing organ become known, it may be possible to grow cells of the organ in an artificial matrix identical to its in situ counterpart; the addition, replacement, or deletion of selected matrix components in a controlled manner could then provide insight into their morphogenetic roles during organogenesis.

The extracellular matrix has been clearly implicated in the branching morphology of many tubular type organs, such as the salivary glands, lungs, and kidneys. Branching involves the epithelial component of such an organ. It is repetitive and continues until the complex, definitive shape of the organ is achieved. Generally, the particular pattern of branching is characteristic of any given organ. Since the morphogenesis of the salivary gland has received considerable attention, let us consider it now as an example of a branching organ requiring the participation of both collagen and glycosaminoglycans.

Each salivary gland arises as a small epithelial bud in the oral cavity that pushes into the underlying mesodermal mesenchyme (Fig. 13–6). The early primordium consists of a bulbous distal tip that joins the oral cavity by a stalk or primary duct. Subsequently, the epithelium undergoes substantial growth and repetitive branching. Branching is initiated by an invagination that deepens into a distinct cleft at the bulbous tip of the primordium, thereby separating it into two lobes. A cleft then appears in the tip of each lobe, and the level of branching is increased. The dual processes of growth, supported by mitosis, and cleft formation

**FIG. 13–6.** A, a schematic drawing of mouse salivary gland epithelial morphogenesis. The in vivo development of the gland takes about three days from the time of the initial appearance of the rudiment (After B. Spooner, 1973. Am. Zool. 13:1007); B, photograph of a 13½-day-old submandibular salivary gland of the mouse showing glycosaminoglycans on epithelial surface (dark line at base of arrow). C, interlobular clefts; E, epithelium; M, mesenchyme. (From M. Bernfield, et al., 1973. Am. Zool. 13:1067.)

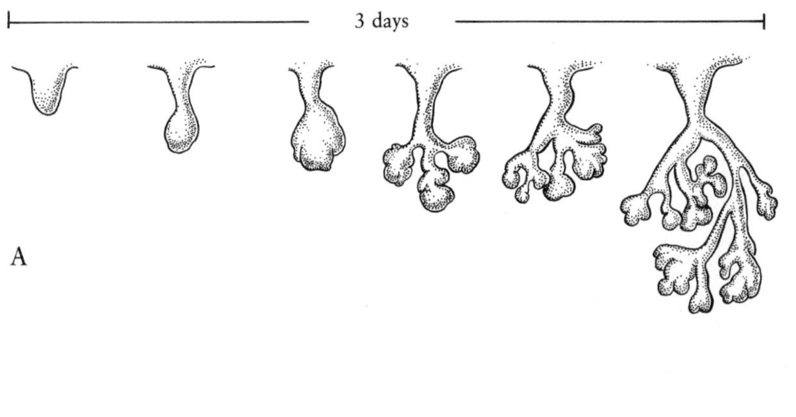


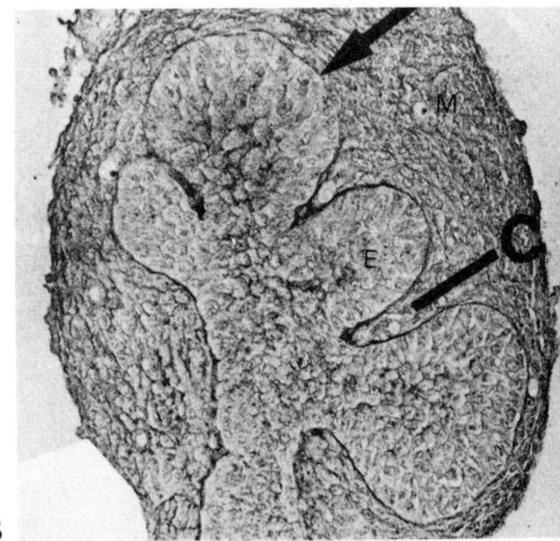

establish the branching pattern of the salivary gland. The process of cleft formation is accompanied by a change in cell shape (i.e., cells become wedge-shaped). This alteration in cell shape during epithelial branching is presumably due to contraction of a system of filaments. Treatment of cultured salivary glands with cytochalasin B stops cleft formation and blocks further branching of the epithelial component (Fig. 13–7). Studies by Spooner with the electron microscope have shown bundles of microfilaments concentrated in basal ends of epithelial cells or in the outer surface of each distal branch. Although microtubules are also present in the epithelial cells, they are not required for salivary cleft formation. Treatment of cultured salivary glands with colchicine or colcemid stunts their growth but does not interfere with the clefts present in the epithelium.

The importance of salivary gland mesoderm in epithelial tube morphogenesis has been demonstrated in several types of experiments by Spooner (Fig. 13-8). If the salivary epithelium is enzymatically isolated from the underlying mesoderm and then grown in culture, it rounds up and fails to branch (Fig. 13-8 A). When the epithelial primordium is combined in organ culture with salivary mesoderm on one side and lung mesoderm on the other side (Fig. 13-8 B) branching morphogenesis occurs only in the portion of the epithelium associated with the salivary mesoderm. Although the epithelium does not branch where it is in contact with the lung mesoderm, a basal lamina and deposited GAGs can be identified at the interface between the two tissues. Also, microfilaments are identified in the epithelial cells associated with the lung mesoderm, suggesting that salivary mesoderm is not specifically required for the continued presence of these organelles. However, salivary mesoderm is absolutely required for epithelial morphogenesis. It induces a very specific pattern of branching in the epithelium by controlling microfilament activity.

Extracellular materials between the salivary gland epithelium and the salivary gland mesoderm appear to mediate the interaction between these two tissues. Histochemical and autoradiographic localization techniques show the matrix between the epithelial rudiment and the mesenchyme to consist of collagen and GAGs. Although the mesenchyme labels heavily with precursors of collagen, such as [$^3$H]proline, very little label is localized in the epithelial basal lamina. These observations suggest that collagen is not a major component of the basal lamina of the embryonic salivary gland. If the salivary gland is cultured in L-azetidine-2-carboxylic acid, a proline analog that interferes with collagen production, branching is inhibited. Treatment of salivary epithelia undergoing morphogenesis with *collagenase,* an enzyme that digests away collagen, also stops the branching process. Although these data suggest that collagen is crucial to branching, its precise role is still not well established. There is speculation that the collagen fibrils function to stabilize the older formed clefts and branches.

With other kinds of enzymatic treatments, it is possible to remove specific GAG components selec-

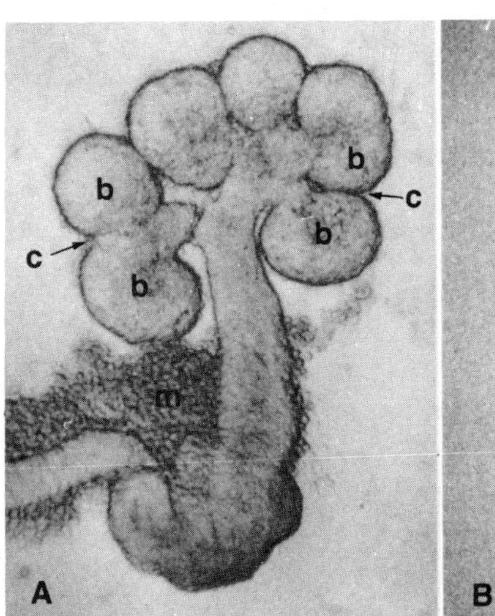


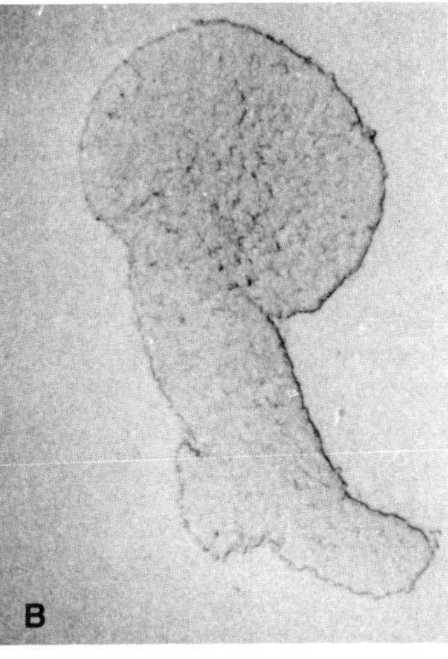


**FIG. 13–7.** A, photomicrograph of living salivary gland epithelium after removal of the surrounding mesoderm (m). Note the clefts (c) and branches (b). B, photomicrograph of salivary gland epithelium that has been treated with cytochalasin B. Note the absence of clefts. (From B. Spooner, 1973. Am. Zool. 13:1007.)

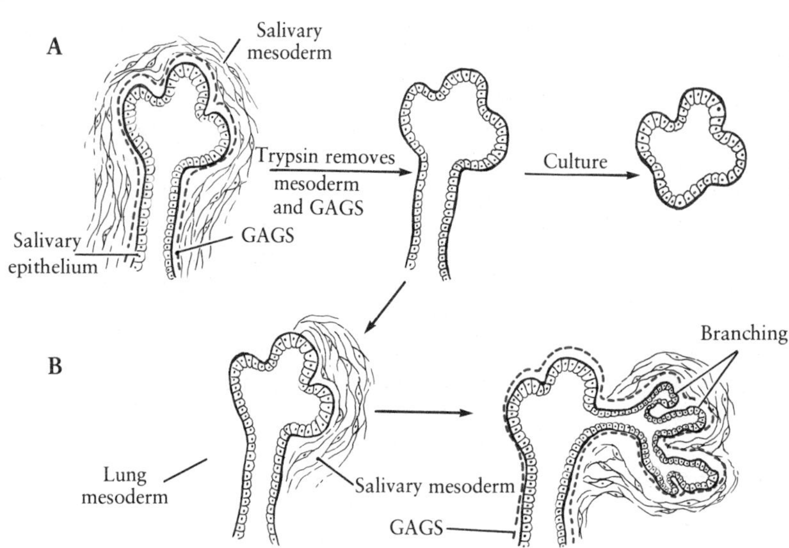


**FIG. 13–8.** Schematic illustration of experiments by Spooner showing that isolated salivary epithelium fails to branch when divested of its own mesoderm (A) or associated with a foreign mesoderm such as lung mesoderm (B). GAGs, glycosaminoglycans. (After B. Spooner, 1973. Am. Zool. 13:1007.)

tively from the extracellular matrix in order to determine which are important to the branching process (Fig. 13–9). When GAGs are removed from the basal surface of the epithelial rudiment by trypsinization, branching morphogenesis ceases; immediate recombination with mesenchyme fails to restore epithelial morphology (Fig. 13–9 B,E), Morphogenesis resumes some 48 hours after the epithelium is combined with salivary mesenchyme and new GAGs have been resynthesized (Fig. 13–9 B,E). If epithelial rudiments are first treated with low concentrations of pure collagenase (which does not remove GAGs) and then exposed to either hyaluronidase or chondroitinase, branching is inhibited. All of these treatments disrupt the basal lamina. The subsequent addition of an investing layer of mesenchyme restores branching of the epithelium (Fig. 13–9 C–E). The structural integrity of the basal lamina appears to be an important requirement for branching. Labeling studies show the basal lamina of the salivary gland to contain hyaluronic acid, chondroitin sulfate, and heparan sulfate. These GAGs are present in highly ordered arrays intimately associated with the plasmalemma.

What is the mode of action of these GAG molecules in controlling epithelial morphogenesis? A hypothetical model constructed by Bernfield and his colleagues from their studies explaining the relationship between the extracellular materials and branching is shown in Figure 13–10. A continuous basal lamina containing newly synthesized GAGs lies beneath the basal side of an epithelial bud or primary lobule (Fig. 13–10 A). On either side of such an epithelial bud, GAGs accumulate at a very low rate, but well-defined bundles of collagen fibers are visible. All cells of the epithelium contain apical (inner) and basal (outer) microfilaments. Contraction of the basal microfilaments at the tip of the bud produces a localized group of wedge-shaped cells, thereby generating an infolding or cleft (Fig. 13–10 B). As the cleft deepens, the primary bud is converted into two branches or secondary lobules. Collagen fibers, whose precursors are synthesized in part from the adjacent mesenchyme, are drawn into the cleft where they act to stabilize the branched morphology. Continued cleft formation is due to contraction of the microfilaments and a high level of mitotic activity in the cells of the secondary lobules (Fig. 13–10 C). The entire sequence is repeated as each new cleft is formed. The disappearance of GAGs from the clefts is an essential step in the branching process, and it may be related to the release of hydrolytic enzymes from the salivary mesenchyme cells.

The relationship between the contraction of microfilaments and the GAGs of the basal lamina during branching morphogenesis is very speculative. Calcium ions are considered to have a regulatory role in the contractile systems of both muscle and nonmuscle cells. *Papaverine,* a drug that blocks smooth muscle contraction by interfering with the flow of calcium ions, inhibits the morphogenesis of salivary glands in organ culture. It is postulated, therefore, that GAGs, which can bind calcium ions, control the morphogenetic events at the tips of branching epithelial tubules by regulating the availability of this cation.

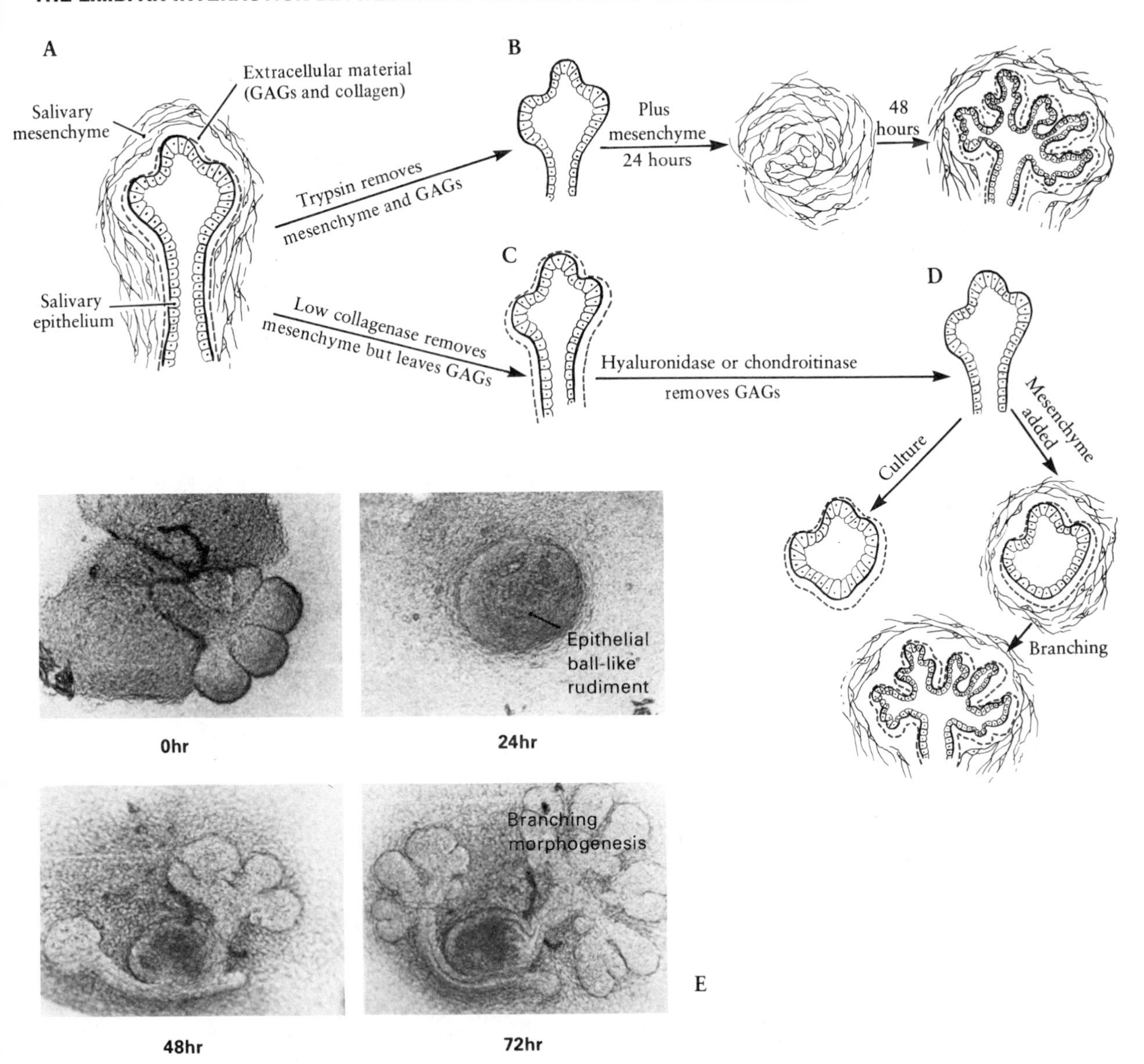


FIG. 13–9. The role of extracellular materials in the branching morphogenesis of salivary epithelium. A, isolated gland rudiment consists of an epithelial bud surrounded by mesenchyme; B, trypsin treatment of rudiment removes the mesenchyme and the glycosaminoglycans (GAGs). The epithelial bud initially rounds up, but it will resume branching some 48 hours after being reassociated with salivary mesenchyme. C, low concentrations of collagenase remove the mesenchyme but leave the layer of GAGs intact. Without the addition of mesenchyme, the epithelium treated in this way loses its branching pattern. D, if C is followed by treatment with hyaluronidase or chondroitinase, all GAGs are removed and the epithelium rounds up unless combined with salivary mesenchyme; E, photograph of trypsin-isolated living salivary epithelium showing its response to contact with fresh mesenchyme (0 hr) after 24, 48, and 72 hours. GAGs are resynthesized, but there is no branching without the mesenchyme. (From M. Bernfield et al., 1973. Am. Zool. 13:1067.)

## The Limb: An Interaction Between an Epithelial Sheet and Mesoderm

During the past several decades of analyzing the development of organs, it has become evident that many organs during their formation require complex inductive interactions between an epithelial tissue, generally of ectodermal or endodermal origin, and a mesenchymal tissue. An exchange of signals between these two types of tissues takes place in such varied organs as the salivary glands, lungs, lenses, liver, pancreas, skin, and limbs.

The vertebrate limbs have received particular attention from investigators because their develop-

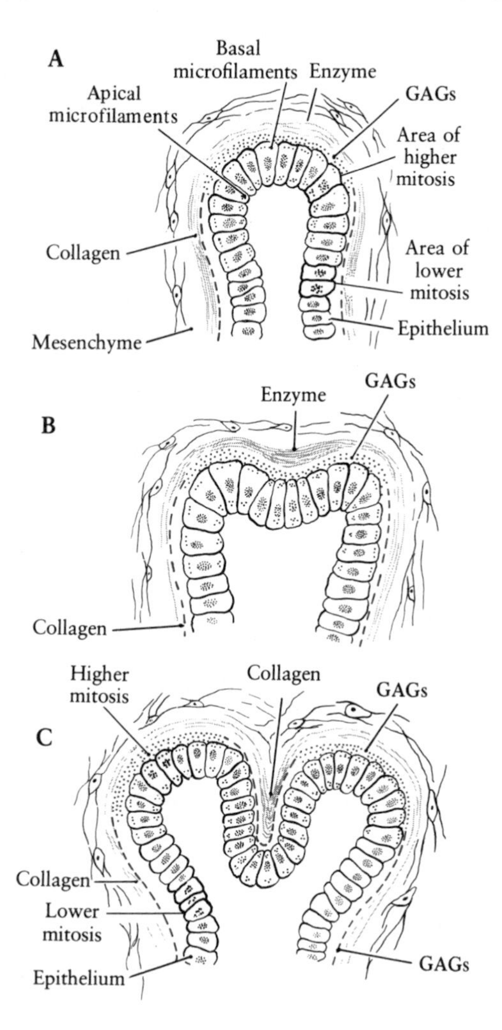


**FIG. 13–10.** Schematic model according to Bernfield and colleagues depicting the relationships between extracellular material and cleft formation in a branching organ. A, the tip of an epithelial bud is a major site of GAG synthesis. Newly synthesized GAGs accumulate in the basal lamina region but turn over rapidly, perhaps owing to "remodeling" by hydrolytic enzymes. B, early cleft formation is due to the contraction of microfilaments in the epithelial cells at the tip of the bud. Collagen fibers begin to form at the top of the cleft because of the regional synthesis of tropocollagen by mesenchymal cells. C, deepening of the cleft and the formation of secondary buds or lobules. The branch points are now stabilized by bundles of collagen. The tips of the secondary lobules show the bulk of newly synthesized GAGs. These are the sites for the next clefts. (From M. Bernfield et al., 1973. Am. Zool. 13:1067.)

ment exhibits all of the classical phenomena associated with organ formation: morphogenesis and growth, histodifferentiation, cytodifferentiation, sequential inductions, gene control, and self-regulation. Also, the limb, particularly of the chick, is a system chosen by many to examine the process of pattern formation. *Pattern formation* refers to the spatial relationships of the various parts of the developing form of an organ. For the limb, this focuses on the three-dimensional arrangement of the bones, muscles, nerves, and blood vessels. The primordia of the limbs are rather large, at least in avian embryos, and special techniques permit their manipulation for study of the interrelationships among differentiation, morphogenesis, and pattern formation. We will examine the development of the paired limbs in some detail here and not return to them again.

## The Limb Field and Limb Bud

The initial signs of paired limb development in vertebrates are internal and visible in the lateral plate

**FIG. 13–11.** Diagrammatic representation of the origin of the limb mesenchyme from the lateral plate mesoderm (A) and its association with the ectoderm to form the limb bud (B) in a typical vertebrate embryo.

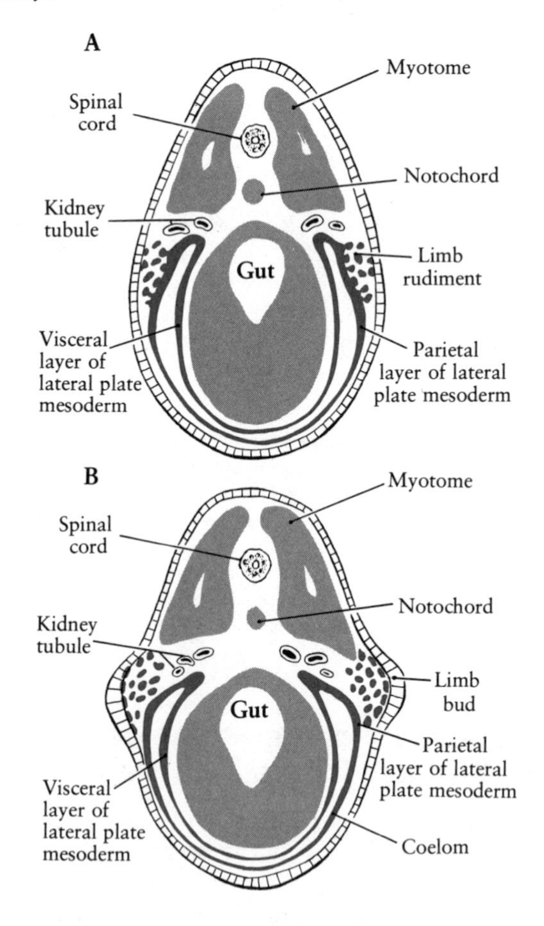

mesoderm on either side of the embryo (Fig. 13–11). The upper edge of the somatic layer of the lateral plate mesoderm thickens. Gradually, many of these cells lose their association with the mesodermal epithelium and migrate laterally beneath the overlying ectoderm. In the case of amphibian embryos, the limb mesenchyme proliferates locally and consequently produces two pairs of discrete, disclike masses *(limb discs)* at sites representing the future pectoral and pelvic appendages. By contrast, the length of the lateral place mesoderm thickens on either side of the amniote embryo and its mesenchyme cells accumulate beneath the ectoderm as a distinct horizontal ridge *(Wolffian ridge)*. Enlargement of each ridge in the pectoral and pelvic positions, coupled with the gradual disappearance of the intermediate portion of the ridge, results in the formation of the definitive *limb buds* (Fig. 13–12). Each limb bud, therefore, consists of a central core of condensed mesenchyme surrounded by an overlying cap of ectoderm.

The limb bud quickly loses its low, swollen appearance. Rapid multiplication of the mesodermal cells transforms the limb bud into an elevated mass projecting from the ventrolateral body wall of the embryo. As the base of the bud narrows and the distal end become flattened, the whole structure appears semicircular in outline (i.e., paddle-shaped). The paddle-shaped part of the limb rudiment will differentiate as the hand or foot (Fig. 13–13). The mesoderm of the bud at this time consists of an outer, compact layer and an inner spongy mass rich in blood sinuses. As the bud continues its outgrowth, the mesodermal cells condense and aggregate in the central core region to block out discrete masses that represent precursors for the various skeletal and muscular parts of the appendage. The mesodermal condensations appear in a fixed order within the limb; proximal ones become organized earlier than distal ones. There has been some controversy over the source of the mesodermal condensations producing muscle tissue in the tetrapod limb. Many authorities have suggested that somites contribute nothing to the muscle masses and that the latter arise solely from nonsomitic sources. However, results of several experiments lead us to believe now that somites contribute presumptive muscle tissue to limb buds in all amniotes. When a block of quail somites, whose cells have a large body of heterochromatin not present in chick cells, is grafted into the axial region of a chick embryo from which the somites have been excised, quail cells are subsequently detected in myogenic condensations but not in the skeletal (precartilaginous) condensations of the limb. Definition of the final contour of the limb is dependent on the interplay of several morphogenetic events, including proliferation and differentiation, cell movements and cell death (see below).

Of the two embryonic tissues that contribute to the construction of the limb, the mesoderm appears to possess the property of "limbness" and is determined very early in development or well in advance of any visible sign of the limb rudiment. This can be demonstrated by excising a fragment of presumptive limb mesoderm from the lateral plate and transplanting it beneath the flank ectoderm. When the mesodermal fragment is taken from embryos shortly after the stage of neural tube formation, its transplantation results in the development of a supernumerary limb. Ectoderm from any region of the embryo at this stage appears fully competent to contribute to the formation of a limb. If the ectoderm of the limb region is excised and the

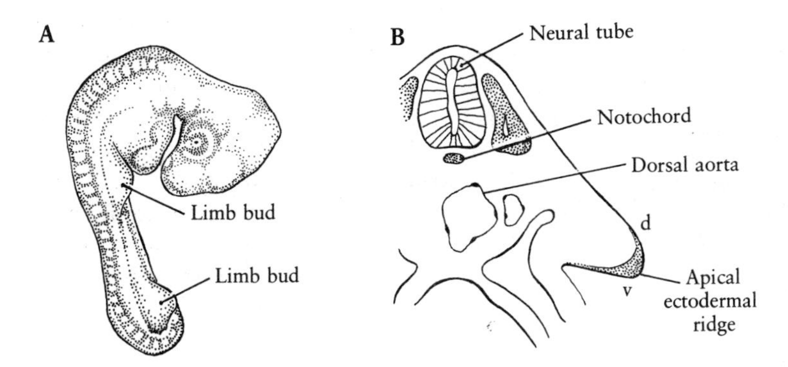


**FIG. 13–12.** A, the limb buds appear as prominent swellings on the body wall of the chick embryo (stage 21); B, cross section through the wing bud of a stage 19 chick embryo. Note the differences in the thickness of the wing bud ectoderm between dorsal (d), ventral (v), and lateral surfaces.

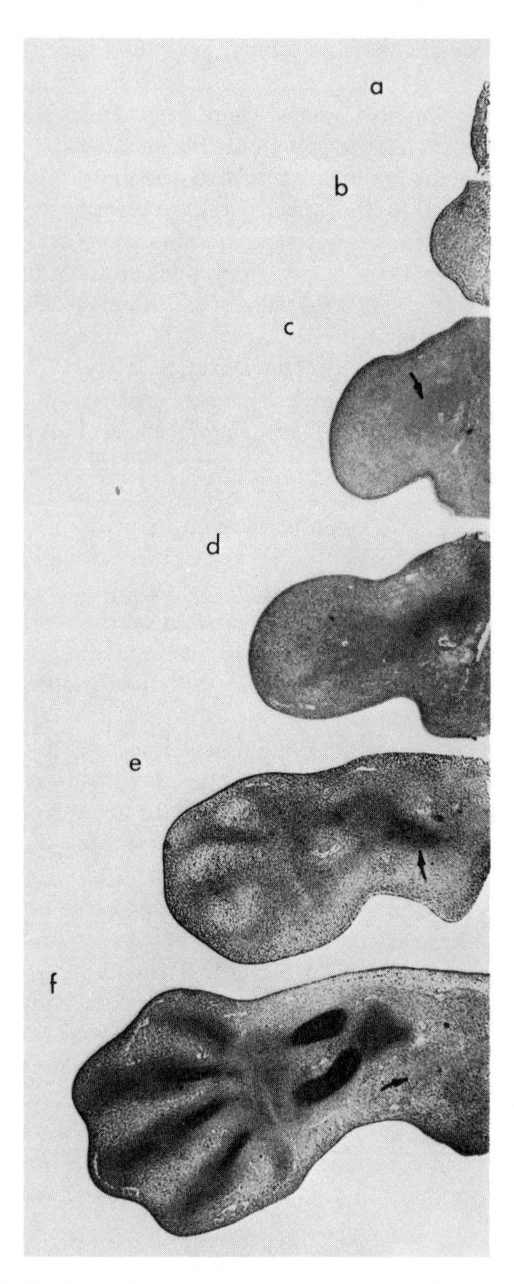


**FIG. 13–13.** A series of stages in the development of the mouse forelimb. A, at day 10 of gestation the limb mesenchyme appears homogeneous; B, at day 11 of gestation the condensation of the humerus is just beginning to form; however, there is no evidence of cartilage matrix as demonstrated by an absence of Alcian blue staining; C, at day 11½ of gestation, the three proximal elements of the forelimb are visible; the humerus contains some Alcian blue staining; D, by day 12 of gestation, the radius and ulna condensations show the beginnings of cartilage formation; E, by day 13 of gestation, cartilage is present in all major skeletal primordia. (From E. Owens and M. Solursh, 1981. Dev. Biol. 88:297.)

presumptive limb mesoderm is then covered by a piece of ectoderm taken from the head or trunk, a normal limb bud will form.

As in the case of other organs, such as the lens and the inner ear, the area of tissues capable of giving rise to the vertebrate limb bud (i.e., prospective limb bud potential) is larger than that area normally contributing to the rudiment of this organ. For example, extirpation of the area normally giving rise to the limb disc in an early-stage salamander embryo is followed after a short delay by the appearance of a limb. Presumably, ectodermal and mesodermal cells adjacent to the site of the extirpated tissues move in and reconstitute the limb disc. If the extirpation is enlarged to include the presumptive limb bud as well as the surrounding cells, no limb will subsequently develop. This larger region, which represents the whole prospective limb bud potential, is referred to as the *limb field*.

The term *primary field* is often employed to describe the general region in an embryo from which a particular organ will develop. Just as in the case of the egg, which can be separated mechanically into two parts with each part capable of forming a miniaturized whole, a primary organ field can be split into two halves and each half will develop into a complete organ. Up until the limb bud stage in amniote embryos, the limb rudiment can be divided into two halves and each half, following transplantation to an appropriate site, will develop into a whole limb. Cells within a field, therefore, appear to recognize their position within the whole. If the field is disrupted by either the removal or addition of cells, there is the property of regulation by the constituent cells such that a normal organ still forms. Another basic property of a primary field is the gradual and progressive determination of its parts. Initially, the parts of a primary field are totipotent and each can form the entire organ. As the potency of a part becomes restricted to its prospective significance, it can then only form a part of the whole organ. Successive determinations subdivide the primary organ field into secondary fields, tertiary fields, and so on until all parts of an organ are fully fixed with respect to their fate. The self-differentiating property of the limb resides in the mesoderm; the acquisition of this property appears to depend on earlier influences from the adjacent somitic mesoderm.

## The Roles of Ectoderm and Mesoderm in Limb Differentiation

During the early growth of the limb bud in amniotes, a sharply defined epidermal thickening appears distally along the edge of the flattened bud (Fig. 13-16 A). The constituent cells of this *apical ectodermal ridge* (AER) originate from the in situ proliferation of the ectoderm as well as from ectodermal cells migrating distally over the dorsal and ventral surfaces of the limb bud. The cells of the ridge are organized as a pseudostratified, columnar epithelium, an organization in distinct contrast to the cuboidal shape of the adjacent, nonapical ridge cells. Ridge cells contain numerous pinocytotic vesicles and oriented microtubules. Observations with the scanning electron microscope now indicate that apical ridges are also characteristic of amphibian limb rudiments.

The AER appears to be induced by the underlying mesenchymal component of the limb bud. When the prospective limb mesoderm from chick embryos (stages 12–17) is isolated and transplanted beneath flank epidermis, it will induce cells of the latter to form an AER and subsequently a supernumerary limb bud. Mesoderm transplanted before stage 12 fails to initiate the development of an AER. It has been suggested by some investigators that an association with the somites at the level of the limb bud may be required for the acquisition of inductive properties by the limb mesoderm. The mesoderm loses this property to induce an AER around stage 17, or soon after the initial appearance of the AER. Shortly after the loss of the inductive capacity by the mesoderm to form AER, the flank ectoderm is no longer competent to respond to the limb-inductive stimulus and form an AER.

Progress in the analysis of limb development can be correlated with a series of advances in techniques for handling and manipulating the limb ectoderm and mesoderm. A particularly useful technique in examining the processes and cellular interactions during limb morphogenesis was developed by Zwilling and is shown in Figure 13–14 A. By treating chick limb buds with a trypsin (or versene) solution he was able to separate the limb bud into its principal components, an outer ectodermal layer and the inner mesenchymal mass. Since the mesoderm cells do not remain viable after trypsinization, other limb buds were treated with an EDTA solution. Ethylenediaminetetraacetic acid acts by disrupting the epithelial integrity of the ectodermal cells, but it leaves the mesoderm intact and in a healthy state. An in vitro system is thus made available to test the importance of the two limb tissues. The isolated ectoderm can be rolled into a sleeve and then packed with mesodermal tissues obtained from different sources or from different-aged em-

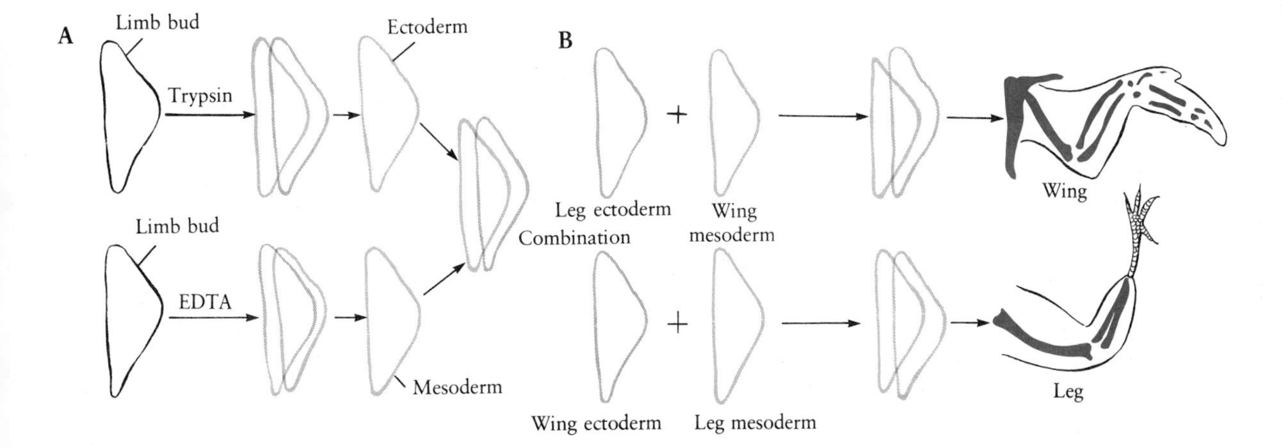


**FIG. 13-14.** The importance of the mesoderm in limb morphogenesis. A, diagram to show the procedure developed by Zwilling for the separation of intact ectoderm and mesoderm from either wing bud or leg bud. The mesoderm is then stuffed into the jacket of ectoderm and the recombinant graft placed on the flank or chorioallantoic membrane of a host embryo. B, wing bud mesoderm surrounded by leg bud ectoderm grows out as a wing. Leg bud mesoderm surrounded by wing bud ectoderm grows out as a leg structure.

bryos. The tissues of the composite limb bud stick together, and the bud itself can then be implanted onto the flank or chorioallantoic membrane of another embryo. For example, suppose the recombinant limb implant is formed by mesoderm tissue of the leg bud and a strip of ectoderm from the wing bud (Fig. 13–14 B). The transplant clearly differentiates as a leg. If wing bud mesoderm is combined with ectoderm from the leg bud, the graft grows out as a wing. The morphological structure of the developing limb is determined by the source of the mesodermal component. The source of the ectoderm does not affect the fundamental nature of the limb. However, the importance of the ectoderm in limb morphogenesis is shown by the fact that isolated mesoderm cannot differentiate a limb in the absence of this superficial layer of cells.

Limb buds appear to produce material for their prospective parts in a proximodistal or base-to-tip sequence. Hence, very young limb buds would only have prospective material for the girdle. Material for successively more distal parts of the limb are added apically as the bud undergoes elongation. Evidence that the AER is indispensable for the proximodistal outgrowth of the limb and the sequential differentiation of its mesodermal components comes from surgical and grafting experiments utilizing chick embryos, particularly from the laboratory of John Saunders and his colleagues (Figs. 13–15, 13–16 B). First, surgical removal of the AER results in limbs lacking some of those parts normally present in control embryos. The extent of the limb deficiencies is dependent on the time at which the AER extirpation is carried out (Fig. 13–15). For example, the removal of the AER from the wing bud of a three-day-old chick embryo (stages 19 and 20) will produce a limb with an absence of skeletal parts distal to the humerus (Fig. 13–16 B). The proximal skeletal parts, such as girdle and humerus, are unaffected by the removal of the AER at this time. Separation of the AER from the wing bud at successively later times will result in fewer deficiencies of the distal skeletal parts. Summerbell conducted several experiments to estimate more precisely the quantitative effects of AER excision. Removal of the AER at stage 20 prevented the formation of skeletal structures distal to the wrist. If the AER was excised at stages 21 to 24, the limb became truncated at the wrist. Only a single phalangeal element was missing when the AER was removed at stage 27. Second, grafting of a

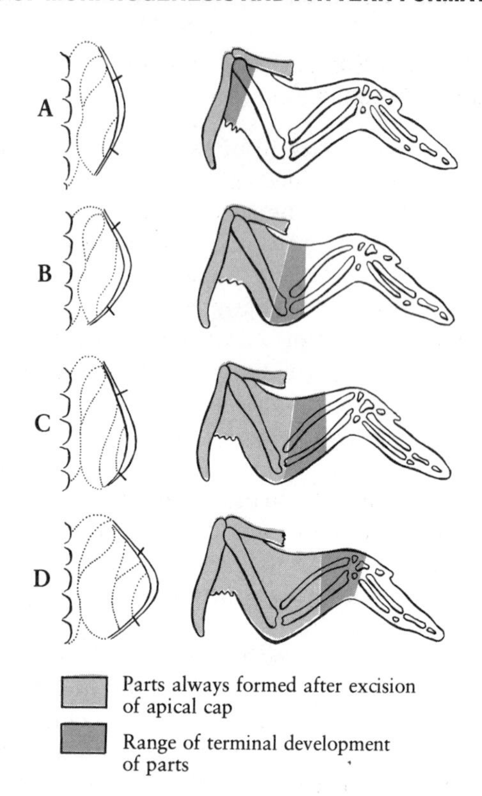


**FIG. 13–15.** Results of a study by Saunders following the excision of the apical ectodermal ridge from wing buds at successively later stages of limb development (A–D). The stippled outlines show the approximate boundaries of the future wing areas in the stages operated. (From J. Saunders, 1948. J. Exp. Zool. 108:363.)

supernumerary AER to the dorsal surface of a limb bud stimulates the outgrowth of a corresponding supernumerary limb. Third, proximal limb bud mesoderm, when grafted subadjacently to the AER, will form distal limb parts under the influence of the ridge. Fourth, further evidence supporting a role for the AER in limb development comes from experiments with the wingless chick mutation. The wingless condition in the chicken is due to a homozygous recessive trait. Although wing buds appear in embryos with this trait, the AER regresses on the third day and there is little further growth of the appendages. If three-day-old ectoderm with an AER from a genetically normal strain is combined with wingless mesoderm, the AER survives for two to three days and then regresses as it does in the wingless phenotype. When wingless ectoderm is combined with normal mesoderm, AER and limb development are inhibited. Interpretation of these results has led to the widely accepted view that the AER "induces" the proximodistal outgrowth of limb structure. Over a period of time, therefore, the successive parts of the limb are added and mapped out as the bud elongates under the influence of the AER (Fig. 13–16).

The nature of the influence or signal of the AER on the underlying mesoderm is not known. A number of experiments, however, have provided some insight into its mechanism of action. Although the AER is required for the outgrowth and differentiation of the limb bud, the ridge does not apparently specify in the underlying mesoderm level-specific structures along the proximodistal axis. If the AER issued specific instructive stimuli at each successive limb level, then limb recombinant grafts of early limb bud mesoderm and late-stage ectoderm with an AER should be deficient in only certain distal skeletal parts. Rubin and Saunders observed, however, that any AER was capable of inducing the normal, proximodistal outgrowth of limb parts in the underlying mesoderm, regardless of the ages of the ectoderm and the mesoderm. Hence, the inductive stimulus released from the AER would appear to remain qualitatively constant from stage to stage and from level to level during limb bud outgrowth. Also, AERs can be exchanged between wing and leg buds without affecting the wing or leg pattern of the developing mesoderm. When duck leg mesoderm is enclosed in a jacket of chick wing ectoderm and grafted to an appropriate site, a duck leg develops. A well-formed limb develops with a chick wing skeleton if 13-day-old rat ectoderm is combined with 3-day-old chick limb mesoderm. The action of the AER, therefore, appears to be permissive and noninstructive. The AER functions to maintain the apical mesoderm cells in a state of developmental plasticity and to elicit the proper sequencing of level-specific patterns in the limb mesoderm. These patterns must be programmed intrinsically within the mesoderm.

Outgrowth of the limb is a consequence of reciprocal interplay between the AER and the limb mesoderm. Specific characteristics of the limb mesoderm include not only its ability to respond to the outgrowth-inducing stimulus of the AER, but also to its abiilty to maintain the integrity of the AER through the production of an *apical ectodermal maintenance factor* (AEMF; also known as the mesodermal maintenance factor). The evidence for the presence of such a factor has been demonstrated in several ways. When a physical barrier such as a piece of mica is inserted between the AER and the apical mesoderm, the AER after several hours loses its structural integrity and becomes flattened. Similar behavior occurs in the ridge if nonlimb mesoderm is stuffed into a jacket of ectoderm with an

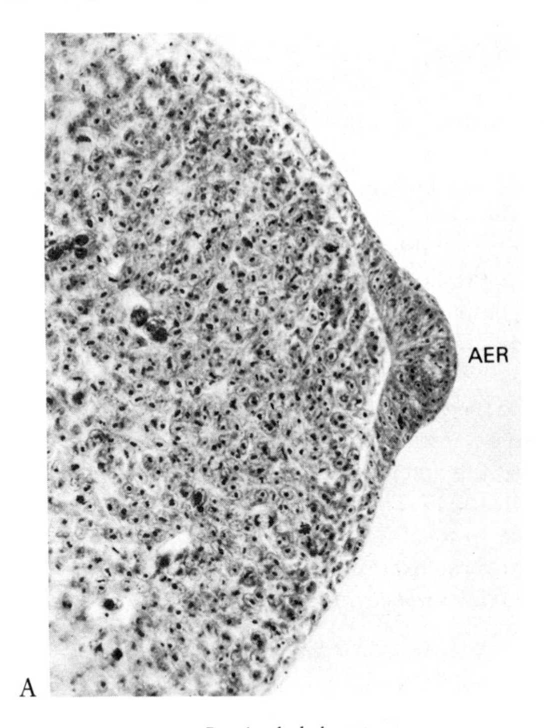


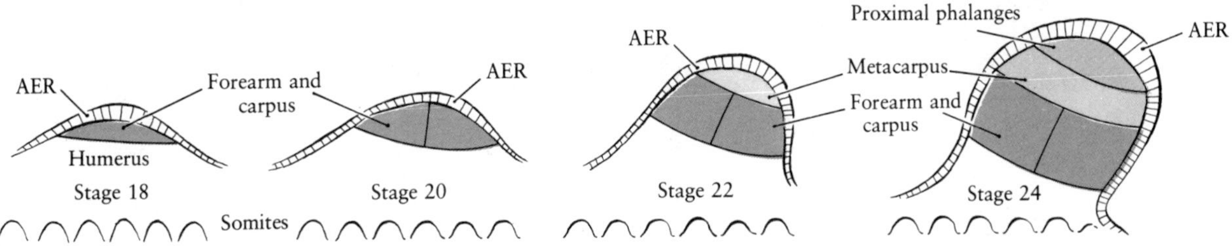


**FIG. 13–16.** A, apical ectodermal ridge (AER) at the apex of the left wing bud in a chick embryo. (From J. Saunders, 1948. J. Exp. Zool. 108:363.) B, the apical ectodermal ridge (AER) induces in proximodistal sequence the various mesodermal parts of the forelimb. Shown here are the areas of the prospective wing at four successive stages of development. (From R. Amprino, 1965. In: Organogenesis, R. H. DeHaan and H. Ursprung, eds. Holt, Rinehart and Winston, New York.)

AER. Although the properties of the AEMF have not been elucidated, experiments by Zwilling using a wingless mutant strain of chickens point to its transmissible nature. Recall that an AER from a normal strain will degenerate when coupled with mesoderm from the mutant strain; however, if the same type of recombinant graft is transplanted to the dorsal surface of a genetically normal limb bud, the AER of the graft remains for a longer period of time and the limb mesoderm responds by outgrowth. In the first experiment, the mesoderm appears to lack the factor (i.e., AEMF) necessary to maintain the AER; in the second experiment, the AEMF of the host mesoderm probably moved through the mutant mesoderm to sustain the AER of the graft.

In the normal chick limb, the position of the AER becomes asymmetrical as the bud elongates. Gradually, the AER along the anterior margin of the bud degenerates because the mesoderm in this region no longer produces AEMF. Hence, according to the Saunders–Zwilling model of limb morphogenesis, the AEMF is present in a limited portion of the mesoderm (largely postaxially), distributed asymmetrically, and transmissible in a proximodistal direction (Fig. 13–17). The distribution of AEMF along the anteroposterior axis of the bud appears to be the condition that ultimately determines the craniocaudal length of the apical ridge, the thickness of the ridge, and probably the number and arrangement of developing limb elements. Early studies by Saunders and others, in which the tip of an early wing bud was rotated through 180 degrees, led to the production of duplicated distal wing parts with reversed symmetry anteriorly. In the normal wing, the anteroposterior sequence of the digits is II, III, and IV. The sequence produced in the rotated wing tip was IV, III, II, III, and IV. The conclusion from these experiments was that AEFM from the postaxial mesoderm maintained the AER preaxially, which, in turn, elicited the outgrowth and differentiation of postaxial structures in the preaxial mesoderm.

Other experiments with polydactylous mutant chicken strains suggest that the anterior or preaxial portion of the limb mesoderm does not normally produce AEMF. Extra preaxial digits are characteristic of polydactylous mutants. If ectoderm isolated from the limb bud of a normal strain is combined with the limb mesoderm of a polydactylous strain, the preaxial section of the apical ridge increases in thickness and induces supernumerary preaxial digits. It is assumed that the production of AEMF is extended to the preaxial portion of the limb mesoderm in polydactylous forms.

The various portions of the overlying apical ectodermal ridge respond to AEMF by promoting the outgrowth of specific regions of the subadjacent mesoderm. Figure 13–18 shows the consequences of removing the cranial or the caudal half of the apical ectodermal ridge of a wing bud. Outgrowth and development of distal skeletal structures are apparent only in the bud mesoderm that is covered by the remaining part of the ridge.

Control of the proximodistal axis of the developing limb is under the influence of the AER. Other experimental studies have provided insight into specification of parts along the anteroposterior axis of the limb bud. It has already been mentioned that polarized duplication of the forelimb is produced if the tip of the wing bud is rotated 180 degrees (Fig. 13–19). When mesoderm from the posterior margin of the wing bud is grafted to the preaxial margin, the AER thickens and a supernumerary wing with polarized limb structures is again induced (Fig. 13–20). Experiments of this type led to the concept of a *zone of polarizing activity* (ZPA). In the chick, ZPA activity is strongest between stages 19 and 28. The polarizing effect can also be demonstrated to be present in the leg bud. As with AERs,

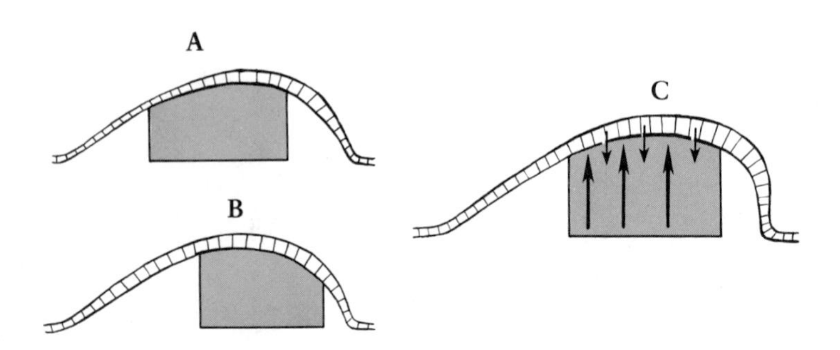


**FIG. 13–17.** Distribution of the mesodermal maintenance factor in wing-bud mesoderm according to Zwilling (A) and Saunders (B). C, the Zwilling-Saunders model of limb development proposes that the mesodermal maintenance factor (large arrows) acts on the apical ectodermal ridge, and the apical ectodermal ridge (small arrows) exerts an influence on the mesoderm by controlling its outgrowth activity. (From R. Amprino, 1965. In: Organogenesis. R. H. DeHaan and H. Ursprung. eds. Holt, Rinehart and Winston, New York.)

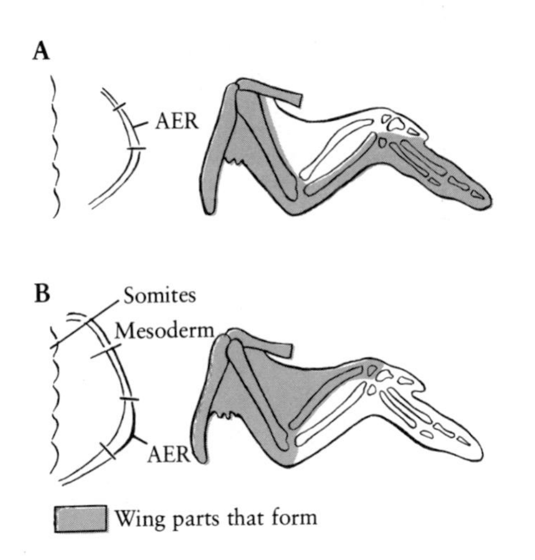


**FIG. 13-18.** Consequences of the removal of the cranial (A) or the caudal (B) half of the apical ectodermal ridge on the development of the wing bud. (From J. Saunders, 1948. J. Exp. Zool. 108:363.)

ZPAs can be interchanged between wing and leg; the response of the recombinant limb bud to the signal generated by the ZPA depends on the origin of the mesoderm. For example, leg bud ZPA grafted to the preaxial border of the chick wing stimulates growth of a supernumerary wing. Similar zones of ZPA activity have been identified in limb buds of the frog, salamander, mouse, hamster, and even the snapping turtle. Fallon and Crosby list a range of species in which their ZPAs all induce outgrowth when transplanted to the chick wing, suggesting that there is a fundamental similarity in the action of the polarizing zone. Another feature of the ZPA is that it organizes the host mesoderm into a reduplicated pattern without necessarily contributing to the resultant structure. For example, a lethally irradiated quail polarizing region has been shown to stimulate extra limb structures while not providing any cells itself.

What is the role of the ZPA in situ? The simple experiment in this analysis requires removal of the ZPA and examination of the development of the ZPA-less wing or leg. When such an experiment is carried out at a stage when the chick limb shows maximal ZPA activity, the resultant limb appears rather normal. If the ZPA is removed from the wing bud between stages 17 and 24, about 50 percent of the embryos show normal wings and 50 percent wings with postaxial defects. Studies of this type have led some investigators to doubt that the ZPA plays a polarizing role in normal development.

The nature of the ZPA action remains to be determined. No subcellular fraction prepared from ZPAs demonstrates polarizing activity in vivo, and no morphogen has been isolated and identified. Treatment of ZPAs with classical inhibitors of metabolism and of protein, RNA, and DNA syntheses has provided little insight into any special cell function that might underlie polarizing activity. Calandra and MacCabe suggest that the morphogen is diffusible. They have demonstrated that there is a gradient of polarizing activity in the chick wing bud, ranging from no detectable activity near the anterior margin to marked activity at the posterior border (i.e., site of the ZPA). Insertion of an impermeable barrier of Mylar film blocks the spread of ZPA activity in the anterior direction, but not if a 0.45-micron Millipore filter is used. Results of techniques using other barrier systems point to the ZPA as an organizer of the anteroposterior axis during normal development. In the chick, digit IV is specified by a high morphogen level and digit II by a low morphogen level.

The vertebrate limb, as pointed out previously, is an elegantly constructed organ with a very precise three-dimensional arrangement of bones, muscles, nerves, blood vessels, feather germs, and/or scales. By 10 to 12 days of incubation in the chick (stages

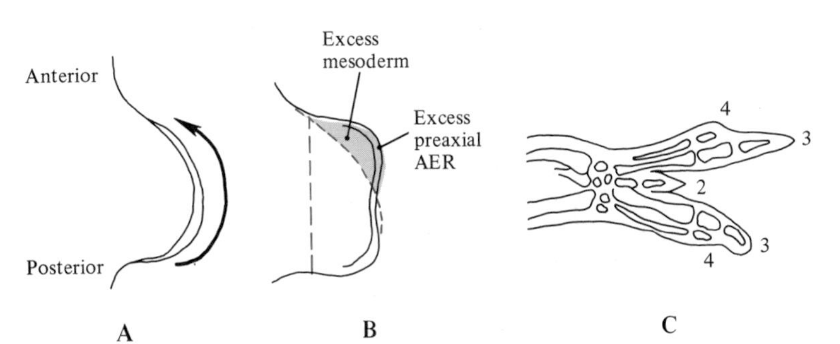


**FIG. 13-19.** A diagrammatic summary of studies by Saunders and colleagues showing that rotation of the tip of the wing bud through 180 degrees (A) results in an abnormal elongation of the AER and an excess of preaxial limb mesoderm (B). The wing becomes duplicated, with the extra digits a mirror image of the normal digits (C).

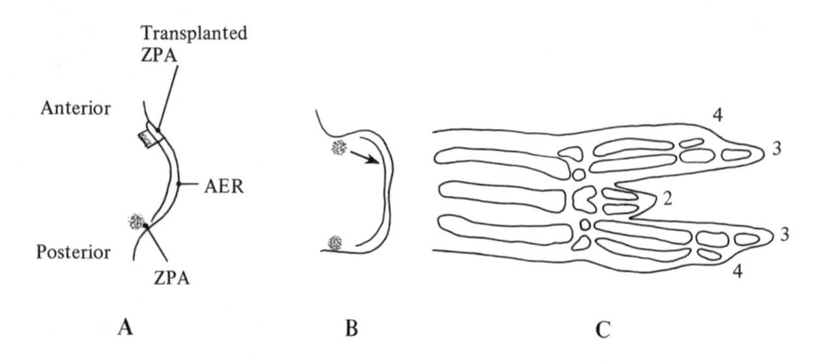


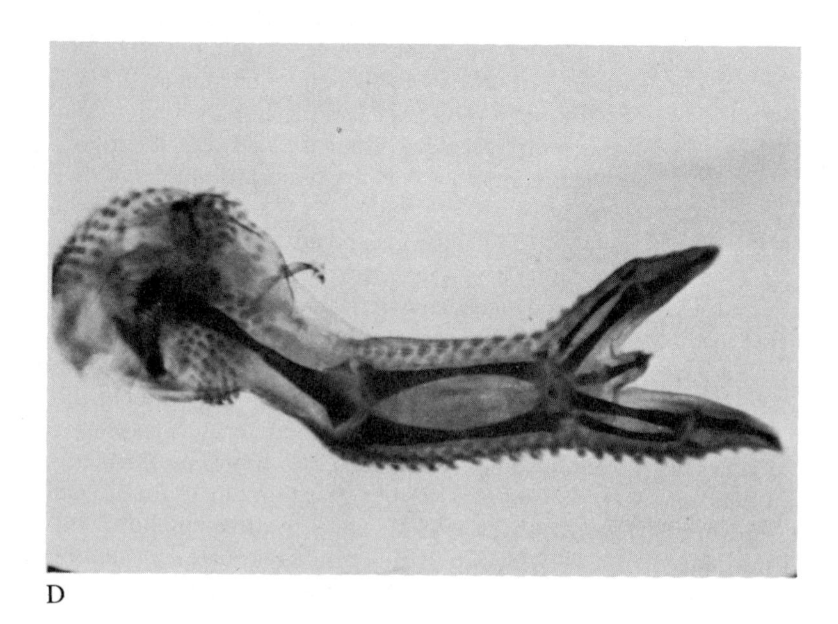


**FIG. 13–20.** Grafting of ZPA from a donor wing bud to the anterior margin of a host wing bud (A) stimulates the outgrowth of the adjacent mesoderm (B). The anteroposterior axis of the duplicated wing tip (C) is always determined by the position of the transplanted ZPA. D, a photograph showing reduplication of the wing tip with digit pattern 4 3 2 3 4. (From D. Summerbell and L. Honig, 1982. Am. Zool. 22:105.)

36–38), the undifferentiated cells of the limb bud have given rise to these various structures that characterize the pattern of the adult wing or leg. Developmental biologists concerned with the limb are interested in knowing how this intricate pattern develops from the undifferentiated bud mass. For example, how does a mesenchymal cell in the limb bud "know" to differentiate into a muscle cell rather than a cartilage cell, or vice versa? Perhaps the mesenchymal cell differentiates into the muscle cell because it is predetermined to do so. Alternatively, however, the location of the mesenchymal cell in the limb bud might dispose it to differentiate into a muscle cell rather than a cartilage cell. The differentiated elements that comprise the limb bear special relationships to each other. During limb pattern development, therefore, the proper number of the correct cell types must be present at the same time, and these cells must be properly distributed in space.

As one might guess, analysis of limb development from several different perspectives (i.e., regulation, growth, regeneration) has resulted in the formulation of several pattern formation models. No attempt will be made to review all of them; we will focus only on those that are current and appear to fit experimental data best. An excellent review of this subject can be found in a symposium, entitled "Principles and Problems of Pattern Formation in Animals," sponsored by the American Society of Zoologists and published in the *American Zoologist* (1982).

One possibility is that pattern determination of the limb is either intrinsic to the cells (i.e., preprogrammed) or generated as a result of instructions impressed on the cells during outgrowth. We have already seen from experiments with quail-chick chimeras that most of the limb musculature appears to be derived from somite mesoderm; hence, somites appear to make a clonal contribution to limb development, supporting the view that myogenic and chondrogenic precursor cell types are preprogrammed to differentiate into muscle and cartilage, respectively. These two cell lines, somite–myogenic and somatopleural chondrogenic–fibroblastic, are presumably distinct from the beginning of limb development. Somites may also be important to specifying pattern. For example, few skeletal elements differentiate in a mouse forelimb primordium unless grafted with its adjacent somites. Without these somites, only a cartilaginous mass develops.

The more conventional models discount predetermined limb pattern and postulate that *positional information* and chemical morphogens, generated during limb outgrowth, are important. There are two major models based on positional information: (1) the ZPA/progress zone model proposed by Wolpert, Summerbell, and others, and (2) the polar coordinate model proposed by French, Bryant, and others. The concept of positional information, initially developed by Hans Driesch, was first applied to the vertebrate limb by Lewis Wolpert. He proposes that pattern formation consists of two processes. First, cells acquire information about their relative position in the limb field; in some fashion, the cells possess a "memory" of their *positional value*. Second, each cell or group of cells uses this positional value to trigger particular information from its genome for the expression of the differentiated state. Positional values for both the proximodistal and anteroposterior geometrical axes are assigned by a specialized area of mesoderm beneath the AER, known as the *progress zone,* as the limb bud grows. Presumably, within the progress zone, the positional value of each cell changes autonomously over time in relation to some metabolic variable. Hence, the positional value of a cell is determined by the length of time that it remains in the progress zone (i.e, the value changes with the number of cell divisions it passes through). Cells leaving the progress zone early have proximal positional values, and cells leaving later have more distal positional values. Once cells leave the progress zone, their positional value is fixed; they undergo differentiation only in accordance with this value. The function of the AER would appear to be to determine when cells are released from the progress zone, thereby specifying the position of cells in the whole. These tenets of the model fit nicely observations on the proximodistal outgrowth of the chick wing (i.e., difference between the humerus and a digit). The primordia of proximal limb parts are laid down before those of distal ones; indeed, the proximal end of each primordium is established before its distal end. Also, there is a mesodermal zone of rapidly dividing and undifferentiated cells just beneath the AER. The Wolpert model further proposes that the ZPA acts as a signaling region to specify the anteroposterior postitional coordinates of limb bud cells (i.e., difference between digit II and digit IV). It is postulated that the ZPA at the posterior margin of the limb bud functions as a source of a morphogen that diffuses through the rest of the limb field. The morphogen breaks down in the cells and produces a concentration gradient with the highest activity at the level of the ZPA itself. The anteroposterior positional value of a cell is specified by the concentration of the morphogen to which it is exposed while in the progress zone. As in the case of its proximodistal positional value, the anteroposterior positional value is fixed when the cell leaves the progress zone. Hence, cells nearest the ZPA differentiate into more posterior limb structures.

The concept that the limb is organized by the progressive distal addition of new cells has been challenged, because results of grafting experiments employing the AER and its associated progress zone do not always agree with the predictions of the model. The best test of the model would be to graft cells of prospective distal structures (i.e., wrist) onto a limb bud stump whose cells are already specified for proximal structures only (i.e., humerus or upper arm). One would expect that the recombinant graft would lack the intercalated structure or forearm (i.e., radius, ulna). Such experiments are subject to variable interpretation, but in the hands of some investigators, forearm structures are present in the grafts. The skeletal structures under these conditions are formed by both stump and graft cells, suggesting that graft cells are reprogrammed to form parts more proximal that their normal fate.

The polar coordinate model is based largely on studies of insect (larval cockroaches) and urodele limb regeneration. It utilizes the concept of positional information, but does not recognize a special signaling region whose action depends on diffusion of a morphogen over a distance.With this model, the growing limb bud is viewed as a three-dimensional cone with a base (proximal) and an apex (distal). Any point on the surface of the cone is specified by two coordinates, one along the major longitudinal axis of the cone (A–D, Fig. 13–21) and

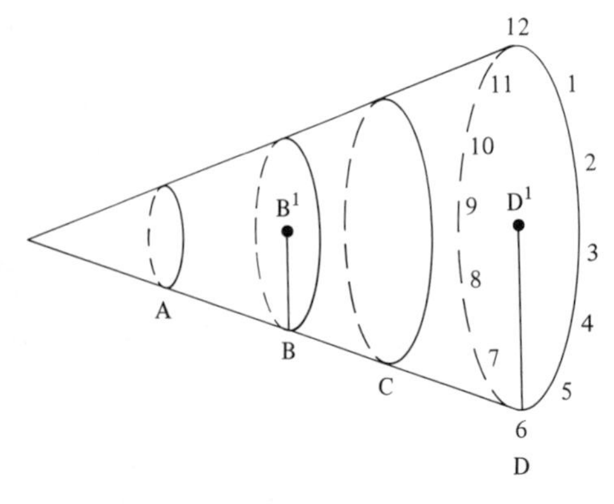


**FIG. 13–21.** Diagrammatic representation of the polar coordinate model of limb development as proposed by Bryant, Bryant, and French (1977). The limb is viewed as a cone with a base (proximal) and an apex (distal). The position of a cell is specified by its level along the major axis (A–D) and by its location on the clock face at that level. For example, the point at B has a level value B and a clock face value of 6. All points along lines B B′ and D D′ have the same positional value. Therefore, one component of a cell's positional information is a value corresponding to position on a circle and the second to position on a radius.

one around its circumference (1–12, Fig. 13–21). At any given cross-sectional level in the cone, the numbers are arranged like a clock face for reference (hence, this is often referred to as the clock face model). Since no differences are postulated to exist along the major axis (i.e., all points have the same positional value), pattern formation in the limb is essentially two-dimensional and dictated by activities at the circumference, presumably in the subdermal mesoderm. The position of a cell, therefore, is specified by its level along the major axis and its location in the clock face at that level. The correct spatial relationships will be established if cells follow two rules. First, if cells with nonadjacent positional values come to lie side by side, cells must be induced to multiply at the junction until cells with appropriate positional values have been intercalated. For example, if a cell with a positional value 5 finds itself next to a cell with a positional value 8, then cells interact and produce cells with values 6 and 7. Second, cells at any proximodistal level can only give rise to new cells with a more distal positional value, provided a complete series of positional values has been exposed (which is the case in studies using amputation). This model depends on short-range neighbor-to-neighbor cell interactions and some earlier unknown process that gives fixed positional values to the cells.

Which is the correct model? Unfortunately, none of the models currently before us satisfies all of the available data generated by limb studies to account for the normal pattern of limb structures formed during development. However, many of these models have been beneficial in stimulating a great deal of exciting research on an individually and collectively complex organ.

## Determination of the Limb Axes

The growth and differentiation of the limb skeleton normally proceed in an ordered, proximodistal sequence. In amniote embryos, the limb girdles, which develop from the same mass of mesoderm as the limb skeletal parts, differentiate at the same time as the *stylopodium* (femur, humerus). Successively the *zeugopodium* (radius and ulna of forelimb, tibia and fibula of hindlimb) and *autopodium* (manus of forelimb, pes of hindlimb) are laid down. Each skeletal element is initially blocked out in mesoderm, which subsequently becomes converted to cartilage. The muscles that will attach to the skeleton of the limb originate from mesenchyme cells of the myotomes.

It is apparent that the organization of the various components of the normally developed limb is complex and asymmetrical. Ross Harrison (1921), in a classical study, analyzed the origin of the axial polarity of the embryo by transplanting limb primordia. Using the salamander, rudiments of the limb were excised at different stages following neurulation, rotated or inverted in such a way that one or several of its axes were altered with respect to the axial polarity of the host embryo, and transplanted (Fig. 13–22). For example, transplantation of the left limb primordium shortly after neurulation can be accomplished by moving the transplant over the back and onto the right flank. Such a transplant reverses the dorsoventral axis with respect to the host, but the anteroposterior axes of host and graft coincide (Fig. 13–22 A). The graft transplanted in this manner develops normally with the limb growing caudally (Fig. 13–22 B). If, however, the same graft is rotated 180 degrees on the right flank such that the anteroposterior axis is reversed with respect to the host (Fig. 13–22 C), the limb grows forward instead of caudad and subsequently shows the posterior digits placed anteriorly (Fig. 13–22 E). Hence, the anteroposterior axis is determined and irreversibly fixed at the time of transplantation.

The dorsal and ventral surfaces were undisturbed in these two experiments (i.e., the dorsal part of the limb rudiment differentiated ventral structures). Transplantation of a right-forelimb primordium to the right flank after rotation of 180 degrees produces a similar limb (Fig. 13–22 D). When the limb disc area was excised from an embryo whose tail rudiment was beginning to elongate and transplanted as in Figure 13–22 A, the dorsoventral axis was observed to be fully determined at this time. Such a graft showed no disturbances in the development of the proximodistal axis. The proximodistal axis is determined at about the time that the limb bud becomes visible morphologically.

**FIG. 13-22.** Specification of the axes of the amphibian limb. A, transplantation of a left limb primordium with inversion of the dorsoventral axis; B, the normal limb resulting from this operation; C, transplantation of a left limb primordium to the right flank with inversion of the anteroposterior axis; D, transplantation of a right forelimb primordium to the right flank after rotation of 180 degrees; E, the limb resulting from the operations in C and D grows forward, indicating that the anteroposterior axis is irreversibly fixed. A, anterior; D, dorsal; P, posterior; V, ventral. (After F. Swett, 1937. Q. Rev. Biol. 12:322.)

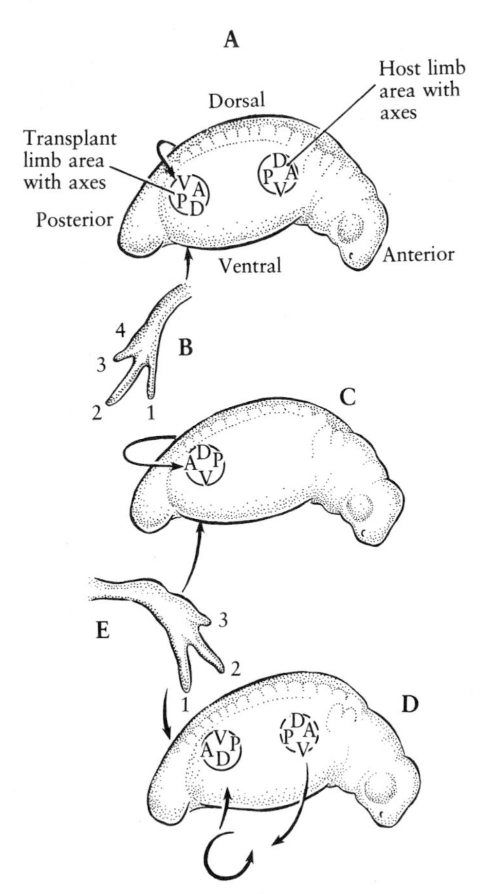


The gradual stabilization of the properties of the limb field as shown by the sequential determination of the axes has also been demonstrated in grafting experiments with the chick embryo. The anteroposterior axis is the first to be determined (five-somite stage). Subsequently, the dorsoventral and the proximodistal axes become fixed to establish the complete axial relationships of the adult limb. The determination of the axes of the chick limb resides in the mesoderm rather than the ectoderm. For example, if prospective wing bud mesoderm is grafted to the flank and the ectoderm is allowed to heal over it, outgrowth occurs and a wing develops; however, the axes are always determined by the intial orientation of the grafted mesoderm rather than by the host ectoderm.

## The Differentiation of the Limb Cells

During the course of its development, the limb initially passes through a morphogenetic phase when axes are determined and shape is acquired (Fig. 13–23). The multiplication of cells, the movement of cells, and adhesion between cells play important roles in giving shape to the limb bud. Although the limb bud from the beginning consists of two cell lineages (myogenic, chondrogenic–fibroblastic), all of the limb mesenchyme cells are homogeneous in

**FIG. 13-23.** Relationships of the relative activities of the morphogenetic and cytodifferentiative phases in the development of the chick limb. Note that chondroitin sulfate, the major constituent of cartilage cells, is initially synthesized during the morphogenetic phase. (From E. Zwilling, 1968. In: The Emergence of Order in Developing Systems. M. Locke, ed. 27th Symposium of the Society for Developmental Biology. Academic Press, New York.)

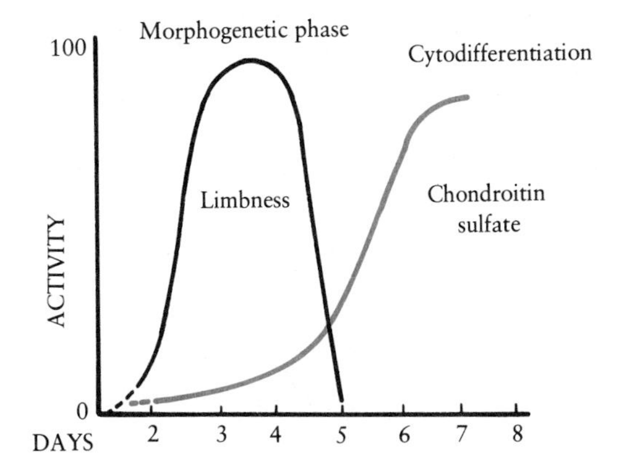

appearance and difficult to distinguish between. Indeed, chondroitin sulfate, a sulfated GAG characteristic of the extracellular matrix of the mature cartilage cell, is synthesized by all mesodermal cells during early limb morphogenesis (Fig. 13–23).

The mechanisms by which embryonic limb mesenchyme differentiates into the various definitive cell types of the limb have been the subject of both in vitro and in vivo analyses for a number of years. An initial step in the differentiation of limb cartilage and muscle cells is the spatial organization of their precursors into condensations or aggregates. It has been shown experimentally that the spatial arrangement of the precursor condensations for both muscle and cartilage tissues are affected by the surgical removal of the AER. How the AER interacts with these mesodermal populations to regulate their spatial distribution is unclear. Additional insight into the behavior of cells during the normal formation of chondrogenic aggregates is provided by numerous mutant strains in which single genetic loci affect the early stages of chondrogenesis. The talpid mutant is produced by a recessive lethal gene that, if in the homozygous state, leads to the formation of a very broad limb bud. Within the limb, the precartilage condensations are typically poorly defined and often result in the fusion of proximal skeletal elements and the presence of excessive distal digits (Fig. 13–24). Talpid mesenchyme cells appear to be more adhesive and less mobile than normal mesodermal cells. Reduced movements of mesodermal cells in the mutant might prevent separation between neighboring precartilage rudiments and thereby result in abnormal patterns of aggregation. The formation of normal precartilage condensations would appear to require changes in cell-to-cell contact, particularly in surface properties controlling adhesiveness.

Studies principally from the laboratory of Solursh and his colleagues suggest that cartilage formation in the limb buds of both chick and mouse embryos follow a very similar developmental pattern (Table 13–1) An early step in chondrogenesis occurs when mesoderm cells acquire the predisposition to form areas of high cell density. Limb mouse mesenchyme cells show this property by 9½ days of gestation; however, these same cells do not have the ability to form cartilage spontaneously. The skeletal condensation for the humerus is clearly recognizable by day 11 of gestation, and the first visible signs of cartilage matrix by day 11½ (Fig. 13–13). Hence, the capacity to form cartilage is acquired shortly after the ability for mesodermal cells to form aggregates. This in turn depends on the acquisition of the ability to interact with neighboring cells. Several studies suggest that chondrocytes synthesize a factor that enhances their differentiation from the mesenchyme. In the chick embryo, cartilage differentiation is indicated by a noticeable

**FIG. 13–24.** Diagram showing areas of precartilage condensation in normal (A) and talpid (B) mouse limb buds.

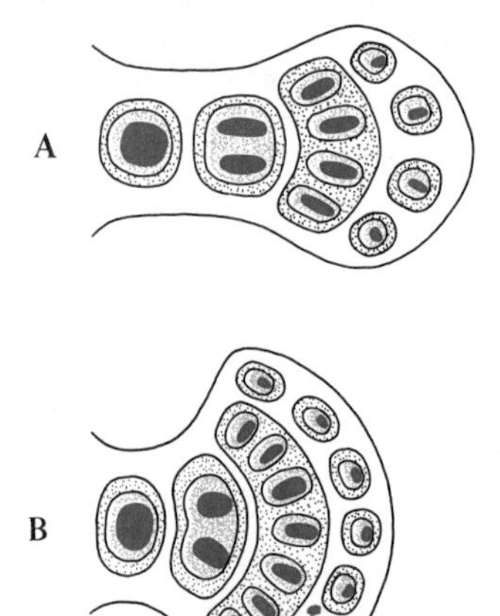


Table 13–1 A Comparison between Mouse and Chick in the Morphological Structures Observed during the Course of Forelimb Development. (From E. Owens and M. Solursh, 1981. Dev. Biol. 88:297.)

| *Mouse Limb Bud Stages* | *Chick Wing Bud Stages* | *Morphological Structures Observed in Cross Section* |
|---|---|---|
| 15 | 17–19 | Homogenous mesenchyme |
| 16–17 | 20–21 | Homogenous mesenchyme |
| 18 | 22–23 | First signs of long bone condensations |
| | 24 | Major condensations formed; no cartilage |
| 19 | 25 | First cartilage observed in humerus |
| 20–Early 21 | 26 | Cartilage in long bones; condensations of other skeletal primordia well formed |
| Middle–late 21 | $26^{+}$ | Cartilage in most skeletal primordia |

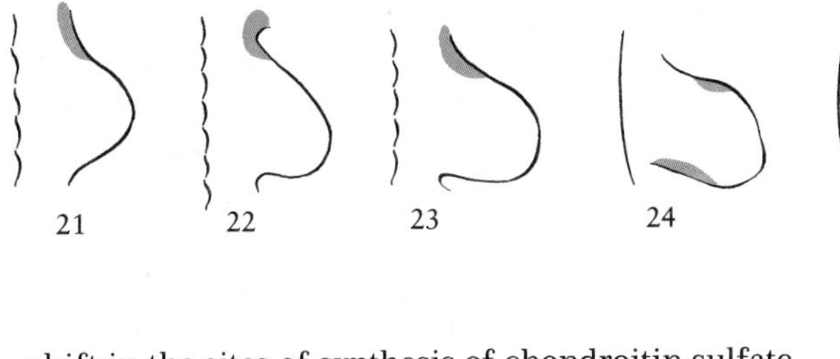


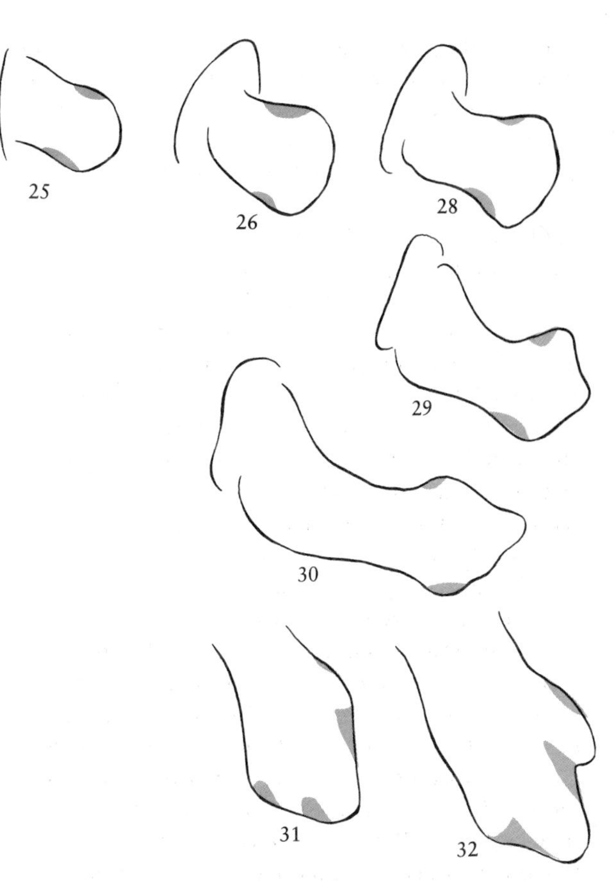


shift in the sites of synthesis of chondroitin sulfate after stage 22. Mesodermal cells in the dorsal and ventral regions of the limb bud (presumptive muscle cells) show a decrease in chondroitin sulfate synthesis, while those in the core of the bud (presumptive fibroblasts and cartilage cells) show a sharp rise in synthesis. By stage 25, cartilage is first observed in the humerus (Fig. 13–13 and Table 13–1).

## Cell Death: A Morphogenetic Agent

One can conclude from the previous sections of this chapter that many factors interact to account for the morphogenesis of tissues and organs in multicellular animal embryos. A process that is common and probably necessary in many cases of organ development is that of cellular degeneration or differentiated cell death. The destruction of cells during the development of an embryo may be massive and quite dramatic. Examples include the removal of many larval tissues at the time of metamorphosis in amphibians and insects (e.g., tadpole tail, intersegmental muscles of pupating insects). More typically, cell death is regional in character where it is employed in the definition of pattern of an organ. Cell necrosis is often observed to precede changes in the shape of an epithelial organ, such as during the invagination of the optic cup (eye) and the olfactory pit (nose). The separation of the lens rudiment from the ectoderm requires cell degeneration. The removal of unwanted or superfluous cells after the union of parts, the closure of openings, and the formation of lumens is frequently effected through cell death.

Although cell death is a prominent occurrence in several embryonic systems of the vertebrate organism, it has been analyzed most extensively as a morphogenetic process in the development of the limb. Figure 13–25 illustrates the areas of necrosis in the superficial mesoderm of the chick wing as detected by supravital staining with Nile blue sulfate. Most of the cell deaths between stages 21 and 23 occur along the anterior edge of the wing bud and the adjacent body wall. This degeneration of cells in the *anterior necrotic zone* (ANZ) assists in shaping the contours of the future shoulder region. By stage 24, cells begin to die in large numbers at the posterior margin of the wing bud and the body wall. This *posterior necrotic zone* (PNZ), which contains approximately 1500 to 2000 cells in the process of degeneration at one time, results in the separation of materials that form the distal part of the scapula from those that contribute to the elbow region and the posterior portion of the upper arm. Its cells are quickly phagocytized by macrophages. At later stages, there is a distinct correlation between the topographical distribution of cell deaths and the emergence of the definition of the manus or hand with its three major digits.

**FIG. 13–25.** The wing bud in late paddle stages and stages of contour formation to show regions of massive necrosis in the superficial mesoderm. (From J. Saunders et al., 1962. Dev. Biol. 5:147.)

A comparison between the leg primordia of the chick and the duck clearly shows the relationship between the distribution of necrotic areas and the modeling of the appendage (Fig. 13–26). In the chick, massive destruction of cells in the interdigital zones brings about the separation of the individual hindlimb digits. In the duck, areas of necrosis are limited to the peripheral regions between the digits. Most of the epidermis remains intact to contribute to the webbing of the foot. There appear to be no areas in either the forelimb or the hindlimb of the mouse or rat that correspond to the anterior and posterior necrotic zones of the chick. Since mammals have the full number of digits (five), the function of the ANZ and the PNZ may be to reduce the availability of mesenchyme for digit formation in the birds and therefore to regulate the spatial distribution of skeletal elements.

A similar area of necrosis, termed the *opaque patch,* is also clearly important to the patterning of the forearm. This region of cell death appears in the wing between stages 24 and 25 and effectively separates the condensations of mesoderm that will give rise to the radius and ulnar bones. In contrast to the cells of the PNZ, cells of the opaque patch show distinct signs of death, such as condensed chromatin and vacuolated cytoplasm, prior to being phagocytized. All cells of the central mesoderm of the forearm actively produce chondroitin sulfate until stage 24. At this time, however, the opaque cells suddenly cease synthesizing chondroitin sulfate, lose their cell-to-cell adhesiveness, and change their programming from presumptive cartilage cells to cells endowed with a death clock. Other zones of necrosis appear between the digits in amniotes without webbed appendages. Cell death in each *interdigital necrotic zone* follows a very precise developmental program. For example, in the chick, necrosis begins proximally between adjacent digits II and III, and then an area of death appears distally under the AER; the two areas of cell death then coalesce. By contrast, cell death between digits III and IV begins proximally and spreads distally. This zone then overlaps the PNZ posterior to digit IV, thus eliminating mesenchyme distal to the terminal phalangeal element of digit IV.

The concept that destructive processes (i.e., cell death), particularly in the interdigital regions, help to shape the limb has immeasurably assisted in our understanding of the development of this organ. But, how are areas of cell death determined? Is this a part of the differentiative process? Saunders and his colleagues have analyzed this problem using the PNZ of the chick wing bud. Evidence from transplantation experiments indicates that the death of the cells of the necrotic zone appears to be set by stage 17. If the PNZ is excised between stages 17 and 22 and then transplanted to a suitable environment, such as the somite region of the host embryo, the cells of the graft die on schedule or at stage 24. Control grafts, prepared from the dorsal side of wing mesoderm, when transplanted into the somite area remain healthy. Also, cells that migrate into and replace the excised PNZ do not undergo cell death. These experiments allow the conclusion that the degeneration of PNZ cells can be attributed to an intrinsic "death clock."

Although the death clock is set by stage 17 in the wing bud, other grafting experiments show that the death program can be reversed prior to stage 22. If

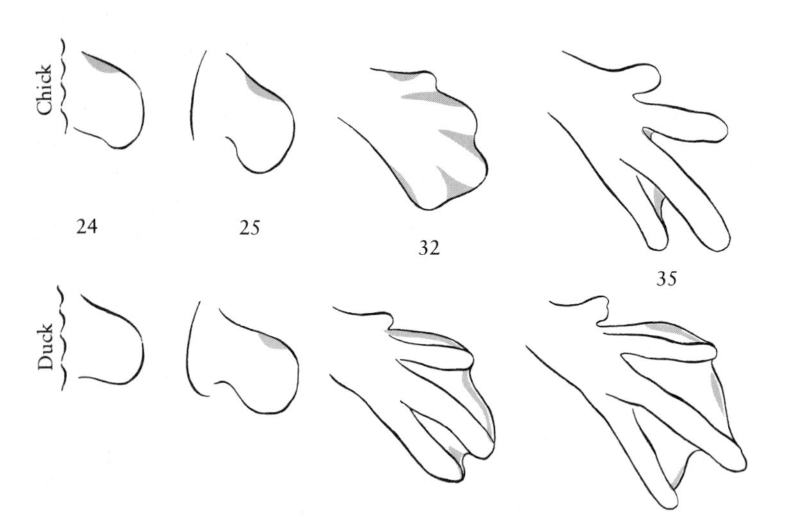


**FIG. 13–26.** Patterns of necrosis at different stages in the leg primordia of the chick and the duck. (From J. Saunders and J. Fallon, 1966. In: Major Problems in Developmental Biology. M. Locke, ed. 25th Symposium of the Society for Developmental Biology. Academic Press, New York.)

the PNZ of stage 21 embryos is removed and grafted to the dorsal mesoderm of the wing bud, there is no triggering of the death clock and the PNZ cells remain alive. After stage 22, however, transplants of the PNZ unequivocally demonstrate that its cells are irreversibly directed toward massive necrosis at stage 24. Additional in vitro experiments on the ability of the dorsal wing mesoderm to prevent scheduled PNZ necrosis suggest that this portion of the wing produces a diffusible substance that in some way sustains the PNZ cells. Presumably, the mesodermal cells of the PNZ in the intact embryo lose the ability to synthesize this substance and die. Additional evidence from studies on certain talpid mutants tends to confirm that areas of cell death, such as the PNZ, are under genetic control. For example, in the talpid$^2$ mutant the ANZ and PNZ are not present. The result is excessive elongation of the AER and the production of many digits.

It has not been possible to detect in the PNZ cells prior to stage 24 morphological changes that would overtly indicate the operation of the death clock. However, alterations in the synthesis of particular macromolecules have been identified in PNZ cells before stage 24 and the onset of phagocytosis. If chick embryos between stages 19 and 23 are injected with tritiated thymidine, the PNZ cells at stage 22 (when they are irreversibly committed to death) show a marked reduction in thymidine incorporation and hence in DNA synthesis. Shortly after the depression in the rate of DNA synthesis, neighboring cells begin to engulf PNZ cells. Substantially fewer labeled amino acids are incorporated into protein in the PNZ after stage 22. Since these changes in patterns of DNA and protein synthesis occur after the commitment of PNZ cells to death, they are symptomatic of death rather than determinants of cell death differentiation.

Patterns of cell death in the chick limb probably involve inductive interactions between the central mesoderm and the overlying epidermis. When the mesoderm of a duck leg is isolated and covered by a strip of epidermis from the wing of a chick, the recombinant limb graft grows and displays a pattern of necrosis similar to that of the duck (i.e., the limb is webbed). The reciprocal recombinant limb graft, or mesoderm of a chicken leg covered by wing epidermis of the duck, shows a pattern of interdigital necrosis that is similar to that of the chick (Fig. 13–27 A,B). Hence, specificity of patterns of necrosis in the limbs appears to be established in the mesoderm. By some unknown mechanism, the mesoderm is able to elicit localized patterns of cell death in the epidermis.

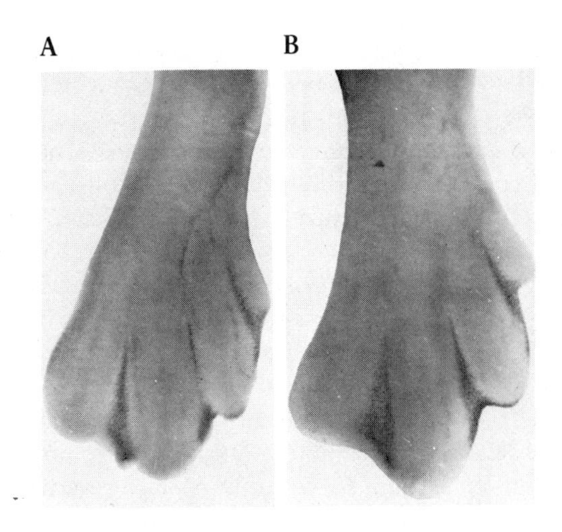

**FIG. 13–27.** Patterns of necrosis revealed between the digits by Nile blue staining. A, chimeric limb bud grown as a flank graft composed of a core of mesoderm from chick leg bud and a jacket of ectoderm from the wing of a duck embryo; B, foot of a control chick embryo. Note the extent of necrosis in A is considerably less than in B. (From J. Saunders and J. Fallon, 1966. In: Major Problems in Developmental Biology. M. Locke, ed. 25th Symposium of the Society for Developmental Biology. Academic Press, New York.)

Obligatory cellular necrosis is also required in the remodeling and differentiation of portions of the digestive tract in a number of vertebrates, including amphibians, birds, and mammals. In the chick embryo, for example, the esophagus characteristically becomes occluded anteriorly at approximately five days of incubation. A period of intense epithelial proliferation then follows for the next several days. The degeneration of epithelial cells in this part of the gut is initiated at about eight days of incubation. Numerous intercellular vesicles appear between the moribund epithelial cells and through coalescence produce a new, definitive esophageal lumen. As in other systems showing necrosis (e.g., developing limbs, chick oviduct), biochemical data indicate that there is a marked increase in acid hydrolase acitvity, particularly of acid phosphatase and $\beta$-glucuronidase, during the remodeling process. The increases in the specific activities of these hydrolytic enzymes between days seven and twelve link them temporally with cell death and remodeling. Failure of the epithelial cells to degenerate results in the formation of a highly vesiculated and partially obstructed lumen, a condition present in homozygous crooked neck dwarf mutant chick embryos. The mechanism controlling the death of these gut cells remains to be determined. There is the possibility that the hydrolytic enzymes are released intracellularly from a lysosomal system to effect degeneration.

A dramatic example of cell death in insects is the degeneration of the intersegmental muscles in the silkmoth *(Antheraea)* following ecdysis from the pupa to the adult stage. Within 48 hours of the emergence of the pupa from the old cuticle, these muscles have degenerated. They appear to provide the motive force for ecdysis itself. Presumably, their contraction drives hemolymph or tissue fluid into the thorax of the pupa and thus assists in the rupture of the cuticular encasement . After ecdysis into the adult stage, these same muscles propel hemolymph into the wings in order to expand them. Developmentally, therefore, it is imperative that these muscles remain functional until the moth has emerged from the pupa stage.

Since hormonal balance is critical in insect metamorphosis (Chapter 11), the role of the endocrine system in setting the death clock for the intersegmental muscle breakdown has been examined by Lockshin and Williams in a series of studies. They found that three weeks before ecdysis the combination of a high level of ecdysone and a low level of juvenile hormone acted as a potentiator for the degeneration of the muscle cells after the emergence of the moth. If juvenile hormone was experimentally injected into the thorax of the pupa at this time, metamorphosis into the adult form was blocked and the breakdown of the intersegmental muscles prevented.

Ecdysone by itself does not appear to initiate directly the actual degeneration process in the muscles. Additional experimental studies by Lockshin and Williams have shown that the stimulus that triggers the differentiated death mechanism is neural in nature. There is observed a marked cessation of motor impulses to the intersegmental muscles shortly after ecdysis. The absence of a neuromuscular transmitter substance in some way then causes lysosome-like particles to release cytolytic enzymes that destroy the muscle cells. Cessation of nervous stimulation as a cause of cell death has been observed in other embryonic systems.

As in insects, hormones play an essential role in the destruction of specific larval tissues, such as tail and gut, during the metamorphosis of amphibians. Amphibian metamorphosis is an important postembryonic event during which there are spectacular changes in larval structures in preparation for transformation to a lung-breathing, terrestrial animal. Almost every tissue and organ system of the larval frog undergoes alterations. A dramatic part of the metamorphic process is the regression of the tail.

The significance of the thyroid gland in metamorphosis was initially discovered by Gudernatsch in 1912. By feeding mammalian thyroid gland tissue to tadpoles, it was shown that the tadpoles exhibited precocious metamorphosis. Conversely, the elimination of iodine (an elemental constituent of *thyroxine*) or of thyroxine by thyroidectomy prevents metamorphosis from taking place. Regression of the tail depends on stimulation by thyroxine. Morphologically, experimental data indicate that the thyroid hormone acts on the tail tissues and induces their lysis. Histolysis of tail tissues always begins at the tip and proceeds toward the base of the tail. A typical pattern of destruction includes thickening of the epidermis, migration of pigment cells, and involution of notochord, neural tube, and muscle cells. Tissues undergoing regression have high activity levels of such hydrolytic enzymes as cathepsin, acid phosphatase, and collagenase. Visible tail resorption is accomplished by phagocytic digestion.

How does thyroxine act at the level of the cell to stimulate cell destruction? The process of tail resorption has been studied on the biochemical level using amputated larval tails cultured in an artificial medium. When such tails are exposed to low concentrations of triiodothyronine (a thyroid hormone related to thyroxine), they undergo resorption at a rate comparable to that observed in control larvae. The effect of the hormone can be completely eliminated by adding actinomycin D (to inhibit RNA synthesis) or cycloheximide or puromycin (to inhibit protein synthesis) to cultures of hormone-stimulated tails. This indicates that mRNA synthesis and protein synthesis are required for cell death and tail regression. It would appear that the thyroid hormone in the tail turns on a specific genetic program of transcription and translation, the result of which is the production of destructive hydrolytic enzymes.

A particularly fascinating feature of the hormonal control of amphibian metamorphosis is that thyroxine elicits a wide range of responses in different larval tissues. Depending on the target tissue, some of these alterations are constructive while others are destructive. Some tissue cells die while other tissue cells proliferate. For example, the same thyroid hormone accelerates the destruction of larval blood cells and stimulates the proliferation of blood cells characteristic of the adult frog. A major and largely unresolved problem is: What properties of the thyroid hormone or of the target cells are responsible for these different responses? An attractive hypothesis is that the response of a given tissue at a given time may relate to the presence of mature thyroxine-binding receptors on its cell surfaces.

## Cell Sorting, Cell Recognition, and Organ Construction

Embryogenesis in vertebrates proceeds through a complicated series of cellular interactions in which different cells become spatially sorted out, segregated, and assembled into specific multicellular groupings that give rise to tissues and organs. Many tissues and organs, such as the gonads, heart, various nervous ganglia, and the adrenal cortex, originate in the embryo from cells that migrate over substantial distances to their final destinations. These cells must move toward the target site and then begin the process of tissue and organ construction. As seen from examination of the behavior of neural crest cells, a single cell or group of migrating cells must "realize" in some manner that it has reached the final target site; presumably this information is provided by the microenvironment. Organs are not haphazard collections of cells and tissues. Rather, each organ has a distinctive, characteristic form, a precisely ordered arrangement of cells, and specific relationships to other organs. Topographical stability of cells and tissues lies at the basis of the integrity of organs. The performance of normal function could not proceed without it.

Embryonic cells are capable of positional rearrangements and assembly into specific morphogenetic patterns because of their surface properties. It is now a generally held view that molecular events occurring at the cell surface are of importance in controlling the capacity of cells to recognize and adhere to each other during tissue formation. The twin processes of *cell recognition* and *selective cell adhesion*, therefore, are essential to the construction and stability of an organ. We have previously noted the role of cell-to-cell recognition and selective cell affinities in the formation of the germ layers (Chapter 8). Holtfreter, several decades ago, dissociated the germ layers of early amphibian embryos and then observed that the cultured single cells were able in many cases to reaggregate to form a tissue of remarkable likeness to the one from which they came. This was an important demonstration that cells of the same type have a way of recognizing each other; they prefer to interact with cells of their own kind. If cells from two different germ layers are dissociated and then intermingled in the same suspension, the resulting aggregate incorporates both types of cells. However, in the course of further development of heterotypic aggregates, the diverse cell types sort out according to kind and form distinct, histogenetically uniform groupings. For obvious technical reasons, the mechanisms underlying embryonic cell associations are difficult to explore in the living embryo. Since the early investigations of Holtfreter and others, therefore, in vitro experimental systems have been developed for practically every embryonic tissue. These systems are based on the reaggregation in vitro of cell suspensions prepared by trypsinization of either chick or mouse embryonic tissues. The dispersed cells are then cultured and allowed to reaggregate into their characteristic histological patterns. The method assumes that the interactions observed between particular cell types in vitro are similar to those taking place during in vivo organogenesis.

The ability of cells to recognize each other and to sort out into distinct, cellular fabrics appears early in development. Sea urchin blastula cells, for example, form more stable adhesions than blastomeres from the cleavage stages. Surface components of the blastomeres, therefore, appear to become altered as the blastular epithelium is organized. Recognition is clearly shown in vertebrate embryos at gastrulation as observed from results of studies in which tissue fragments from different germ layers are combined in vitro. The cells of each germ layer possess their own adhesive properties and germ layer specificity. During the course of further development, different populations of cells within the same germ layer acquire characteristic recognition specificities. Hence, the capacity of cell populations derived from the same germ layer to sort out from one another would appear to depend on the acquisition and maintenance of unique surface properties.

What is the nature of the surface of an embryonic cell that allows it to associate with similar cell types? When during the process of organ formation are these properties acquired? How do these properties relate to the assembly of tissues and their stabilization in an organ? The capacity of cells from different germ layers as well as of different populations of cells from the same germ layer to recognize each other and sort out is apparent during the early morphogenetic phase of organ formation. For example, nonlimb mesodermal cells cannot participate with limb bud mesodermal cells to construct an integrated chimeric organ. Suppose dissociated mesodermal cells from chick leg bud (stages 18 and 19) and mesodermal cells from chick somites (stages 13–15) are mixed randomly and then placed as a pellet into a jacket of limb bud ectoderm. The ectodermal–mesodermal "assembly" is then grafted onto the dorsal wing of an appropriate host embryo. Within a period of 18 to 20 hours there is unequivocal evidence that the mesodermal cells are sorted out according to kind. Typically, the somitic cells ususally move to a central position within the graft, while the limb mesodermal cells occupy a more peripheral position. Similar observations are made on grafts formed by combining limb mesodermal cells with flank mesodermal cells isolated from chick embryos between stages 14 and 18. Despite the fact that somitic mesoderm and limb bud mesodermal cells have the capacity to differentiate into the same cell types (i.e., cartilage and bone cells), they segregate on the basis of their rudiment of origin within the mesodermal germ layer.

It is also clear that the surface properties of the same cell type during different stages of organogenesis may become altered; the differentiation of new and unique recognition properties would then affect behavior toward neighboring cells. For example, suppose the chondrocytes from limb cartilages (derivatives of the limb mesoderm) and the chondrocytes of the vertebral cartilages (derivatives of the somitic mesoderm) are digested from their matrices (eight-day chick embryo) and then randomly mixed within limb bud ectoderm Several days before the transplantation, the resultant limb graft shows internally a single chimeric mass of differentiated cartilage (i.e., limb and vertebral chondrocytes did not segregate but remained randomized). The intrinsic surface properties that probably cause these cells to segregate from each other during the morphogenetic phase of limb development are apparently no longer present. This suggests that when the morphogenetic phase of an organ is over and the cytodifferentiation of specialized cell types is under way, the surface properties of cells likewise become altered to permit them to associate into highly organized cellular fabrics.

Sorting-out studies in mixed aggregates using tissues from older chick and mouse embryos have provided most of what is known about the mechanism by which individual cells of organs are organized and held together in their specific groupings and ordered arrays. A variety of cell combinations have been used, including retina and heart cells, cartilage and heart cells, and heart ånd liver cells. Figure 13–28 illustrates two types of sorting-out patterns that are commonly exhibited by a mixture of embryonic cells. When dissociated, four-day limb bud chondrogenic cells are randomly mixed with five-day liver cells, the cultured heterotypic aggregate within several days shows the formation of a discrete outer liver tissue completely enveloping the inner, chondrogenic tissue. A different pattern of segregation is observed when heart cells of five-day-old embryos are mixed with retinal cells of seven-day-old embryos. Following the mixing of the two types of embryonic cells, the heart cells withdraw from the surface of the aggregate. Gradually, the heart cells organize as a series of clusters throughout the larger, continuous association of retinal cells. The number of coherent masses of heart tissue is dependent on the proportion of heart cells in the original cell aggregate.

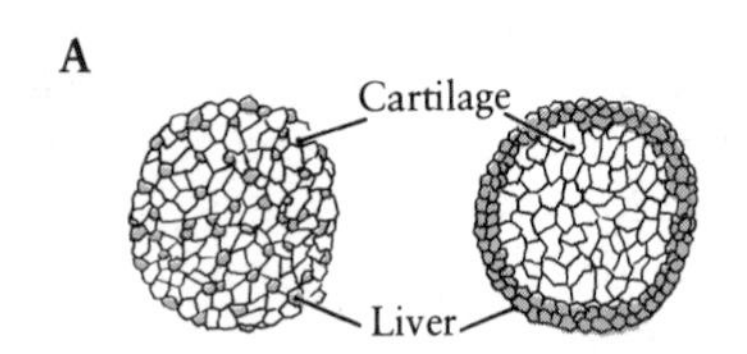


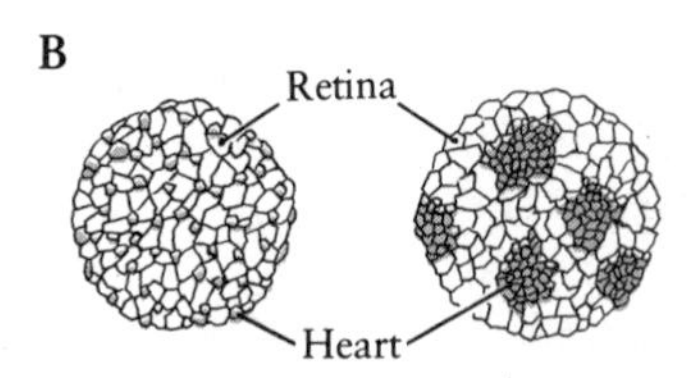


**FIG. 13–28.** Two types of sorting-out patterns as observed in mixtures of embryonic cells. A, liver and cartilage cells; B, heart and retinal cells.

Examination of a large number of such binary embryonic cell combinations leads to the general conclusion that heterologous or different cells segregate into distinct groupings, while like or homologous cells (i.e., cells belonging to the same tissue) associate with one another. For this reason, Moscona and colleagues find it useful to distinguish two categories of embryonic cell recognition during organ formation: (1) tissue-specific cell recognition, which refers to cells belonging to different tissues that segregate from one another and assemble preferentially into tissue-forming groups; and (2) cell type-specific cell recognition, which refers generally to the affinities between cells from the same tissues (i.e., with identical phenotypes). Analysis of the mechanisms underlying cell-to-cell recognition, morphogenetic cellular rearrangements, and cell adhesion remains a primary challenge to the developmental biologist. Although still rudimentary, our knowledge of how cells discriminate from one another (recognition specificity) and selectively form stable bonds between each other (cell adhesion) is increasing with the application of new technologies to probe cell surface phenomena. There is general agreement that the specificity of the recognition and adhesion processes involve interactions of cell surface constituents that function as *intercellular ligands* or cell-to-cell receptors. However, opinions differ as to the nature of these constituents and how they are organized to effect cell recognition and association into tissues.

Moscona and his co-workers have provided evidence using several different experimental approaches that cells from different embryonic tissues possess qualitatively different surface constituents. Such tissue-specific, membrane-associated molecules are known as *cognins*. Most of the work on cognins has focused on a single tissue type, the chick embryo neural retina. Retina cognin is extracted from the membranes of retina cells of embryos younger than 13 days of development. When this factor is combined with suspensions of dissociated pre-13-day retinal cells, there is rapid aggregation of the cells into a recognizable tissue. The cognin appears to be available only during the period when the neural retina undergoes its histological organization (i.e., it is stage specific). The retina-specific, cell-aggregating factor appears to be a glycoprotein with a molecular weight of 50,000 to 60,000; it consists of some 400 amino acids and various sugar residues (less that 20% by weight). The mode of action of retina cognin in bringing about tissue-specific cell affinities remains to be elucidated. Experimentally, the retinal cell-aggregating activity is sensitive to trypsin but not to enzymes that preferentially oxidize sugar residues, suggesting a critical role for the peptide moiety. Chemical "variants" of the tissue-specific cognin or differences in the topographical arrangement of cognin arrays on the cell surface may be possible mechanisms underlying the capacity of different cells within a single tissue to become histologically organized into a unified structure. An alternative view is put forth by Steinberg and colleagues. Although acknowledging the importance of cell surface components, he proposes that differences in tissue-specific recognition and adhesion between cells relate primarily to the number and/or distribution of adhesive sites over the cell surface.

The normal architecture of an organ requires that cells of a specific function be organized into a given tissue and that these tissues become spatially segregated into separate, distinct groupings. Morphogenetic processes involved in organ formation include cellular movements, recognition of a terminal cellular distribution, and stability of the final cellular configuration The organization of cells into multicellular tissue patterns requires interactions between cells. The identification of the cellular components and their role in mediating such interactions is necessary for an understanding of normal embryonic development as well as abnormal development.

## References


Amprino, R. 1965. Aspects of limb morphogenesis in the chicken. In: Organogenesis. R. DeHaan and H. Ursprung, eds. New York: Holt, Rinehart and Winston.

Bard, J. and E. Hay. 1975. The behavior of fibroblasts from the developing avian cornea. Morphology and movement *in situ* and *in vitro*. J. Cell Biol. 67:400–418.

Ben-Shaul, Y., R. Hausman, and A. Moscona. 1979. Visualization of a cell surface glycoprotein, the retina cognin, on embryonic cells by immuno-latex labeling and scanning electron microscopy. Dev. Bio. 72:89–101.

Bernfield, M., R. Cohn, and S. Banerjee. 1973. Glycosaminoglycans and epithelial organ formation. Am. Zool. 13:1067–1083.

Bryant, S., P. Bryant, and V. French. 1977. Biological regeneration and pattern formation. Sci. Am. 237:67–81.

Bryant, S., N. Holder, and P. Tank. 1982. Cell–cell interactions and distal outgrowth in amphibian limbs. Am. Zool. 22:143–151.

Calandra, A. and J. MacCabe. 1978. The *in vitro* maintenance of the limb bud apical ridge by cell-free preparations. Dev. Bio. 62:258–269.

Ede, D. 1971. Control of form and pattern in the vertebrate limb. In: Control Mechanisms of Growth and Differentiation. 25th Symposium of the Society for Experimental Biology, pp. 235–254. D. Davies and M. Balls, eds. New York: Academic Press.

Fallon, J. and G. Crosby. 1975. Normal development of the chick wing following removal of the polarizing zone. J. Exp. Zool. 193:449–455.

French, V., P. Bryant, and S. Bryant. 1976. Pattern regulation in epimorphic fields. Science 193:969–981.

Gipson, I. and T. Kiorpes. 1982. Epithelial sheet movement: protein and glycoprotein synthesis. Dev. Biol. 92:259–262.

Greensburg, J., S. Seppa, H. Seppa, and A. Hewitt. 1981. Role of collagen and fibronectin in neural crest cell adhesion and migration. Dev. Biol. 87:259–266.

Harrison, R. G. 1921. On relations of symmetry in transplanted limbs. J. Exp. Zool. 32:1–136.

Hilfer, S. and E. Hilfer. 1983. Computer simulation of organogenesis: an approach to the analysis of shape changes in epithelial organs. Dev. Biol. 97:444–453.

Hinchliffe, J. and D. Johnson. 1980. The Development of the Vertebrate Limb. New York: Oxford University Press.

Hornig, L. 1983. Does anterior (nonpolarizing region) tissue signal in the developing chick limb. Dev. Biol. 97:424–432.

Iten, L. 1982. Pattern specification and pattern regulation in the embryonic chick limb bud. Am. Zool. 22:117–129.

Lockshin, R. A. 1971. Programmed cell death. Nature of the nervous signal controlling breakdown of intersegmental muscles. J. Insect Physiol. 17:149–158.

Lockshin, R. A. and C. M. Williams. 1965. Programmed cell death. III. Neural control of the breakdown of intersegmental muscles of silkmoths. J. Insect Physiol. 11:601–610.

Manasek, F. 1975. The extracellular matrix: a dynamic component of the developing embryo. Curr. Top. Dev. Biol. 10:35–102.

Meier, S. and E. Hay. 1974. Control of corneal differentiation by extracellular materials. Collagen as a promoter and stabilizer of epithelial stroma production. Dev. Biol. 38:249–270.

Moscona, A. A. 1974. Surface specificities and embryonic cells: lectin receptors, cell recognition, and specific cell ligands. In: The Cell Surface in Development. A. A. Moscona, ed., New York: John Wiley & Sons.

Odell, G., G. Oster, P. Albrech, and B. Burnside. 1981. The mechanical basis of morphogenesis. I. Epithelial folding and invagination. Dev. Biol. 85:446–482.

Owens, E. and M. Solursh. 1981. In vitro histogenic capacities of limb mesenchyme from various stage mouse embryos. Dev. Biol. 88:297–311.

Reiter, R. and M. Solursh. 1982. Mitogenic property of the apical ectodermal ridge. Dev. Biol. 93:28–35.

Rowe, D., J. Cairns, and J. Fallon. 1982. Spatial and temporal patterns of cell death in limb bud mesoderm after apical ectodermal ridge removal. Dev. Biol. 93:83–91.

Rubin, L. and J. Saunders, 1972. Ectodermal–mesodermal interactions in the growth of limb buds in the chick embryo: constancy and temporal limits of ectodermal induction. Dev. Biol. 28:94–112.

Rutter, W., J. Kemp, W. Bradshaw, W. Clark, R. Ronzio, and T. Sanders. 1968. Regulation of specific protein synthesis in cytodifferentiation. J. Cell. Physiol. 72:1–18.

Rutz, R. and S. Hauschka. 1983. Spatial analysis of limb bud myogenesis: elaboration of the proximodistal gradient of myoblasts requires the continuing presence of apical ecodermal ridge. Dev. Biol. 96:366–374.

Saunders, J. 1948. The proximodistal sequence of origin of the parts of the chick wing and the role of the ectoderm. J. Exp. Zool. 108:363–403.

Saunders, J. 1966. Death in embryonic systems. Science 154:604–612.

Saunders, J. and J. Fallon. 1966. Cell death in morphogenesis. In: Major Problems in Developmental Biology, pp. 289–314. M. Locke, ed. New York: Academic Press.

Saunders, J. and C. Reuss. 1974. Inductive and axial properties of prospective wing bud mesoderm in the chick embryo. Dev. Biol. 38:41–50.

Saunders, J., M. Gasseling, and L. Saunders. 1962. Cellular death in morphogenesis of the avian wing. Dev. Biol. 5:147–178.

Smith, J. and L. Wolpert. 1981. Pattern formation along the anteroposterior axis of the chick wing: the increase in width following a polarizing region graft and the effect of X irradiation. J. Embryol. Exp. Morphol. 63:127–144.

Smith, R. and M. Bernfield. 1982. Mesenchyme cells degrade epithelial basal lamina glycosaminoglycan. Dev. Biol. 94:378–380.

Spooner, B. 1973. Microfilaments, cell shape changes, and morphogenesis of salivary epithelium. Am. Zool. 13:1007–1022.

Stark, R. and R. Searls. 1973. A description of chick wing bud development and a model of limb morphogenesis. Dev. Biol. 33:138–153.

Steinberg, M. 1970. Does differential adhesion govern self-assembly processes in histogenesis? Equilibrium configurations and the emergence of a hierachy among populations of embryonic cells. J. Exp. Zool. 173:395–434.

Steinberg, M. 1978. Cell-cell recognition in multicellular assembly levels of specificity. In: Cell-Cell Recognition. 32nd Symposium of the Society for Experimental Biology, pp. 25–49. A. Curtis, ed. New York: Academic Press.

Steinberg, M. and T. Poole. 1982. Cellular adhesive differentials as determinants of morphogenetic movements and organ segregation. In: Developmental Order: its Origin and Regulation. 40th Symposium of the Society for Experimental Biology, pp. 351–378. S. Subtelny and P. Green, eds. New York: Alan R. Liss.

Sugrue, S. and E. Hay. 1982. Interaction of embryonic corneal epithelium with exogenous collagen, laminin, and fibronectin: role of endogenous protein synthesis. Dev. Biol. 92:97–106.

Summerbell, D. 1981. The control of growth and the development of pattern across the anteroposterior axis of the chick limb bud. J. Embryol. Exp. Morphol. 63:161–180.

Summerbell, D. and L. Honig. 1982. The control of pattern across the anteroposterior axis of the chick limb bud by a unique signaling region. Am. Zool. 22:105–116.

Tarin, D. and A. Sturdee. 1971. Early limb development in *Xenopus laevis*. J. Embryol. Exp. Morphol. 26:169–179.

Thiery, J., J. Duband, and A. Delouvée. 1982. Pathways and mechanisms of avian trunk neural crest cell migration and localization. Dev. Biol. 93:324–343.

Thompson, H. and B. Spooner. 1982. Inhibition of branching morphogenesis and alteration of glycosaminoglycan biosynthesis in salivary glands treated with β-D-xyloside. Dev. Biol. 89:417–424.

Tomasek, J., E. Hay, and K. Fugiwara. 1982. Collagen modulates cell shape and cytoskeleton of embryonic corneal and fibroma fibroblasts: distribution of actin, α-actinin, and myosin. Dev. Biol. 92:107–122.

Tosney, K. 1982. The segregation and early migration of cranial neural crest cells in the avian embryo. Dev. Biol. 38:13–24.

Trinkaus, J. 1976. On the mechanism of metazoan cell movements. In: The Cell Surface in Animal Embryogenesis and Development, Vol. 1., pp. 227–329. G. Poste and G. Nicolson, eds., New York: North-Holland.

Watanabe, M., D. Bertolini, D. Kew, and R. Turner. 1982. Changes in the nature of the cell adhesions of the sea urchin embryo. Dev. Biol. 91:278–285.

Wolpert, L. and A. Hornbruch. 1981. Positional signaling along the anteroposterior axis of the chick wing. The effect of multiple polarizing region grafts. J. Embryol. Exp. Morphol. 63:145–159.

Zwilling, E. 1961. Limb morphogenesis. Adv. Morphog. 1:301–330.

Zwilling, E. 1968. Morphogenetic phases in development. In: The Emergence of Order in Developing Systems, pp. 184–207. M. Locke, ed. New York: Academic Press.

Zwilling, E. 1974. Effects of contact between mutant (wingless) limb buds and those of genetically normal chick embryos: confirmation of a hypothesis. Dev. Biol. 39:37–48.

# 14

# DEVELOPMENT OF THE FACE, PALATE, ORAL CAVITY, AND PHARYNX

The ventrolateral aspect of the head of the early human embryo shows a number of barlike processes separated from each other by a series of grooves. The ridges are the *visceral arches* and the grooves are the *visceral furrows* (Fig. 14–1). Four well-defined arches are seen at the end of the first month (Fig. 14–1 A). The fifth arch always remains rudimentary. A lateral view of a slightly older embryo shows the formation of the *mandibular* and *maxillary processes* from the first arch (Fig. 14–1 B).

## The Face

At the beginning of the second month when the pharyngeal membrane ruptures to establish connection between the ectodermal stomodeum and the endodermal foregut, the lateral boundaries of the stomodeum are formed by the mandibular and the maxillary processes (Fig. 14–2 A,B). In a frontal view of the face at this time, most of the structures that will contribute to the formation of the face are distinguishable (Fig. 14–2 A,B). The lower jaw, the caudal boundary of the stomodeum, is formed by the mandibular processes whose origin from a pair of lateral primordia is evident in this figure. The maxillary processes, which form a large part of the upper jaw, are present only as small rudiments at this time. The maxillary and mandibular processes, both derivatives of the first visceral arch, merge with each other at the lateral boundary of the stomodeum. The transverse diameter of the stomodeum is considerably greater than that of the definitive oral opening. Present also are a pair of nasal placodes that represent the beginning of the development of the nose. Between them is the unpaired *frontonasal prominence.* During the next two weeks (Fig, 14–2 B–D) the maxillary processes become more prominent and the olfactory placodes invaginate to form the *olfactory pits.* The rim of each olfactory pit is deficient where it communicates with the oral cavity, and the olfactory pit might well be termed the olfactory groove at this time. *Lateral nasal processes* and *medial nasal processes* form the lateral and medial ridges of each olfactory pit.

The outline of the face emerges as the structures named above continue to grow (Fig. 14–2 E,F). A deepening of the nasal pits and the stomodeum is brought about by the forward growth of the mesodermal structures surrounding these orifices. The olfactory pits approach each other in the midline, gradually squeezing out the frontonasal prominence. The maxillary processes grow medially and fuse with an extension of the medial nasal process.

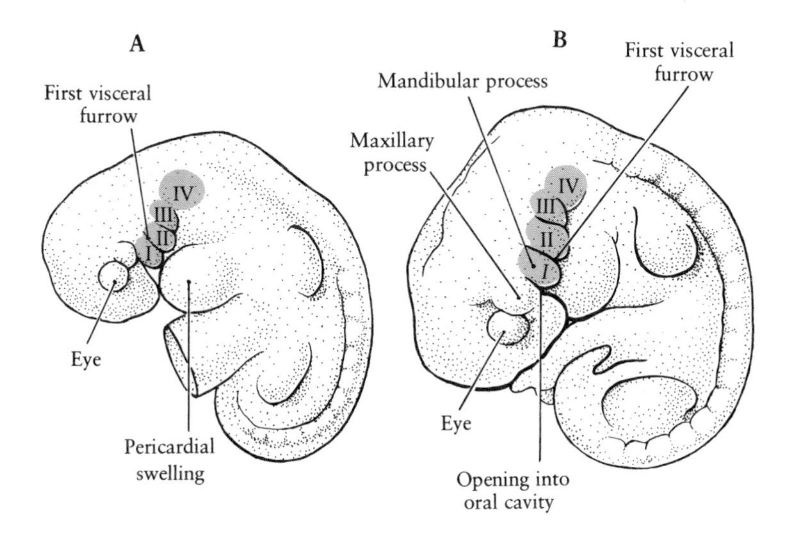


**FIG. 14–1.** External appearance of the visceral arch system. A, 4.2 mm (week 5); B, 6.3 mm (week 6). (From E. Blechschmidt, 1961. The Stages of Human Development before Birth. W. B. Saunders Company, Philadelphia.)

The junction of the maxillary processes and the median nasal processes forms the *philtrum* of the adult lip. Fusion of these processes closes the externally observed connection between the olfactory grooves and the oral cavity and converts the olfactory grooves into blind passageways whose external openings are the nostrils *(external nares)*. The median nasal processes fuse to form the *median nasal septum* separating the two nasal cavities. The external nares thus are brought closer together from their original lateral positions. A transverse furrow develops between the nasal region of the frontonasal prominence and the frontal region of the skull, setting off the nose as a separate structure.

The developing eyes and ears are also seen in a frontal view of the face during the second month (Fig. 14–2 D,E). A groove extends between the maxillary process and the lateral nasal process from the corner of the eye. This is the *nasolacrimal furrow*. It marks the beginning of the formation of the nasolacrimal canal, which connects the orbital and nasal cavities in the adult. The first visceral furrow between the first and second visceral arches marks the position of the future external auditory meatus. Contributions from the areas of the first and second arches surrounding the first visceral furrow will form the external ear.

These external modifications occur during the second month and are followed by the development of the underlying bony structures of the jaws. The most medial part of the upper jaw forms from an extension inward of the fused median nasal processes and is homologous to the premaxillary process and bone of lower forms. It bears the incisor teeth. The remainder of the upper jaw, carrying all of the other upper teeth, develops from the maxillary portions of the first arch, while the tooth-bearing portions of the lower jaw are formed from the mandibular division.

## The Palate

The deep end of each olfactory pit is separated from the underlying oral cavity by an epithelial plate, the *oronasal membrane* (Fig. 14–3 A), and thus at first each olfactory pit ends blindly. When the oronasal membrane perforates late in the second month, the nasal cavity now opens into the anterior part of the oral cavity by way of a pair of foramina, the *primitive choanae* (Fig. 14–3 B,C). The most anterior part of the nasal cavity is separated from the oral cavity by the *median palatine process (primary palate)*, which is the inner extension of the fused median nasal processes mentioned above as being homologous to the premaxillary bone. The short passageway from the anterior nares to the primitive choanae is extended during the third month by the formation of a much larger roof over the oral cavity, the *secondary palate*. Only a small anterior part of the definitive palate is formed by the median palatine process. The major part of the palate is formed by the medial growth of a pair of *lateral palatine processes*, shelflike outgrowths from the inferior surface of the maxillary processes (Fig. 14–4). The portion of the palate derived from the lateral palatine processes is termed the secondary palate. At first the palatine shelves hang vertically downward along the sides of the tongue at the time when

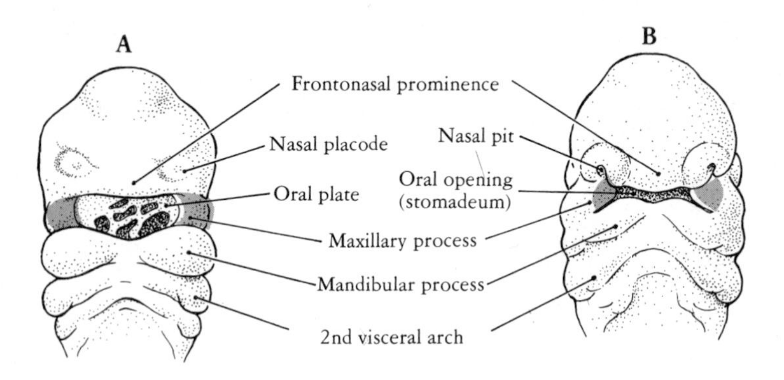


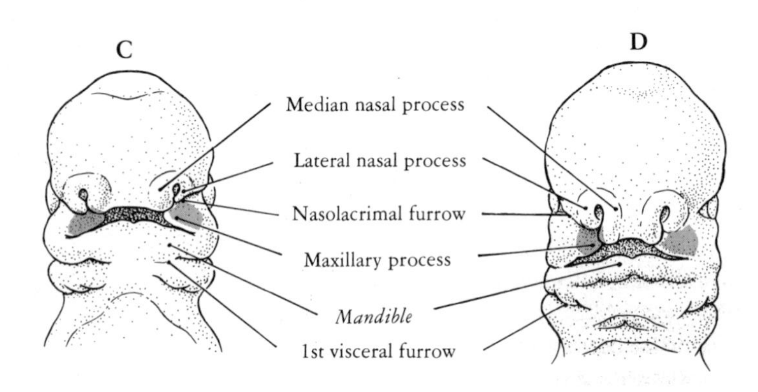


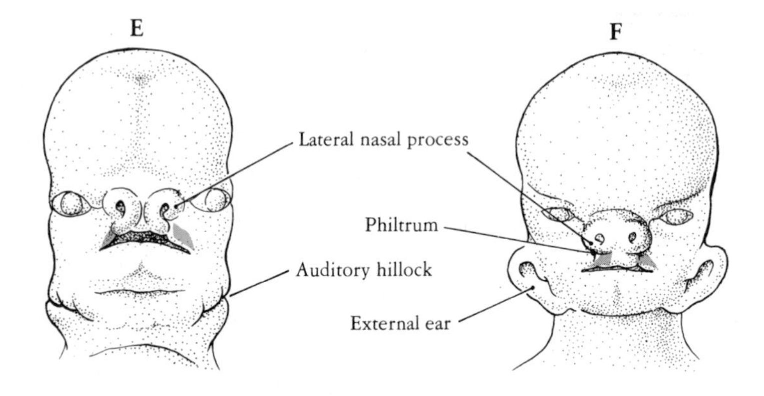


**FIG. 14–2.** Development of the human face as seen from the frontal aspect. A, four weeks; B, five weeks; C, five and a half weeks; D, six weeks; E, seven weeks; F, eight weeks.

the developing tongue occupies a large portion of the oronasal cavity (Fig. 14–4 C). Subsequently they move from a vertical to a horizontal position above the tongue—a process known as shelf elevation—and shortly thereafter grow toward the midline and make contact with each other and with the medial palatine process (Fig. 14–4 E–H). The preliminary adhesion is between the epithelial coverings of the apposing palatal shelves, but the epithelial layers soon break down, establishing a continuity and a definitive fusion between the mesenchymal tissue of the two shelves.

Although the entire process of palate formation has been studied extensively, attention most recently has been focused on the mechanism of shelf elevation and the mechanism of the breakdown of the midline epithelial seam. The reorientation of the palatal shelves from a vertical to a horizontal position has been variously attributed either to extrinsic "forces" lying outside of the palatal tissues or to intrinsic "forces" involving macromolecular synthesis and cellular reorganization within the palatal tissues themselves.

Tongue regression and the differentation of the

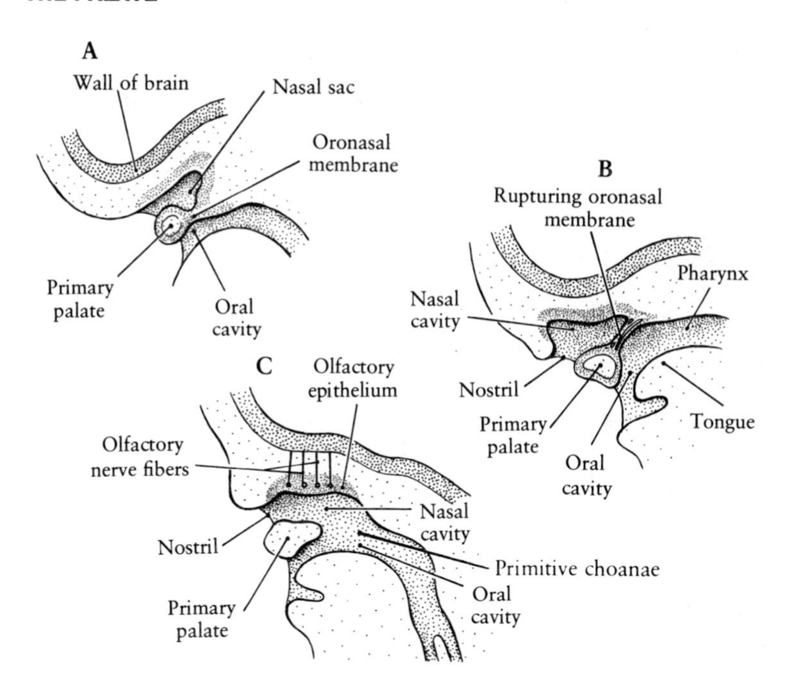


**FIG. 14–3.** Sagittal sections of the early human embryo showing the formation of the nasal cavities and the primary choanae. A, five weeks; B, six weeks; C, seven weeks.

tongue myoneural apparatus both occur at the time of shelf elevation, and it is tempting to assign a cause-and-effect relationship to these readily observable events. However, shelf elevation in the absence of a tongue (aglossia) and in the presence of a tongue of markedly reduced size (microglossia) has been observed in many cases, and it has been concluded that neither morphological nor biochemical changes in the developing tongue are involved in the mechanism of palatal shelf elevation.

Thus, we must look to intrinsic forces to account for shelf reorientation. It has been observed that at the time of shelf elevation there is an increase in the acid mucopolysaccharide content of the ground substance of the palatal shelves. This could produce an elevating force due to the turgor associated with the strong water-binding affinities of GAGs. A buildup of the other important molecular component of the ground substance, collagen, during shelf elevation has also been reported, but what precise role collagen fibers may play in the process of shelf elevation is not known.

Another hypothesis suggests that the mechanism may lie within the cells themselves. An increased synthesis of the contractile proteins actin and myosin and the presence of a system of microfilaments—presumably containing the actin–myosin complex—have been observed just prior to shelf elevation. These changes may then be associated with morphological movements of the cells themselves that would control shelf reorientation. The concomitant increase in GAGs might then serve to facilitate the cellular movement activated by contractile proteins.

Once the epithelia of the palatal shelves have made contact to form a midline epithelial seam, the epithelial cells must then be removed to establish mesenchymal cell continuity and permanent shelf fusion. It is now generally established that intracellular autolytic processes are responsible for the degeneration of the epithelial cells. Hydrolytic enzymes have been shown to be localized in the epithelial cells at the time of seam formation and subsequent cellular breakdown. Cyclic AMP increases in amount in the palatal shelf cells during shelf elevation and epithelial adhesion, and is predominantly localized in the epithelium. The midline epithelial seam stains most heavily at the time of epithelial adhesion and cell death. Since cell death is an active event requiring the synthesis of specific proteins, it has been suggested that cAMP is important to the synthesis or activation of the proteins required for epithelial cell autolysis (Greene et al., 1980).

When the lateral palatine processes fuse in the midline, they roof over the oral cavity and extend the posterior openings of the nasal cavities back to the region of the pharynx (Fig. 14–4 F,H). The re-

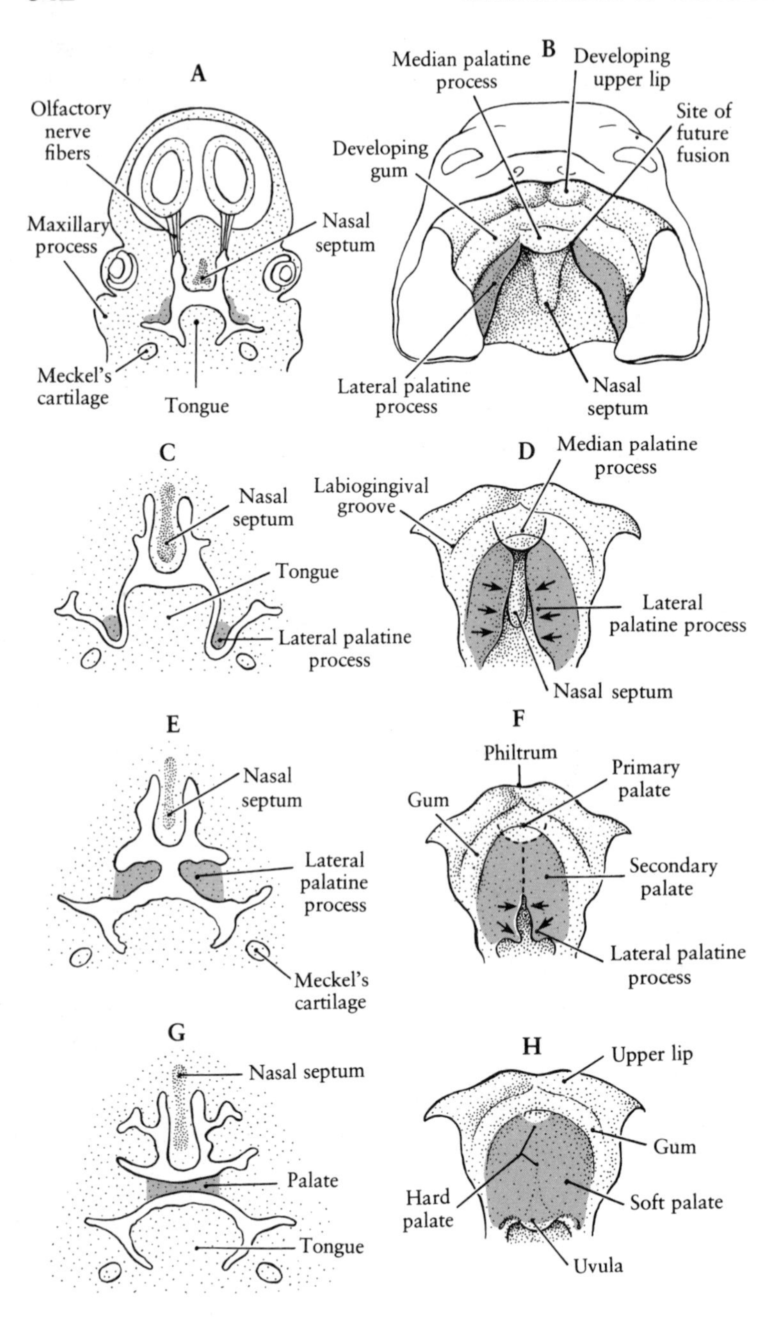


**FIG. 14–4.** Drawings of frontal sections (A, C, E, and G) and of the roof of the oral cavity (B, D, F, and H) illustrating the development of the palate from week 6 to week 12.

spiratory and alimentary passageways thus establish separate external entrances but open in common posteriorly into the pharynx. The two nasal cavities are completely separated from each other by the *nasal septum,* which fuses with the palate (Fig. 14–4, C,E,G). The nasal septum differentiates from the medial mass of the cartilaginous model of the ethmoid bone. The lower portion of the septum that joins the palate remains cartilaginous. The upper portion ossifies to form the perpendicular plate of the ethmoid bone, thus completing the nasal septum. The anterior part of the palate ossifies to form the *hard palate.* The most posterior part, posterior to the nasal septum, however, contains no bone and becomes the *soft palate,* its most posterior portion forming the triangular *uvula* (Fig. 14–4 H).

### Cleft Palate and Cleft Lip

Failure of fusion of any part of the palatine processes results in a gap in the roof of the mouth, known as *cleft palate*. In the soft palate, the gap is commonly in the midline. However, in the anterior part of the hard palate, the cleft is located on one or both sides of the midline, resulting from the failure of the fusion of one or both of the lateral palatine processes with the medial palatine process.

Cleft palate is one of the most common congenital malformations in humans, its incidence varying from 1 in 600 to 1 in 1000 in different countries. Since nongenetic exogenous causes are probably responsible for a large proportion (as much as 80%) of palatal clefts, this abnormality has been extensively studied. Of particular importance to these studies was the discovery by Baxter and Fraser (1950) that cortisone administration to mice early in pregnancy produced a high incidence of cleft palate, thus providing a reliable experimental system. Nevertheless, the mechanism of the production of cleft palate in experimental animals is still undetermined, and any evidence of the mechanism through which it occurs in the human palate is conflicting and inconclusive.

Cleft palate may result either from a failure in the initial contact of the palatal shelves with each other or with the medial palatine process, or—after normal epithelial adhesion has taken place—a failure to follow through with the subsequent events of epithelial breakdown and fusion of the mesenchymal tissues. Palatal cleft abnormalities produced by cortisone injections may involve either or both of these processes. Cortisone injection decreases the synthesis in the palatal shelf tissues of mucopolysaccharides and collagen, molecules that have both been implicated as of possible importance in shelf elevation and medial growth. Glucocorticoids also inhibit cell proliferation, and the decreased mitosis noted in the palatal mesenchyme following cortisone administration could result in cleft formation due to growth-inhibiting effects. Both of these cortisone-produced effects could produce palatal clefts due to the failure of normal medial growth and epithelial contact.

In addition, it has been shown that cortisone administration also results in the retardation of protein synthesis in the epithelial cells of the palatal shelves. Inhibition of protein synthesis could result in the failure of the formation of the lysosomal enzymes necessary for cell autolysis and thus interfere with the controlled epithelial cell degeneration necessary for shelf fusion.

Agents other than cortisone (hypervitaminosis A, radiation, aminopropionitrile, and other teratogens) may also result in the production of cleft palate, but the use of these agents has been no more informative. The mechanism of the development of palatal shelf abnormalities is still unknown, and in fact it has not yet even been determined whether the primary teratological effect is on the epithelium, the mesenchyme, or both (Shah, 1979).

Failure of the fusion of the maxillary process with the median nasal process results in a common anomaly known as *cleft lip*. The gap in the lip is at one side of the midline at the philtrum. The name "harelip" is thus somewhat of a misnomer, since in the rabbit the gap is in the midline. A cleft lip may occur on one or on both sides and may be continuous posteriorly with a cleft palate.

## The Oral Cavity and the Pharynx

### The Teeth

A sagittal section through the primitive jaws shows that up until the sixth week they are solid masses. At this time a thickened plate of epithelium, the *labial lamina*, appears in the midline and spreads outwardly in a semicircle in both directions around each jaw (Fig. 14–5). As the labial lamina sinks into the underlying mesoderm, the central cells disappear and a groove, the *labial groove*, is formed (Fig. 14–5 B). The labial groove becomes the *vestibule* separating the lips from the gums (Fig. 14–5 C).

Projecting inwardly from each labial lamina is a second semicircular shelf of tissue, the *dental lamina* (Fig. 14–5 A,B). Since the dental lamina is an ingrowth of the oral epithelium, it must be appreciated that it contains the rapidly proliferating cellular layers of normal surface epithelia. Through the mitotic activity of these cells, the dental lamina increases rapidly in size. During the third month, specialized regions, the *enamel organs*, develop (Fig. 14–5 B,C). They form as cup-shaped structures that sink into the mesoderm but remain connected to the gum epithelium by way of the dental lamina, now a constricted cord of cells. Five of these structures develop on each side in both the upper and

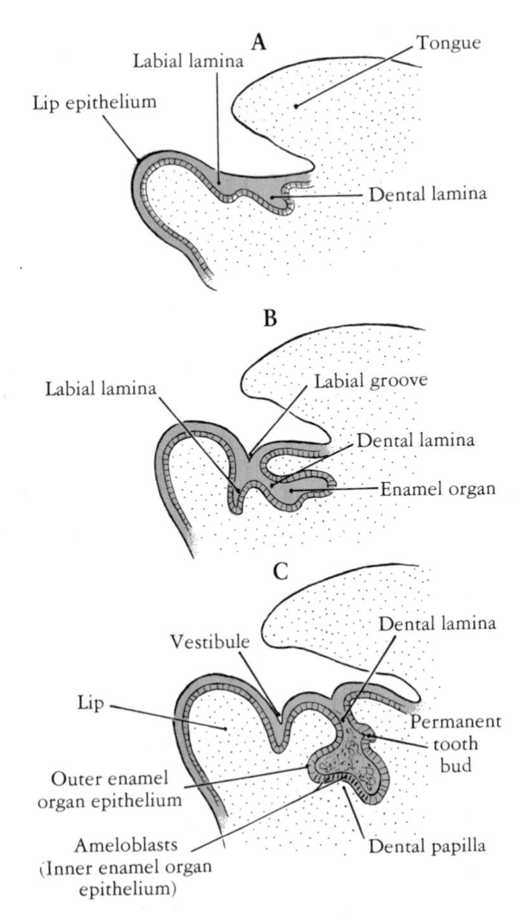


FIG. 14–5. Sagittal sections of the lip and the jaw showing the separation of the lip and the gum by the labial groove and the early development of a tooth. A, 7 weeks; B, 8 weeks; C, 11 weeks.

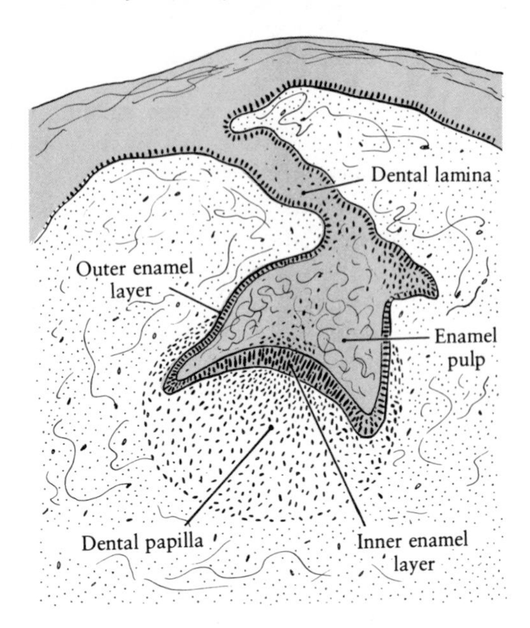


FIG. 14–6. Developing human tooth at three months.

FIG. 14–7. Developing human incisor at seven months. A, sagittal section; B, detail of rectangle in A.

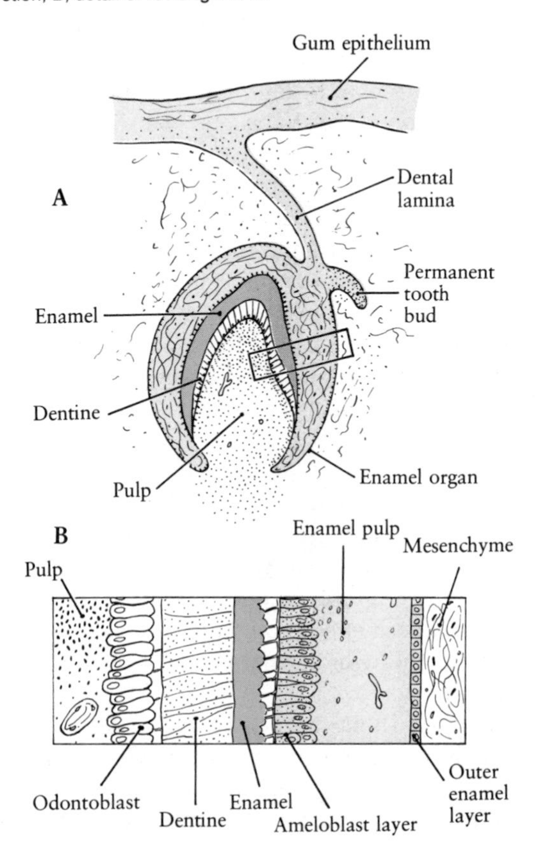


the lower jaws. They will form the deciduous or milk teeth. On its mesenchymal side each enamel organ becomes indented by a core of mesoderm, the *dental papilla* (Figs. 14–5 C and 14–6). Both the enamel organ and the dental papilla are surrounded by a connective tissue sheath, the *dental sac*. The entire structure is known as the *tooth germ* or *tooth bud*.

The enamel organ consists of an inner concave group of columnar cells, the *inner enamel organ epithelium*, which is continuous with an outer convex layer of cuboidal cells, the *outer enamel organ epithelium*. Between these two epithelia, the ectodermal matrix forms a stellate reticulum that contains large amounts of intercellular fluid rich in mucopolysaccharides, morphologically resembling embryonic mucuous connective tissue (Figs. 14–6 and

14–7 A). The indented inner epithelium gives rise to the *enamel* of the tooth. Its cells are termed *ameloblasts*. The cells of the enclosed dental papilla closest to the ameloblast layer differentiate into a *dentine*-forming layer of cells called *odontoblasts* (Fig. 14–7 B). The remainder of the mesodermal core of the dental papilla differentiates into the *pulp* of the tooth. The pulp consists of a framework of reticular tissue binding together the blood vessels, lymphatics, and nerves of the tooth. Enamel and dentine are formed simultaneously by continued secretions from the ameloblasts and odontoblasts, the oldest enamel and dentine being in apposition in the center of the developing tooth. Enamel, dentine, and bone have similar characteristics consisting of an organic framework in which are deposited inorganic salts. Bone contains approximately 33 percent organic material, dentine somewhat less, and enamel only about 5 percent. The root of the tooth begins to form after the crown is almost completed, and it has not yet completed development even at the time the tooth erupts.

*Epithelial-Mesenchymal Interactions*

Differentiation of the teeth is still another excellent example of how, in many organs, development depends on an orderly sequence of interactions between epithelial and mesenchymal components. In the development of the teeth, the initial event in the sequence takes place when the migrating mesenchymal cells (of neural crest origin) that will form the dental papilla contact the stomodeal epithelium. This interaction is necessary for the later differentiation of the odontoblasts. Then, interaction between the dental mesenchyme and the oral ectoderm results in the formation of the tooth bud. Within the tooth bud, odontoblast differentiation is the result of a signal from the inner enamel ectoderm. In turn, the odontoblasts secrete predentine and thereby convey a signal back to the inner enamel ectoderm that results in the differentiation of ameloblasts and the subsequent formation of enamel.

The last two steps of this process that occur in what is termed the bell stage of the tooth bud (Fig. 14–7) have been studied extensively. Interaction between the dental papilla mesoderm and enamel organ ectoderm is prerequisite for the differentiation of odontoblasts. However, only dental papilla mesoderm may be induced to form odontoblasts. They cannot be made to form in any other mesoderm, not even in the oral mesoderm of the rodent in the region of the *diastema* (the gap between the canines and the incisors where no teeth develop). The dental papilla is thus predetermined, and the signal from the dental lamina is merely a trigger. It is a permissive induction. However, the reciprocal induction of ameloblasts by the differentiated odontoblasts is quite different. It is directive. In contrast to the lack of effect of dental lamina ectoderm on the mesoderm of the diastema, dental papilla mesoderm will readily induce the differentiation of ameloblasts in the diastema ectoderm. In fact, the mesoderm of the dental papilla can induce the development of ameloblasts in a variety of epithelia including even the skin of the foot.

Electron-microscopic examination of the tooth bud at the time of odontoblast differentiation shows tufts of fibrils of the mesoderm of the dental papilla impinging on the inner enamel epithelium whose basal lamina is still intact. The triggering impulse from the enamel ectoderm thus does not require direct cell-to-cell contact but is mediated through the external cell matrix—in this instance the basal lamina. Once the odontoblasts differentiate and begin to secrete predentine, which appears as an accumulation of collagen fibers containing matrix vesicles at the epithelial–mesenchymal interface, the basal lamina of the dental epithelium breaks down and ameloblasts differentiate. A cell-to-cell contact that includes the formation of desmosomes between the processes of the odontoblasts and the dental epithelium is thus necessary for the induction of ameloblasts.

Transfilter induction experiments support these findings. When filters whose pore size is in the range of 0.6 to 0.8 $\mu$m are placed between the dental mesoderm and ectoderm, the normal sequence of events takes place. In three days the mesenchymal cells line up in rows, and on the following day odontoblasts differentiate and begin to secrete predentine. At the end of a week, the ectodermal cells on the other side of the filter become polarized, differentiate into ameloblasts, and begin to secrete enamel. Electron-microscopic examination of the filter always shows penetration of cytoplasmic processes into the pores from the mesodermal side of the filter. If the pore size is reduced to 0.2 micron, the differentiation of the odontoblasts is delayed for three days, and their differentiation at six days correlates with the slower penetration of the cytoplasmic processes into the smaller pores. Again, following the differentiation of the odontoblasts, ameloblasts form on the ectodermal side of the filter four days later. With pores 0.1 $\mu$m or less in diameter, no cytoplasmic penetration occurs and odontoblasts do not differentiate. In this case, neither do ameloblasts.

Thus, since penetration of mesodermal cytoplasmic processes into the pore canals is necessary for odontoblast differentiation, the induction process is not the result of what might be called a long-range signal. However, since EM studies show that the basal lamina of the enamel ectoderm is still intact at the time of odontoblast induction, neither is there a need for heterotypic cell contact. Only close contact between the cytoplasmic processes of the mesenchymal component and the external cell matrix of the epithelial component is necessary to allow the triggering signal to be transmitted. The induction of ameloblast formation, however, is preceded by the breakdown of the basal lamina of the epithelium, allowing cell-to-cell contact before the directive inductive signal is transmitted. The nature of the signal is not known.

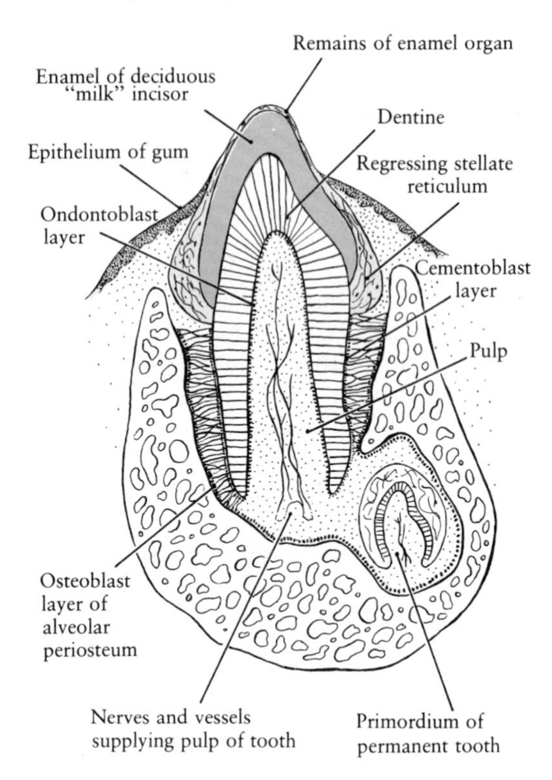


**FIG. 14–8.** Drawing of an erupting tooth showing its relation to its bony alveolar socket in the jaw.

*Tooth Eruption*

Part of the dental sac has an important function in the late development of the tooth. As the tooth erupts, both the dental sac and the enamel organ are sloughed off in the part that has erupted. However, the dental sac that surrounds the root differentiates into an organ closely applied to the dentine of the tooth on one side and the bony alveolar socket of the jaw on the other. Both of these layers of periosteal tissue together constitute the *periodontal membrane.* The cells of the dental sac closest to the dentine differentiate into *cementoblasts.* The cementoblasts secrete a substance, *cementum,* histologically and chemically similar to bone, around the root of the tooth. The remaining cells of the dental sac form fibers that hold the tooth in place by embedding themselves in both the dentine of the tooth and the bone of the alveolar socket (Fig. 14–8).

Each deciduous tooth will eventually be replaced by a permanent tooth. The permanent teeth develop from lingual extensions of the dental lamina in the same manner as the deciduous teeth (Figs. 14–7 and 14–8). Those permanent teeth that have no deciduous precursors, the molars, develop from backward-growing extensions of the ends of the semicircular dental lamina of each jaw. The permanent teeth grow slowly at first, but gradually their growth exerts a pressure on the deciduous teeth. This pressure, along with a partial absorption of their roots, results in the milk teeth being shed.

The order in which the deciduous teeth erupt is generally from the midline laterally. However, their replacement does not follow this pattern. The approximate times of eruption of the deciduous and permanent teeth are as follows:

| *Deciduous Teeth* | | *Permanent Teeth* | |
|---|---|---|---|
| central incisors | 6–8 months | | 7 years |
| lateral incisors | 7–10 months | | 8 years |
| canines | 14–20 months | | 12 years |
| first molars | 12–16 months | first bicuspids | 11 years |
| second molars | 20–30 months | second bicuspids | 11 years |
| | | first molars | 6 years |
| | | second molars | 12 years |
| | | third molars | 17–25 years (often later—or never) |

## The Hypophysis

The hypophysis is entirely an ectodermal derivative. However, it develops from two separate primordia, *Rathke's pouch,* an outgrowth of the stomodeum, and the *infundibulum,* an outgrowth of the diencephalon.

Rathke's pouch may be seen during the fourth week of development as a hollow evagination of the roof of the stomodeum just in front of the region of the pharyngeal membrane (Fig. 14–9 A), extending toward the floor of the overlying brain. When Rathke's pouch meets the infundibulum, its oral attachment becomes constricted (Fig. 14–9 B,C), and it then completely loses its connection to the roof of the oral cavity (Fig. 14–9 D). By week 12 the buccal anlagen is completely separated from the oral cavity by the developing sphenoid bone, which forms a bony depression, the *sella turcica,* which partially encloses the pituitary anlagen (Fig. 14–9 E). Rathke's pouch develops into the *anterior lobe* of the pituitary gland.

After its contact with the infundibulum, the anterior lobe undergoes marked changes. The cells of its anterior wall proliferate rapidly to form the *pars distalis.* In so doing they impose on the lumen of Rathke's pouch until this space is reduced to a narrow cleft, the *residual lumen.* The wall of Rathke's pouch in contact with the infundibulum forms the *pars intermedia* of the anterior lobe. This part of the pituitary may be well defined in the fetus and the infant, but in the adult, it merges with the *pars nervosa* of the *posterior lobe* and becomes obscure. Paired lateral extensions of the pars distalis grow out, fuse, and form a collar around the infundibular stalk. This is the *pars tuberalis.*

The pars distalis becomes highly vascularized, and its cells become arranged in interlaced columns forming a network around the blood vessels. Most of the blood vessels are branches of the internal carotid artery. During the third month of development, there is evidence of cell specialization when secretory granules appear in a number of cells, indicating their differentiation into cells that are destined to synthesize specific pituitary hormones. These functional fetal pituitary cells may secrete trophic hormones that are necessary for the complete development and function of the thyroid and adrenal glands. However, the majority of the cells of the pars distalis remain chromophobic during fetal life and do not show accumulations of the characteristic acidophilic and basophilic granules found in the adult gland.

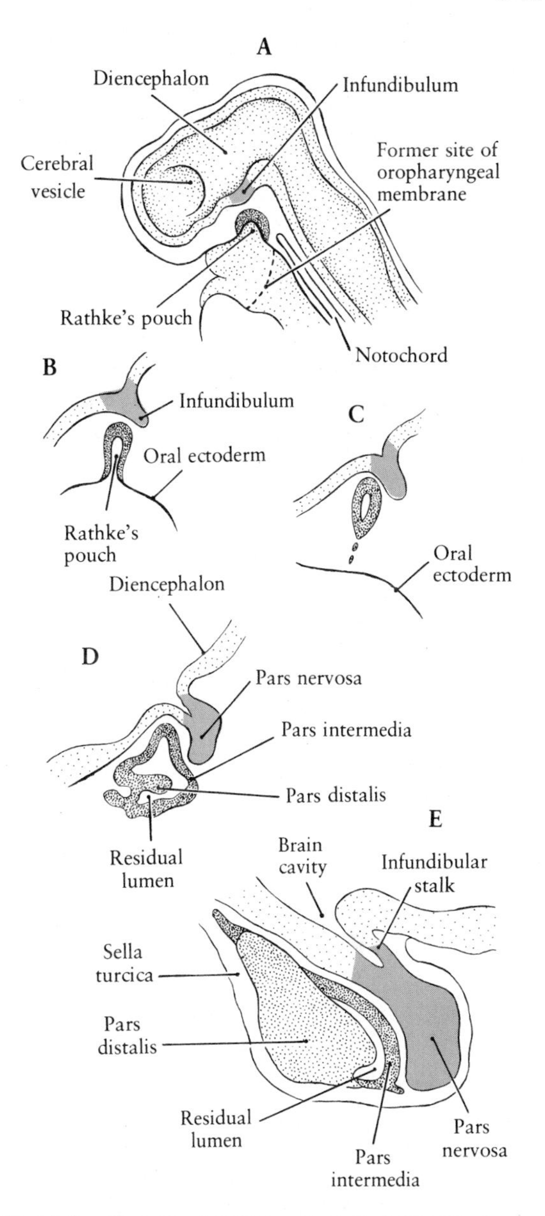


**FIG. 14–9.** Diagrams of sagittal sections illustrating the development of the hypophysis. A, end of the first month showing the dual origin of the gland from the roof of the oral cavity (Rathke's pouch) and the floor of the diencephalon (infundibulum); B, C, meeting and fusion of two primordia during month 2; D, 8 weeks; E, 11 weeks.

The infundibulum forms the posterior lobe of the pituitary. Its distal end enlarges to become the pars nervosa, which remains connected to the hypothalamus by way of the mostly solid *infundibular stalk* (Fig. 14–9 E). The cells of the pars nervosa differentiate into cells known as *pituicytes,* which are modified neuroglia cells. Nerve cells are not found in the pars nervosa, but many nerve fibers grow into this area from nuclei in the hypothalamus. These are the neurosecretory fibers whose function was described in Chapter 5.

## Derivatives of the Visceral Arch System

A section through the pharyngeal region at the end of the first month (Fig. 14–10 A) shows that behind the oral cavity the pharynx has flattened and broadened and developed a series of endodermal diverticula, *pharyngeal pouches,* growing laterally toward the corresponding ectodermal visceral furrows. In fishes the pharyngeal pouches and the visceral furrows break through and join to form *visceral clefts.* These visceral clefts are also termed gill slits, since the visceral (branchial) arches bear respiratory structures, the gills. The gill slits provide exit for the respiratory water, which enters through the mouth. In amniotes, visceral clefts usually do not develop and, if they do, are only temporary and never serve any respiratory function. Thus, it is quire incorrect to say that the amniote embryo—despite the fact that it develops in an aquatic environment—is ever a fish (or even fishlike). Although

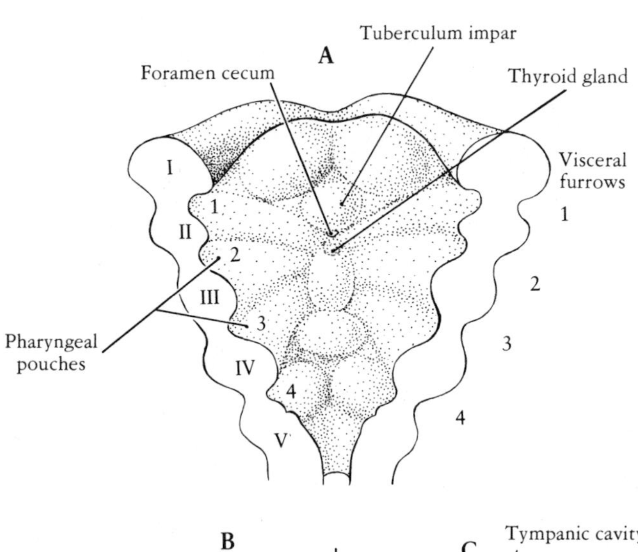


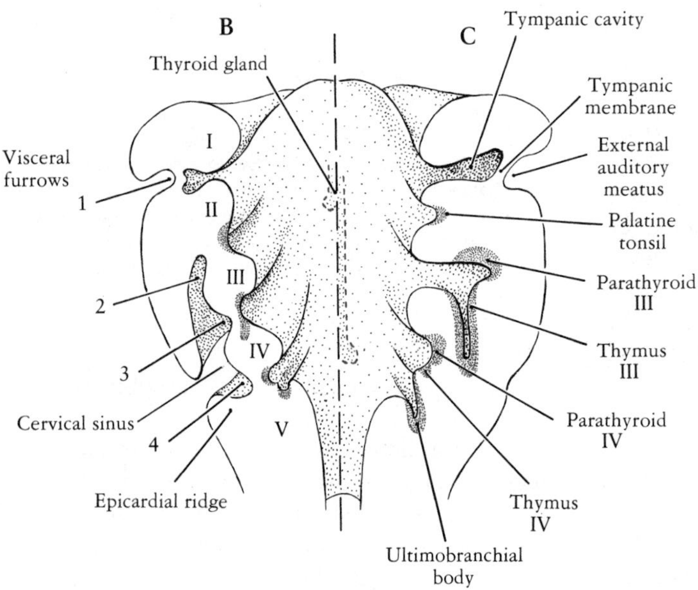


FIG. 14–10. Diagrams of three stages in the differentiation of the visceral arch system. A, about four weeks; B, formation of the cervical sinus (week 6); C, differentiation of the pouches (week 6).

the visceral arch system in the amniotes gives up its respiratory function, the furrows, arches, and pouches develop into a wide variety of other specialized structures.

## Visceral Furrows

The first visceral furrow forms the external auditory meatus. It is the only visceral furrow that forms an adult structure, the rest forming indistinguishable parts of the neck contour.

## Visceral Arches

At the end of week 6, the second arch grows over the more caudal ones, sinking them into a depression known as the *cervical sinus* (Fig. 14–10 B). When the second arch fuses with the epipericardial ridge, the cervical sinus is cut off from the surface and is eventually obliterated (Fig. 14–10 C). The caudal growth of the second arch may be likened to the same event in the bony fishes, which forms the flaplike operculum over the posterior gill arches. Thus, after a short existence of only about two weeks, any resemblance to the gill-bearing branchial arch system of the teleosts is lost.

Each visceral arch has three components: (1) an artery, (2) a nerve, and (3) mesenchyme—which will form cartilage, bone, and muscle.

The visceral arches at one time or another provide a pathway for the paired aortic arches from the aortic sac to the dorsal aorta, although many of these connections are only transitory. The transformation of the aortic arches into the adult pattern will be described in the chapter on the circulatory system.

The motor fibers of cranial nerves V, VII, and IX supply the muscles derived from visceral arches one, two, and three, respectively. The Xth nerve supplies muscles derived from the remaining arches. This one-to-one relationship between the muscles derived from specific arches and their motor innervation does not hold true for their sensory supply. This is particularly true for the exteroceptors, almost all of which are supplied by the trigeminal (Vth) cranial nerve. However, there does seem to be a relationshp between visceral arch origin and sensory nerve supply to visceral structures (mucous membranes, taste buds).

The differentiation of the mandibular and the maxillary processes into the jaws has already been described. In addition, the first and second arches also form the middle ear ossicles as will be described in the section on the ear. The second arch mesenchyme also forms the lesser cornu and the upper part of the body of the hyoid bone. The third arch forms the remainder of the hyoid bone—the greater cornu and the lower half of the body. The fourth and fifth arches form the laryngeal cartilages (Fig. 14–11).

The first arch mesoderm forms the muscles of mastication and also the anterior belly of the digastric, the mylohyoid, and the tensor tympani. The muscles of facial expression are all derived from the second arch mesoderm, which also forms the posterior belly of the digastric, the stylohyoid, and the stapedius. The third arch mesoderm forms the more cranial pharyngeal muscles, and the fourth and fifth arches form the lower pharyngeal and the laryngeal muscles.

## Pharyngeal Pouches

The first pharyngeal pouch forms the tympanic cavity of the middle ear and its connection to the pharynx, the Eustachian tube. Again, their development will be discussed more completely in the section on the ear.

The dorsal and most of the ventral portions of the second pharyngeal pouch become obliterated by the proliferation of the endoderm. However, a small recess persists, the *tonsillar fossa,* the primordium of the *palatine tonsil* (Fig. 14–10). It is invaded

**FIG. 14–11.** Diagram of the contributions of the five visceral arches to the skeletal system.

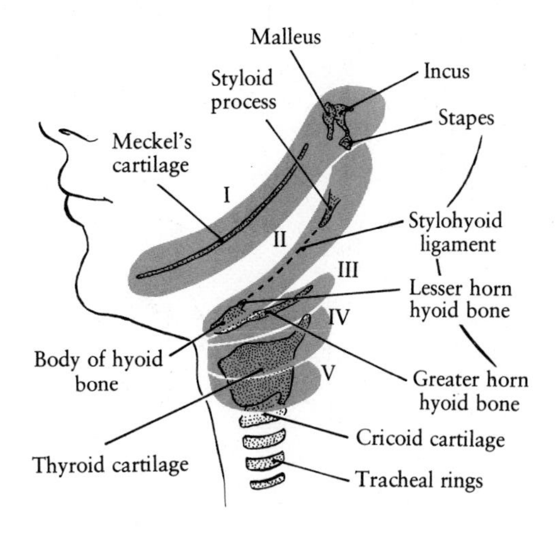

by mesenchyme and at about the fifth month begins to develop aggregations of lymphatic material. Similar aggregations of lymphatic tissue (but not of second pouch origin) occur in the *lingual tonsil* and the dorsal wall of the nasopharynx to form the *pharyngeal tonsil (adenoids)*.

In the third pouch, communication to the pharynx becomes constricted during the third month. The lateral end of the pouch, however, continues to expand and forms a large pouchlike region with dorsal and ventral sacculations (Fig. 14–10 B,C). The dorsal sacculations give rise to parts of the *parathyroid glands,* and the ventral sacculations form the major portion of the *thymus gland* (Fig. 14–10 C). In the 13-millimeter embryo, during week 6, the constricted connection to the pharynx is lost and the developing thymic and parathyroid tissues become free from any pharyngeal attachments. The third pouch now shows an accelerated rate of growth, particularly in the region that will form the thymus (Fig. 14–12). This region pushes caudally and its lumen is gradually eliminated (Fig. 14–12 D). During the early caudal movement of the thymic rudiment, the parathyroid remains attached to it and is pulled caudally past the position of the part of the fourth pouch, which is also developing a part of the parathyroid gland. This explains the definitive position of parathyroid III caudal to that of parathyroid IV.

At about the 20-millimeter stage, toward the end of month 2, the rudiments of the thymus and the parathyroid of pouch 3 separate. The thymic rudiment continues its caudal migration and its caudal end extends into the thoracic cavity where it meets and fuses with its counterpart of the other side. The cranial end remains in the cervical region and some cords persist up to or higher than the level of the thyroid gland.

Each fourth pharyngeal pouch develops a parathyroid rudiment, parathyroid IV, from its dorsal portion, in the same manner as the third (Fig. 14–10). The pair from the fourth pouch does not become associated with the caudally growing thymus and thus retains its position and ends up at the level of the cranial end of the thyroid, becoming the *superior parathyroid* of the adult. Both pairs of parathyroids become embedded in the thyroid tissue (Fig. 14–13).

The fate of the ventral region of the fourth pouch

**FIG. 14–12.** Later development of the parathyroid and thymus. A, parathyroid III and thymus III separated from the pharynx; B, section through A; C, D, caudal elongation of the thymus and the beginning of the obliteration of its cavity. (From G. L. Weller, 1932.)

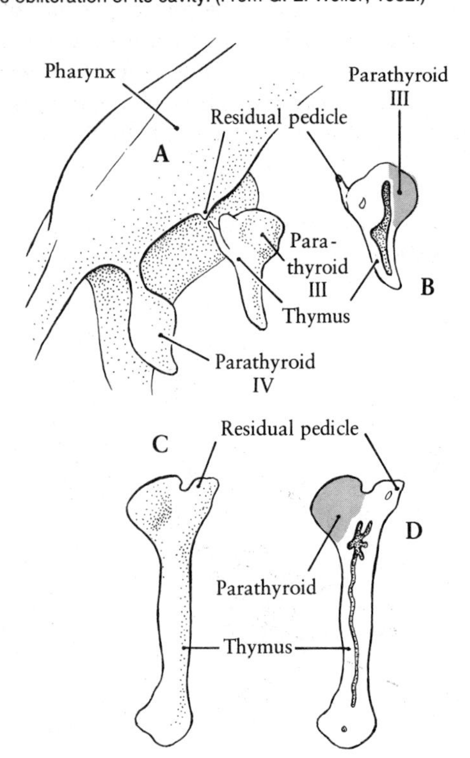


**FIG. 14–13.** Diagram of the caudal migration of the thyroid, thymus, parathyroids, and ultimobranchial bodies.

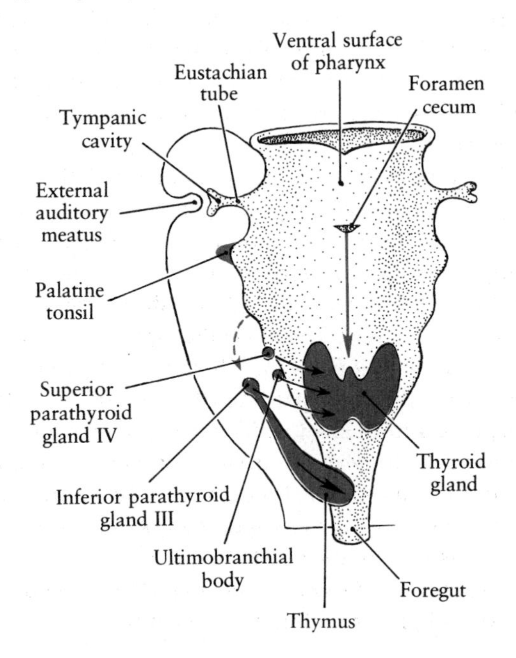

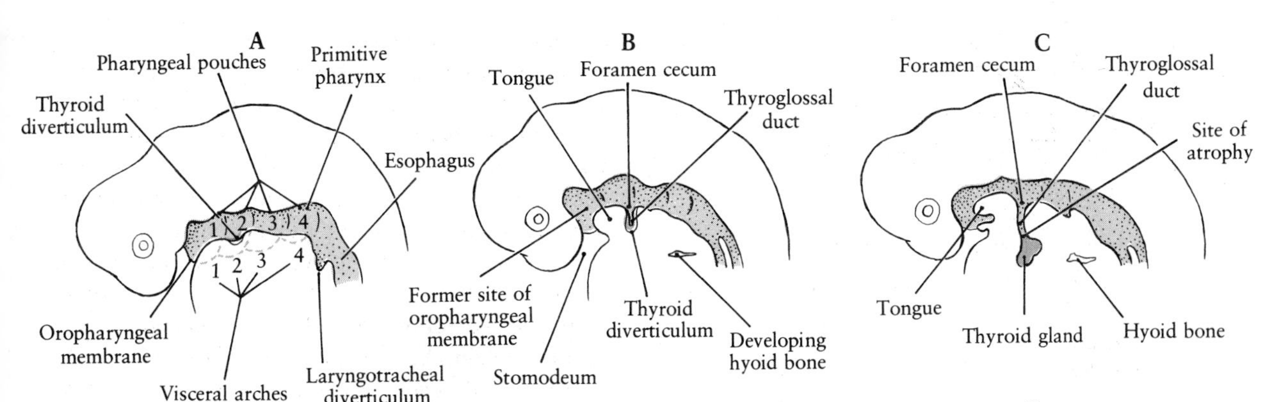


is in some doubt. A small portion may contribute to the thymus gland. A pair of large *ultimobranchial bodies* is found in this region in the sixth week (Fig. 14–10). The ultimobranchial bodies have variously been considered as a part of the fourth pouch or as rudimentary fifth pouches. The ultimobranchial bodies lose their connection with the pharynx and become embedded in the thyroid gland as it grows caudally. The ultimate fate of the tissue of these bodies has been considered to be (1) nothing, (2) thyroid tissue, or (3) special tissue in the thyroid gland, parafollicular cells, responsible for the secretion of thyrocalcitonin.

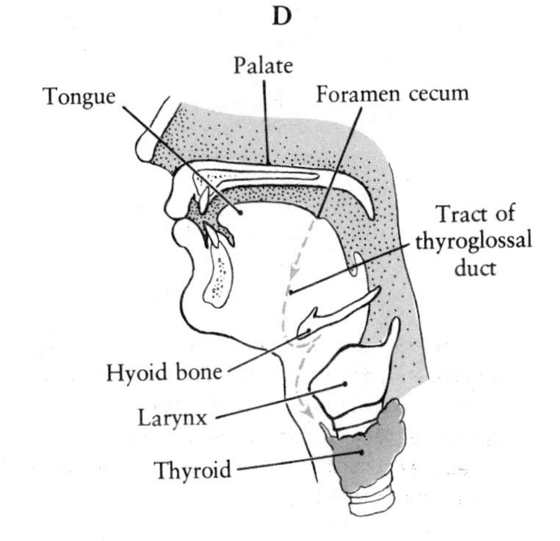


**FIG. 14–14.** Diagrams of sagittal sections of the head showing successive stages in the development of the thyroid gland at four, five, and six weeks (A, B, and C); D, adult head showing path of the thyroid gland to its position anterior to the larynx.

## The Floor of the Oral Cavity

### *The Thyroid Gland in Man*

The thyroid gland develops in close association with the pharyngeal pouches. It is recognizable in early somite embryos as a thickening in the floor of the pharynx between the first and second pouches just caudad of the region that will form the tuberculum impar of the tongue (Fig. 14–10 A). In man, at 16–17 days, the thickening forms a ventral outpocketing that becomes closely associated with the underlying endothelium of the developing heart. The evagination develops into a flask-shaped vesicle attached to its origin by a narrow neck, the *thyroglossal duct* (Figs. 14–14 C and 14–15 A,B). The vesicle quickly becomes bilobed. The lumen of the thyroglossal duct soon becomes occluded, and early in the second month the stalk breaks up and the bilobed terminal portion loses its connection to the pharynx (Figs. 14–14 D and 14–15 C). The lumen of the vesicle also disappears at this time, and the gland becomes a solid mass of expanding tissue.

As the heart moves caudally, the thyroid gland moves along with it, and by the end of the seventh week it becomes located in its adult position in the anterior lower neck region where it lies ventral to the developing larynx as a bilobed U-shaped structure, its two lobes connected by a narrow *isthmus* (Fig. 14–15 D). Subsequent morphological development involves a progressive increase in size.

The histological and biochemical differentiation of the thyroid gland has been studied in a number of mammals, including man. In man, there are three histological stages: (1) a precolloid stage (from 47 to 72 days), (2) a beginning colloid stage (from 73 to 80 days), and (3) a stage of follicular growth (from 80 days to birth). Colloid appears in canaliculi between the developing follicles. The canaliculi open into a central clover-shaped colloid space that is surrounded by the developing follicles.

There are some species differences in the time of the development of an iodine-concentrating mech-

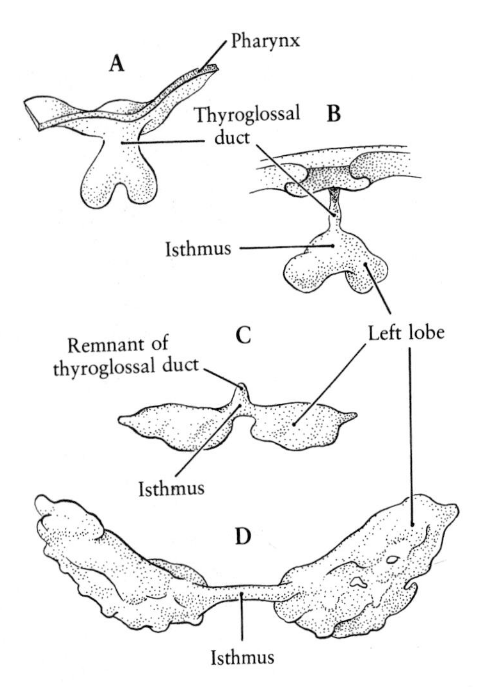


**FIG. 14–15.** Development of the human thyroid gland. A, formation of bilobed rudiment showing its connection to the pharynx by the hollow thyroglossal duct; B, thyroglossal duct elongates as gland grows caudally; C, connection to the pharynx is lost and the two lobes are connected by an isthmus; D, large right and left lobes connected by a thin isthmus.

anism and in the sequence of the synthesis of the active thyroid hormone, thyroxine (triiodothyronine, $T_4$). In the mouse, organic binding of iodine and formation of colloid occur on days 15–16, a day before the appearance of follicles and the production of $T_4$. In most subprimate mammals there is a stepwise development of synthetic activity that begins with the ability to trap iodine; next develops the mechanism to synthesize iodotyrosine, and then develops the mechanism to couple iodotyrosine to form iodothyronine—the same steps that take place during the synthesis of the hormone in the adult gland. However, in man and in the monkey, the ability to concentrate iodine and the mechanism for synthesizing $T_4$ appear at the same time—about day 74 in man—when the follicles are forming lumens and the colloid is present centrally in the cloverleaf pattern.

*The Thyroid Gland in the Chick*

The development of the thyroid gland in the chick embryo follows the same morphological pattern described above. It can be divided into five stages: (1) primordium formation, (2) vesicle formation, (3) stalk formation and detachment, (4) bilateral division, and (5) mesenchyme invasion. Up until 48 hours of incubation (stage 12), the floor of the pharynx consists of a single layer of loosely connected cuboidal cells all having the same appearance. At stage 12, the thyroid primordium is first seen when a small group of cells between the second pair of arches develop differences that distinguish them from the surrounding cells. They become more tightly packed; the cell surface open to the pharynx becomes scalloped, large characteristically staining droplets appear in the apical ctyoplasm, and an indication of a fibrillar band stretching across several cells appears in the apical cytoplasm.

In the chick, $T_4$ can be detected chromatographically at stage 12, the time the thyroid primodium can first be distinguished morphologically. This appearance of an organ-specific product at the time the organ can first be distinguished morphologically has been reported so far in only one other organ, the pancreas, where insulin can be detected in the early primordium.

$T_4$ accumulation in the cells of the developing glands does not show a steady increase with time, but shows at first a rapid logarithmic increase to about day 4½ (stage 24), a plateau until about day 6½ (stage 29), and then a second slower rise to a second plateau at day 12, which level is maintained until hatching (Fig. 14–16 A). The number of cells per gland follows almost the same pattern, rising to a plateau between days 5 and 6½ and then steadily increasing until hatching (Fig. 14–16 B). It should be noted that plateaus in $T_4$ accumulation and cell number occur at a time of complex morphogenetic activity—formation of a bilobed structure, detachment of the lobes, and mesenchymal invasion. The second rise in $T_4$ accumulation and cell number is correlated with the formation of the thyroid follicles. It has been suggested that the cessation of cell division during an important morphogenetic (or biochemical) phase of differentiation may be important to the later development of the thyroid gland. This stage has been compared to the "protodifferentiation" period described in the development of the pancreas.

*The Tongue*

The tongue is derived from the floor of the oral cavity and the pharynx. At the end of the first month of development, in the 6-millimeter embryo, the

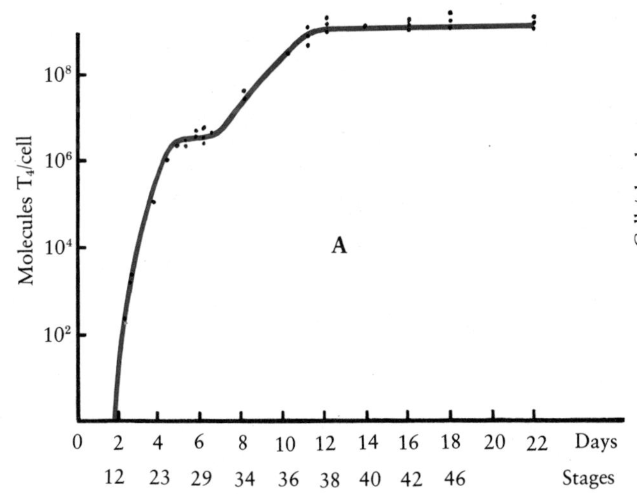


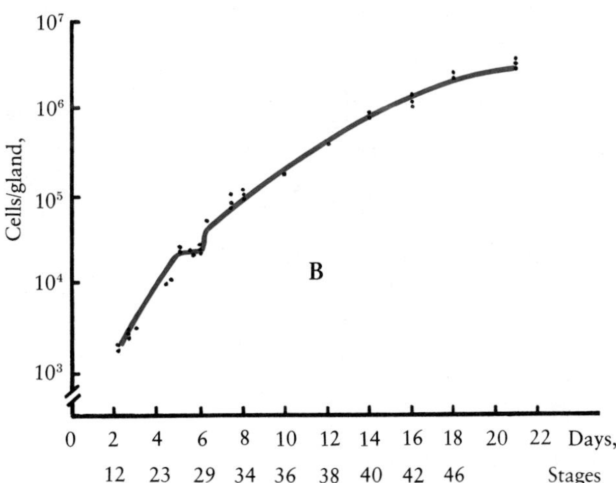


**FIG. 14–16.** A, changes in the amount of thyroid hormone per cell during the development of the chick; B, changes in the number of cells per thyroid gland during the development of the chick. (From W. G. Shain et al., 1972. Dev. Biol. 28:202.)

visceral arches are well defined and the ventral ends of the pharyngeal grooves and the arches between them extend toward the midline (Fig. 14–17 A). Between and caudal to the first arches is a small median elevation, the *tuberculum impar*. From it and the adjacent mandibular region the anterior two thirds of the tongue, the *body*, will be formed. The posterior third of the tongue, the *root*, will be formed by the union of the second visceral arches and also will receive some contributions from the third and fourth arches. The *foramen cecum* is an important landmark located just posterior to the tuberculum impar. It marks the point of origin of the thyroid evagination and also separates the body from the root of the tongue.

The major part of the body of the tongue arises from paired *lateral lingual swellings* of the first arch (Fig. 14–17 A,B). The lateral lingual swellings increase rapidly in size and unite with the tuberculum impar, which itself lags in development, is obscured by the lingual swellings, and contributes little of significance to the definitive organ. The fusion of the lingual swellings is indicated on the dorsum of the tongue as the *median sulcus* (Fig. 14–17 C,D) and appears internally as the *septum* of the tongue. The foramen cecum persists as a small depression on the dorsal surface of the tongue at the apex of a V-shaped sulcus, the *sulcus limitans*. Just anterior to the sulcus limitans are located the large circumvallate papillae.

The formation of the root of the tongue is somewhat more complicated. Between and uniting the second and third visceral arches, posterior to the tuberculum impar, is a swelling known as the *copula* (Fig. 14–17 A,B). The root of the tongue is formed by the copula and from the ventral parts of the second, third, and fourth arches. In its development, the material of the third arch migrates anteriorly and encroaches on the territory of the second, and a large part of the sensory innervation to the root of the tongue is by way of cranial nerve IX, the nerve of the third visceral arch. The vagus nerve, the nerve of the fourth arch, also supplies the root of the tongue, indicating the contribution of fourth arch material.

The main sensory supply to the body of the tongue is by way of branches of cranial nerves V and VII, nerves of the first and second arches, respectively. However, a small part of the tongue anterior to the sulcus limitans is supplied by the glossopharyngeal nerve, indicating some third arch contribution to the body.

The tissue beneath the epithelial membrane of the tongue consists mainly of skeletal muscle and connective tissue. The connective tissue is formed from the mesoderm of the visceral arches. However, the mesoderm of the visceral arches probably does not form the tongue muscles. On the basis of both comparative anatomy and the fact that the hypoglossal nerve supplies the tongue muscles, it appears that this tissue is derived from more caudally located somatic muscle of the occipital myotomes.

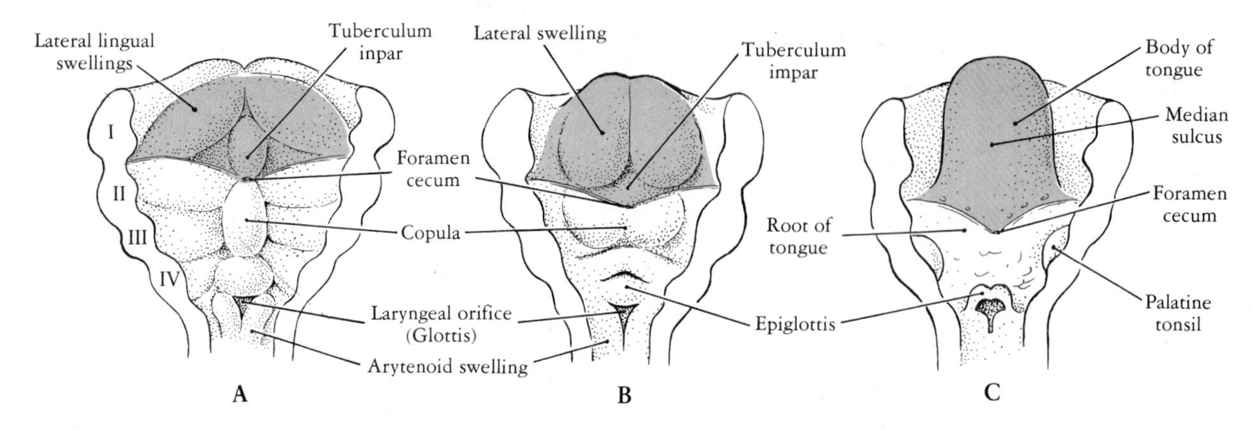


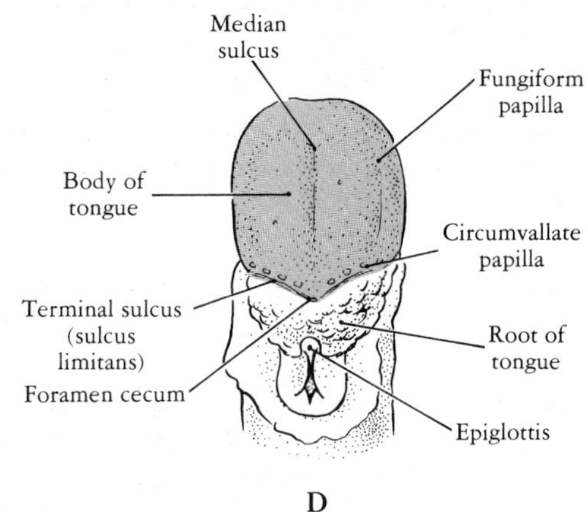


FIG. 14–17. Development of the human tongue seen from above. A, five weeks; B, six weeks; C, five months; D, adult.

The oral surface of the tongue develops specialized gustatory receptors, the taste buds. In addition to the circumvallate papillae already mentioned, foliate papillae appear during the third month, and fungiform and filliform papillae appear still later. All types develop as elevations of the surface epithelium.

During the fifth month, lymphocytes invade the root of the tongue, marking the beginning of the development of the lingual tonsil. Its crypts, however, do not develop until after birth.

*The Epiglottis*

The slit-shaped opening into the larynx, the *glottis*, is apparent in the fifth week between the fourth and fifth visceral arches just caudad of the root of the tongue (Fig. 14–17 A). Anterior to the glottis, between it and the copula, a swelling marks the beginning of the formation of the *epiglottis*, and a pair of *arytenoid swellings*, one on either side of the glottis, mark the beginning of the development of these laryngeal cartilages (Fig. 14–17 A). The arytenoid swellings migrate rostrally toward the epiglottis and in so doing form a transverse component on the upper end of the glottis so that the opening now becomes T-shaped (Fig. 14–17 C). Later, the upper part of the opening becomes more ovoid, although a deep interarytenoid notch persists in the sagittal plane.

*Salivary Glands*

The epithelium of the oral cavity gives rise to a number of solid evaginations that will develop into the salivary glands. The *submandibular gland* develops during the third month as an outgrowth from the floor of the oral cavity between the tongue and the gum. The *parotid gland* appears somewhat later as an outgrowth from the cheek immediately posterior to the angle of the jaw, and the primordium of the *sublingual gland* develops lateral to it. The openings of the ducts of the salivary glands in the adult are somewhat more anterior than the point of the original evagination, owing to the closure of the posterior parts of the gutterlike grooves from which they are derived. The original solid epithelial outgrowths undergo repeated branching, and by the sixth month the numerous branches become completely canalized. The morphogenesis of the salivary glands and the epithelial–mesenchymal interactions that take place at this time have been described in Chapter 13.

## References

Baxter, H. and F. C. Fraser. 1950. Production of congenital defects of offsprings of female mice treated with cortisone. McGill Med. J. 19:245–251.

Burnet, F. M. 1962. The thymus gland. Sci Am. 207:50–57.
Fisher, D. A. and J. H. Dussault. 1976. Development of the mammalian thyroid gland. Handbook of Physiology, Section 7, Vol. 3, pp. 21–38. R. O. Greep and E. B. Astwood, eds. Baltimore: Williams and Wilkins.
Greene, R. M., J. L. Shanfeld, Z. Davidovitch, and R. M. Pratt. 1980. Immunochemical localization of cyclic AMP in the developing rodent secondary palate. J. Embryol. exp. Morphol. 60:271–280.
Shah, R. M. 1979. Current concepts on the mechanisms of normal and abnormal secondary palate formation. In: Advances in the Study of Birth Defects, Vol. 1. Teratogenic Mechanisms. T. V. N. Persaud, ed. Baltimore: University Park Press.
Shain, W. G., S. R. Hilfer, and V. G. Fonte, 1972. Early organogenesis of the embryonic chick thyroid. Dev. Biol. 28:202–218.
Weller, G. L. 1932. Development of the thyroid, parathyroid and thymus glands in man. Carnegie Contrib. Embryol. 24:93–139.

# 15

# THE DIGESTIVE TUBE AND ASSOCIATED GLANDS

## The Digestive Tube

### The Pharynx

The *pharynx* is connected cranially with the oral and the nasal cavities and caudally with the *larynx* and the *esophagus*. Following the development of the nasal cavities and the palate, the pharynx may be divided into three areas. The *oropharynx* communicates with the oral cavity from which it is marked off by an indistinct fold of mucous membrane, the *palatoglossal fold* (anterior pillar of the fauces). A second fold of mucous membrane connects the palate to the floor of the pharynx, the *palatopharyngeal fold* (posterior pillar of the fauces). The palatine tonsil develops between these two sheets of mucous membrane. The *nasopharynx* is the portion of the pharynx that communicates with the nasal cavities by way of the posterior nares. The Eustachian tubes open into the lateral walls of the nasopharynx. The *laryngopharynx,* the most caudal part of the pharynx, communicates posteriorly with the esophagus and, by the way of the glottis, with the larynx.

### The Esophagus, Stomach, and Intestine

The alimentary tract caudal to the pharynx differentiates into the esophagus, *stomach,* and the *small* and *large intestines,* and the *rectum.* The alimentary tract shows a basic similarity of development throughout its length, and differences exist only in the degree and extent of changes in size, shape, and position and in the production of a number of specialized glandular outgrowths. The basic histological differentiation of the tract will be described after a consideration of the gross morphological changes.

The esophagus is the portion of the foregut into which the pharynx opens. At first only a very short connection between the pharynx and the stomach, the esophagus increases rapidly in length with the development of the lungs and the caudal shift in position of the heart and the stomach (Fig. 15–1). It remains, however, as a relatively straight, unspecialized connection between the pharynx and the stomach. It is the most constricted part of the alimentary tract.

The stomach is the caudal continuation of the

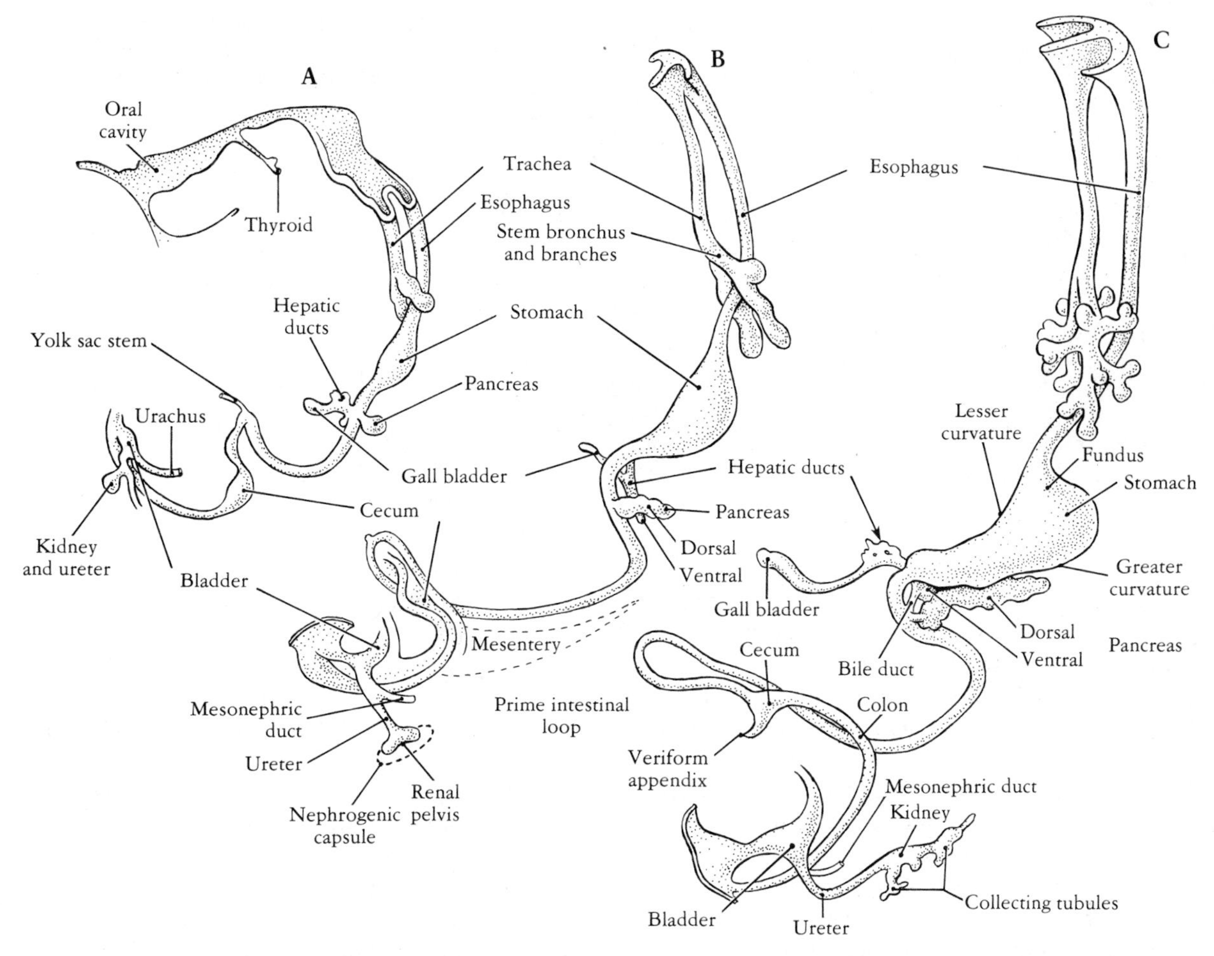


**FIG. 15–1.** Drawings of reconstructions of three stages in the development of the digestive tube in man. A, 7–8 mm, 31–32 days; B, 8–11 mm, 32–34 days; C, 11–13 mm, 34–35 days. (From G. L. Streeter, 1948. Carnegie Contributions to Embryology 32:133.)

esophagus. It shows a fusiform dilation as early as the end of the first month, which forecasts its future expansion about a week later (Fig. 15–1 A). Continued growth develops a greatly enlarged organ that already shows the adult configuration with a cranial *lesser curvature,* a caudal *greater curvature,* and a bulge near the esophageal junction, the *fundus* (Fig. 15–1 B,C). Dorsal and ventral mesenteries (mesogastria) are attached to the greater and lesser curvatures, respectively, indicating that during this stage of development the stomach goes through a 90-degree rotation to the right.

The intestine is the portion of the digestive tube between the stomach and the cloaca. The major portion of both the small and the large intestine develops from the roof of the open midgut between the anterior and the posterior intestinal portals. The midgut is at first in open and broad communciation with the yolk sac, and the entire tract lies in the sagittal plane of the body (Fig. 15–2). As the body folds develop and undercut the embryo, the connection between the embryonic and extraembryonic structures is by way of the body stalk. At the end of the first month, the embryonic intestine, which has begun to increase rapidly in length, begins to herniate into the body stalk, and its original broad connection to the yolk sac is marked by the point of attachment of the yolk stalk (vitelline duct) at the end of the hairpin-shaped herniation (Fig. 15–3 A). This small diverticulum may be used as a landmark in respect to which the midgut may be divided into cephalic (proximal) and caudal (distal) limbs. The cephalic limb will form the caudal part of the *duodenum,* the *jejunum* and the greater part

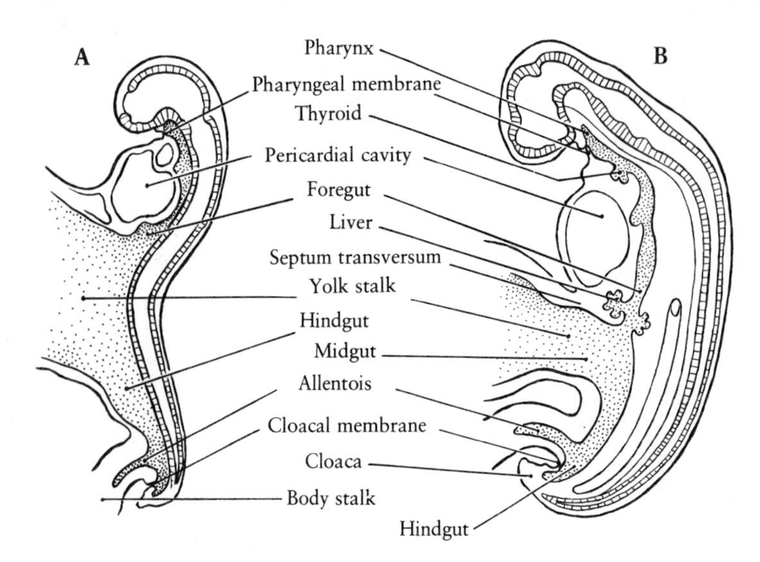


**FIG. 15–2.** Saggital sections of early human embryos showing the digestive tract. A, 2.5 mm, 18 somites; C, 2.5 mm, 25 somites.

of the *ileum.* The caudal limb will form the caudal part of the ileum, the *cecum,* the *ascending colon,* and about half of the *transverse colon.* The cranial part of the duodenum develops from the foregut. The caudal part of the colon and the rectum develop from the hindgut. The cecum marks the point of transition from the small to the large intestine. It appears early in the second month as a dilation caudal to the point of attachment of the yolk stalk (Figs. 15–1 A and 15–3 B,C).

The positional relationship between the small and large intestines in the adult may readily be understood by a consideration of the rotation of the herniated midgut and the sequence of its return into the embryonic abdominal cavity as diagrammed in Figure 15–3. The midgut moves into the extraembryonic coelom owing to a lack of space in the abdominal cavity, which is usurped by the relatively massive embryonic liver and mesonephros. The cephalic limb now elongates rapidly and is thrown into coils. The caudal limb does not coil and shows little change except the development of the cecum (Fig. 15–3 B).

The entire midgut, within the body stalk, starts a counterclockwise rotation around the superior mesenteric artery (a branch of the dorsal aorta that will supply all of the midgut derivatives). About 90 degrees of the entire rotation is completed while the midgut is herniated (Fig. 15–3 B).

At about the tenth week, the relative size of the liver and the mesonephros decreases, and the abdominal cavity expands and the herniated midgut begins to return to the abdominal cavity (Fig. 15–3 C). As it does so, the rotation continues and eventually goes through about 270 degrees (Fig. 15–3 D). The small intestine, the cephalic limb of the midgut, moves out of the body stalk first and, in so doing, pushes the caudal part of the large intestine (hindgut) to the left of the body cavity. The jejunum is the first part to return, and its cranial coils come to be located on the left side, the more caudal folds of the jejunum and the ileum gradually filling up the remainder of the cavity. The caudal limb of the midgut is then withdrawn, the cecal region being the last part to leave the body stalk to occupy a position just below the liver in the upper right quadrant of the abdominal cavity (Fig. 15–3 D). The large intestine then occupies the left and cranial parts of the abdominal cavity, and the much coiled small intestine occupies the central part. Further changes involve a descent of the cecal region so that the small intestine becomes completely enclosed in the C-shaped large intestine, the latter consisting of ascending, transverse, and descending portions (Fig. 15–3 E).

The cecum continues to mark the transition between the small and the large intestine. As the cecum increases in size, the original end-to-end connection between these two areas is changed, and the small intestine enters the large intestine at a right angle (Fig. 15–4). The cecum then appears as a pouch below this *ileocolic junction.* The growth of the distal end of the cecum lags behind the other parts of the organ from the third month on, result-

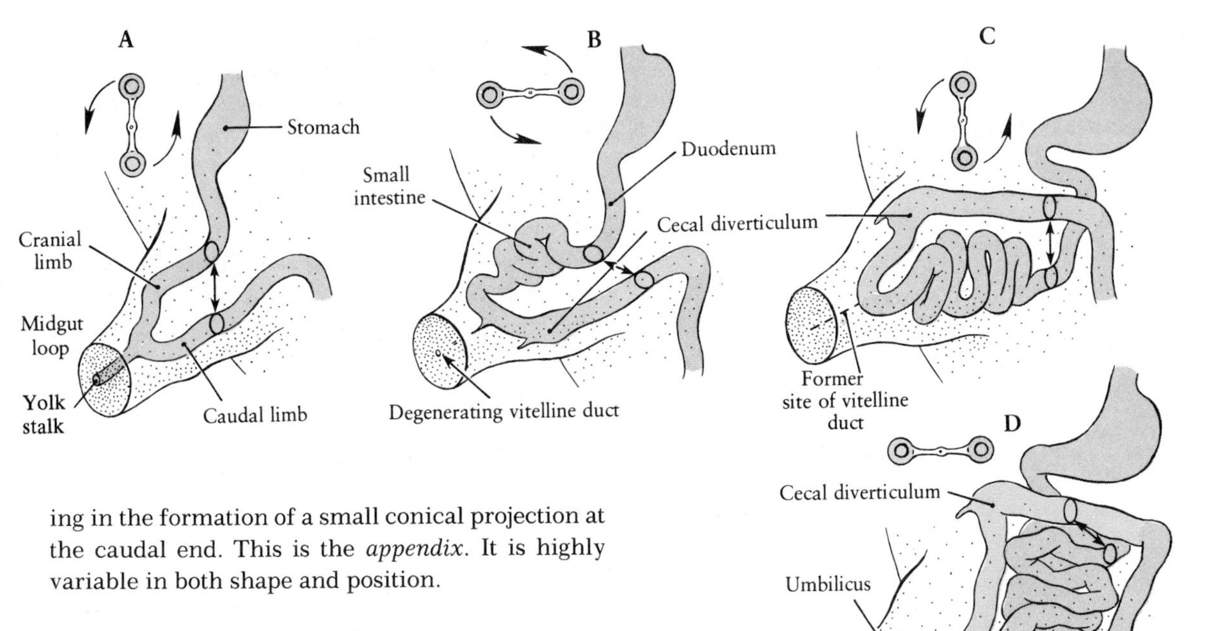


ing in the formation of a small conical projection at the caudal end. This is the *appendix*. It is highly variable in both shape and position.

*Histology of the Alimentary Tract*

The characteristic histological structure of the adult alimentary tract is that of an epithelium lined tube surrounded by layers of muscle and connective tissue. The lining of the tract and the glands developed from it are the only parts derived from embryonic endoderm. The connective tissue and muscle are the products of the splanchnic mesoderm that becomes associated with the endodermal lining. The general pattern consists of four layers: (1) mucosa, (2) submucosa, (3) muscularis, and (4) serosa—named in order from the inside to the outside (Fig. 15–5).

The mucosa consists of the epithelial lining plus a thin layer of connective tissue and in some areas a thin layer of muscle. The epithelium shows some variation throughout the tract and, although it is generally simple columnar, in the esophagus it is stratified squamous. In the stomach and the intestine, it forms characteristic glands with various kinds of secretory and absorptive cells that begin to function at different times of development in different regions of the tract. Proteolytic enzymes and digestive ferments of the small intestine have been detected during the fourth month; and although some enzymes such as amylase may not be present until birth, it is apparent that the alimentary tract is ready to commence functional activity well in advance of the time it will be called on to do so.

For the first few months of development, the tract is usually devoid of solid material. After this time,

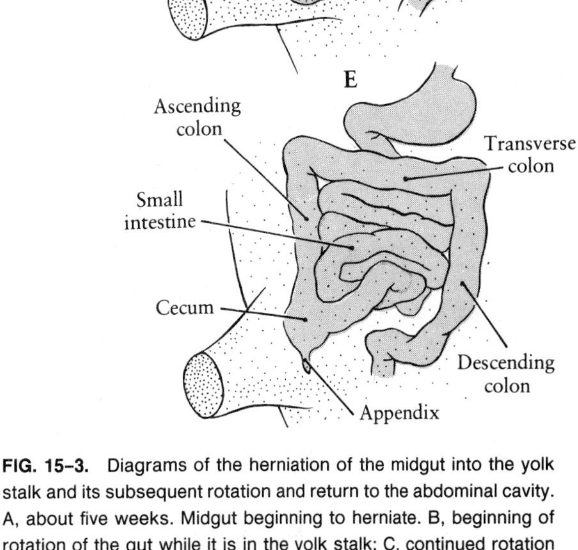


**FIG. 15–3.** Diagrams of the herniation of the midgut into the yolk stalk and its subsequent rotation and return to the abdominal cavity. A, about five weeks. Midgut beginning to herniate. B, beginning of rotation of the gut while it is in the yolk stalk; C, continued rotation as the gut begins to return into the abdominal cavity; D, midgut completely within the abdominal cavity after rotation of 270 degrees; E, final positional changes completed as the cecum descends into the lower right quadrant. The insets, which indicate the rotation of the intestine, represent cross sections through the cranial and caudal limbs at the levels indicated by the circles and arrows in the diagrams. (After K. L. Moore, 1977. The Developing Human. W. B. Saunders Company, Philadelphia.)

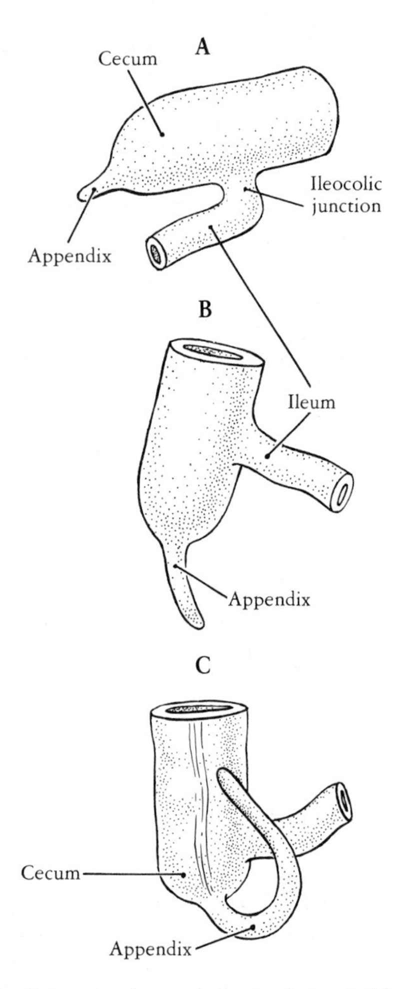


FIG. 15-4. Successive changes in the development of the cecum, the appendix, and the small intestine. A, about 7 weeks; B, about 12 weeks; C, at birth.

an increasing amount of material begins to accumulate in the lumen. Swallowed amniotic fluid contributes sloughed epithelial cells and lanugo hairs. The tract itself adds more epithelial cells, mucus, and bile secretions. The contents of the neonatal gut are called *meconium* and, owing to their long retention in combination with bile secretions, they are greenish in color. The green color of the stool quickly changes during the first few days after birth to the normal yellowish color.

## Associated Glands

### The Liver and the Gallbladder

The liver is the most precocious of the glandular derivatives of the alimentary tract. It is first indicated in the third week as a thickening of the endodermal lining of the ventral wall of the foregut in the region of the anterior intestinal portal (Fig. 15–2 A). This is the region of the tract that will become the duodenum. The thickening soon gives rise to an outgrowth, the *hepatic diverticulum,* which grows into the septum transversum (Fig. 15–2 B). The septum transversum is a mesodermal partition lying between the pericardial cavity and the yolk stalk. It will form the major part of the diaphragm separating the pericardial and abdominal cavities (Chapter 17). The rapidly proliferating cells divide into two components, a more cranial, larger, *pars hepatica,* which will form the liver tissue and its duct system, and a more caudal, smaller, *pars*

FIG. 15-5. Two stages in the histogenesis of the intestine. A, month 2; B, month 5.

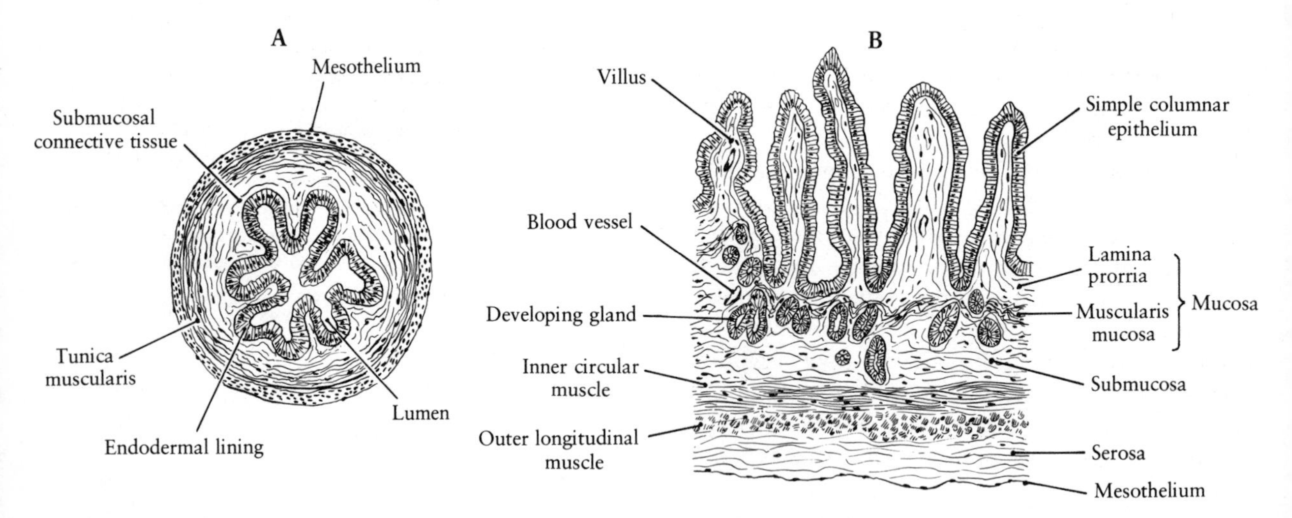

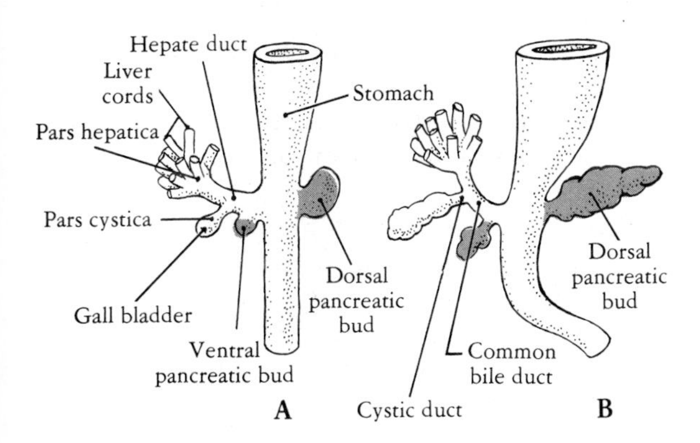


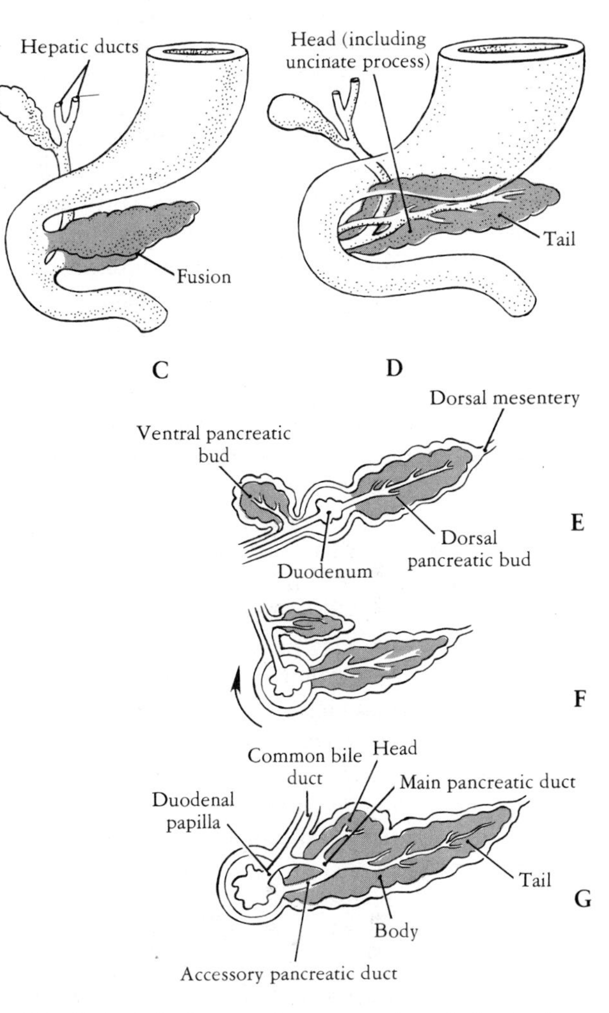


**FIG. 15-6.** A–D, diagrams of successive stages in the development of the liver, the gallbladder, and the pancreas from the fifth to seventh weeks; E–G, cross sections of B, C, and D. (After K. L. Moore, 1977. The Developing Human. W. B. Saunders Company, Philadelphia.)

*cystica,* which will form the gallbladder and the cystic duct (Fig. 15–6 A,B).

The pars hepatica develops a maze of anastomosing cords of epithelial cells that continue to proliferate as they invade the septum transversum. The cords break up the vitelline and umbilical veins, transforming these vessels into hepatic sinusoids. This results in an intermingling of hepatic cords and sinusoidal channels. The sinusoids form a network interspersed between the veins bringing blood to the liver (the vitelline and umbilical veins—the paired vitelline veins later differentiating into the single hepatic portal vein) and the veins draining the liver (the hepatic veins). The parenchyma of the liver develops from the hepatic cords, forming what are actually anastomosing sheets of epithelium, each consisting of a single layer of cuboidal cells. Bile canaliculi develop between the hepatic cords. They connect to intrahepatic bile ducts. These, in turn, connect to the larger hepatic ducts that drain into the common bile duct (Fig. 15–6 B).

The mesoderm of the septum transversum differentiates into the connective tissue, the hematopoietic tissue, and the Kupffer cells of the liver. Blood cells are differentiated in the second to the seventh month, but at birth the hematopoietic activity of the liver has stopped. The blood-forming function of the embryonic (fetal) liver is largely responsible for the rapid increase in the size of this organ, which in the third month makes up about 10 percent of the total weight of the fetus.

The gallbladder begins its development at the distal end of the pars cystica as a solid cylindrically shaped group of cells, connected to the common bile duct by a stem, the *cystic duct* (Fig. 15–6). The gallbladder grows ventrally but does not invade the tissue of the septum transversum.

Owing to the rotation of the duodenum, the opening of the common bile duct into the duodenum shifts from an anterior to a posterior position (Fig. 15–6, C,D). Consequently, the common bile duct will be found running posterior to the duodenum.

### *Epithelial-Mesenchymal Interaction in the Development of the Chick Liver*

The localization of the liver-forming areas of the chick embryo has been well documented by the work of Rudnick and Rawles (1937). Early in development, hepatic and cardiac potencies are closely associated in the region just anterior to the primitive streak. Later, in the 10- to 15-somite stage, the hepatic region covers a large part of the blastoderm on either side of the somites, the bilateral areas merging anteriorly in the floor of the foregut and the cardiac fold. The presumptive liver tissue, which is located in the region of the anterior intestinal portal, represents presumptive hepatic endoderm, and the remainder represents presumptive hepatic mesoderm. What is the relationship between the presumptive endoderm, which forms the proliferating epithelial cords of the liver, and the mesoderm into which the cords grow? If the early hepatic bud from which the underlying mesenchyme has been removed is cultured in vitro or transplanted into the coelom of a host chick embryo, it degenerates or forms small masses of undifferentiated cells. Mesoderm is thus necessary for the differentiation of the hepatic cells. In fact, it has been demonstrated that for complete differentiation the endoderm must be successively subjected to the action of two mesodermal tissues: (1) the midventral cardiac mesoderm and (2) the lateral hepatic mesoderm. Presumptive hepatic endoderm from which the mesoderm has been removed will not differentiate even when grafted into the area of the hepatic mesoderm, if the operation is performed before the five-somite stage. At the five- to six-somite stage, the hepatic endoderm and the precardiac mesoderm meet in the cardiac fold, and only after the endoderm has been exposed to the cardiac mesoderm will it form hepatic tissue when transplanted to the hepatic mesoderm area. The first step in liver differentiation, then, is the interaction between the hepatic endoderm and the mesoderm of the cardiac area; and it is necessary before the second step, the interaction with the hepatic mesoderm, can exert its effect. Both interactions must take place. Neither one, by itself, will result in the normal differentiation of the hepatic tissue.

The first induction is apparently specific. It can only be exerted by the mesoderm of the cardiac region, and only hepatic endoderm will respond to it. The second is less specific, since the hepatic endoderm will form well-differentiated hepatic cords when associated with mesenteric or metanephric mesoderm. Although these cords appear normal, the cells are unable to synthesize glycogen. However, if they are now put into association with hepatic mesoderm, they will begin to synthesize it. Apparently the inductive action of the foreign mesoderm is an incomplete one, and the secondary induction itself may be considered to consist of two steps: (1) the formation of hepatic epithelial cords (histological differentiation) and (2) the onset of specific metabolic activity (functional differentiation). Only hepatic mesoderm can bring about the latter.

Neither the mode of action nor the chemical nature of the liver-inducing substances is known. However, there is little doubt that they differ chemically from the inducing substances playing roles in the differentiation of other organs such as the lung, the limb, and the kidney.

## The Pancreas

The pancreas is the second largest gland in the body. It has a dual function and consists of an exocrine portion that synthesizes digestive enzymes, which pass by way of a duct system into the duodenum, and an endocrine portion that synthesizes insulin and glucagon, two hormones essential to the regulation of carbohydrate metabolism.

### *The Morphological Development of the Pancreas*

The pancreas develops from the duodenal epithelium in the region of the liver as a pair of evaginations, one dorsal and one ventral, which later fuse with each other. The *dorsal pancreas* arises from the dorsal wall of the duodenum opposite and slightly cranial to the liver diverticulum and pushes into the dorsal mesentery. The *ventral pancreas* arises from the ventral region of the duodenum in the angle of the hepatic diverticulum, but as the hepatic diverticulum grows, the ventral pancreas establishes connection with it and opens into the common bile duct (Fig. 15–6 A,B). When the duodenum rotates to the right, the ventral pancreatic diverticulum is carried into the dorsal mesentery lying below and behind the dorsal pancreas and separated from it by the part of the left vitelline vein that will contribute to the formation of the hepatic portal vein. The two pancreatic primordia fuse (Fig. 15–6 C,D). The ventral pancreas contributes the tissue that will form the lower portion of the head; and the dorsal pancreas forms the remainder of the head, the tail, and the body of the adult pancreas. When the two primordia fuse, their duct systems also become interconnected. Then, the proximal part of the dorsal duct degenerates, and the pancreas empties into the persistent ventral pancreatic duct, a tributary of the common bile duct. Within the pancreatic tissue, the duct of the dorsal pancreas persists and drains the tail and the body of the organ. Although in man the ventral duct (duct of Wirsung) persists as the only definitive opening into the duodenum, in other species (e.g., the pig and the cow) only the dorsal duct (duct of Santorini) persists. In the horse and the dog, both ducts persist.

*The Exocrine Pancreas in the Rodent*

The development of the pancreas in the mouse and the rat has been thoroughly studied, and some interesting facts on the biochemical and ultrastructural aspects of development have been described. In these rodents the pancreas also develops from a dorsal and a ventral diverticulum that later fuse. The dorsal pancreas appears first, in the 20-somite embryo about the middle of the gestation period (9½ days in the mouse, 11 days in the rat). Its original broad connection to the gut soon narrows, and the pancreas develops as a ramified system of tubules whose lumens form a connecting network throughout the pancreatic tissue. Each tubule consists of a single layer of polarized cuboidal cells whose apical ends are joined together by junctional complexes. Microvilli project into the narrow lumen. The basal end of the cells is covered by a common basal lamina continuous with that of the gut. Early in its formation, mesodermal cells accumulate around the pancreatic diverticulum. This contact with the mesoderm or with some factor passing from the mesoderm to the pancreatic cells is necessary for further differentiation. The tubules grow rapidly for several days, and acinar, duct, and endocrine tissues develop from the epithelium.

*The Biochemical and Cytological Differentiation of the Exocrine Pancreas.* The pancreas goes through a number of stages in its differentiation. If cells of the nine-day mouse embryo from the region of the gut where the pancreatic diverticulum will form are grown in vitro, they will develop pancreatic tissue and enzymes, provided mesoderm is also present. Before this time, they will not. Thus, even before the pancreatic diverticulum forms, some of the endoderm cells of the gut have acquired the capability of forming pancreatic tissue. At this time there is no morphological, cytological, or biochemical difference between these cells and any other cells of the gut.

When the pancreatic diverticulum first develops, its cells pass through what has been termed a *primary transition phase* when small amounts of the characteristic digestive enzymes can be detected in the cells (Fig. 15–7 A). After the primary transition, the cells enter a *protodifferentiation* state in which a constant low level of enzymes is present, cell division is rapid, and no cytodifferentiation has occurred. The protodifferentiation state lasts for three to four days. The low level of enzyme is probably not due to the fact that a few of the pancreatic cells have differentiated completely to a point at which they are synthesizing the adult level of enzyme.

**FIG. 15–7.** Model of the differentiation of the epithelial cells of the pancreas. A, exocrine cells and insulin-secreting cells. Concentrations of specific cell products rise from an undetectable level to a detectable low level during the primary transition phase as the pancreatic diverticulum forms and becomes associated with the mesoderm. Products remain at a constant low level in the protodifferentiated state and then rise to high levels during the secondary transition phase, which leads to the differentiated stage. The appearance of characteristic granules occurs in the secondary transition stage; B, glucagon secretion. Levels of glucagon rise rapidly during the primary transition stage of the cells diagrammed in A. Primary transition and protodifferentiated stages have not been identified. (From R. Pictet and W. J. Rutter, 1972. Handbook of Physiology, Section 7, Vol. 1. R. Greep and E. B. Astwood, eds. Williams and Wilkins, Baltimore.)

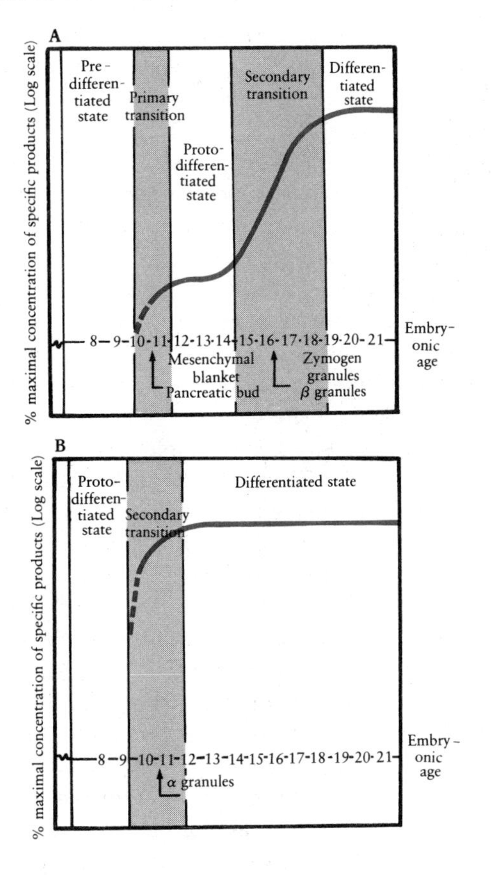

Rather it is due to all of the acinar cells of the pancreatic diverticulum synthesizing enzymes at a constant low level.

A *secondary transition* then occurs in which the rate of enzyme synthesis increases, and there is a dramatic increase by 1000 times or more of the concentration within the cells. Cytochemical changes now appear during this transition phase. First, there is an increase in the amount of rough endoplasmic reticulum, the site of enzyme synthesis, and this increase is followed by the appearance of zymogen granules in which the enzymes are stored. After the secondary transition, the acinar cells are considered to be fully differentiated.

*The Mesodermal Factor.* The need for a *mesodermal factor* (MF) to induce the differentiation of pancreatic tissue has been mentioned. MF is neither tissue nor species specific. Kidney or salivary gland mesoderm as well as mesoderm from the chick embryo will all support differentiation of the epithelial cells of the pancreas.

If pancreatic epithelium and mesoderm are cultured on opposite surfaces of a porous filter, the mesoderm can still produce its effect. Although the amount of differentiation is affected by the thickness of the filter and the size of the pores, contact between the two tissues is probably not necessary, as has been demonstrated in a number of other mesenchymal–epithelial-reacting systems. The mesodermal factor has been extracted and partially purified. It is trypsin sensitive and therefore proteinaceous. Its activity is also destroyed by sodium periodate, indicating it has a carbohydrate moiety.

At the time the cells of the gut first develop the capability of forming pancreatic derivatives, mesoderm must be present to achieve this result. However, late in the protodifferentiation state, about a half a day before the secondary transition, the inductive process has been completed and MF is no longer needed. The inductive influence of the mesoderm thus takes place over a relatively short span of three to four days. What is happening during the time the mesoderm is necessary? The major event is the rapid increase in DNA synthesis and cell division. This is apparently necessary for differentiation, since inhibition of DNA synthesis results in a failure of differentiation. Inhibition of mRNA synthesis during the protodifferentiation state also results in a failure of differentiation. It may then be concluded that new mRNA is being synthesized at this time as a preliminary to enzyme synthesis, and this RNA synthesis in turn depends on DNA synthesis. When cell division ceases, differentiation takes place. This is easily demonstrated, since cell division shows a gradient of activity toward the end of the protodifferentiation state, with the peripheral cells continuing to divide after the central cells have stopped. It is in the nondividing central cells that zymogen granules first appear. Also, inhibitors of DNA synthesis no longer have any effect on the differentiation of the central cells but still inhibit differentiation in the peripheral cells. Thus, a loss of proliferative activity appears to be associated with the attainment of the secondary transition stage leading to the final differentiation stage.

The mechanism by which MF exerts its influence is speculative. There is good evidence that it acts at the cell surface. If MF is covalently bound to large insoluble beads of Sepharose (a derivative of agar) that are then placed among a culture of pancreatic epithelial cells, the cells attach to the beads, become oriented, synthesize DNA, and divide. It has been proposed that MF promotes close contacts between cells. In the absence of MF, epithelial cells in culture tend to spread out in a loose arrangement—something that does not happen if MF is present. Changes in membrane permeability and cell shape may also be produced and, in turn, stimulate cell division.

*The Endocrine Pancreas in the Rodent*

The epithelium of the developing pancreatic tubules consists of a single layer of cuboidal cells whose apical ends are closely connected by junctional complexes. The orientation of the mitotic spindle and the resulting cleavage plane are apparently instrumental in maintaining this single-layered configuration. The spindles are oriented parallel to the cell surface, and the cleavage planes thus develop perpendicular to the lumen. The cleavage furrow first appears at the basal region of the cell and moves apically. Each daughter cell remains firmly attached by its junctional complex to its neighbor cell and, in turn, the daughter cells develop apical junctional complexes between themselves before the cleavage plane is completed.

At the time of the early formation of the pancreatic diverticulum, the endocrine cells, which can be distinguished histologically from the exocrine cells, may be present in small numbers as a part of the single-layered tubule epithelium linked to the exocrine cells. Later formation of endocrine cells probably takes place by an orientation of the mitotic spindle perpendicular to the apical surface; the cell division then results in the formation of one apical daughter cell still linked to its neighbors and another separated from its neighbors and "escaped" from the single-layered epithelium—although still retained within the exocrine complex beneath the basal lamina (Fig. 15–8 A). There is a marked increase in the number of endocrine cells during the protodifferentiation period. Since it has been demonstrated that there is very little DNA synthesis or cell division among the endocrine islet cells, this increase must be the result of a continued supply from the dividing exocrine cells. Islet cells accumulate in small groups but still beneath the basal lamina (Fig. 15–8 B). Some of these groups may join to form larger outpocketings of islet cells. The islets later separate completely from the exocrine complex by the fusion of the basal lamina at the point of the outpocketing (Fig. 15–8 C). Capillaries invade the islets, and the endocrine secretions of the islet cells are removed by way of this vascular supply.

**FIG. 15–8.** Histogenesis of the islet tissue. A, islet cells develop and some become located between the cells lining the lumen and the basal lamina; B, islet cells increase in number and form small clusters; C, islet cells become completely separated from exocrine cells by pinching off (at arrows) of the basal lamina at the point of outpocketing. Islets are penetrated by capillaries. (From R. Pictet and W. J. Rutter, 1972. In: Handbook of Physiology, Section 7, Vol. 1, R. Greep and E. B. Astwood, eds. Williams and Wilkins, Baltimore.)

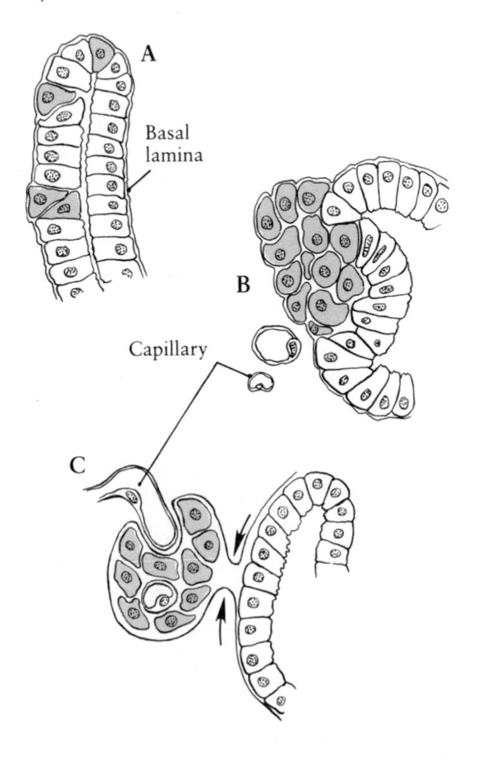


*Biochemical and Cytological Differentiation of the Endocrine Pancreas.* The biochemical and cytological differentiation of the insulin-secreting B cells of the islets follows the same pattern just described for the exocrine pancreas (Fig. 15–7 A). Following a preliminary initiating event leading to a primary transition phase, the synthesis of insulin takes place in the early pancreatic diverticulum and continues at low levels during the protodifferentiation state unaccompanied by any cytodifferentiation. Increasing quantities of insulin are synthesized during the secondary transition phase, and this event is accompanied by cytodifferentiation ending in the appearance of $\beta$ granules and the attainment of the differentiated state. The A cells, which synthesize glucagon, present a different picture. At the time of the formation of the pancreatic diverticulum during the primary transition phase, glucagon is already present at high concentrations—some 1000-fold higher than insulin (Fig. 15–7 B). This is correlated with the presence of A cells containing $\alpha$ granules. Glucagon levels and cytodifferentiation of the A cells reach the completely differentiated state while the exocrine pancreas and the insulin-secreting islet cells are still in the protodifferentiation state. The lack of a method for assaying low levels of glucagon does not at present allow an answer to the question of whether or not A cells also pass through a primary transition and a protodifferentiation state. If they do, this would occur before the formation of any morphologically or histologically recognizable pancreatic tissue.

## References

Pictet R. and W. J. Rutter. 1972. Development of the embryonic endocrine pancreas. In: Handbook of Physiology, Section 7, Endocrinology, I. R. Greep and E. B. Astwood, eds. Baltimore: Williams and Wilkins.

Rawles, M. E. 1936. A study of the localization of organ-forming areas in the chick blastoderm of the head-process stage. J. Exp. Zool. 72:271–315.

Rudnick, D. and M. E. Rawles, 1937. Differentiation of the gut in chorioallantoic grafts from chick blastoderms. Physiol. Zool. 10:381–395.

Shephard, T. H. 1965. The thyroid. In: Organogenesis. R. L. DeHann and H. Ursprung, eds. New York: Holt, Rinehart and Winston.

Streeter, G. L. 1942, 1945, 1948, 1949, 1951. Developmental horizons in human embryos. Carnegie Contrib. Embryol. 30:211–245; 31:29–63; 32:133–203; 33:149–167; 34:165–196.

# 16

# THE RESPIRATORY SYSTEM

The tissues of the embryo, as well as those of the adult organism, require energy to carry out their various activities. The bulk of this energy is derived from a series of complex chemical reactions within the cell that involve the utilization of oxygen *(internal respiration)*. An end product of pathways designed to capture this energy (in the form of ATP) is carbon dioxide, a gas that is either modified or eliminated directly from the internal environment of the organism.

For survival, both embryos and adult organisms have developed specialized systems for the supply of oxygen to and the elimination of carbon dioxide from their cells and tissues. The organs exchanging gases with the external environment *(external respiration)* in adult vertebrates are chiefly *internal gills* (fishes) and *lungs* (tetrapods). Both gills and lungs represent specialized structural derivatives of the embryonic pharynx. Other sites where external respiration may occur include the integument (amphibians), oral cavity (amphibians), and cloaca (reptiles).

The supply of oxygen and the elimination of carbon dioxide in most young developing embryos take place by simple diffusion and exchange with the external environment. However, the energy requirements in rapidly growing embryos are such that specialized organs must be developed to promote these vital functions. In the placental mammal, the fetal and maternal tissues of the placenta serve as the site where carbon dioxide produced by the embryo diffuses into the maternal blood, and oxygen carried by the maternal circulation diffuses into the embryonic circulation. In reptiles and birds, the allantois with its extensive system of blood vessels acts as a respiratory organ. The allantois continues to function in this capacity until the embryo hatches from the eggshell and begins to breathe the surrounding air. The frog tadpole utilizes the moist skin and a series of pharyngeal gills.

Both gills and lungs are laid down during embryogenesis. To anyone who has dissected and examined these organs of respiration, the morphology of gills and lungs appears to be complex and quite different. In several respects, however, the structural differences between gills and lungs are superficial and primarily relate to the "conducting portion" of the respiratory organ. The basic requirement of any respiratory organ is a surface area across which gaseous exchange between the blood vascular system and the oxygen-containing medium can take place. The efficiency of the exchange is enhanced if the surface area is large enough to permit sufficient flow and a diffusion distance short enough to allow rapid movement of molecules. Large surface areas are achieved by folding or branching of the respiratory surface. Short diffusion distances are created by making the cells of the blood–air (or blood–water) barrier thin. Whether the organ of respiration is a gill or a lung, therefore, the physical principles governing gaseous exchange dictate a common structural design at the site of external respiration.

## Development of the Organs of Respiration: Gills

Recall that the pharyngeal region of the embryonic foregut consists in all vertebrates of a series of laterally directed *pharyngeal* or *visceral pouches*. The pairs of visceral pouches develop sequentially, beginning with the first pair lying behind the mandibular arch. Ectodermal folds opposite to the visceral pouches push inward, producing a series of *visceral grooves* on the surface of the embryo; each groove corresponds in position to a visceral pouch. There is experimental evidence to suggest that the visceral grooves are induced to formation by the visceral pouches as the latter contact or approximate the ectoderm. Temporarily, a *pharyngeal membrane* is formed by the lateral endodermal wall of the visceral pouch and the adjacent inner wall of the ectodermal groove. A *gill cleft* is formed when the pharyngeal membrane ruptures or becomes perforated, thus affording an open channel between the cavity of the pharynx and the external environment.

Most aquatic vertebrates use *gill lamellae* or *gill filaments* as sites of gaseous exchange (i.e., adult sharks and teleost fishes; larval frogs) (Fig. 16–1). These structures arise as the result of the covering epithelium of the visceral pouches being thrown into a complex labyrinth of primary and secondary foldings. Although the number of pouches participating in gill lamellae or filament formation varies among vertebrate species, the gills formed in this fashion are of the *internal type*, since they are specializations of the inner or endodermal germ layer. Typically, gill lamellae are anchored onto the anterior and posterior surfaces of the *gill septum*, the latter being a flattened, lateral extension of the gill arch. The gills on each surface of the gill arch constitute a *demibranch*, while the two surfaces on either side of a gill arch form a complete gill or *holobranch*. Internal changes in the gill arches stimulate the vascularization of the lamellae or filaments. A large blood vessel beneath the pharynx (i.e., the ventral aorta) supplies a *branchial artery* to each gill arch. With the formation of the gill lamellae or filaments, each branchial artery eventually becomes divided into an *afferent branchial artery* and an *efferent branchial artery*. An extensive capillary network with a large surface area joins together the afferent and efferent vessels. The affer-

**FIG. 16–1.** Development of gills in the frog tadpole. A, the gill area in a 6-mm tadpole; B, external gills in a 9-mm tadpole; C, a reconstruction of a frog tadpole to show the topographical relationships between internal and external gills. (From R. Rugh, 1977. A Guide to Vertebrate Development. Burgess Publishing Company, Minneapolis.)

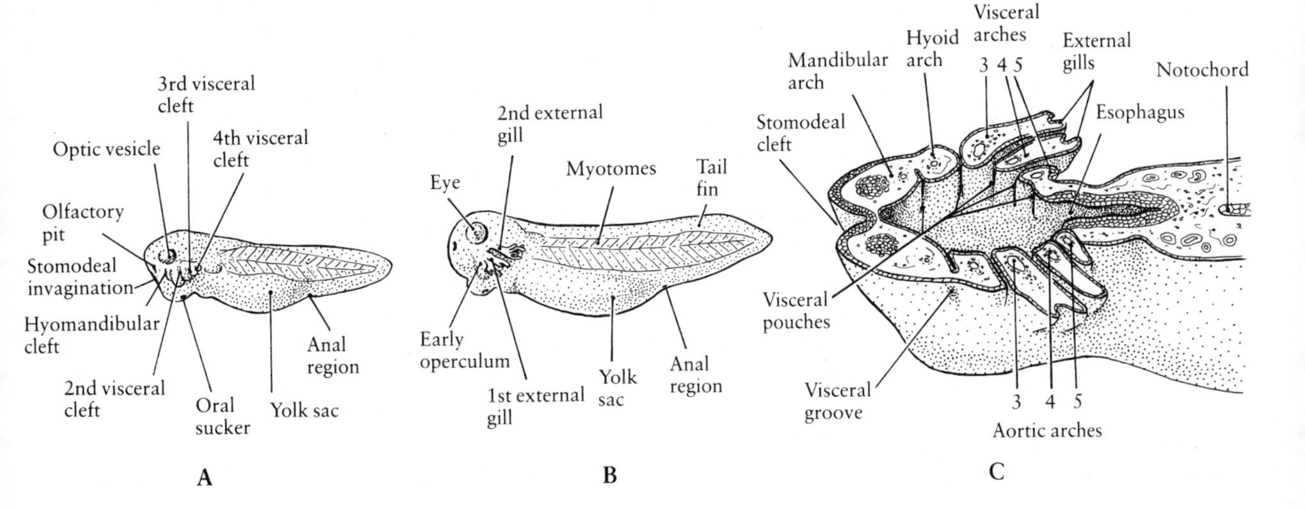

ent artery carries blood from the ventral aorta to the thin-walled capillaries; the efferent artery returns blood from the gills toward the dorsal aorta.

True *external gills* are primarily an embryonic or larval respiratory organ and occur in such forms as lungfishes, frogs, and salamanders. However, in some adult amphibians that continue a completely aquatic existence, such as the mudpuppy (*Necturus*), external gills are retained as the major organ of respiration. External gills are formed during the larval stage in all amphibians and in the case of frogs, they are replaced by gills of the internal type. Experimental studies with salamander embryos have shown that the endodermal pouches induce external gill development upon reaching the ectoderm. Gills of the external type originate as fleshy outgrowths from the dorsal ends of the visceral arches, particularly from the third, fourth, and fifth visceral arches. From these conelike structures, finger-shaped projections or gill filaments extend outward. As in the case of internal gills, the vasculature of the external gill filaments is supplied by the branchial arteries of the visceral arches. External respiration is achieved by the movement of the external gill filaments in the surrounding medium.

The development of gill filaments does not, of course, occur in organisms using lungs as primary sites of internal respiration. In amniotes, for example, four pairs of visceral pouches develop, but only the first three pairs show corresponding visceral grooves and clefts; the fourth pair of pouches rarely opens to the outside. The visceral clefts of the first three pairs of pouches eventually become closed over by folds of the ectoderm.

## Development of the Organs of Respiration: Lungs

Lungs or homologous structures are present in most major classes of vertebrates, from bony fishes, to amphibians, to mammals. Without exception, the lungs arise from a ventral pharyngeal diverticulum that appears relatively early in embryonic life. The pocketlike diverticulum typically divides at its tip into two branches; the latter are the primordia to the two bronchi and the lungs proper. The unpaired portion of the rudiment remains as the trachea. Both pharyngeal endoderm and pharyngeal mesoderm will participate in the construction of the postpharyngeal parts of the respiratory system, including the lungs proper.

Among vertebrates, the lungs show a range in the complexity of their structural organization. In the amphibians, particularly the salamanders, the lungs are a pair of simple, smooth-lined sacs that develop as expansions at the distal ends of an initially single respiratory primordium. By contrast, the lungs in mammals consist of many lobules composed in turn of a complicated system of branching tubes that terminate in thin-walled sacs or alveoli. The association of special air-storage compartments or *air sacs* with the air passageways distinguishes the lungs of birds from those of amphibians and mammals. Despite these dissimilarities, the lung in all vertebrate species basically develops in similar fashion, but to different stages of complexity. The degree of complexity can be directly correlated with the extent of internal subdivision and compartmentalization.

### The Mammal—Human

The primordium of the postpharyngeal components of the respiratory system—the larynx, the trachea, the primary bronchi, and lungs—initially appears in the form of a midventral trough or furrow in the endoderm of the pharynx. Seen in the human embryo at apprxoimately four weeks of development, this *laryngotracheal ridge* (or *tracheobronchiolar ridge*) is bluntly rounded and has an extensive communication with the ventrocaudal part of the pharynx (Fig. 16–2 A). Beginning posteriorly at the junction of the laryngotracheal ridge and the future esophagus, a pair of lateral grooves, one on either side, pinches inward and produces a pair of lateral folds (Fig. 16–2 B). The paired folds grow toward each other and fuse to form a temporary *tracheoesophageal septum*. As the lateral grooves deepen and gradually extend in the cranial direction, the septum is split, thereby separating a *laryngotracheal* (or *tracheobronchial*) *tube* from the rest of the gut tract except at the level of the future glottis or *laryngeal aditus* (approximately at the level of the fourth pair of visceral arches) (Fig. 16–2 C,D). The laryngotracheal tube elongates in the caudal direction, ventral to and approximately parallel to the esophagus. While this is occurring, the distal end of the tube enlarges and divides to form a pair of rounded swellings known as *lung buds*. These are probably better termed *primary bronchial buds*, since each is a rudiment for the major branch to each future lung. By approximately the 4-millime-

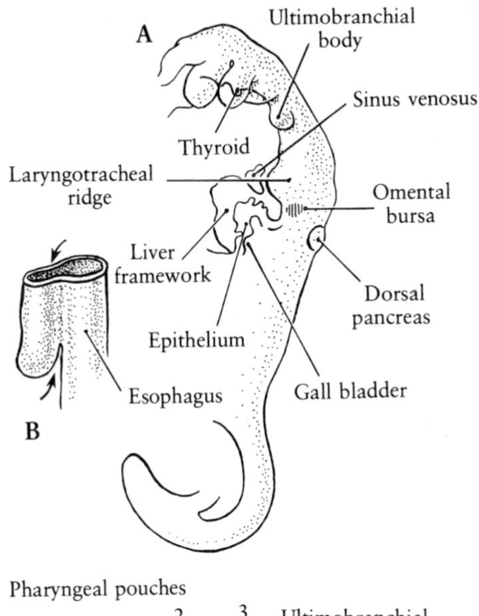


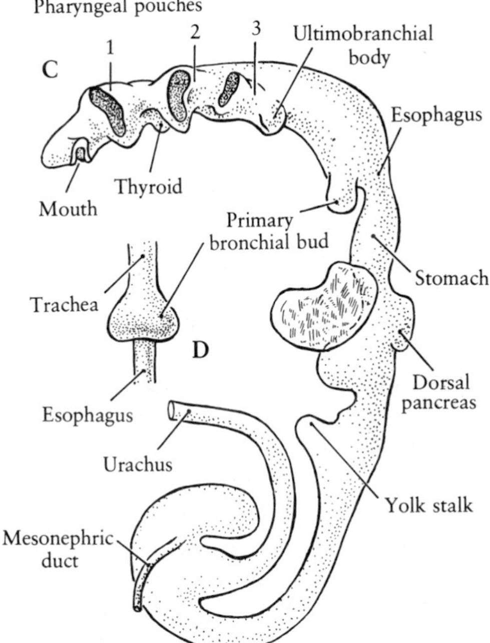


**FIG. 16–2.** A, outline drawing of the gut endoderm to show the laryngotracheal primordium in a 3-mm human embryo; B, a lateral view of the laryngotracheal primordium at the same stage to show the lateral furrows as they pinch in between the future esophagus and trachea; C, outline drawing of the gut endoderm to show the respiratory primordium in a 4-mm embryo; D, ventral view of the respiratory primordium in a 4-mm embryo. (A, C, from G. Streeter, 1945. Carnegie Contrib. Embryol. 31:27.)

ter stage, the components of the future respiratory system are morphologically represented by the laryngeal region, the tubular trachea, and the paired primary bronchial buds (Fig. 16–2 D).

*The Larynx*

The larynx is a complicated organ of elastic and muscular tissues, supported by several cartilages, whose foundations are laid down very early in development (Fig. 16–3). It develops primarily around the site of the evagination of the primary respiratory primordium.

The slit that opens from the floor of the pharynx into the trachea is the laryngeal aditus or laryngeal orifice. Located between the bases of the third and fourth visceral arches, it is a narrow, longitudinal slit bounded cranially by a rounded pronounced thickening of the floor of the pharynx known as the *hypobranchial eminence*. The hypobranchial eminence will gradually become modified in shape, assume the form of a transverse flap, and give rise to the *epiglottis*. (Fig. 16–3 A). The epiglottis will become muscular and act to protect against the entrance of foreign objects into the larynx and respiratory passageways during swallowing.

Concurrently, two additional thickenings, the *arytenoid swellings*, form on either side just caudal to the laryngeal aditus (Fig. 16–3 A). They arise as thickenings of mesodermal tissues originating from the fourth and fifth pairs of visceral arches. Each arytenoid swelling then thickens and lengthens toward the hypobranchial eminence. The two ridges of the tissue formed in this fashion, the *aryepiglottic folds*, convert the laryngeal aditus into a T-shaped orifice (Fig. 16–3 B).

For a short period of time (between 7 and 10 weeks of development), the entrance into the larynx ends blindly because the epithelial lining of the upper part of the larynx becomes fused together. The epithelial union then breaks down, and the lumen of this part of the respiratory system recanalizes, leaving an enlarged, oval laryngeal orifice and a pair of lateral recesses termed the *laryngeal ventricles*. Anteroposterior folds of the respiratory epithelium along the cranial and caudal walls of each laryngeal ventricle form the *vestibular folds* (false vocal cords) and the *vocal cords* (true vocal cords), respectively.

As with the rest of the postpharyngeal parts of the respiratory system, the inner or endodermal layer differentiates as the epithelial lining of the larynx.

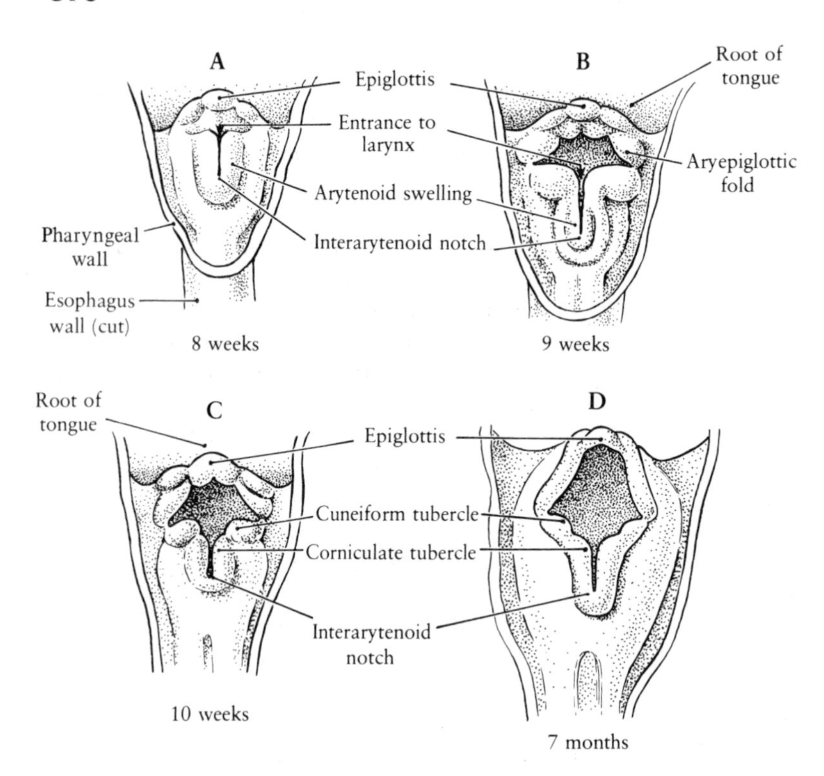


**FIG. 16–3.** Development of the human larynx. A, 8 weeks; B, 9 weeks; C, 10 weeks; D, 7 months.

Dense mesenchyme from the fourth and fifth pairs of visceral arches supports the epithelium. The skeletal tissue of the larynx, chiefly in the form of several *laryngeal cartilages*, develops during the seventh week as the localized arytenoid swellings. Shortly thereafter, the primordia of the elastic cartilage plates, the *cuneiform* and *corniculate tubercles*, are clearly visible along the margins of the lumen of the larynx (Fig. 16–3 C,D). Later in fetal life the epiglottis becomes reinforced with a plate of cartilage. The *laryngeal muscles* originate from the same mesenchymal masses that form the laryngeal cartilages and hence are innervated by branches of the vagus nerve. The definitive topography of the larynx is assumed during the last third of gestation (Fig. 16–3 D).

*The Trachea, Primary Bronchi, and Lungs*

Following its separation from the foregut, the tracheal portion of the respiratory primordium grows rapidly and carries the primary bronchial buds caudally until they reach their definitive position in the thorax (Fig. 16–4). A cross section through the lung region at this time shows that each primary bronchial bud consists of an innermost epithelial layer surrounded by vascularized mesenchyme (Fig. 16–5). Initially, the primary bronchial buds tend to be symmetrically arranged (Fig. 16–6 A). Quickly, however, the lung buds become noticeably different in size and orientation. The right primary bronchial bud becomes larger in appearance and tends to be less sharply directed to the side (Fig. 16–6 B). This particular pattern of the primary bronchial tubes readily explains why foreign objects, postnatally, more frequently enter the right main bronchus than the left main bronchus. During the fifth week, each endodermal lung bud divides *monopodially* (see below) to give rise to a lateral diverticulum or bud; subsequently, the right lung bud gives origin on its craniodorsal side to a second monopodial diverticulum (Fig. 16–6 C). By about the beginning of the second month, the right lung bud has differentiated into an undivided, proximal *right primary* or *main bronchus* and distally three *stem bronchi;* the left lung bud has differentiated into an undivided proximal *left primary* or *main bronchus* and distally two stem bronchi. Each stem bronchus is destined to branch and rebranch and, with the surrounding pulmonary mesenchyme, will give origin to the definitive *pulmonary lobes* that characterize adult lung organization. Hence, the right lung typically has three lobes (upper, middle, and

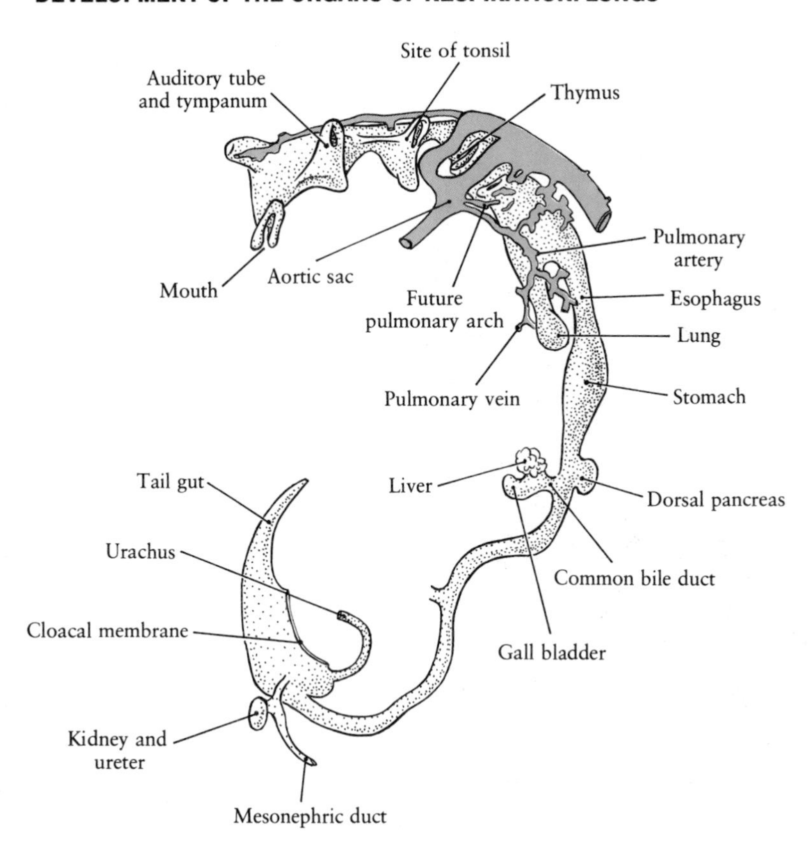


**FIG. 16–4.** Outline drawing of the gut to show the caudal growth of the respiratory primordium in the 5.7-mm human embryo. (From G. Streeter, 1945. Carnegie Contrib. Embryol. 31:27.)

lower) and the left lung two lobes (upper and lower).

The early branching of the primary bronchial buds tends to be monopodial. That is, a branch or diverticulum is formed on one side while the main branch continues to grow beyond the point of branching without any significant change in direction. The subsequent branching pattern of the stem bronchi tends to be *dichotomous*, with a given branch being bifurcated into two symmetrically

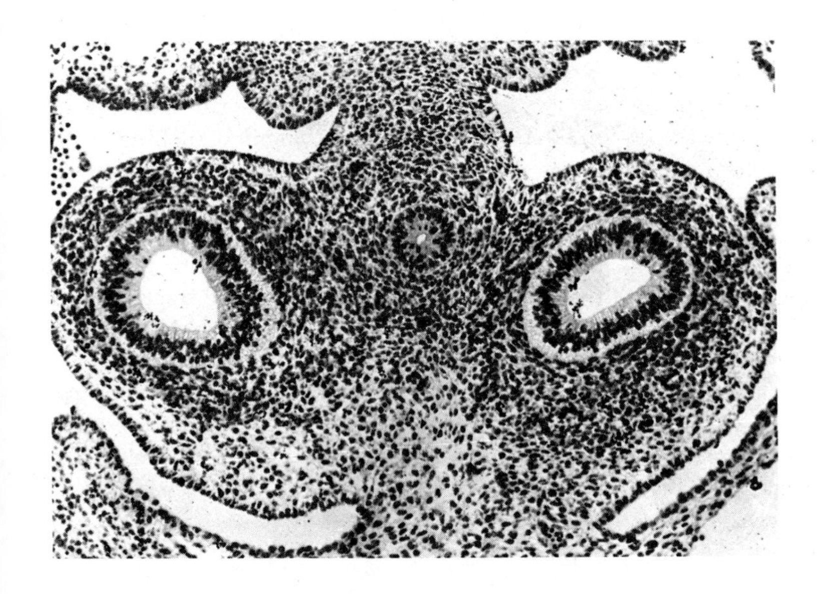

**FIG. 16–5.** A photomicrograph of a cross section through the 6.7-mm human embryo showing the paired primary bronchi and the esophagus. (From G. Streeter, 1945. Carnegie Contrib. Embryol. 31:27.)

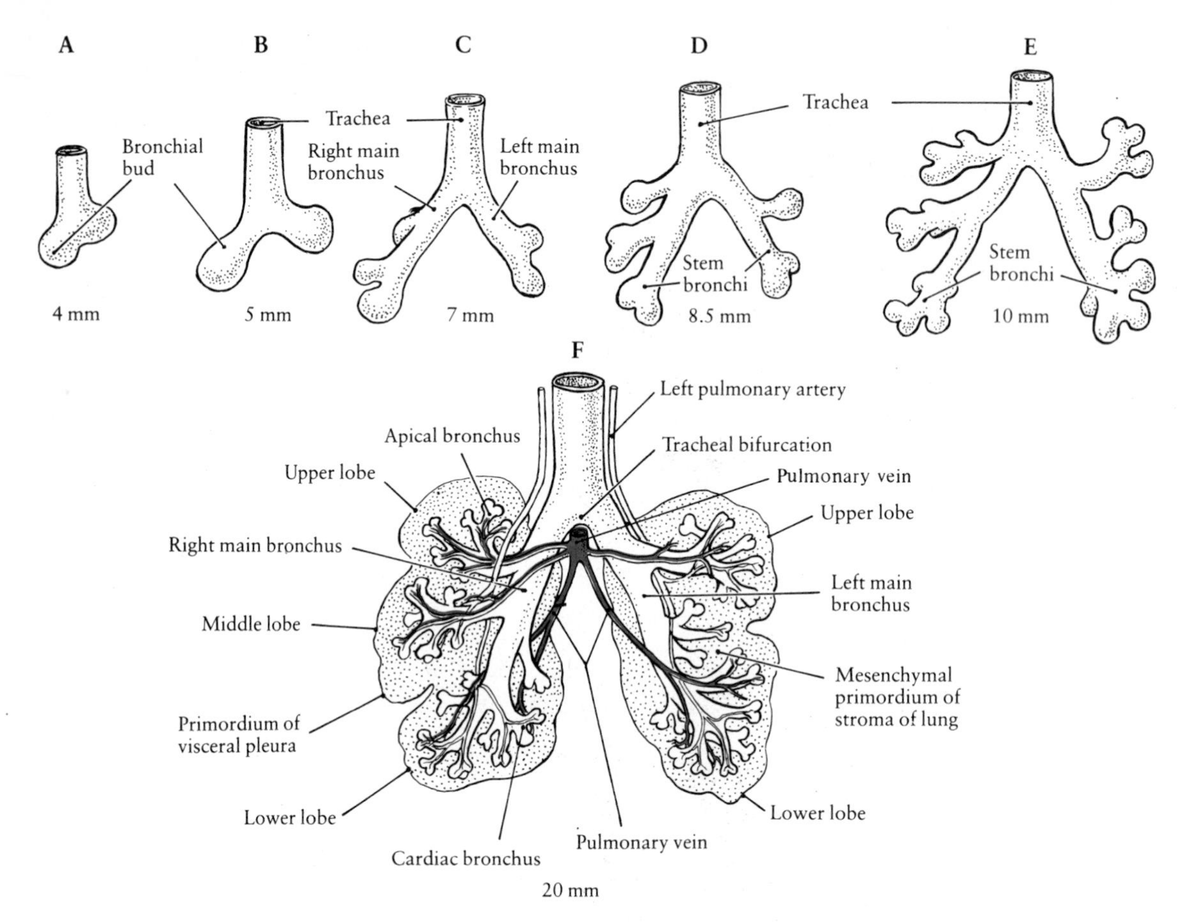


**FIG. 16–6.** Diagrams showing the progressive development of the major bronchi of human lungs (ventral views). (After L. Arey, 1965. Developmental Anatomy. W. B. Saunders Company, Philadelphia.)

placed branches (Fig. 16–6 E). By these processes a large and arboreous system of tubes forms, which is termed the early bronchial or *respiratory tree*. With enlargement of the bronchial tree, the pulmonary mesenchyme surrounding the stem bronchi and their descendant tubes becomes furrowed, thereby yielding the principal lobes of the lung (Fig. 16–6 F). By the seventh week, 10 bronchi of the third order of branching have appeared in the right lung and 8 for the left lung. Each of these bronchial tubes supplies the remaining branches for the clinically important *bronchopulmonary segments* or lobules. These lobules are separated from each other by connective tissue partitions derived from mesenchyme.

It is now generally recognized that human lung development can be subdivided into five stages: (1) *embryonic period* from conception to week 5 in utero; (2) *pseudoglandular period* from week 5 to 17 in utero; (3) *canalicular period* from week 16 to 24 in utero; (4) *terminal sac period* from week 24 until birth; and (5) *alveolar period* from late fetal life and after birth. Following the appearance of the respiratory primordium from the floor of the foregut, each lung through dichotomous branching establishes the major air passageways and treelike network of narrow tubes lined by a thick columnar or cuboidal epithelium. By 16 weeks of development, all of the branches of the air-conducting portion of the tracheobronchial tree, from the trachea up to and including the terminal bronchioles, are formed. The lung resembles an exocrine gland in its construction (hence the name pseudoglandular period). During the canalicular period, the functionally important respiratory or gas-exchanging portion of the lung becomes delineated with the appearance of new tubular branches, the *respiratory bronchioles*. Each respiratory bronchiole terminates in two or three thin-walled dilations called *terminal sacs* or *primitive alveoli*. At this time there is also an increase in the size of the lumens of the bronchi and bronchioles, and a gradual thinning of their epithelial walls. The terminal sac period is characterized by further differentiation of the respiratory portion of the lung; respiratory bronchioles rapidly subdivide into an array of thin-walled primitive alveolar ducts and primitive alveoli. Primitive alveoli are larger and lack the smooth outline so typical of true alveoli. The immature alveoli are lined by a continuous vascularized epithelium that becomes progressively attenuated. There is sufficient respiratory surface in the fetus at this time for gaseous exchange, represented by the terminal sac epithelium, to permit survival in case of premature birth. By the end of the terminal sac period, the organizational pattern of the respiratory portion of the lung is complete.

Current information indicates that *definitive alveolar ducts* and *mature alveoli* probably do not form until very late in fetal life and after birth (alveolar period). The alveoli form initially by the attenuation of prenatal primitive alveoli and subsequently by the segmentation of existing alveoli. They also have been observed to form from alveolar ducts and respiratory bronchioles. Mature alveoli are lined by a simple squamous epithelium. Increase in the size of the lung after birth is largely attributable to additional branching of the fetal respiratory tree and the multiplication of alveoli. New respiratory bronchioles and alveoli are produced postnatally for six or seven generations of branchings. Precise information on the extent of alveolar multiplication is limited. It has been estimated that there are about 20 million air spaces (presumably primitive alveoli) in the fetal lung at birth; by the age of eight years, the adult number of 300 million has been reached.

In summary, the pattern of both conducting and respiratory passageways of the lung is established by birth. For the next several years following birth, there is substantial growth of the lung; this growth is manifest in an increase in the number and size of alveoli as well as in the number of respiratory bronchioles.

Development of an adequate pulmonary vasculature is critical to the proper functioning of the lung at birth. Vascularization of the embryonic lung is initially evident at the time of primary bronchial bud formation (Figs. 16–4, 16–5). Each bronchial bud becomes invested with a fine network or plexus of capillary-like blood vessels arising from the *aortic sac* (see Chapter 18). Subsequently, branches descending from the sixth pair of aortic arches, the rudiments of the *pulmonary arteries*, join with the lung bud capillaries. The pulmonary artery to each lung will branch and rebranch, tending to follow the pattern of bronchial tube branching. Extensive proliferation of pulmonary capillaries is particularly noticeable during the canalicular period of lung development. These capillaries press tightly against the thin epithelial walls of the respiratory bronchioles and terminal sacs. With attenuation of the terminal sacs, the underlying capillaries bulge into the lumens of the alveoli. The air–blood barrier, therefore, is formed by two adjacent and tightly apposed epithelial membranes, one contributed by the alveoli and the other by the pulmonary capillaries. The pulmonary veins will develop from pulmonary mesenchyme lying between pulmonary lobules.

Development of the pulmonary veins occurs after that of the arteries. Similar to the pulmonary arterial branches, the architecture and branching of the venous system closely follows the pattern of the bronchial tree (Fig. 16-6 F).

## General Mechanisms of Lung Development

The dual origin of the lung—pulmonary epithelial endoderm and pulmonary mesenchyme—permits an analysis of the contributions made by each embryonic layer to the morphogenesis and cellular structure of the adult organ. As pointed out above, the epithelial component of the mammalian lung branches into a complex network of bronchial tubes. The pulmonary mesenchyme invests and condenses around the bronchial tubules as the respiratory tree takes form and shape.

Since early studies by Rudnick (1933), the fetal lung has always been considered to be capable of self-development or self-differentiation. That is, the primordium of the lung when removed and cultured under proper experimental conditions will develop to form an organ with a remarkable likeness to the lung in vivo. The lung explant can differentiate most of the cellular elements found in the adult lung. Also, many of the patterns of chemical change within differentiating cells are faithfully repeated in the lung explant. Since the lung is a self-developing entity, its epithelial and mesenchymal components can be manipulated in culture to determine their role in lung morphogenesis and the relationship between morphogenesis and the differentiation of various cell types.

Presumably, the laryngotracheal ridge appears during embryogenesis in response to some initiating induction in the foregut. However, the nature and the source of the inducing stimulus remain unknown. In the laryngotracheal area there are several chemical changes (such as increases in glycogen, alkaline phosphatase, and ribonucleoprotein) that accompany the formation of the early lung primordium, but their significance, if any, to the induction process has not been eludicated.

Concrete evidence that the epithelial and mesenchymal layers of the pulmonary primordium interact in an inductive relationship and that they are indispensable for lung morphogenesis and cytodifferentiation come from organ culture studies utilizing the more advanced fetal lung. Much attention has been given to the dependence of pulmonary epithelial branching on the presence of the surrounding mesenchyme. Rudnick first demonstrated that branching of the bronchial tree fails to occur if the pulmonary epithelium is deprived of its investing mesenchyme. Also, removal of the mesenchyme during the epithelial branching will immediately interrupt the latter process. Other investigators have also shown that epithelial morphogenesis is dependent on underlying mesenchyme by grafting a piece of *bronchial mesoderm* next to the tracheal endoderm (Fig. 16–7 A). After approximately six hours in culture, a supernumerary bud appears in the explant at the original grafting site (Fig. 16–7 B). The extra bronchial bud is then observed to grow and branch profusely in a pattern reminiscent to that seen in vivo. Note that the left primary lung bud in the absence of investing pulmonary mesenchyme fails to subdivide. If *tracheal mesoderm* is excised and grafted to the base of one of the primary lung buds (Fig. 16–7 C), the lung bud with the investment of tracheal mesoderm fails to branch (Fig. 16–7 D). Hence, the pulmonary mesenchyme appears to exhibit very different capacities to stimulate epithelial budding in the developing lung. The proximal, older, or more mature tracheal mesoderm prevents or inhibits epithelial budding and thus accounts for the undivided nature of the structural trachea. The distal, younger mesenchyme stimulates the epithelial budding to form the respiratory tree. Electron-microscopic studies by Wessells (1970) have shown that the tracheal mesoderm is structurally different from the bronchial mesoderm. Cells of the former are highly ordered and form a tight investment around the tracheal endoderm. A layer of highly oriented collagen fibers is situated along the tracheal endoderm. At the tip of active bronchial buds, these same collagen fibers are randomly oriented (Fig. 16–8). Such an arrangement of fibers along the trachea could conceivably effect an inhibiting action by preventing the pas-

**FIG. 16–7.** Grafting a piece of bronchial mesoderm next to the tracheal endoderm (A) will induce a supernumerary lung bud (B). A lung bud with an adjacent graft of tracheal mesoderm (C) fails to subdivide (D).

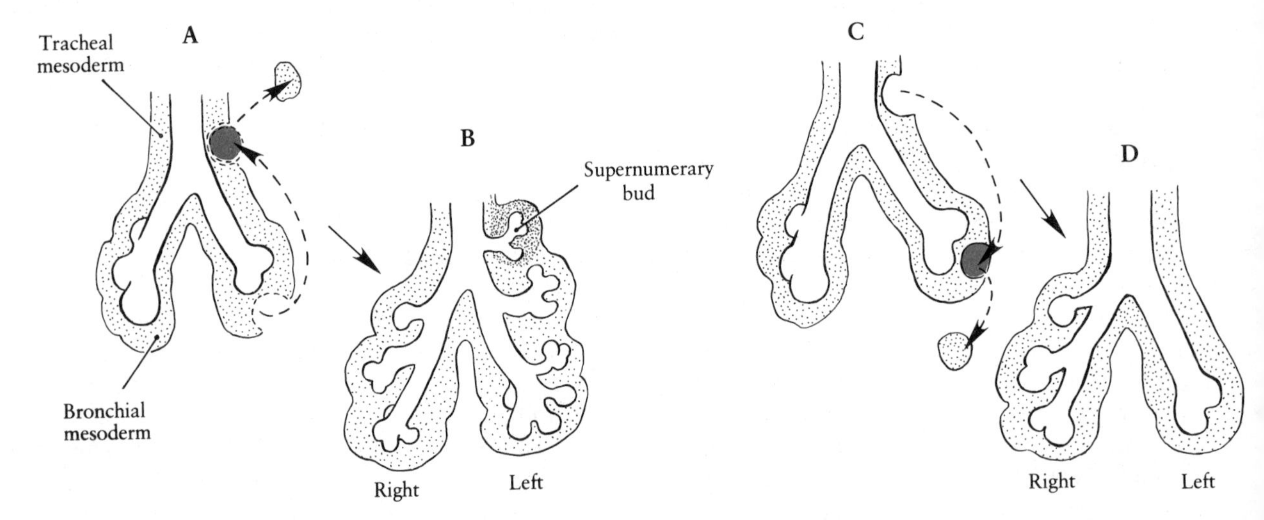

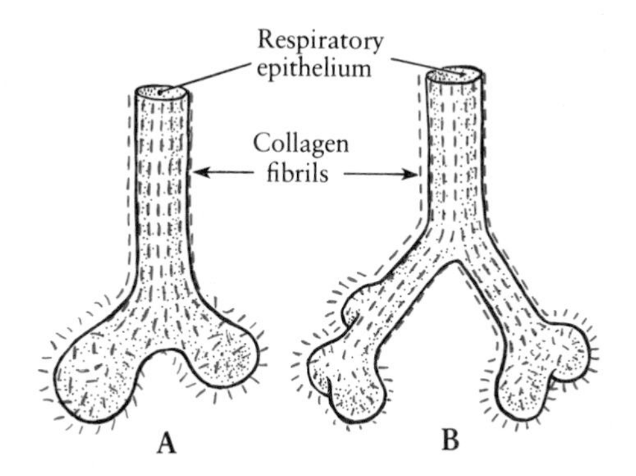


**FIG. 16–8.** A diagram after studies by Wessells and others showing the distribution of collagen along the tracheobronchial epithelium of a mammalian embryo at two successive steps in lung morphogenesis. Note that branching occurs where there is a random arrangement of collagen fibrils.

sage of an inducing stimulus from the pulmonary mesenchyme.

How specific is the pulmonary mesenchyme in inducing epithelial budding? A variety of foreign mesoderms, such as those excised from gut or salivary gland, will elicit formation of the laryngotracheal or primary lung bud primordium. However, the branching of the primary lung buds is always dependent on an appropriate interaction with bronchial mesoderm. Although the first step in lung formation occurs in the presence of what we might term *nonspecific mesoderm*, the shape and form of the respiratory tree require a very specific or *homologous mesoderm*.

Whether the morphogenetic factors controlling the species-specific pattern of branching are initially resident in the bronchial mesenchyme or bronchial epithelium is still a major unresolved question. For example, during normal chick lung morphogenesis, monopodial branching leads to a system of tubes and expanded air sacs. By contrast, mouse lung epithelium is more active and exhibits both dichotomous and monopodial branching in constructing a complex system of bronchial tubes. Taderera (1967) has observed that the in vitro combination of mouse lung epithelium and chick lung mesenchyme *(a chimeric explant)* initially exhibits a bronchial branching pattern characteristic of intact mouse lung explants. However, after 5 days in culture, a few of the terminal branches displayed a typical chick lung morphology. This suggests that the mesoderm may play an important role in determining branching pattern. Other heterotypic recombinations (such as salivary epithelium and lung mesenchyme) point to the epithelium as contributing to the control of a branching pattern.

The question of whether cellular contact is required between the lung mesenchyme and the lung epithelial tissue for induction is still unresolved. Earlier studies, in which Millipore filters (25 $\mu$m thick; 0.45 $\mu$m pore size) were placed between the two pulmonary tissues, suggested that direct cell contact was not necessary for induction to occur. However, the pulmonary mesenchyme in experiments utilizing the transfilter technique remains simple in organization and generally shows few histological cell types. Normally, in the complete absence of pulmonary epithelium, the pulmonary mesenchyme fails to differentiate any of the characteristic connective tissue and smooth muscle cell types, thus showing that the mesenchyme and epithelium are reciprocally dependent on each other for their differentiation. The failure of the pulmonary mesenchyme to differentiate smooth muscle across the filter may mean that either contact between the interacting tissues is necessary or that inductively active materials are impeded by the filter owing to molecular size or lowered mobility. More recently, Wessells and other have examined lung morphogenesis in the mouse through induction of supernumerary buds in normally unbranched tracheal epithelium by the transplantation of bronchial mesenchyme. They maintain that 11- and 12-day-old tracheal epithelium forms supernumerary buds only at points of contact with grafted pulmonary mesenchyme.

Although several explanations have been advanced, the precise role of the epithelium and the mesoderm in controlling lung morphogenesis remains poorly understood. Because of the suspected role of extracellular materials in mediating tissue interactions in the construction of other organs, some attention has been given to characterizing the macromolecular constituents between lung epithelium and lung mesenchyme, and assessing their function in the branching process. Both collagen and glycosaminoglycans (GAGs) can be identified by a variety of cytochemical techniques between the interacting tissues. Evidence suggesting that collagen, produced by the mesenchyme, is required for branching has come principally from experiments in which the morphogenesis of lung rudiments in culture is interrupted upon treatment with collagenase.

More recently, Spooner and his colleagues have explored the possible involvement of collagen in branching morphogenesis by employing drugs known to interfere with collagen metabolism. Treatment of 11-day-old mouse lungs in culture for 72 hours with compounds known to block collagen synthesis and processing, such as L-azetidine-2-carboxylic acid (LACA), results in failure of the lung primordia to continue branching and in general retardation of epithelial growth (Fig. 16–9). At the electron-microscopic level, normal lung rudiments are characterized by an abundance of collagen fibers between the mesenchyme cells (Fig. 16–10); however, there is a noticeable reduction in collagen and other extracellular materials in LACA-treated lung tissue. Basal laminae of the lung epithelia are unaffected by LACA treatment, an observation consistent with reports that basal laminae of embryonic epithelia are rich in GAGs. Unfortunately, data do not provide insight as to how collagen is involved in controlling the branching pattern. It is postulated that collagen may initiate or stabilize branch points in the lung epithelium. There are other studies suggesting that the bronchial mesoderm acts to regulate epithelial branching through initiating localized increases in mitotic activity. This might involve the passage of a mitotic stimulating factor from the mesenchyme.

**Fig. 16–9.** Effects of L-azetidine-2-carboxylic acid (LACA) on lung morphogenesis. A control lung is shown successively at (A) 24 hours, (B) 48 hours, and (C) 72 hours of culture. A LACA-treated (60 $\mu$g/ml) lung is shown successively at (D) 24 hours, (E) 48 hours, and (F) 72 hours. Between 48 and 72 hours of culture, LACA-treated lung fails to continue branching. (From B. Spooner and J. Faubion, 1980. Dev. Biol. 77:84.)

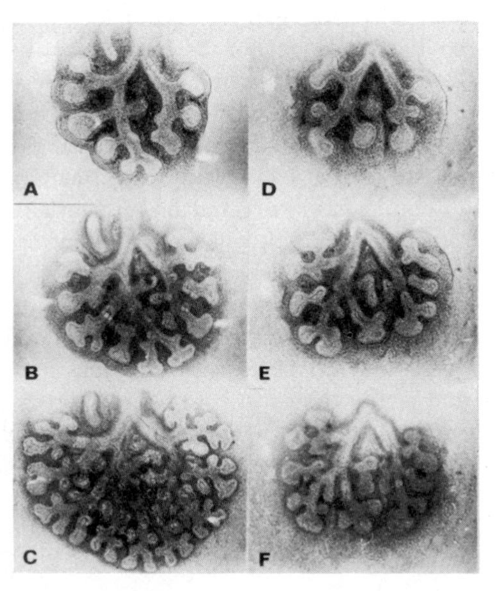


The division of an epithelial lung bud into two branches is probably dependent on a sequence of events. An actively dividing bud shows a high rate of mitosis at its tip. Additionally, its epithelial cells are observed to synthesize and secrete GAGs into the extracellular space. In contrast to nondividing portions of the respiratory epithelium, there is a rapid turnover of the newly synthesized GAGs at the tip of a dividing epithelial bud. It has been proposed that the GAG turnover is induced by hydrolytic enzymes released from the pulmonary mesenchyme. These enzymes appear to digest cell surface GAG of the basal lamina at the morphogenetically active regions. The tip region is thus maintained in an unstable state. Stabiliziation of the branching of the pulmonary epithelium is probably accompanied by reduced turnover in extracellular glycosaminoglycans and by modification of the nature of the extracellular substances.

Finally, all current evidence tends to indicate that the differentiation of the cell types of the lung and the expression of shape and form of the lung are coupled events in the development of this organ.The differentiation of cell types in both tissues begins at the trachea and proceeds toward the respiratory portion of the lung. In the human lung, the epithelium gradually decreases in thickness so that the proximal air passages are lined by a pseudostratified columnar epithelium, while the distal air passages are lined by a squamous epithelium. Of the eight cell types that characterize the epithelium of the adult human lung, four have begun to differentiate by 16 weeks of gestation; these are ciliated, nonciliated (pre-Clara), secretory, and basal cells. By birth, the two characteristic cell types of the alveoli (type I and type II) are differentiated. The mesenchyme differentiates into cartilage, connective tissue (particularly elastic tissue), and smooth muscle. The differentiation of these various mesenchymal-derived cell types tends to lag slightly behind the development of the epithelial cells.

## Respiratory Movements

As long as the fetus remains in the aquatic uterine environment, respiratory exchange occurs in the placenta, and the lung does not perform any respiratory function. Nevertheless, it has been known

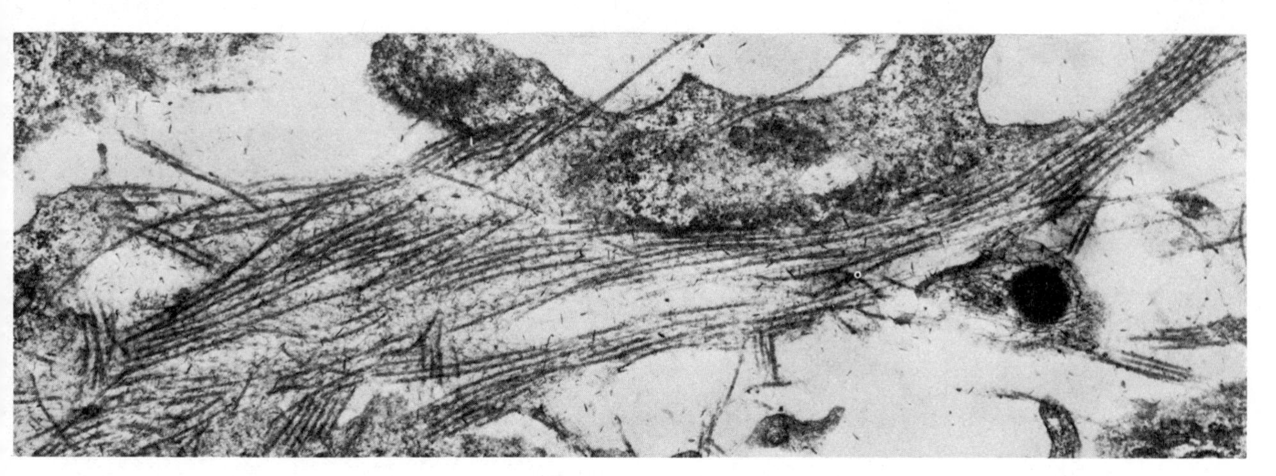

**FIG. 16–10.** Electron-micrographic demonstration of extracellular collagen in the mesenchymal cells of cultured embryonic lung (72 hours). (From B. Spooner and J. Faubion, 1980. Dev. Biol. 77:84.)

for many years that the fetus may show contractions of the thoracic musculature and diaphragm that resemble normal respiratory movements. The majority of investigators report that the respiratory movements of the fetus are irregular and not constant, thus casting serious doubt on the view that respiratory movements in the newborn are merely a resumption of the normal physiological, intrauterine movements that had been interrupted by the birth process.

Fetal respiratory movements can be elicited in human embryos from the third month onward. Although opinions are divided, a generally accepted view is that respiratory movements occur only when the fetus fails to receive an adequate supply of oxygen by way of the placental circulation. It is known, for example, that in experimental animals artifically induced *anoxia* of the fetus stimulates vigorous respiratory movements. The amount of amniotic fluid that finds its way into the lungs is normally not of any consequence. Experimentally, the presence of amniotic fluid in the lungs has been demonstrated by the detection of labeled amniotic fluid (produced by injecting ink or thorotrast, a radioopaque substance, into the amniotic cavity) in the trachea and bronchi. During late fetal life, there appears to be a state of respiratory movement inhibition. Near birth when the placenta becomes increasingly less efficient as a center of respiratory exchange and when some anoxia probably exists, there are surprisingly few respiratory movements. The neuromuscular architecture for initiating and controlling these movements is fully functional at this time.

Most of the spaces within the fetal lung are filled with fluid derived from the lung itself, the glands of the trachea, and the amniotic cavity. Aeration of the lungs, therefore, is due to the rapid replacement of fluid by air and not due to the simple inflation of a collapsed, empty lung. The fluid from the lungs is eliminated by several routes. Most of it enters the networks of pulmonary capillaries and lymphatic vessels that surround the pulmonary alveoli. Some fluid is also probably cleared from the lungs through the mouth and nose as the result of pressure exerted on the pleural cavities during the birth process.

The newborn animal apparently expands the soggy lungs by increasing the size of the pleural cavities. The original expansion must overcome the cohesion of the wet adhesive walls of the air passageways as well as the pressure exerted by the fluid of the pleural cavities. This is a remarkable feat when one considers that the muscular system at birth is weak and incompletely developed. It is not surprising that for as long as 10 days after birth, portions of the lungs are likely to remain uninflated *(atelectasis)*.

## Abnormalities of the Respiratory System

As one might expect, there are departures from the normal pattern of size and shape of various parts of the respiratory system (i.e., larynx, bronchial tubes), but these generally have little serious effect on the individual. Occasionally, an entire lung or lung lobe may be absent *(agenesis)*.

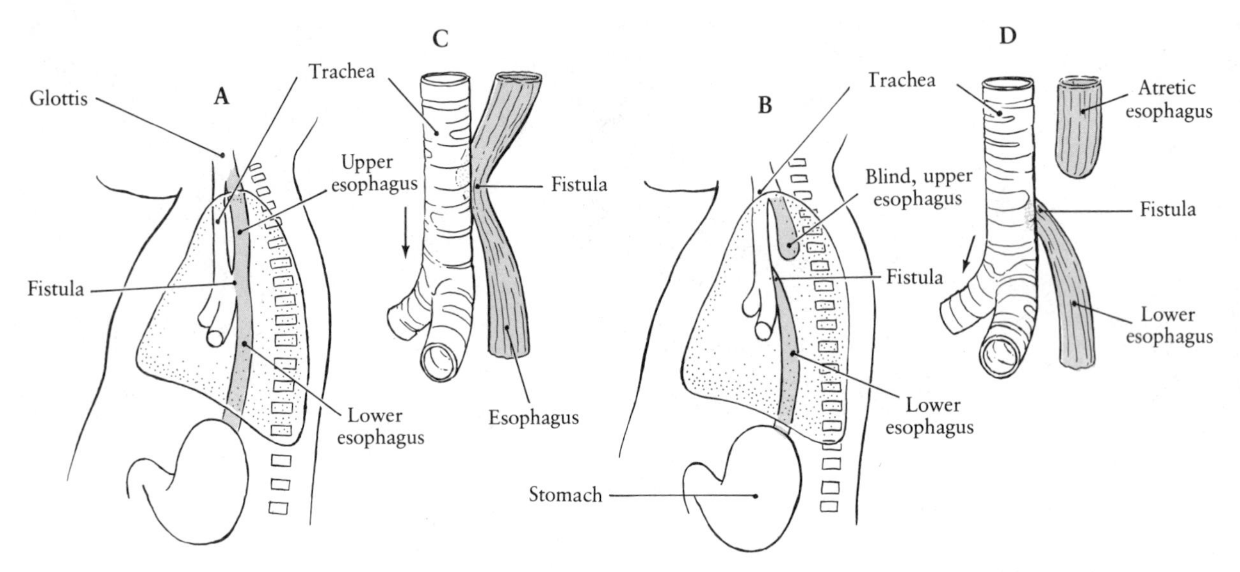


**FIG. 16–11.** A and B are sketches illustrating conditions encountered in two cases of tracheoesophageal fistula (After C. Haight, 1944. Ann. Surg. 120:623); C and D illustrate the suggested mechanical basis for the origin of the fistula in A and B, respectively. The rate of tracheal elongation is excessive (arrows) and determines the severity of the fistula syndrome. (After P. Gruenwald, 1940. Anat. Rec. 78:293.)

A serious congenital anomaly that is detected in approximately 1 in 3000 to 1 in 4000 births is *tracheoesophageal fistula*. The fistula is commonly visible at the level of the seventh cervical or first thoracic vertebra and appears as an opening between the trachea and esophagus below the level of the larynx (Fig. 16–11 A). Typically, the opening between the trachea and esophagus is large, with the upper part of the esophagus terminating as a blind tube (*esophageal atresia;* Fig. 16–11 B). The difficulties for the nursing infant are self-evident. Milk is regurgitated from the atretic esophagus and, if passed into the lungs, may result in pneumonia. The fistula arises during the fourth week when the laryngotracheal primordium is separating from the esophagus.

The basis for the origin of fistulas is largely speculative. Presumably, the laryngotracheal primordium grows more rapidly than the adjacent digestive tube, a process that interferes with the normal development of the paired lateral grooves. There may result a posterior connection between the esophagus and the trachea, typically in the vicinity of the paired primary bronchi (Fig. 16–11 C). If the rate of tracheal elongation is excessive, the lower portion of the esophagus is drawn out into a narrow muscular ridge and the upper portion of the esophagus into a distended blind sac (Fig. 16–11 D). For many infants, the fistula can be surgically ligated and an anastomosis effected between the upper and lower esophageal components.

*Bronchiectasis* is a congenital anomaly of the bronchial tubes, particularly the terminal bronchi. It is manifest in irregular saccular enlargements or evaginations from the bronchi. Postnatally, they are subject to chronic infection because they fail to drain properly. Prenatally, some of these sacs may become completely blocked to form fluid-filled *bronchial cysts*.

*Hyaline membrane disease* is a serious affliction of the respiratory system that is caused by a lack of *surfactant*, a detergent-like substance that is normally produced by alveolar cells and acts to reduce surface tension forces within the alveoli. Alveoli tend to collapse without surfactant. Some investigators support the hypothesis that disturbances in the development of the pulmonary vasculature, resulting in anoxia, affect the capacity of alveolar cells to produce surfactant. Obstetric techniques *(amniocentesis)* now make it possible to measure the levels of fetal surfactant.

## References

Amy, R., D. Bowes, P. Burri, J. Haines, and W. Thurlbeck. 1977. Postnatal growth of the mouse lung. J. Anat. 124:131–151.

Inselman, L. and R. Mellins. 1981. Growth and development of the lung J. Pediatr. 98:1–15.

Corliss, C. E. 1976. Patten's Human Embryology, pp. 296–306. New York: McGraw-Hill Book Co.

Gruenwald, P. 1940. A case of atresia of the esophagus combined with tracheo-esophageal fistula in a 9-mm human embryo, and its embryological explanation. Anat. Rec. 78:293–302.

Haight, C. 1914. Congenital atresia of the esophagus with tracheoesophageal fistula. Reconstruction of esophageal continuity by primary anastomosis. Ann. Surg. 120:623–655.

Hamilton, W. J. and H. W. Mossman. 1972. Human Embryology, 4th ed., pp. 291–376. Baltimore: Williams and Wilkins.

Rudnick, D. 1933. Development capacities of the chick lung in chorioallantoic grafts. J. Exp. Zool. 66:125–154.

Smolich, J., B. Stratfor, J. Maloney, and B. Ritchie. 1967. Postnatal development of the epithelium of larynx and trachea in the rat: scanning electron microscopy. J. Anat. 124:657–673.

Sorokin, S. 1965. Recent work on developing lungs. In: Organogenesis, pp. 467–491. R. L. DeHaan and H. Ursprung, eds. New York: Holt, Rinehart and Winston.

Spooner, B. and J. Faubion. 1980. Collagen involvement in branching morphogenesis of embryonic lung and salivary gland. Dev. Biol. 77:84–102.

Spooner, B. and N. Wessells. 1970. Mammalian lung development: interactions in primordium formation and bronchial morphogenesis. J. Exp. Zool. 175:445–454.

Streeter, G. L. 1945. Developmental horizons in human embryos. Description of age group XIII, embryos about 4 or 5 millimeters long, and age group XIV, period of indentation of the lens vesicle. Carnegie Contrib. Embryol. 31:27–63.

Streeter, G. L. 1948. Developmental horizons in human embryos. Description of age groups XV, XVI, XVII and XVIII, being the third issue of a survey of the Carnegie Collection. Carnegie Contrib. Embryol. 32:133–203.

Taderera, J. V. 1967. Control of lung differentiation *in vitro*, Dev. Biol. 16:489–512.

Wessells, N. 1970. Mammalian lung development: interactions in formation and morphogenesis of tracheal buds. J. Exp. Zool. 175:455–466.

Wessells, N. and J. Cohen. 1968. Effects of collagenase on developing epithelia *in vitro:* lung, ureteric bud, and pancreas. Dev. Biol. 18:294–309.

# 17

# THE COELOM AND MESENTERIES

Coelomic cavities are fluid-filled spaces that come to surround the various viscera of the vertebrate body during the course of development. These include the *pericardial cavity* around the heart, the *pleural cavities* surrounding the lungs, and the *peritoneal cavity* in which lie the stomach, intestines, pancreas, and other organs. Only the pericardial cavity and the peritoneal cavity are common to all vertebrates. Lungs present in such forms as amphibians, reptiles, and birds are located in the peritoneal cavity, which is more appropriately termed the *pleuroperitoneal cavity*.

The coelomic cavities of the mammal (pericardial, paired pleural, peritoneal) arise through the subdivision of the early *intraembryonic coelom* by the following partitions or membranes: the unpaired *transverse septum*, which effects an initial separation between pericardial and peritoneal cavities; the paired *pleuropericardial folds*, which fuse with the transverse septum and separate pericardial from pleural cavities, and the paired *pleuroperitoneal folds*, which also join with the transverse septum and complete the separation of each pleural cavity from the peritoneal cavity. These various membranes unite to contribute to the formation of the *diaphragm*, a muscular partition between pleural and abdominal cavities found only in mammals.

## The Intraembryonic Coelom

The early intraembryonic coelom in all vertebrates originates on either side of the body as the result of the confluence of small isolated spaces that appear in the lateral plate mesoderm (Fig. 17–1). In higher vertebrates, such as birds and mammals, the coelom extends within the lateral plate mesoderm beyond the confines of the developing body as the *extraembryonic coelom*. At the embryonic disc stage in the mammal, these cavities extend forward on either side to fuse anteriorly in the region of the future heart mesoderm *(cardiogenic mesoderm)* (Fig. 17–2). The cranial portion of the horseshoe-shaped intraembryonic coelom represents the primordium of the pericardial cavity. The right and left limbs represent the presumptive pleural cavities, as each will subsequently receive a developing lung bud, and the presumptive peritoneal cavity. Note that at approximately the level of the first pair of somites the intraembryonic coelom is continuous on either side with the extraembryonic coelom. As the body of the embryo is progressively folded off from the extraembryonic membranes, the extraembryonic and intraembryonic portions of the coelom are thereby separated from each other. It is only the intraembryonic coelom that is partitioned to accommodate the embryonic viscera.

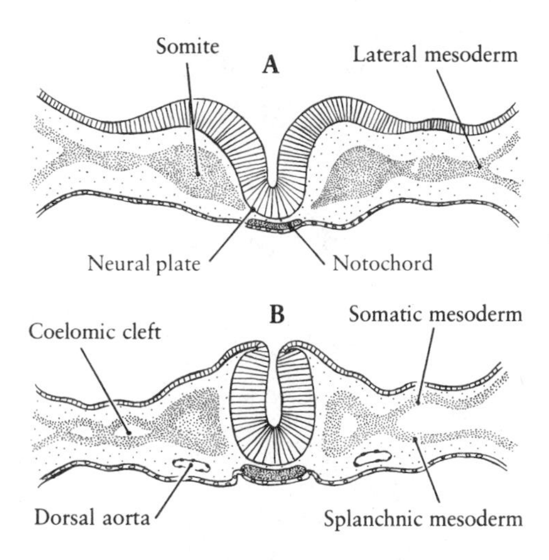


**FIG. 17–1.** Origin of the vertebrate coelom by confluence of spaces in the lateral plate mesoderm. A, human embryo, two-somite stage; B, human embryo, seven-somite stage.

The intraembryonic coelom divides the lateral plate mesoderm into a *somatic (parietal) layer*, continuous with the extraembryonic mesoderm of the amnion, and a *visceral (splanchnic) layer* continuous with the extraembryonic mesoderm of the yolk sac (Figs. 17–1 and 17–2). The somatic mesoderm and the embryonic ectoderm form the body wall or *somatopleure*, and the visceral mesoderm and the endoderm form the gut wall or *splanchnopleure*.

**FIG. 17–2.** A diagrammatic reconstruction of a young human embryo (2.5-mm stage) to show the relationships between the intraembryonic coelom, the extraembryonic coelom, and the mesoderm. The arrow indicates the site of union between intraembryonic and extraembryonic coeloms. (From W. Hamilton and H. Mossman, 1972. Human Embryology. Macmillan, London.)

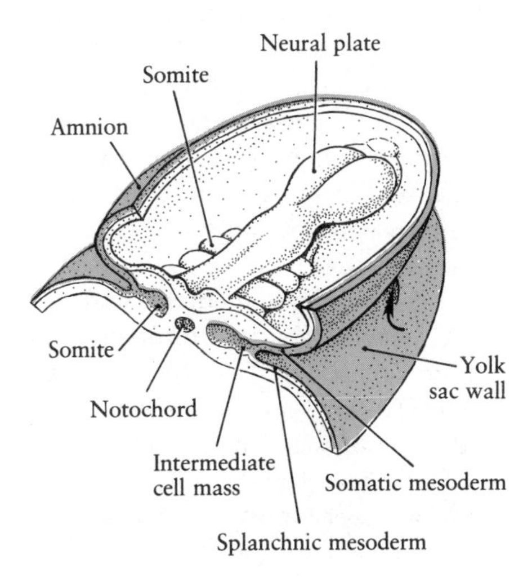


## The Pericardial Cavity

The presumptive pericardial cavity is initially located anterior to the level of the neural plate (Fig. 17–3). With the formation of the head fold, the heart primordium, in the form of a pair of developing *endothelial tubes*, and the coelomic space in which it resides are bent ventrally and caudally beneath the foregut (Fig. 17–3 A,B). As a consequence of this reversal of position of the presumptive pericardial cavity, the original cranial wall of this chamber now becomes its definitive caudal wall. The mass of mesoderm, representing fused somatic and splanchnic layers of mesoderm, which occupies the space between the gut, yolk stalk, and ventral body wall constitutes the transverse septum (Fig. 17–3 C,D). However, the septum transversum, extending dorsally from the ventral body wall to the floor of the foregut, only effects a partial separation between the pericardial and peritoneal portions of the intraembryonic coelom. Caudally, the dorsolateral corners of the pericardial cavity are still connected to the remaining portion of the intraembryonic coelom by somewhat restricted passageways, now termed the *pericardioperitoneal canals* (Figs. 17–3, 17–4, 17–5). Hence, the transverse septum never extends all the way to the dorsal body wall.

The transverse septum represents the initial step in diaphragm formation and clearly foreshadows the division of the intraembryonic coelom into thoracic and abdominal regions. Only the cranial portion of the original transverse septum will continue in its role as a partition. The rapidly growing liver primordium penetrates the more caudal part of the septum and, as the liver increases in size and withdraws, this portion of the partition is drawn out as part of the ventral mesentery. Since both heart and liver abut against the transverse septum, the stems of all major embryonic and extraembryonic veins pass through the transverse septum.

## The Pleural Cavities

Dorsal to the transverse septum, the region of the pericardial cavity is continuous with the pericardioperitoneal canals (Fig. 17–4). Beginning at about four weeks of human development, each endoder-

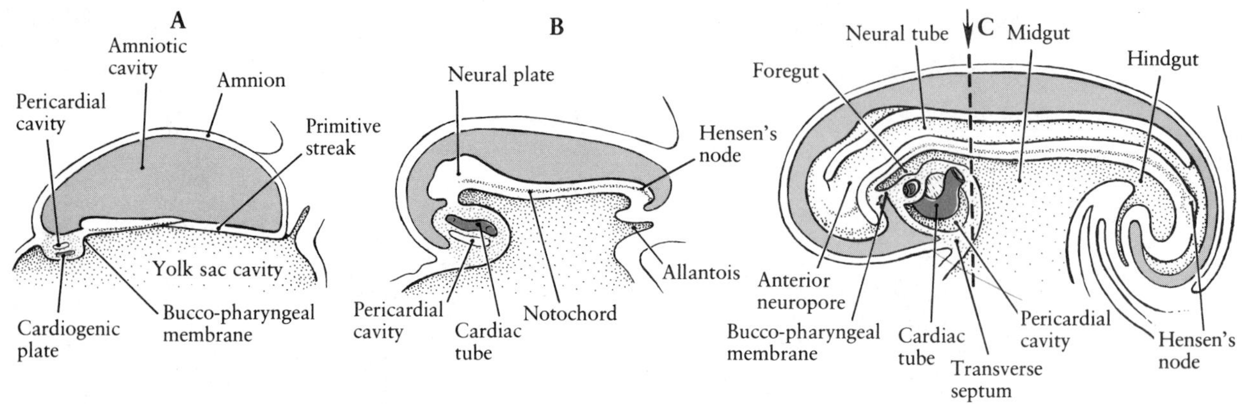


mal lung bud with its surrounding mass of pulmonary mesenchyme pushes into the pericardioperitoneal canal. The splanchnic mesoderm forms a covering for the lung primordium. At this time, the pericardioperitoneal canals are better termed the pleural cavities. The communication between the pericardial cavity and the pleural cavity on either side is the *pericardiopleural opening;* each pleural cavity communicates with the peritoneal cavity by the *pleuroperitoneal opening* (Figs. 17–4, 17–5 A). The pleural cavities are initially very narrow and slitlike, but they soon become greatly enlarged to accommodate the expanding respiratory tree.

The separation of the pericardial cavity from each pleural cavity is effected by the closure of the pericardiopleural opening through the development of the *pleuropericardial membrane.* The steps in this closure process can be traced in Figure 17–5 A–D. In the dorsal margin of the transverse septum, the *common cardinal vein* or *duct of Cuvier* passes transversely and medially to enter the sinus venosus (Fig. 17–5 B). Laterally, each vein tends to course in a crescent-shaped ridge of somatic mesoderm, termed the *pulmonary ridge,* as it passes from the lateral body wall to the transverse septum. As the heart descends caudally, the common cardinal vein is forced to pass obliquely and ventromedially to enter the sinus venosus. This shift in the course of the common cardinal vein causes the pulmonary ridge to be drawn out into a curtainlike partition, the pleuropericardial membrane (Figs. 17–5 C–D, 17–6 A,B). By about the sixth week of development, the free edge of the pleuropericardial membrane fuses with the median mass of esophageal mesenchyme (the *primitive mediastinum)* to close off the pleuropericardial opening (Fig. 17–6 C).

Closure of the pleuroperitoneal openings to separate the pleural cavities from the peritoneal cavity occurs during the seventh week of development when the lungs show extensive growth and lateral expansion. On either side, a pleuroperitoneal membrane arises in the dorsolateral aspect of the caudalmost part of the pleuroperitoneal canal as a crescent-shaped fold of mesoderm extending between the cephalic portion of the kidney *(mesonephros)* and the transverse septum (Figs. 17–5 B, 17–6 A). The membranes extend medially and somewhat caudally, leaving a rapidly diminishing angle between the dorsal body wall and the esophageal mesentery (Figs. 17–5 C, 17–6 B). Fusion of their free edges with the esophageal mesentery and the transverse septum closes the communication between the pleural and peritoneal cavities (Figs. 17–5 D, 17–6 C).

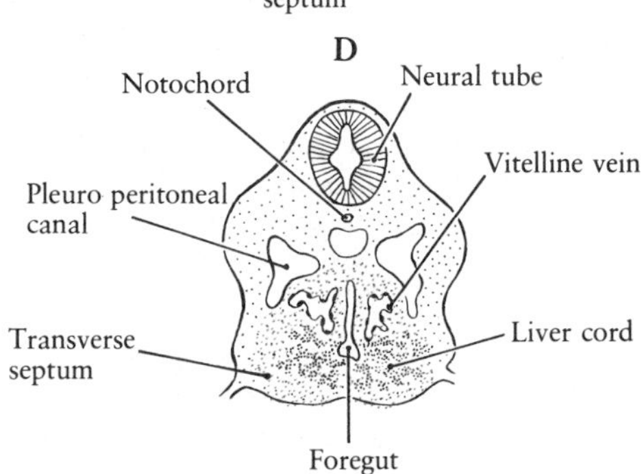


**FIG. 17–3.** A series of schematic drawings in sagittal view (A–C) to show the formation of the pericardial cavity and the transverse septum in the human embryo. A, presomite embryo illustrating the anterior position of the pericardial cavity; B, formation of the head fold and the beginning of the rotation of the heart and pericardial cavity in a 7-somite human embryo; C, formation of the transverse septum in a 14-somite stage embryo; D, a cross section in the plane indicated in C (arrow) to show position and relationship of the transverse septum.

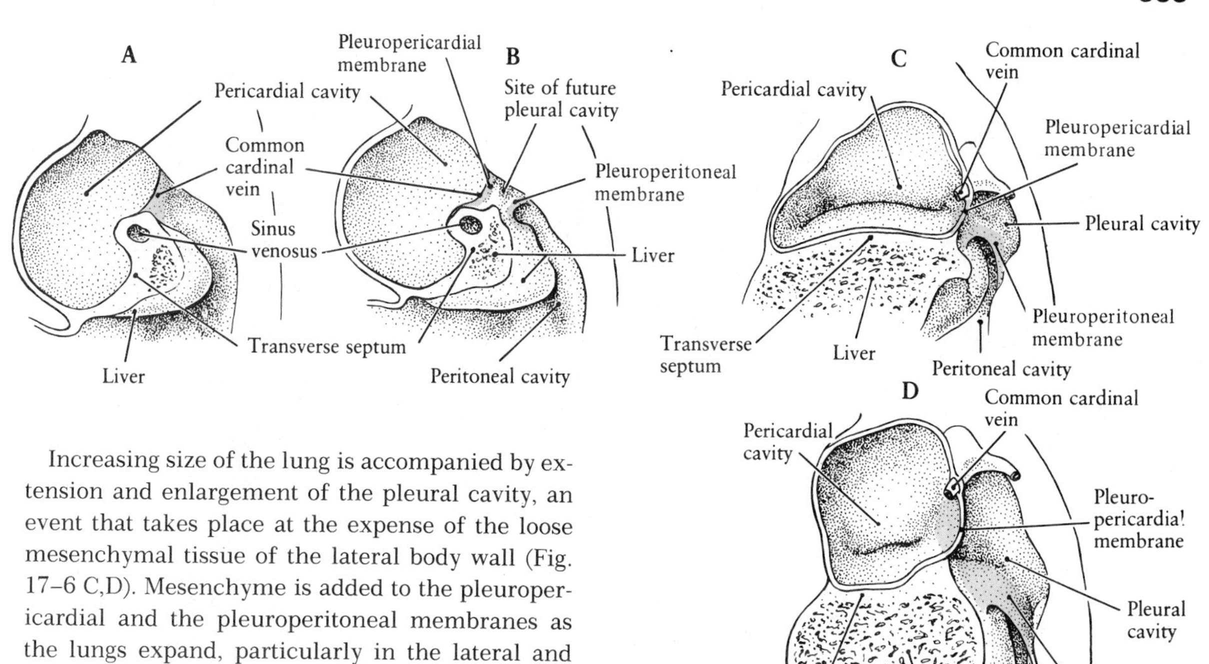


Increasing size of the lung is accompanied by extension and enlargement of the pleural cavity, an event that takes place at the expense of the loose mesenchymal tissue of the lateral body wall (Fig. 17–6 C,D). Mesenchyme is added to the pleuropericardial and the pleuroperitoneal membranes as the lungs expand, particularly in the lateral and ventral directions, and separates the body wall tissue. Gradually, the lungs and the pleuropericardial membranes come to lie on either side of the heart in their typical adult relationships (Fig. 17–6 D). The original cardiac surface of each pleuropericardial membrane now constitutes the lateral wall of the pericardial cavity. The partition separating heart from lung represents the original pleuropericardial membrane plus mesenchymal additions captured from the body wall. The fibrous partition formed in this fashion encloses the heart like a sac and is termed the *pericardium* (Fig. 17–6 D).

**FIG. 17–4.** A reconstruction of a model to illustrate the continuity (arrows) between pericardial cavity, pericardioperitoneal canals, and the peritoneal cavity in a late somite human embryo. (From W. Hamilton and H. Mossman, 1972. Human Embryology. Macmillan, London.)

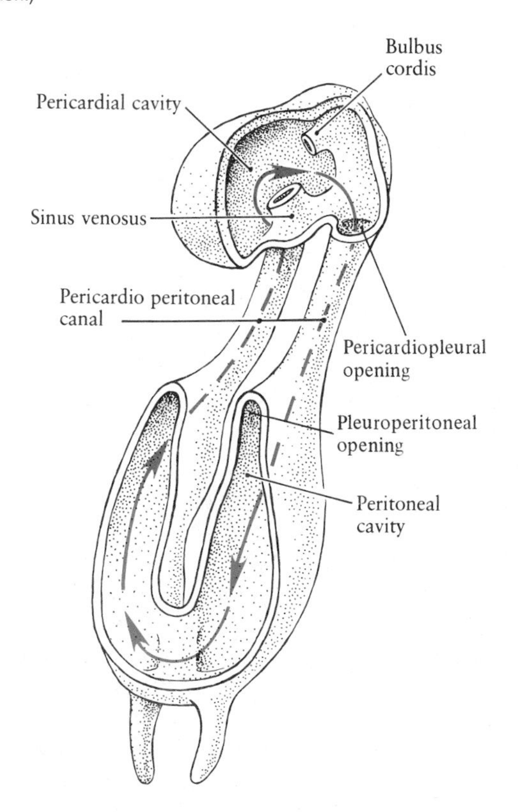


**FIG. 17–5.** Drawings showing partitioning of the human intraembryonic coelom. A, various regions of the coelom (right-side view) before partitioning; B, a 5-mm embryo, cut longitudinally near the midline, showing origin of the pleuropericardial and pleuroperitoneal membranes; C, complete separation of the pericardial cavity from the right pleural cavity by the right pleuropericardial membrane in a 13-mm embryo; D, complete separation of the peritoneal cavity from the right pleural cavity by the right pleuroperitoneal membrane in a 15-mm embryo.

## The Diaphragm

The complete separation of the pleural cavities from the peritoneal cavity is effected by a compos-

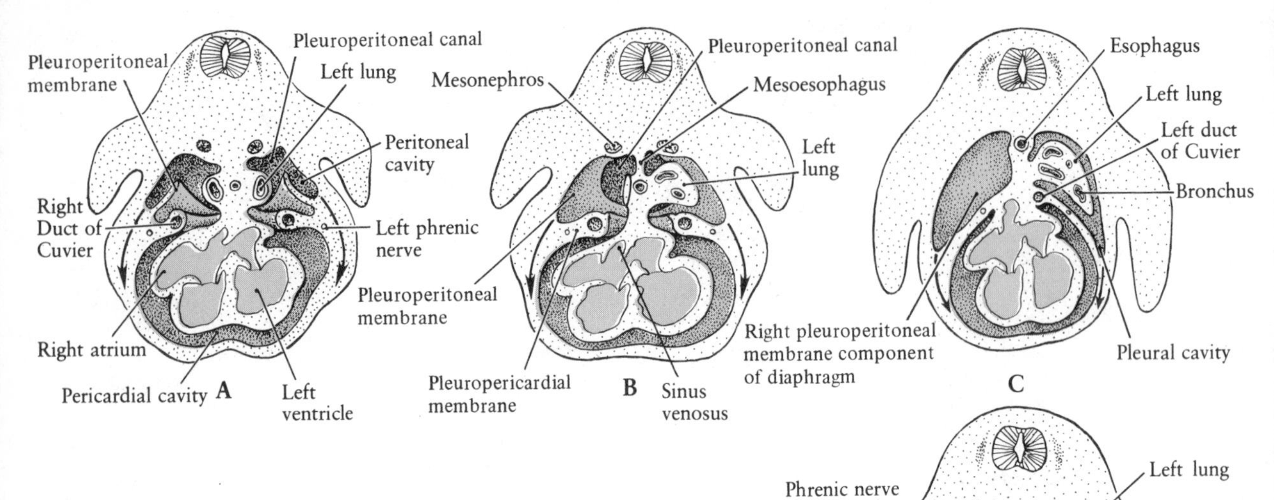


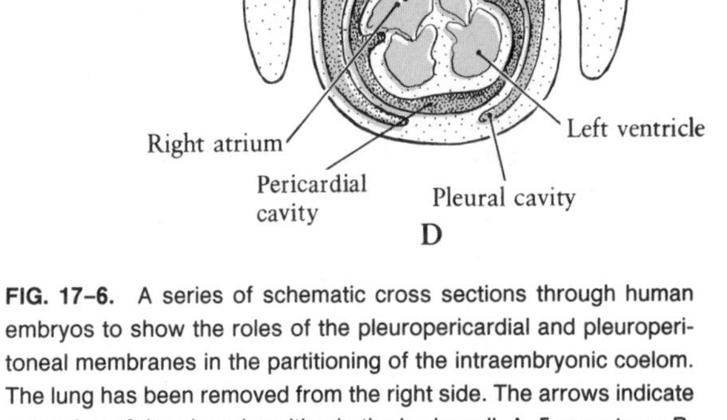


FIG. 17–6. A series of schematic cross sections through human embryos to show the roles of the pleuropericardial and pleuroperitoneal membranes in the partitioning of the intraembryonic coelom. The lung has been removed from the right side. The arrows indicate extension of the pleural cavities in the body wall. A, 5-mm stage; B, 9-mm stage; C, 22-mm stage; D, approximately 25 mm. (A–C, from W. Hamilton and H. Mossman, 1972. Human Embryology. Macmillan, London.)

ite partition known as the diaphragm. Its origin is complex, but its chief components have been referred to above and include the following: (1) an anterior, central portion that represents the greater part of the cranial aspect of the transverse septum; this becomes the *central tendon* of the diaphragm; (2) dorsal, paired pleuroperitoneal membranes whose anterior margins become continuous with the posterodorsal edges of the central tendon; (3) a dorsal, unpaired portion from the dorsal esophageal mesentery that constitutes the medial part of the diaphragm; (4) circumferential portions derived from the lateral body wall that are added to the periphery of the pleuroperitoneal membranes as the lungs expand.

The muscular portion of the diaphragm is largely derived from the early migration (shortly after the early limb bud stage) of *myoblasts* or primitive muscle cells into the transverse septum and the pleuroperitoneal membranes. Presumably the muscle cells originate from the hypaxial portions of the third, fourth, and fifth cervical myotomes, since the motor and in part the sensory *(phrenic nerve)* innervations of the diaphragm are from the third, fourth, and fifth cervical nerves. There is also evidence that some muscle fibers may originate from mesenchyme cells in the transverse septum itself.

During the descent of the heart into the thoracic region and with the rapid enlargement of the lungs, the diaphragm undergoes an extensive movement in the caudal direction. This is clearly indicated by positional changes in the cervical nerves that pass to the muscular tissue of this partition. For example, the diaphragm lies opposite the third, fourth, and fifth cervical myotomes at approximately three weeks of development. By eight weeks, the dorsal parts of the diaphragm have moved far caudally, giving this partition a strong dome-shaped contour. The diaphragm eventually lies at the level of the lower thoracic or upper lumbar segments of the body.

## The Peritoneal Cavity

The peritoneal cavity is formed from that part of the intraembryonic coelom lying caudal to the transverse septum. Initially, the gut tube with its suspending mesentery separates the intraembryonic coelom into right and left halves (Fig. 17–7). With rupture of the ventral part of the mesentery, there results a large embryonic peritoneal cavity that extends from the thoracic to the pelvic region of the embryo.

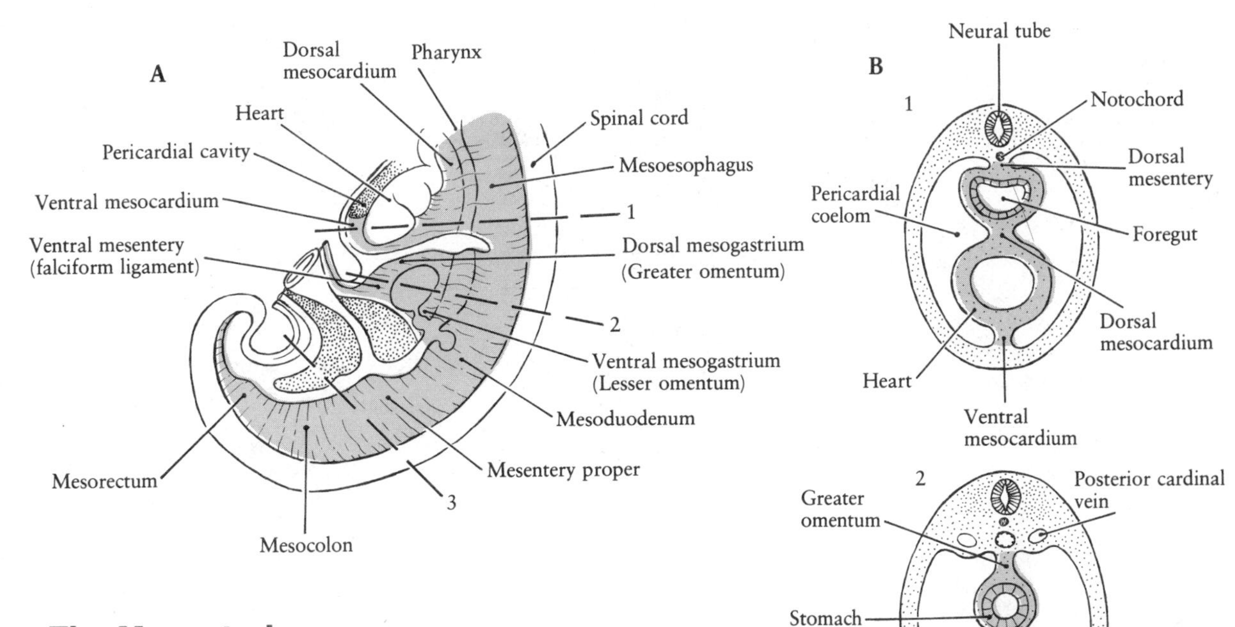


## The Mesenteries

During the transverse folding process that separates the embryo from the extraembryonic membranes (Fig. 17–7 A), the sheet of embryonic endoderm becomes rolled and fashioned into the gut tube (Fig. 17–7 B). Concurrently the splanchnic

**FIG. 17–7.** Schematic drawings to illustrate successive stages in the formation of the primitive mesentery in human embryos. The arrows indicate sites at which somatic and visceral layers of mesoderm join together. A, 2 mm; B, 4 mm; C, 8 mm.

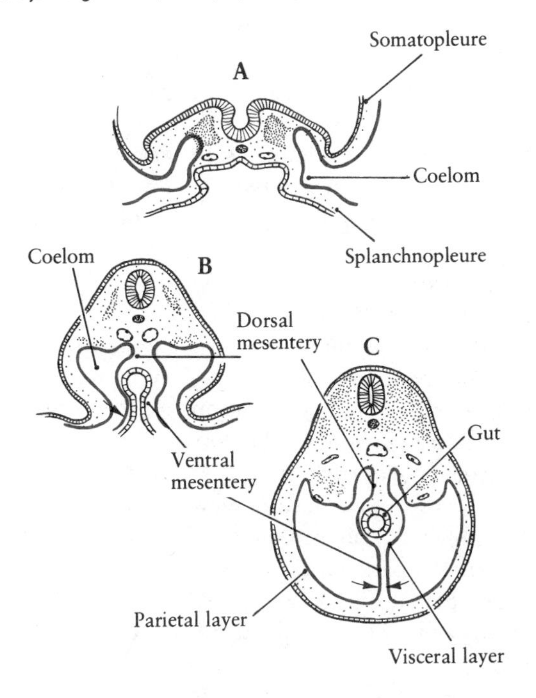


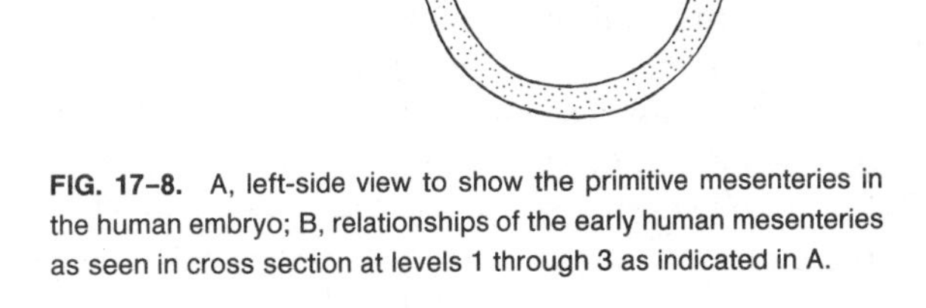


**FIG. 17–8.** A, left-side view to show the primitive mesenteries in the human embryo; B, relationships of the early human mesenteries as seen in cross section at levels 1 through 3 as indicated in A.

mesoderm from either side swings toward the midline and wraps around the endodermal tube to give rise to a double-layered partition known as the *primitive mesentery* (Fig. 17–7 C). Extending from the roof of the intraembryonic coelom to the midventral body wall, the primitive mesentery is interrupted by the early, straight gut tube to form upper *(dorsal mesentery)* and lower *(ventral mesentery)* components (Fig. 17–7 C). The dorsal mesentery tends to persist, but the ventral mesentery, as mentioned above, is quite temporary, and its degenera-

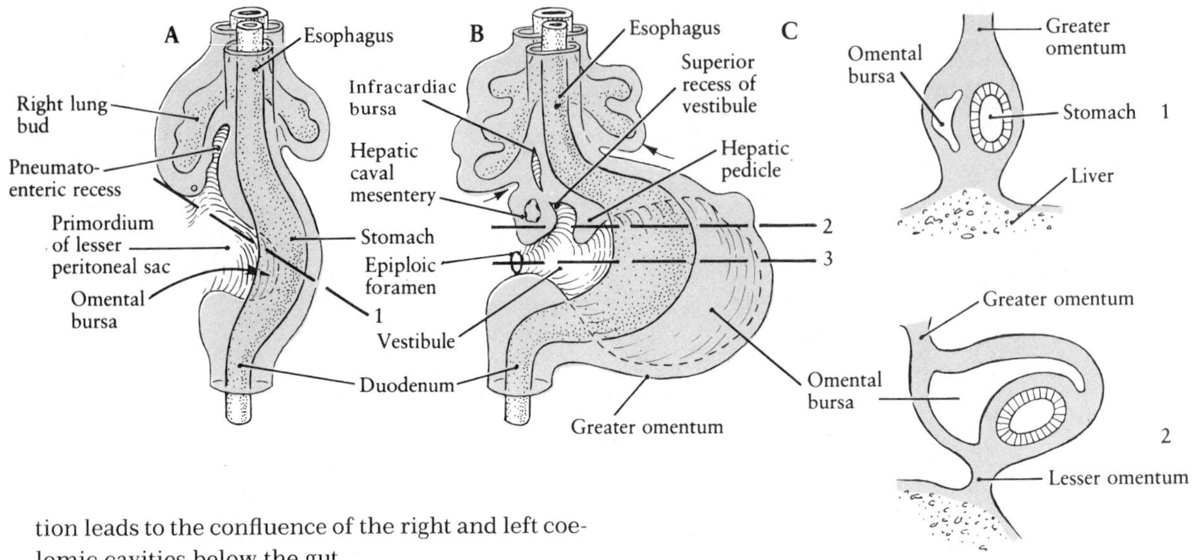


tion leads to the confluence of the right and left coelomic cavities below the gut.

The double-layered dorsal mesentery will differentiate into a variety of structures. It will form connective tissue and a covering epithelium *(mesothelium)* that lines the peritoneal cavity. Where the mesentery continues around the endodermal tube of the gut, it will differentiate as the *serosa.* It will also contribute to the connective tissue, vascular tissue, and smooth muscle tissue of the gut wall.

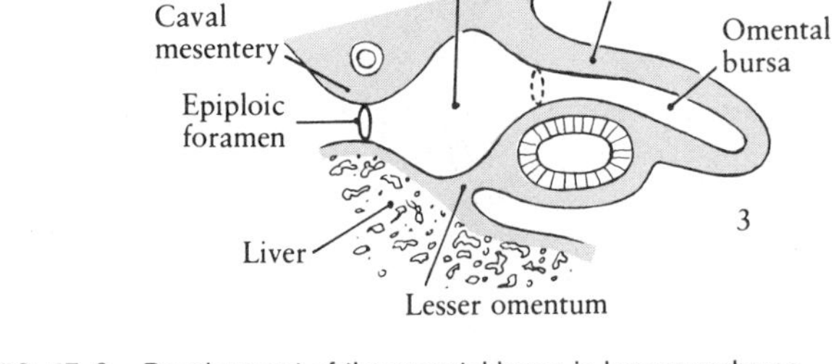


**FIG. 17–9.** Development of the omental bursa in human embryos at four (A) and six (B) weeks, ventral views. C, transverse views at levels indicated in A and B. The large arrows indicate approximate level of transverse septum.

## Derivatives of the Dorsal Mesentery

The gut tube is initially suspended throughout most of its length by a definitive dorsal mesentery. The mesentery extends in the midplane from the roof of the peritoneal cavity to the gastrointestinal tract and serves as a supporting vehicle for blood vessels and nerves passing to and from the gut. Commonly, distinctive names are given to the parts of the dorsal mesentery that support the different regions of the gut. Thus, there is the *dorsal mesogastrium* or *greater omentum* of the stomach, the *mesoduodenum* of the duodenum, the *dorsal mesentery proper* of jejunum and ileum, the *mesocolon,* and the *mesorectum.* The early relationships of some of these mesenteries are shown at several levels of the human embryo in Figure 17–8 A,B.

The pharynx and the upper portion of the esophagus lack a dorsal mesentery, since the intraembryonic coelom does not normally extend this far cranially. The remainder of most of the esophagus, however, is supported by a thick, dorsal mesentery known as the *mesoesophagus* (Figs. 17–6 B, 17–8 A). It will contribute to a thick, specialized medial septum in the adult termed the *mediastinum.* The mediastinum supports the early primary lung bud rudiments and the esophagus in its transit through the pleural cavities. Near its junction with the stomach, the mesoesophagus thins out into a typical mesentery.

The primitive, simple relationships of the dorsal mesentery behind the esophagus are quickly complicated as the gut undergoes growth, elongation, and folding to produce its adult configuration. Part of the dorsal mesentery becomes greatly exaggerated; other portions of the dorsal mesentery disappear or secondarily fuse with each other. Modification of the dorsal mesentery of the stomach in particular establishes new relationships for the spleen, the pancreas, and the duodenum.

The dorsal mesogastrium is that part of the dorsal mesentery that suspends the stomach (Figs. 17–

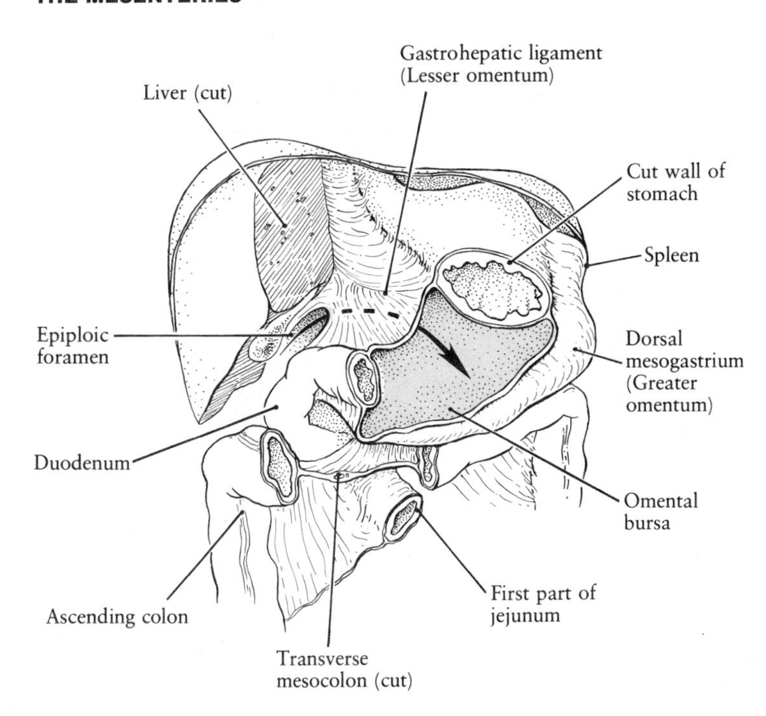


**FIG. 17–10.** Model showing the relationships of the omental bursa to the peritoneal cavity and the surrounding viscera. (From L. Arey, 1974. Developmental Anatomy. W. B. Saunders Company, Philadelphia.)

8, 17–9 and 17–12). Beginning at about four weeks of human development, a sacculation appears in the mesogastrium known as the *omental bursa* (Fig. 17–9 A). Although the formation of the bursa sac is often described as being an invagination dependent on the rotation of the stomach, it more likely forms as the result of the coalescence of separate clefts that initially appear in the right surface of the dorsal mesogastrium (Fig. 17–9 C). Gradually, the saccular recess deepens toward the left behind the stomach, bringing about a shift in the attachment of the dorsal mesogastrium (Fig. 17–9 C). The finger-shaped projection termed the *pneumoenteric recess* is continuous with the bursa and extends cranially between the right lung and the esophagus (Fig. 17–9 A). Its anterior end is interrupted by the growing diaphragm to form a blind sac that often persists in the adult as the *infracardiac bursa* (Fig. 17–9 B).

As the stomach undergoes its clockwise rotation to bring the cardiac end of this organ to the left and the pyloric end to the right (Figs. 17–10, 17–11 A–C), the bursa expands transversely and comes to lie dorsal to the stomach and to the right of the esophagus (Fig. 17–9 C). Accompanying these axial changes in the orientation of the stomach is a marked extension of the dorsal mesogastrium to a point well beyond the greater curvature of the stomach (Fig. 17–11 A–E). Indeed, the dorsal mesogastrium appears as an apronlike, double-folded mesentery, the *greater omentum,* which sprawls over the small intestine. It is clearly distinguishable from the *lesser omentum,* which is the ventral mesentery passing between stomach and liver (Figs. 17–9 C, 17–10).

The entire portion of the peritoneal cavity that becomes captured above the stomach and delineated above by the folded dorsal mesogastrium and below by the lesser omentum is called the *lesser peritoneal space* or *sac* (Fig. 17–9 A,B). Its organization can be dissected as follows. The omental bursa is that part of the lesser peritoneal space entirely bounded by the greater omentum of the stomach (Fig. 17–9 C). It opens into the *vestibule,* a chamber outlined below by the lesser omentum and above by the peritoneal wall and liplike fold of the dorsal mesentery *(caval mesentery).* The vestibule in turn communicates with the general peritoneal cavity by a slitlike aperture termed the *epiploic foramen* or the *foramen of Winslow* (Figs. 17–9 C, 17–10).

The caudal extension of the omental bursa establishes secondary attachments and greatly influences the growth and position of such structures as the spleen, pancreas, and duodenum. As the bursa expands caudally beyond the greater curvature of the stomach, it meets, adheres to, and fuses with

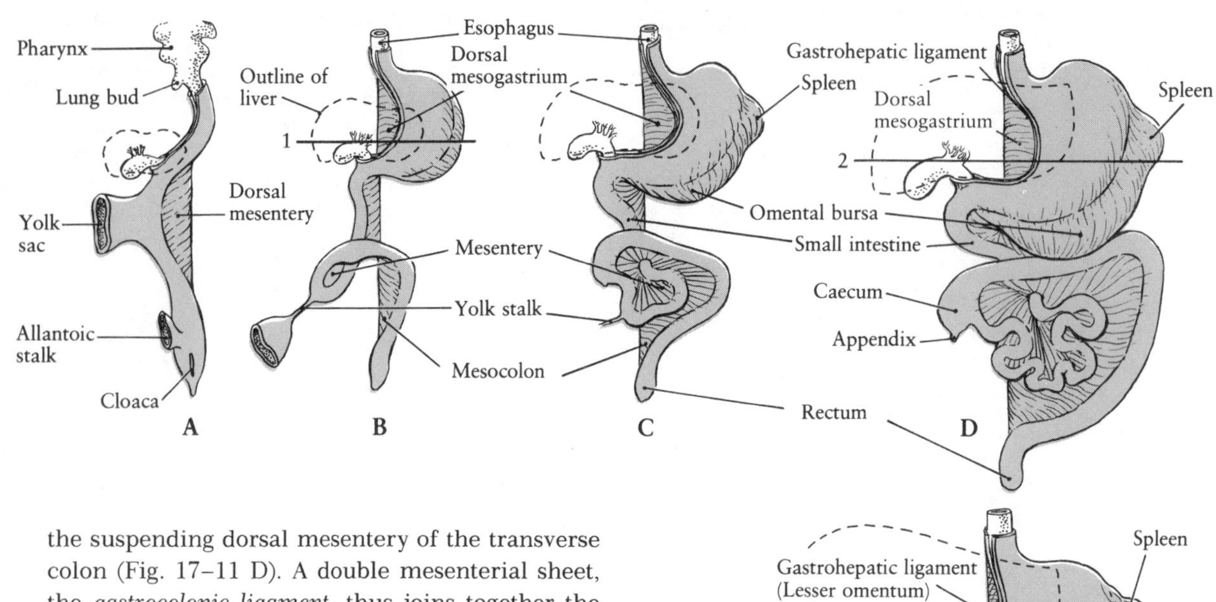


the suspending dorsal mesentery of the transverse colon (Fig. 17–11 D). A double mesenterial sheet, the *gastrocolonic ligament*, thus joins together the stomach and the colon (Fig. 17–13). Beyond the colonic attachment, the walls of the bursa collapse and unite so that its cavity is oblitered (Fig. 17–13).

The spleen appears within the dorsal mesogastrium shortly after the initial phases of the axial rotation of the stomach (Fig. 17–12 B). As the spleen enlarges and bulges from the left face of the dorsal mesogastrium, it reaches the dorsolateral body wall and is pressed against it. That portion of the greater omentum between the stomach and the spleen is the *gastrosplenic ligament* (Fig. 17–12 C).

As pointed out in the chapter on the gastrointestinal tract, the pancreas initially begins to form between the two layers of the dorsal mesentery suspending the duodenum. Rather quickly the proliferating pancreatic tissue pushes into the greater omentum. As in the case of the spleen, the pancreas is carried dorsad against the body wall where the greater omentum contacts and fuses with the parietal peritoneum (Figs. 17–12 C, 17–13). Gradually, the fusion between the dorsal mesentery and the body wall becomes more extensive, leaving the pancreas tightly adherent to the dorsal body wall. The original right face of the dorsal mesogastrium now covers the surface of the pancreas.

Similar to the rest of the enteric tract, the duodenum initially has its own dorsal mesentery or *mesoduodenum*. Rotation of the stomach brings the duodenum closer to the dorsal body wall, with consequent shortening of the mesoduodenum. By approximately the third month of human development, the duodenum comes to lie against the body

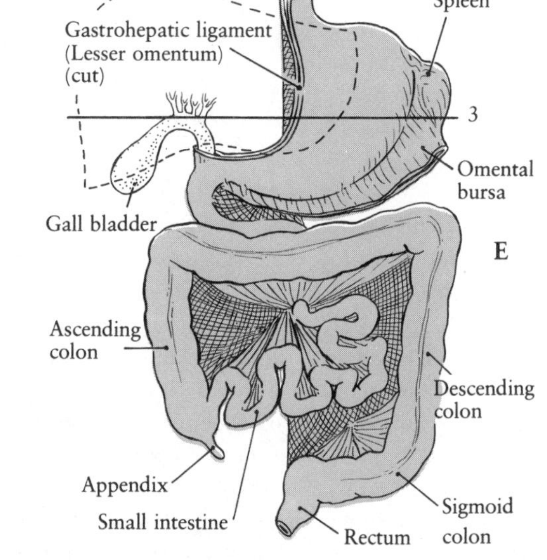


**FIG. 17–11.** A series of schematic drawings in frontal view showing the major developmental changes in the position of the enteric tract and its associated mesenteries. The cross-hatched areas in E indicate the part of the dorsal mesentery of the duodenum and segments of the large intestine that become fused to the dorsal body wall. The heavy lines marked 1, 2, and 3 represent the locations of transverse sections shown in Figure 17–12. (From Human Embryology by B. M. Patten. Copyright © 1968 by McGraw-Hill, Inc. Used with permission of McGraw-Hill Book Company.)

wall and its own mesentery is completely resorbed (Fig. 17–13). In this new fixed, retroperitoneal position, most of the duodenum is situated between the transverse mesocolon and the more dorsal part of the mesogastrium (Fig. 17–11 E). These mesenteries, especially the mesocolon, secondarily form the peritoneal covering for the duodenum.

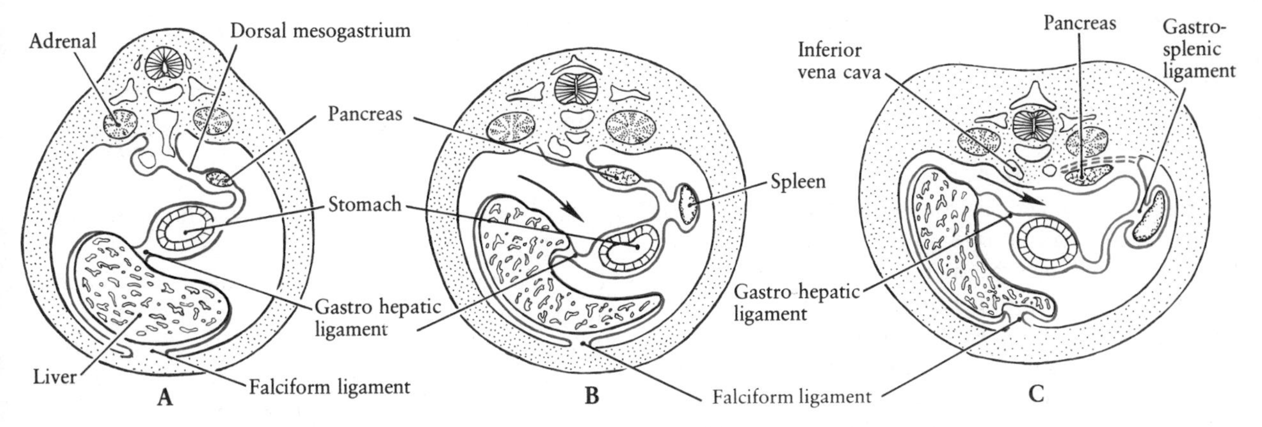


**FIG. 17–12.** Transverse sections through the region of the stomach to show the changes in the relationships between the mesenteries. Note the changes in the position of the pancreas. The arrows in B and C indicate position of the epiploic foramen. (From Human Embryology by B. M. Patten. Copyright © 1968 by McGraw-Hill, Inc. Used with permission of McGraw-Hill Book Company.)

The dorsal mesentery of the remainder of the intestine is greatly affected by the rapid growth of this part of the enteric tract and its herniation into the umbilical cord. The result is the production of an elongate, rather fan-shaped appearing mesentery. In contrast to the fixed position of the duodenum, the jejunum and the ileum portions of the small intestine remain freely movable within the peritoneal cavity and are supported by the folded *mesentery proper*.

Much of the embryonic suspending mesentery of the large intestine is lost as portions of this part of the gut become fixed to the body wall. The free and obliterated portions of this mesentery are illustrated in Figure 17–11 E. For example, the *ascending* and *descending mesocolons* become pressed against the dorsal body wall, shorten, and progressively fuse with the adjacent peritoneum. Consequently, the ascending and descending segments of the colon are fixed in this position. The *transverse mesocolon* remains largely free, although it secondarily covers the duodenum as previously mentioned. The *sigmoid mesocolon* remains free, but the mesorectum disappears as the rectum becomes fixed against the body wall at the level of the sacrum.

**FIG. 17–13.** Schematic longitudinal sections of the body to show the secondary associations of the omental bursa in human embryos.

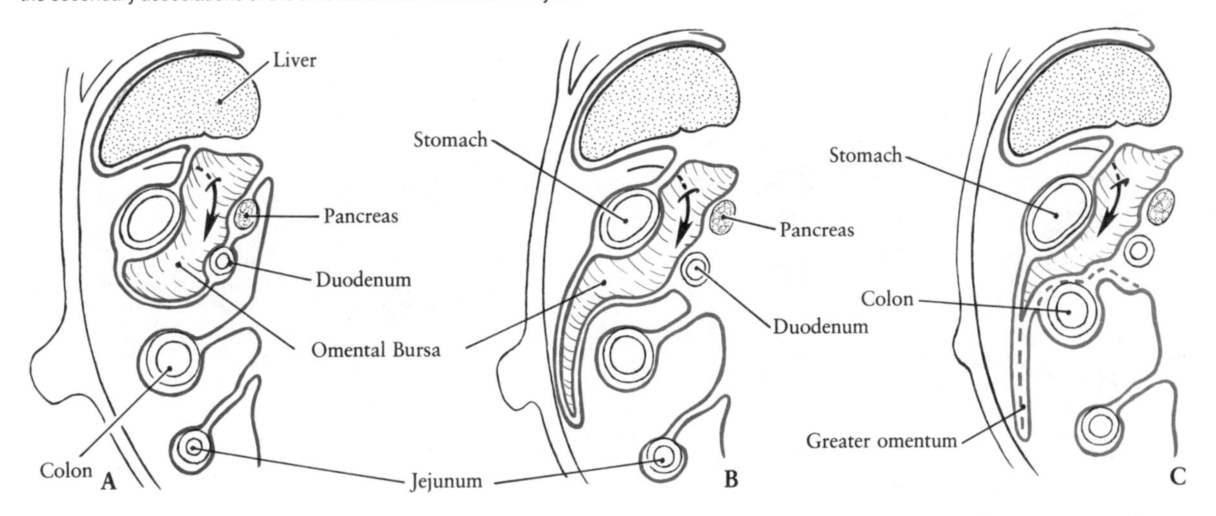

## Derivatives of the Ventral Mesentery

The two layers of the splanchnic mesoderm that meet below the gut and contribute to the formation of the primitive mesentery (Fig. 17–7) constitute the *ventral mesentery*. For most of the length of the intraembryonic coelom, the ventral mesentery is quite temporary and soon becomes obliterated. At the level of the pleural cavity, this leaves the dorsal mesentery to enclose each enlarging lung as the *visceral pleura*. In the pericardial cavity, the splanchnic mesoderm beneath the foregut folds around the tubular heart and as such forms a specialized region of the ventral mesentery. The portion of the ventral mesentery between the heart and the foregut constitutes the *dorsal mesocardium* (Fig. 17–8 B). It soon disappears, leaving the heart without any permanent supporting mesentery. The portion of the ventral mesentery between the heart and the floor of the pericardial cavity is the *ventral mesocardium* (Fig. 17–8 B). This mesentery is also transitory.

The ventral mesentery persists in the vicinity of the stomach, upper duodenum, and liver, and contributes to several special mesenterial supports called *ligaments*. A permanent ventral mesentery in this general region of the gut appears to arise secondarily as the result of the growth of the liver bud into the mesoderm of the transverse septum. As the stomach and liver draw away caudally, the splanchnic mesoderm is drawn into a definitive ventral mesentery that is continuous from the lesser curvature of the stomach to the ventral body wall. It can be divided into three parts: (1) a portion between the diaphragm and the liver *(coronary ligament)* and a portion between the ventral body wall and the liver *(falciform ligament)* (Figs. 17–8, 17–14); (2) a portion extending from the stomach and duodenum to the liver (lesser omentum) (Figs. 17–9, 17–10, 17–14); this in turn can be regionalized into a cranial *gastrohepatic ligament* and a caudal *hepatoduodenal ligament;* (3) a portion that becomes the enveloping capsule of the liver (Glisson's capsule).

## Abnormalities in the Development of Coelomic Cavities

The formation of the musculotendinous diaphragm is dependent on rigidly determined movements of the pleuropericardial and pleuroperitoneal membranes between six and seven weeks of development in the human embryo. *Diaphragmatic hernias* occur in approximately 0.08 percent of births and are abnormalities of the diaphragm re-

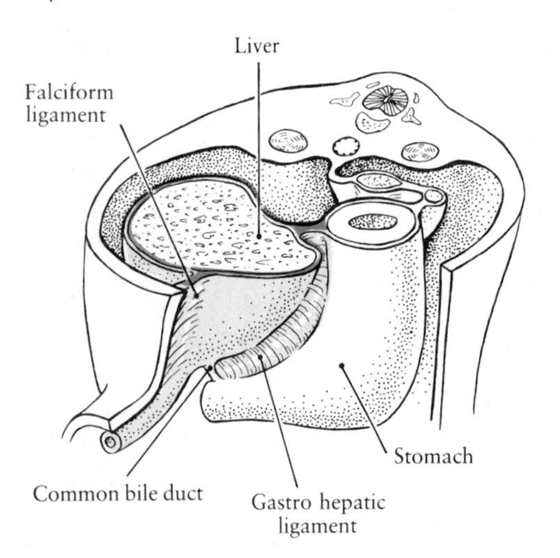


**FIG. 17–14.** A drawing to show several derivatives of the ventral mesentery in the vicinity of the liver and the stomach in a 17-mm human embryo. A portion of the liver has been removed. (From W. Hamilton and H. Mossman, 1972. Human Embryology. Macmillan, London.)

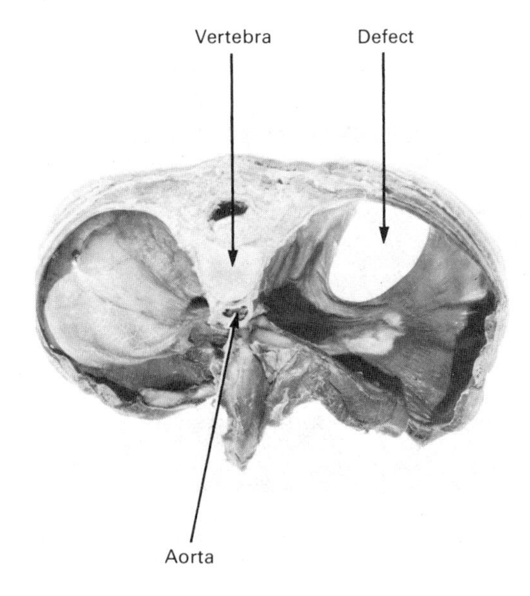


**FIG. 17–15.** Photograph of a transverse section through the thoracic region of a newborn infant to show a large left lateral posterolateral defect in the muscular diaphragm. (From K. L. Moore, 1977. The Developing Human: Clinically Oriented Embryology, 2nd ed. Courtesy of W. B. Saunders Company, Philadelphia.)

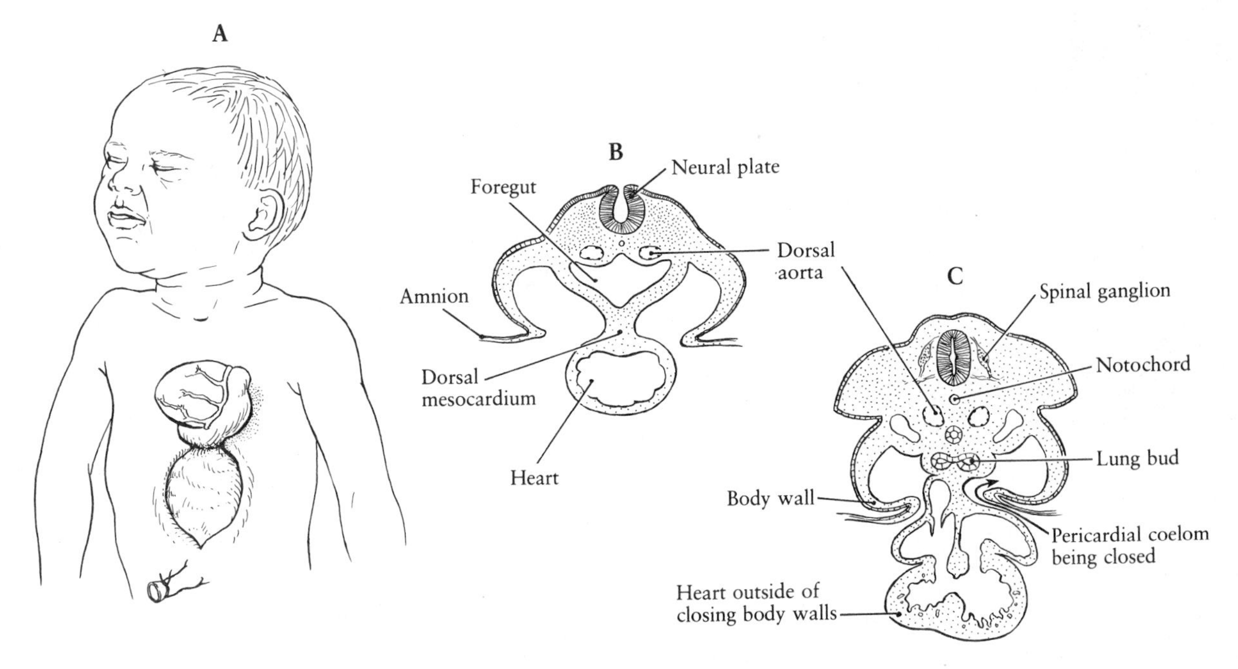


**FIG. 17–16.** Ectopia cordis. A, sketch of infant showing heart on incompletely closed ventral body walls; B, C, hypothetical cross sections of embryos to show probable manner in which this abnormality arises. (From Human Embryology by B. M. Patten. Copyright © 1968 by McGraw-Hill, Inc. Used with permission of McGraw-Hill Book Company.)

sulting from defective formation, movement, and/or fusion of the pleuroperitoneal membranes.

Posterolateral defect of the diaphragm, which occurs about once in 2200 births, appears as a large opening (the so-called foramen of Bochdalek) in the diaphragm because of a failure in closure of the pleuroperitoneal opening (Fig. 17–15). The defect appears to be more prevalent on the left side of the diaphragm than on the right side. The abdominal viscera, such as spleen, stomach, and small intestine, tend to herniate into the pleural cavity and thereby interfere with the normal function of both lungs and heart. If the herniation takes place before complete growth of the lung, the lung is often hypoplastic or greatly reduced in size. Although mortality is high in infants if diaphragmatic abnormalities are left untreated, surgical procedures are available that can be employed to restore normal relationships between the body cavities.

Defects in the separation of the pericardial cavity from the pleural cavities by the pleuropericardial membranes are less common than diaphragmatic hernias. Failure to close the pleuropericardial opening, if present, is usually on the left side. Occasionally, the left atrium may herniate into the left pleural cavity upon atrial systole.

A rare but severe abnormality that usually results in postnatal death is *ectopia cordis* (Fig. 17–16). The heart lies outside of the body, a condition presumably arising because of faulty separation between extraembryonic and intraembryonic regions in the cardiac area as the ventral body walls of the embryo are closing (Fig. 17–16).

## References

Arey, L. 1974. Developmental Anatomy. Philadelphia: W. B. Saunders Company.

Corliss, C. E. 1976. Patten's Human Embryology. Elements of Clinical Development, pp. 307–324. New York: McGraw-Hill Book Co.

Hamilton, W. and H. Mossman. 1972. Human Embryology. London: Macmillan.

Moore, K. L. 1977. The Developing Human: Clinically-Oriented Embryology, 2nd edition, pp. 145–155. Philadelphia: W. B. Saunders Company.

Wells, L. J. 1954. Development of the human diaphragm and pleural sacs. Carnegie Contrib. Embryol. 35:107–143.

# 18

# THE CARDIOVASCULAR SYSTEM

One of the first organ systems to become functional in the embryo is the cardiovascular system. This is not surprising when one bears in mind that the embryo cannot grow beyond a volume of a few cubic millimeters using processes of simple diffusion to meet critical metabolic requirements. When an embryo has reached a certain small size, therefore, an elaborate system of vascular channels is fashioned to provide for the nutritional, respiratory, and excretory needs of developing tissues and cells. As in the adult, the main blood vessels in the embryo tend to be associated with centers of intense metabolic activity such as the yolk and placenta. The circulating blood carries nutrients and oxygen from organs of absorption to sites of growing and differentiating cells. In turn, waste materials are picked up and transported to organs facilitating elimination or storage. By necessity, the topographical arrangement of blood vessels in the embryo (particularly the amniote embryo) is quite different from that of the adult, because centers of metabolic activity shift during the development of the organism.

The embryo then is faced with a difficult task in the design of its cardiovascular system. It must not only construct a vascular system associated with the functional metabolic centers of the embryo, but it must also provide an arrangement of vessels that anticipates a shift in the sites where these metabolic activities are to be carried out in the posthatch or postnatal organism. In the case of the mammalian embryo, the activities of the placenta are at birth passed on to the digestive tract, lungs, and kidneys. Although accessory fetal organs such as the yolk sac and placenta are temporary, they possess an enormous circulating blood supply. Indeed, in order for the embryo proper to circulate blood to the extensive extraembryonic circulation, its heart and blood vessels must be many times larger, relative to body size, than the adult heart and vessels to adult body size. Greater absolute increase in fetal growth reduces this discrepancy after about the fifth month of gestation in the human embryo.

Beyond the physiological significance of the developing circulation, there is evidence, particularly from the vertebrate limb, that the vascular architecture of the embryo may play an important morphogenetic role in the formation of certain organs by establishing and maintaining discrete nutrient microenvironments. Uncommitted cells, for example, might respond to a particular nutrient environment by expressing an appropriate developmental program leading to the differentiation of a specific cell type. Alternatively, the vascular environment could conceivably provide some key nutrient that permits committed cells to express a distinctive phenotype. The vascular system of any developing organ, therefore, may serve as a scaffolding within and around which cells are either directed or permitted to develop along particular phenotypic pathways.

For purposes of convenience, the cardiovascular system can be divided into three major components: (1) the blood vascular system, including heart, arteries, and veins; (2) the hematopoietic or blood-forming organs; and (3) the lymphatic system. We will consider each of these, but place most of our emphasis on the blood vascular system, since regional and temporal alterations in the pattern of developing blood vessels are more dramatic and interesting.

## Angiogenesis and Hematopoiesis

Blood cells and blood vessels are specializations of mesenchyme. This important embryonic tissue fashions the epithelial lining, the collagenous and elastic tissues, and the smooth muscle that participate in the construction of arteries and veins.

The earliest formative vascular tissue to appear in higher vertebrate embryos is termed *angioblast* (or *hemagioblast*); the process of primitive blood vessel development is called *angiogenesis (hemangiogenesis)*. The tissue is in the form of solid masses of cells located in the splanchnic mesoderm of the yolk sac (Fig. 18–1). These clusters of *blood islands* differentiate as a consequence of an interaction between the splanchnic mesoderm and the underlying endoderm. During early vessel development, each blood island becomes hollowed out (Fig. 18–1 A–C). The more peripheral cells of the blood island become organized as a flattened, vascular *endothelium*. The central cells become hematopoietic stem cells that differentiate into the *primitive blood cells*. Within the endothelial-lined channels, fluid or primitive blood plasma accumulates and suspends the blood cells. Because these mesenchyme cells can differentiate into either primitive blood cells or vascular endothelium, they have been termed angioblasts or hemangioblasts. By growth and union of these hollowed-out cords of cells, the originally solid, isolated clusters of hemangioblastic tissue are converted into plexuses of blood vessels. These are present on the yolk sac, the body stalk, and the chorion of human embryos as early as the head process stage. Once the system of closed vessels is established, new extraembryonic vessels arise as outgrowths from preexisting vessels.

The first vessels within the embryo proper are detected during the period of early somite formation. These develop from islands of hemangioblastic tissue localized in the mesenchyme of the early organ primordia and form by the same processes as previously described for extraembryonic blood vessels. It was initially thought that the source of intraembryonic vessels was angioblastic tissue spreading from the yolk sac into the embryo. Experimental evidence now tends to support the concept that the intraembryonic endothelium differentiates in situ from intraembryonic mesenchyme. Only secondarily does the network of intraembryonic vessels join with that over the yolk sac. Once a primitive,

FIG. 18–1. The development of primitive blood vessels and blood cells from yolk sac blood islands. A, aggregation of cells to form blood islands in human embryo in the fourth week; B, beginning of the differentiation of the endothelium and the primitive blood cells; C, a more advanced condition showing organized endothelium and primitive blood cells suspended in plasma; D, the Corner 10-somite embryo showing blood islands on the yolk sac.

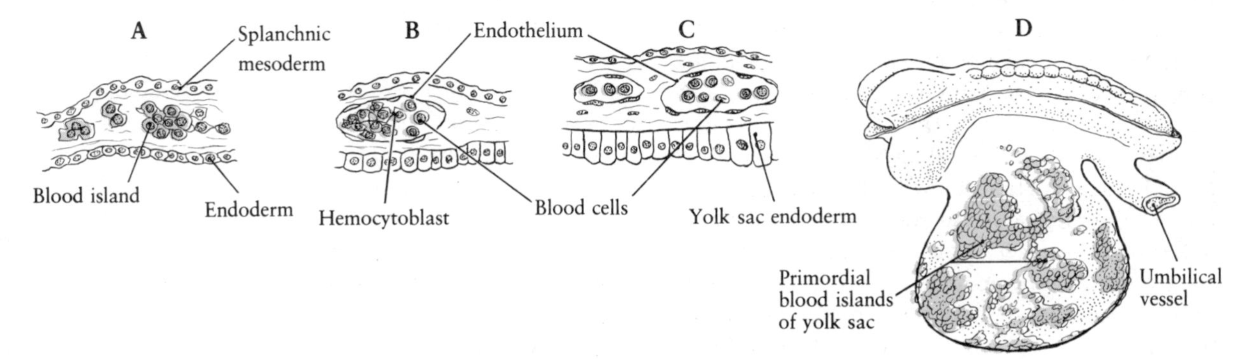

closed system of embryonic vessels is established, new blood vessels arise primarily by sprouting from preexisting vessels.

*Hematopoiesis* or *hemopoiesis* is a term referring to the development of various blood cell types. In species that have been studied to date, including frog tadpole, chick, mouse, and human, blood cells form by the differentiation of mesenchyme tissue at different locations (particularly for red blood cells) during embryogenesis. In the frog embryo, for example, red blood cells develop and mature in the mesonephric kidney and the liver. Following metamorphosis of the tadpole, the spleen becomes the center of red blood cell maturation. Hematopoiesis first appears in the yolk sac blood islands on day 7 or 8 of gestation of the mouse embryo. By day 10, the fetal liver starts to produce red blood cells; the liver will remain the major site of erythrocyte formation until near birth. The sequence and time of appearance of hematopoietic centers in the human embryo and fetus are as follows: yolk sac (week 4), body mesenchyme and blood vessels (week 5), liver (week 6), spleen, thymus, and lymph glands (weeks 8–16), and bone marrow (week 16). At any of these sites, the mesenchyme cells round up, lose their typical mesenchymal intercellular junctional complexes, proliferate, and become free basophilic progenitors of the various blood cell types.

The question of the origin of the hematopoietic cells that give rise to the definitive blood cells has been widely debated. According to some investigators, extrinsic precursor cells migrate from the yolk sac by way of the vitelline veins to populate the lymphoid and other blood-forming organs of the embryo. The other viewpoint holds that the blood cells develop locally or in situ by the transformation of mesenchymal cells. Recent experiments utilizing tissue grafts of blood-producing organs from mouse and quail implanted into chick (i.e., yielding a chimeric avian embryo) suggest strongly that there are two distinct populations of hematopoietic precursor cells in the embryo. For example, if the liver primordium taken from 8- to 16-day mouse embryos is grafted into a 3-day chick (or quail) embryo, hematopoietic cells will develop in the liver only if the graft is obtained from embryos at the 28- to 32-somite stage. This clearly suggests that mouse fetal hepatic erythropoiesis cannot develop autonomously because of an absence of stem cells. Several investigators, using unique nuclear and chromosomal cell markers, have demonstrated in the chick embryo that the hematopoietic cells that differentiate in the thymus, the bursa of Fabricius, the spleen, and the bone marrow are of extraembryonic origin. Hence, it appears that the first population of hematopoietic cells arises in the yolk sac and subsequently colonizes the liver, spleen, and lymph organs of the embryo. The derivatives of these colonies of hematopoietic cells of extrinsic origin are not well known. Subsequently, therefore, it is believed that a new population of hematopoietic stem cells, perhaps arising from either the body mesenchyme of the embryo or primitive intraembryonic blood vessels, seeds the liver and spleen, and replaces those stem cells of yolk sac origin. Most of the adult avian hematopoietic system is derived predominantly from stem cells that arise intraembryonically.

There is now an extensive body of literature that tends to support the concept that a single pluripotential stem cell (the hematopoietic stem cell or *hemocytoblast*) has the capacity to differentiate into erythrocytes (red blood cells), granulocytes (granular white blood cells), and megakaryocytes (nongranular white blood cells, such as lymphocytes and monocytes). Yolk sacs isolated from 7- to 13-day mouse embryos contain stem cells that are capable of differentiating into all of these cell types. Direct evidence for a single stem cell for both myeloid and lymphoid cell lines in adult mice has been obtained by injecting suspensions of bone marrow cells with characteristic radiation-induced chromosomal abnormalities into lethally irradiated recipient mice. The irradiation destroys the host's capacity to form hematopoietic cells. When the blood-forming tissues of the irradiated animal are examined, the same chromosomal markers can be detected in both red blood and lymphoid cell lines. Although the precise morphological identity of this stem cell has yet to be fully ascertained, it is thought to be large and lymphocyte-like, possessing a granular, basophilic cytoplasm, and very mobile. How these pluripotential stem cells are able to differentiate into discrete blood cell lineages is a major unresolved question. It is suspected that somehow the stem cells selectively acquire a responsiveness to only those regulatory mechanisms that characterize and control the differentiated cell line.

*Erythropoiesis* is the process by which hemocytoblastic stem cells become transformed into mature red blood cells containing hemoglobin. The study of erythroid cell differentiation in vertebrates has contributed substantially to our current understanding of the basic processes and regulatory mechanisms that prevail during normal and abnormal cell differentiation. There are several unique features of erythropoiesis that make it a suitable system for studies in cell differentiation. First, relatively large numbers of blood cells can be isolated from accessible sites in the embryo, cultured in vitro, and subjected to various manipulations, Second, hemoglobin, a metalloprotein constituting about 90 percent of the protein synthesized in the red blood cell, is biochemically and genetically well characterized; it is a useful marker in examining the relationships between cellular morphology and biochemistry during erythroid cell differentiation. Third, erythroid cell production can be stimulated by a glycoprotein substance called *erythropoietin*.

Advances in in vitro culture techniques have enabled analysis of the importance of erythropoietin during development in the differentiation of red blood cells. In the adult mammal, the number of red blood cells and the amount of hemoglobin synthesized is regulated by the titer or level of circulating erythropoietin. Presumably, in response to low tissue levels of oxygenation, the kidney either elaborates and releases erythropoietin itself or a renal activator of a plasma (liver) erythropoietin precursor. Several models have been proposed to show how erythropoietin stimulates red blood cell formation. One of these models is shown in Figure 18–2. It recognizes that there are two major compartments in the red blood cell system: the *stem cell compartment* and the circulating *erythrocyte cell compartment*. In the former compartment, erythropoietin acts to trigger sensitized, nondividing hemocytoblasts to differentiate into erythroid cells. This substance appears to act only on those cells already committed to the red blood cell line. Sudden removal of a segment of the stem cell population stimulates other stem cells to proliferate in order to replace those removed. Erythropoietin production is controlled by the number of cells in the circulating compartment and the amount of oxygen they transport. The circulating erythrocyte compartment contains cells with a finite lifespan (110–120 days) and incapable of self-renewal.

Red blood cells developing within hematopoietic body organs of the embryo form from so-called late erythroid precursors (often designated *CFU-E cells*) and appear to be dependent on the presence of erythropoietin. By contrast, early erythroid precursor cells (i.e., those of the yolk sac and designated *BFU-E cells*) do not respond to erythropoietin treatment and apparently undergo differentiation independent of its control. Erythropoietin normally stimulates cell division and proliferation of erythroid cells. Because these cells become organized in the form of clusters or colonies, a convenient biological assay is available to measure the sensitivity of the various hematopoietic organs to erythropoietin during ontogeny. Interestingly, the sensitivity of the same hematopoietic organs to erythropoietin changes during development. Fetal livers (13–14 or 16–17 days of gestation) can be stimulated by relatively small doses of this hormone. Newborn livers and adult blood-forming organs require higher

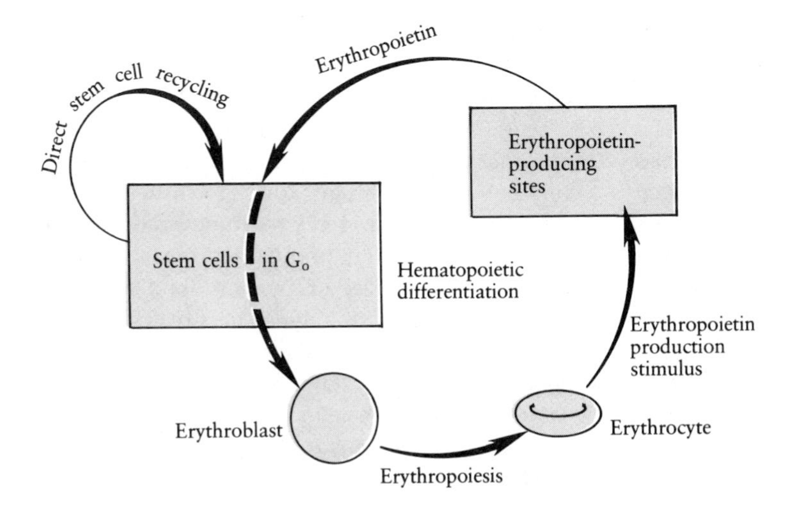


**FIG. 18–2.** A diagrammatic representation of a model showing the relationship between erythropoietin and red blood cell formation. (After J. Okunewick, 1970. Cell Differentiation, ed. by Ole A. Scheide and Jean De Vellis. © 1970 by Litton Educational Publishing, Inc. Reprinted by permission of Van Nostrand Reinhold Company.)

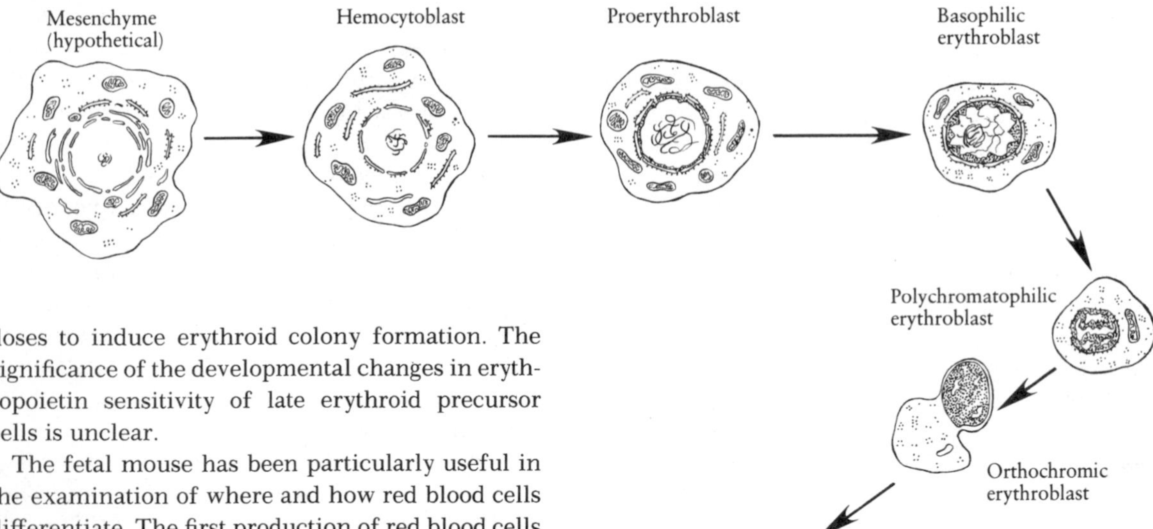


**FIG. 18–3.** A suggested scheme for the differentiation of erythropoietic cells in the embryonic liver and in all adult erythropoietic tissues of a mammal. (After R. Rifkind, 1974. Concepts of Development. J. Lash and J. Whittaker, eds. Sinauer Associates, Sunderland, Mass.)

doses to induce erythroid colony formation. The significance of the developmental changes in erythropoietin sensitivity of late erythroid precursor cells is unclear.

The fetal mouse has been particularly useful in the examination of where and how red blood cells differentiate. The first production of red blood cells appears in the yolk sac during day 7 of gestation. These cells enter the circulation as *erythroblasts* several days later. Here the erythroblasts proliferate and in synchronous fashion complete their maturation by day 15. These *primitive erythroid cells* are large and nucleated with a very short lifespan. By day 21 of gestation (birth), most of these cells have been phagocytized by the reticuloendothelial system of the fetus.

In all vertebrates studied so far, a *definitive erythroid cell population* derived from intraembryonic precursor cells gradually replaces the yolk sac-derived primitive erythroid cell line. In contrast to primitive erythrocytes, the definitive erythrocytes are smaller and more adultlike; also, the definitive cell population is self-renewing.

The fetal liver is the first major site for the production of the definitive red blood cell type. Differentiation of the definitive erythroid cell is characterized by a series of systematic changes in cell morphology and the production of hemoglobin (Fig 18–3). Transformation of the hemocytoblast into the *proerythroblast* begins around day 11 in the mouse and is marked by a slight basophilia in the cytoplasm. Proerythroblasts have large nuclei and uncondensed chromatin; they lack detectable hemoglobin but show intense synthesis of ribosomal RNA. The accumulation of RNA in the cytoplasm is responsible for the intense basophilia of the next stage of cellular differentiation known as the *basophilic erythroblast*. Although it is difficult to detect hemoglobin cytochemically, the synthesis of this protein is probably initiated in this cell type. Hemoglobin is clearly detectable by staining in the *polychromatic erythroblast*. Polychromatophilic cells are nucleated, and they show marked reduction in the rate of RNA synthesis and intense accumulation of hemoglobin. In the *orthochromic erythroblast*, the nucleus becomes pycnotic and eventually is eliminated from the cell with a small amount of cytoplasm. The cell now enters the circulation on day 12 as an *anucleated reticulocyte*. Anucleated reticulocytes continue to make hemoglobin for a day or two. When hemoglobin synthesis is terminated, the cell becomes a *mature erythrocyte*. Adult erythrocytes, therefore, are terminally differentiated cells with a relatively simple structure. Their differentiation involves the gradual inactivation of chromatin material and the increasing use of the intracellular machinery for the synthesis of cytoplasmic (globin) and plasma membrane proteins. An important protein of the erythrocyte membrane is *spectrin*, composed of two nonidentical polypeptides. Spectrin forms a dense meshwork on the cytoplasmic side of the plasma membrane. Its interaction with actin and several other proteins establishes a cytoskeletal framework that functions to maintain the biconcave shape of the mammalian erythrocyte and restrict the mobility of certain transmembrane polypeptides through the membrane.

The hemoglobin molecule is constructed of globin, a protein containing four polypeptide chains, and four iron-containing heme groups. During the development of the vertebrate embryo, there are distinct changes in the type of hemoglobin being synthesized. Human hemoglobins, for example, are heterogeneous at all stages of development (Fig. 18–4). Yolk sac-derived erythrocytes are found in embryos up to eight weeks and consist of a mixture of *embryonic hemoglobins.* The predominant embryonic hemoglobins (Hb Gower 1 and Hb Gower 2) are formed by $\xi$, $\epsilon$, and $\alpha$ chains. *Fetal hemoglobin* (Hb F; $\alpha_2\gamma_2$) appears with the onset of erythropoiesis in the liver and remains in the circulation until birth. Finally, by the end of the first trimester, the final type of hemoglobin, adult hemoglobin (Hb A; predominantly $\alpha_2\beta_2$), is detected in the circulation. The $\beta$ chains constitute about 10 percent of the non-$\alpha$ chains up to about 36 weeks of gestation; their rate of production increases with a decline in $\gamma$-chain synthesis of Hb F. A small amount of Hb F persists into adult life in a limited population of red blood cells called *F cells.*

Very little is known about the factors involved in regulating the switch from one hemoglobin type to another. It would be convenient to believe that the temporal shifts in the type of hemoglobin being synthesized are correlated with the site of erythropoiesis; that is, the switch from embryonic to fetal hemoglobin reflects the shift from yolk sac to liver red blood cell formation. Such does not appear to be the case. Yolk sac-derived erythrocytes synthesize Hb F in addition to embryonic hemoglobins, and the first erythrocytes originating from the liver tissue are known to synthesize embryonic hemoglobins. Therefore, transitions between hemoglobin type would appear to be related primarily to gestational stage of the developing organism and not to the site of erythropoiesis. However, such may not be the case in all vertebrate embryos. In *Rana* (bullfrog) tadpoles, two erythropoietic sites (kidney and liver) produce different types of hemoglobin. The choice of Hb type to be synthesized is made at an early step in the differentiation of the red blood cell and appears to be effected by the milieu of the blood-forming organ. When larval liver is cocultured with kidney tissue, there is a switch in hemoglobin pattern; part of the shift is toward the Hb type produced by the kidney. Insight into the regulation of hemoglobin switching may come from newly developed techniques that allow for the growth of erythroid cell colonies in vitro and analysis of the hemoglobin type produced in colonies prepared from red blood cell precursors at varying stages of differentiation.

In view of the fact that erythroid cells become specialized for the synthesis of a well-characterized molecule (hemoglobin), the red blood cell system (both embryonic and adult) has been used by a number of investigators for study of the genetic mechanisms controlling the production of a protein during the differentiative process. Efforts have been directed at determining the relative contributions of selective DNA transcriptional and posttranscriptional level controls of gene expression leading to the maturation of red blood cells. Unfortunately, a unified and comprehensive picture of regulation during erythrocyte differentiation is still to be developed. This in part relates to ever-changing experimental approaches to the question and to differences in the interpretation of data in light of the increased sophistication of instrumentation for detecting hemoglobin. A good case in point is the early development of red blood cells in the chick embryo. The blood islands of the yolk sac produce a single, synchronous population of erythrocytes; they appear between the definitive primitive streak and head fold stages of development (about 18–24 hours of incubation). Hemoglobin is detected by routine staining methods at the six- to eight-somite stage, or some 10 hours after the formation of the head fold. If chick embryos are cultured with actinomycin D at selected stages of development and then examined for hemoglobin production, it can be demonstrated that addition of the drug prior to the head fold stage inhibits the normal temporal expression of hemoglobin. This means that there is a distinct delay in the known time of globin mRNA synthesis and in its translation into the polypeptide chains of the globin molecule. The traditional view has been that the globin mRNA exists in an inactive form until the activation of a translational control mechanism in the cytoplasm at the six- to eight-somite stage. We now suspect that this may not be the primary level of regulation. Sensitive tracer methods show that globin mRNA, once transferred into the cytoplasm, can be found in two compartments: (1) as actively translated on RNA in polyribosomes to yield small quantities of newly synthesized hemoglobin, and

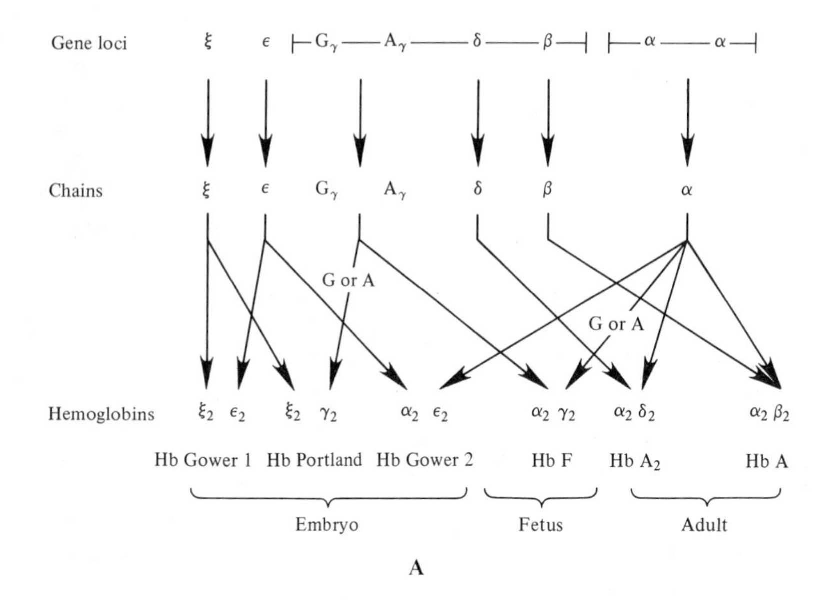


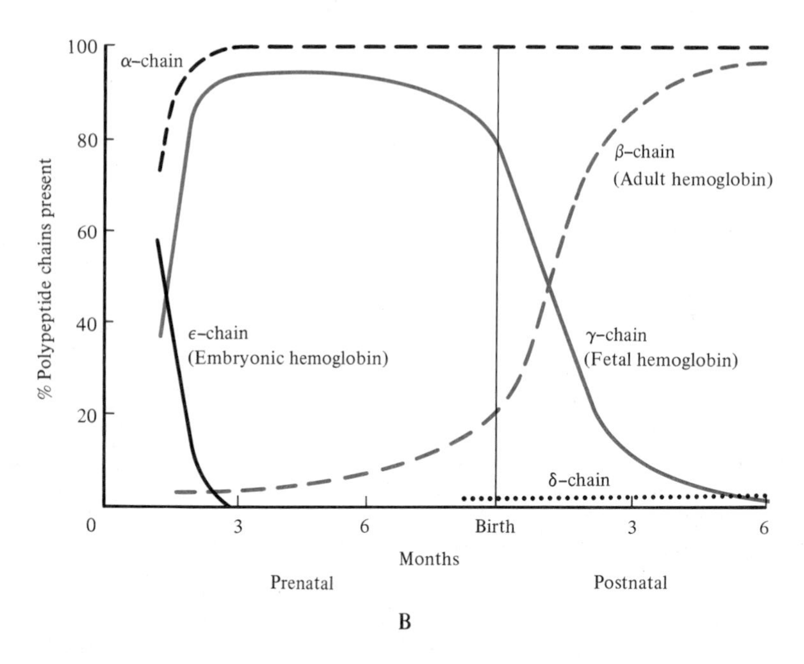


**FIG. 18–4.** Structure and periods of synthesis of human hemoglobins. A, composition of the different types of hemoglobins. Upper row illustrates the gene loci that code for the hemoglobin chains. The lower row shows the various polypeptide combinations and the different hemoglobins formed from them. (From D. Weatherell and J. Clegg, 1979. Cell 16:467.) B, developmental changes in human polypeptide chains. (From E. Huehn et al., 1964. Cold Spring Harbor Symp. Quant. Biol. 29:327.)

(2) in an inactive form associated with free mRNP complexes. The translation of some globin mRNA shortly after transcription clearly indicates transcriptional-level control during erythropoiesis. Recently, Imaizumi-Scherrer and colleagues have assayed the complexity of four distinct RNA populations in differentiating duck red blood cells. Their data show that approximately 1200 mRNA species, including mRNAs for the alpha and beta polypeptide chains of hemoglobin, are not represented at all in the polyribosomal mRNA population, suggesting that posttranscriptional controls must also be important in the differentiation of the erythrocyte.

The 13-day mouse liver contains the complete morphological series of red cell types. It has been used to examine the relationship between the various stages of cellular differentiation and the expression of gene products. The addition of erythropoietin to the fetal liver in vitro results initially in the synthesis of 4S, 5S, and ribosomal RNA. No globin mRNA is detectable at this time. With continued exposure to the hormone, DNA synthesis is initiated and the globin genes are transcribed.

These activities take place in the proerythroblast stage. By the basophilic erythroblast stage of differentiation, there is intense accumulation of globin mRNA and the various globin polypeptides can be detected. Hemoglobin synthesis and DNA production are both inhibited by actinomycin D and puromycin. This suggests that, under hormonal stimulation, there is an early transcription resulting in the production of a protein. This protein is necessary for DNA replication. Following DNA replication, there is a second, later phase of transcription that yields the globin mRNA required for hemoglobin synthesis. While the synthesis of globin protein increases in erythroid cells, there is a reduction in the synthesis of many nonglobin proteins.

## The Primitive Vascular System

Diffuse capillary plexuses always precede the formation of arteries and veins in any given region of the embryo. Arteries and veins gradually become differentiated through the enlargement of individual capillaries and the fusion and confluence of adjacent ones. Capillaries from which the flow of blood is diverted during this process undergo regression and atrophy. Presumably, genetic background and local hemodynamic influences, such as the direction and velocity of the blood, control not only the selection of those channels that are to enlarge, but also the structural characteristics of their walls (i.e., whether they are to be arteries or veins).

The primitive vascular system in all vertebrate embryos consists of simple, paired, symmetrically arranged endothelial tubes. Initially, arteries and veins cannot be structurally distinguished; however, they are named in terms of their fate and relationship to the heart. Human embryos of about 12 somites clearly show the initial arrangement of definitive blood vessels (Fig. 18–5). Directly beneath the notochord are the paired *dorsal aortae.* These are continued around the tip of the pharynx as the first pair of *aortic arches* where they then join the anterior end of the heart. Paired dorsal *intersegmental arteries* spring from the dorsal aortae and pass between successive pairs of somites. Vessels passing to and from the yolk sac *(vitelline arteries* and *veins)* and placenta *(umbilical* or *allantoic arteries* and *veins)* are established and circulate blood extraembryonically shortly after the onset of contractions of the heart (about 26 days). A pair of *anterior cardinal veins* returns blood from the capillary plexuses of the head end of the embryo to the heart.

At slightly later stages of development (Fig. 18–6), additional aortic arches are sequentially added, the dorsal aortae fuse into a single median vessel as far forward as the pharynx, and the posterior end of the embryo is drained by a pair of *posterior cardinal veins.* Each posterior cardinal vein joins with an anterior cardinal vein to form a short common vessel, the *common cardinal vein (duct of Cuvier),* which empties into the heart.

The arrangement of blood vessels described above undergoes substantial alteration during the course of development. Blood vessels fuse, hypertrophy, regress, or atrophy with shifts in patterns of blood flow and internal changes in the organization of the heart. For purposes of convenience, we will

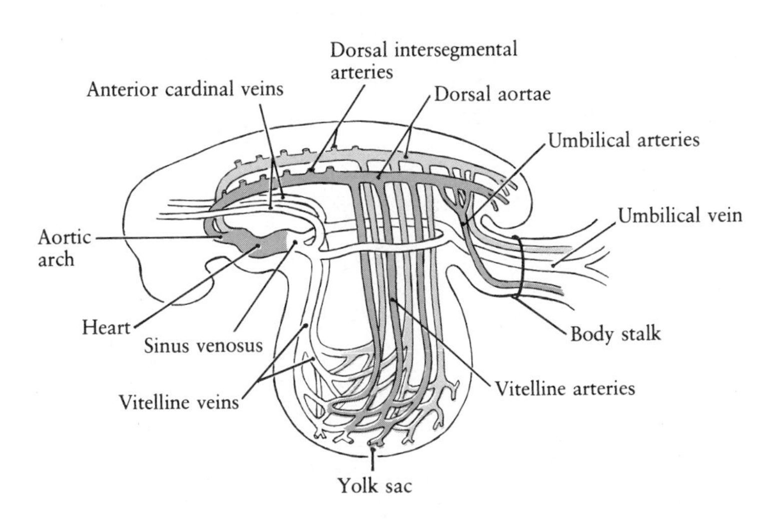


**FIG. 18–5.** A diagram of the early vascular system in the human embryo as seen from the left side.

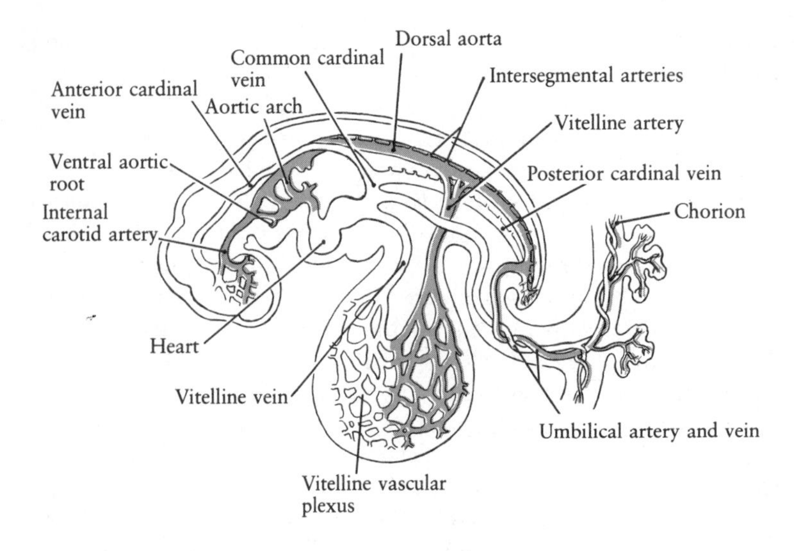


**FIG. 18–6.** A drawing to show the arrangement of the heart and blood vessels at the end of the first month in the human embryo. Left-side view.

describe under separate subheadings the changes in the heart, arteries, and veins that result in the adult arrangement of the cardiovascular system. The student should keep in mind, however, that alterations in the organization of the heart and in the patterns of arteries and veins occur simultaneously and are interdependent.

## The Heart

### Localization and Induction of Heart-Forming Tissue

The primitive heart in most vertebrates is tubular and arises as the result of the fusion of a pair of cardiac primordia. This is particularly evident in forms such as the bony fishes, reptiles, and birds where the developing embryo is flattened and pressed against the yolk. Long before the heart is recognizable as a tubular structure, cells with heart-forming capacity can be localized using techniques of vital staining, extirpation, and transplantation.

The approximate locations of heart-forming mesodermal cells in the urodele at successive stages of development are shown in Fig. 18–7. At the onset of gastrulation, the areas of presumptive heart tissue are located on either side of the dorsal lip of the blastopore (Fig. 18–7 A,B). After passing through the dorsolateral lips of the blastopore, each cardiac area is located in the edge of the mesodermal mantle, rather high on the flank of the embryo adjacent to the presumptive hindbrain of the neural plate (Fig. 18–7 C). If one of these cardiac

**FIG. 18–7.** Localization of heart-forming tissue at various stages in a urodele embryo. A, gastrula, dorsal view; B, gastrula, left lateral view; C, neurula, left lateral view; D, tail bud, right lateral view. (After W. Copenhaver, 1955. Analyses of Development. B. Willier, P. Weiss, and V. Hamburger, eds. W. B. Saunders Company, Philadelphia.)

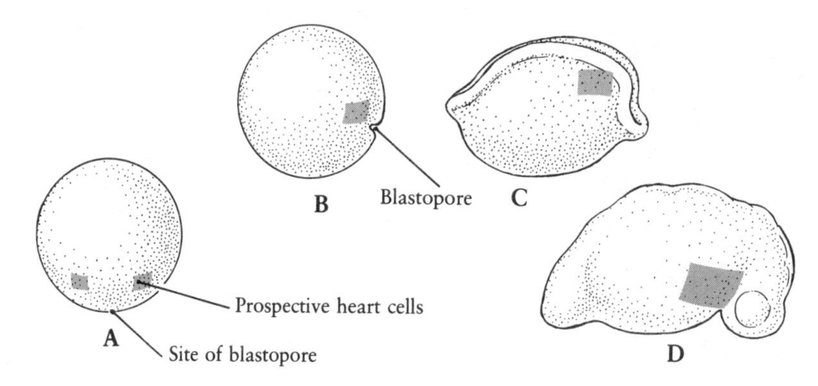

regions is dissected out at the tail bud stage (Fig. 18–7 D) and cultured in the proper medium, it will differentiate into an S-shaped, pulsating heart tube.

The position and extent of the presumptive cardiac regions at various stages in the chick embryo have been rather precisely mapped out by explanting fragments of the blastoderm to the chorioallantoic membrane of eight- to nine-day host embryos (Fig. 18–8). Explants with heart tissue will differentiate into cellular vesicles of contractile activity. Hence, the capacity of these tissue fragments to self-differentiate into cardiac tissue has been a useful tool in constructing a fate map for the heart. At about the time that the primitive streak forms, the presumptive cardiac tissue is located in the mesoderm just lateral to the tip of the streak (Fig. 18–8 B,C). By the definitive to late primitive streak stage, the cardiac regions are paired and appear anterolaterally to Hensen's node (Fig. 18–8 E,F).

There is some evidence that the heart-forming regions are determined in their cardiogenic potency by interactions with neighboring tissues, particularly the endoderm. Precardiac mesoderm appears to require the presence of endoderm for successful cardiac muscle differentiation. In the amphibian, for example, the presumptive heart mesoderm is in contact with the foregut endoderm during its movement after migration through the blastopore. Balinsky and others have shown that the extirpation of the entire endoderm from embryos at the neurula stage results in the complete absence of the heart. Also, fragments of presumptive cardiac mesoderm from salamander embryos at gastrula or early neurula stages typically produce well-formed hearts in culture only if explanted with endoderm. Jacobson and Duncan (1968) have shown that a particular fraction of endoderm, prepared by passing homogenized endoderm through a Sephadex column, can partially substitute for intact endoderm in culture. Hence, the formation of heart tissue would appear to be dependent on inductive influences emanating from the endoderm.

**FIG. 18–8.** Localization of heart-forming tissues at progressively later stages in the chick embryo.

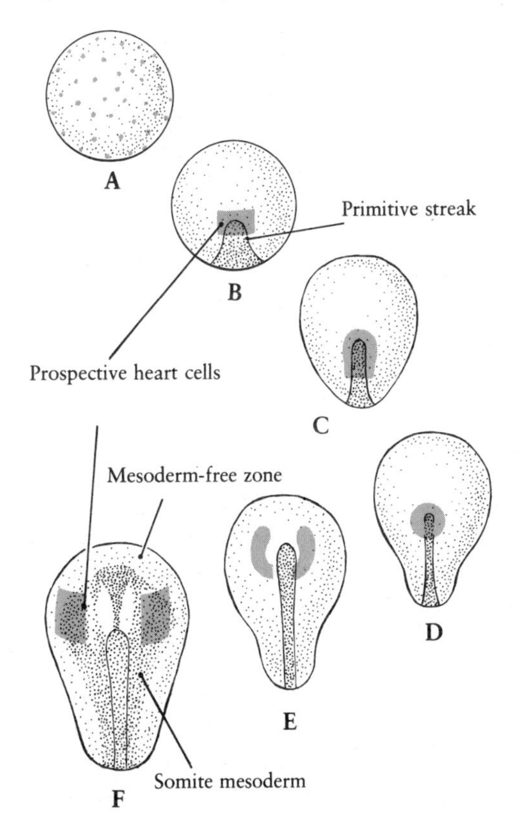


The importance of the endoderm in the determination of heart-forming tissue is less clear in higher vertebrates. Studies by Waddington and his colleagues are inconclusive. Cultured chick embryos, following removal of the entire area pellucida endoderm (hypoblast) at the definitive primitive streak stage, generally showed an absence of well-formed hearts. However, one case was reported in which the embryo displayed an organized heart in the absence of endoderm. DeHaan removed the endoderm overlying the presumptive cardiac region from only one side of the primitive streak stage chick embryo. No heart appeared on the operated side of the embryo. When he cultured the fragments removed from the operated side of the embryo, approximately 70 percent of these developed masses of beating heart tissue. This demonstrated that extirpation techniques may remove not only the endoderm but also the mesoderm that contains presumptive heart tissue as well. The absence of heart tissue in Waddington's studies is probably related to the fact that the presumptive heart-forming tissue has been removed with the presumed inducer. It has also been demonstrated that the avian precardiac mesoderm can differentiate in vitro to some extent in the absence of endoderm, but the development of mature cardiac cells is clearly retarded in the absence of this germ layer. Generally, it appears that the differentiating mesoderm will develop into a tubular organ only if endoderm is present in the culture. Manasek proposes that the normal morphogenesis of the heart can only take place if there is normal cytodifferentiation and maturation of cardiac tissue. He suggests that the basal lamina of the endoderm, perhaps through the synthesis of small molecules such as collagen or glycoproteins, mediates the interaction between the precardiac mesoderm and the endoderm.

### Formation of the Primitive Tubular Heart

The formation of the heart tube and its subsequent looping are basic but complex processes of cardiac morphogenesis. As the following discussion will illustrate, the visible, descriptive events of heart formation in vertebrate embryos are rather well known. However, the forces underlying the morphogenetic changes and the mechanisms controlling events of heart formation are only gradually being elucidated.

The heart can be viewed as a highly specialized blood vessel with very thick muscular walls. The heart follows a generally similar pattern of early embryonic development in most vertebrates. The formation of this organ in the amphibian embryo is rather simple and perhaps should be considered before that of other vertebrates (Fig. 18–9). Recall that by the end of neurulation the lateral plate mesoderm, lying on either side of the tubular gut, is split into parietal and visceral layers. At the end of neurulation, the free, ventromedial edges of the mesodermal mantle swing toward the midline and become noticeably thickened in the heart-forming regions (Fig. 18–9 A,B). Cells proliferate from the heart-forming regions, migrate beneath the gut, and become organized as a longitudinal, vascular strand of tissue. A lumen develops within the vascular strand, thus converting a solid cord of cells into an endothelial-lined tube (Fig. 18–9 C). The endothelial lining of the future heart cavity constitutes the rudiment of the *endocardium.*

As the endocardial tube is being formed, the edges of the lateral plate mesoderm continue to push toward the midline. Subsequently, the visceral layers of mesoderm from either side meet below and above the endocardial tube to form suspending partitions, the *ventral* and *dorsal mesocardia* (Fig. 18–9 C–E). The visceral mesoderm enveloping the endocardial tube will thicken and form the *epimyocardium* layer (Fig. 18–9 D,E). The epimyocardial rudiment will later differentiate into

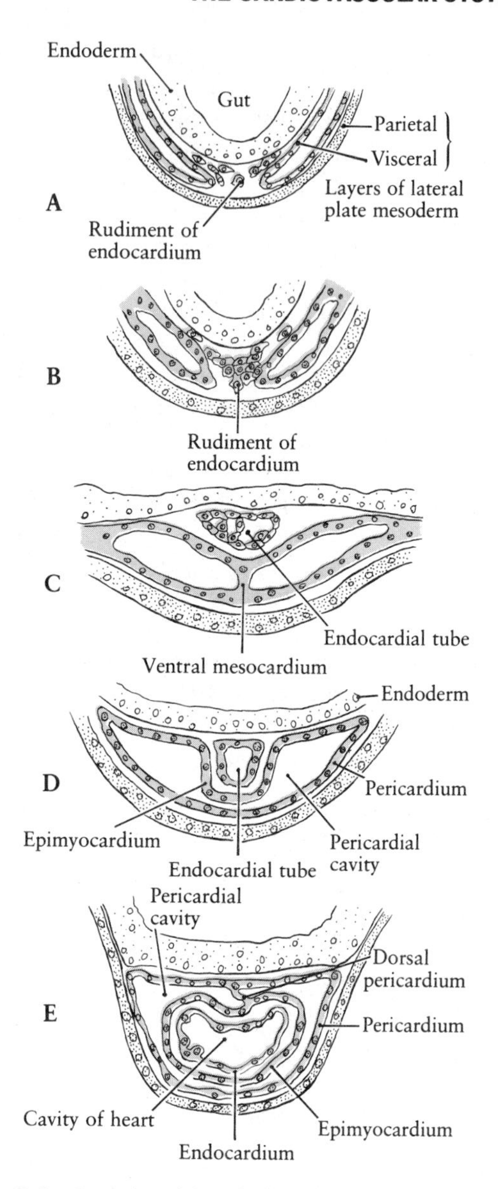


**FIG. 18–9.** Formation of the primitive tubular heart in amphibians as seen in a series of transverse sections.

the cardiac muscle tissue (myocardium) and covering layer of the heart (epicardium). The coelomic cavities on either side of the tubular heart will expand. With the disappearance of the ventral mesocardium, they join together as the *pericardial cavity* (Fig. 18–9 D,E).

The early stages in the development of the heart in bony fishes, reptiles, and birds are considerably more complicated in that the postgastrulative embryo, organized as flat sheets of cells, rests upon a large yolk mass. Consequently, the unpaired heart

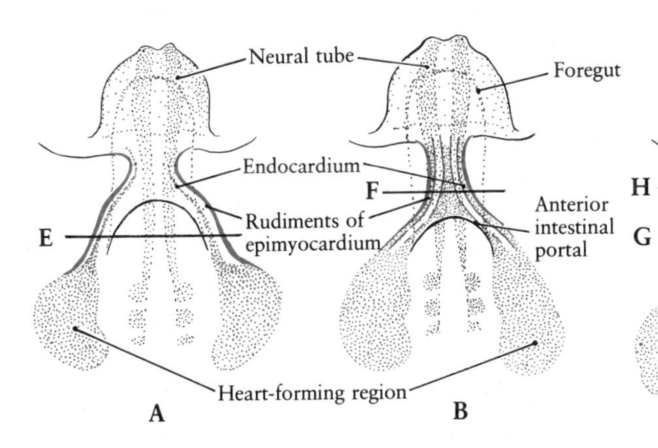


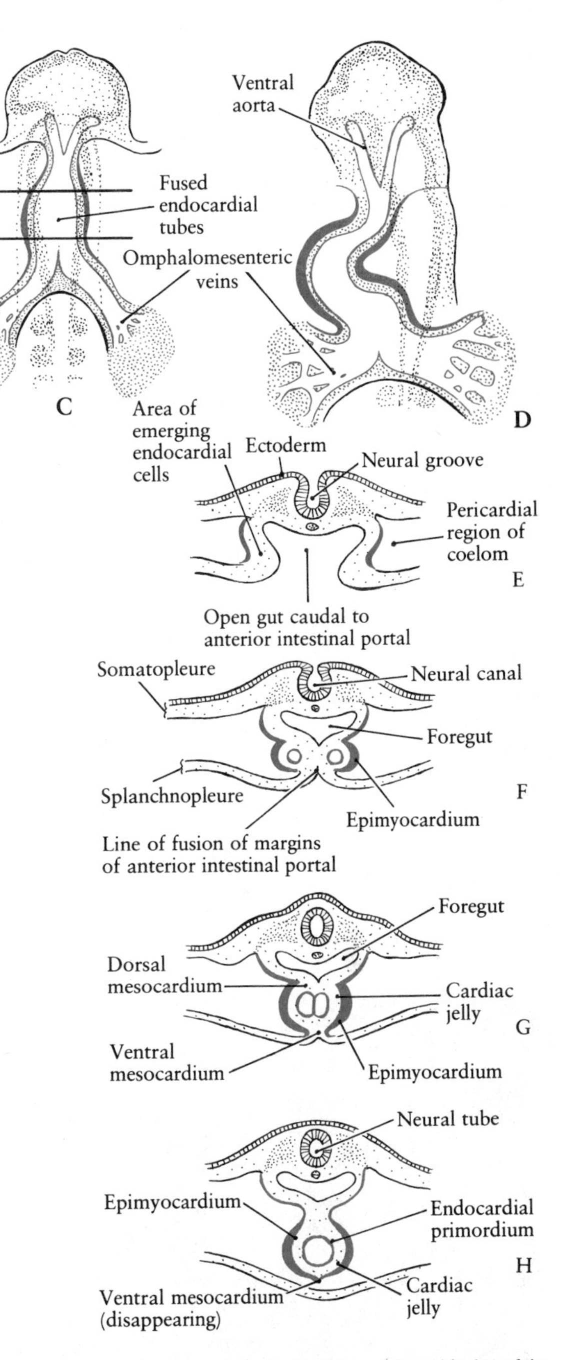


**FIG. 18–10.** A–D, ventral views to show the origin and fusion of the paired cardiac primordia in the chick embryo; E–H, transverse sections through the heart region at locations indicated in A–C above. (A–D, after R. DeHaan, 1965. In: Organogenesis, R. L. DeHaan and H. Ursprung, eds. Holt, Rinehart and Winston, New York.)

arises gradually as the two widely separated, lateral cardiac primordia swing toward the ventral midline with the lateral plate mesoderm and fuse below the foregut.

The heart begins to form in the chick embryo shortly after the head process stage (Fig. 18–10 A). At this time the heart-forming regions are indicated as thickenings in the splanchnic mesoderm on either side of the anterior intestinal portal (Fig. 18–10 E). Cells detach themselves from the mesial surfaces of the splanchnic mesoderm and become organized on either side as a pair of thin-walled endothelial tubes (Fig. 18–10 B,F). These endocardial primordia, destined to give rise to the internal lining of the heart, are invested by the thickened, intact splanchnic mesoderm or epimyocardial primordia (Fig. 18–10 E,F).

Traditionally, the epicardium has been considered to be a derivative of the epimyocardium in the chick embryo. More recent studies using light and electron microscopy suggest that the outer splanchnic layer of the early heart consists only of myocardial cells. The epicardial cells appear to originate from the mesothelium adjacent to the sinus venosus and migrate as a flattened monolayer of cells over the entire heart surface (Fig. 18–22 B). The term epimyocardium, therefore, should perhaps be considered a misnomer and discarded. For purposes of convenience, however, we will retain this term in our illustrations of chick cardiogenesis to indicate the site where myocardial and epicardial layers will differentiate. The reader should keep in mind that nothing is implied as to the pricise origin of the epicardium. As the body folds undercut the embryo and separate it from the yolk, the ventral wall of the gut and the ventral body wall of the embryo are completed. In this process, the endocardial tubes from the right and left sides are brought together and fuse in the midline below the gut (Fig. 18–10 C,D,F,G). Simultaneously, the epimyocardial rudiments close around the endocardial vessels, meeting initially below and then above them (Fig. 18–10 G). The double layer of splanchnic mesoderm meeting above the endocardial tubes constitutes the *dorsal mesocardium* and that below the endocardial tubes the *ventral mesocardium*. The ventral mesocardium is very transitory and disappears shortly after it forms (Fig. 18–10 H), leaving the primitive tubular heart suspended in an unpaired pericardial cavity by the dorsal mesocardium.

As in other vertebrate embryos, a distinct extracellular space separates the endocardium and the epimyocardium of the primitive tubular heart in the chick (Fig. 18–10 C,D). This space is filled with a loose, fibrillar reticulum known as the *cardiac jelly* (Fig. 18–11). The cardiac jelly is a gel-like extracellular matrix with a unique set of biomechanical properties. It is highly viscous and rich in glycosaminoglycans (hyaluronate, chondroitin sulfate), collagen, and glycoproteins. Some of these glycoproteins are known to contain fucose residues. These particular glycoproteins are probably synthesized by the endoderm and may represent the endodermal requirement for cardiac differentiation. Most of the macromolecules of the cardiac jelly are synthesized by the developing myocardium. The production of these extracellular substances decreases as the heart matures.

In the human embryo, the primordium for both the heart and the pericardial cavity initially lies as a crescent-shaped zone of mesoderm cephalic to the embryonic disc (Fig. 18–12). It is visible as early as the primitive streak stage (15 days). At a slightly later stage (17–18 days), the thickened mesoderm becomes split into parietal and splanchnic layers through the appearance and coalescence of many vesicular spaces, a process leading to the formation of the pericardial cavity (Fig 18–13 A,B). A U-shaped *cardiogenic plate* or heart primordium now lies below the pericardial cavity (Fig. 18–13 B).

As the head of the embryo pushes rapidly forward, the limbs of the cardiogenic plate are swept

**FIG. 18–11.** A, cross-sectional view of the early embryonic chick heart showing the large expanse of extracellular matrix (CJ, cardiac jelly) between the endocardium (E) and the myocardium (M). (From T. Fitzharris and R. Markwald, 1982. Dev. Biol. 92:315.) B, SEM of the early chick heart extracellular matrix showing the randomized distribution of the microfibrils. (From R. Markwald et al., 1979. Dev. Biol. 69:634.)

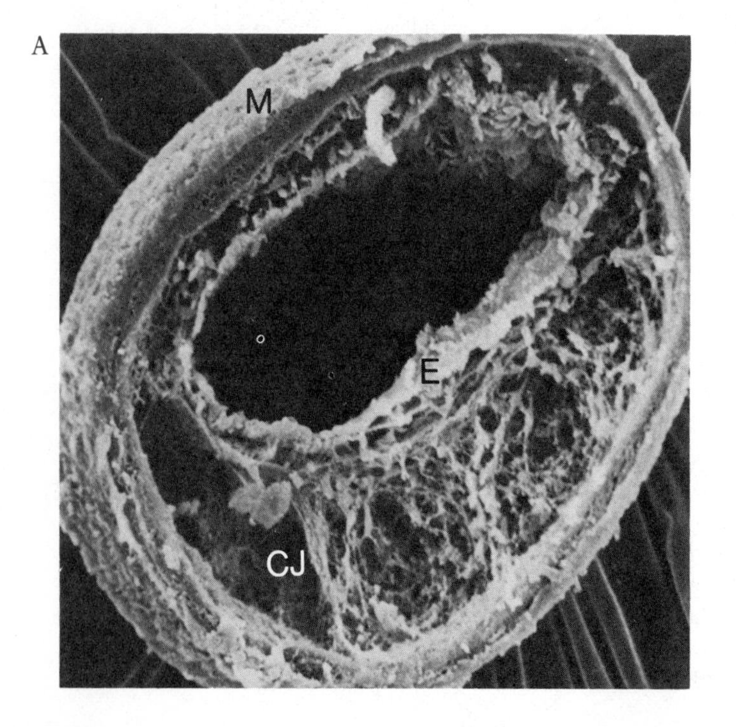


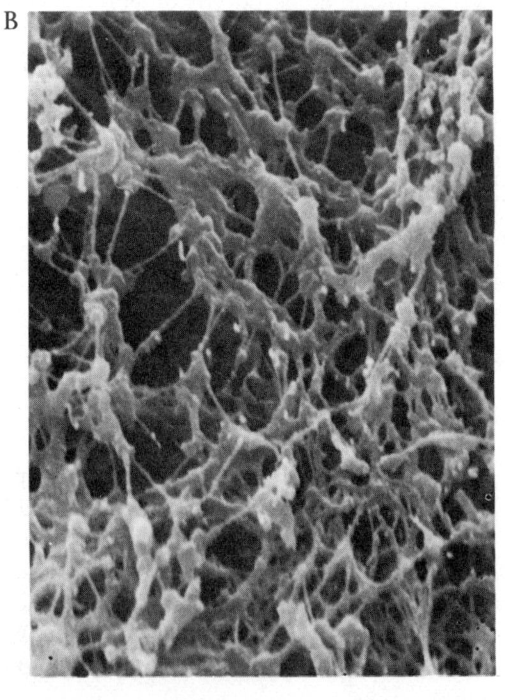

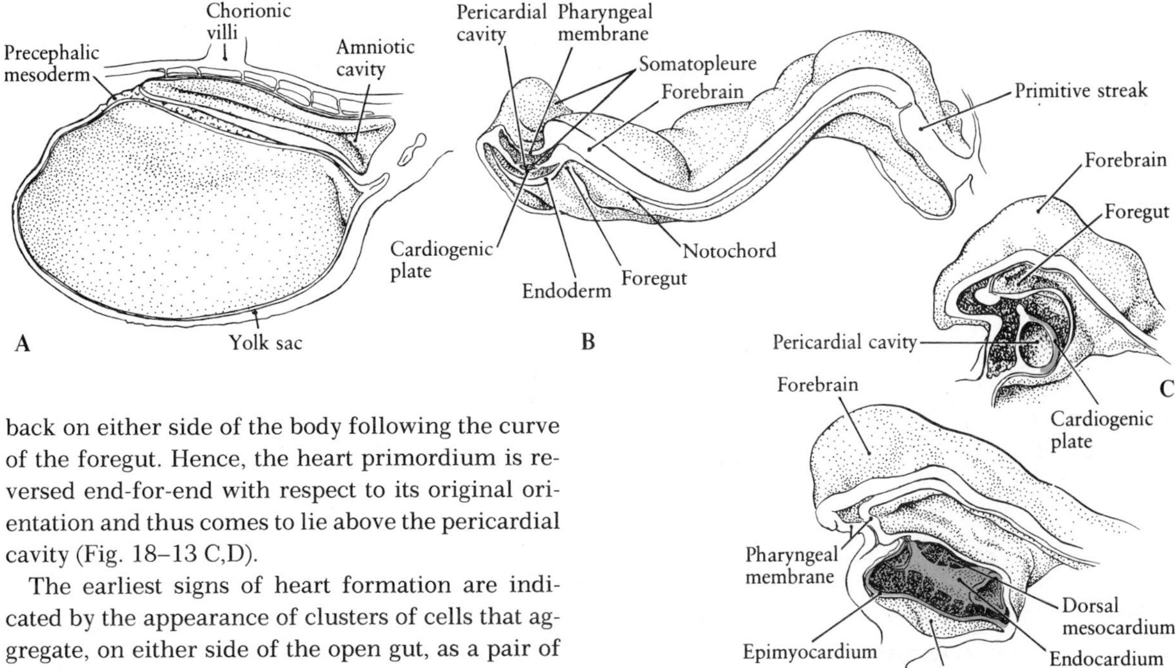


FIG. 18–13. A series of drawings to show the early transformation of the cardiogenic plate in the human embryo. (From C. Davis, 1927. Carnegie Contrib. Embryol. 19:245.)

back on either side of the body following the curve of the foregut. Hence, the heart primordium is reversed end-for-end with respect to its original orientation and thus comes to lie above the pericardial cavity (Fig. 18–13 C,D).

The earliest signs of heart formation are indicated by the appearance of clusters of cells that aggregate, on either side of the open gut, as a pair of elongated strands (cardiogenic cords) between the endoderm and the splanchnic mesoderm of the cardiogenic plate (Fig. 18–14 A). Each vascular cord then quickly becomes canalized or hollowed out to produce a thin-walled endocardial tube (Figs. 18–13 D, 18–14 B). As in the chick embryo, the splanchnic mesoderm on either side thickens as the epimyocardial rudiment where it lies adjacent to the endocardial tube (Fig. 18–14 B).

While these changes occur in the splanchnic mesoderm, the folding off of the embryonic body progresses concurrently with closure of the floor of the foregut. The paired endocardial tubes are brought closer together as the process of embryonic separation from extraembryonic tissues reaches the level of the heart (Fig. 18–15 A). The two tubes then fuse to form a single, endocardial tube lying in the midline (Figs. 18–14 C,D, 18–15 B,C). It is enveloped by a single troughlike fold formed by the right and left epimyocardial rudiments (Fig. 18–14 D). The apposed layers of the splanchnic mesoderm meet above the endocardial tube as the dorsal mesentery of the heart or dorsal mesocardium (Fig. 18–14 D). No comparable structure is produced below the heart so that the originally paired right and left coelomic cavities become immediately confluent to form the definitive, unpaired pericar-

FIG. 18–12. A dorsal view of the late presomite human embryo to show angioblastic tissue in the splanchnic mesoderm. The ectoderm and somatopleure have been removed. (From C. Davis, 1927. Carnegie Contrib. Embryol. 19:245.)

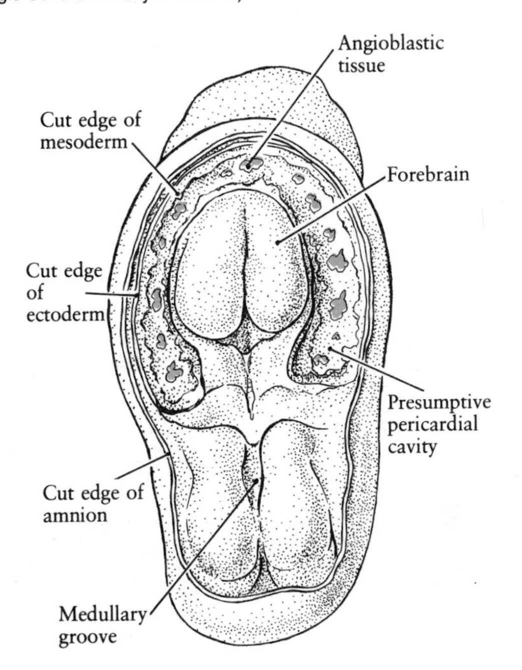

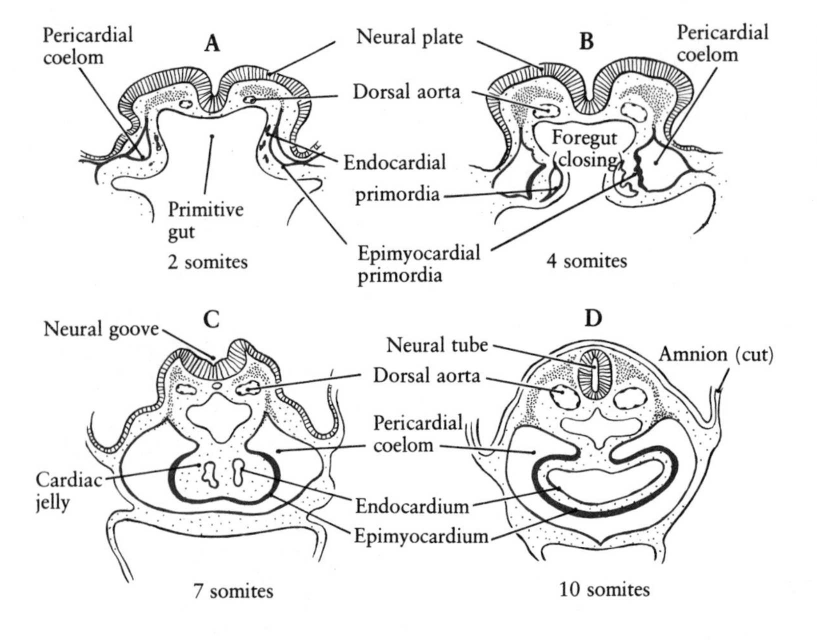


**FIG. 18–14.** Four stages in the fusion of the paired cardiac primordia of the human heart as seen in transverse section. (From Human Embryology by B. M. Patten. Copyright © 1968 by McGraw-Hill, Inc. Used with permission of McGraw-Hill Book Company.)

dial cavity. A distinct extracellular space, filled with a loose, gelatinous reticulum or cardiac jelly, separates the endocardium and the epimyocardium of the primitive tubular heart (Fig. 18–14 C). Continued retreat of the anterior intestinal portal joins the paired cardiac primordia to the portion of the heart already formed. Fusion continues until the entire heart is a single organ.

Even before the paired cardiac halves have started to merge in the midline, each endocardial tube, partially invested by its epimyocardial mantle, shows a sequence of dilations that foreshadow the future chambers of the tubular heart (Fig. 18–16). Named in the order in which they transport blood through the heart, the primary divisions of the early embryonic heart are *sinus venosus, atrium, ventricle, and bulbus cordis* (Fig. 18–17).

The bulbus cordis, also known as the *conus arteriosus* in lower vertebrates, is often referred to as the outflow tract of the heart. By approximately the eight-somite stage of the human embryo, the bulbar and ventricular cardiac rudiments have completely merged to form single, median chambers, while the atrial region is still paired and unfused (Fig. 18–15 B). The right and left sinus venosus at this time lie in the loose mesenchyme of the septum transversum. Each sinus venosus serves as the center of confluence for the cardinal, umbilical,

**FIG. 18–15.** A–D, four stages in the formation of the human heart as exposed by dissection. Ventral views. (From C. Davis, 1927. Carnegie Contrib. Embryol. 19:245.)

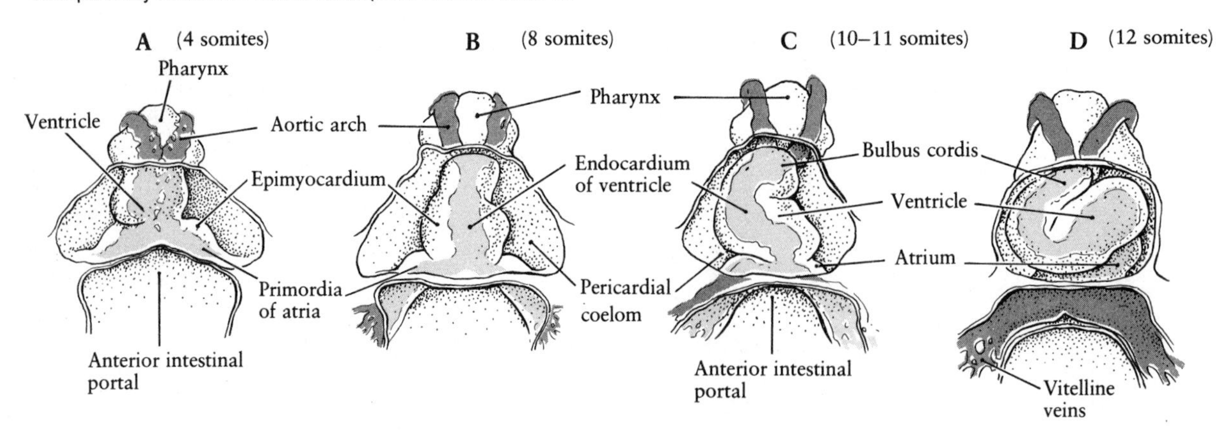

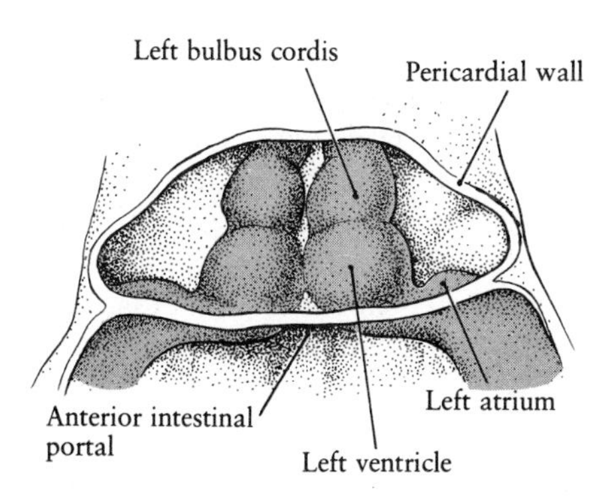


**FIG. 18–16.** The paired cardiac tubes showing regionalization into chambers in the six-somite human embryo. Ventral view. (From C. Davis, 1927. Carnegie Contrib. Embryol. 19:245.)

and vitelline veins. Cephalically, the bulbus cordis continues into a short *ventral aorta* or *truncus arteriosus* (Fig. 18–17). The ventral aorta is formed by the fusion of paired endothelial tubes that are fashioned from two strands of mesenchyme anterior to the level of the heart. Within the next several days, the paired atria and sinus venosuses will merge to complete formation of the primitive tubular heart (Figs. 18–14 D, 18–20 A,B, 18–21 A,B). In short, the early, roughly symmetrical tubular heart of all vertebrate embryos is formed by fusion of bilateral precardiac splanchnic mesodermal cell masses. It initially consists of two layers (endocardium and "epimyocardium") separated by a proteoglycan-rich extracellular matrix. Morphological changes result in the gradual appearance of four anteroposterior ordered chambers. Concurrently, stellate cells gradually migrate into the cardiac jelly (Fig. 18–18). These cells originate from the endocardium, proliferate in the extracellular matrix, and give rise to thickened pads of mesenchymatous tissue in the atrioventricular canal (*dorsal* and *ventral endocardial cushions*) and along the walls of the bulbus cordis (*bulbar ridges*) (Figs. 18–17, 18–28, 18–30, 18–32). These ridges of endocardial connective tissue are the earliest signs of the future septa of the heart.

The convergence of the two lateral cardiac primordia in the ventral midline to produce a single heart tube is a complex event. It appears to involve a series of different, but simultaneously occurring and coordinated morphogenetic cell movements. DeHaan and his colleagues have analyzed some of these movements in chick embryos using several techniques, including that of time-lapse cinematography. The movements that result in the union of the paired cardiac regions include (1) folding movements of the endoderm to form the crescent-shaped pouch of the anterior intestinal portal, (2) rapid, anteromesial migration of special clusters of cells (*precardiac clusters*) within the cardiac regions, and (3) ventral emigration movements of cells that lose their association with the splanchnic mesoderm to form the hemangioblast or presumptive endocardial layer.

The movements of the endoderm are of great im-

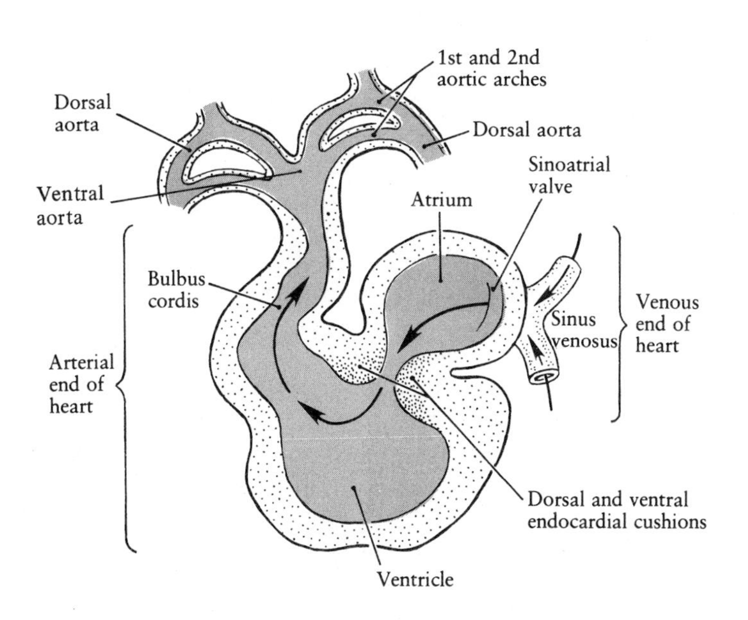


**FIG. 18–17.** A diagrammatic section through the human heart (12-somite stage) showing the beginning of the formation of the bulbar ridges and endocardial cushions. Arrows indicate the direction of blood flow through the heart.

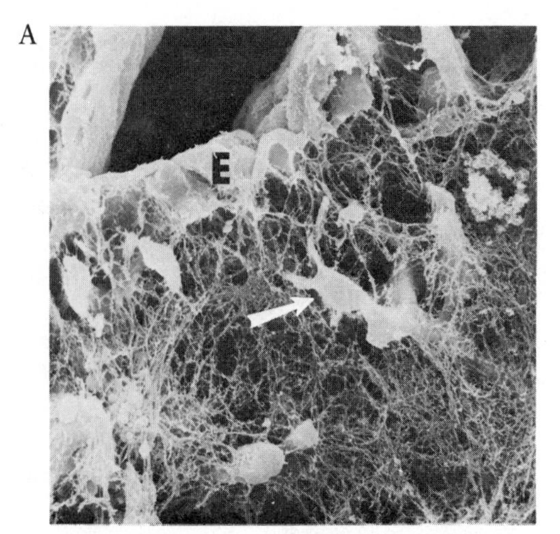


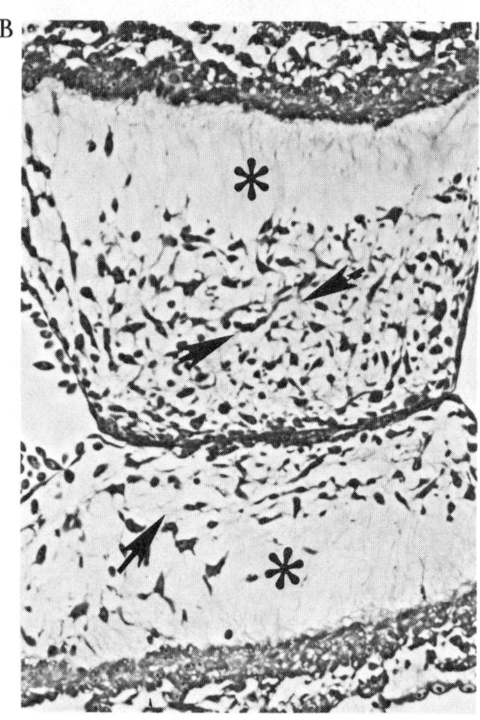


**FIG. 18–18.** A, the beginning of the formation of the endocardial cushions by proliferation of cells (arrow) from the endocardium (E). (From T. Fitzharris and R. Markwald, 1982. Dev. Biol. 92:315.) B, the film of the chick embryo to show that the formation of the heart tube from paired heart-forming areas is accompanied by directed movements of clusters of precardiac cells. The precardiac cell clusters are numbered 1–4 and will give rise to the endocardium. (After R. DeHaan, 1965. In: Organogenesis. R. L. DeHaan and H. Ursprung, eds. Holt, Rinehart and Winston, New York.)

portance in heart tube formation. Because of its cohesive properties, the endoderm adheres tightly to the overlying cardiac primordia. As the endoderm moves medially and obliquely backwards to complete the roof and floor of the gut, the heart-forming regions are dragged passively toward the midline.

The heart-forming areas of the chick embryo are initially rather broad and diffuse, consisting of small clusters of tightly packed cells within an intact mesodermal epithelium. Both electron and light microscope observations show that the translocation of the cells of the cardiac epithelium is collective; that it, the mesodermal epithelium moves as a sheet. The cells of the cardiac epithelium will give rise to the cardiac muscle. By analyzing tracings of photographs made from time-lapse films, DeHaan has shown that the precardiac cell clusters actively migrate toward the anterior intestinal portal during heart tube formation and give rise to the endocardial cells (Fig. 18–19). Initially, the clusters exhibit random movements with no apparent relation to their "goal" of the anterior intestinal portal (Fig. 18–19 A,B). At a later stage, however, when the anterior ends of the cardiac regions begin to move forward and mesiad, the clusters become arranged in a stable configuration (Fig. 18–19 C). Subsequently, each cluster migrates anteromesially and joins the forming heart in the order of its position along the anteroposterior axis (Fig. 18–19 D,E). The sudden change in the migratory behavior of the precardiac clusters from random to oriented movements is probably related to influences emanating from the underlying endoderm. DeHaan has noted that the early embryonic endoderm consists of flattened squamous cells organized as an epithelium. Shortly thereafter, a band of lunate, column-shaped cells differentiates in the endoderm, running in a crescent-shaped arc, directly beneath the cardiac regions. It is presumed that the precardiac clusters use the endoderm as a substratum and are oriented by the altered cell shape in this layer. Since glycoproteins are synthesized by the endoderm and accumulate in the basal lamina during the period of precardiac cell translocation, it is conceivable that these macromolecules provide directional cues to the precardiac cell clusters.

There is ample evidence from tissue culture studies that the heart-forming areas undergo biochemical (functional) differentiation before the visible appearance of the tubular heart. If the heart-forming area is trisected at different stages and the pieces are then cultured separately, each explant differentiates into a number of heartlike, pulsating vesicles. The posteriormost piece differentiates earlier than the other tissue fragments, and its vesicles show very rapid beating or pacemaker activity. The anteriormost explant of the heart-forming area shows the slowest pacemaker activity. The concept

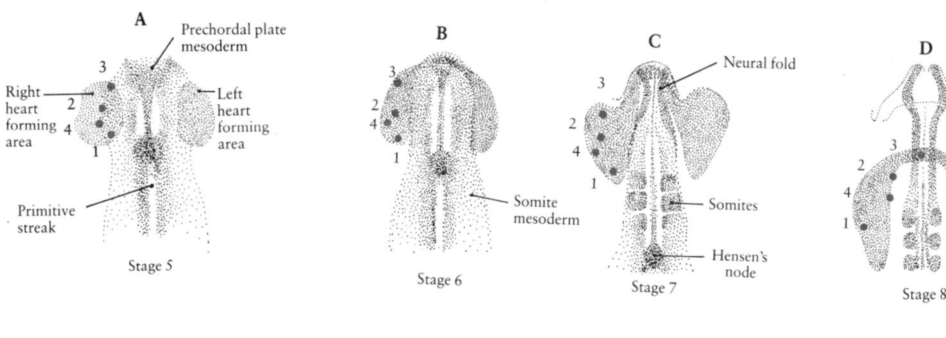


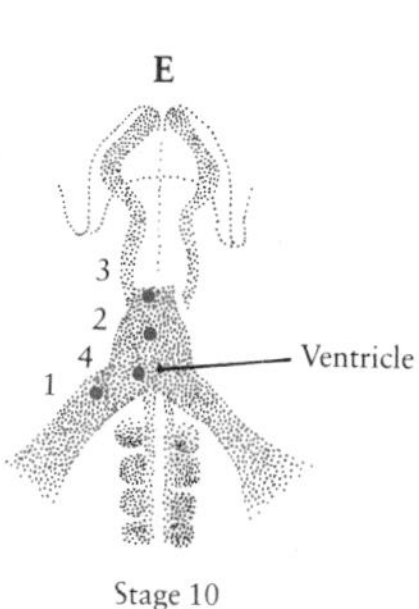


FIG. 18–19. A–E, tracings of photographs made from time-lapse film of the chick embryo to show that the formation of the heart tube from paired heart-forming areas is accompanied by directed movements of clusters of precardiac cells. The precardiac cell clusters are numbered 1–4 and will give rise to the endocardium. (After R. DeHaan, 1965. In: Organogenesis. R. L. DeHaan and H. Ursprung, eds. Holt, Rinehart and Winston, New York.)

emerges that there is an early regionalization of pacemaker activity along an anterior–posterior gradient in the cardiac primordia. Differences in the rate of contractility between the early cardiac cells determines the pattern of beating in the various chambers of the tubular heart.

## Looping and the Establishment of External Form for the Heart

The endocardial tube and its epimyocardial mantle are suspended from the roof of the pericardial cavity by the dorsal mesocardium (Fig. 18–14). By the 16-somite stage (day 24), this curtain of tissue has disappeared leaving most of the heart free and movable within the pericardial cavity. The arterial (cranial) and venous (caudal) ends of the heart tube remain fixed in place by the aortic arches and the major veins in the septum transversum, respectively. This arrangement allows the originally straight heart tube to change shape and position as it grows into an adult organ.

During its development, the tubular heart undergoes characteristic changes in its shape that tend to be shared in common by most vertebrate embryos. Initially, the heart tube simultaneously bends and rotates toward the right side of the body axis; this is known as *D-looping* (Fig. 18–20 A). It is primarily the midportion of the cardiac tube that experiences extensive alteration in position. A consequence of looping is the beginning of the regionalization of the tubular heart into the cardiac chambers. Subsequently, the cardiac tube is thrown into a U-shaped bend (bulboventricular loop; Figs. 18–20 B, 18–21 B, 18–22 A) and then into a compact S-shaped loop that nearly fills the pericardial cavity (Figs. 18–15 D, 18–20 C, 18–22 C). As the heart tube bends, the bulboventricular portion moves ventrad and caudad so that the ventricle, formerly situated cephalic to the common atrium, is brought to its characteristic adult position posterior to the atrium (Fig. 18–20 C,D). Because the common atrium is bounded below by the bulbus cordis and above by the sinus venosus, it enlarges primarily in the lateral and ventrolateral directions. The result is a pair of sacculations that foreshadows the future, definitive right and left atria. The undilated narrowed portion of the cardiac tube between the atrium and the ventricle is the *common atrioventricular canal* (Fig. 18–23 A). The sinus venosus remains anchored to the transverse septum during cardiac loop formation. It is a thin-walled chamber into which empty the major systemic, vitelline, and umbilical veins (Fig. 18–21 B,C). The sinus quickly becomes differentiated into three regions as its right side, owing to important shifts in the pattern of blood flow returning from the liver, undergoes enlargement (Fig. 18–21 C,D). These are the *right* and *left horns* of the *sinus venosus* and a narrow in-

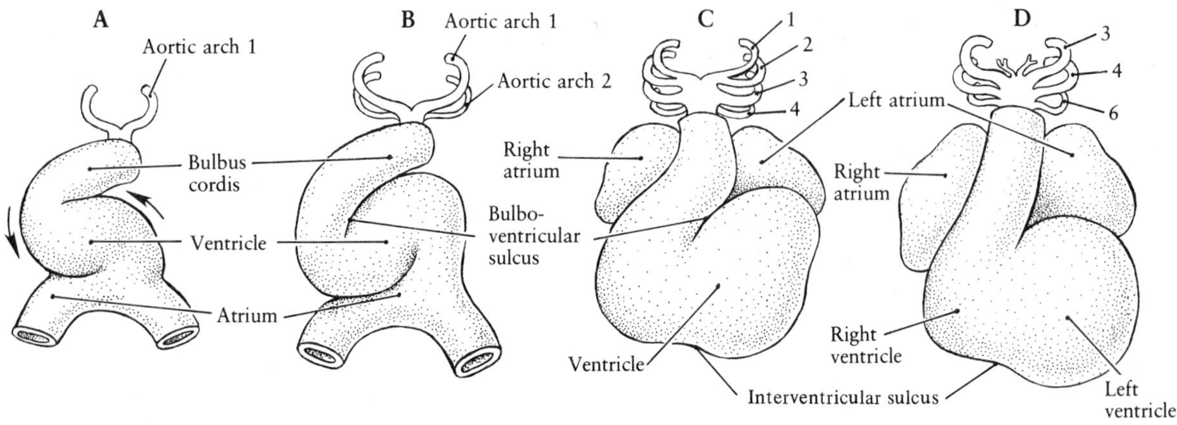


tervening *transverse portion* of the *sinus venosus*. The opening of the sinus venosus into the atrium by the way of the *sinoatrial orifice* is concentrated on the right side of the common atrium (Fig. 18–23 A).

Growth of the cardiac tube appears to stimulate several additional alterations in external form. The proximal part of the bulbus cordis is absorbed into the right side of the common primitive ventricle. Presumably, this is related to a lag in the development of the wall of the bulboventricular loop marked by the bulboventricular sulcus (Fig. 18–20 B,C). At about the same time the ventral surface of the ventricle shows a distinct, median longitudinal groove or *interventricular sulcus* (Figs. 18–20 C,D). The sulcus marks the position of an internal, muscular partition beginning to divide the common ventricle into two chambers (Fig. 18–23 A).

The S-shaped cardiac loop brings about an arrangement of chambers that is observed during cardiogenesis in most vertebrate embryos. In adult cartilaginous and bony fishes, the heart retains this configuration and functions as a single, tubular organ pumping venous blood to the gills for oxygenation. Substitution of lungs for internal gills in air-breathing vertebrates is associated with major internal alterations in heart structure and in the organization of the aortic arch arteries. The introduction of paired lungs into the vertebrate plan of organization requires the heart to act as a double pump, pumping one bloodstream (pulmonary) from the right side of the heart to the lungs and another bloodstream (systemic) from the left side of the heart to the general body circulation by way of the aorta. Conversion of the S-shaped, tubular heart into an elaborately valved, four-chambered, partitioned organ, which effects a complete separation of pulmonary and systemic bloodstreams, is

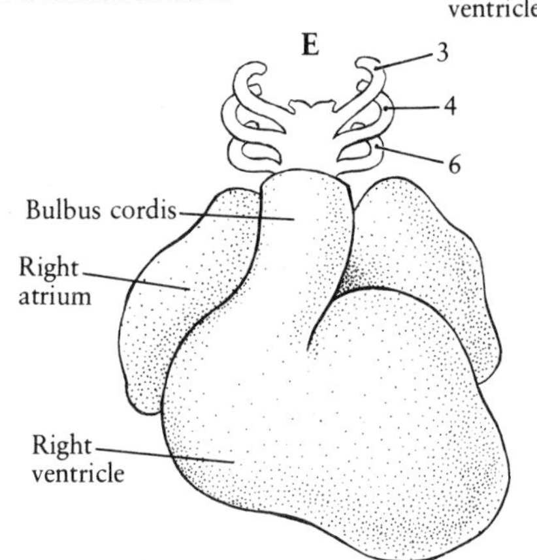


**FIG. 18–20.** Ventral views of reconstructions of the hearts of young human embryos during successive stages in cardiac loop formation. (From T. Kramer, 1942. Am. J. Anat. 71:343.)

achieved only in birds, mammals, and crocodiles. The amphibians and reptiles have advanced beyond the fishes in the sense that there is some internal subdivision of the atrium and ventricle.

Little is known about the morphoregulatory mechanisms controlling cardiac looping and the development of the shape of the vertebrate heart. For a number of years, it was assumed that the velocity and pressure of blood flowing through the cardiac tube were responsible for changes in external shape and in determination of where cardiac septa were located. Unfortunately, the design and execution of meaningful experiments on embryos to test the importance of this hypothesis have proved difficult. Early studies by Bacon using the frog showed that the embryonic heart tube when isolated in vitro was capable of curvature despite the absence of circulating blood. Llorca and Gill claimed that in the chick embryo curvature of the

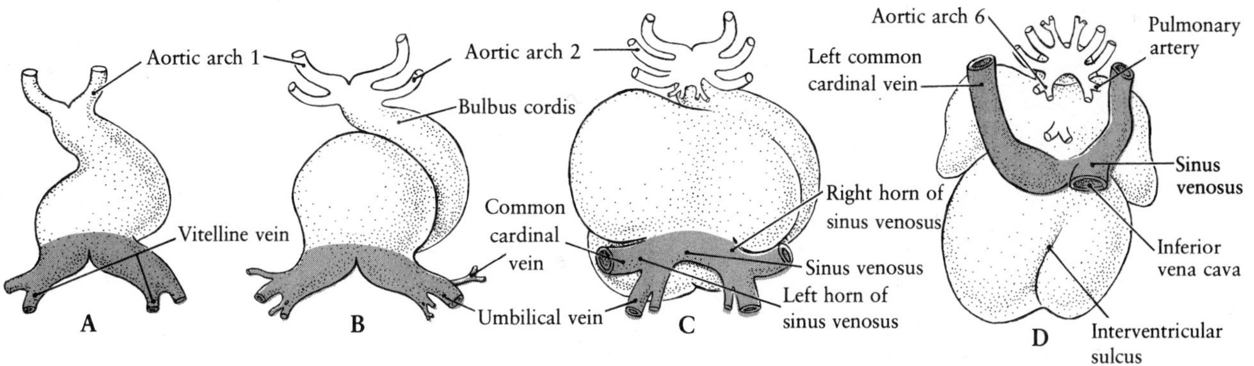


isolated heart could only take place in a coelomic cavity, and even then its morphogenesis was always imperfect when compared with control embryos. Differences in mitotic rate (or growth rate) between two sides of the tubular heart are frequently cited as the basis for looping. Available experimental evidence to date suggests that growth and physical deformation are dominant factors in effecting heart shape. Early cardiac shape changes, including looping, appear to be brought about primarily by physical forces intrinsic to the heart itself. Studies, particularly by Manasek and colleagues, have attempted to correlate cardiac looping with changes in myocardial cell shape and in structural and compositional modifications of the cardiac jelly. For example, the shapes of the myocardial cells on the two sides of the presumptive ventricle of the early cardiac loop are quite different. The cells of the greater curvature change from being cuboidal to a more flattened, squamous shape, and with a consequent increase in their apical surface area; the cells of the lesser curvature do not undergo this shape change. Since the myocardial cells do not change positions relative to one another, the sudden increase in surface area would appear to be accommodated by a change in the shape of the organ. However, it is still uncertain if these changes in cell shape cause the heart to loop or if the cells change shape in response to the bending. Increasingly, it appears that mechanical forces organized by biochemical events in the cardiac jelly, such as synthesis and turnover, underlie the looping process. Injection of selected enzymes, such as pronase and testicular hyaluronidase, into the cardiac jelly matrix of the early chick embryo is followed by marked cardiac shape changes. Nonspecific proteolytic enzymes in particular cause dramatic loss of heart shape, indicating that extracellular proteins

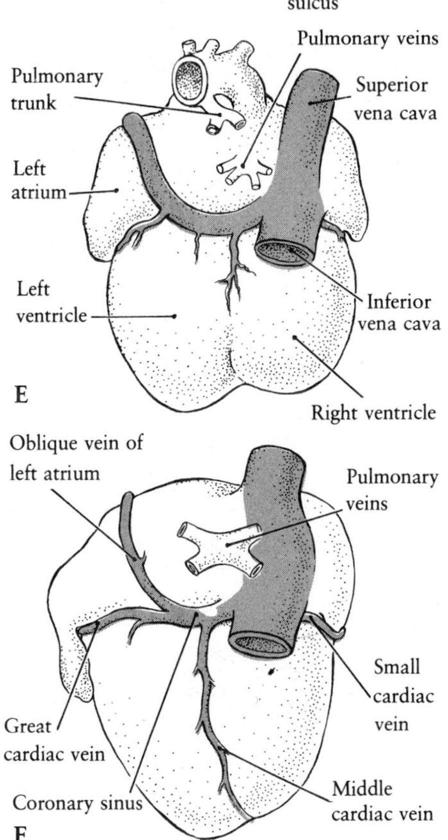


**FIG. 18–21.** A–F, dorsal views of successive stages in the development of the human heart, showing particularly the changing relations of the sinus venosus. (From Foundations of Embryology by B. Patten and B. Carlson. Copyright © 1974 by McGraw-Hill, Inc. Used with permission of the McGraw-Hill Book Company.)

are essential in maintaining the shape of the early cardiac tube. The cardiac jelly is a fibrillar system holding the endocardial and myocardial tissue layers in their relative positions. Selective synthesis or removal of matrix components could lead to the bending forces required for looping.

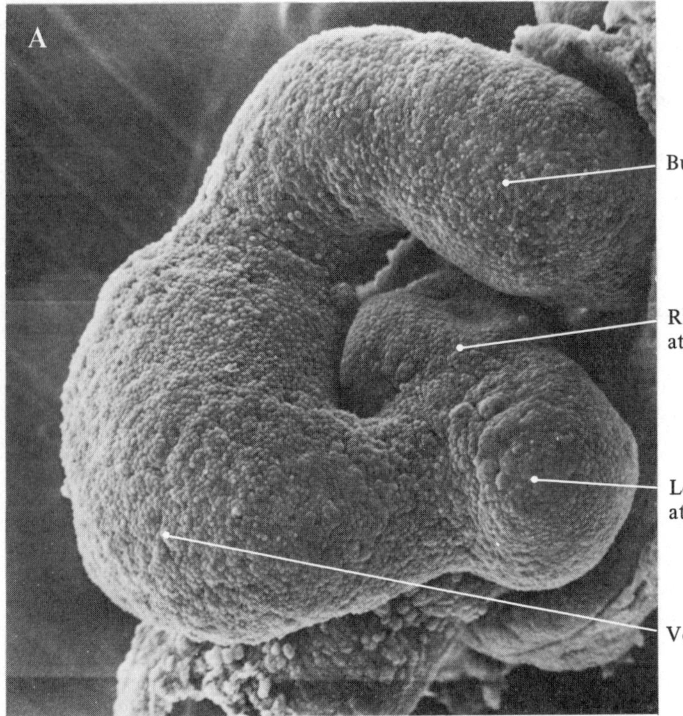


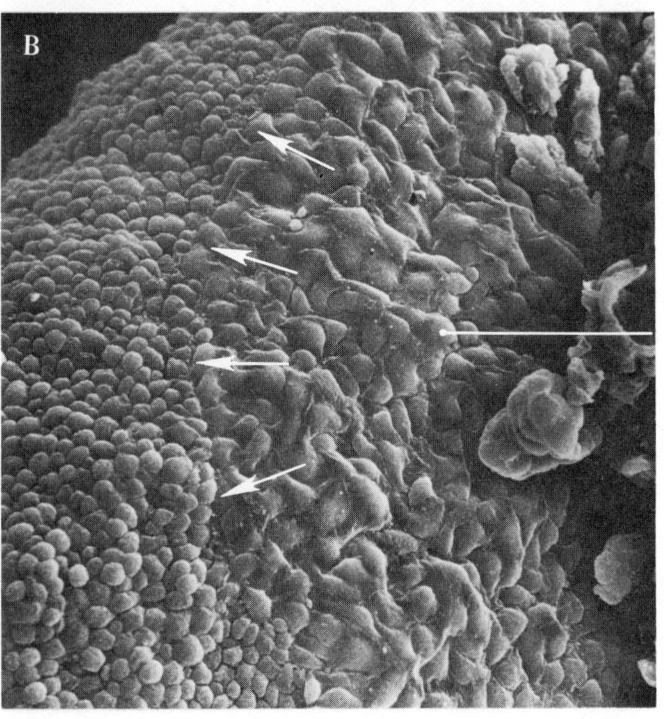


**FIG. 18–22.** A, an SEM ventral view of the chick embryo to show the U-shaped configuration of the early cardiac tube; B, a higher magnification of a portion of the dorsal surface of the U-shaped heart showing the leading edge (arrows) of the epicardium advancing over the myocardium. (From E. Ho and Y. Shimada, 1978. Dev. Biol. 66:579); C (facing page), an SEM view of the anterior aspect of a rat heart at 13 days of gestation to show the outflow tract of the heart. (From Thompson et. al., 1983. Anat. Rec. 206:207.)

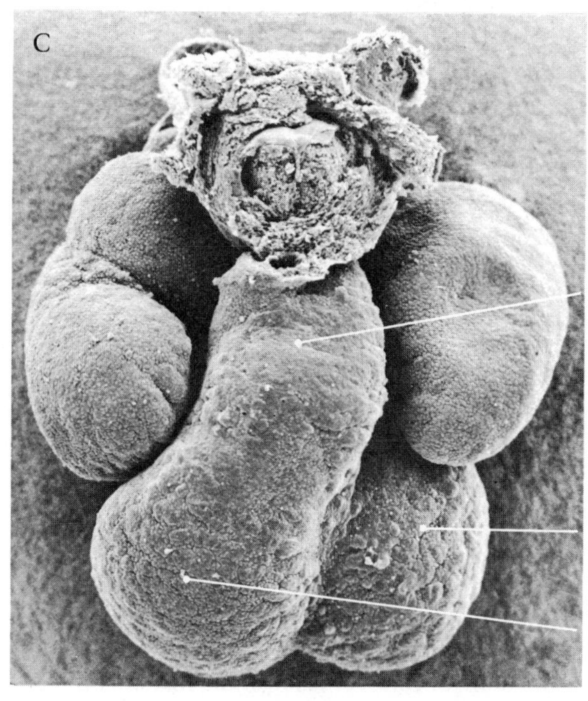


## Partitioning of the Venous End of the Heart

The partitioning or septation of the cardiac tube is certainly one of the most important and dramatic changes observed in the developing heart. Division of the primitive heart into chambers is a cardiovascular event that is most susceptible to maldevelopment. Cardiac septa initially appear in the venous end and subsequently in the arterial end of the heart. They migrate toward the base of the ventri-

**FIG. 18–23.** Semischematic drawings of the interior of the heart to show initial steps in partitioning. A, cardiac septa as they appear in human embryo at five weeks; B, cardiac septa as they appear at six weeks. Note that the interatrial foramen primum is nearly closed and the interatrial foramen secundum is beginning to appear; C, the interior of the heart at a stage when the foramen primum is closed and the septum secundum is established. (From B. Patten, 1960. Am. J. Anat. 107:271.)

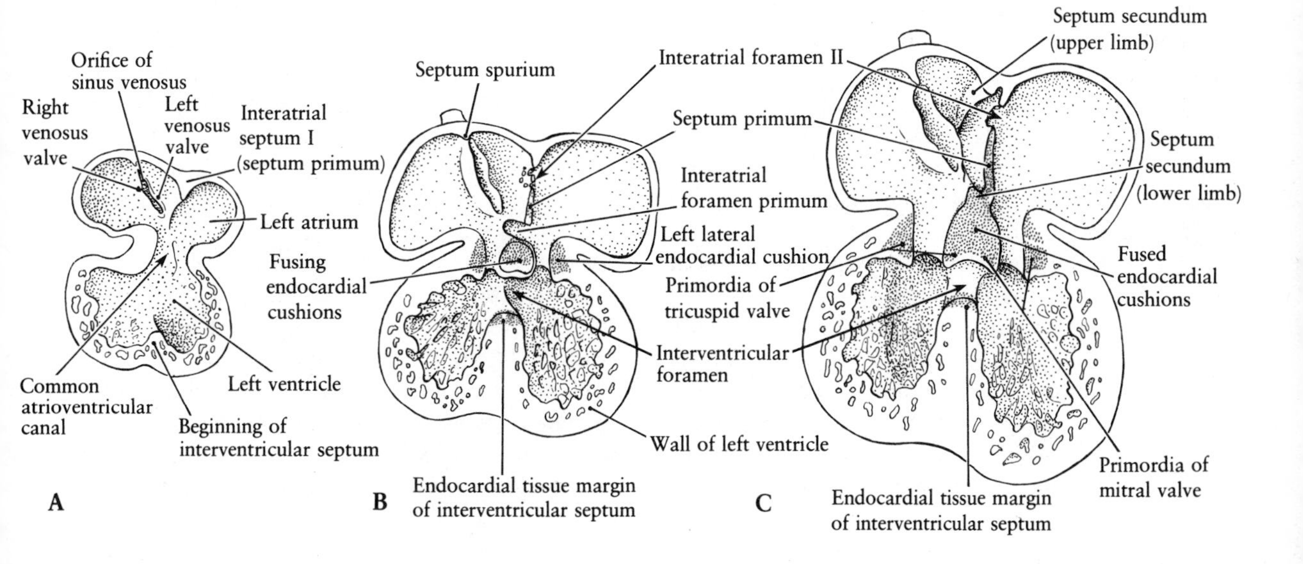

cle, where their fusion assures proper internal division into right and left functional compartments. Precise timing in the formation and movement of septa is therefore critical to normal subdivision. The venous end of the embryonic heart consists of the sinus venosus, the common atrium, and the common atrioventricular canal (Fig. 18–17). Initially, the sinus venosus opens into the center of the primitive atrium, and the right and left horns are about of equal size (Fig. 18–21 C). Progressive enlargement of the right horn results from two left-to-right shunts of blood, which appear during the fourth week of development in human embryos. As this occurs, the sinoatrial opening moves to the right side of the common atrium (Figs. 18–21 D, 18–23 A). Where the orifice opens into the atrium it is bounded by a pair of thickened ridges termed the *right* and *left venosus valves* or sinoatrial valves (Fig. 18–23 A). These ridges typically merge on the cephalodorsal wall of the atrium to form a projection known as the *septum spurium* (Fig. 18–23 B). As its names implies, the septum spurium plays no direct role in septation of the heart.

At approximately the 5-millimeter stage (32 days), a sickle-shaped, sagittal, muscular fold appears in the cranial wall of the atrium (Fig. 18–23 A). The *primary interatrial septum (septum primum; interatrial septum I)* grows downward toward the common atrioventricular canal, thereby separating the common atrium into *right* and *left atria (auricles)*. The space between the free edge of the septum primum and the common atrioventricular canal constitutes the *interatrial foramen primum or primary interatrial foramen* (Figs. 18–23 B, 18–30 A). It is normally obliterated as the septum primum reaches the dorsal and ventral endocardial cushions (Fig. 18–30 B).

By the time of the appearance of the primary interatrial septum, massive swellings rich in glycosaminoglycans, collagen (types I and II), and fibronectin appear in the atrioventricular canal. The development of these cushions (as well as the ridges of the bulbus cordis) has been the subject of considerable study, in part because they offer a unique model system for examining the three-dimensional migration of cells. The myocardium secretes the extracellular macromolecules of the cardiac jelly matrix. The matrix is initially organized as a loose latticework of collagen-like microfibrils coated with a fine filamentous and granular material (Figs. 18–11 B, 18–18 A). Subsequently, the endocardium, noticeably thicker in the atrioventricular canal, seeds mesenchymal cells into the matrix (Fig. 18–18 A). This is followed by a wave-like centrifugal translocation of the cushion cells through the matrix toward the myocardium. Active cell movement appears to require changes in the alignment and composition of the matrix; these are presumed to be coupled to the passage of the first cushion cells through the matrix. The increase in the size of the dorsal and ventral endocardial cushions is due to the interstitial growth of cushion cells during and following active migration (Fig. 18–18 B). These cushions contact and fuse together in human embryos of 38 days to form a sagittal partition known as the septum intermedium (Figs. 18–23 B,C, 18–24 B,C). As a consequence the common atrioventricular canal is split into the narrow *right* and *left atrioventricular canals* (Figs. 18–23 C, 18–24 C,D).

A very critical event occurs in the development of the human heart just prior to the fusion of the primary interatrial septum with the septum intermedium (Fig. 18–23 C). If the primary interatrial septum were to remain intact after fusion with the cushion complex, the right half of the venous end of the heart would be the only one to receive a substantial return of blood. The left half of the venous end of the heart would receive little, because return from the lungs is scanty. To assure that the left atrium, and, indeed, the future systemic portion of the heart, receives adequate blood flow during fetal life, there is considerable tissue remodeling of the primary interatrial septum. A number of small perforations appear in the cranial portion of the primary interatrial septum (Fig. 18–23 B). Scanning electron microscopy has shown that these foramina do not seem to form as a result of ruptures or "blowouts" in the primary interatrial septum. Neither does cell death nor degeneration appear to play a role in forming the *foramen secundum* or *secondary interatrial foramen* (Figs. 18–23 C, 18–30 B). Rather, specialized rounded endocardial cells appear to extend thin cytoplasmic processes into the septal core from both the right and left auricular surfaces, thereby separating cells of the core and isolating a portion of septal tissue. Narrow, shallow indentations are the first indications of the developing foramen secundum. The foramina enlarge and coalesce as numerous cellular processes divide

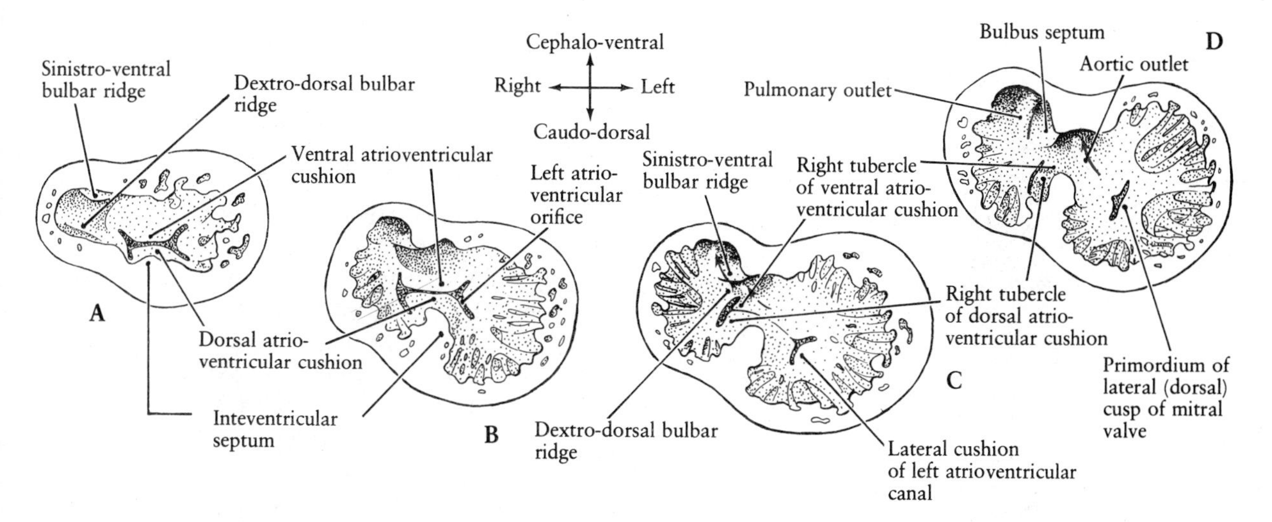


**FIG. 18–24.** A–D, formation of the endocardial cushions and their fusion to form the septum intermedium in human embryos. Transverse views posterior to the atrioventricular canal. A, 8.8 mm; B, 11 mm; C, 13 mm; D, 14.5 mm. (From T. Kramer, 1942. Am. J. Anat. 71:343.)

the septal core into many isolated islands of tissue. These same endocardial cells may secrete hyaluronidase to assist in the remodeling of this part of the heart. The foramen secundum is established at about the time of obliteration of the foramen primum. The timing of the formation of the foramen secundum and the fusion of the primary interatrial septum with the dorsal and ventral endocardial cushions is critical. The acceleration of septa fusion or retardation in the formation of the secondary interatrial opening could severely disturb the distribution of blood within the cardiac tube and the eventual division of the common atrium.

Closure of the primary interatrial foramen is followed by the formation of the second interatrial septum *(secondary interatrial septum* or *septum secundum)* along the right atrial wall between the attachments of the left venosus valve and the primary interatrial septum (Fig. 18–23 C). This fold from the ventrocranial wall is also crescent-shaped, but its free edge tends to be directed toward the base of the sinoatrial orifice. The difference in the direction of growth of the septum secundum is compared with the septum primum in Figure 18–25. As the septum secundum migrates toward the septum intermedium, its free edge extends beyond the foramen secundum of the primary interatrial septum. A permanent opening, termed the *foramen ovale,* remains in the secondary interatrial septum (Fig. 18–32). Although the foramen ovale becomes smaller during the course of development, it will remain open until after birth (Fig. 18–26).

The structural relationships between the primary interatrial septum and the foramen ovale are important to the one-way passage of blood from the right atrium to the left atrium. Carefully examine Figures 18–26 and 18–32, and note that the free edge of the primary interatrial septum is situated opposite to the gap formed by the foramen ovale. The growth and orientation of the free margin of the secondary interatrial septum are such that the septum partially overrides the opening of the *inferior vena cava.* Functionally, the blood from this vessel becomes split into two streams, one passing directly through the foramen ovale into the left atrium and other into the right atrial chamber. Because of this function, the margin of the septum se-

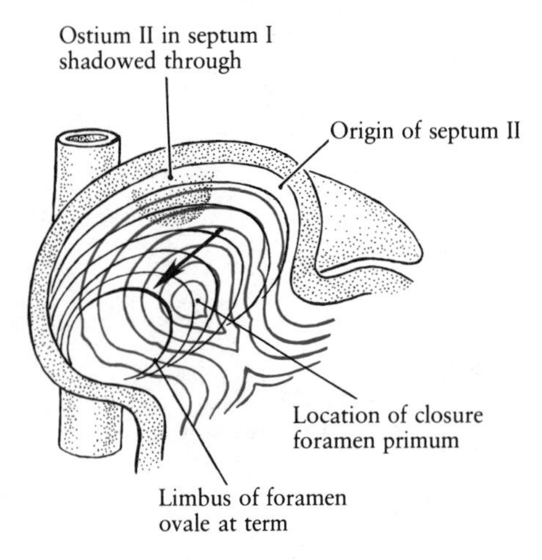


**FIG. 18-25.** Schematic drawing showing by a series of contour lines the growth of the septum secundum. The direction of growth of the septum primum is indicated by curved solid black lines. (From B. Patten, 1960. Am. J. Anat. 107:271.)

cundum if often called the *crista dividens*. As blood moves through the foramen ovale, the free, thin edge of the primary interatrial septum is forced laterally. However, upon atrial contraction, this flap is forced over the foramen ovale and thus prevents the regurgitation of blood back into the right atrium. The lower part of the primary interatrial septum is often referred to as the *valve of the foramen ovale*. After birth the cranial edge of the valve of the foramen ovale is pressed against the secondary interatrial septum owing in great measure to the progressive increase in the return of blood from the paired lungs. It fuses with the cephalic portion of the secondary interatrial septum to form the definitive interatrial septum. The foramen ovale is thereby obliterated.

During the partitioning of the common atrium and the common atrioventricular canal, much of the right horn of the sinus venosus is absorbed into the wall of the right atrium because it fails to keep pace with the rapid growth of the rest of the cardiac tube (Fig. 18–27 A,B). The smooth part of the right atrium or *sinus venarum* is derived from this part of the sinus venosus and is the site into which the major systemic veins empty (Fig. 18–27 B). The left horn dwindles and generally persists only as the stem of the *oblique vein* of the left atrium (Fig. 18–27 B). The transverse portion of the sinus remains as the *coronary sinus*, receiving the oblique vein and other cardiac veins. The loss of the functional importance of the sinus venosus causes the major systemic veins, the *superior vena cava* and the *inferior vena cava*, to open independently into the right atrium.

Important changes also occur in the right and left venosus valves guarding the sinoatrial orifice. The left valve gradually approaches and fuses with the cephalic extremity of the secondary interatrial septum. After becoming a prominent ridge in the right atrium, the right venosus valve becomes reduced

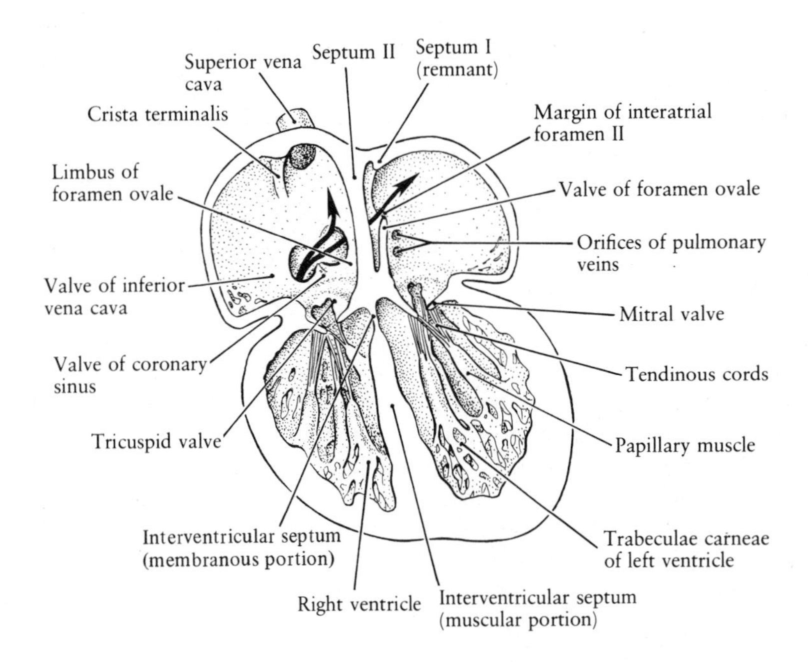


**FIG. 18-26.** Semischematic drawing to show the relationships between the two interatrial septa during the latter part of fetal life. Note particularly the arrangement of the lower part of the septum primum. (From Foundations of Embryology by B. Patten and B. Carlson. Copyright © 1974 by McGraw-Hill, Inc. Used with permission of the McGraw-Hill Book Company.)

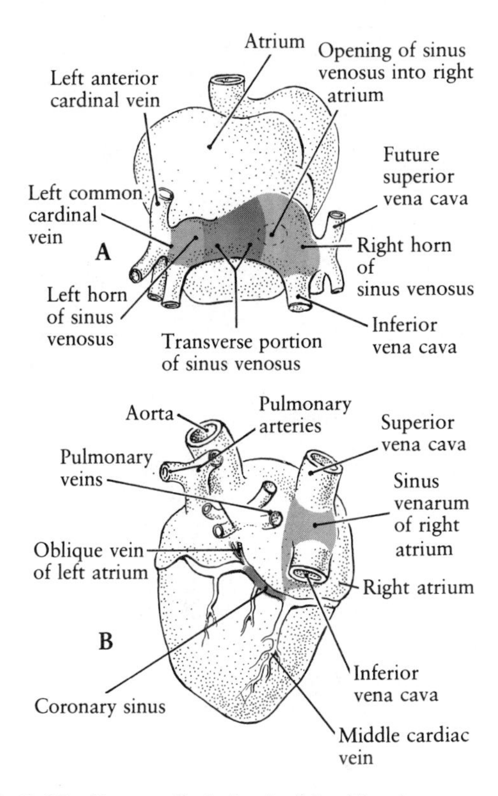


**FIG. 18–27.** Diagrams illustrating the fate of the sinus venosus in the human embryo. A, dorsal view of the sinus venosus at about one month; B, dorsal view of the heart after transformation of the sinus venosus. The extraembryonic veins are not shown.

and divided into two major components: (1) a cranial part known as the *crista terminalis,* a vertical ridge that separates the sinus venarum from the primitive atrium (Fig. 18–26), and (2) a caudal part that is divided into the *valve of the inferior vena cava,* located to the right of the ostium of the inferior vena cava, and a smaller, *caudal valve of the coronary sinus* (Fig. 18–26).

Thus the venous limb of the primitive cardiac tube is greatly modified to produce the right and left atria, chambers separated in the adult heart by a continuous muscular interatrial septum fashioned from the embryonic septum primum and septum secundum. The right atrium is formed from the right half of the common atrium, the right half of the right atrioventricular canal, and a portion of the sinus venosus. The left atrium is formed from the left half of the common atrium, the left half of the common atrioventicular canal, and portions of the stems of the pulmonary veins. Most of the inner surface of both atria has a rough, trabeculated appearance due to the presence of thin, muscular projections known as the *pectinate muscles.*

## The Atrioventricular Valves

The openings of the right and left atria into their corresponding ventricles are guarded by the *atrioventricular valves (tricuspid valve* on the right and *mitral valve* on the left). The flaps comprising these valves arise from localized proliferations of endocardial tissue on the lateral walls of the atrioventricular canal *(lateral endocardial cushions)* and from the dorsal and ventral endocardial cushions (Fig. 18–23 B,C). These soft cushions become hollowed out on their ventricular sides by the action of blood during ventricular contraction. Each flap differentiates as a mass of fibrous tissue and becomes connected to special muscles of the ventricle *(papillary muscles)* by cords of connective tissue *(chordae tendineae)* (Fig. 18–26). Three flaps or valvular cusps are formed around the right atrioventricular canal and two around the left.

## Partitioning of the Arterial Limb of the Heart

The arterial limb of the embryonic heart consists of the ventricle and an outflow tract variously referred to as bulbus cordis, conus arteriosus, or truncus arteriosus. To avoid confusion in nomenclature, we will retain the term bulbus cordis to designate this part of the heart and use truncus arteriosus to refer to the vessel giving rise to the aortic arches. In this context, truncus arteriosus is synonymous with ventral aorta and *aortic sac.* Partitioning of the arterial limb into pulmonary and systemic streams occurs concurrently with the internal separation of the venous limb.

As previously mentioned, the proximal portion of the bulbus cordis is incorporated into the wall of the right ventricle because of the laggard growth of this part of the heart. This part of the embryonic heart is represented in the adult organ by the *infundibulum* of the right ventricle. According to the classical descriptions of vertebrate cardiac morphogenesis, the remainder of the bulbus cordis is divided into two channels, the *aorta* and the *pulmonary trunk,* by a *spiral aorticopulmonary* or *bulbar septum.* The rudiments of this septum are visible throughout the bulbus in human embryos of 5 millimeters in length. They appear as two prominent, opposed thickenings of subendocardial tissue

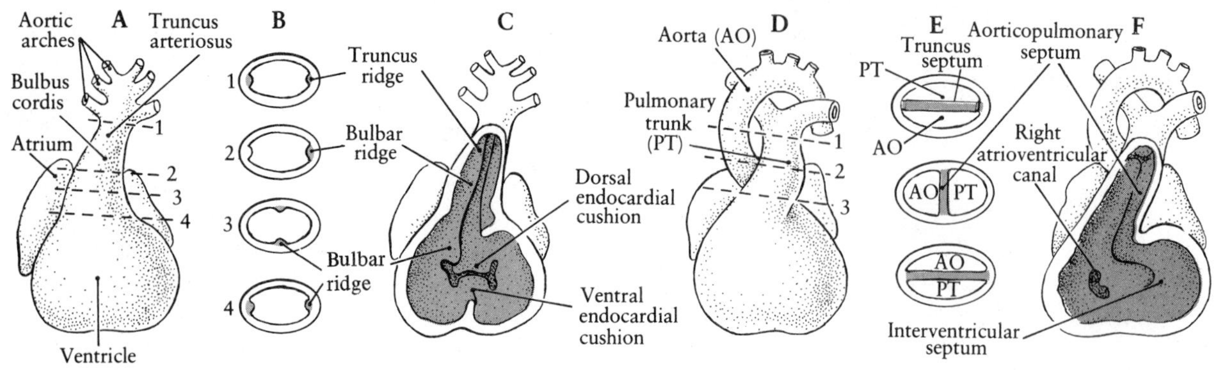


termed the bulbar ridges (Figs. 18–28, 18–31 A). The bulbar ridges, similar to the cushion pads of the atrioventicular canal, are complex structures composed of two epithelia, the original endocardium and the peripheral myocardium, separated by mesenchyme cells embedded in an abundant extracellular matrix. Similar ridges of tissue are found in the truncus arteriosus and are directly continuous with those of the bulbus cordis. Figure 18–28 B illustrates that the bulbar ridges show a complex spiral arrangement from the distal end to the proximal end of the bulbus cordis. The rotation of the ridges is presumably a reflection of the pattern of bloodstream flow through this part of the heart (Fig. 18–28 G). Proximally, the *right bulbar ridge* projects into the lumen of the ventricle just above the right atrioventricular canal. The *left bulbar ridge* lies opposite to the right one and adjacent to the forming septum of the ventricle (Figs. 18–30 A,B, 18–31 A).

The bulbar ridges by the eighth week have enlarged and fused distally to form the aorticopulmonary septum. It divides the lumen of the bulbus into a pulmonary trunk (dorsal) and a systemic trunk or aorta (ventral) (Figs. 18–28 F, 18–31 B,C). Partitioning of the bulbus cordis continues a process initiated in the truncus arteriosus. Here, a *truncus septum* forms through fusion of pads of endocardial tissue and bifurcates the single lumen between aortic arches IV and VI (Fig. 18–28 A–F). Fusion of the bulbar ridges along the free edge of the aorticopulmonary septum then continues toward the ventricle with the septum following the same spiral course as the rudiments from which it is formed. The result is that the aorta is continuous anteriorly with the third and fourth pair of aortic arches and the pulmonary trunk with the sixth pair of aortic arches (Fig. 18–28 H).

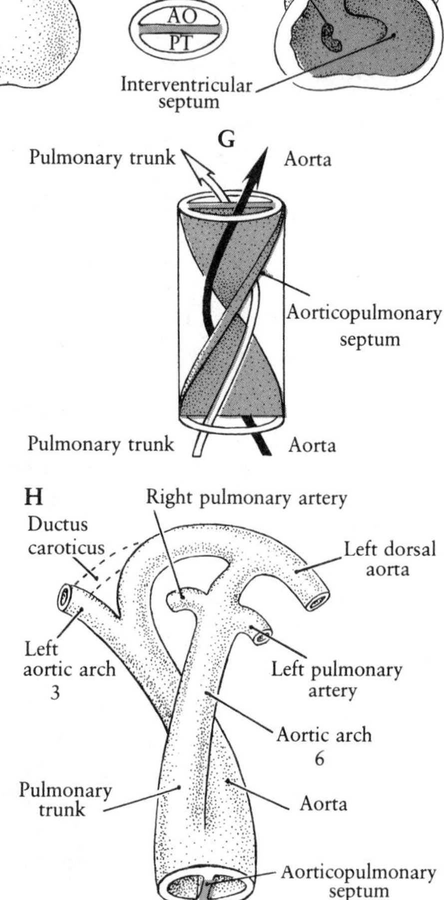


**FIG. 18–28.** A series of diagrammatic drawings illustrating the division of the bulbus cordis and the truncus arteriosus. A, ventral view of the human heart at five weeks; B, transverse sections through the levels of the heart as indicated in A to show the truncal and bulbar ridges; C, ventral view of the heart to show the endocardial cushions; D, ventral view of the heart after the initiation of the division of the arterial limb; E, transverse sections through the bulbus (D) to show relationships between the aorticopulmonary septum, the aorta, and the pulmonary trunk; F, ventral view of the heart demonstrating the aorticopulmonary septum; G, diagram illustrating the spiral of the aorticopulmonary septum; H, diagram showing the two major arteries twisting around each other as they leave the heart.

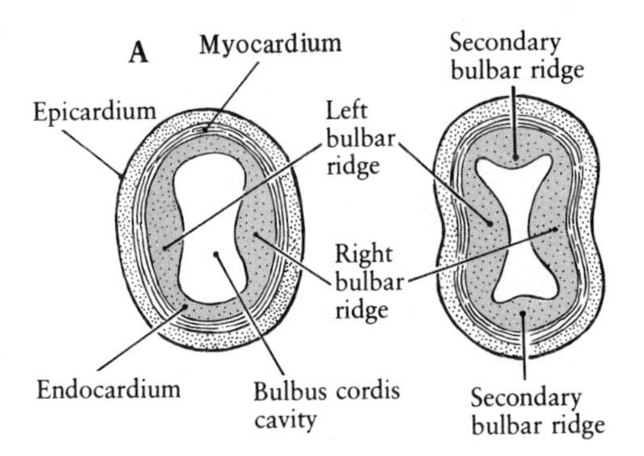


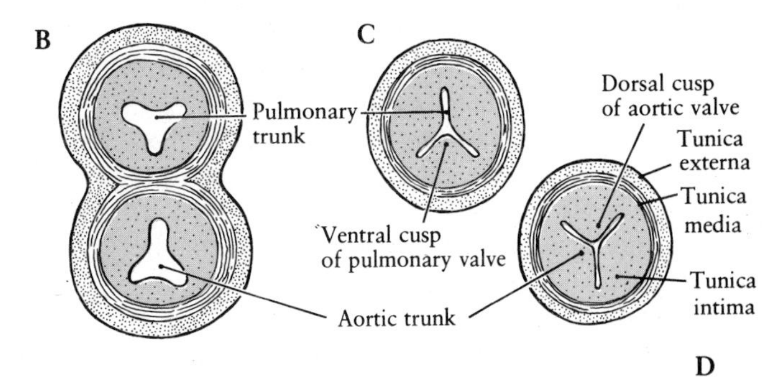


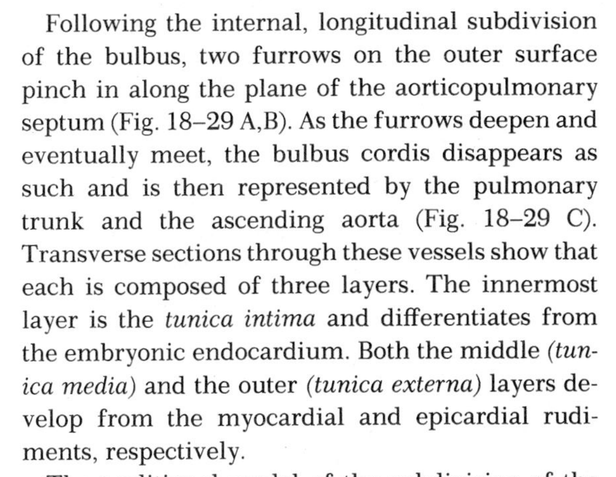


**FIG. 18–29.** Drawings of sections to show the localization and modification of the primordia of the semilunar valves at the bases of the pulmonary trunk and the aorta. A, transverse sections through the bulbus cordis; B, transverse section through the base of the bulbus cordis after formation of the aorticopulmonary septum; C, the isolated aorta and pulmonary trunk at the time of valve formation; D, longitudinal sections through the base of the major vessel to show hollowing out (arrows) of the bulbar ridges to form valve cusps. (From T. Kramer, 1942. Am. J. Anat. 71:343.)

Following the internal, longitudinal subdivision of the bulbus, two furrows on the outer surface pinch in along the plane of the aorticopulmonary septum (Fig. 18–29 A,B). As the furrows deepen and eventually meet, the bulbus cordis disappears as such and is then represented by the pulmonary trunk and the ascending aorta (Fig. 18–29 C). Transverse sections through these vessels show that each is composed of three layers. The innermost layer is the *tunica intima* and differentiates from the embryonic endocardium. Both the middle *(tunica media)* and the outer *(tunica externa)* layers develop from the myocardial and epicardial rudiments, respectively.

The traditional model of the subdivision of the outflow tract is based largely on reconstructions from two-dimensional histological sections. Attempts to reconcile this model with other observations have led to considerable confusion concerning the partitioning of this part of the heart. For example, the base of the pulmonary trunk and the aorta contains substantial smooth muscle. It always has been assumed that the bulbar myocardium dedifferentiates into dense mesenchymal tissue, which then serves as the source for the smooth muscle of the two major blood vessels. Recent studies with rat and chick embryos have employed computer graphic methods to reconstruct in three dimensions the positional and kinetic relationships of structures thought to be involved in the septation of the outflow tract. Results of these investigations are more consistent with the view that the entire aorticopulmonary septum is somehow translocated toward the ventricles by retraction of the myocardial layer of the bulbus, rather than by a zipperlike longitudinal fusion of the two bulbar ridges. The mechanics of the forces that effect this apparent movement against the vigorous flow of blood have not been specified. The smooth muscle tissue of the pulmonary trunk and aorta appears to arise from the bulbar ridges just "downstream" of the advancing aorticopulmonary septum as the bulbar myocardium recedes toward the ventricles.

Division of the primitive ventricle into right and left chambers is first indicated by a ridge of loosely woven muscular fibers or *trabeculae carneae*. Gradually, these muscle fibers become consolidated into a crescent-shaped muscular fold, the *interventricular septum* which projects inward from the apex of the ventricle (Fig. 18–23 A,B). Most of the initial increase in length of the muscular interventricular septum results from the dilation of the ventricles on either side of it, a process that produces a furrow on the external surface known as the interventricular sulcus (Figs. 18–20 C,D, 18–23 A,B). The dorsal limb of the interventricular septum extends toward the atrioventricular canal where it fuses with the *right tubercle* of the *dorsal endocardial cushion* (just to the left of the right atrioventricular canal) (Figs. 18–24 C,D, 18–30 A,B). The ventral limb of the partition merges with the *right tubercle* of the *ventral endocardial cushion* (just to the right of the left atrioventricular canal) (Figs. 18–24 C,D, 18–30 A,B). The space bounded by the free edge of the interventricular septum and the fused endocardial cushions of the atrioventricular canal is the *interventricular foramen* (Fig. 18–30 A,B).

In contrast to the atrium, where closure between the right and left sides is delayed until birth, the interventricular foramen is rapidly obliterated so that by eight weeks there is little evidence of communication between the right and left ventricles. Closure of the space along the anterior margin of the interventricular septum is a complex process. It apparently results from the rapid growth of the ventricular cavities and the proliferation and active migration of pliable endocardial tissues from several sources, including the margin of the interventricular septum itself.

The details of the final steps in the partitioning of the ventricle are illustrated in Figure 18–31. Cells proliferated from the right side of the fused dorsal and ventral endocardial cushions move ventrally and caudally along the free margin of the interventricular septum, thus reducing the craniocaudal extent of the interventricular foramen. Simultaneously, the right bulbar ridge enlarges, fuses with the right tubercle of the ventral endocardial cu-

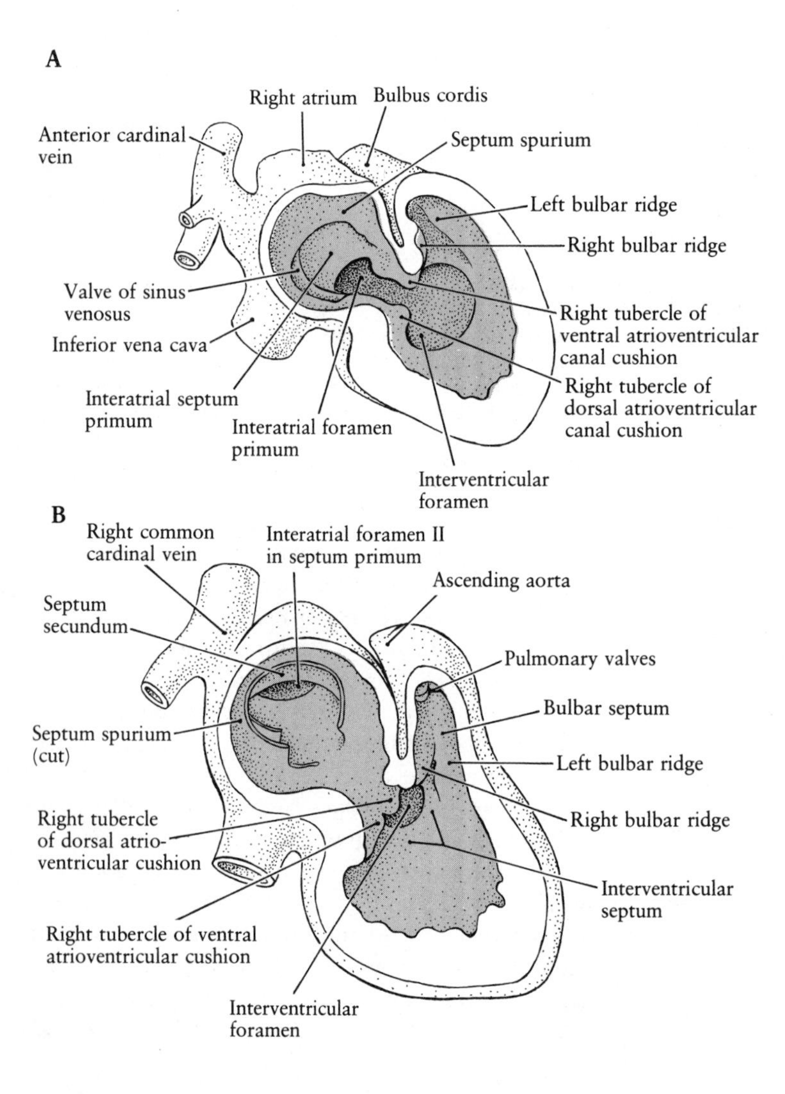


**FIG. 18–30.** Semischematic drawings in sagittal view to show the relationships between the various cardiac septa in the human embryonic heart. A, 8 to 10 mm; B, 12 to 14 mm. (From T. Kramer, 1942. Am. J. Anat. 71:343.)

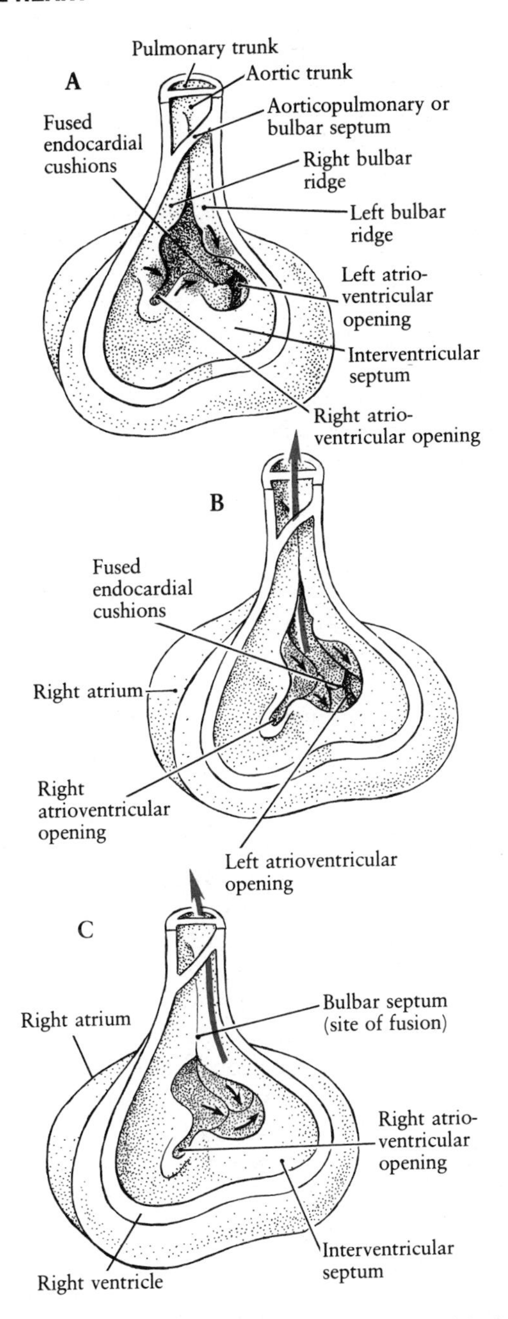


**FIG. 18–31.** A series of drawings illustrating closure of the interventricular foramen and formation of the membranous part of the interventricular septum. A, division of the bulbus cordis and fusion of the atrioventricular cushions; B, reduction in the size of the interventricular foramen by the proliferation of endocardial tissues; C, completion of the bulbar and interventricular septa and closure of the interventricular foramen. (After W. Hamilton and H. Mossman, 1972. Human Embryology. Macmillan, London.)

shion, and then migrates ventrocaudally and to the left.

This effectively occludes the communication between the right atrium and the common ventricle through the ventral part of the atrioventricular canal. Similarly, the left bulbar ridge enlarges, fuses with the right tubercle of the dorsal cushion, and then migrates dorsocaudally and to the right. This effectively occludes the communication between the left atrium and the common ventricle through the dorsal part of the atrioventricular canal. The two bulbar ridges then fuse with each other and with the endocardial tissue along the margin of the interventricular septum to close the interventricular foramen. These activities join the right atrioventricular canal with the right ventricle and the pulmonary trunk, and the left atrioventricular canal with the left ventricle and the ascending aorta.

The definitive interventricular septum is thus composed of two sections. The muscular part originates from the coalescence of numerous muscular trabeculae. The base of the interventricular partition is initially formed by the soft endocardial tissue of the bulbar ridges and the endocardial cushions. This endocardial tissue gradually becomes converted into fibrous connective tissue to form the membranous part of the interventricular septum (the *pars membranacea septi*).

As with the bulbus cordis, much of our knowledge about the septation of the ventricle is based on manual and artistic reconstructions rendered from extensive tissue sections. Techniques using internal relief castings, scanning electron microscopy, and computer-assisted imaging should clarify many of the complexities of septation of this part of the heart.

## Semilunar Valves of the Aorta and Pulmonary Trunk

Valves develop in the arterial limb of the heart between the ventricles and the bases of their associated arterial vessels. These arise as specializations of bulbar endocardial tissue. In addition to the main bulbar ridges, two smaller accessory ridges appear beneath the endocardium to extend throughout most of the bulbus (Fig. 18–29 A). After the physical separation of the bulbus along the aorticopulmonary septum, the base of each vessel contains one of the accessory ridges and half of each of the larger ridges (Fig. 18–29 B,C). Thus, three endocardial swellings guard the orifice of both the aorta and pulmonary artery. They soon become hollowed out on their distal sides to form the three cusps of the *semilunar valves* (Fig. 18–29 D).

## The Cardiac Wall and the Conducting System

The primitive cardiac tube is initially composed of two distinct layers, an inner endocardium and an outer myocardium, separated by a large amount of extracellular matrix. As the heart becomes increasingly more convoluted, the cardiac jelly becomes thinner and thereby permits a closer approximation between the endocardium and the myocardium. Indeed, the cells of the endocardium even appear eventually to penetrate into the innermost cellular layer of the developing myocardium.

In most vertebrate embryos, cardiac function begins shortly after gastrulation or the primitive streak stage. Regular periodic contractions of the mouse heart, for example, are observed by day 9 postcoitum. The initial differentiation of the myocardium into functional cardiac myocytes, therefore, occurs rather early in embryonic life. Contractility has been shown to be associated with the acquisition of organized myofilaments and myofibrils. Hence, the biochemical and morphological changes that transform mesenchymal cells into functionally active heart muscle must take place very rapidly. Mesenchymal cells do not spontaneously transform into myocytes; they do so only when in contact with previously differentiated cardiac muscle cells, suggesting that young muscle cells might induce the process of differentiation. It is not clear whether this process is mediated by intercellular contact or by the transmission of a biochemical signal(s). All other noncardiac muscle cell types, such as those contributing to the connective tissue framework of the cardiac skeleton (fibroblasts), the epithelial and connective tissues of the epicardium, and the vascular smooth muscle tissue, are added during the course of heart development after the cytodifferentiation of cardiac muscle. Cell heterogeneity, therefore, in the outer embryonic heart layer increases during ontogeny as nonmuscle cell types assist in the architectural design of the heart. The cardiac muscle is initially a single, continuous layer throughout the heart tube. Gradually, however, it differentiates into a thin, superficial layer of dense muscle and a thick, spongy layer of loosely arranged muscular trabeculae. Spaces between the muscle trabeculae are lined by endocardium and constitute a significant part of the ventricular cavities.

The conducting system of the adult vertebrate heart controls the rate of cardiac rhythm by determining the time at which the muscle cells of the different chambers are to contract. It consists of concentrations of specialized muscle cells and fibers beneath the endocardium, and is best developed in the mammalian heart. Embryonically, the initial pacemaker of the cardiac muscle is located in the caudal part of the left cardiac tube. Following complete formation of the tubular heart, the right horn of the sinus venosus, beating at the highest rate of any of the cardiac chambers, acts as the pacemaker for the rest of the organ. In more primitive vertebrates, such as fish and amphibians, the pacemaker property remains in the sinus venosus, which also functions as a contractile chamber. However, in mammalian embryos when the right horn of the sinus venosus is incorporated into the right atrium, the pacemaker function is also transferred to this part of the heart in the form of the *sinoatrial node*. In the mouse the primordial sinoatrial node is first visible as an aggregate of compact cells in the medial wall of the *right common cardinal vein* at day 11. Histologically, these cells are indistinguishable from the early sinus musculature and are therefore presumably derived from it. Within two days, the sinoatrial node reaches its full fetal size. The origin and formation of the remainder of the conducting system, particularly the *atrioventricular node* and the *bundle of His* (Fig. 18–32 E,F) have not been satisfactorily resolved. The best light- and electron-microscopic data available suggest that the atrioventricular node and the bundle of His arise along the inner dorsal wall of the atrioventricular canal; they appear simultaneously and in continuity from the beginning. The primordium of the atrioventricular portion of the conducting system appears between days 9 and 10 postcoitum in the mouse embryo. Transformation of the primordial atrioventricular node into its compact form occurs during the second half of gestation. The early morphological cytodifferentiation of the conducting fibers is similar to that of the cardiac myocytes.

Nerve fibers from both vagus nerves and sympathetic chains subsequently innervate the sinoatrial

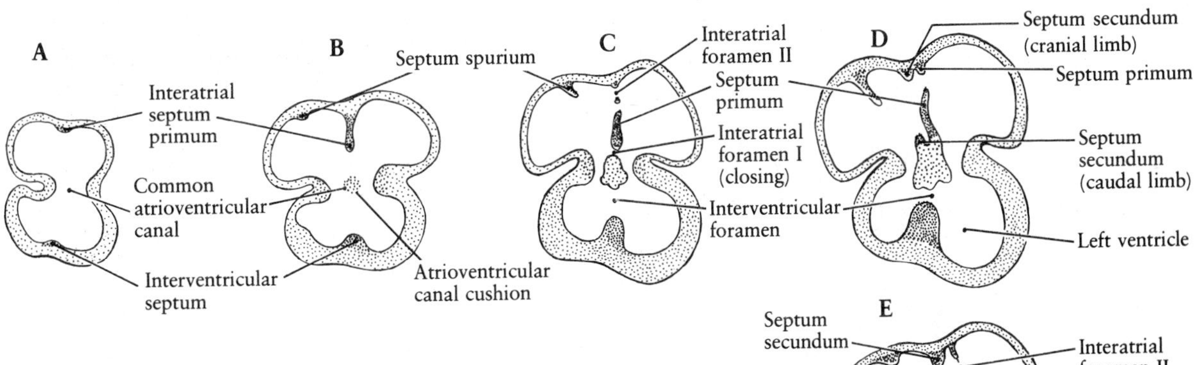


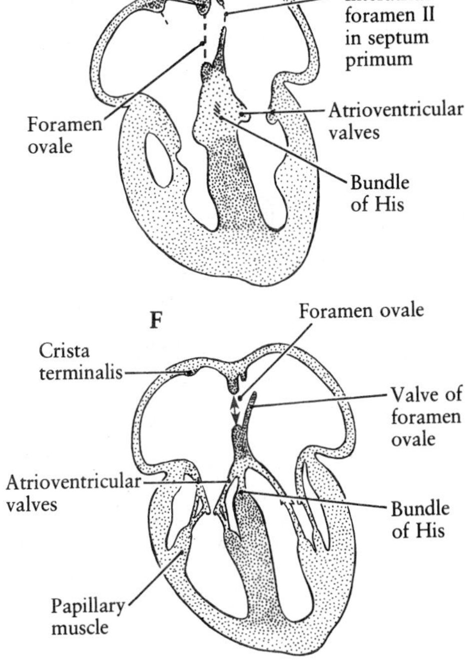


**FIG. 18–32.** A–F, a series of drawings of human hearts to show the rate of progress of the cardiac septa. Frontal views. (From Foundations of Embryology by B. Patten and B. Carlson. Copyright © 1974 by McGraw-Hill, Inc. Used with permission of McGraw-Hill Book Company.)

node, the atrioventricular node, and the bundle of His. Generally, all histological features of these various components of the conducting system are delineated by birth.

## The Arteries

The primary arteries of the early vertebrate embryo are the *right* and the *left primitive aortae.* Each of these vessels can be readily divided into three regions in early somite embryos (Fig. 18–5): (1) a short ventral segment continuous caudally with the endocardial tube of the heart, (2) a primitive branchial or aortic arch artery, and (3) a long dorsal segment *(dorsal aortae),* which distributes blood from the heart to various organs chiefly by way of the intersegmental arteries, the vitelline arteries, and the umbilical arteries.

Following the formation of the primitive tubular heart, the short ventral sections of the primitive aortae fuse to form the midventral truncus arteriosus. The truncus at its cranial end later becomes substantially dilated into the aortic sac (Fig. 18–33 A). From the cephalic portion of the aortic sac, the first pair of aortic arches bends around the rostral part of the pharynx to join the paired dorsal aortae. Successive pairs of aortic arches will differentiate in the mesenchyme of the visceral arches as the embryo continues to develop (Fig. 18–33 A,B).

The complicated adult plan of arteries is derived from this arrangement of truncus arteriosus, paired aortic arches, and paired dorsal aortae. Vertebrates such as teleost and elasmobranch fishes, in which respiration is at the level of the pharynx (i.e., gills), retain this simple arrangement of blood vessels. Each aortic arch artery, interrupted by the capillary network of the gill, becomes divided into *afferent* and *efferent branchial arteries.* Behind the pharynx, the dorsal aortae fuse into a single, median vessel, the *dorsal aorta.* The introduction of pulmonary respiration and the consequent partitioning of the heart results in major changes in the symmetrical pattern of embryonic arteries, particularly in higher vertebrates.

### The Pharyngeal (Aortic) Arch Arteries

Six pairs of aortic arches, connecting the truncus arteriosus (or ventral aorta) with the dorsal aortae, develop in most vertebrate embryos. However, the entire set of aortic arch arteries are not all present

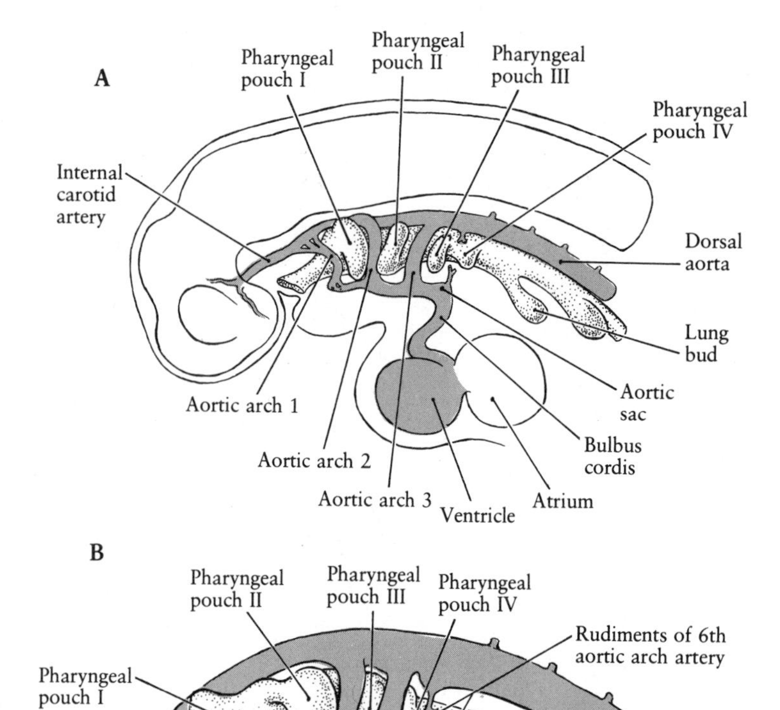


**FIG. 18–33.** Drawings from reconstructed models of human embryos to show the development and transformation of the aortic arches. A, 4 mm, left-side view; B, 5–11 mm, left-side view. (After W. Hamilton and H. Mossman, 1972. Human Embryology. Macmillan, London.)

at the same time. Generally, the first two pairs of arches involute and disappear rather quickly.

In human embryos, the primitive aortic arch pattern is laid down within the first four weeks of development. It is transformed into the basic adult arrangement of arteries during the period between six to eight weeks. The first and second pairs of aortic arches appear between the third and fourth weeks, but subsequently undergo regression and involution. Their atrophy is probably associated with the relative caudal migration of the heart and the aortic sac, making them less directly accessible to blood flow than the third and fourth pairs of aortic arches immediately behind them (Fig. 18–33 B). There is some evidence that parts of the first pair of arteries contribute to the *maxillary arteries* and the second pair to the *stapedial arteries*. The ventral portions of the first two aortic arches may also contribute to the *external carotid arteries* (Fig. 18–33 B).

The early degeneration of the first two pairs of aortic arches, as well as that of the fifth pair of arches (these appear in about 50 percent of embryos), leaves only the third, fourth, and sixth pairs along with their dorsal and ventral roots to play important roles in the construction of adult vessels (Figs. 18–33 B, 18–34). On either side, the portion of the dorsal aorta between the third and fourth arches (called the *ductus caroticus*) eventually disappears (Fig. 18–33 B). The third pair of aortic arches with contributions from the short cephalic extensions of the dorsal aortae form the primitive *internal carotid arteries* (Fig. 18–34 A,B). The *external carotid arteries* originate as vascular sprouts from the bases of the third arches. Contributions to these arteries are probably made by the vascular channnels produced by the degeneration of the first two pairs of arches. The common stem of each third aortic arch, caudal to the origin of the exter-

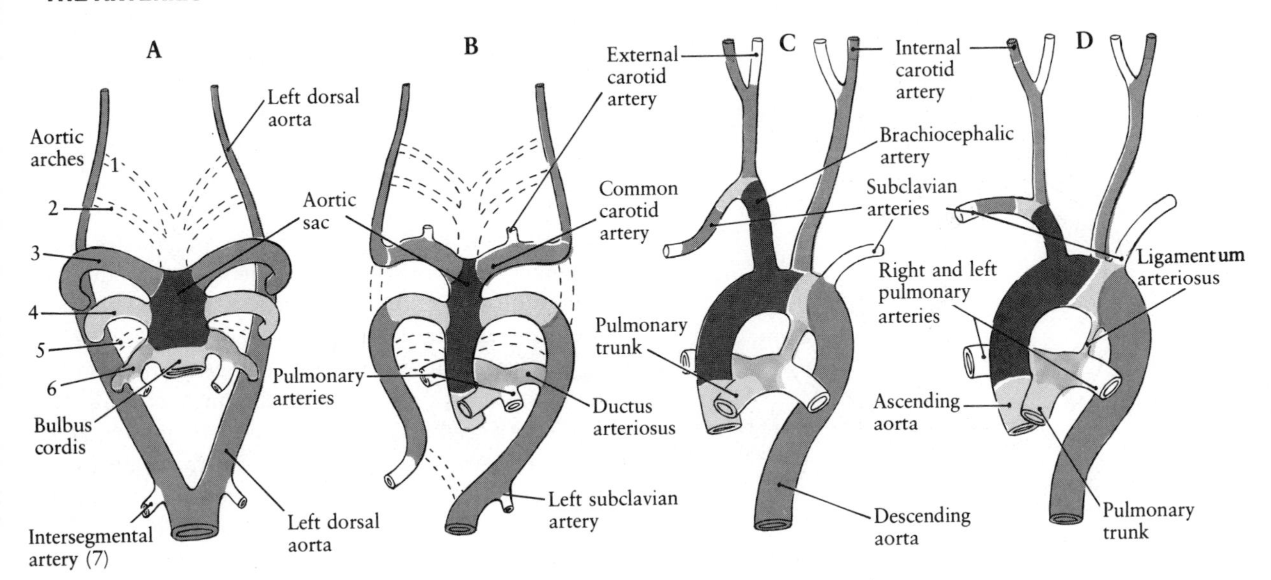


**FIG. 18-34.** Diagram summarizing the transformation of the aortic sac, bulbus cordis, aortic arches, and dorsal aortae into the adult arterial pattern. A, aortic arches at six weeks; B, seven weeks; C, eight weeks; D, the arterial vessels of a six-month-old infant. Broken lines indicate vessels that normally disappear.

nal carotid artery, remains as the *common carotid artery* (Fig. 18–34 B).

Both of the fourth pairs of aortic arches persist, but their contribution to the adult organization of arteries is quite different. On the right, a ventral portion of the aortic sac elongates into the *brachiocephalic* or *innominate artery;* the latter serves as a common trunk for the right common carotid and right subclavian vessels (Fig. 18–34 C). The *right subclavian artery* is formed chiefly from the right fourth aortic arch and a short segment of the dorsal aorta (Fig.18–34 B,C). Therefore, much of the right dorsal aorta, from its point of junction with the left dorsal aorta to the subclavian artery, undergoes regression and disappears. Concurrently, the portion of the right sixth aortic arch beyond the origin of the right pulmonary artery drops out. On the left side, the fourth aortic arch is retained as a major component of the systemic channel leading from the heart to the dorsal aorta. With contributions from the ascending aorta (i.e., divided truncus and bulbus) and the left dorsal aorta, it persists as the *arch of the aorta.* It is interesting to note that the right half of the fourth arch is retained in the bird as the main route to the dorsal aorta. As discussed later in this chapter, the *left subclavian artery* originates as an enlargement of a segmental branch of the left dorsal aorta near its union with the left fourth aortic arch. It is equivalent to only the distalmost part of the right subclavian artery.

The pulmonary or sixth arches begin as vascular sprouts from the dorsal aorta and the aortic sac (Fig. 18–33 B). The dorsal rudiment taps the extension from the aortic sac in such a way that a distal segment of the latter remains as the *pulmonary artery.* The right sixth arch loses its connection with the dorsal aorta, leaving its proximal part as the stem of the right pulmonary artery.The distal part of the left sixth arch persists as a shunt (the *ductus arteriosus* or *duct of Botallo*) until its atrophy at birth (Fig. 18–34 C,D). The remainder of the arch then constitutes the stem of the left pulmonary artery (Fig. 18–34 D).

The transformation of the aortic arches occurs simultaneously with the internal subdivision of the truncus arteriosus and bulbus cordis. The result is that the aorta opens into the third and fourth aortic arches (or derivatives) while the pulmonary trunk is continuous with the left sixth aortic arch.

## Branches of the Dorsal Aorta

Even before the dorsal aorta becomes a single, median vessel behind the pharynx, each of the endothelial tubes that contribute to its formation gives rise to several major sets of tributaries. A schematic arrangement of these vessels through the trunk of the embryo is shown in Figure 18–35. The most conspicuous of these vessels are the *dorsal, somatic intersegmental arteries.* These pass between adjacent

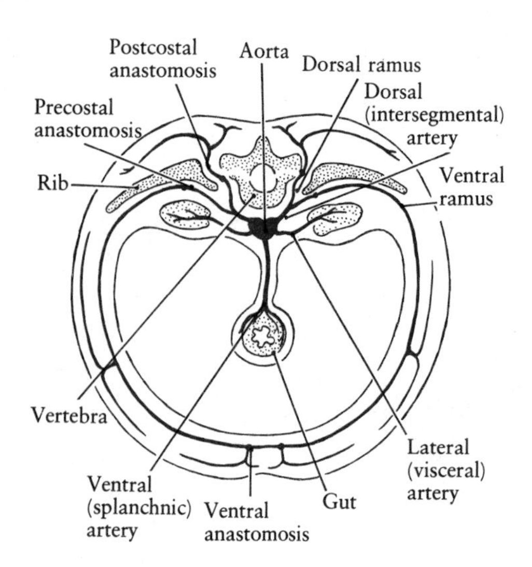


**FIG. 18–35.** A schematic drawing in transverse view to illustrate the arrangement of the branches of the dorsal aorta. Longitudinal anastomoses are indicated as beadlike enlargements.

or successive somites to become continuous with vascular plexuses developing in the somites, the lateral body wall, and the neural tube. *Lateral arteries* arise from the sides of the descending aorta and supply structures associated with the intermediate mesoderm. *Ventral, splanchnic arteries* are distributed to the gut and its accessory organs such as the yolk sac and allantois.

The dorsal intersegmental arteries initially arise as a series of dorsolateral, symmetrically arranged vessels extending from the level of the occipital somites to the somites of the sacral region. Although each artery is primitively associated with the neural tube, it gradually sends secondary branches out as adjacent structures grow and are added. For convenience, two tributaries of the original artery can be identified (Fig. 18–35). A dorsal tributary (or *dorsal ramus*) passes caudally between successive ribs and transverse processes supplying vessels to the spinal cord, the dorsal musculature, and the skin. A ventrolateral tributary (the *ventral ramus*) curves around the body wall, courses along the primary branches of the spinal nerves, and terminates ventrally in an anastomosis with the same vessel of the opposite side. The ventral branches of the intersegmental arteries persist in the thoracic and lumbar regions of the embryo as the serially arranged *intercostal* and *lumbar arteries* (Fig. 18–36).

At the occipital, cervical, and sacral regions of the embryo, there is an extensive reorganization of the intersegmental arteries. With definition of the neck and the forelimbs and with the loss of the ductus caroticus, longitudinal vascular connections appear just dorsal to the ribs *(postcostal anastomoses)* to join together the intersegmental arteries of the cervical region (Fig. 18–36). The roots of the first six cervical arteries then degenerate, leaving the dorsal branch of the seventh intersegmental artery and its longitudinal anastomosis on either side as the *vertebral artery* (Fig. 18–36 B).

The vertebral arteries then extend cranially and bend mesially just beneath the mesencephalon. Here the vertebrals fuse to form the single, *basilar artery* (Figs. 18–36 A, 18–37). At approximately the level of the diencephalon, branches of the basilar artery unite with the internal carotid arteries to form the *circular arteriosus of Willis*. Hence, the carotid and the vertebral-basilar circulations are two separate routes of blood supply to the brain.

With the caudal descent of the heart and the elongation of the neck, the seventh intersegmental artery comes to lie opposite the dorsal end of the fourth pair of aortic arches (Fig. 18–34 B). Conveniently, this is the level at which the anterior limb bud begins to take shape during the fifth week of human development. It is not surprising, therefore, to find that the seventh intersegmental artery on either side enlarges to become the subclavian artery. Beyond the vertebral artery, the subclavian continues into the forelimb as the *axillary artery*.

The striking dissimilarity in the origin of the right and left subclavian arteries has been referred to under the aortic arches. It is directly related to the fate of the fourth pair of aortic arches. Since the left fourth arch is retained, the left subclavian artery is formed solely from the seventh intersegmental artery. On the right, the aorta drops out behind the seventh intersegmental artery, leaving the right subclavian artery to be fashioned from the latter plus the portion annexed from the right fourth aortic arch.

Several additional branches from the subclavian arteries provide insight into the construction of other arteries. Longitudinal anastomoses between the tips of the ventral branches of intersegmental arteries in the thoracic and lumbar regions bring about the formation of the *internal thoracic (internal mammary)*, *superior epigastric*, and *inferior epigastric arteries* (Fig. 18–36 A). The *thyrocervical arteries* develop on either side from longitudinal anastomoses below the ribs *(precostal anastomoses)* and

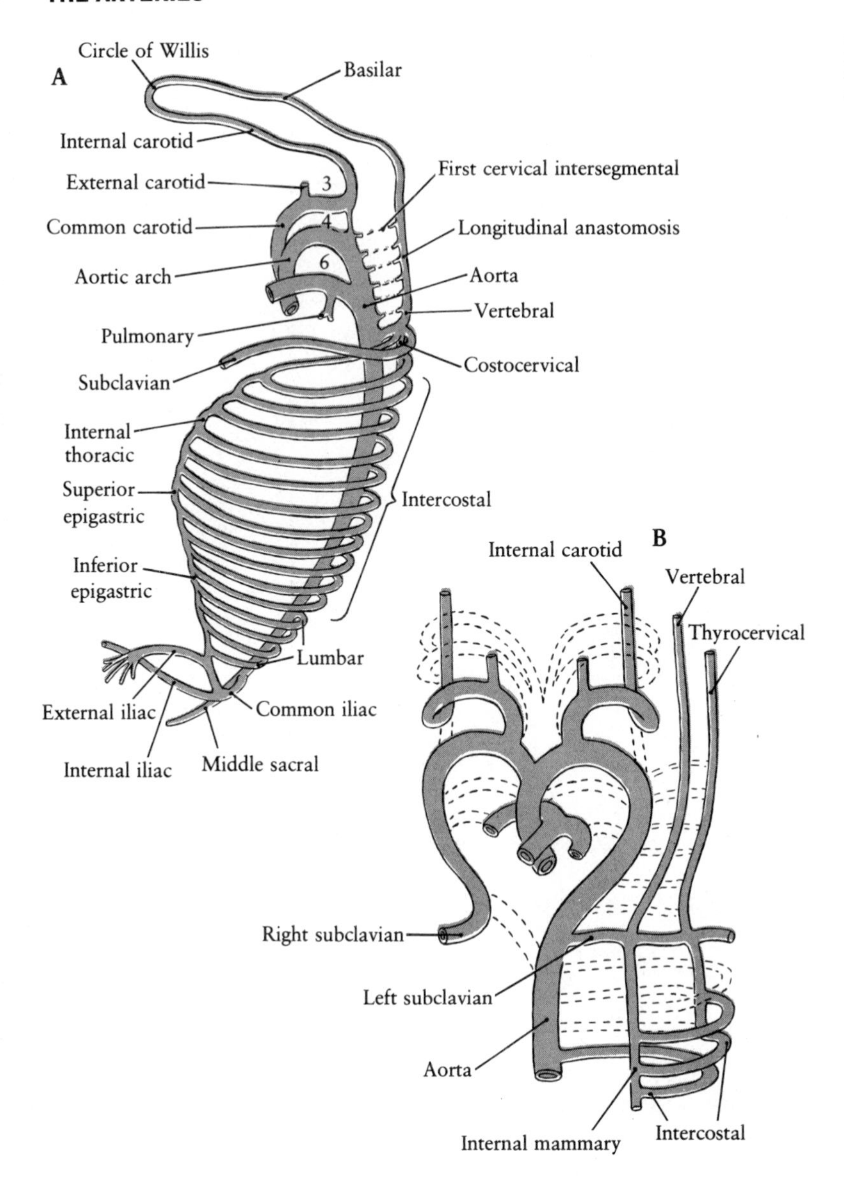


**FIG. 18–36.** Derivatives of the branches of the human dorsal aorta. A, left-side view; B, ventral view showing origin of several cranial arteries.

between ventral rami of intersegmental arteries anterior to the subclavian arteries. The *costocervical arteries* develop in similar fashion but from the three ventral rami immediately caudal to the subclavian arteries.

The lateral branches of the dorsal aorta supply the organ derivatives of the intermediate mesoderm. Initially, these blood vessels develop in relationship to the paired mesonephric kidneys and are quite numerous (Fig. 18–38 A). Regression of the mesonephros reduces the number of lateral arteries. They are represented in the adult by the *inferior phrenic, suprarenal, renal,* and *internal spermatic* or *internal ovarian arteries* (Fig. 18–38 B).

Of the three major types of branches of the dorsal aorta, the ventral ones are the least segmental in nature. They are primitively represented by the paired *vitelline (omphalomesenteric) arteries* to the yolk sac and the paired *umbilical (allantoic) arteries* to the allantois (Fig. 18–39 A). Fusion of the dorsal aortae into a single vessel stimulates the consolidation of vitelline vessels into single, ventral blood vessels. Subsequently, three major arterial trunks, which pass by way of the dorsal mesentery to the gut, are visible. These are the *celiac artery* (to stomach, duodenum, liver, pancreas, and spleen), the *superior mesenteric artery* (to small intestine and part of large intestine), and the *inferior mesenteric*

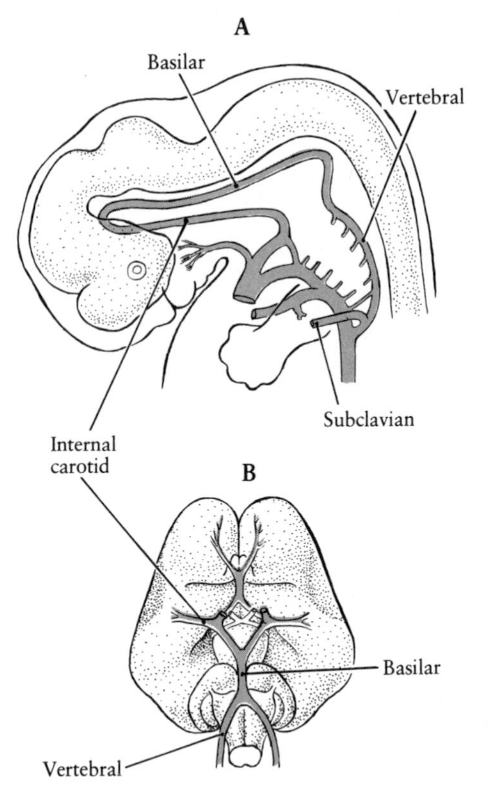


**FIG. 18–37.** Drawings to show the relationships between the internal carotid and vertebral arteries. A, 6 weeks, left-side view; B, 14 weeks, ventral view of brain. (After L. Arey, 1974. Developmental Anatomy. W. B. Saunders Company, Philadelphia.)

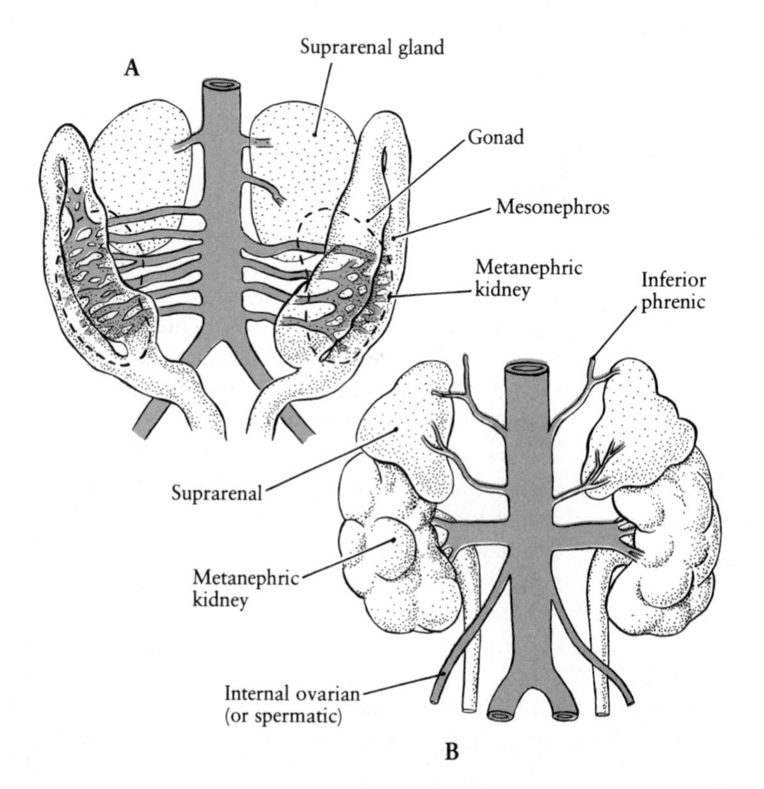


**FIG. 18–38.** Ventral views of the lateral branches of the human aorta at seven weeks (A) and birth (B).

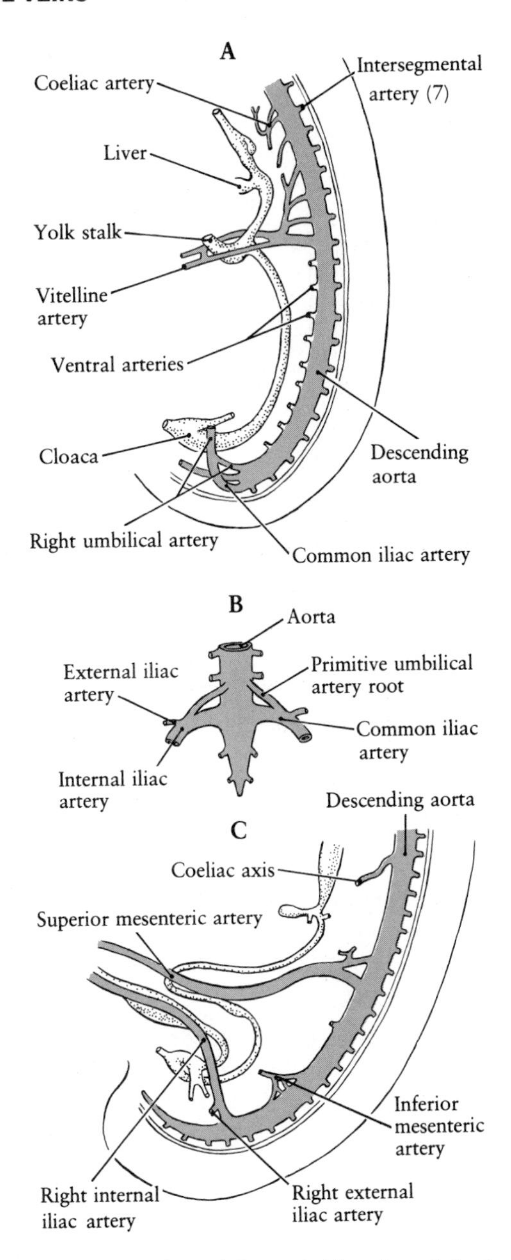


FIG. 18-39. Derivatives of the ventral branches of the human aorta. A, 5 mm, right-side view; B, 5 mm, ventral view of the terminal end of the aorta; C, 9 mm, right-side view.

*artery* (to the descending colon, sigmoid colon, and rectum) (Fig. 18–39 C). It has been suggested that these three arteries probably represent the original splanchnic vessels of the seventh cervical, third thoracic, and fifth thoracic segments, respectively.

The umbilical arteries are established very early in development and accompany the outgrowth of the allantois. As the embryo increases in length, these blood vessels, passing through the umbilical cord to the placenta, migrate caudally to the lower lumbar region. Here, on either side, an anastomosis forms between the umbilical artery and the fifth lumbar intersegmental artery.

The root of each umbilical artery then regresses, leaving the lumbar intersegmental artery to form the stem of the definitive umbilical artery. When at birth the placental circulation ceases, the umbilical arteries beyond the embryo proper begin to atrophy. Within the embryo, the stem of each umbilical artery is retained as the *common iliac artery* (Fig. 18–39 B). Beyond the origin of the *external iliac artery,* which is a new vessel supplying the hindlimb bud, the umbilical artery persists as the *internal iliac artery* (Fig. 18–39 B).

Caudal to the level of the umbilical arteries, the dorsal aorta becomes diminished in size. It continues toward the tail as the slender *caudal artery* or *middle sacral artery.*

Both the anterior and posterior limb buds are well established in human embryos of three days. Each limb bud is a simple swelling of ectoderm that encloses closely packed cells originating from the adjacent mesoderm. As each appendage bud enlarges, it acquires a vascular plexus supplied by several intersegmental branches of the dorsal aorta. In the case of the forelimb, the lateral branch of the seventh cervical intersegmental artery forms the *axillary artery.* The axillary artery is initially continued into the upper arm as the brachial artery and into the forearm as the *interosseous artery* (Fig. 18–40 A). The most prominent vessels of the forearm, the *radial* and *ulnar arteries,* arise as branches of the brachial artery.

The main artery of the early hindlimb is the *sciatic artery* (Fig. 18–40 B). It accompanies the sciatic nerve of the lower extremity and originates as a branch from the future internal iliac artery. Subsequently, this vessel is largely replaced by the *femoral artery,* which springs from the external iliac artery. The sciatic artery is reduced to the *inferior gluteal artery* (proximally) and the *popliteal* and *peroneal arteries* (distally). The major vessels of the foot, the *anterior* and *posterior tibial arteries,* are specializations of the sciatic and femoral arteries, respectively.

## The Veins

The veins of the early amniote embryo can easily be grouped into three systems (Fig. 18–6): (1) the vitelline system consisting of the right and left *vi-*

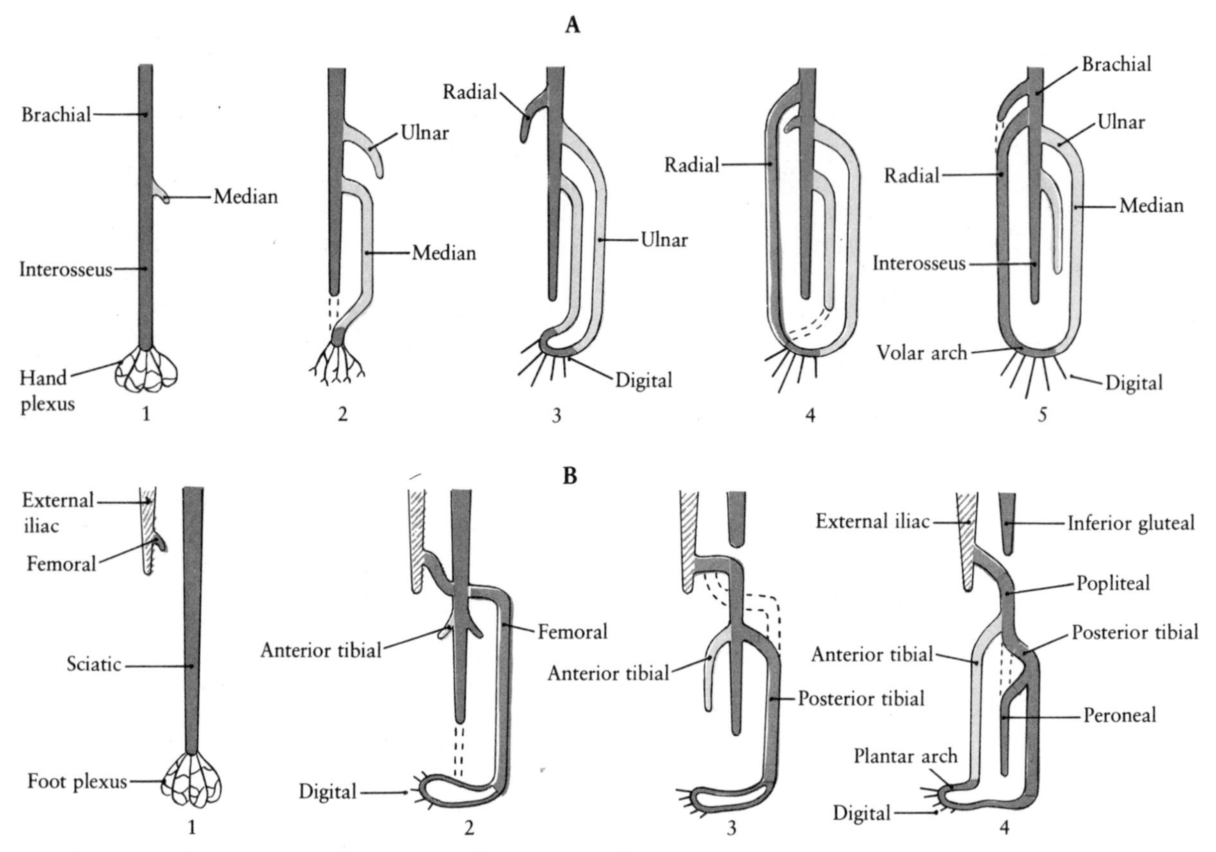


**FIG. 18-40.** Stages in the development of the arteries of the human arm (A) and leg (B). (From L. Arey, 1974. Developmental Anatomy. W. B. Saunders Company, Philadelphia.)

*telline veins (omphalomesenteric veins)*, which arise in the splanchnic mesoderm of the yolk sac; (2) the umbilical system consisting of the right and left *umbilical veins (allantoic veins)*, which drain the capillaries of the chorionic villi or allantois; and (3) the cardinal system, which initially consists of the *anterior* and *posterior cardinal veins*. Subsequently, paired *subcardinal* and *supracardinal veins* appear to supplement and gradually replace the postcardinal veins by participating in the formation of the *inferior vena cava*.

Most of the veins of the embryo arise as capillary plexuses that increase in complexity by sprouting and anastomosing with adjacent plexuses. Fusion and enlargement give rise to fewer but larger channels. Initially, the major vessels of the venous system show a symmetrical arrangement. Striking alterations, particularly in the trunk region of the embryo, substantially modify this primitive organization of the veins. Such changes in the vascular pattern are easy to understand if one remembers that the young vessels are plexiform and that the natural tendency for the blood is to seek the most direct route back to the heart. Central to giving shape to the final, asymmetrical venous pattern in mammalian embryos are two dramatic shifts in the position and direction of blood flow. The first is a left-to-right shunt of blood that results from the tranformation of the vitelline and umbilical veins within the liver. The second is a left-to-right shunt of blood brought about by an oblique cross-connection between the left and right anterior cardinal veins. With the right half of the atrium designed to receive blood of the systemic circulation, there is a distinct tendency to emphasize those venous channels on the right side of the embryo.

## Transformation of the Vitelline Veins

The fate of the vitelline veins, as with the umbilical veins, is intimately tied to the growth and differentiation of the liver. The paired vitelline veins ini-

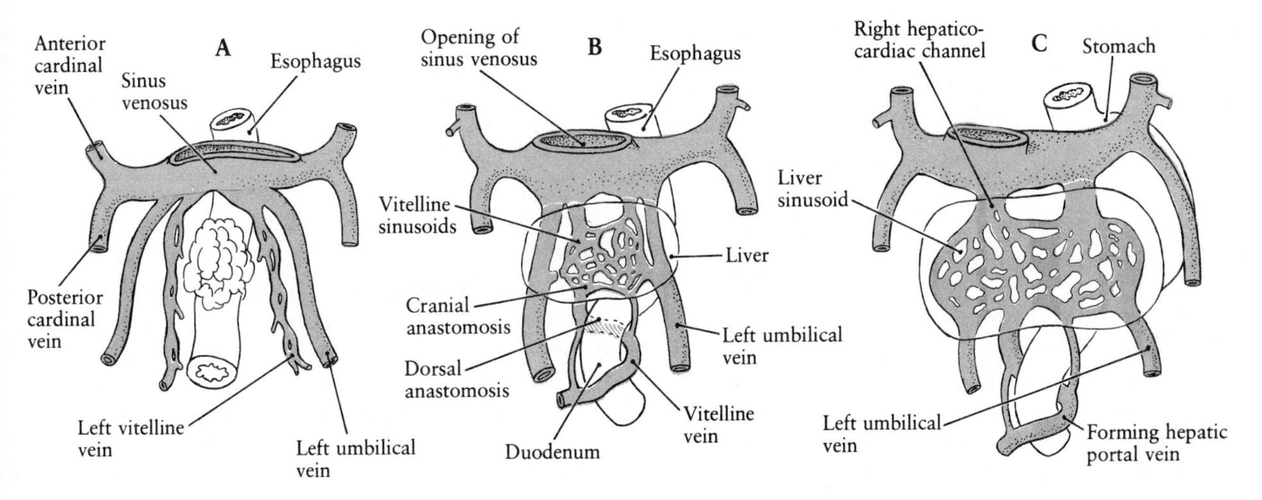


tially pass on either side of the anterior intestinal portal and course through the septum transversum to empty into the sinus venosus (Fig. 18–41 A). The continuity is interrupted by the growth ventrally of the liver primordium. The endothelium of the two vessels breaks up into the primitive *hepatic sinusoids* of the liver (Fig. 18–41 B). The stems of the vitelline vessels between the liver and the sinus venosus are then retained as the stems of the right and left vitelline veins *(hepaticocardiac channels)* (Fig. 18–41 B,C). The right vitelline vein will eventually enlarge and contribute to the proximal end of the inferior vena cava.

The distal segments of the paired vitelline veins become joined together by three cross anastomoses (cranial, middle, caudal) with the middle connection lying dorsal to the duodenum (Fig. 18–41 B). When the stomach and duodenum elongate and rotate from their original midsagittal position, the most direct route to the liver for the venous blood is not by way of the original right and left vitelline veins, but rather from one vessel to the other via the interconnecting anastomoses (Fig. 18–41 C). The blood from the right vitelline vein tends to flow across the caudal (ventral) anastomosis to the left vitelline vein, and then across the middle interconnecting plexus to the hepatic end of the persistent part of the right vitelline vein. The intervening portions of the two vitelline vessels drop out, leaving a composite, S-shaped vessel known as the *hepatic portal vein* (Fig. 18–41 D). Caudally, it extends to the union of the splenic and superior mesenteric veins. Although the *superior mesenteric vein* would appear to be represented by the distal part of the left vitelline vein, it is a new vessel that develops in situ in

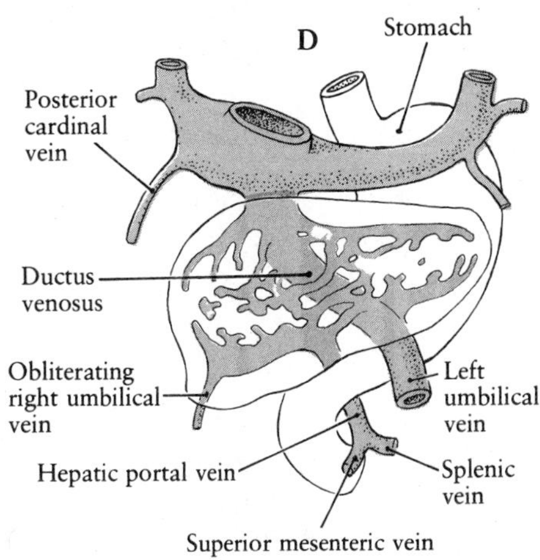


**FIG. 18–41.** A series of schematic drawings (A–D) to show the transformation of the vitelline and umbilical veins within the developing liver.

the dorsal mesentery of the intestinal loop. The distalmost segments of the vitelline veins regress with decline of the yolk sac.

## Transformation of the Umbilical Veins

As the primitive right and left lobes of the liver expand laterally, the umbilical veins become enmeshed and broken up into sinusoids (Fig. 18–41 B,C). The stems of the umbilical veins persist for only a short time as blood returning from the placenta is progressively diverted through the hepatic sinusoids in its return to the heart. By approximately the fifth week of human development, the entire right umbilical vein and the proximal stem of the left umbilical vein have dropped out (Fig. 18–41 D). The left umbilical vein then continues throughout fetal life as the major blood vessel carrying oxygenated blood from the placenta toward the heart.

Initially, the blood from the left umbilical vein crosses the liver into the sinusoids formed by the right proximal vitelline vein. This route becomes very circuitous as the right lobe of the liver continues to enlarge. Gradually, therefore, the hepatic sinusoids between the point of entry of the left vitelline vein into the liver and the stem of the right vitelline vein enlarge and merge to form a large intrahepatic channel known as the *ductus venosus* (Fig. 18–41 D). The ductus venosus passes into the enlarging right horn of the sinus venosus by way of the right hepaticocardiac channel. A sphincter mechanism in the ductus venosus regulates the flow of umbilical blood into the liver.

## Formation of the Superior Vena Cava

The right and left anterior cardinal veins are the main venous drainage channels from the cranial and neck regions of the early embryo (Fig. 18–6). Each runs caudally to join the corresponding posterior cardinal vein at the level of the *common cardinal vein.* Frequently, the cephalic portion of the anterior cardinal vein is termed the *primary head vein* because it eventually receives smaller vessels from the deeper *(cerebral veins)* and more superficial *(dural sinuses)* parts of the brain.

With definition of the neck region and descent of the heart, the anterior cardinal veins become elongated. At approximately eight weeks of development, an oblique, transverse anastomosis joins the left anterior cardinal to the base of the right anterior cardinal vein (Fig. 18–42 A,B). The proximal part of the left anterior cardinal then regresses to persist as part of the *highest intercostal vein* (Fig. 18–42 C). The left common cardinal vein, no longer an important vessel in the return of blood, contributes to the formation of the *oblique vein* of the left atrium. The intercardinal anastomosis itself forms the *left brachiocephalic* (or *innominate) vein* (Fig. 18–42 C).

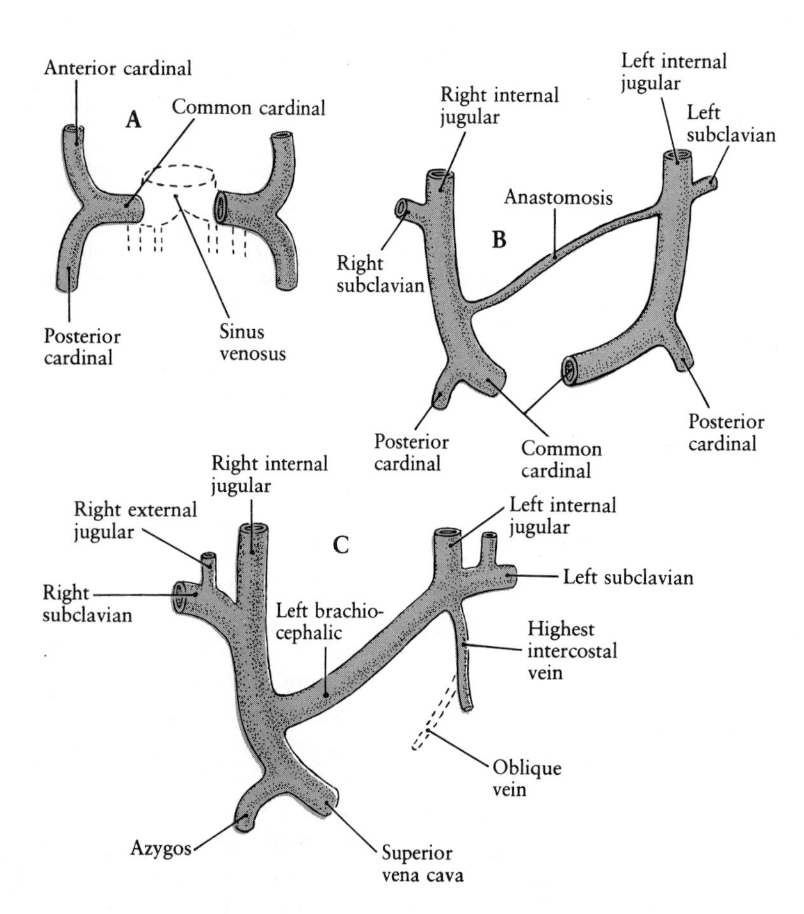


**FIG. 18–42.** Transformation of the anterior and posterior cardinal veins in the human embryo. A, six weeks; B, eight weeks; C, adult.

The right common cardinal vein and the base of the right anterior cardinal vein (to the union with the left brachiocephalic vein) constitute the *superior vena cava,* a vessel destined to serve as the major channel draining the head, neck, and anterior appendages (Fig. 18–42 C). Each anterior cardinal vein beyond the level of the external jugular and subclavian veins becomes the *internal jugular vein.* The *external jugular vein* on either side originates as a secondary channel from a capillary plexus in the facial region.

## Formation of the Inferior Vena Cava

The formation of the inferior vena cava or *postcaval vein* is very complex. It involves major alterations in the three pairs of cardinal veins that appear in succession to drain the body wall, the viscera, and the lower limbs. The posterior cardinal veins appear slightly later than the anterior cardinals and arise as two longitudinal vessels dorsolateral to the urogenital fold (Fig. 18–43). They are concerned primarily with the drainage of the lateral body wall, the mesonephric kidneys, and the hindlimb buds. Subsequently, the paired *subcardinal veins* differentiate on the ventromedial surfaces of the mesonephroi (Fig. 18–43). They terminate cranially in the posterior cardinal veins. Numerous transverse anastomoses through the vascular sinus of the kidneys also join the posterior and subcardinal vessels. Finally, a third set of longitudinal veins, the *supracardinal veins,* develops dorsomedial to the posterior cardinal complex. On either side, each supracardinal is continuous anteriorly and posteriorly with the posterior cardinal vein (Fig. 18–44).

Between the sixth and eighth weeks of development, enlargement, atrophy, and fusion between these cardinal vessels lead to the emergence of a single, unpaired vessel or inferior vena cava. The origin of this vessel requires additional description (Fig. 18–44). Initially the systemic circulation on either side of the embryo is drained toward the common cardinal veins by both the posterior and subcardinal vessels (Fig. 18–44 A). Shortly thereafter, the right subcardinal vein becomes connected to the proximal right vitelline vein by a vascular plexus that develops in the *caval mesentery.* The caval mesentery is a thickening of the dorsal body wall just to the right of the dorsal mesentery; it effectively acts as a bridge between the right lobe of the liver and the right mesonephros. Since this is a very direct route to the heart, enlargement of this new channel proceeds rapidly. The right proximal vitelline vein forms the *hepatic segment* of the inferior vena cava. The cephalic end of the right subcardinal vein becomes the *prerenal* or *mesenteric segment* of the inferior vena cava (Fig. 18–44 B,C).

The consolidation and enlargement of the hepatic and prerenal portions of the inferior vena cava indirectly hasten the degeneration of the posterior cardinal veins. On the right, a portion of the right posterior cardinal vein persists to contribute to the *azygous vein* (Figs. 18–42 C, 18–44 D).

As more blood from the caudal part of the body is collected and passed through the mesonephroi,

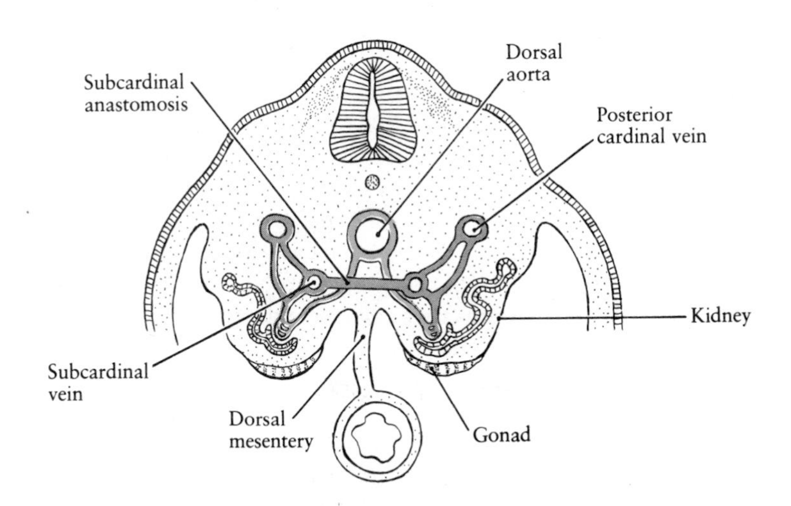


**FIG. 18–43.** A schematic drawing of a transverse section through the posterior abdominal wall of the early human embryo to show the relationships between the posterior and subcardinal veins.

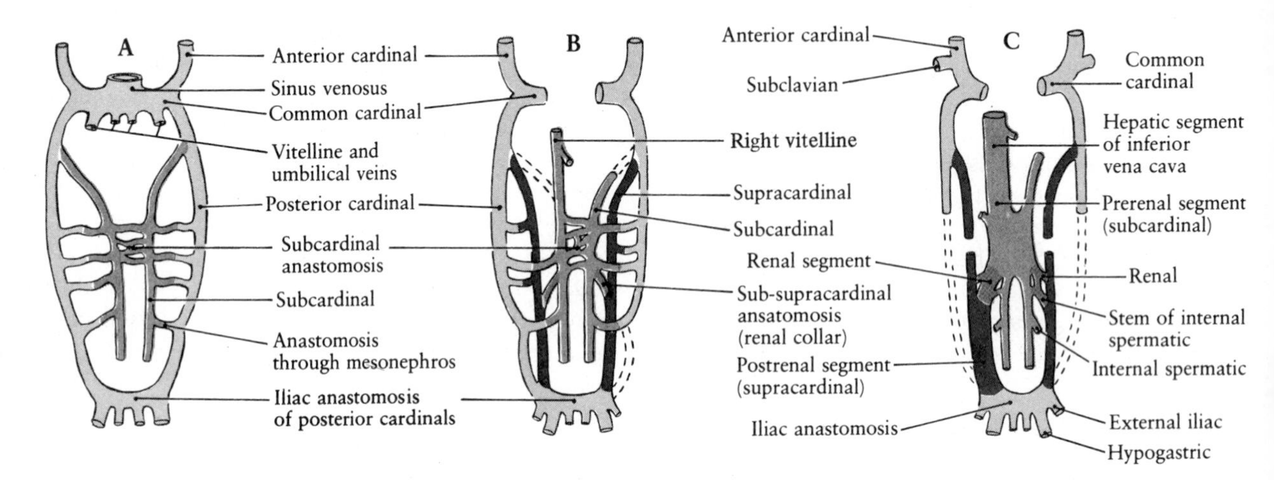


an extensive *subcardinal anastomosis* develops between the kidneys (Fig. 18–44 B). Gradually, a single, main channel emerges within this anastomosis of irregular venous spaces. Since this channel comes to lie between the metanephric kidneys, it contributes to the *interrenal segment* or *renal segment* of the inferior vena cava (Fig. 18–44 C,D).

Elimination of the need for passing blood through the kidney on its return to the heart is effected by a special connection (known as the *renal collar* or *renal anastomosis*) between the right supracardinal vein and the interrenal segment of the inferior vena cava (Fig. 18–44 B). Behind this connection, the right supracardinal vein forms the *postrenal segment* of the inferior vena cava. The chief vessels opening into this part of the inferior vena cava are the *common, external,* and *internal iliac veins*. All of these vessels are fashioned from an early *iliac anastomosis* between the right and left posterior cardinal veins at the level of the hindlimbs (Fig. 18–44 B–D).

Most of the left supracardinal vein drops out, although a prerenal portion may persist as the *hemiazygous vein* (Fig. 18–44 D). A short cross-connection unites it with the azygous vein. A short anterior portion of the right supracardinal vein contributes to the formation of the azygous vein.

Several additional tributaries of the inferior vena cava should be mentioned because they reflect the differences that take place in vascular reorganization on the right and left sides of the embryo. The metanephric kidneys are drained by the *renal veins*. The origin of the right renal vein is simple; it represents a consolidation of vascular channels that joins directly the renal anastomosis (Fig. 18–44 C,D). The left renal vein is more complicated because the renal anastomosis on the left is not incor-

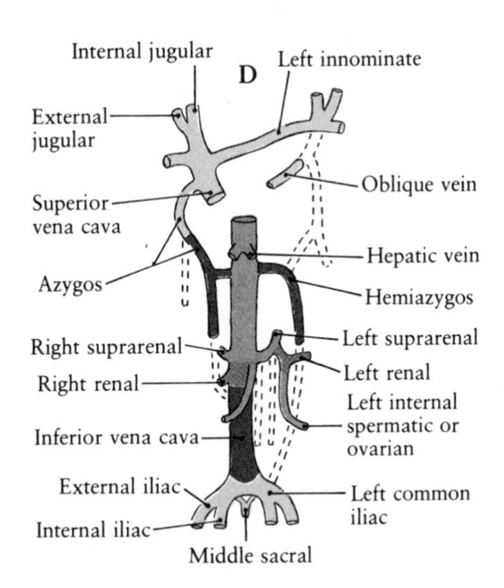


**FIG. 18–44.** Transformation of the primitive veins of the trunk of the human embryo to show formation of the inferior vena cava. Ventral views. A, six weeks; B, seven weeks; C, eight weeks; D, adult. (From L. Arey, 1974. Developmental Anatomy. W. B. Saunders Company, Philadelphia.)

porated into the inferior vena cava. Hence, it is a consolidation of a primitive renal vein, the left renal anastomosis, and a fused portion of the subcardinal anastomosis (Fig. 18–44 C,D). The *right* and *left suprarenal veins* are also not homologous vessels as Figure 18–44 illustrates. The *spermatic* and *ovarian veins* represent portions of the subcardinal veins that persist caudal to the kidneys.

## The Pulmonary Veins

From the left atrium, the primitive pulmonary vein arises as a sprout of endothelium that joins the pulmonary capillary plexus on the endodermal pri-

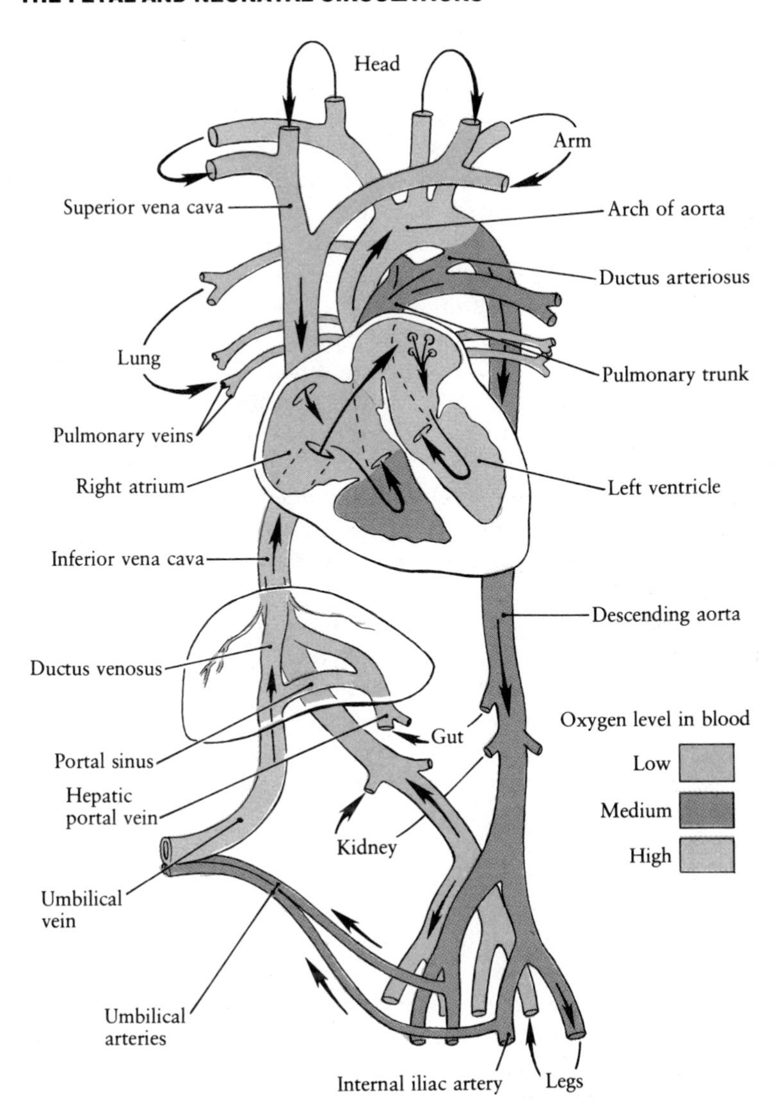


**FIG. 18–45.** The pattern of circulation in the human fetus. Arrows indicate the direction of blood flow and colors the approximate levels of oxygen saturation.

mordia of the lungs. As the sinus venosus is being absorbed into the right atrium, the stem of this single pulmonary vein, as well as the bases of the right and left pulmonary veins, are absorbed into the wall of the left atrium. Four separate pulmonary venous orifices come to open into the dorsal wall of the left atrium (Figs. 18–26, 18–45).

### The Veins of the Limbs

From the capillary plexus that develops in the mesoderm of the limb buds, there emerges a *border vein* that acts as the early drainage channel of blood brought in by the axillary artery. In the forelimb, the border vein initially opens into the posterior cardinal vein. Secondarily, however, it is transferred to the anterior cardinal vein as a result of the caudal descent of the heart. The *subclavian veins*, the *axillary veins*, and the *basilar veins* develop from the embryonic border vein. In the hindlimb, the major vein *(femoral vein)* develops as a vessel from the posterior cardinal vein. Within the hindlimb proper, the border vein differentiates into the *anterior tibial, small saphenous,* and *inferior gluteal veins.*

## The Fetal and Neonatal Circulations

The fetal cardiovascular system is designed primarily to serve prenatal requirements and to permit those modifications at birth that quickly establish the postnatal circulatory pattern. Although birth requires immediate changes in the flow of blood, the particular arrangement of the fetal blood vessels allows the neonate to accomplish this transition with relative ease.

As pointed out previously, the placenta acts as an organ of transfer of oxygen and nutritive material from the maternal bloodstream to the fetal circulation, and of carbon dioxide and nitrogenous wastes of metabolism from the fetal to the maternal circulation. Some knowledge of the fetal circulation is necessary in order to appreciate the abrupt changes that occur with the shift at birth in the sites where these various activities are carried out.

A simplified scheme of the fetal circulation is diagrammed in Figure 18–45). Blood, approximately 80 percent saturated with oxygen, returns from the placenta to the liver by way of the umbilical vein. Using special techniques *(angiocardiography)* that permit analysis of the distribution patterns and oxygenation of the fetal blood, it has been demonstrated that about half of the placental return is diverted through the hepatic sinusoids, whereas the other half bypasses the liver and courses through the ductus venosus into the inferior vena cava. A muscular sphincter at the junction of the ductus venosus and the umbilical vein regulates the flow into each of these vascular pathways.

Blood from both the ductus venosus and the hepatic portal circulation is collected by the inferior vena cava and emptied into the right atrium. Since the inferior vena cava already contains deoxygenated blood from the caudal regions of the fetus, its level of oxygenation is not as high as that in the umbilical vein. Most of the blood flow from the inferior vena cava is diverted by the lower margin of the septum secundum (i.e., the crista dividens) through the foramen ovale into the left atrium (Fig. 18–26). Here there is some mixture with the deoxygenated blood returning from the lungs. In this way, most of the blood returning from the placenta bypasses the lungs and is shunted to the left side of the heart. The vessels of the heart, head, and neck thus receive blood with a high level of oxygen tension.

A small stream from the inferior vena cava passes through the right atrium and mixes with a large volume of deoxygenated blood from the superior vena cava and the coronary sinus. After entering the right ventricle, this blood courses by way of the pulmonary trunk through the ductus arteriosus and into the descending aorta. Initially, probably less than 10 percent of the right ventricular blood flow goes to the lungs. Flow to the lungs does appear to increase steadily in later stages of fetal life. A large portion of the mixed blood in the dorsal aorta passes out into the umbilical arteries for reoxygenation in the capillaries of the chorionic villi.

Major adjustments occur in the pattern of circulation within minutes of the transfer of respiration function from the placenta to the lungs (Fig. 18–46). An immediate response to the first breath of the neonate is a contraction of the two umbilical arteries to prevent further blood from circulating to the placenta. This is followed by constriction of the umbilical vein and the ductus venosus. If the umbilical cord is not tied for several minutes, there is some transfer of blood from the placenta to the neonate.

Alterations in the umbilical arteries, the umbilical vein, and the ductus venosus occur simultaneously with a narrowing of the ductus arteriosus. Closure of the ductus arteriosus is an essential step in the shift from the fetal to the neonatal pattern of circulation. In contrast to earlier views, it is now agreed that complete closure of this shunt takes place over a period of several months. Initially, however, there is a *functional closure* of the ductus arteriosus within the first several minutes of birth. This involves the contraction of the muscular tissue in the wall of the ductus, a process presumably mediated under the influence of a substance *(bradykinin)* released by the lungs upon their initial inflation. A consequence of this reduction of the ductus arteriosus is marked increase in blood flow to the lungs. *Anatomical closure* of the ductus is achieved by the proliferation of the tunica intima.

The sudden increase in the volume of the lungs is associated with a sharp reduction in pulmonary vascular resistance. The increase in pulmonary blood flow leads to a sharp rise in the pressure of the left atrium. Since the pressure in the right atrium is decreased with the sudden occlusion of the placental circulation, there is a cessation of blood flow through the foramen ovale *(transatrial flow)* as the valve of the foramen is forced against the septum secundum. Hence, the foramen ovale may be regarded as functionally closed. Anatomical closure of the foramen ovale is effected through the proliferation and hypertrophy of endothelial and connective tissues around the opening, a process that also requires several months for completion.

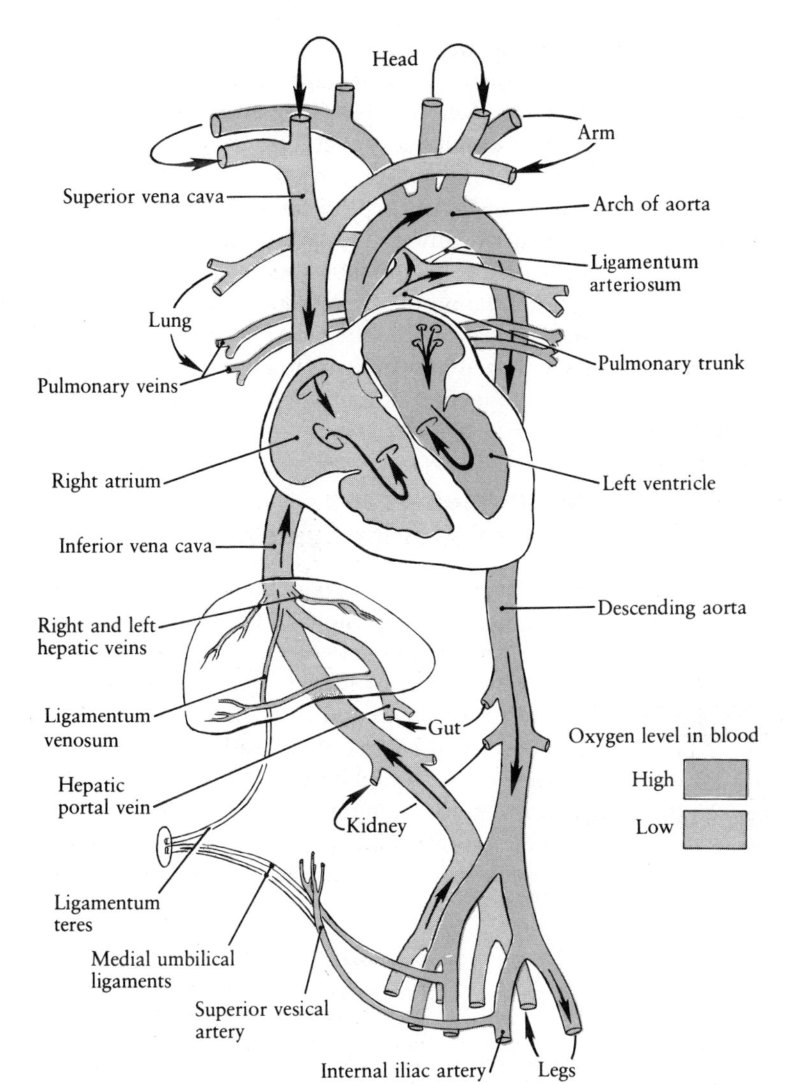


**FIG. 18–46.** The pattern of circulation in the human neonate. Arrows indicate the direction of blood flow. The derivatives of fetal vessels are also shown.

As long as pulmonary blood flow is normal and the pressure in the left atrium is at least equal to that in the right atrium, the fact that there is a structural opening between the two chambers for some time is of little importance.

Vessels no longer required in the postnatal pattern of circulation eventually become obliterated (i.e., their lumens disappear due to hypertrophy of the tunica intima). However, the fibrous connective tissue of the walls of these vessels commonly persists as *ligaments* in the adult (Fig. 18–46). The intra-abdominal portion of the umbilical vein forms the *ligamentum teres,* which extends from the umbilicus to the hepatic portal vein. The *ligamentum venosum,* a remnant of the ductus venosus, passes through the liver and joins the inferior vena cava and the hepatic portal vein. The ductus arteriosus becomes the *ligamentum arteriosum,* which passes from the left pulmonary artery to the arch of the aorta. Much of the intra-abdominal portion of the umbilical arteries remains as the *medial umbilical ligaments.*

## The Lymphatic System

The lymphatic system is a network of terminally closed vessels that return tissue fluids to the venous system. It begins to develop in the human embryo shortly after the appearance of the primitive cardiovascular system.

There are differences of opinion on the precise method of origin of the lymph vessels. The older view is that the lymph channels arise in a manner similar to that previously described for blood vessels (i.e., progressive coalescence of mesenchymal-lined spaces). A current view is that the earliest lymph vessels originate as capillary sprouts from the endothelium of embryonic veins.

The early lymph channels tend to be distributed along the main venous trunks. Dilations of these vessels give rise to six lymph sacs (Fig. 18–47 A): (1) two *jugular lymph sacs* near the union of the subclavian and anterior cardinal veins, (2) two *iliac lymph sacs* near the union of iliac and posterior cardinal veins, (3) an unpaired *retroperitoneal sac* in the base of the dorsal mesentery near the suprarenal glands, and (4) the *cisterna chyli* dorsal to the retroperitoneal sac. By continuous elongation, centrifugal growth, and branching, lymph vessels extend out from these lymph sacs to most of the tissues of the body.

At a slightly later stage in development, two large lymph channels (the *right* and *left thoracic ducts*) join the cephalic jugular sacs with the caudal cisterna chyli (Fig. 18–47 B). The adult *thoracic duct* originates from the cranial end of the left lymph duct, the interconnecting anastomosis between the two lymph ducts, and the caudal part of the right lymph duct (Fig. 18–47 C). It opens into the venous system by way of the left jugular sac at the junction of the internal jugular and subclavian veins. The *right lymphatic duct* is derived from the cranial part of the right thoracic duct (Fig. 18–47 C).

Most of the lymph sacs become transformed into lymph nodes during the early fetal period. Mesenchyme cells invade the sac and subdivide its cavity into a network of *lymph sinuses*. Proliferation and differentiation of the mesenchymal tissue lead to the formation of clusters of lymphoid masses. Lymph nodules and germinal centers of lymphocyte production do not appear in these lymphoid masses until just before or shortly after birth.

## Anomalies of the Cardiovascular System

Disturbances in the development of the heart and in the usual arrangement of the blood vessels are among the commonest of developmental anomalies. The overall incidence of congenital malformations of the heart and of its associated major vessels is about 0.7 percent of live births in humans. In view of the complexity of the events leading to the formation of the four-chambered heart and the striking rearrangements in the vascular pathways that occur with the definition of the final vascular

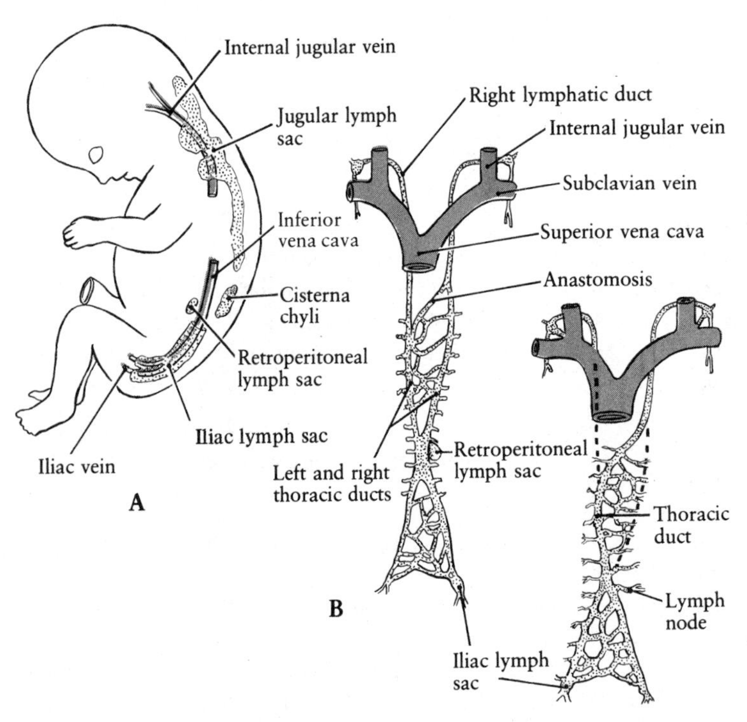


**FIG. 18–47.** Diagrams showing the development of the human lymphatic system. A, lymphatic system at seven weeks, left lateral view; B, ventral view of the lymphatic system at nine weeks showing the paired thoracic ducts; C, formation of the adult thoracic and right lymphatic ducts. (After K. L. Moore, 1977. The Developing Human: Clinically Oriented Embryology, 2nd ed. Courtesy of W. B. Saunders Company, Philadelphia.)

pattern of vessels, it is indeed surprising that malformations of the cardiovascular system are not more common.

Congenital defects, particularly those of the heart, are interesting embryologically. Clinically, these disturbances run the gamut from variations in vascular routes that are functionally insignificant to abnormalities that are serious and life threatening. The gravity of defects is generally determined by the extent to which they interfere with the pulmonary circulation, particularly with respect to the efficiency of the oxygenation of the blood and the sufficiency of the blood being returned from the lungs for delivery into the systemic circulation from the left ventricle. A few of the more common disturbances in the cardiovascular system are described below.

## Heterotaxis (Reverse Rotation) and Displacement of the Heart

Normally, the cardiac tube is thrown into a loop to the right (Fig. 18–48 A). If the tube bends to the left, there is a transposition in which the heart and its two major vessels are reversed left-to-right (Fig. 18–48 B). This so-called left looping results in a positional abnormality of the heart known as *dextrocardia*. Interestingly, if dextrocardia is accompanied by transposition of the viscera *(situs inversus)*, there are few associated cardiac defects and the heart can function in a normal way.

In *ectopia cordis*, the heart lies outside of the pericardial activity owing to defective developments of the mediastinum and the pericardium. *Extrathoracic ectopia cordis* is an extreme condition in which the heart protrudes through the chest wall because of a failure of the lateral body folds to meet and fuse properly.

Failure of the aorticopulmonary septum to pursue its normal spiral course can result in complete transposition of the major vessels (Fig. 18–49). In typical cases, the aorta lies anterior to the pulmonary trunk and springs from the right ventricle; the pulmonary trunk originates from the left ventricle. For survival, there must be associated defects in the cardiac septa and persistence of the ductus arteriosus so that some exchange between the pulmonary and systemic circuits can take place.

**FIG. 18–48.** Sketches of the cardiac tube to show normal right looping (A) and left looping resulting in dextrocardia (B).

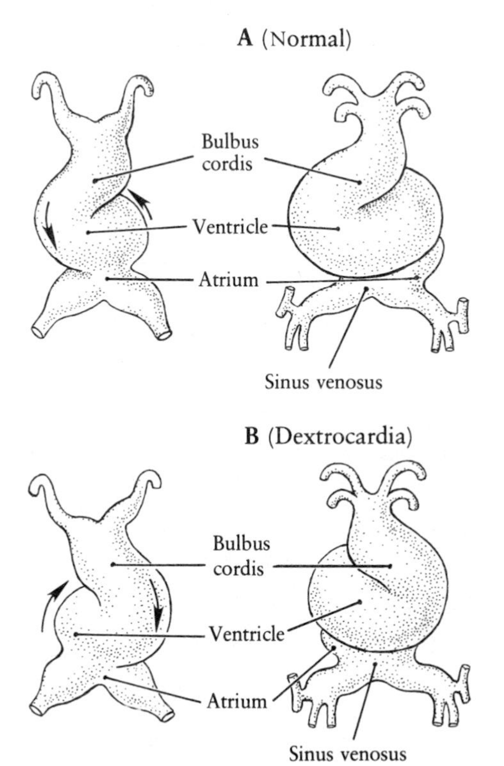


## Malformations Due to Arrest of Development

A number of cardiac defects can be traced to failures in processes relating to the growth, migration, and/or degeneration of septa or their primordia.

**FIG. 18–49.** A sketch showing complete transposition of the aorta and pulmonary trunk.

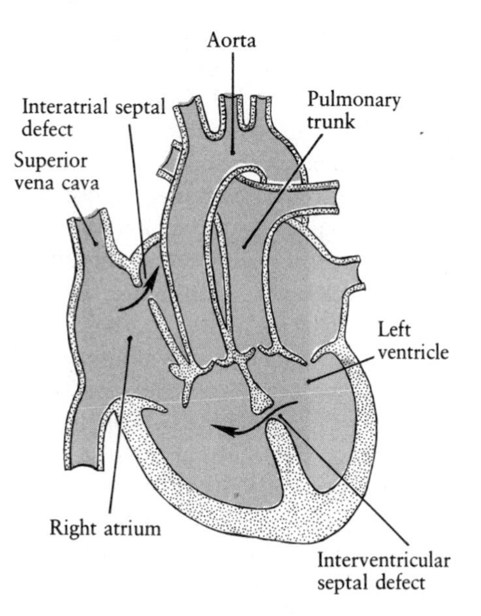

This is well illustrated with examples of the defective closure of fetal openings in the heart.

Atrial septal and ventricular septal defects are among the most common congenital heart anomalies. The most frequent type of atrial septal defects is *persistent* or *patent foramen ovale* (Fig. 18–50). Patent foramen ovale usually arises from an abnormal resorption of the septum primum (valve of the foramen ovale) during the formation of the foramen secundum. For example, if the septum primum is excessively resorbed, the future valve of the foramen ovale is too short to blanket the foramen ovale (Fig. 18–50 B). Also, if the septum secundum is too short, resulting in an abnormally large foramen ovale, the septum primum is unable to close at birth (Fig. 18–50 C). A large percentage (25%) of persons have a condition known as *probe patent foramen ovale*. The presence of this functional opening in the interatrial septum can be demonstrated by passing a probe obliquely from one atrium to the other atrium. Unless the patent foramen ovale is forced open because of other defects in the heart, this condition is not considered to be pathological.

Other clinically significant atrial septal defects include *patent foramen primum, persistent atrioventricular canal* (owing to failure of the fusion of the dorsal and ventral endocardial cushions), and *common atrium* (owing to failure of the septum primum and septum secundum to develop). Generally, interatrial septal defects at the level of the foramen ovale or the foramen primum are compatible with life, provided the atrioventricular canal is normally partitioned and guarded by functional valves. Commonly, these hearts show an increase in size on the right side because of the extra load imposed upon it by blood moving from the left to right with each atrial contraction.

Isolated ventricular septal defects are the most frequent of all cardiac anomalies. As one might expect, these defects typically involve the formation of the membranous portion of the interventricular septum and the closure of the interventricular foramen. Endocardial tissues of the bulbar ridges, the cushions of the atrioventricular canal, and the free edge of the muscular interventricular septum must meet at the right time and in the right place to complete division of the ventricle. Failure to accomplish this task is commonly associated with disturbances in the partitioning of the bulbus cordis. Complete absence of the interventricular septum is

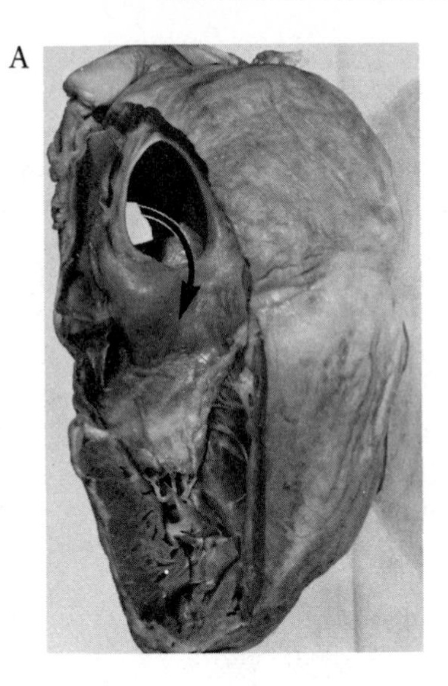


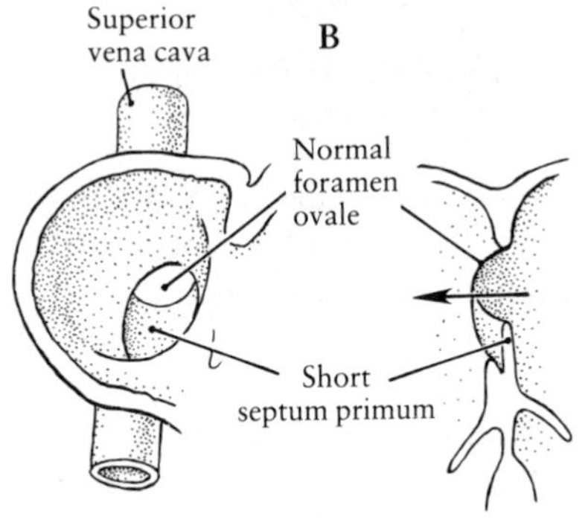


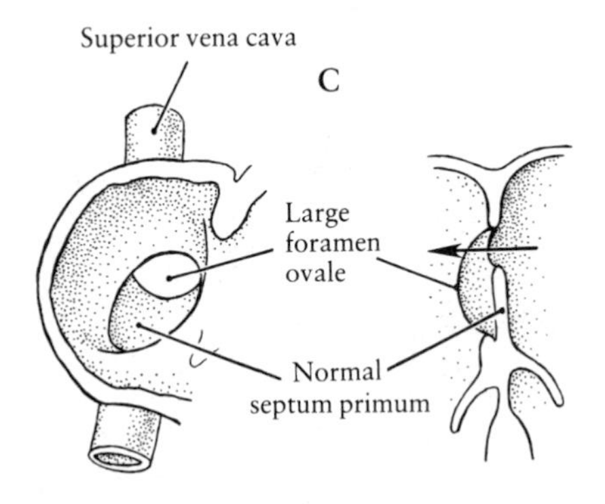


**FIG. 18–50.** Patent foramen ovale. A, photograph of an adult heart as seen from the right side; B, a sketch from the right aspect of the interatrial septum showing defect due to excessive resorption of the septum primum; C, a sketch from the right aspect of the interatrial septum showing defect due to a short septum secundum. (From K. L. Moore, 1977. The Developing Human: Clinically Oriented Embryology, 2nd ed. Courtesy of W. B. Saunders Company, Philadelphia.)

very rare and results in a three-chambered heart *(cor triloculare biatriatum).*

The bulbus cordis is the site of several anomalous conditions, most of which can be traced to disturbances in its partitioning by the aorticopulmonary septum. *Persistent bulbus cordis* results from the failure of the aorticopulmonary septum to develop and divide the bulbus into pulmonary trunk and the aorta. Both ventricles then pump blood into a common outlet with free mingling of the blood from the two sides of the heart. The inefficiency of this pattern of circulation is obvious and must be corrected if there is to be postnatal survival.

Typically, the bulbus cordis is equally subdivided into an aorta and pulmonary trunk. Occasionally, this part of the arterial outlet is partitioned unequally, presumably owing to disturbances in the position of the bulbar ridges that fuse to form the aorticopulmonary septum. According to the direction of the malplacement, what results is either a large pulmonary trunk and a small aorta or a small pulmonary trunk and a large aorta (Fig. 18–51 A). If the aorta is narrowed, the condition is *aortic stenosis;* if the pulmonary trunk is constricted, it is *pulmonary stenosis*. In either case, the aorticopulmonary septum cannot properly fuse with the interventricular septum, and a ventricular septal defect results. Commonly, the larger of the two vessels will override the septal defect in the ventricle (Fig. 18–51 B). The complete absence of a lumen in the aorta or pulmonary trunk leads to a condition known as *atresia.*

Several of the above examples indicate that disturbances in the development of one part of the heart are invariably accompanied by the formation of defects in another part of the heart. A classic example of this is the *tetralogy of Fallot*. Four defects constitute this condition (Fig. 18–51 B): (1) pulmonary stenosis, (2) ventricular septal defect, (3) aorta overriding the septal defect, and (4) hypertrophy of the right ventricle. These conditions result in little blood reaching the lungs for aeration. The lack of oxygen in the peripheral circulation gives a bluish tinge (cyanosis) to the skin and the lips. If the ductus arteriosus remains open, blood flow from the aorta to the pulmonary artery can supply sufficient blood to the lungs to support life. Several surgical methods are available to correct deficiencies in the pulmonary circulation if the ductus arteriosus closes. The Blalock and Taussig technique involves anastomosing the brachiocephalic artery to the right pulmonary artery or the left subclavian artery to the left pulmonary artery. Another techique involves the coupling of the end of the right pulmonary artery to the side of the superior vena cava. These vascular connections are designed as shunts to increase pulmonary flow.

## Aortic Arch Anomalies

The transformation of the embryonic arrangement of the aortic arches can be accompanied by several types of variations. Disturbances include the persistence of vessels that normally disappear (i.e., right aortic arch) or the disappearance of a vessel that is normally retained (i.e., left aortic arch). A common malformation is *coarctation of the aorta* in which there is stenosis of the aorta just above or below the

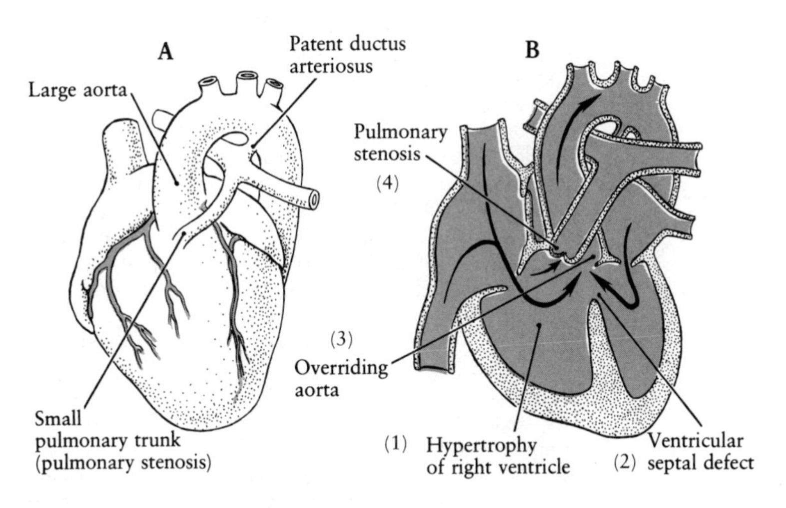


**FIG. 18–51.** A, a sketch of an infant's heart to show a narrowed pulmonary trunk (pulmonary stenosis) and a large aorta resulting from unequal partitioning of the bulbus cordis; B, a frontal section through a heart to show the tetralogy of Fallot.

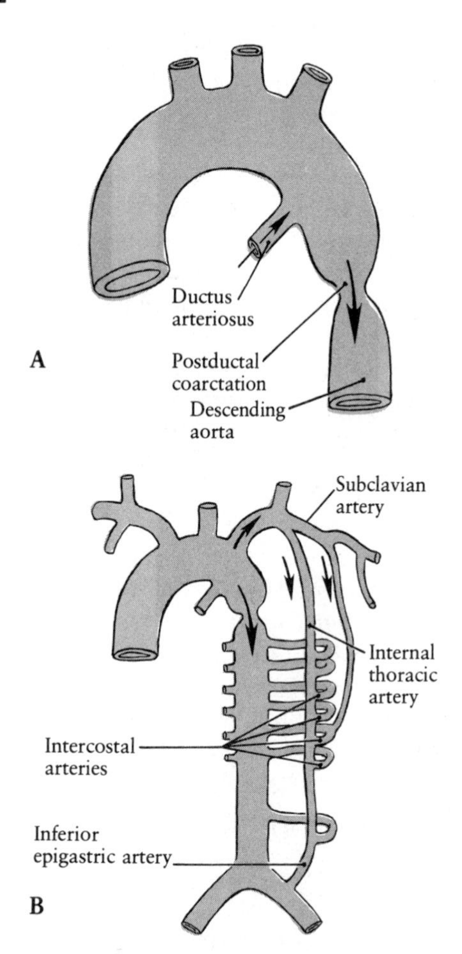


**FIG. 18-52.** Coarctation of the aorta below the ductus arteriosus (A) and the common routes of collateral circulation that develop in association with the defect (B).

ductus arteriosus (Fig. 18–52 A). When the aorta is constricted below the ductus, an extensive collateral circulation often develops to assist with the movement of blood to the peripheral parts of the body (Fig. 18–52 B).

Failure of the distal portion of the sixth aortic arch to involute at birth to form the ligamentum arteriosum leads to *patent ductus arteriosus*. The primary cause of patency is failure of the contraction of the musculature surrounding the ductus. It is a defect associated with maternal rubella infection during early pregnancy.

## References

Bernanke, D. and R. Markwald, 1982. Migratory behavior of cardiac cushion tissue cells in a collagen-lattice culture system. Dev. Biol. 91:235–245.

Broyles, R., G. Johnson, P. Maples, and G. Kindell. 1981. Two erythropoietic microenvironments and two larval red cell lines in bullfrog tadpoles. Dev. Biol. 81:299–314.

Chapman, B. and A. Tobin. 1979. Distribution of developmentally regulated hemoglobins in embryonic erythroid populations. Dev. Biol. 69:375–387.

Cole, R. and T. Regan. 1977. Regulation of prenatal haemopoiesis: evidence for negative feedback control of erythropoiesis in the foetal mouse. J. Embryol. Exp. Morphol. 37:237–249.

Davis, C. L., 1927. Development of the human heart from its first appearance to the stage found in embryos of twenty paired somites. Carnegie Contrib. Embryol. 19:245–284.

DeHaan, R. L., 1963a. Organization of the cardiogenic plate in the early chick embryo. Acta Embryol. Morphol. Exp. 6:26–38.

DeHaan, R. L. 1963b. Regional organization of prepacemaker cells in the cardiac primordia of the early chick embryo. J. Embryol. Exp. Morphol. 11:65–76.

DeHaan, R. L. 1965. Morphogenesis of the vertebrate heart. In: Organogenesis. R. L. DeHaan and H. Ursprung, eds. New York: Holt, Rinehart and Winston.

DeVries, P. A. and J. B. de C. M. Saunders. 1982. Development of the ventricle and spiral outflow tract in the human heart. A contribution to the development of the human heart from age group IX to age group XV. Carnegie Contrib. Embryol. 37:87–114.

Fitzharris, T. and R. Markwald. 1982. Cellular migration through the cardiac jelly matrix: a stereoanalysis by high-voltage electron microscopy. Dev. Biol. 92:315–329.

Hamilton, W. and H. Mossman. 1972. Human Embryology. London: Macmillan

Ho, E. and Y. Shimada. 1978. Formation of the epicardium studied with the scanning electron microscope. Dev. Biol. 66:578–585.

Houssaint, E. 1981. Differentiation of the mouse hepatic primordium. II. Extrinsic origin of the haemopoietic cell line. Cell Differ. 10:243–252.

Huehns, E., N. Dance, G. Beavens, F. Hecht, and A. Motulsky. 1964. Human embryonic hemoglobins. Cold Spring Harbor Symp. Quant. Biol. 29:327–331.

Hurle, J., J. Icardo, and J. Ojeda. 1980. Compositional and structural heterogeneity of the cardiac jelly of the chick embryo tubular heart: a TEM, SEM and histochemical study. J. Embryol. Exp. Morphol. 56:211–223.

Icardo, J., J. Ojeda, and J. Hurle. 1982. Endocardial cell polarity during looping of the heart in the chick embryo. Dev. Biol. 90:203–209.

Imaizumi-Scherrer, M., K. Maundrell, O. Civelli, and K. Scherrer. 1982. Transcriptional and post-transcriptional regulation in duck erythroblasts. Dev. Biol. 93:126–138.

Jacobson, A. G. and J. T. Duncan. 1968. Heart induction in salamanders. J. Exp. Zool. 167:79–103.

Jargiello, D. and A. Caplan. 1983. The establishment of vascular-derived microenvironments in the developing chick wing. Dev. Biol. 97:364–374.

Kramer, T. C. 1942. The partitioning of the truncus and conus and the formation of the membranous portion of the interventricular septum in the human heart. Am. J. Anat. 71:343–370.

Llorca, F. and D. Gill. 1967. A causal analysis of the heart curvature in the chick embryo. Wilhelm Roux' Arch. Entwicklungsmech. Org. 158:52–63.

Manasek, F. 1976. Heart development: interactions involved in cardiac morphogenesis. In: The Cell Surface in Animal Embryogenesis and Development, pp. 545–598. G. Poste and G. L. Nicolson, eds. New York: Elsevier/North-Holland Biomedical Press.

Manasek, F. 1981. Determinants of heart shape in early embryos. Fed. Proc. 40:2011–2016.

Manasek, F. J., M. Burnside, and R. Waterman. 1972. Myocardial cell shape change as a mechanism of embryonic heart looping. Dev. Biol. 29:349–371.

Markwald, R., T. Fitzharris, H. Bank, and D. Bernanke. 1978. Structural analyses on the matrical organization of glycosaminoglycans in developing endocardial cushions. Dev. Biol. 62:292–316.

Markwald, R., T. Fitzharris, D. Bolender, and D. Bernanke. 1979. Structural analysis of cell:matrix association during the morphogenesis of atrioventricular cushion tissue. Dev. Biol. 69:634–654.

Moore, K. 1977. The Developing Human: Clinically Oriented Embryology. Philadelphia: W. B. Saunders Company.

Morse, D. and M. Hendrix. 1980. Atrial septation. II. Formation of the foramina secunda in the chick. Dev. Biol. 78:25–36.

Nakamura, A. and F. Manasek. 1981. An experimental study of the relation of cardiac jelly to the shape of the early chick embryonic heart. J. Embryol. Exp. Morphol. 65:235–256.

Orts-Llorca, F., J. Fonolla, and J. Sobrado. 1982. The formation, septation and fate of the truncus arteriosus in man. J. Anat. 134:41–56.

Patten, B. 1960. Persistent interatrial foramen primum. Am. J. Anat. 107:271–280.

Rawles, M. E. 1943. The heart-forming areas of the early chick blastoderm. Physiol. Zool. 16:22–43.

Rich, I. and B. Kubanek. 1980. The ontogeny of erythropoiesis in the mouse detected by the erythroid colony-forming technique. II. Transition in erythropoietin sensitivity during development. J. Embryol. Exp. Morphol. 58:143–155.

Rifkind, R. 1974. Erythroid cell differentiation. In: Concepts of Development. J. Lash and J. R. Whittaker, eds. Sunderland, Mass.: Sinauer Associates, Inc.

Rychter, Z. 1962. Experimental morphology of the aortic arches and the heart loop in chick embryos. Adv. Morphog. 2:333–371.

Thompson, R., Y. Wong, and T. Fitzharris. 1983. A computer graphic study of cardiac truncal septation. Anat. Rec. 206:207–214.

Turpen, J., C. Knudson, and P. Hoefen. 1981. The early ontogeny of hematopoietic cells studied by grafting cytogenetically labeled tissue analgen: localization of a prospective stem cell compartment. Dev. Biol. 85:99–112.

Viŕagh, S. and C. Challice. 1982. The development of the conducting system in the mouse embryo heart. IV. Differentiation of the atrioventricular conducting system. Dev. Biol. 89:25–40.

Waddington, C. H. 1932. Experiments on the development of chick and duck embryos cultivated *in vitro*. Philos. Trans. R. Soc. Lond. (Biol.) 221:179–230.

Weatherall, D. and J. Clegg. 1979. Recent developments in the molecular genetics of human hemoglobin. Cell 16:467–479.

Zagris, N. 1980. Erythroid cell differentiation in unincubated chick blastoderm in culture. J. Embryol. Exp. Morphol. 58:209–216.

# 19

# THE UROGENITAL SYSTEM

The excretory and the reproductive systems are closely associated anatomically, developmentally, and functionally—a fact implied by the use of the term urogenital system. Both systems develop from a common mesodermal ridge that runs longitudinally along the posterior abdominal wall, lateral to the dorsal mesentery of the gut. The ducts of both systems open into the same cavity, the cloaca. In addition, in the male, parts of the functional excretory system of the embryo, when they lose their excretory responsibilities, do not degenerate but are converted into parts of the functional reproductive system of the adult. Both systems in the adult male share a common excretory pathway, the urethra.

However, although they develop in close association with each other, for descriptive purposes we will consider the two systems separately.

## The Excretory System

### The Intermediate Mesoderm

When the embryonic mesoderm first appears, it is concentrated around the notochordal process where it is known as the paraxial mesoderm. From here it spreads out laterally between the ectoderm and the endoderm as the lateral plate mesoderm (Fig. 19–1 A). Shortly after its appearance, the paraxial mesoderm begins to form segmentally arranged somites proceeding in a craniocaudal direction. The lateral mesoderm splits into two layers to surround the coelom: somatic mesoderm associated with the ectoderm (together termed the somatopleure), and splanchnic mesoderm associated with the endoderm (together termed the splanchnopleure). The region of the lateral mesoderm between the somite and the coelom is the intermediate mesoderm (Fig. 19–1 B). The intermediate mesoderm will give rise to the excretory system.

During the development of the mammalian excretory system, three different types of kidneys appear. These are the *pronephros*, the *mesonephros*, and the *metanephros*. They develop in a temporal sequence as well as in a spatial craniocaudal sequence (Fig. 19–2). Since all three types develop from the same longitudinal ridge of nephrogenic tissue—often with no definite boundaries between them—there is some justification for considering all three as parts of a single unit that develop in different regions and at different times, as a progressive differentiation along a continuum.

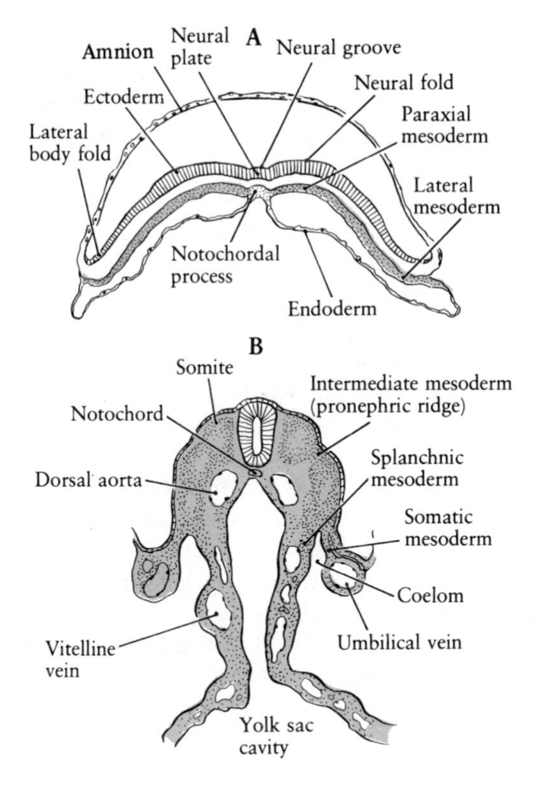


**FIG. 19–1.** A, section through the neural plate of an 18-day-old presomite human embryo. Mesoderm lateral to the notochord subdivided into paraxial and lateral portions. B, section through the somite region of a 22-day-old 10-somite human embryo. The intermediate mesoderm, here differentiating into pronephros and called the pronephric ridge, is located between the somite mesoderm and the coelom.

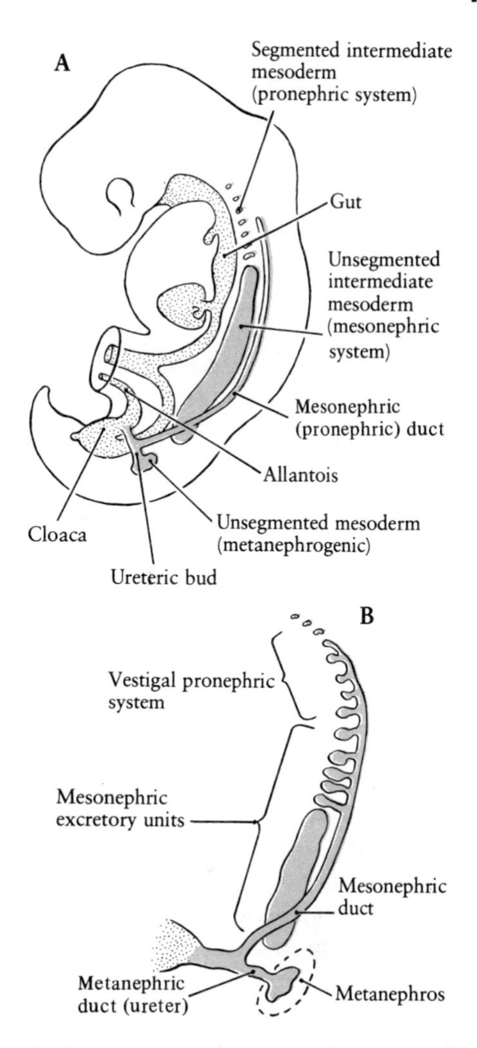


**FIG. 19–2.** Diagram showing the development of the pronephros, the mesonephros, and the metanephros from the intermediate mesoderm (nephrogenic ridge) in a craniocaudal sequence.

## The Pronephros

The pronephros or head kidney, never functional in man, is the functional kidney of some lower vertebrates. A functional pronephros (Fig. 19–3) consists of paired pronephric tubules developed from the nephrogenous tissue. One end of each tubule opens into the coelom, the other into a longitudinally running duct, the *pronephric duct,* uniting the lateral ends of successive pronephric tubules. A branch of the dorsal aorta indents either the pronephric tubule or the coelom close to the point where the pronephric tubule opens into it. The former is termed an *internal glomerulus,* the latter an *external glomerulus.* The glomerulus is covered by a thin layer of epithelium through which wastes are filtered either into the tubule or into the coelom. Wastes in the coelom move into the tubule through the *peritoneal funnel.* Wastes that are filtered through the internal glomerulus enter an expanded portion of the pronephric tubule, the *nephrocoele,* from which they pass into a more lateral portion of the tubule through the *nephrostome.* The tubules drain into the pronephric duct, the excretory duct opening into the cloaca.

In the human embryo, the pronephros is made up of 7 to 10 pairs of solid "tubules" that develop from the nephrogenic cord in the cervical region during week 3. The exact caudal limit of the pronephros is not well defined, since it overlaps the territory of the mesonephros. Pronephric tubules are present for only a short period of time, the more anterior degenerating while the more posterior are still developing. By the 5-millimeter stage, all of the

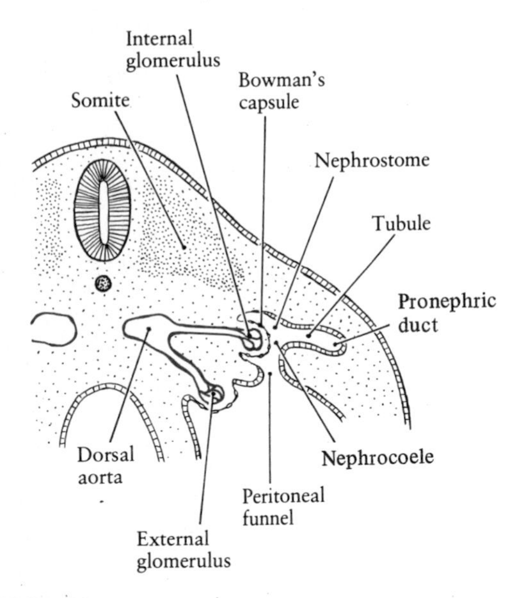


**FIG. 19–3.** Schematic drawing of a functional pronephros showing the relationship of the two types of glomeruli to the coelom.

tubules have degenerated. As a part of the pronephros, a pronephric duct is formed by joining of segmental delaminations from the nephrogenic ridge. Below the region of the pronephros, the blind end of the pronephric duct grows caudally as a solid rod of cells that subsequently hollows out. It opens into the cloaca. The rudimentary pronephric tubules do not establish any connection with the pronephric duct. Although the tubules quickly degenerate, the pronephric duct persists and becomes the functional excretory duct of the mesonephros when this structure develops (Fig. 19–2 B). In the male, it eventually forms a part of the reproductive tract when the mesonephros degenerates.

## The Mesonephros

The mesonephros in the human embryo extends from about the 10th to the 26th somite (fourth lumbar). The *mesonephric tubules* arise from the nephrogenic cord in a caudally running progression. Mesonephric tubules are not segmentally arranged, and as many as three to four per segment may be present. When this organ reaches its maximum development during month 2, approximately 35 to 49 tubules are present. Each tubule connects laterally to what was originally the pronephric duct but which is now termed the *mesonephric* or *Wolffian duct* (Fig. 19–4 A). The medial end of each tubule becomes invaginated by a glomerulus around which forms a thin-walled double epithelial membrane, *Bowman's capsule.* The capsule and the glomerulus together make up the mesonephric or *renal corpuscle* (Fig. 19–4 B). The tubule becomes S-shaped, and a secretory region near the capsule and an excretory region toward the duct may be recognized. The size of the human mesonephros is somewhat small compared to the large organ developed in the pig or the rabbit, but is large in comparison to that of the rat or the mouse. The size of the mesonephros appears to bear some relation to the permeability of the placenta.

As the mesonephric tubules increase in size, the mesonephros cannot be accommodated in the body wall, and it bulges into the coelom as a longitudinal ridge lateral to the dorsal mesentery. Since the gonad also develops from this tissue, it is called the *urogenital ridge* (Fig. 19–4 A,B). It is suspended from the dorsal body wall by the *urogenital mesentery.* The urogenital ridge soon divides into a medial *genital ridge* and a lateral *mesonephric ridge* (Fig. 19–4 C).

## The Metanephros

The metanephros is the last of the three types of excretory organs to appear, and it will develop into the functional kidney of the adult. It has a dual origin: (1) the collecting system develops from an outgrowth of the mesonephric duct; (2) the excretory system develops from the nephrogenic ridge caudal to the mesonephros.

The *metanephric diverticulum (ureteric bud)* appears as an outgrowth from the dorsomesial wall of the mesonephric duct just before it enters the cloaca (Fig. 19–5 A). The diverticulum grows first dorsally and then cranially, forming an elongated duct, the *ureter.* The distal blind end of this duct expands to form the *renal pelvis.* It pushes into the caudal end of the nephrogenic ridge, which then becomes molded about it as the *metanephric cap* (Fig. 19–5 B). The renal pelvis divides to form two or three primary tubules that are the future *major calyces* (Fig. 19–5 C). Each primary tubule further subdivides and forms secondary tubules that represent the future *minor calyces* (Fig. 19–5 D). Two to four open into each major calyx. The secondary tubules continue to subdivide until approximately 13 to 14 generations of tubules are formed. As each tubule divides, the metanephric cap is also subdi-

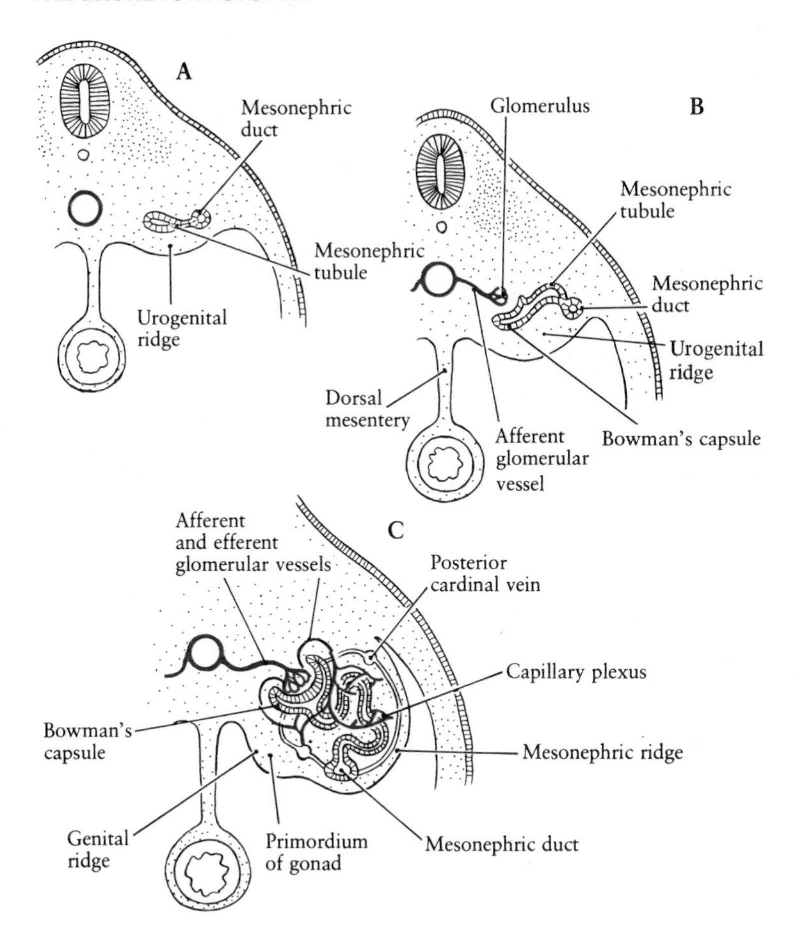


**FIG. 19–4.** Successive stages in the development of the human mesonephros during the fifth and sixth weeks. A, connection of a mesonephric tubule to the mesonephric duct; B, medial end of a mesonephric tubule being invaginated by a glomerulus to form a renal corpuscle; C, renal corpuscle, mesonephric tubule, and mesonephric duct differentiated. Genital ridge forming medial to the mesonephric tissue.

vided so that the end of each tubule always retains an individual covering of metanephrogenic tissue. As the secondary tubules enlarge and develop into the minor calyces, they absorb the third- and fourth-generation tubules so that the tubules of the fifth order, some 10 to 25 in number, open into each minor calyx. These are the *papillary ducts* of the adult kidney. Tubules of higher orders form the collecting tubules that converge on the papillary ducts forming the renal pyramids (Fig. 19–5 E).

The metanephric cap moves laterally and forms clusters of tissue on each side of the collecting tubule. These metanephric vesicles appear first in the 18- to 20-millimeter embryo during month 2 (Fig. 19–6 A,B). The vesicle enlarges and becomes S-shaped. The limb of the S away from the collecting tubule is invaginated by a tuft of capillaries, the glomerulus, to form the renal corpuscle (Fig. 19–6 C). Both ends of the S-shaped vesicle remain relatively fixed in position, and secondary curvatures and histological modification of the tissue between these points produce the complicated twisting nephric tubule. Convolutions near the glomerulus and near the collecting tubule form the *proximal* and *distal convoluted tubules,* respectively (Fig. 19–6 D,E). Between these two regions, the remaining portion of the nephron develops into the thin-walled *loop of Henle* (Fig. 19–6 D,E). The glomerulus and the convoluted tubules become part of the *cortex* of the kidney, and the collecting tubules draining into the minor calyces form a part of the *medulla.* The loops of Henle dip down into the medulla (Fig. 19–7).

The confluence of all of the papillary ducts that open into a minor calyx forms the apex of a renal unit, the *pyramid,* whose base is in the cortex (Fig. 19–7). In the embryo and the newborn, the surface of the kidney is grooved, forming lobules indicating the location of the primary pyramids. These grooves gradually disappear, and lobulations are not seen in the adult kidney.

The human kidney begins to function during the third month of development, in the sense that it be-

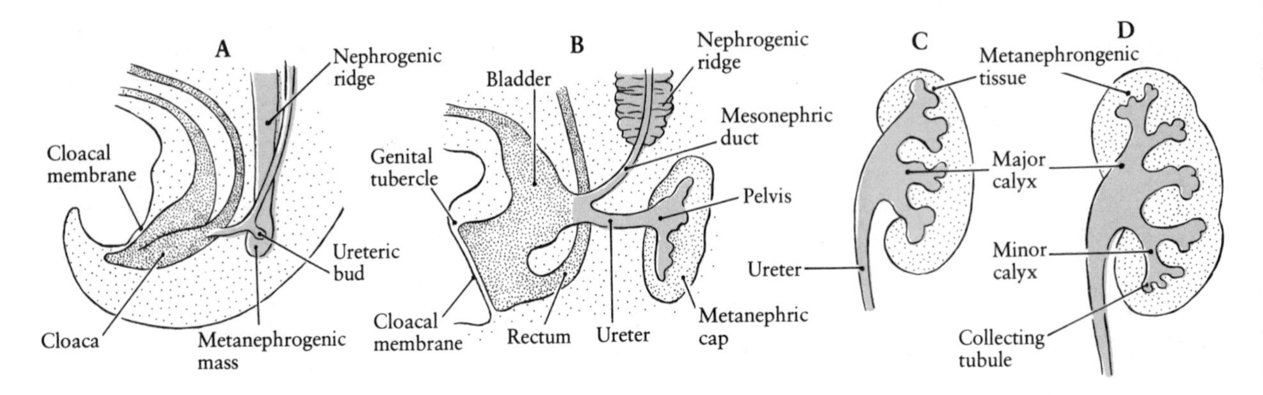


gins to secrete urine and excrete it through its duct system into the bladder. This, of course, serves no necessary excretory function, since the placenta serves in this capacity. The small amount of urine formed by the embryonic kidney is passed from the bladder into the amniotic sac where it is swallowed along with other amniotic fluids.

The kidney is unusual in that it undergoes a cranial migration, from its original pelvic location to a position opposite the first lumbar vertebra. This positional change, which brings the cranial pole of the kidney into contact with the caudally migrating suprarenal gland, is due in part to an active migration and in part to the straightening of the lumbar curvature.

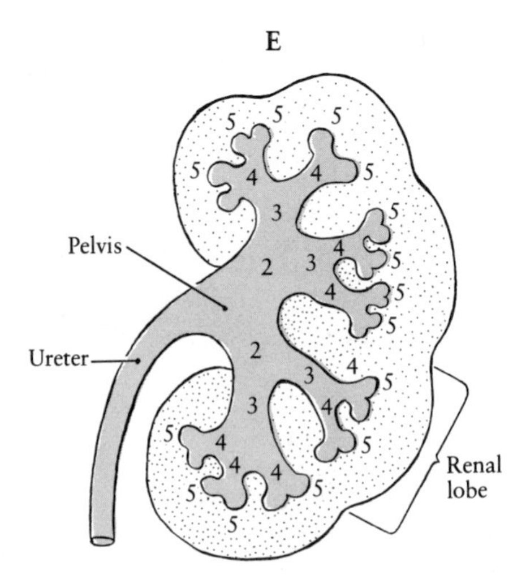


FIG. 19–5. Origin of the human metanephric duct (ureteric bud) from the mesonephric duct and its combination with the nephrogenic ridge tissue (metanephrogenic mass) to form the kidney. A, 5 mm; B, 11 mm; C, D, development of the distal end of the ureteric bud into the renal pelvis, calyces, and the duct system of the kidney. C, 15 mm; D, 20 mm; E, 30 mm: further branching of the collecting system of the kidney. The numbers designate the subdivisions of the diverticula from the pelvis: 2, major calyces; 3, minor calyces.

## Dependent Differentiation

Neither the mesonephric nor the metanephric tubules are self-differentiating, the former needing the presence of the mesonephric (pronephric) duct, the latter needing the presence of the ureteric bud. If the pronephric duct is transected at any level, its further caudal extension is effectively blocked (Fig. 19–8 A). If the caudal extension of the pronephric duct is prevented from reaching the mesonephric level of the nephrogenic cord, mesonephric tubules do not differentiate (Fig. 19–8 B). In addition, if the ureteric bud is prevented from reaching the metanephrogenic region of the nephrogenic cord, a metanephrogenic blastema may form, but tubules will not differentiate. However, in both the chick and the mammal, central nervous system tissue will promote tubule formation in mesonephrogenic and metanephrogenic mesenchyme; the action is thus more comparable to evocation and is not the result of a specific induction. An entirely normal mesonephros is never formed in the presence of nervous tissue only, and thus some of the determination of the structure must be attributed to the influence of the pronephric duct beyond its action as an evocator. In the light of the interrelationship of the kidney and the mesonephric duct, it is not surprising that, in the human male, kidney agenesis on one side is accompanied by the absence of the vas deferens on the same side, since the vas deferens develops from the mesonephric duct.

The differentiation of the metanephros is an excellent example of one of a number of secondary inductions that take place during development. Many of the results obtained from studies on primary induction have been found to apply also to

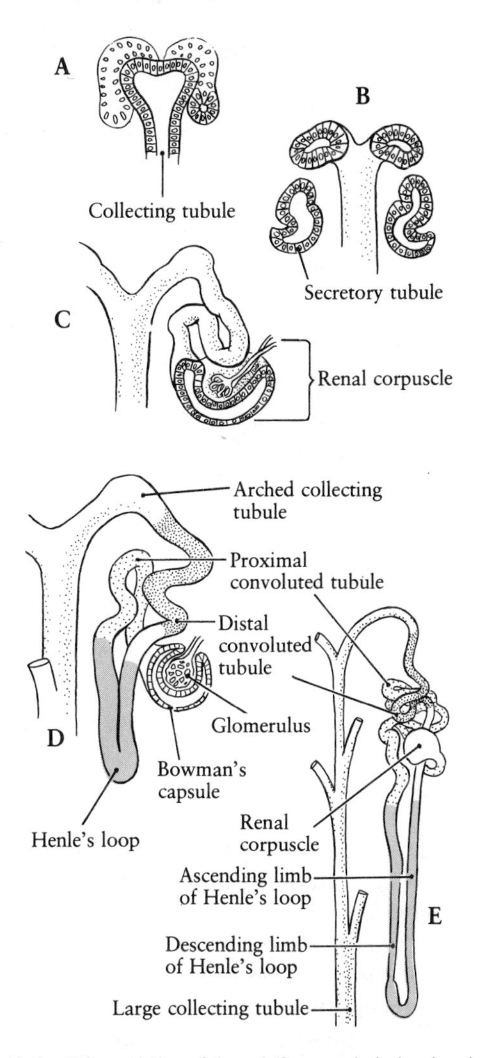


**FIG. 19–6.** Differentiation of the uriniferous tubule (nephron) from the metanephrogenic tissue.

secondary inductions. It was early determined that the morphogenetic signal for kidney tubule induction could pass through various types of membrane filters. However, the percentage of inductions decreased as the pore size of the membranes decreased. While filters with pore sizes of 0.2 mm or greater regularly permitted the inductive signal to pass, when filters with a pore size of 0.1 mm were interposed between the inducer and the metanephrogenic tissue, tubules were formed only occasionally. Examination of the filters by light and electron microscopy gave a good correlation between the ingrowth of cytoplasmic processes into the pores of the filer and the transmission of the inductive signal. Filters with a pore size of 0.2 mm or larger uniformly showed ingrowths of cytoplasmic processes from both the inducing and the responding tissues. In addition to pore size, pore density is important. The greater the number of pores per unit area, the greater the number of tubules formed. Thus, contact area is also of importance in inductive activity.

Although close apposition of cellular extensions is necessary for kidney tubule induction, the mechanism of the process and the nature of the signal remain speculative. It is possible that apposition might be needed for the formation of specialized junctions between the cell membranes allowing the passage of the inducer into the target cell. However, such specialized junctions have not been found between cytoplasmic processes within the filter pores, although this does not rule out the possibility of their presence, since their identification within the filter is difficult.

Experiments using metabolic inhibitors have shed some light on the nature of the stimulus. Treatment of the inducing tissue with DNA inhibitors does not affect its capacity for induction. However, tubules will not form in kidney mesenchyme that has been treated with the same inhibitors. One would of course expect that cell division would be necessary during tubule formation. Inhibitors of both RNA and protein synthesis will abolish the inductive capacity of the inducing tissue. If the explant (inducer and kidney mesenchyme) is cultured in 6-diazo-5-oxo-L-norleucine (DON), a compound that inhibits glycosaminoglycan synthesis, tubule formation does not take place. The inhibitory effect of DON may be negated by adding glucosamine to the culture. This then suggests that GAGs may be involved either in the establishment of cell contact between the two systems or in the interchange of messages between the inducing and reacting tissues.

## The Bladder and the Urethra

The caudal portion of the hindgut to which the allantois is connected and into which the pronephric duct originally opens is the *cloaca* (Fig. 19–9). A cloacal membrane separates the cloaca from the proctodeum. At about the 5-millimeter stage, near the end of the first month, a mesodermal partition, the *urorectal septum*, begins fo form at the cranial angle between the allantois and the hindgut (Fig. 19–9 B). During the second month, the urorectal

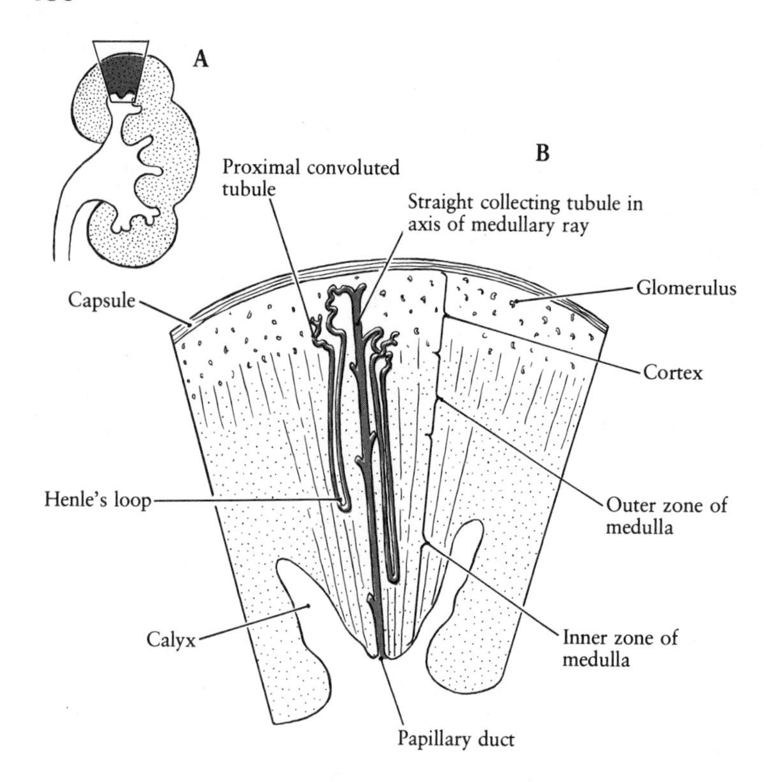


**FIG. 19–7.** A, Drawing of a pyramid of the kidney of a six-month-old embryo showing the relationship of the uriniferous tubules and glomeruli to the medulla and cortex. Inset, B, relationship of lobule drawn to the entire kidney.

**FIG. 19–8.** A, absence of the nephric duct on the right side of a chick embryo following transection of the duct at a more cranial level; B, failure of the mesonephric tubules to form on the left side of a chick embryo in the absence of the left nephric duct. Only a formless blastema is present. (After C. H. Waddington, 1938.)

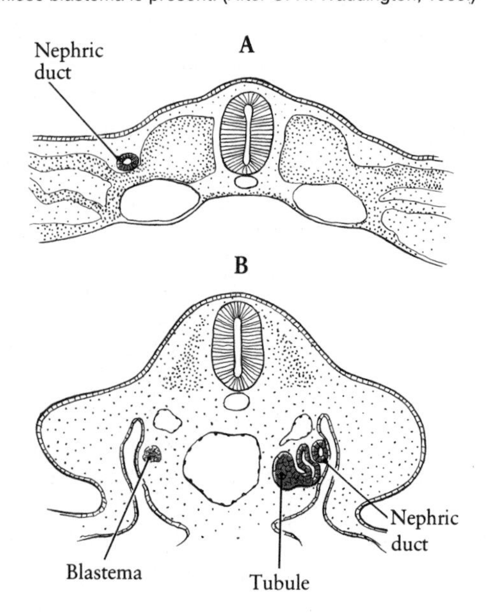


septum grows toward and fuses with the cloacal membrane, partitioning the cloaca into an anterior *urogenital sinus* and *bladder* and a posterior *rectum* (Fig. 19–9 B,C). At the point of fusion of the urorectal septum and the cloacal membrane, the *perineal body* is formed; the cloacal membrane is now divided into a urogenital membrane and an anal membrane, indicating the separate openings to the outside of the urogenital and the gastrointestinal systems. Between them is the *perineum.* In most vertebrates the cloaca is not subdivided and remains as a common outlet for both the digestive and the urogenital systems.

As the urorectal septum develops, the mesonephric ducts, from which the ureteric buds are developing, now open into the urogenital sinus (Fig. 19–9 B). Growth of the region of the urogenital sinus, into which the mesonephric ducts open, imposes marked changes on the relationship of the mesonephric and metanephric ducts (Fig. 19–10). The caudal ends of the mesonephric ducts are gradually absorbed by the expanding urogenital sinus, and the mesonephric and metanephric ducts each now obtain their own individual openings into the urogenital sinus (Fig. 19–10 B). A cranial movement of the metanephric duct openings and a movement of

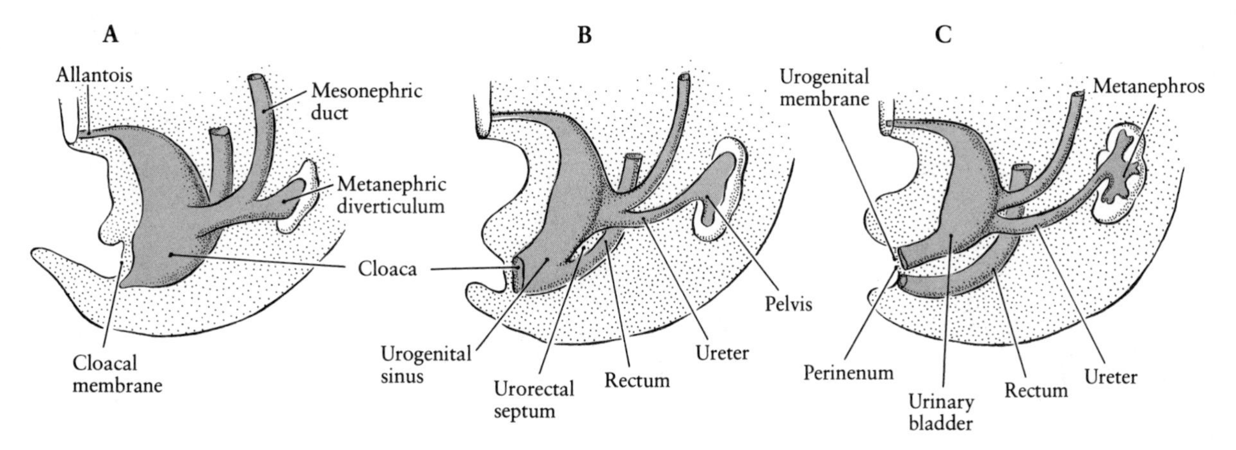


**FIG. 19–9.** Diagrams of the formation of the urorectal septum and the separation of the cloaca into the urogenital sinus and the rectum. A, four weeks; B, five weeks; C, six weeks.

the openings of the mesonephric ducts medially toward each other result in the definitive configuration of a triangle, the base of which is craniad and is represented by the openings of the metanephric ducts while the apex is caudal and represented by the openings of the mesonephric ducts (Fig. 19–10 C,D). This is the *trigone* of the adult. The mesonephric ducts open on an elevation on the posterior wall of the urogenital sinus, which is called *Muller's tubercle*.

**FIG. 19–10.** Dorsal view of the bladder showing the relationship of the mesonephric ducts and the ureter. The two systems attain separate openings, and the ureters move craniad and the mesonephric ducts caudad.

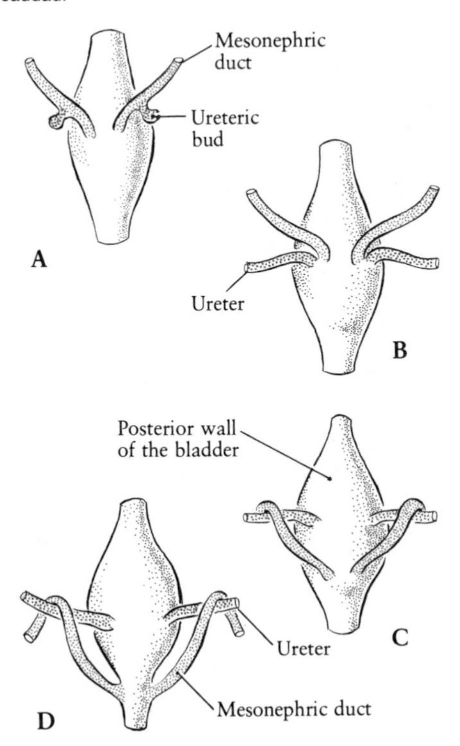


The bladder develops from the anterior part of the former cloaca, craniad of the openings of the mesonephric ducts. It is thus mainly formed by the expansion of the proximal part of the allantois, although the urogenital sinus makes some contribution. The allantois distal to the part that forms the bladder becomes reduced in size, and its lumen is occluded. It remains as a cord connecting the cranial end of the bladder to the umbilicus. In the fetus it is called the *urachus* (Fig. 19–11 A). In the adult it is the *middle umbilical ligament*. Uncommonly the distal portion of the allantois, between the bladder and the umbilicus, remains patent, establishing a *urachal fistula* through which urine may escape.

The urogenital sinus may be separated into a cranial (pelvic) portion nearest to the bladder, receiving the openings of the mesonephric ducts, and a more caudal (phallic) portion. Depending on the sex of the individual, the fates of these two parts of the urogenital sinus differ, as will be described more completely in the section on the reproductive system. In brief, in the male, the urogenital sinus forms a long *urethra* whose channel is extended to an opening at the tip of the penis (Fig. 19–11 A). In the female, a short urethra forms corresponding only to the extent of the male urethra from the base of the bladder to Muller's tubercle. The remainder

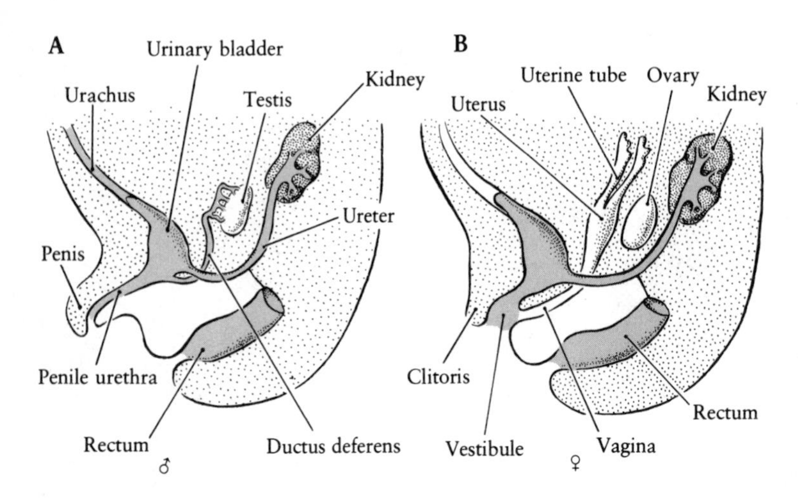


**FIG. 19–11.** Diagrams of 12-week-old male and female urogenital systems. In the male, the urethra is long and extends to the tip of the penis. The short female urethra opens into the vestibule.

of the pelvic (plus all of the phallic) portions unite to form a shallow *vestibule* into which both the urinary and the genital ducts open (Fig. 19–11 B).

## The Genital System

As mentioned in the previous section, the gonads develop in close association with the mesonephros as a part of a longitudinally running thickened ridge, the urogenital ridge, developed from the intermediate mesoderm and located on the dorsal body wall lateral to the dorsal mesentery. During the fifth week, the common ridge begins to separate into two, and a ventromedial thickening, the genital ridge, represents the portion of the structure that will develop into the gonad (Fig. 19–4).

### The Primordial Germ Cells

As the following sections in this chapter will illustrate, there is now conclusive evidence that the germ cells of the gonads are of extragonadal origin and are set aside early in development to become the stem cells, *primordial germ cells,* of the germ line. Descriptive studies using the light and the electron microscope, as well as cytochemical, biochemical, and immunological analysis have clearly shown easily recognizable differences between primordial germ cells and somatic cells in the blastula and early gastrula stages in many species, and other studies have been able to trace these cells from their extragonadal locations into the developing gonad. Transplantation operations and the destruction of the primordial germ cells by irradiation or other methods have shown that without these cells, although a gonad may develop, it will always lack germ cells. The primordial germ cells are thus the precursors of the germ line and indispensable for the development of the entire future population of sperm and ova.

That a specialized cytoplasmic substance could direct the differentiation of cells into the germ line was first discovered in studies on insects. At about the middle of the 19th century, it was known that the cells that occupy the posterior pole of the insect egg are those destined to become the germ cells. Shortly after the turn of the century, Hegnar proposed that the granules observed in the pole plasm are, in fact, the germinal determinants. Since that time, numerous experiments have shown that the cytoplasm of the posterior pole of the dipterans and the coleopterans possesses some factor that determines that the nuclei that colonize it will differentiate into germ cells while those that do not will become somatic cells. A brief discussion of some of the experimental analyses of the role of the pole plasm has been presented in Chapter 11.

#### *The Germ Plasm and the Primordial Germ Cells in Anurans*

Studies on the germ plasm and the germ cells of the anurans have yielded strikingly similar results to those on the insects. In *Xenopus* the primordial germ cells may be distinguished on the basis of microscopically recognizable cytoplasmic inclusions, the germ plasm. The germ plasm is found at the vegetal pole of the egg, and in the two-cell stage it is aggregated in small patches extending several millimeters away from the pole (Fig. 19–12 A). As the two-cell stage divides, the aggregates move to-

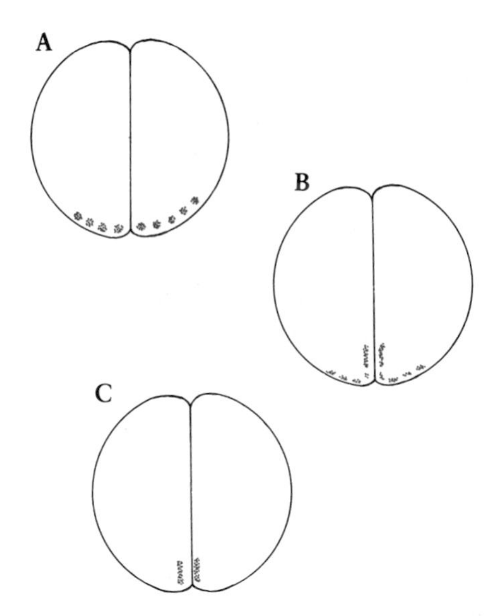


**FIG. 19–12.** Diagrams of the location and state of aggregation of the germ plasm granules during the first two cleavages in *Xenopus*. A, two-celled stage showing individual aggregates in the vegetal pole region; B, beginning of the second cleavage. Aggregates moving toward the cleavage furrow and coalescing; C, section of a four-celled embryo. Granules in a single aggregate in each cell along the cleavage furrow. (After P. McD. Whitington and K. E. Dixon, 1975. J. Embryol. exp. Morphol. 33:57.)

ward the cleavage furrow, coalescing as they do so (Fig. 19–12 B), and in the majority of the eggs each of the four cells of the four-cell stage has a single aggregate located close to the cleavage furrow (Fig. 19–12 C). The average number of cells containing germ plasm granules in all embryos studied increases only slightly to about five in the blastula stage. This is not because cells containing germ plasm do not divide. They do, as can be seen by the progressive decrease in size of the cells that are found to have germ plasm granules. However, the germ plasm is located at the polar region of each cell in which it appears, and therefore, in the process of normal cleavage it is passed on to only one of the two daughter cells. An increase in the number of cells containing pole plasm in some embryos is merely due to the inherent variability of the cleavage pattern. The number of germ plasm-containing cells in the early gastrula depends solely on the relative positions of the cleavage planes and the germ plasm. During gastrulation the germ plasm migrates from the polar region of the cell to a position capping or ringing the nucleus. From this time on until the early tadpole stage, when the primordial germ cells begin their migration out of the endoderm, each goes through two to three divisions. These, as are all other successive divisions, are now cloning, and each of the daughter cells receives some germ plasm, although not in the exact same amount.

Because of the original position of the germ plasm, the primordial germ cells are located above and around the vegetal pole through the blastula stage (Fig. 19–13 A). During gastrulation they participate in the same morphogenetic movements of the surrounding endoderm cells as a part of the invaginating endoderm. At the end of gastrulation and during the tail bud stages, they are located in the deep floor of the archenteron near the blastopore (Fig. 19–13 B,C). In the early tadpole stage they undergo a dramatic change in position and are now found in the most dorsal region of the endoderm of the gut (Fig. 19–13 D). When the two sheets of lateral mesoderm that are moving dorsally to surround the gut meet, the primordial germ cells leave the endoderm and lie below this sheet of mesoderm (Fig. 19–13 E). When this mesoderm then forms the dorsal mesentery, the primordial germ cells become embedded in it, (Fig. 19–14 A), and from here they migrate dorsally through the dorsal mesoderm and then laterally into the developing genital ridges (Fig. 19–14 B,C). During their migration the germ cells are always completely covered by the cytoplasm of the adjacent somatic cells. Specialized junctions (desmosomes) are formed between the somatic cells, but they are not found between the germ cells themselves or between germ cells and somatic cells. Once the genital ridge has developed and been populated by germ cells, a few junctions between germ cells and somatic cells may be formed, indicating a more permanent relationship once the migratory process is completed. It is probable that the migration of the germ cells is due mainly to their own active amoeboid movement. This type of movement has been demonstrated by germ cells cultured in vitro.

The cells of the splanchnic mesoderm of the body wall lateral to the dorsal mesentery proliferate and differentiate into an organized band of cells, the genital ridge, covering and containing the germ cells. These mesodermal cells will form the germinal epithelium of the gonad. During their rather extensive migration, the germ cells rarely show any mitotic activity. However, once established in the developing gonad, they begin to divide again.

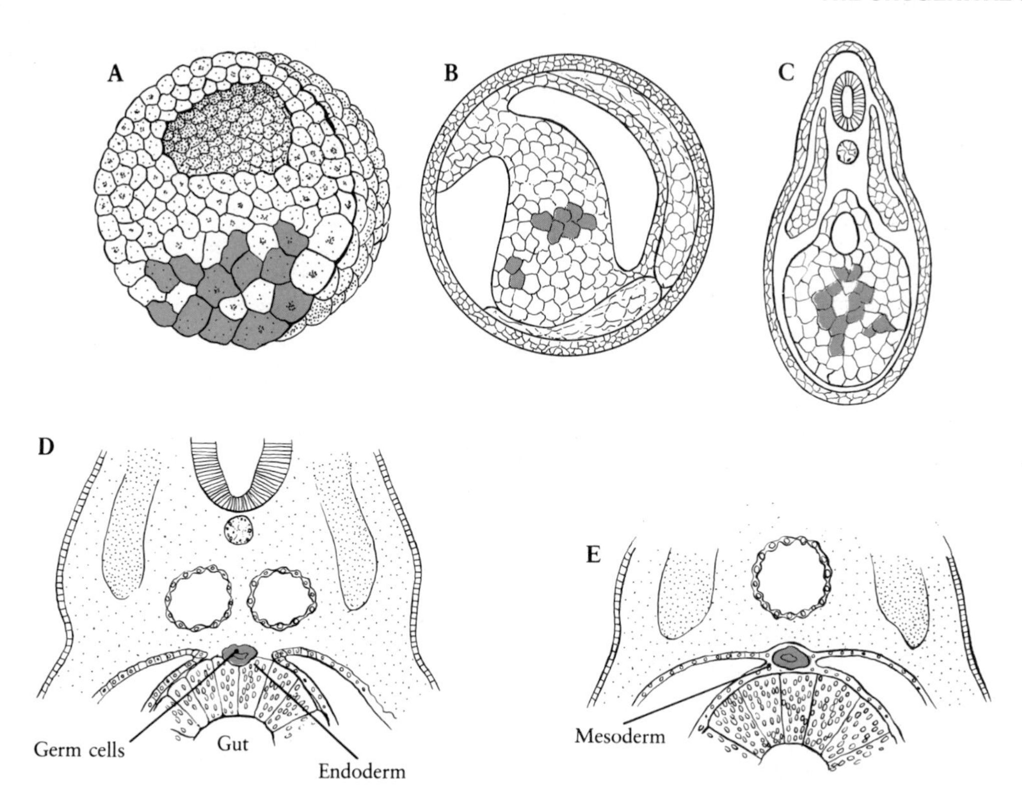


**FIG. 19–13.** Diagrams to represent the position of the primordial germ cells in *Xenopus* embryos of different ages. A, blastula; B, gastrula; C, tail bud; D, localization of the germ cells at the most dorsal region of the endoderm of the gut before the two sheets of splanchnic mesoderm meet in the dorsal midline to form the dorsal mesentery; E, the germ cells have left the endoderm and lie between the sheets of the dorsal mesentery. Germ cells may be located at any of the colored areas. (After P. McD. Whitington and K. E. Dixon, 1974. J. Embryol. exp. Morphol. 33, 57.)

That the primordial germ cells represent the progenitors of the entire germ line has been proven in a number of different experiments. In one, the belly endoderm of one species of *Xenopus* in which the nuclei contain two nucleoli may be replaced by that of another strain in which the nuclei have only a single nucleolus. Following this, the germ cells of the host will then all contain only one nucleolus, and this characteristic of the graft will also be found in all of the host's offspring. Thus, the primordial germ cells grafted into the host in the belly endoderm are the sole basis for the establishment of the germ line.

### *Primordial Germ Cells in the Urodeles*

A similar type of experiment has been performed in the urodele amphibians. In this class of amphibians, germ plasm cannot be seen at the ventral pole of the cleaving egg, and primordial germ cells are not delineated until the gastrula stage, when they may be distinguished among the cells of the lateral mesoderm. Neither removal of cytoplasm at the vegetal pole of the undivided egg nor removal or transplantation of blastomeres at the vegetal pole of the blastula has any effect on the number of germ cells in the larva. Nor does removal in the gastrula stage of the presumptive endoderm that will form the caudal part of the gut have any effect on the germ cells. The transplantation of belly endoderm between two species of urodeles whose germ cells are distinguishable on the basis of pigment granule content has the opposite result from the experiment just described using *Xenopus*. The germ cells are always characteristic of the host. These experiments all demonstrate that neither the germ plasm nor the primordial germ cells are associated with the endoderm in these species. However, if the lateral mesoderm is removed, almost all of the larvae develop gonads lacking germ cells. Grafts of lateral and ventral lip mesoderm (which will form the lat-

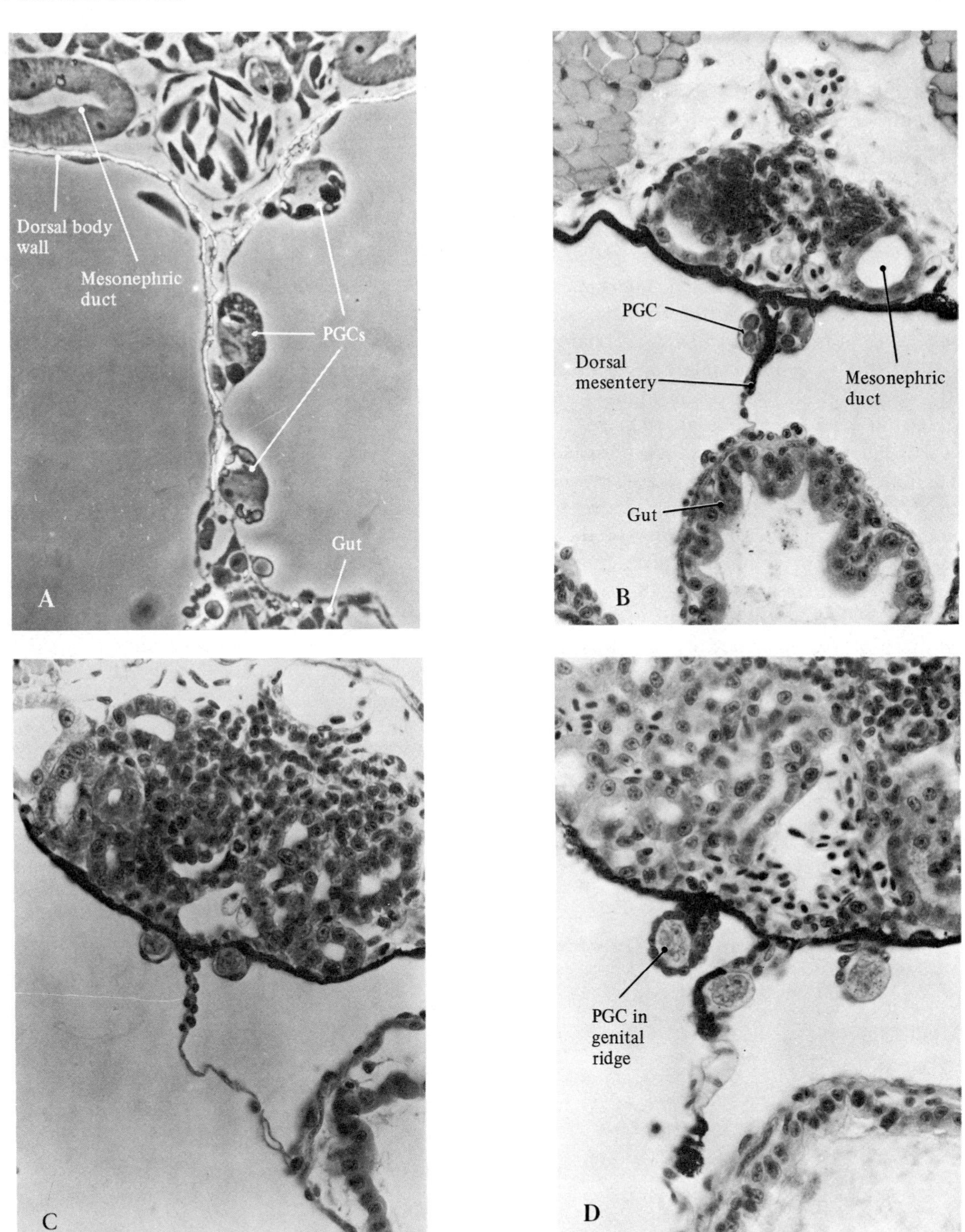


**FIG. 19-14.** Migration of the primordial germ cells in *Xenopus* from the dorsal mesentery into the posterior abdominal wall and the formation of the genital ridges. A, 3 primordial germ cells in the dorsal mesentery; B, primordial germ cells at the root of the dorsal mesentery; C, the germ cells after leaving the dorsal mesentery are in the abdominal wall; D, germ cells in the newly formed genital ridges. (From C. C. Wylee and J. Heasam, 1976. J. Embryol. exp. Morphol. 35, 125.)

eral mesoderm) between *Triturus cristatus* and *Ambystoma mexicanum,* where differences in pigment granules distinguish the cells of the two species, result in the formation of graft-type germ cells. Again, the primordial germ cells, although they appear later and in a different germ layer, are shown to be the forerunners of the entire germ line.

*Primordial Germ Cells in the Chick*

In the chick embryo, about 20 primordial germ cells may be distinguished uniformly distributed in the blastodisc of the unincubated egg, presumably separated from the somatic cells, during the cleavage stages. These cells increase in number and in later blastula stages become localized along the anterior and lateral borders of the blastodisc. In the primitive streak stage, they form a crescent-shaped group of cells anterior to the primitive streak in the extraembryonic area at the juncture of the area opaca and the area pellucida, forming what is known as the *germinal crescent* (Fig. 19–15). Depriving the embryo of any interaction with this germinal crescent, either by destroying it by cauterization or radiation or by explanting the embryo and not the germinal crescent, results in the formation of sterile gonads. Here again we have clear proof of the origin of all of the germ cells from a group of primordial germ cells segregated from the somatic cells early in development.

Interestingly, in the chick the primordial germ cells move to the developing gonads by the way of the vitelline circulation, which they enter by active migration between the endoderm cells at approximately 33 to 38 hours of incubation. Two embryos may be explanted side by side, one of which has been "sterilized" by the removal of the germinal crescent. Unless circulatory connection is established between the explants, the operated embryo will remain sterile. But if a connection does develop, primordial germ cells pass from the normal to the operated embryo and colonize its gonads and produce germ cells. The unoperated embryo, of course, always develops a normal complement of germ cells, since its own primordial germ cells pass to the gonads in the embryo's own circulatory system.

**FIG. 19–15.** Position of the primordial germ cells in the chick embryo between the area opaca and area pellucida forming the germinal crescent craniad of the head process.

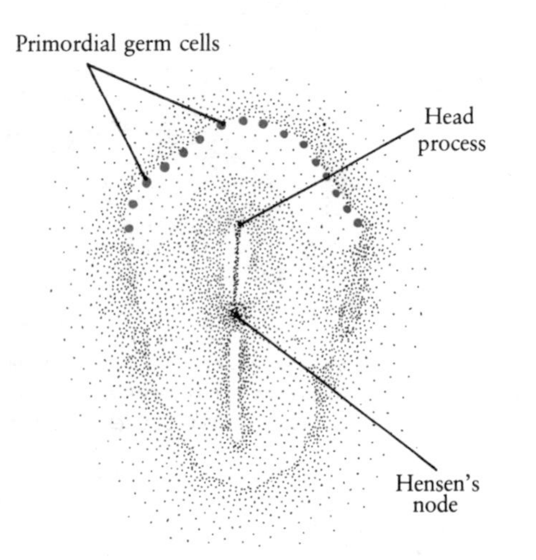


*Relation of the Germ Plasm to the Primordial Germ Cells*

From the above experiments we may conclude that there is clear evidence that cells with specific distinguishing characteristics are set aside early in the development of many species, cells that will later migrate into and populate the genital ridges. Experimental analysis by destruction and transplantation supplements the descriptive work on their migration and establishes the fact that these cells represent the stem cells of the entire germ line. But what is the relation of the germ plasm to the primordial germ cells?

In the anurans, where the germ plasm is situated at a specific location in the cleaving egg, we have the opportunity to perform the same kinds of experiments that we have described in the insects. Irradiation of the vegetal hemisphere of the frog egg *(Rana pipiens)* at the beginning of the first cleavage with a UV dose of 8000 ergs per square millimeter results in the development of larvae lacking germ cells completely. Lower doses have less of an effect, but even a dose as low as 2600 ergs per square millimeter results in an easily observable reduction in the number of germ cells. Irradiation of the animal hemisphere has no effect on germ cell production. UV irradiation of *Xenopus* eggs at the beginning of the first cleavage has the same effect. With a UV dose rate of 150 ergs per square millimeter per second, as the time of irradiation increases the number of germ cells decrease and no primordial germ cells develop when the total dose is 4500 ergs per square millimeter or higher (Fig. 19–16).

Microsurgical operations, where a portion of the cytoplasm at the vegetal pole of the embryo is re-

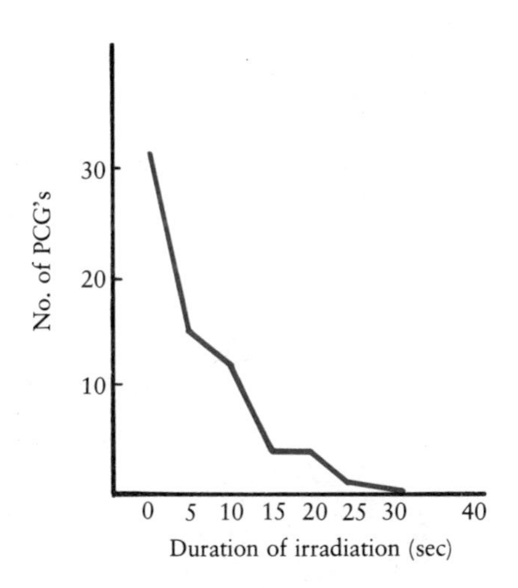


**FIG. 19–16.** Average number of primordial germ cells (PGCs) per tadpole following UV irradiation of the *Xenopus* egg at the time of the first cleavage with 150 ergs/mm$^2$/sec for increasing lengths of time. (From K. Tanabe and M. Kotani, 1974. J. Embryol. exp. Morphol. 31:89.)

moved, also often result in a total or partial absence of germ cells. Although such experiments obviously indicate the importance of the germ plasm, it is hard to define either qualitatively or quantitatively exactly what was removed.

If we sterilize a frog embryo by irradiating it at the beginning of the first cleavage, what will happen if we inject cytoplasm from the vegetal pole of an unirradiated egg into the irradiated egg? In one such experiment, when vegetal cytoplasm from an unirradiated fertilized egg was injected into four-cell embryos that had been irradiated at the beginning of the first cleavage, about 50 percent of the eggs developed into larvae with germ cells, although in smaller numbers than normal. Transfer of animal pole cytoplasm into irradiated eggs had no effect on germ cell development and all of the irradiated eggs developed into larvae without germ cells.

There seems to be little doubt then that the substance at the vegetal pole that can be histologically identified in the cleaving anuran egg and traced into cells that will develop into primordial germ cells is, in fact, responsible for inducing these cells to differentiate into germ line stem cells. Can we define the nature of this material? Unfortunately, as is the case in another important factor in the development of the amphibian, the organizer, we cannot do so precisely. There are, however, a number of indications that it may be a nucleic acid. There is good correspondence between the UV absorption spectrum for nucleic acids and the relative efficiency of UV irradiation of different wavelengths in germ cell reduction. Since DNA and RNA absorption spectra are the same, this correlation does not distinguish between them. However, the cluster of dense granules found in the vegetal cytoplasm and considered to be a part of the germ plasm stain positively for RNA. In addition, these granules resemble those found in the pole plasm of the insects that also stain positively for RNA.

Despite all of the evidence relating the germ plasm to the primordial germ cells, the exact role of the germ plasm in controlling germ cell differentiation is poorly defined. In the anurans the few cells that contain the germ plasm behave exactly as do the other endodermal cells up to gastrulation. They divide and undergo the same morphogenetic movements. The first indication of any difference is seen when they begin to show an active migration from the deep endoderm to the roof of the gut. Interestingly, this migration coincides with the movement of the germ plasm from the pole of the cell to the region of the nucleus. This intracellular shifting has led to the suggestion that the germ plasm may well be inactive in early development, and its first role in germ cell formation may be to direct the migration of the primordial germ cells from the deep endoderm into the genital ridges. It may also play a role in cell division, inhibiting mitotic activity during the migration stages, initiating mitotic activity once the germ cells are established in the developing gonad, and initiating meiotic activity in the gonial cells.

*Primordial Germ Cells in the Mammal*

In mammals, including man, cells that segregate early in development and later migrate into the genital ridges have also been described and identified as primordial germ cells. They show various distinguishing features depending on the staining techniques employed. When stained with hematoxylin and eosin, they have a clear cytoplasm, but when stained by the azur A method for nucleic acids, the cytoplasm is strongly basophilic. Numerous studies have used the high alkaline phosphatase activity of the primordial germ cells to determine the site of their first appearance and the route of their migration. In the mouse, they have also been shown to have at the EM level of observation a number of features that distinguish them from other cells. These include many ribosomes and polysomes, dense granulofibrillar bodies, and annulate cisternae. Also, their nuclei contain compact masses of granules and fibrils, evidence that nuclear metabolic activity is high, perhaps related to the large numbers of ribosomes and polysomes in the cytoplasm. The cells are also distinguished by the small amount of endoplasmic reticulum and Golgi complex that they contain.

In the mouse, primordial germ cells have been located in the 8–9-day-old embryo in the endoderm of the allantoic diverticulum, the yolk sac, and the hindgut. However, it has been suggested that the primordial germ cells enter the endoderm secondarily and actually originate in the splanchnic mesoderm in the region of the hindgut. The primordial germ cells of the 8–9-day-old embryo on the basis of both light and electron microscopy studies, show more resemblances to splanchnic mesoderm cells than they do to endoderm cells. In addition, they have been located partially in the endoderm and partially in the mesoderm layers, suggesting that they may be migrating from a mesodermal to an endodermal site. Thus, the precise time of appearance and the origin of the primordial germ cells in the mammal have not been determined. The primordial germ cells seen in the endoderm of the 8–9-day mouse embryo may well be cells that have developed from stem cells set aside at some other location in the 6–7-day-old embryo.

Germ plasm material similar to that described in cleavage stages in invertebrates and amphibians has not been found in the mammal. However, structures similar to the germ plasm of these lower forms consisting of fibrogranular material lacking a surrounding membrane have been described in the germ cells of mammals. The term *nuage* (French for cloud) has been applied to these structures. Nuage has been detected in the primordial germ cells of the mouse and the rat but is more easily demonstrated and particularly abundant in the hamster oocyte (Fig. 19–17).

In the 10- to 11-day mouse embryo, gonadal ridges develop medial to the mesonephros as small bulges on the posterior coelomic wall, and at this time the primordial germ cells leave the gut endoderm and migrate into the dorsal mesentery. By days 12 and 13, the primordial germ cells have moved into the genital ridges, a migration that is thus essentially the same as that described for the amphibians. The morphology of the primordial

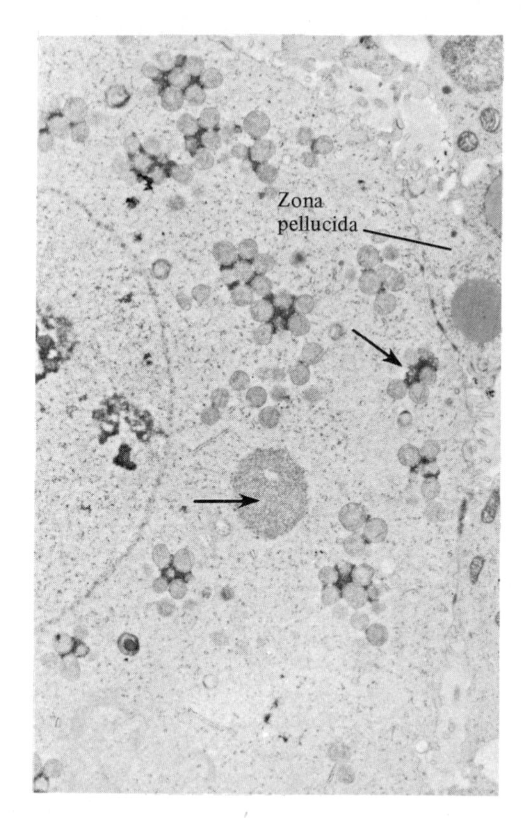


**FIG. 19–17.** EM of an oocyte from a 12-day-old hamster newborn. Nuage (arrows) present in the cytoplasm in two forms: 1, a large fibrogranular body and 2, electron-dense substance between clusters of mitochondria. (From E. M. Eddy et al., Gamete Res. 4:333.)

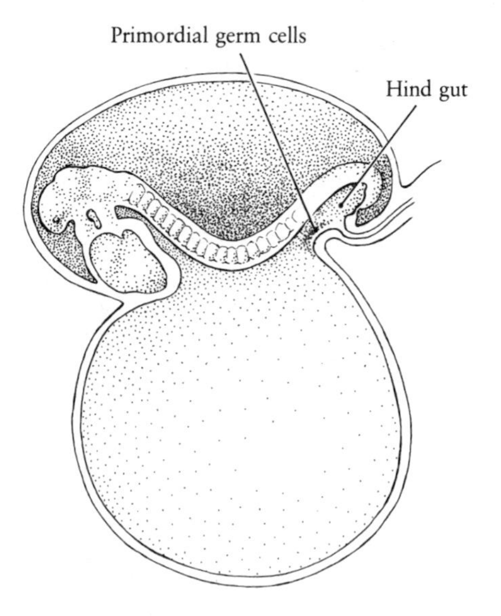


**FIG. 19–18.** Diagram of a 16-somite human embryo showing the location of the primordial germ cells in the endoderm of the yolk sac and hindgut. (After E. Witschi, 1948. Carnegie Contrib. Embryol. 32:67.)

germ cells changes during this migratory period. When first seen in the yolk sac endoderm, they have the appearance of sedentary cells and are probably carried into the hindgut passively by the invagination of the surrounding endoderm. In contrast, the primordial germ cells in the hindgut and the dorsal mesentery have the morphology of actively moving cells, elongated polarized cells with blunt pseudopodia at one end and an indented nucleus at the other, with most of the cytoplasmic organelles clustered at one side of the nucleus.

In man, large primordial germ cells with round and extremely large nuclei, large nucleoli, and clear cytoplasm are seen in the endoderm of the yolk sac of the 13-somite embryo. From here they migrate into the endoderm of the hindgut (Fig. 19–18) and at the 25-somite stage start to leave the hindgut to move to the genital ridges by way of the dorsal mesentery (Fig. 19–19 A,B). This migration takes place during the fifth week, at which time the primordial germ cells form lobate and filiform pseudopodia. As in the mouse, their morphology suggests that their migration is due largely to their own amoeboid movements.

It has not as yet been determined what controls the specificity of the time and the pattern of the migration of the primordial germ cells. It has been proposed that contact guidance exerted by the surrounding endodermal and mesodermal tissues might be a factor. In addition, it has long been

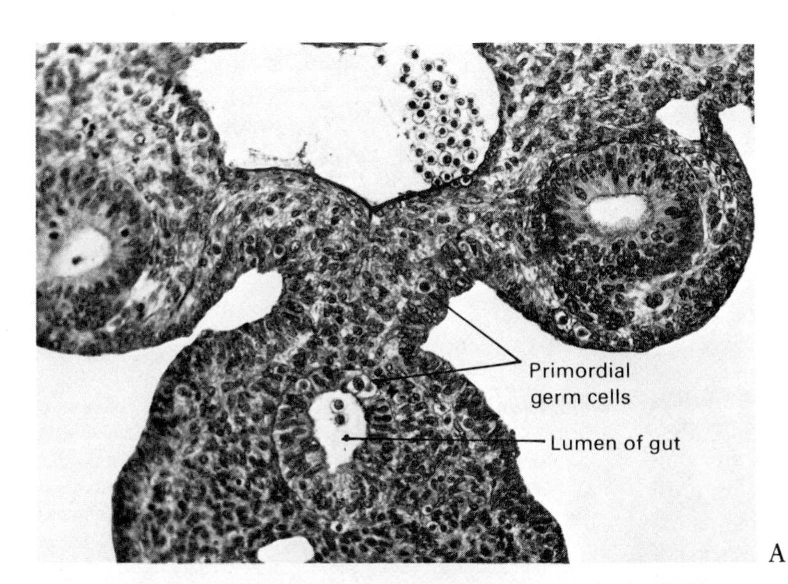


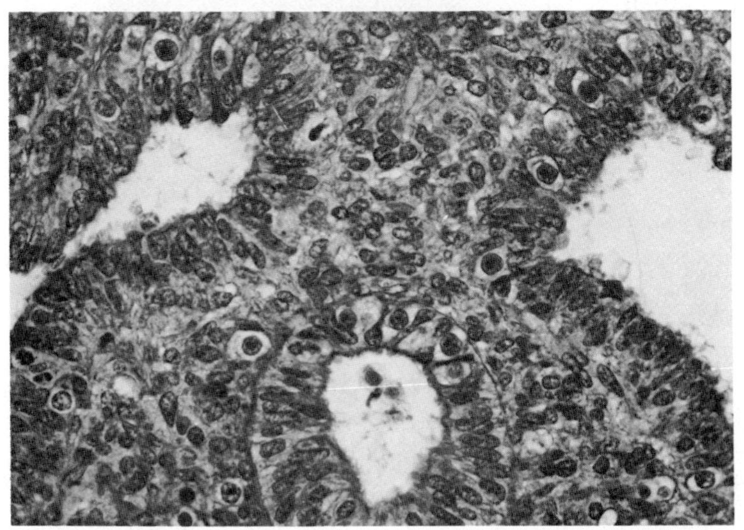


FIG. 19–19. A, cross section through the lower part of the gut of a 32-somite human embryo with primordial germ cells in the dorsal endodermal epithelium, the gut mesenchyme, and the coelomic angles; B, higher power view of two primordial germ cells in the gut epithelium. The cell on the right has broken down the basement membrane and is migrating into the mesenchyme. (From E. Witschi, 1948. Courtesy of the Carnegie Institution of Washington, Davis Division.)

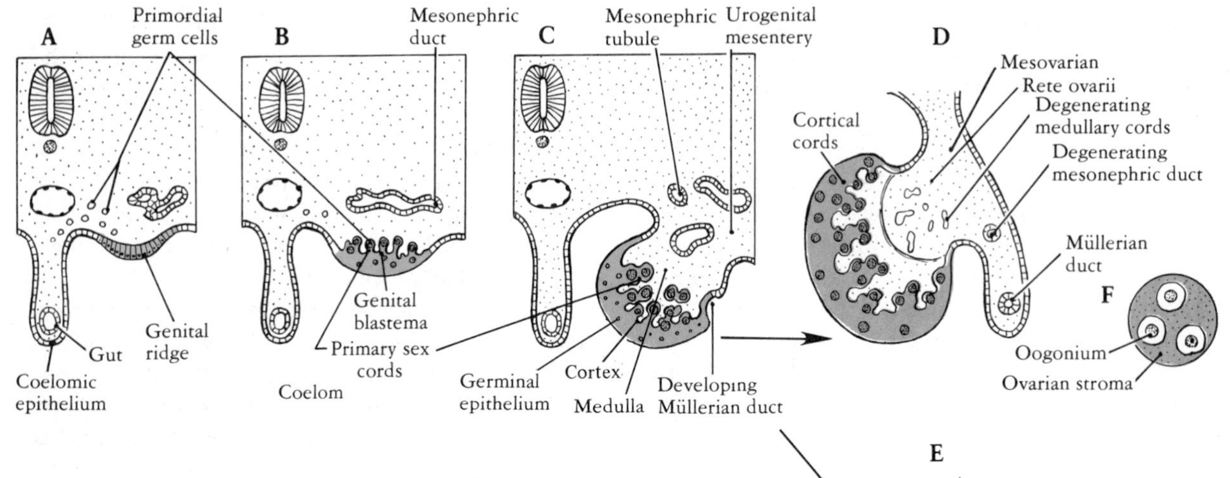


thought that a chemotactic effect of the gonadal ridges may also play a role. There is some experimental evidence to support the latter. When the hindgut from a 9½-day mouse embryo is transplanted into the coelom of a 60-hour chick embryo, the mouse primordial germ cells accumulate on the side of the graft nearest the chick mesonephros and genital ridge. Some primordial germ cells may even leave the mouse tissue and locate in the chick gonad.

## The Indifferent Gonad

Although the sex of the individual is determined genetically at the time of fertilization, it is not possible to determine the sex of the embryo morphologically until after the age of six weeks. Up until this time, differentiation of the genital ridge is exactly the same in the male and female, and the potential testis and ovary are indistinguishable from each other. The gonad may thus be called, up to this stage of its development, the indifferent gonad (Fig. 19–20 A–C).

During the fifth week, the thickening *germinal epithelium* proliferates cells into the underlying mesenchyme. The first cells to be proliferated occupy the deepest position—next to the mesonephros—and form the *rete blastema*. Further proliferation of the germinal epithelium forms irregular cords of cells, the *primary sex cords*. They surround the migrating primordial germ cells and constitute the *genital blastema* (Fig. 19–20 B). The indifferent gonad may be divided into a *primary cortex*, the germinal epithelium, and a *primary medulla*, the *genital blastema* (primary sex cords) and the rete blastema. However, there is no actual separation between any of these parts of the gonad.

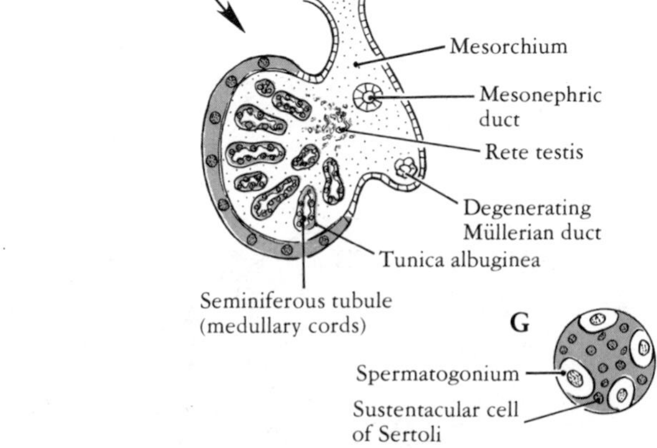


**FIG. 19–20.** Early differentiation of the ovary and the testis. A, B, primordial germ cells migrating from the dorsal mesentery into the germinal blastema; C, the indifferent gonad at the age of six weeks; D, differentiation of the ovary by the proliferation of the secondary sex cords (cortical cords); E, differentiation of the testis by the proliferation of the primary sex cords (medullary cords); F, section through the ovarian cortex showing three primary follicles; G, section through a seminiferous tubule.

The indifferent gonad bulges into the coelom, as the medial part of the urogenital ridge, connected to the dorsal body wall by a mesentery that it shares in common with the mesonephros, the urogenital mesentery (Fig. 19–20 C). As the gonad increases in size, it separates from the mesonephros and acquires its own separate mesentery—the *mesorchium* in the male, and *mesovarium* in the female (Fig. 19–20 D,E).

## Development of the Testis

During the seventh week, two events occur to distinguish a testis from an ovary. Both of these events occur in the indifferent gonad destined to become a testis. Their failure to occur at this time thereby distinguishes an ovary. The changes that take place in the prospective testis are: first, a continuing proliferation of the primary sex cords; and second, the formation of a connective tissue layer, the *tunica albuginea,* between the cortex and the medulla (Fig. 19–20 E).

The primary sex cords of the genital blastema continue to proliferate and form branched strands of cells, the *medullary sex cords.* These will form the *seminiferous tubules* of the testis. The seminiferous tubules are lined with an epithelium consisting of two types of cells: primordial germ cells that will differentiate into spermatogonia, and cells derived from the germinal epithelium that will form the sustentacular or *Sertoli cells* (Fig. 19–20 G). The early sex cords are composed primarily of Sertoli cells.

The rete blastema differentiates into a network of channels, the *rete testis,* which establishes connection with the seminiferous tubules through the straight tubules, *tubuli recti.* The general topography of the testis is well marked by the middle of pregnancy, and the duct system is well organized (Fig. 19–21). However, the duct system is not completely canalized, and the sperm are not differentiated until puberty.

The mesenchymal portion of the medulla forms the connective tissue components. Between the seminiferous tubules, it forms septa that partition the testis into over 200 lobules. The septa converge toward the mesorchium to form the *mediastinum* where the rete testis is located. Peripherally, the septa extend to the tunica albuginea, which by about 10 weeks completely envelops the testis. The tunica albuginea separates the cortex from the medulla and effectively prevents any further participation of the original germinal epithelium in the differentiation of the testis. The germinal epithelium flattens out to form the mesothelium, and the tunica albuginea forms the underlying capsule.

In addition to connective tissue, the mesenchymal cells of the medulla also differentiate into interstitial or *Leydig cells.* Leydig cells reach a maximum number in the fetus at about the end of the fifth month; after this period they decline. As they fall off in numbers, the rate of the differentiation of the seminiferous tubules increases. However, a causal relationship between these two events has not been established.

## Development of the Ovary

In the indifferent gonad destined to become an ovary, the primary sex cords do not continue to proliferate after the sixth week, nor does a tunica albuginea develop at this time. Rather, the primary sex cords remain indistinct and are intermingled

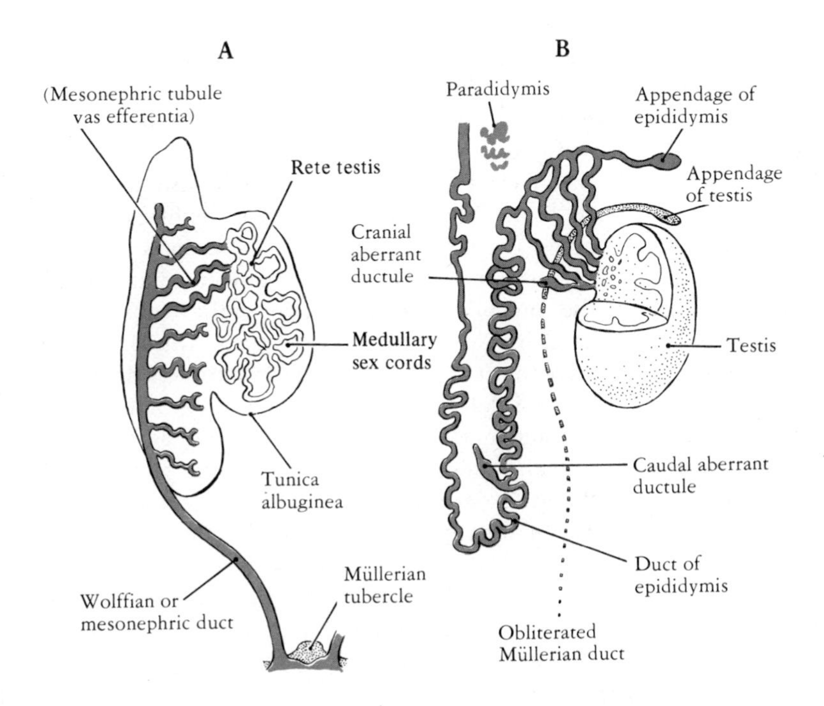


**FIG. 19–21.** A, diagram of the male genital ducts in the fourth month of development. The Müllerian system has regressed. B, diagram of the male genital ducts showing the vestigial remnants of the Wolffian and Müllerian systems. (After L. B. Arey, 1974. Developmental Anatomy. W. B. Saunders Company, Philadelphia.)

with connective tissue that does not form septa. Their fate is to form a rudimentary *rete ovarii* (homologous to the rete testis) and the medullary tissue of the ovary.

The development of the cortical portion of the gonad distinguishes the ovary from the testis. At about 10 weeks the coelomic germinal epithelium undergoes a secondary proliferation (Fig. 19–20 D). This proliferation forms cortical sex cords or *secondary sex cords,* a process that has no counterpart in the male. These cords penetrate the underlying mesenchyme but remain close to the surface. During the fourth month, isolated masses of the cortical sex cords surround individual primordial germ cells forming primordial follicles. A primordial (primary) follicle consists of a single layer of flattened cells, follicular cells, derived from the secondary sex cords, surrounding an oogonium, which is the derivative of a primordial germ cell (Fig. 19–20 G). Ovarian follicle cells and testicular Sertoli cells are homologous to each other. The mitotic activity of the primordial germ cells continues throughout most of fetal life, and the proliferation of the germinal epithelium also continues until a thin tunica albuginea is formed during the eighth month. The primary (medullary) sex cords and the primordial germ cells associated with them in the indifferent gonad do not contribute to the formation of ovarian follicles. They degenerate and are replaced by a vascular, fibrous connective tissue stroma that forms the definitive medulla of the ovary.

Proliferative activity in the ovary ceases before birth, and the formation of new or additional germ cells does not ever take place again. Thus, the human female at birth contains in her ovaries all of the germ cells that she will ever produce. Contrast this to the male where the major function of the postpuberal testis is the continual production of many millions of new spermatogonia.

Although some of the primary follicles in the fetal ovary may respond to increased estrogen secretion in late fetal life by forming antra, these follicles do not mature but become atretic. Most of the primary follicles remain in their undifferentiated stage until, at puberty, they respond to hormonal stimulation by developing into mature follicles.

## The Reproductive Duct System

The sex ducts and the sex accessory glands develop from discrete primordia in each sex. However, in all embryos, no matter what the sex, there is a time when the primordia of the duct systems of both sexes are present. The male appropriates as its major sex duct the mesonephric (Wolffian) ducts. The major portion of the female duct system develops from the *paramesonephric (Müllerian)* ducts, which first appear as infoldings of the coelomic epithelium of the urogenital ridge lateral to the mesonephric ducts.

The development of the duct system of both sexes is complicated by the fact that in each sex remnants of the system of the opposite sex persist and form nonfunctional vestigial structures.

### *The Male*

The germinal epithelium of the seminiferous tubules is the site of sperm production and from this location, the sperm must find their way to the outside. The development of the passageways for the first stage in this journey, from the seminiferous tubules to tubuli recti to rete testis—all derivatives of the primary sex cords—has already been described.

It has been noted that the male appropriates the mesonephric duct to form the vas deferens. A connection between the vas deferens and the rete testis must then be established. This connection is made through the mesonephric tubules at the level of the developing testis. As the mesonephros degenerates, the tubules in the region of the testis become shortened and somewhat disorganized, and the glomeruli become fibrous and are extruded from the glomerular capsule. The rete testis is in close contact with the blind ends of these tubules. The rete testis establishes connections with them, and they become canalized and form the *vas efferentia,* the passage from the testis to the mesonephric duct (Fig. 19–21 A). The region of the mesonephric duct just below the point where the vas efferentia open becomes elongated and highly convoluted and forms the *epididymis.* The remainder of the Wolffian duct caudad of the epididymis acquires a thick coat of smooth muscle and develops into the vas deferens. The vas deferens opens into the urogenital sinus at a point marked by Muller's tubercle as has been previously described. Just before the point where the vas deferens opens into the urogenital sinus, a sacculation appears at about three months, representing the beginning of the development of the *seminal vesicles.* The short portion of the vas deferens between the point of origin of the seminal vesicles and its entrance into the urogenital sinus

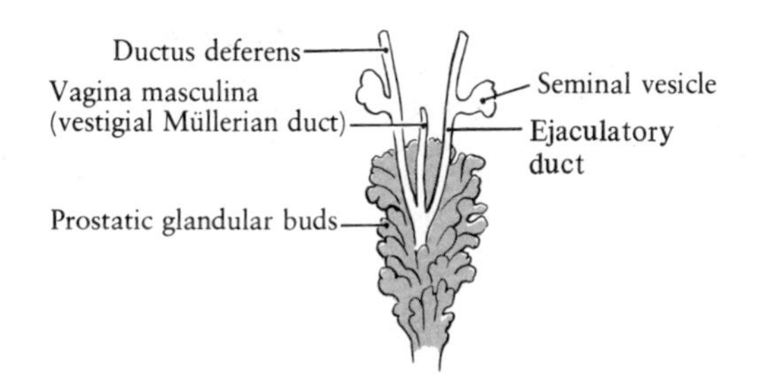


**FIG. 19–22.** Appearance of the seminal vesicles and the prostate gland at about four months.

(urethra) is known as the *ejaculatory duct* (Fig. 19–22).

Some parts of the mesonephros not forming functional sections of the male duct system fail to degenerate completely and persist as vestigial structures (Fig. 19–21 B). The blind cranial end of the mesonephric duct forms the *appendix epididymis*. Some of the cranial group of mesonephric tubules, which form the vas efferens, end blindly and become the *cranial aberrant ductules*. A caudal group of tubules, which never form any functional component of the system, also persist as the *paradidymis* and the *caudal aberrant ductule*.

*The Female*

The paramesonephric (Müllerian) ducts form the *uterine tubes* (oviducts, Fallopian tubes), the *uterus*, and part of the *vagina*. These ducts first appear during the fifth week as bilateral thickenings of the coelomic epithelium on the anterolateral surface of the urogenital ridge near its cranial pole (Figs. 19–20 C and 19–23 A).

**FIG. 19–23.** A, early development of the Müllerian duct as a thickening of the coelomic epithelium lateral to the Wolffian duct; B, the plate invaginates to form a groove.

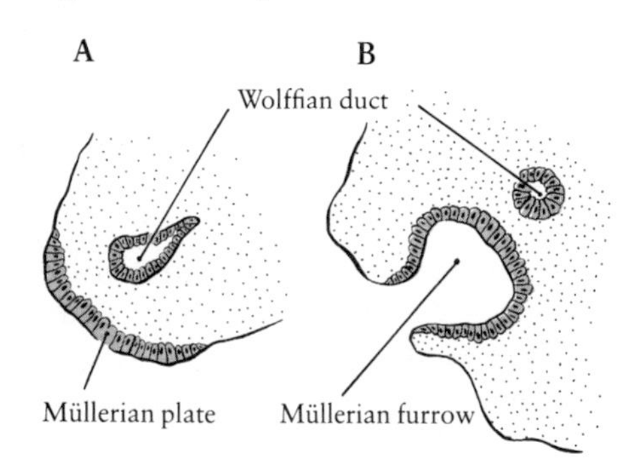


Each thickening sinks in to form a groove (Fig. 19–23 B) whose lips then close over to form a duct, which, however, remains open at its cranial end. The Müllerian ducts grow caudally in close relationship with the mesonephric ducts. As they approach the cloaca, they move medially, anterior to the mesonephric ducts, and fuse into a common *uterovaginal canal* (Fig. 19–24). Fusion of the two Müllerian ducts also brings together the two folds of peritoneum in which they lie. The fused uterovaginal canal is thus connected on either side by a broad sheet of tissue to the posterior body wall. This sheet is the *broad ligament*. The uterovaginal canal and the broad ligament divide the peritoneal cavity in this region into an anterior *vesicouterine pouch* and a posterior *rectouterine pouch* (of Douglas; Figs. 19–24 C and 19–25). The uterovaginal canal ends blindly in contact with the urogenital sinus (Figs. 19–24 and 19–25), and at this site is formed the elevation on the posterior inner wall of the sinus known as *Müller's tubercle* (Fig. 19–26).

The unfused cranial portions of the Mullerian ducts form the paired uterine tubes. The funnel-shaped opening at the cranial end of the tube becomes the abdominal ostium and develops fringes or fimbriae surrounding the opening into the uterine tube.

The fused portion of the ducts forms the *corpus* and the *cervix* of the uterus. The juncture between the two Müllerian ducts is at first Y-shaped, and the uterus is thus bicornuate up until the third month (Fig. 19–26 A). Thereafter, a secondary cranial extension of this area occurs so that the original angular junction of the Müllerian tubes now becomes dome-shaped, forming the *fundus* (Fig. 19–26 B,C). The mesenchyme surrounding the uterus becomes greatly thickened to form the uterine myometrium, the muscular wall of the uterus.

At the point where the blind end of the uterovaginal canal approaches the urogenital sinus, a pair of solid masses of tissue grows out of the urogenital sinus into the caudal end of the uterovaginal canal. These are the *sinovaginal bulbs* (Figs. 19–25 A,B and 19–26). They form a solid plate of cells, the vaginal plate, which proliferates rapidly and increases the distance between the urogenital sinus and the primordium of the uterus. Later, the central core of this plate breaks down and forms the vaginal lumen. By the fifth month, the vagina is entirely canalized, and winglike extensions surround the uterine cervix forming the vaginal *fornices*. Al-

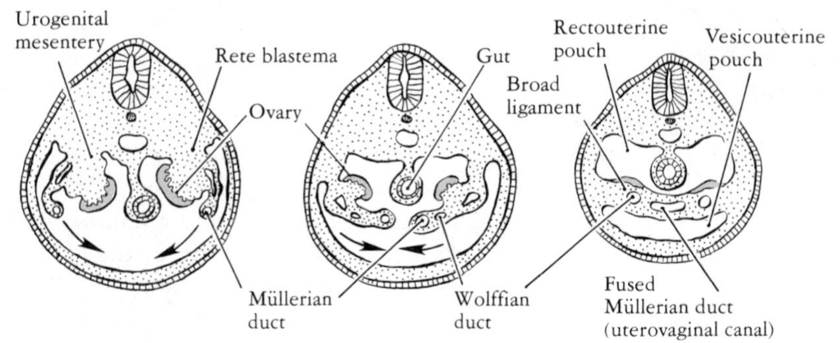


**FIG. 19–24.** Anteromedial movement and fusion of the Müllerian ducts represented by diagrams drawn of successively more caudal levels. Ovary becomes positioned on the posterior aspect of the two folds of mesentery (future broad ligaments) in which the Müllerian ducts lie.

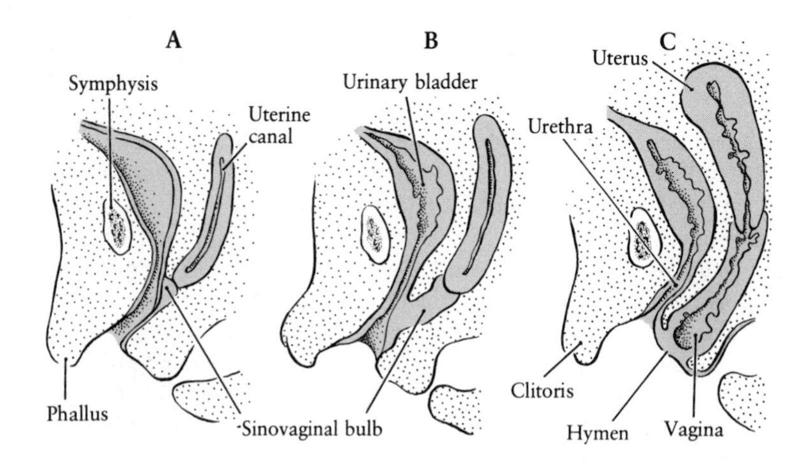


**FIG. 19–25.** Diagram of the development of the uterus and the vagina. A, nine weeks; B, end of the third month. Sinovaginal bulbs between the urogenital sinus and the uterus. Solid at first, they begin to hollow out; C, at birth. The vagina is separated from the vestibule by an incomplete hymen.

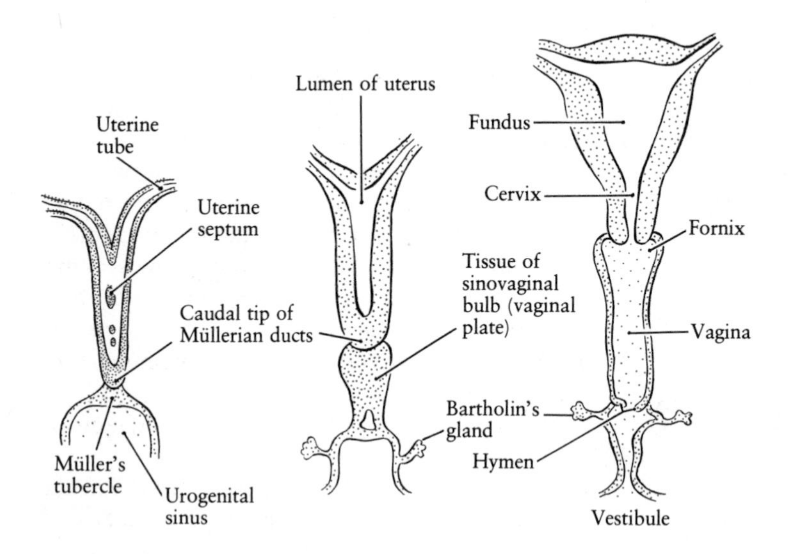


**FIG. 19–26.** Diagrams of frontal sections of the stages shown in Figure 19–25. Bartholin's glands develop as outgrowths of the urogenital sinus. They open into the vestibule.

though the sinovaginal bulb forms the vaginal epithelium, the fibromuscular coat of the organ develops from the uterovaginal primordium. The vagina is separated from the urogenital sinus by a fold of tissue, the *hymen*, which persists to varying degrees postnatally as an incomplete partition over the entrance to the vagina.

## Heterosexual Remnants of the Duct System

*The Male*

The Müllerian system in the male atrophies during the third month of development with the exception of cranial and caudal remnants. The cranial remnant becomes the *appendix testis*. The caudal remnant persists as a small pouch on the dorsal wall of the urethra, the *prostatic utricle* or *vagina masculina* (Fig. 19–21 B).

*The Female*

Portions of both the cranial and the caudal groups of mesonephric tubules persist as vestigial structures in the female (Fig. 19–27 B). The cranial tubules form a number of tiny canals attached to a short remnant of the mesonephric duct, which in the male forms the epididymis. Collectively, these tubules, located within the broad ligament, form the *epoöphoron*. A few of the cranial tubules become the cystic *aberrant tubules*. Remnants of the caudal group of tubules form the *paroöphoron*. Isolated remnants of the mesonephric duct, corresponding to regions that form the vas deferens in the male, may persist in the broad ligament lateral to the uterus or in the walls of the vagina as the *duct of Gartner*. The cranial end of the Wolffian duct persists as the *vesicular appendage*, the homolog of the appendage of the epididymis in the male.

## Genital Ligaments and the Descent of the Gonads

*The Male*

The developing gonad and the mesonephros are attached to the posterior body wall by the urogenital mesentery (Fig. 19–20 C). As the mesonephros degenerates, this broad connection narrows and becomes the mesentery of the gonad, which in the case of the testis is called the *mesorchium* (Fig. 19–20 E). The cranial pole of the testis is continuous with the mesonephros and the urogenital mesentery that extends up to the diaphragm. Both of these structures degenerate completely, after which the cranial pole of the testis does not have any ligamentous attachment. However, the peritoneal fold attached to the caudal pole of the testis persists and, reinforced by mesoderm and parts of the degenerating mesonephros, becomes the *caudal genital ligament* (Fig. 19–28 A). The caudal genital ligament extends from the caudal pole of the testis to the region where the urogenital ridge bends toward the midline. From this bend, a new ligament makes connection with the adjacent body wall and is, in turn, continuous with a mesenchymal condensation extending through the abdominal wall into the genital (scrotal) swellings. The passage through the abdominal wall is the *inguinal canal*. Together these three tissue cords connecting the caudal pole of the testis to the scrotum constitute the *gubernaculum testis*. As the abdominal cavity increases in cross-sectional and longitudinal directions, the testis remains attached to the scrotum by the gubernaculum. The gubernaculum does not keep pace with the growth of the abdominal region, and thus the

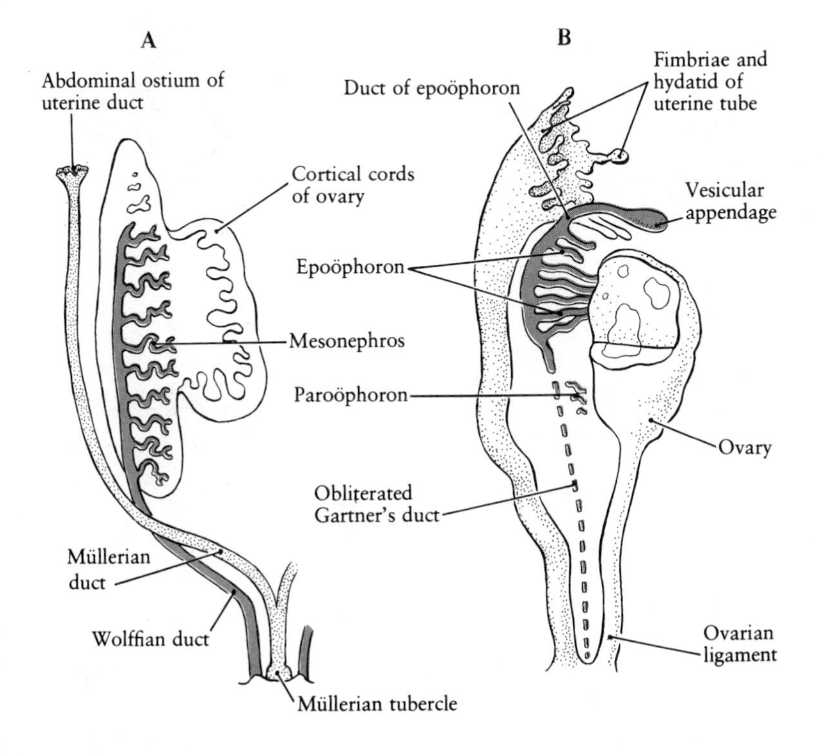


**FIG. 19–27** A, diagram of the duct system of the female in the second month of development. Fused Müllerian ducts meet the urogenital sinus at a point marked by Müller's tubercle. B, vestigial remnants of the mesonephric tubules and mesonephros in the female. (After L. B. Arey, 1974. Developmental Anatomy. W. B. Saunders Company, Philadelphia.)

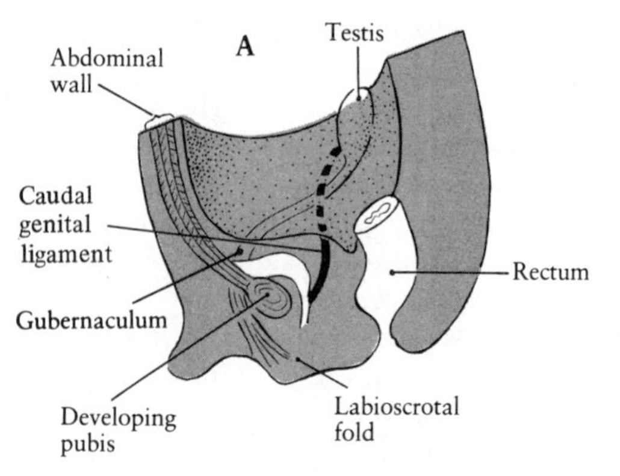


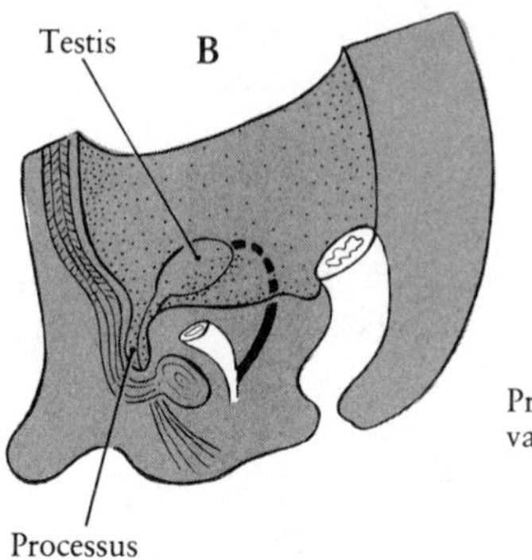


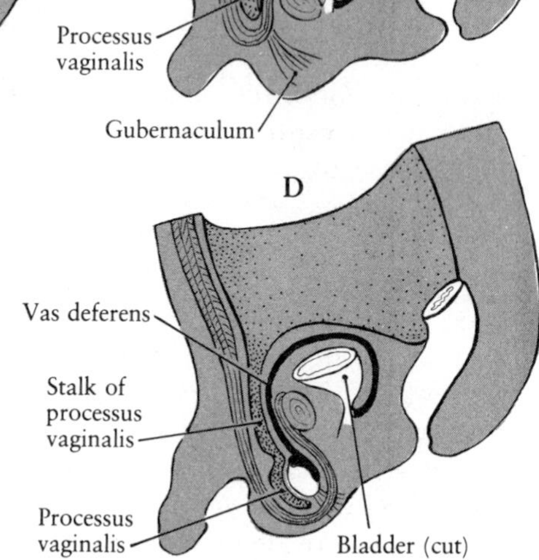


FIG. 19–28. Descent of the testis shown in sagittal section. A, before descent at about seven weeks; B, C, passage into the processus vaginalis during the seventh month; D, fully descended but with the processus vaginalis still in communication with the abdominal cavity; E, one month postnatally. Processus vaginalis closed off. All of the muscle layers of the anterior abdominal wall are represented by muscle and fascia of the scrotal sac; F, enlargement of testes and ducts in scrotum.



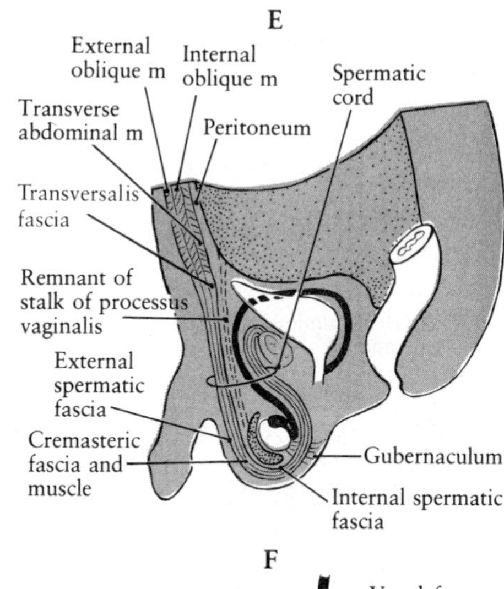


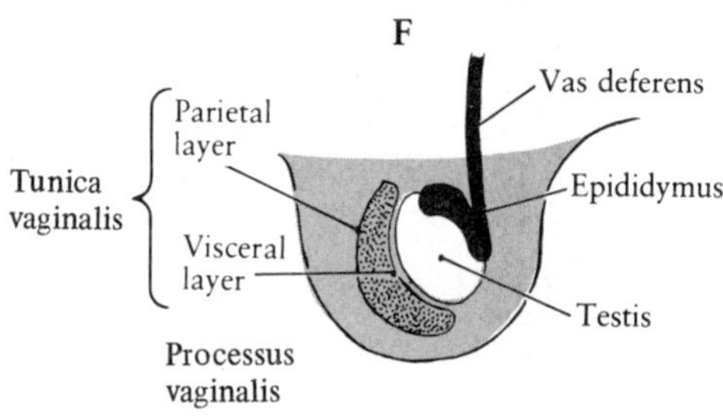


testes undergoes an apparent caudal migration, so that by the end of the third month they no longer are found in the abdominal cavity but are in the pelvic cavity close to the area of the inguinal canals.

At the beginning of the third month, a bilateral evagination of the coelom forms just anterior to each gubernaculum. Each peritoneally lined extension of the coelom is known as a *processus vaginalis*. Each sac evaginates into the anterior abdominal wall and follows the course of the gubernaculum through the inguinal canal into the scrotal swelling (Fig. 19–28). The herniation of the vaginal process is covered by the muscular and connective tissue components found in this region of the anterior abdominal wall. They will form coverings of the testis, each of which can be related to its corresponding layer in the abdominal wall.

The testes remain in the pelvic cavity in close proximity to the processus vaginalis until the seventh month. At this time, the testes descend through the inguinal canal into the scrotum (Fig. 19–28 D). The testes and the gubernaculum always lie beneath the peritoneal lining of the processus vaginalis. The epididymis, vas deferens, and the blood vessels and nerves supplying the testes are carried into the vaginal process with the testes. These ducts, vessels, and nerves passing back into the peritoneal cavity through the inguinal canal collectively constitute the *spermatic cord*. The original broad opening between the vaginal process and the peritoneal cavity becomes narrower and is obliterated before birth. Within the scrotum, a remnant of the vaginal process persists. The lining of this cavity is the parietal layer of the *tunica vaginalis* that becomes, where it covers the testis, the visceral layer of the tunica vaginalis (Fig. 19–28 F).

The final descent of the testes from the peritoneal cavity into the scrotal sacs is accompanied by a marked shortening of the gubernaculum. The role the gubernaculum plays in the descent of the testis is questionable. Although it ceases to grow during the seventh month and even undergoes an actual shortening, this process probably does not serve to pull the testis through the inguinal canal to its ultimate position in the scrotal sac. The gubernaculum does initially form a pathway through the abdominal wall for the processus vaginalis to follow during the formation of the inguinal canal, but by the time of the descent of the testis, the gubernaculum, which never does attach to the skin of the scrotum, has been converted into a soft mucoid tissue. Descent may be aided by an increase in intra-abdominal pressure resulting both from the growth of the abdominal viscera and contractions of the fetal abdominal muscles. There is good evidence that androgens and gonadotrophins assist in the process. Clinically, injection of these hormones induces descent in many cases of cryptorchidism.

*The Female*

In the female, both the cranial and the caudal poles of the ovary are connected to ligamentous remnants of the urogenital region, forming the *suspensory ligament* of the ovary and *ovarian (proper) ligaments*, respectively. The ovarian ligament attaches the ovary to the broad ligament. A caudal extension connects the broad ligament and the uterus to the labia majora. This is the *round ligament*. Thus, the ovarian ligament and the round ligament together are homologous to the gubernaculum of the male. The male has no homolog of the suspensory ligament. The ovary undergoes a descent but does not normally pass through the inguinal canal into the genital swellings although a small processus vaginalis is formed. It is usually obliterated but may persist as a small diverticulum of the peritoneum, the *canal of Nuck*.

The ovary is connected to the posterior wall of the broad ligament by the *mesovarium*, a persistent part of the original urogenital ligament. The uterine tubes are connected to the broad ligament by the mesosalpinx, connective tissue remnants of the mesonephros.

## External Genitalia

The primordia of the external genitalia represent, as do those of the gonads, a true case of ambisexuality in which there is present at an early stage a set of structures that have the potentiality of differentiating in either a male or a female direction.

During the third week of development, a pair of elevated folds of mesoderm develop on either side of the cloacal membrane. These are the *genital (cloacal) folds*. During the fourth week, these folds meet anteriorly at the site of a median elevation, the *genital tubercle* (Fig. 19–29 A,B). At the same time, a second pair of rather indistinct elevations, the *genital (labioscrotal) swellings*, appear (Fig. 19–29 A,B). They surround both the genital tubercle and the cloacal folds. When the urorectal septum fuses with the cloacal membrane, it divides it into an anterior urogenital membrane and a posterior anal membrane surrounded by the urogenital folds and the anal folds, respectively. Up to this time, six weeks, the appearance of the external genitalia in the male and the female is identical.

*The Male*

During the third month, a male is characterized by a marked enlargement and elongation of the genital tubercle (phallus) as it begins to develop into the *penis* (Fig. 19–29 C). As the phallus elongates, it pulls the urogenital folds anteriorly, and they form an elongated pair of folds stretching along the underside of the developing penis. The urogenital sinus has now broken through and appears as a deep groove, the *urogenital (urethral) groove*, between the urogenital folds but not reaching all the way to the tip of the penis (Fig. 19–29 C). The urethral groove is linked by the endoderm of the phallic portion of the urogenital sinus.

The genital folds grow toward each other and at the end of the third month they fuse along the midventral axis of the penis, forming the penoscrotal raphe (Fig. 19–29 E). The urethral groove is thus converted into a canal, the *penile urethra*. The most distal part of the definitive penile urethra, however, is not a derivative of the urogenital groove, but is formed by canalization of a cord of ectodermal cells, which grows inward from the tip of the glans to meet the portion of the urethra formed from the

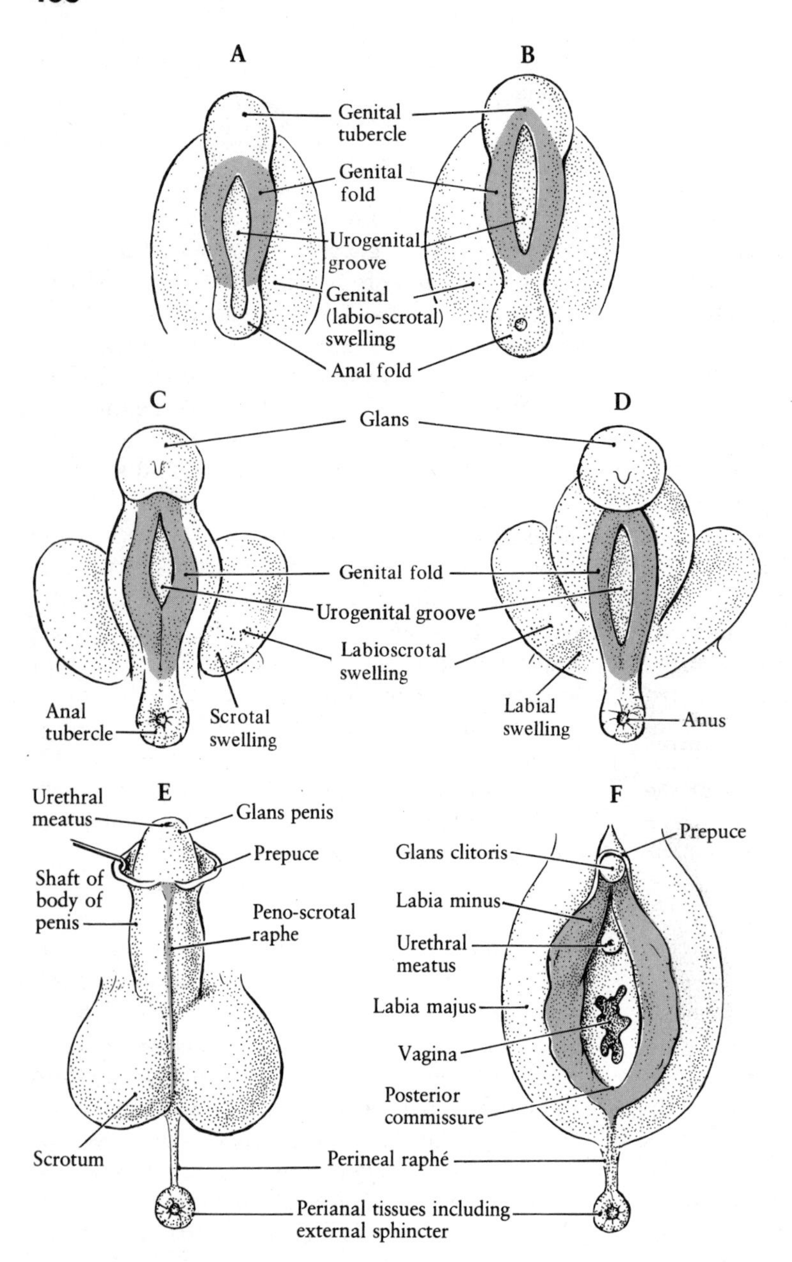


**FIG. 19–29.** Development of the male and female genitalia from a perineal view. A, B, indifferent stages at 4 and 7 weeks; C, E, male at about 9 and 12 weeks; D, F, female at about 9 and 12 weeks.

urethral groove (Fig. 19–30 A). The *external urethral meatus* and the *navicular fossa* (glandular part of the urethra) are derivatives of the ectodermal cord (Fig. 19–30 B). The skin at the distal margin of the penis grows forward over the *glans penis* to form the *prepuce*. Initially the prepuce fuses with the glans but separates from it during infancy.

The genital swellings become the *scrotal swellings* and grow toward and fuse with each other to form the scrotum. The line of fusion is marked by the *scrotal raphe* (Fig. 19–29 E).

The *corpus cavernosum urethra (corpus spongiosum)* surrounding the penile urethra and the paired *corpora cavernosa penis* develop from the mesenchymal tissue of the shaft of the penis.

*The Female*

The changes that occur in the female are not so profound as those in the male. The phallus enlarges only slightly and bends caudally to form the *clitoris*. It has a *glans clitoris* and a *prepuce*. The urethral folds do not fuse, and they form the *labia minora*.

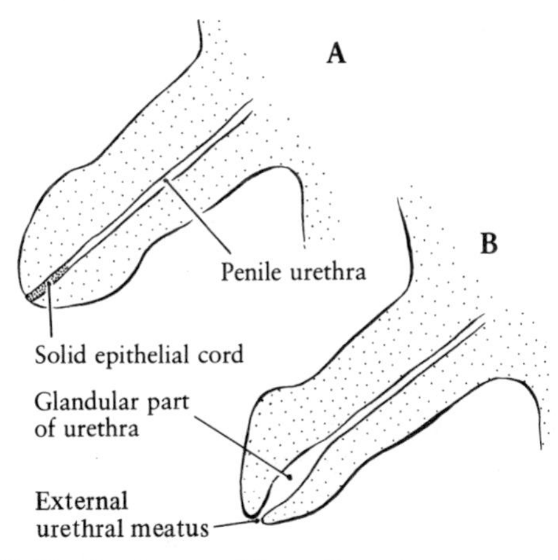


**FIG. 19–30.** Development of the penile urethra.

They flank a shallow cavity, the *vestibule,* which represents the urethral portion of the urogenital sinus (Fig. 19–29 D). Nor do the genital swellings fuse, except posteriorly where they form the posterior commissure. They enlarge to become the horseshoe-shaped *labia majora* (Fig. 19–29 F).

## The Accessory Sex Glands

### *The Male*

Accessory sex glands are more highly developed in the male. The origin of the seminal vesicles from the lower ends of the mesonephric ducts has been described. The mesoderm into which they grow forms the muscle of their walls.

Three other glands, all derivatives of the urethral epithelium, also develop; the *prostate,* the *bulbourethral* (Cowper's) glands, and the *urethral glands* (of Littré). The large prostate develops as a multiple outgrowth of the urethra above and below the openings of the mesonephric ducts (Figs. 19–22 and 19–31). These differentiate into the glandular epithelium of the organ, while the surrounding mesoderm forms their muscle and connective tissue components. The bulbourethral glands arise as a pair of solid outgrowths at the beginning of the penile urethra, from the portion that will form the membranous urethra. They extend a short distance posteriorly paralleling the urethra, and pick up their muscle and connective tissue coats from the surrounding mesoderm. The numerous small glands of Littré develop as buds from the cavernous urethra.

### *The Female*

Since the seminal vesicles are outgrowths of the mesonephric ducts, it is understandable that the female, in which these ducts degenerate, forms no homolog of these glands. However, homologs of the three derivatives of the urethral epithelium do develop. In the female, the sex accessory glands are all considerably smaller than the male homologs. The *urethral* and *paraurethral glands* (of Skene) correspond to the prostate. Outgrowths of the urogenital sinus form the *major vestibular glands* (of Bartholin) and the *minor vestibular glands,* both opening into the vestibule. They are the homologs of the bulbourethral glands and the urethral glands of Littré, respectively.

# Control of the Differentiation of the Reproductive System

As described in the previous sections, parts of both the male and the female reproductive systems differentiate from embryonic structures that at one stage in development are common to both sexes (ambisexuality—gonad, external genitalia). However, other parts develop from discrete primordia

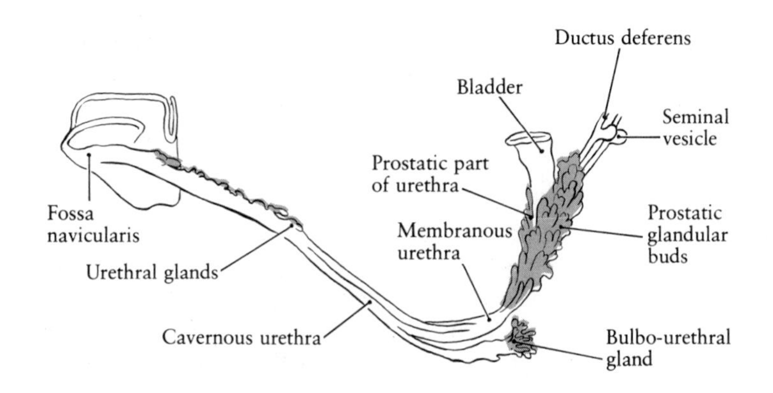


**FIG. 19–31** Vas deferens, urethra, and related structures at about midterm. A, lateral view; B, posterior view.

(bisexuality—duct system). The presence of both ambisexual and bisexual components of the reproductive organs in the embryo makes the problem of what controls the characteristic differentiation of these organs in the respective sexes an interesting one. Is the control of differentiation of the gonad the same as that for the differentiation of the genital tract? How large a role is played by the sex chromosomes, and how do they produce their effects? In veiw of the fact that the gonads of the adult secrete a number of powerful hormones that control the maturation and maintenance of the reproductive system, is there any such control over differentiation of the system in the embryo?

## Genetic Control of Differentiation

The genetic sex of an individual is determined at the time of fertilization. Generally one sex is heterogametic and the other is homogametic in respect to the sex chromosomes. In man, the male is the heterogametic sex with 22 pairs of autosomes and a pair of unmatched sex chromosomes (XY). The female has the same number of autosomes but a matched pair of sex chromosomes (XX). Normally, the genetic sex expresses itself in the development of either a typical male or a typical female.

Genetic control of the development of the reproductive system, however, is not absolute. In fact, it can be reversed. Generally, the lower one goes on the evolutionary scale, the more easily can sexual development be modified. In fact, so labile is genetic sex determination in the platyfish (a favorite viviparous fish of aquarium fanciers) that about 1 percent of these fish normally develop into functional adults of a sex opposite to their genotype. Experimental sex reversal, using sex hormones, can easily be accomplished in many teleosts. In the amphibians, while there are differences between species, sex reversal following experimental exposure of the embryos to steroid hormones of the opposite sex has been reported often. In *Rana* and *Hyla*, the female is homogametic and may be completely masculinized by androgen treatment. In *Xenopus*, lifelong sex reversal of the male toad may be induced by adding estrogens for only three days to the aquarium water containing the larvae. In *Xenopus*, the male is the homogametic sex (ZZ). Mating these sex-reversed ZZ females to normal males should, and does, result in entirely male offspring.

Birds and toads present somewhat of a special situation in which genetic sex may be reversed surgically owing to the presence in the adult of gonadal remnants of the opposite sex that are stimulated to differentiate when the normal gonads are removed. Bidder's organ in the male toad is a cortical remnant, a rudimentary ovary, located at the cranial pole of each testis. Following removal of the testes, Bidder's organ enlarges and develops into a functional ovary. In the female chicken, the right gonad remains rudimentary. It consists mainly of medullary tissue and, if the functional left ovary is removed, the right rudimentary gonad will develop into a testis in which spermatogenesis proceeds normally. However, since in the female the Wolffian duct system has regressed, there is no way for the sperm to reach the outside.

In mammals, where sex reversal does not occur, the genetic control of sex would appear to be quite straightforward and simple, with XX–XY chromosomal dimorphism expressing itself in female–male sexual dimorphism. What actually do we know about the sex chromosomes and their genes, and the nature of the genetic control of development?

### *The Y Chromosome*

For many years after the sex chromosomes were discovered there was some question whether it was the X or the Y chromosome, or both, that carried the determining factors for sex. It is now known that the sex chromosomes function in the control of the differentiation of the gonads, with the Y chromosome providing the determining factors. In the absence of a Y chromosome (XX or XO genotype), an ovary develops. In the presence of a Y chromosome, a testis develops. Thus, XXY genotypes are males, and humans with as many as four X chromosomes show unequivocal testicular gonad histology—provided they have a Y chromosome also.

It has also been proposed that the Y chromosome is necessary for spermatogenesis. However, the evidence for this is indirect. In some species of marsupials where some or all of the somatic cells are XO, the spermatogonia are always XY. In a mosaic mouse where the possibility for the production of XYY and XO spermatogonia exists, only the former are present. One human male with a Y chromosome with a small deletion on one arm of the Y chromosome developed normal testicular tissue, but the testes had no spermatogonia. The deletion apparently did not involve the gene loci controlling testicular differentation but did involve the loci controlling spermatogenesis.

*The H-Y Histocompatibility Antigen.* Once it was determined that the Y chromosome was responsible for the differentiation of the testis, investigators began to search for a testicular-organizing substance mediating this effect. The search for such a substance took a substantial step forward in 1955 when it was found that female mice unexpectedly rejected skin grafts from male mice of the same inbred strain and that, in addition, the serum from these female mice had cytotoxic activity for male mouse cells. This reaction was determined to be due to the presence in the male of a histoincompatibility antigen associated with the Y chromosome, and the term H-Y antigen was applied to it. The production of a humoral antibody by sensitized female mice provides a mechanism to test for the presence of H-Y antigen in other species by absorbing the antibody with cells of the other species and then testing for residual cytotoxic activity. Use of this technique produced the important results that male cells of all the mammalian species tested, including man, specifically absorbed the H-Y antibody while female cells did not. The occurrence of this antigen in all male species and its absence in all female species tested suggested that it might, in fact, be the testis-organizing protein through which the Y chromosome exerted its effect on gonadal differentiation.

The experimental evidence supports this proposal. If a suspension of cells from a mouse testis is allowed to reaggregate in culture, the cells will form numerous short, seminiferous tubule-like structures. However, if these same cells, before they are allowed to reaggregate, are first subjected to an excess of H-Y antibody—which effectively strips them of their H-Y antigen—they now form follicle-like structures in which a single male primordial germ cell is surrounded by what appear to be follicle cells. Thus, tubular organization is the function of the H-Y antigen, and in its absence testis cells that presumably would have become Sertoli cells now form follicle cells.

The differentiation of the indifferent gonad into a testis thus depends on the expression of H-Y antigen, which in turn is usually dependent on the presence of a Y chromosome. Thus, the Y chromosome, which for many years was considered to be largely inert, is certainly not, since its genes function actively in the control of maleness. Attempts have been made to map the structural and regulatory loci for H-Y antigen expression on the Y chromosome. Regulator control has been mapped to the short arm of the Y, but the structural locus (loci) has not been precisely determined. In certain cases, maleness may be expressed in the absence of a Y chromosome. XX males occur in a frequency of about one to several thousand and have been shown to express H-Y antigen. In fact, in the mole-vole, *Eliobus lutescens*, the Y chromosome has been permanently eliminated, and the sex chromosome constitution is XO in both sexes. The males, however, all express the H-Y antigen and the females do not. In this species, then, the H-Y gene is probably linked to the X chromosome. The direction of the differentiation of the fetal gonad then should be considered to depend solely on the presence or absence of the expression of the H-Y antigen.

Conversely, some individuals with a Y chromosome may show a female phenotype. In one instance, this has been found to be due to a mutation of the X-linked gene which specifies the production of the androgen-receptor protein. In the absence of a receptor protein, the tissues fail to respond to any androgen present. This illustrates that the sole function of the H-Y antigen in determining maleness is to organize testicular material. In this mutant the antigen fulfills its testis-determining role, but other factors intervene to prevent the action of the male hormone. A different situation occurs in the wood lemming (*Mypopus schistocolor*). In this species, some females produce all female progeny when mated to normal XY males. This has been determined to be due to a mutation on the X chromosome of the female. In the presence of this mutation XY lemmings develop into completely normal females that, in turn, never produce anything but female progeny. Here, the H-Y antigen is itself repressed and never gets the opportunity to exercise its function.

The H-Y antigen has also been shown to be present in species other than mammals, and evidence for the presence of an antigen identical or showing cross-reaction with mouse H-Y has been presented both in species in which the male is the heterogametic species (the frog, *Rana pipiens*) and in which the female is the heterogametic sex (the white leghorn chicken, *Gallus domesticus*, and the South African clawed frog, *Xenopus laevis*). When the female is heterogametic, the H-Y (H-W) antigen is found in the female. Thus, the mouse H-Y antigen is a cell surface component, which over an evolutionary span of several hundred million years, has functioned to direct the indifferent fetal gonad to differentiate in the direction typified by the heterogametic sex of the particular species in question (testes in XY males and ovaries in ZW females).

*The X Chromoxome*

The production of viable YY males in amphibians shows that the sex chromosomes of the homogametic sex are not required for viability in these forms. The sex chromosomes in the lower forms code for common enzymatic functions, strongly suggesting their origin from a common ancestral chromosome. However, during the evolution of the higher forms, the X and Y chromosomes became much more specialized and the Y lost most of its structural loci, making the X the sole chromosome containing most of the sex-linked genes. Since all of the cells of the female contain two X chromosomes and those of the male contain only one, it would appear that there might be an imbalance in either the female or the male in the enzymatic functions controlled by the genes of the X chromosome. However, there is a mechanism that prevents this, a phenomenon termed *X-chromosome inactivation*. The story of X-chromosome inactivation started with a report by Barr and Bertram (1949) that the neurons of female cats contained a piece of heterochromatin adjacent to the nuclear membrane that was not found in male neurons. It was later proposed that this piece of heterochromatin represented one of the X chromosomes of the female and that the inactivation of one of the X chromosomes equalized the dosage of X-chromosomal gene–controlled functions between the sexes. In fact, where the normal XX female retains one active X chromosome and inactivates the other showing a single Barr body in her somatic cells (Fig. 19–32 A), an XXXX female, in order to retain only a single active X, inactivates three X chromosomes and correspondingly shows three Barr bodies (Fig. 19–32 B). Evidence that this heterochromatin is indeed inactive comes from several sources. It has been demonstrated that the *sex chromatin (Barr body)* does not incorporate RNA precursors. Also, translocation of an autosomal gene affecting coat color to the inactive sex chromatin results in the inactivation of the autosomal gene. This experiment also nicely demonstrates that the sex chromatin does represent one of the X chromosomes, since cytological examination of pigmented cells from a pigment mosaic shows a single, enlarged X (containing translocated material), while unpigmented cells show a single, normal-sized X. Biochemical evidence for X-chromosome inactivation is demonstrated by the fact that clones of cells from humans heterozygous for allelic variants of glucose-6-

**FIG. 19–32.** Barr bodies from human female buccal mucosal cells. A, sex chromatin-positive nucleus with a single Barr body at the nuclear membrane; B, three Barr bodies seen in a single cell from a 48XXXX female. (From L. R. Shapiro, 1982. In: Pathology of the Female Genital Tract. A. Blaustein, ed. Springer-Verlag, New York.)

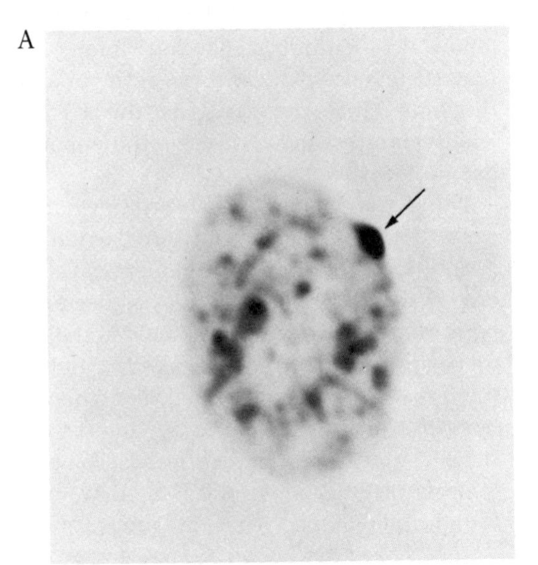


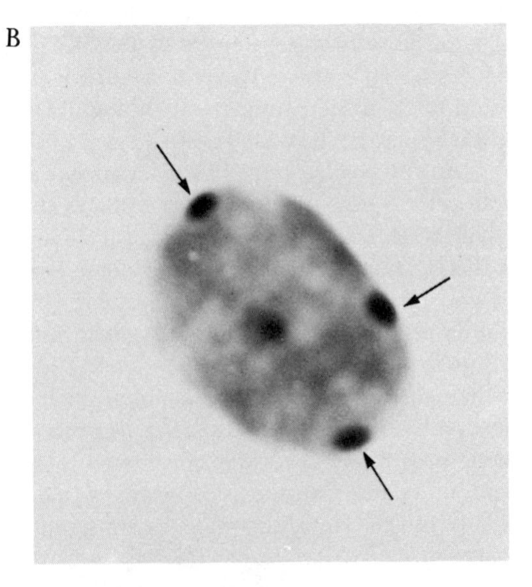

cells of the female regularly turn off one of their X chromosomes, ensuring a uniformity in the male and the female of the enzymatic activity controlled by the genes of the X chromosome.

The cells of the germ line in the female, however, present a different story. In the germ cells, both X chromosomes are active. XO mice that are fertile for a short period of time, but whose oocytes show the effect of having only a single X by impaired development of the offspring following fertilization, have only half of the activity of the sex-linked enzymes of normal XX mice. Oocytes of human females with allelic variants of glucose-6-phosphate, in contrast to somatic cells, show the hybrid enzyme. Thus, the second X is active in the germ line of the female and in fact is required for normal development of the germ cells.

In the male germ line we find still another variation: The spermatocyte inactivates its single X. Heterochromatin typical of that seen in female somatic cells is seen in spermatocytes, where it is also shows transcriptional inactivity. X-Chromosome inactivation in the male germ line is necessary if meiosis is to occur. Retention of an active X results in meiotic failure.

The number of functional X chromosomes in the body is thus characteristic of both the sex of the individual and the tissue.

It ranges from two in the female germ line to one in the somatic cells of both males and females and none in the male germ line.

## Hormonal Control of Differentiation

Since the adult system is so markedly influenced by the sex hormones, it is not surprising that there has been considerable investigation of the role of sex hormones in the differentiation of the reproductive system in the embryo. These investigations have included experiments on the hormonal control of the differentiation of the gonads, sex ducts, and the external genitalia, employing a wide variety of methods such as castration, transplantation of gonads, injection of steroid hormones into the fetus, and injection of steroid hormones into the pregnant female.

### *The Gonads*

We have seen that the H-Y antigen associated with the Y chromosome is responsible for testicular differentiation, and that in its absence an ovary develops. Once the embryonic gonads begin to secrete hormones do these hormones have any influence on the later differentiation of the gonad? One method of studying the effect of hormones on the development of the gonad is through parabiosis. Parabiotic union of salamander larvae before the sex of the larvae can be determined will result in about 50 percent of the unions being between individuals of the opposite sex. In such heterosexual pairs, the testes develop normally but the ovaries are inhibited. The parabiosed females are sterile and may later develop testicular nodules in the sterile gonad that show spermatogenesis and produce mature sperm. From the results of such parabiosis experiments, it has been concluded that the testes differentiate earlier than the ovaries and secrete male hormones that pass across to the female larva and inhibit the development of its gonad. Although union of larvae of the opposite sex of the same age results in male dominance, the opposite effect—the female inhibiting male differentiation—may result from the union of older or more rapidly developing females with younger or more slowly developing males.

Other experiments in fish, amphibians, and birds have been reported, including hormone administration and gonad transplantation, which demonstrate that steroid hormones may influence the differentiation of the gonads in these phyla. With few exceptions, the situation in the mammal is not the same, and the gonad not only develops according to the genetic sex without any stimulation from the sex hormones but also is generally refractory to hormonal influence whether it occurs naturally or by experimental design. One exception is the freemartin, a masculinized heifer that develops as a twin of a normal male. During development, the fetal membranes of the twins are united in such a manner that they provide cross-circulation between the male and the female fetuses. In the freemartin, the ovary is inhibited to a greater or lesser degree, the most masculinizing effect being the development of sterile seminiferous tubules.

Numerous attempts have been made to duplicate nature's freemartin "experiment" and to demonstrate an inhibitory effect of sex hormones on the differentiation of the gonads in mammals. The results have almost always been negative. Injection of the mother or the fetus with steroid hormones does not result in any effect on the developing gonad, although, as will be discussed in the next section, effects on the duct system and external genitalia do occur. One exception is the oppossum. The oppossum is born at an early stage of development, when the reproductive system is still undifferentiated, and spends the last part of what should be its intrauterine life attached to a nipple in its mother's pouch. Injection of estradiol into newborn male opossums stimulates the development of the cortical region of the testis, forming an ovatestis in which oocytes may differentiate.

Although steroid hormone injections usually do not modify gonad differentiation, if fetal gonads of the opposite sex are placed in close proximity to each other, some changes in gonad differentiation may occur. Transplants of heterosexual rat gonads where the ovary is taken from a fetus older (16 days) than that which provides the testis transplant (13 days) have shown that the older fetal ovary can suppress, but not reverse, the differentiation of the younger testis. However, in heterosexual transplants of gonads from embryos of the same age, the testis appears to exert a greater influence than the ovary. When ovaries and testes from mouse embryos of the same age are transplanted together below the kidney capsule, the ovaries develop into ovatestes. In this situation, potential female germ cells form spermatogonia.

It has been suggested that the fetal testis secretes a hormone different from that of the adult testis, one that can inhibit the differentiation of the cortical tissue of the developing gonad. However, such a hormone has never been identified nor synthesized, and the concept remains controversial. Despite a small number of cases in which it has been shown that sex hormones may influence the differentiation of the gonad, the conclusion is generally reached that under normal conditions the gonad in mammals differentiates according to its genetic constitution and is not influenced by hormonal factors.

*The Duct System and the External Genitalia*

The development of the duct system and the external genitalia presents a different story; here it has been shown that the secretions of the developing gonads—in particular the testes—control the direction of differentiation. One method of determining the effect of the gonad is to remove it. In the castrated male fetus, the Wolffian duct system either does not differentiate or regresses, the Müllerian system persists, and the external genitalia remain undifferentiated and femalelike. In the castrated female fetus, the Müllerian system and the external genitalia develop normally, and the Wolffian system either does not differentiate or regresses. These effects indicate that the male reproductive tract requires the stimulation of hormones from the developing testes, while the Müllerian derivatives can develop normally without any hormonal influence. This conclusion is supported by tissue culture experiments. The Wolffian duct will develop and differentiate in tissue culture only if testicular material is present. Contrary to this, the Müllerian ducts develop whether ovarian tissue is present or not. The concept of this fundamental mechanism of sequential sexual differentiation—in which the male is the induced phenotype and carries a Y chromosome that calls for the differentiation of a testis, which then secretes fetal hormones that in turn regulate the differentiation of the male sex ducts and external genitalia—evolved between 1947 and 1952 mainly through the studies of Jost (1972). Testosterone, the principal steroid of the adult testis, is also the androgen secreted by the fetal testis.

If the male system differentiates only in the presence of testicular hormones and the female system needs no hormonal stimulation, this readily explains why, in the female fetus, the Wolffian system regresses and the Müllerian system differentiates. But why, in the male fetus, does the Müllerian system regress? It would be logical to expect that substances from the fetal testes, in addition to stimulating the male system, might also inhibit the female. This is not the case. In the normal male fetus and in the castrated male fetus into which a fetal testis has been transplanted, the Wolffian system differentiates and the Müllerian system regresses. However, in the castrated male fetus injected with androgens, although the Wolffian system persists, so does the Müllerian. It would appear that the testicular substance responsible for the regression of the Müllerian system is not an androgen. This is further supported by the fact that if an androgen antagonist is given to a normal male fetus, the Wolffian ducts will regress, but so will the Müllerian. The Müllerian-inhibiting substance is a glycoprotein of about 70,000 molecular weight and is synthesized by the seminiferous tubules.

The general conclusion, then, is that in the male two different substances secreted by the developing testes are needed for the differentiation of the male reproductive tract and the regression of the female tract, respectively. The first is testosterone. The second is not an androgen but has not been completely identified. In the female, in the absence of any stimulating androgenic effect, the Wolffian system regresses and in the absence of an inhibitory signal from testicular tissue, the Müllerian system differentiates.

Figure 19–33 diagrams the relationship among the histological differentiation of the gonads, testosterone and estradiol secretion, and the anatomical differentiation of the genital systems of the male and female in the human. In both sexes the migration of the primordial germ cells occurs at the same time, but while the histological differentiation of the testis—formation of the spermatic cords—follows immediately and even takes place at the same time as germ cell migration, the histological differentiation of the ovary does not start until the second trimester. Both the testis and the ovary synthesize their respective androgenic (testosterone) and estrogenic (estradiol) steroid hormones at the same time, starting near the end of the second month. Again, in the male, androgen synthesis is accompanied by further histological differentiation of the gonad—Leydig cell development—but in the female, estrogen synthesis occurs without any histological differentiation in the ovary. In the male, regression of the Müllerian system is the first indication of the differentiation of the sex ducts. It begins before any significant testosterone synthesis is detected, thus further supporting the fact that the Müllerian-inhibiting substance is not testosterone. Differentiation of the Wolffian ducts and the external genitalia is started and completed during the

**FIG. 19–33.** Diagram of the time of occurrence and the relationship between steroid synthesis, histological differentiation of the gonads, and the differentiation of the genital ducts and the external genitalia in the human male and female. (From J. D. Wilson et al., 1981. Science 211:1278.)

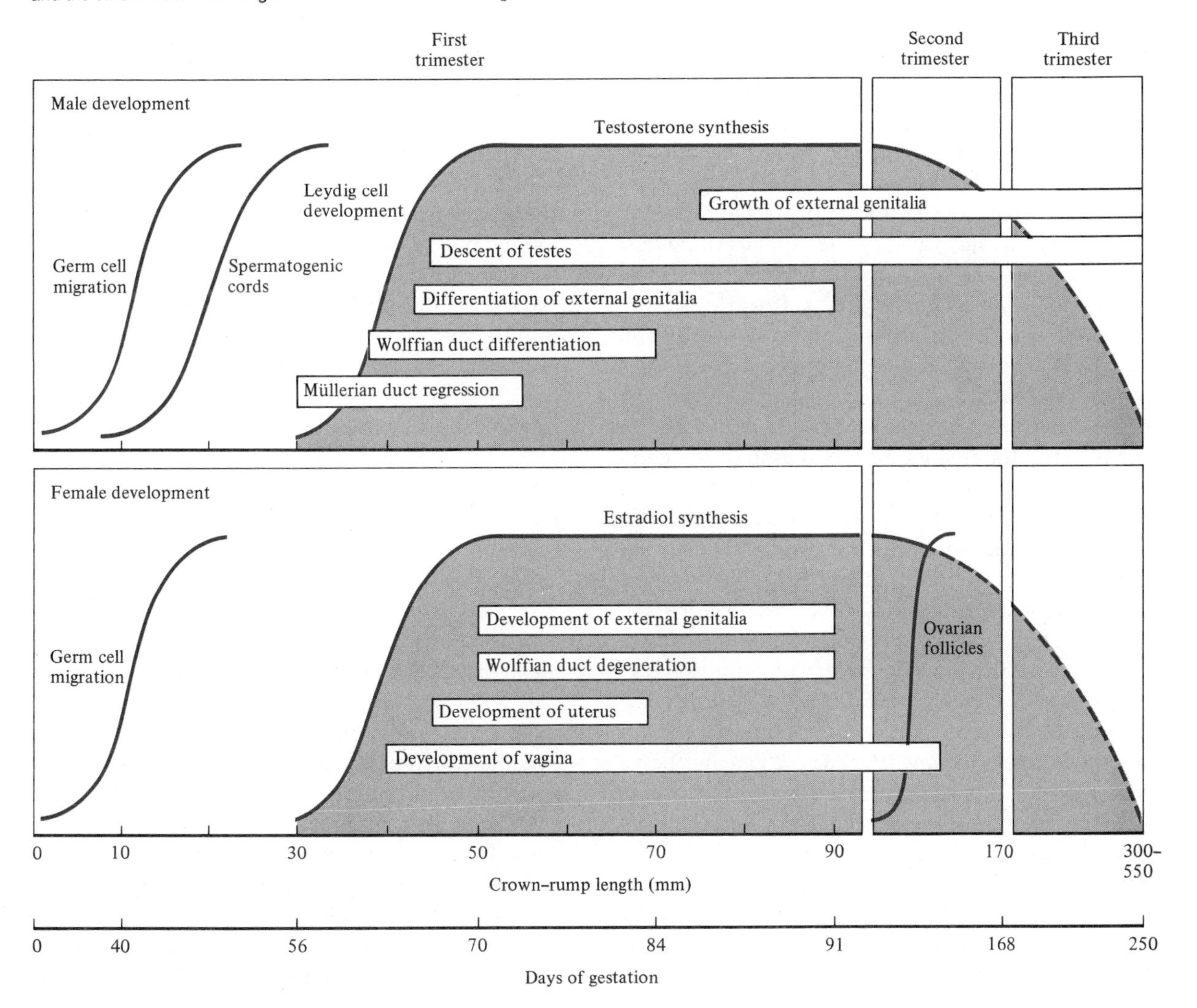

third month. Descent of the testis also starts during the third month, but the final descent through the inguinal canal into the scrotum may not be completed in some instances until after birth. Growth of the external genitalia is also not completely accomplished until after birth. In the female, the regression of the Wolffian duct system and the differentiation of the Müllerian ducts and the external genitalia take place over the same period of time as the corresponding differentiation of the ovarian follicles. Although there is a high level of estrogen synthesis during this time, its role in differentiation is questionable, since, as we have seen, a normal female system will develop in the absence of any hormones.

The genetic determinants that ultimately control the differentiation of the ovary and the testis, which is then followed by the synthesis of the appropriate sex hormone, do so by influencing the rates of only a limited number of reactions in the pathway of steroid synthesis. In the rabbit at the time when histological differentiation of the testis is well advanced, there are only two differences in the enzymes controlling steroid hormone synthesis (Wilson et al., 1981). First, the testis has about 50 times as much 3$\beta$-hydroxysteroid dehydrogenase, an enzyme that is rate limiting in testosterone synthesis. Second, the ovary has the capacity to convert the testosterone it synthesizes into estradiol. All of the other enzymes acting in the pathway of steroid hormone synthesis are the same in the testis and the ovary. Thus, differences in only a few enzymes at a critical time in differentiation dictate the type of steroid synthesis in the embryonic gonad and in turn control the further differentiation of the genital system.

The period of time during which a testis can exert its influence is limited to that period during which the reproductive structures are differentiating. There is then what may be called a "critical period," and only during this time can testicular hormones be effective. Once any part of the Wolffian system has developed, it escapes from the influence of the male hormone and will not regress if the androgenic influence is withdrawn. The same holds true for the Müllerian system. Once it has differentiated, it will not regress. Equally important, once either system has regressed, it can no longer be called back into existence and is lost forever. It follows that, if the Wolffian system is stimulated by the male hormone during only part of the critical period and then the stimulus is removed, its differentiation will be incomplete, stopping at the time the male hormone is withdrawn but not regressing to a more undifferentiated level.

## Intersexuality

On the basis of the above description of the control of the differentation of the reproductive system, it is apparent that serious abnormalities of this system result either from the failure of the male gonad to exert its normal influence on the male fetus or from the influence of androgens on the female fetus during the development of its reproductive system. Timing is critical: When a stimulus is applied or is lacking is of utmost importance. The strength of the stimulus is also a factor. In the human, a large number of reproductive system abnormalities present evidence of the variety of effects that may result when the normal hormonal levels are not maintained.

In studying abnormalities of the reproductive system, it is often difficult to determine the actual (genetic) sex of the individual by a physical examination, so complex are the abnormalities that present themselves. However, the presence or absence of sex chromatin may be used for this purpose. The specific genetic sex may be determined by the examination of leukocytes obtained from blood smears.

Genital abnormalities that result in intersexual individuals present a bewildering variety of differences that are often difficult to classify. Indeed, new classifications are introduced frequently. These are important for clinical purposes, but we will not attempt to present any system of classification; instead, we will examine only a few of the types of known intersexual variations.

### *Hermaphrodism*

A *true hermaphrodite* possesses both testicular and ovarian tissues. This is a rare condition in humans and, since the time of the first description of a true hermaphrodite in the medical literature in 1899 until 1974, only 302 cases had been reported. The gonadal tissue may appear as a separate testis and ovary or as a combined ovatestis. Sterility is the rule, and only in rare cases is germ cell maturation completed. An ovatestis is the combination most characteristic. Approximately one third of the cases show a testis on one side and an ovary on the other. The latter condition is usually accompanied by the presence of sex ducts that are different on each

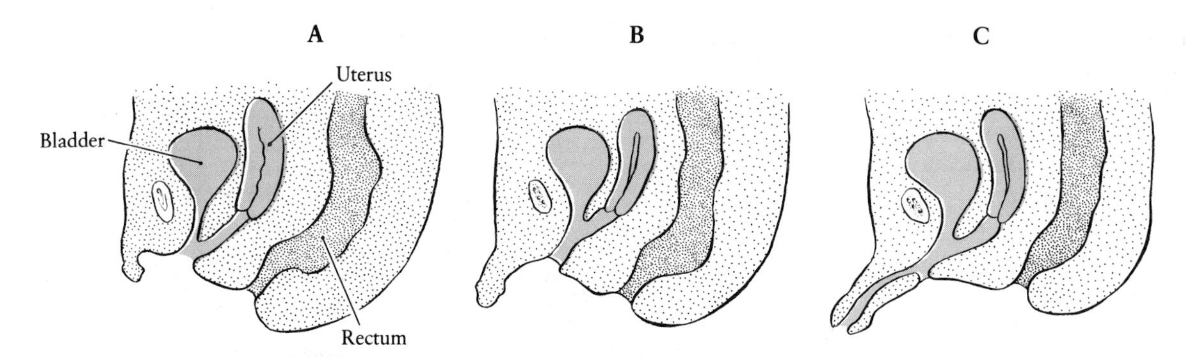


**FIG. 19–34.** Some types of urogenital structures developed in hermaphrodites ranging from A, external genitalia that are femalelike with a common external opening for the urethra and the vagina, to B, an opening into a urogenital sinus, and C, development of a penis with hypospadia.

side, conforming to the type of gonad found there. The external genitalia range from malelike to femalelike (Fig. 19–34). A common condition is *hypospadia* where, owing to hypoplasia of the inferior portion of the phallus, the urethra does not extend to the tip of the penis but opens somewhere along the inferior aspect (Fig. 19–34 C).

Although externally apparent malformations may be noted from the time of birth, hermaphrodites are generally raised as being members of one particular sex—that which most closely fits the appearance of the genitalia. The failure to undergo the normal morphological changes that occur at the time of puberty then results in a medical examination and a determination of the true status of the individual.

*Pseudohermaphrodism*

*Male* and *female pseudohermaphrodism* are conditions much more common than true hermaphrodism. A male pseudohermaphrodite is an individual whose gonad is a testis but whose sex ducts, genitalia, and secondary sexual characteristics resemble more closely those of the female sex. In agreement with the presence of only testicular tissue, these individuals have no Barr bodies and an XY chromosome configuration. Although many male pseudohermaphrodites may have malelike genitalia, hypospadia is common. In addition, the sex ducts show a much greater development of the Müllerian derivatives, usually with a well-developed uterus and uterine tubes, with the testes often occupying the position where the ovaries would be expected (Fig. 19–35).

Female pseudohermaphrodism is a much rarer condition in which the individual is genetically female, is chromatin positive, and has ovarian tissue, although she shows modifications of the urogenital system in a male direction. Not to be confused with this condition are others, including the *adrenogenital syndrome*, where genetic females develop malelike structures whose cause is extragonadal. Most commonly, female pseudohermaphrodites have an enlarged phallus resembling a penis, which, however, is hypospadic. Otherwise, the external genitalia are femalelike. In many cases, the vagina is short and narrow and opens into a urogenital sinus, as does the urethra. The uterus and the uterine tubes are more or less normal.

*Gonadal Dysgenesis*

As the name implies, this is a condition in which the gonad fails to develop. As would then be expected, since the genital system develops without any influence from the gonad, the sex ducts and genitalia are normally and completely female, al-

**FIG. 19–35.** Urogenital system in a male pseudohermaphrodite.

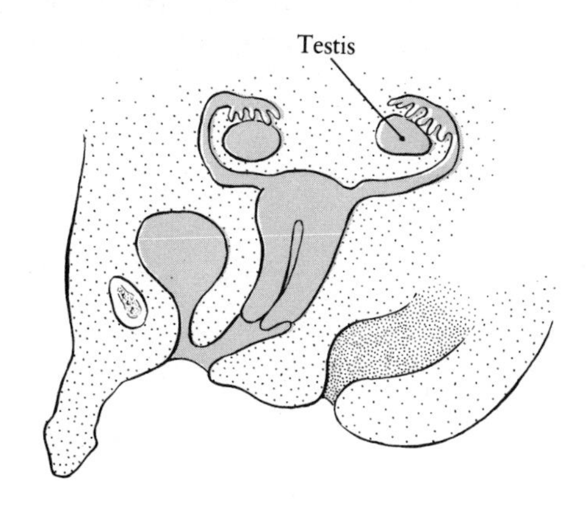

though immature. The genetic sex may be XX, XY, or XO. Their stature is generally small, and a webbed neck is typical. Gonadal remnants are present and distinguish this condition from true agonadism, a very rare occurrence in which the gonad never develops at all.

*Klinefelter's Syndrome*

True Klinefelter's syndrome is applied to chromatin-positive but phenotypic males, with an XXY chromosomal configuration. They resemble normal males with hypogonadism, are often mentally retarded, and show enlarged breasts (gynecomastia). It is estimated that the incidence of this condition is one in every thousand males.

*The Adrenogenital Syndrome*

This syndrome is characterized by a masculinization of the genitalia and the physique owing to an oversecretion of androgens from the adrenal glands. A variety of conditions are seen in this syndrome, depending on the genetic sex of the individual and the time when the increased secretion of androgens occurs. If the individual is male and the condition is congenital, the reproductive system develops normally although the penis may be slightly enlarged. On the other hand, congenital occurrence in a female results in masculinization of the genital organs. The external genitalia show a wide variety of conditions depending on the amount of androgen secreted by the adrenal glands. Postnatal development in males with the syndrome is characterized by precocious puberty, strong masculature, and short stature. In females, virilization of the external genitalia continues, and they also show precocious development of pubic and axillary hair and are short and muscular. Owing to the large amount of circulating androgen, pituitary secretion of gondotrophins is suppressed, and in both sexes the gonads do not mature.

## References


Barr, M. L. and E. G. Bertram. 1949. A morphological distinction between neurons of the male and female, and the behavior of the nucleolar satellite during accelerated nucleoprotein synthesis. Nature 163:676–677.

Boczkowski, K. 1968. The syndrome of pure gonadal dysgenesis. In: Medical Gynaecology and Sociology. Oxford: Pergamon Press.

Bruent, J., R. R. Mowbray, and P. M. F. Bishop. 1968. Management of the undescended testis. Br. Med. J. 1:1367–1372.

Bulmar, G. 1957. The development of the human vagina. J. Anat. 91:490–502.

Gillman, J. 1948. The development of the gonads in man, with a consideration of the role of the fetal endocrines and histogenesis of ovarian tumors. Carnegie Contrib. Embryol. 32:81–131.

Glenister, T. W. A. 1954. The origin of the urethral plate in man. J. Anat. 88:413–425.

Gruenwald, P. 1952. Development of the excretory system. Ann. N.Y. Acad. Sci. 55:142–146.

Jost, A. 1972. A new look at the mechanism controlling sex differentiation in mammals. Johns Hopkins Med. J. 130:38–53.

O'Connor, R. J. 1939. Experiments on the development of the amphibian metanephros. J. Anat. 74:34–44.

Ohno, S. 1978. The role of H-Y antigen in primary sex determination. J.A.M.A. 239:217–220.

Tanabe, K. and M. Kotani. 1974. Relationship between the amount of the "germinal plasm" and the number of primordial germ cells in *Xenopus laevis*. J. Embryol. exp. Morphol. 31:89–98.

Torrey, T. W. 1954. The early development of the human nephros. Carnegie Contrib. Embryol. 34:175–197.

Torrey, W. 1965. Morphogenesis of the vertebrate kidney. In: Organogenesis. R. L. DeHaan and H. Ursprung, eds. New York: Holt, Rinehart and Winston.

Waddington, C. H. 1938. The morphogenic function of a vestigial organ in the chick. J. Exp. Zool. 15:371–377.

Whittington, P. McD. and K. E. Dixon. 1975. Quantitative studies of germ plasm and germ cells during early embryogenesis of *Xenopus laevis*. J. Embryol. exp. Morphol. 33:57–74.

Wilson, J. D., F. W. George, and J. E. Griffin. 1981. The hormonal control of sexual development. Science 211:1278–1284.

Witschi, E. 1948. Migrations of germ cells of human embryos from the yolk sac to the primitive folds. Carnegie Contrib. Embryol. 32:67–80.

# 20

# THE NERVOUS SYSTEM

The nervous system may be divided into the *central nervous system*—the brain and its caudal continuation, the spinal cord—and the *peripheral nervous system* made up of millions of nerve fibers that connect the central nervous system to all parts of the body. The part of the nervous system that innervates visceral structures—smooth muscle and glands of the viscera and the integument—is called the *autonomic nervous system*. the autonomic nervous system is not a separate system but merely that part of the nervous system that functions in the involuntary control of visceral organs. The autonomic nervous system is, in turn, subdivided into *sympathetic* and *parasympathetic* divisions. The sympathetic division has the cell bodies of its motor fibers located in the thoracic and upper lumbar regions of the spinal cord. The parasympathetic division has nerve fibers that are associated with cranial nerves III, VII, IX, X, and XI and with cell bodies in the sacral region of the spinal cord. Because of the distinct anatomical locations of those divisions, they are also termed the *thoracolumbar* and the *craniosacral* divisions.

The basic unit of the nervous system is the nerve cell or *neuron*, a cell that is specialized in the properties of excitation and conduction. A neuron consists of a cell body containing the nucleus and a number of processes that transmit the nerve impulse. *Dendrites* transmit impulses toward the cell body, and *axons* transmit impulses away from the cell body. Each neuron has a single axon but may have a number of dendrites.

Each neuron is a structural and functional unit of the nervous system and makes up one of a number of links in a chain through which the nerve impulse passes. Each neuron is in close contact with one or a number of other neurons. The point at which the impulse passes from one neuron to another is the *synapse*, where the axon of one neuron terminates on the dendrite(s) of another. Although they are in close contact at the synapse, there is no cytoplasmic continuity between the processes, merely an interface where they are functionally related to allow the passage of the nerve impulse from cell to cell.

The structures that we call nerves, as in the frog nerve–muscle preparation we use to study muscle contraction, are actually bundles of neuronal processes connecting the central nervous system with peripheral structures. Often the cell bodies of neurons form aggregations at some point along a nerve, forming *ganglia* such as the sensory ganglia of the spinal nerves. Corresponding accumulations of nerve cell bodies within the central nervous system are themselves called *nuclei*—a somewhat confusing terminology in that these central nervous system nuclei are in turn made up of the individual nuclei and cytoplasm of a number of neurons.

The early development of the nervous system, which involves the formation of the neural tube, has been described previously. Further development consists of considerable modification of this originally more or less straight tubular structure, the most marked changes and the greatest growth taking place at the cranial (rostral) end, resulting in the formation of the brain. Caudad of the brain, the neural tube retains a fairly uniform diameter, and since the developmental changes that occur here are much less complicated than those that occur in the brain, the histogenesis and morphogenesis of the nervous system will first be examined in the spinal cord.

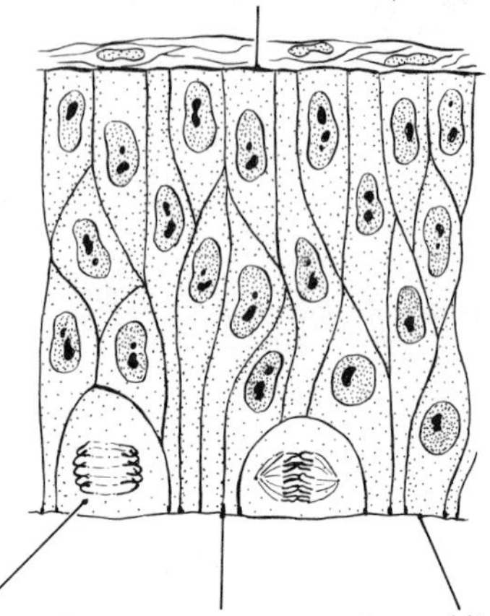

**FIG. 20–1.** Drawing of a cross section of the neural tube of an early chick embryo. Although the position of the nuclei presents a layered appearance, the neural tube at this stage is a pseudostratified columnar epithelium. (After J. Langman et al., 1966. J. Comp. Neurol. 127:399.)

# THE SPINAL CORD

## Histogenesis of the Spinal Cord—Neurogenesis

Neurogenesis involves three processes: proliferation, migration, and maturation. When the neural plate first forms, it consists of a single layer of columnar cells. These cells proliferate so rapidly that by the time the neural tube is formed, it consists of many more cells, which appear to be arranged in a number of layers (Fig. 20–1). A basement membrane, the *external limiting membrane,* covers the surface of the neural tube, and the lumen of the tube, the central canal, is lined by a thin *internal limiting membrane.* At this stage of development, the cells of the tube are not actually arranged in layers but are pseudostratified. The nuclei appear to be located at essentially two different levels. One level consists of oval nuclei below the external limiting membrane contained in wedge-shaped cells with slender cytoplasmic processes extending to the internal limiting membrane, where they are interconnected by terminal bars (Fig. 20–1). This is the layer that is usually termed the mantle layer, but recently the name *intermediate layer* has been suggested for it. The second layer consists of round cells in various stages of mitosis with broad contact with the internal limiting membrane. Since mitotic figures are seen only in this region close to the lumen of the tube, it has been termed the germinal layer on the assumption that this layer of proliferating cells functions to supply cells to the outer layers. However, more accurate studies of histological preparations, reported in the mid 1930s and confirmed experimentally in the 1950s and 1960s by colchicine treatment, by tritiated thymidine uptake, and by electron-microscopic studies, have presented a different interpretation. What appear to be two layers of different kinds of cells actually represent different stages in the cell cycle of the same type of cell, the *neuroepithelial cell* (Fig. 20–2 A). During the synthetic phase of the cell cycle, the nuclei are found in the wedge-shaped cells at different levels below the external limiting membrane. The nuclei, as they enter the mitotic stage of the cell cycle, now move toward the lumen to take up a position in what is now termed the *ventricular layer.* The broad dividing cells squeeze the cytoplasmic processes of the wedge-shaped cells into slender strips. In the early stages of neurogenesis, the mitotic spindles are arranged parallel to the internal limiting membrane. Division completed, the daughter nuclei move away from the lumen to occupy their positions in the intermediate layer. They repeat the migration when they divide again.

Sometime after the closure of the neural tube, there appear close to the external limiting membrane cells that are histologically different from the

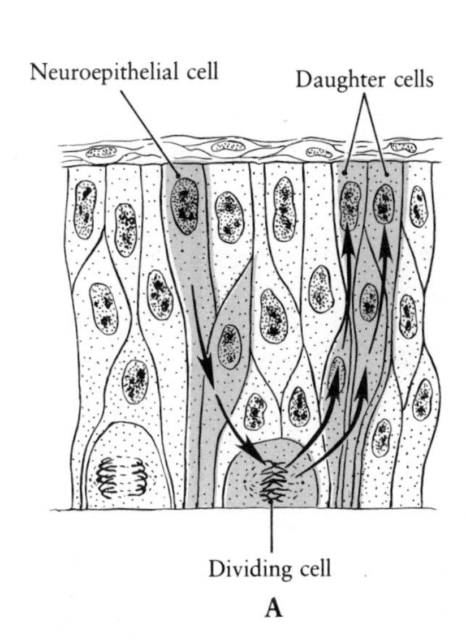


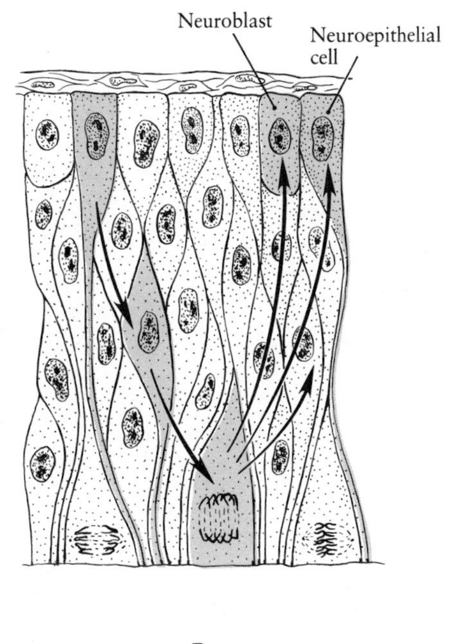


**FIG. 20–2.** A, schematic drawing of the wall of the chick neural fold. The nuclei of the cells synthesizing DNA are located in a region at a distance from the internal limiting membrane but move toward it when they start to divide. Following division, the daughter nuclei again move away from the lumen; B, schematic drawing of the chick neural tube. The tube has increased in thickness over A. Some mitotic figures in the ventricular zone are now oriented perpendicular to the lumen. One daughter cell from such a dividing neuroepithelial cell loses its terminal bar connections and migrates into the outer portion of the intermediate zone as a primitive neuroblast. (After J. Langman, et al., 1966. J. Comp. Neurol. 127:399.)

neuroepithelial cells and that, in addition, do not synthesize DNA. These cells are *neuroblasts* and represent the first stage in the differentiation of the neuron. At the time the first neuroblasts appear, it has been observed that some of the dividing neuroepithelial cells have their spindles oriented perpendicular to the internal limiting membrane. It has been proposed that this would represent a possible mechanism for the release of one of the daughter cells. The one away from the lumen would be freed from its terminal bar connections and thus able to migrate to the outermost level of the neural tube (Fig. 20–2 B).

During its further differentiation, the neuoblast passes successively through apolar, bipolar, and unipolar to multipolar stages (Fig. 20–3). The persisting process of the unipolar stage grows out of the intermediate layer to become a part of the *marginal* layer below the external limiting membrane. This process becomes the axon of the nerve cell. On the side of the cell opposite the axon, the appearance of a number of small, branched processes marks the beginning of dendrite formation. Further differentiation involves the appearance of fine neurofibrils in the cytoplasm of the cell and its processes and, considerably later, the appearance of RNA-rich Nissl bodies in the cell. The dendrites make synaptic connections with the axons of adjacent neurons. The axons running in the marginal zone may synapse with dendrites of nerve cells within the spinal cord, forming association neurons; or they may pierce the external limiting membrane and leave the central nervous system, many axons at each segmental level forming the *motor (anterior, ventral) root* of a spinal nerve that becomes associated with the myotome at that level.

## Histogenesis in the Spinal Cord: Gliogenesis

The next cells to differentiate from the neuroepithelial cells are the *glioblasts,* forerunners of the *neuroglia,* the supportive cells of the nervous system. Some supportive elements can be recognized early in the development of the neural tube. These are *ependymal* cells whose nuclei lie in the ventricular layer and whose process extend from the in-

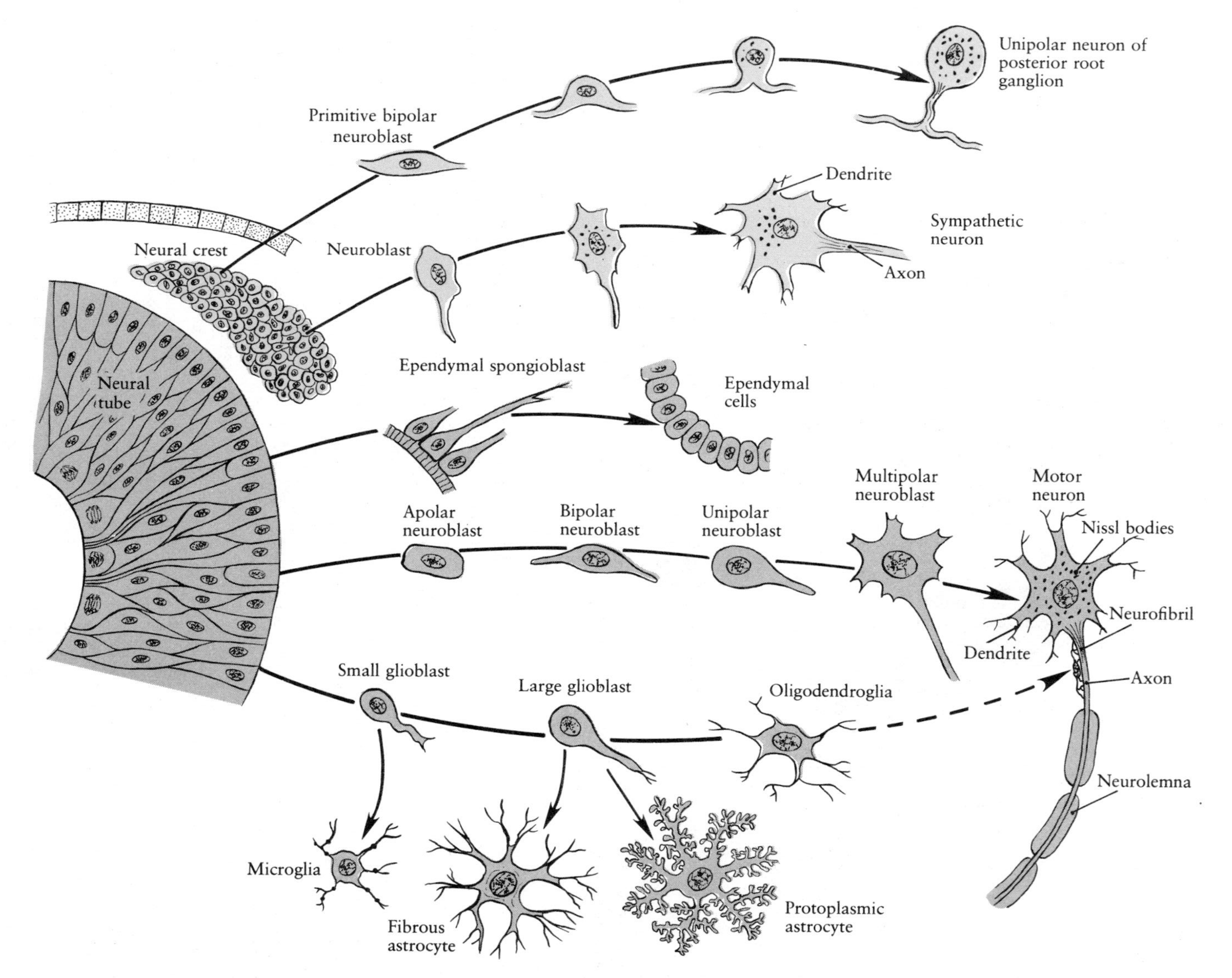


**FIG. 20–3.** Diagram showing the types of neurons and neurologia developed from the neuroepithelial cells of the neural tube and from the neural crest.

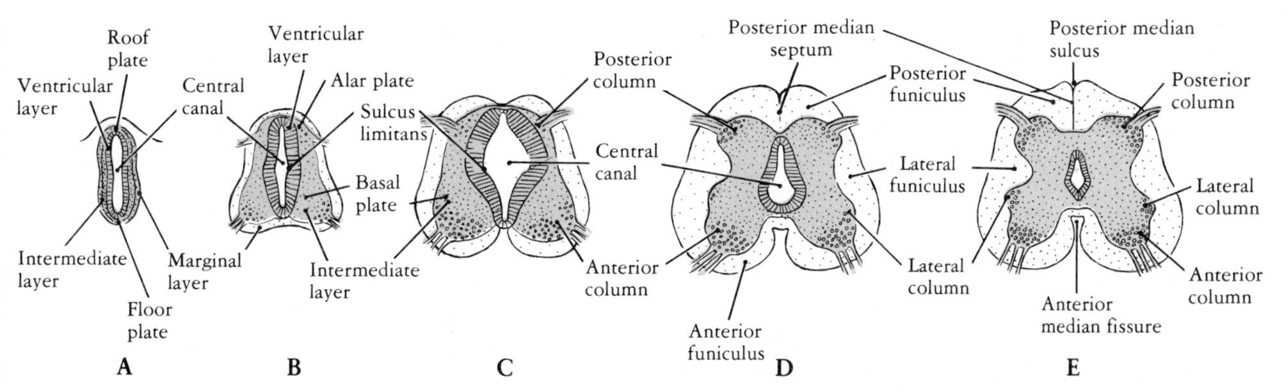


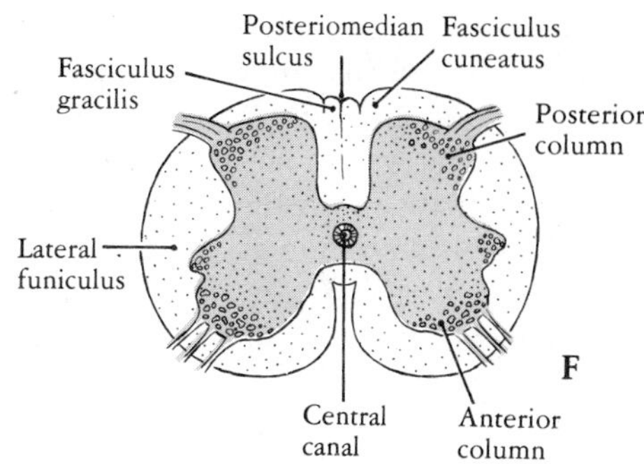


**FIG. 20–4.** Six stages in the development of the human spinal cord. A, 3.7-mm embryo; B, 10-mm embryo; C, 11-mm embryo; D, 30-mm embryo; E, 45-mm embryo; F, 80-mm embryo.

ternal to the external limiting membrane. After the disappearance of the neuroepithelial cells—as they differentiate into the forerunners of the neurons and the neuroglia—ependymal cells persist as the definitive epithelial lining of the neural canal.

The glioblasts are the precursors of *astrocytes* and *oligodendroglia* cells (Fig. 20–3). Astrocytes may be either *fibros* or *protoplamic,* the former most prevalent in the white matter, the latter in the gray. The ends of the processes of the astrocytes often develop expansions that become closely associated with nervous system membranes and blood vessels. In addition to their supportive function, astrocytes are also concerned with the supply of nutrients to the nervous elements. The oligodendroglia develop fewer and less delicate processes than the astrocytes. They are found in both the white and gray matter of the nervous system but are more common in the white, where they function in the formation of myelin sheaths around the axons of the nerve cells. Delicate extensions of these cells are seen closely related to the myelin sheath.

Another type of glial cell, *microglia,* appears late in development; its origin and identification are controversial. Since it appears at the time that the blood vessels are invading the nervous tissue, it is often suggested that the mesodermal connective tissue of the walls of the blood vessels may be the precursors of the microglia. A different interpretation has been offered. It recognizes a small glioblast, a derivative of the neuroepithelial cells, which may differentiate directly into a microglial cell or, by growth, become a precursor of the larger glial cells, the astrocytes and oligodendroglia. Microglia cells generally appear inactive, but following injury to nervous tissue, they proliferate rapidly and become phagocytic.

Neuroglia cells differ from the connective tissue cells of mesodermal origin in two respects. They are not stained by the usual connective tissue staining methods but do show up clearly following silver impregnation, as is characteristic of all nerve cells and their processes. In addition, they do not form intercellular fibrous elements.

## Further Development of the Spinal Cord

In man, by the end of the first month of development, the spinal cord shows a thick intermediate zone outside of which is seen a thin marginal layer (Fig. 20–4 A). The central canal is elongated in an anteroposterior direction and covered anteriorly by a thin *floor plate* and posteriorly by a thin *roof plate.* Shortly thereafter (Fig. 20–4 B), the central canal shows an indentation on each of its lateral walls. This is the *sulcus limitans,* which serves as a marker by which the neural tube may be divided into a posterior *alar plate* and an anterior *basal plate.* The alar plate neurons will form the sensory and coordinating parts of the cord, and the basal plate cells will form the motor elements. By the middle of the second month, the differentiating alar and basal plates are well formed (Fig. 20–4 C).

As development progresses (Fig. 20–4 D–F), the size of the central canal is decreased by the approximation and fusion of the walls of the alar plate, forming a seam, the *posterior median septum*. The shallow groove above this septum is the *posterior median sulcus*. The basal plate regions develop rapidly and overlap the floor plate to form a deep fissure below it, the *anterior median fissure*.

The cells of the intermediate layer gradually assume the characteristic butterfly or H-shaped configuration of the adult cord; the two posterior and the two anterior limbs of the H are often termed the dorsal and ventral horns, respectively. However, in conformance with their location in respect to the anatomical position and also to emphasize the fact that they are, in reality, longitudinally oriented columns of cells, we prefer to label them the *posterior* and *anterior columns*.

The neurons of the posterior column are associated with sensory (afferent) impulses. Fibers from the sensory receptors (exteroceptors in the integument, proprioceptors in muscles and tendons, and interoceptors in the visceral organs) return to the spinal cord where they make synaptic connections with the dendrites of the neurons in the posterior column.

The neurons of the anterior column are associated with motor (efferent) impulses. Their axons grow out of the spinal cord to make peripheral connections with effectors that are both somatic (skeletal muscle) and visceral in nature.

The marginal region of the cord consists mainly of longitudinally running fiber tracts carrying sensory impulses upward from the spinal cord to the brain and motor impulses downward from the brain to the cord. On the basis of their location, three regions of the marginal zone are designated as the *posterior (dorsal)*, *lateral*, and *anterior (ventral) funiculi* (Fig. 20–4 D,E).

## The Spinal Nerves

Thirty-one pairs of segmentally arranged spinal nerves develop. Each spinal nerve is formed by the juncture of an anterior (ventral) motor root made up of axons carrying nerve impulses away from the spinal cord and a posterior (dorsal) sensory root made up of processes from neurons located in the posterior root ganglia carrying impulses toward the

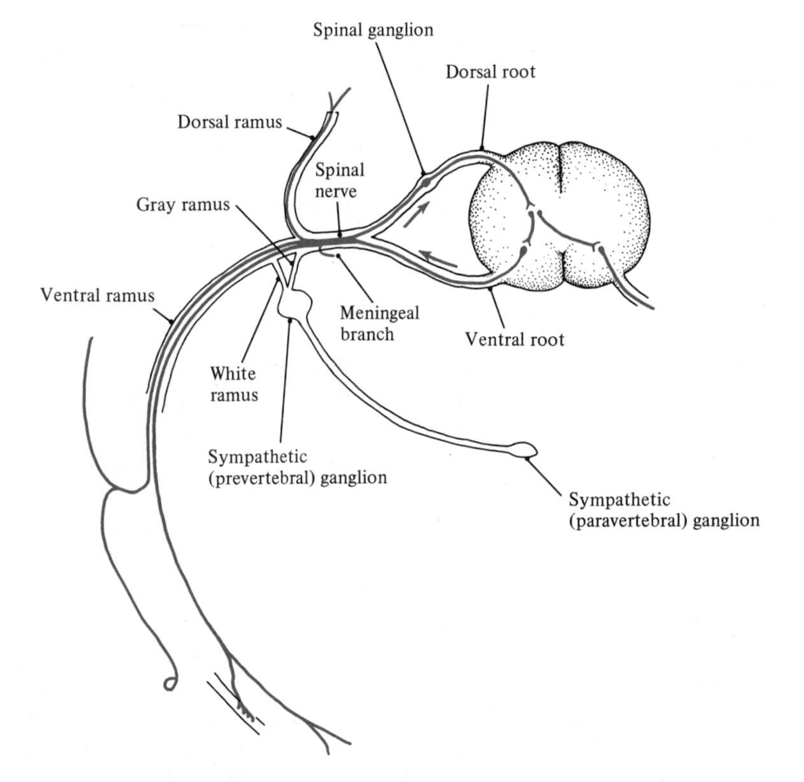


**FIG. 20–5.** Diagram of a typical spinal nerve.

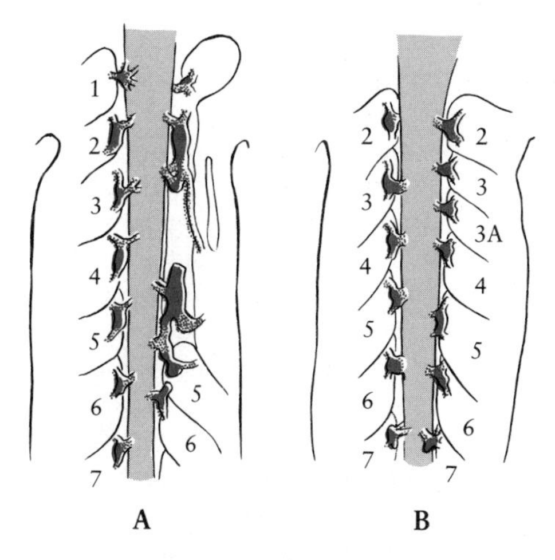


**FIG. 20–6.** The effect of somite excision and somite transplantation on the development of the spinal ganglia and nerve roots. A, incomplete segmentation of ganglia on the right side after excision of somites 2, 3, 4, and 5; B, three somites (3, 4, and 5) on the right side were excised and replaced with four somites. Note the presence of a supernumerary ganglion and nerve (3A) associated with the additional somite. (After P. Weiss, 1955. In: Analysis of Development, B. J. Willier, P. Weiss, and V. Hamburger, eds. W. B. Saunders Company, Philadelphia.)

spinal cord (Fig. 20–5). A short distance from the union of the anterior and posterior roots, the spinal nerve gives off a dorsal ramus that supplies the muscles and skin of the back. Further on, the spinal nerve sends branches termed *rami communicans* to autonomic nervous system ganglia and then continues as the large ventral ramus to supply the ventral regions of the body and the extremities.

The pattern of 31 pairs of spinal nerves is another example of metamerism. However, this segmentation is not intrinsic to the nervous system. It has been known for some time that this pattern is secondarily imposed through the influence of the segmentally arranged somites. If the somites on one side of a salamander embryo are removed, the spinal nerves and ganglia on that side will show an incomplete haphazard segmentation (Fig. 20–6 A). If additional somites are transplanted to an area, supernumerary nerves and ganglia will develop (Fig. 20–6 B).

## The Anterior Roots of the Spinal Nerves

Most of the fibers that make up the anterior roots of the spinal nerves are the axons of neurons, which develop in the anterior column of the spinal cord. These are motor fibers that will innervate the large skeletal muscle masses, and for this reason they are classified functionally as *somatic efferent fibers* (Fig. 20–7).

In the thoracic and upper lumbar regions of the cord, a third column develops, a small *lateral column*, between the anterior and posterior columns. The neurons in this column develop axons that also leave the cord to make up a part of the anterior root of the spinal lerve, but soon after becoming a part

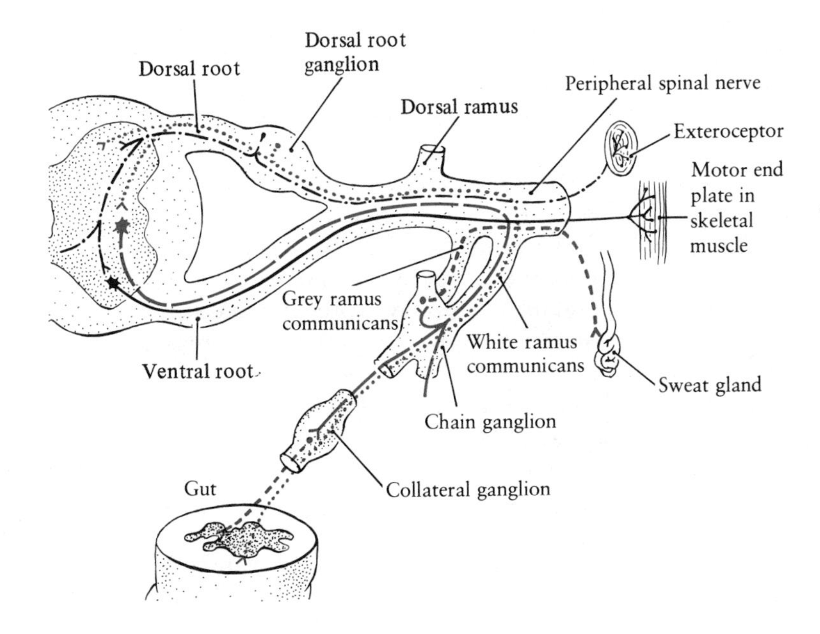


**FIG. 20–7.** Diagram of a spinal nerve and its ventral or (motor) and dorsal (sensory) roots. The chain ganglia of the sympathetic division of the autonomic nervous system are connected to the spinal cord by the gray ramus communicans and the white ramus communicans. The functional types of fibers, their location, and their central and peripheral connections are shown. Somatic efferent —, visceral efferent preganglionic ———, visceral efferent postganglionic ------, somatic afferent —·—·—, visceral afferent · · · · · ·. The autonomic nerve fibers are colored.

of the spinal nerve, they leave it and grow toward one of a number of different ganglia (accumulations of nerve cell bodies and supportive cells) (Fig. 20–7) located in a number of different positions in the trunk area. Here, they synapse. The neurons with which they synapse now send their axons to visceral effectors. The axons of the spinal cord neurons are thus termed *preganglionic* fibers and those of the ganglia are termed *postganglionic*. This two-neuron chain, with the first neuron in the spinal cord and the second in a ganglion, represents the sympathetic division of the autonomic nervous system. Similarly, the neurons of the parasympathetic division of the autonomic nervous system, which are located in the brain stem and the sacral region of the spinal cord, develop preganglionic fibers that synapse in different ganglia. Postganglionic fibers from the neurons in these ganglia then send their axons to the appropriate visceral effectors. The motor ganglia in which all of the preganglionic fibers of the autonomic nervous system synapse are derivatives of the neural crest.

## The Neural Crest

The early development of the neural crest was described in Chapter 8. When first formed, the neural crest appears as two longitudinally running bands of cells on either side of the spinal cord, between it and the somites. The cells of the neural crest soon migrate away from their original locations and become widely distributed throughout the body, contributing to the formation of a remarkable variety of different structures. Its major contribution is to the peripheral nervous system, all of which, with the exception of the motor fibers from central nervous system neurons and some of the cranial nerve sensory ganglion components, is derived from the neural crest. In the head region it forms the root ganglia of the VIIth (geniculate ganglion), IXth (superior ganglion), and Xth (jugular ganglion) cranial nerves. The trunk ganglia of these nerves, part of the geniculate, the petrosal, and the nodose, respectively, are formed from ectodermal placodes in the head region. The neural crest also forms the autonomic nervous system ganglia associated with cranial nerves III, VII, IX, and X. The spinal neural crest forms all of the spinal sensory ganglia and the ganglia of the autonomic nervous system in the trunk. The supportive cells of peripheral ganglia are also of neural crest origin. In addition, neural crest cells differentiate into a wide variety of nonneural structures. In the head region they contribute to the head mesenchyme (which is then called *mesectoderm*) that forms the bones and cartilages of the facial and visceral skeleton, connective tissue of the salivary, thyroid, and thymus glands, the ciliary muscles of the eye, the melanophores of the iris, and the odontoblasts of the teeth. The neural crest also contributes to the endocrine system, forming the medulla of the adrenal gland, the cells of the carotid body, and the calcitonin-producing cells of the ultimobranchial bodies. The trunk pigment cells of the skin and enteric structures are all neural crest derivatives. The spinal neural crest forms the posterior sensory root ganglia of the spinal nerves, the trunk and collateral ganglia of the autonomic nervous system, the cells of the adrenal medulla, the Schwann cells, and the pigment cells.

Cells migrate away from the neural crest in two streams (Fig. 13–2). The first is dorsolateral into the superficial ectoderm dorsal to the neural tube and the somites. Most of these cells will form pigment cells. The second is ventral between the neural tube and the somites. These cells will form both the sensory ganglia of the cranial nerves and of the posterior roots of the spinal nerves, and the ganglia of the autonomic nervous system.

### *The Autonomic Nervous System Ganglia*

Generally, the first cells to leave the neural crest end up in locations the farthest away from their original position. The three groups of cells that migrate the greatest distances all become ganglia of the autonomic nervous system, whose axons (postganglionic) carry impulses to smooth muscles and glands of the viscera.

One group consists of a chain of segmentally arranged ganglia located ventrolaterad of the vertebral column. These *chain (paravertebral) ganglia* are interconnected by longitudinally running fibers making up the sympathetic trunks of the autonomic nervous system. Efferent preganglionic fibers of the neurons of the lateral column of the spinal cord run to the chain ganglia by way of the white rami communicans (Fig. 20–7). Although the chain ganglia are located at all levels of the spinal cord, the lateral columns— and, correspondingly, the white rami communicans—are found only in the thoracic and upper three lumbar segments. Most of the postganglionic fibers of the chain ganglia neurons return to the spinal nerves by way of the gray rami communicans supplying motor innervation to peripheral visceral organs such as blood vessels, arrector pili muscles, and sweat glands.

Some of the preganglionic fibers from lateral column spinal cord neurons may pass through the chain ganglia without synapsing to reach more deeply located *collateral ganglia* (Fig. 20–7), irregular aggregations of cells associated with the three main trunks of the dorsal aorta (celiac, superior mesenteric, inferior mesenteric). Postganglionic fibers from the collateral ganglia neurons that are, of course, derivatives of neural crest cells that migrate to these locations supply motor innervation to abdominal and pelvic visceral organs. These spinal cord preganglionic fibers, chain and collateral ganglia, and their postganglionic fibers together make up the *sympathetic division* of the autonomic nervous system. In the chick it has been demonstrated that the neural crest from the 8th to the 28th somites is responsible for the formation of the ganglia that make up this system. Crest cells from these spinal levels do not make any contributions to the parasympathetic division of the autonomic nervous system (Fig. 20–8).

The neural crest cells that migrate the greatest distances are those of the vagal region of the neural axis (somites 1–7 in the chick; Fig. 20–8). They eventually reach locations within the walls of the thoracic and abdominal viscera forming *terminal* or *enteric plexuses*. Preganglionic fibers reach these plexuses by way of the Xth cranial nerve, and the postganglionic fibers supply motor innervation to the smooth muscle of these regions. Other crest cells of the head region differentiate into autonomic ganglia (ciliary, otic, parotid) supplied by preganglionic fibers in other cranial nerves. Postganglionic fibers supply the salivary glands and the intrinsic eye muscles. These ganglia and nerves constitute the cranial component of the *parasympathetic division* of the autonomic nervous system. Enteric plexuses that develop in the pelvic viscera are derived from neural crest cells that migrate out from the sacral region of the spinal cord (somites caudal to the 28th in the chick, Fig. 20–8). The preganglionic fibers that supply these plexuses and the short postganglionic fibers that run from them to the smooth muscle of the pelvic viscera make up the sacral component of the parasympathetic division of the autonomic nervous system.

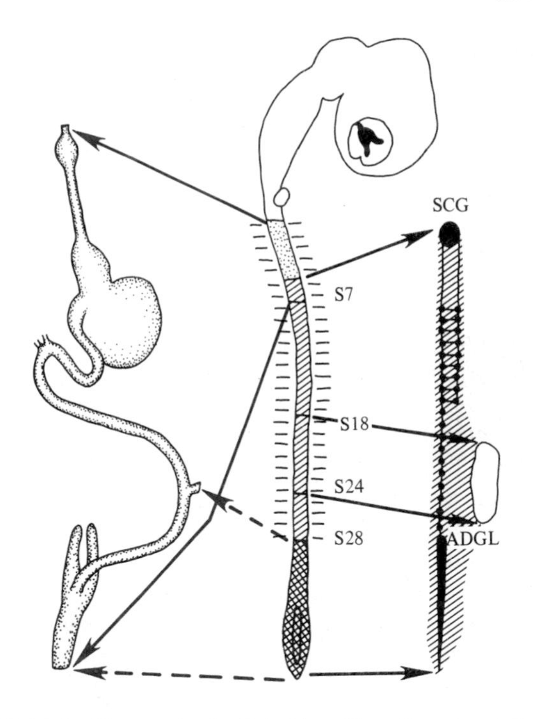


**FIG. 20–8.** Diagram showing the origins of the autonomic ganglion cells and the adrenomedullary cells. The sympathetic ganglia cells (SGC) are derived from the neural crest from all levels of the spinal cord below somite 7. The vagal neural crest, somite levels 1 through 7, gives rise to parasympathetic enteric cells of the entire alimentary tract. Spinal levels below somite 28 supply enteric cells for the postumbilical gut. The neural crest from the region of somites 8 to 28 does not form any enteric cells. Neural crest from somite levels 18 to 24 supplies the adrenomedullary cells (ADGL).

*Posterior Root Ganglia*

The group of neural crest cells that remain closest to the spinal cord form the sensory *spinal ganglia (posterior root ganglia)* (Fig. 20–7). The segmental arrangement of the somites is reflected in the segmental arrangement of the posterior root ganglia. Each of the 31 paris of spinal nerves develops a spinal ganglion.

The axons and dendrites that grow out of the spinal ganglia form the fibers making up the posterior roots of the spinal nerves. The neurons in the spinal ganglia at first show two processes leaving the cell. These later fuse to form a single process that branches at some distance from the cell body (Fig. 20–3). One branch of this process grows peripherally and becomes funcitonally linked with a sensory receptor. The other branch grows centrally into the spinal cord. The fibers that make connections with *exteroceptors* (touch, pain, temperature receptors) and *proprioceptors* (neuromuscular and neurotendinous receptors) are classified functionally as *somatic afferent*. Those that connect to *enteroceptors* in the visceral organs are classified as *visceral afferent*. Thus, the cells of the spinal ganglia develop the sensory or afferent components of the spinal nerves carrying impulses form the numerous somatic and visceral sensory receptors into the central nervous system.

*Schwann Cells*

All peripheral nerves are surrounded by a delicate sheath or *neurolemma* made up of what are known as *Schwann cells*. These cells are neural crest derivatives. The Schwann cells migrate from the neural crest to the growing nerve fiber and then grow peripherally along with the fiber, continuing to divide as they do so.

The one known function of the Schwann cells is the formation of the myelin sheath of some of the peripheral nerve fibers. During the process of myelinization, the naked nerve fiber settles into a groove in the surface of the Schwann cell and gradually the Schwann cell encircles it (Fig. 20–9 A, B). The point where the two folds of the Schwann cell meet is called the *mesaxon*. Envelopment continues, and the Schwann cell forms scroll-like layers around the nerve fiber. At first these layers contain Schwann cell cytoplasm, but this then disappears and is found only at the innermost and outermost regions of the cell. The loss of the cytoplasm brings into contact the two inner layers of the Schwann cell membrane to form a dense line, the *major dense line* (Fig. 20–9 C). Contact of the outer surfaces of the Schwann cell membrane forms a thinner line, the *intraperiod line*. Thus, myelin is nothing more than the apposed membranes of the lamellae of the Schwann cell wrapped around the nerve fiber.

Whether or not a nerve fiber will be myelinated depends on its size. In mammals, peripheral nerve fibers less than 1 micron in diameter are unmyelinated, although they still have a Schwann cell sheath. In myelinated fibers, the number of lamellae increases the diameter of the fiber. Myelinization is not necessary for nerve impulse conduction—which occurs in the embryo before the formation of myelin—but, after myelinization, the velocity of the conduction of the nerve impulse increases.

Schwann cells are not found within the central nervous system, although many fibers in the brain and spinal cord are myelinated. It was logical to propose that one of the glia cells was responsible for myelinization in the central nervous system, although exactly which one it was at first was not known. Recently, EM studies have shown a continuity between the oligodendroglia cell and the myelin sheath, thus proving conclusively that this cell is the one responsible. A single oligodendroglia cell may form myelin lamellae around more than one nerve fiber, but the process of myelinization and the end result is essentially the same as in the peripheral nerve fibers.

*Adrenal Gland*

Neural crest cells also form a part of the adrenal (suprarenal) gland. The adrenal glands of higher vertebrates are formed by two primordia of widely

**FIG. 20–9.** Diagram showing the formation of the myelin sheath. A, axon being enveloped by a Schwann cell; B, axon completely enveloped. The mesaxon is the channel formed where the folds of a Schwann cell meet, opening to the extracellular space; C, lamellae of a Schwann cell wrapped around the axon forming the myelin. A dense line results from the loss of cytoplasm and the apposition of the inner surfaces of the lamellae, and an intraperiod line results from the apposition of the outer surfaces.

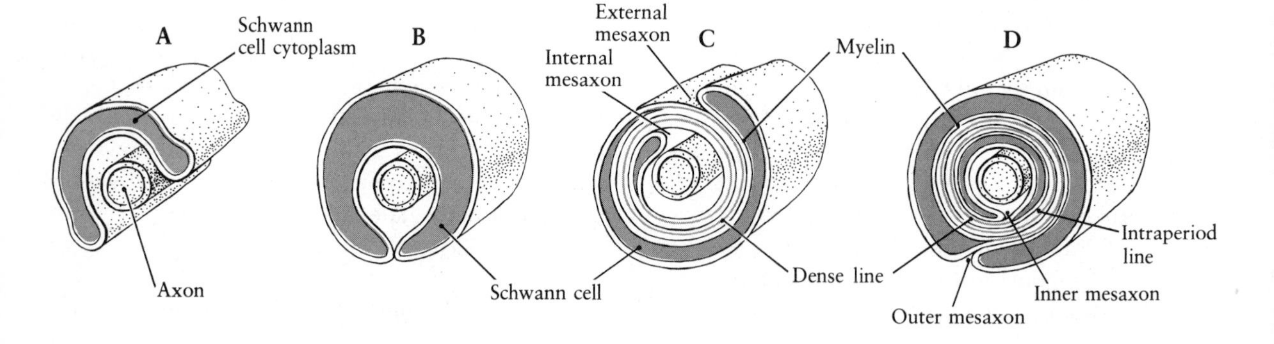

different origin, which combine secondarily into a single organ. The adrenal cortex is of mesodermal origin, and it encloses a medulla derived from ectodermal neural crest tissue. In some lower vertebrates the cortical and medullary parts of the gland normally remain as separate organs.

The adrenal cortex first appears as a group of cells in the angle of the posterior abdominal wall between the dorsal mesentery and the developing genital ridge at the level of the cranial end of the mesonephros (Fig. 20–10 A, B). The cells are derived from the coelomic mesoderm as are those of the gonad. The adrenal cortical cells separate from the mesothelium and proliferate to form, by the end of the second month, a cluster of large acidophilic cells enclosed in a connective tissue capsule. A second proliferation of cells from the coelomic mesothelium is added to the original mass at a later stage, but the main portion of the adrenal cortex at birth is made up of the original group of cells, termed the fetal cortex.

During the second month of development, migrating neural crest cells reach the medial aspect of the adrenal anlagen by way of the sympathetic ganglia. In the chick these cells migrate out from the neural crest at levels from the 18th to the 24th somites (Fig. 20–8). These cells, which will form the medulla of the gland, form a mass on the medial aspect of the cortical cells (Fig. 20–10 C). They are gradually encapsulated by the cortex and give up their function as nerve cells and differentiate into endocrine elements. Owing to the presence of the hormones epinephrine and norepinephrine in these cells, they stain brown with chrome salts and are thus part of what is known as the chromaffin system, of which the adrenal medulla represents the major portion. Other chromaffin cells are found as clusters (paraganglia) associated with autonomic nervous system ganglia and also as clusters located along the course of the dorsal aorta.

At birth the adrenal gland is a comparatively large organ, almost one third the size of the kidney. However, a large part of the adrenal at this time consists of the primary or fetal cortex located between the definitive cortex and the medulla. Postnatally, the fetal cortex involutes and the gland decreases both actually and relatively in size. In the adult it is about 1/30th of the size of the kidney. The function of the fetal cortex is not known, although studies have shown that it is capable of secreting steroid hormones. The possible role of the adrenal in parturition has been discussed in Chapter 5.

The double origin of the cortex and the medulla is reflected in differences in function of these two parts of the gland in the adult. The cells of the medulla elaborate epinephrine or norepinephrine and function in relation to the sympathetic nervous system, being primarily concerned with emotional adjustments. The adrenal cortex secretes steroid hormones that function in carbohydrate and protein metabolism and in the control of water and electrolyte balance. In addition, both androgens and estrogens are synthesized by the adrenal cortex.

**FIG. 20–10.** Diagram showing the origin of the cortex and the medulla of the adrenal gland from the mesothelium and the neural crest, respectively.

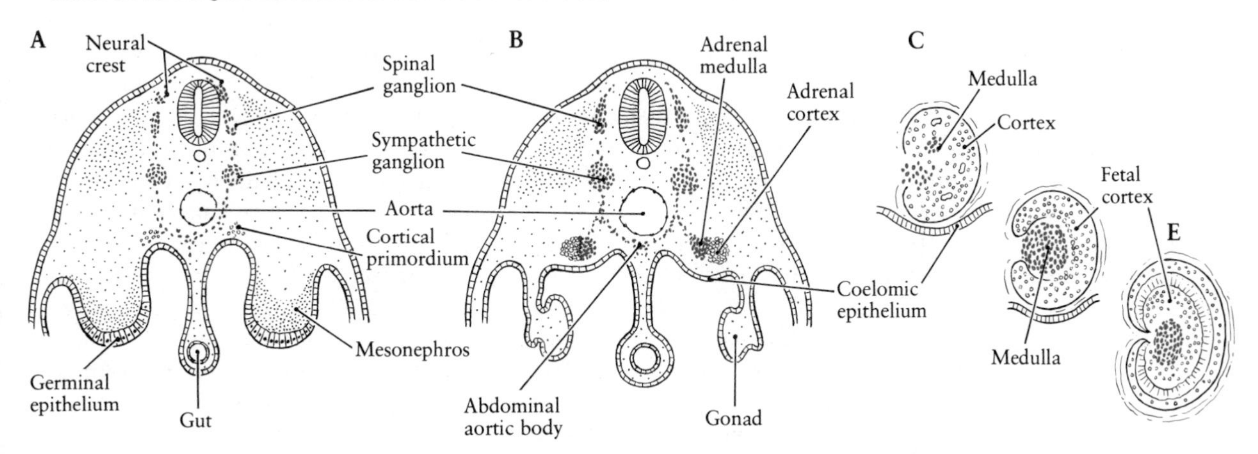

### Control of Neural Crest Migration and Differentiation

The physiological and morphological diversity of the neural crest derivatives and their widespread distribution have provoked many studies on the possible factors controlling their migration and differentiation. Most of the studies involve transplantation techniques. Since neural crest cells, once they leave their point of origin, are indistinguishable from the cells of the tissues through which they migrate, it is necessary to mark the cells of the graft in some manner so as to follow their development in the host. Thus, the host neural tube is excised and replaced by a transplant whose cells may be distinguished from the host cells. The graft may be labeled with tritiated thymidine and cell migration determined by autoradiography. This method of labeling imposes a time limit on the study due to the dilution of the label with successive cell division. Another method is to use two different species whose cells are distinctive. Thus, transplants of Japanese quail neural tube may be made to chick hosts and the quail cells followed because they are easily distinguished from the chick cells on the basis of their nucleolus-associated heterochromatin (Le Douarin, 1980). However, transplantation operations are traumatic and cause a delay in the onset of cell migration. This problem may be overcome by transplanting marked neural crest cells into a host by injecting them directly into the somitic cavity (Bronner-Fraser and Cohen, 1980). This technique offers the additional bonus that cloned cells may be used. Individual crest cells from in vitro cultures may be isolated and themselves cultured, and all of the transplants from these clones are the direct descendants of a single neural crest precursor cell.

Neural crest cell migration occurs in a temporospatial (craniocaudad) sequence. Migration begins in the cranial regions of the embryo and moves progressively caudally. While cells of the more cranial regions are actively migrating, cells at the more caudal levels have not as yet left their point of origin. We have already seen that crest cells at different levels of the embryo normally migrate to different localities and differentiate into different structures (Fig. 20–8). At any one spinal level, neural crest cells migrating over the ventral route may eventually form any one of three structures: spinal sensory ganglia, autonomic nervous system ganglia, adrenal medulla. Migration may take place over a period of days and carry the cells long distances from their source, ending in a precise distribution of the cells in their definitive structures at specific localities. A number of questions immediately come to mind: How is the migration initiated? How is it maintained? What factors control the precise orientation of the movement and the final localization of the cells?

A major consideration in attempting to answer these questions in analyzing neural crest cell migration is to ask also, at what time in the differentiation of the neural crest are developmental limitations imposed? Are the cells already determined prior to or early in migration to differentiate into specific structures? If so, they could presumably follow directional clues specific to their predetermined constitution, and the pattern of migration would reside within the cells themselves. If, however, the cells were not yet determined before or early in migration, then their pattern of distribution could be governed by nonspecific environmental factors to which all cells would react similarly. We have seen that the first cells to leave the spinal cord neural crest migrate the greatest distances and form autonomic nervous system ganglia, whereas those that leave later end up close to the neural tube and form spinal ganglia. This could mean that the ultimate distribution could be due to predetermined cells leaving the neural tube in a specific order. A number of experiments have been conducted to test this possibility. Neural tube segments from regions of the chick spinal cord from which many neural crest cells had already migrated were grafted into regions of a host embryo where neural crest migration had not yet started (younger embryos or more caudal levels of the same embryo). Since normally the first cells to leave the crest form autonomic neuroblasts, these structures would be lacking in grafts from cranial levels of older embryos if there were a perdetermination of migratory pathways. The cells determined to form autonomic ganglia would have already migrated out before the graft was made. Then, the only cells left in the "older" neural crest grafts would be those destined to form spinal ganglia. Such is not the case. The cells from the older grafts have the same range of migratory capabilities as the cells from the younger grafts; in addition, they differentiate into sympathetic neuroblasts showing the same pattern of localization as the normal embryo.

We may also interchange cranial and spinal neural tubes. If chick neural tube from spinal levels is grafted into vagal levels, the neural crest cells that in the spinal region would have migrated no further than the adrenal gland now undergo the extended migration characteristic of vagal levels and end up in the abdominal viscera differentiating into enteric parasympathetic ganglia. Cranial neural crest cells grafted to spinal levels also migrate and differentiate according to their implanation site rather than their origin. However, the results with cranial neural crest grafts introduce a slight complication in that same vagal level cells, in addition to populating spinal level locations, may find their way into enteric locations, and some grafts from cranial levels may form mesectoderm derivatives at the host spinal level such as cartilage and parts of the mesonephros.

Despite the few discrepancies encountered when vagal crest is grafted to spinal levels, these experiments support the conclusion that the environment produced by the tissues in the early embryo plays the major role in determining the migratory route and eventually the differentiation of the neural crest cells. Specific pathways are produced during embryogenesis as the environment continually changes its ability to act as a substrate for cell migration. Preferential pathways produced by the environmental substrate guide the undifferentiated crest cells to their destinations. The environmental influence on the control of cell migration may be exerted even on postmitotic nonmigratory cells. Melanocytes that have ceased migration in vitro, if injected back into embryonic regions of active neural crest cell migration, reinitiate migration out over the ventral pathway. Even cells of parasympathetic ganglia localized and already undergoing cytodifferentiation, when implanted in young chick embryos, will reinitiate migration over the ventral pathway. In addition, they now participate in the formation of normal sympathetic ganglia illustrating not only that nonmigratory localized cells may be induced to renew migration under the proper environmental conditions, but also, even at this late stage of differentiation their fate may be changed by the environment in which they finally localize.

The potentialities of crest cells are also nicely illustrated by experiments on cloned chick cells. These clones—all descendants of a single cultured cell—may contain all pigmented, all nonpigmented, or a mixture of pigmented and nonpigmented cells. If cells of a mixed clone (containing pigmented and nonpigmented cells) are injected into chick embryos, they are capable of giving rise to both pigment cells and adrenergic neurons. That descendants of a single crest cell can differentiate into different phenotypes is unequivocal proof that at least some if not most of the population of premigratory neural crest cells are bipotential and probably pluripotential.

It is interesting to note that interchanges between cranial and spinal levels of the neural crest result in changes in the type of neurotransmitter substances secreted at the nerve terminals of the transposed neurons. Autonomic nervous system neurons secrete two different types of neurotransmitter substances—(1) adrenaline (epinephrine) or noradrenaline (norepinephrine) and (2) acetylcholine—and are thus termed *adrenergic (noradrenergic)* or *cholinergic* neurons, respectively. All of the preganglionic fibers of the autonomic nervous system—fibers whose cell bodies lie within the central nervous system—are noradrenergic. Sympathetic ganglia neurons are also noradrenergic (with the exception of those supplying the sweat glands), but parasympathetic ganglia neurons are cholinergic. Thus, the presumptive sympathetic ganglia neurons, which—when transplanted to vagal regions of the embryo—become enteric ganglia, now change the nature of their transmitter substances from noradrenergic to cholinergic. The reverse is of course true for vagal region neural crest cells transplanted to spinal regions. Actually, even in normal autonomic nervous system embryogenesis, there is for the parasympathetic ganglia cells a change in the type of transmitter substances elaborated by the developing neurons. All autonomic nervous system neurons at first express noradrenergic characteristics, regardless of the type of neurotransmitter they will elaborate at maturity. In the case of the sympathetic ganglia neurons, the noradrenergic phenotype persists; however, in the case of the parasympathetic ganglia the noradrenergic phenotype later changes to the cholinergic.

The early expression of a noradrenergic phenotype was considered to be dependent on the embryonic environment through which the neural crest cells migrated, specifically an interaction with somitic mesenchyme that had previously interacted with ventral neural tube ectoderm. However, more recent investigations have show that cells cloned in vitro will secrete noradrenaline and thus require no specific embryonic environmental stimuli for this manifestation. The early acquisition of noradrenergic characteristics may be an expression of programmed intracellular events that are merely triggered by the embryonic environment. Triggering does not interfere with subsequent migration, following which a change in expression occurs if the cells differentiate into enteric ganglia. Nor does it preclude subsequent DNA synthesis and cell division.

One might consider that the long migratory path that the presumptive enteric ganglia cells follow might be a factor in the switch from noradrenergic to cholinergic expression. This is not the case. Rather, it can be shown that cholinergic differentiation is dependent on the definitive localization of the cells. Quail spinal region neural crest cells cultured in vitro and then transplanted into the chick colorectum (never having undergone any migration at all) differentiate into normal parasympathetic ganglia containing significant levels of acetylcholinesterase activity. Localization within the gut, rather than moving over any particular pathway, is apparently the controlling factor in the differentiation of the enteric ganglia cells.

We have not by any means attempted to consider all of the questions bearing on the process of neural crest differentiation. Topics such as what initiates the original migration of the neural crest cells away from the neural tube, what is the nature of the guidance effect during migration, and what are the factors involved in stopping migration and promoting cell aggregation and subsequent differentiation in specific locations are all important problems whose investigation is underway in a number of laboratories (Bronner-Fraser and Cohen, 1980; Le Douarin, 1980).

## Establishment of Pattern in the Nervous System

In no other system in the animal are the cell patterns as complicated and the interconnections between cells as important as in the nervous system. The adult nervous system presents an amazingly complex network of billions of cells whose synaptic interconnections must be so patterned that they can function in the control of activities, which range from the seemingly uncomplicated spinal reflex through the initiation and coordination of muscular activity to the little-understood process of thought and learning. The development of the nervous system involves an orderly succession of events and changes in the multiplication, differentiation, migration, and interconnection of its cellular elements whose proper functioning is the result of a strict topographical organization.

We may divide the establishment of the pattern of the nervous system into three phases. The first is the determination of the gross morphological pattern of the central nervous system, which sets off the brain, with its major subdivisions, rostrally from the spinal cord caudally. Histogenesis within the central nervous system, with the development of characteristic types of neurons in different locations along the central axis, follows as a part of this stage of development. The second phase is the outgrowth of nerve fibers from the central nervous system and from the autonomic nervous system ganglia toward the vicinity of the peripheral structures that they will innervate. The final phase is the establishment of the specific connections with the correct target organs and the development of the correct central synaptic connections that will allow this intricate network to function in the coordination of all of those activities controlled by the nervous system.

We may ask many questions about the neural ontogeny, but unfortunatley, our experiments give us very few specific answers. How do nerve fibers grow? What determines the direction in which they grow? What controls the connection of peripheral nerve endings to specific muscle fibers and sensory receptors? What determines the formation of the synaptic connections within the central nervous system, which results in the correlation between sensory input and motor output?

### Axon Growth

The mechanism of axon growth was still in dispute at the end of the 19th century. A number of possibilities had been suggested. The noted cell biologist, Schwann, proposed that a chain of individual cells, the Schwann cells, which form the axon sheath, also fused together to form the nerve fiber. This theory was readily disproved when it was shown that, following the removal of the neural crest, the axon developed normally in the complete absence of Schwann cells. In fact, as described previously, the axon develops before the Schwann cells reach it.

Another theory, based on faulty histology, stated that the early nervous system was a syncytium interconnected by filaments that later developed into the nerve fibers. More accurate histology prevailed against this proposal.

A third theory, that of the neuroanatomist Ramón y Cajal, proposed that the axon developed as an outgrowth of the cytoplasm of the young neuron. The actual outgrowth of nerve processes from the neuron was first described by Harrison in 1907 using the tissue culture technique he invented. Harrison pictured the outgrowth of nerve processes from neurons of the spinal cord of the frog as being ameboid in nature, the fibers putting out and retracting pseudopodia until one became dominant and then repeating the process (Fig. 20–11).

The axon grows out of the neuron in a consistent direction, that is, toward the periphery of the spinal cord. This orientation appears to be due to some innate polarization of the neuron rather than to any external influence exerted at the time of outgrowth. The polarization may be the result of its original attachment to the inner surface of the neural tube, and it can be recognized by a change in the ultrastructure of the neuron before any outgrowth appears. Golgi bodies and rough endoplasmic reticulum, which are scanty in the young neuron, accumulate in increasing quantities in the region of the future outgrowth. This implies a timed gene activation switching on the synthetic activity necessary for the formation of the materials needed for axon growth followed by their differential distribution to the region of the cell where they will be utilized.

**FIG. 20–11.** Diagram of five stages in the ameboid-like growth of a nerve fiber. Dotted areas represent extensions that were put out and then withdrawn.

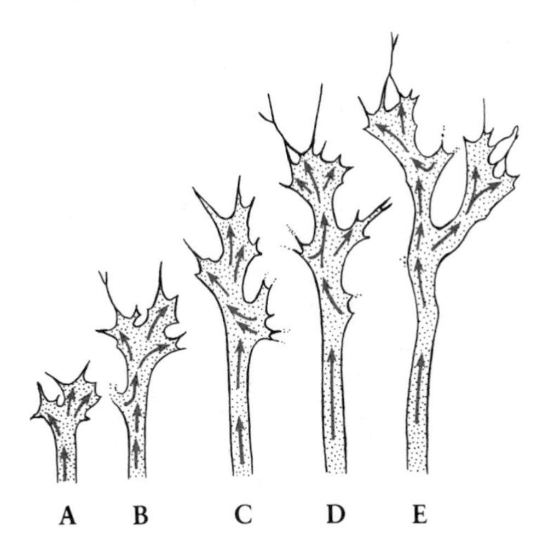


After the initial outgrowth, new materials for continued growth and maintenance are synthesized in the cell body and transported to the axon, and are also synthesized in the axon itself. Although the cell body has large numbers of ribosomes, none are found in the axoplasm. The many mitochondria located in the axon, then, must be the site for RNA and protein synthesis in the nerve fiber itself.

## Influence of Peripheral Structures on Neuroblast Proliferation and Fiber Outgrowth

Numerous experiments involving removal, addition, and transplanation of organs have shown that the size of the peripheral area to be innervated has a direct effect on the number of neurons in the areas supplying these structures. If the size of the peripheral field is reduced or increased, the number of neurons whose fibers supply the field is correspondingly reduced or increased. This applies to motor neurons located in the anterior column of the spinal cord and to sensory neurons located in the posterior root ganglia.

### *Motor Neurons and Peripheral Organs*

If a forelimb bud is removed early in development, a hypoplasia of the spinal cord motor column and the posterior root ganglia at the level of the spinal cord normally supplying the forelimb results (Fig. 20–12 A,B). Conversely, a limb bud transplanted to a location opposite a region of the spinal cord that does not normally supply nerves to a limb will cause a hyperplasia of this region of the cord (Fig. 20–12 C,D). Most of these types of experiments have been done on amphibians and chicks, but the few examples for mammals support the same conclusions. Hypoplasia of corresponding nerve centers has been reported in man in cases of abrachia and other congenital limb abnormalities.

The reduction in the number of neurons following limb bud removal is due not to a failure of the normal proliferation of neurons but to an abnormal amount of cell death following normal proliferation. Peripheral structures apparently have no effect on the proliferation and differentiation of neuroblasts or on the initial outgrowth of their axons. However, the absence of the peripheral or-

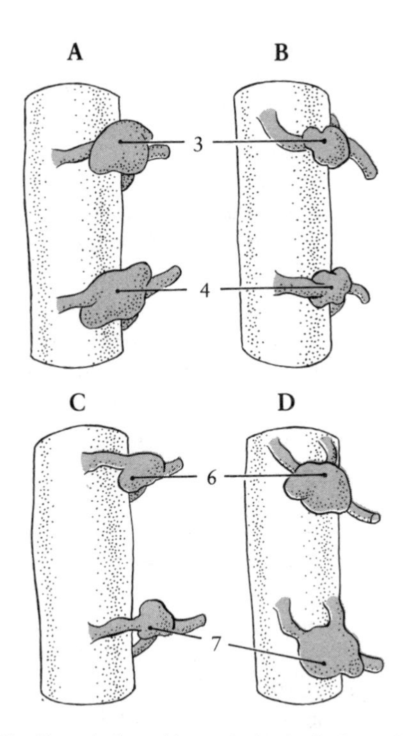


FIG. 20–12. Hypoplasia and hyperplasia of spinal ganglia associated with limb excision and limb transplantation. A, right third and fourth spinal ganglia connected with a normal limb; B, hypoplasia of right third and fourth spinal ganglia following excision of the right forelimb primordium; C, normal sixth and seventh spinal ganglia, which have no connection with a limb; D, hyperplasia of right sixth and seventh spinal ganglia following connection with a transplanted limb bud. (From S. R. Detwiler, 1936. Neuroembryology. Hafner Publishing Company.)

gans that the nerve fibers would normally supply results in a massive death of the neurons. The motor column opposite the region where a limb bud has been removed shows only about 10 to 20 percent of the normal number of neurons.

The peak of cell death coincides with the time of establishment of neuromuscular connections and the onset of limb movement. This leads to the conclusion that the establishment of a connection to a muscle fiber and the formation of a motor endplate are essential for the continued growth and maintenance of the neuron. In fact, death of neurons in the anterior column is a part of normal development, in that there is always an overproduction of neurons that differentiate and send out axons toward the periphery. This is then followed by the death of those neurons whose axons fail to make connections with muscle fibers. In *Xenopus* it has been estimated that from three to eight neurons die for each one that survives. All of the motor neurons in the embryo whose axons are growing out of the anterior column of the spinal cord are in active competition for the establishment of a permanent connection with a muscle fiber. A struggle for survival exists. The number surviving, then, represents the number that are sufficient to "saturate" the periphery. The reason the periphery can be saturated is that each muscle fiber will accept only a single motor axon terminal in mammals and only a limited number in submammalian species. Only the uninnervated membrane of the muscle fiber is receptive to a nerve fiber. Once innervated, the muscle fiber membrane is no longer receptive. This phenomenon, which reminds one of the change in the egg membrane that prevents polyspermy, may be due to the release of acetylcholine by the motor endplate. If the muscle is poisoned with botulinus toxin, which prevents the release of acetylcholine, muscle fibers will now accept additional axon fiber connections.

Thus, autonomous proliferaiton, differentiation, and axon outgrowth are followed by the selective death of those neurons whose axons fail to establish functional connection with a muscle fiber. Theories about the mechanism causing the death of the neurons are speculative. It has been proposed that, once a functional connection has been made, this information is transferred from the muscle to the axon and then relayed back to the nerve cell body. The information could be in the form of some "material" that is necessary for the continued growth of the neuron.

As nerves grow into the target areas, they show increased branching or sprouting. These arborations on the nerve endings are referred to as *terminal fields*. It has been proposed that sprouting is the result of some stimulus from the target tissue. Once a nerve ending has made connection with the target, the sprouting stimulus disappears. Here we see another example of nerve–target interaction. Apparently the sprouting factor is masked, once the target is innervated, by a neuronal factor that acts continuously since, if the target is denervated, sprouting is induced from adjacent neurons presumably by a renewed release of a sprouting factor from the target tissue. The inhibition of the release of a sprouting factor by the action of the neuron can be blocked by treating the nerves with colchicine, which blocks the flow of material from the nerve cell body out along the axon. When nerves to a target area are treated with colchicine, the result is the same as if the target area were denervated. Adjacent nerve fibers sprout to supply the treated area, and hyperinnervation of individual muscle fibers occurs.

*Guidance of Axon Growth and the Establishment of Neuromuscular Specificity*

In the adult, the path of a motor fiber from its central origin to its peripheral connection may be long and tortuous, leading to the consideration that indeed some extremely potent force must have been at work to guide the fiber along the proper path. However, we must consider that axon outgrowth takes place in the early embryo where distances are shorter and paths more direct. The first fibers that grow into the limb bud pass over these relatively short paths to connect to developing muscle. These "pioneering" fibers may now be passively towed along as the muscle with which they connect differentiates, grows, and migrates. In addition, the pioneering fibers now provide a pathway that fibers developing later may follow.

Nevertheless, some sort of directional guidance is probably exerted by the developing peripheral organs, and that it does occur is nicely illustrated from the results of transplantation experiments. If a barrier such as a small sliver of mica is placed between the spinal cord and the developing limb bud of a salamander, the brachial nerve fibers from segments three, four, and five, which normally supply the limb, will grow toward the barrier and then deviate from their course and grow around the barrier to reach the limb bud. If a frog limb bud is removed and transplanted a short distance caudal from its normal site, the brachial nerve fibers from segments three, four, and five will deviate from their usual routes and grow toward the transplant. However, if the limb bud is transplanted too far away from its normal site, it can no longer attract motor fibers from the normal spinal level of supply. Instead, fibers from the cord at the level of the transplant now grow toward the limb. Although the limb may be innervated by some fibers from levels other than its normal source of supply, it will not now show coordinated movements.

We should note, that it has been known since the early 1920s that only those nerves from the spinal cord that normally supply motor fibers to the limb musculature, brachial, and lumbosacral cord regions, will support coordinated limb movements. A grafted limb innervated by nerves from the trunk (thoracic) level of the cord may show muscle contraction but not coordinated movement. This control of limb movement by the appropriate cord levels is established in the embryo. The capacity to control limb movement is already present shortly after the closure of the neural tube, since segments of the brachial region of the spinal cord transplanted to trunk levels will support normal movements of a supernumerary limb grafted into an area close to the brachial cord graft. An interesting result from this type of experiment was seen when a grafted brachial segment of the cord supplied the innervation for a supernumerary hindlimb graft or when grafted lumbosacral cord supplied a supernumerary forelimb graft. In both of these cases, limbs developed from the grafts moved in synchrony with the normal host forelimb if controlled by a graft from the brachial cord level or in synchrony with the host hindlimb if controlled by a graft from the lumbosacral cord, irrespective of whether they were forelimb or hindlimb grafts.

In addition, in heterotopic grafts, when a forelimb was controlled by a lumbosacral cord graft, its movements were those characteristic of a hindlimb with the elbow remaining relatively motionless as does the knee of the hindlimb. This is even more apparent in grafts in chick embryos where wing grafts controlled by lumbosacral nerves were unable to perform typical winglike movements of the joints and hindlimb grafts controlled by brachial nerves showed winglike movements. We would have to conclude from the results of these experiments that the synchrony and the characteristics of the movement of the grafted limbs were controlled by the level of the cord that supplied the innervation to the graft.

The results are complicated by the fact that grafted limbs that appear to move in synchrony with normal ones when viewed with the naked eye are actually out of synchrony when studied by electromyography (relating electrical records from individual muscles to limb position during limb movement). Supernumerary limb muscles show a slight delay in their time of contraction.

These observations on the control and the synchronization of muscle movement in supernumerary limbs do not support the concept of *myotopic modulation* (or specification) proposed by Weiss (1936) and Sperry (1965). Weiss' conclusions were based on results of experiments on transplanting individual muscles to a region near a normal limb in an amphibian tadpole and directing nerves to the transplant from the nerve supply to the normal limb. The transplanted muscle contracted in synchrony with the corresponding muscle in the host limb. The same result applied for several supernumerary muscles or for the entire limb, each transplanted muscle acting in concert with the corresponding muscle of the normal limb. Under the conditions of these experiments, the same nerve fibers did not always innervate the same supernumerary muscle because they were diverted to the transplant as a matter of chance or at the choice of the experimenter. Thus, the conclusion was reached that no matter what nerve fiber establishes connection with a muscle, that muscle in some way is able to confer its name to the nerve fiber, imprinting it with the specific character (biochemical?) of the particular muscle. Once the axon is modulated by the specific character of its peripheral connection, this is conveyed back to the motoneuron cell body in the spinal cord and in turn to the premotor cell neuron with which the motoneuron establishes synaptic connection. Thus, all nerve fibers reaching a biceps msucle are stamped with the name "biceps," and their central connections carry this name also. A different explanation of the results of these experiments has been proposed (Szekely and Czeh, 1967). Examination of the spinal cord representation of the forelimb muscles in *Ambystoma* (the location of the nerve cell bodies that supply motor fibers to specific muscles) shows considerable overlap along the cord to such an extent that muscles of supernumerary limbs are quite likely to be supplied by a complete set of appropriate motoneurons without assuming any kind of specification. However, no studies have yet been made on the location of the motoneurons supplying supernumerary muscles.

Sperry, in a series of experiments beginning in the 1940s, proposed a somewhat different hypothesis to explain neuronal specificity based mainly on a series of behavioral and anatomical studies in adult animals. He first determined that the pattern of muscle innervation that was developed in the embryo established a nervous system pattern of the control of muscle movement that could not be modified in the adult. If the nerves to the muscles of the hindlimb of the rat that control flexion and extension of the ankle joint were switched so that the nerves that originally supplied extensor muscles were now directed to supply flexor muscles, flexion and extension movements of the joint were reversed. When the rat tried to lift its foot, it instead pressed downward; when it tried to press downward on the ball of the foot, the toes curled upward and the animal fell back on its heels. It is not surprising that the same result occured if the nerve supply to the muscle was left intact but the insertions of the flexor and extensor tendons were interchanged. Extensive training in an effort to overcome the reversal of foot movements was to no avail. The effect was permanent. Other similar experiments also demonstrated a similar lack of plasticity in afferent neuronal circuits as we will describe in the section on sensory nerve receptors. These results were interpreted by Sperry to mean that the lack of plasticity in the adult was due to the establishment of specific permanent nerve circuits in the embryo that were the result of biochemical affinities of the neurons. The biochemical specificity of the peripheral fiber is conferred on it by the specific target with which it makes connection. Central (premotor) neurons whose fibers synapse with the motoneurons innervating peripheral targets contain inherent biochemical "flavors." The establishment of central connections then occurs only between those fibers whose central biochemical flavor matches the flavor of the particular muscle and in turn the motoneuron supplying it. Sperry's hypothesis is thus similar to that of Weiss in that the biochemical flavor of the peripheral nerve fiber is conveyed to it by its muscle fiber connection. It differs in the proposal that the biochemical makeup of the premotor neuron is inherent, allowing it to synapse only with specific matching motoneurons. Both concepts, when applied to the establishment of neuromuscular specificity during development, propose that motoneuron fibers would grow into the limb bud in a more or less haphazard or random fashion originally lacking any identity and have identity conferred on them by the chance contact they make with specific muscle fibers.

However, any consideration of the mechanism of neuronal specificity should take into account the fact that the pattern of innervation of the vertebrate limb is as constant and repeatable as the muscular and skeletal elements of the limb themselves. In any species there is usually only slight variation from one individual to another. In the chick there is a characteristic pattern of nerve pathways in both the dorsal and ventral regions of the wing supplying the extensor and flexor musculature, respectively. The growth of motor fibers into the limb bud is neither diffuse nor random, and bundles of fibers form the same specific patterns as they grow toward the premuscle masses. The outgrowing fibers appear to follow consistent "tracks" within the limb bud mesenchyme. These tracks seem to be set up and controlled by the developing limb bud, since the nerve fiber outgrowth from the lumbosacral cord to a transplanted wing bud replacing the normal leg bud follows tracks similar to those found in the innervation of the normal wing. Thus, although nerve outgrowth may be attracted to any nonspecific rapidly growing structure, the attraction of the limb buds for their nerve supply also dictates a pattern of consistent tracks characteristic of the particular limb. Also a part of this pattern is the distribution of the cell bodies of the motoneurons within the ventral column of the spinal cord. Those whose fibers innervate the ventral (flexor) musculature are located at more rostral levels than those supplying the dorsal (extensor) musculature. Also the flexor motoneurons are found in the dorsomedial region of the column, and the extensor motoneurons are located in the ventrolateral area. The location of the motoneurons within the spinal cord may specify the direction of the initial outgrowth of the axons from these cell bodies. The tracks that the growing nerves follow toward the developing limb, however, are not toward specific muscles but toward either dorsal or ventral muscle masses. Fibers from motoneurons in the dorsomedial region of the cord will orient toward the dorsal muscle mass, and those from the ventrolateral neurons will orient toward the ventral muscle mass, the forerunners of the extensor and flexor muscles, respectively.

It is possible to consider that the axons growing out from the spinal cord motoneurons are subject to a succession of influences that in sum eventually control the final pattern. The first level of control would direct them from the spinal cord to the base of the differentiating limb bud. A second level of control would then guide the fibers through the limb mesenchyme over specific pathways to muscles within either the dorsal or the ventral muscle masses. The definitive synapse between individual axons and muscle fibers would represent the third and final step in the process.

Horseradish peroxidase (HRP), when injected into specific regions of the embryonic spinal cord, labels the axons derived from the motoneurons of these segments so that their pathways can be followed from the time of their initial outgrowth. Early growth shows very little deviation along the craniocaudal axis of the cord during the time the nerves are growing toward the base of the limb bud. That is, axons from cranial motoneurons retain their original position cranial to axons from more caudal motoneurons. At the base of the limb bud, axons for different muscle masses form nerve bundles that diverge from other bundles, often crossing over each other and thus forming the characteristic pattern of the brachial or lumbosacral plexus. A major observation at this time is that axons from the medial regions of the spinal cord project toward the ventral muscle mass and those from lateral regions project toward the dorsal muscle mass. At any craniocaudal segment of the spinal cord, each spinal nerve would contain axons from both medially and laterally located motoneurons. Plexus formation would be the result of sorting out and projecting axons from the medial and lateral regions of the spinal cord from all levels to the ventral and dorsal muscle masses, respectively. This selective process would result in the formation of a network of peripheral nerves in a specific pattern between the spinal cord and the base of the limb. The pattern would be determined by the topographical position of the motoneurons within the spinal cord. An early determination of medial and lateral neurons to project to ventral and dorsal muscle masses, respectively, may be demonstrated by reversing the craniocaudal axis of the chick lumbosacral spinal cord before axon outgrowth. Following this, the axons change their pathways within the plexus so that they reach their normal muscle masses. Although the individual nerves pursue other than normal pathways to reach their appropriate destinations, the lumbosacral plexus still develops its normal pattern.

Early experiments on the transplantation of limb buds so that their dorsoventral (DV) axes were reversed also supported the concept of early specification of motoneurons in the medial and lateral locations along the cord axis. In these transplants the original ventral flexor muscles are now located in a dorsal position in the limb and the dorsal extensor muscles are located in a ventral position. Nevertheless, the motoneuron specificity is preserved, and medially and laterally located neurons send out axons that change their pathways to supply the normal innervation to flexor and extensor muscles, respectively, even though the dorsal and ventral locations of these muscles are now reversed. However, the results are not always 100 percent perfect, and in a number of cases axons do not correct their paths to cross over to reach their normal muscles but pass through the plexus and major nerve trunks along a more or less direct path and end up innervating "inappropriate" muscles.

The results from experiments on the innervation of transplanted supernumerary limbs in the chick also support a concept of early specification of motoneurons according to their medial and lateral positions in the developing spinal cord. In the case where the muscles of the supernumerary limb are not supplied by nerves from the normal levels of the cord, the nerves that do supply the ventral and dorsal muscle masses are still those that are outgrowths of medial and lateral regions of the cord, respectively, although the nerves are derived from different levels of the cord than the normal ones. Even if the supernumerary limb is a leg transplanted next to a wing, the mediolateral selectivity for ventral and dorsal muscle masses is expressed.

It has been suggested (Hollyday, 1980) that the mediolateral specificity might be explained on the basis of chemoaffinity labels present on the growth cones of the axons and the muscle masses toward which they grow, ensuring that the axons would grow into the appropriate muscle masses. This initial specificity between dorsal and ventral muscle masses and their nerves would require only two labels—one for dorsal and one for ventral—which would be shared by all muscles differentiating in the respective muscle masses.

Evidence that does not support the concept of the early specification of medially and laterally located motoneurons comes from more recent experiments on the rotation of chick limb buds (Summerbell and Stirling, 1982). If the limb bud is rotated 180 degrees—thus reversing both the anteroposterior (AP) and dorsoventral (DV) axes—axons from the anterior and posterior levels of the cord grow toward the base of the limb bud and enter it in an order that is reversed as far as the original axes of the limb are concerned. That is, axons from the anterior segments of the cord grow into what is now the anterior region of the reversed limb, the region of the limb that is normally posterior, and motoneurons from the posterior cord segments grow into normally anterior regions of the limb. Once in the limb, the axons retain their AP positions relative to the host axis and to each other, positions that are reversed relative to the graft axis. Most of these axons now supply muscles different from those they normally would. There is no evidence that they alter their pathways to seek out predetermined targets, although in a few cases axons do make connections with both inappropriate and appropriate muscles.

In addition Summerbell and Stirling (1980), using HRP labeling, demonstrated, contrary to earlier experiments, that following reversal of the DV axis of the chick limb, the flexor muscles were now innervated by motoneurons from lateral regions of the cord and extensor muscles by motoneurons from medial regions of the cord, just the opposite from conditions found in the normal limb.

These last-mentioned experiments would support what is called a passive-deployment hypothesis. It does not invoke any predetermined matching of motoneurons and muscles but is based on the observation that, in normal development, the outgrowing nerve fibers follow the same pattern of tracks within the limb mesenchyme, resulting in a consistent and repeatable pattern of nerve pathways. We have already described nerve fiber growth as ameboid in nature, the nerve fiber extending and retracting pseudopodia-like processes as it progresses. However, the analogy of nerve fiber growth to ameboid movement may give the erroneous impression that nerve fibers can move freely through tissue spaces and fluids. They cannot. Fiber growth can proceed only along interfaces such as solid–liquid, liquid–gas, or two immiscible liquids. The major interfaces available to the growing nerve fiber in the embryo are apparently those formed by the numerous fibrous units that make up a large part of the amorphous ground substance of the tissues. It seems probable that the nerve fiber shows directional growth because the tissue units are not arranged haphazardly but have a definite orientation. Weiss (1934) was able to obtain directional growth of nerve fibers in tissue culture by stretching the blood clot substratum. The previous random growth of the fibers now became oriented in the direction of the stretching of the fibrous units of the substratum. Nerve fibers growing in vitro on a mica coverslip scored with a pattern of scratches will also orient along these pathways exclusively. Weiss termed this type of behavior *contact guidance*. To what extent contact guidance can explain the growth of the nerve fiber in the embryo is problematical. In what way could an oriented ground substance develop in the embryo in such a pattern that the nerve fibers could use it as a pathway? The fibrous units in the embryo might be oriented through stretching as a result of differential growth. It is possible that the dehydration around rapidly growing organs could stretch the ground substance fibers in their direction and provide a guidance system directing the nerve fibers toward rapdily proliferating structures.

Obviously there is no simple hypothesis that explains all of the experimental results and observations. There is evidence that there may be an early specification of motoneurons based on their medial or lateral position within the neural tube. The early specification that medial and lateral motoneurons project to dorsal and ventral muscle masses may be the factor controlling the pattern of the nerve pathways to the base of the limb. In addition, there is good evidence that the consistent pathways that the nerves to the limb muscles show may depend on the limb mesenchyme itself and its ability to specify, by some means or another, particular tracks through the developing musculature so that the routes taken by the pioneering fibers are consistent, with branching occurring at defined and predictable points along the pathway. The tracks may be determined by simple physical factors or may involve more subtle and as yet undefined trophic factors. This establishment of specific pathways would consistently bring the axons of the motoneurons into the area of the development of specific muscles. The final step, the establishment of the connection between individual nerve and muscle fibers, probably involves some type of as yet undetermined recognition process.

## Sensory Nerves and Receptors

During development, the sensory fibers growing out of sensory ganglia always grow into the sensory field before the development of any sensory receptors, suggesting that the differentiation of the receptors may depend on some stimulus brought in by the sensory fibers. There are many different types of receptors responding to different kinds of chemical, thermal, and mechanical stimuli. Each receptor can respond only to a single kind of stimulus, and the question then arises whether, in addition to inducing the development of the receptor, the nerve fiber also controls its specificity. Most of the experiments on this subject have been done on postembryonic stages, but the results have obvious implications for the role of the sensory nerve fiber in the differentiation of its receptor.

Most types of receptors, including taste buds, Pacinian corpuscles, and neuromuscular spindles, degenerate when they are denervated and regenerate after their nerve supply is restored. In addition, their differentiation is dependent on the arrival of nerve terminals. A classic example of this trophic effect is seen in the dependence of taste buds on their sensory innervation. In the rat and the rabbit, a reduction in the size of the taste buds is seen eight hours after denervation, and they completely disappear in five to seven days. The taste buds on the posterior part of the tongue are supplied by cranial nerve IX and degenerate when this nerve is cut. They will regenerate if supplied by fibers from cranial nerve IX, but will also regenerate if supplied by fibers from cranial nerves VII or X. These are the three cranial nerves that normally supply sensory fibers to taste buds: VII to those on the anterior region of the tongue, IX to those on the posterior region, and X to buds in the larynx and pharynx. The taste buds will not regenerate if they are supplied by cranial nerve XII, which is a purely motor nerve to the skeletal muscle of the tongue, thus establishing a specificity at least between sensory fibers and sensory receptors. However, the specificity goes even further than this. Denervated taste buds will not regenerate if supplied by sensory fibers from branches of cranial nerve V. Thus, the trophic stimulus for the maintenance of taste buds is supplied only by sensory nerves, but only by sensory nerves that contain gustatory fibers.

A further aspect of this relationship is seen in the results from another experiment. The taste buds on the anterior and posterior regions of the tongue, which we have seen are supplied by cranial nerves VII and IX, respectively differ in their responses to different chemicals, and these differences can be detected on the basis of electrical response measured in the nerves that supply these regions. If cranial nerves VII and IX are switched so that VII supplies the posterior and IX the anterior region of the tongue, the electrical responses in the switched nerves are reversed and correspond to the region of the tongue the nerves supply. Thus, the precise functional specificity of the taste receptors is an inherent property of the epithelium in which they develop, and the nerves exert a nonspecific morphogenetic action that calls for the maintenance of gustatory receptors in general, but does not determine their exact specificity.

Another example of epithelial control over the specificity of the receptor is seen when the skin of the beak and tongue of the duck, which contains certain specific types of mechanoreceptors, is grafted to the foot. These receptors, which are normally supplied by a branch of the Vth cranial nerve, degenerate when grafted but regenerate when sensory nerve fibers of the foot innervate the graft. Here, the sensory nerve fibers from the spinal cord are capable of stimulating the development of head sensory receptors normally supplied by cranial nerve fibers, but the specific nature of the receptors is inherent in the head epithelium.

### Central Connections of Sensory Nerves

How are the central connections of sensory fibers determined? Are they specified by the function and location of the receptor or does the sensory fiber receive its specificity from the center and then make the appropriate peripheral connections? The evidence seems to be in favor of the first supposition.

If a hindlimb bud is grafted into the back region of a frog tadpole, it will develop, and the adult frog now carries an extra hindlimb in this abnormal position. Under the conditions of this experiment, motor nerves do not innervate the graft, which does serve, however, as a sensory field for sensory nerves growing out of the adjacent spinal ganglia. If the limb is stimulated, it does not move; instead, the corresponding normal hindlimb responds to the stimulus, as if it had been stimulated itself. The explanation is that the trunk sensory nerves supplying the grafted limb have received the imprint of the hindlimb sensory field and have thus made the appropriate central connections for sensory fibers from hindlimb receptors.

Another experiment that illustrates the same principle involves removing a strip of skin from the belly and the back of a frog tadpole and transplanting it with its dorsoventral axes reversed. Belly skin is now found on the back and back skin on the belly. After metamorphosis, when the grafted area now on the back is stimulated, the frog responds by wiping at the belly area with its limb, and, when the grafted belly skin is stimulated, the frog scratches its back. After being transplanted in a new position, the sensory receptors of the skin still signal their old position to the central nervous system. Some quality in the skin determines the pattern of the reflexes set up in the central nervous system. Once determined, this pattern is permanent.

A final example of the specificity of peripheral to central connections involves experiments done mainly on frogs and fish on the growth of the optic tract nerve fibers centrally from the retina to their connections with the tectum of the midbrain—the visual center in these forms. Electrophysiological experiments in which the regional electrical activity of the tectum is recorded following stimulation of the retina with a spot of light show that there is a specific point-to-point projection of the retina back to the optic tectum. If the optic tract is cut, it will regenerate, and when it does normal visuomotor coordination is restored and the pattern of the point-to-point electrophysiological mapping is the same as it was before the tract was severed. The regenerating optic fibers thus made their connections with the original positions in the tectum.

The specificity of these connections is nicely illustrated by the introduction of a number of surgical rearrangements of the eye after the optic tract has been cut. If the eye is rotated 180 degrees, the optic tract fibers will still regenerate and vision will be restored. However, the visuomotor reflexes are now reversed, and objects in the nasal field of vision are now seen as being in the temporal field of vision, since they are now seen by the part of the retina that, before rotation, saw the temporal field of vision, which now still projects back to its original temporal field of the tectum. The frog strikes at a fly as if it is in the direction just opposite to the one it actually is. This defect is permanent, and no amount of experience ever corrects it. The optic fibers grow back to their original specific points of connection in the tectum, no matter what the orientation of the eye from which they originate.

It is not known why the fibers grow back to their original positions during regeneration nor, in fact, how the retinotectal connections form in the embryo in the first place. Some kind of guidance must be necessary for the fibers as they leave the retina: first, to get into the correct position in the optic tract that will take them to the proper tectum; second, to get into the proper branch that will end up in the correct general region of the tectum; and, third, to make the correct connections with the tectal cells.

Although the mechanism for the guidance of the

fibers back to the tectum is not known, there is some indication that the final connections are specified by the retinal cells. Rotation of the optic cup in the adult, which produces a reversal of the visuomotor reflexes, results in perfectly normal vision if the operation is performed in *Ambystoma* any time before embryonic stage 34 (a time when the forelimb bud is just forming) in the tadpole. If the eye is reversed in stages 34 to 36, the operation results in greater and greater mixups of the visuomotor reflexes, and, if performed after stage 36, complete reversal results. Thus, there is a time in development when the retinal cells change from an unspecified to a permanently specified condition. This specificity occurs in *Ambystoma* just before the tectal connections are made (stage 38). In *Xenopus,* specification occurs earlier, some 20 hours before the outgrowth of axons from the retinal cells. In both species, specification does not take place throughout the entire retina at the same time but, as in the limb, the AP axis is determined before the DV axis. This may be shown by comparing experiments that rotate the AP and DV axes independently with experiments that rotate both axes at the same time. Contralateral "over the back" transplants that invert the DV axis but do not change the AP axis (see Fig. 13–16) show that there is a stage in the development of the salamander when this type of transplant will result in normal vision. At the same stage of development, transplants that simply rotate the eye 90 degrees, so that the AP axis becomes the DV axis and the DV axis becomes the AP axis, result in normal vision in the new AP field of vision but abnormal vision in the DV field. Thus, we may conclude that at this stage of development the original DV axis was as yet undetermined and could regulate to function normally in either an inverted position (first experiment) or as an AP axis (second experiment). However, at the same time, the AP axis was already fixed and, when put into the position of the DV axis, could no longer adjust. Specification of the retinal cells is not accompanied by any as yet recognizable morphological or ultrastructural changes in the cells. However, DNA synthesis stops just before specificity appears.

Experiments in which the optic cups from two different *Xenopus* embryos were cut in half and then re-fused so that the resulting eyes consisted of two nasal or two temporal halves have added more information to the problem of retinotectal connections. Normally, fibers from the nasal and temporal parts of the retina, which represent the caudal and cranial fields of vision, project to the caudal and cranial regions of the tectum, respectively (Fig. 20–13 A). Following operations that produce double-nasal or double-temporal eyes, electrophysiological mapping shows that the corresponding halves of double-nasal or double-temporal eyes both project back to the same tectal positions—as one might expect. However, unexpectedly, each nasal half or temporal half projected not to half of the tectum, as it normally would, but its fibers spread out over the entire tectum (Fig. 20–13). In addition, when the optic chiasma was uncrossed in these operated animals so that the optic fibers from a double eye were directed to a tectum previously connected to a normal eye and vice versa, the projection from each half of the double eye still spread out to occupy all of the previously normal tectum, and the fibers from the normal eye took up their normal projections to the tectum previously connected with the double eye. Thus, parts of the tectum that had previously been connected to the temporal half of the normal eye were still capable of forming connections to double-nasal eyes having no temporal fibers at all. Also, parts of the tectum previously connected to the nasal half of a double-nasal eye were still capable of forming connections to the tem-

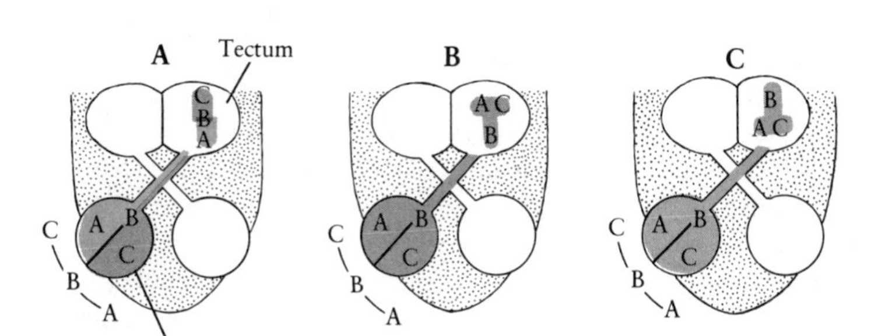


**FIG. 20–13.** The relationship between the visual field, the retina, and the projection of the retina to the tectum in *Xenopus* in normal, double-nasal, and double-temporal eyes. (From R. M. Gaze, 1963. J. Physiol. 165:484.)

poral part of a normal eye. Thus, the retinotectal connections do not direct that the tectum assume a permanent and fixed map of specifications. Retinal fibers can establish connections with any part of the tectum under certain experimental conditions. The normal nasal retina to caudal tectum and temporal retina to cranial tectum specificity does not imply a strict specific cell-to-cell pattern between the retina and the tectum. What is important is the topographical relation between the optic tract fibers as they project to the tectum. If it is the relative position of the optic tract fibers that is important, it is possible that the timing of the development of fibers from different parts of the retina could be translated into a corresponding orderly formation of tectal connections. If axonal outgrowth occurs in a spatiotemporal order, this would determine a spatiotemporal pattern of arrival of axons in the tectum and could be the mechanism determining the pattern of tectal connections. Development of a permanent specificity following these connections could result from the later formation of biochemical affinities between the connected cells.

As stated in the beginning of this section, much of the search for the reason for the establishment of the proper nervous system interconnections and the development of coordinated activity do not really give us many specific answers to the questions raised. In view of the tremendous complexity of this system, this lack of detailed knowledge is not surprising. Yet, we have been dealing with such relatively simple aspects as the connections between axons and muscle fibers and between sensory receptors and their nerve fibers, as well as the establishment of simple central synapses. One cannot fail to appreciate how much more bewildering a problem is presented in an attempt to understand the connections and mechanisms controlling memory, learning, and thought processes.

## The Brain

From the beginning, the rostral part of the neural plate—the part that will form the brain—is the largest part of the developing central nervous system. Within the brain region itself, the further rostrally one goes, the more complicated the differentiation becomes and the further the structures depart from the relatively simple organization of the spinal cord.

### External Form

The cranial end of the neural tube at about day 25 shows three regional enlargements separated by two constrictions. These are the primary brain vesicles; the *prosencephalon* (forebrain), the *mesencephalon* (midbrain), and the *rhombencephalon* (hindbrain) (Fig. 20–14 A). At the time the three primary vesicles are formed, the originally straight neural tube, owing largely to the greater growth of the posterior portion of the mesencephalon, shows a bending or flexure in the middle of the brain. This is the *cephalic flexure* (Fig. 20–14 A). Shortly after the appearance of the cephalic flexure, a second bend develops approximately in the region of the juncture of the brain and the spinal cord. This is the *cervical flexure* (Fig. 20–14 B).

Two of the three primary brain vesicles, the prosencephalon and the rhombencephalon, subdivide so that the 9-millimeter embryo, early in the second month, shows five major brain regions (Fig. 20–14 C). These were named by His in 1893 the *telencephalon, diencephalon, mesencephalon, metencephalon,* and *myelencephalon* proceeding in a rostral-to-caudal direction. Shortly after the five divisions are recognizable, a third flexure appears between the myelencephalon and the metencephalon. This is the *pontine flexure* (Fig. 20–14 C). The pontine flexure bends in a direction opposite to the first two flexures. In the 11-millimeter embryo (Fig. 20–14 D), the five brain divisions and the three flexures are easily recognized. A well-defined posterior constriction separates the telencephalon from the diencephalon. The developing optic cup and the infundibulum are two landmarks in the diencephalon. The cephalic flexure marks the middle of the mesencephalon, and a constriction called the *isthmus* separates the mesencephalon and the metencephalon. The three flexures gradually become less pronounced, and eventually the pontine and cervical flexures completely disappear. The cephalic flexure, however, persists, and because this occurs in the region of the midbrain, the diencephalon and the telencephalon are permanently set at a slight angle to the rest of the brain.

The central canal of the spinal cord runs the length of the neural tube, and in the part of the tube that will form the brain, it undergoes modifications in accordance with the regional specializations of the brain to form a number of local enlargements

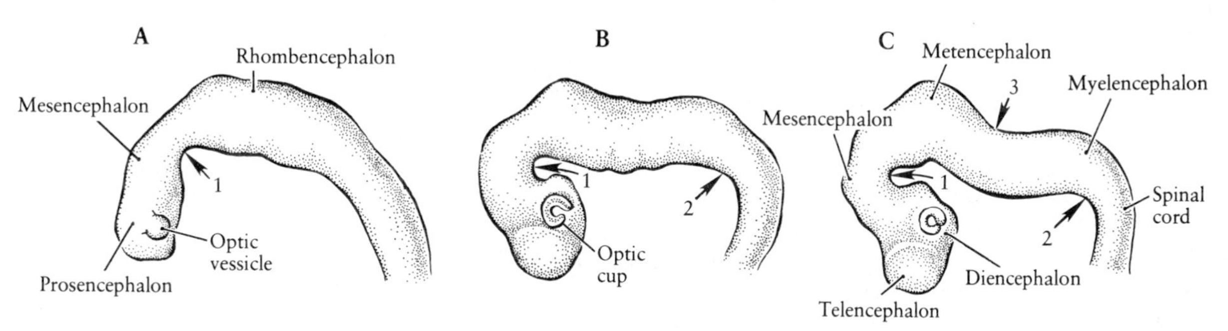


known as *ventricles* (Fig. 20–15). The first and second ventricles lie within the telencephalon and become greatly enlarged as the cerebral hemispheres expand. The third lies within the diencephalon and is connected to the ventricles of the telencephalon by a narrow canal, the *foramen of Monro (interventricular foramen).* The fourth ventricle develops in the rhombencephalon and later shows a wide lateral expansion. It is connected to the third by a narrow passage through the mesencephalon, the *aqueduct of Sylvius (cerebral aqueduct).*

**FIG. 20–14.** Four stages showing the external configuration of the brain during its early development. A, about 25 days showing the three primary divisions and the cephalic flexure (3.5 mm); B, about 30 days (5 mm); C, about 36 days (9 mm); D, about 39 days (11 mm). 1, cephalic flexure; 2, cervical flexure; 3, pontine flexure.

## Internal Configuration

In the brain, the characteristic H-shaped pattern formed by the cell bodies of the neurons in the spinal cord is found only in the most caudal region of the myelencephalon. Elsewhere, the histological appearance of the brain bears little resemblance to that of the spinal cord. This is the result of a number of factors: (1) the development of important fiber tracts that impinge on the cellular area that represents the intermediate area of the cord, (2) the migration of neurons from the intermediate layer into the outer marginal layer of the brain, and (3) the accumulation of groups of neurons in particular regions of the brain to form what are known as *cranial nuclei.*

**FIG. 20–15.** Expansion of the central canal in the region of the brain to form the ventricular system. (After E. L. House and B. Pansky, 1967. A Functional Approach to Neuroanatomy. McGraw-Hill Book Co., New York.)

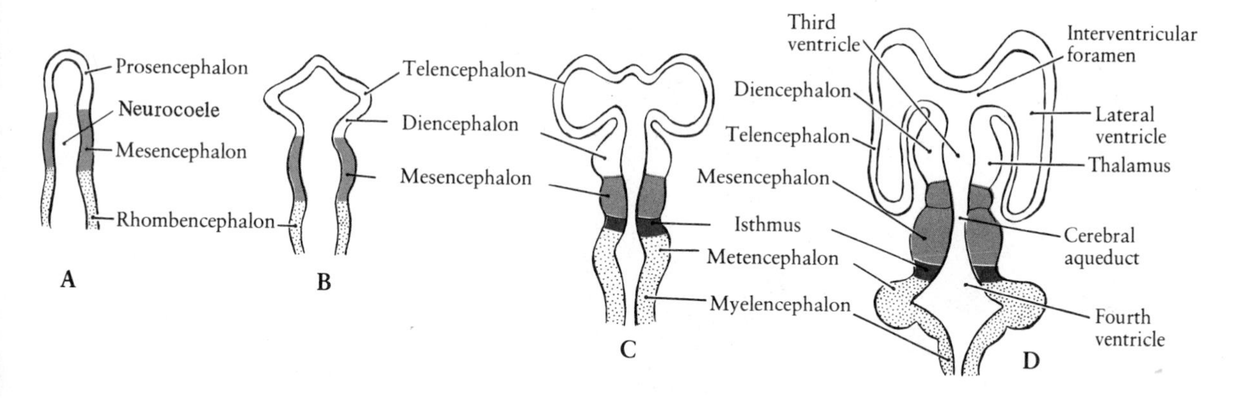

*The Functional Classification of Cranial Nerve Nuclei and Fibers*

Many of the nuclei in the brain are associated with the cranial nerves. The neurons of cranial nerve nuclei form groups of cells, each group serving a specific function. Thus, all of the motor fibers of the cranial nerves are axons from specific cranial nuclei sending their efferent impulses away from the brain stem. At the same time, the sensory fibers in the various cranial nerves, whose cell bodies are located in cranial nerve sensory ganglia outside of the brain stem, send their afferent impulses into the brain where they make central synapses with specific sensory nuclei. Very often a single cranial nucleus, be it motor or sensory in function, may be associated with fibers in more than one cranial nerve. This is particularly true for the sensory cranial nerve nuclei where, for instance, all of the visceral afferent fibers, no matter which cranial nerve they are found in, make their central synapses with the same nucleus. However, any single cranial nerve nucleus is always associated with only one functional type of nerve fiber.

We have seen that we can group the spinal nerve fibers into four classes or types on the basis of their function. These same four functional types are also found in the cranial nerves. However, the cranial nerves have three additional (special) types of fibers, and we are then concerned with seven different types of fibers in the cranial nerves. Any one cranial nerve may have one or more than one type of fiber associated with it. The efferent fibers are outgrowths of nuclei derived from the basal plate of different regions of the brain, and the sensory fibers make their central connections with nuclei, which are alar plate derivatives. The alar and the basal plates each give rise to three functional types of cranial nerve nuclei. Some of these nuclei are found at only one level of the brain, while others may be present as long bands of cell bodies stretching through many levels. In the early embryo, these six types of nuclei show a regular arrangement (Fig. 20–16), but later migrations and disturbances from developing fiber tracts result in considerable rearrangement. The functional types of the cranial nerve nuclei are as follows:

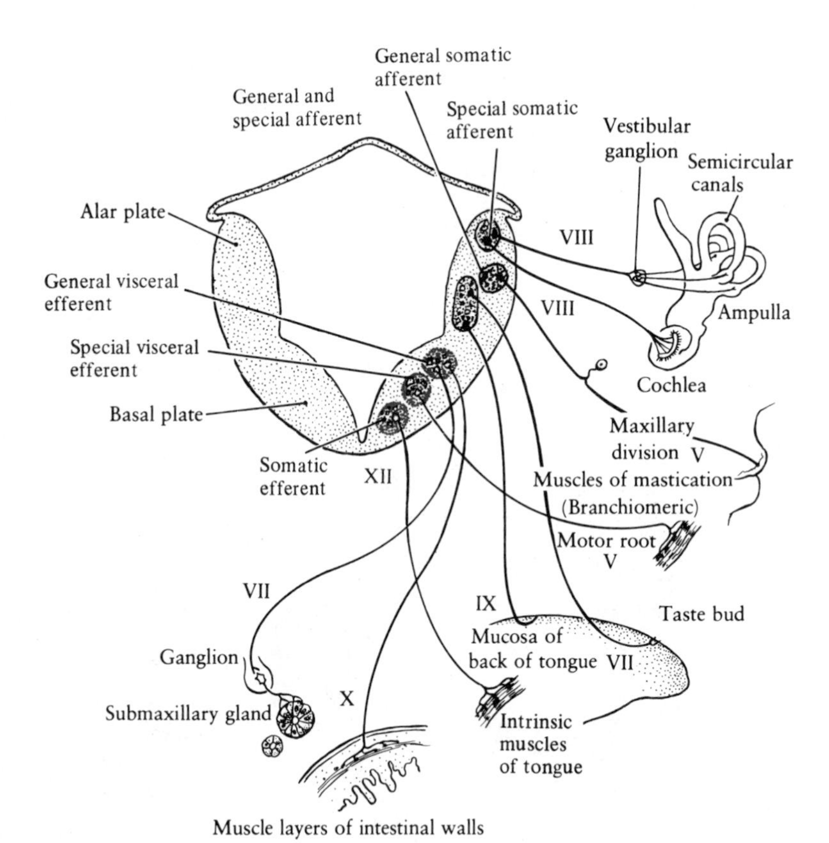


**FIG. 20–16.** General location in early development of the alar and basal plate nuclei associated with the different functional classifications. (See text for description of functional classifications.)

1. *Somatic Efferent (SE)*—as in the spinal cord, motor neurons whose fibers supply skeletal muscle derived from somatic mesoderm.

2. *General Visceral Efferent (GVE)*—as in the spinal cord, motor nuclei whose fibers are preganglionic to structures of visceral origin such as the lungs, the heart, and the alimentary tract—a part of the parasympathetic division of the autonomic nervous system. In this division, the ganglia are located close to the structures they supply (even within the walls, as is the case with the enteric ganglia previously considered), and the postganglionic fibers are correspondingly short.

3. *Special Visceral Efferent (SVE)*—contrary to the implication of the name, the fibers of these nuclei are not a special part of the autonomic nervous system supplying visceral organs. Instead, they are efferent fibers to skeletal muscle. However, the muscles these fibers supply are derived from head or visceral arch mesoderm. The skeletal muscles formed from this type of mesoderm are histologically and functionally exactly the same as those formed from somatic mesoderm. They differ in only one thing—their embryonic origin. Because of this origin, their cranial nerve motor fibers are given this particular functional classification.

4. and 5. *General and Special Visceral Afferent (GVA, SVA)*—all of the visceral afferent fibers in the cranial nerves (with the exception of those in the olfactory nerve) make their central connections with the same cranial sensory nucleus. The general visceral fibers are connected peripherally to general visceral receptors. The special visceral fibers are connected to what are considered special visceral receptors—taste buds and olfactory receptors.

6. *General Somatic Afferent (GSA)*—nuclei receiving sensory input from peripheral somatic sensory receptors in the head region.

7. *Special Somatic Afferent (SSA)*—nuclei receiving sensory input from special sense organs of ectodermal origin—the eye and the ear.

## Development of the Myelencephalon

The myelencephalon shows the least variation from the spinal cord. Early in development a change in the H-shaped pattern of the cord is seen. It is due primarily to two events: (1) the expansion of the fourth ventricle and, as a consequence of this, the separation of the alar plates, and (2) the development of the brain nuclei.

The fourth ventricle is already a large cavity in the early embryo. When the pontine flexure develops, the fourth ventricle is expanded laterally. In this process the alar plates become separated and the roof plate stretches to form a thin ependymal layer roofing over the fourth ventricle and joining the right and left alar plates (Fig. 20–17). When the pia mater—the innermost meninx of the brain—develops, it forms a vascular plexus, the *tela choroidea,* in a T-shaped pattern over the roof of the ventricle. This plexus of blood vessels, covered by the ependymal epithelium, then invaginates into the fourth ventricle forming the *choroid plexus* of that ventricle, whose function is the secretion of cerebrospinal fluid into the ventricular system.

The lateral walls of the fourth ventricle show well-defined thick alar and basal plates marked off from each other by a distinct groove, the sulcus limitans (Fig. 20–17 A). The cell bodies of neurons serving a common function move to specific areas of the alar and basal plates where they accumulate

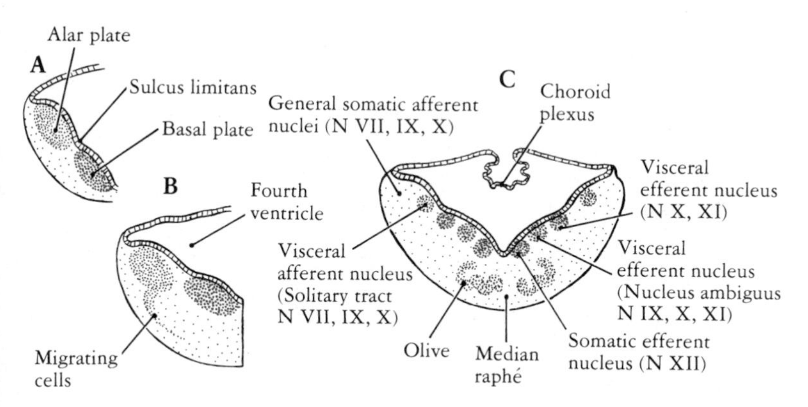


**FIG. 20–17.** Three stages in the early development of the myelencephalon showing the lateral expansion of the fourth ventricle, the formation of the alar and basal plate nuclei, and the development of the choroid plexus.

as groups of cells collectively called a nucleus. The pattern made by these nuclei is characteristic of the region of the brain examined. Often the ultimate position of a nucleus may be the result of extensive migration of its individual neurons. One example of this is the migration of alar plate neurons of the myelencephalon into the basal plate to form the sensory way station, the *olivary nucleus* (Fig. 20–17).

As in the spinal cord, the neuroblasts derived from the basal plate send out fibers to form the efferent components of the cranial nerves, and those from the alar plate form the central nervous system connections (or way stations) for afferent impulses. Some derivatives of the basal plate of the myelencephalon are the *nucleus ambiguus,* which is the point of origin of the special visceral efferent fibers running in cranial nerves IX, X, and XI, the *dorsal motor nucleus* of the Xth nerve (GVE), and the motor nucleus of the XIIth (hypoglossal) nerve (SE) (Fig. 20–17 C). From the alar plates are derived the *solitary nucleus,* which receives the visceral sensory fibers from cranial nerves VII, IX, and X; the *nucleus gracilis* and *nucleus cuneatus,* which are relay stations for proprioceptive impulses ascending from the spinal cord to the cerebellum and cerebrum.

The floor plate of the myelencephalon forms a non-nervous median raphe.

## Development of the Metencephalon

The metencephalon consists of three parts: (1) a primary axial portion, the *tegmentum,* a continuation rostrally of the general structure of the myelencephalon; (2) a specialized expansion of the most posterior regions of the alar plates to form the *cerebellum;* and (3) a basal portion, the *pons*, containing a group of nuclei, the *pontine nuclei,* and the fibers connecting these nuclei to other parts of the central nervous system, particularly the cerebrum and the cerebellum. Regions 2 and 3 are both phylogenetically newer acquisitions and develop later than the tegmental region of the metencephalon.

The primary axial portion forms the floor of the metencephalic part of the fourth ventricle and shows the characteristic alar and basal plates separated by the sulcus limitans (Fig. 20–18 A). The lateral expansion of the middle of the fourth ventricle gives it a diamond-shaped appearance, and the floor of the fourth ventricle, including the part formed by the myelencephalon, is known as the *rhomboid fossa.* The basal plates of the metencephalon form motor nuclei associated with cranial nerves V (SVE), VI (SE), and VII (GVE, SVE). Nuclei developing in the alar plates provide sensory connections for cranial nerves V (GSA), VII (GVA, SVA), and VIII (SSA) (Fig. 20–18 C).

### *The Cerebellum*

The most posterior parts of the alar plates rostral to the lateral recesses which are separated from each other medially by the roof plate, become thickened to form the *rhombic lips* (Fig. 20–18 B). Progressing rostrally, the fourth ventricle narrows and the rhombic lips approach and fuse with each other. Proliferation of the rhombic lips gives rise to the *cerebellar swellings* or *plates,* which are recognizable in the 7-to 8-millimeter embryo early in the second month (Fig. 20–19). During the second month each cerebellar plate thickens rapidly and forms a pair of bulges projecting into the fourth ventricle, which soon join medially to form a single dumbbell-shaped cerebellum (Fig. 20–20).

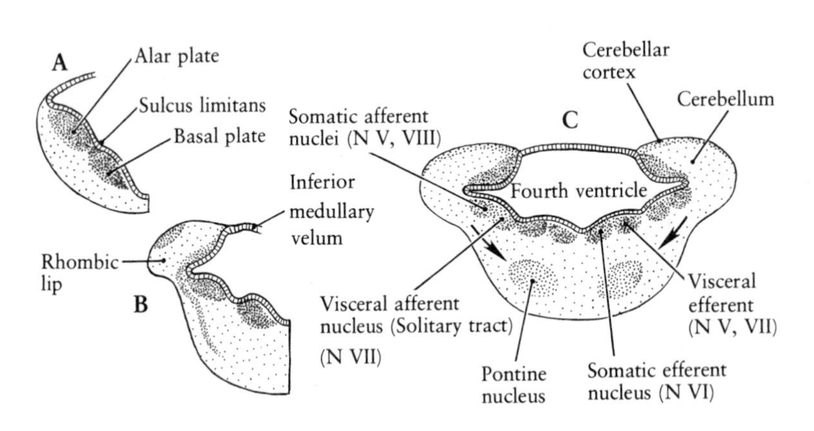


**FIG. 20–18.** Three stages in the early development of the metencephalon showing the differentiation of the alar and basal plates and the formation of the rhombic lip. Arrows represent movement of alar plate cells into the basilar portion of the pons.

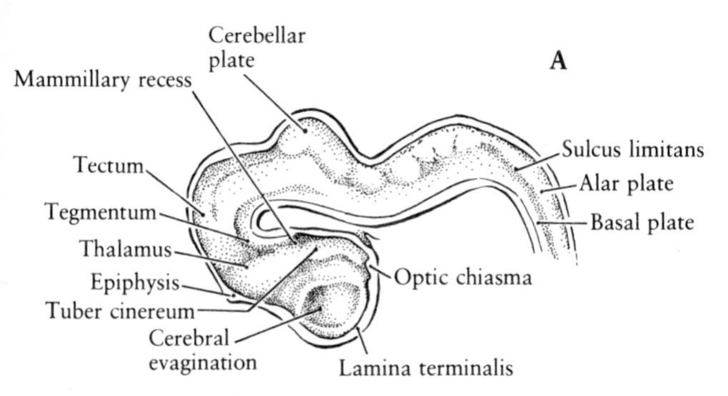


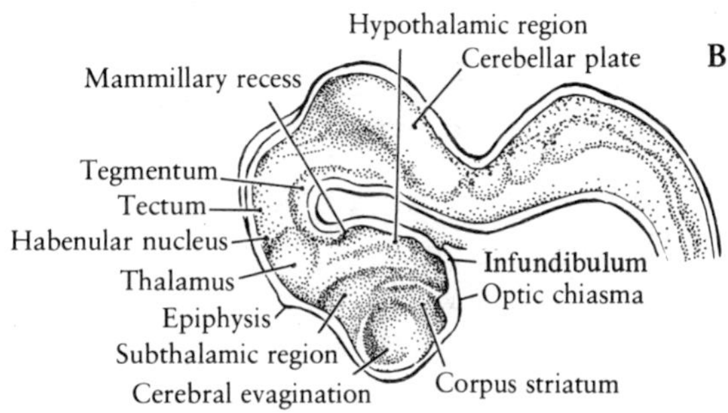


During the third and fourth months, there is a rapid growth of the cerebellum and fissures appear on its surface that serve as landmarks to delineate its principal lobes. The first to appear is the *posterolateral fissure*. This sets off a region caudal to this fissure, the *flocculonodular* lobe, made up of a central *nodulus* and two lateral arms, the *flocculi* (Fig. 20–21). The part of the cerebellum rostral to the fissure is called the *corpus cerebelli* (Fig. 20–21 A). The median nodulus and the paired lateral floccular lobes are thus the most caudal part of the cerebellum, lying just cranial to the lateral recesses, and they are the first part of the cerebellum to differentiate. As is the general rule, the first part of a system to differentiate is the oldest part phylogenetically. Functionally, the flocculonodular lobe is concerned with the maintenance of equilibrium. The coordinating functions of the cerebellum, functions that are phylogenetically more recently acquired, develop in the corpus cerebelli.

The *fissure prima* develops late in the fourth month, dividing the corpus cerebelli into anterior and posterior lobes (Fig. 20–21 B), and shortly thereafter a *fissure secunda* appears in the posterior lobe (Fig. 20–21 C). The part of the cerebellum just rostral to the fissure secunda, particularly its lateral aspects, is the last part to develop. In man, this undergoes a marked expansion to form the lateral hemispheres. This newest part of the cerebellum receives the name *neocerebellum*. Functionally it is concerned with the integration of the muscular activity initiated by the phylogenetically newest part of the cerebrum, the neopallium. Both of these parts of the brain show their greatest development in the primates.

As development proceeds many more subdivisions appear, and the adult cerebellum shows a complex pattern of fissures and lobules to which cumbersome and often fanciful names have been

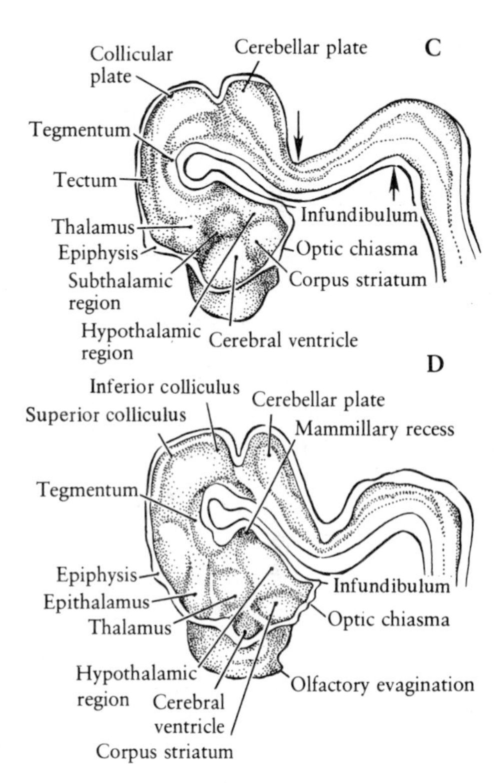


**FIG. 20–19.** Drawings of reconstructions of the brain of four human embryos early in the second month. A, about 31 days (7–8 mm); B, about 33 days (9–10 mm); C, about 35 days (12 mm); D, about 37 days (14.6 mm). (From G. L. Streeter, 1948. Carnegie Contrib. Embryol. 32:133.)

given. However, the fundamental divisions of embryological and functional significance are those outlined above. In their order of development, we may thus recognize three regions whose major functions differ, although there is certainly some overlap among them: (1) the flocculonodular lobe, termed the *archicerebellum,* which functions to maintain equilibrium; (2) the part of the body of the cerebellum anterior to the primary fissure, the

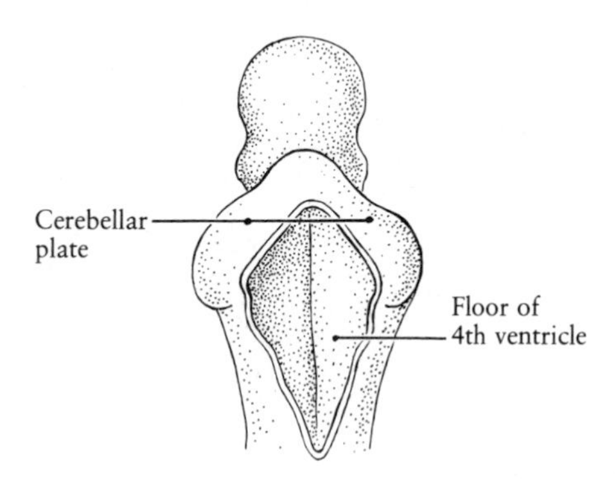


**FIG. 20–20.** Fusion of the right and left cerebellar plates.

anterior lobe, and the part of the body of the cerebellum between the posterolateral fissure and the fissure secunda; collectively these parts are termed the *paleocerebellum* and function to maintain muscle tone and posture; and (3) the region between the fissure prima and the fissure secunda, termed the *neocerebellum,* which functions to coordinate muscular activity. All three regions have essentially different connections to the medullary nuclei of the cerebellum and to other parts of the brain in accordance with the functions they serve.

*Histogenesis of the Cerebellum.* The migration of the neuroepithelial cells and the neuroblasts of the cerebellum differs from that described for the spinal cord. The cerebellar plate at first shows the same characteristic three-layered (ventricular, intermediate, marginal) pattern as the spinal cord. However, as the cerebellar plate thickens, neuroepithelial cells move out of their position next to the lumen of the fourth ventricle and migrate to the periphery of the marginal zone to form what is termed the *external granular layer* (Fig. 20–22 A). Later, an *internal granular layer* develops in the marginal layer through the outward migration of other neuroepithelial derivatives and also by the inward migration of cells from the external granular layer (Fig. 20–22 B). Large, flask-shaped cells, *Purkinje cells,* mark the border between these two granular layers (Fig. 20–22 C). In later development, most of the cells of the outer granular layer migrate inward and end up in the inner granular layer, leaving a cell-poor layer, now called the *molecular layer,* as the most superficial part of the cerebellar cortex. The molecular layer is occupied by the dendrites of the Purkinje cells and the axons of the cells of the inner granular layer.

Other neuroepithelial derivatives remain in the deeper part of the cerebellum. They form four pairs of *medullary nuclei* and through their dendrites make synaptic connections with the axons of

**FIG. 20–21.** Three stages in the development of the human cerebellum. Drawings on the bottom are views of the posterior surface, and those on the top represent sagittal sections through the cerebellar cortex. A, formation of the posterolateral fissure setting off the body of the cerebellum from the flocculonodular node; B, formation of the primary fissure in the body of the cerebellum; C, later development of the lobes of the cerebellum (about five months).

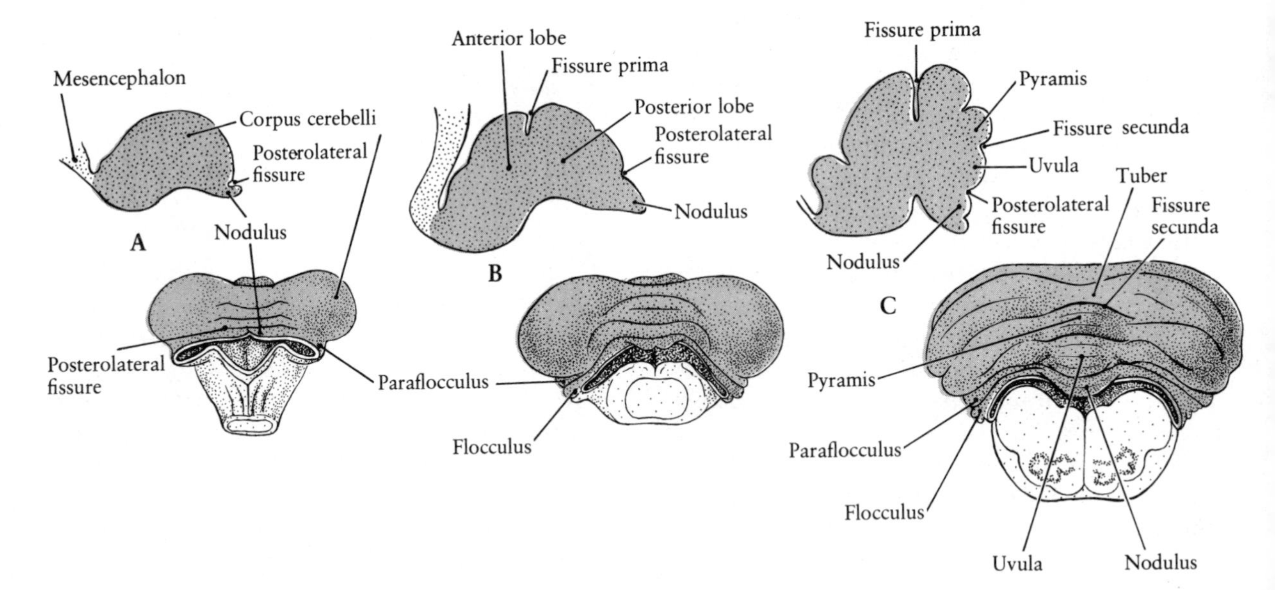

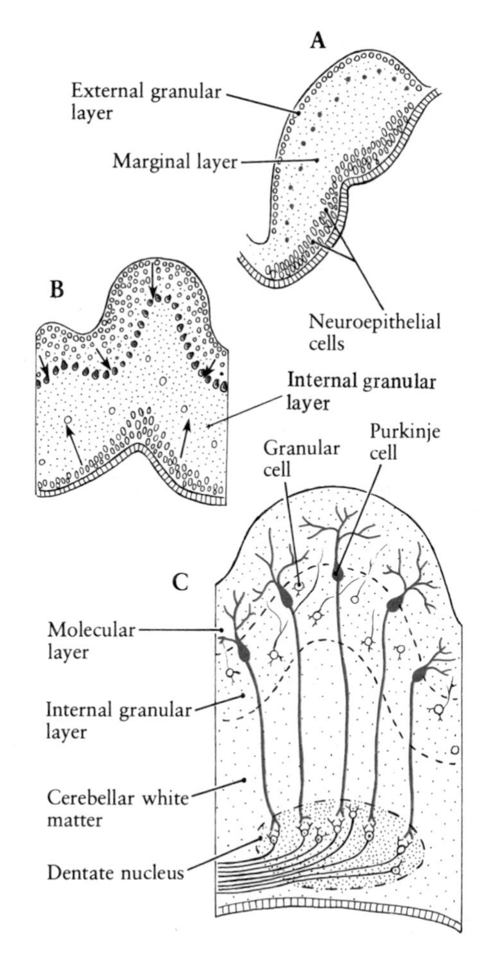


FIG. 20–22. Histogenesis of the cerebellar cortex. Arrows indicate the direction of migration of the cells.

the Purkinje cells. The medullary nuclei relay impulses from the cerebellar cortex to other parts of the central nervous system. The first medullary nuclei to differentiate are the most medially located, and their connections are with the oldest part of the cortex, the flocculonodular lobe. The most lateral, which are the largest and the last to develop, are the *dentate nuclei* (Fig. 20–23), whose connections are with the Purkinje cells of the neocerebellum. Efferent impulses from the dentate nuclei are carried cranially into the mesencephalon by fibers that make up the major portions of the *superior cerebellar peduncles*, one of the three pathways to and from the cerebellar cortex.

The roof plate in the region between the cerebellar plates is absorbed in the expansion of the cortex, but cranially and caudally it persists as the *superior* and *inferior medullary velum*, respectively (Fig. 20–23).

*The Basilar Portion of the Pons*

The basilar portion of the metencephalon develops in connection with a group of nuclei, the pontine nuclei. Although these nuclei are located in the most anterior (basal) portion of the pons, they are derived from neuroblasts that have migrated to this position and are actually of alar plate origin (Fig. 20–18). The pontine nuclei are associated with the neocerebellum, to which they are connected by fibers that cross over in the most basilar region of the pons and ascend to the cerebellum by way of the

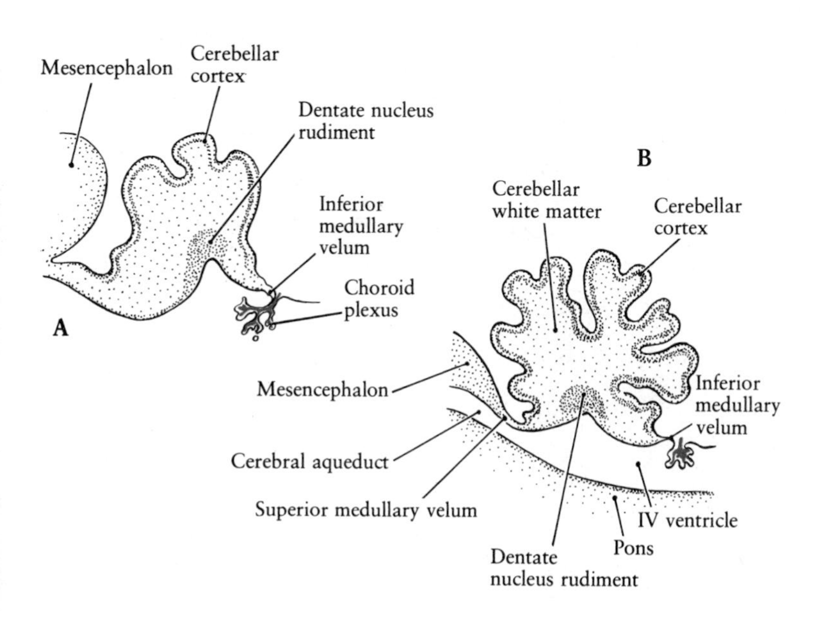


FIG. 20–23. Sagittal sections of the developing human cerebellum. A, during the fourth month (100 mm); B, during the fifth month (150 mm). (After W. J. Hamilton and H. W. Mossman, 1972. Human Embryology. Macmillan Press, London.)

*middle cerebellar peduncle (brachium pontis).* The pontine nuclei receive impulses from the cerebral cortex and thus form a link by way of which the newest part of the cerebellar cortex may interact with the newest part of the cerebral cortex in the coordination of muscular activity.

## Development of the Mesencephalon

The mesencephalon does not show any striking modifications or expansions and is soon overshadowed by the cerebral cortex and the cerebellum. Three regions of the mesencephalon may be recognized: (1) the *tegmentum,* developed from the basal plates; (2) the *tectum,* developed from the alar plates; and (3) the *cerebral peduncles* consisting of fiber tracts mainly from the cerebrum (Fig. 20–24).

The development of the tegmentum begins late in month 1 (3- to 5-mm embryo) as the basal plate begins to thicken. As is the general rule, the basal plate begins its differentiation before the alar plate. During the second month it gives rise to the somatic efferent nuclei of cranial nerves III and IV. Associated with the nucleus of the IIIrd nerve is a cranial component whose fibers are preganglionic fibers of the parasympathetic division of the autonomic nervous system (Fig. 20–24 C).

At the time the tegmentum is developing, the tectum is still undifferential. Not until the beginning of month 3 (24- to 27-mm embryos) does the alar plate region show any significant expansion. At this time the alar plates thicken and form two longitudinally running ridges on the posterior aspect of the mesencephalon, the *collicular plates* (Figs. 20–19 C and 20–25). Later, each ridge is divided transversely forming a rostral and a caudal pair of swellings, the *superior* and *inferior colliculi,* respectively (Fig. 20–25 B). Collectively the colliculi are called the *corpora quadrigemina.* As they increase in size, they obscure the roof plate. In the development of the tectum, neuroblasts migrate toward the surface, where they form peripheral tectal nuclei (Fig. 20–24 B,C); the colliculi thus resemble the cerebellar cortex in the reversal of white and gray matter. The superior colliculi are centers for visual reflexes; the inferior colliculi are centers for auditory reflexes.

Two pairs of prominent nuclei develop in the tegmental area: the *red nucleus* and the *substantia nigra* (Fig. 20–24 C). Although their origin is not completely agreed on, they are generally considered to be alar plate derivatives that migrate into the tegmental region.

The cerebral peduncles *(basis pedunculi)* make up the most anterior region of the mesencephalon. They consist of nerve fibers running longitudinally through the marginal layer of the tegmentum. They contain fibers from cerebral cortex nuclei passing to lower brain and spinal regions, *corticopontine, corticobulbar,* and *corticospinal* tracts.

As the basal and alar plates expand, they bulge into the ventricle of the mesencephalon and the mesocele becomes progressively smaller ending up as a narrow channel, the *aqueduct of Sylvius,* connecting the fourth ventricle of the hindbrain to the third ventricle of the diencephalon. The floor plate is lacking. It is considered to terminate at the level of the rostral end of the metencephalon.

## Development of the Diencephalon

Early in the development of the prosencephalon a lateral diverticulum appears on either side. These are the *optic vesicles* (Figs. 20–14 A and 20–26) that will develop into the major portions of the eye. The

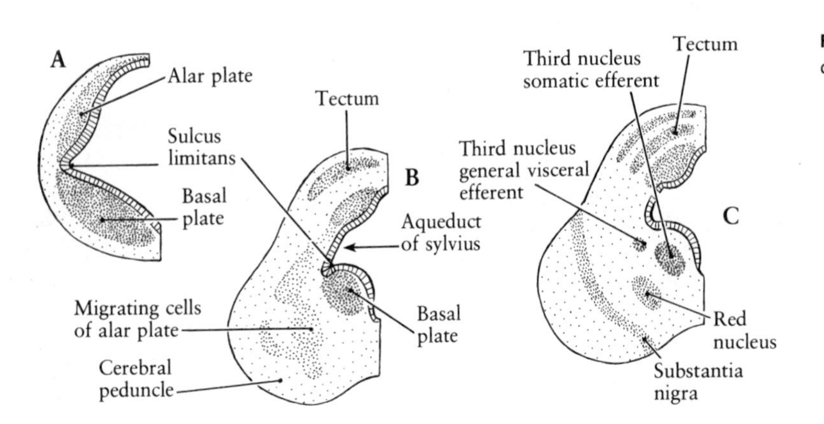


**FIG. 20–24.** Three stages in the differentiation of the mesencephalon.

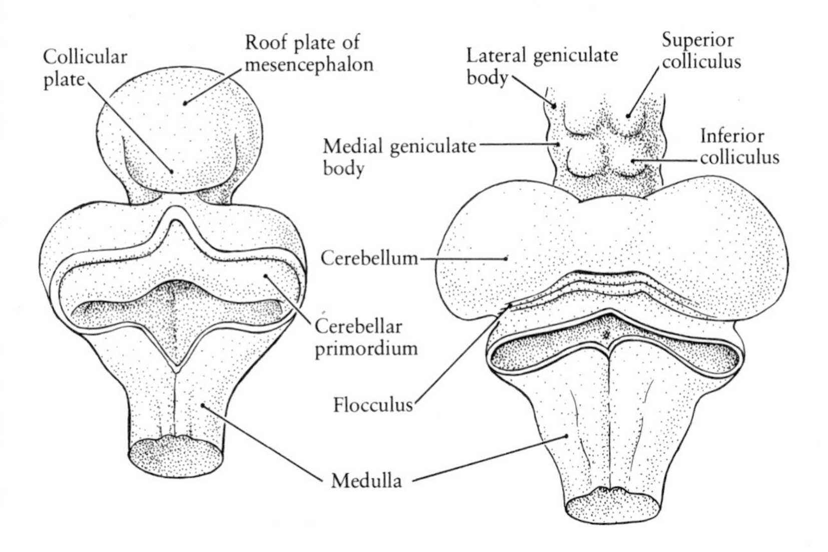


**FIG. 20–25.** A, dorsal view of the hindbrain and the midbrain in a two-month old human embryo; B, same view in a four-month-old embryo.

optic vesicles are connected to the ventricle of the diencephalon by the hollow *optic stalk* (Fig. 20–26 B). They mark the rostral end of the diencephalon. The diencephalon is prominent during the middle of the second month, at which time the boundaries between it and the telencephalon are easily distinguished. Later it becomes overshadowed by the cerebral hemispheres, and the walls of the diencephalon and the telencephalon become closely connected.

Figure 20–27 shows the developing forebrain during the seventh week, at which time all of the primordia of the future diencephalic structures are recognizable. The diencephalon consists of three major subdivisions: the *epithalamus,* the *thalamus,* and the *hypothalamus.* The most anterior and the first to develop, the hypothalamus, is marked off from the thalamus by the *hypothalamic sulcus.* The hypothalamic sulcus is not a continuation of the sulcus limitans, which stops in the mesencephalon, and thus does not divide the diencephalon into alar and basal plates. All of the forebrain, with the exception of roof plate derivatives, is formed from alar plate material. The *optic chiasma,* the *infundi-*

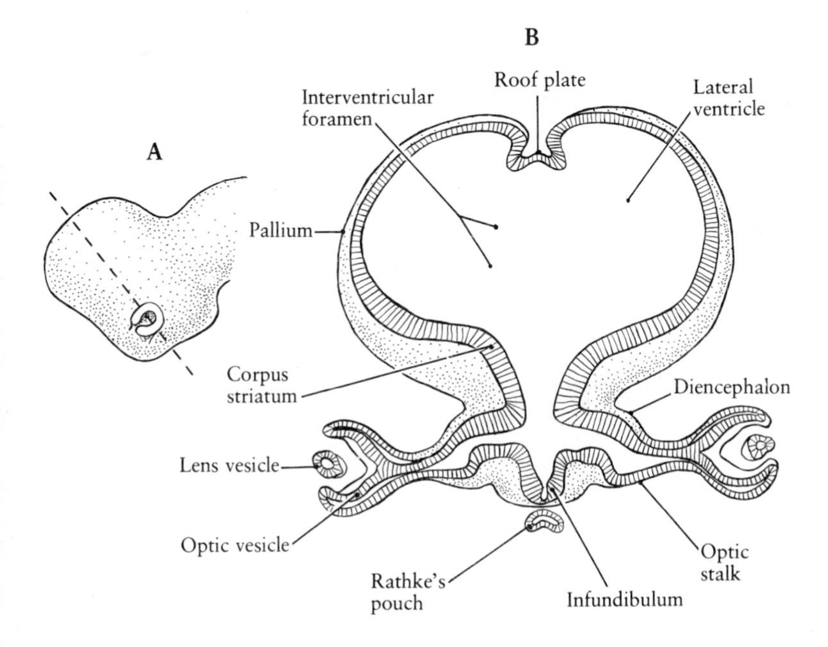


**FIG. 20–26.** Human diencephalon (10 mm). A, lateral view showing developing optic vesicle; B, cross section at level indicated in A.

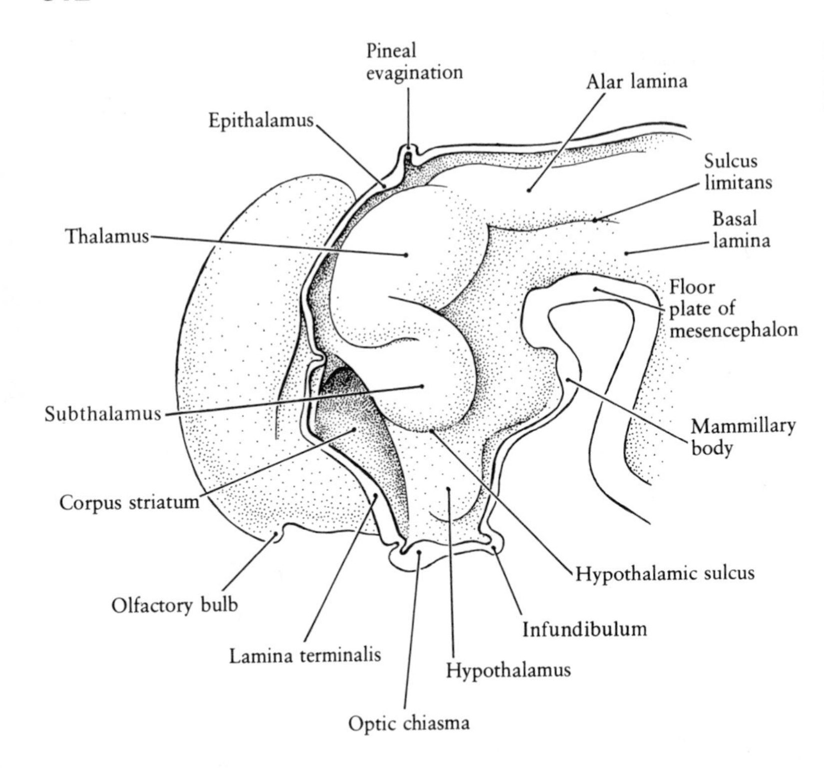


**FIG. 20–27.** View of the medial surface of a reconstruction of the forebrain of a 19-mm human embryo.

*bulum*, and the *mammillary bodies* are hypothalamic derivatives (Figs. 20–19 and 20–27). The optic chiasma represents the crossing over of parts of the optic tracts from one side to the other, the infundibulum will develop into the neural parts of the pituitary glands and the mammillary bodies will develop nuclei functioning as olfactory way stations. The neuroblasts of the intermediate layer of the hypothalmus will differentiate into a series of nuclei concerned with visceral functions, the brain center of the autonomic nervous system.

The epithalamus is the next area to differentiate. The cells of the intermediate layer in this region develop into the *habenular nuclei*, which are associated with olfactory impulses and lie adjacent to a doral evagination of the roof of the diencephalon, the *epiphysis* (Figs. 20–19 B and 20–27). The epiphysis will form the *pineal body*. Two commissures develop in the region; the *habenular commissure* cranial to the epiphysis, and the *posterior commissure* caudal to the epiphysis (Fig. 20–29 B).

The thalamus consists of an older portion, the *ventral thalamus (subthalamus)*, which is an important transitional zone from the mesencephalon. A new portion, the *dorsal thalamus*, which reaches its highest development in man, differentiates important nuclear masses that become association and relay centers for impulses to and from the cerebral

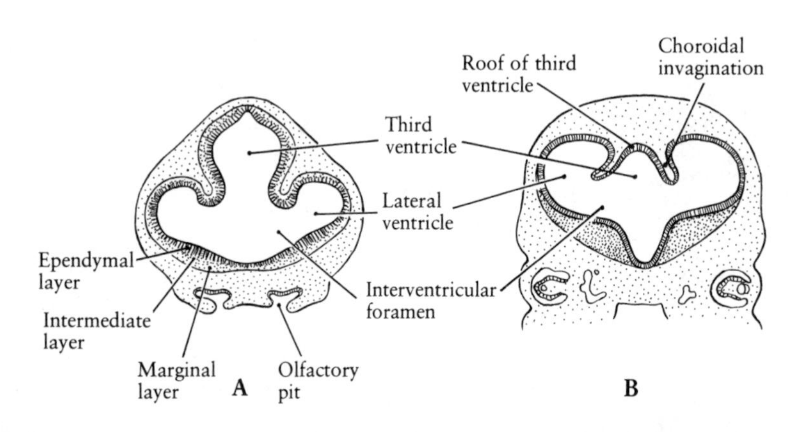


**FIG. 20–28.** Transverse sections through the forebrain of human embryos. A, 13 mm; B, 15 mm.

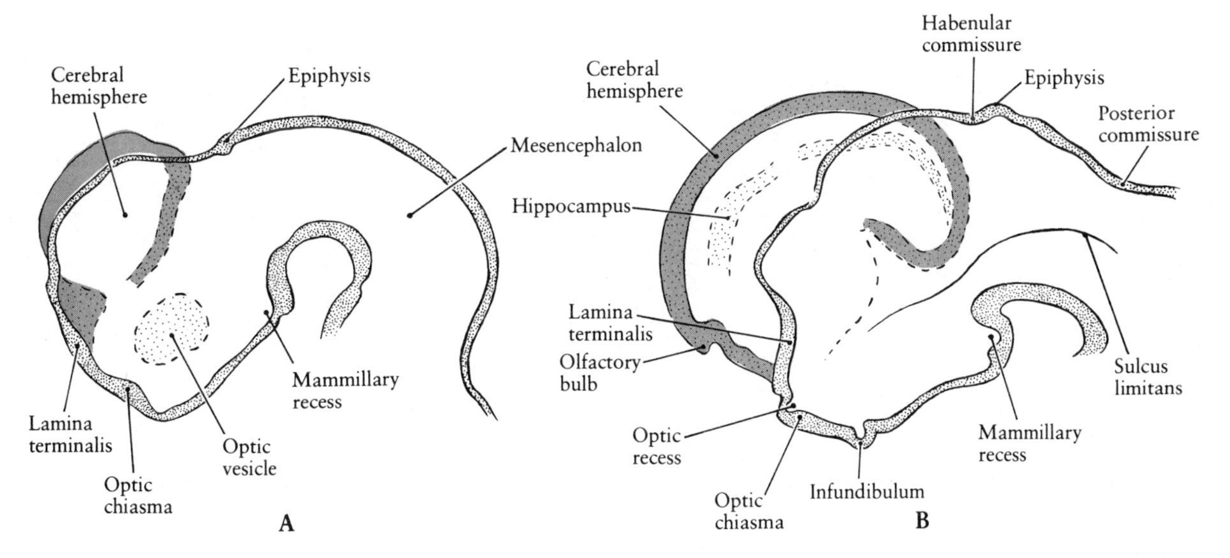


**FIG. 20–29.** Diagrams of midsagittal sections of the brains of human embryos during the fifth and sixth weeks of development. A, 7.5 mm; B, 19 mm. (After G. W. Bartelmez and A. S. Dekeban, 1962. Carnegie Contrib. Embryol. 37:13.)

cortex. The late-developing thalami grow rapidly, bulge into the third ventricle, and join to form the *intermediate mass.*

The most posterior part of the thalamus differentiates the *lateral* and *medial geniculate bodies,* the former associated with the superior and the latter with the inferior colliculi of the mesencephalon (Fig. 20–25). This region of the thalamus is sometimes called the *metathalamus.*

In additional to forming the epiphysis, the roof plate also forms the thin ependymal tela choroidea, which becomes a part of the choroid plexus of the third ventricle (Fig. 20–28 B).

## Development of the Telencephalon

The differentiation of the telencephalon begins later than that of other regions of the brain, but its rapid development, once started, gives rise to a structure that far overshadows the rest of the brain. This is accomplished by the tremendous expansion of the lateral and posterior walls of the prosencephalic cavity. At first the expanded lateral ventricles retain a wide connection with the cavity of the third ventricle (Figs. 20–26 and 20–28), but this cavity becomes progressively smaller as development proceeds. During its early expansion, the cerebrum extends laterally, posteriorly, and caudally but only slightly cranially.

During the second month, extension in a cranial direction begins; by the middle of the month, the *lamina terminalis,* which marks the most rostral end of the original brain stem and was still located at the most rostral position in the brain at the end of the first month (Figs. 20–29 A and 20–30 B), is now found to be about in the middle of the expanding hemispheres (Figs. 20–29 B and 20–30 C). Caudally, the lamina terminalis is set off from the optic chiasma by the *optic recess,* which marks the boundary between the telencephalon and the diencephalon (Fig. 20–29 B). At this time, the optic chiasma is one of three important commissures found in this region of the brain. The other two are the *anterior commissure* and the *hippocampal commissure,* which are located in the lamina terminalis (Fig. 20–30). Since the lamina terminalis retains its original position while the cerebral hemispheres expand rostrally to it, these two commissures appear to be located in the middle of the cerebrum in later stages of development (Fig. 20–30). Posteriorly and caudally, the lamina terminalis connects to the roof of the diencephalon.

Further development during the second month consists of the continued expansion of the cerebral hemispheres, which will eventually extend far enough caudally to cover the lateral aspects of parts of the metencephalon, and of the beginning of the differentiation in its walls. The caudal expansion of the cerebral vesicles brings their medial walls into contact with the walls of the diencephalon with which they fuse. The early differentiation of the walls of the cerebral vesicles is seen in Figure 20–

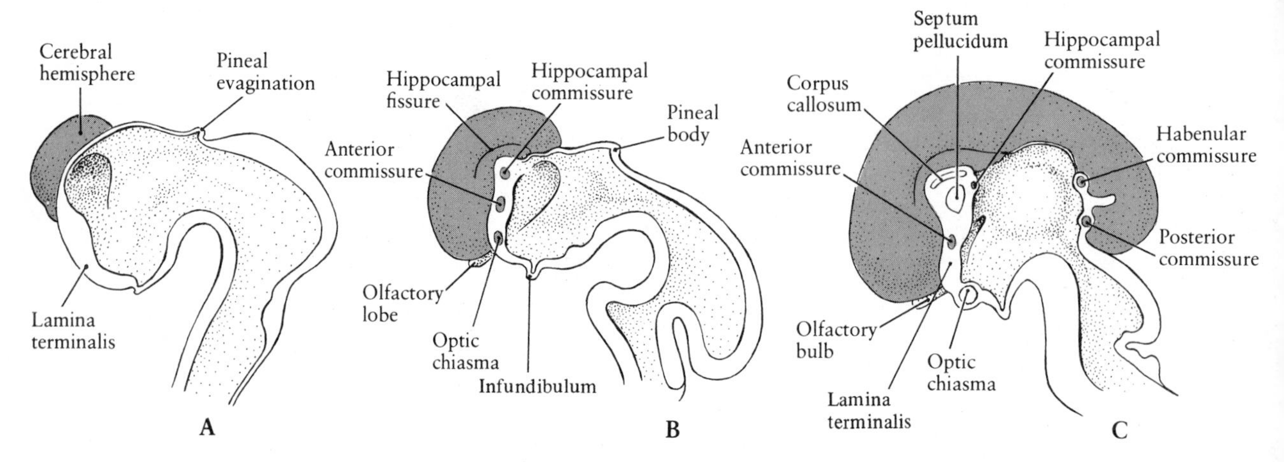


**FIG. 20-30.** Sagittal sections through the brain stem of three human embryos showing the development of the commissures. A, 9 mm; B, 25 mm; C, 65 mm. (After W. J. Hamilton and H. W. Mossman, 1972. Human Embryology. Macmillan Press, London.)

31 B. A marked thickening of the anterolateral wall differentiates first and represents the beginning of the formation of the *corpus striatum.* Next, the medial wall differentiates to form the *hippocampus.* The medial wall posterior to the hippocampus, the posterior wall, and the suprastrial region of the wall (i.e., all of the wall of the cerebral vesicle between the hippocampus and the corpus striatum) is the primordium of the *neopallium,* the newest part of the cerebral cortex.

Medially, along its posteromedial attachment to the diencephalon, the ependymal layer of the roof plate projects laterally into each ventricle to contribute to the formation of the choroid plexus of the first and second ventricles. It is convenient to consider separately the development of the three parts of the cerebrum: the corpus striatum, the hippocampus, and the neopallium.

*Corpus Striatum*

The corpus striatum can be separated into a medial accumulation of cells that retains its proximity to the ventricle and a more lateral region that gradually moves away from the ventricle. The former will develop into the *caudate nucleus,* the latter into the *lentiform nucleus* (Fig. 20–32).

As the cerebral cortex continues to expand caudally and then bends anteriorly, the corpus striatum is drawn out into a longitudinally running, comma-shaped ridge that is closely associated with the thalamus of the diencephalon (Fig. 20–33). The arched tail of the caudate nucleus ends in an accumulation of nerve cells termed the *amygdaloid nucleus.* Fiber tracts running in both directions between the developing neopallium and the thalamus appear on both sides of the lentiform nucleus. This thick bundle of fibers, in the form of a V that is

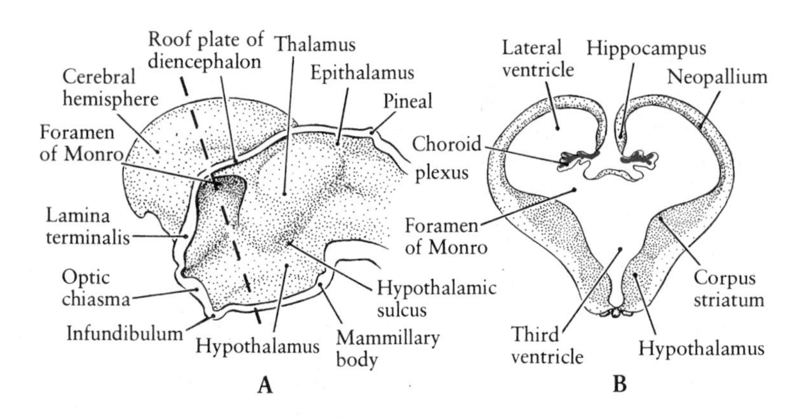


**FIG. 20-31.** A, drawing of the medial surface of a reconstruction of the forebrain of a seven-week-old human embryo; B, cross section through A at the level indicated. (From J. Langman, 1963. Medical Embryology. Williams and Wilkins, Baltimore.)

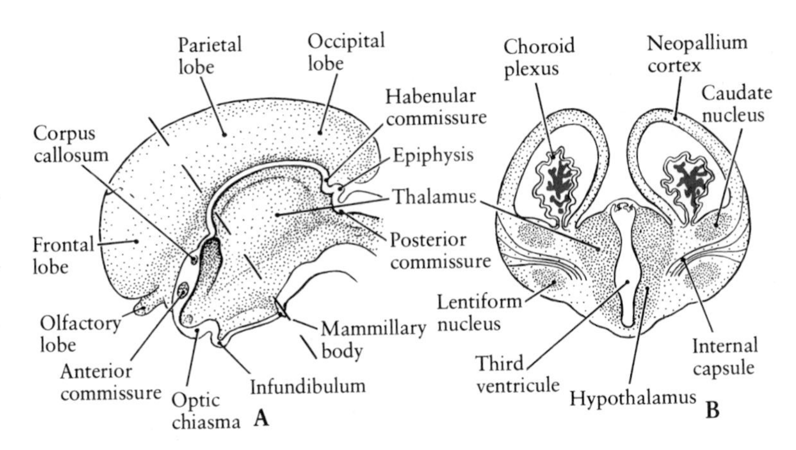


**FIG. 20–32.** A, diagram of a sagittal section of the cerebrum at about 10 weeks; B, drawing of a cross section through A at the level indicated.

open laterally, is called the *internal capsule*. The lentiform nucleus lies in the cavity of the V, so that it is separated by the posterior limb of the internal capsule from the caudate nucleus and by the anterior limb from the thalamus (Figs. 20–32 B and 20–33). The caudate nucleus and the thalamus thus come into direct contact. The lentiform nucleus divides into a lateral group of closely packed neurons, the *putamen*, and a medial portion, the *globus pallidus*. Another group of neurons develops lateral to the putamen to form the *claustrum*, the most superficial of the striatal nuclei (Fig. 20–33 B). The claustrum lies beneath an area of the cerebral cortex called the *insula*, a part of the cortex that develops more slowly than other regions and is overgrown by them so that it forms no part of the visible surface of the cortex but is hidden under the other lobes. Collectively, in the adult, the caudate nucleus, the lentiform nucleus, and the internal capsule are known as the *corpus striatum*, the highest center of the brain to develop in the lower vertebrates. Even in the birds, no substantial neopallium develops, and the corpus striatum is the center for correlation of sensory impulses and control of motor activity.

### *Hippocampal Cortex and Olfactory Complex*

The hippocampal pallium is carried caudally and anterolaterally with the growth of the temporal lobes of the telencephalon. It appears on the medial wall of the cerebral vesicle as a curved area parallel

**FIG. 20–33.** A, diagram of a sagittal section through a human cerebrum of approximately 21 weeks; B, cross section through A at the level indicated. (From J. Langman, 1963. Medical Embryology. Williams and Wilkins, Baltimore.)

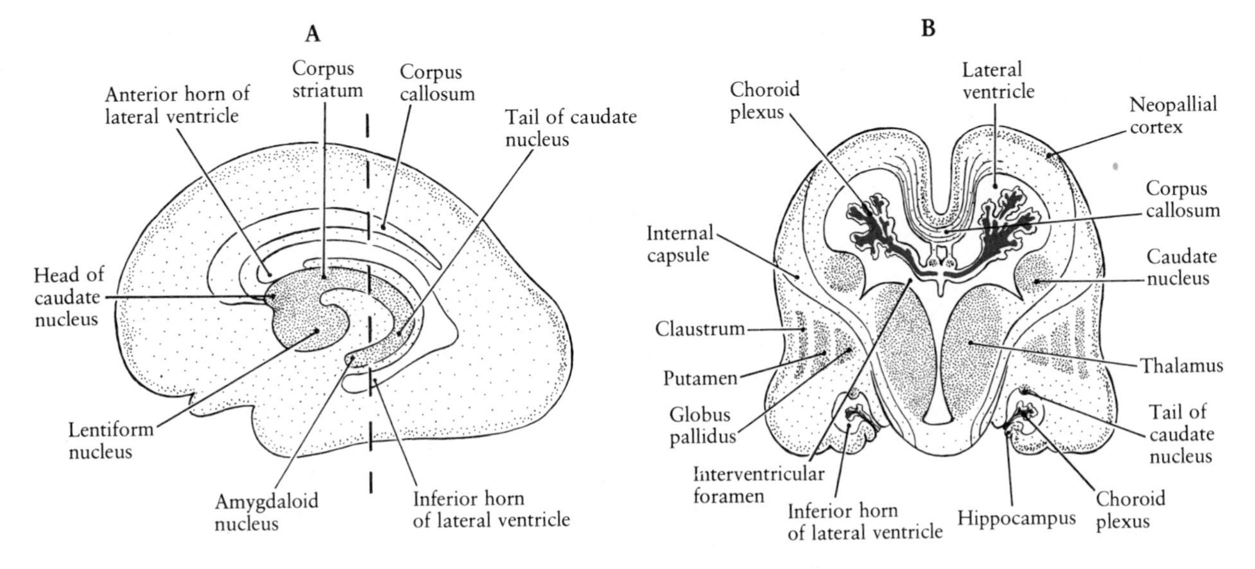

to the choroidal fissure between it and the neopallium (Fig. 20–31 B). Much of the hippocampal structure undergoes retrogressive changes. The rostral parts of the hippocampal region are interconnected by a commissure, the *hippocampal commissure,* which develops in the posterior portion of the lamina terminalis (Fig. 20–30). The hippocampal regions of the brain are generally considered to be part of the *rhinencephalon,* the parts of the brain concerned with olfaction.

The major part of the rhinencephalon begins to form at about four months as a projection, the *olfactory lobe,* on the anterior rostral surface of each cerebral hemisphere (Figs. 20–30 and 20–32). The olfactory lobe divides into a rostral part, the *olfactory bulb,* and the *olfactory stalk.* Fibers from the neurons in the olfactory bulb pass back through the olfactory stalk to centers developed from the caudal part of the olfactory lobe.

*Neopallium*

By far the most extensive changes in the walls of the cerebral vesicles occur in its phylogenetically newest part, the neopallium. The spinal cord configuration of ventricular, intermediate, and marginal layers is interrupted during the third month by the migration of neurons into the mariginal layer to form outer cellular layers as in the cerebellum and the corpora quadrigemina. As the neopallial walls increase in thickness, these outer cellular layers also become much thicker, and its neurons become stratified. By the end of the seventh month, six layers of neurons make up the basic pattern of the neopallium. This pattern is modified in different regions of the cortex, and about 100 different variations of the basic pattern have been described, each characteristic of a particular region.

The area of the cortex becomes greatly increased by the formation of numerous folds *(gyri)* and grooves *(sulci)* (Fig. 20–34). The gyri and sulci form a pattern on the surface of the cerebral hemispheres by means of which it may be divided into a number of different lobes. The *central sulcus* separates the *frontal* and *parietal lobes.* A *parietooccipital sulcus,* more prominent on the medial surface, and a line connecting it to the preoccipital notch defines the parietal and *occipital lobes.* The *lateral* (Sylvian) *fissure* separates the frontal and *temporal lobes,* and a line connecting the lateral fissue to the parietooccipital sulcus further defines the temporal lobe. The slower growing insula lies under the juncture of the temporal, parietal, and frontal lobes.

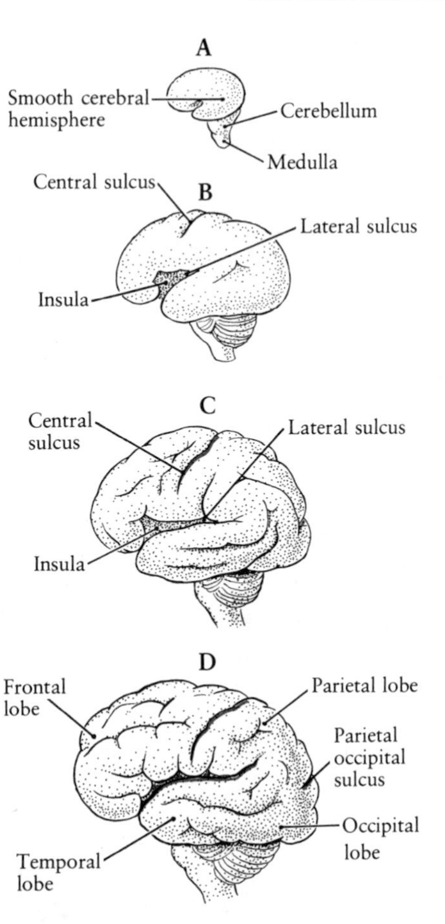


**FIG. 20–34.** Diagram of successive stages in the development of the surface of the cerebrum. A, 13 weeks; B, 26 weeks; C, 35 weeks; D, newborn.

*Commissures*

The two sides of the cerebrum are interconnected by three commissures that develop in relation to the thickened area of the lamina terminalis known as the commissural plate (Figs. 20–30 and 20–35). The first to develop is the anterior commissure, followed shortly by the hippocampal commissure as has already been described. The anterior commissure contains fibers connecting olfactory bulb derivatives. The last commissure to develop, and the largest, is the *corpus callosum* (Fig. 20–30 C and 20–35). Its fibers are from the neopallium, crossing over close to the hippocampal commissure. With the increase in the size of the neopallium and its continued expansion, the corpus callosum also increases rapidly in size and expands both cranially and caudally to form a large C-shaped structure in the center of the cerebrum (Fig. 20–35 C,D). The hippocampal commissure is carried caudally by

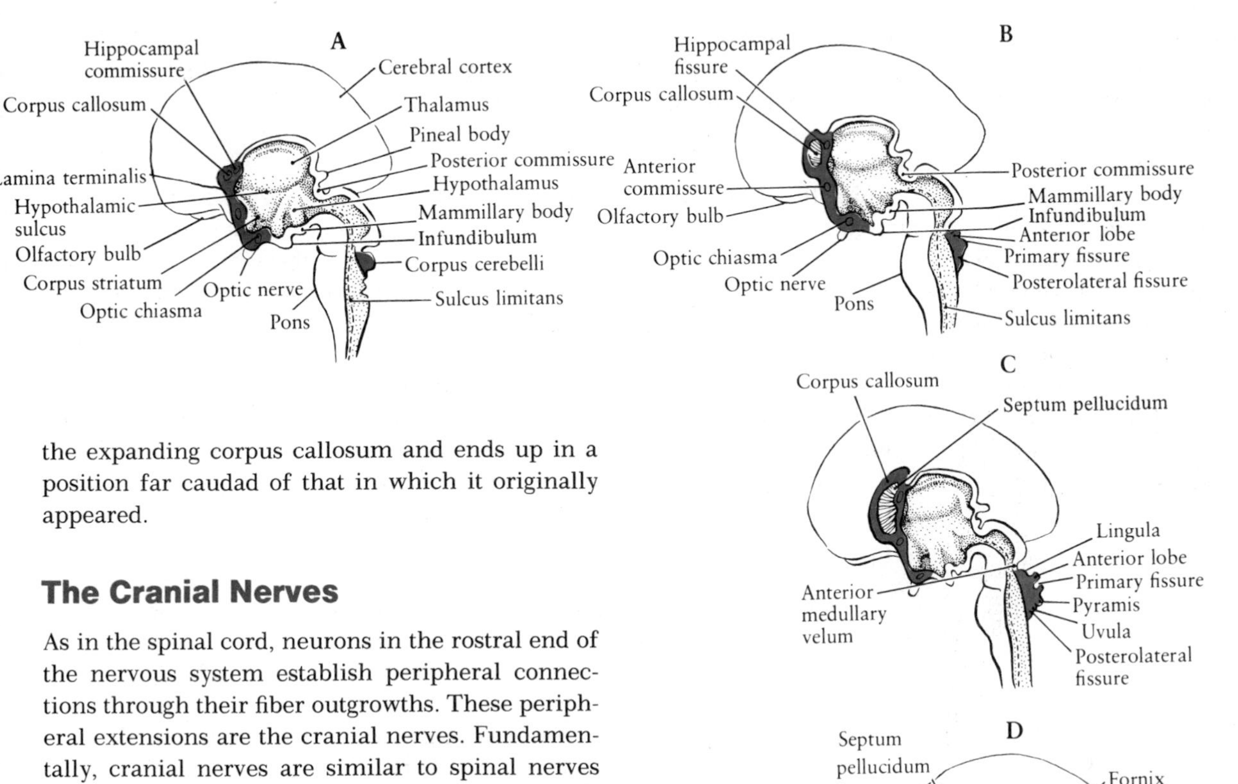


**FIG. 20–35.** Diagrams of sagittal sections through the brains of human embryos during the third to the fifth months of development. A, 11 weeks; B, 4 months; C, 4½ months; D, 5 months. (After E. L. House and B. Pansky, 1967. A Functional Approach to Neuroanatomy. McGraw-Hill Book Co., New York.)

the expanding corpus callosum and ends up in a position far caudad of that in which it originally appeared.

## The Cranial Nerves

As in the spinal cord, neurons in the rostral end of the nervous system establish peripheral connections through their fiber outgrowths. These peripheral extensions are the cranial nerves. Fundamentally, cranial nerves are similar to spinal nerves already discussed. Cranial motor fibers, which carry efferent impulses peripherally, are the axons of nerve cells in the brain stem. Cranial sensory fibers, which carry impulses centrally from sensory receptors, are the dendrites of nerve cells. The cell bodies of the afferent nerve fibers are not located within the brain stem but are found in cranial nerve ganglia, just as the afferent fibers in the spinal nerve are parts of neurons located in the posterior root ganglia. Fibers growing centrally from the cranial nerve ganglia make connections with specific cranial nuclei within the central nervous system. The cranial nerves differ from the spinal nerves in that they are not segmentally arranged; nor do they have anterior and posterior roots.

There are 12 pairs of cranial nerves, any one of which may carry one or more kinds of nerve fibers with different functional classifications. It is appropriate to look at the large nerve trunks that constitute the cranial nerves as convenient pathways over which nerve fibers associated with the cranial nuclei arrive at their appropriate terminations. Although fibers of different functional classification may use the same convenient route to their peripheral destination, when we examine their central connections we find that within the central nervous system there is a more strict separation into functional components, and central nervous system nuclei or tracts are segregated to serve only one or at the most two functional classifications. By central connections for efferent fibers we mean the cranial nerve nuclei from which the efferent nerve fibers take origin. Central connections of afferent fibers are those nuclei to which sensory fibers return, carrying impulses into the central nervous system from their peripheral receptors.

In the human embryo, the cranial nerves, their sensory ganglia, and their central connections are apparent early in the second month of development in the 10-millimeter embryo (Fig. 20–36). In discussing the cranial nerves, we will not take them up sequentially according to their number, but will consider them in relation to the functional classification of the fibers they carry. The peripheral pathways over which these fibers pass are of obvious importance and must be thoroughly learned in neuroanatomy. However, a study of the early development of the fibers and the establishment of their central and peripheral connections allows the embryologist to appreciate the functional aspect of the cranial nerves at a time when these connections are relatively unobscured.

The names and numbers of the cranial nerves and their sensory ganglia, when they are present, are listed below for reference:

| *Nerve Number* | *Name* | *Ganglion* |
|---|---|---|
| I | Olfactory | |
| II | Optic | |
| III | Oculomotor | |
| IV | Trochlear | |
| V | Trigeminal | Semilunar (Gasserian) |
| VI | Abducens | |
| VII | Facial | Geniculate |
| VIII | Acoustic | Spiral and Vestibular |
| IX | Glossopharyngeal | Superior and Petrosal |
| X | Vagus | Jugular and Nodose |
| XI | Accessory | |
| XII | Hypoglossal | |

Figure 20–36, to which continual reference should be made while studying the cranial nerves, is a diagrammatic representation of the cranial nerves in the 10-millimeter human embryo.

## General Somatic Efferent

There are four cranial nerves that carry general somatic efferent fibers: III, IV, VI, and XII. The nuclei of origin of the somatic efferent fibers may be considered to represent a rostral extension of the anterior column of the spinal cord into the brain stem. This is particularly evident when considering the hypoglossal nerve, whose rootlets can be seen emerging from the anterolateral sulcus in a position corresponding to the emergence of the anterior roots of the spinal nerves. The cranial nerve nuclei of the somatic efferent fibers, although they are considerably separated in their longitudinal aspect, are found to occupy similar positions in the brain stem close to the midline in the basal plate near the ventricular cavities.

*Oculomoter Nerve* (III)
The nucleus of origin of the third nerve is found in the basal plate of the mesencephalon at the level of the superior colliculus. Fibers emerge from the base of the mesencephalon in the region of the cephalic flexure and form a single trunk that passes to the developing muscle mass in the region of the orbit. As its name implies, the oculomotor nerve innervates the extrinsic eye muscles. Of the six eye muscles, it supplies the superior, inferior, and medial recti and the inferior oblique.

*Trochlear Nerve* (IV)
The nucleus of origin of the trochlear nerve also develops in the basal plate of the mesencephalon, but caudal to the motor nucleus of the third nerve—at the level of the inferior colliculus. The fibers from motor nucleus IV do not leave the brain stem directly but pass dorsally around the cerebral aqueduct, decussate in the dorsal part of the mesencephalon, and leave the midbrain in the region of the isthmus. The trochlear nerve is a slender nerve that innervates the superior oblique eye muscle.

*Abducens Nerve* (VI)
The abducens nucleus is the most caudal of those nuclei giving rise to fibers that pass to the extrinsic eye muscles. It lies in the floor of the fourth ventricle, and its fibers form a nerve that emerges just caudad of the pons and passes to the lateral rectus eye muscle. The name of the nerve indicates the action of this muscle.

*Hypoglossal Nerve* (XII)
The XIIth cranial nerve arises from a long slender nucleus located close to the midline stretching all through the myelencephalon. The rootlets of the hypoglossal nerve emerge all along the sides of the medulla to form the main trunk of the nerve, which supplies all of the intrinsic and most of the extrinsic muscles of the tongue.

## Special Visceral Efferent

The cranial nerve nuclei, which in the embryo are just lateral to the somatic efferent nuclei, form a broken longitudinal column lying in the anterolat-

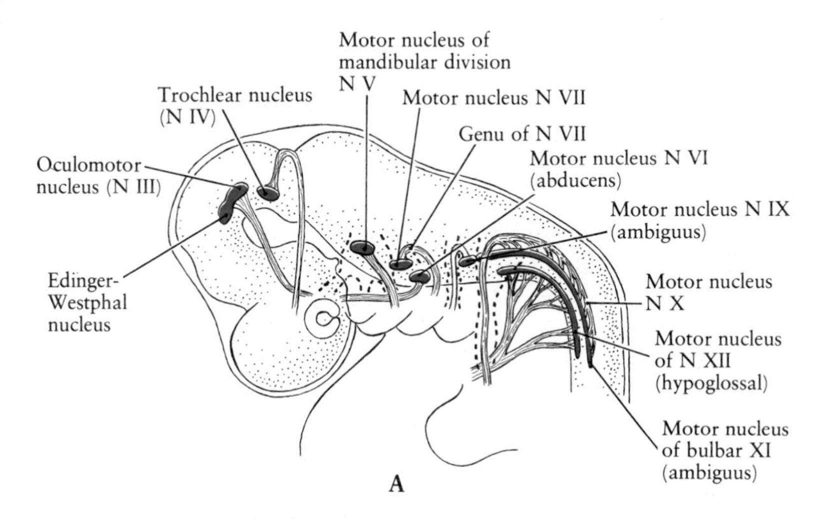


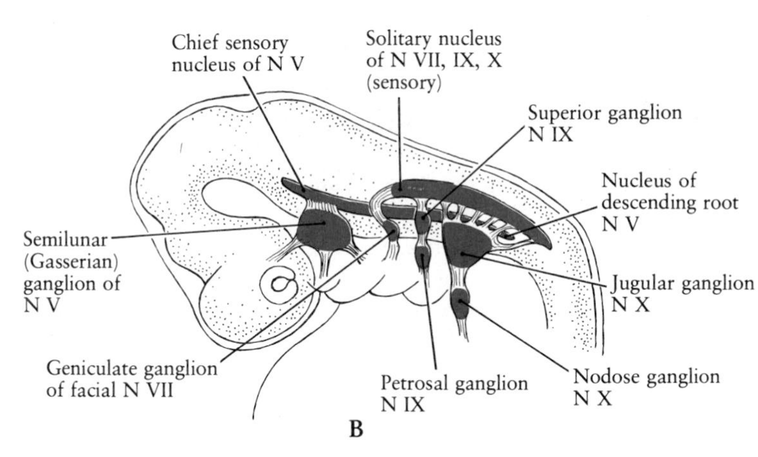


**FIG. 20–36.** The cranial nerves, their central connections, and their sensory ganglia in a human embryo of about six weeks of age. A, nuclei giving rise to motor fibers; B, nuclei with which sensory fibers make central connections. (After B. Patten, 1947. Human Embryology. McGraw-Hill Book Co., New York.)

eral region of the pons and the medulla. They are the nuclei of origin of the special visceral efferent fibers found in cranial nerves V, VII, IX, X, and XI.

*Trigeminal Nerve* (V)
Cranial nerve V supplies motor fibers to the muscles derived from the first visceral arch. These muscles will develop into the muscles of mastication; in the adult, all of the masticator muscles are supplied by the Vth nerve. The special visceral efferent fibers of this nerve arise in the motor nucleus of the trigeminal nerve developed in the reticular formation of the pons. They run in the mandibular division of the nerve.

*Facial Nerve* (VII)
Cranial nerve VII supplies motor fibers to muscles formed from the second visceral arch. Collectively they form the muscles of facial expression. The motor nucleus of the VIIth cranial nerve is located in the caudal region of the pons. The nucleus of the facial nerve first develops caudad of the nucleus of the abducens nerve but then undergoes a change in position relative to the abducens nucleus, moving dorsally and cranially. As a result of this migration, the motor fibers of the facial nerve pursue an arched course around the abducens nucleus, forming what is known as the *genu* of the facial nerve.

*Glossopharyngeal, Vagus, and Spinal Accessory Nerves* (IX, X, and XI)
The special visceral efferent fibers of these three nerves arise from a single nucleus, the *nucleus ambiguus*, extending throughout the length of the myelencephalon. Fibers running in the glossopharyngeal nerve supply muscles of the pharynx derived from the third visceral arch. Fibers of the vagus nerve supply muscles of the pharynx and larynx

derived from the fourth and fifth arches, and the spinal accessory nerve supplies certain neck and shoulder muscles.

The vagus and spinal accessory nerves, considered separate nerves in the adult, form part of a single complex in the embryo. The accessory nerve has both a bulbar and a spinal portion. Fibers from the bulbar portion arise from the nucleus ambiguus and are distributed with the fibers of the vagus nerve to muscles derived from the caudal visceral arches. Fibers of the spinal portion of the nerve arise in the lateral part of the anterior gray column of the first five or six cervical segments of the spinal cord. The fibers from these segments emerge from the lateral surface of the cord, ascend alongside the cord through the foramen magnum, and, joining the fibers of the bulbar division, emerge with them from the cranial cavity. However, they soon leave the common trunk as the external branch of the spinal accessory nerve and supply the sternocleidomastoid and trapezius muscles. These muscles are considered to be of branchiomeric origin and, therefore, the fibers in the accessory nerve supplying them are classified as special visceral efferent.

### General Visceral Efferent

General visceral efferent fibers in the cranial nerves are a part of the parasympathetic division of the autonomic nervous system. As do the autonomic nervous system fibers of the spinal nerves, these fibers arising in the brain represent the first or preganglionic fibers of a two-neuron relay to the structure innervated. Preganglionic fibers of the parasympathetic nervous system are found in cranial nerves III, VII, IX, X, and XI.

*Oculomotor Nerve* (III)
The origin of the preganglionic fibers is in a special nucleus (the Edinger–Westphal nucleus) that lies in the rostral part of the motor nucleus of the oculomotor nerve in the basal plate of the mesencephalon. These fibers run in the IIIrd nerve to the *ciliary ganglion* from which postganglionic fibers supply the circular muscles of the iris and the ciliary muscle of the lens.

*Facial Nerve* (VII)
These fibers originate in the *superior salivatory nucleus* and supply the lacrimal gland and the submaxillary and sublingual salivary glands. The *sphenopalatine ganglion* contains the cell bodies of the postganglionic fibers to the lacrimal gland, and the *submaxillary ganglion* contains those whose fibers supply the salivary glands.

*Glossopharngeal Nerve* (IX)
These fibers supply the parotid gland. Preganglionic fibers arising in the *inferior salivatory nucleus* in the basal plate of the myelencephalon pass to the *otic ganglion,* whose postganglionic fibers then run to the salivary gland.

*Vagus and Spinal Accessory Nerves* (X, XI)
The dorsal motor nucleus of the vagus nerve in the basal plate of the myelencephalon is the point of origin for the preganglionic fibers running in these nerves to be distributed to the thoracic and abdominal viscera.

### General and Special Visceral Afferent

The general and special visceral afferents are sensory fibers from visceral organs, and they run in cranial nerves I, VII, IX, and X. Visceral afferent fibers in nerves VII, IX, and X include (1) fibers from the taste buds, mainly in VII and IX but a few in X, classified as special visceral afferent; and (2) fibers from the alimentary tract and other thoracic and abdominal viscera, classified as general visceral afferent. All of the visceral afferent fibers in these three nerves (VII, IX, and X) enter the solitary tract in the rhombencephalon and terminate in the nucleus of the solitary tract. The nucleus of the solitary tract starts in the myelencephalon and extends as far rostrad as the motor nucleus of the Vth nerve.

*Facial Nerve* (VII)
The sensory ganglion of the seventh nerve is the *geniculate* ganglion. Special visceral afferent fibers from it supply the taste buds on the anterior two thirds of the tongue. General visceral afferent fibers from the geniculate ganglion neurons supply the oral and pharyngeal mucosa.

*Glossopharyngeal Nerve* (IX)
Cranial nerve IX has two sensory ganglia. The one lying closer to the brain is the *superior* and the more distal one is the *petrosal.* The latter contains the cell bodies of origin of the special visceral afferent fibers to the taste buds on the posterior third of the tongue and the general visceral afferent fi-

bers from visceral receptors in the pharyngeal mucosa. The central connections of the VIIth and IXth nerves are with the most rostral portions of the nucleus of the solitary tract.

*Vagus Nerve* (X)
The vagus nerve also has two ganglia along its trunk, a proximal *jugular* and a more distal *nodose*, of which the latter contains the cell bodies of origin of the special visceral afferent fibers. A few fibers of the vagus nerve supply the taste buds of the pharyngeal mucosa and the epiglottis (special visceral afferent), but the vast majority of the vagus fibers convey general visceral afferent fibers from the alimentary tract and the thoracic and abdominal viscera. All central connections are to the solitary tract and its nucleus.

*Olfactory Nerve* (I)
Although the nasal placode from which the olfactory organ develops is of ectodermal origin, the Ist nerve is classified as special visceral afferent because of the general connection of the sense of smell with other mainly alimentary functions. However, the central terminations of the olfactory fibers are not with the nucleus of the solitary tract, but are in the olfactory bulb in the most rostral part of the telencephalon. Unlike other sensory nerves, the olfactory nerve does not have a ganglion. Its fibers develop from bipolar cells in the olfactory epithelium of the nose. The distal processes of these fibers protrude as bristles above the epithelial surface, and the proximal processes pass through the ethmoid bone to the olfactory bulb where they synapse with the dendrites of *mitral cells* (Fig. 20–37). Mitral cells are the second neurons in a chain conveying olfactory stimuli. Their axons pass through the olfactory tract to various terminations within the rhinencephalon.

**FIG. 20–37.** Diagram of the fibers from the olfactory epithelium passing through the cribiform plate of the ethmoid bone to synapse, in the olfactory bulb, with dendrites of the mitral cells. The axons of the mitral cells form the olfactory tract.

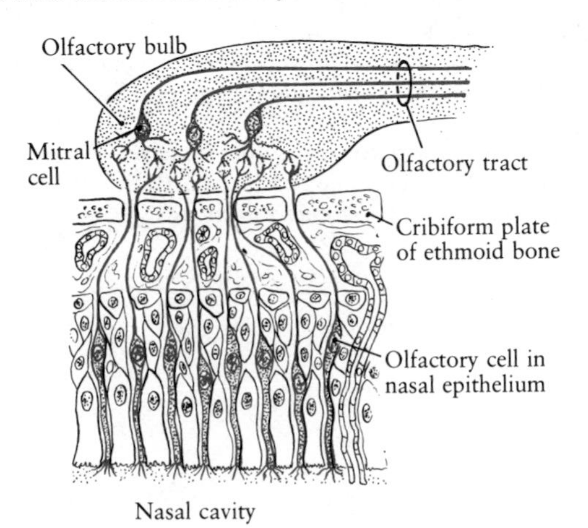


## General Somatic Afferent

General somatic afferent nerve fibers carry sensations from somatic receptors of the head region to their central connections within the brain stem. These fibers are found in cranial nerves V, VII, IX, and X. The general somatic afferent fibers in all four of these cranial nerves terminate in the sensory nucleus of the Vth nerve.

*Trigeminal* (V)
The trigeminal is the main general somatic sensory cranial nerve. It has a large ganglion, the *trigeminal*, distad of which the nerve splits into its three divisions, the *mandibular*, the *maxillary*, and the *ophthalmic*. Centrally, the nerve fibers from the trigeminal ganglion terminate either in the main sensory nucleus of the Vth nerve in the pons or, after descending in the spinal tract of the Vth nerve, in the nucleus of the spinal tract, which is continuous cranially with the main sensory nucleus in the brain. Caudally, the nucleus of the spinal tract is continuous with the part of the posterior column of the spinal cord that receives somatic sensations from the neck and body regions.

*Facial, Glossopharyngeal, and Vagus Nerves* (VII, IX, and X)
These three cranial nerves each carry a few general somatic afferent fibers from the region of the external auditory meatus and the ear. The ganglia in which their cell bodies are located are the geniculate, the superior, and the jugular, respectively.

## Special Somatic Afferent

These fibers convey sensations from the organs of special sense of the head region, which are of ectodermal origin. Their fibers are found in cranial nerves II and VIII.

*Optic Tract* (II)

Since the sensory receptor of the eye, the retina, is actually an extension of the brain, the optic nerve is really a tract within the brain itself. Fibers in the optic tract are those from the third set of neurons in the visual chain, and they carry visual impulses from the rods and cones to terminations mainly in the lateral geniculate body and the superior colliculi. The development of the eye and its visual receptors will be considered in the next chapter.

*Acoustic Nerve* (VIII)

The acoustic nerve carries fibers both from the organ of hearing, the *cochlea,* and the organ of equilibrium, the *semicircular canals.* Fibers from the cochlea have their cells located in the *acoustic (spiral) ganglion* closely associated with the cochlear duct and make central connections with the cochlear nucleus in the hindbrain. Those nerve fibers that carry impulses from the semicircular canals arise in the *vestibular ganglion* located in the internal auditory meatus and make central connections with a number of vestibular nuclei located in the myelencephalon and the metencephalon. Further consideration will be given to this nerve in the discussion of the development of the ear and the semicircular canals.

## References


Angevine, J. B., Jr., D. Bodian, A. J. Coulombre, M. V. Edds, Jr., V. Hamburger, M. Jacobson, K. M. Lyser, M. C. Prestige, R. L. Sidman, S. Varon, and P. A. Weiss. 1970. Embryonic vertebrate central nervous system: revised terminology. Anat. Rec. 166:257–262.

Bronner-Fraser, M. E. and A. M. Cohen, 1980. The neural crest: What can it tell us about cell migration and determination? In: Current Topics in Developmental Biology, Vol. 16. A. A. Moscona and A. Monroy, eds. New York: Academic Press.

Detwiler, S. R. 1936. Neuroembryology. New York: Macmillan.

Duncan, D. 1957. An electron microscope study of the embryonic neural tube and notochord. Texas Rep. Biol. Med. 15:367–377.

Harrison, R. G. 1907. Observations on the living developing nerve fiber. Anat. Rec. 1:116–118.

Hollyday, M. 1980. Motoneuron histogenesis and the development of limb innervation. In: Current Topics in Developmental Biology, Vol. 16. A. A. Moscona and A. Monroy, eds. New York: Academic Press.

Hunt, R. K. and M. Jacobson. 1974. Neural specificity revisited. Curr. Topics Dev. Biol. 8:203–259.

Jacobson, M. and R. K. Hunt. 1973. The origin of nerve cell specificity. Sci. Am. 228:26–35.

Langman, J., R. L. Guerront, and B. G. Freeman. 1966. Behavior of neuroepithelial cells during the closure of the neural tube. J. Comp. neurol. 127:399–412.

Le Douarin, N. 1980. Migration and differentiation of neural crest cells. In: Current Topics in Developmental Biology, Vol. 16. A. A. Moscona and A. Monroy, eds. New York: Academic Press.

Sauer, F. C. 1935. Mitosis in the neural tube. J. Comp. Neurol. 62:377–405.

Sperry, R. W. 1959. The growth of nerve circuits. Sci. Am. 201:68–75.

Sperry, R. W. 1965. Embryogenesis of behavioral nerve nets. In: Organogenesis. R. L. DeHaan and H. Ursprung, eds. New York: Holt, Rinehart and Winston.

Summerbell, D. and R. V. Sterling. 1980. The innervation of dorso-ventrally reversed chick wings: evidence that motor axons do not actively seek out their appropriate targets. J. Embryol. exp. Morphol. 61:233–247.

Summerbell, D. and R. V. Sterling. 1982. Development of the pattern of innervation of the chick limb. Am. Zool. 22:172–184.

Szekely, G. and G. Czeh. 1967. Localization of motoneurons in the limb-moving spinal cord segments of *Amblystoma.* Acta Physiol. Acad. Sci. Hung. 32:3–18.

Vaughn, J. E. and A. Peters. 1971. The morphology and development of neuroglia cells. In: Cellular Aspects of Neural Growth and Differentiation, pp. 103–140. D. C. Pease, ed. Los Angeles: University of California Press.

Watterson, R. L., P. Veneziano, and A. Barth. 1956. Absence of a true germinal zone in neural tube of young chick embryos as demonstrated by the colchicine technique. Anat. Rec. 124:379.

Weiss, P. 1934. In vitro experiments on the factors determining the course of the outgrowing nerve fiber. J. Exp. Zool. 68:393–448.

Weiss, P. 1936. Selectivity controlling the central-peripheral relations in the nervous system. Biol. Rev. 11:494–531.

Weiss, P. 1955. Neurogenesis. In: Analysis of Development. B. H. Willier, P. Weiss, and V. Hamburger, eds. Philadelphia: W. B. Saunders Company.

# 21

# THE SENSE ORGANS

## The Eye

The development of the eye presents a fascinating but at the same time a complicated picture in that its component parts are of diverse origin. The optic cup and its sensory receptor area, the *retina,* are actually a part of the brain (an evagination of the forebrain), and as was mentioned in the discussion of the cranial nerves, the optic nerve then is really a tract within the brain. The *lens,* which focuses the light rays on the retina, is formed as an ingrowth from the ectoderm overlying the optic cup. Head mesoderm forms the tough fibrous coat, the *sclera* (which gives the eyeball its shape) and the vascular *choroid coat* as well as the transparent *cornea.* Mesoderm, which corresponds to head somite mesoderm of lower forms, develops the extrinsic eye muscles that attach to the sclera and move the eyeball. Finally, the eyelids are formed from folds of the skin after most of the other parts of the eye have been established. These heterogeneous components must be integrated into a definitive visual organ whose parts must match precisely. It is not surprising that during this process we will find some excellent samples of dependent differentiation, which we will discuss after we have described the morphogenesis of the eye.

### Optic Vesicle and Early Optic Cup

The neural part of the eye first appears early in the third week as a pair of *optic grooves* on either side of the midline at the expanded cranial end of the still-open neural folds (Fig. 21–1 A). As the neural folds close, the optic grooves deepen to form the *optic vesicles* as lateral evaginations of the brain wall, each with a cavity continuous with the central canal of the neural tube (Fig. 21–1 B).

During the fourth week, the optic vesicle closely approaches the surface ectoderm and its outer (distal) wall thickens and begins to invaginate, the first step in the conversion of the optic vesicle into the *optic cup* (Fig. 21–1 C). The distal portion of the optic vesicle as it invaginates then becomes the inner layer of the optic cup, the portion that will form the *retinal layer* of the eye. A constriction develops between the optic cup and the brain marking the beginning of the *optic stalk* (Fig. 21–1 C,D).

When the optic vesicle approaches the surface, the ectoderm overlying it thickens to form the *lens placode.* This thickening occurs early in month 2 (4.5 mm), at which time the lens placode consists of a number of layers of cells (Fig. 21–1 C). When the optic cup forms, the lens placode invaginates into it to form the *lens vesicle* (Fig. 21–1 D). Shortly

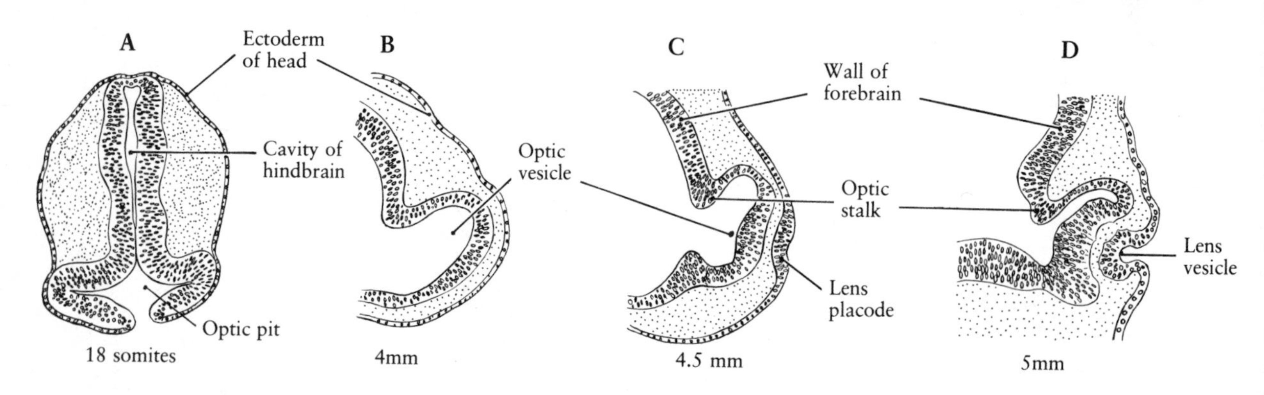


thereafter the lens vesicle becomes closed (Fig. 21–1 E); by the sixth week it breaks away from the surface epithelium and appears as a rounded vesicle lying within the cavity of the optic cup. By the end of the sixth week, the cavity of the lens vesicle is excentrally placed owing to the beginning of the differentiation of the deeper (inner) region of the lens vesicle (Fig. 21–1 F).

The invagination that converts the optic vesicle into the optic cup does not occur in the exact center but is also extended to the midventral line. Because this invagination is excentric and because the distal ventral portion of the optic vesicle stops growing while the other margins continue to expand, the midventral wall of the optic cup shows a defect, the *optic (retinal) fissure* (Fig. 21–2). Blood vessels that develop in the nearby mesenchyme form the *hyaloid artery,* which enters the optic fissure to supply the inner surface of the optic cup and the lens. The hyaloid vein drains these areas. These vessels provide an intraocular vascular system for the developing eye (Fig. 21–3). However, this system atrophies completely at a later time, and a new intraocular supply develops. In addition to allowing entrance to blood vessels, the optic fissure provides a shortcut for the return of the axons of the retinal neurons to the diencephalon. Without benefit of this ventral defect, the nerves would have to reach the optic stalk by running out to the margin of the optic cup (Fig. 21–4). The optic fissure is not a permanent fixture but is obliterated during the second month of development by the overgrowth of its margins, first in the optic cup and later in the optic stalk.

## Differentiation of the Optic Cup

At the end of the fifth week, the double-walled optic cup extends only a short distance past the equator

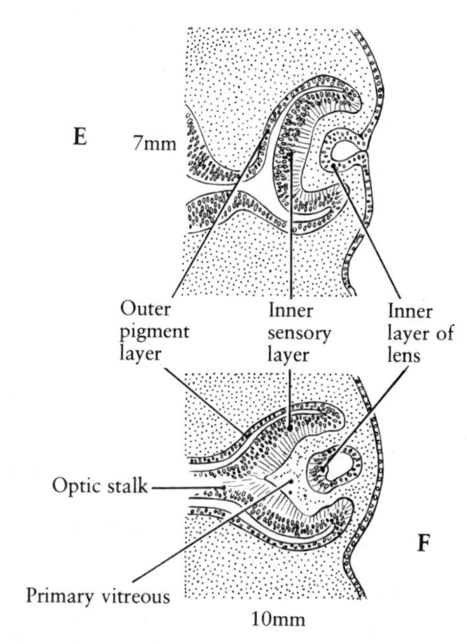


**FIG. 21–1.** Early development of the lens and the optic cup. (From I. Mann, 1964. The Development of the Human Eye. Grune & Stratton, New York.)

of the lens (Fig. 21–2). The inner layer differentiates just like the wall of the neural tube—which, of course, it is. Owing to the invagination that has taken place, however, its layers are reversed, the ependymal layer being outermost and the mantle layer innermost. Further forward growth of the outer margin of the optic cup over the lens marks the beginning of the formation of a circumferential, marginal, non-nervous part of the cup that will overlie the outer region of the lens. This thinner non-nervous area is called the *pars caeca retinae* and it is separated from the nervous, light-sensitive region of the retina, the *parts optica retinae,* by a depression, the *ora serrata* (Fig. 21–5). The gap made by the outer circumference of the optic cup is the *pupil.*

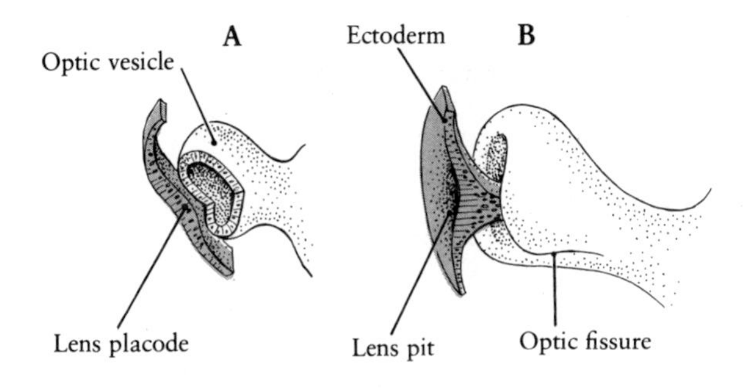


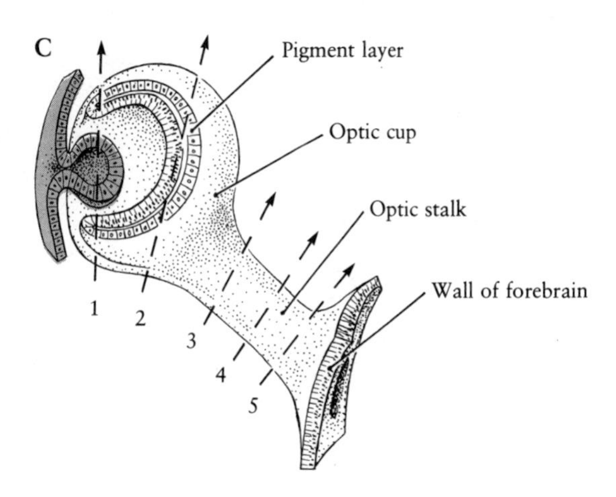


FIG. 21-2. Models of the lens and the optic cup showing the formation of the optic fissure. The lens is shown as sectioned. In A and C part of the optic cup has been cut away. A, 4.5 mm; B, 5.5 mm; C, 7.5 mm. Arrows and numbers refer to Figure 21-3. (From I. Mann, 1964. The Development of the Human Eye. Grune & Stratton, New York.)

The outer layer of the optic cup is always much thinner than the inner layer; early in its development it acquires granules and soon becomes densely pigmented. This pigmented layer extends all the way to the pupillary margin of the cup covering both the pars optica retinae and the pars caeca retinae (Fig. 21–5).

## Histogenesis of the Pars Optica Retinae

The inner layer of the pars optica retinae is surrounded by a cavity that corresponds to the lumen of the neural tube; thus, its outer layer represents the ependymal and ventricular layers of the neural tube, and its inner layer corresponds to the marginal layer of the tube. Toward the end of month 2 (17 mm), cells from the outer layer start to migrate into the inner layer to form an *inner neuroblastic layer*. This migration begins first in the region of the posterior pole; the cells that remain in the outer layer form the *outer neuroblastic layer* (Fig. 21–6). A basement membrane (external limiting membrane) is found outside of the outer neuroblastic layer. The cells of the outer neuroblastic layer will differentiate into cells whose outer ends form the *rods* and *cones,* the visual receptors. The protoplasmic processes of these cells extend beyond the limits of the external limiting membrane in the direction of the pigment layer. By the end of the seventh month, the shape of these processes serves to distinguish between the two types of cells (Fig. 21–7).

Some cells of the inner nuclear layer migrate still further inward to form a layer of ganglion cells. The axons of these ganglion cells form a fibrous layer over the inner surface of the retina, the fibers converging from their points of origin toward the optic stalk. The cells remaining in the inner nuclear layer become bipolar neurons that conduct impulses from the rods and cones to the ganglion cells. A schematic representation of this system is diagrammed in Figure 21–8. The region of the retina that lies in the direct visual axis is known as the *macula*. In the macula, only cones are present. The macula lies in the center of a shallow depression, the *fovea centralis*.

The cavity of the original optic vesicle, between the inner and outer layers, is gradually obliterated.

FIG. 21-3. Cross sections of the optic cup and lens at the levels indicated by the arrows in Figure 21-2. (From I. Mann, 1964. The Development of the Human Eye. Grune & Stratton, New York.)

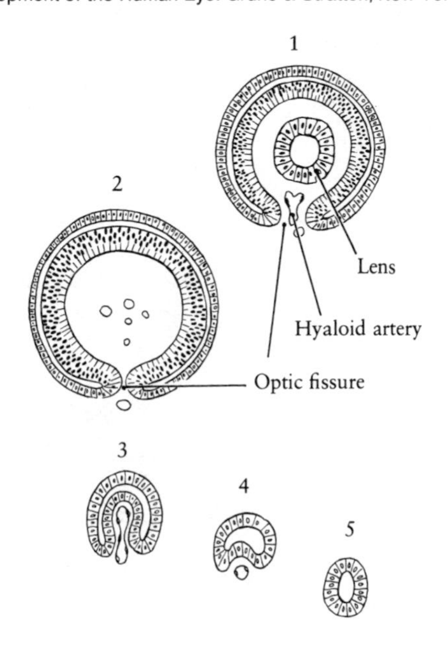

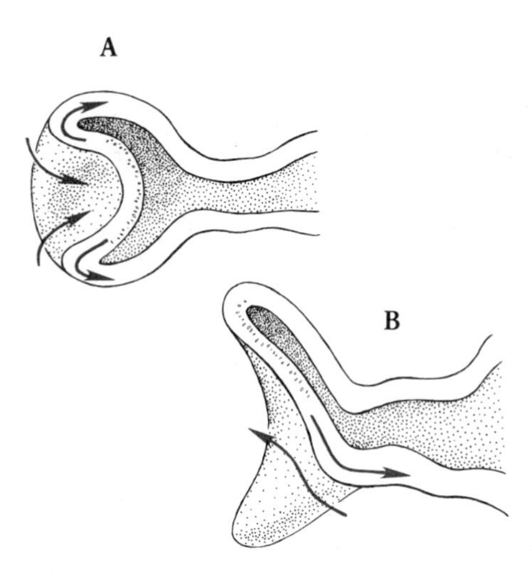


FIG. 21-4. A, course the optic nerve fibers and blood vessels would have to take in the absence of an optic fissure. Both would have to run around the margin of the optic cup; B, when the optic fissure develops, the nerve fibers and the blood vessels can take the shortcut indicated by the arrows.

These layers become closely associated with each other, with protoplasmic processes from the pigment layer interspersed about the processes of the rods and cones. The pigment granules migrate into and out of the processes of the pigment cells, to screen the photoreceptors from bright light in the former case and to allow maximum sensitivity in dim light in the latter. Despite this close association, the two layers do not fuse and a potential intraretinal space persists as a remnant of the ventricular system representing a potential plane of cleavage in which retinal separation can occur following injury to the eye.

## Development of the Pars Caeca Retinae

### *The Pars Ciliaris Retinae*

The double-layered *pars ciliaris retinae* is the region of the pars caeca retinae just peripheral to the ora serrata (Fig. 21–5). During the third month, as it is invaded by the overlying mesoderm during the formation of the *ciliary body,* it becomes thrown into folds (Fig. 21–9). These folds, each consisting of an inner unpigmented and an outer pigmented layer of simple columnar cells, become arranged radially projecting toward the lens to form the *ciliary processes* (Fig. 21–10). The mesoderm of the ciliary body develops the ciliary muscle fibers. Contraction of the ciliary muscle acts to modify (thicken) the lens by relaxing the tension of the *suspensory ligament,* which connects the tips of the ciliary processes to the capsule of the lens.

### *The Pars Iridica Retinae*

Peripheral to the ciliary body is the *pars iridica retinae.* Both of its layers become pigmented, and the pars iridica retinae, plus the mesoderm associated with it, differentiates into the *iris.* A thin layer of mesoderm adherent to the outer layer of the pars iridica retinae extends beyond the inner margin of the iris over the surface of the lens forming the *pupillary membrane* (Fig. 21–9). This membrane is

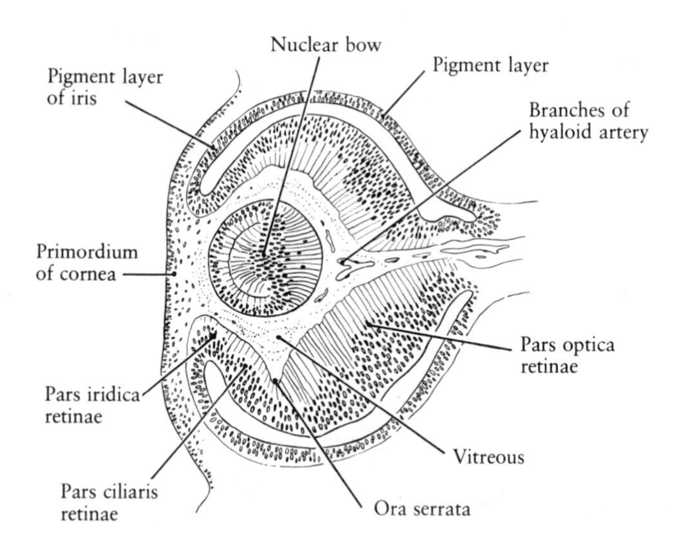


FIG. 21-5. Section of the eye of a human embryo during the sixth week of development (11.5 mm). The ora serrata divides the nervous from the non-nervous region of the optic cup. (After A. Fischel, 1929.)

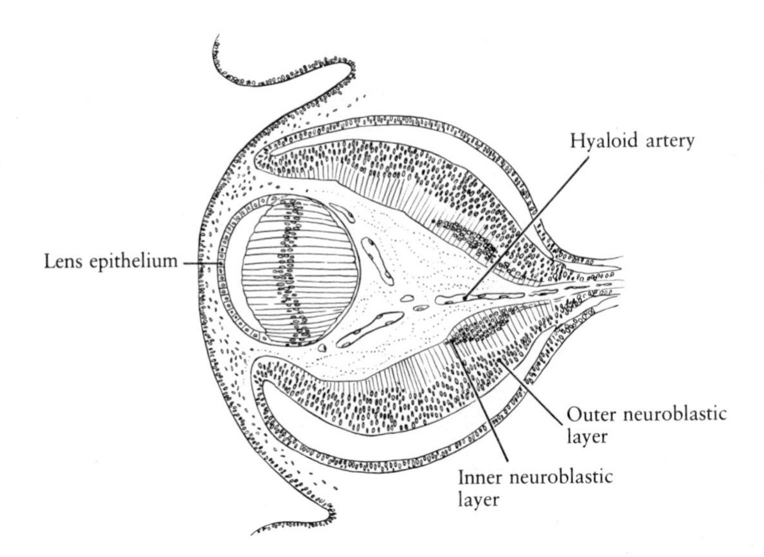


**FIG. 21–6.** Section through the eye of a human embryo at seven weeks of development (17 mm). The inner neuroblastic layer of the retina is beginning to form in the region of the optic stalk. (From I. Mann, 1964. The Development of the Human Eye. Grune & Stratton, New York.

resorbed before birth. Smooth muscle fibers develop outside of the pars iridica retinae in the mesoderm covering it. Eventually, they differentiate into the dilator and constrictor muscles of the iris (Fig. 21–10 B). They have long been considered to be formed from the outer layer of the pars iridica retinae rather than from the overlying mesoderm. In this case, this constitutes one of the few exceptions to the general condition that muscle tissue is of mesodermal origin. The sphincter muscles form a group close to the margin of the iris, and the dilator muscles form a group of radially oriented muscles closer to the ciliary body (Fig. 21–10).

**FIG. 21–7.** Section showing developing rods and cones at about seven months. (After I. Mann, 1964. The Development of the Human Eye. Grune & Stratton, New York.

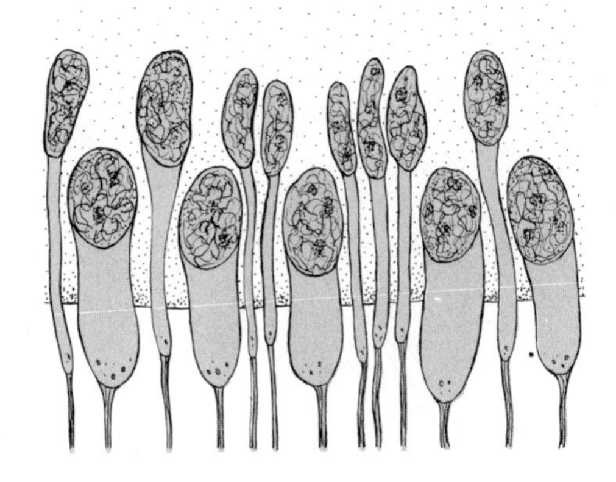

## Development of the Lens

We have followed the development of the lens from the formation of the lens placode up to the time when it forms a lens vesicle cut off from its ectodermal origin, lying within the opening of the optic cup. The beginning of the differentiation of the lens fibers was indicated as taking place during the sixth week when the cells of the inner wall of the lens vesicle grow toward the outer wall, gradually obliterating the cavity of the lens vesicle (Fig. 21–1 F). Only the cells of the inner portion of the lens vesicle undergo this differentiation to form the *primary lens fibers*. The cells of the outer wall remain as a simple cuboidal epithelium forming the *anterior lens epithelium*. The nuclei of the primary lens fibers move to the equator of the lens, and form a line that is convex outwardly, the *nuclear bow* (Fig. 21–11). Each fiber is a single cell that stretches from the inner to the outer pole of the lens, with its nucleus resting approximately on the equator. These primary lens fibers form the core of the lens. Further development consists of multiplication of nuclei in the equatorial region, which forms the *secondary lens fibers*. As each fiber forms, it grows meridionally in both directions from its equatorial nucleus toward either pole. However, new fibers fail to grow the complete distance toward each pole, and, as a result, the meeting place of the lens fibers is not at the two opposite poles but in an irregular line known as the *lens suture* (Fig. 21–9). New lens fibers continue to be added to the

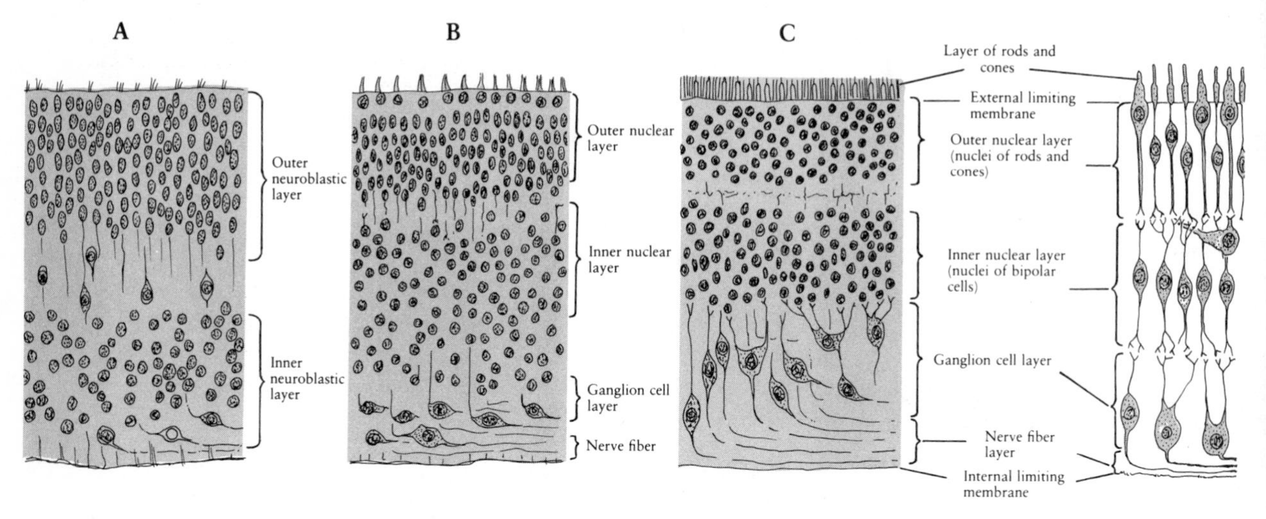


**FIG. 21–8.** Stages in the histogenesis of the retina. A, about 7 weeks (17 mm); B, about 11 weeks (65 mm); C, about 27 weeks (250 mm). (From I. Mann, 1964, The Development of the Human Eye. Grune and Stratton, New York.)

lens up until about the 20th year of life. The lens sutures change as new fibers are added and show a complex pattern with increasing age.

## Accessory Structures

*The Choroid and Sclera*

When the optic vesicle first forms, the mesoderm surrounding it is loosely arranged as a network of mesenchymal cells showing no differentiation into cartilage, muscle, or connective tissue. The first sign of differentiation is the appearance of a vascular network formed as a continuation of the developing internal ophthalmic artery (a branch of the internal carotid). A part of the vascular bed develops into the hyaloid artery, already described as entering the optic fissure forming the intraocular vascular system. An extraocular system also develops and during the second month shows conspicuous blood vessels. This vascular layer, completely covering the optic cup, constitutes the *choroid coat* (Fig. 21–12).

Outside of the vascular choroid, the *sclera* develops as a fibrous connective tissue covering condensed from the surrounding mesenchyme. The sclera is continuous with the dura mater of the

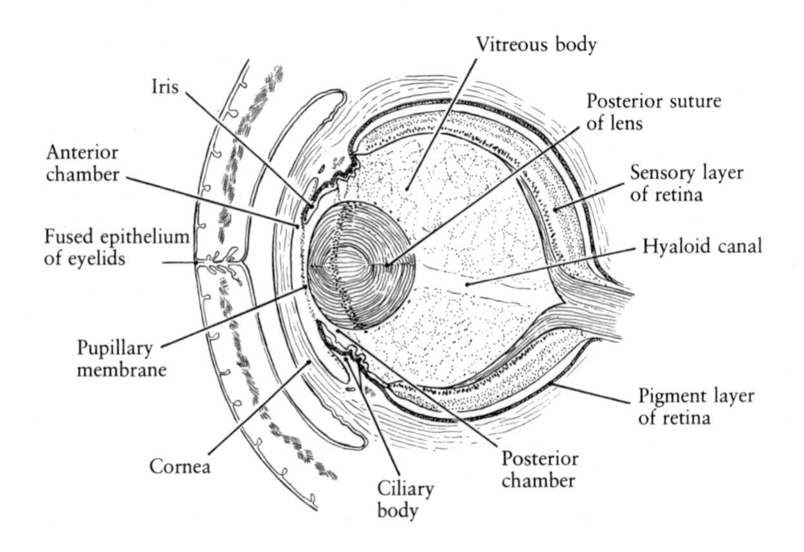


**FIG. 21–9.** Cross section of an eye from a fetus of about 19 weeks (174 mm) showing fused eyelids and developing eyelashes.

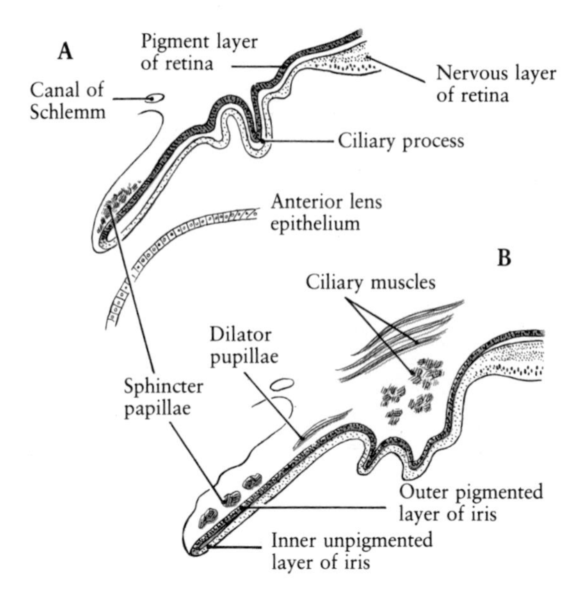


**FIG. 21–10.** Development of the pars caeca and the muscles of the iris and the ciliary body.

brain centrally over the optic tract. This tough connective tissue layer molds the eyeball and also provides attachment for the extrinsic eye muscles (Fig. 21–12).

*The Vitreous Body*

The area between the lens and the inner wall of the optic cup is occupied in the adult by a network of fibers whose interstitial spaces are filled with a gelatinous material. This is the *vitreous body* (Fig. 21–12). Very early in the formation of the lens placode and the optic cup, the space between them is occupied by a few mesodermal cells (Fig. 21–1 D). As the lens vesicle and the optic cup develop further, the space between them becomes larger and contains a few mesodermal cells interspersed between fibrils connecting the lens and the optic cup. Later, when blood vessels appear in this area, the ectodermal lens fibrils and the mesodermal cells become associated with the blood vessels to form the primary vitreous (Fig. 21–1 D,E). It is not possible to distinguish whether the primary vitreous is of ectodermal or mesodermal origin, or both. Later, when the space between the lens and the optic cup enlarges further, the definitive secondary vitreous forms. Again, it appears that it is partially of ectodermal origin, consisting of fibrous contributions from the inner retinal wall as well as contributions from the atrophy of the hyaloid artery.

**FIG. 21–11.** Eye of a human embryo of about seven weeks (17 mm) at the time the eyelids are beginning to develop. (After A. Fischel, 1929.)

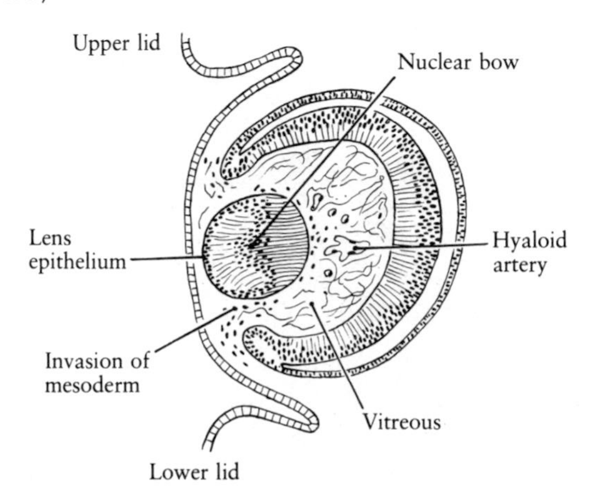


*The Cornea*

The cornea is a continuation of the sclera over the outer surface of the lens (Figs. 21–9 and 21–12). Proliferation of the mesoderm over the outer surface of the lens leaves a cavity between the cornea and the lens. This is the *anterior chamber* of the eye. The anterior chamber of the eye is separated from the outer lens epithelium by the pupillary membrane (Fig. 21–9).

Of course, both the cornea and the lens must be transparent in order to transmit the light rays through to the retina. It is generally stated that the cells of these structures can carry out their function only after they acquire transparency during development. Since embryonic tissues are generally already transparent, it should rather be stated that these particular tissues retain their embryonic condition of transparency.

The pupillary membrane, as mentioned previously, is a temporary structure that is normally resorbed during the sixth month. Following its resorption, the anterior chamber then becomes confluent with the space between the iris, the lens, and the suspensory ligament. This space is the *posterior chamber* of the eye (Fig. 21–9). Both the anterior and the posterior chambers are filled with a fluid *aqueous humor*.

The cornea and the sclera show the same radius of curvature in early stages. However, beginning at four months, the radius of curvature of the cornea decreases and, as a result, the corneoscleral bound-

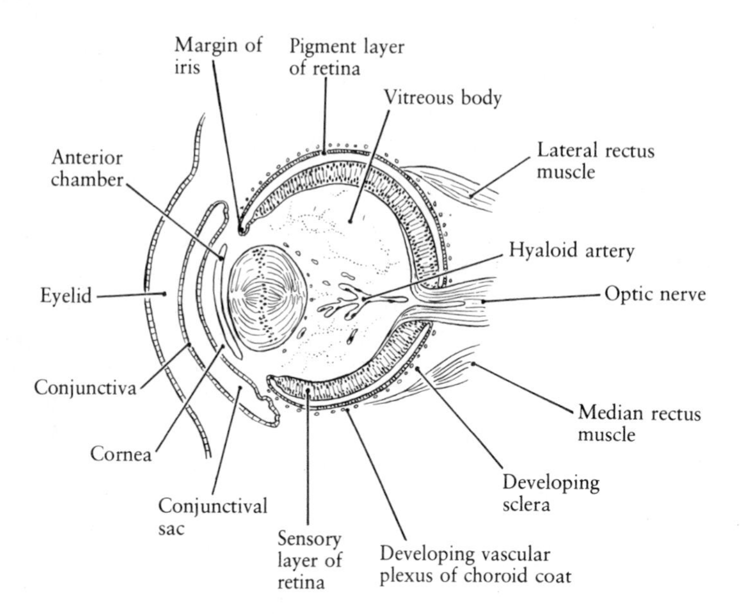


**FIG. 21–12.** Section of the eye of a human fetus at about 11 weeks (48 mm).

ary is easily distinguished. Failure of development of the correct curvature of the cornea results in improper focusing of the image on the retina. The entire focusing apparatus is, of course, dependent on the position, shape, and curvature of the cornea, the lens, and the retina, all of which play a part in the formation of the point retinal image.

*The Eyelids*

The eyelids develop as folds of ectoderm that grow toward each other over the surface of the cornea during the second month (Fig. 21–11). These folds meet and their epithelial layers fuse (Figs. 21–9 and 21–12). Very soon after this fusion, invagination of the common epithelium indicates the beginning of the formation of the hair follicles of the eyelashes, those of the upper lid developing slightly in advance of the lower. The eyelashes differentiate exactly as hair on any other part of the ectodermal surface. Three types of glands develop in the eyelids (Fig. 21–13). Associated with and opening into the lumen of the hair follicles are sebaceous glands *(glands of Zeis)* and the modified sweat *glands of Moll*. In addition, about 30 large sebaceous glands develop from the epithelium just anterior to the posterior lid margin. These are the *tarsal* or *Meibomian* glands. By the end of the fifth month, the adhesion between the upper and lower lids begins to break down and the lids are separated completely during the seventh month.

The stratified columnar epithelium covering the

**FIG. 21–13.** Development of the eyelashes and associated glands during the time the eyelids are fused. (From I. Mann, 1964. The Development of the Human Eye. Grune & Stratton, New York.)

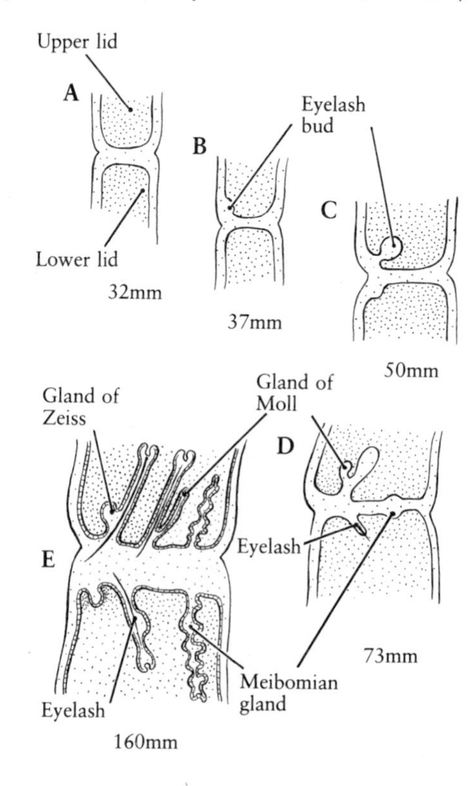

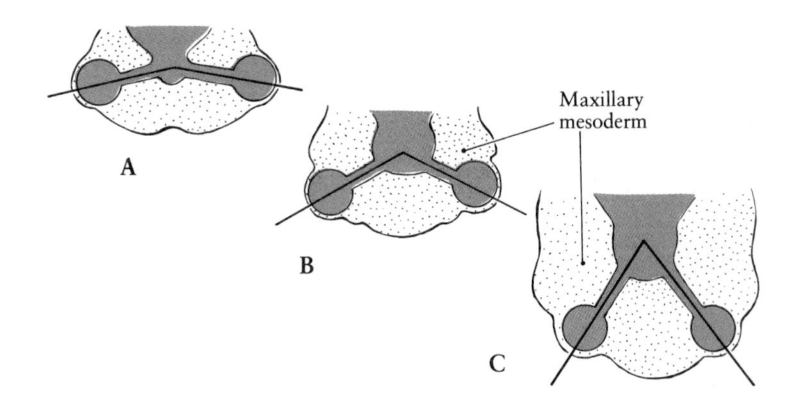


FIG. 21–14. Decrease in the angle of the optic axis during development. A, 8 mm, 160°; B, 16 mm, 120°; C, 40 mm, 72°. (After I. Mann, 1964. The Development of the Human Eye. Grune & Stratton, New York.)

inner surface of the eyelids, which is continuous with the corneal epithelium, is the *conjunctiva*. The *conjunctival sac* is the space between the cornea and the conjunctiva (Fig. 21–12).

### Positional Changes of the Eyes

At the time of the outgrowth of the optic cup, the right and left optic stalks lie in the same straight line forming an angle of 180 degrees. By the middle of the second month, the angle has been reduced to 160 degrees (Fig. 21–14). As a wedge of maxillary mesoderm pushes upward posterior to the eyes, the optical axes continue to converge, the highest rate of convergence occurring during the third month when the angle is reduced to about 105 degrees. At birth it has been reduced to about 71 degrees, only slightly less acute than in the adult (Fig. 21–14). This decrease in the angle of the optical axis allows for the overlapping of the visual fields and the development of binocular vision characteristic of mammalian species.

### Abnormalities in the Development of the Eye

*Coloboma* is a term applied to the presence of a notch or gap in the retina, the choroid, or the iris, or in several of these structures. Most colobomata may be presumed to be the result of a failure of complete closure of the optic fissure. The notch may be extensive, involving all three of the above structures; or, it may involve the iris alone. Coloboma of the iris alone occurs as often as not in some plane other than that of the optic fissure and in these cases may have a different developmental origin from defects that involve both the iris and the choroid.

Opacity of the lens may occur at any time in fetal or postnatal life. The high incidence of congenital cataracts in children whose mothers had contracted German measles early in pregnancy has already been discussed.

Many bizarre abnormalities of the eye are known, but they are fortunately rather rare. *Cyclopia* is a condition in which only a single median eye (or two eyes in various degrees of fusion) are present. *Anophthalmia* is the complete absence of the eye and may be the result either of the failure of the optic vesicle to evaginate from the forebrain or of its atrophy, after its evagination. *Microphthalmia* is a reduction in the size of the eye and is often accompanied by the formation of optic cysts—herniations of the retina into the scleral sac—which, in some cases, may be larger than the microphthalmic eye itself.

## The Ear

As does the eye, the ear also represents a composite organ to which contributions are made from a variety of sources. The ear consists of three separate parts, the *inner* ear, the *middle* ear, and the *outer* ear, each from a different origin. The major portion of the inner ear, which represents the actual organ of hearing, the *cochlea,* and the organ of equilibrium, the *semicircular canals,* the *sacculus,* and the *utriculus,* is derived from an ectodermal *otic placode*. The inner ear thus serves two functions: equilibrium and hearing. In the lower vertebrates, the inner ear is solely an organ of equilibrium to which the higher vertebrates have appended the organ of hearing. The middle ear, the transmitting apparatus, consists of three ear bones, formed from mesoderm of the first and second visceral arches,

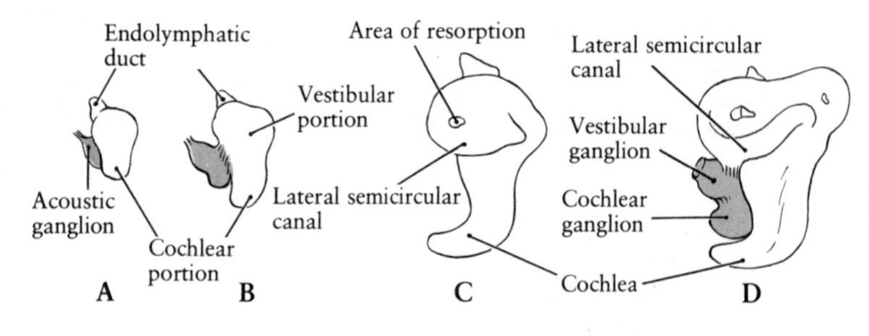


enclosed in a cavity that is an extension of the first pharyngeal pouch. The external ear is a fleshy cartilaginous structure derived from visceral arch mesoderm surrounding the first visceral furrow, which persists as the external auditory meatus.

## The Inner Ear

The otic placode is the first sensory placode to develop. It is externally apparent in the early-somite embryo as a thickening of the ectoderm at the level of the middle of the hindbrain. Before the end of the third week, in the nine-somite embryo, it is quite prominent (Fig. 21–15 A). The auditory placode invaginates to form the *auditory pit* during the fourth week (Fig. 21–15 B) and by the end of the first month has sunk completely below the surface to form the *auditory vesicle (otocyst)* (Fig. 21–15 C).

**FIG. 21–15.** Formation of the auditory vesicle in the early human embryo. A, 9 somites, 20 days; B, 16 somites, 23 days; C, 30 somites, 30 days.

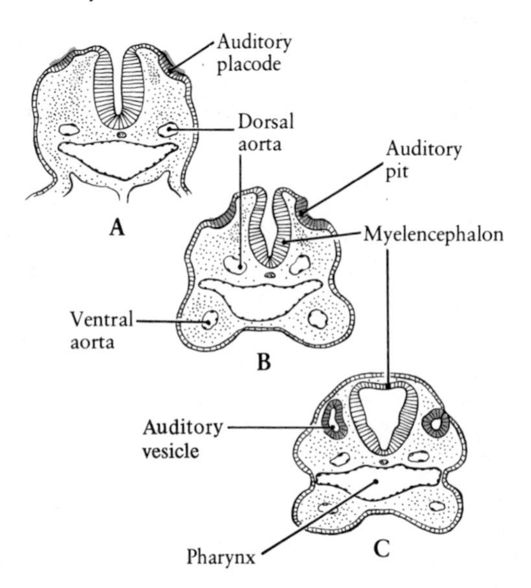


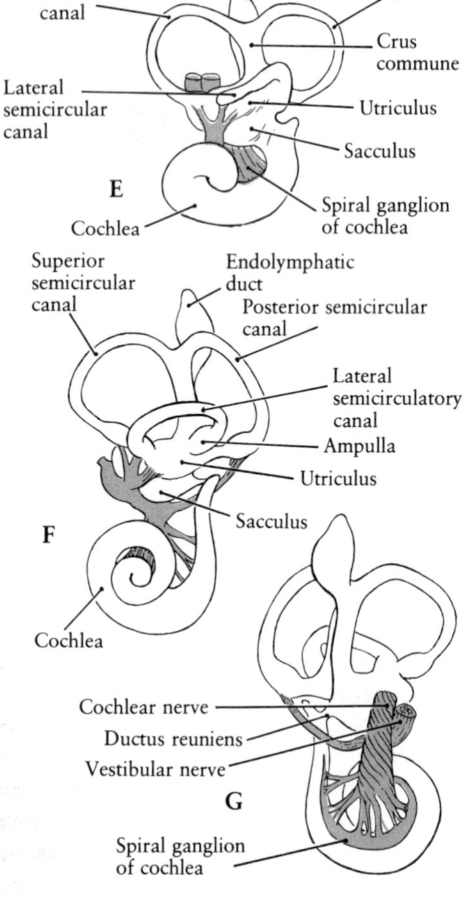


**FIG. 21–16.** Development of the inner ear. A, 6 mm; B, 9 mm; C, 11 mm; D, 13 mm; E, 20 mm; F, 30 mm; G, 30 mm. A–F are lateral views. G is a medial view.

The otocyst lies directly caudad of and in contact with the facioacoustic ganglion.

During the fifth week, the otocyst begins to elongate dorsoventrally, and from the part of the vesicle just dorsad of its point of detachment from the epidermis, an appendage forms. This rounded protuberance is the forerunner of the *endolymphatic duct* (Fig. 21–16 A,B). It develops rapidly, its dorsal tip expanding and its ventral portion contracting. By the end of month 2 (20–30 mm), the tip has widened to form a flat pouch, the *endolymphatic sac*, while the narrow ventral portion forms a tube, the endolymphatic duct, opening into the part of the otocyst that will develop into the *sacculus* and the *utriculus* (Fig. 21–16 E–G).

From the beginning of its dorsoventral expansion

during the fifth week, the otocyst may be divided into two areas. One is a large triangular dorsal portion to which the endolymphatic duct is attached, the *vestibular pouch*. It will form the semicircular canals. The other, ventral to the vestibular pouch, is more slender and flattened. It is the *cochlear pouch,* and it will form the cochlea (Fig. 21–16 A,B). The meeting place of these two regions is called the *atrium* and will give rise to the sacculus and the utriculus.

During the sixth week, the vestibular portion develops three disklike expansions oriented perpendicularly to each other, one in the superior, one in the posterior, and one in the lateral plane (Fig. 21–16 D). With the absorption of their central portions, these disks are converted into canals, the semicircular canals (Fig. 21–16 E). Each semicircular canal develops one bulbar end, the *ampulla.* The superior and posterior canals share an arm in common, the *crus commune,* and each develops its ampulla at the other end. The lateral canal develops its ampulla at its rostral end (Fig. 21–16 F). Sensory receptors, called *cristae,* develop within the ampullae. They respond to the movements of the fluid within the semicircular canals.

The utriculus and the sacculus begin their differentiation later than the semicircular canals, during the middle of the second month (Fig. 21–16 E). Each develops sensory receptors, *maculae,* that initiate impulses recording the position of the head. Both the cristae and the maculae are innervated by nerve fibers from the vestibular division of the acoustic ganglion (Fig. 21–16 D–F).

The cochlear portion of the inner ear also begins its development during the middle of the second month at a time when the semicircular canals are almost completely formed. At six weeks, the cochlear duct appears as an elongated tube that bends rostrally at its ventral end (Fig. 21–16 C). During the second half of the month, the duct elongates at a rapid rate, and the original bending is continued to produce a spiral duct with two and a quarter turns. The original broad connection to the vestibular portion of the otocyst is gradually reduced to a small canal, the *ductus reuniens,* uniting the dorsal portion of the cochlear duct to the sacculus (Fig. 21–16 G).

The cochlear duct represents only a part of the definitive organ of hearing. The mesenchyme surrounding the duct differentiates into a cartilaginous capsule enclosing the membranous cochlear duct. During the third month, the cartilaginous capsule undergoes resorption; the cochlear duct is now surrounded by an open space, the *perilymphatic space,* filled with a fluid, the *perilymph.* The cochlear duct is triangular, with the base of the triangle attached to one wall of the perilymphatic space and the apex stretching across the space to attach to the opposite wall (Fig. 21–17 B). The perilymphatic space is thus divided into two passages by the cochlear duct. One of these passages, the *scala vestibuli,* ends at the *oval window* between the inner and middle ear, the region to which the malleus bone of the middle ear abuts. The other, the *scala tympani,* ends at the *round window* between the inner and the middle ear. The structure that supports the cochlear duct and separates the scala vestibuli and the scala tympani is the *basilar membrane.* The organ of hearing (the *organ of Corti*) rests on this membrane (Figs. 21–17 and 21–18). The sensory receptors of this organ consist of a number of rows of neuroepithelial elements called *hair cells,* which are covered by

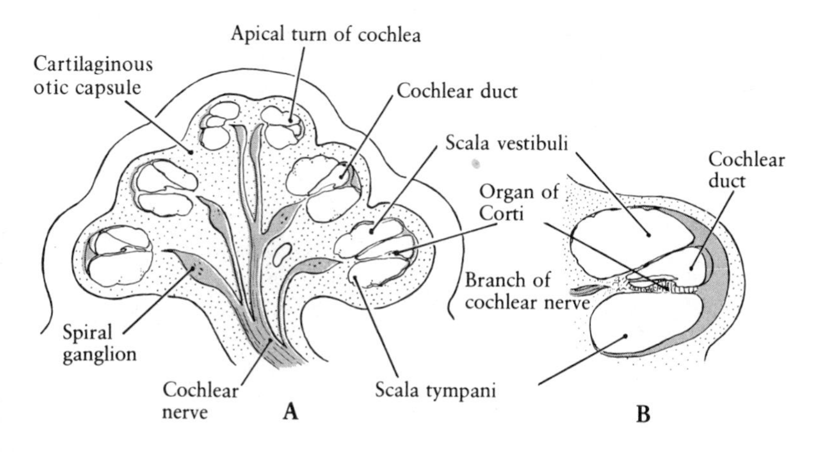


**FIG. 21–17.** A, section through the cochlea of a four-month-old fetus; B, enlargement to show the relation of the cochlear duct to the scala vestibuli and the scala tympani.

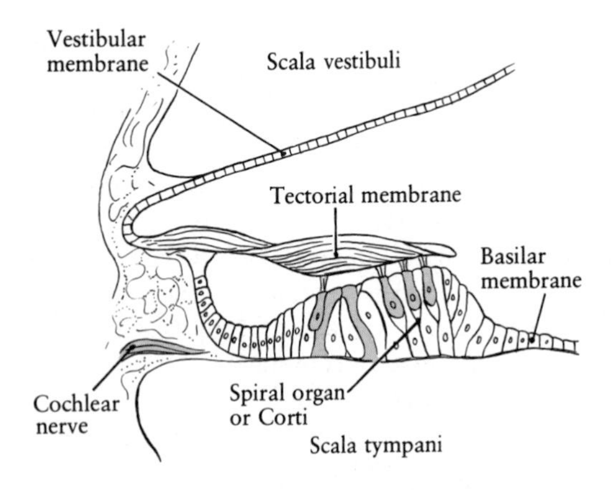


**FIG. 21–18.** Section through the spiral organ of Corti of a 250-mm fetus.

a *tectorial membrane* adhering to the tips of the hair cells (Fig. 21–18). Wavelike motion of the perilymph imparted to it by the ear ossicle at the oval window results in vibrations of the basilar membrane and consequent stimulation of the hair cells. Nerve impulses are then initiated in the nerve fibers surrounding the bases of the hair cells.

The sensory receptors of the inner ear are supplied by nerve fibers from the acoustic division of the facioacoustic ganglion. The more caudal portion of this ganglion is the acoustic division that is seen in contact with the rostral portion of the developing inner ear during the fourth week (Fig. 21–16). The superior portion of the acoustic ganglion supplies fibers to the semicircular canals, the sacculus, and the utriculus. It consists of two divisions, a superior one supplying the ampullae of the superior and lateral semicircular canals and the utriculus, and an inferior one supplying the sacculus and the ampulla of the posterior semicircular canal (Fig. 21–19). These fibers converge to form the vestibular part of the acoustic nerve.

The inferior portion of the acoustic ganglion differentiates into the *spiral ganglion,* whose fibers supply the hair cells of the spiral organ of Corti. As the cochlear duct develops its typical spiral pattern, the spiral ganglion conforms to this same pattern (Fig. 21–19 C,D). The centrally directed fibers from the spiral ganglion form the cochlear division of the acoustic nerve.

## The Middle Ear and the External Ear

At the time that the inner ear is developing as the sound-receiving element of the organ, the middle-ear cavity *(tympanic cavity)* and the middle-ear ossicles are developing as the sound-transmitting element of the system. The tympanic cavity is derived from the first pharyngeal pouch. This pouch, which appears at about three weeks, grows laterally and makes temporary contact with the inpocketing that is the first visceral furrow (Fig. 21–20). Shortly, this contact with the ectoderm is lost and the distal portion of the first pouch expands, forming the primordium of the tympanic cavity, the *tubotympanic recess*. The more proximal region of the pouch remains narrow, forming the *auditory (Eustachian) tube* (Fig. 21–20 B), the channel by which the middle ear is connected to the pharynx. At the same

**FIG. 21–19.** Differentiation of the left acoustic ganglion and nerve. The vestibular ganglion is finely stippled and the spiral ganglion is coarsely stippled. A, 4 mm; B, 7 mm; C, 20 mm; D, 30 mm.

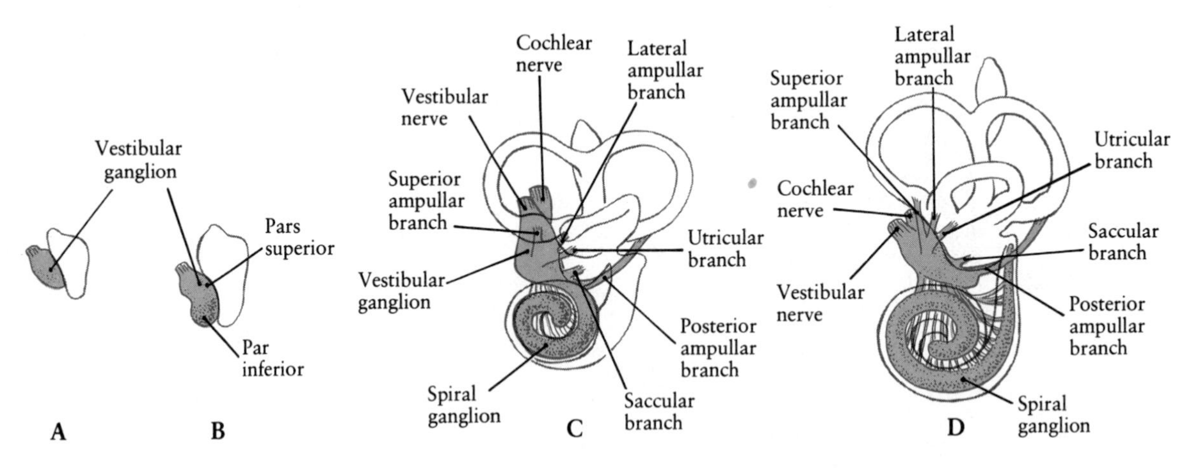

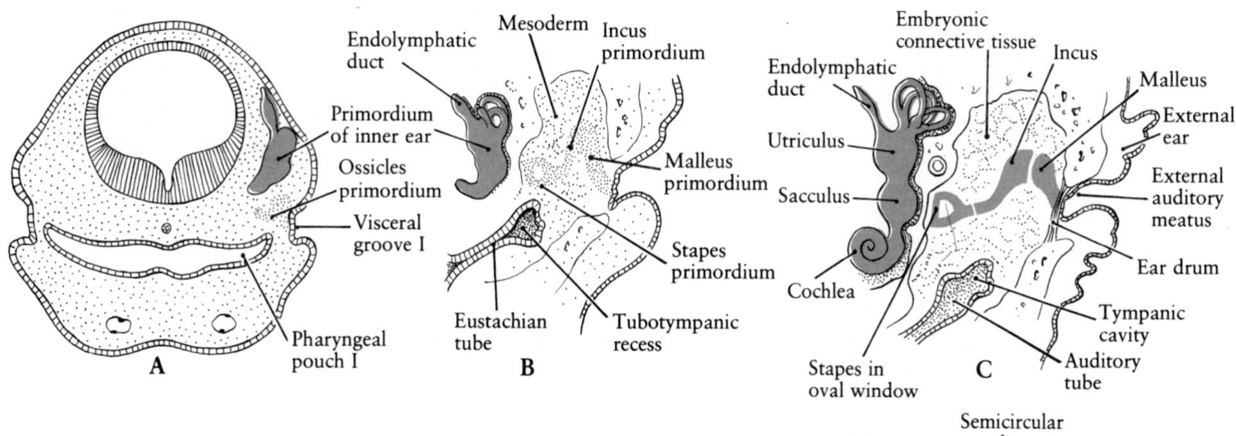


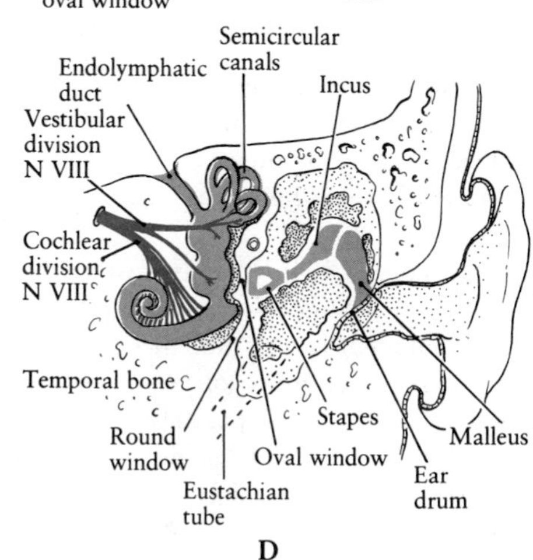


FIG. 21–20. Schematic diagrams of four stages in the development of the middle ear. (From B. Patten, 1968. McGraw-Hill Book Co., New York.)

time, the mesenchyme just above the tympanic cavity lateral to the developing otic vesicle shows three areas of condensation resulting from the proliferation of the mesenchyme of the first and second visceral arches (Fig. 21–20 A,B). These represent the precursors of the three middle-ear bones, the *malleus*, the *incus*, and the *stapes*. The first two are derivatives of the first arch, and the stapes is a derivative of the second arch. The ear ossicles remain embedded in mesoderm until the eighth month, when the connective tissue surrounding them begins to undergo resorption and the tympanic cavity enlarges to envelop them (Fig. 21–20 C). At birth, there is still some unresorbed mesoderm around the ear ossicles (Fig. 21–20 D).

The *tympanic membrane (eardrum)* is a derivative of the closing plate between the first pharyngeal pouch and the first visceral furrow. The first visceral furrow thus becomes the *external auditory meatus* (Fig. 21–20).

The *auricle (pinna)* develops from swellings formed as mesenchymal proliferations of the first and second visceral arches surrounding the first visceral furrow (Fig. 21–21). Three *auditory hillocks* are first-arch derivatives, and three are second-arch derivatives. Fusion of these individual hillocks and their further growth forms the auricle. In view of the number of growth centers involved, it is not surprising that there is great individual difference in the shape and the size of the external ear.

## Analysis of Sensory Organ Development

As we have seen in other chapters, most organs form as a result of the gradual and cumulative effects of interaction among embryonic tissues. Of the three primary germ layers in vertebrate embryos, the ectoderm is probably the most dependent on the presence of adjacent tissues for the emergence of the various organs that it forms. We have seen in Chapter 10 that the central nervous system is the first major ectodermal organ to appear, dependent on an induction from the underlying chordamesodermal tissue. Functional components of vertebrate eyes, olfactory organs, and inner ear structures are also specializations of the embryonic head ectoderm. Rudiments of the nasal sac, the lens of the eye, and the inner ear are morphologically detectable as thickenings or outpocketings of the ectoderm shortly after gastrulation. However, determination of all of these organs is probably initiated just before or during the process of gastrulation itself. The developmental histories and differentiation of nose, lens, and inner ear are quite

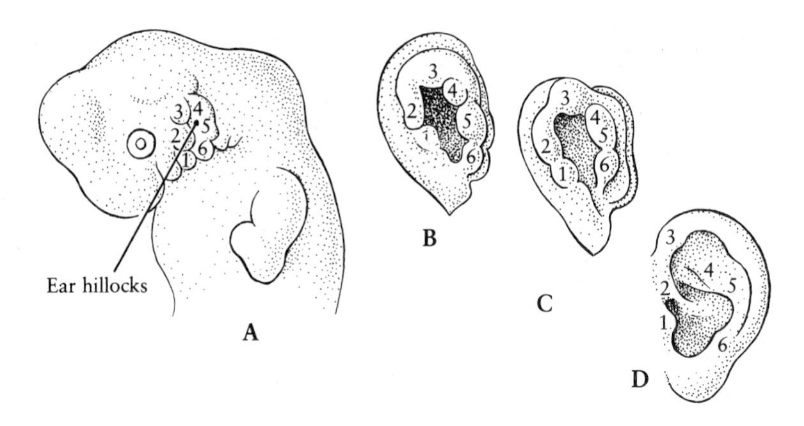


FIG. 21–21. Development of the human auricle. A, 11 mm; B, 13.5 mm; C, 15 mm; D, adult. 1–6, auditory hillocks of the first and second visceral arches, which become 1, tragus; 2, 3, helix; 4, 5, antihelix; 6, antitragus. (From L. Arey, 1974. Developmental Anatomy. W. B. Saunders Company, Philadelphia.)

similar and are dependent on a temporal sequence of inductions by tissues acting synergistically. Indeed, a part of an organ developing as a result of an induction may then subsequently act as an inducing stimulus.

By comparison with the eye, analysis of the mechanisms underlying the developing of the nose and inner ear is very limited. Our knowledge in this area is based almost solely on a series of elaborate extirpation and explantation experiments conducted by Jacobson using amphibian embryos. During the gastrulation process, the heart mesoderm and the deuterencephalic portion of the chordamesoderm stimulate the overlying prospective ear ectoderm to "predispose" it to form the inner ear. Unlike the nose, the endoderm does not appear to be an important inductor of the inner ear. Complete removal of the endoderm at the neurula stage has little effect on ear production. The process of inner ear determination is continued as neurulation is completed. The hindbrain (medulla) acts as a secondary inductor and induces the "pre-ear" ectoderm to thicken into an otic or auditory placode. The influence of the hindbrain appears to be strongest during late neurulation. Prospective ear ectoderm of the early neurula is not sufficiently determined to form this organ when explanted by itself. Indeed, before the stimulus of the rhombencephalon, the ear ectoderm tissue is equipotential and labile; that is, any of its parts can give rise to any part of the inner ear depending on its position in the whole. Similar to the limb, the axial organization and compartmentalization of the inner ear are determined in steps. Under the inductive influence of the developing hindbrain, the anteroposterior axis of the ear in *Ambystoma* embryos is fixed by the end of neurulation and the dorsoventral axis by the early tail bud stage. A summary of inner ear induction utilizing diverse inductor tissues is shown in Figure 21–22.

The hindbrain appears to induce the overlying ectoderm to form an inner ear in an instructive manner. How messenger or inductor substances might be transmitted to the ear ectoderm is not clear, but there is little evidence that these are diffusible molecules that pass from the inducing tissue across the interspace to the reacting tissue. A fine-structural analysis of inner ear development by Model and others suggests that inductor substances might be transmitted through gap junctions between cells of the interacting tissues. During the known period of inner ear determination in the axolotl embryo, many fingerlike processes from both hindbrain and inner ear cells extend into the interspace. Specialized (focal) junctions connect the two cell types. These focal contacts are rather small (10–30 nm in diameter), and they exhibit the septilaminar appearance so characteristic of gap junctions (Chapter 7). The sudden disappearance of these contacts and the interdigitating processes suggests that the gap junctions could provide the structural basis for cell-to-cell communication between medulla and inner ear cells during the inductive interaction. As pointed out in Chapter 7, gap junctions in other cell systems have been observed to be capable of the intercellular transfer of molecules.

Similarly, induction of the nose rudiment involves the succession of diverse inductor tissues. There is an early and significant primary induction by the anterior end of the endoderm shortly after the onset of gastrulation. Later inductions emanate from the prechordal plate mesoderm and the prosencephalon of the forebrain. As in the case of the inner ear, although direct causal relationships

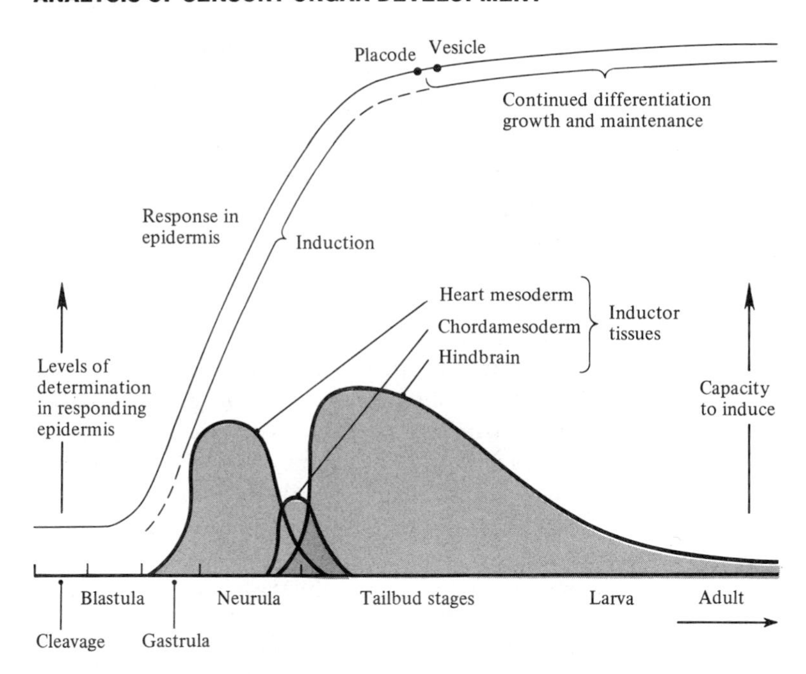


**FIG. 21-22.** Normal course of events during inner ear determination. Induction is gradual and involves diverse inductor tissues. Continued induction after placode formation is essential for complete differentiation and growth of this organ. (From A. G. Jacobson, 1966. Science 152:25.)

can be experimentally demonstrated in the development of the olfactory organ, little is known about the mechanisms controlling the inductions and the bases underlying the differentiations of the sensory cells.

Next to the brain, the vertebrate eye is probably the most complex organ with respect to its development. Originating from several tissue sources that are dependent on a complicated chain of inductions, a number of cell types are functionally unified into an organ of unusual efficiency. The early development of the eye, particularly the retina–lens complex, provides a useful model system for investigating such fundamental processes as induction, morphogenesis, cell differentiation, growth, and aging.

Recall that the early rudiments of the whole eye are the primary optic vesicle and the lens placode. Vital staining experiments have shown that the cells destined to form the optic vesicle lie well forward in the neural plate (Fig. 21–23 A). Presumptive lens material is localized in the embryonic head ectoderm lateral to and slightly anterior to the optic vesicle rudiments (Fig. 21–23 B). As the neural plate rolls into a tube, the lateral walls of the prosencephalon evaginate into the pair of optic vesicles.

Elevation of the neural folds brings the future retina cells of the optic vesicle into contact with the presumptive lens cells. The lens then develops as a disc shaped thickening in the skin ectoderm overlying the primary optic vesicle. Embryologists have been aware for many years that the normal development of the whole eye is dependent on the association of several embryonic tissues. In the case of the eye vesicle rudiment, its expression is dependent on an inductive influence originating from the anterior end of the archenteron (i.e., archencephalic inductor). The determination of the optic vesicle appears to be irreversibly set by the end of the neural plate stage. If the eye rudiment is excised at this time and transplanted with some surrounding mesenchyme, it will differentiate in normal fashion into retina, iris, and so on.

As in the case of many organ primordia (Chapter 13), the early eye rudiment is labile and shows the property of self-regulation. That is, part of the rudiment has the ability to develop into a whole eye. For example, the eye rudiment at the neural plate stage or the optic vesicle stage can be dissected into two halves in the frog embryo. Each half will develop into a complete eye, albeit smaller than normal. Even at the early optic cup stage, a piece of presumptive pigmented epithelium, following its removal and transplantation into the vicinity of an eye in another embryo, will form a whole eye (Fig. 21–24). These experiments also indicate that the individual parts of the eye are not determined at the same time that the eye as a whole is determined.

One of the best-known examples of induction is

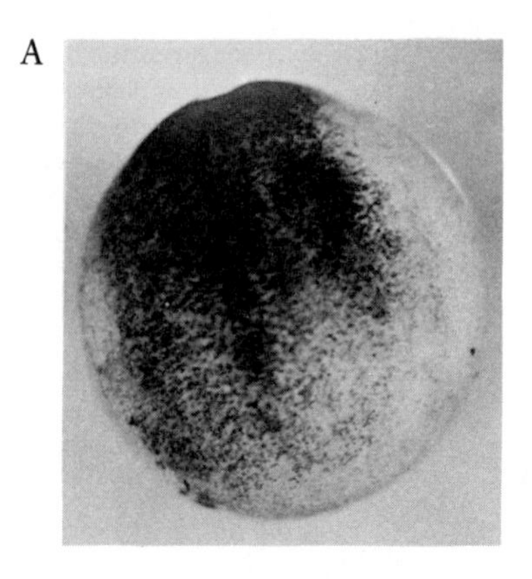


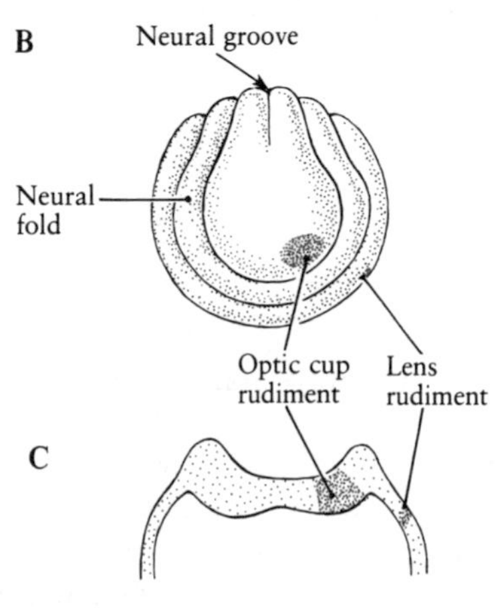


**FIG. 21–23.** A, photograph showing the neural plate during an early stage of neurulation in an amphibian. (Courtesy of R. Brun.) B (dorsal view) and C (transverse view) indicate the location of the presumptive optic cup and presumptive lens in the neurula stage of an amphibian. (After H. Spemann, 1938. Embryonic Development and Induction. Yale University Press, New Haven.)

**FIG. 21–24.** When a large piece of pigmented epithelium is transplanted into the vicinity of a normal eye of another embryo it develops into a complete eye with pigment coat and retina. (After N. Dragomirow, 1933. Wilhelm Roux' Arch. Entwicklungsmech. Org. 129:522.)

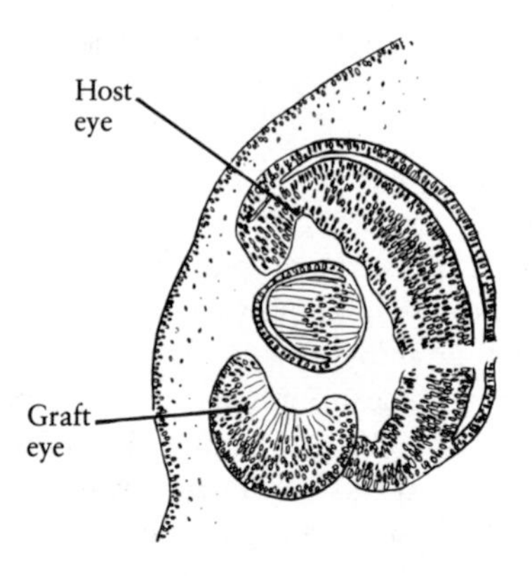


the differentiation of the lens after an interaction with the primary optic vesicle. The direct causal relationship between the presence of the primary optic vesicle and lens development has been experimentally demonstrated in several ways (Fig. 21–25). If the optic vesicle is removed, the skin ectoderm does not form a lens (Fig. 21–25 B). Excision and transplantation of the optic vesicle to a site beneath the flank of an embryo stimulates the formation of a lens in an abnormal position (Fig. 21–25 C). Further support for the dependence of the differentiation of the lens on a prolonged inductive influence from the primary optic vesicle comes from studies with mutant strains of mice. For example, in a certain strain of mice, the optic vesicle does not grow normally because the rate of cellular mitosis is reduced. Consequently, either the optic vesicle does not make contact with the ectoderm or contact is delayed. Such mice show either no lenses (anophthalmia) or very small lenses (microphthalmia). In addition to being smaller in size, the lens may be improperly centered in the optic cup. The period of inductive interaction by the optic vesicle, therefore, would appear to ensure the proper size, shape, and placement of the lens in relation to the retina.

Results of other experiments have clearly indicated that tissues other than the primary optic vesicle are also required for normal induction of the lens. In *Rana esculenta,* for example, a lenslike structure develops from ectoderm in the normal lens position following ablation of the primary optic vesicle. Jacobson has conducted a number of experiments on lens induction using the salamander embryo by culturing presumptive lens ectoderm with one or a combination of other tissues. He concludes that lens differentiation, as with the olfactory sac and inner ear, results from a sequence of synergistic tissue interactions. The inductor tissues are the pharyngeal endoderm, which moves beneath the presumptive lens ectoderm during gastrulation, the presumptive cardiac mesoderm, and the neural tissue of the primary optic vesicle. Similar tissue interactions are now recognized as being important during lens development in avian and mammalian embryos.

As in other organ systems that require embryonic inductions, a major question in eye development is whether there is a diffusible substance that specifically induces lens differentiation. Earlier studies attempted to resolve this question by placing cello-

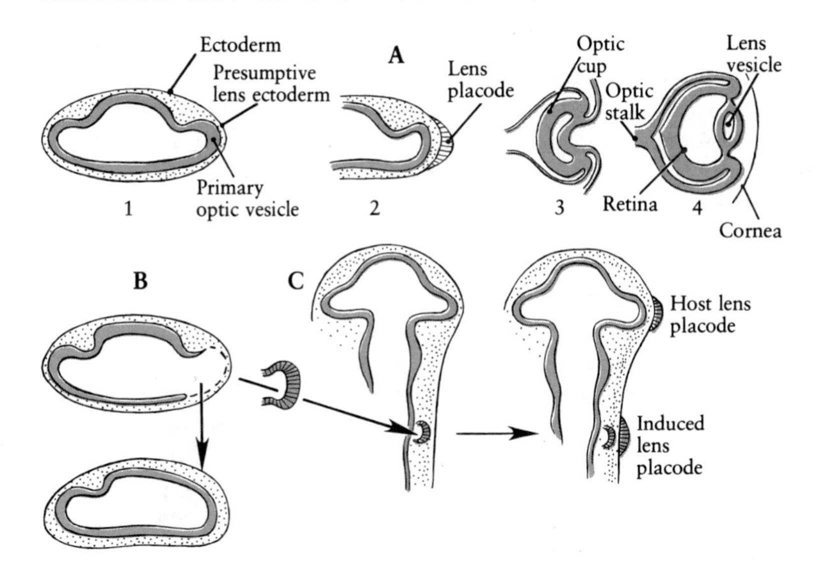


**FIG. 21-25.** Diagram to show the importance of the optic vesicle in lens induction. A, 1–4, steps in the normal development of the optic cup, lens, and cornea. Transverse views through the prosencephalon. B, extirpation of the primary optic vesicle before it approximates the presumptive lens ectoderm results in the absence of lens development; C, transplantation of the primary optic vesicle beneath the flank ectoderm of a second embryo at a comparable stage of development induces an accessory lens. A, 1–4 and B, transverse views; C, dorsal view.

phane sheets, porous membranes, or agar slices between the two interacting tissues. The results were inconclusive and subject to varying interpretations. Support for the long-range extracellular diffusion concept has come from transfilter studies using chick optic vesicle and trunk ectoderm. Uncommitted ectoderm forms distinct lenslike bodies capable of synthesizing lens proteins when separated from the optic vesicle by Millipore filters of different thicknesses and pore sizes. Furthermore, although several investigators have drawn attention to the strong adherence of the optic vesicle and the overlying ectoderm during the inductive period, electron microscopic images clearly show an interspace, up to 100 microns in width, between the interacting tissues (Fig. 21–26). This space in the chick embryo is acellular, is lined by the basement membranes on the basal surfaces of the optic vesicle and ectoderm, and contains a fibrous, hyaluronate material closely resembling the composition of the vitreous body. Hence, direct cellular contact between the presumptive neural retina and presumptive lens ectoderm does not appear to be required for a successful lens induction to occur. Since the influence of the retina on the lens is not restricted to the initial inductive period but is also required for full differentiation and maintenance of the lens, the signal molecule must be small to enable it to penetrate the capsule surrounding the lens.

There is other evidence that such an instructive lens molecule is unlikely and that a general alteration in the microenvironment of the reacting cells is necessary for lens induction. This concept implies that all head ectodermal cells have the potential to form lens cells; however, the realization of lens potential can only occur under certain environmental cues. A variety of tissues under specified conditions can form lenses. In *Xenopus* larvae, removal of the normal lens (lentectomy) is followed by regeneration of lens from cells of the cornea. Several eye tissues, including the pigmented iris and the retina itself, can form lenslike structures under given cell culture conditions. Results from cell culture studies in particular support the view

**FIG. 21-26.** Drawing of an electron micrograph showing the interspace (IS) and adjacent parts of presumptive lens (PL) and presumptive retina (PR) cells. Note the basement membranes and the fibrous nature of the matrix between the presumptive lens and presumptive retina. (After P. Silver and J. Wakely, 1974. J. Anat. 118:19.)

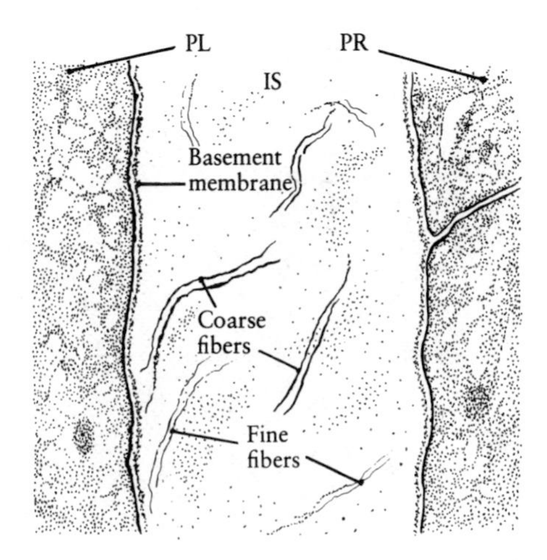

that many cells have a tendency to differentiate into lens cells, particularly those of ectodermal origin. The possibility has been raised that the primary optic vesicle–lens interaction in the intact embryo may be mediated by specific alterations in the structure and composition of the extracellular matrix between the tissues. In the chick embryo, histochemical and autoradiographic studies have shown that the region between the primary optic vesicle and the presumptive lens ectoderm is rich in glycoproteins and glycosaminoglycans (Fig. 21–27). These matrix molecules appear to be synthesized and released by the epithelia of both interacting tissues. There appears to be a striking correlation between the onset of lens development (i.e., the elongation of ectodermal cells and the synthesis of lens proteins) and an increase in the quantity of the matrix glycoproteins during the period of lens induction. However, how the extracellular matrix facilitates lens morphogenesis is unclear. Quantitative changes in the density of the matrix per se might provide a specific stimulus leading to spatial specificity and initiation of lens cell specialization. If induction involves the production of an inducing factor by the primary optic vesicle, the extracellular matrix may be important in modifying its composition or regulating its delivery to the overlying ectodermal tissue.

The lens is a favorite organ for the study of morphogenesis and cell differentiation, primarily because it is accessible for manipulation and experimentation. The first morphological sign of lens differentiation is the formation of the lens placode. As with the neural plate, the placode cells contain microtubules oriented parallel to the direction of elongation. Cell elongation is a striking feature of lens development and growth, but it is still not clear whether microtubules are the only factor underlying the change in cell shape. Shortly after the lens placode becomes invaginated to form the lens vesicle, the epithelial cells in the inner or posterior wall of the hollow vesicle begin to elongate. These soon fill the lumen of the vesicle and are called the *primary lens fibers*. The same induction responsible for determination of the lens as a whole is also involved in the differentiation of the lens epithelial cells into lens fiber cells. Epithelial cells of the lens vesicle, when grafted in the vicinity of either presumptive retina or the epithelium of the inner ear vesicle, may develop lens fibers (Fig. 21–28). Details on the morphological, cytological, and biochemical transformation of lens epithelial cells into lens fiber cells have to a large extent been worked out using the adult lens.

The adult lens grows throughout the whole lifespan of the organism. It is solid, avascular, and composed of the following (Fig. 21–29): a single layer of epithelial cells, a zone of cellular elongation (or equatorial region) made up of cells that are in the process of becoming fiber cells, and inner fiber cells. The cells that differentiate from the epithelium are laid down in layers at the periphery of the organ. Fiber cells develop continuously in the zone of cellular elongation after the embryonic lens has formed. The fiber (primary) cells formed during embryonic growth constitute the central or nucleus region, while the newly formed fiber (secondary) cells are located in the cortex of the lens. The mechanism of cell elongation during lens fiber cell differentiation has been the subject of considerable

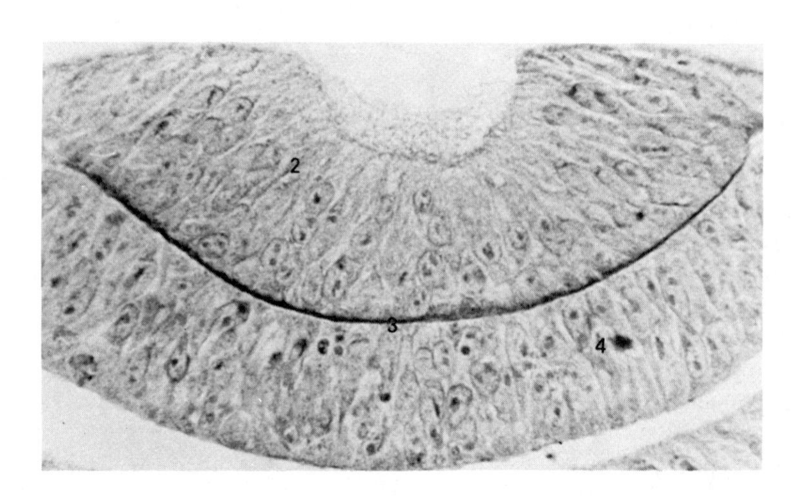


**FIG. 21–27.** Photomicrograph of a section of an invaginating chick lens rudiment showing PAS-positive extracellular material (3) between the lens (2) and optic vesicle (4). The extracellular matrix is composed of acidic glycosaminoglycans and glycoproteins. (From R. W. Hendrix and J. Zwaan, 1974. Differentiation 2:357.)

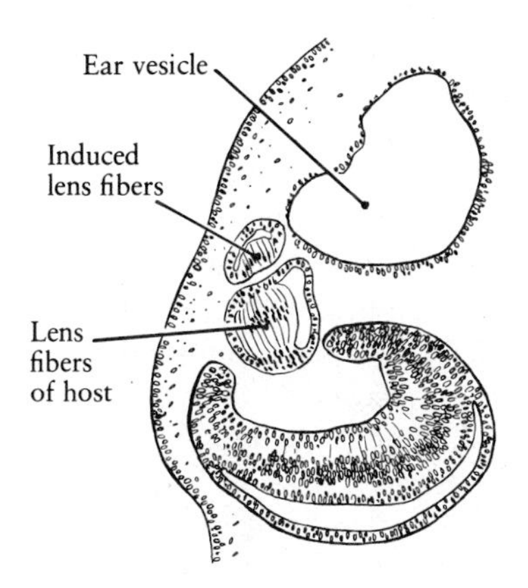


**FIG. 21–28.** Epithelial cells of the lens vesicle when grafted in the vicinity of the inner ear vesicle may be induced to form an additional mass of lens fibers. (After N. Dragomirow, 1929. Wilhelm Roux' Arch. Entwicklungsmech. Org. 116:633.)

study and controversy. Although several studies suggest that microtubules are involved in lens fiber formation, other mechanisms of elongation have been proposed. Beebe and colleagues have demonstrated that cultured six-day embryonic lens epithelial cells in the presence of *nicodazole* (a synthetic inhibitor of microtubules) can still elongate in the absence of microtubules. They propose that lens cell elongation is directly caused by an increase in cell volume. Following the formation of a lens fiber, the cell loses its replicative ability and enters a stationary phase of the cell cycle. Since the part of the lens pit-vesicle that elongates is held by the optic cup, it would appear that the optic cup environment induces primary fiber formation and maintains lens growth patterns. Hence, the overall lens morphology is governed by morphogenetic events of a totally separate tissue.

**FIG. 21–29.** Diagram of the structure of the lens of a typical adult vertebrate. (See text for additional details.) (From J. Papaconstantinou, 1967. Science 156:338. Copyright 1967 by the American Association for the Advancement of Science.)

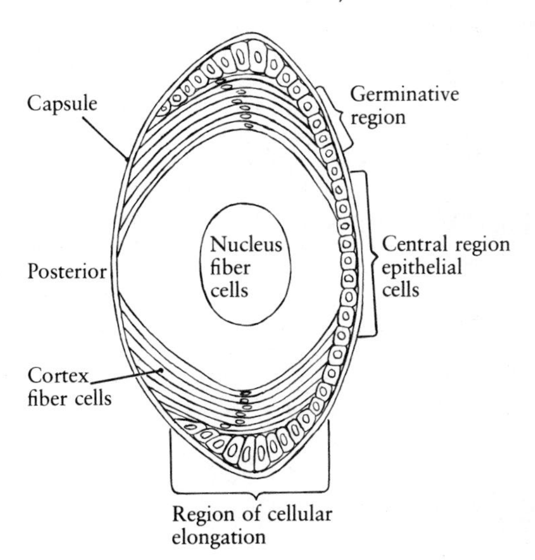


Since fiber cell formation is a terminal step in the differentiation of a specific cell type, the lens is an ideal system for the study of the interrelationships among morphogenesis, cellular differentiation, and protein synthesis. Lens epithelial cells are characterized by their cuboidal shape, their basophilic staining properties, and their ability to undergo mitosis. Their transformation into fiber cells is accompanied by physical changes in shape, an increase in the population of ribosomes, and the synthesis of structural proteins termed *lens crystallins.* Crystallins are water soluble and make up more than 90 percent of the total protein of the vertebrate lens. There are four classes of crystallins ($\alpha$, $\beta$, $\gamma$, and $\delta$) that are immunologically distinct; each crystallin is composed of multiple polypeptides. $\alpha$, $\beta$, and $\gamma$ crystallins are formed in the lenses of a variety of vertebrates, including amphibians and mammals; $\alpha$, $\beta$, and $\delta$ crystallins are found only in avian and reptilian lenses. Techniques of column chromatography, electrophoresis, and immunofluorescence have established that crystallins are normally restricted to the lens cells (i.e., they are tissue specific).

The crystallins are useful markers for the differential expression of genes during the development and maturation of the lens. They show a complex sequence of appearance and accumulation that is temporally and spatially specific in the lens cells. The site of the first appearance of crystallins in the lens strongly suggests that their synthesis is a direct response to induction of the ectoderm. In the chick embryo, the $\delta$ crystallin, or the primary lens fiber protein of the adult bird lens, is the first to appear and does so at about the time that the lens placode begins to invaginate from the ectoderm. $\beta$ Crystallin is detected at approximately 56 hours of development, but it is not present in substantial amounts until the lens vesicle stage (80 hours). It accumulates in the lens following hatching. $\alpha$ Crystallin can first be identified at the lens vesicle stage. The $\beta$ and $\alpha$ crystallins are the principle proteins of fiber cell differentiation. $\delta$-Crystallin synthesis decreases in lens epithelial cells during the latter part of embryonic development and in fiber cells shortly after hatching. The regulation of this change in $\delta$-crystallin synthesis has been rather extensively studied in the chick. When 6-day-old and 19-day-old embryonic lens epithelial cells are compared, the rate of $\delta$-crystallin synthesis is substantially less in the older tissue. However, the lens epithelial cells of both stages have equivalent amounts of $\delta$-crystallin mRNA sequences. Therefore, the drop in the synthesis of this protein cannot be correlated with a decrease in the amount of available mRNA. It is now known that the 19-day-old embryonic lens epithelial cells appear to utilize the $\delta$-crystallin mRNA less efficiently than 6-day-old cells. The molecular mechanism underlying this reduced efficiency of mRNA utilization has yet to be clarified, although possibilities include the partial degradation of or structural changes in the mRNA molecules. Hence, this is another example of a translational control being used to regulate selectively the synthesis of a protein during embryonic development. By contrast, the lack of synthesis of $\delta$ crystallin in adult epithelial cells is clearly due to an absence of detectable amounts of $\delta$-crystallin mRNA.

Elegant immunofluorescence methods have also shown that crystallins in the mouse and the rat are initially detected in those cells of the invaginating lens placode and newly formed lens vesicle that lie directly opposite to the central area of the presumptive neural retina (Fig. 21–30). $\alpha$ Crystallin is the first crystallin to be detected in the rat embryo and is localized in some of the lens pit cells at 12 days of development. By 14 days all lens cells contain this protein. At about 12.5 days, $\beta$ and $\gamma$ crystallins are initially observed in some cells located in the posterior portion of the lens vesicle (i.e., presumptive primary fiber cells). Subsequently, these two crystallins are restricted to the elongating cells and the fiber cells, or those lens cells situated in the optic cup. All three types of crystallins are synthesized in the lens fiber cells.

What are the factors that control or regulate the temporal sequence and spatial distribution of these lens proteins? Unfortunately, little is known about how the crystallin genes of the lens cells are sequentially activated to produce their specific proteins. One possibility is that the presumptive retina over a period of time releases a sequence of specific crystallin inductor substances to which the lens epithelium responds. However, we have already seen that the evidence for lens-specific inductors is weak. Alternatively, the coordination of expression of the genes coding for specific crystallins may be a programmed response of the lens cells, initiated by the same generalized, nonspecific signal that starts the lens differentiation process.

The spatial pattern and distribution of the lens crystallins appear to be controlled by the position of

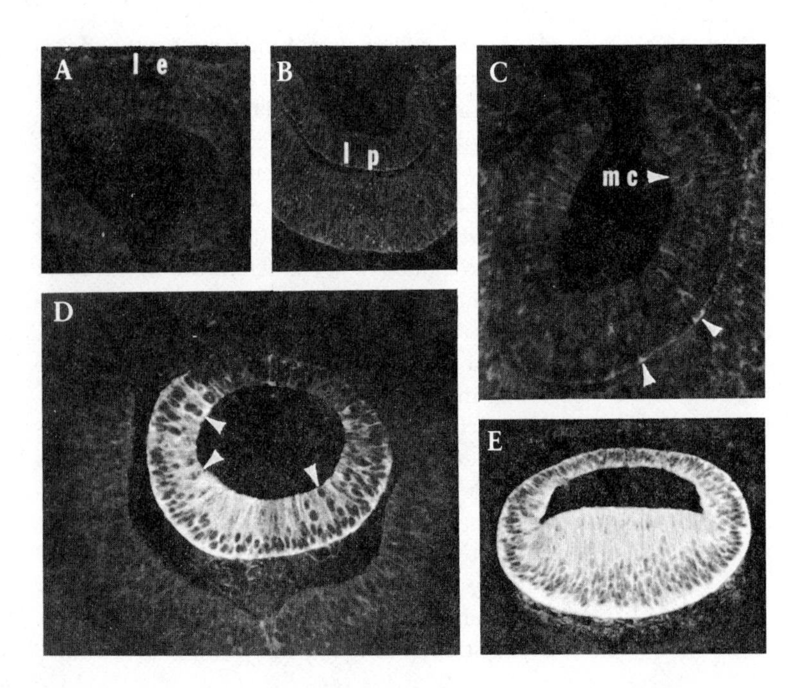


**FIG. 21–30.** Immunofluorescence is a technique in which a specific tissue protein is visually localized by reacting it with a fluorescein-labeled antibody made against the antigen. A, optic vesicle and its area of contact with the presumptive lens ectoderm (le) in a 9.25-day-old mouse embryo. Note that no lens crystallins are detected. B, section through the lens placode (lp) and optic cup of a 10.5-day-old mouse embryo; C, invaginated lens of an 11.0-day-old embryo. Note that crystallins are present in the cytoplasm of a few of the innermost cells (arrows) but not in apically situated mitotic cells (mc). D, 11.5-day-old embryonic mouse eye showing strong fluorescence in cells along the interior wall of the lens vesicle and in mitotic cells (arrows); E, all lens cells fluoresce in a 12.5-day-old mouse eye. (From M. van de Kemp and J. Zwaan, 1973. J. Exp. Zool. 186:23.)

the optic cup. In the mouse, $\alpha$ crystallin is present in all cells of the lens, while $\beta$ and $\gamma$ crystallins are detected only in elongating and fiber cells. Elongating and fiber cells develop in that part of the lens primordium adjacent to and held in the optic cup. There is strong evidence that the spatial organization of the two major forms of lens cells depends on the optic cup. If the lens of a five-day chick embryo is rotated 180 degrees so that the epithelium facing the cornea is opposite to the neural retina, the cells of this epithelium will elongate and form fiber cells. Removal of the neural epithelium will inhibit these changes in cell shape. Experimentally, it has been shown that the neural retina also induces the overlying epithelial cells to synthesize the $\beta$ and $\gamma$ crystallins. This means that lens cells outside or beyond the influence of the optic cup can only synthesize $\alpha$ crystallin. Recently, Ostrer and his colleagues have found a 35,000 molecular weight $\beta$-crystallin polypeptide to be a specific marker for lens cell elongation; the appearance of this polypeptide correlates with the appearance of its mRNA in the elongating cells.

Lens differentiation, then, is characterized by the simultaneous appearance of spatially distinct cell populations and spatial differences in the expression of genes coding for specific crystallins. In view of the complexity of the differentiation, it is not surprising that the mechanism(s) coordinating this sequence of morphogenesis and cytodifferentiation needs further clarification. For example, are lens cell elongation and the synthesis of $\beta$ and $\gamma$ crystallins induced by the same retinal substance(s)? Are these events interdependent or independently controlled?

The lens also exerts an important influence over the development of the whole eye, particularly the cornea and the neural retina. Acting with the primary optic vesicle, the lens is primarily responsible for the induction and maintenance of the cornea. This can be demonstrated by transplanting the retina–lens complex beneath the skin in another part of the embryo and observing the development of an accessory cornea; it can also be demonstrated by replacing the normal cornea with skin from another part of the embryo and observing transformation of the graft into cornea. In contrast to the neural plate and lens ectoderm, the competence to form corneal tissue remains for a long period of time.

The close association between the invaginating lens placode and the wall of the primary optic vesicle has led to the conclusion that the lens is instrumental in the formation of the optic cup. Although this process has been described in many vertebrate embryos, little is known about how the optic cup invagination is initiated and how it is regulated. Distinct changes in the extracellular materials between lens and optic vesicle occur during invagination to form the retina. Before lens placode formation in the chick embryo, a thick fibrillar layer separates the basal surfaces of the primary optic vesicle and the ectoderm. However, when the lens placode forms, the two tissues are no longer separate; the basal laminae of placode and optic vesicle fuse to construct a thick basement membrane. Fusion occurs only at the margin of the lens placode. During invagination, therefore, the margin of the optic cup is attached at the edge of the lens vesicle (Fig. 21–31). Experiments with drugs that inhibit the synthesis of glycoconjugates suggest that materials associated with the optic cup may be important to the invagination process. Both acidic glycosaminoglycans and glycoproteins can be identified along the apical (inner) and basal (outer) surfaces of the optic vesicle. There is a substantial increase in the matrix of the apical surface during lens placode formation and invagination, particularly at the margin of the presumptive retina. If chick embryos are treated with drugs such as *tunicamycin* (an inhibitor of the N-glycosylation of glycoproteins) before the optic cup stage, there is a marked decrease

**FIG. 21–31.** Scanning electron micrograph of an optic primordium (P) of the chick embryo showing that the invaginating lens epithelium (L) is closely apposed to the presumptive retina (R). The basement membrane that holds the two layers together has been exposed (arrow). (From J. Yang and S. R. Hilfer, 1982. Dev. Biol. 92:41.)

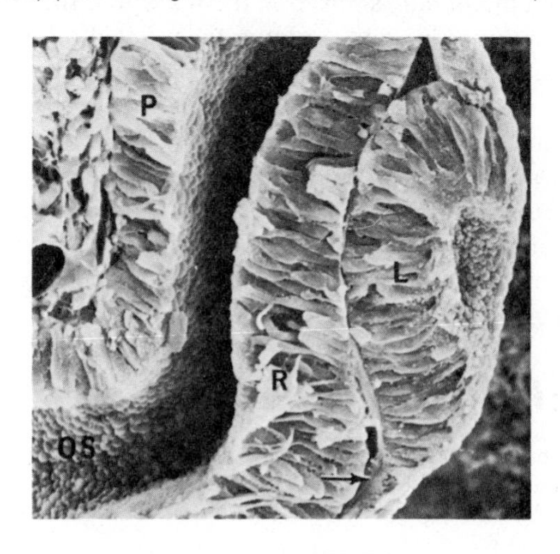

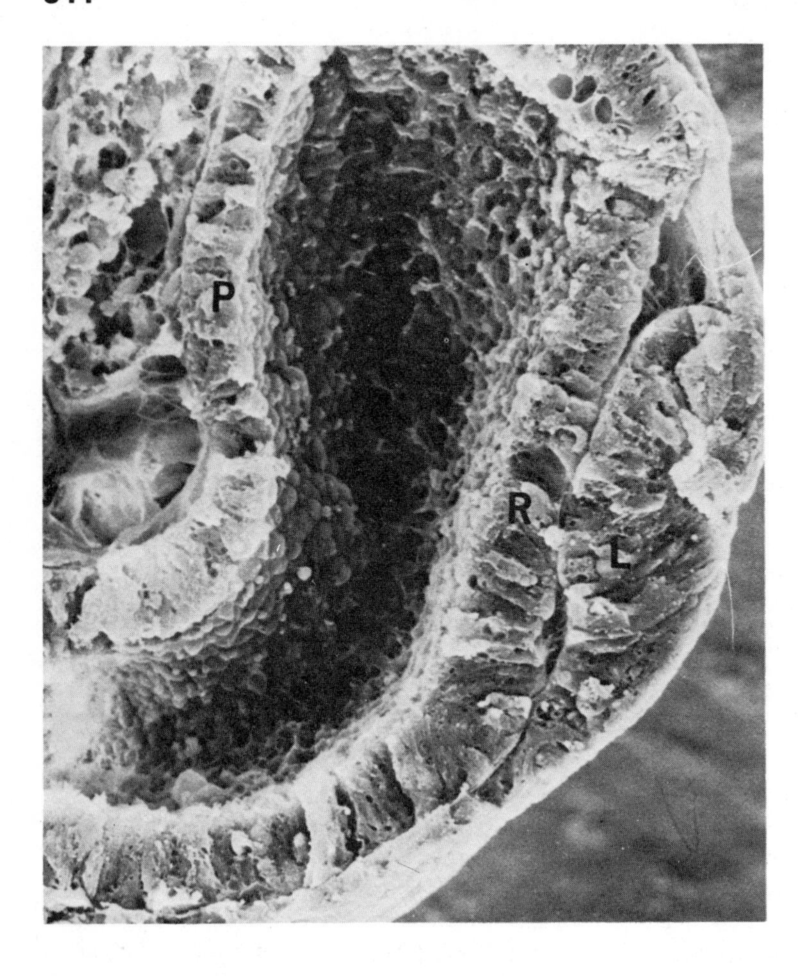


**FIG. 21–32.** Scanning electron micrograph of the eye of a chick embryo that was treated in ovo with tunicamycin before invagination of the optic cup. A flattened lens (L) has formed, but the optic vesicle shows little evidence of invagination. P, primary optic vesicle; R, presumptive retina. (From J. Yang and S. R. Hilfer, 1982. Dev. Biol. 92:41.)

of extracellular materials; this is very evident along the apical surface. In most cases, these embryos show complete inhibition of optic cup formation (Fig. 21–32). When the basal surface of an optic cup is examined, the attachment to the margin of the lens epithelium is loose; consequently, the basement membrane slides past the margin of the presumptive retina and the latter fails to fold in properly. Hence, the matrix along the basal surface of the optic vesicle serves to affix this part of the eye primordium to the lens epithelium. The role of the extracellular matrix along the apical surface is subject to speculation. It is likely that this material may stabilize the invaginated retina against the pigmented layer during later stages of optic cup formation.

## References

Beebe, D. C., P. J. Compart, M. C. Johnson, D. E. Feagans, and R. N. Feinberg. 1982. The mechanism of cell elongation during lens fiber cell differentiation. Dev. Biol. 92:54–59.

Bloemendal, H. 1977. The vertebrate eye lens. Science 198:127–138.

Coulombre, A. J. 1965. The eye. In: Organogenesis. R. L. DeHaan and H. Ursprung, eds. New York: Holt, Rinehart and Winston.

Hendrix, R. W. and J. Zwaan. 1974. Changes in the glycoprotein concentration of the extracellular matrix between lens and optic vesicle associated with early lens differentiation. Differentiation. 2:357–362.

Jacobson, A. G. 1963. The determination of nose, lens, and ear. I. Interactions within the ectoderm and between the ectoderm and underlying tissues. J. Exp. Zool. 154:273–283.

Jacobson, A. G. 1966. Inductive processes in embryonic development. Science 152:25–34.

Karkinen-Jääskeläinen, M. 1978a. Permissive and directive interactions in lens induction. J. Embryol. Exp. Morphol. 44:167–179.

Karkinen-Jääskeläinen, M. 1978b. Transfilter lens induction in avian embryo. Differentiation. 12:31–37.

Mann, I. 1964. The Development of the Human Eye. New York: Grune & Stratton.

McAvoy, J. W. 1980. Induction of the eye lens. Differentiation. 17:137–149.

McDevitt, D. S. and S. K. Brahama. 1981. Ontogeny and localization of the $\alpha$, $\beta$ and $\gamma$ crystallins in newt eye lens development. Dev. Biol. 84:449–454.

Model, P., L. Jarrett, and R. Bonazzoli. 1981. Cellular contacts between hindbrain and prospective ear during inductive interaction in the axolotl embryo. J. Embryol. Exp. Morphol. 66:27–41.

Ostrer, H., D. Beebe, and J. Piatigorsky. 1981. β-Crystallin mRNAs: differential distribution in the developing chicken lens. Dev. Biol. 86:403–408.

Papaconstantinou, J. 1967. Molecular aspects of cell differentiation. Science 156:338–346.

Silver, P. H. S. and J. Wakely. 1974. Fine structure, origin, and fate of extracellular materials in the interspace between the presumptive lens and presumptive retina of the chick embryo. J. Anat. 118:19–31.

Topashov, G. V. and O. G. Stroeva, 1961. Morphogenesis of the vertebrate eye. Adv. Morphog. 1:331–378.

Treton, J., T. Shinohara, and J. Piatigorsky. 1982. Degradation of δ-crystallin mRNA in the lens fiber cells of the chicken. Dev. Biol. 92:60–65.

Twitty, J. 1955. Eye. In: Analysis of Development. B. H. Willier, P. A. Weiss, and J. Hamburger, eds. Philadelphia: W. B. Saunders Company.

van de Kemp, M. and J. Zwaan. 1973. Intracellular localization of lens antigens in the developing eye of the mouse embryo. J. Exp. Zool. 186:23–32.

Yang, J. and S. R. Hilfer. 1982. The effect of inhibitors of glycoconjugate synthesis on optic cup formation in the chick embryo. Dev. Biol. 92:41–53.

Yntema, C. L. 1955. Ear and nose. In: Analysis of Development. B. H. Willier, P. A. Weiss, and J. Hamburger, eds., Philadelphia: W. B. Saunders Company.

# 22

# DEVELOPMENT OF THE SOMITES

We have described the development of the somites into segmentally arranged masses of mesodermal cells connected laterally to the lateral mesoderm by way of the intermediate mesoderm (Chapter 8). In man, the first somite appears during week 3 of development, and early in week 5, when somite formation is complete, some 42 to 44 pairs of somites have formed. Although the number of somites that develop is subject to some variation, somite count in these early stages provides a reliable measure of developmental progress.

In the differentiation of each somite from the paraxial mesoderm, the cells at first appear very similar (Fig. 22–1 A) but soon exhibit a wide range of developmental potentialities. Shortly after it first forms, each somite's boundaries become more regular and its cells increase rapidly in number and assume a radial orientation around a small central lumen, the *myocoele* (Fig. 22–1 B). Further development consists of a lengthening of the somite in a dorsoventral direction and a flattening in a lateral direction accompanied by the formation of a slit-shaped myocoele (Fig. 22–1 C,D). At this time, three regions of the somite may be recognized and named on the basis of their prospective fate (Fig. 22–1 C,D). In the most medial and ventral region, the cells lose their epithelial appearance and migrate away from the main body of the somite toward the notochord and the neural tube as a mass of mesenchyme. This region, which will give rise to skeletal material, is the *sclerotome*. A dorsal intersegmental artery runs between adjacent sclerotomes. The most superficial cells lying in a ventrolateral position beneath the ectoderm make up the *dermatome*, which will develop into the connective tissue of the epidermis and the subcutaneous tissue beneath it. In all probability some of the cells of the region marked off as the dermatome may also contribute to the formation of skeletal muscle. The third region is the *myotome*, which forms a platelike group of spindle-shaped cells medial and dorsal to the dermatome. It will form skeletal muscle. The further development of these three regions will be considered separately.

## Muscle Development

The muscular tissue of the body, with the exception of that found in the iris and the muscles associated with the sweat and mammary glands, is derived from mesoderm. Muscle tissue is developed from primitive myoblasts, specialized cells that will differentiate fibers in which the function of contractility is highly developed. On the basis of their his-

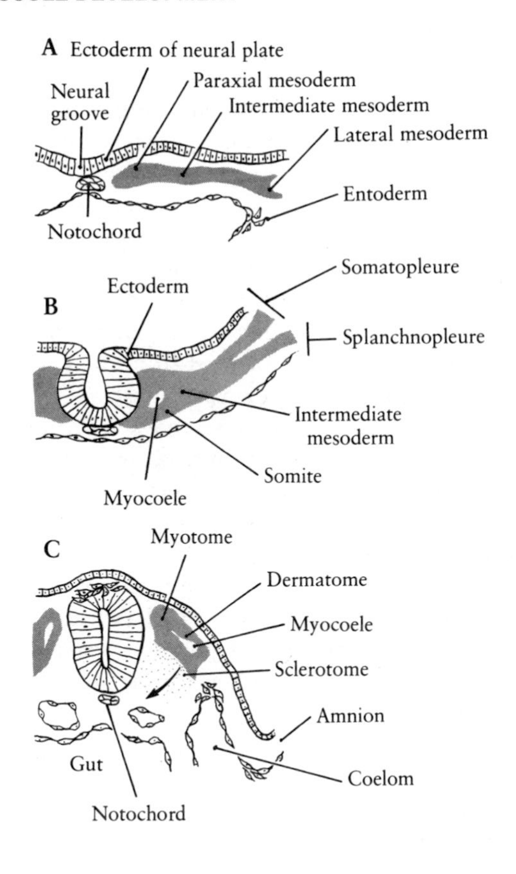


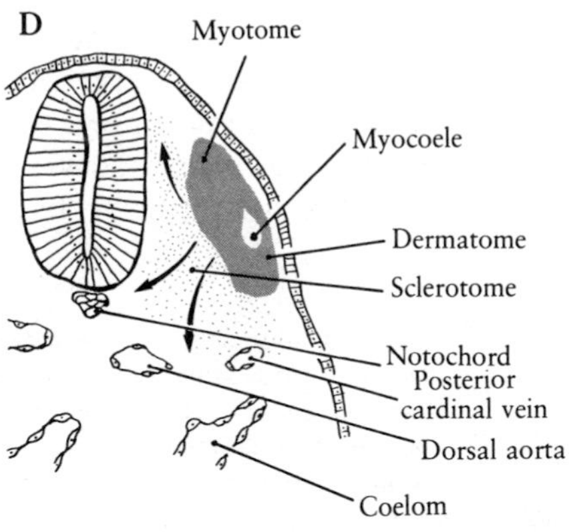


**FIG. 22–1.** Diagrams of the early differentiation of a somite. A, beginning of somite formation from the paraxial mesoderm; B, formation of the myocoele; C, beginning of the migration of sclerotome cells toward the notochord; D, further migration of the sclerotome masses.

tology, origin, and ultimate location, three types of muscle tissue are recognized: (1) *skeletal* muscle, which is attached to and serves to move the bones of the body; (2) *smooth* muscle, which is found in the alimentary, respiratory, and urogenital tracts and in the walls of blood vessels and the ducts of glands; and (3) *cardiac* muscle, found in the walls of the heart. We will consider mainly the development of the first two types in this section. Cardiac muscle develops from the splanchnic mesoderm that surrounds the heart tube. The development of the heart was described in Chapter 18.

## Skeletal Muscle

Skeletal muscle develops from three different regions of mesoderm. The axial (trunk) muscles, and probably the muscles of the limb girdles, develop from the myotomes of the somites. Most of the muscles of the head region develop from the visceral mesoderm of the visceral arches, and the muscles of the limbs develop from local accumulations of cells of the lateral mesoderm.

### *The Development of a Myotome*

Each myotome increases in size and, at five weeks, divides into a smaller group of cells located dorsally, the *epimere*, and a larger ventral group of cells, the *hypomere*, thus establishing an *epaxial* and *hypaxial* column of myotomes (Fig. 22–2). At the same time, each spinal nerve growing into its respective somite divides into a *posterior primary division* and an *anterior primary division* that estab-

**FIG. 22–2.** Division of the myotome into a dorsal epimere and a ventral hypomere in the five-week-old embryo. The posterior primary division and the anterior primary division of the spinal nerve supply the epimere and the hypomere, respectively.

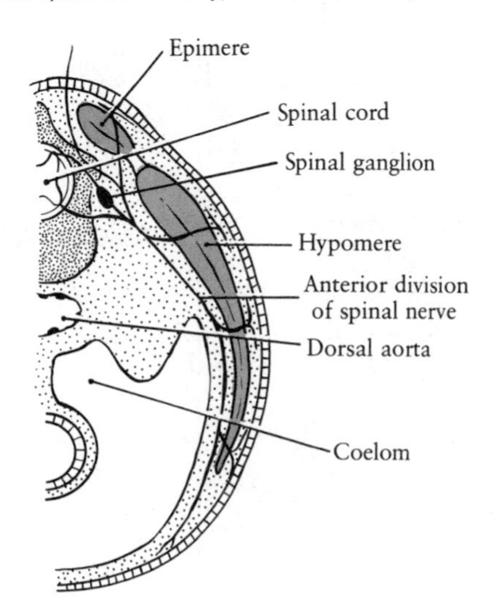

lish permanent connections with the epimere and the hypomere, respectively (Fig. 22–2).

The epaxial musculature subdivides further into deep and superficial portions. The deep portion may fuse over a few consecutive somites but, in general, retains its original segmental arrangement giving rise to short intervertebral muscles. The superficial portion fuses over a large number of consecutive segments and then, by longitudinal splitting, develops the long extensor muscles of the head and trunk (Fig. 22–3).

The hypaxial musculature extends into the lateral and ventral body wall. The myoblasts of the more cranial hypomeres give rise to the scalenes, shoulder girdle muscles, and intercostal muscles (Fig. 22–3 A). The more caudal hypomeres form the broad, coatlike-layered muscles of the thoracic and abdominal walls and the longitudinally running rectus abdominis (Fig. 22–3 B). The hypomeres of the lumbar somites form the muscles of the pelvic girdle.

*Muscles of the Limbs*

The limb buds in man form during the second month as ectodermal outpocketings of the body wall filled with mesenchyme. Condensations of the limb bud mesenchyme form premuscle masses near the base of each limb bud. The superior limb buds form opposite the last six cervical and the first two thoracic myotomes, and the inferior limb buds develop opposite the second to the fifth lumbar and the first three sacral myotomes. Branches of the spinal nerves supplying these myotomes extend into the premuscle masses and, as the limb elongates and the muscles differentiate, the nerves retain these early connections. The muscles develop into a ventral limb flexor group and a dorsal limb extensor group (Fig. 22–3 A). Anterior and posterior branches of the spinal nerves supply the flexor and extensor groups, respectively. Thus, in the superior extremity, the median and ulnar nerves, from the anterior division, innervate the flexor muscles while the radial nerve, derived from the posterior division, innervates the extensor muscles.

Although it has been shown that the fin musculature of fishes is derived from extensions of the myotomes into the developing fins, evidence for a myotomal origin of the limb musculature in the tetrapods is not convincing. Most embryologists consider that these muscles differentiate in situ from limb bud mesenchyme, probably of lateral plate origin. An analysis of the morphogenesis of the vertebrate limb has already been presented in Chapter 13.

*Tongue Muscles*

Cranial to the cervical somites, four *occipital somites* develop. The first occipital somite then disappears, and the myotomes of the remaining three differentiate into the intrinsic muscles of the tongue. The migration of the myoblasts from the occipital myotomes to the floor of the oral cavity to form the tongue muscles has not been observed directly. However, the hypoglossal nerve has been

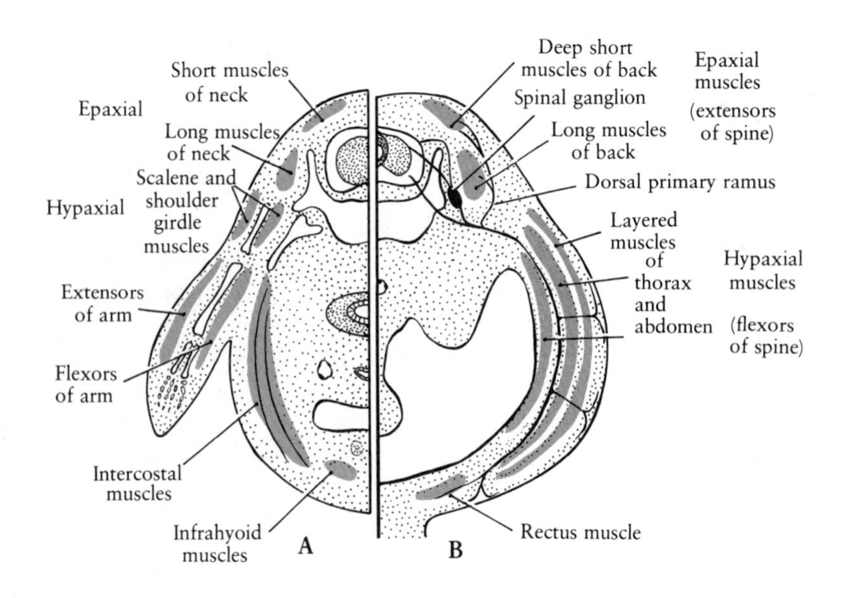


**FIG. 22–3.** Diagrams of the arrangement of the primitive muscle masses derived from the epimeres and the hypomeres. A, section through the shoulder region; B, section through the trunk region.

shown to grow into the occipital somites, and the innervation of the tongue muscles by this nerve indicates the origin of these muscles from the myoblasts of the occipital somites.

*Eye Muscles*
The intrinsic muscles of the eye appear to develop in situ from head mesenchyme. The three muscle masses from which they develop can be considered as three pairs of *head (preotic) somites*, even though typical somites do not develop in the head region of any mammalian embryo. In the shark embryo, in which head somites do differentiate, the myotome portions of the three somites in this region develop into the extrinsic eye muscles and are supplied by cranial nerves III, IV, and VI. Thus, in the mammal the three mesenchymal masses that give rise to the eye muscles and that are, in turn, supplied by the same three cranial nerves, represent the differentiation of what is, at least phylogenetically, head somite material.

*Visceral Arch Muscles*
The development of the muscles derived from the visceral arches has been described in Chapter 14.

## Smooth Muscle

The smooth muscle of the alimentary tract arises independently of the somites as a differentiation of the splanchnic mesoderm covering the alimentary tract. From it are developed the circular and longitudinal muscle coats of the tract as well as the muscularis mucosa, the thin layer of muscle found in the mucosa of many parts of the gut. The smooth muscle of the trachea and the bronchi develop in the same manner.

Other smooth muscle is found as part of the coat of the vessels of the circulatory system. In those vessels that develop in the splanchnopleure, such as the allantoic and vitelline vessels, the muscle is also of splanchnopleuric origin. However, it is probable that, as the blood vessels extend into the somatopleure of the body wall and limb buds, the muscles developing in the walls of these vessels arise from the surrounding somatic mesoderm. The smooth muscle of some parts of the urogenital system is also derived from somatic mesoderm, and it is not unreasonable to consider that mesenchyme anywhere in the body is capable of forming this type of muscle.

## Myogenesis

Muscle cells and tissues have been a favored material in cell differentiation studies, because they possess a highly specific morphology and identifiable markers of their differentiated state, such as myosin and actin. Also, the differentiation of skeletal muscle cells in cell culture from isolated embryonic myogenic cells closely parallels the developmental sequence observed in vivo. The cell culture system for analyzing muscle development offers several advantages over the study of muscle in the intact embryo. In contrast to the embryo in which muscle tissue is complicated by many distinct cell types, including nonmyogenic cells, cell suspensions of embryonic muscle can be prepared that are relatively homogeneous. These cells can be manipulated experimentally to give a greater degree of synchrony of development than in the embryo. In this section we will consider primarily the events leading to the differentiation of skeletal muscle, because this cell type, in contrast to smooth and cardiac muscle cells, has been rather extensively studied. The salient features of skeletal myogenesis can be arranged in the following chronological sequence: (1) the formation by proliferation of presumptive myoblast and myoblast cells, (2) the synthesis of myosin, actin, and other associated muscle-specific molecules, and (3) the fusion of myoblasts into multinucleated cells termed *myotubes* that continue synthesis of contractile proteins and their assembly into myofibrils.

The differentiation of the skeletal muscle cell can be considered to consist of two primary events. There is a beginning period when presumptive myoblast cells, which originate chiefly from the mesoderm of the lateral plate or somites, increase by cell mitoses (Fig. 22–4 A). These cells are observed to increase at a constant exponential rate in culture. At a rather predictable time, the mononucleated myogenic cells cease to divide and fuse to form multinucleated cells containing many nuclei with a common cytoplasm (Fig. 22–4 B–D). The number and size of these cellular syncytia increase very rapidly for the first 48 hours after the fusion process begins. Although details of the signal or signals that initiate the fusion between myoblasts, or between myoblasts and myotubes, are still unclear, observations on the behavior of muscle cells in culture suggest that fusion is a two-step process. There is an initial cell recognition by primed homotypic

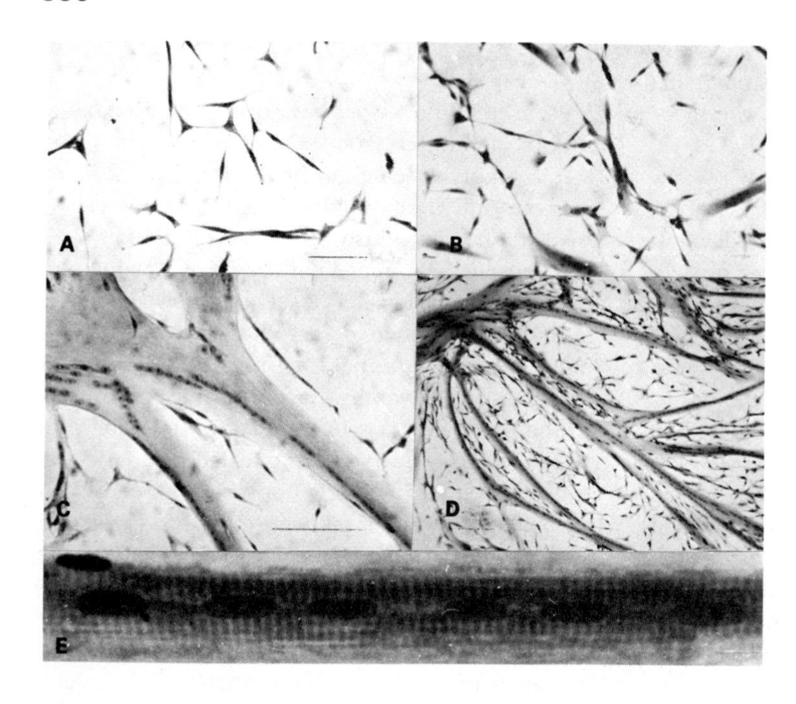


**FIG. 22–4.** Photomicrographs of quail muscle cells in culture. A, mononucleated myoblasts recognizable by their characteristic bipolar shape; B, the beginnings of the formation of myotubes from fusing myoblasts; C, a higher magnification of the fusion process; D, extensive formation of myotubes; E, cross-striations in a multinucleated myotube. (From I. Konigsberg and P. A. Buckley, 1974. In: Concepts of Development. J. Lash and J. Whittaker, eds. Sinauer Associates, Stanford, Conn.)

(similar) cells. This has been shown by labeling fibroblasts, chondroblasts, and liver cells with tritiated thymidine and then mixing these nonmyogenic cells with unlabeled myoblast cells that are in the process of fusion. Labeled nonmyogenic cells are never incorporated into myotubes, indicating that there is a very precise system of recognition sites on the surfaces of myogenic cells about to fuse. The second step appears to involve rearrangements in the membrane surfaces of myogenic cells, for several collisions precede a period of stationary contact before myoblasts fuse either with each other or with myotubes.

Skeletal muscle is an unusual tissue in that it is formed by the physical fusion of its constituent cells. Initially it was thought that the myotubes were formed by mitosis of a single muscle cell without subsequent division of its cyotplasm. A subject of considerable controversy has been the relationship between the event of fusion, the cell cycle, and cytodifferentiation. Fusion of the mononucleated myogenic cells appears to be tightly coupled to the mitotic cycle. Bischoff and Holtzer (1969) have determined from analysis of myogenic cell cultures that myoblasts fuse only when they are in the $G_1$ phase (before DNA synthesis) of the cell cycle. Once fusion has taken place, the nuclei of myotubes do not synthesize DNA and hence do not divide, except under certain abnormal conditions, such as during repair processes. That fusion results in the cessation of DNA synthesis and cell division may be easily demonstrated by the fact that prefused myoblasts contain DNA in two amounts corresponding to 2N and 4N, while the multinucleated cells or fused myoblasts show only a single amount corresponding to 2N. The latter is the amount of DNA present in the $G_1$ or postmitotic phase of the cell cycle. Also, myotubes never take up tritiated thymidine, indicating that they are no longer synthesizing DNA.

The second phase of the differentiation of the skeletal muscle cell takes place after the fusion of myoblasts into myotubes. Myosin, actin, and other proteins characteristic of this tissue type accumulate rapidly after fusion. Myosin and actin self-assemble into the thick and thin myofilaments, respectively, which characterize the myofibrils of muscle tissue (Figs. 22–4 E; 22–5). There is evidence that myosin is synthesized on polysomes containing 50 to 60 ribosomes. Approximately 200 molecules of myosin self-assemble to form a thick filament about 150 Å in diameter and 1.6 microns in length. Considerably less information is available on the formation of the thin or actin filament. Once assembled, however, thin filaments form a hexagonal array around each myosin filament (Fig. 22–5). The earliest myofibrils have very few myofilaments. Additional filaments are added at the cir-

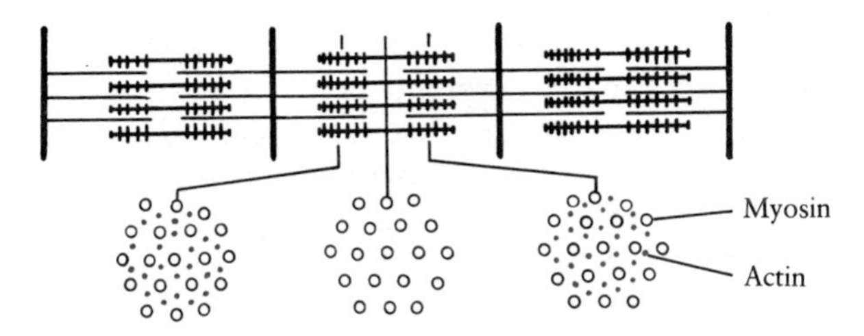


**FIG. 22-5.** Diagrammatic representation of a part of a myofibril showing the arrangement of actin- and myosin-containing filaments. Note the hexagonal array of actin filaments around the myosin filament. (From A. Huxley, 1969. Science 164:1356. Copyright 1969 by the American Association for the Advancement of Science.)

cumference of a growing myofibril until the fully mature state is reached.

Since fusion inhibits DNA synthesis, one might logically ask whether this step in muscle differentiation initiates the synthesis of the muscle-specific contractile proteins. Two particularly useful tools in detecting minute quantities of myosin are the electron microscope and use of fluorescein-labeled antibodies prepared against myosin. Analysis of various types of muscle cells using these techniques suggests that the synthesis of myosin is initiated at variable times during the development of a muscle cell. For example, the mononucleate myoblasts of early chick somite tissue not only accumulate myosin before their fusion, but organize the protein into recognizable filaments. By contrast, leg muscle cells from older chick embryos clearly do not begin the production of myosin until after the fusion of myoblasts. The messenger RNAs coding for myosin are present in the prefused rat myoblast, but their translation is delayed until myotube formation. This suggests that specific biosynthesis of myosin is regulated at the translational level and linked to myogenic fusion.

The relationship between cell division and the differentiation of cell types has been and continues to be the subject of heated debate between investigators. During the differentiation of a cell type, there is an initial period in which cell numbers increase by the ordinary mitoses of stem cells. Subsequently, cytodifferentiation follows in which the proteins or so-called *luxury molecules* that characterize the terminal cell type are synthesized. Cells then are differentiated into types on the basis of the types of luxury molecules they produce (i.e., myosin in muscle, hemoglobin in red blood cells, etc.). It is generally held that in this type of system the differentiated state is accompanied by a loss in capacity for the synthesis of DNA and for further proliferative activity.

Sharp differences of opinion exist about the mechanism involved in the transformation of proliferating myoblasts into nonproliferating, differentiated myotubules. On the basis of studies with erythrogenic, myogenic, and chondrogenic cells, Holtzer and his collaborators (1972) have proposed that the cell cycle plays an obligatory role in organizing and channeling the differentiation of these cell types. They consider that there is a causal relationship between division and differentiation and that DNA synthesis and nuclear division are essential prerequisites for the formation of myotubules. Holtzer distinguishes two types of cell cycles in cell populations. First, there is a proliferative cycle that results in daughter cells whose patterns of protein synthesis are identical to those of the parent cell. The sole function of proliferative cell cycles is to increase cell numbers. Second, there is a *quantal cell cycle* that produces daughter cells capable of synthesizing protein patterns very different from the parent cell. That is, there is a reprogramming of genes in the progeny cells. With regard to muscle tissue, the quantal scheme would postulate that the presumptive myoblast passes through an obligatory S phase (DNA synthesis), at which time the genome is reprogrammed. The daughter or myoblast cells are then "postmitotic" and capable of overt differentiation (i.e., fusion and synthesis of muscle-specific proteins). The transition to the terminal myoblast capable of cytodifferentiation is coupled to this single, critical round of DNA synthesis. According to the quantal concept, a cell cannot synthesize cell-specific proteins until the terminal quantal mitosis has occurred.

A model for the establishment of the muscle cell lineage, shown in Figure 22–6, proposes the existence of several discrete, discontinuous populations of cells that act as precursors to the myoblast cell. One or two quantal divisions interspersed with proliferative cycles are envisaged as setting up the broad determinations of ectoderm, endoderm, and mesoderm. Further quantal mitoses within the mesodermal poulation create lineages leading to cartilage, muscle cells, and so on. Within the myogenic line of cells, $\alpha$-myogenic and $\beta$-myogenic cells would be early precursor cells to the stem or presumptive myoblast cells. These are transient populations and cease to perpetuate their own phe-

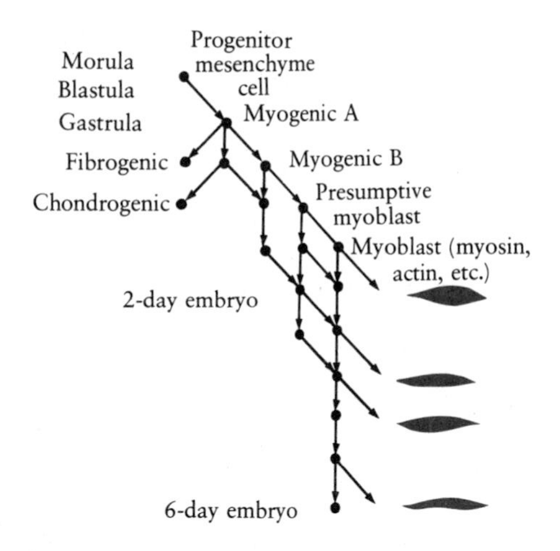


**FIG. 22-6.** Holtzer's concept of the role of quantal mitoses in the origin of cell lineages and the differentiation of the myoblast cell. A vertical mitosis is proliferative and perpetuates the phenotype. A horizontal mitosis is quantal and alters the phenotype of the daughter cells. (From H. Holtzer, 1970. In: Cell Differentiation. O. A. Scheide and J. de Vellis, eds. © 1970 by Litton Educational Publishing, Inc. Reprinted by permission of Van Nostrand Reinhold Company.)

notypes because they give rise to higher levels of differentiation. The presumptive myoblasts can either proliferate into more presumptive myoblasts or undergo a terminal quantal mitosis to initiate the terminal differentiative step. All horizontal shifts between cell "compartments" presumably involve the selective derepression of batteries of genes.

The concept of a quantal cell cycle has not been conclusively proved for any type of tissue, but the results of a number of experiments on muscle differentiation may be cited in its support. The fusion of myoblasts to form myotubules can be prevented by inhibiting or delaying what is considered to be the quantal cell cycle. Chick myoblasts cultured under conditions in which DNA synthesis is inhibited—high population density or the presence of specific inhibitors of DNA synthesis—form fewer myotubules than those cultured under normal conditions. Also, if myoblasts are cultured in a medium that allows them to divide normally but inhibits fusion to form myotubules—a so-called nonpermissive medium—they will fuse when put into a permissive medium, but only after a period of time sufficient to allow for DNA synthesis and cell division. They thus seem to need to complete one cell cycle, or at least DNA synthesis and cell division, under conditions that are permissive for further differentiation before they are actually able to accomplish this differentiation.

If myoblasts in culture are pulse labeled (exposed for about half an hour) with tritiated thymidine and then immediately radioautographed and also tested for the presence of myosin by their ability to bind fluorescent antimyosin, individual cells may show either labeled DNA or the presence of myosin, but not both. However, if the cells are labeled and tested at timed intervals following labeling, some cells will show both labeled DNA and the presence of myosin, but only beginning at ten hours after labeling. Ten hours is the time it would take cells labeled in S to pass through $G_2$ and M (3 to 4 hours) and then spend 6 to 7 hours in $G_1$ of the next cycle before beginning to synthesize myosin. These are cells that have fused into myotubules about 5 hours into $G_1$ of the new cycle and have begun to synthesize myosin shortly after fusion. The nuclei of the myotubules remain in $G_1$ (more correctly, $G_0$) permanently unless needed for muscle repair following injury.

However, not all investigators agree with the concept that myotubule formation is dependent on a program intrinsic to the cell and unique to the differentiation process. Konigsberg and his co-workers (1974, 1978) proposed that the controlling factors were to be found in the environment, noting that dense cultures of cells promoted myotubule formation. As cells are cultured over a period of time, their density increases and so does their fusion rate. Also, if the number of cells in an inoculum used to set up a culture is varied, those cultures with the larger number of cells show fusion earlier than the less-dense cultures. Equal aliquots from the same culture will either continue to proliferate when a large volume of culture medium is added or initiate fusion when cultured in a smaller volume. Since it has been known for some time that fusion always takes place in $G_1$, it is interesting to note that in high-density fusing populations, the $G_1$ is prolonged and averages almost twice the length of the $G_1$ in proliferating populations. However, protraction of $G_1$ is not a programmed event intrinsic to the cell, but occurs as a response to the environment. In cell cultures followed over a period of time, cell cycle measurements indicate that cell density increases continually and $G_1$ begins to lengthen significantly on day 3. Day 3 is also the first day that cell fusion appears. Thus, it is concluded that increasing cell density brings about a protraction of $G_1$ and an initiation of cell fusion. However, this does not mean that all of the cells in which $G_1$ has been prolonged and that are capable of fusion, but have not as yet done so, have withdrawn from the cell cycle. Time-lapse cinematography has shown that when two daughter cells were followed, in about half of the daughter pairs one cell started fusion and the other went on to divide again. The commitments of daughter cells are thus not necessarily the same.

It is possible to induce prolongation of $G_1$ and myotubule formation in proliferating cells by culturing them in media collected from older fusing cells. This led some investigators to propose that the media from the older fusing cultures contained a fusion-promoting factor, an addition of cell type-specific material to the media by the fusing myotubules. Konigsberg, in contrast, proposed that the older media promote fusion because they have been depleted of the high-molecular-weight growth-promoting components of the original media. If this is the case, the addition of such factors to the older media should decrease the amount of fusion. When the serum and the high-molecular-weight fraction of the embryo extract of the original medium are added to the older cultures, they do just that. Conversely, decreasing the amount of serum and embryo extract in the original medium results in precocious fusion. Thus, the continued proliferation of myoblasts conditions the medium by depleting it of growth-promoting materials, and the result is a prolongation of the $G_1$ phase of the cell cycle and an initiation of fusion. However, some cells may remain in $G_1$ for well over twice the normal length of time and still reenter S. It is thus proposed that the changes in the cells' environment resulting in the lengthening of $G_1$ are neither accompanied by an obligatory withdrawal from the cell cycle nor preceded by an intrinsically programmed quantal cell cycle.

Another investigator (Nadal-Ginard, 1978) studied cell fusion using a permanent cell line ($L_6E_9$) of rat myoblasts. These myoblasts can be induced to undergo fusion and myosin synthesis in a high percentage of cases by altering the culture medium. In normal growth medium, the line is purely proliferative and composed of cycling nondifferentiating cells. When placed into a "differentiation medium" after first being pulse labeled, they fuse after undergoing two to three divisions. At the time of fusion, 98 percent of the nuclei were labeled, and thus, practically all of the cells had divided before fusing. DNA synthesis in the proliferating population can be inhibited by a number of drugs that do not affect cell viability. If these inhibitors are added to the differentiating medium at the same time that the proliferating cells are added, a large percentage ($>$ 90%) of the cells neither synthesize DNA nor divide, yet a high percentage (66–75%) form myotubules and do so at an earlier time than the controls. Thus, in this cell line DNA synthesis is not required to switch from a proliferating to a differentiating program, and a quantal type of cell division is not needed. However, the $L_6E_9$ lines consists of genetically altered permanent myoblasts, and the results using these cells may not apply to normal myoblasts.

As another example of environmental influence on muscle differentiation, it has been proposed that environmental clues may play a part in determining the specific direction in which limb mesenchymal cells will differentiate. These cells have the capacity to differentiate into only a few different types of tissues: muscle, cartilage, bone, and connective tissue. One of the environmental factors that directs a mesenchymal cell into either cartilage or muscle differentiation may be related to an effect on the size of the internal pool of nicotinamide adenine dinucleotide (NAD). High internal NAD pools are associated with myogenic and low-NAD pools with chondrogenic differentiation. The size of the internal NAD pool is in turn regulated by the external concentration of nicotinamide. The cells are permeable to nicotinamide but not to NAD. Once inside the cell, nicotinamide is converted to NAD, and thus, high external levels of nicotinamide dictate high internal levels of NAD and myogenic differentiation. Chick limb mesodermal cells respond to the level of nicotinamide in the culture medium in a manner that supports this hypothesis. Lowering levels favors chondrogenesis, and raising levels favors myogenesis. Once a cell has been committed to a certain pathway, it continues along this line and does not have the option of reverting and starting over again. It is interesting to note that cartilage develops in the relatively avascular regions of the limb, while muscle develops in the more vascular areas. Since nicotinamide is stored in the yolk, its relative tissue concentration is determined by the density of the vascular supply.

Skeletal muscle and cardiac muscle cells share several common features during the myogenesis.

The myoblasts of both cell types undergo a mitosis to produce daughter cells that cease to synthesize DNA and commence to translate rapidly for actin, myosin, and other muscle-specific proteins. While these programs are apparently initiated concurrently in the $G_1$ phase of daughter cells of skeletal tissue, the decision to shut down DNA synthesis is delayed for one or two division cycles in cardiac muscle cells.

## Connective and Supportive Tissue

The connective and supportive tissues of the body are derived from embryonic mesoderm. Part of this mesoderm appears in the segmentally arranged somites that give rise to the axial skeleton and its muscles. However, everywhere between the epithelial layers of the endoderm and the ectoderm, another kind of mesodermal tissue is present as a loose aggregation of cells. This is the mesenchyme. The mesenchymal cells in the embryo are stellate in shape and form a loose network, the interstices of which are occupied by a structureless intercellular matrix or ground substance. The branching processes of the mesenchymal cells are adherent rather than continuous, as has been shown by extensive observations of tissue cultures. This loosely arranged material of the early embryo soon differentiates in a number of different directions that will result in the formation of connective and supportive tissue. Its development in any direction is distinguished not so much by the differentiation of the cells as by what takes place in the intercellular matrix. The matrix is characterized by the presence of fibers of three different kinds; *recticular, collagenous*, and *elastic*. The cells that form these fibers are called *fibroblasts*. Fiber formation occurs as an organization of the ground substance in close proximity to the cell surface. Intercellular material, including the fibers, is nonliving and dependent for its continuous existence on the presence of the cells of the connective tissue.

The type of connective tissue that remains most like the embryonic mesenchyme is the reticular connective tissue, which forms the supporting framework in many lymphoid organs. *Tendons* and *ligaments* are connective tissue developments in which the collagenous fibers are closely packed in a parallel orientation. *Fascia* is connective tissue with a loosely arranged network of both collagenous and elastic fibers. *Supportive tissue*, cartilage and bone, is also an important derivative of the embryonic mesoderm. Although cartilage and bone contain the same kinds of fibers present in other types of connective tissue, the major change in the formation of these tissues is an increase in the hardening and strengthening of the intercellular matrix.

### The Differentiation of Cartilage

Cartilage begins to develop during the second month when, in dense aggregations of mesenchyme cells, *centers of chondrification* appear. In these aggregations the cells, *chondroblasts*, are closely packed and their processes are withdrawn (Fig. 22–7 A,B). This precartilage mass increases in size and is molded into the shape that the cartilage will eventually take. Increasing amounts of matrix now appear and the cells are forced farther and far-

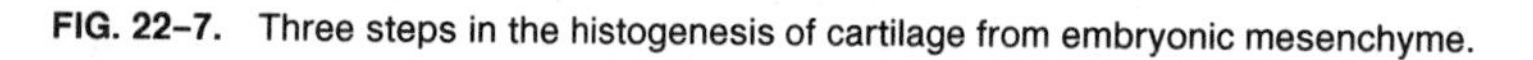
**FIG. 22–7.** Three steps in the histogenesis of cartilage from embryonic mesenchyme.

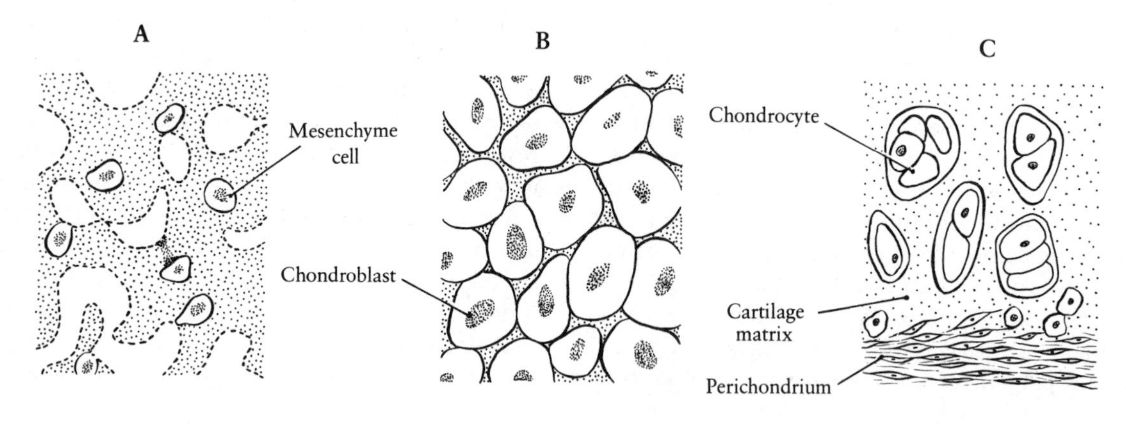

ther apart, each cell surrounded by matrix and enclosed in its own separate compartment. Young chondroblasts may divide once or twice and, in mature cartilage, the division products then occur as isolated groups of two to four cells embedded in the matrix and separated by it from other similar groups (Fig. 22–7 C). Each cartilage is covered by a tough connective tissue sheet of *perichondrium*. Continued growth of cartilage may occur by either appositional or interstitial growth. In the former, new chondroblasts are differentiated from the connective tissue of the perichondrium, and these cells add new matrix to the surface. A small amount of interstitial growth may also occur by the addition of new material on the inside of the structure, thus resulting in internal expansion. The definitive cartilage is avascular, and the *chondrocytes* are supplied by the diffusion of substances through the matrix.

During fetal life the major portion of the skeleton is cartilaginous, but cartilage persists in the adult only in certain restricted areas such as the articular surfaces of the long bones and in the respiratory tract.

*The Role of the Notochord in Cartilage Differentiation*

The cartilage of the vertebral column is formed from somite mesoderm. Early experiments indicated that vertebral chondrogenesis is dependent on an inductive influence from the embryonic notochord or spinal chord. When grown alone, somites from early chick embryos do not form cartilage, but they do so if cultured with either embryonic notochord or spinal cord. More recent experiments, however, have shown that under the proper culture conditions embryonic somites can form cartilage under their own power. The mass of the cultured material is important. Individual somites or small numbers of somites seldom form cartilage matrix. Even mature chondrocytes that are actively synthesizing cartilage will lose their typical shape and become fibroblast-like when cultured in small groups of cells, and although they continue to synthesize DNA and divide, they no longer form cartilage matrix. However, if entire rows of somites from a number of embryos are cultured together in a tightly packed mass, they will synthesize cartilage without any outside influence. The cells in the densely packed center of the culture aggregate, form connections with each other, and deposit large quantities of matrix. Cells on the periphery tend to flatten out and avoid forming close associations with each other. Hence, they fail to synthesize matrix. Thus, the establishment of a dense aggregate of sufficient size is enough of a stimulus to trigger matrix formation. Indeed, the formation of a dense mass of precartilage cells is the first step in normal chondrogenesis.

However, under normal conditions vertebral cartilage is formed by an interaction between the somite mesoderm and embryonic notochord and spinal cord. This interaction depends on a diffusible substance, since it can take place through a Millipore filter and can be produced using tissue extracts. The nature of the substance has not been defined. Its action should be considered as a trigger mechanism rather than as an inductive one. Precartilage cells are already biased toward the synthesis of matrix, and the notochord acts as a stimulus for the stabilization and enhancement of already existing pathways of differentiation. It is not an inducer of new activity. The stimulus could be in the form of the release of a block, which would then allow the accumulation of terminal products; or it might act as a repressor of some other of the cells' potentialities such as myogenesis.

*Biochemical Differentiation*

One of the components of cartilage is chondroitin sulfate, repeating units of sulfated *N*-acetylgalactosamine and uronic acid, which, when occurring as side chains attached to a protein backbone, forms proteochondroitin sulfate. In the differentiation of cartilage, it would be interesting to determine at what time the metabolic pathways for the synthesis of chondroitin sulfate are established. The steps in its synthesis involve compounds that are reasonably specific and for which assays are available. All of the evidence indicates that low levels of chondroitin sulfate synthesis occur in the earliest somites formed as well as, surprisingly, in other embryonic tissues, including extraembryonic membranes. Levels of chondroitin sulfate increase during cartilage differentiation, and it is found in large quantities only in mature cartilage. Cartilage differentiation is thus often defined by the increasing amount of chondroitin sulfate measured by the uptake of radioactive sulfur.

Recently it has been determined that the synthesis of chondroitin sulfate and proteochondroitin sulfate may be more specific to cells that will differentiate into chondrocytes than had been reported previously. Analysis of proteochondroitin sulfate synthesized by chick chondrocytes in vitro has shown it to be chromatographically heterogeneous with two distinct peaks. The first represents about 90 percent of the total, the second about 10 percent. It was then suggested that the proteochondroitin sulfate seen in the first peak represented a cartilage-specific molecule, while that seen in the second was a more widespread nonspecific molecule. One would then expect that the chondroitin sulfate seen in the early somite would be the latter, nonspecific kind and that, as the cartilage differentiated, the specific variety would appear and increase. In fact, analysis of explants from stages 23 and 24 of chick limb buds cultured for different lengths of time shows a striking change in the elution profile of proteochondroitin sulfate over a period from 18 hours to nine days (Fig. 22–8). In the analysis, three peaks appear, peaks Ia and Ib representing monomer and aggregate forms of the cartilage-specific molecule and peak II representing the widespread molecule. In the material cultured for 18 hours, peaks Ia and Ib represent about 22 percent of the total and peak II the remainder. After two and nine days of culture, the peaks remain qualitatively alike, but there is a quantitative increase in peaks Ia and Ib and a decrease in peak II so that at nine days peak II contains only about 11 percent of the total. Thus, cartilage differentiation is marked by a disproportionate increase of a particular molecule of proteochondroitin sulfate, which is specific for cartilage matrix. The presence of small amounts of this material in the precartilaginous mesoderm may represent the same situation seen in the presence of myosin in the limb bud at stages before any myoblasts can be identified morphologically. Both cases would involve an initial biochemical differentiation, the establishment of specific metabolic pathways, which underlies and presages the final morphological differentiation.

**FIG. 22–8.** Elution profiles of proteochondroitin sulfate obtained from cell cultures of stage 23–24 chick limb buds. Abscissa: effluent volume; ordinate: percentage of total dpm. A, after 18 hours of culture. Only 22 percent of the activity is found in the cartilage-specific molecule (Ia and Ib). B, after two days of culture. Activity is shifting from the cartilage-nonspecific to the cartilage-specific molecule; C, after nine days of culture. About 90 percent of the activity is now found in the cartilage-specific molecule. (From P. F. Goetinck et al., 1974. Exp. Cell Res. 87:241).

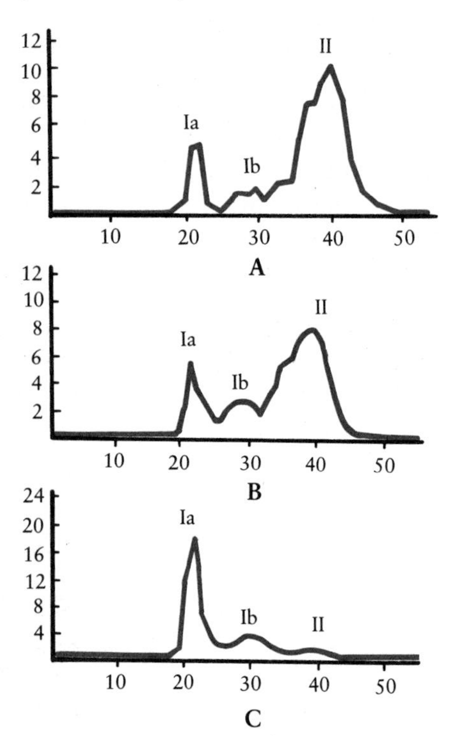


Similarly, the type of collagen synthesized in cartilage is unique to that tissue. It is composed of three $\alpha$ chains that have a different type of structure, known as type II, from the $\alpha$ chains of collagen, known as type I, found in skin and bone. The early precartilaginous limb mesoderm synthesizes type I collagen. As development progresses there is a shift of collagen synthesis to type II in the chondrogenic areas while the other limb mesoderm continues to synthesize type I collagen.

Studies on chondrogenesis have been aided by the discovery of a mutant in the chicken known as *nanomelia* which affects cartilage synthesis. Double recessive embryos develop severe micromelia and the limb cartilage contains only 10 percent of the amount of sulfated proteoglycans of the normal. In the *nanomelia* embryos the proteoglycan elution curve does not change from the precartilaginous pattern containing only small amounts of type I proteoglycan (Fig. 22–9 A,C). The elution pattern of cartilage in the limb of the normal chick shows the characteristic change to an increase in the synthesis of type I proteoglycan (Fig. 22–9 C). However, analysis of collagen synthesis in nanomelic cartilage shows that the nanomelic embryo synthesizes normal amounts of cartilage-specific collagen. The *nanomelia* has no effect on collagen synthesis. Thus, the mechanism responsible for the change in

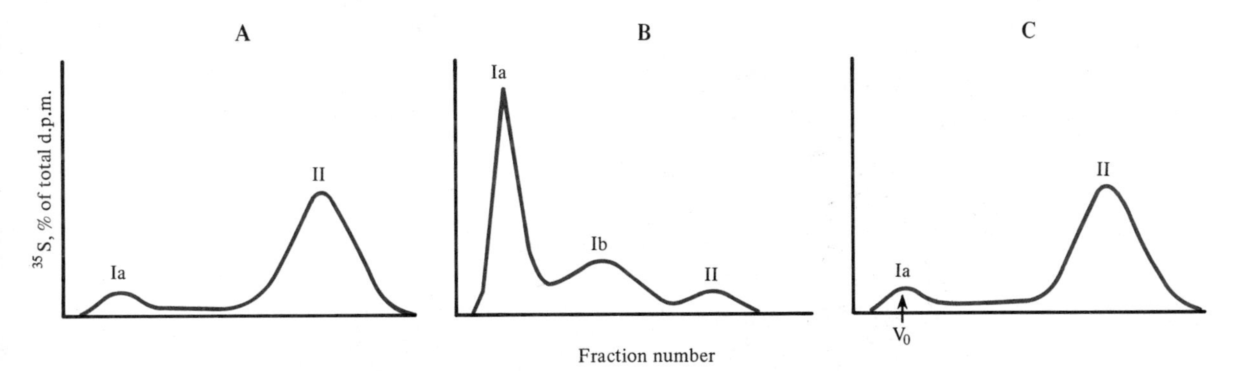


**FIG. 22–9.** Diagrammatic representation of elution profiles of proteoglycans in the chick embryo. A, normal precartilaginous limb bud, stage 18-19; B, normal cartilage; C, *nanomelia* cartilage. (From P. F. Goetinck and J. P. Pennypacker, 1977. In Vertebrate Limb and Somite Morphogenesis, pp. 421–431. D. A. Ede, J. R. Hinchkliffe, and M. Balls, eds. Cambridge University Press, New York.)

the synthesis of sulfated proteoglycans from type II in the precartilaginous mesoderm to type I in chondrogenis mesoderm, which is deficient in the *nanomelia* mutant, must be different from that responsible for the synthesis of cartilage-specific collagen, which is not affected in the mutant.

## The Differentiation of Bone

The differentiation of bone proceeds by two methods: *intramembranous* and *endochondral ossification.* However, the microscopic organization of bone formed by both of these methods is the same. Intramembranous ossification occurs mainly in the development of the bones of the skull, particularly in those bones that form the vault of the cranium. Intramembranous bone differentiates from loose aggregates of embryonic connective tissue. In its formation the matrix between the cells first accumulates bundles of collagenous fibers, and then the cells become oriented in epithelial-like layers along these fibers (Fig. 22–10). These cells are *osteoblasts* (bone-forming cells), whose function is to lay down the intercellular matrix that is first known as *osteoid.* Later, this matrix ia hardened by the deposition of calcium salts, presumably through the activity of the osteoblasts. As the bone matrix is formed, the osteoblasts become bone cells, *osteocytes,* trapped within the matrix (Fig. 22–10). They are retained in small spaces called *lacunae.* The connective tissue covering the bone, *periosteum,* continues to supply new osteoblasts, which replace those that have been engulfed, as the bone increases in size. Since materials cannot diffuse through the calcified bone matrix, the osteoblasts depend for their existence on an intricate network of small channels, *canaliculi,* which interconnect their lacunae and provide a passage to the main central canals, the *Haversian canals* (Fig. 22–11). Bone is laid down in the form of sheets called *lamellae.* The major part of the bone is made up of lamellae arranged concentrically around a central Haversian canal containing the nerves and blood vessels that supply the bone. *Endosteal* and *periosteal* lamellae are those that are found parallel to the inner surface of the bone surrounding the marrow cavity and those that are found parallel to the outer surface of the bone, respectively (Fig. 22–11).

The largest part of the skeleton develops by the replacement by bone of a previously existing cartilaginous model; endochondral ossification (Fig. 22–

**FIG. 22–10.** Differentiation of membrane bone.

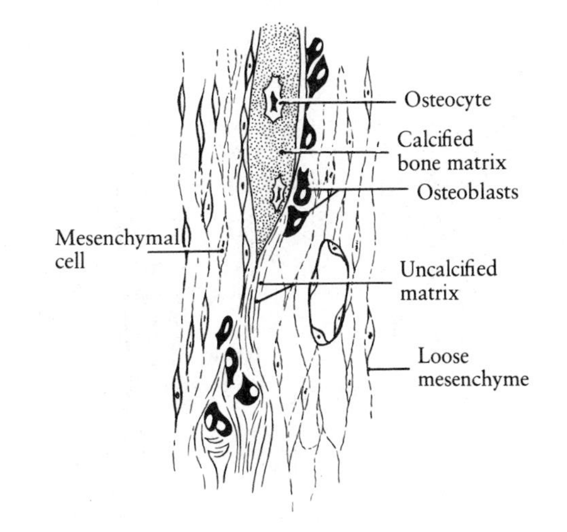

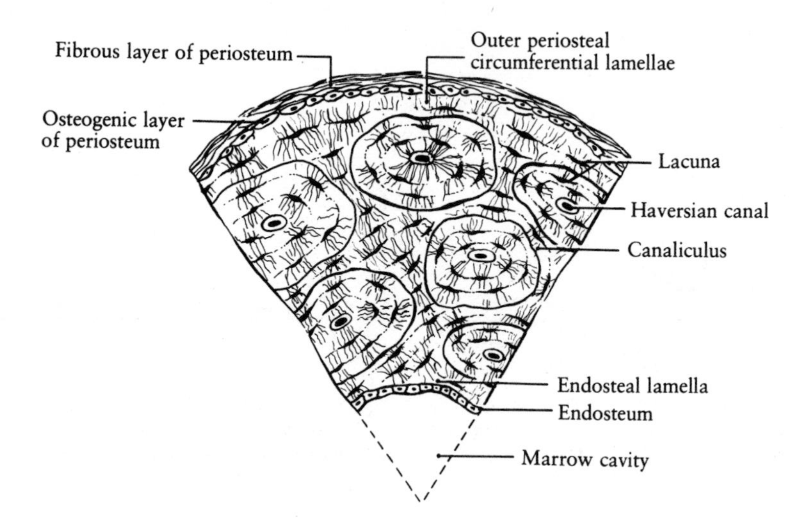


**FIG. 22–11.** Cross section of the shaft of an adult long bone showing the arrangement of the Haversian, endosteal, and periosteal lamellae.

12). In a typical long bone, *primary centers of ossification* arise in the center of the shaft. Here, the cartilage cells hypertrophy, much of the matrix is resorbed, and calcium salts are deposited in the remaining matrix. The result is the formation of branching trabeculae of calcified cartilage. Connective tissue of perichondrial origin and blood vessels grow into the center of ossification to begin the formation of the marrow cavity. These connective tissue cells become osteoblasts, which then deposit bone on the spicules of calcified cartilage.

The spongy bone of the shaft *(diaphysis)* of a long bone is continually being remodelled both by the formation of new trabeculae and the resorption and removal of old ones. The removal of trabeculae is brought about by the action of large multinucleate cells called *osteoclasts*. The general trend is toward the removal of trabeculae rather than the formation of new ones. In this process the individual marrow spaces become confluent and eventually a large central marrow cavity is formed.

One of the most common bone dysplasias is *osteopetrosis* (marble bone) which is characterized by the persistence of primary spongy bone. Cartilage formation, calcification, and the formation of trabeculae appear normal, but the resorption of the trabeculae is impaired. One would expect that the osteoclasts should be in some way involved in this condition and, indeed, in osteopetrotic rats it has been demonstrated that the osteoclasts do not function normally. They are unable to increase the synthesis and release of the lysosomal enzymes needed for normal bone resorption.

The ends of the long bones, the *epiphyses*, generally do not develop centers of ossification until after birth. When they do, the same processes that

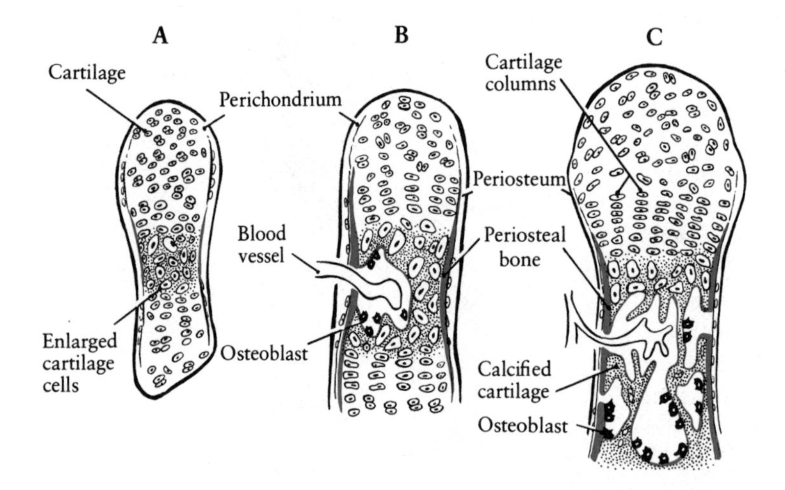


**FIG. 22–12.** Diagram of endochondral ossification in the diaphysis of a long bone.

have previously taken place in the diaphysis of the bone now occur in the epiphyses. Between each epiphysis and the diaphysis, a plate of cartilage persists that allows for the further increase in the length of the bone during postnatal life. Closure (ossification) of this *epiphyseal-diaphyseal plate* of cartilage precludes any further elongation of the bone. However, a layer of cartilage persists over the ends of the bones, providing a surface for articulation.

*Development of the Axial Skeleton*

The axial skeleton of the embryo is simply the notochord. This primary skeleton persists as the only adult axial skeleton in *Amphioxus* and also makes up a large part of the adult skeleton of the cyclostomes. In all other vertebrates, the notochord is replaced by a stiffer, either cartilaginous or bony, skeleton. In mammals, the axial skeleton consists of the skull, the vertebral column, the ribs, and the sternum.

*The Skull:* The skull is formed by a series of flat bones and irregular bones which, with one exception, the mandible, are immovably fused together. Anatomically, in the human, it is made up of eight bones, the *cranium,* which directly surround the brain, and 14 bones which make up the *facial* portion of the skull.

During development there is a sequential progression in the type of tissue which makes up the skull. The earliest indication of skull formation is a mass of dense mesenchyme enveloping the cranial end of the notochord during the 5th and 6th weeks. This is the *desmocranium.* During the 7th week the mesenchyme in the basal region of the skull chondrifies forming the *chondrocranium.* At the time the base of the skull is cartilaginous, the roof and the sides remain as a dense connective tissue capsule in which membrane bone will later develop. A period of ossification then ensues, in which both endochondral and membrane bone are formed, and results in the differentiation of the *osteocranium.*

On the basis of the origin of the individual bones, the skull may also be divided into a *neurocranium* and a *viscerocranium (splanchnocranium).* The viscerocranium consists of bones which are derived from visceral arch mesoderm, as described in Chapter 14: the upper and lower jaws, the palate, the middle ear bones, and the hyoid bone. The neurocranium makes up the remainder and the major part of the skull.

The basal region of the neurocranium is formed from plates of cartilage that surround the cranial end of the notochord. The notochord originally extends to the pharyngeal membrane but stops at the sella turcica after the development of the hypophysis. A plate of cartilage surrounds the cranial end of the notochord and extends caudally to the occipital somites (Fig. 22–13 A). This *parachordal cartilage* (Fig. 22–13 A) plus contributions from the sclerotomes of the caudal 3 of the 4 occipital sclerotomes (the first one disappears) form the *occipital* bone which surrounds the foramen magnum (Fig. 22–13 B). Rostral to the parachordal plate are the *hypophyseal cartilages,* and another pair of cartilages, the *trabeculae cranii,* develop rostral to them. These will form the *body* of the *sphenoid* bone and the *ethmoid* bone, respectively. Three pairs of cartilages develop from mesenchymal condensations outside of the median cartilaginous plates (Fig. 22–13 A). The most rostral, the *ala orbitalis,* forms the *lesser wing* of the *sphenoid.* Caudal to them the *ala temporalis* cartilages form the *greater wings* of the *sphenoid.* The components of the sphenoid bones later fuse with each other and with the basal plate except at points where openings persist for the passage of the cranial nerves to the eye muscles and the first two divisions of the trigeminal nerve (Fig. 22–13 B).

A pair of cartilaginous *periotic capsules* develops in association with the otic capsule. They will form the *petrous* and *mastoid* portions of the *temporal* bone and then fuse with the ala temporalis and parachordal plate derivatives, leaving, however, a large jugular foramen for the passage of the 9th, 10th, and 11th cranial nerves (Fig. 22–13 B).

In contrast to the basal region of the skull, which is composed of the irregular bones formed from the above-described cartilaginous condensations, the walls and the roof of the skull are made up of flat plates of bone which develop by intramembranous ossification. At birth these flat bones are still incompletely formed and are separated from each other by connective tissue bands or sutures (Fig. 22–14). Where two sutures meet, fairly large areas of the brain are covered only by a connective tissue membrane. The spaces are termed *fontanelles.* They are present at all four corners of the *parietal* bones, the largest being the *anterior* fontanelle *(bregma)* at the juncture of the *coronal* and *sagittal* sutures.

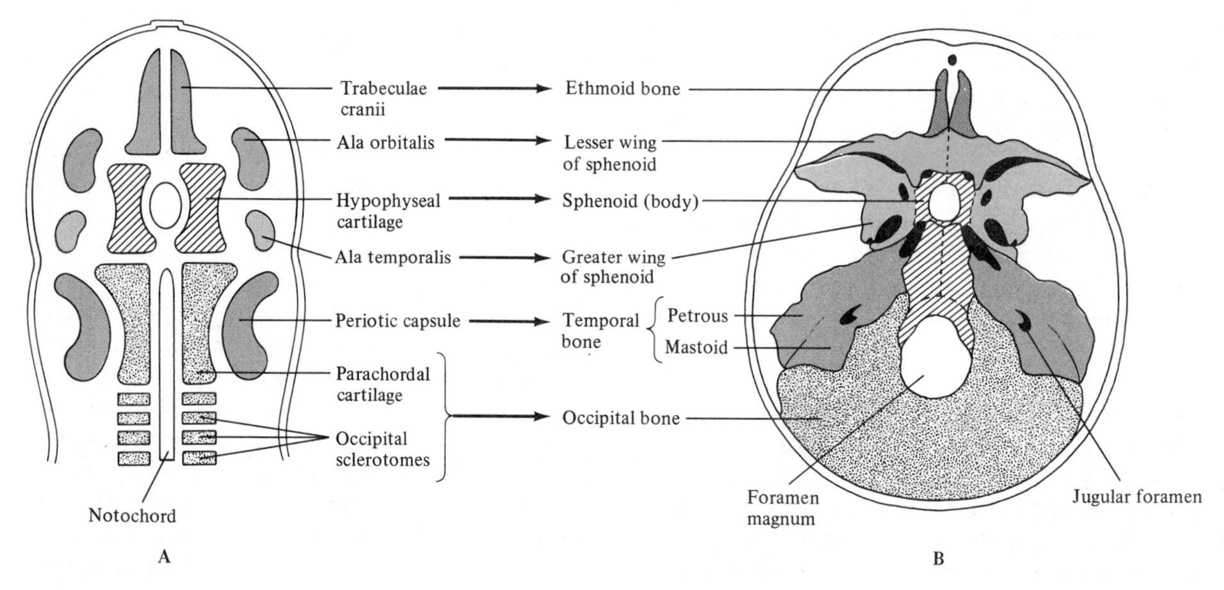


**FIG. 22–13.** A, Diagrammatic representation of the cartilaginous components that make up the chondrocranium; B, dorsal view of the base of the skull showing the bones derived from the components in A.

*The Vertebrae and the Ribs:* The vertebrae and the ribs develop from the sclerotomes of the somites. During the fourth week, the cells from each sclerotome migrate in three different directions (Fig. 22–15 A): (1) ventromedially to surround the notochord, (2) dorsally to cover the neural tube, and (3) ventrally into the ventral body wall.

At each segmental level, the sclerotomal mass surrounding the notochord consists of a loosely arranged cranial portion and a densely arranged caudal portion (Fig. 22–15 B,D). These two portions of the individual sclerotome then undergo a rearrangement in which the denser caudal part of one sclerotome joins the looser cranial half of the sclerotome just caudal to it. The fused cranial and caudal parts of the adjacent sclerotomes will form the body of one vertebra. This then explains why the intersegmental arteries (which originally ran between the adjacent segmentally arranged sclerotomes) now pass over the center of the bodies of the vertebrae, and why the spinal nerves (which originally grew toward the center of each somite) now run between the bodies of adjacent vertebrae (Fig. 22–15 F).

**FIG. 22–14.** Skull of the newborn viewed from above.

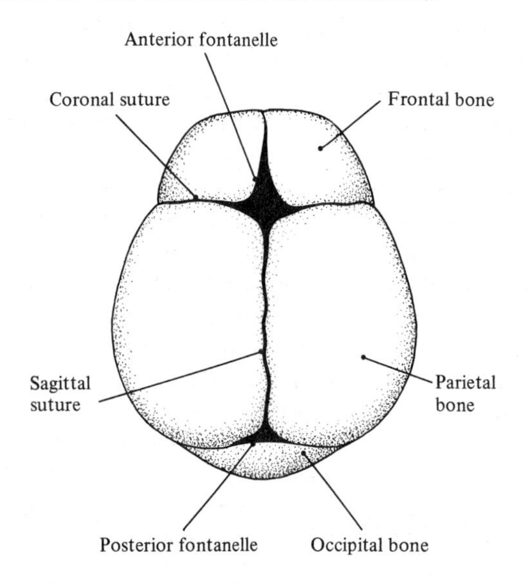


A small cranial part of the dense caudal portion of the original sclerotome moves craniad and becomes located opposite the center of the adjacent myotome. This part of each sclerotome differentiates into an *intervertebral disk.* The only parts of the notochord that persist in the adult are the remnants of this structure, which are incorporated into the intervertebral disks where they form the gelatinous center of the disk, the *nucleus pulposus.*

The cells of the sclerotome that migrate dorsally to surround the developing neural tube differentiate into the *neural arch,* with its central canal through which the spinal cord passes, and into the spinous and transverse processes of the vertebrae.

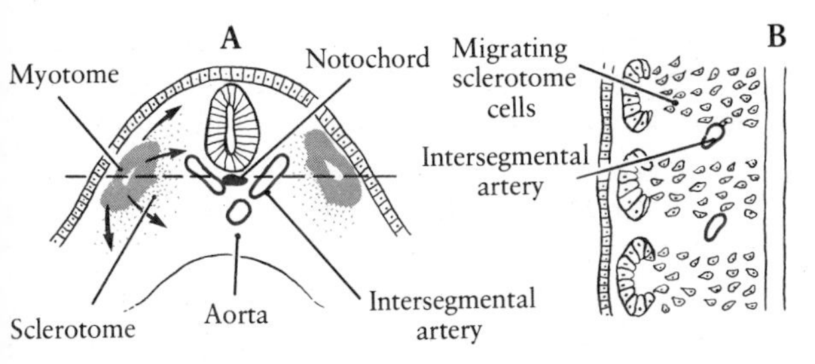


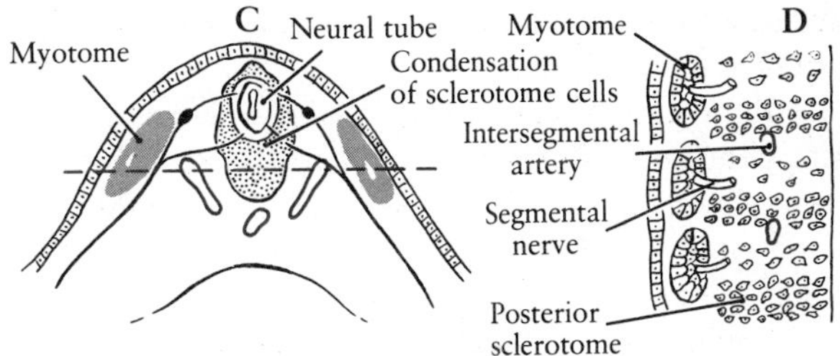


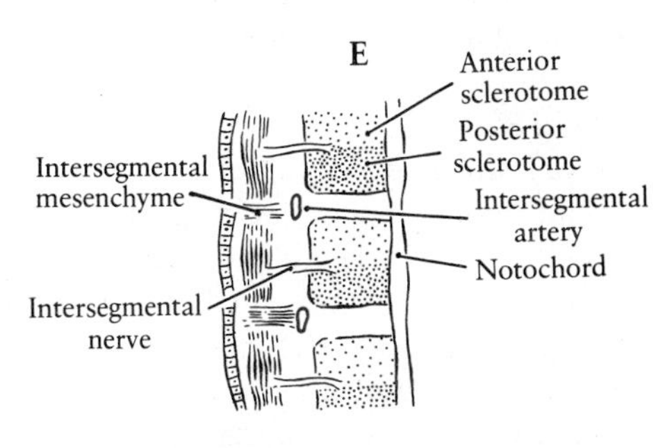


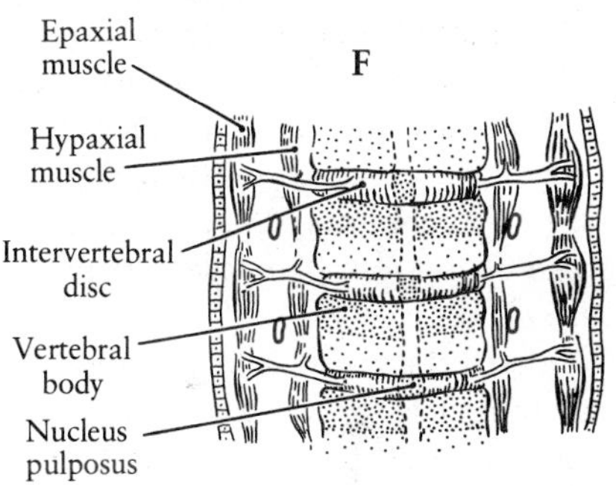


**FIG. 22–15.** Diagrams of four stages in the formation of the bodies of vertebrae (B, D, E, F). A, diagram of a transverse section through a four-week-old embryo; C, diagram of a transverse section through a five-week-old embryo. Dotted lines in A and C represent the level of the frontal sections diagrammed in B, D, E, and F. Since the vertebrae form by the fusion of caudal and cranial parts of adjacent sclerotomes, the spinal nerves, orginally oriented toward the center of a somite, now pass between the bodies of the vertebrae.

The cells of the sclerotome that migrate ventrally form the costal processes of the vertebrae and, in the thoracic region, the ribs.

The mesenchymal masses that are differentiating into the vertebrae begin cartilage formation during the second month, and the cartilaginous models begin endochondral ossification in the late embryonic period. Ossification is not complete until about the mid-twenties of postnatal life.

*The Sternum.* The sternum, the midventral line of bones to which the ribs attach, develops in the second month from local condensations of mesenchyme as a longitudinally oriented pair of sternal primordia, which are at first widely separated from one another and have no connections with the ribs. The ribs soon attach to the developing paired sternal bars that then unite with each other progressively in a craniocaudal direction. The cartilaginous model begins ossification during the fifth month, but all of the centers of ossification are not present until after birth.

*Development of the Appendicular Skeleton*

The appendicular skeleton consists of the shoulder (pectoral) and hip (pelvic) girdles and the bones of the limbs. The bones of the appendicular skeleton develop in association with the limb buds. As do the muscles of the limbs, the bones also develop some distance from the somites; and although the bones of the respective girdles probably develop from somite sclerotome, the limb bones differentiate from the unsegmented lateral mesoderm.

The mesenchymal condensations that will form the cartilaginous models of the girdles and the limbs are present early in the second month, begin to calcify at about seven weeks, and show centers of ossification by eight weeks. Differentiation in each limb proceeds in a proximodistal direction, and the superior extremity develops somewhat in advance of the inferior. In general, the larger bones are the first to chondrify and the first to ossify. Most of the limb bones show primary centers of ossification before birth, but in most of them ossification in the epiphysis does not begin until after birth.

## Congenital Anomalies

There are numerous types of skeletal abnormalities but fortunately most of them are rare.

In addition to osteopetrosis, another general dysplasia involving the differentiation of long bones is *achondroplasia*, the common cause of dwarfism. It results from a decreased rate of growth of the cartilage plate between the epiphysis and the diaphysis, and the bones of the extremities are correspondingly shortened.

Both the pituitary and the thyroid glands are necessary for normal growth, including the growth of the skeleton, and deficiencies in either of the hormonal products of these glands results in an overall inhibition of growth. Thyroid deficiency results in *cretinism*, a rare condition today, characterized by dwarfism, mental deficiency, arrested sexual development and, skeletal abnormalities.

*Axial Skeleton*

*Cranioschisis* is a condition in which the roof of the skull remains open. In severe cases when the brain is almost totally lacking it is known as *acrania*. The skull may be either small *(microcephaly)* or enlarged *(macrocephaly)*, both conditions being secondary and resulting from the development of abnormally small or enlarged brains.

It is not surprising when we consider the complicated mechanism of rearrangement of adjacent sclerotomes to form the vertebrae, to find that successive vertebrae may fuse asymmetrically or that half a vertebra may be missing or not joined to its mate. Increase or decrease in the total number may also occur.

One of the most serious defects of vertebral development is the failure of the two halves of the vertebral arches to fuse, resulting in *spina bifida*. The occurrence of this defect in only one vertebra, usually in the lumbar or sacral region is farily common but not serious. However, the rare involvement of a series of vertebrae leaves a large gap in the roof of the neural canal and is usually accompanied by abnormal development of the spinal cord with herniation through the gap.

*Appendicular Skeleton*

Minor defects of the limbs which involve the skeleton are relatively common. These may involve fusion of the fingers or toes (more commonly the toes) *(syndactyly)* or the development of supernumerary fingers or toes *(polydactyly)*. More serious limb deformities are *amelia*, complete absence of a limb or limbs, and *meromelia*, partial absence of a limb or limbs. *Phocomelia* is a type of meromelia in which the hands and feet are attached to the axial skeleton by only a small irregularly shaped bone and appear to spring directly from the trunk. The relationship between the tranquilizer thalidimide and phocomelia was considered in Chapter 12.

## References


Bischoff, R. and H. Holtzer. 1969. Mitosis and the processes of differentiation of myogenic cells *in vitro*. J. Cell Biol. 41:188–200.

Holtzer, A. 1970. Myogenesis. In: Cell Differentiation, pp. 476–503. O. A. Schjeide and J. de Vellis, eds. New York: Van Nostrand Reinhold.

Holtzer, H., H. Weintraub, R. Mayer, and B. Mochran. 1972. The cell cycle, cell lineages, and cell differentiation. Curr. Topics Dev. Biol. 7:229–256.

Konigsberg, I. R. 1971. Diffusion-mediated control of myoblast fusion. Dev. Biol. 26:133–152.

Konigsberg, I. R. and P. A. Buckley. 1974. Regulation of the cell cycle and myogenesis by cell-medium interaction. In: Concepts of Development, pp. 179–193. J. Lash and J. Whittaker, eds. Stamford, Conn.: Sinauer Associates.

Konigsberg, I. R., P. A. Sollman, and L. O. Mixter. 1978. The duration of terminal $G_1$ of fusing myoblasts. Dev. Biol. 63:11–26.

Levitt, D. and A. Dorfman, 1974. Concepts and mechanisms of cartilage development. Curr. Topics Dev. Biol. 8:103–149.

Lipton, B. H. and A. G. Jacobson. 1974a. Analysis of normal somite development. Dev. Biol. 38:73–90.

Lipton, B. H. and A. G. Jacobson, 1974b. Experimental analysis of the mechanisms of somite morphogenesis. Dev. Biol. 38:91–103.

Nadal-Ginard, B. 1978. Commitment, fusion and biochemical differentiation of a myogenic cell line in the absence of DNA synthesis. Cell 15:855–864.

O'Neill, M. C. and F. E. Stockdale. 1972. A kinetic analysis of myogenesis *in vitro*. J. Cell Biol. 25:52–65.

Yaffe, E. and H. Dym. 1973. Gene expression during differentiation of contractile muscle fibers. Cold Spring Harbor Symp. Quant. Biol. 37:543–547.

# 23

# THE INTEGUMENTARY SYSTEM

The integumentary system of the adult vertebrate consists of two morphologically distinct layers and their associated appendages, such as hairs, feathers, scales, and glands. The superficial layer, or *epidermis,* is a stratified squamous epithelium that develops from the surface ectoderm. The deeper layer, or *dermis (corium),* is a connective tissue originating from mesoderm. An acellular basement membrane lies between the epidermis and the dermis.

Study of the integument and its maturation is particularly intriguing. As a tissue with a permanent germinal population that produces differentiating cells, the epidermis can provide a model system for examining complicated processes of differentiation involving sudden and marked changes in cell morphology. One approach has been to use the human fetal skin in an effort to resolve key questions regarding the postnatal skin, since conditions in the fetus are in slower motion, so to speak, and therefore capable of closer analysis. More permissive legislation relating to the termination of pregnancy has greatly increased the availability of human fetal material for this purpose. As with events seen during the development of other organs, there is increasing evidence that interactions between epidermis and dermis play important roles in the differentiation of the skin and its appendages. The possibility has been raised that interference with these interactions may be a causative factor in epidermal abnormality and disease.

## The Epidermis

In early embryos of most vertebrates, the cells of the ectoderm proliferate and form a protective, transitory epithelium known as the *periderm* or covering layer (Fig. 23–1 A,B). Cells in the basal region of the ectoderm will become the *generative* or *germinative (Malpighian) layer,* a zone destined to give rise to the stratified epithelium of the adult epidermis (Fig. 23–1 B–H).

The developing epidermis of human embryos and fetuses has recently received a significant amount of attention using new techniques for analysis, including transmission and scanning electron microscopy. The ectoderm of very young human embryos (less than 36 days old) is a simple epithelium whose plasma membranes facing the amniotic fluid exhibit occasional microvilli (Fig. 23–1 A). Between 5 and 20 weeks, there develops a clear distinction between the periderm and the germinative layer of the epidermis (Fig. 23–1 B,C). There

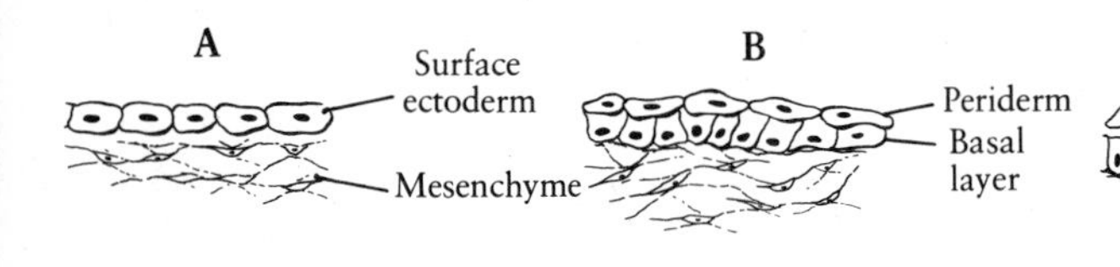


is an extensive elaboration of special intercellular contacts known as *desmosomes* during this time. The initial indication of desmosome differentiation is a localized increase in the density of apposed segments of the plasma membranes of adjacent cells.

Until recently, the periderm was generally viewed as a passive, protective covering for the rest of the epidermis while the latter was in varying stages of keratinization. Ultrastructural observations now tend to indicate that the periderm, over a limited period of time, probably functions actively in providing for and maintaining the well-being of the fetus. Cells of the young periderm are low cuboidal in sectional view, and flat and polygonal in surface view. However, between 9 and 16 weeks of development, the cells of the periderm become very tall and elevated (Figs. 23–1 C–E; 23–2). The amniotic surface of each cell shows microvilli, globular projections, and various clefts and infoldings; these are all specializations designed to provide maximum surface area of the epidermis to the amniotic fluid. The microvilli in particular are currently thought to assist in the uptake and transport of glucose from the amniotic cavity into the fetus. Subsequently, the periderm undergoes regression, with its cells becoming flattened, altered with respect to internal morphology (i.e., pycnotic nuclei), and sloughed off or desquamated into the amniotic cavity. The periderm is completely absent in embryos of approximately 23 weeks (Fig. 23–1 F,G).

A major question regarding the periderm has been whether it undergoes keratinization (i.e., cornification) before the final stages of regression. Most studies tend to indicate that the specific alterations in periderm cells are quite different from those involving the transformation of the underlying epidermal cells into keratinized squames. Additionally, cells of the periderm have never been observed to contain *keratohyalin granules.*

The epidermis becomes increasingly stratified as cells of the stratum germinativum actively divide and their daughter cells crowd toward the surface. By 12 weeks in the human embryo, for example, sections through the epidermis show that it consists of one layer of stratum germinativum, one to three

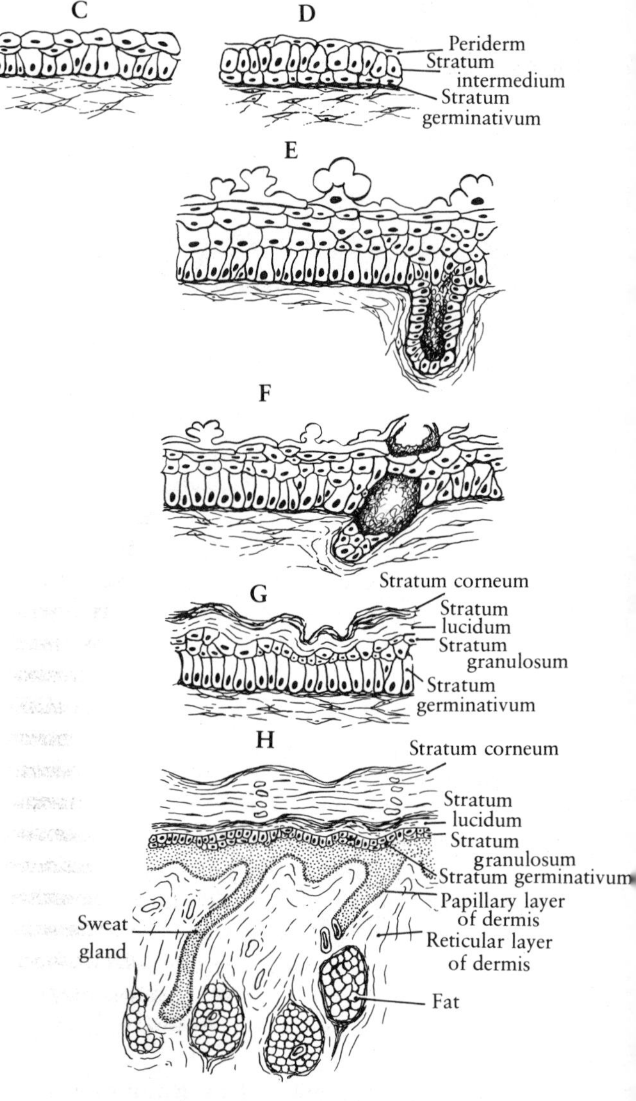


**FIG. 23–1.** Schematic diagrams illustrating the development of the periderm, epidermis, and dermis in human embryos. A, indifferent ectoderm and mesenchyme (36 days); B, the early bilaminar epidermis with periderm flattened (36–55 days); C, periderm layer consists of elevated, domelike cells (55–75 days); D, appearance of intermediate layer in epidermis (65–96 days); E, periderm modified to form large clusters of simple and complex blebs (95–120 days); F, regression of the periderm (108–160 days); G, epidermis has characteristics of adult tissue type (160 days); H, the integument at birth. (A–G, after K. Holbrook and G. Odland, 1975. J. Invest. Dermatol. 65:16.)

layers of a *stratum intermedium,* and a simple periderm (Fig. 23–1 D). By about 25 weeks, the fetal epidermis is completely keratinized and similar in organization to that of the adult, consisting of a basal, germinative layer, a *stratum granulosum*

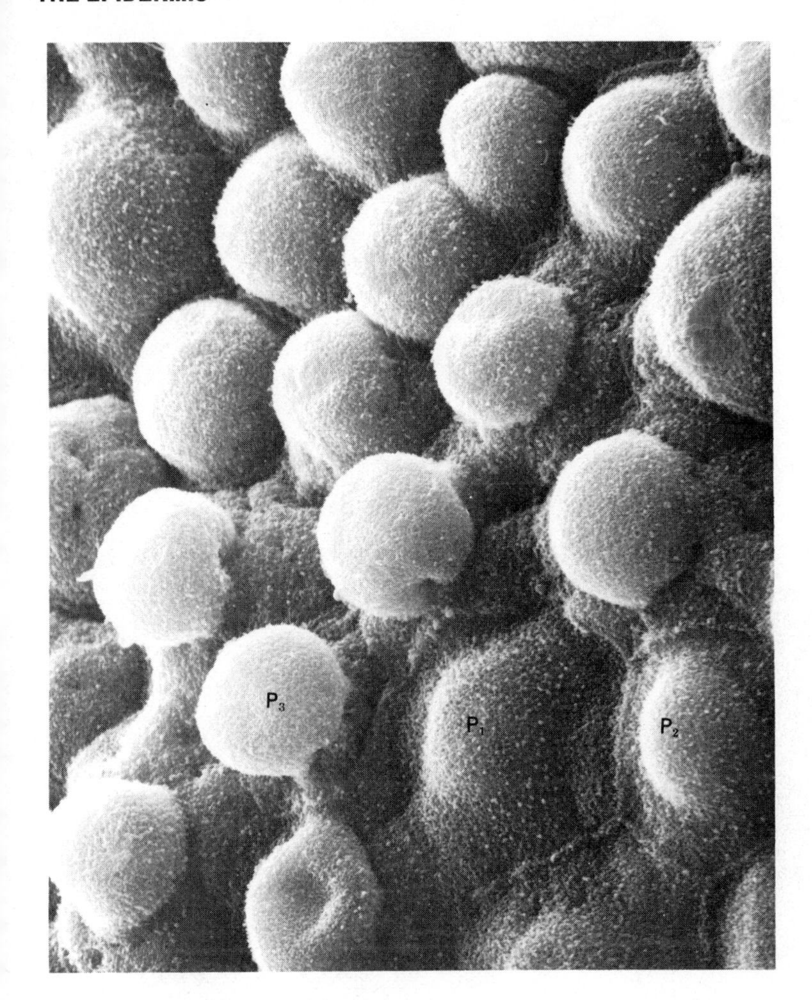


**FIG. 23–2.** A scanning electron microscope view of the fetal skin at 88 days. Many of the peridermal cells are tall and elevated. Note the flat-surfaced cells ($P_1$), elevated cells ($P_2$), and cells with formed blebs ($P_3$). (From K. Holbrook and G. Odland, 1975. J. Invest. Dermatol. 65:16.)

whose cells have distinct keratohyalin granules, a *stratum lucidum* (two to three layers thick) whose cells contain variable amounts of glycogen, and five or six layers of *stratum corneum* (Figs. 23–1 G, H; 23–3).

Stratification of the epidermal epithelium is accompanied by cytomorphic and differentiative changes that progressively transform cells of basal layer origin through a series of stages into the flattened, avital squames of the stratum corneum. The fully keratinized cell of the stratum corneum is one with a matrix of filaments embedded in an amorphous substance and arranged in bundles or fibrils. Whether epidermal keratinization is regarded as a process of differentiation or disintegration, it is a complex process involving synthesis of keratin proteins, keratohyalin granules, alterations in the cell surface and the junctional complexes between cells, dehydration, and death of cells. The dynamic aspects of keratinization are poorly understood. Most investigators tend to believe that cytoplasmic filaments of the basal cells are prekeratin, fibrous proteins that become chemically and morphologically altered to yield the keratin bundles characteristic of the stratum corneum cells. There remain unresolved questions relating to the nature, origin, and fate of keratohyalin granules as well as their role in the cornification process. At the level of the light microscope, these granules are very distinct particles in the cytoplasm of cells comprising the stratum granulosum. Although they were initially believed to be precursors to the keratinous proteins, it is now suggested that keratohyalin granules may function in organizing and arranging the substance of the stratum corneum cell into its final keratin pattern.

Several other cells, which are generally termed *epidermal nonkeratinocytes*, can be identified in the

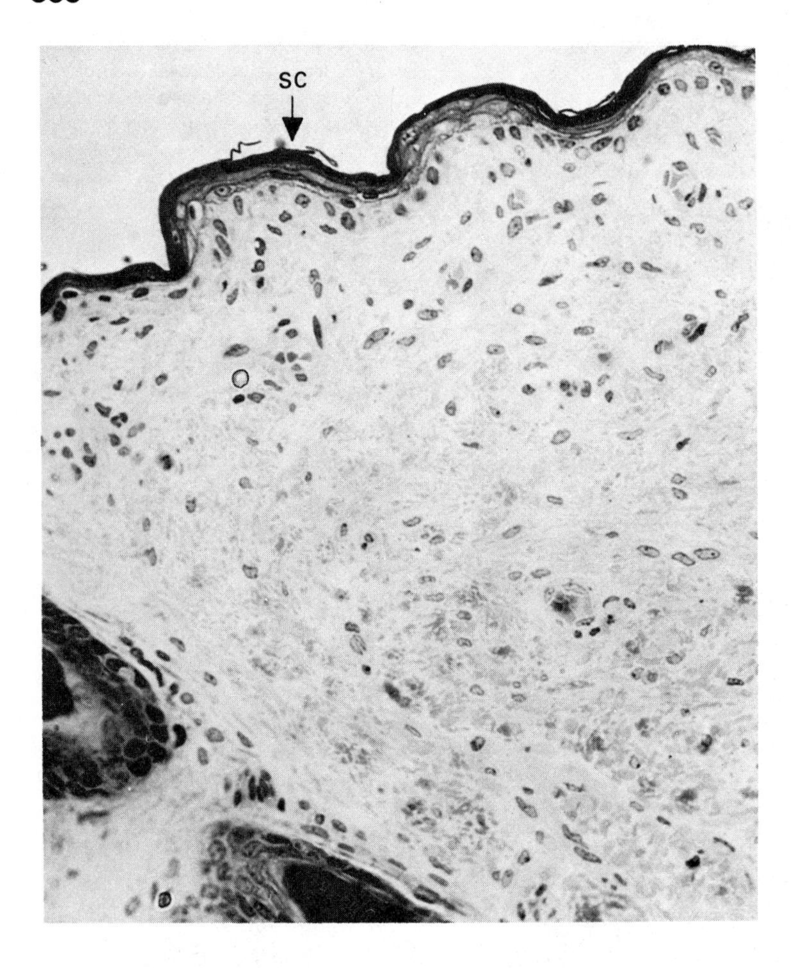


**FIG. 23–3.** The human fetal skin at 185 days as shown by the light microscope. All layers of the epidermis are present. Note the absence of all periderm cells. SC, stratum corneum. (From K. Holbrook and G. Odland, 1975. J. Invest. Dermatol. 65:16.)

developing human epidermis. Neural crest cells are known to migrate into the dermis and differentiate as *melanoblasts*. Subsequently, these invade the basal layer of the epidermis and, as *melanocytes*, specialize in pigment formation (Fig. 23–4). Brown and black melanin pigments (the *eumelanins*), which commonly make up the different body color patterns observed in vertebrates, are produced in special organelles *(melanosomes)* of the melanocyte cytoplasm by the oxidation of L-tyrosine in the presence of the enzyme *tyrosinase*. Active melanocytes have been observed as early as eight weeks in the human epidermis.

**FIG. 23–4.** Early melanocytes entering the human epidermis from the dermis.

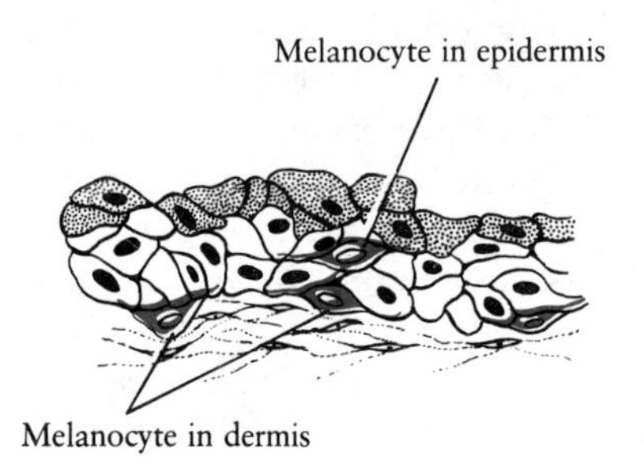


*Merkel cells* are intraepidermal elements generally considered to be integumentary mechanoreceptors (Fig. 23–5). Although they were once thought to be an epidermal keratinocyte, most workers now support the view that the Merkel cells, probably of neural crest origin, migrate through the dermis into the epidermis.

A particularly interesting cell type is the *Langerhans cell,* a squamouslike cell with distinctive granules *(Langerhans granules)* and present within the epidermis in a fully differentiated state by 14 weeks. The general function, the origin, and the fate of this cell remain a mystery. They have been implicated in immune reactions as well as in organizing the highly ordered, columnlike arrangement of epidermal keratinocytes.

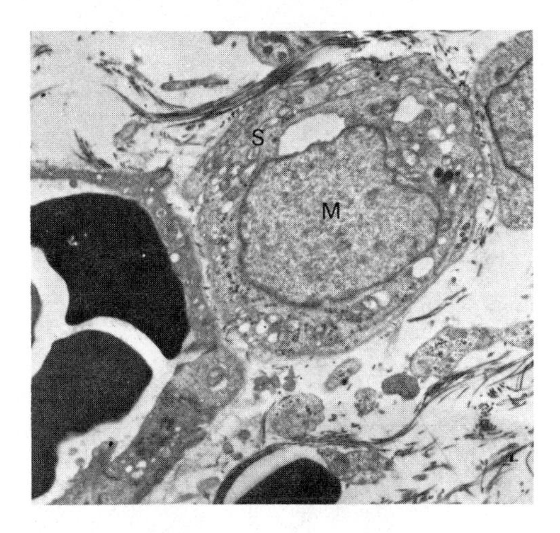


**FIG. 23-5.** A Merkel cell (M) in the dermis of a 21-week-old human fetus. It is closely associated with peripheral axonal–Schwann cell complexes (S) in the skin. (From A. S. Breathnach, 1971. J. Invest. Dermatol. 57:133.)

## The Dermis

The dermis is derived from mesenchyme tissue beneath the surface ectoderm (Fig. 23–1). It is generally recognized that there are two sources for the mesenchyme. Most of the dermis differentiates from the somatic plate mesoderm. The remainder originates from the dermatome of the somite.

When fully developed, the dermis consists of a highly vascularized, fibroelastic connective tissue divided into *papillary* and *reticular layers* (Fig. 23–1 H). Differentiation of collagenous fibers and elastic fibers from mesenchyme is well established in human embryos of approximately 24 weeks. Dome-shaped thickenings of the dermis *(dermal papillae)* project into the basal layer of the epidermis; these alternate with downgrowths of the stratum germinativum termed *epidermal ridges* (Fig. 23–1 H). The unevenness of the dermoepidermal boundary is an effective structural adaptation that permits maximum resistance to forces of shear, thereby acting to maintain the structural integrity of the integument.

## The Cutaneous Appendages

The vertebrate skin displays an astounding degree of functional and morphological diversity. This is expressed by the presence of discrete, highly specialized structures or appendages, including a variety of glands (ranging from the mammary, sweat, and sebaceous glands of mammals to the preen or uropygial glands of birds), hairs, feathers, terminal phalangeal coverings (nails, hoofs, claws), and scales. The primordia of most of these appendages arise initially as localized thickenings of the epidermis. An exception is the scales of fishes, which are specializations of the dermal mesenchyme. Although the development of these integumentary derivatives has been rather carefully studied at the anatomical level, our knowledge regarding mechanisms underlying their differentiation is fragmentary.

### The Nails

Each nail is initially foreshadowed by a thickened area of the epidermis (the *primary nail field*) on the dorsal side of the tip of each digit (Fig. 23–6 A). The nail field soon becomes more sharply defined with the appearance of elevated folds of the epidermis known as the *proximal* and *lateral nail folds* (Fig. 23–6 A,B). Although some keratinization of the nail field does take place, forming the so-called false

**FIG. 23-6.** Development of the human nail. A, dorsal view of the tip of the fingernail at about 10 weeks; B, sagittal section through the fingernail at about 14 weeks; C, sagittal section through the fingernail at birth. (After L. Arey, 1974. Developmental Anatomy. W. B. Saunders Company, Philadelphia.)

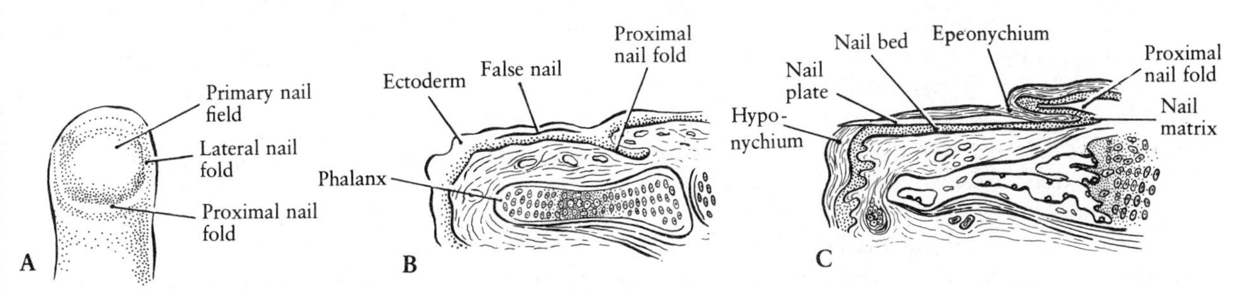

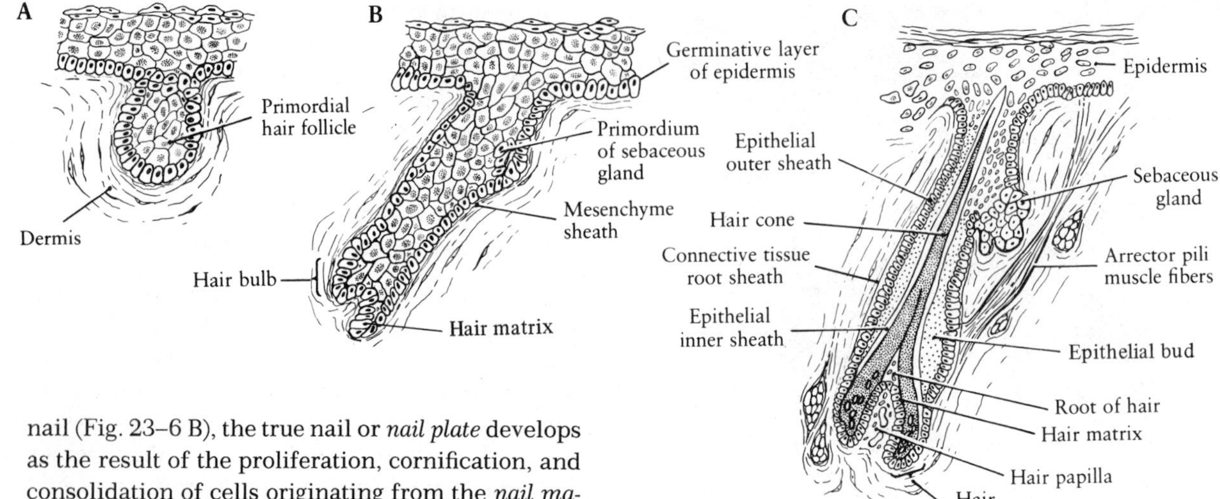


FIG. 23–7. Drawings to illustrate the successive stages in the development of hair. A, the early primordial hair follicle; B, beginning of the formation of the hair matrix of the hair bulb; C, hair cone formation. (From B. M. Patten, 1968. Human Embryology. Copyright © 1968 by McGraw-Hill, Inc. Used with permission of McGraw-Hill Book Co.)

nail (Fig. 23–6 B), the true nail or *nail plate* develops as the result of the proliferation, cornification, and consolidation of cells originating from the *nail matrix* or the germinative layer of the proximal epidermal nail fold (Fig. 23–6 C). The nail plate grows forward over the *nail bed* from the base of the terminal phalanx to reach the tip of the digit just before birth.

The periderm and the stratum corneum for a time constitute a covering layer, termed the *eponychium* (Fig. 23–6 C), for each nail plate. By the latter part of fetal life, this layer is lost except for fragmentary, keratinized portions near the margins of the nail (cuticle of the nail).

## The Hairs

Although hairs begin to develop from the epidermis at the end of eight weeks in the eyebrows, lips, and chin of human embryos, those of the general integument do not begin to appear until about the fourth fetal month. The primordium of each hair originates as a solid column of cells *(hair follicle)*, produced by localized proliferation of the stratum germinativum, which pushes down into the underlying mesenchyme (Fig. 23–7 A). The deepest part of the hair follicle *(hair bulb)* quickly becomes enlarged into a club-shaped mass and then invaginated by a domelike mass of mesenchyme tissue (Fig. 23–7 B,C). A detailed view through the hair follicle at this time shows the moundlike mesenchyme, or *hair papilla,* capped by epithelial cells that, because they later give rise to the hair proper, constitute the *hair matrix* (Fig. 23–7 B,C). The tissue of the hair papilla is continuous with mesenchyme cells that invest the rest of the hair follicle; these mesenchyme cells will differentiate into the *dermal* or *connective tissue root sheath* (Fig. 23–7 C).

Proliferation of cells by the hair matrix produces a young, cone-shaped mass of cells *(hair cone)*, which pushes toward the surface through the central cells of the hair follicle (Fig. 23–7 C). The more peripherally situated cells of this axial core will give rise to the *inner epithelial root sheath* while the remaining, centrally located cells will form the *hair shaft*. Above the region of the hair matrix, the cells of the hair shaft become keratinized and topographically organized into an *outer cuticle*, a *middle cortex*, and a *central medulla*. The peripheral cells of the original follicle wall will differentiate into the *outer epithelial root sheath* (Fig. 23–7 C).

Two thickenings of the outer root sheath appear on the lower side of the obliquely directed hair follicle (Fig. 23–7 C). The lower one is the *epithelial bud*, a region of rapid cell proliferation that contributes to the growth of the hair follicle. The upper swelling is the primordium of the *sebaceous gland*. Mesenchyme below the epithelial bud aggregates to form the *arrector pili muscle*, a bundle of smooth muscle fibers attached to the connective tissue root sheath of the follicle and the papillary layer of the dermis.

The first hairs to emerge in the fetal skin are rather slender and spatially close together (Fig. 23–8). They form a downy coat commonly termed *la-*

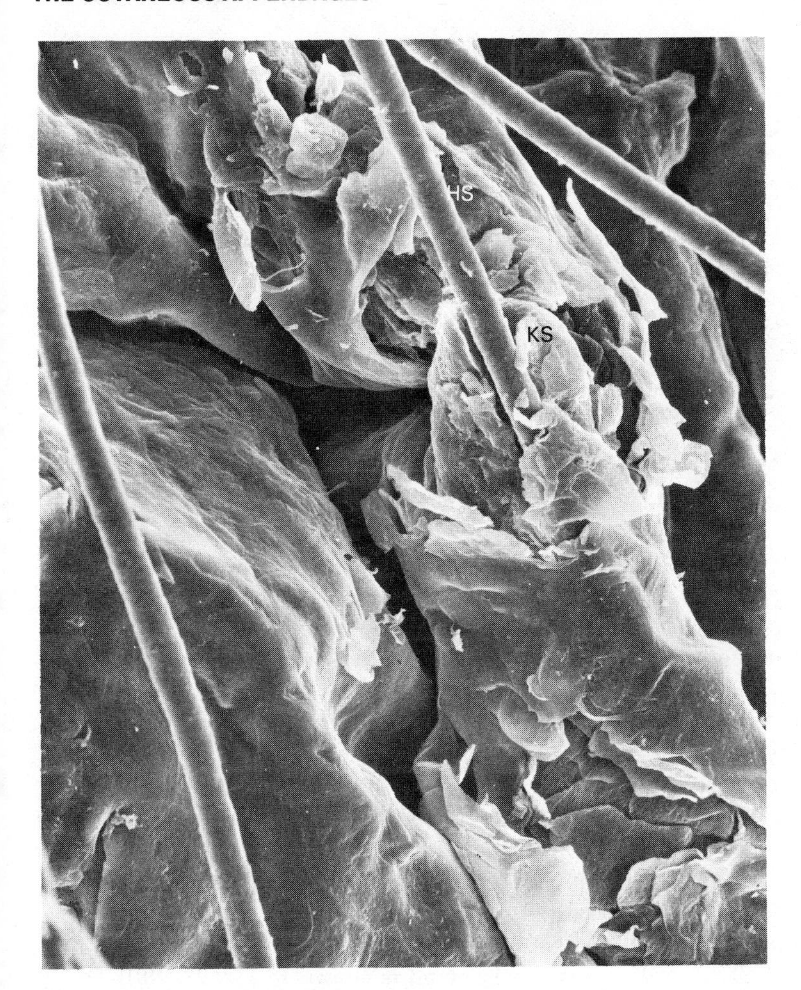


**FIG. 23–8.** A surface view of the fetal skin at 23 weeks showing hair shafts (HS) and keratinized epidermal cells (KS). (From K. Holbrook and G. Odland, 1975. J. Invest. Dermatol. 65:16.)

*nugo*. Hairs of this type are typically shed into the amniotic fluid by birth and replaced by fine *vellus hairs* characteristic of the prepuberal skin. Vellus hairs appear to be at least in part derived from new hair follicles.

## The Sebaceous Glands

Most of the sebaceous glands arise as lateral evaginations from the outer epithelial root sheath of the hair follicle (Fig. 23–7 B,C). The solid swelling becomes lobulated to form the several alveoli that characterize the adult gland. The lumen of the sebaceous gland is formed through the breakdown of central sebaceous cells with the resultant oily secretion *(sebum)* passing into the amniotic fluid by way of the hair canal. The sebum mixes with desquamated peridermal cells to form *vernix caseosa,* a whitish, cheeselike substance that acts as a protective coating for the fetal skin. Since the sebaceous gland is *holocrine* (i.e., the secretion consists of disintegrated gland cells), periodic replacement of secretory cells is required.

Sebaceous glands independent of hair follicles are found in the upper eyelids, external genitalia, and around the anus. They arise in a similar manner from epithelial buds of the epidermis.

## The Sweat Glands

The ordinary eccrine sweat glands develop as solid, cylindrical downgrowths of the generative layer of the epidermis (Fig. 23–9 A). As each bud pushes into the underlying mesenchyme, the distal segment of the primordium becomes coiled to form the secretory portion of the gland. Recent studies with the electron microscope show that lumen formation in sweat glands appears to be a complex

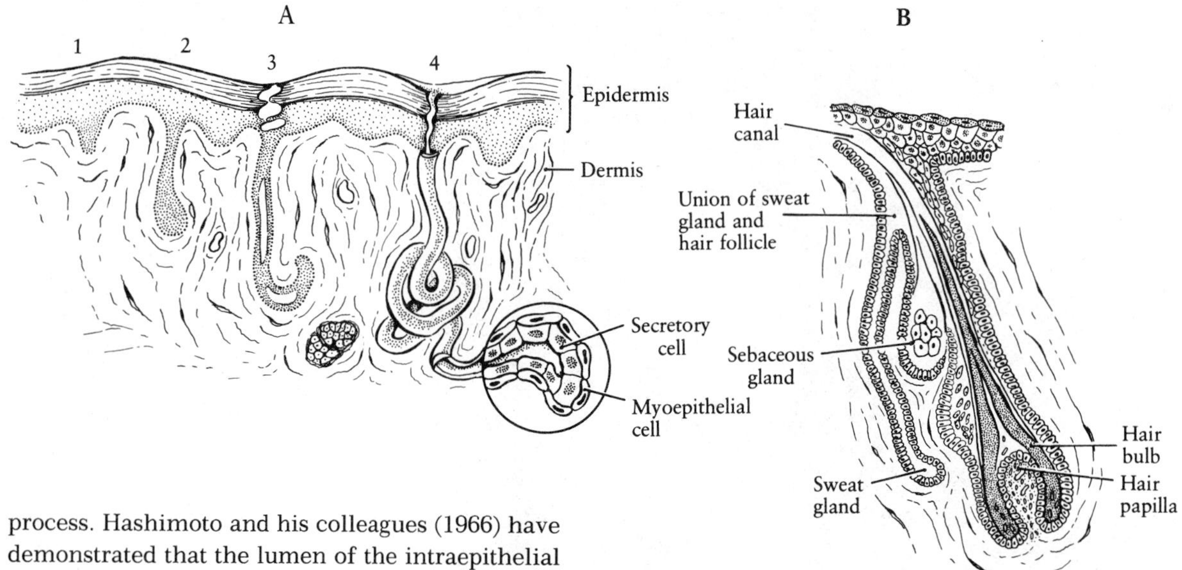


**FIG. 23–9.** A, successive stages in the development of the common eccrine sweat gland; B, an apocrine sweat gland developing in association with a hair follicle.

process. Hashimoto and his colleagues (1966) have demonstrated that the lumen of the intraepithelial portion of the duct forms extracellularly by the separation of cells, while the lumen of the intradermal portion of the duct arises through the formation of cytoplasmic vesicles within the primordial cells. These vesicles break through the plasma membranes and coalesce. Epithelial cells in the secretory segment of the gland differentiate into *secretory* and *myoepithelial cells* (Fig. 23–9 A). Myoepithelial cells are of special interest because they are considered to be specialized smooth muscle cells, derived from ectoderm, which aid in the expulsion of sweat from the gland.

Aprocrine sweat glands develop in association with hair follicles and have a distribution largely limited to the axilla, the pubic region, and the areola of the mammary glands. They develop from downgrowths of the stratum germinativum and as such open into the hair canal above the sebaceous glands (Fig. 23–9 B).

Actual secretion by sweat glands is probably negligible before birth.

## The Mammary Glands

Although mammary glands do not normally function until adulthood, their primordia appear relatively early in mammalian development. The first visible evidence of mammary gland development appears in the form of a pair of epidermal thickenings along the ventrolateral body walls from the axillary to the inguinal regions (Fig. 23–10 A). These so-called *mammary ridges* or *milk lines* are particularly prominent in mammals with serially arranged mammary glands, such as the rat and cow. In humans, the milk ridges disappear rather quickly except in the pectoral region where the paired glands will develop.

The progressive development of the human mammary gland is shown in Fig. 23–10 B–D. Localized proliferation within the mammary ridge produces, on either side, a lenticular-shaped epithelial mass of cells that extends down into the underlying dermis. Each mass continues to enlarge and gradually becomes globular in shape. During the fifth fetal month, 20 to 25 solid epithelial cords bud off and push deeper into the mesenchyme. These cords of cells are the primordia of the *lactiferous ducts* (milk ducts). Each slowly acquires a lumen, branches at its distal end, and will serve as a focal point around which a lobe of the mammary gland is organized (Fig. 23–10 C). Subsequently, the epidermis at the site of the origin of the gland becomes keratinized and hollowed out to form a shallow mammary pit into which the ducts open (Fig. 23–10 D). Shortly after birth, this same area will elevate into the *mammary nipple.* The areola of the gland will develop from the circular area of the epidermis around the nipple (Fig. 23–10 D).

Until pregnancy occurs, the mammary gland remains incompletely developed (Fig. 23–11). Most of

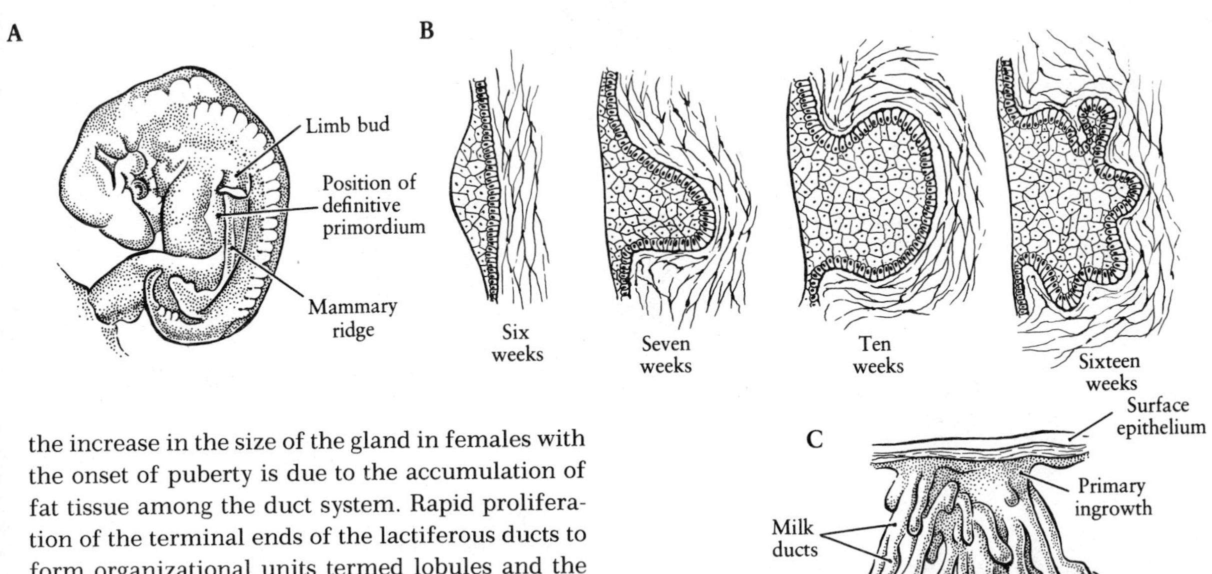


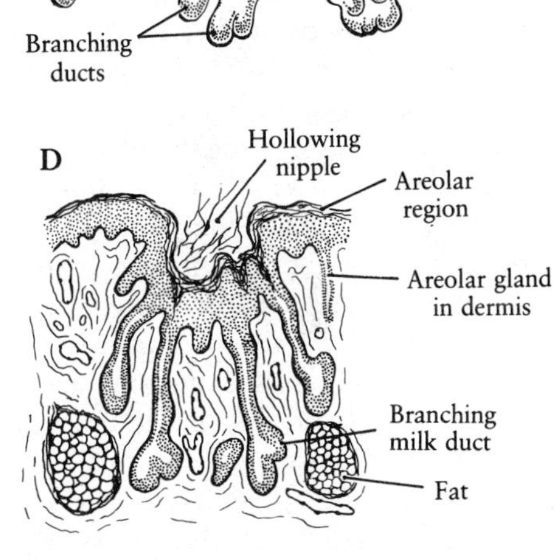


FIG. 23–10. Development of the human mammary gland. A, the position of the mammary ridge at 6 weeks; B, a series of vertical sections through the epidermal gland primordium from 6 to 16 weeks; C, the appearance of the gland at 6 months; D, the mammary gland in vertical section at 8 months.

the increase in the size of the gland in females with the onset of puberty is due to the accumulation of fat tissue among the duct system. Rapid proliferation of the terminal ends of the lactiferous ducts to form organizational units termed lobules and the secretory portions of the gland occurs during the last third of gestation. The mammary glands in males normally undergo little postnatal development.

## Dermal–Epidermal Interactions and Their Analysis

At an early stage of development in all vertebrates, the general integument consists of a flat, two-layered epidermis, originating from embryonic ectoderm, and a dermis of uniform constitution derived from somatic mesenchyme. From this relatively undifferentiated state, an epidermis emerges whose cells are stratified, rigidly organized, and in varying stages of keratinization. Closer examination of the epidermis reveals that there are regional differences in its microanatomy, expressed primarily in terms of its thickness and degree of cellular stratification. Also, there are variations in the extent to which the epidermis is interrupted by the appearance of discrete, highly specialized cutaneous appendages, such as hairs and glands. Because the skin is the largest organ in the body and fulfills a spectrum of important physiological functions, there has been an intense interest in the processes leading to the differentiation of the embryonic skin, particularly those responsible for the origin and maintenance of the various structural specializations associated with the epidermis.

As pointed out in previous chapters, one of the most fundamental processes in embryonic development is the interaction between populations of cells and tissues of diverse ontogenetic origins (i.e., presumptive neural ectoderm and chordamesoderm). The interaction typically leads to the expression of new, differentiated cell and tissue types. Based largely on the results of the studies with the chicken embryo, it is now generally agreed that (1) reciprocal interactions between epidermis and dermis are required for normal skin development, and (2) regionally distinctive differentiations of the epi-

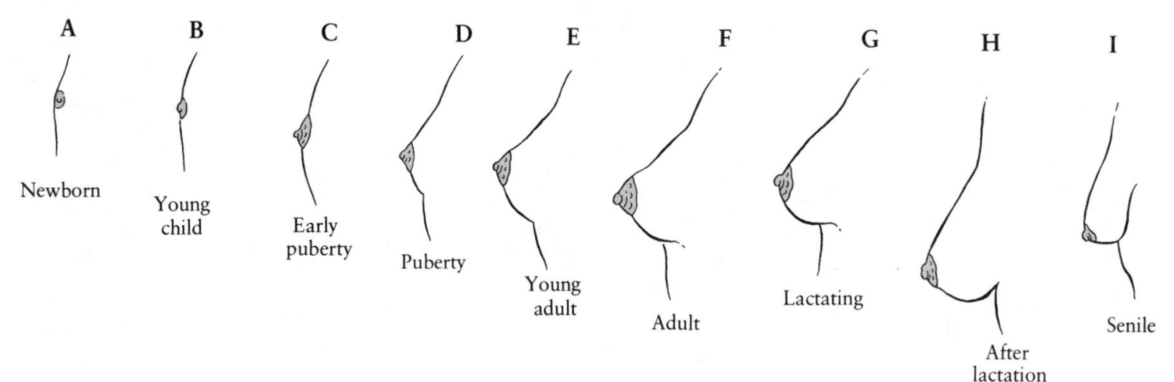


**FIG. 23–11.** Sketches of the human breast in profile to show changes in contour with age and states of functional activity.

dermis are determined under the influence of the dermal mesenchyme.

Techniques available for the study of dermoepidermal interactions in embryonic and adult tissues are similar to those employed to examine epitheliomesenchymal relationships in general (Chapter 13). Thin sections of different kinds of skin can be grafted into wounds that are prepared in the integument of genetically compatible hosts. For example, one can observe the in vivo response of epidermal epithelia excised from different sites (ear epidermis or cornea epidermis) as they grow over the surface of common mesenchyme that invades the wound bed. A particularly useful method for determining the interrelationships between epidermis and dermis is illustrated in the diagram of Figure 23–12. A section of skin is dissected out and cut into smaller fragments. The epidermal and dermal components of these fragments are then completely separated from each other by chemical (EDTA) or enzymatic (trypsin) treatment (Fig. 23–12 A). Heterotypic recombinant grafts, formed by bringing together the two tissues from different sources, are then transplanted to heterotopic sites in the same host or to the chorioallantoic membrane of an embryonated chicken egg (Fig. 23–12 B).

When the whole skin of embryonic chickens or fetal mammals is cultured, growth and differentiation of both epidermis and dermis generally resemble patterns observed in vivo. The criteria for epidermal growth and differentiation are the maintenance of a healthy stratum germinativum, indicated by mitotic figures in its cells, the formation of keratin, and the development of an organized stratum corneum. However, isolated grafts of epidermis (i.e., without dermis) in organ culture tend to curl up, remain largely undifferentiated, and eventually degenerate. Thus, epidermal tissue needs a substrate over which to spread before it is capable of differentiation. Although this is normally provided by the underlying living dermis, dead freeze-thawed dermis, collagen gel, and Millipore filters all will provide adequate substitutes. A

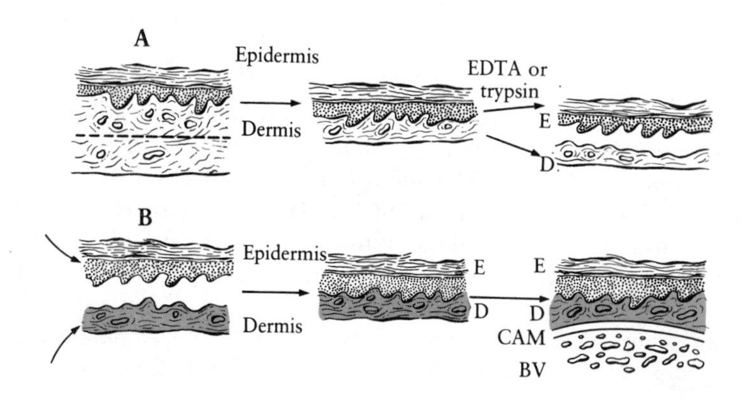


**FIG. 23–12.** Diagram illustrating the steps in the recombination of skin components and the transplantation of resulting grafts to the chorioallantoic membrane. A, separation of skin components by EDTA or trypsin into isolates of epidermis (E) and dermis (D); B, recombinants of epidermal and dermal isolates are assembled under a microscope and placed on a portion of the chorioallantoic membrane (CAM) rich in blood vessels (BV). (After R. Briggaman and C. Wheeler, 1968. J. Invest. Dermatol. 51:454.)

living dermal substrate is not necessary for epidermal proliferation. Epidermis from isolates of very young chicken embryos has been shown to become completely keratinized; however, these same grafts show very few mitoses and little evidence of the tissue organization characteristic of normal epidermis. Isolated explants of dermis undergo necrosis rather rapidly without significant differentiation. When epidermis is recombined with its own dermis in culture, both layers survive and differentiate to produce a graft of normal whole skin.

Extensive studies have been conducted on the respective roles of the dermis and the epidermis in the differentiation of epidermal appendages. Chick and duck embryos have been most widely studied, so that more data are available on the mechanism of feather development than on any other amniote appendage. Experiments on reptiles and mammals have also been conducted, and they have demonstrated that the same principles apply to the development of scales and hairs as to the development of feathers.

Epidermis and dermis may be combined in a number of different ways. Using the same species but combining dermis and epidermis from different regions of the embryo is known as heterotopic recombination. The association of epidermis and dermis from two different species constitutes a heterospecific recombination. Heterotopic recombinations have very effectively made us aware of the importance of the dermis in the differentiation of specific epidermal appendages. Saunders has demonstrated, for example, that grafts of ectoderm-free prospective thigh, mesoderm, following transplantation into a prepared site on the wing bud of a four-day-old chick embryo (Fig. 23–13 A), become covered with wing ectoderm. Feathers subsequently develop in the wing epidermis above the mesemchymal graft that display structural and organizational characteristics typical of feathers appearing on the thigh. Also, when transplanted to the wing bud, mesoderm from the distal end of the leg bud causes the wing epidermis to form scales and claws, epidermal specializations associated with the leg (Fig. 23–13 B). The specific character of the appendage is thus controlled by the dermis in heterotopic recombinations.

Not all of the regions of the chick skin develop appendages. Some areas are bare (glabrous). What will happen if glabrous epidermis is combined with dermis from a feather-forming region? Conversely, does the dermis in glabrous regions have the potential to induce appendages when combined with feather- or scale-forming epidermis? Do certain epidermal regions not develop appendages because of a lack of a proper inductive influence from the dermis or because the epidermis is unable to respond to that stimulus? The answers are clear. Dermis from glabrous regions will not induce the development of either scales or feathers in epidermis from regions that normally develop these appendages. Conversely, dermis from scale- and feather-forming regions induces the formation of its own specific type of appendage when combined with epidermis from glabrous regions. Even the corneal epidermis will form feather buds when combined with wing dermis.

The age of both the dermis and the epidermis is important in these interactions. The inductive capacity is acquired over a period of time, and if der-

**FIG. 23–13.** Diagram illustrating mesodermal control of the development of the epidermis in the chick. A, thigh mesoderm grafted into wing bud; B, apical mesoderm of leg bud grafted into wing bud. (From J. Cairns and J. Saunders, 1954. J. Exp. Zool. 127:221.)

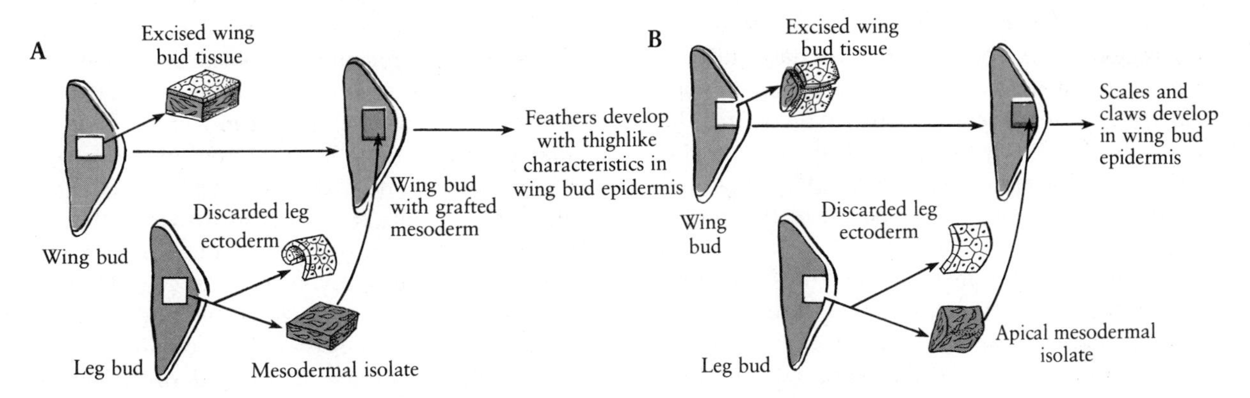

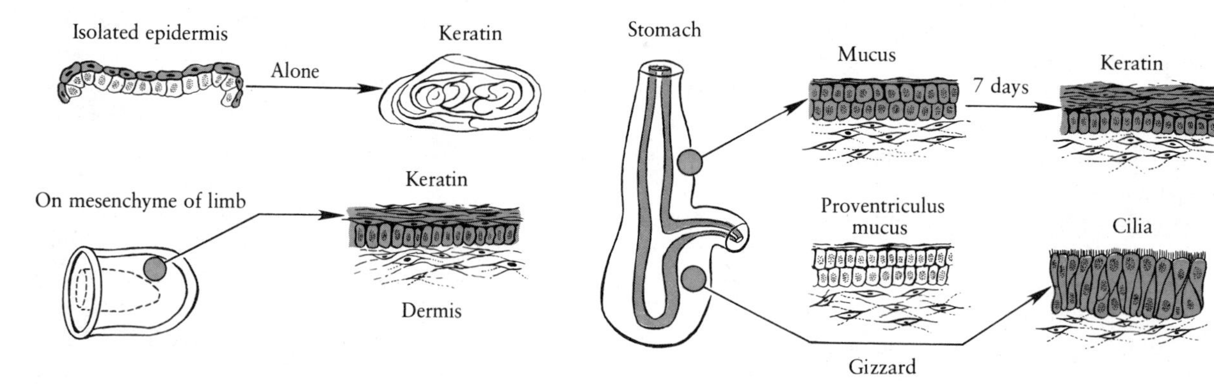


**FIG. 23–14.** Drawings to show the differentiation of isolated embryonic chick limb epidermis (A) when combined with mesenchymes of different sources (B). (After C. B. McLoughlin, 1963. In: Cell Differentiation, Symposia of the Society for Experimental Biology, No. 17.)

mis is taken from too early a stage it will not induce the formation of appendages. Similarly, the epidermis over a period of time loses the capacity to change the type of appendage from what it would normally form to that induced by heterotopic dermis. Leg epidermis from chick embryos older than 8.5 days will not form feathers when combined with wing dermis. By this time the leg epidermis has responded to the inductive influence of its own dermis to such an extent that it has lost the competence to respond to a foreign stimulus. Thus, the differentiation of specific epidermal appendages is due to the response of competent epidermis to a specific inducing signal from the underlying dermis that determines the type of appendage to be developed.

In the mouse, the same results have been reported. Dermis from the upper lip region where vibrissae develop will induce the differentiation of vibrissae in the dorsal body epidermis, which normally forms pelage hair, and vice versa. Dermis from the sole of the foot—a glabrous region—will not support the differentiation of hair in normal body epidermis. Conversely, pelage hair may be induced in the epidermis of the sole of the foot when this epidermis is combined with dorsal dermis. As in the chick, the dermis contains the information for region-specific epidermal appendage differentiation and can induce the differentiation of its specific type of appendage not only in epidermis that normally forms appendages but also in glabrous epidermis. The epidermis appears to be indifferent and merely responds to the specific instructions of the dermis.

The importance of mesenchymal factors in influencing the expression of the epidermis has been clearly shown in a series of classic experiments by McLoughlin (1961). Isolated five-day-old chick limb epidermis was explanted in vitro with mesenchyme excised from several sources, including gizzard, proventriculus, and heart (Fig. 23–14). The response of the epidermis to stimulation by the mesenchyme was different and very specific in each of these heterotopic recombinants. With gizzard mesenchyme the epidermis failed to keratinize, but its cells were induced to secrete mucus and sometimes form cilia. This is a marked alteration in the prospective fate of the epidermis, which normally would produce only keratinous proteins. Explants of heart mesenchyme became subdivided into two zones: a central region of myoblast cells surrounded by a region of fibroblast cells. The epidermis was prevented from keratinizing on the region of myoblasts and spread into a simple squamous epithelium; however, the epidermis keratinized heavily on heart fibroblasts. On proventriculus mesenchyme, the epidermis was initially prevented from keratinizing and secreted mucus, but after approximately seven days its cells became keratinized. The latter suggests that the mesodermal influence must be continuous to maintain the modified differentiation of the epidermis. Studies by Briggaman and Wheeler (1968) using recombinant grafts of human skin have demonstrated that the dermis is continuously required to conserve and maintain the stable adult features of the integument. Hence, dermoepidermal interactions are not restricted to the embryonic period.

Heterospecific recombinations have provided other interesting results. What will happen if mouse epidermis is combined with chick dermis? Can the chick wing dermis induce the mouse epidermis to form feathers? It cannot. Neither can mouse dermis induce hair formation in the chick epidermis. Thus, foreign dermis cannot force epidermis to differentiate appendages that are not a part of the vocabulary of that particular epidermis. However, foreign dermis can induce the formation of appendages in dermis of a different species, but the nature of the appendages—scales, hair, or feathers—is epidermally controlled even though it is dependent on the dermis for its expression. This type of interaction was reported some years ago, in the 1930s, when ectodermal grafts were made to the ventral head region of frogs and salamanders. The oral ectoderm in this region in the frog forms epithelial pads, called suckers, while that of the salamander forms rodlike outgrowths, called balancers. Frog trunk epidermis can be transplanted to the oral region of the salamander, where it will respond to the inductive influence of the salamander dermis and form oral appendages. These appendages, however, are suckers, not balancers. The converse experiment yields the same results, frogs with balancers. The inductive influence of the host mesoderm will act across species lines and induce the graft to form oral appendages, but the type of appendage developed depends on the reacting tissue. Within a species, as demonstrated by heterotopic recombinations, all of the epidermal appendages characteristic of that species can be induced by the dermis, and the type of appendage is determined exclusively by the dermis. In interspecific recombinations, the dermis can still exert its inductive influence but the epidermis is limited in its response to forming only that type of appendage that is part of its genetic vocabulary. In this case, the specific nature of the appendage is determined by the epidermis.

Usually the appendages formed in epidermis combined with heterospecific dermis do not differentiate completely but form "arrested" appendages. The foreign dermis can start the process of differentiation, but it cannot carry it to completion. It provides the first step in the differentiation process, the trigger, a non-class-specific induction. But there is a second step leading to complete differentiation, and it has been shown that this step depends on the presence of dermis that is the same type as the epidermis. The first step has been called the "nonspecific dermal message," to which the epidermis responds according to its own class-specific properties by forming a rudimentary appendage. A second "class-specific dermal message" is then needed to complete the differentiation of the rudimentary appendage, controlling the morphogenetic events that determine the final architecture of the appendage.

Thus, in heterospecific recombinations of chick epidermis and mouse dermis, if a small amount of chick dermis is added to the explant (making the dermis of the explant about 90% mouse and 10% chick), most of the explants will still form arrested feather buds but a small number will go on to form feathers. It is significant that chick dermis from glabrous regions—which we have seen cannot induce feather formation even in epidermis that normally forms feathers—when added to the explant will support feather development. This supports the proposal that feather differentiation depends on a two-step induction process. Although glabrous dermis is unable to supply the signal for the first step, it is capable of supplying the second class-specific signal.

How does the dermal mesenchyme act to enable the epidermis to acquire its specific properties? If mesenchyme tissues are the inducers of epidermal specificities, how are their instructions transferred to the epidermis? Unfortunately, answers to these questions are incomplete and far from being satisfactory. This is undoubtedly related to the fact that responses by the epidermis to mesenchymal stimuli involve a complex of processes, including determination, mitosis, morphogenesis, and cytodifferentiation.

Electron-microscopic studies of the developing hair follicle in the mouse have shown that there are readily observable differences in the structure of the basal lamina over different parts of the hair follicle. At 14 days, the basal lamina of the lateral follicle walls is intact and beginning to form hemidesmosomes (structures functioning to attach adjacent cells to each other). There is a progressive maturation on days 17 to 20, with the basal lamina remaining intact and showing an increase in both the number and the completeness of the hemidesmosomes. The picture is consistent with the concept of the development of a barrier between the epidermal hair matrix cells and the surrounding dermal cells over the entire lateral surface of the hair follicle. However, examination of the basal lamina in the region of the dermal papilla reveals that at 14 days, although it is intact, it is not as mature as that of the lateral follicle wall. From 17 to 20 days the basal lamina does not mature, develops no hemidesmosomes, becomes irregular and less dense, and shows gaps through which the processes of mesenchymal cells make contact with the hair matrix cells. This heterotypic cell contact is similar to that which has been reported in a number of other developing organs such as lung, tooth, and salivary glands at the site and the time of known tissue interaction and differentiation. In the regions of tissue interaction, the basal lamina shows a different morphology, a morphology that apparently facilitates the exchange of materials or information between two different cell populations.

In line with the concept that gaps in the basal lamina are important in allowing tissue interactions leading to changes in differentiation of epithelial cells is the effect resulting from the addition of vitamin A to the culture medium of hair follicles of the rat grown in vitro (Hardy et al., 1983). The basal lamina of follicles after one day in standard medium remain intact around the lateral follicle walls, and gaps only occur in the region of the dermal papilla (Fig. 23–15 A). However, after one day in medium with added vitamin A, gaps appear, randomly distributed over the entire lateral wall of the follicle (Fig. 23–15 A). After three days in standard medium, the basal lamina still shows gaps only over the dermal papilla. The hair shaft and inner root sheath have started to develop, and some of the dermal cells have aligned with the lateral walls of the follicle, beginning to form a dermal sheath (Fig. 23–15 B). However, after three days in vitamin A–enriched medium, the basal lamina of the lateral walls of the follicle show gaps through which epithelial cells extend, forming buds (Fig. 23–15 C). The basal lamina is absent over the surface of the buds and continuous only in the regions between the buds. There is a very limited differentiation of the hair matrix. The epithelial cells of the buds later form mucus-secreting cells. Although vitamin A acts on the lateral walls of the developing hair follicle in a manner resembling an inductor, it is not known whether it is itself an inductor molecule, a releaser of enzymes that break down the basal lamina, or a modifier of dermal tissue. Various other observations suggest that the mesenchyme exerts its influence through the basement membrane and the extracellular matrix between epidermis and dermis. For example, at a specific time in development, the capacity to synthesize DNA and to divide becomes restricted to the stratum germinativum of the epidermis. The cells above this layer are no longer in contact with the basement membrane and undergo cytodifferentiation. It has been suggested

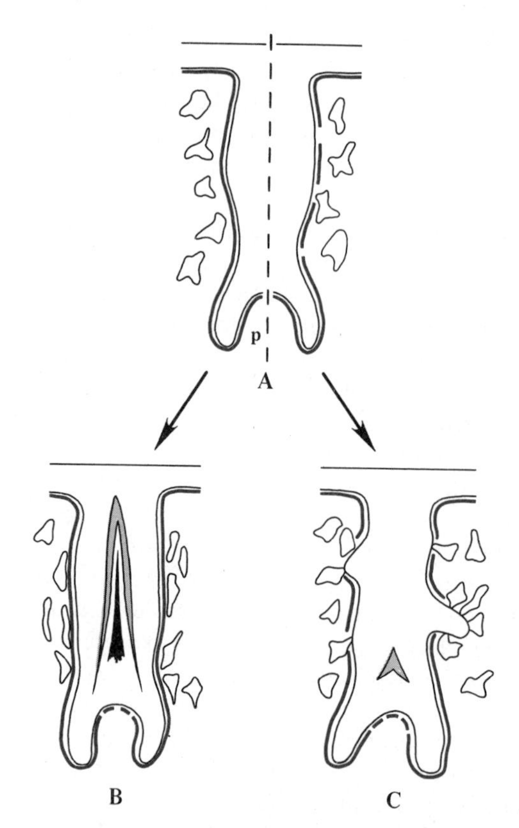


**FIG. 23–15.** Schematic diagram of the effect of adding vitamin A to the culture medium of developing mouse vibrissae follicles. A, after one day in vitro. The left side represents a follicle in standard medium. The basal lamina (thick line) is intact except over the dermal papilla (p). The right side represents a follicle in vitamin A–enriched medium. Gaps in the basal lamina are found all along the lateral walls of the follicle. B, after three days in standard medium. Gaps only over the dermal papilla, epithelium of hair matrix developing, dermal cells beginning to form a dermal sheath alongside of the follicle. C, after three days in vitamin A–enriched medium. Epithelial buds have developed along the lateral walls not covered by a basal lamina, basal lamina intact only between buds, hair matrix undeveloped. (After M. H. Hardy et al., 1983. J. Invest. Dermatol. 80:27.)

that the basal layer retains its competence to divide in response to a growth factor that passes through the basal lamina.

## Malformations of the Integumentary System

### Disturbances in Keratinization

There are several integumentary disorders that represent departures from normal levels of keratinization in the epidermis and its appendages. A particularly interesting, though rare, hereditary disorder is *congenital ectodermal dysplasia.* It is characterized in severe cases by the absence or hypoplasia of the eccrine sweat glands, the sebaceous glands, and mucus glands; hair and teeth may be deffective or completely absent. Early diagnosis of the malformation is essential, since the absence of sweat glands severely disturbs thermoregulation.

Occasionally the skin may become overkeratinized *(hyperkeratosis)* during early infancy. Frequently under these conditions, the stratum corneum cracks into numerous scalelike thickenings. This disorder, termed *ichthyosis simplex,* is transmitted by a single autosomal gene.

*Hypertrichosis* is a condition in which there is an overabundance of hair in regions where hair is normally sparse. It can be traced to the development of supernumerary hair follicles and/or the abnormal persistance of fetal hair follicles. *Hypotrichosis* is the converse disorder.

### Abnormalities of the Mammary Glands

Absence of the mammary glands *(amastia)* or mammary nipples *(athelia)* is rare. More common are supernumerary breasts *(polymastia)* and supernumerary nipples *(polythelia).* Extra breasts and nipples develop from accessory mammary epithelial buds that develop along the mammary ridges, particularly in the vicinity of the normal pair of glands. Aberrant mammary glands and nipples have been reported at sites far from the conventional positions of the paired mammary ridges (Fig. 23–16).

**FIG. 23–16.** Abnormalities in the development of breasts and nipples. Schematic diagrams summarize, on a single individual, abnormal locations where supernumerary gland development has been reported in the literature.

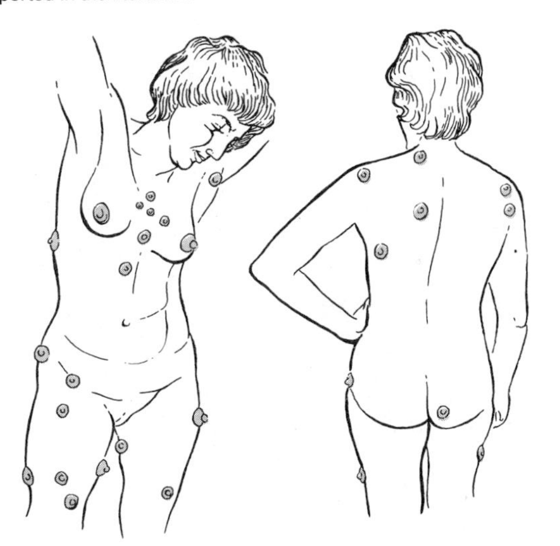

## References


Breathnach, A. S. 1971. Embryology of the human skin. A review of ultrastructural studies. J. Invest. Dermatol. 57:133–143.

Briggaman, R. A. and C. Wheeler. 1968. Epidermal–dermal interactions in adult human skin: role of dermis in epidermal maintenance. J. Invest. Dermatol. 51:454–465.

Cairns, J. and J. Saunders. 1954. The influence of embryonic mesoderm on the regional specification of epidermal derivatives in the chick. J. exp. Zool. 127:221–248.

Dodson, J. W. 1967. The differentiation of epidermis. I. The interrelationship of epidermis and dermis in embryonic chicken skin. J. Embryol. exp. Morphol. 17:83–105.

Hardy, M. H., R. J. Van Exan, K. S. Sonstegard, and P. R. Sweeny. 1983. Basal lamina changes during tissue interactions in hair follicles—an in vitro study of normal dermal papillae and vitamin A–induced glandular morphogenesis. J. Invest. Dermatol. 80:27–34.

Hashimoto, K., B. Gross, R. DiBella, and W. Lever. 1966. The ultrastructure of the skin of human embryos. J. Invest. Dermatol. 47:317–335.

Holbrook, K. and G. Odland. 1975. The fine structure of developing human epidermis: light, scanning and transmission electron microscopy of the periderm. J. Invest. Dermatol. 65:16–38.

Lawrence, I. 1971. Timed reciprocal dermal–epidermal interactions between comb, mid-dorsal and tarsometatarsal skin components. J. Exp. Zool. 178:195–210.

McLoughlin, C. B. 1961. The importance of mesenchymal factors in the differentiation of chick epidermis. II. Modification of epidermal differentiation by contact with different types of mesenchyme. J. Embryol. exp. Morphol. 9:385–409.

Rawles, M. 1963. Tissue interactions in scale and feather development as studies in dermal–epidermal recombinations. J. Embryol. exp. Morphol. 11:765–789.

# INDEX